10th National Physics Seminar (SNF 2021)

AA003750

Journal of Physics: Conference Series Volume 2019

Jakarta, Indonesia
19 June 2021

ISBN: 978-1-7138-4990-2
ISSN: 1742-6588

Printed from e-media with permission by:

Curran Associates, Inc.
57 Morehouse Lane
Red Hook, NY 12571

Some format issues inherent in the e-media version may also appear in this print version.

Institute of Physics
Dirac House, Temple Back
Bristol BS1 6BE UK

Phone: 44 1 17 929 7481
Fax: 44 1 17 920 0979

techtracking@iop.org

Additional copies of this publication are available from:

Curran Associates, Inc.
57 Morehouse Lane
Red Hook, NY 12571 USA
Phone: 845-758-0400
Fax: 845-758-2633
Email: curran@proceedings.com
Web: www.proceedings.com

TABLE OF CONTENTS

PREFACE

It is a great honour that the 10th Seminar Nasional Fisika (SNF) 2021 has been successfully conducted by the Department of Physics Universitas Negeri Jakarta collaborated with the Physical Society of Indonesia, Chapter of Jakarta and Banten. Due to the pandemic of Covid-19, the seminar is virtually conducted in Jakarta, Indonesia, through the platform of zoom meeting on June 19, 2021. In this time, the seminar takes a theme "Independent Learning of Physics for 21st Century through Collaboration Research".

As a scientific forum, the seminar is attended by 268 participants coming from the students, lecturers, engineers, and practitioners who participate either as presenters or as participants. The main aim of the seminar is to share ideas and knowledge, and to build collaboration with each other. We receive 130 papers which will be selected to be published in the **Journal of Physics: Conference Series** with the scope of physics and its related area. In this volume, we accept 117 papers for the publication after meeting the scope, checking originality, and performing peer review.

On behalf of the committee of SNF 2021, we would like to fully respect to keynote speakers and all the presenters that contribute to the papers. We would like also to thank the teamwork of undergraduate students who support us behind the scene without whom we are surely nothing. Many thanks are addressed to reviewers who invest the time to decide if the paper is suitable for the publication. Finally, we would like also to thank someone as a funder who does not want to mention his name.

At last, we expect we can enhance the knowledge and insights of physics and its application in technology and make collaboration in the future time. See you all in the next seminar.

The Chairman

Dr. Teguh Budi Prayitno

Journal of Physics: Conference Series **2019** (2021) 011001 doi:10.1088/1742-6596/2019/1/011001

COMMITTEE

Advisory Board

Prof. Dr. Komarudin, M.Si	(Rector of Universitas Negeri Jakarta)
Prof. Dr. Suyono, M.Si	(Vice Rektor I of Universitas Negeri Jakarta)
Dr. Adisyahputra, MS	(Dean of Faculty of Mathematics and Science, Universitas Negeri Jakarta)
Ariadne L. Juwono, M.Eng, Ph.D	(Chairperson of Physical Society of Indonesia, Chapter of Jakarta and Banten)

Scientific Commitee:

Prof. Dr. Anto Sulaksono	(Universitas Indonesia)
Dr. Budhy Kurniawan	(Universitas Indonesia)
Dr. Acep Purqon	(Institut Teknologi Bandung)
Dr. Ing. R. Wahyu Widanarto	(Universitas Jenderal Soedirman)
Prof. Dr. Putut Marwoto	(Universitas Negeri Semarang)
Dr. Artoto Arkundato	(Universitas Negeri Jember)
Prof. Dr. Agus Setyo Budi, M.Si	(Universitas Negeri Jakarta)
Prof. Dr. I Made Astra, M.Si	(Universitas Negeri Jakarta)
Prof. Dr. Yetty Supriyati, M.Pd	(Universitas Negeri Jakarta)
Prof. Dr. Mangasi A. Marpaung, M.Si	(Universitas Negeri Jakarta)
Prof. Dr. Erfan Handoko, M.Si	(Universitas Negeri Jakarta)
Prof. Dr. Sunaryo, M.Si	(Universitas Negeri Jakarta)
Dr. Widyaningrum Indrasari, M.Si	(Universitas Negeri Jakarta)
Dr. Umiatin, M.Si	(Universitas Negeri Jakarta)
Dr. Esmar Budi, M. T	(Universitas Negeri Jakarta)
Dr. Iwan Sugihartono, M.Si	(Universitas Negeri Jakarta)
Dr. Hadi Nasbey, M.Si	(Universitas Negeri Jakarta)
Dr. rer. nat. Bambang Heru, M.Si	(Universitas Negeri Jakarta)
Dr. Anggoro Budi Susilo, M.Si	(Universitas Negeri Jakarta)
Fauzi Bakri, M.Si	(Universitas Negeri Jakarta)
Riser Fahdiran, M.Si	(Universitas Negeri Jakarta)
Handjoko Permana, M.Si	(Universitas Negeri Jakarta)
Dewi Muliyati, M.Si	(Universitas Negeri Jakarta)

Organizing Commitee:

Dr. Teguh Budi Prayitno	(Chairman, Universitas Negeri Jakarta)
Dr. Firmanul Catur Wibowo	(Vice Chairman, Universitas Negeri Jakarta)
Lari Andres Sanjaya, M.Pd	(Secretary, Universitas Negeri Jakarta)
Dwi Susanti, M.Pd	(Treasurer, Universitas Negeri Jakarta)
Dr. Mutia Delina	(Moderator, Universitas Negeri Jakarta)

VIRTUAL DOCUMENTATION

ABS-45_Rena...
ABS-048_Mu...
ABS-098_Ima...
Recording
UNJ

Recording
View
Mutia Delina (UNJ)
Unmute
Stop Video
220
Participants
Chat
Share Screen
Record
Reactions
Leave

Recording
View
Mutia Delina (UNJ)
Unmute
Stop Video
Participants
Chat
Share Screen
Record
Reactions
Leave

Erfan Handoko
Prof. Dr. Erfan Handoko, M.Si
Achievement
UNIVERSITAS NEGERI JA
Unmute
Stop Video
Participants
Chat
Share Screen
Record
Reactions
Leave

Handjoko
N
Unmute
Stop Video
Participants
Chat
Share Screen
Record
Reactions
Leave

HOW PRESENTED THE OIL DROP EXPERIMENT IN TEXTBOOKS? A HISTORY AN PHILOSOPHY OF SCIENCE PERSPECTIVE
Prof. Dr. Bayram COŞTU
Yildiz Technical University
Istanbul, TURKEY

Public Relations and Documentation
• Dr. Hadi Nasbey, M.Si
• Upik Rahma Fitri, M.Pd.
• Muhamad Fajrin, ST.
Promotion and Sponsorship
• Prof. Dr. Agus Setyo Budi
• Prof. Dr. Yetty Supriyati, M.Pd
• Prof. Dr. Sunaryo, M.Si
• Dr.rer.nat. Bambang Heru I., M.Si
• Dr. Iwan Sugihartono, M.Si
• Prof. Dr. I Made Astra, M.Si
• Drs. Siswoyo, M.Pd
Transportation and Accommodation
• Riser Fahdiran, M.Si
Handjoko
Recording
thank u prof
Leave

The 10th National Physics Seminar

Journal of Physics: Conference Series **2019** (2021) 011002 doi:10.1088/1742-6596/2019/1/011002

Peer review declaration

All papers published in this volume of Journal of Physics: Conference Series have been peer reviewed through processes administered by the Editors. Reviews were conducted by expert referees to the professional and scientific standards expected of a proceedings journal published by IOP Publishing.

- **Type of peer review: Single-blind / Double-blind / Triple-blind / Open / Other**
  The review process in SNF-2021 was conducted by a single-blind process, which means that only referees know the identity of authors, but the authors do not.

- **Conference submission management system:**
  Open Journal System

- **Number of submissions received:**
  157

- **Number of submissions sent for review:**
  130

- **Number of submissions accepted:**
  117

- **Acceptance Rate (Number of Submissions Accepted / Number of Submissions Received X 100):**
  $$\frac{117}{157} \times 100\% = 74.5\%$$

- **Average number of reviews per paper:**
  10

- **Total number of reviewers involved:**
  13

- **Any additional info on review process:**
  None

- **Contact person for queries:**
  Name : Dr. Teguh Budi Prayitno
  Affiliation: State University of Jakarta
  Email : teguh-budi@unj.ac.id

E-Book Static Fluid and Dynamic Fluid Web-Based with a Problem-Based Learning Model to Improve Students Physics Problem-Solving Skills

Z Ramli[1], Sunaryo[1] and V Serevina[1]
[1]Magister of Physics Education, Faculty of Mathematics and Science, Jakarta State University, Rawamangun Muka Street No. 1, Jakarta, Indonesia., 13220
Email: zulkifliramli53@gmail.com, vserevina@unj.ac.id, sunaryo@unj.ac.id

Abstract. This study aims to develop physics learning media in the form a static fluid and dynamic fluids e-book based on website with Problem Based Learning for students learning. The research subjects were 43 5th semester student of Jakarta State University. The research method used is Research and Development (R&D) research using the ADDIE development model (Analysis, Design, Development, Implementation, and Evaluation). The result of the instrument validation showed 90.31% for the validation of media experts, 94.94% for the validation of material experts, and 77.09% for the validation of learning experts. Based on the results of the validation of media experts, the material, and learning can show that this e-book in terms of several indicators used for validations has very decent criteria. Based on the result of trials to students, the results obtained an average score of all aspects of 88.39% with a very feasible interpretation and the impact on students Physics Problem Solving was measured. Based on the result of the effectiveness test show that there are differences in physics problem solving among students who use static fluids and dynamic fluids e-book based on Problem Based Learning and static fluids and dynamic fluids e-book based on Non-Problem Based Learning. Based on the D'Cohens test, the use of static fluids and dynamic fluids e-book based on Problem Based Learning made an effective contribution to increasing students Physics Problem Solving by 2.51 in the medium category, so it can be concluded that the developed e-book can increase the Physics Problem Solving students.

1. Introduction

Considering the rapid technological advance of the Internet in particular has brought benefits and a powerful influence to everyone connected with education. With these benefits, it is also helpful in curriculum 13 that emphasizes technology-based learning in accordance with the 2014 dictation of curriculum 2013, which in general the 2013 curriculum developed based on three factors: internal challenges, external challenges, and an improvement over mindset [1].

The books are the learning component necessary for the study of a particular material, where students can learn independently. In addition to the communications networks of technology harvesting such as the website, there should be a tremendous range of tools available to teachers and students with a view to building the skills, knowledge, and expertise needed in the work and life of the 21st century [2]. So, utilizing network technology, an e-book is combined with the website.

The media teaching of the e-book-based website is a media that has a purpose to facilitate communication in learning [3]. The teaching material that can make learning more interesting, interactive, and efficient is e-book [4], a converted text book into digital format, the e-book also has a sense as a learning environment that has an application that contains a multimedia, instructional resource database that stores a multimedia presentation on the topic in a book [5].

Further highlighting the excelling value of e-book learning is supported by Darlen research with the interactive e-book development theme for junior physics showing the product development of an interactive e-book interactive grade VIII junior high school physics resource [6]. Illa also tells of the interactive e-book that the use of e-book can interact between teachers and students in remote learning and the students are more interested in using the e-book in learning [7]. It corresponds to what will be developed as an interactive e-book source for learning physics. The focus to be achieved on the developed e-book is to increase interaction between teachers and students in close-or remote learning, wherever and whenever, such as e-learning. Astrini with the theme of developing interactive e-book media on the structure and function of plant organ tissues stated that the interactive e-book media developed was feasible with an overall average of 94% and the videos presented could clarify understanding of the concept [8]. This is important to develop through e-books, which display interactive videos that can guide students to understand a material through visualizations such as images, audio, video or animation of physics learning.

Some of the results of web-based learning make no significant difference to traditional (face to face) and positive responses from web-based learning users because one can be accessed at any time or place [9]. Web-based use of learning media has a higher metacognitive capacity than those using web-based learning media [10]. Regarding the discussion of fluid materials is found in Nisrina research which reveals the mastery of fluid concepts in students taught with interactive multimedia-based learning higher than those taught with conventional learning [11]. With this web-based e-book combination that contains hyperlink's menu of discussion rooms, it is expected that students will be able to write their criticisms.

In addition to the use of Internet networks, and regardless of the importance of the use of teaching materials in learning, the model of learning is an essential component that can sustain the achievement of the purpose of learning [12]. According to the purpose of learning, good educators should begin with an improvement in the learning process done by offering an approach or strategy that can improve learners' ability and results. One strategy has been Problem-Based Learning (PBL) model. This PBL study is a problem-based learning model that first appeared during the learning process [13]. This was consistent with the problem that occurred in the class, where the use of the PBL changed the condition of passive learning to be active and creative or that was said teacher oriented to student oriented.

Research conducted on Aji states on the theme development of pbs-based physics learning modules to increase the ability to solve physics results that an PBL physics-based learning module on the topic of balance and oscillation can improve the student physics solving capability. This can be known by the assessment of the material experts and the physics teachers on the is worth component, which is that the issues presented can improve the problem-solving ability of learners who get a percentage of 91.7% [14]. Mahayukti also wrote about the development of a problem-oriented e-book to improve a student's critical thinking skill with a problem-based e-module use will direct learners to solve the problem on their own and this will be a real experience in solving the problem [15]. It relates to the research developed that is with websites based PBL models. The e-book model PBL is complemented with interactive videos, visual images/pictures that adjust with physical problems so that students can have critical thinking ability and problem-solving ability to acquire knowledge concepts. The e-book also has a report card for the summative and formative and experimental evaluations so that the student can visualize and pour out creative ideas and find solutions for the project.

Based on the above, it will be necessary to study the development of a steady, steady, and dynamic website-based e-book development with a PBL model to enhance the problem-solving capabilities of student physics at the Jakarta state university physics education.

2. Research Methodology

The research method used Research and Development (R&D) method which is the method used to develop and validate a product used in learning. The development model to be used is the ADDIE model (Analysis, Design, Development, Implementation, and Evaluation).

This begins the gradual stage, identifying problems in research and knowing what students and educators need in learning physics. Early studies were analysis of needs and a study of literature. The design phase, the design plan e-book needs, the analysis of the students' characteristics that will use e-book, divide and conceptualize matter into sub materials, determine the content in the e-book (text, picture or graph, video, and evaluation), transcribe e-book scripts, specify e-book instruments. The e-book container is a website that requires designing web links, websites, website template, and hosting websites. The stage of development, incorporating material content such as title-writing, writing text using the Microsoft Office Word 2019, creating a front and back cover of e-book and creating pictures using a Corel Draw X5 and is stored in PNG format, editing a supporting video of materials using inshot software, making an formative evaluation using Google Form software, combining all of the content of the e-book (text, image or graph, video, animation, and evaluations) by using a professional PDF flip and Posting it on the website. The intrinsic stage, the resulting data is derived from the validation of media experts, materials experts, learners, graphic design experts, the physics faculty test, and students.

2.1. Data Analysis Techniques

Assessments of validation tests and product tests are calculated based on scores of acquisitions on each aspect, in the following ways:

$$P = \frac{S}{N} \times 100\% \qquad (1)$$

P = The percentage of success (%)
S = The total score
N = The maximum score

Data obtained, then quantifies the interpretation of the score, as follows (Riduwan, 2007).

Table 1. Scale Likert Interpretation

Percentage (%)	Interpretation
76 – 100	Well Deserved
51 – 75	Worthy
26 – 50	Unfit
0 – 25	So Unseemly

To see the difference between the value of physics problem solving ability of students in static fluid and dynamic fluid using e-books with PBL models and e-books without PBL models, so that the gain value t test can be used, with notes $N_1 \neq N_2$ and $t_{count} > t_{table}$ can be written the following equation (Farida, 2017).

$$t = \frac{\overline{X_1} - \overline{X_2}}{\sqrt{\frac{SD_1^2}{N_1 - 2} + \frac{SD_2^2}{N_2 - 2}}} \qquad (2)$$

t = The number or degree coefficient of the mean of the two groups
$\overline{X_1}$ = The average distribution value of the experimental class
$\overline{X_2}$ = The average distribution value of the control class
N_1 = Number of experimental class samples
N_2 = Number of control class samples
SD_1^2 = The value of the distribution variance of the experimental class sample
SD_2^2 = The value of the variance of the control class sample distribution

Next, the D'Cohens test was carried out to test whether there was an effect of the e-book with the PBL model on the ability to solve physics problems, which D'Cohens' interpretation was as follows (Santoso, 2010).

Table 2. Effect Size Criteria

Effect Size	Information
D < 0,8	Large
0,2 < D < 0,8	Medium
D < 0,2	Small

3. Result and Discussion

The e-book media draft that is made include the cover, the introduction, the usage guide, basic competence, the table of contents, on the e-book, glossary, history of scientists, the library list. Here will show you the developed e-book draft:

(a) (b)

Figure 1. E-book display (a) Cover (b) Concept map

The e-book media based websites with PBL models can be viewed at www.physicse-bookfluid.com. Here will be shown a website based media e-book with an email model:

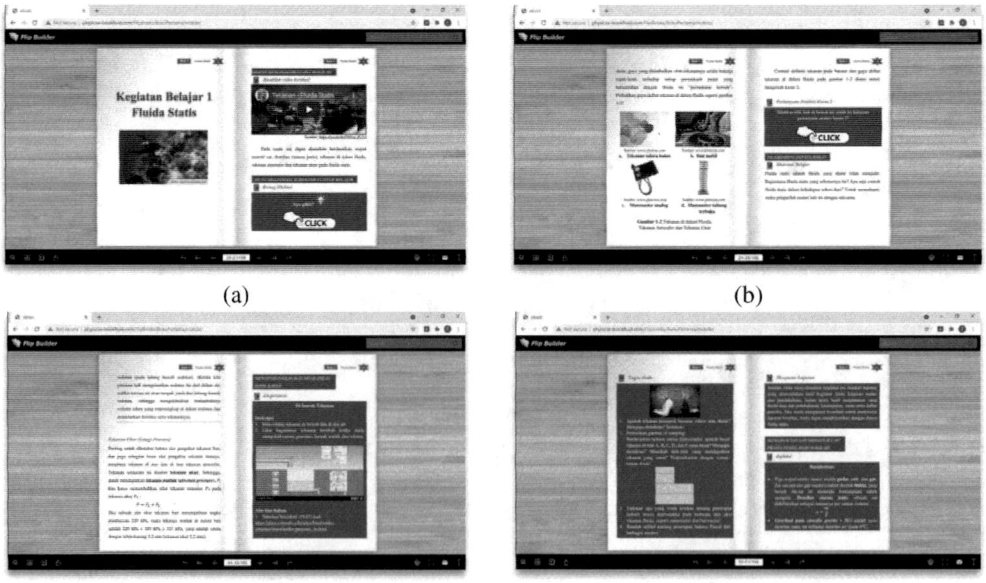

(a) (b)

(c) (d)

Figure 2. E-book web-based with PBL on display
(a) Student orientation on problems & Organizing students to study (b) Guiding individual/group investigations
(c) Develop and present the work (d) Analyze and evaluate the problem-solving process

The static fluid and dynamic fluid e-book media are arranged based on the problem-based learning model as follows:

Table 3. Stages of Problem-Solving Skills

Stages of Problem-Solving Skills	Indicator
Visualize the problem	Visualizing the problem into a visual representation
	Identify problems based on basic concepts
Describe the problem in physics description	Turning visual representations into physical descriptions
	Create a free body diagram/sketch that describes the problem
	Make a list of known quantities according to the diagram/sketch
	Determine the quantity in question which is known according to the diagram/sketch
Plan the solution	Converts physical descriptions into mathematical representations
	Determine the right equation for problem solving
Execute the plan	Substituting the value of a known quantity into the equation
	Perform calculations using the selected equation
Check and evaluate	Evaluating conformity with the concept
	Evaluating units

The asset-based e-book worthiness test using PBL models is done using methods of spreading questionnaire to media experts (professors), materials (professors), teachers (professors), and users (professors and students).

Validation of media experts is to test the feasibility and get the opinion of the developed e-book and to obtain information of corrections, Suggestions, and critics to evaluate and revise e-book. According to media expert tests, the average percentage of assessments given by media experts is 90.31% with very decent interpretation. This suggests that the developed e-book is already worth using in the study of physics.

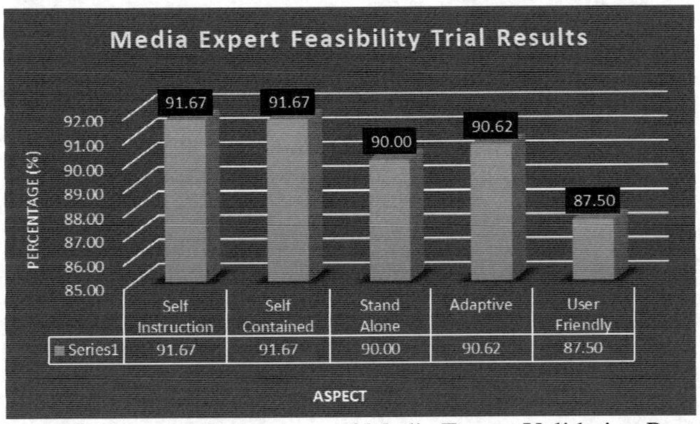

Figure 3. Diagram Percentage of Media Expert Validation Result

Validation of the material experts aims to know the worthiness of the material in the e-book developed for static and dynamic fluid materials and obtain information of improvements, Suggestions, and criticisms to evaluate and revise the material to the learning media. Based on the results of the validation of the material experts, the average percentage of assessments given by the material expert is 94.94 with very reasonable interpretation. This suggests that the developed e-book is already worth using in learning.

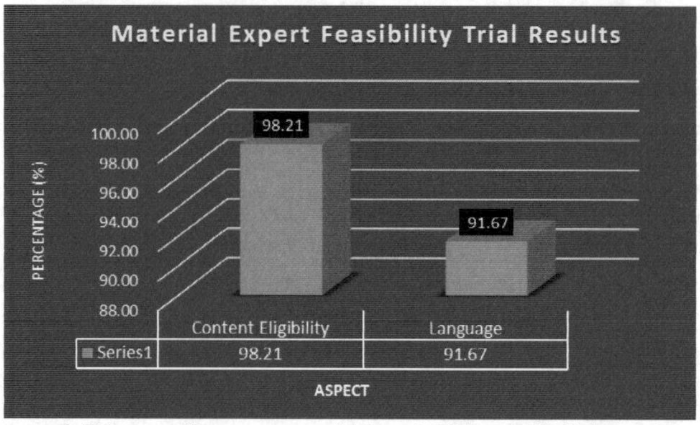

Figure 4. Diagram Percentage of Material Expert Validation Result

Validation by the learner is intended to know the worthiness of the stages of learning presented in the form of the developed e-book and to obtain information of improvements, Suggestions, and criticisms to evaluate and revise the stages of the learning process on the e-book. Based on the researchers' validation test, the average percentage of assessments given by the learner is 77.09% with a perfectly feasible interpretation. This suggests that the developed e-book is already worth using in learning.

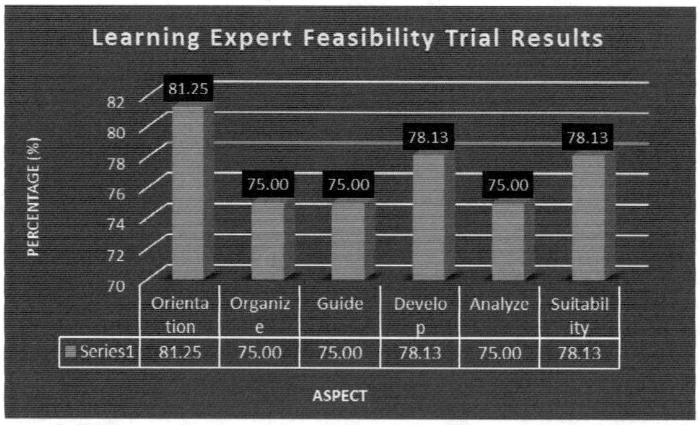

Figure 5. Diagram Percentage of Learning Expert Validation Result

College's trial was based on feasibility on content, presentation, language, and grafting. According to the professor's test, the average percentage of the score given by the professor of physics is 96.93% with a very decent interpretation. This suggests that the developed e-book is already worth using in learning.

Table 4. Test Result of Physics Educator's

Aspect	Result
Content Eligibility	98.96
Presentation	95.00
Language	95.85
Graphic	97.92
Overall Percentage	96.93

The developed e-book is validated by media experts, material experts, and learning experts. The results of the validation are used as material for evaluation and improvement of e-book development.

Next, a small group trial was conducted with 5 (five) students. Based on the results of small group trials of students, the average percentage of assessments given by students is 91.04%.

Table 5. Test Result of Small Groups of Students

Aspect	Result
Content Eligibility	91.67
Presentation	87.50
Language	90.00
Graphic	95.00
Overall Percentage	91.04

The results of the evaluation of the large group of students test on e-book obtained results that the developed e-book can be a good learning resource for students to understand physics materials. After a revision of development, the e-book was tested as a lesson in the Jakarta state university physics study program 2020/2021. The assessment given by students consists of the feasibility of content, presentation, language, and grafting.

Table 6. Test Result of Large Groups of Students

Aspect	Result
Content Eligibility	87.79
Presentation	85.83
Language	91.25
Graphic	89.72
Overall Percentage	88.65

Based on the results of the large group test results for students acquired an average percentage of assessment of 88.65% with an interpretation that is perfectly feasible. This suggests that the developed e-book has been worth using on the study of physics.

To measure the effect of website-based static fluid and dynamic fluid e-books with the PBL model on improving students' physics problem solving abilities, the t-test gain value was used in both the experimental class and the control class. Based on the t-test, the gain value shows the results of the $t_{count} = 8.58 > t_{table} = 2.01$ so it can be said that the results of the experiments carried out have an influence on the experimental class.

The level of contribution (D'Cohens test) of the e-book with the PBL model is 2.51 so it can be said that the e-book with the PBL model on improving students' physics problem solving skills has a contribution to the high category level. This makes a student's involvement in problem-based learning that can listen to a variety of perspectives and can improve high level thinking ability [16].

4. Conclusions

Based on the validation tests by materials, media experts, and learning experts, and field tests, as well as field tests, it can be concluded that the e-book static fluid and dynamic fluid website based models with advanced PBL models are worthy of independent teaching to enhance student physics problem solving.

Acknowledgments

Thanks to the UNJ professor of physical education and to the professors who have guided to research this development. Thanks also to Mr. Dr. Esmar Budi and mother Pertiwi, S.Pd., M.Pd. was a material expert; Father Dr. Iwan Sugihartono, M.Si. and father Iwan Permana Suwarna, M.Pd. as media expert; Prof. Dr. I Made Astra, M.Si. and Dr. Hadi Nasbey, M.Si. the teacher who has been willing to validate and give advice for improved products..

References

[1] Kemdikbud 2014 *Materi Pelatihan Guru Implementasi Kurikulum 2013* (Jakarta)
[2] Roblyer, M. D. 2013 *Integrating Educational Technology Into Teaching-6th ed* (USA: Person Education)
[3] Smaldino, S. E., Lowther, D. L., & Russel, J. D. 2011 *Instructional Technology And Media For Learning* (Jakarta: Kencana)
[4] Woo, T. K. 2011 *Asian Association of Open Universities Journal,* 6(1) *p. 51-58*
[5] Shiratuddin, N., Landoni, M., Gibb, F., & Hassan, S. 2004 *Journal of Digital Information* 4(3)
[6] Darlen, R. F., Sjarkawi, S., & Lukman, A. 2015 *Jurnal Tekno-Pedagogi* 5(1) p. 13-23
[7] Restiyowati, I., & Sanjaya, I. G. 2012 *Journal Of Chemical Education* 1(1) p. 130-135
[8] Astrini, W., Djoko, B. J., & Pratiwi, P. R. 2012 *Unesa Jurnal Bio Edu* 1(3)
[9] Nam, C. S., & Jackson, S. T. 2007 *Journal of Information Technology Education* 6
[10] Kalatting, S., Serevina, V., & Astra, I. M. 2015 *Journal Penelitian dan Pengembangan Pendidikan Fisika* 1(1)
[11] Nisrina, N., Gunawan, G., & Harjono, A. 2016 *Jurnal Pendidikan Fisika dan Teknologi* 2(2)
[12] Huda, M. 2014 *Model-Model Pengajaran dan Pembelajaran* (Yogyakarta: Pustaka Pelajar)
[13] Abidin, Y. 2014 *Desain Sistem Pembelajaran dalam Konteks Kurikulum 2013* (Bandung: Refika Aditama)
[14] Aji, S. D., Hudha, M. N., & Rismawati, A. Y. 2017 *Science Education Journal* 1(1) p.36-51
[15] Mahayukti, I. S. 2013 *Jurnal Pendidikan Indonesia* p. *264-275*
[16] Smith, R. O. 2014 *Journal of Research and Practice for Adult Literacy, Secondary and Basic Education* p. 50-56

Journal of Physics: Conference Series **2019** (2021) 012002 doi:10.1088/1742-6596/2019/1/012002

Development of E-Module with A Scientific Approach To Improve The Student's Critical Thinking Skills At Class Xi Student High School In Optical Tools Material

A Haryanto[1], Sunaryo[1] and C E Rustana[1]
[1]Department of Magister Physics Education, Universitas Negeri Jakarta
Jl. Rawamangun Muka, Jakarta13220, Indonesia
Email: yantoadiyanto42@gmail.com, sunaryo@unj.ac.id, crustana@unj.ac.id

Abstract. This research aims to produce a physics learning media in the form of e-module with a scientific approach to learning physics on the optical instrument material for SMA / MA students. The subjects of this research were 46 students of class XI SMA Muhammadiyah Kota Bima. This developmental research refers to the method developed by ADDIE (Analysis, Design, Development, Implementation, Evaluation). Quasi-experimental method used in this research to test the effectiveness of the product with a research focus on e-module optical tools with a scientific approach to help improve students' critical thinking skills. The results of e-module validation by media experts showed 87.5%, material expert validation 82.78%, and learning expert validation 83.84%. Based on validation by media, material, and learning experts, it shows that this e-module in terms of several indicators used for validation has very feasible outcome criteria. Based on the results of trials in small and large groups of students, it was obtained that the average total aspect score of 88.78% and 86.6% was categorized as very feasible and the impact was measured on students' critical thinking. Based on the results of the D'Cohens test, users of e-module optical tools can provide effectiveness in improving students' critical thinking skills by 0.83 in the moderate category. So, it can be concluded that the e-module that has been developed by researcher can improve the students' critical thinking skills at SMA / MA.

1. Introduction

Education in schools is directed not only at mastering and understanding scientific concepts, but also at improving students' thinking abilities and skills, especially higher-order thinking skills, namely critical thinking skills, teachers need to teach their students to learn and think (teaching of thinking). [1].

The learning process that is often carried out by teachers is still dominant in the aspects of knowledge and understanding the concepts, so as a result, students do not develop critical thinking skills as desired. The lack of critical thinking skills of students can be seen from the 2015 PISA (Program for International Student Assessment) report which shows that Indonesia can only be ranked 62 out of 69 countries in the field of science [2]. When compared with the other countries.

One of the students' thinking skills can be developed by using assistive media in the learning process and the teacher can choose the right media. According to [3], Students 'critical thinking skills can be developed with the use of appropriate learning media, the selection of appropriate learning media can determine the success of developing students' critical thinking skills.

E-Module Assisted Guided Inquiry to Improve Critical Thinking Skills, the results of the analysis of increasing critical thinking skills showed that both classes were able to improve students' critical thinking skills, but the experimental class was more effective than the control class. The conclusion of this study is that learning with e-module assisted inquiry guide can improve students' critical thinking skills [4]. The development of web-based e-modules to improve the competence attainment of physics knowledge in high school static and dynamic electricity material, this research shows that the development of web-based e-modules on static electricity and dynamic electricity can be used as multimedia learning and can increase the achievement of students' knowledge competencies. [5]. The effect of the Problem Based Learning (PBL) learning model and the ability to think critically on the learning outcomes of junior high school students 'ecosystem material, it is obtained information that there is an influence of the Problem Based Learning model and the ability to think critically on the learning outcomes of junior high school students' ecosystem, there is an interaction between Problem Based Learning learning models and the ability to think critically on the learning outcomes of junior high school students' ecosystem material [6]. The development of a physics e-module based on Problem Based Learning (PBL) on temperature and heat material to improve the science process skills of high school students "shows that this e-module in terms of the indicators used for trials has very good criteria and is suitable for use as learning teaching materials [7].

Based on the research results of Ismi Rahayu, Sukardi with the title "The Development Of E-Modules Project Based Learning for Students of Computer and Basic Networks at Vocational School"The effectiveness of e-module project-based learning is stated to be effective in improving learning outcomes [8]. The results of research by Imron Hamzah and Sriyani Mentari "Development of Accounting E-Module to Support the Scientific Approach of Students, Grade X Vocational High School" from the result, it can be concluded that the scientific approach to supporting accounting e-module is appropriate and able to improve students learning motivation and result [9].

Based on the description above, motivated research examines the reality conditions faced by students in physics lessons. The research with the title "the Development of e-Module with a Scientific Approach to Help the Critical improve Thinking Skills of Senior High School Students class XI in Optical Instrument Material".

2. Method

The research method used is the research and development (R&D) method, which is a method used to develop and validate a product used in learning. The development model that will be used is the ADDIE (Analysis, Design, Development, Implementation, and Evaluation) model. ADDIE model contains learning activities that are deliberately designed and developed to create a learning process in students through, in this process, students will have the ability include the knowledge, skills, and attitudes which needed to carry out tasks and jobs [10].

This stage begins with the Analyze stage by identifying the problems in research and knowing what students and educators need in learning physics, knowing the problems that occur in the field, both students and educators, especially learning resources and learning media that are appropriate, easy to use, anytime and anywhere. Initial research is carried out, namely needs analysis.

Design Phase Planning for the development of e-module physics with a scientific approach to increase critical thinking is prepared by compiling an e-module requirement plan, namely determining the e-module material, analyzing the characteristics of students who will use the e-module, dividing and conceptualizing the material into sub-material, determining content in the e-module (text, pictures or graphics, video, animation, and evaluation), write the e-module script, determining the e-module media instrument (media validation sheet and rubric, material validation sheet and rubric, student response questionnaire, response questionnaire teacher).

Table 1. Steps of scientific approach and delivery of E-module material

Learning Steps	Summary
Observe	This stage begins with basic questions related to the lesson to be learned. In the e-module, this section is displayed in the form of images, videos or animations related to the examples or events and is followed by several questions.
Ask	Teachers and students will provide each other with information or questions according to the problems given at the observing stage
Gathering Information	At this stage, students collect a number of information or facts in order to answer questions and problems posed by previous students. You do this by reading the material in the e-module, or an experiment that will be monitored by the teacher.
Associate	At this stage, students process information or problems that have been collected by discussing them with friends in a group.
Communicating	At this stage, each group conveys the results of learning activities to others clearly both orally and in writing

Development stage This stage carries out development activities, namely combining material content such as writing titles, writing text and creating pictures using Microsoft Office Word 2007 and Corel Draw X5, making the front and back covers of e-modules using Microsoft Office Word 2007 and storing them in PNG format, Editing video and pictures supporting material using AVS Video editor 6.5 software, Creating questions using Google Form software, Making formative evaluations, Combining all e-module content (text, pictures or graphics, video, animation, and evaluation) using Flip PDF Professional and include it in the 3D Pageflip Professional Software.

The revised e-module implementation phase, according to the results of the validation and small group testing will then be tested on large groups, namely students of SMA Muhammadiyah Kota Bima. Large group trials were carried out during the e-module implementation process in the experimental class. Furthermore, students will get learning using e-modules. After learning is complete, students are given a large group trial questionnaire.

Evaluation Stage In the evaluation stage, which aims to determine the feasibility of the product and the effectiveness of the product that has been developed in helping to increase critical thinking. The resulting data obtained from the validation of media experts, material experts, and learning experts.

The d'cohens test aims to test whether the e-module with a scientific approach is effective for improving critical thinking skills, according to the Effect Size Criteria. From the results of the evaluation and data analysis, d'Cohens is categorized as moderate. So it can be concluded that the developed e-module is effective for learning physics.

3. Result and Discussion

This research was conducted to develop a product in the form of an e-module with a scientific approach to the subject matter of optical devices. e-module developed has been tested for its feasibility by media experts, material experts, and learning experts. The results of the developed e-module validation are as follows:

Table 2. The Result Of The Media Expert Validation Test

Aspect Assessment	Result
Display Aspects	87.5 %
Programming Aspects	87.5 %

Table 3. The Result Of The Material Expert Validation Test

Aspect Assessment	Result
Self Instructional	84.78 %
Self Contained	87.5 %
Stand Alone	75 %
Adaptive	79.16 %
User Friendly	87.5 %

Table 4. The Result Of The Learning Expert Validation Test

Aspect Assessment	Result
Contents	81.73 %
Strategy	84.37 %
Evaluation	85.42 %

Based on the results of expert testing, the average percentage of assessment given by material experts in all aspects is 87.5% as shown in table 1 above, while the average percentage of assessment by media experts and learning experts as shown in tables 2 and 3 is 82. , 78% and 83.84% with very good interpretation. This shows that the e-module developed is suitable for use in learning physics. The following image shows a page view showing of the final developed e-module.

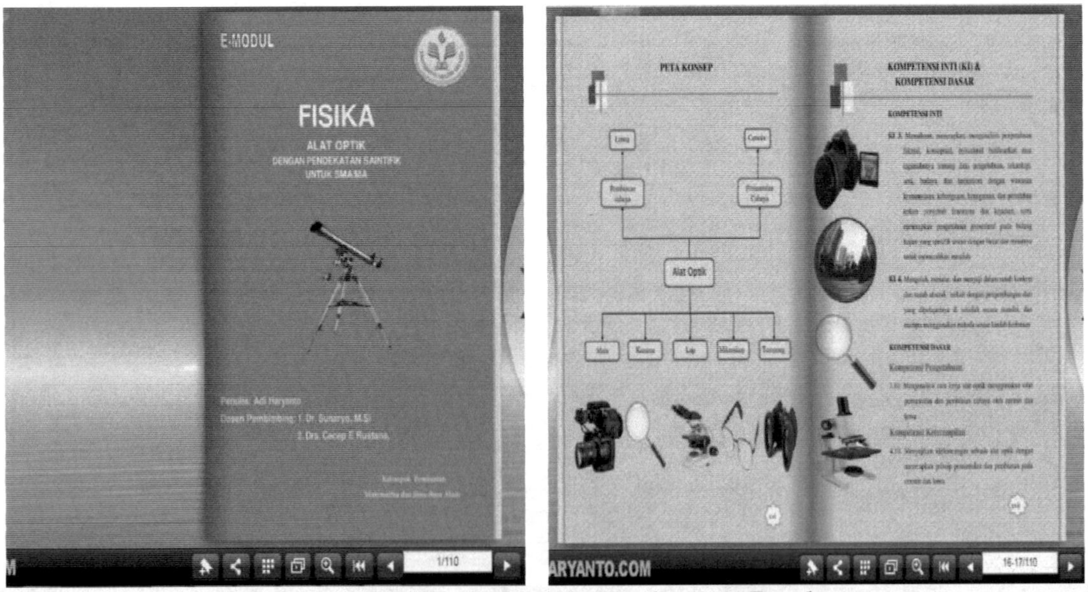

Figure 1. Display of e-module Development Results

Table 5. Indicators of Critical Thinking Skills

Critical thinking indicator	Critical thinking sub indicator
Give a simple explanation (elementary clarification)	Focusing the question, Analyzing the argument, Answering an explanation or challenge.
basic support	Adapting to the source
Inference	Induce and consider the results of induction
advanced clarification	Define terms and consider them
strategy and tactics	decide/determine actions and Interact with others

Based on the results of trials on physics teachers and students, data is obtained as shown in the graph as follows.

Figure 2. Percentage of Physics Teacher Testing Results

Based on the results of the teacher trials, the average percentage of the assessment given by the physics teacher was 87.1% with a very decent interpretation. This shows that the e-module developed is suitable for use in learning. All input from the teacher is accepted, then the e-module is revised and added according to the advice of the teacher.

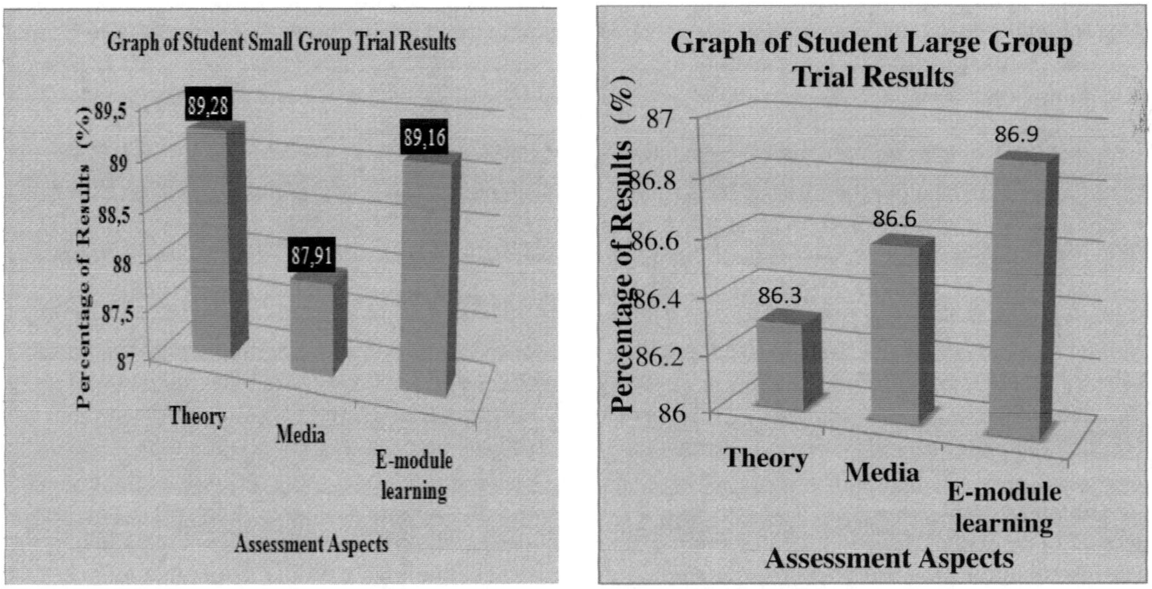

(a) (b)

Figure 3. a) Percentage of Student Small Group Trial Results. b) Percentage of Student Large Group Trial Results

An average of 87.1% in all aspects including the feasibility of material, media, e-module learning and graphics is calculated as the result of the physics teacher trial as shown in Figure 2.

In general, the total average of 88.78% and 86.6% for all related aspects is calculated from all the results of the validation of teacher and student validators, it is stated that e-module with a scientific approach can be interpreted as a suitable learning medium for use in the learning process. teaching on optical instrument material.

To find out the increase in differences in students who use e-module with a scientific approach with students who do not use e-module with a scientific approach to the optical instrument material. The t test is used to determine the hypothesis whether H_0 is rejected or accepted. The hypothesis is:

- H_0 =There is no increase between the gain value of the control class and the experimental class.
- H1 =There is an increase between the gain value of the control class and the experimental class.

From the N-Gain t test, it was found that the value of t count = 2.94 > t table = 2.01 with a significant level of 5%. Then H_0 = rejected and H_1 = accepted. So it can be said that the results of the experiment carried out have an influence on the experimental class.

D'Cohens' test was carried out to test its effectiveness. The d'Cohens test result is 0.83, categorized as moderate according to the following standards [11]. As shown in table 6 below.

Table 6. Effect size criteria

Effect Size	Category
D< 0,2	Small
0,2 < D < 0,8	Medium
D > 0,8	Big

Then the e-module with the scientific approach is interpreted in the medium category. This e-module can be used as a source of independent learning for students so that students can develop critical thinking skills independently. Based on the description above, it can be concluded that there has been an increase in students' critical thinking skills using teaching materials in the form of e-modules with a scientific approach.

4. Conclusion

Based on the results of the research and discussion of the development of optical e-module with a scientific approach, the following conclusions can be drawn:

1. The results of the validation of material experts obtained an average percentage of all aspects of 82.78%, media experts 87.5%, and learning experts 83.84%. While the overall average percentage of these experts was obtained at 84.70%. The results of the feasibility test indicate that the optical instrument e-module developed is categorized is very feasible to use as a learning resource in learning physics.
2. The results of the D'cohens test using e-module optical tools provide an effective contribution to students facilitate in improving critical thinking skills by 0.83 in the moderate category, it can be concluded that the e-module that was developed can improve the critical thinking of Senior high school students.

Acknowledgments

The authors would like to express their deepest gratitude to the supervisor, head of the physics education study program, validator of media, material, and learning experts, principals and physics teachers of SMA Muhammadiyah.

References

[1] Turmuji A. 2015 *Kompasiana.com/Turmuji Ahmad*
[2] Iswadi, S. 2016 *Sekelimut dari Hasil Pisa 2015*
[3] Mayer, R. E. 2009 *Multimedia Learning* (Yogyakarta: Pustaka Pelajar)
[4] Budiarti, S., Nuswowati, M., & Cahyono, E. 2016 *Journal of Innovative Science Education* 5 (2): 144-151
[5] Taufik, S. JH. 2018 *Wahana Pendidikan Fisika* 3(2) p. 51-61
[6] Fidiawati R, Komala R & Rusdi 2018 *Jurnal Biologi dan Pembelajarannya* 5 (2) p. 8-14
[7] Mayanty, S. 2018 *Pengembangan e-modul fisika berbasis Problem Based Learning (PBL) pada materi suhu dan kalor untuk meningkatkan keterampilan proses sains siswa SMA* (Jakarta: Universitas Negeri jakarta)
[8] Hamzah, Imron & Mentari, S. 2017 *Journal of Accounting and Business Education* 2 (1) p.78-88.
[9] Rahayu, I & Sukardi 2020 *Journal of Education Technology* 4(4) p. 398-403
[10] Pribadi, B. A. 2014 *Desain dan pengembangan program pelatihan berbasis kompetensi: implementasi model ADDIE* (Jakarta: prenada media group)
[11] Kotrlik J W and Atherton J C 2011 *Reporting and interpreting effect size in quantitative Agricultural Education Research* 52 p. 132-42

Application of CTL-Based Stop Motion Learning Videos for Understanding Elementary School Science Concepts in Lebak Regency

D K Adiputra[1, a)], and A Apriyansa[2, b)]
[1]STKIP Setia Budhi Rangkasbitung.
Jln. Budi Utomo Rangkasbitung, Banten – Indonesia
[2]Universitas Negeri Jakarta,
Jln. Rawamangun Muka, Jakarta Timur - Indonesia
Email: [a)]dedemadridista57@gmail.com, [b)]anak.rangkas@gmail.com

Abstract. The impact of the Covid-19 pandemic has made it difficult for elementary education institutions to implement online learning. In the application of audio-visual stop motion learning media based on CTL in elementary schools, it will become a reference material for teachers in implementing online learning, especially in science learning which adheres to scientific concepts through observation and experiments on a contextual basis. Thus, this study aims to apply audio-visual stop motion learning media based on CTL in elementary schools and to measure children's understanding of the concept of science. The specific goal is to make online learning more meaningful by using CTL-based audio-visual stop motion video learning media in science learning in elementary schools. The benefit of teachers and students is that there are learning media that become references for online learning designs.

1. Introduction

The current positive case for Covid-19 as of October 21, 2020, according to the Ministry of Health, which was conveyed through the Covid-19 Task Force on the www.covid19.go.id page, there were an additional 4267 confirmed positive for the Corona virus with a total of 373,109 positive cases in Indonesia. This causes all the habits of people's activities begin to change. The community must limit social interaction during activities outside the home. This also applies in the world of education, especially in the implementation of learning activities.

The government issued Circular Letter Number 36962/MPK.A/HK/2020 which was made in order to minimize the spread of Covid-19 by directing, teaching and learning activities both at schools and college campuses using the online method (in the network). This decision confused many parties. The unpreparedness of schools in implementing online learning programs is one of the main factors of the difficulties faced. Based on the preliminary study, information was found that one of the unpreparedness in carrying out online learning is the preparation of learning tools, such as Learning Implementation Plans, learning media, and evaluation of learning with online methods.

In the implementation of online learning, the teacher said that they only used the ability to lecture and give questions. That way, students become bored, no focused, and not motivated in learning. In addition, learning becomes unattractive because the learning process is not face-to-face. Even in the assessment process, students get less than optimal results, especially in science learning. So we need a solution that can make it easier for teachers to carry out online learning and attract students' attention so that learning objectives can be achieved even though learning must be carried out online.

Science learning will not be conveyed properly if it is only presented with the lecture method. Must use at least pictures so that students can visualize the science material. The use of learning videos can be a solution to the problems encountered. This is supported by Agustiningsih's research [1] Video is a form of learning media that can accommodate student learning needs in accordance with the competencies mandated in the curriculum. This research is an experimental research type with the research subjects being fourth grade students at SDN Ajung 03 Jember. Data collection methods used are test, observation and interview methods. The results of the study are shown from the results of calculating the difference in the pre-test and post-test scores of the experimental and control classes which show that 7.8 and 1.998, then > that is 7.8> 1.998 from db = 65 at a significance level of 5% so that it can be stated that learning using video media is better than not using video media or in other words rejected and accepted. Referring to the results of key research that the effect of the application of video on student learning outcomes so that the video can be used as an alternative learning media in order to support the implementation of the 2013 curriculum. Furthermore, based on research by Guswiani, et al. [2] This study aims to determine the use of video media on students' motivation and learning outcomes. The research approach used is a quantitative approach with Quasi Experimental Design research methods, namely experimental research designs, with data collection techniques using tests and observations. Based on the results of data analysis, it was concluded that the use of learning videos was very effective in increasing student motivation and learning outcomes in Front Office learning in Class XI Hospitality Accommodation at SMKN 3 Garut. This is indicated by the difference between motivation and student learning outcomes between classes that apply video media and classes that are given learning video media. The existence of media that can present images as well as sound, can help students divert their boredom and will be more interested in participating in learning that uses multimedia than just following conventional learning through lectures conducted by teachers [3], especially in online learning. Science material requires visualization of concepts in its delivery to students. The effectiveness and efficiency of online learning can be increased by using innovative learning media, one of which is CTL-based stop motion audio-visual video media.

Stop motion can help teachers visualize science material with interesting concepts. In line with Sumardi's research [4] Based on the results presented, inquiry-based stop-motion learning media can be used as learning media in schools for students. This video is also designed based on the components of contextual learning, namely constructivism, inquiry, questioning, community learning, modeling, reflection, and authentic assessment. These components are put together in the form of videos so that the learning process can be more meaningful. Learning will be more meaningful if the process is related to what students learn and what they experience directly in real life. That way, students will not only respond to learning that is only limited to remembering a concept, but a fact that can be useful in students' lives [5]. So, the solution to this problem is to apply CTL-based stop motion audio-visual video media in elementary schools.

2. Literature Review

2.1. Stop Motion

Stop motion animation is an easy-to-use technique for taking still images one-by-one with the camera and producing playable video clips [6]. Then, stop motion also shows adaptability to support teachers in learning various science concepts and knowledge of technology pedagogical content. In the development of this type of animation with objects being moved, clay is often used, so stop motion is often also referred to as claymotion [7]. But now available various types of stop motion animation. The International Design School explained that there are seven types of stop motion animation, which are as follows:

- Cut Out Animation
- Clay animation
- Puppet/Doll Animation
- Pixilation Animation
- Graphic Animation
- Object Animation
- Silhouette Animation

2.2. Contextual Teaching and Learning

Contextual learning is based on the understanding of progressivism by John Dewey, an educational philosopher, namely learning whose methods are based on children's experiences. When students accept learning and relate the information they know to what is being studied, they will be actively learning on their own well [8]. Contextual learning certainly requires a design that is reflected in the ideas and principles of Contextual Teaching and Learning (CTL). The teacher then develops is as follows [9].

2.2.1. Constructivism

Building contextual learning by using a constructivist foundation where knowledge is built by students themselves, discovering their own concepts, seeking their own meanings, concluding their own understanding through knowledge and findings that are owned or discovered by themselves.

2.2.2. Inquiry

Learning that starts with questions so that it can raise students' curiosity, motivation is a learning process based on discovery, namely inquiry

2.2.3. Questions

Knowledge will be obtained when someone starts asking questions. This questioning activity can encourage students' curiosity. By asking, they can seek other knowledge, motivate, criticize a finding, identify the information received lead students to find conclusions about something.

2.2.4. Learning Society

This activity is carried out so that students are accustomed to doing work together and looking for more benefits of learning. Students can carry out this component of the learning community during the learning process. For example, criticizing each other, exchanging information, exchanging ideas, exchanging experiences, and also exchanging knowledge between students with other students.

2.2.5. Modeling

The teacher can give a job by displaying something so that students can imitate, imitate and or do something that is shown by the teacher according to the model shown.

2.2.6. Reflection

Students who have carried out a series of lessons can review or provide responses that have the aim of making known and unidentifiable knowledge so that students are able to take an action to develop knowledge.

2.2.7. Authentic Rating

Authentic assessment is a procedure for providing grades that display student competencies in real terms. Learning must be able to lead students to gain knowledge and understanding of an information. Learning improvement is assessed not only on the results obtained, but also on the process that has been passed by students.

2.3. Understanding of Science Concepts

The location of a concept in science learning is part of a product that includes science facts. Understanding science concepts is the ability of students to be able to understand a concept or fact and answer it using their own sentences without changing the meaning of the intended concept. Understanding the concept of science is defined as the process of explaining a fact or science concept in detail, through observation and experiment [10].

2.4. Science Concept Understanding Indicators

Cognitive processes, including: interpreting (interpreting), giving examples (exemplifying), classifying (classifying), summarizing (summarizing), drawing inferences / inferring (inferring), comparing (comparing), and explaining (explaining) [11].

- Interpreting, changing from one form of information to another, for example from words to graphics or pictures, or vice versa, from words to numbers, or vice versa, as well as from words to words, for example summarizing or paraphrasing;
- Exemplifying provides an example of a general concept or principle. Giving examples require the ability to identify the characteristics of a concept and then use these characteristics to create examples;
- Classifying, recognizing that something (object or phenomenon) falls into a certain category;
- Summarizing, making a statement that represents all the information or making an abstract of an article;
- Inferring, finding a pattern from a series of examples or facts;
- Comparing, detecting similarities and differences between two objects, ideas or situations;
- Explaining, constructing and using a causal model in a system.

3. Method

This research method is descriptive qualitative with data collection techniques in the form of literature studies from various relevant references and by using interview techniques and google forms to determine the level of effectiveness and motivation in learning CTL-based Stop Motion Learning Videos. The data collected were analyzed descriptively and qualitatively. The goal is to see the effectiveness of learning from home during the Covid-19 pandemic.

4. Results

Based on the results of this study, it obtained very significant results based on the scientific understanding test on the topic of energy and its changes with the following results:

Table 1. Recapitulation of formative test results.

No	Description	Results
1	Number of Completed Students	22
2	Number of students who did not complete	4
3	The highest score	8.7
4	Lowest value	6
5	Formative test mean	7.73
6	Percentage of learning completeness	92.31%

Source: The results of the formative test of students' speaking skills

The table can be explained that by applying CTL-based stop motion video media to improve learning outcomes on the concept of energy and its changes, the average value of student achievement is 7.73 and learning completeness reaches 92.31% or there are 22 participants from 26 students who have complete in learning to improve science learning outcomes on the concept of energy and its changes. These results indicate that students are quite good in learning even though there are some students who have not finished, because students who get a score of 7.00 are only 92.31% of science learning outcomes on the concept of energy and its changes. In addition to the results of tests conducted online, this research is strengthened by the results of interviews and google form questionnaires, where all students want science learning with online learning media, which can be linked to everyday life, and consider CTL-based stop motion learning videos to be effective for used during online learning.

5. Conclusion

Based on all the discussions and analyzes that have been carried out, it can be concluded that using CTL-Based Stop Motion Learning Videos can improve understanding of science learning on the topic of animal metamorphosis in grade IV SDN Muara Ciujung Timur, Lebak Regency. So the use of CTL-Based Stop Motion Learning Videos is highly recommended in science learning to support online learning to make it more interesting, and students are motivated in science learning in elementary schools.

References

[1] Agustiningsih. (2015). Video Sebagai alternatif MediaPembelajaran dalam Rangka Mendukung Keberhasilan Penerapan Kurikulum 2013 di Sekolah Dasar Madrasah. Jurnal PGSD Universitas Jember, Pancaran, 4 (1), hal 55-68

[2] Guswiani, W., (dkk). Efektivitas Penggunaan Video Pembelajaran Dalam Meningkatkan Motivasi Dan Hasil Belajar Siswa Dalam Pembelajaran Front Office Di Kelas XI Akomodasi Perhotelan SMKN 3 Garut. JTEP-Jurnal Teknologi Pendidikan dan Pembelajaran 3 (2) 2018.

[3] Sumardi.,(dkk). Efektifitas Media Pembelajaran Stop Motion berbasis Inquiry dalam Pembelajaran Sejarah di Era Revolusi Industri 4.0. Jurnal Pendidikan Sejarah Universitas Negeri Padang (4), hal 418-427.

[4] K. Maryanti, "Pengembangan Media Pembelajaran Video Animasi Stop Motion untuk Pembelajaran Biologi dengan Aplikasi Picpac," *E-Jurnal Prograł Stud. Pendidik. Biol.*, vol. 8, no. 1, pp. 26–33, 2017.

[5] Sari P, "Efektivitas Hasil Belajar Matematika Melalui Model CTL Berbasis Masalah Terbuka Siswa Kelas IV SD," *J. Ilm. Sekol. Dasar*, vol. 4, no. 2, pp. 248–256, 2020.

[6] Sun, "Stop-motion to Foster Digital Literacy in Elementary School," *Media Educ. Res. J.*, vol. 51, no. 25, pp. 93–102, 2017.

[7] Ranang, *Animasi Kartun dari Analog sampai Digital*. Jakarta: Indeks, 2010.

[8] Leviudin, *Belajar dan Pembelajaran*. Yogyakarta: CV. Budi Utama, 2017.

[9] Trianto, *Model Pembelajaran Terpadu*. Jakarta: PT. Bumi Aksara, 2014.

[10] A. Susanto, *Teori Belajar & Pembelajaran di SD*. Jakarta: Prenadamedia Grup, 2013.

[11] K. Anderson., *Kerangka Landasan untuk Pembelajaran, Pengajaran, dan Assesmen*. Yogyakarta: Pustaka Belajar, 2010.

Augmented Reality Geometrical Optics (AR-GiOs) for Physics Learning in High Schools

H P Kencana, B H Iswanto and F C Wibowo
Physics Education Master Program, Universitas Negeri Jakarta, Jl. Rawamangun Muka No.1, Jakarta 13220, Indonesia
Email : hapsaripradakencana_1310819001@mhs.unj.ac.id, bhi@unj.ac.id, fcwibowo@unj.ac.id

Abstract. This study aims to produce "Augmented Reality Geometrical Optics (AR-GiOs)" as a medium for physics learning for geometrical optics in high school. The method used in this study is the Research and Development method with the ADDIE model (Analyze, Design, Develop, Implement, and Evaluate). AR-GiOs was developed to facilitate students in learning the concept of image formation and image properties of mirrors and lenses. AR-GiOs can display four 3D simulations, including 3D simulations of the image formation process by a concave mirror, a convex mirror, a concave lens, and a convex lens. In addition to AR, in this study, a worksheet was also developed to guide student learning activities and valuable as a marker. The results of product validation covering the material and media aspects got a very good average score (84% and 90%). Based on the results of the expert validation test, it can be concluded that AR-GiOs is suitable for use as a medium for learning physics in high school.

1. Introduction

Optical image formation is one of the abstract physics materials, so it requires critical thinking skills to understand the theories and compare them with symptoms in everyday life [1]. Most students have misconceptions about analyzing image formation by mirrors and optical devices [2]. Students also have difficulty understanding the concept of image formation, such as real images, virtual images, and magnification [3]. Other researchers also stated that students had difficulty in determining the location of the image, the type of image (real or virtual) [4]. Some of the factors that can lead to student misunderstandings include students' daily experiences, language used, teachers, and textbooks [3].

Mobile technology can be an interesting learning medium that helps students overcome conceptual difficulties and learn more enthusiastically [5, 6]. The initiatives which transform mobile devices into tools for learning, and which support equitable opportunities for students who cannot afford them, generally provide affordable solutions to educational challenges [7]. Augmented Reality (AR) is one of the solutions needed to be present in learning activities. Augmented Reality (AR) can synergize with smartphones so that it becomes an effective, efficient, and useful learning medium in the learning environment. Augmented Reality (AR) improves digital-age literacy, creative thinking, communication, collaboration, and problem-solving ability, which constitute the so-called twenty-first-century skills, necessary to transform information rather than just receive it [8]. Especially in the field of physics, it can provide a more realistic expression through various physics-based simulations that allow students to achieve high achievements [9]. Augmented Reality (AR) has many advantages for learning, including concretizing abstract concepts [10, 11], increasing mastery of concepts [10, 11, 12, 13, 14, 15], and improving critical thinking skills [16].

Previous research has proposed the use of Augmented Reality 3D techniques for physics experiments, in particular, to produce a convex lens image formation, experimental device integrated with Augmented Reality (AR) to stimulate student interest in learning and increase activity levels [17]. However, this study only produced experimental devices for the convex lens. Other research has also developed an Augmented Reality (AR) based magnetic field learning application that can emphasize natural interactions. The weakness of this application is that the system is not always stable and the resulting animations overlap [18]. Other research has also produced MagAR, a magnetism teaching material using augmented reality and sensing technology [19]. However, these two studies did not produce Augmented Reality for geometric optical materials.

Several previous studies have produced Augmented Reality for Physics learning media but have not provided a solution for students' problems, specifically the process of forming images on geometric optical material. So that the researchers aim to produce "Augmented Reality Geometrical Optics (AR-GiOs)" as a medium for learning physics in high school on the subject of geometric optical image formation.

2. Methods

This research used research and development method with the ADDIE model. The ADDIE model consists of five stages, including Analyze, Design, Develop, Implement, and Evaluate [20]. However, this research only reached the third stage of the ADDIE model. The analysis was conducted to determine the needs of learning Physics in schools. The analysis carried out includes literature study, analysis of student needs, and analysis of teacher needs. Design is the creation of storyboards for AR-GiOs media. At the development stage, AR-GiOs application development and student worksheets were carried out. The software used to develop AR-GiOs is Blender, Unity 3D, and Vuforia. Meanwhile, student worksheets were created using Microsoft Word and Paint 3D.

3. Result and Discussion

3.1. AR-GiOs

AR-GiOs facilitates students in learning the concept of image formation and image properties by mirrors and lenses. AR-GiOs aims to visualize abstract concepts in the formation of images completely and well for students. In addition, AR-GiOs's ability to display three-dimensional objects will improve the quality of learning.

AR-GiOs can display four 3D simulations, including 3D simulations of the image formation process by a concave mirror, convex mirror, concave lens, and convex lens. The user can change the object distance and object height values in the 3D simulation so that the focal distance, image position, image height, and magnification values will be obtained. The object used in the 3D simulation is a flashlight with the GO logo. The orange lines in the figure show special rays.

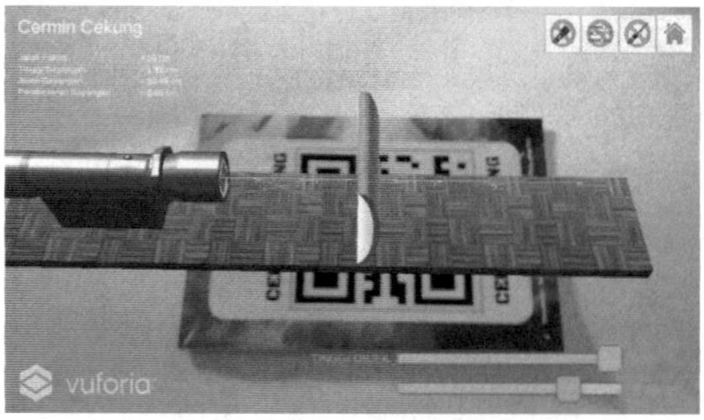

Figure 1. The process of forming an image by a concave mirror.

Figure 2. Image of an object by a concave mirror.

Figure 1 shows a 3D simulation of the image formation process by a concave mirror. Figure 2 shows the results of the image of an object by a concave mirror. In picture 2, objects are 25.42 cm from the mirror, then the image of the object is 16.49 cm. So based on these experiments, the image of

Journal of Physics: Conference Series **2019** (2021) 012004 doi:10.1088/1742-6596/2019/1/012004

the object is reduced, real, and inverted. If the object is placed in front of the focus then the image of the object will be enlarged, virtual, upright. If an object is placed between the focus and the radius of curvature of a concave mirror, the image of the object will be enlarged, real, and inverted [21].

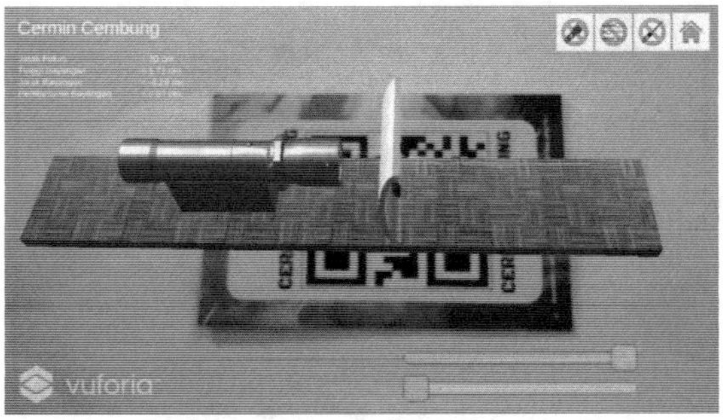

Figure 3. The process of forming an image of a convex mirror.

Figure 4. Image of an object by a convex mirror.

Figure 3 shows a 3D simulation of the image formation process by a convex mirror. Figure 4 shows the results of the image of an object by a concave mirror. In figure 4, objects are 7.5 cm from the mirror, then the image of the object is -4.29 cm. So based on these experiments, the image of the object is virtual, upright, and reduced. When the user changes the position of objects in a 3D simulation of a convex mirror, the image will always be behind the mirror, so that the disposition of the image formed by the convex mirror is always virtual, upright, and reduced [21].

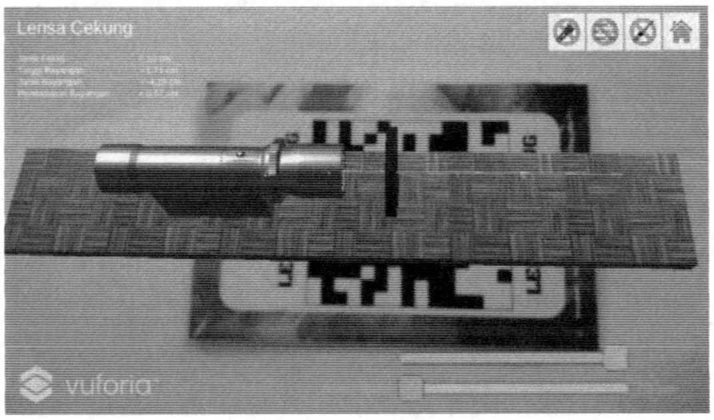

Figure 5. The process of forming an image by a concave lens.

Figure 6. Image of an object by a concave lens.

Figure 5 shows a 3D simulation of the image formation process by a concave lens. Figure 6 shows the image of an object by a concave lens. In figure 6 objects are 7.5 cm from the lens, then the image of the object is -4.29 cm. So that based on these experiments, the properties of the image formed are virtual, upright, and reduced. The disposition of the image formed by a concave lens is always virtual, upright, and reduced [21].

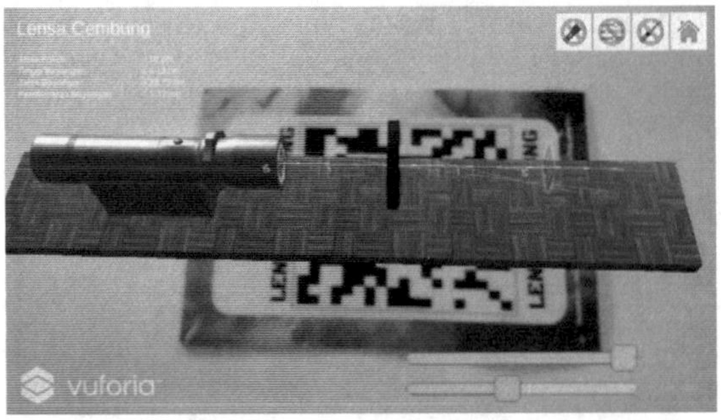

Figure 7. The process of forming an image by a convex lens.

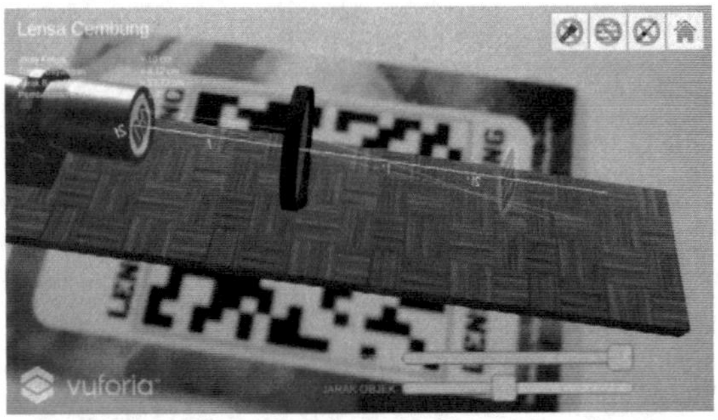

Figure 8. Image of an object by a convex lens.

Figure 7 shows a 3D simulation of the image formation process by a convex lens. Figure 8 shows the results of the image of an object by a convex lens. In figure 8 objects are 17.28 cm from the lens, then the image of the object is 23.73 cm. So based on these experiments, the image of the object is enlarged, real, and inverted. If the object is placed in front of the focus, the image of the object will be enlarged, virtual, and upright. If the object is placed behind the radius of curvature of a convex lens, the image of the object will be enlarged, real, and inverted [21].

3.2. Student Worksheet

In making learning activities complete, student worksheets are developed to guide student learning activities. Student worksheets also serve as markers. Student worksheets are integrated into the AR-GiOs application so that before starting learning students can download and print student worksheets. AR-GiOs student worksheet components, including cover pages, introduction, table of contents, information on "AR-GiOs Worksheet", instructions for use, AR-GiOs Markers, core competencies, basic competencies and indicators, experimental activities, and bibliography.

The student worksheet contains four experimental activities, including the experimental activity of forming an image on a concave mirror, a convex mirror, a concave lens, and a convex lens. Experimental activities contain (1) experimental objectives, (2) basic theory, (3) initial questions, (4) tools and materials, (5) experimental steps, (6) observation tables, (7) data analysis and discussion, and (8) conclusion.

PERCOBAAN PEMBENTUKAN BAYANGAN PADA CERMIN CEKUNG

TUJUAN PERCOBAAN

1. Setelah melakukan percobaan cermin cekung secara virtual menggunakan AR-GO, diharapkan peserta didik mampu menganalisis proses pembentukan bayangan percobaan cermin cekung.
2. Setelah mendapatkan data hasil percobaan cermin cekung secara virtual menggunakan AR-GO, diharapkan peserta didik mampu menganalisis data hasil percobaan cermin cekung yaitu fokus lensa, posisi benda, posisi bayangan, tinggi benda, dan tinggi bayangan.
3. Setelah menganalisis data hasil percobaan cermin cekung secara virtual menggunakan AR-GO, diharapkan peserta didik mampu menganalisis sifat bayangan dari hasil pembentukan bayangan pada cermin cekung.

1 | LEMBAR KERJA PESERTA DIDIK AR-GO

Figure 9. Experiment front page and experiment objectives.

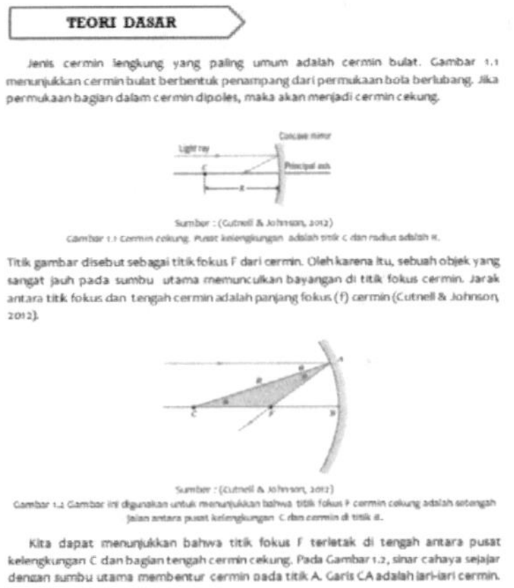

TEORI DASAR

Jenis cermin lengkung yang paling umum adalah cermin bulat. Gambar 1.1 menunjukkan cermin bulat berbentuk penampang dari permukaan bola berlubang. Jika permukaan bagian dalam cermin dipoles, maka akan menjadi cermin cekung.

Sumber : (Cutnell & Johnson, 2012)

Titik gambar disebut sebagai titik fokus F dari cermin. Oleh karena itu, sebuah objek yang sangat jauh pada sumbu utama memunculkan bayangan di titik fokus cermin. Jarak antara titik fokus dan tengah cermin adalah panjang fokus (f) cermin (Cutnell & Johnson, 2012).

Sumber : (Cutnell & Johnson, 2012)

Kita dapat menunjukkan bahwa titik fokus F terletak di tengah antara pusat kelengkungan C dan bagian tengah cermin cekung. Pada Gambar 1.2, sinar cahaya sejajar dengan sumbu utama membentur cermin pada titik A. Garis CA adalah jari-jari cermin.

Figure 10. Basic theory.

Figure 9 shows the front page of the experiment and the objectives of the experiment. The purpose of the experiment informs the knowledge and skills that students will acquire after learning. Students can identify optical concepts from the basic theory in Figure 10. Before starting the experiment, students answer initial questions to determine initial abilities. Virtual tools and materials used in virtual experiments include concave mirrors, convex mirrors, concave lenses, convex lenses, flashlights with the GO logo, and experimental tables.

LANGKAH PERCOBAAN

1. Sebelum memulai percobaan, diharapkan peserta didik membaca panduan penggunaan terlebih dahulu agar tidak kesulitan dalam melakukan percobaan.
2. Jalankan aplikasi AR-GO maka akan muncul pemindai kamera. Pindai gambar berlogo AR dibawah ini. Pindai dengan baik maka simulasi cermin cekung akan muncul.

AR-GO
CERMIN CEKUNG
AR
CERMIN CEKUNG

3. Simulasi cermin cekung akan muncul seperti gambar dibawah ini.

Figure 11. Experiment step.

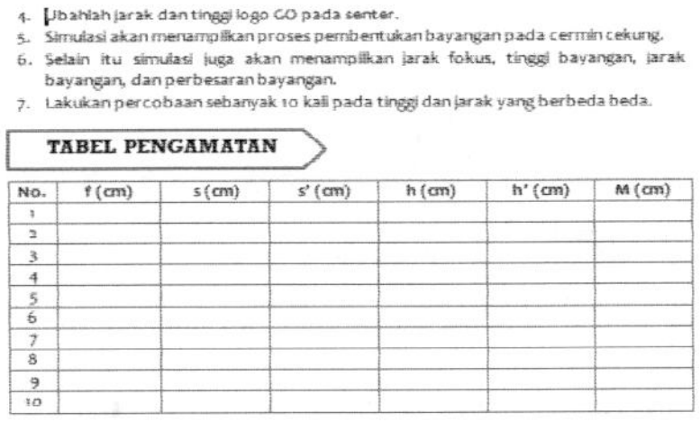

4. Ubahlah jarak dan tinggi logo GO pada senter.
5. Simulasi akan menampilkan proses pembentukan bayangan pada cermin cekung.
6. Selain itu simulasi juga akan menampilkan jarak fokus, tinggi bayangan, jarak bayangan, dan perbesaran bayangan.
7. Lakukan percobaan sebanyak 10 kali pada tinggi dan jarak yang berbeda beda.

TABEL PENGAMATAN

No.	f (cm)	s (cm)	s' (cm)	h (cm)	h' (cm)	M (cm)
1						
2						
3						
4						
5						
6						
7						
8						
9						
10						

Figure 12. Observation table.

Students conduct experiments following the experimental steps in Figure 11. AR-GiOs uses 4 markers in figure 11, including convex mirror marker, concave mirror marker, convex lens marker, and concave lens marker. When the user selects the start, learning menu, then scans one of the markers. Then a 3D simulation will appear on the cell phone screen. The AR-GiOs application will be able to display 3D animation. Students write down experimental data in the observation table that has been prepared in Figure 12. The data obtained include focus, object distance, object image distance, object height, image height, and magnification. After the experimental data is obtained, students are expected to conduct data analysis and discussion. The last stage of the experiment is the students write the conclusion of the experiment.

3.3. Validation Test Results

AR-GiOs has been tested for validity by media experts and material experts. The following are the results of AR validation and student worksheets. As shown in Table 1, the results of the validation of media experts in this research product obtained an average value of 90%, which was categorized as very well. The results of the validation of material experts get an average score of 84%, which was categorized as very well. So it can be concluded that the AR-GiOs meets the requirements and is feasible as a medium for Physics learning on geometric optics for high school students.

Table 1. Validation results

Aspects Measured	Presentation Scale	Interpretation
Media Expert Validation		
Effectiveness of AR	87%	Very Good
Efficiency of AR	92%	Very Good
Satisfaction of AR	91%	Very Good
Cover Layout of Worksheet	87%	Very Good
Component of the student worksheet	90%	Very Good
Content of the student worksheet	94%	Very Good
Average of all aspects	90%	Very Good
Material Expert Validation		
3D Simulation Concept	90%	Very Good
Material Content of Worksheet	83%	Very Good
Writing Language of Worksheet	80%	Good
Average of all aspects	84%	Very Good

3.4. Discussion

Based on the results of the validation that has been done by media experts and material experts, it was found that "Augmented Reality Geometrical Optics (AR-GiOs)" has met the requirements and is feasible as a learning medium for the physics of geometric optical material on the concept of forming an image for class XI SMA. AR-GiOs solves the problems experienced by students, including: (1) the process of forming images, (2) the nature of the image from the image formed, and (3) the position of the image from the image formation on mirrors and lenses. AR-GiOs can visualize and concretize abstract concepts of image formation processes on mirrors and lenses, this is following the results of previous studies [9], [10], [11]. In addition, the ability of AR-GiOs that can display three-dimensional objects can improve the quality of learning, this is following the results of previous studies [8], [9].

AR-GiOs has overcome the limitations of previous studies, only produced experimental devices with convex lenses [17]. In addition, the experimental device developed has not discussed the formation of virtual and real images. AR-GiOs succeeded in visualizing the image formation process by a concave mirror, concave mirror, concave lens, and convex lens. In addition to this, users can vary the distance of objects and the height of objects in the 3D simulation. So users will get data, including focus distance, shadow position, image height, and magnification. AR-GiOs has succeeded in providing a satisfying experience and interaction satisfaction to students.

4. Conclusion

AR-GiOs has succeeded in simulating the image formation process in a concave mirror, convex mirror, concave lens, and convex lens. AR-GiOs can concretize abstract concepts in the process of forming real and virtual images. By using 3-dimensional visualization, AR-GiOs can provide a more realistic expression of the abstract shadow formation process. AR-GiOs can facilitate students in learning the concept of image formation and image properties of mirrors and lenses. Augmented Reality Geometrical Optics (AR-GiOs) has been worthy as a medium for learning physics on geometric optical material.

References

[1] Nurhayati N and Angraeni L 2017 Analisis Kemampuan Berpikir Tingkat Tinggi Mahasiswa (Higher Order Thinking) dalam Menyelesaikan Soal Konsep Optika melalui Model Problem Based Learning *Jurnal Penelitian & Pengembangan Pendidikan Fisika* 3(2) 119-126

[2] Kaniawati I, Rahmadani S, Fratiwi N J, Suyana I, Danawan A, Samsudin A, and Suhendi E 2020 An Analysis of Students' Misconceptions About the Implementation of Active Learning of Optics and Photonics Approach Assisted by Computer Simulation *International Journal of Emerging Technologies in Learning (iJET)* 15(9) 76-93

[3] Widiyatmoko A and Shimizu K 2018 Literature Review of Factors Contributing to Students'

Misconceptions in Light and Optical Instruments *International Journal Of Environmental & Science Education* 13(10) 853-863

[4] Gurel, Eryılmaz, and McDermott 2015 A review and comparison of diagnostic instruments to identify students' misconceptions in science *EURASIA J. Math. Sci. Technol. Educ.* 11(5) 989-1008

[5] Vieyra R E and Vieyra C 2014 Analyzing forces on amusement park rides with mobile devices *The Physics Teacher* 52(3) 149-51

[6] González M, Martín M, Llamas C, Martínez Ó, Vegas J, Herguedas M, and Hernández C 2015 Teaching and Learning Physics with Smartphones *J. Cases Inf. Technol* 17(1) 31-50

[7] UNESCO 2013a *Policy Guidelines for Mobile Learning* (Paris: UNESCO)

[8] Papanastasiou G, Drigas A, Skianis C, Lytras M, and Papanastasiou E 2018 Virtual and augmented reality effects on K-12 higher and tertiary education students' twenty-first century skills *Springer-Verlag London Ltd., part of Springer Nature 2018*

[9] Sung N J S, Ma J, Choi Y J, and Hong M 2019 Real-Time Augmented Reality Physics Simulator for Education *Appl. Sci. 2019* 9(19) 4019

[10] Radu I and Schneider B 2019 What Can We Learn from Augmented Reality (AR)? Benefits and Drawbacks of AR for Inquiry-based Learning of Physics *Computing Systems Proceedings (CHI 2019)* 544

[11] Ivanova M and Ivanov G 2011 Enhancement of learning and teaching in computer graphics through marker augmented reality technology *International Journal of New Computer Architectures and Their Applications (IJNCAA)* 1(1) 176-184

[12] Aprinaldi A, Rahmawati Y, and Komaro M 2019 Implementation of Augmented Reality (AR) android based in learning *Journal of Physics: Conference Series* 1402

[13] Fidan M and Tuncel M 2019 Integrating augmented reality into problem based learning: The effects on learning achievement and attitude in physics education *Computers & Education* 142

[14] Ismail A, Festiana I, Hartini T I, Yusal Y, and Malik A 2019 Enhancing students' conceptual understanding of electricity using learning media-based augmented reality *Journal of Physics: Conf. Series* 1157

[15] Yoon S, Anderson E, Lin J, and Elinich K 2017 How augmented reality enables conceptual understanding of challenging science content *Educational Technology & Society* 20 (1) 156-168

[16] Anggraini S, Setyaningrum W, Retnawati H, and Marsigit 2019 How to improve critical thinking skills and spatial reasoning with augmented reality in mathematics learning *Journal of Physics: Conference series* 1581

[17] Cai S, Chiang F K, and Wang X 2013 Using the Augmented Reality 3D Technique for a Convex Imaging Experiment in a Physics Course *International Journal of Engineering Education* 29(4) 856–865

[18] Cai S, Chiang F K, Sun Y, Lin C, and Lee J L 2017 Applications of augmented reality-based natural interactive learning in magnetic field instruction *Interactive Learning Environments* 25(6) 778–791

[19] Abdusselam M S and Karal H 2020 The effect of using augmented reality and sensing technology to teach magnetism in high school physics *Technology, Pedagogy, and Education.*

[20] Branch R M 2010 *Instructional Design: The ADDIE Approach* (New York: Springer)

[21] Cutnell J D and Johnson K W 2012 *Physics 9 Edition* (United States of America: John Wiley & Sons, Inc)

Journal of Physics: Conference Series **2019** (2021) 012005 doi:10.1088/1742-6596/2019/1/012005

Four Tier Test (FTT) Development in The Form of Virtualization Static Fluid Test (VSFT) using Rasch Model Analysis to Support Learning During the Covid-19 Pandemic

N Anggraini, B H Iswanto and F C Wibowo
Physics Education Master Program, Universitas Negeri Jakarta
Jl. Rawamangun Muka, Jakarta 13220, Indonesia
Email: neri.anggraini@gmail.com, bhi@unj.ac.id, fcwibowo@unj.ac.id

Abstract. This study aims to develop a valid and reliable test instrument. The test instrument developed is a Four Tier Test (FTT) in a Virtual Static Fluid Test (Test) on Static Fluid material using the Rasch model analysis. The method used in this research is the method of research and development with the 4D model. This research produces an educational product in the form of an FTT in the form of VSFT. The items developed in this study amounted to 20 items. The validation results from several experts gave an average score of 88.13% with very good criteria. The CVR value for each item is 1, while the CVI value is 1 with very good criteria. The validation results from several educational practitioners gave an average score of 89.58% with very good criteria. The results of the Rasch Model analysis showed that the test instrument is accepted, the reliability of all the items has good criteria, all the items are valid and divided into three difficulties, also each item is not biased towards gender differences. The results of the analysis of student misconception showed that only 4.68% of the students understood the whole concept, 23.42% of students only understood partially, 58.95% of students have a misconception, and 12.95% of students did not understand the concept at all. The conclusion of this study is that the FTT instrument in the form of VSFT is valid and reliable as an assessment instrument during the Covid-19 period, and also able to analyze student conceptions.

1. Introduction

The new coronavirus pneumonia triggered by Covid-19 is now spreading around the world [1]. Covid-19 is a global health emergency, and the World Health Organization (WHO) declared it a pandemic. The pandemic has led education providers to decide on expanded education virtualization to remotely conduct online exams and master or doctoral seminars [2]. Online learning and virtual teaching are emerging as new avenues of science and engineering education and have accelerated the spread of educational resources globally, especially during pandemics [3].

During the online learning period, a test tool is needed to help students understand the concepts asked in the questions so that it makes it easier for students to answer questions. One of the test tools that support this convenience is a type of virtual test (virtualization test). The use of the Virtualization test allows question makers to use images, graphics, animations, and videos that can visualize test subject statements that contain abstract concepts and are difficult to explain in words to make it easier to understand questions [4]. The use of virtual tests gives a positive response to teachers and students in terms of ease of use of virtual tests, multimedia, benefits and time of testing [5,6].

Static fluid is a fundamental physics material that examines the physical concept with all its reactions to a given force. However, understanding the concept of static fluids is often seen as complex by students because it requires in-depth reasoning. One of the learning difficulties that is often found is a misconception.

Misconceptions can be a barrier to organizing knowledge, so it needs to be overcome [7]. Therefore, it is crucial to develop an evaluation tool that can detect misconceptions experienced by students [8].

The four-tier test format instrument is one of the instruments used to diagnose students' conception of a physics concept [9]. The four tiers diagnostic test is a four-level multiple-choice test composed of questions with four answer choices, confidence in choosing an answer, the reason for choosing the answer, and the level of confidence in the reason. [10]. The first level is a regular multiple-choice test with a cheat that addresses specific misconceptions. The second level of the test asks for confidence in the first level. The third level of the test asks the reasons for the answer at the first level. The fourth level of the test asks for the confidence of the answers at the third level. However, until now, many test instrument developments have been carried out using classical measurement model analysis.

The Rasch model has been used to analyze various assessment instruments to provide detailed information on several aspects of the quality of the test. The Rasch model is used to analyze instrument items and analyze students [11]. The Rasch model can be used to determine the reliability index to the item analysis for each level, the student's reliability, and dimensionality and detect the bias of the items on the instrument. [12]. Thus, the Rasch model analysis provides several options and ways in which a test instrument is developed.

Using the Rasch model has many advantages. The Rasch Model analysis supports the development and validation of invariance measures in providing empirical evidence and insights for critical psychometric properties of tests and questionnaires. The Rasch model can explain the differences in the difficulty of different levels [13]. The Rasch model can also determine a student's ability, where the likelihood of a student answering an item correctly increases monotonically with the student's abilities. [14].

So that in the context of assessing the Covid-19 pandemic, this study aims to develop a valid and reliable test instrument. The test instrument developed is a Four Tier Test (FTT) in a Virtual Static Fluid Test (Test) on Static Fluid material using the Rasch model analysis.

2. Method

The method used in this research is the method of research and development with the 4D model. 4D development consists of four stages, namely: Defining, Designing, Developing, and Disseminating).

2.1. Defining

The purpose of this stage is to define and define the terms of the virtual test instrument in the four-tier test format, which begins with an objective analysis of the constraints of the material the instrument will develop. This stage includes four main steps: needs analysis, student analysis, concept analysis, and learning objectives.

2.2. Designing

The purpose of this stage is to prepare a Four-Tier Test (FTT) instrument, which will be outlined in the form of a Virtualization Static Fluid Test (VSFT). It consists of three steps, namely designing a VSFT grid in FTT format, compiling VSFT questions in FTT format and compiling a validation instrument.

2.3. Developing

The purpose of this stage is to produce revised learning instruments based on suggestions from experts. This stage includes validation of the tools by experts and revisions.

2.4. Disseminating

The purpose of this stage is to produce tested learning instruments. This stage includes revision and broad implementation to students, data analysis and making conclusions. The data analysis obtained will be analyzed with the Rasch Model measurement model.

3. Result and Discussion

3.1. Four-Tier Test (FTT) in the form of Virtualization Static Fluid Test (VSFT)

This research produces an educational product in the form of a Four Tier Test (FTT) in a Virtualization Static Fluid Test (VSFT). Initially, this educational product had a homepage that contained student identities.

After students fill in their data entirely and start the test, the system will move from the Home page to the next page, which contains many items that students will answer. The items developed in this study amounted to 20 items. Each item is equipped with a stimulus to help students work on it. Each item requires students to give four answers in sequence (four-tier test).

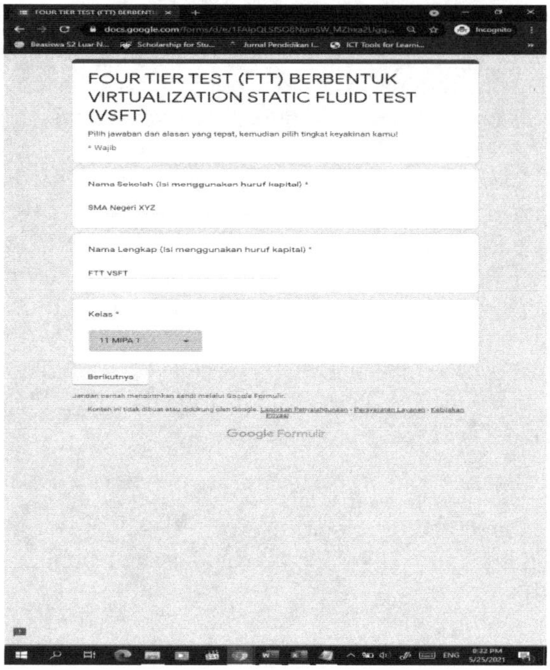

Figure 1. Homepage.

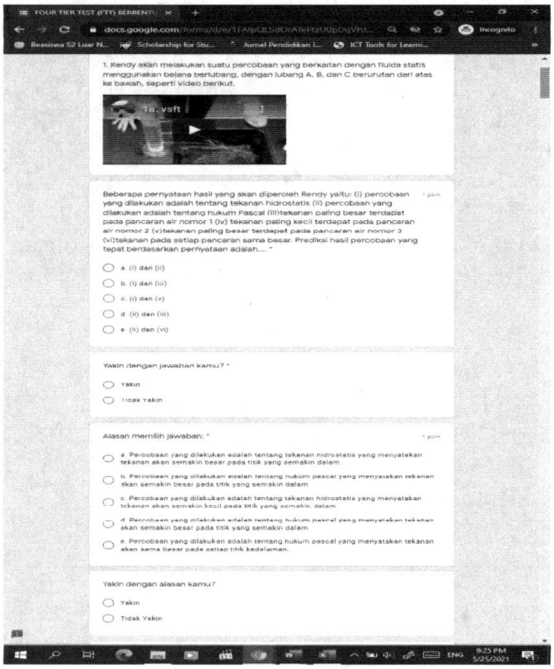

Figure 2. Four Tier Test.

3.2. Expert Validation

The validation results by several experts gave an average value of 88.13% with very good criteria. This informs that the FTT in the form of VSFT has been developed with very good quality and can be used for the test implementation. The value is divided into three aspects: the Physics material aspect of 86.67% with very good criteria, the media aspect of 88.00% with very good criteria, and the assessment aspect of 90.00% with good criteria.

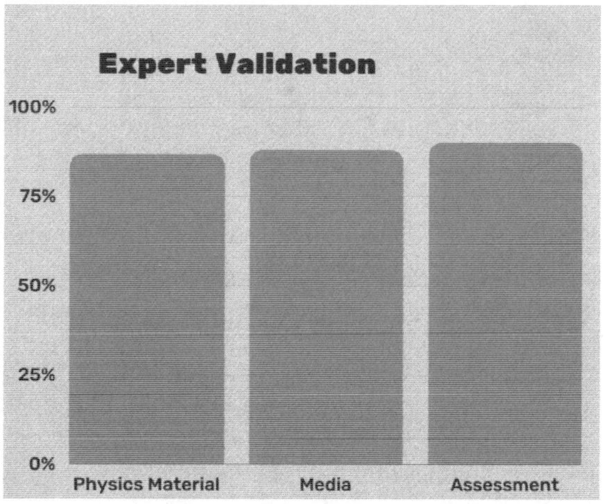

Figure 3. Expert Validation.

3.3. Content Validity Ratio (CVR) and Content Validity Index (CVI)

Each item that has been developed is then validated to determine the validation of the content of each item. After that, the validation results are used to calculate the CVR value for each item. The CVR value for each item is 1. This shows that all items are valid and fit for use. This is in line with previous research regarding developing an instrument that states that content validity (through CVR calculations) is very important and crucial. [15,16].

The CVI value obtained has a value of 1 with very good criteria. This explains that all items that have been developed when viewed as a whole as a test instrument are declared valid and usable. Validity content ensures that items and test instruments that have been designed and developed can measure adequately [17].

3.4. Educational Practitioner Validation

The validation results by several educational practitioners gave an average score of 89.58% with very good criteria. This informs that the FTT in the form of VSFT has been developed with very good quality and can be used for the test implementation. The value is divided into three aspects: the Physics material aspects of 88.39% with very good criteria, the design aspect of 89.84% with very good criteria, and the aspects of function and purpose of 91.41% criteria.

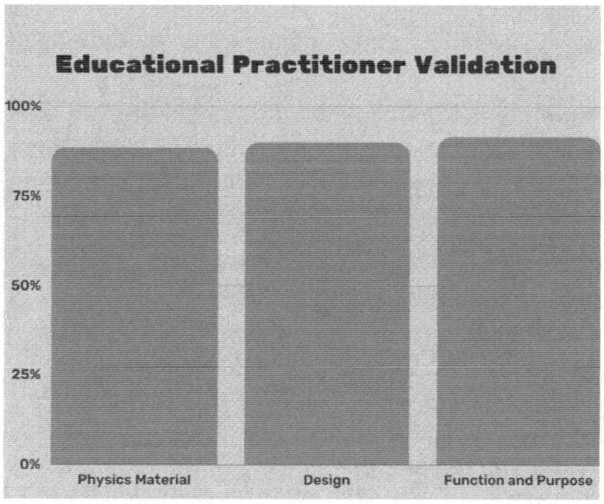

Figure 4. Educational Practitioner Validation.

3.5. Rasch Model Analysis

The items that have been developed and validated theoretically will then be tested on many students to be analyzed with the Rasch Model. This is supported by a statement in a previous study that the Rasch Model analysis provides a solution to the limitations experienced using the classical measurement model [18] so fit items can be gotten [19]. The results of the Rasch Model analysis for each item that has been developed are as follows.

The results of the first Rasch Model analysis showed a unidimensionality value of 22.4%. This explains that the test instrument is accepted and can measure what should be measured. Unidimensionality is the latent property of a measuring instrument [20,21] in verifying the measurement construction of the instrument [22].

The results of the second Rasch Model analysis show a reliability value of 0.92. This explains that the reliability between all the items that have been developed and all the students taking the test has a special criterion. The results of the third Rasch Model analysis show a reliability value of 0.90. This explains that the reliability of all the items that have been developed has good criteria. This is in line with several previous studies which stated that reliability provides more detailed information between the property of the items on the instrument and the individuals who respond to these items [23]. Reliability also informs about consistency, both consistency of items and consistency of person [24].

The fourth Rasch Model analysis results show that each item that has been developed meets at least one of the three item-fit criteria. This explains that all the items that have been developed are valid. This is supported by several studies that state that validity informs the validity of the test content and informs the consequences of using the test score [25]. Using the Rasch model, the analysis results can determine the suitability of the items to the identified constructs [26] against the test instruments in general [27].

The results of the fifth Rasch Model analysis show that all the items that have been developed are divided into three difficulty levels as follows:

Table 1. Difficulty Level.

Difficulty Level	Number of Question
Hard	13, 2, 19
Medium	10, 3, 9, 7, 17, 8, 4, 14, 1, 12, 20, 5
Easy	11, 6, 15, 18, 16

This is in line with other studies which state that the varying levels of difficulty of the questions is another criterion of a good measuring instrument [28,29]. The level of difficulty of the items does not have to be sequential in a test instrument [30].

This study also analyzed the possibility of bias against the existence of gender differences (male and female). The sixth Rasch Model analysis results showed that the DIF value of each item was less than 0.05. This explains that each item is not biased towards gender differences (male and female). Previous studies have shown that the DIF value checks the possibility of item bias caused by responses from different groups [31–33]. Biased questions arise because there are two different groups with the same ability, but there are significant differences when answering the questions [34].

The interpretation using the Rasch measurement model provides more detailed information about the quality of items and persons and their appropriateness in measuring what should be measured [35,36]. Apart from analyzing the quality of the items that have been developed, this study also analyzes the quality of the students taking the test. This is following several previous studies that stated that the Rasch Model analysis was able to explain the quality of the items developed and the quality of the students taking the test [37]. The Rasch model provides information about the consistency of test-takers in answering questions so that there is a possibility that the test taker may be careless in answering, guessing, or cheating [38]. This is, of course, very helpful for teachers in providing evaluations by the abilities of students who take the test [39,40]. This also makes it easy for teachers to check/evaluate and determine the appropriate feedback for each student [41] so that the evaluation of student test-takers by the teacher becomes easier and more precise [42].

The results of the first Rasch Model analysis show a reliability value of 0.86. This explains that the reliability of all students taking the test has good criteria.

The second Rasch Model analysis results showed that the test takers' ability was divided into three levels, namely high, medium, and low abilities. By changing the scores obtained by the students taking the test, the test takers' ability can be determined [43]. There were 52 students who took the test with high ability, 246 students with moderate ability, and 65 students with low ability. This agrees with previous research, which states that a person measure indicates the ability of a person [44].

The third Rasch Model analysis results show that each student who takes the test meets at least one of the three person-fit criteria. This explains that each student who takes the test has an appropriate response pattern. The results of this analysis are important to obtain because if there is an inappropriate response pattern, the Rasch Model analysis will predict the possibility of students taking the test carelessly working on the questions, being lucky in answering, or the possibility of cheating [45].

3.6. Analysis of Student Misconceptions

The analysis results showed that out of 363 students, only 4.68% of the students understood the whole concept of Static Fluid, and 23.42% of students only understood part of the concept of Static Fluid. Meanwhile, 58.95% of students experienced a misconception of Static Fluid material, and 12.95% of students did not understand the concept of Static Fluid at all. This is in line with the statement in previous research that many students still experience misconceptions about physics material [46]. Wherewith the four-tier test, students' misconceptions can be easily identified [47–50].

4. Conclusion

Based on the results of the research that has been obtained, it can be concluded that the Four Tier Test (FTT) instrument in the form of Virtualization Static Fluid Test (VSFT) is valid as an assessment instrument for learning support during the Covid-19 period. In addition, the Four Tier Test (FTT) instrument in the form of a Virtualization Static Fluid Test (VSFT), which has been developed, is also reliable as a learning support assessment instrument during the Covid-19 period. Then, the Four Tier Test (FTT) instrument in the form of a Virtualization Static Fluid Test (VSFT) that has been developed is also able to analyze student conceptions, making it easier for teachers to evaluate and respond appropriately to each student.

References

[1] Feng G, Liu L, Cui W and Wang F 2020 Electron beam irradiation on novel coronavirus (COVID-19): A Monte-Carlo simulation *Chinese Physics B* **29**

[2] Ray S and Srivastava S 2020 Virtualization of science education: a lesson from the COVID-19 pandemic *Journal of Proteins and Proteomics* **11**

[3] Hunter P 2015 The virtual university: Digital tools for e-learning and remote learning are becoming an increasingly important tool for teaching at universities *EMBO Reports* **16** 146–8

[4] Nahadi N, Firman H and Kurniadi H 2018 Development and Validation of Chemistry Virtual Test Based Multiple Representations *Journal of Education and Learning (EduLearn)* **12**

[5] Stavesand J E, Reglitz S and Himmler A 2017 Optimizing the Benefit of Virtual Testing with a Process-Oriented Approach *SAE International Journal of Aerospace* **10**

[6] Mahfira C, Sanjaya Y and Rusyati L 2018 Students' Impression towards Science Virtual Test (SVT) on Digestive System Topic *Journal of Physics: Conference Series* vol 1013

[7] Istighfarin L 2015 Profil Miskonsepsi Siswa Pada Materi Struktur Dan Fungsi Jaringan Tumbuhan *BioEdu* **4**

[8] Fariyani Q, Rusilowati A and Sugianto 2015 Pengembangan Four-Tier Diagnostic Test Untuk Mengungkap Miskonsepsi Fisika Siswa SMA Kelas X *Journal of Innovative Science Education* **4** 41–9

[9] Rawh P, Samsudin A and Nugraha M G 2020 Pengembangan Four-Tier Diagnostic Test untuk Mengidentifikasi Profil Konsepsi Siswa pada Materi Alat-Alat Optik *WaPFi (Wahana Pendidikan Fisika)* **5**

[10] Islami D, Munawaroh F and Hadi W P 2018 Analisis miskonsepsi siswa pada mata pelajaran ipa materi listrik statis menggunakan four tier test *Science Education National Conference*
[11] Zamri A and Nordin 2015 Modeling a Multiple Choice Mathematics Test with the Rasch Model *Indian Journal of Science and Technology* **8** 1–6
[12] Suryani Y E 2018 Aplikasi Rasch Model dalam Mengevaluasi Intelligenz Structure Test *Psikohumaniora: Jurnal Penelitian Psikologi* **3** 73–100
[13] Fulmer G W, Chu H, Treagust D F and Neumann K 2016 Is it harder to know or to reason? Analyzing two-tier science assessment items using the Rasch measurement model *Asia-Pacific Science Education* **1** 1–16
[14] Xiao Y, Han J, Koenig K, Xiong J and Bao L 2018 Multilevel Rasch modeling of two-tier multiple choice test: A case study using Lawson ' s classroom test of scientific reasoning *Physical Review Physics Education Research* **14** 20104
[15] Alizadeh-Siuki H, Tehrani H, Gholian-Aval M, Ebrahimipour H, Jafari A and Vahedian-Shahroodi M 2020 Psychometric properties of a questionnaire on brucellosis prevention behaviors based on the precede model among rural farmers and their family members *Risk Management and Healthcare Policy* **13**
[16] Waltz C F, Strickland O L and Lenz E R 2016 *Measurement in Nursing and Health Research*
[17] Al-Taweel D and Awad A 2020 Development and validation of medication assessment tools to evaluate prescribing adherence to evidence-based guidelines for secondary prevention of coronary heart disease in post-acute coronary syndromes patients in Kuwait *PLoS ONE* **15**
[18] Nurhudaya, Taufik A, Yudha E S and Suryana D 2019 The Raven's advanced progressive matrices in education assessment with a Rasch analysis *Universal Journal of Educational Research* **7**
[19] Azizah A and Wahyuningsih S 2020 Penggunaan Model Rasch Untuk Analisis Instrumen Tes Pada Mata Kuliah Matematika Aktuaria *JUPITEK: Jurnal Pendidikan Matematika* **3**
[20] Al Zoubi F, Mayo N, Rochette A and Thomas A 2018 Applying modern measurement approaches to constructs relevant to evidence-based practice among Canadian physical and occupational therapists *Implementation Science* **13**
[21] Nielsen T and Dammeyer J 2019 Measuring higher education students' perceived stress: An IRT-based construct validity study of the PSS-10 *Studies in Educational Evaluation* **63**
[22] Chang J, Li Y, Song H, Yong L, Luo L, Zhang Z and Song N 2020 Assessment of Validity of Children's Movement Skill Quotient (CMSQ) Based on the Physical Education Classroom Environment *BioMed Research International* **2020**
[23] Bodzin A, Hammond T, Fu Q and Farina W 2020 Development of Instruments to Assess Students' Spatial Learning Attitudes (SLA) and Interest in Science, Technology and Geospatial Technology (STEM-GEO) *International Journal of Educational Methodology* **6** 67–81
[24] Koçak D 2020 Investigation of rater tendencies and reliability in different assessment methods with many facet rasch model *International Electronic Journal of Elementary Education* **12**
[25] Mokshein S E, Ishak H and Ahmad H 2019 The use of rasch measurement model in English testing *Cakrawala Pendidikan* **38**
[26] Md Yunos J, Ibrahim Mukhtar M, Alias M, Lee M F, Tee T K, Rubani S N K, Hamid H, Yunus F A N, Sulaiman J and Sumarwati S 2017 Validity of vocational pedagogy constructs using the rasch measurement model *Journal of Technical Education and Training* **9**
[27] Yasin S N T M, Yunus M F M and Ismail I 2018 The use of rasch measurement model for the validity and reliability *Journal of Counseling and Educational Technology* **1**
[28] Barus G, Sinaga J D and Moron Y D 2019 Testing the quality of film-based creative and innovative character education test with rasch model *Jurnal Konseling dan Pendidikan* **7**
[29] Hamdu G, Fuadi F N, Yulianto A and Akhirani Y S 2020 Items Quality Analysis Using Rasch Model To Measure Elementary School Students' Critical Thinking Skill On Stem Learning *JPI (Jurnal Pendidikan Indonesia)* **9**
[30] Isnani I, Utami W B, Susongko P and Lestiani H T 2019 Estimation of college students' ability on real analysis course using Rasch model *Research and Evaluation in Education* **5**
[31] Xu R H, Wong E L yi, Lu S Y jun, Zhou L M, Chang J H and Wang D 2020 Validation of the

Toronto Empathy Questionnaire (TEQ) Among Medical Students in China: Analyses Using Three Psychometric Methods *Frontiers in Psychology* **11**
[32] Veas A, Castejón J L, Gilar R and Miñano P 2016 Validation of the Spanish adaptation of the School Atitude Assessment Survey-Revised using multidimensional Rasch analysis *Anales de Psicología* **33**
[33] Rahmani B D 2018 Differential Item Functional Analysis on Pedagogic and Content Knowledge (PCK) Questionnaire for Indonesian Teachers using RASCH Model *Journal of Physics: Conference Series*
[34] Alwi I 2017 Sensitivity Of Mantel Haenszel Model And Rasch Model As Viewed From Sample Size *Jurnal Evaluasi Pendidikan* **2**
[35] Bichi A A, Talib R, Atan N A, Ibrahim H and Yusof S M 2019 Validation of a developed university placement test using classical test theory and Rasch measurement approach *International Journal of Advanced and Applied Sciences* **6** 22–9
[36] Tornabene R E, Lavington E and Nehm R H 2018 Testing validity inferences for Genetic Drift Inventory scores using Rasch modeling and item order analyses *Evolution: Education and Outreach* **11**
[37] Maulana S, Supriyati Y and Astra I M 2020 Development of Physics Problems for The Final Assessment in High School Levell with Rasch Model Analysis *Unnes Science Education Journal* **9** 22–9
[38] Susongko P 2016 Validation of Science Achievement Test with the Rasch Model *Jurnal Pendidikan IPA Indonesia* **5** 268–77
[39] Suranata K, Rangka I B, Ifdil I, Ardi A, Susiani K, Prasetyaningtyas W E, Daharnis D, Alizamar A, Erlinda L and Rahim R 2018 Diagnosis of students zone proximal development on math design instruction : A Rasch analysis *Journal of Physics: Conference Series*
[40] Mursidi A and Soeharto 2016 An Introduction: Evaluation of Quality Assurance for Higher Educational Institutions using Rasch Model *Journal of Education, Teaching and Learning* **1** 1–6
[41] Börkan B 2017 Exploring Variability Sources in Student Evaluation of Teaching via Many-Facet Rasch Model *Eğitimde ve Psikolojide Ölçme ve Değerlendirme Dergisi*
[42] Erman Aslanoglu A, Karakaya I and Sata M 2020 Evaluation of university students' rating behaviors in self and peer rating process via many facet rasch model *Eurasian Journal of Educational Research* **2020**
[43] Planinic M, Boone W J, Susac A and Ivanjek L 2019 Rasch analysis in physics education research: Why measurement matters *Physical Review Physics Education Research* **15**
[44] Anselmi P, Colledani D and Robusto E 2019 A Comparison of Classical and Modern Measures of Internal Consistency *Frontiers in Psychology* **10**
[45] Boone W J 2016 Rasch Analysis for Instrument Development: Why, When, and How ? *CBE Life Sciences Education* **15** 1–7
[46] Abbas M L H 2019 Identifikasi Miskonsepsi Mahasiswa Tadris Fisika Menggunakan Four Tier Diagnostic Test pada Mata Kuliah Kalkulus II *Jurnal Matematika dan Pendidikan Matematika* **4**
[47] Fenditasari K, Jumadi, Istiyono E and Hendra 2020 Identification of misconceptions on heat and temperature among physics education students using four-tier diagnostic test *Journal of Physics: Conference Series* vol 1470
[48] Tumanggor A M R, Supahar, Kuswanto H and Ringo E S 2020 Using four-tier diagnostic test instruments to detect physics teacher candidates' misconceptions: Case of mechanical wave concepts *Journal of Physics: Conference Series* vol 1440
[49] Yuberti Y, Suryani Y and Kurniawati I 2020 Four-Tier Diagnostic Test with Certainty of Response Index to Identify Misconception in Physics *Indonesian Journal of Science and Mathematics Education* **3**
[50] Sholihat F N, Samsudin A and Nugraha M G 2017 Identifikasi Miskonsepsi dan Penyebab Miskonsepsi Siswa Menggunakan Four-Tier Diagnostic Test Pada Sub-Materi Fluida Dinamik: Azas Kontinuitas *Jurnal Penelitian & Pengembangan Pendidikan Fisika* **3**

Video Based Experiment to Determine Focal Length of a Positive Lens in Physics Learning

Raida, B H Iswanto and I Sugihartono
Department of Physics Education, State University of Jakarta, Jakarta, Indonesia
Jl. Rawamangun Muka, Jakarta 13220, Indonesia
Email: raida_1310819011@mhs.unj.ac.id , bhi@unj.ac.id

Abstract. During the COVID-19 pandemic, conducting experiments in the lab has become constrained due to physical distancing protocols. Experiments activities are important to help students understand the concepts of physics, one of which is the optical learning of image formation on a positive lens. Video based experiment is one of the solutions that can be used as an alternative to do experiments activities in the laboratory due to the COVID-19 pandemic. This study aimed to design and develop a video based experiment to make students to determine focal length of a positive lens. This study used the R&D method with ADDIE development model reference (Analysis, Design, Development, Implementation, and Evaluation). According to the evaluation of material experts and media experts video based experiment covering aspects of content, material, language and presentation of media experts by 81% and material experts obtained by 84%, categorized as very high. Based on the results of the expert test, it can be concluded that the video based experiment is feasible to be used as an optical learning media for students in high school.

1. Introduction

Physics is a science that deals with facts, processes, theories, concepts and generalizations, which is not only about memorizing, but also processes that allow students to relate one concept to another to express meaningful relationships [1]. Therefore, one of the effective methods that can be used in learning physics is experiments. Experiments or practical activities have benefits in learning physics, namely to train students to understand physics concepts and improve skills in conducting scientific experiments [2]. So experiments are a significant activity in learning physics. Currently cases of transmission of the Covid-19 virus are growing very fast. This virus has spread to countries including Indonesia. The spread of Covid-19 covering all fields, such as social, cultural, economic, and educational, those fields perceives its influence in this situation. In the field of education, the Covid-19 pandemic has had a tremendous impact on schools, teachers, and students [3].

Laboratory-based video is one of solutions that can be used as an alternative to do experiments activities in the laboratory. Research conducted by Rodrriguez [4] video based experiment allows students to have experiments that show students what happens in the laboratory, according to the appropriate theory [4]. Another study states video based experiment may improve students' understanding of concepts [5]. In addition, implementing a video based experiment can improve students' understanding in learning the study of acceleration [6]. However, the use of a video based experiment in determining the focal length of the lens has not been found. To determine the focal length can be obtained in the following equation [7] :

$$\frac{1}{p}+\frac{1}{q}=\frac{1}{f} \tag{1}$$

Information:
p is positive if the object is in front of the lens (real object).
p is negative if the object is behind the lens (virtual object).
q is positive if the image is behind the lens (real image).
q is negative if the image is in front of the lens (virtual image).
f is positive if the lens is converging (convex).
f is negative if the lens is diverging (concave).

This equation, which called the lens equation, can be used to relate the image distance and object distance to the lens. Research by Ceuppens [8] states that students have difficulty in determining optical points such as focus points, objects and shadows so that they cannot determine ray diagrams in image analysis of virtual or real image properties. A video based experiment showing a convex lens practicum will help students understand the image formation of a positive lens.

The main objective of this research is to design and develop a video based experiment. In this paper, the focus of the video is an analysis of the determination focal length of a positive lens. The research questions for this research are: How to design and develop a video based experiment to determine focal length in a positive lens?

2. Method

The model that will be developed using this R&D method is the ADDIE model. The ADDIE model is one of the approach models used in media development, especially in research and development (R&D). Through the ADDIE model (Analysis, Design, Development, Implementation and Evaluation) [9]. The first phase is the analysis phase, in this phase the researcher conducted a needs analysis to determine students' understanding of determining the focus distance on the lens by using laboratory-based video as a substitute for experiments at school. At the design and development phase, researchers designed and developed video based experiment include making storyboard, recording, lighting and video editing. The camera used is a Sony A600 with specifications 24.3 MP APS-C Exmor Sensor, full HD 1080p XAVCS S video at 20/0 fps. Video based experiment was developed with scenarios of camera layout and optical room layout are made. The camera layout is used by using video capture techniques with bird eye view, high angle and view level shooting techniques. This technique is used to get different camera angles so that the video does not look monotonous. The camera layout is shown in Figure 1.

Figure 1. Bird's eye view.

Figure 2. High Angles.

After the video was made, then the video was developed through the editing process using Powtoon software to add animation, music and other tools. In the Implementation Phase, the video product developed was implemented in the physics experiments of optical instruments. After that, the evaluation phase was carried out; showing a video based experiment video to the experts. In fact, the evaluation phase was carried out in each phase to produce a video based experiment based on the assessment and input from experts through a questionnaire.

3. Results and Discussion

3.1 *Design and develop Video Based experiment*

This research produces a product in the form of video based experiment to determine focal length on positive lenses. There are several didactic advantages in using video based experiment for teaching concepts in physics: (1) the videos can be freely distributed on the web and explored with students within the classroom, or as a homework assignment to improve students' understanding of specific content, (2) the students can carry out the analysis of the experiments at any time, as many times as they want, and (3) the students apply and associate physical concepts in real situations [10]. The contents of the video are divided into several schemes that show practical activities with stages up to the end, namely (1) introducing tools and materials presented in figure 3. In tool introduction, video-based experiments showed one by one the tools used in the image formation experiment with the object name according to the displayed object. (2) tool assembly process was presented in Figure 4. The process of assembling the tools in the video-based experiment is shown systematically starting from the assembly of precision rails, placement of clamping stacks, placement of objects, lenses and screens to the process of turning on a 12 Volt lamp as a light source to produce shadows. (3) the practical process of determining the focus of the lens was presented in Figure 5. The practicum was shown in a video-based experiment by changing the object distance according to the optical space I, II and III of the lens.

Figure 3. Introduction to Tools and Materials.

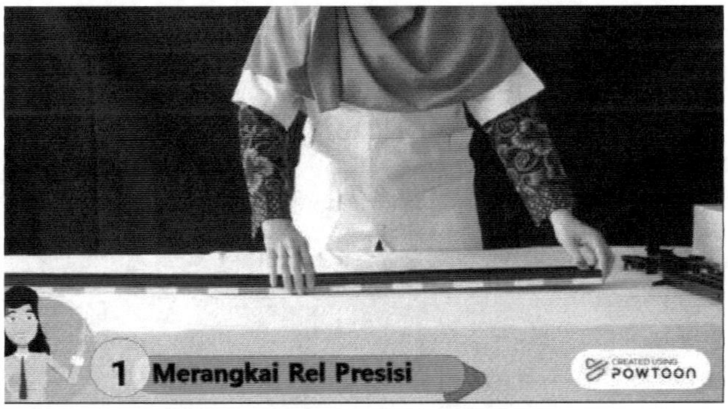

Figure 4. Toolkit.

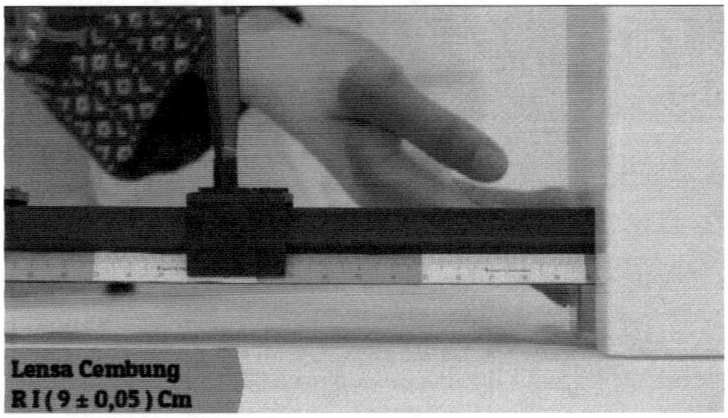

Figure 5. Experiments.

Journal of Physics: Conference Series **2019** (2021) 012006 doi:10.1088/1742-6596/2019/1/012006

The video was made using a +5 cm and +20 cm positive lens focus. The optical rooms used are room I, room II and room III. The video shows the experiments using a positive lens with different object distances according to the optical space so as to produce different image distances as in table 1.

Table 1. object distance and shadow distance used in video based experiment (accuracy: 0,01 cm)

Length of object (cm)	Length of shadow (cm)
5,00	2,50
8,00	13,00
15,00	7, 50
9,00	16,00
25,00	32,00
42,00	25,00

The video editing process used the powtoon application. Powtoon is a freeware service or online program that can be used by internet users [11] the use of video editing with powtoon can be used to make animated videos, and some image manipulations equipped with visual and audio features that can attract students to always pay attention to the videos made presented in figure 6. The results of the determination of focal length were obtained from the results of the analysis of the focus distance, object distance and image distance as well as the characteristics of the image in each optical space which can be seen from the video based experiment presented in Figure 7.

Figure 6. Animation with powtoon app.

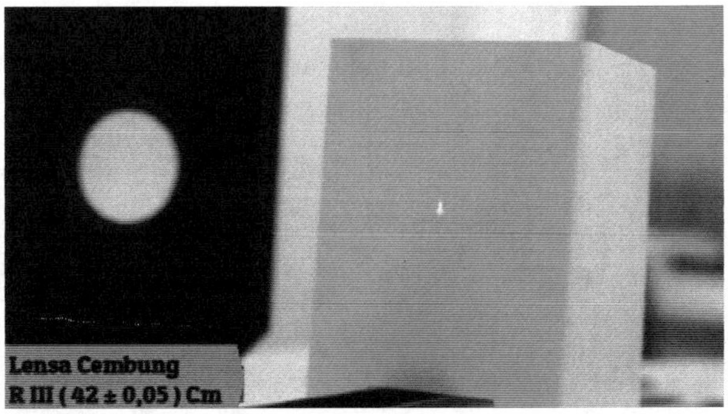

Figure 7. Focus analysis from image distance.

Students can analyze videos by linking video based experiment, the results of calculations and analyzes performed by themselves. To support student analysis materials in knowing the results of the image formation, student worksheets were assisted. Students can stop the video to find out the specific distance of the object and the distance of the image, after that students can determine the focal of the positive lens used in accordance with the experiment shown by video based experiment.

3.2 Video rating by experts

Based on the results of media expert validation, suggestions for improvement or revision were obtained, namely: the video based experiment voice over was quite loud so that the narrator's dubbing was not clear and the light background color was replaced with a dark color.

The results of video based experiment eligibility based on media expert assessments are presented in table 2.

Table 2. The results of the eligibility of media experts

Aspect	V Average	Category
Contents	82%	Very Good
Language	80%	Very Good
Presentation	80%	Very Good

Based on the results of the validation of the material, suggestions for improvement or revision were obtained, namely: the contents of part A on the LKPD questions related to optics and give questions about the results of experiments regarding the effect of object distance, focus point and others on the nature of the image formed. The results of the feasibility video based experiment based on the material expert's assessment were presented in table 3.

Table 3. The results of the eligibility of media experts

Aspect	V Average	Category
Material	92%	Very Good
Language	80%	Very Good
Presentation	80%	Very Good

4. Conclusion

Based on the results and discussion, it can be concluded that this research has produced learning media in the form of video based experiment. Video has gone through a series of stages of development. The video series was combined with several camera layout scenarios, optical room scenarios and student analysis results. Based on the results of the video based experiment expert assessment that has been developed, it has a very good feasibility, so the video was worthy of being used as a media of learning optics in high school.

Acknowledgement

The author would like to thank Mr. Tri Guna Santoso as the physics teacher and the laboratory assistant of SMAN 2 Cileungsi who have helped and supported the author in this research.

References

[1] E. Wiwin and R. Kustijono 2017 The Use Of Physics Practicum To Train Science Process Skills And Its Effect on Scientific Attitude of Vocational High School Students *Journal of Physics* 997(1) 1-8

[2] A. Emda 2017 Laboratorium Sebagai Sarana Pembelajaran Kimia dalam Meningkatkan Pengetahuan dan Keterampilan Kerja Ilmiah *Lantanida Jurnal* 2(2) 83-92

[3] S. Widodo, Y. Wibowo and W. Wagiran 2020 Online Learning Readiness During The Covid-19 Pandemic *Journal of Physics: Conference Series 1700(1) 1742-6596*

[4] F. G. Rodrriguez, J. L. P. Silvia, F. L. Rosano, F. Contrereas and A. I. M. Vitea 2001 A Student Centered Methodology For The Development of A Physics Video Based Laboratory *ELSEVER 13(*5) 527-548

[5] E. S. Nurani , H. Nuroso and J. Siswanto 2014 Kemampuan Analisis Siswa SMP Dalam Mengerjakan Soal Optik Geometris *Semnas Entrepreneurship* 141-147

[6] L. Trudel, A. Metioui and G. Abrez 2016 Student Teachers' Modeling of Acceleration using a Video-Based Laboratory in Physics Education: A Multimodal Case Study *Systemics, Cybernetics And Informatics* 14(3) 23-30

[7] D. Halliday, R. Resnick and J. Walker 2007 Fundamental of physics 8th edition (United state: Willey)

[8] S. Ceuppens, J. Deprez, W. Dehaene and M. D. Cock 2018 Tackling Misconseptions in Geometrical Optics *Physics Education* 53(4) 1-10

[9] R. M. Branch 2009 Instructional Design The ADDIE Approach (New York: Springer)

[10] M. Rodrigues and P. S. Carvalho 2014 Teaching optical phenomena with Tracker *Phys. Educ,* vol. 49(6) 671-677

[11] B. Graham 2015 Power up your powtoon studio project (Mumbai: Packt Publishing)

Journal of Physics: Conference Series **2019** (2021) 012007 doi:10.1088/1742-6596/2019/1/012007

Module Electronic Sound and Light Wave Developed By Scientific Approach For Improving Science Literacy

M Nenchi, A S Budi and C E Rustana
Physics Education Master Program, Universitas Negeri Jakarta
Jl. Rawamangun Muka, Jakarta 13220, Indonesia
Email: melfinenchi14@gmail.com, agussb@unj.ac.id and crustana@unj.ac.id

Abstract. The purpose of this study is to develop a physics learning media for Sound and Light wave physics learning in high school students. The research method used Analyze, Design, Development, Implementation and Evaluate (ADDIE) with one group pretest-posttest design method to improve scientific literacy. The research participants are students in grade eleven of high school students. The E-module was validated by experts and the results showed 84.59% for media expert, 83,49 % for material expert, and 79.75% for learning expert with very proper category. The instrument validation results for high school students were 84.27% with the very proper category. Based on the results of scientific literacy for high school students, the normalized gain score 0.79 with a effective category means E-module can improve students' Science Literacy. So it can be concluded that, there is improving scientific literacy in students using the sound and light wave e-module based on a scientific approach.

1. Introduction

The scientific learning approach provides an opportunity to use logic, discuss, formulate simple questions that provide guidelines for solving main problems, formulate theories or several possible ways to solve a problem or several solutions, to conduct experiments and to draw conclusions based on data collected or experimental results [1]. The utilization of the scientific approach in learning has a moderate effect on student's scientific literacy skills [2]. Therefore necessary the Factors that can influence students' scientific literacy results are the selection of textbooks, worksheets, learning models, and scientific literacy-based evaluation tools [3]. Based on the learning resources are needed. E-modules are non-printed teaching materials that can be used to assist students in the teaching and learning process. These teaching materials can help the role of educators in the classroom, so that students can learn anywhere and anytime [4]. E-modules are needed to complete student handbooks, one of the characteristics of media learning materials that motivated and generated student's curiosity [5]. The implementation of interactive e-modules can improve self-directed learning of students with self-confidence, high motivation, initiative and learning outcomes [6].

Assessment of learning outcomes in Indonesia consists of literacy, numeric and character assessments. Scientific literacy is related to understanding scientific concepts, knowledge and processes for making decision [7]. Scientific literacy is not just an ordinary thinking ability, but an ability that can solve problems using scientific concepts [8]. Scientific literacy gives students instruction to apply their scientific knowledge and recognize to solve the real problem in their life [9]. Scientific literacy shows that science is a contextual factor that underlies the emergence of analysis. Student"s scientific literacy includes several aspects, namely explaining phenomena scientifically, evaluating, interpreting, designing and average scientific literacy [10].

Based on the results of research conducted by Linda, it shows that the increase in the student's self-directed learning processes before and after using e-modules increased with medium category [6]. Elizar's research results from the development, scientific approach module effectiveness have an influence on learning outcomes [11]. Preliminary research shows about 71.6% of students experience learning difficulties. Students difficult to understand their textbooks on sound and light waves in physics learning. Based on the preliminary research, it is necessary for development sound and light wave e-modules based on a scientific approach to improved science literacy.

2. Methods

This study aims to develop the sound and light wave e-module based on a Scientific Approach to improve scientific literacy skill of the students. The subjects used in this study are thirty students of class XI from the same school. The research method used is Analyze, design, Development, Implementation and evaluation (ADDIE) which is evaluation is carried out at each stage before moving on to the next stage. The ADDIE model is one of the systematic instructional designs which has five stages, namely, Analyze, design, Development, Implementation and evaluation, with a diagram in figure 1

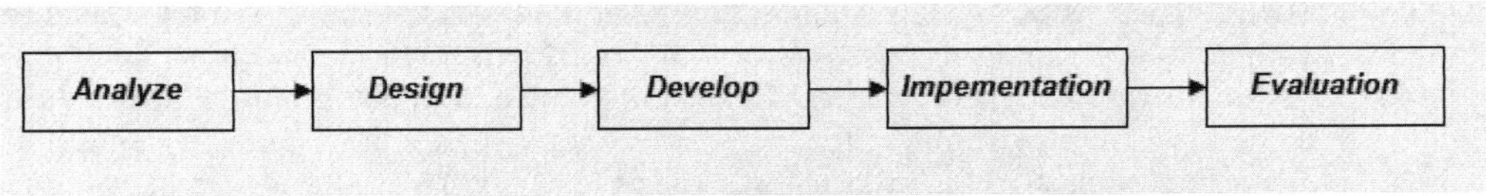

Figure 1. ADDIE Model Flow

Figure 1 is the ADDIE model flow. The explanation of the ADDIE model is as follows:

2.1. Analyze

Analysis in the ADDIE model is to collect data and initial information to identify problems and needs in schools in physics learning, find out the material that is in accordance with the developed media, and find out the opinions of teachers and students about the development of e-modules.

2.2. 2. Design

The design in the ADDIE model is the preparation of e-modules using applications to create e-modules. Then design the e-module as a companion to the teaching materials needed by students in order to demonstrate competence after participating in the learning process.

2.3. 3. Develop

Development in the ADDIE model is developing, namely developing product designs that have been made by making e-modules using software. The developed e-module consists of the material content of sound waves and light waves. After making the e-module, then the product validation test is carried out to media experts, material experts, and learning experts assess the e-module product with the aim of obtaining input and improvement.

2.4. 4. Implementation

The implementation of the ADDIE model is implemented for high school students in class XI by conducting field tests, students fill out student validation of the e-module product that has been developed. The results of the questionnaire are used as the basis for evaluating e-module products.

2.5. 5. Evaluation

The evaluation of the ADDIE model is the evaluation of the e-module product by testing the effectiveness of the e-module product that has been developed in helping to improve the scientific literacy of high school students.

One group pretest-posttest design method to improve scientific literacy, before and after the using of E-modules. The results of the pretest and the posttest literacy data analysis were tested for normality. The data collection was carried out by giving a pretest and posttest, which contents were done after the learning process. the pretest and posttest values obtained were then analyzed using n-gain score was developed by (Hake, 1999) with following formula [12]:

$$\text{N-gain}=\frac{S_{post}-S_{pre}}{S_{max}-S_{pre}} \quad (1)$$

The N-gain obtained is then interpreted based on the criteria in table 1.

Table 1. Criteria for the N-gain Value

N-gain score	Grade Level
0,70 > N-gain	Effective
0,30 ≤ N-gain≤ 0,70	Effective Enough
N-gain< 0,30	Less Effective

3. Result and Discussion

The sound and light wave e-module with a scientific approach can make students learn more independently. This is because the e-module is presented with attractive illustrations. E-modules can be accessed online, so students can study independently. The topic of sound and light waves in the e-module is developed based on the stages of a scientific approach that leads students to think scientifically. The stages of the scientific approach to the sound and light wave e-module are clearly written so that they can help students in the thinking process. Sound and light wave e-module based on a scientific approach applying the five components of a scientific approach that is observing, questioning, experimenting, associating, and communicating. At the observing stage, students observe or watch video shows and relate them to the topic being studied. In the questioning stage, students click on the e-module and ask questions about the video being shown. At the stage of experimenting, students learn the topics that are already in the e-module and make groups to do the practicum contained in the e-module. At the associating stage, students relate the information obtained with the lessons learned. At the communicating stage, students draw conclusions and communicate them directly or students can write on e-modules. There are evaluation questions at the end of each sub-topic that can be used as an evaluation by students after studying sound and sound waves. In addition, the developed e-modules are presented comprehensively in accordance with the basic competencies that students must have. Following are some examples of sound and light wave e-module based on scientific approach shown in Figure 3

Figure 2. Cover

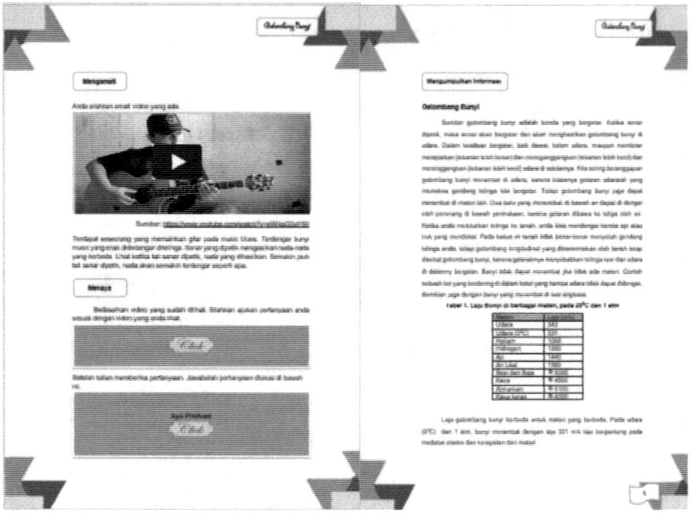

Figure 3. Scientific literacy syntax

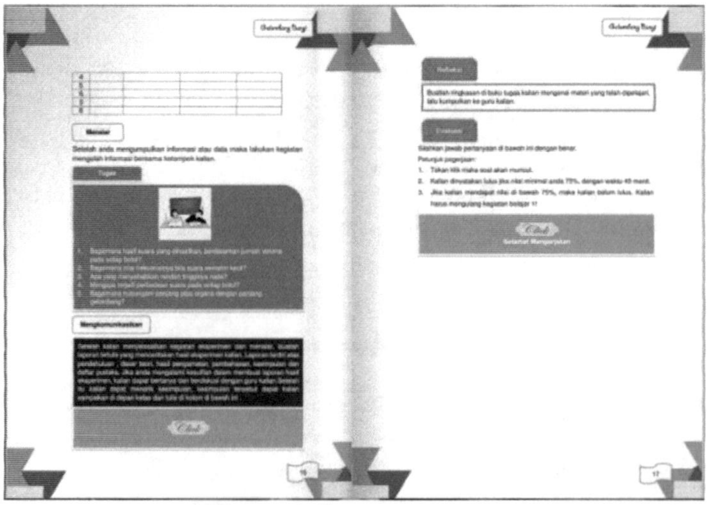

Figure 4. Competence test

Figure 2. is the home page, on the home page there are several tools and covers of the e-module. The tools on the home page are as follows: scrolling the e-module page, zooming in to enlarge and reducing the e-module size, thumbnails to reduce the e-module size, auto flip to move the e-module automatically, Backward, First to return to the e-module, first page, previous page, next page, last to go to the last page, forward, print, full screen, share and select text.

Figure 3. is a learning activity contained in the e-model. In learning activities there is a scientific approach. Figure 3 shows several stages of the scientific approach. At the observing stage, students observe or watch video shows and relate them to the topic being studied. In the questioning stage, students click on the e-module and ask questions about the video being shown. At the stage of experimenting, students learn the topics that are already in the e-module and make groups to do the practice contained in the e-module.

Figure 4. shows several stages of the scientific approach and evaluation questions. The scientific approach in Figure 4 is At the associating stage, students relate the information obtained with the lessons learned. At the communicating stage, students draw conclusions and communicate them directly or students can write on e-modules. There are evaluation questions at the end of each sub-topic that can be used as an evaluation by students after studying sound and sound waves.

The results of the e-module products validation are carried by 3 validator experts included, namely material experts, learning experts and media experts. The results of the material expert of e-module validation can be seen in the table 2.

Table 2. Material Expert Validation Test Results

Assessment Aspect	Results
Self Instruction	80,21%
Self Contained	85%
Adaptive	87,5%
User Friendly	81,25%
Mean	83,49%

The average result of material expert validation is 83.49% on the very proper interpretation.

3.1. Learning Expert Validation Results

Learning validation is carried out to find out how proper the e-module is to be used in learning. The results of the validation of the learning expert e-module can be seen in table 3:

Table 3. Learning Expert Validation Test Results

Assessment Aspect	Results
Contents	93,75%
Usefulness	75%
Scientific approach	75%
Serving Eligibility	75%
Mean	79,75%

The average result of the validation of learning experts is 79.75% on the very proper interpretation.

3.2. Media Expert Validation Results

Media expert validation is carried out to find out how proper e-module is as a medium in learning. The results of the media expert e-module validation can be seen in table 5:

Table 4. Media Expert Validation Test Results

Assessment Aspect	Results
Display	82,81%
Programming	86,36%
Mean	84,59%

The average result of media expert validation is 84.59% on the very proper interpretation.

3.3. Trial

Trial of e-module on small group and big group of high school students. The students used E-Module based on scientific approach was conducted to see how feasible the e-module was used in learning. Trials were carried out by teachers and students. The test results can be seen in table 5:

Table 5. Test Results

Assessment Aspect	Results
Small group of students	91,56%
big group of students	84,27%
Teacher	88,75%

Based on the experimental results of thirty students, pretest and posttest were analyzed by normalization test. Pretest and posttest analysis showed normal results. The pretest and posttest have the average N-gain score was 0.79 with an effective interpretation. So it can be concluded that there is improving scientific literacy in students using the sound and light wave e-module based on a scientific approach.

4. Conclusion

Based on the result and discussion, it can be concluded that e-modules have been developed based on a scientific approach. The sound and light wave e-module has been developed in the proper categories for materials, media, and self-directed learning. Overall the development of the light and sound wave E-Module based on a scientific approach can improve the scientific literacy of students in high school.

Acknowledgments

The author would like to thank the principals and physics teachers of SMA LABSCHOOL Cibubur for granting permission and providing assistance in conducting this research.

References

[1] A. Nofrianto dan Jumrawarsi, "Teacher's Problem and Scientific Learning Approach: An Investigation on Teacher's Problem-Posing Ability," *Journal of Physics: Conf. Series* , vol. 1155, pp. 1-9, 2019.

[2] N. Shofiyah , I. Afrilia dan F. E. Wulandari, "Scientific Approach and The Effect on Students Scientific Literacy," *Journal of Physics: Conference Series* , vol. 1594, pp. 1-6, 2020.

[3] A. Rusilowati, B. Astuti dan N. A. Rahman, "How to improve student's scientific literacy," *Journal of Physic: Conference Series,* vol. 1170, pp. 1-5, 2019.

[4] C. D. T. Haspen dan Syafriani, "The preliminary study in the development of e-Physics module integrated ethnoscience," *Journal of Physics: Conference Series* , vol. 1481, pp. 1-5, 2020.

[5] F. A. Perdana, Sarwanto, Sukarmin dan I. Sujadi, "Development of e-module combining science process skills and dynamics motion material to increasing critical thinking skills and improve

student learning motivation senior high school," *International Journal of Science and Applied Science: Conference Series* , vol. 1, pp. 45-54, 2016.

[6] R. Linda, H. Nufus dan Susilawati, "The implementation of chemistry interactive e-module based on Kvisoft Flipbook Maker to improve student' self-learning," *AIP Conference Proceedings* , vol. 2243, pp. 1-7, 2020.

[7] R. M. Vieira dan C. T. Vieira, "Fostering Scientific Literacy and Critical Thinking in Elementary Science Education," *Int J of Sci and Math Educ,* 2014.

[8] Suryanti, M. Ibrahim dan N. S. Lede, "Process skills approach to develop primary students' scientific literacy: A case study with low achieving students on water cycle," *Materials Science and Engineering* , vol. 296, pp. 1-7, 2018.

[9] Y. Stefanova, M. Minevska and S. Evtimova, "SCIENTIFIC LITERACY: PROBLEMS OF SCIENCE EDUCATION IN BULGARIAN SCHOOL," *PROBLEMS OF EDUCATION IN THE 21st CENTURY,* p. 113, 2010.

[10] P. W. Hastuti, W. Setianingsih dan E. Widodo, "Integrating Inquiry Based Learning and Ethnoscience To Chance Students's Scientific Skills and Science Literacy," *Journal of Physics: Conference Series* , vol. 1387, pp. 1-7, 2019.

[11] E. Ellizar, H. Hardeli, S. Beltris dan R. Suharni, "Development of Scientific Approach Based on Discovery Learning Module," *Materials Science and Engineering,* vol. Volume 335, p. 8, 2018.

[12] J. R. Milda, S. L. Rahayu dan H. Susilo, "Implementation of E-module Endocrine System Based on Problem Based Learning (PBL) to Improve Scientific Literacy and Cognitive Learning Outcome," *AIP Conference Proceedings,* vol. 2330, pp. 1-5, 2021.

Using Accelerometer Smartphone Sensor and Phyphyox for Friction Experiment in High School

Y F Ilmi, A B Susila and B H Iswanto
Department of Physics Education, Universitas Negeri Jakarta
Jl. Rawamangun Muka, Jakarta 13220, Indonesia
Email: yusinafadla@gmail.com, anggarabs@unj.ac.id and bhi@unj.ac.id

Abstract. Experiments related to friction in schools are mostly done by direct observation, and it becomes more difficult to determine the value of the coefficients of static and kinetic friction due to the low level of accuracy. This study aims to develop a friction coefficient practicum tool using the accelerometer sensor on a smartphone with the *Phyphox* application. The research was conducted using the 4D Thiagarajan method (Define, Design, Develop, Disseminate). This tool demonstrates experimental friction on an inclined plane with a variety of surface materials (aluminium, wood, glass, acrylic). The results recorded by *Phyphox* are spreadsheets of object acceleration data per unit time in x, y, and z coordinate graphs. The sliding plane frame is designed using an Arduino microcontroller which can make it easier for students to automatically form inclined plane angles. This tool has been tested by experts with an average score are 85.25% and has been tested on physics teachers with an average score of 82.5% in the good category. This friction coefficient practicum tool is feasible and can help students understand the concept of friction

1. Introduction

Physics is one of the practical experimental subjects, some of the concepts of physics are formed through mathematical analysis, but in the end the theories that are formed must be tested through experiments [1]. Physics studies the facts that exist then packaged into physical concepts and developed into laws or theories of physics through an experiment [2]. Experimental activities can help students acquire, integrate and build knowledge in a friendly way [3]. Therefore, experimental/ laboratory activities have an important role at all levels of education.

Practicum based on mobile application technology (*Phyphox*) is an application that provides direct access to sensors on smartphones in which there are many simulations that can be used as physics learning media [4]. *Phyphox* application can be obtained for free and can be downloaded on *Google Playstore* and *Appstore* [5]. The use of the functions of the *Phyphox* application produces valid and reliable data that can be presented to students in real time, without the need to analysed raw data files given in spreadsheet form [6]. *Phyphox* makes experiments more accessible and expands the tools available to students with simple methods for controlling experiments remotely and with data in the field [5].

With the *Phyphox* application as a physics experiment application, it is considered capable of solving problems in physics learning. With *Phyphox* in a smartphone, a teacher does not need a long time to complete the material. Most of the sensors are legible and their measurement data is displayed graphically and equipped with many other suitable innovative features, so *Phyphox* is very good for use in schools [7]. Thus, smartphones have provided opportunities for new learning perspectives,

Journal of Physics: Conference Series **2019** (2021) 012008 doi:10.1088/1742-6596/2019/1/012008

including relationships between teachers, students, and learning material [8]. Hence, smartphones can be used as experimental tools in the world of education, including physics [9].

Newton's law is one of the fundamental physics materials in the dynamics of motion. The force of friction is an essential element in motion dynamics. It is essential to teach this topic in physics learning [10]. Friction is one of the interesting topics to study in physics learning. However, the frictional force material is one of the physics materials that is prone to misconceptions, one of which is due to the lack of demonstrations about the frictional force [11]. Many schools conduct experiments related to friction by direct observation, and it becomes more difficult when it is done to determine the value of the coefficient of friction due to the low level of accuracy.

Based on observations, it was found that 41.2% of the 17 teachers only used the lecture learning method in delivering Newton's Law material and 68.8% answered the use of learning media in the frictional force material that was often used, namely power point media. This is due to the lack of complete and inadequate laboratory equipment and materials in schools. By utilizing the *Phyphox* application on a smartphone, the experimental observation process that has been recorded in the form of a spreadsheet can then be processed more accurately until finally information is obtained in the form of quantities to determine the value of the friction coefficient of objects. Furthermore, by utilizing the *Phyphox* application in physics learning, it can improve students' analytical skills and creative thinking skills [12].

2. Method

The purpose of this research is to develop a friction coefficient practicum tool on an inclined plane using the accelerometer sensor on a smartphone with the help of the *Phyphox* application which is feasible as a learning medium. This research is a research and development (R&D) and uses the 4D Thiagarajan method, namely Define, Design, Develop, and Disseminate, with the flow chart in Figure 1.

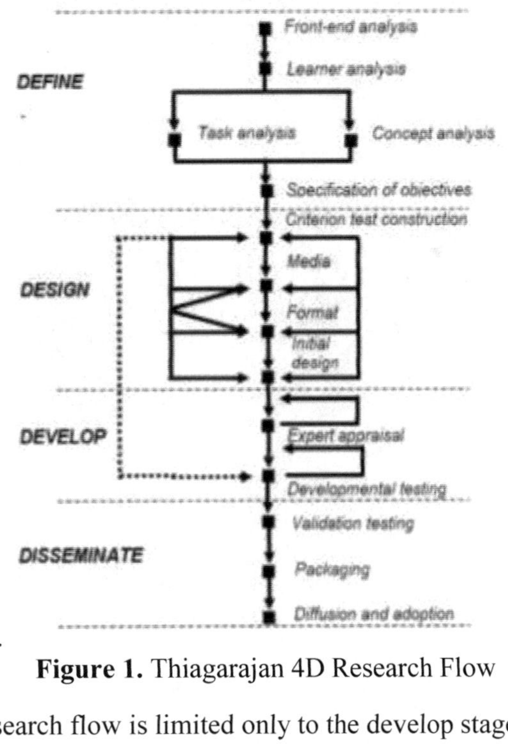

Figure 1. Thiagarajan 4D Research Flow

The limitation of this research flow is limited only to the develop stage. The details are as follows:

1. Define stage. A preliminary study starts from analysing the media needed in schools, data are obtained from observations in several schools, this is done to obtain information about the learning process and the availability of practicum equipment in schools.

2. Design stage. The initial design of the practical tool was developed. At this stage, a feasibility test instrument, a product trial instrument was also designed.
3. Development Stage. At this stage, the device framework begins to be developed based on the initial design and development of practicum tools needed by schools.

As for the data collection technique, namely by using a Likert scale test with an assessment weight of (4) strongly agree/ very good, (3) agree/ good, (2) disagree/ less good, (1) strongly disagree/ bad. The following is a presentation of the Likert scale interpretation

Table 1. Likert Scale Interpretation

Percentage	Interpretation
76 – 100 %	Very Good
51 – 75 %	Good
26 – 50 %	Less Good
0 – 25 %	Bad

3. Result and Discussion

The developed media are a practicum tool to determine the coefficient of friction of objects on an inclined plane, which is used for two practical activities, namely determining the coefficient of static and kinetic friction on various kinds of surfaces. The measurement results, use the accelerometer sensor on the smartphone with the help of the *Phyphox* application.

3.1. Design

The tools and materials needed to design the coefficient of friction practicum tools include a smartphone with the *Phyphox* application installed, an Arduino microcontroller, a servo motor, a wooden board in an inclined plane frame with a size of 60x20 cm, as shown in Figure 2.

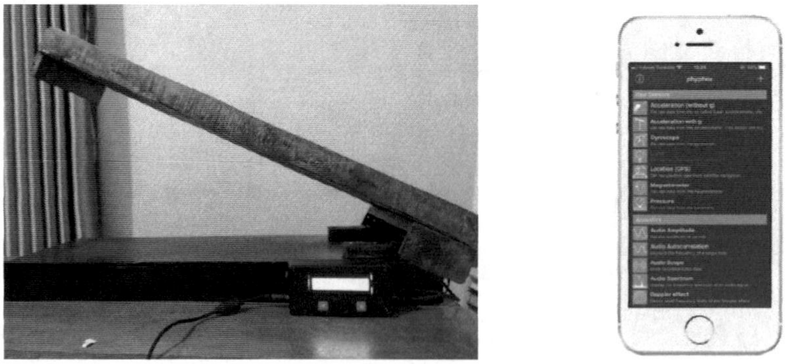

(a) *(b)*

Figure 2. Tools and Materials, (a) Inclined plane, (b) *Phyphox* in smartphone

The inclined plane frame is designed using an Arduino microcontroller, IC LM2596, 16x20 LCD and a servo motor that can make it easier for users to automatically form angles with a circuit as shown in Figure 3.

Journal of Physics: Conference Series **2019** (2021) 012008 doi:10.1088/1742-6596/2019/1/012008

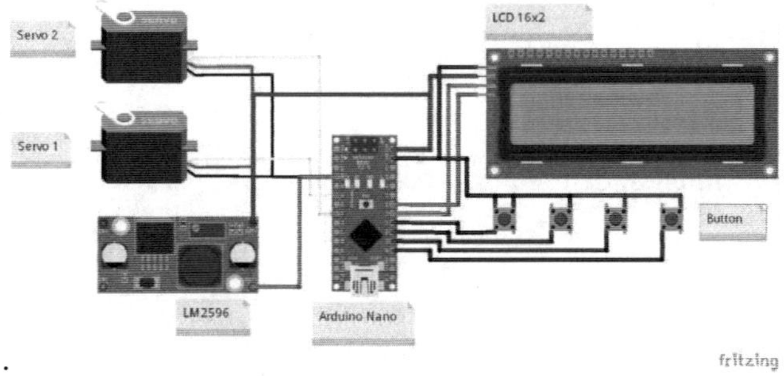

Figure 3 Inclined Plane Circuit

3.2. Experimental data

This experiment uses an iPhone 7 with software version 14.1 in which there is an accelerometer sensor inside. How to use the coefficient of friction practicum on inclined planes, which are:

1) Static coefficient measurement
The smartphone is placed on a flat surface with an angle of 0°. The glide plane is slowly raised by the object begins to move. The experiment was repeated until 5 data were obtained with different combinations of surface types of materials and bases. The resulting data from *Phyphox* is then exported for analysis to obtain a static coefficient value.
2) Kinetic coefficient measurement.
The smartphone is placed with the slope of the glide plane at three variations of angles, namely 30°, 45°, and 60°. The experiment was carried out by changing the type of load and the plane of the slide. The results of the data from *Phyphox* were then exported, to be analysed to obtain the value of the kinetic coefficient.

The results displayed in the *Phyphox* application are graphs of acceleration in x, y, and z coordinates per unit of time. The data recorded by *Phyphox* are in the form of a spreadsheet that can be exported from a smartphone to a laptop in CSV or Microsoft Excel format. The following is an example of the display generated by the *Phyphox* application:

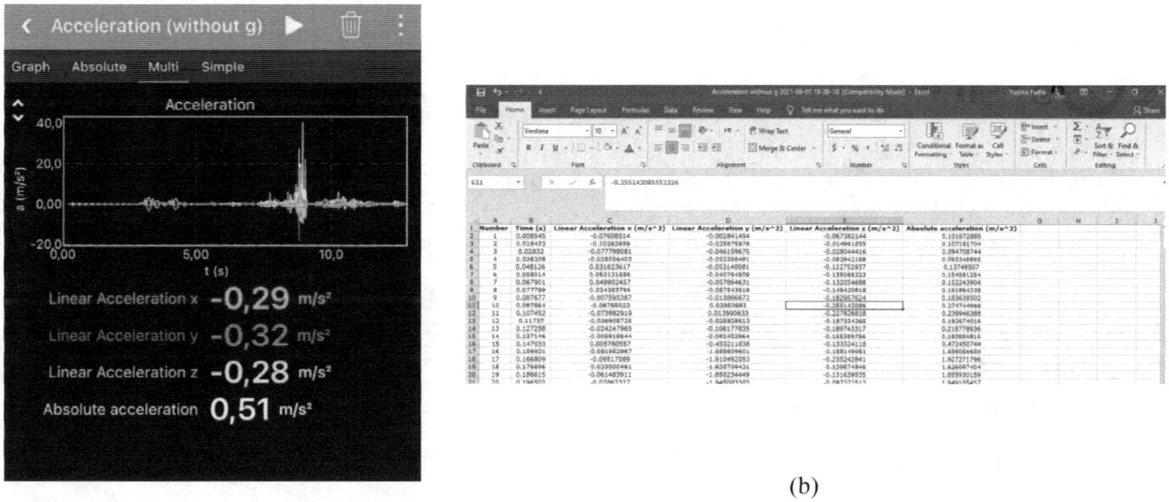

(b)

(a)

Figure 4. Display of data recorded by *phyphox*, (a) Graphical form and average values displayed on a smartphone, (b) Spreadsheet form in Microsoft excel

The results of the data that have been exported to Microsoft Excel are then analyzed to determine the value of the coefficient. The experiment was carried out by the researcher. The data obtained were as follows:

Table 2. Example of experimental data on the coefficient of kinetic friction with wood–aluminium at an angle 30°

Trial of-	θ	t (m.s^{-1})	a (m.s^{-2})	μ_k
1	30	1.23	2.76	0.252
2	30	1.24	2.77	0.25
3	30	1.21	2.76	0.252
4	30	1.24	2.77	0.25
5	30	1.18	2.87	0.24
Average	30	1.22	2.78	0.248

The results of the sample measurement data with wood - aluminium surfaces with the help of the *Phyphox* application, the coefficient of kinetic friction at an angle of 30° is 0.248 $\pm$ 0.005

Table 3. The value of the coefficient of kinetic friction for the entire surface area at an angle 30°

Plane surface/ Load base	Wood	Aluminium	Glass	Acrylic
Wood	0.345 $\pm$ 0.002	0.323 $\pm$ 0.002	0.254 $\pm$ 0.002	0.285 $\pm$ 0.002
Aluminium	0.248 $\pm$ 0.005	0.280 $\pm$ 0.003	0.204 $\pm$ 0.003	0.242 $\pm$ 0.001
Glass	0.235 $\pm$ 0.003	0.279 $\pm$ 0.003	0.195 $\pm$ 0.002	0.202 $\pm$ 0.002
Acrylic	0.283 $\pm$ 0.002	0.266 $\pm$ 0.002	0.215 $\pm$ 0.002	0.250 $\pm$ 0.001

Based on table 3, the value of the coefficient of kinetic friction in various pairs of a material at an angle of 30° tends to be constant. The order of the largest to smallest friction coefficient values is wood-wood > aluminium-wood > acrylic-wood > wood-acrylic > aluminium-aluminium > aluminium-glass > aluminium-acrylic > glass-wood > acrylic-acrylic > wood-aluminium > acrylic-aluminium > wood-glass > glass-acrylic > glass-aluminium >acrylic-glass > glass-glass.

Table 4. Example of static friction coefficient experimental data with wood–aluminium.

Trial of-	θ	t	a (m.s^{-2})	μ_k
1	21	1.20	0.544	0.38
2	20	1.21	0.562	0.36
3	21	1.18	0.543	0.38
4	19	1.24	0.482	0.34
5	21	1.21	0.521	0.38
Average	20.4	1.20	0.53	0.368

An example of experimental data measuring static coefficients with wood-aluminium material obtained the average angle when the object starts to move is 20.4° with a static coefficient of 0.368 $\pm$ 0.017.

Table 5. The value of the coefficient of static friction for the entire surface area

Plane surface/ Load base	Wood	Aluminium	Glass	Acrylic
Wood	0.564 ± 0.006	0.420 ±0.008	0.518 ±0.005	0.429 ±0.008
Aluminium	0.368 ± 0.017	0.371 ±0.008	0.342 ±0.007	0.369 ±0.083
Glass	0.341 ±0.006	0.304 ±0.009	0.301 ±0.007	0.411 ±0.008
Acrylic	0.411 ±0.010	0.396 ±0.005	0.358 ±0.011	0.432 ±0.010

The data in the table above shows the value of the coefficient of static friction on various pairs of materials, sorted from the largest to the smallest values, namely, wood-wood > glass-wood > acrylic-acrylic > acrylic-wood > aluminium-wood > acrylic-glass = wood -glass > aluminium -acrylic >aluminum-aluminum > acrylic-aluminum > wood-aluminium > glass-acrylic > glass-aluminum > wood-glass > aluminium-glass > glass-glass.

By comparing the friction coefficient values in table 3 and table 5, for the combination of various pairs of the same material, the static coefficient value has a value that tends to be greater than the kinetic coefficient. On the coefficient of static and kinetic friction, the pairs of materials with the largest and smallest order have the same order. The analytical method using the *Phyphox* application in experiments to determine the coefficient of static and kinetic friction of various pairs of materials is very helpful for students in understanding the concept of frictional force. In addition, the information provided will be more accurate when compared to direct observation.

The use of sensors on smartphones can be a reference for learning media, particularly in the experimental activities in the classroom. In this research, the kinetic and static friction coefficient between four surfaces can be determined through the acceleration sensor by utilizing the *Phyphox* application. It is simple, and the devices can be easily found. This experiment is immensely practical, so the teacher and students can carry it out in the classroom without requiring a laboratory [13]. The use of various materials that are easily found can ease learning the experimental activities particularly. Besides, smartphones as the facilities in collecting data can be directly used without any complicated stages [14].

3.3. Implementation

The appropriateness and quality of the friction coefficient practicum tool developed is known through a test by experts. The appropriateness test in this study consisted of a material expert test, a media expert test and a learning expert test. With the results of the feasibility test in table 6.

Table 6. The expert test results

Expert test of-	Score	Interpretation
Media	80.4%	Very Good
Material	90.84%	Good
instructional	84.5%	Good

Based on the results of the table above, all aspects of the assessment that have been tested by experts are in the very good category with an average score of 85.25%. Furthermore, this practicum tool was tested by 8 physics teachers by giving a questionnaire. With the results of the score in table 7.

Table 7. Teacher test results

Aspect	Score	Interpretation
Structure	85.0	Good
Design	80.6	Good
Concept	83.4	Good
Content	84.3	Good
Interactivity	79.2	Good

Based on these results, all aspects assessed by the teachers were in the very good category. With a total average in all aspects of 82.5%.

4. Conclusion

Based on the results of the effectiveness test to the teacher, it was found that the use of the *Phyphox* application on a smartphone was considered effective for learning physics, especially for experimental activities in the classroom so that it helped understanding the frictional force material with a total score of 82.5 in the good category. In addition, experimental activities like this can save time. Learning with *Phyphox* also provides opportunities for students to actively and directly participate in understanding the material and prove it with real evidence. Through the acceleration value of the sliding object obtained on an inclined plane using various materials, the coefficients of static and kinetic friction on each material can be easily determined. These results can provide further development of the use of light sensors on smartphones, spring systems or fluids for physics experiments and block on inclined planes. Further development in this research can be done by analysing theory analytically.

References

[1] Nurfadhilah, I. H. R and N. E, 2019 Pengembangan Panduan Eksperimen Fisika Menggunakan Smarthphone dengan Aplikasi Phyphox pada Materi Tumbukan," *Jurnal Penelitian Pembelajaran Fisika,* **Vol. 10 No.2**, pp. 101-108.

[2] J. Hughes and F. , 1986 *Experimental Physics With A Rotation Table* (London: Phys Education)

[3] S. Catalina , E. Carlos and P. David, 2017 Teaching Physics through Eksperimental Projects *24th International Congress of Theoretical and Applied Mechanics* **Vol. 20** , pp. 189-194

[4] K. J and V. P, 2013 Smartphones as Experimental Tools: Different Methods to Determine the Gravitational Acceleration in Classroom Physics by Using Everyday Devices *European Journal of Physics Education* **Vol. 4 No. 1** pp. 16-27

[5] S. S, H. S, H. H and S. C 2018 Advanced tools for smartphonebased experiments: phyphox *IOP Science,* pp. 1-5

[6] T. Pierratos and P. M. Hariton 2020 Utilizing the phyphox app for measuring kinematics variables with a smartphone *IOP Science* **Vol. 5 No. 22**

[7] B. Götze 2017 Smartphone-Experimente zu harmonischen Pendelschwingungen mit der App phyphox *Didaktik der Physik Frühjahrstagung – Dresden* p. 233

[8] D. Sulisworo and M. Toifur 2016 The Role of Mobile Learning on the Environment Shifting at High School in Indonesia, *Int. J. Mobile Learning and Organisation,* **Vol. 10, no. 3**, p. 168

[9] Y. Pramudya and E. P. Raharja 2019 The Influence of Inclined Plane Angle to the Oscillation Period of Spring and Block Systems *Advances in Engineering Research,* **Vol. 189**, p. 9

[10] M. L and S. D 2018 Measuring the coefficients of kinetic and rolling friction by exploring decaying mass-spring oscillations *Phys. Educ,* **Vol. 53 No.1**, p. 2

[11] A. A. Arsyad and D. Sartika 2016 Identifikasi Miskonsepsi pada Materi Gaya Gesek *Jurnal Saintifik* **Vol.2 No. 2**, pp. 101-104

[12] F. D. Fatmala 2020 The Effect of GIL Assisted Phyphox in Physics Learning towards Creative Thinking *Jurnal Pembelajaran Fisika,* **Vol. 8 (2)**, p. 148

[13] M. E. Wahyuni, D. Sulisworo and Ishafit 2020 The Utilization of Sensors on Smartphone to Determine the Coefficient of Kinetic Friction with the Inclined Plane in Supporting Physics Learning," *International Journal of Advanced Science and Technology,* **Vol. 29, No. 5**, p. 5350

[14] K. S, 2018 A simple experiment to measure the maximum coefficient of static friction with a smartphone *Phys. Educ,* **Vol. 53 No. 5**

Virtual Test Instruments to Measure Scientific Literacy of High School Students on Work and Energy

Y Oktalia, A B Susila and B H Iswanto
Physics Education Master Program, Universitas Negeri Jakarta
Jl. Rawamangun Muka, Jakarta 13220, Indonesia
Email: yesyoktalia6@gmail.com, anggarabs@unj.ac.id, and bhi@unj.ac.id

Abstract. Science literation is one aspect of assessment that is required in the K-13 curriculum. This study aims to describe the science literacy skills of high school students in the matter of effort and energy in learning physics and to determine the characteristics and quality of scientific literacy instruments using virtual test. The research method used is the 4D development method: Define, Design, Development, Disseminete. The scientific literacy test was validated by expert lecturers and physics teachers with CVI results in the very good category and CVR analysis in the valid category. The results of the analysis of the Rasch model show that the test instrument is accepted and can measure what it should measure. Research participants are tenth grade who has studied the concepts of work and energy. The virtual test contains scientific literacy skills which contain physics questions on the concepts of work and energy. The results showed that the scientific literacy skills of high school students were 65,2% for the contextual aspect, 65,1% for the knowledge aspect, and 69,5% for the competence aspect. Based on the research results, students' scientific literacy skills are in the good category.

1. Introduction

Scientific literacy is an important and needed skill in today's digital era. The importance of scientific literacy because the problems are related to knowledge and technology. Science education aims to improve students' knowledge, competence in an effort to meet the needs of life in various situations such as to understand the environment, health, economy, modern social, and technology [1] [2]. Systematic and scientific education really supports sustainable education [3]. Deep and broad scientific abilities allow students to understand the world [4]. Scientific literacy plays an important role in solving real problems [5] involving scientific thinking skills and acting in addressing existing social issues such as the development of life skills, where the need for reasoning skills in a social context and emphasizes that scientific literacy is for everyone [6].

The effectiveness and efficiency of educational teaching cannot be separated from the learning assessment instruments used [7]. Teachers must be provided with assessment instruments to support their duties, namely evaluating student learning outcomes. Assessment documents the educational process carried out on how students achieve learning objectives in the form of scores, degrees, or grades [8]. The use of computer technology to conduct student assessments has succeeded in encouraging a high level of validity, being more interactive and in accordance with the objectives to be achieved [9]. One form of assessment using technology is a virtual test. Virtual tests can measure competence more effectively than paper and pencil tests [10]. Basyari found that learning using multimedia makes learning more meaningful [11] and virtual exam assessment (ASVITE) based on interactive lecture demonstrations (ILD) supports 21st century employment skills competency Marwanti et al al [12].

Journal of Physics: Conference Series **2019** (2021) 012009 doi:10.1088/1742-6596/2019/1/012009

Work and energy are basic physics materials that examine physics concepts about motion which are often seen as complex if the assessment is in the form of text so that students need a stimulus in the form of videos or pictures. Therefore, it is very important to develop more effective virtual evaluation [10].

Instruments developed literacy test contains three aspects: context, knowledge and competence [13]. Each developed question is equipped with a stimulus in the form of videos or pictures to avoid misconceptions. Currently, the development of evaluation instruments is mostly done with classical models.

So it takes the development of virtual test instrument to measure the scientific literacy of high school students on the work and energy.

2. Method

The method used in this research is the Research & Development method. The development of the test instrument was carried out with reference to the 4D model developed by Thiagarajan. The 4D model consists of four stages:

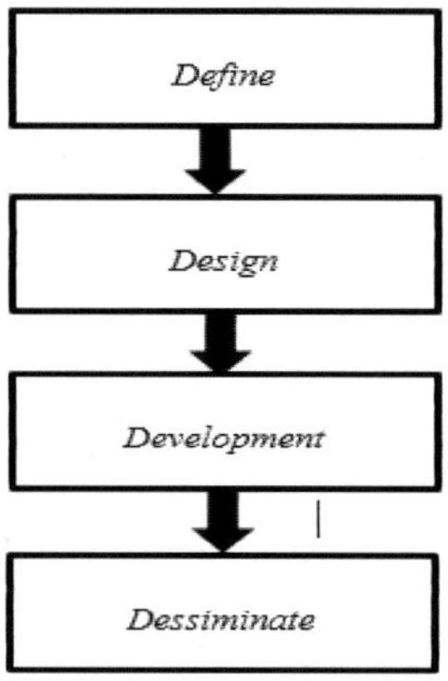

Figure 1. 4D Model Test Development Procedure

2.1. Define

The purpose at this stage is to define the terms of the virtual test instrument in the format of a scientific literacy test, which begins with an objective analysis of the material constraints to be developed by the instrument. This stage includes four main steps: needs analysis, student analysis, curriculum analysis and learning objectives.

2.2. Design

The purpose of this stage is to prepare a literacy test instrument that will be virtual. This stage includes three steps: compiling a test grid, determining the form and length of the test, and developing a validation instrument.

2.3. Development

The purpose of this stage is to produce an instrument that is suitable for use based on expert opinion and statistical analysis. This stage includes four steps: writing tests, reviewing tests, expert validation and revising based on expert suggestions. Data from the results of expert validation instruments will be analyzed for validity and reliability using CVR and CVI analysis.

2.4. Disseminate

The purpose of this stage is to produce scientific literacy instruments that are tested. This stage includes implementation to students, data analysis and conclusion. Analysis of the data obtained will be analyzed using the Rasch Model.

3. Results and Discussion

3.1. Scientific Literacy virtual test

The virtual instrument-based literacy test contains one aspect of scientific literacy. The following are some examples of literacy questions presented in the video and the picture shown in Figure 2.

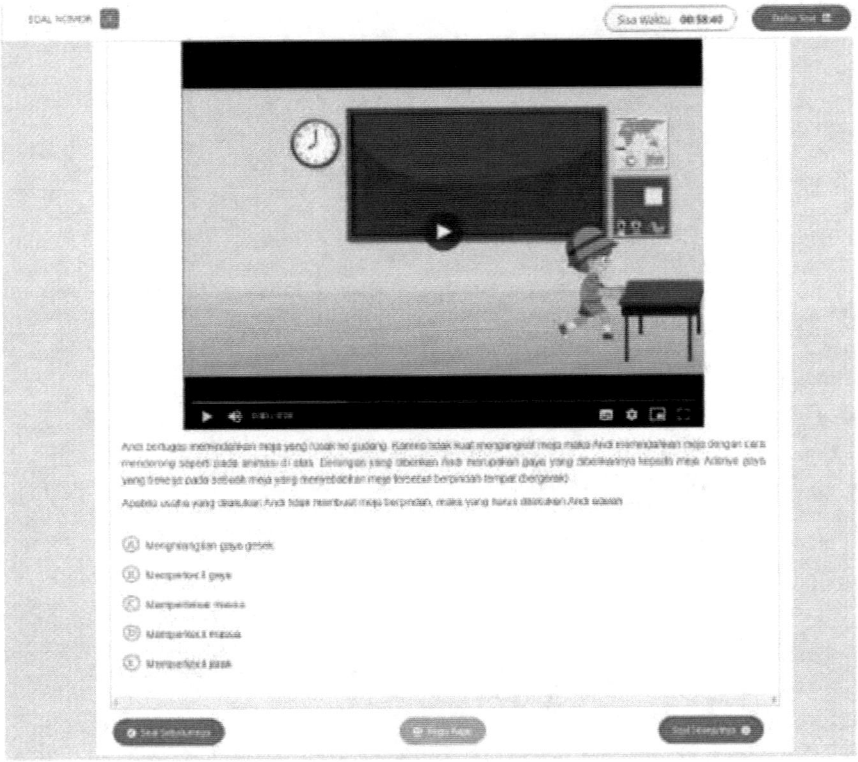

(a) context aspect

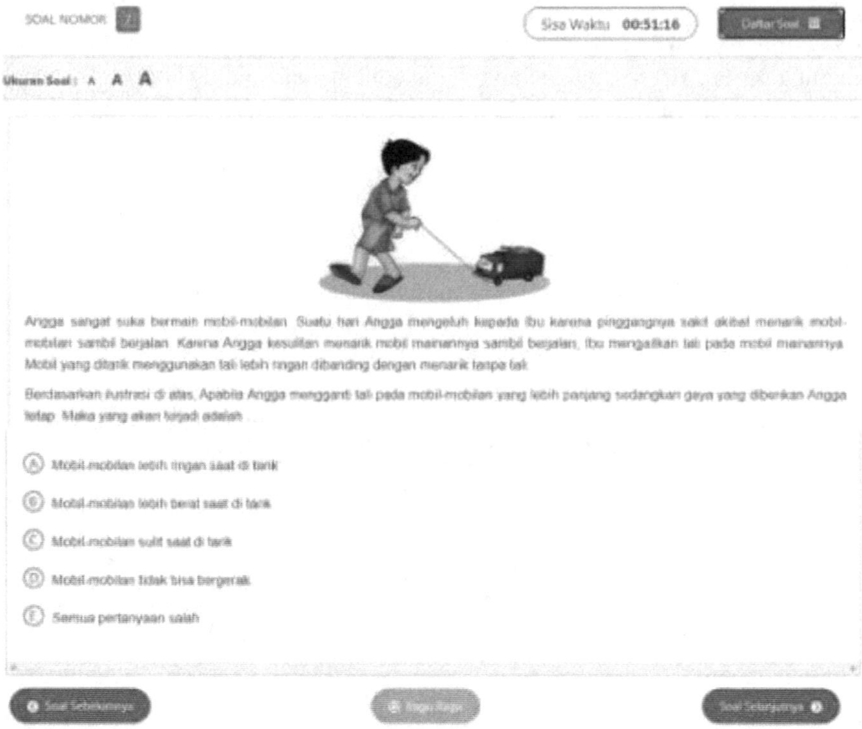

(b) Knowledge indicators

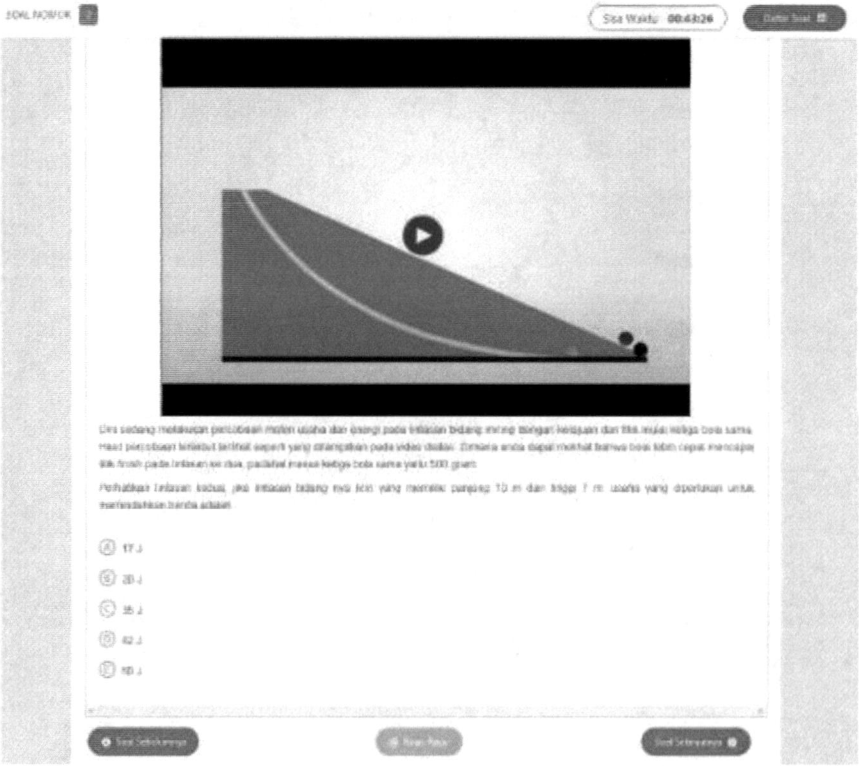

(c) Competence indicators

The stimulus questions used video or animation tested to determine the student's scientific literacy consisted of the three indicators. It can be seen that an example of a problem in Figure 2 (a) shows a problem with context aspect indicators, students are asked to understand the application of work and energy with approach scientific literacy test (b) Knowledge indicators where at that stage students designing and evaluating investigations, (c) Competence indicators which means being able to interpretation data and use scientific evidence.

Based on the results of the validation, scientific literacy test on work and energy learning based on the indicators. The results of the analysis of the Rasch Model show that The average percentage of material validation by the student was 31,5%, with very good interpretation and 0,97 for reliability. The results of expert validation tests consisted of three aspects, namely substantance, a construction and language which were analyzed using the Content Validty Ratio (CVR). The average percentage of the CVR was 0.905 with a very good interpretation.

Table 1. CVR and CVI Values

	Ne	CVR	Criteria	CVI	Criteria
question 1	375	0.88	Valid	0.94	very good
question 2	371	0.86	Valid	0.93	very good
question 3	362	0.81	Valid	0.91	very good
question 4	371	0.86	Valid	0.93	very good
question 5	372	0.86	Valid	0.93	very good
question 6	376	0.88	Valid	0.94	very good
question 7	379	0.90	Valid	0.95	very good
question 8	380	0.90	Valid	0.95	very good
question 9	373	0.87	Valid	0.93	very good
question 10	377	0.89	Valid	0.94	very good
question 11	373	0.87	Valid	0.93	very good
question 12	378	0.89	Valid	0.95	very good
question 13	377	0.89	Valid	0.94	very good
question 14	378	0.89	Valid	0.95	very good
question 15	380	0.90	Valid	0.95	very good
question 16	378	0.89	Valid	0.95	very good
question 17	379	0.90	Valid	0.95	very good
question 18	371	0.86	Valid	0.93	very good
question 19	366	0.83	Valid	0.92	very good
question 20	379	0.90	Valid	0.95	very good

3.2 Analysis of scientific literacy results

Based on table the results of the implementation science literacy test show that 65,2% of students are able to understand the matter of scientific literacy in the matter of work and energy in the context aspect Contain that students are able to answer questions related to application on work and energy. 65,1% the knowledge indicator, students are able to answer questions in designing and evaluating investigations and 69,5% of students are able to answer questions on the aspect of competence which means being able to interpret data use scientific evidence.

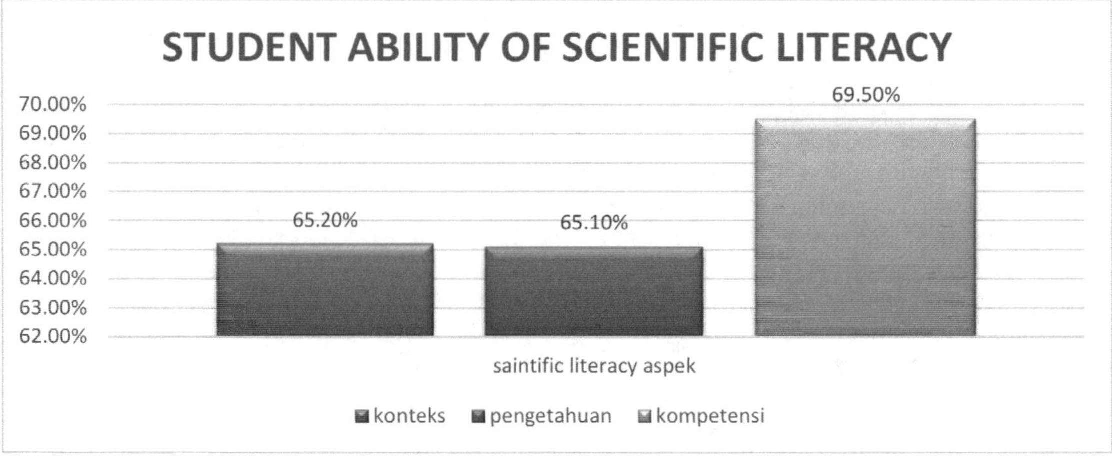

Figure 3. Data on science literacy skills of high school students

The average of all aspects of the feasibility test instrument is possible to be used as an assessment to measure scientific literacy of work and energy in physics learning. The instrument is stated to be able to measure what it is supposed to measure. Virtual instruments also make it easier for teachers to check or evaluate and determine appropriate feedback for each student [14].

4 Conclusion

Based on the research results that have been obtained, it can be concluded that the virtual test instrument for scientific literacy of work and energy learning is valid as an assessment instrument that show 31,5%, with very good interpretation and 0,97 for reliability. And based on the test results of students, it was found that the scientific literacy skills of high school students' scientific literacy skills 66,6% were in the good category. Virtual test instrument of scientific literacy makes it easier for students and teachers to use it and can create students' abilities in scientific literacy

References

[1] Bahcivan E and Kapucu S 2014 Turkish Preservice Elementary Science Teachers' Conception of Learning Science and Science *International Journal of Environmental & Science Education* pp 9(4) 429-442

[2] Pratiwi S N, Cari C and Aminah S 2019 Pembelajaran IPA Abad 21 dengan Literasi Sains Siswa *Jurnal Materi dan Pembelajaran Fisika (JMPF)* vol 9 no 1 2019 ISSN : 2089-6158

[3] Wamsler C 2020 Education for sustainability Fostering a more conscious society and transformation towards sustainability *International Journal of Sustainability in Higher Education* vol 21 no 1

[4] Dani D 2009 Scientific Literacy and Purposes for Teaching Science: A Case Study of Lebanese Private School Teachers *International Journal of Enviromental & Sains Education vol 4 no 3* pp. 289-299

[5] Widowati A, Widodo E, P Anjarsari and Setuju 2017 The Development of Scientific Literacy through Nature of Science (NoS) within Inquiry Based Learning Approach *ournal of Physics: Conf Series 909*

[6] Rychen S L 2003 Key competencies for a successful life and a well functioning society (Cambridge: MA: Hogrefe & Huber)

[7] Atta H B, Vlorensius, Arras I and Iksanudin 2019 Developing an instrument for students scientific literacy *Journal of Physics: Conference Series 1422 012019*

[8] Kuznetsova O, Palferova S and Sherstobitova A 2019 Application of Multivariate Statistical Methods for Assessment of Educational Competencies *Smart Education and e-Learning, vol. 144 pp,* pp. 608-618

[9] Scalise K d G 2016 Computer-Based Assessment in E-Learning: A Framework for Constructing-Intermediate Constraint Questions and Tasks for Technology Platforms *The Journal of Technology, Learning, and Assessment,* pp. 4(6) 356-368

[10] Valantika L, Firman H and Nahadi 2017 Development virtual test of reaction rate based visual perceptual skills to measure students' mastery concept *IOP Conf. Series*

[11] Basyari A, Sunaryo and Iswanto B H 2012 Pengembangan Media Pembelajaran Fisika Berbasis Adobe Flash untuk Menjelaskan Fisika Inti dan Radioaktivitas untuk SMA Kelas XII in *Seminar Nasional Fisika*

[12] Marwanti K, Suherman, Wibowo F. C, Darman D . and Guntara Y 2020 Assessment Virtual Test (ASVITE): Assessment Virtual *JPPPF (Jurnal Penelitian dan Pengembangan Pendidikan Fisika) vol 6 (1)*

[13] OECD 2017 Assessment and Analytical Framework: Science, Reading, Mathematic, Financial Literacy and Collaborative Problem Solving Revised Edition (Paris: OECD Publishing)

[14] B B 2017 Exploring Variability Sources in Student Evaluation of Teaching via Many-Facet Rasch Model *Eğitimde ve Psikolojide Ölçme ve Değerlendirme Dergisi*

[15] Ayre C and Scally A J 2014 Critical values for Lawshe's content validity ratio: Revisiting the original methods of calculation. Measurement and Evaluation in Counseling and Development vol. 47(1) p. 79–86

The effect of learning models and divergent thinking on higher-order thinking skills

R Hidayati, Y Supriyati and A S Budi
Department of Physics Education, Universitas Negeri Jakarta
Jl. Rawamangun Muka, Jakarta 13220, Indonesia
Email: riahidayati2727@gmail.com, yetti.supriyati@unj.ac.id, agussb@unj.ac.id

Abstract. This study aims to determine the effect of interaction between Problem Based Learning models, Creative Problem-Solving models, and divergent thinking on high order thinking skills. The method used in this study is a quasi-experimental design with treatment by level 2x2. The sample number of this study is as many as 120 students of grade X high school. The sample was divided into two, namely the control class and the experimental class. The control class was given learning using the Problem Based Learning model while the experimental class used the Creative Problem Solving model by having high and low divergent thinking skills. Data about students' high order thinking skills were obtained using a test in the form of an essay consisting of 5 questions and analyzed using the two-way ANOVA statistical test and the Tukey test as a follow-up test. The results showed that 1) there is a difference in high order thinking students who use the creative problem-solving model with students who use the problem-based learning model, 2) there is a difference in high order thinking skills between students with high divergent thinking ability with low divergent ability students and 3) there is an influence of interaction between learning model and divergent thinking ability to high order thinking skill.

1. Introduction

Physics is one of the subjects that is considered important to be taught as a vehicle to develop higher proficiency to develop science and technology [1]. Entering the 21st century, science and technology are developing very rapidly. The rapid pace of development of science and technology certainly impacts challenges and competition by every country globally [2]. To be able to play a role in the global world, every country is absolutely to prepare a generation that has 21st Century Skills. The best way that can be done to make it happen is through education. According to the National Education Association (2002) states that 18 kinds of 21st Century Skills need to be supplied to each individual, one of which is Learning and Innovation Skills which consists of 4 aspects, namely critical thinking, communication, collaboration, and creativity [3]. Critical thinking is one of the skills of high-order thinking skills which is a skill that students must have in solving a problem. One's thinking ability can affect learning ability, speed, and effectiveness of learning [4]. The ability to think is needed in solving problems, especially in physics [5]. Higher-order thinking skills are an important aspect of teaching and learning. High-order thinking skills include the ability to analyze, evaluate, and create [4]. While divergent thinking is the ability of individuals to find various alternative answers to a

Journal of Physics: Conference Series **2019** (2021) 012010 doi:10.1088/1742-6596/2019/1/012010

problem [6]. The ability of students in solving high-level questions is still low, especially on questions C4 and C5 [7]. The lack of students' ability to solve high-level questions is due to the lack of developing high-level students' thinking skills in the learning process [8]. Learning models that are seen as being able to help and facilitate to facilitate students in mastering physics concepts and practice developing higher-order thinking skills, one of which is the Creative Problem Solving (CPS) learning model because with the Creative Problem Solving learning model students will be more active in the learning process and students will be accustomed to solving and developing their mindset in dealing with a problem [9]. In addition to the reasons above, several studies have proven that problem-solving learning models can also improve higher-order thinking skills according to research results which state that problem-based learning models are effective for improving students' higher-order thinking skills [10].

2. Method

This research is research in the form of a quasi-experimental design (quasi-experimental) with treatment by level 2 x 2. In this study, there are two independent variables, namely the learning model (treatment variable) and divergent thinking ability (attribute variable), and one dependent variable, namely High Order Thinking Skills. The independent variables are grouped into two types, namely Problem Based Learning models and Creative Problem-Solving models, while divergent thinking skills consist of high divergent thinking skills and low divergent thinking abilities.

Table 1. Research designs

Divergent Thinking	Learning Models	
	Creative Problem Solving	Problem Based Learning
High	A_1B_1	A_2B_1
Low	A_1B_2	A_2B_2

This design has a control group, but cannot function fully to control external variables that affect the implementation of the experiment [11]. This study used two classes, namely, the control class was treated using the Problem Based Learning model while the experimental class was treated using the Creative Problem Solving model. The focus of this research is to analyze the differences in High Order Thinking Skills on the material of Elasticity and Hooke's Law of class X students based on the differences in the level of students' divergent thinking abilities.

3. Results and Discussion

The summary of data regarding students' high-order thinking skills can be seen in Table 2 below. Based on the results of the high-order thinking skill test of students who took part in learning the creative problem-solving model, the students' maximum score = 100, minimum score = 35, and average = 74.58. Meanwhile, for the results of the high order thinking skill test of students who took part in the problem-based learning model, the maximum score of students = 100, the minimum score of students = 30, and the average = 72.50. Furthermore, for the test results of high order thinking skills of students who have high divergent thinking skills, the maximum score of students = 96, minimum value = 70, and average = 79.72. Meanwhile, for the results of the high order thinking skill test for students who have low divergent thinking skills, the maximum score of students = 70, the minimum score of students = 20, and the average = 59.02.

Table 2. The Summary of Data High Order Thinking Skill

	A_1	A_2	B_1	B_2	$A_1 B_1$	$A_1 B_2$	$A_2 B_1$	$A_2 B_2$
Average	74.58	72.50	79.72	59.02	71.17	78.00	71.50	73.33
Variance	285.24	145.41	50.14	142.25	346.14	201.00	125.25	163.89
Standard Deviation	16.89	12.06	7.08	11.93	18.60	14.18	11.19	12.80
Maximum Value	100	100	96	70	100	95	95	100
Minimum Value	35	30	70	20	35	40	45	30

To test the normality of the data in this study used SPSS analysis Kolmogorov-Smirnov analysis test. Where a data is called normally distributed if the significance level > from the specified significance level is 0.05.

Tabel 3. The Normality Test with Kolmogorov-Smirnov

No	Sampel	Statistic Kolmogorov - Smirvon	Df	Description
1	A_1	0.080	60	Normal
2	A_2	0.061	60	Normal
3	B_1	0.099	60	Normal
4	B_2	0.066	60	Normal
5	$A_1 B_1$	0.338	30	Normal
6	$A_1 B_2$	0.373	30	Normal
7	$A_2 B_1$	0.265	30	Normal
8	$A_2 B_2$	0.169	30	Normal

Based on the results of the normality analysis of the data obtained, it shows that the research data are normally distributed.

Furthermore, the homogeneity test in this study also used Levene's SPSS test. Where a data has a homogeneous variance if the significance level > of the specified significance level is 0.05.

Tabel 4. The Homogeneity Test

df	Mean Square	F	Sig
1	563.333	2.615	0.109

Based on the results of the homogeneity analysis of the data obtained, it shows that all data groups have homogeneous variants.

Based on the results of the normality test and the homogeneity of variance test, it can be concluded that the data from all groups came from a population that was normally distributed and had a homogeneous variance. Therefore, hypothesis testing with ANOVA can be done. Furthermore, if there is an interaction between the influence of the learning model of divergent thinking, then a further test is carried out using the Tukey test.

The results of this study indicate that students' high-order thinking skills are better when using creative problem-solving models than problem-based learning models. High order thinking skills of students are also influenced by divergent thinking skills and there is also an interaction effect between learning models and divergent thinking on high order thinking skills.

The Creative Problem Solving learning model is a problem-centered learning model that emphasizes the balance between divergent thinking and convergent thinking. In addition, the Creative Problem Solving learning model can also increase students' activities and creative thinking and think critically in the learning process [12]. During the learning process, students do not only listen and take notes. Expressing opinions, asking friends during discussions, and other activities both mentally, physically, and socially so that students can use various ways with their creative power to solve a problem.

Many factors influence the success of learning, one of which is the use of learning models. Creative problem solving is solving problems by developing their thinking by using various alternatives in solving problems. Variety of learning models that may be applied to solve problems are creative problem-solving, learning models (Creative Problem Solving Model) which are variations of Problem Solving learning by solving problems through systematic techniques in organizing creative ideas to solve problems [13]. A person's creative way of thinking that can help solve a problem and achieve a certain goal using various ways or ideas that are not only one-way is called divergent thinking. This helps students practice higher-order thinking skills where students can analyze, evaluate and produce a new solution to the problems at hand, not just knowing and memorizing a concept. The CPS model is a learning model to solve problems creatively. According to Osborn, almost all problem-solving efforts always involve the six characteristics which are abbreviated as OFPISA: Objective, Finding, Fact Finding, Problem Finding, Solution Finding, and Acceptance Finding [14]. In addition, the teacher's role in CPS is to direct creative problem-solving efforts. The teacher is also tasked with providing subject matter or discussion topics that can stimulate students to think creatively in solving problems [15].

Problem-based learning is one of the innovative learning models that can provide active learning conditions for students. This learning model is designed with real problems that are useful for improving students' high-level thinking skills. The role of teachers in this model is as a facilitator so that learning is not centered on teachers but students. This problem-based learning model can develop students' thinking processes because learning begins with giving problems to students. Students must therefore study or understand the problem to solve it. Guided by the teacher, students can solve the problem well according to their creativity and understanding so that students will gain new knowledge from their own experience.

4. Conclusion

Based on the results and discussion, it can be concluded that there are differences in high order thinking skills of students who use creative problem-solving models and students who use problem-based learning models, there are differences in high order thinking skills between students with high divergent thinking abilities and students with low divergent abilities and there is an interaction effect between learning models and divergent thinking skills to high order thinking skills.

Acknowledgments

We would like to take this opportunity to thank the entire committee of The 10th National Physics Seminar (SNF 2021) organized by the Physics Department, Faculty of Mathematics and Natural Sciences, Universitas Negeri Jakarta (UNJ), supported by the Physical Society of Indonesia (PSI) Jakarta and Banten Chapter and for your effort and expertise as a reviewer. Your help enabled us to meet the scheduled time and to maintain the standards of peer-reviewed journals. Your dedication is most appreciated as the number of manuscripts is increasing every year.

References

[1] Soros P, Pinkham K, and Ekkapim S 2018 *AIP Conference Proceeding 1923*

[2] Hariharasudan A and Sebastian Kot 2018 *Social Sciences*

[3] Sandro 2020 *Advances in Social Science, Education and Humanities Research, volume 566*

[4] Yee M H, Widad B O and Jailani BM 2011 *International Journal of Social Science and Humanity*

[5] Rothman J, Marzuki, Wahyudi, and Petrie S D 2019 *Journal of Turkish Science Education Volume 16 Issue 4*

[6] Colzato L S, Szapora A, Lippelt D et al 2017 *Mindfulness 8*, 10-16

[7] Yusuf I and Widyaningsih S W 2018 *Jurnal Komunikasi Pendidikan* 42-49

[8] Nurhayati and Angraeni L 2017 *JPPPF – Jurnal Penelitian & Pengembangan Pendidikan Fisika*, 119-126

[9] Fauziah M, Marmoah S, Murwaningsih T, and Saddhono K 2019 *European Journal of Educational Research Volume* 9, Issue 2, 537-568

[10] Sucipto 2017 *Jurnal Pendidikan Volume 2 Nomor 1*, 63-71

[11] Sugiyono 2017 *Metode Penelitian Kuantitatif, Kualitatif, dan R&D* Bandung: Alfabeta, CV

[12] Hariawan H, Kamaluddin K, and Cahyono U 2014 *Ejurnal Pendidikan Fisika Tadulako*, 48-54

[13] Totiana F, Susanti E, and Redjeki T 2012 *Jurnal Pendidikan Kimia (JPK)*

[14] Huda M 2013 *Model-Model Pengajaran dan Pembelajaran* Yogyakarta: Pustaka Pelajar

[15] Muti'ah U, Waluya S B, and Mulyono 2019 *Prosiding Seminar Nasional Pascasarjana UNNES*, 88-93

Journal of Physics: Conference Series **2019** (2021) 012011 doi:10.1088/1742-6596/2019/1/012011

Virtual Microscopic Simulation (VMS) Design on Light Waves: Interference and Diffraction

D Anggraini [1], F C Wibowo[2] and M Delina[2]
[1]Physics Education Master Program, Universitas Negeri Jakarta, Jl. Rawamagun Muka, Rawamangun, 13220, Indonesia
[2]Department of Physics Education, Universitas Negeri Jakarta, Jl. Rawamagun Muka, Rawamangun, 13220, Indonesia
Email: dewi25308@gmail.com, fcwibowo@unj.ac.id, Mutia_Delina@unj.ac.id

Abstract. The properties of light waves are only looking indirectly through physic phenomena, such as interference and diffraction of light. This causes light waves to become of the physics materials that are difficult for students to understand. This study aims to produce a Virtual Microscopic Simulation (VMS) design on light wave material. The research method used in this research is research and development using the ADDIE development model. However, the ADDIE development model used is only to Design stage. The instrument used to measure the feasibility of VMS media is a validation test by media and materials experts. The results showed that characteristics of VMS design in light waves consist of two topics, namely interference and diffraction of light. VMS has a positive impact on students in the physics learning, especially about microscopic materials. VMS expected to effective in helping students to understand the light wave material, especially about the concept of interference and diffraction of light.

1. Introduction

Visible light is a form of electromagnetic radiation, so the light has properties like a wave. The properties of light waves are only looking indirectly through physic phenomena, such as interference and diffraction of light. In physics learning, often when studying the concept of interference and diffraction of light in the classroom, students directed to solve problems with the use of complicated math and then to prove for real carried out experiments of interference and diffraction light in the laboratory [1]. This causes most students have difficulty in learning the concept of interference and diffraction light correctly [2]. In addition, the student's greatest difficulty in understanding light waves lies in the fact that students are unable to interpret of interference and diffraction concept in terms of basic wave models as well as their lack of understanding de Broglie wavelengths [3]. Many students fail to observe important features of particular patterns and identify the differences among similar patterns [4]. Even students who have studied light waves at the introductory level and beyond are often unable to explain the patterns of interference and diffraction produced on the screen, whether in a single slit or double slit. Many don't know whether to apply geometric optics or physical optics to certain situations and make it possible to improperly combine the elements of the two. Meanwhile, the symptoms of interference and diffraction of light are often met in everyday life, such as oil scattered on a paved road reflects the rainbow colours that seem after the rain. The colours seen in the oil are due to interference between the light reflected on the front surface and the thin back surface of the oil. If many sources of waves present in the oil result in diffraction [5]. In some other cases, most students have difficulty recognizing double slit interference patterns and diffraction lattice patterns of light [6]. Therefore, teachers and students must show attention to observing and distinguishing patterns of interference and diffraction in light waves.

Based on this description, needed approach that facilitates development of a relatively deep idea of light waves, where the tools used can visualize light waves in real [7]. One of the efforts to solve this problem is to develop new media that can describe and visualize the physical phenomenon of interference and diffraction light which are then integrated with existing theories. The right medium to help students in visualizing light waves is computer simulation.

Computer simulations usually model abstract concepts and involve mathematical models [8]. Computer simulation is a simulation run by a single computer used to show a behaviour of a system. The computer simulations developed are adapted to the characteristics of the system. Computer simulations are designed to be effective in conceptual change [9]. Computer simulations offer representations ideal, dynamics and visual of physical phenomena and experiments dangerous, expensive or unfit in school laboratory [10]. Computer simulations can be used to demonstrate or practice phenomena that are difficult to be shown in the classroom [11]. The advantage of simulation when compared to other media are simulations can present simplified external visualizations of phenomena that are not easily observed by students and direct them to think interconnection between theory and its visualization [12] [13].

Computer simulations commonly used in classroom learning are virtual simulations. Virtual simulation is a learning media that displays a physical model of an object or an abstract phenomenon. Virtual simulations available in physics learning are still limited to certain physics concepts such as force, Hooke's Law, Faraday's Law, heat transfer, dynamic electricity, transition phase, and sound waves. In the concept of light waves, simulations are already available for interference and diffraction of light topics across several platforms as in PhET simulation. Based on the results of the analysis of media availability, it was found that the available simulations only show the visualization of how the light source passing through the slit experiences interference and diffraction of properties, where these properties described by light and dark patterns captured on the screen. Available simulation media have not been able to visualize microscopically how the process of interference and diffraction of light occurs both in a single slit and double slits so that students are still quite difficult to recognize and distinguish interference patterns and light diffraction correctly. From these findings can give opportunity to conduct research by designing VMS on light wave topics. Advantages of using VMS by users can clearly visualize the abstract phenomena microscopically and manifestly; reduce the occurrence of misunderstanding; improve students' understanding; and build a view that more scientific [14] [15]. This study will discuss the design of VMS media, which has an overview of the process of interference, and diffraction of light microscopically and real.

2. Method

The method used in this research is the Research & Development method. Media development made reference to the ADDIE development model. However, the ADDIE development model only to the design stage. Analysis stage is a stage to know the needs of physics learning media in schools. The analysis includes literature studies in the form of journal analysis and availability of simulation media, needs student analysis, and needs teacher analysis. While the design stage is the stage that determines how a learning program designed according to the needs and goals of learning. The design includes the storyboard of Virtual Microscopic Simulation (VMS). The instrument used to measure the feasibility of VMS media is a media and material validation test by several experts. The scheme of ADDIE model development can see in Figure 1.

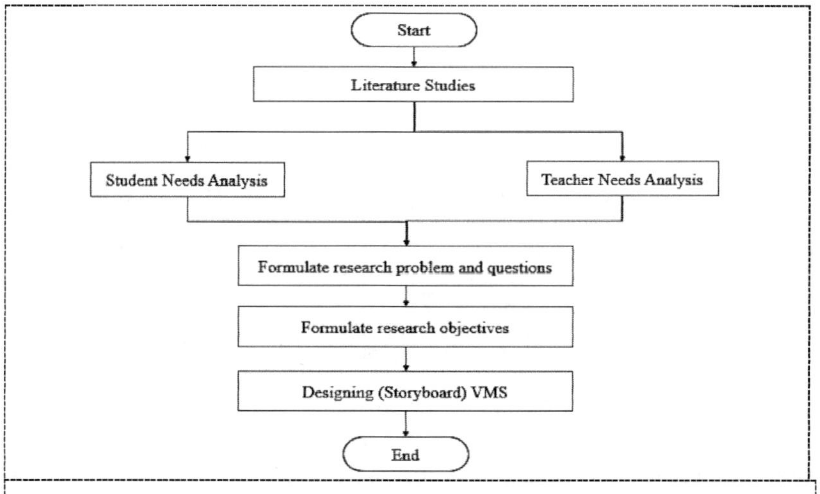

Figure 1. ADDIE development model scheme

3. Results and Discussion

3.1 Main Menu

Virtual Microscopic Simulation (VMS) is a tool used to visually imitate an abstract phenomenon to make it look more real with the help of a computer. The design characteristics of VMS on light wave material consists of two topics, namely interference and diffraction of light. The design of the VMS on light wave material can see in Figure 2.

Figure 2. Main Menu

Figure 2 shows the main menu page on VMS, which contains the developer profile menu, double slit interference menu, a single slit diffraction menu, diffraction grating menu, previous button and exit button. Each interference and diffraction of light menu contain a simulation and an explanatory material of the simulation display.

3.2 Double Slit Interference

The simulation of double slit interference as shown in Figure 3.

Journal of Physics: Conference Series **2019** (2021) 012011 doi:10.1088/1742-6596/2019/1/012011

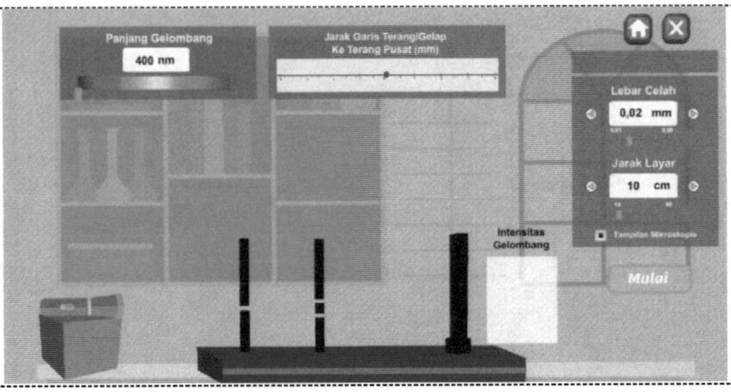

Figure 3. Double slit interference simulation

Figure 3 shows a simulation of double slit interference selected by the user. On this screen there several settings menus that can be changed according to the experiments by the user. The settings menu in this simulation consists of a wavelength setting button, a wide setting button between two slits, a screen to slit spacing setting button, a microscopic display button that serves to microscopic display the process of light interference in a double slit. In addition, if the user wants to return to the main menu can click the home button and to exit the simulation the user can click the exit button. The microscopic simulation display of double-slit interference is illustrated in Figure 4.

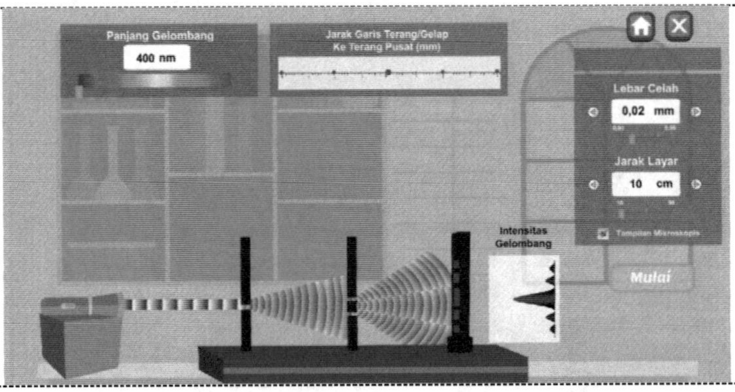

Figure 4. The interference pattern for a double slit

The output in this simulation will present a display of the light propagation process that through two narrow slits to create a light/dark pattern captured on the screen. When a light/dark pattern captured on the screen, a display box will appear in the form of a graph of wave intensity and the distance between the light/dark lines to the central light.

3.3 Single Slit Diffraction

Light diffraction is the bending of light waves through a narrow slit. The simulation of single slit diffraction as shown in Figure 5.

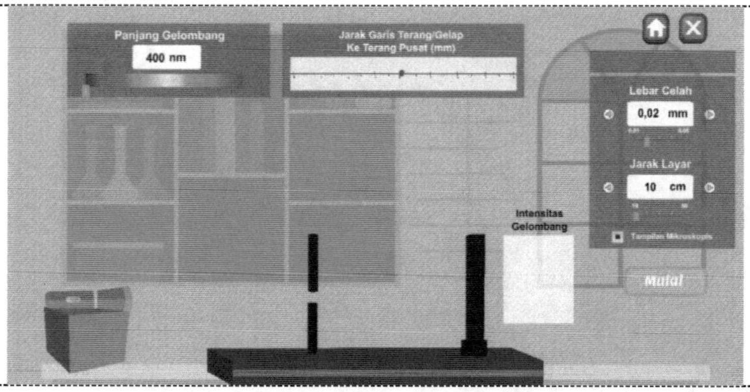

Figure 5. Single slit diffraction simulation

Figure 5 shows a simulation of a single slit diffraction selected by the user. On this screen there several settings menus that can change according to the experiments by the user. The settings menu in this simulation includes a wavelength setting button, a wide setting button between slits, a screen to slit distance setting button, a microscopic display button that serves to microscopic display the process of light diffraction performing at a single slit. If the user wants to return to the main menu can click the home button and to exit the simulation click the exit button. The microscopic simulation display of single-slit diffraction is illustrated in Figure 6.

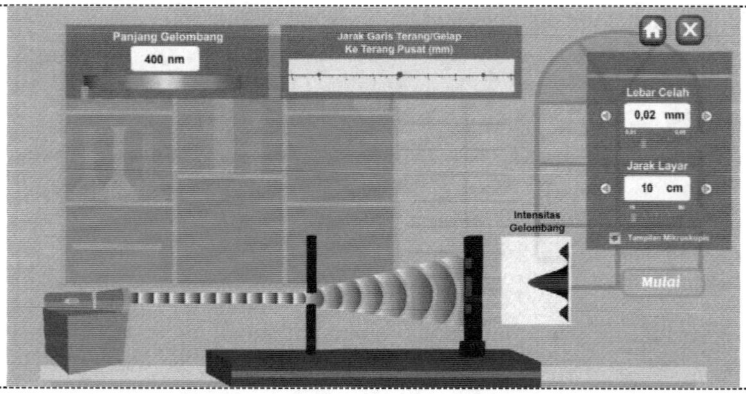

Figure 6. The diffraction pattern for a single slit

The output in this simulation will present a display of the light propagation process that through a narrow slit resulting in a light/dark pattern captured on the screen. If a light/dark pattern is captured on the screen, a display box will appear in the form of a graph of wave intensity and the distance between the light/dark lines to the central light.

3.4 Diffraction grating

The simulation of diffraction grating as shown in Figure 7.

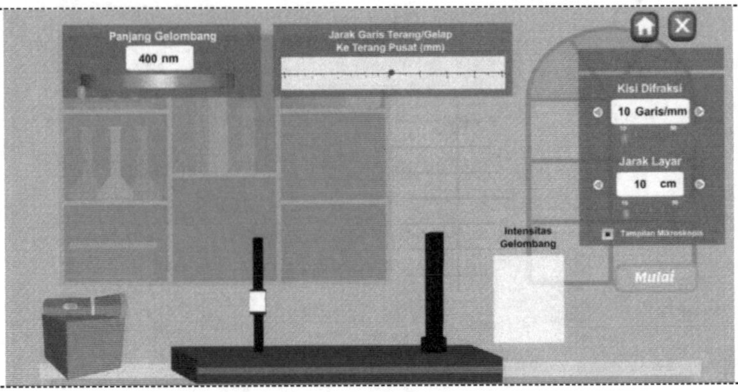

Figure 7. Diffraction grating simulation

Figure 7 shows a simulation of the diffraction grating selected by the user. On this screen there several settings menus that can change according to the experiments by users. The settings menu in this simulation consists of wavelength setting buttons, lots of grids setting button, the screen in a grid spacing setting button, a microscopic display button that serves to microscopic display the process of light diffraction occurrence on the grid. On the screen there is also a home button that serves to return to the main menu, and for users who want to exit the simulation can click the exit button. The microscopic simulation display of a diffraction grating is illustrated in Figure 8.

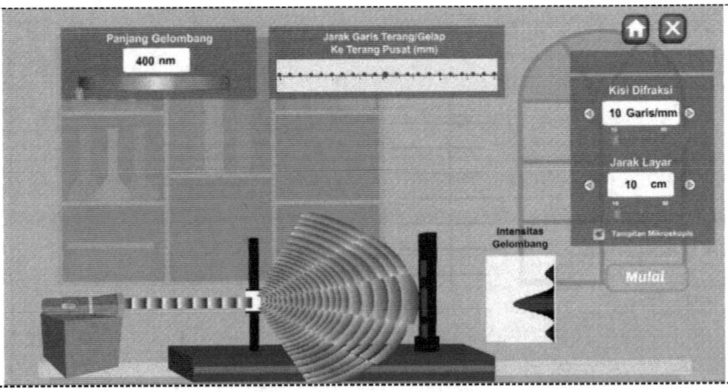

Figure 8. The diffraction pattern for a grating

The output in this simulation will present a display of the light propagation process that through a grid, resulting in a light/dark pattern captured on the screen. If a light/dark pattern captured on the screen, a display box will appear in the form of a graph of wave intensity and the distance between the light/dark lines to the central light. The design characteristics of VMS on light wave material consists of two topics, namely interference and diffraction of light. The result of VMS design will produce VMS media used by students to see, recognize, and understand how the process of interference and diffraction of light so that patterns of interference and diffraction of light form in the form of light and dark lines. The advantages of the developed VMS product are (1) the VMS visualizes microscopically how light propagates so that light to have light interference and diffraction properties, (2) VMS displays the relationship between the distance between the slits and the intensity of light waves. This helps students to know that the smaller the distance between the slits, the greater intensity of the waves. (3) VMS displays how the relationship between wavelength and intensity of light waves. This helps students to know that the greater the wavelength, the greater intensity of the wave. (4) VMS is easy to access and use anytime anywhere. VMS has a positive impact on students in the study of physics, especially of microscopic material. With the development of VMS expected to effectively help students to reduce misconceptions and improve their understanding in studying light wave material, especially on the topic of interference and diffraction of light.

4. Conclusion

Virtual Microscopic Simulation (VMS) is a tool used to visually imitate an abstract phenomenon to make it look more real with the help of a computer. The design characteristics of VMS on light wave material consists of two topics, namely interference and diffraction of light. VMS has a positive impact on students in the study of physics, especially on microscopic material. With the development of VMS expected to effectively help students to understand the light wave material, especially about the concept of interference and diffraction of light.

References

[1] L Maurer 2013 Simulating interference and diffraction in instructional laboratories *Phys. Educ* **48** 227
[2] V Mesic, K Neumann, I Aviani, E Hasović, W J Boone, N Erceg, V Grubelnik, A Sušac, D S Glamočić, M Karuza, A Vidak, A Alihodžić, R Repnik 2019 Measuring students' conceptual understanding of wave optics: A Rasch modeling approach *Phys. Rev. Phys. Educ. Res* **15** 010115
[3] S Vokos, P S Shaffer, B S Ambrose, L C McDermott Y 2000 Student understanding of the wave nature of matter: Diffraction and interference of particles *American Journal of Physics* **68** S42
[4] A Susac, M Planinic, A Bubic, K Jelicic, L Ivanjek, K M Cvenic, M Palmovic 2021 Effect of Students' investigative experiments on students' recognition of interference and diffraction patterns: An eye-tracking study *Phys. Rev. Phys. Educ. Res* **17** 010110
[5] H D Young and R A Freedman 2001 Fisika Universitas Edisi Kesepuluh Bandung: Erlangga
[6] A Susac, M Planinic, A Bubic, L Ivanjek and M Palmovic 2020 Student recognition of interference and diffraction patterns: An eye-tracking study *Phys. Rev. Phys. Educ. Res* **16** 020133
[7] V Mesic, E Hajder, K Neumann and N Erceg 2016 Comparing different approaches to visualizing light waves : An experimental study on teaching wave optics *Phys. Rev. Phys. Educ. Res* **12** 010135
[8] D R Darman, F C Wibowo, A Suhandi, W Setiawan, H Abizar, S Nurhaji, S Nulhakim, L Nulhakim and A Istiandaru 2019 Virtual media simulation technology on mathematical representation of sound waves *Journal of Physics: Conf. Series* **1188** 012092
[9] K C Trundle and R L Bell 2010 The use of a computer simulation to promote conceptual change: A quasi-experimental study *Computers and Education* **54** 1078
[10] C Sarabando, J P Cravino and A A Soares 2014 Contribution of a computer simulation to students' learning of the physics concepts of weight and mass *Procedia Technology* **13** 112
[11] G Triyani, A Danawan, I Suyana, I Kaniawati 2019 An investigation of students' misconceptions about momentum and impulse through interactive conceptual Instruction (ICI) with computer simulation *Journal of Physics: Conf. Series* **1280** 052008
[12] A Budi, D Muliyati 2018 Discovering and understanding the vector field using simulation in android app *Journal of Physics: Conf. Series* **1013** 012062
[13] S Moser, J Zumbach, I Deibl 2017 The effect of metacognitive training and prompting on learning success in simulation-based physics learning *Science Education* **101** 944
[14] F C Wibowo, A Suhandi, D Rusdiana, Y Ruhiat, D R Darman, A Samsudin 2017 Effectiveness of Microscopic Virtual Simulation (MVS) for conceptualizing students' conceptions on phase transitions *American Scientific Publishers* **23** 839-843
[15] F C Wibowo, B H Iswanto 2019 Designing MOOCS with Virtual Microscopic Simulation (VMS) for increasing of student's level of understanding *Journal of Physics: Conference Series* **1402** 066094

Design of Massive Online Simulation (MOS) On Concept Archimedes' Principle

N Khofiya, F C Wibowo and M Delina
Physics Education Master Program, Universitas Negeri Jakarta,
Jl. Rawamangun Muka, Jakarta 13220, Indonesia
Email: nidaankhofiya_1310819008@mhs.unj.ac.id, fcwibowo@unj.ac.id, mutia_delina@unj.ac.id

Abstract. In learning physics, Archimedes' law is a physics material that is quite difficult and requires a deep understanding of concepts based on study files from several schools in Banten. As many as 96.4% of students need online simulation media using computers or androids to help study Archimedes Law material. The Massive Online Simulation (MOS) design is divided into six parts, namely the dashboard, home, materials, videos, simulations and quizzes. This study aims to design a media in the form of Massive Online Simulation (MOS) on Archimedes' legal material. The method used is ADDIE (Analysis, Design, Development, Implementation and Evaluation). Research and development of the ADDIE model is carried out only up to the Design stage. This research can produce an attractive Massive Online Simulation (MOS) design that can be accessed anytime and anywhere.

1. Introduction

Digital information and communication skills of learners have increased rapidly and are considered important skills in the 21st century, which refers to several aspects related to knowledge, beliefs, attitudes, and values regarding information and communication technology (ICT) [1]. The development of science and technology that is increasing rapidly in the field of information and communication technology has an influence in various fields including the field of education. Advances in Information Technology have encouraged many changes, including in the field of education, one of which is the emergence of multimedia as a learning medium. Various kinds of diverse technological advances can increase the interaction between teachers and students [2]. Learning media that are currently developing and are often used by teachers and students during the covid-19 pandemic, such as edupage, google meet, schoology, moodle and so on are examples of Massive Online [3]. Massive Online is a distance learning platform, generally contains free courses on a topic, and is characterized by a very high number of registrations [4]. There is a study that states that MOOCS media can increase academic self-efficacy, teaching attendance, and perceived benefits have a significant direct effect on learning engagement but there is no learning mechanism [5].

Based on the results of observations in schools that have been carried out 72% of students learn by using laptops / computers and telephone. As many as 85.5% of students like simulation media when applied in the learning process. However, about 88% of students have difficulty in understanding Archimedes' Law material. As many as 96.4% of students actually need online simulation media using a computer or android to help study Archimedes' Law material. In overcoming this, an online simulation media is needed that is attractive and easy to access anywhere and anytime to be able to assist students in understanding Archimedes' Law material.

Massive Online Simulation shows that academic, teaching attendance, and perceived usefulness have a significant direct effect on learning engagement [6]. In physics learning, Archimedes' law is one of the physics materials that are quite difficult and requires an in-depth understanding of the concept because it is difficult to observe directly or invisible or unobservable, and students experience serious confusion and difficulties [7]. In addition, current teachers are required to change the direct learning system into online learning (distance) due to the spread of Covid-19. The development of various physical phenomena of visualization media and has not been observed, until now, is mostly done for the sake of learning physics, but the developed simulation does not have a learning mechanism, cannot display various objects in the simulation, is less user friendly, and is only in the form of videos, and cannot be said virtual simulation media. The novelty in this research is to develop a Massive Online design which contains Archimedes law simulation material that can facilitate students in learning physics, so that students can use MOS as a fun physical learning resource anywhere and anytime. As well as the novelties in this research is to develop a design based on the shortcomings of the media availability analysis and MOOCS shown in the results and discussion. Therefore, referring to the findings and results of the research above, it is deemed necessary to develop an MOS design, namely a media development design to obtain a product in the form of a complete virtual simulation with the use of simulation mechanisms, materials, videos, simulations, quizzes and user friendly.

2. Method

The research method used is ADDIE (Analysis, Design, Development, Implementation and Evaluation). Research on the development of the ADDIE model was carried out only at the design stage, because the purpose of this study was to design media in the form of Massive Online Simulation (MOS) on interesting and accessible Archimedes law materials anywhere and anytime, with 80 high school students as research subjects in several schools in the province. Banten, Indonesia. The research flow for the development of Massive Online Simulation (MOS) is shown in Figure 1.

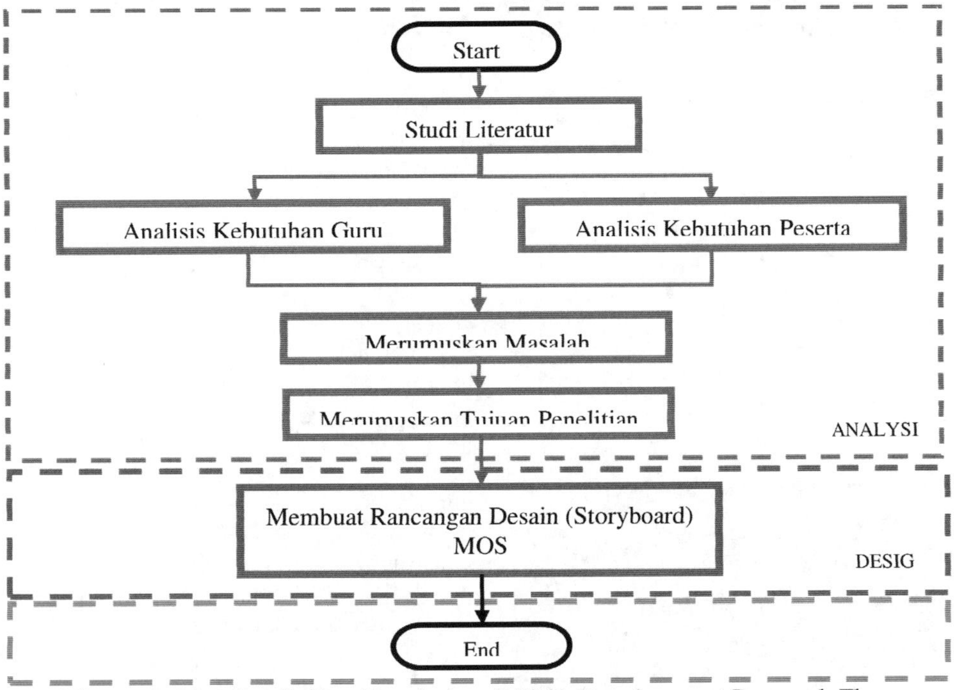

Figure 1. Massive Online Simulation (MOS) Development Research Flow

As shown in Figure 1. The research flow consists of literature studies, field studies, namely analyzing the needs of teachers and students, formulating problems, formulating research objectives, and making Massive Online Simulation (MOS) designs.

3. Result and Discussion

3.1. Media availability analysis

Virtual simulation has advantages in attracting the attention of students. Simulations can help students understand the meaning of scientific knowledge, concepts and facts for students. Virtual simulations contribute to teachers and students' professional development, and provide an understanding of science in science teaching [8]. Virtual simulations give students the opportunity to collaborate with learning and are able to strengthen knowledge about interprofessional competencies [9]. Virtual simulation has advantages in attracting the attention of students [10]. Virtual simulations can contribute to the teaching and learning process by giving students the opportunity to learn by doing, providing interesting and fun activities to encourage students to discover, and ensuring active classroom interaction through discussion in the learning process [11]. The following is a platform analysis of the availability of simulation media on Archimedes' legal material, shown in Table 2 which has been developed by previous researchers and has been published.

Table 1. Analysis of Platform Availability of Virtual Simulation Media on Archimedes' Legal Materials

Virtual Simulation Media has been Developed	Source	Information
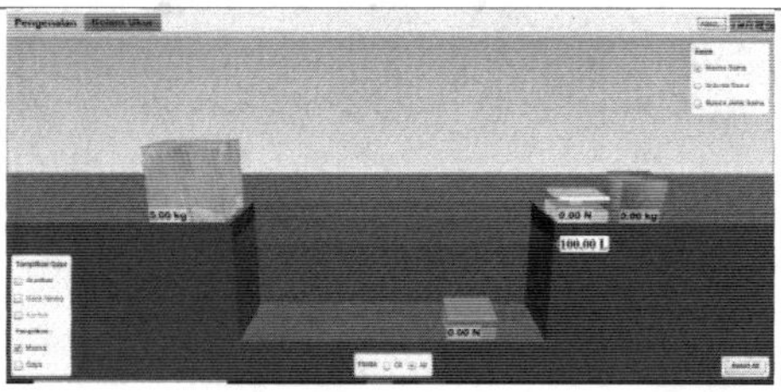	https://phet.colorado.edu/sims/density-and-buoyancy/buoyancy_in.html	This Archimedes law simulation can display various kinds of objects, but cannot display learning mechanisms and the data is only a description of physical phenomena.
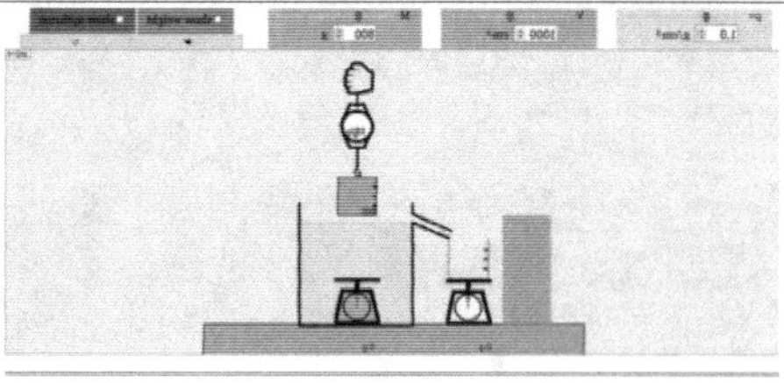	https://iwant2study.org/lookangejss/02_newtonianmechanics_6pressure/ejss_model_archimedes/archimedes_Simulation.xhtml	This Archimedes law simulation runs very slowly and cannot display various kinds of objects such as iron or aluminum.
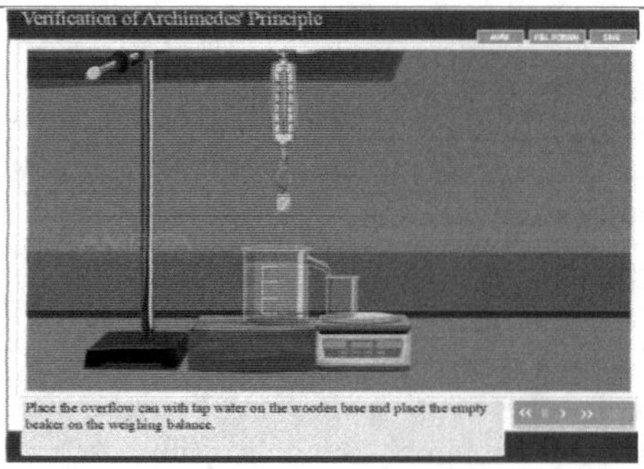	https://amrita.olabs.edu.in/?sub=1&brch=1&sim=72&cnt=131	This Archimedes law simulation cannot display various kinds of objects to determine floating, floating and sinking objects. However, there are conclusions after doing the simulation.

Journal of Physics: Conference Series **2019** (2021) 012012 doi:10.1088/1742-6596/2019/1/012012

Virtual Simulation Media has been Developed	**Source**	**Information**
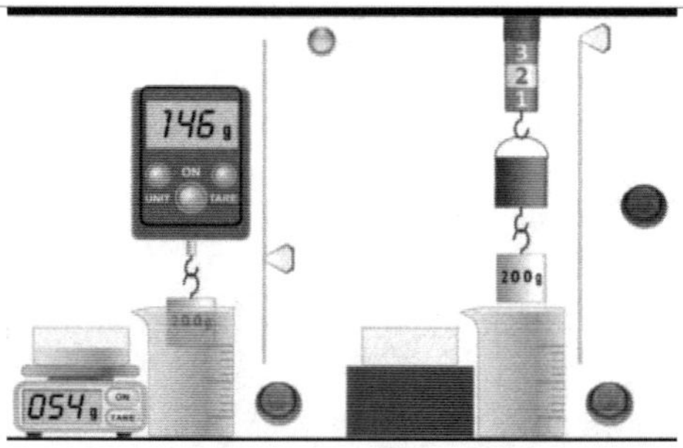	https://www.vascak.cz/data/android/physicsatschool/template.php?s=mech_archimedes&l=en	This Archimedes law simulation can be easily operated, but does not display physical phenomena
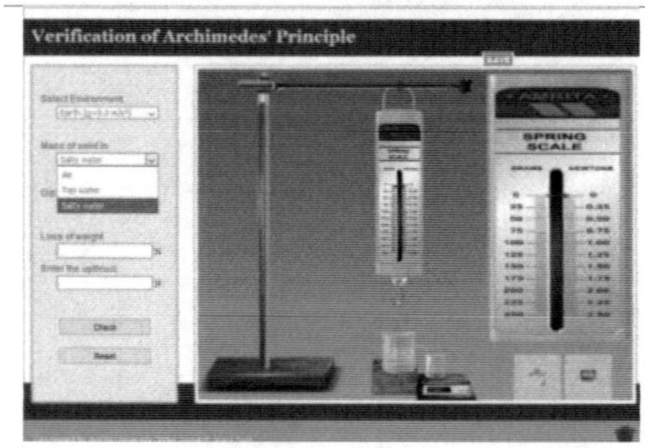	https://www.golabz.eu/lab/verification-of-archimedes-principle	Archimedes' law simulation provides a variety of objects but is not user friendly
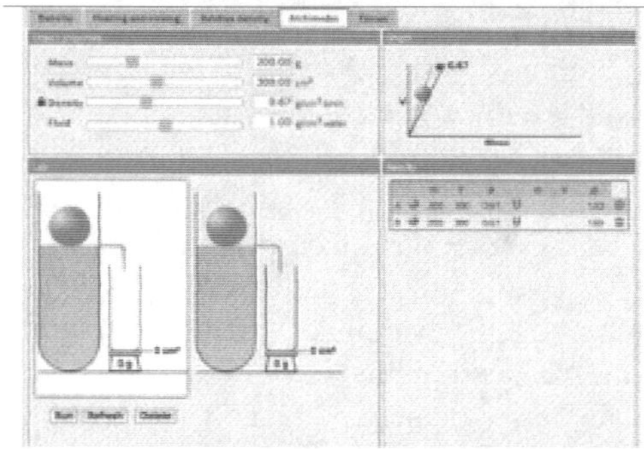	https://www.golabz.eu/lab/splash-virtual-buoyancy-laboratory	This simulation can display data, but has not been able to manipulate models and expand learning
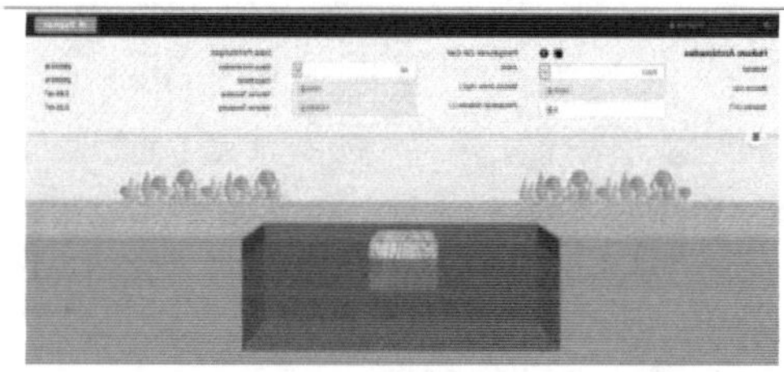	https://vlab.belajar.kemdikbud.go.id/Experiments/hukumarchimedes/#/	This Archimedes simulation is almost similar to phet, but the objects provided are more varied, the data are only limited to a description of physical phenomena.
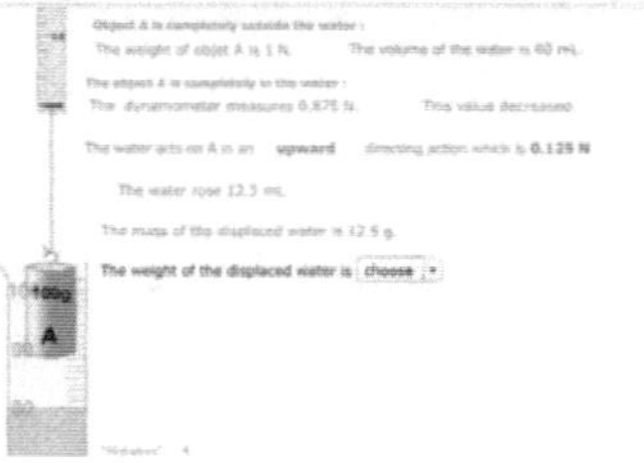	ww.physics-chemistry-interactive-flash-animation.com/mechanics_forces_gravitation_energy_interactive/buoyancy_	This Archimedes simulation has not been able to manipulate the model and expand learning

Virtual Simulation Media has been Developed	Source	Information
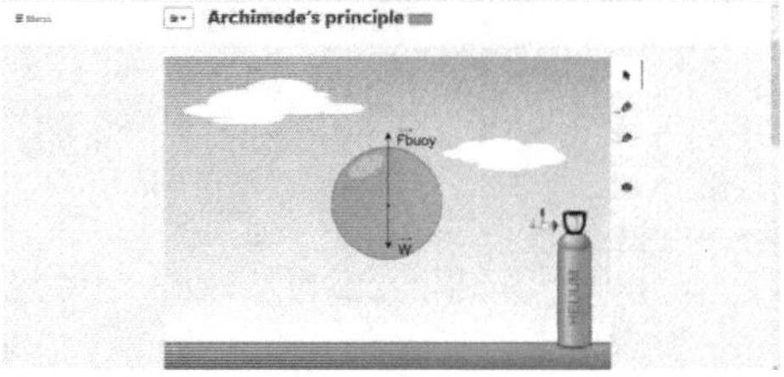	https://www.edumedia-sciences.com/en/media/365-archimedes-principle	This Archimedes simulation has not been able to manipulate the model and expand learning
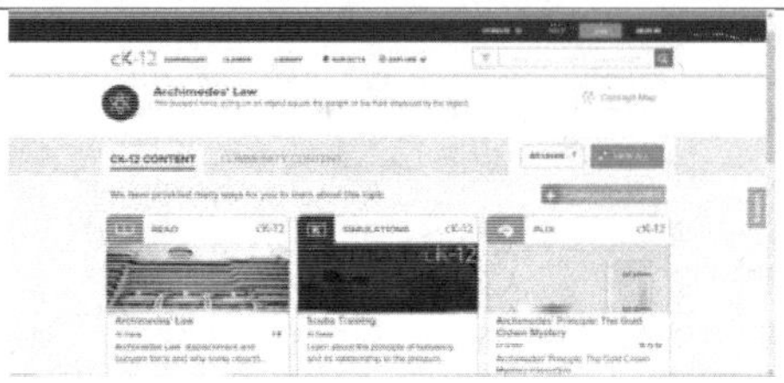	https://www.ck12.org/physics/archimedes-law/	This platform is quite complete like the Moodle platform, but when it is opened, one of the simulations is only a video

Based on Table 1, information is obtained that some of the simulation media in Archimedes' legal material still have shortcomings, namely, there is no learning mechanism, cannot display various kinds of objects in the simulation, is less user friendly, and is only in the form of video, and cannot be said to be virtual simulation media because Therefore, a media development design is needed to get a product in the form of a virtual simulation complete with a simulation use mechanism, a sequence of material on Archimedes' law material and user friendly.

Massive Online Simulation (MOS) is an adoption of the word MOOCs. Massive Online Simulation (MOS) is learning that is carried out online which has the nature of open learning, free of charge or free, with online media as the channel, referring to students who are given objects in the form of virtual representations of real objects or commonly called simulations, occurring through the website. And courses, structured around a set of learning objectives within a defined field of study. The following is an analysis of the Massive Online Simulation (MOS) availability platform shown in Table 2 which has been developed by previous researchers and has been published.

Table 2. Analysis of Massive Online Simulation (MOS) Availability Platforms

Massive Online Simulation that has been developed	Source	Information
	https://www.my-mooc.com/en/	This MOOCS has a good design and complete courses, but does not show the use of MOOCS.

Massive Online Simulation that has been developed	Source	Information
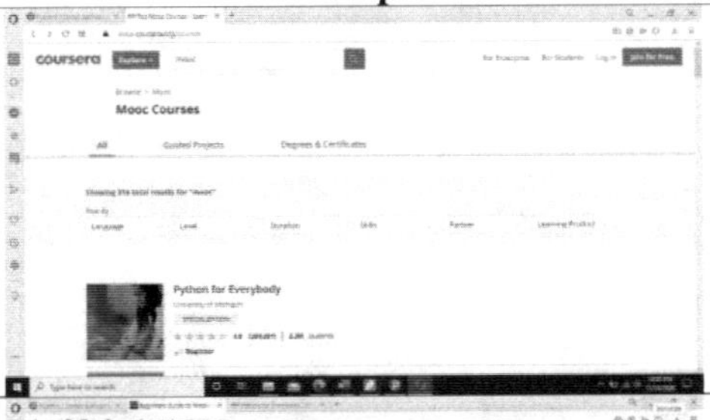	https://www.coursera.org/courses?query=mooc	This MOOCS features several accessible, but paid courses.
	https://www.classcentral.com/help/moocs	This MOOCS displays the use of MOOCS and learning mechanisms, but does not display interesting images or features.
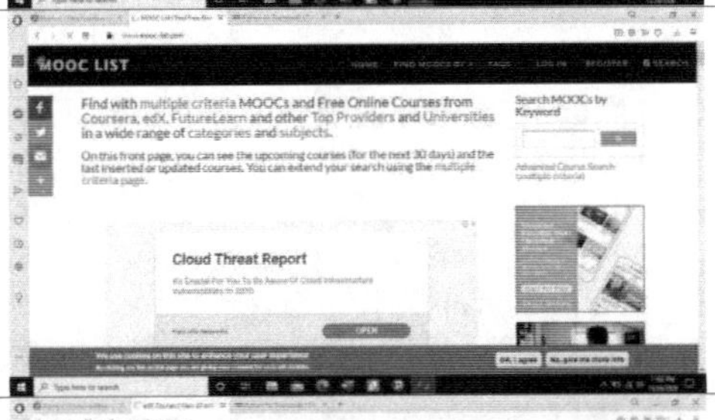	https://www.mooc-list.com	This MOOCS is one of the providers of free courses, but there is a free limitation in use for only 30 days.
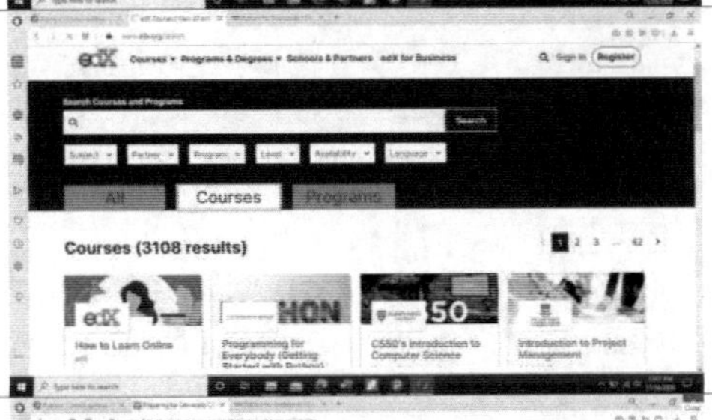	https://www.edx.org/search	This course service provider is very complete and has an attractive design.
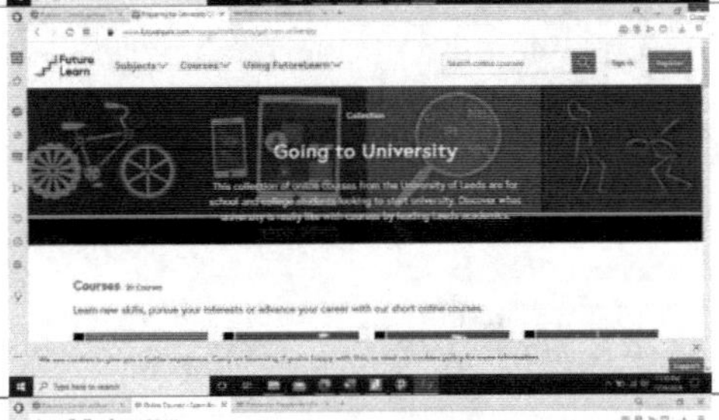	https://www.futurelearn.com/courses/collections/get-into-university	This MOOCS features a very attractive design and appearance, but there are no mechanics or tutorials for using media.
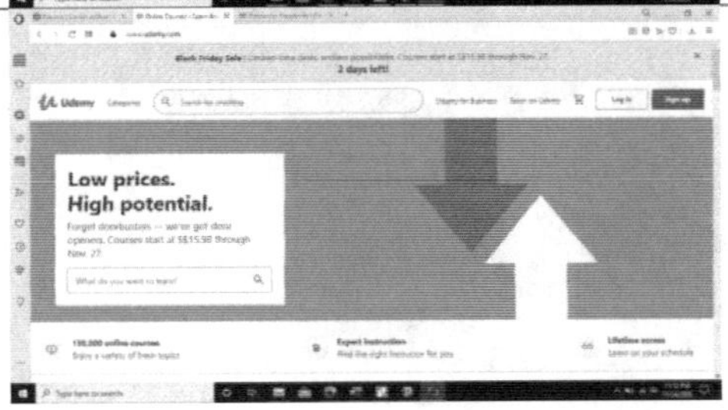	https://www.udemy.com	This MOOCS features multiple learnings and has a unique design.

Massive Online Simulation that has been developed	Source	Information
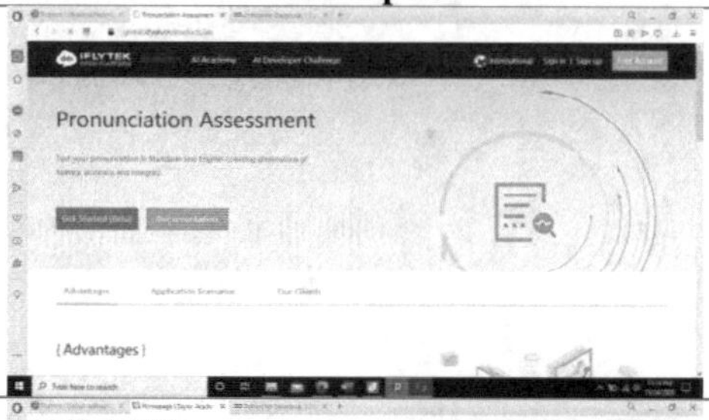	https://global.xfyun.cn/products/ise	This MOOCS has a free account feature, so it can make it easier for students to have an account.
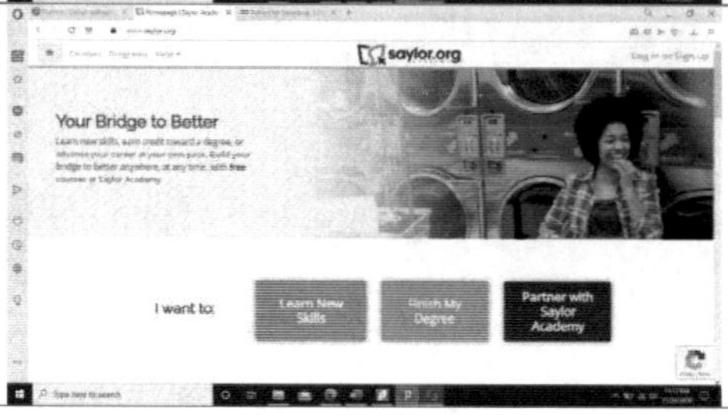	https://www.saylor.com	The design used is attractive and good, but the features used are few.

Based on Table 2, information is obtained that some MOOCS media still have shortcomings, namely they do not display the use of MOOCS, are paid, do not display interesting pictures or features, there is no learning mechanism, therefore, a MOOCS development design that displays the use of media, is free (free), displays interesting pictures or features, there is a learning mechanism and by adding simulations in the massive online to get a product in the form of Massive Online Simulation (MOS). The advantages of this study are that the MOS design developed can display the use of media, it is free, displays interesting images or features, there is a learning mechanism by adding online simulations to get products in the form of Massive Online Simulation (MOS), user friendly and media designed to be attractive and interesting, and can be accessed anywhere and anytime.

3.2. MOS development

Massive Online Simulation (MOS) is learning that is carried out online which has the nature of open learning, free of charge, with online media as the channel, referring to students who are given objects in the form of virtual representations of real objects or commonly called simulations, occurring via the web, and courses, structured around a set of learning objectives within a defined field of study. This media were developed using the ISpring Suite platform. In addition, the media are also made interactive, Massive Online has several components, namely home, dashboard, materials, videos, simulations accompanied by calculation data, and quizzes. The Massive Online Simulation (MOS) design developed in this study is as follows.

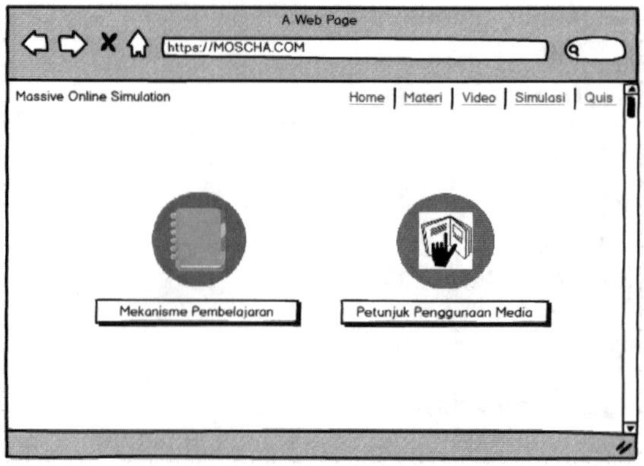

Figure 2. MOS display at home

Based on Figure 2, this screen appears when students have successfully logged in to Massive Online Simulation (MOS). This display is the home screen for Massive Online Simulation (MOS).

There are several main menus on the home screen, namely the home menu, materials, videos, simulations and quizzes, as well as a menu of learning mechanisms and instructions for using media.

1. Home, the main screen when logged in to each account. And there is a menu of learning mechanisms and instructions for using media so that students are guided in using massive online simulation.
2. Material, on the material menu, of course, there is Archimedes' law material.
3. Video, on the interesting video menu there is a video of the application of Archimedes' law in life which can be witnessed in real and up close.
4. Simulation, in the simulation menu, there is a simulation of Archimedes' law with several kinds of objects.
5. Quiz, on the quiz menu there are 2 quizzes to find out ICT literacy in students when using massive online simulation of Archimedes legal materials.

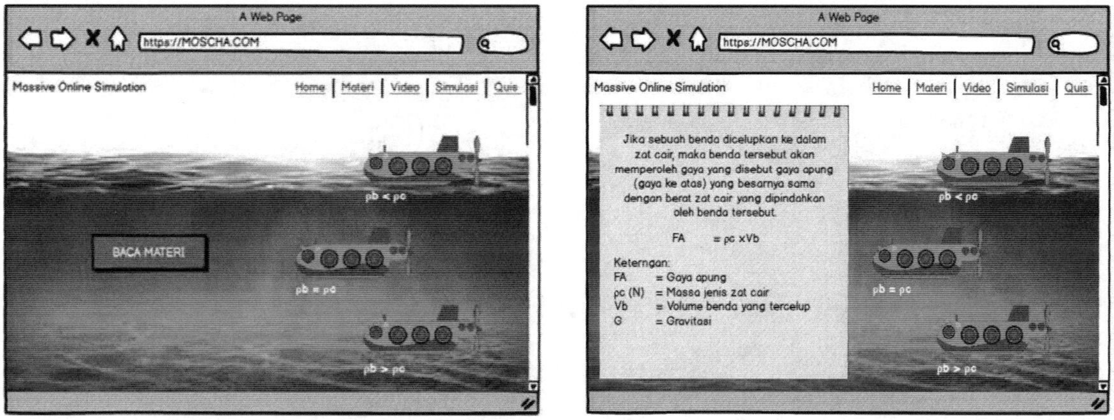

Figure 3. MOS Display on Material

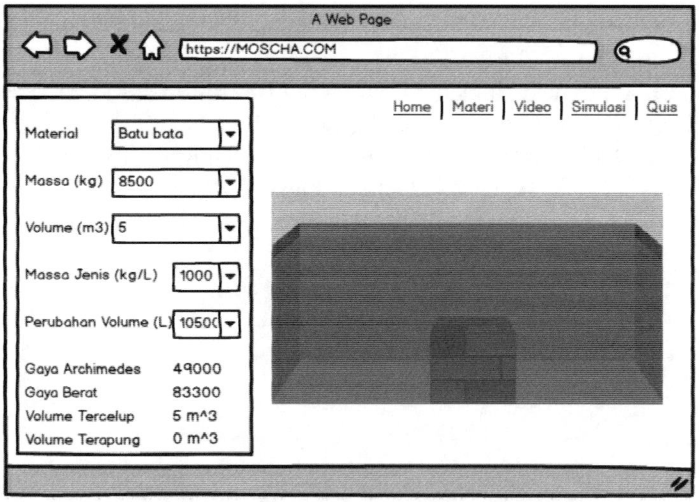

Figure 4. MOS Display in Simulation

Based on Figures 3 and 4, the display of material and simulation in Massive Online Simulation (MOS). On the material display, there are several main menus, namely the home menu, materials, videos, simulations and quizzes and there is a "read material" button to go to the Archimedes law sequence display. In the simulation display there are several components and calculation data, namely material, mass, volume, density and volume change and data for calculating Archimedes force, gravity, submerged volume and floating volume are provided. The material in the simulation display can be selected which consists of several types of objects, namely bricks, floating objects and Styrofoam in accordance with Archimedes' law concepts, namely floating, floating, and sinking. When students select an object on the material or click on a brick material. The simulation screen will display a simulation of bricks submerged in water and the calculated data.

4. Conclusion

Students are interested and like simulation media when applied in the learning process. From the research it was stated that students experienced serious confusion and difficulties in Archimedes' law material. So a Massive Online Simulation (MOS) design was made for Archimedes' legal materials, by analyzing the availability of media on Archimedes' legal materials and analyzing the availability of MOS. Based on the analysis of the availability of media on Archimedes' legal material and massive online, it is necessary to design media to get a product in the form of a virtual simulation complete with learning mechanisms, simulations, materials, videos and quizzes on Archimedes' legal materials. Display images or features that are attractive, free, and can be accessed anywhere and anytime to get a product in the form of Massive Online Simulation (MOS) which will later be developed in the form of Massive Online Simulation which will be applied in a learning process.

References

[1] Schmid, R., & Petko, D. (2019). *Computers and Education, 136* (September 2018), 75–86.
[2] Sun, A., & Chen, X. (2016). *Journal of Information Technology Education: Research, 15*(2016), 157–190.
[3] Basilaia, G., & Kvavadze, D. (2020). *Pedagogical Research, 5*(4).
[4] Sciarrone, F., & Temperini, M. (2020). *Soft Computing, 24*(15), 11121–11134.
[5] Jung, Y., & Lee, J. (2018). Computers & Education, Vol. 122, 9–22. https://doi.org/10.1016/j.compedu.2018.02.013
[6] Lee, Y. (2019). *Journal of Educational Computing Research, 57*(2), 471–490.
[7] Loverude, M. E., Kautz, C. H., & Heron, P. R. L. (2003). *American Journal of Physics, 71*(May 2014), 1178–1187.
[8] Wibowo, F. C., Suhandi, A ., Rusdiana, D., Ruhiat, Y., Darman, D. R., & Samsudin, A. (2017). Advanced Science Letters, 23(2), 839–843.
[9] Aleksandrov, E., & Levitskaya, A. (2018). Media Education (Mediaobrazovanie), 58(4), 1–147.
[10] Martín-Gutiérrez, J., Mora, C. E., Añorbe-Díaz, B., & González-Marrero, A. (2017). Eurasia Journal of Mathematics, Science and Technology Education, 13(2), 469–486.
[11] Faour, M. A., Ayoubi, Z., & The, Z. (2018). Journal of Education in Science, Environment and Health, 4(1), 54–68.

Journal of Physics: Conference Series **2019** (2021) 012013 doi:10.1088/1742-6596/2019/1/012013

Virtual Microscopic Simulation (VMS) for physics learning of the photoelectric effect in high school

A F Mandagi, B H Iswanto and I Sugihartono
Physics Education Master Program, Universitas Negeri Jakarta, Jl. Rawamangun Muka, Jakarta 13220, Indonesia
Email: anisamandagi202426@gmail.com; bhi@unj.ac.id; iwan-sugihartono@unj.ac.id

Abstract. The photoelectric effect is a physics concept that is often considered difficult by students because it contains abstract and microscopic concepts. So we need a medium that can visualize the concept of particle wave dualism on the photoelectric effect phenomenon microscopically. This study aims to produce "Virtual Microscopic Simulation (VMS)" as a medium for learning physics on the topic of the photoelectric effect in high school. The research method used is Research and Development with a 4D model (Define, Design, Development, and Disseminate). Validation of media feasibility has been carried out by several experts covering aspects of content and media with an average score of 77.2% and 75.7% with good categories. Based on the results of the expert validation test, it can be concluded that VMS can visualize the photoelectric effect process that is suitable for use as a medium for learning physics for students in high school.

1. Introduction

Physics is a fundamental science because it is the basis for the development of science and technology. Given the importance of the role of physics in human life, physics should be understood with the correct concept. The concept of physics consists of macroscopic and microscopic properties [1]. Macroscopic properties can be observed and measured easily, so it is generally well understood. On the other hand, microscopic properties cannot be understood properly because they are abstract and cannot be observed directly (unobservable) [2]. If microscopic properties can be understood, then macroscopic properties will also be easy to understand [3].

The photoelectric effect is included in the microscopic properties due to take place at the microscopic level that cannot be observed directly [4]. Students find it difficult to understand the concept of the photoelectric effect, for example, in the process of ejection of electrons and the flow of electrons in a circuit [5]. The characteristics of the abstract photoelectric effect concept make the potential for students to have an incomplete understanding. In addition, the phenomenon of the photoelectric effect is not easy to understand because of its complex nature and requires a large amount of time and money if done in the classroom [5].

The use of appropriate learning media can visualize microscopic concepts, making it easier for students to understand the concept of physics as a whole [6]. The use of instructional media in learning activities becomes an important part that should be well prepared and planned. Simulations can help students visualize aspects of science that are too big or too small to see [7]. Simulation can help the student to demonstrate physical phenomena, abstract and difficult to teach that can be delivered in a more clear and simple for students [8].

From a preliminary study that has been carried out by distributing questionnaires to one high school in Jakarta. A total of 35 students of class XII contributed in filling out the questionnaire. The results of the analysis of student's needs regarding learning of the photoelectric effect show that physics learning of the photoelectric effect material has not been maximized, especially in the use of learning media. As many as 65% of students are not interested in studying physics and as many as 83% of students find it difficult to understand the material of the photoelectric effect. The results of the teacher's needs analysis of learning the photoelectric effect show the teacher's responses through the distributed questionnaires. Aspects of limited learning media and low concept understanding results indicate that students have difficulty understanding the photoelectric effect material through the media used during learning. As many as 65% of teachers agree that the media used does not support the understanding of the photoelectric effect concept.

Several developments of abstract and microscopic photoelectric effect simulation media have been found in previous studies, including: photoelectric effect simulation combined with games [5] and photoelectric effect PhET simulation [9, 10]. However, from several photoelectric effect simulations that have been developed, there are several weaknesses, including: the simulation does not properly assess the purpose of the battery in the circuit. In addition, to determine the response to the use of PhET simulation on the topic of the photoelectric effect, a response to the use of PhET simulation media was carried out. It was found that 31% of students stated that they could not open the PhET simulation, this is because the PhET simulation has a .Jar format so that the applications run are very limited.

Therefore, the media to be developed will visualize the phenomenon of the Photoelectric Effect microscopically with virtual simulation, with additional features in the form of a voltage slider that represents the battery in the simulation with a value of -1V to +1V (smallest scale 0.1 V), and file type media in flash format (.swf) so that it is expected to make it easier for students to use the media. Based on this description, it becomes the basis for developing "Virtual Microscopic Simulation (VMS)" for physics learning that can visualize the concept of the photoelectric effect at the microscopic level in high school.

2. Method

This type of research is a R&D method (Research and Development). Developed learning media in the form of Virtual Microscopic Simulation (VMS) of the photoelectric effect created using Adobe Flash CS6 with a 4-D model of development [11]. The 4-D development model consists of four main phases, namely: Define, Design, Develop, and Desseminet. But this research was carried out until the Development stage, namely expert validation of the content and media aspects developed. Broadly speaking, the research procedures consisted of field studies, literature studies, instrument development, VMS development, and expert judgement. Validation instrument to media experts and content in the form of a questionnaire sheet is used to measure the feasibility of the resulting product

2.1 Virtual Microscopic Simulation (VMS) Media Evaluation Questionnaire for Media Experts

To measure the feasibility of the VMS media that has been developed, it is necessary to have a media research instrument that is tested by media experts. The making of the instrument is based on interactive multimedia criteria put forward by several experts and reference sources. The criteria for assessing multimedia-based learning categorized as good are as follows:
a. Ease of navigation, i.e. the developed program should be as simple as possible.
b. Media integration between aspects of language and skills to be learned.
c. The media developed must provide the learning desired by students.
d. Cognition, namely the program developed must meet the learning needs of students.

The following are the aspects used to assess VMS media, which are shown in Table 1.

Table 1. Aspects of the Media Expert Questionnaire

No	Assessment Aspect	Indicator Aspect Assessment
1	Media design quality	• Reliable • Maintainable
2	Media Interactivity	• Communicative • Creative (layout design, typography, color) • Mobile media (animation, interactive layout, navigation icon)

2.2 Virtual Microscopic Simulation (VMS) Media Evaluation Questionnaire for Content Experts

The research instrument in the form of a questionnaire was also made to obtain data on the validation of the VMS media, which was developed through examiners by material experts listed in Table 2.

Table 2. Aspects of the Questionnaire for Content Experts

No	Assessment Aspect	Indicator Aspect Assessment
1	Media compatibility with curriculum	• The suitability of the material with basic competencies • Material depth • Relevance of learning objectives to basic competencies
2	Media content	• Easy to understand, systematic, coherent and clear logic flow • The suitability of the material with the learning objectives • Presentation of logical and systematic material
3	Language	• Language effectiveness and efficiency • The language used is in accordance with the rules of the Indonesian language

2.3 Research procedure

The research procedure is concrete and detailed steps that describe the development model used, namely the 4-D model. This research is limited only to the Develop stage. According to the 4-D model, the procedure of this study is shown in Figure 1.

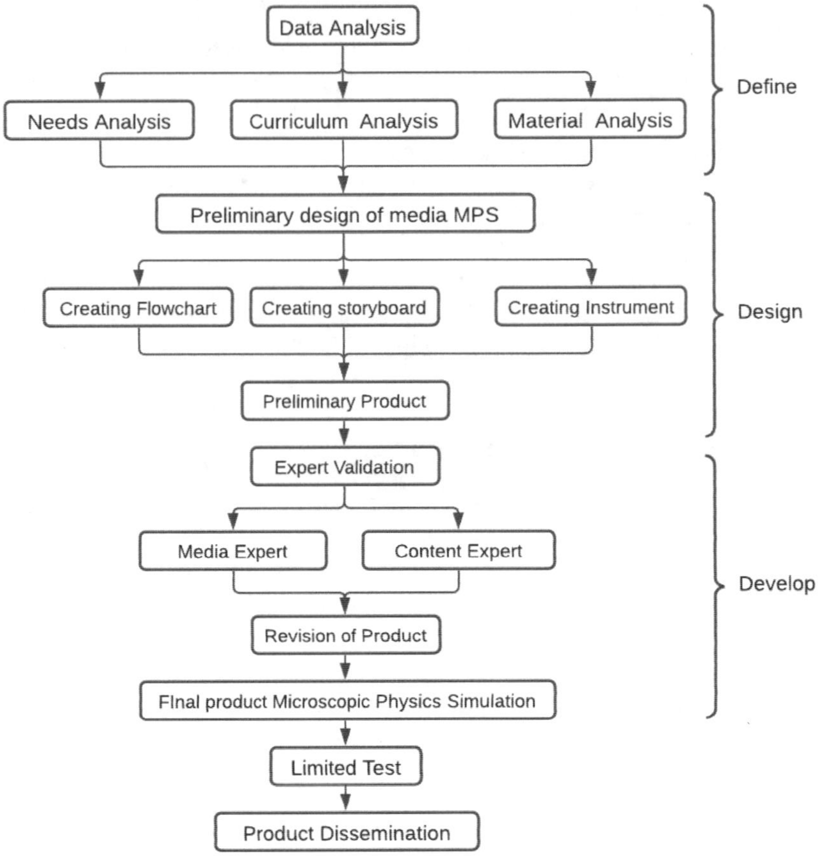

Figure 1. Research procedure using the 4-D model

Based on Figure 1, the first stage is the definition which is done to collect information from curriculum analysis. Furthermore, a needs analysis is carried out to see the actual situation in schools. Then, the material analysis is also called learning design. Learning design consists of: (1) identification of basic competencies; (2) formulating success indicators, (3) making test instrument indicators; (4) and development of learning strategies. The second step, the design of the development media is carried out with the following stages: (1) making a flowchart of the schema used as navigation; (2) collecting supporting concepts for media creation such as images, clip art, etc.; (3) create a storyboard of the developed media; (4) make an initial draft of the product of the developed media, according to the storyboard and flowchart; and (5) create an expert validation instrument to assess the media developed. The third step is evaluating VMS products which consist of validating VMS media through content experts and media experts, followed by data analysis and product revision. The last step is Deisseminate, which is the stage of promoting the media to users so that it can be accepted, both individually and in groups. Before the Desseminate stage is carried out, the media must be tested on a limited scale. However, from this stage to the Desseminate stage it cannot be done due to time constraints. Therefore, the development of VMS media, according to the 4-D model only reached the development phase or the expert validation stage.

Data analysis used quantitative data analysis methods obtained from the results of questionnaires given to media experts and content experts about the assessment of the developed media. Additional media also came from interviews with physics teachers, and criticisms and suggestions from experts regarding the quality of products developed for physics learning.

3. Result and Discussion

3.1 Validation Media of Virtual Microscopic Simulation (VMS)

Products that have been created are then validated by media experts and content experts. This assessment aims to obtain suggestions, criticisms, opinions, and evaluations from the media that have been made. In this validation there are 6 validators consisting of 3 lecturers from media experts and 3 lecturers from content experts. The results of the research by media experts and content experts can be seen in the recapitulation of the validation results in Table 3 and Table 4.

Table 3. VMS media validation results by Media Experts

No	Assessment Aspect	Media Expert Assessment Score		
		I	II	III
Media Design Quality				
1	Font, font size, font color and font layout of the letters used	4	4	4
2	Navigation buttons are easy to find and attractive	4	4	3
3	Simulation clarity, layout and size	4	3	4
4	VMS media background	4	4	4
5	The color composition used from VMS	4	4	3
Total		20	19	18
Average (percentage)		19 (76%)		
Category		Good		
Media Interactivity				
1	Easy-to-use navigation buttons	4	3	4
2	Easy-to-use menu button	4	4	3
3	Use of clear and simple language	4	4	4
Total		12	11	11
Average (percentage)		11,3 (75,6%)		
Category		Good		

Table 4. VMS media validation results by Content Expert

No	Assessment Aspect	Media Expert Assessment Score		
		I	II	III
Media compatibility with curriculum				
1	The relationship between content and basic competencies	4	4	4
2	Systematic content presentation	4	4	4
3	Considering the difficulty level of the content	4	3	4
4	Presentation of content to achieve learning objectives	3	4	4
5	The relevance of the content to the given simulation	4	4	4
Total		19	19	20
Average (percentage)		19,3 (77,3%)		
Category		Good		
Media Content				
1	Easy to understand, systematic, coherent and clear logic flow	4	4	4
2	The suitability of the material with the learning objectives	4	4	3
3	Presentation of logical and systematic material	4	4	4
Total		12	12	11
Average (percentage)		11,7 (77,8%)		
Category		Good		
Language				
1	Language effectiveness and efficiency	4	4	4
2	The language used is in accordance with the rules of the Indonesian language	4	4	3
Total		8	8	7
Average (percentage)		7,7 (77,7%)		
Category		Good		

From Table 3 and Table 4, information is obtained that the appearance and interactivity of the developed VMS media is quite good and the developed VMS media is in accordance with the photoelectric effect content. Based on the results of the VMS media validation assessment, an average of 76.45% was obtained and was declared good. The average results of the feasibility assessment of each expert on the VMS media are shown in Figure 2.

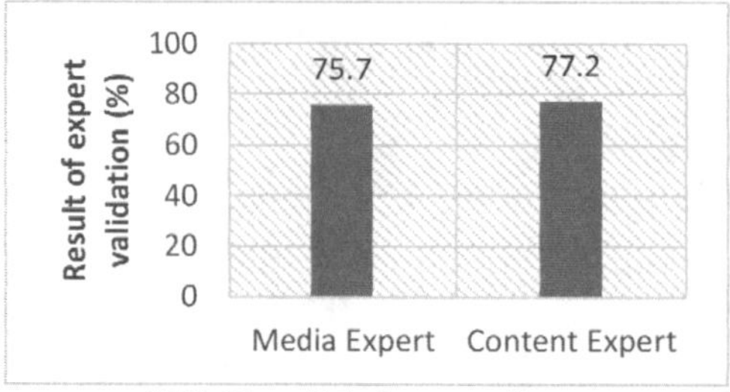

Figure 2. Percentage of VMS validation media expert and content expert

Based on the description, the media can already be used in research after the improvements made based on suggestions from experts, so it is feasible to use in physics learning, especially in the concept of the photoelectric effect.

3.2 Characteristics of Virtual Microscopic Simulation (VMS) Media

VMS media is developed using Adobe Flash CS6 which can combine text, sound, color, and animation. These advantages can be used to visualize the phenomenon of the photoelectric effect microscopically. The output file from adobe flash CS6 has a shockwafe flash extension file (.swf) which can be animated directly if there is already a flash player plugin on the computer.

The VMS media design that has been validated and revised based on suggestions and notes from experts contains an initial display, main menu, info menu, simulation menu, graph menu, and table menu. The initial appearance of the media is shown in Figure 3. The initial display consists of the name of the media, namely "Virtual Microscopic Simulation (VMS)". In the main menu there are four submenu buttons, which consist of: (1) The theory menu which shows a brief theory of the photoelectric effect; (2) Simulation menu that shows the simulation of the photoelectric effect phenomenon displayed by microscopic visualization; (3) The Graph menu shows the variables that affect the photoelectric effect; and (4) The exercise menu contains an understanding test of the photoelectric effect. The developed simulation menu is shown in Figure 4.

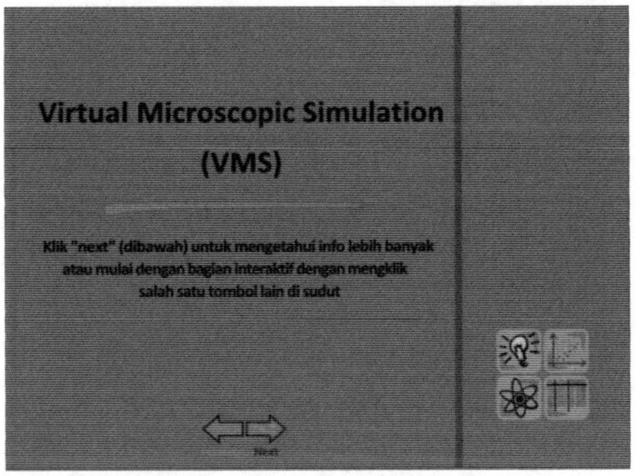

Figure 3. VMS preview

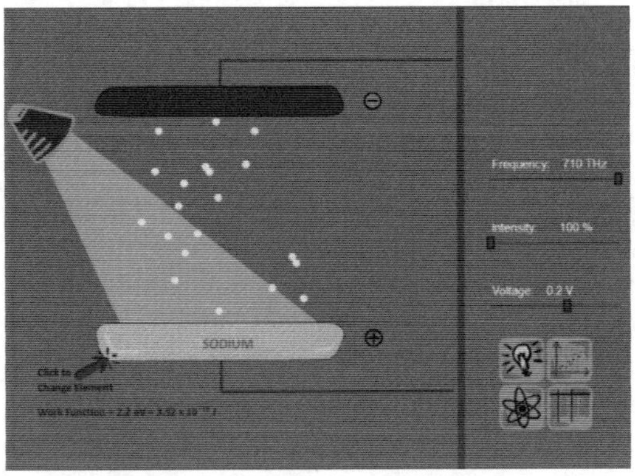

Figure 4. VMS simulation menu

Figure 3 shows the simulation menu. In the simulation menu there are several variables, such as frequency, intensity, voltage, and type of metal. The variable frequency, intensity, and battery voltage are in the form of a slider that can be moved to the right or left according to the value. The target metal type variable is a plate in a circuit that can change if the plate is clicked. In the simulation circuit there is a light source that will emit photons, the value of which depends on the frequency of the light which is indicated by the color gradation of the light particles. If the frequency of light is greater than the threshold frequency of the target metal, the electrons in the target metal will exit or move from the positive pole to the negative pole. The number of electrons that come out of the target metal is influenced by the value of the light intensity, the greater the number of electrons that come out, the more electrons will come out. In the simulation menu there is an addition, namely a voltage slider that has a value of -1v to +1v with the smallest scale of 0.1 v, to find the stopping voltage on the photoelectric effect.

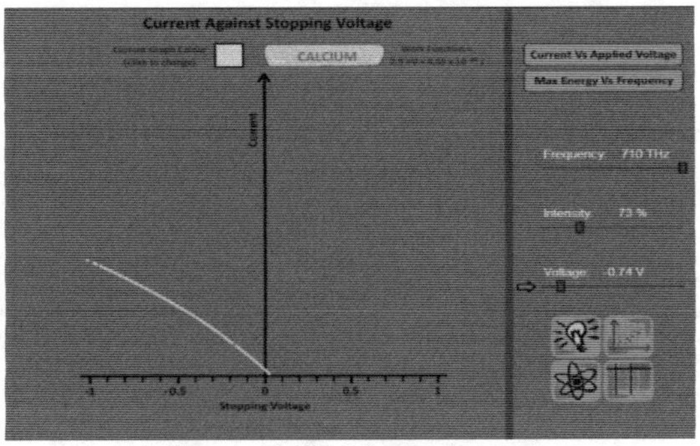

Figure 5. VMS graph menu

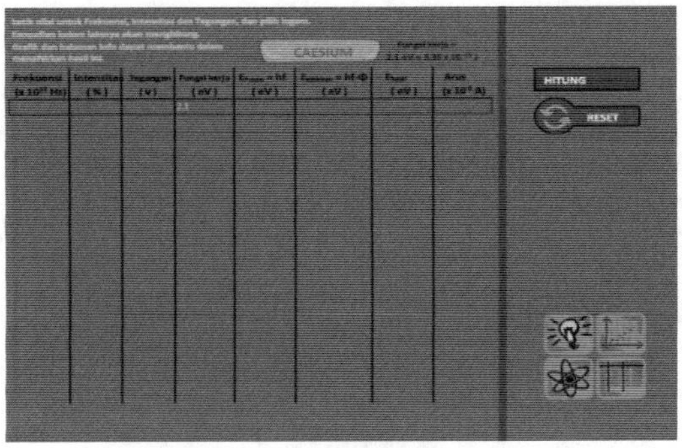

Figure 6. VMS table menu

The novelty of VMS is a graphical menu that is displayed on a full screen shown in Figure 5 with the aim that users can see more clearly the graph that is formed. The graph menu explains the relationship "Current vs Applied Voltage" and "Max Energy vs Frequency". The target metal used consisted of 4 different metals, namely sodium, potassium, calcium, and cesium. The metal type will change when the user presses the target metal. The frequency range used is 430 THz - 710 THz from red to purple. The voltage used is within -1V to +1V with a difference of 0.01V. Figure 6 shows another novelty of VMS, namely the table menu that is used to calculate the value of energy and current in the Photoelectric Effect based on the values entered into the table.

4. Conclusion

Based on the results of research and analysis, it can be concluded that the developed Virtual Microscopic Simulation (VMS) media can be a means of learning physics for student in high school, especially on the concept of the photoelectric effect with better categories, with novelty: (1) simulation displays the correct purpose of the battery on the circuit; and (2) The VMS file type is in .swf format so that it can be run immediately without a long loading time if the computer has a flash player installed.

References

[1] H. Gould and J. Tobochnik, *Princeton University Press,* vol. **79**, no. 4, pp. 431-432, 2010.

[2] F. C. Wibowo, A. Suhandi, Nahadi, A. Samsudin, D. R. Darman, Z. Suherli, A. Hasani, S. M. Leksono, A. Hendrayana, Suherman, S. Hidayat, D. Hamdani and B. Costu. *Asia-Pacific Forum on Science Learning and Teaching (APFSLT),* vol. **18**, no. 2, p. 2, 2017.

[3] F. C. Wibowo, A. Suhandi, D. Rusdiana and Y. Ruhiat. *Advanced Science Letters,* vol. **23**, no. 2, pp. 839-843, 2017.

[4] S. Klassen, *Science and Education,* vol. **20**, no. 7, pp. 719-731, 2011.

[5] T. Namgyel and K. Buaraphan, *Asia-Pacific Forum on Science Learning and Teaching,* vol. **18**, no. 2, pp. 1-30, 2017.

[6] F. C. Wibowo, A. Setiawan, U. Alizkan, D. R. Darmawan and E. Budi, *Universal Journal of Educational Research,* vol. 7, no. **12**, pp. 2867-2882, 2019.

[7] S. Khan, *Journal of Science Education and Technology,* vol. **20**, no. 3, pp. 215-232, 2011.

[8] E. Bozkurt and A. Ilik, *Procedia - Social and Behavioral Sciences,* pp. 4587-4591, 2010.

[9] Supowoko, Sarwanto, Sukarmin and Suparmi, *International Journal of Science and Applied Science: Conference Series,* vol. **33**, no. 1, 2017.

[10] A. Sokolowski, *Physics Education,* vol. **48**, no. 1, pp. 35-41, 2013.

[11] A. F. Rahmadani, H. Hidayat and E. Syahmaidi, *International Journal of Scientific Research and Management,* vol. **6**, no. 1, 2018.

[12] R. Estriégana, J. A. Medina and R. P. Barchino, *Computers & Education* , vol. **135**, pp. 1-14, 2019.

[13] F. C. Wibowo, A. Suhandi, D. Rusdiana, A. Samsudin, D. R. Darman, M. N. Faizin, Wiyanto, Supriyatman and A. Permanasari, *Journal of Physics: Conference Series,* vol. **877**, 2017.

[14] K. G. Nelson, A. F. McKenna, S. Brem and J. C. Hilpert, *Journal of Engineering Education* , vol. **106**, no. 2, pp. 218-244, 2017.

[15] V. Potkonjak, M. Gardner, V. Callagan, P. Mattila, C. Guetl, V. M. Petrovic and K. Jovanovi, *Computer & Education,* vol. **95**, pp. 309-327, 2016.

[16] A. Ballu, X. Yan, A. Blachard and T. Clet, *Procedia CIRP,* vol. **43**, pp. 148-153, 2016.

Journal of Physics: Conference Series **2019** (2021) 012014 doi:10.1088/1742-6596/2019/1/012014

The Development of Digital Comic As Learning Media Based On Picture-And-Picture Learning Model On Global Warming Materials During Distance Learning

V Serevina, I M Astra and A F Syahida
Department of Physics Education, Universitas Negeri Jakarta
Jl. Rawamangun Muka, Jakarta 13220, Indonesia
Email: vserevina@unj.ac.id, imadeastra@gmail.com, azzahra.fadya.s@gmail.com

Abstract. Global warming is an environmental physics material that studies the science of rising Earth's temperature, which includes causes, effects and remedies. Digital comics as a learning media are simple media but can involve the material as a whole with interesting delivery. This study aims to develop learning media in the form of digital comics based on picture and picture learning models on global warming materials that are suitable for use during distance learning at the high school level. The method used in this research is Research and Development (R&D) with a 4D research model (define, design, develop, disseminate). The resulting product is in the form of a digital comic that is suitable for use in physics learning during distance learning and tested on 62 students. The feasibility test for learning media on global warming material was validated by media experts, material experts and learning experts using a validation test questionnaire. Then the trial was conducted with a number of students at the high school level. The expected result is that this learning media is feasible and effective for use in physics subjects at the high school level.

1. Introduction

The Ministry of Education and Culture of Indonesia issued Circular Letter Number 15 of 2020 concerning Guidelines for the Implementation of Learning from Home in an Emergency Period for the Spread of COVID-19. In this circular, it is stated that the purpose of implementing learning from home is to ensure the fulfillment of the rights of students to obtain educational services during the COVID-19 emergency, protect education unit residents from the adverse effects of COVID-19, prevent the spread and transmission of COVID-19 in the education unit and ensure the fulfillment of psychosocial support for educators, students and parents. To address this matter. To respond to these instructions, educational institutions in Indonesia from elementary to high school levels have begun to implement distance learning online [1]. Distance learning means the learning process is carried out separately between students and teachers [2]. Although relatively new in Indonesia, according to several studies, online distance learning can be as effective as school-based education. The United States Department of Education analyzed more than 1,000 cases. From this research, it is proven that students who take online classes can outperform students who take offline classes in most subjects [3]. Therefore, online distance learning can be an alternative form of learning during the COVID-19 emergency [4].

One of the learning models that can create an interactive learning process is the picture and picture learning model [5]. The picture and picture learning model rely on images as learning media so that students can remember the subject matter through visual media [6]. Picture and picture learning model is one type of cooperative learning model that uses pictures that are paired or sequenced so that they become a logical picture arrangement [7]. From Serevina and Lulu's research, virtual media experimentation makes it easier for students to remember and understand the concept of circular motion through illustrated images [8], that means learning with picture media can help students understand the material with a more interesting way. One of the learning media that is suitable for this learning model is digital comic. This is due to the millennial generation's interest in digital media, which has one function as a tool that makes it easier for someone to access information and entertainment, including stories in the form of digital comics [9].

Basically, comics are usually printed using paper. However, in this modern world, the use of digital comics is becoming more popular. Digital comics are facilitated by devices such as computers, tablets or smartphones [10]. The use of digital comics as a learning medium is expected to be an interesting, creative and innovative learning media that can also make it easier for students to understand the subject matter. The use of online technology can make learning more active, creative and fun as well as cover a wider range of knowledge. The use of digital comics for online learning can improve the quality of the student learning process [11].

The subject chosen is global warming for 11th grade high school science majors. The following is a concept map regarding global warming.

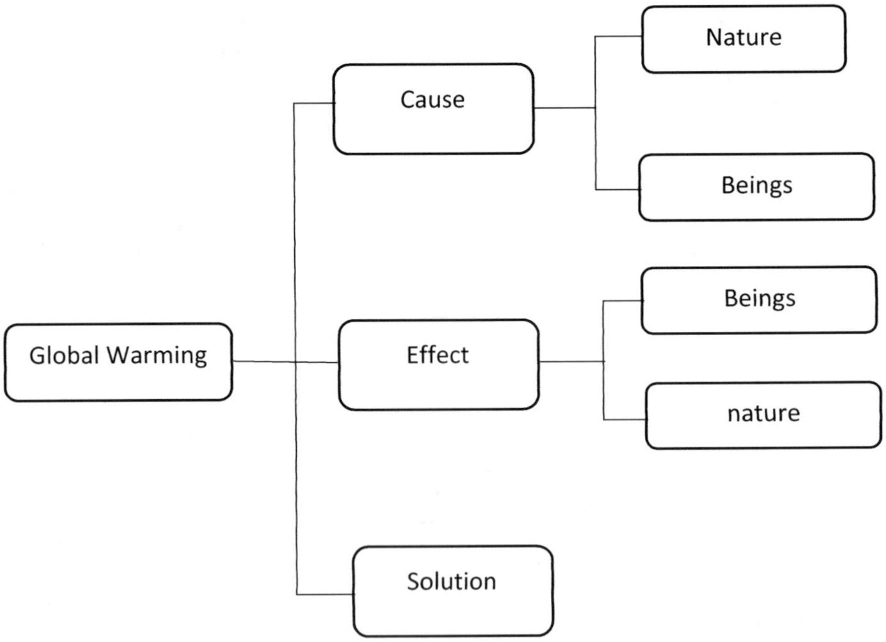

Table 1. Global Warming Concept Map

Journal of Physics: Conference Series **2019** (2021) 012014 doi:10.1088/1742-6596/2019/1/012014

2. Method

The research method used in this research is research and development methods with a 4D as the model. In the 4D model, there are four stages which include define, design, develop and disseminate [12]. At the define stage, an analysis will be carried out on the needs of students. Furthermore, at the design stage, a design will be made for the learning media under study. Then at the development stage, the learning media will be developed with validation by experts and tested on a limited basis with students and teachers. If the product is feasible and has been tested, the final stage will be carried out which is the disseminate stage where the product will be widely produced.

The data in this study were obtained through the results of validity questionnaires by expert lecturers and trials by educators and students. The data obtained from the test results were conducted on teachers and students by observing the development and using questionnaires. The validation data are divided into three data, namely the validation of the feasibility of the material by the material expert, the validation of the feasibility of the media by the media expert and the validation of the feasibility of learning from the learning expert. While the trial data were obtained from teachers and students, which were then analyzed with a score of 1-5 choices.

Table 2. Score weight

Alternative Answer	Score Weight
Totally agree	5
Agree	4
Enough	3
Disagree	2
Strongly disagree	1

Source: [17]

After that, the percentage of success is calculated using the formula:

$$Percentage\ score = \frac{Total\ score}{Max\ total\ score} \times 100\%$$

The percentage score obtained is then used to measure the quality of the developed learning media. The quality of the media can be known by using the interpretation of scores, namely:

Table 3. Interpretation of Scores

Percentage	Interpretation
0%-20%	Very feasible
21%-40%	Less feasible
41%-60%	Enough
61%-80%	Eligible
81%-100%	Very eligible

Source: [17]

3. Result and Discussion

Based on the steps of the 4D research model, the results of the research in accordance with the research and development procedures are as follows.

3.1 Define

Analysis of the literature and characteristics of students has been carried out with the aim of knowing the needs of students for learning media. Therefore, a needs analysis test was carried out on 62 respondents who were high school students majoring in science. The results of the needs analysis that has been carried out are as follows.

3.1.1 Literature Analysis

Distance learning is a new form of the learning process. Distance learning has the same goals as face-to-face learning with the aim of delivering the same material, but has a different form of presentation where students and teachers do not interact directly. Distance learning is also influenced by information technology, the internet and so on [13]. Based on the results of analytical tests that have been carried out, as many as 62.9% of respondents stated that the models and media used in the distance learning process did not make it easy to learn. Meanwhile, 54.8% of respondents stated that the distance learning that had taken place had not been effective and efficient. Therefore, according to students, an interactive and interesting learning model is needed to be implemented in the distance learning process.

3.1.2 Analysis of Student Characteristics

In school, most teachers allow students to use cell phones because cell phones are considered to make students easier in the learning process so that the learning process becomes more effective. However, in reality, this does not work effectively because students instead use their cell phones to open social media, games or even online stores. For this reason, android-based learning media such as digital comics are needed as a medium for learning physics concepts [14]. Digital comic as learning media can facilitate the learning process, especially to understand physics concepts with examples from everyday life [15]. Based on the analysis test results that have been carried out, 87.1% of respondents stated that they like to solve puzzles in the form of pictures such as puzzles and 80.6% of respondents stated that they were more interested in learning by using picture media. Therefore, the use of image media or visual learning media is very important in the student learning process. While 69.3% of respondents stated that they felt happy and interested in reading digital comics, 46.8% of respondents stated that they read digital comics very often and 37.1% of respondents stated that they read digital comics quite often. While 54.8% of respondents stated that they were interested in learning by using digital comic learning media and another 25.8% stated that they were quite interested in using digital comic learning media. Judging from these responses, it can be concluded that respondents who are high school students are enthusiastic about learning by using digital comic media and support the development of digital comics as learning media.

3.2 Design

At this stage, the design of learning media in the form of digital comics is carried out. There are several stages carried out in the design stage of learning media, including: (1) Determining the subject matter to be developed. In this developmental research, the material that will be used is global warming material, (2) determining the indicators for achieving global warming material, (3) compiling a scenario that will become the main storyline in the comic. The storyline used will be in accordance with global warming material, (4) make rough plans or rough sketches, (5) work on illustrations to realize the rough sketches that have been made and (5) arrange illustrations that have been made in the form of panels into an easy comic to understand. The application used is the Canva.

3.3 Develop

After the product has been designed, the product is developed. Validation is done by experts. If the product is declared unfit for use in the learning process, it will be revised. If the product is valid, a limited trial will be conducted on students and teachers.

3.4 Disseminate

At this stage, the product that has been validated and tested will then be disseminated as a suitable learning medium for distance learning for global warming material.

Figure 1. Cover

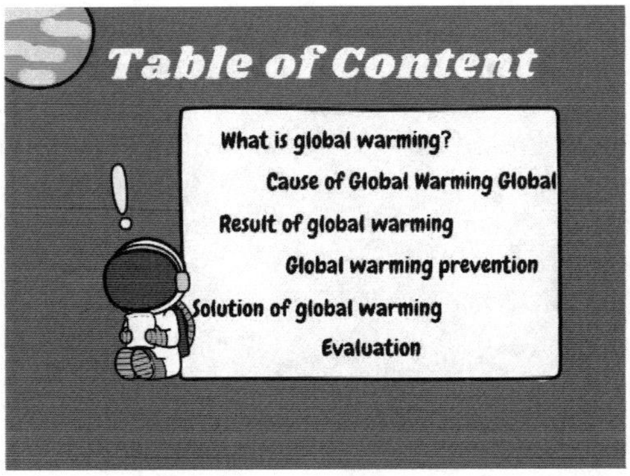

Figure 2. Table of Content

In the process of learning physics, students as focus or center in the learning process while the teacher is the facilitator. So, it takes learning media that can help teachers convey information [16]. Comic is a visual media, that contains fictional and non-fiction stories which are then presented in the form of images. In this digital era, comics that were originally printed on sheets of paper can now be accessed using digital devices such as smart phones or other devices. Comics are also in demand by children and teenagers because comics can convey information in an interesting, short and easy to understand way. In addition, comics also do not require long sentences to convey information so that comics have a light nature.

Digital comics are not only interesting, but can also be a useful learning media in the learning process. Digital comic as a learning media can increase students' interest in learning, it can also make it easier for students to understand a learning material. In addition, digital comics also have other advantages, such as being able to be accessed easily through digital devices owned by students and can be accessed anywhere and whenever students want without having to face to face in class. This advantage is very suitable for use in the distance learning process due to COVID-19 which is spreading throughout the world.

4. Conclusion

The digital comic learning media based on the picture and picture learning model was developed by research and development using the 4D research method. According to the need analysis, students need interesting media and learning models during the pandemic. In addition, students also stated that they were happy and interested in learning to use digital comic learning media. Preliminary research results show that digital comic learning media is needed by students.

References

[1] Nissa, I.C., Febrilia, B.R.A., Astutik, F., Safitri, B.R.A. and Khery, Y 2020 *JPMB: Jurnal Pemberdayaan Masyarakat Berkarakter*, 3(2), pp.185-198

[2] Anshori, Sodiq. 2018 *Civic-Culture: Jurnal Ilmu Pendidikan PKn dan Sosial Budaya 2, no. 1*

[3] Demuyakor, J 2020. *Online Journal of Communication and Media Technologies,* 10(3), p.e2020.

[4] Firman, F. and Rahayu, S 2020 "Pembelajaran Online di Tengah Pandemi Covid-19". *Indonesian Journal of Educational Science (IJES),* 2(2), pp.81-89

[5] De Leoniari, N.L 2018 . *Journal for Lesson and Learning Studies,* 1(2), pp.73-81

[6] Suci, S.H.A., Rosyidah, E., Asitah, N., Aini, N., Murni, A.W., Anam, F., Purnomo, A., Sallu, S., Mulyaningsih, I. and Kuraesin, A.D. 2018. *Journal of Physics: Conference Series,* Vol. 1114, No. 1, p. 012044

[7] Widyawati, Wiwik Yully. 2019. *Jurnal Kredo,* 2(2), pp. 226-241

[8] Serevina, V. and Lulu, M. 2021. *AIP Conference Proceedings (Vol. 2320, No. 1, p. 020052). AIP Publishing LLC.*

[9] Swandi, I Wayan, Arya Pageh Wibawa, Gede Yoga Kharisma Pradana, I Nyoman Suarka. 2020 *International Journal of Innovation, Creativity and Change,* 13(3), 718-732

[10] Nguyen, N.V., Rigaud, C. and Burie, J.C. 2017. *2017 14th IAPR international conference on document analysis and recognition (ICDAR),* Vol. 3, pp. 41-46

[11] Risniawati, M., Serevina, V. and Delina, M. 2020. *Journal of Physics: Conference Series (Vol. 1481, No. 1, p. 012075). IOP Publishing*

[12] Mulyatiningsih, E. 2016. "Pengembangan Model Pembelajaran". Accessed from *http://staff. uny. ac. id/sites/default/files/pengabdian/dra-endang-mulyatiningsih-mpd/7cpengembangan-model-pembelajaran. pdf.*

[13] Yarovaya, Olga, Larisa Yarovaya, and Elena Bogatskaya. 2020. *E3S Web of Conferences,* vol. 210, p. 18051. *EDP Sciences*

[14] Suri, Dewi Atikah, Irnin Agustina Dwi Astuti, Yoga Budi Bhakti, and Ria Asep Sumarni. 2021. In *2nd Annual Conference on Social Science and Humanities (ANCOSH 2020)*, pp. 358-361. Atlantis Press

[15] Rina, Nofha, Jenny Suminar, Ninis Damayani, and Hanny Hafiar. 2020. *International Association of Online Engineering*: 107-127.

[16] Gaol, M.L., Serevina, V. and Supriyati, Y 2019. *Prosiding Seminar Nasional Fisika (E-Journal),* Vol. 8, pp. SNF2019-PE

[17] Maryuliana, M., Subroto, I.M.I. and Haviana, S.F.C 2016. *TRANSISTOR Elektro dan Informatika 1, no. 1 (2016): 1-12.*

Evaluation of Blended Learning Program Physics of Mechanical Wave Materials Using CSE-UCLA Model

Sriwati, M Delina and Y Supriyati
Physics Education Master Program, Universitas Negeri Jakarta
Jl. Rawamangun Muka, Jakarta 13220, Indonesia
Email: sriwati280984@gmail.com, mutia_delina@unj.ac.id, yetti_supriyati@unj.ac.id

Abstract. This study aims to obtain an overview of the blended learning program for physics subjects in mechanical wave material during the covid-19 pandemic, and to find out the supporting factors and obstacles so that they can be projected into solutions that are adaptive to the needs, availability and skills of students. The research method used is qualitative with an evaluative approach to the CSE-UCLA model. The subjects of this study were students of class XI IPA-1 and XI IPA-2, totaling 64 people at SMA Negeri 1 Timpah, Timpah Village, Central Kalimantan, Indonesia. Data were collected through questionnaires, interviews, observation and documentation. The data analysis procedure used Milles & Hubberman model analysis, including data reduction, data presentation, and drawing conclusions. Furthermore, the data and research findings were triangulated to check the validity of the data. The results of the program evaluation research from system assessment components, program planning, program implementation, improvement programs, and certification programs overall show that the physics blended learning program at SMA Negeri 1 Timpah supports the government's policy in responding to the COVID-19 outbreak to learn from home, which is relevant to the supporting factors and obstacles.

1. Introduction

The detection of the Coronavirus or covid-19 in the city of Wuhan, China, which then spread to all countries until it entered Indonesia, also had an impact on education. New coronavirus pneumonia in Wuhan triggered by covid-19 is now spreading all over the world [1]. Due to the highly contagious nature of this virus, social distancing and restrictions have become unavoidable to prevent its rapid spread in the community [2]. The pandemic has made education providers decide to expand education virtualization to conduct online exams and all learning were carried out remotely [3]. Changes in learning patterns are expected to continue to be carried out optimally, for that an evaluation of learning programs can be carried out. Program evaluation aims to determine the achievement of program objectives that have been implemented [4]. Evaluation is very necessary in learning and will determine the success of an organization [5]. Through learning evaluation, educational institutions can determine students' knowledge abilities and competencies based on students' test scores [6].

In principle, evaluation is an activity carried out by evaluators to collect, analyze, and present complete and accurate information about an object / program / service / policy being studied [7]. Model Center for the Study of Evaluation – University of California in Los Angeles (CSE-UCLA) is an evaluation model developed by Alkin which consists of five evaluation components, namely: system assessment, program planning, program implementation, program improvement, and program certification, then this evaluation model is appropriate and suitable to be used to evaluate programs / policies in education and other fields related to service programs that help human life [7]. The evaluation of the CSE-UCLA model has been used in research showing that this evaluation model can be used to evaluate educational services either manually or electronically, namely the Alkin model (or often called the CSE-UCLA model) [8]. Designing an evaluation using the CSE-UCLA model was also carried out by Atmaja, Divayana and Seteman in the research of the Regional Archives and Library Office of Badung Regency to be able to recommend aspects of services that need to be optimized [9]. However, how to effectively evaluate blended learning has not yet formed a mature theory [10].

Weaknesses and advantages of face-to-face learning and web-based learning can be covered with each other by designing for the activities of both, this kind of learning is called blended learning [11]. Blended learning is a learning strategy that integrates face-to-face learning and online learning with a combination of various media [12]. Blended learning has significantly empowered learners and expanded their repertoire of learning experiences [13]. Mixed learning effectively expands learning time, and also expands learning content, giving students more opportunities for independent study and self-expression, and their evaluation is more comprehensive and scientific [14]. However, blended learning designs need to be considered, especially when and what types of learning materials should be delivered online and offline [15].

The material chosen in the evaluation research of this blended learning program is class XI mechanical wave material which is the basic material of waves so that it is considered suitable to be evaluated as an introduction to student knowledge to make it easier for students to follow the next wave of material. One of the main subjects of physics is waves [16]. Students have a false intuition about the relationship between amplitude and frequency, and students have difficulty representing a graph of displacement against time [17]. Research on mechanical waves has been carried out by Afrizon, Dewi, and Dwirida to analyze the physical parameters of Minangkabau aerophone musical instruments as a meaningful learning context [18] stating that Minangkabau musical instruments are one of the tangible local wisdom related to the material. Physics for class XI high school students, namely sound waves as part of mechanical waves. Research on waves was also carried out by Jahirwan Ut Jasron, Sudjito Soeparman, Lilis Yuliati, Djarot B. Darmadi on the characteristics of waves that determine water fluctuations in the water column [19] stating that the steepness of the waves is different, the wave energy also changes. Gallardo Pérez, M Vergel Ortega, and Rojas Suárez have also conducted research for learning on the concept of waves stating that studying the concept of waves sequentially leads students to construct knowledge for use in context, to develop effective reasoning processes, to develop self-directed learning skills and to motivates students to learn, along with developing the capacity to work in teams and supporting the development of physical-mathematical thinking [20].

The originality of this research is the evaluation of blended learning Physics of mechanical wave material for students in areas that are constrained by network and internet signals, namely at Senior High School or SMA Negeri 1 Timpah, Timpah District, Kapuas Regency, Central Kalimantan. This school is situated on Palangka Raya-Buntok Causeway, but electric lighting cannot be enjoyed at night from 16.00-06.00 West Indonesia Time (WIB). During the COVID-19 pandemic, SMA Negeri 1 Timpah implemented a blended learning policy with Learning From Home in accordance with the instructions of the Central Kalimantan Provincial Education Office and while still allowing face-to-face learning to schools according to the needs of teachers and students by alternating between class and still adhere to health protocols. Input from students in this school is not only from the Timpah Junior High School, but also from the One Roof Junior High School in the village which is still separated by forest or community rubber plantations. Student knowledge is built by the student's own experience through a construction process obtained through interaction with objects, phenomena, sources of accurate information and the student's environment [17]. The novelty of this research is the evaluation of blended learning Physics of mechanical wave material using the CSE-UCLE model at SMA Negeri 1 Timpah which has never been evaluated for the mixed learning program that is being implemented in response to the COVID-19 pandemic. The availability of facilities and infrastructure, the distance students travel from home to school and including internet facilities, especially in schools in rural areas need to find solutions or alternative learning activities. The availability of facilities and infrastructure, the distance students travel from home to school and including internet facilities, especially in schools in rural areas need to find solutions or alternative learning activities. Quality education will not be implemented if there are no adequate supporting facilities and infrastructure [7]. The progress of student learning outcomes must be evaluated and monitored [6].

Many educational institutions have opened online classrooms and advocate blended learning models that combine traditional learning methods with computer-assisted learning, but the educational evaluation system and appropriate examination methods have not undergone appropriate reforms [10]. Thus, the formulation of the research problem is: 1) How is the evaluation of the physics blended learning program for wave mechanics using the CSE-UCLA model at SMA Negeri 1 Timpah? 2) What are the supporting factors and obstacles in implementing the evaluation of the physics blended learning program for wave mechanics using the CSE-UCLA model at SMA Negeri 1 Timpah? So the purpose of this study was to determine the evaluation of the blended learning program for physics subjects on wave mechanics and to find out the supporting factors and obstacles to implementing the program using the CSE-UCLA model at SMA Negeri 1 Timpah.

2. Method

This research method is a qualitative research with an evaluative approach using the CSE-UCLA model. The object of this research is blended learning physics of mechanical wave material. While the subjects of this study were 64 students in the second semester of the 2020/2021 academic year, including 32 students in class XI IPA-1, and class XI IPA-2 32 people, at SMA Negeri 1 Timpah, Village / District Timpah, Kapuas Regency, Central Kalimantan, Indonesia. This study involved the principal, 1 physics teacher, 1 chair of the Physics Subject Teacher Consultation (MGMP) in Kapuas Regency and 3 members.

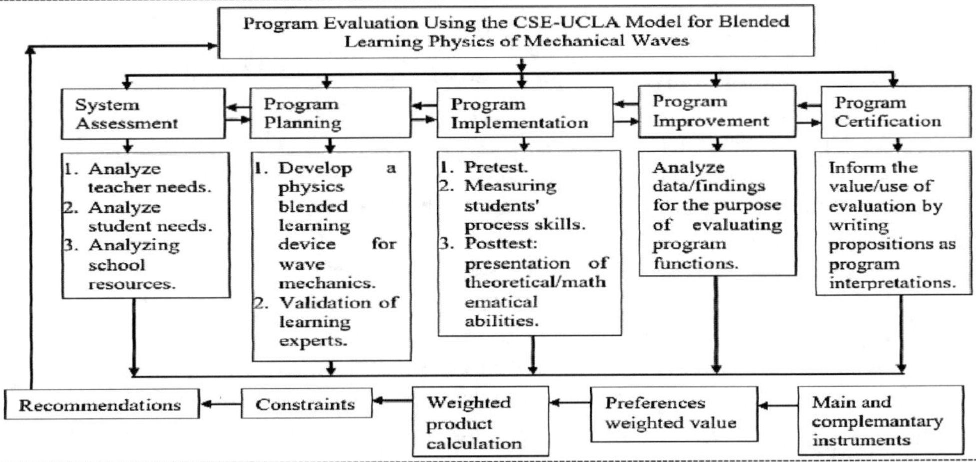

Figure 1. Evaluation of physics blended learning program using the CSE-UCLA model.

Figure 1 above is a series of data collection techniques carried out by observation, interviews, questionnaires and documentation. The results of the analysis are not expressed in numbers but are expressed in the form of words and pictures. The data analysis procedure during the data collection process and after data collection uses a model of qualitative data analysis, namely the Milles & Hubberman model [21], where the activities in this analysis are carried out interactively and last to completion with steps consisting of: data reduction, data display, and conclusion. Then triangulation was carried out for the validity of the data. The data obtained is then organized and tabulated, so that it can be used to compile the necessary evaluation aspects with the CSE-UCLA model [9].

3. Result and Discussion

Based on the research objectives, to determine the implementation of the evaluation of the blended learning program for physics subjects on the wave mechanics material and to determine the supporting factors and their implementation at SMA Negeri 1 Timpah. To achieve this goal, an evaluation instrument using the CSE-UCLA model is used which is described as follows:

3.1. Evaluation of the physics blended learning program for wave mechanics

It is important to evaluate learning to take advantage of each learning objective [22] (Bohori & et al, 2020). In general, the evaluation of the program in blended learning physics of wave mechanics is carried out to find out how prepared, planning, results, and useful it is at SMA Negeri 1 Timpah, in terms of aspects of the CSE-UCLA component model 5.

The System Assessment component of the physics blended learning program as a school policy during the COVID-19 outbreak was carried out by giving questionnaires to teachers, students and interviews with school principals. The results of the assessment of disclosing teacher needs on aspects of: 1) Blended learning interaction tools, 2) Appropriateness of the device, 3) Synchronous and foreign devices, 4) Learning skills 5) Evaluation needs, it was found that 8 out of 12 statements were yes. The results of the assessment of the disclosure of student needs showed that 12 of the 13 statements were yes. The results of the assessment of the principal on the disclosure of school resources, found that 17 of the 20 statements provided information that supported the implementation of the physics blended learning program. This means that the results of the overall assessment state that the blended learning physics program can be applied to support teaching and learning activities that are currently being enforced by social restrictions and Learning From Home. While being given a statement that the answer is no is an obstacle in the blended learning program. So it is necessary to combine e-learning and conventional learning in a class called the blended learning model which is suitable to be applied at the high school level, because the character of competent students still needs assistance as students [12]. Thus, the physics blended learning program at SMA Negeri 1 Timpah is declared to be applicable to support the Central Kalimantan Regional Education Office's policy program regarding studying from home during the recovery from the COVID-19 outbreak. This is a way to meet the needs of teachers, students with limited school resources, blended learning can be the right choice with the use of useful technology in education and web-based learning helps teachers and students organize learning without having to meet face to face [11].

The Planning Program component begins with compiling a physics blended learning device for mechanical wave materials, then evaluated by an evaluator consisting of 1 physics teacher and 4 Physics MGMP members. Basic Competence (KD): 3.8 Analyzing the characteristics of mechanical waves. 4.8 Propose problem solving ideas about mechanical wave characteristics, for example on a rope [23]. In addition, the wavelength increases with the increase in the wave period, and vice versa, the wave height decreases with the increase in the wave period [19]. All of these waves have two main characteristics: They are governed by Newton's laws, and they can only exist in material media, such as water, air, and rock [16]. Evaluation of aspects of this planning program includes: Design of Learning Implementation (RPP), preparation of materials, preparation of exercises, preparation of assignments, and the completion of the assessment rubric. The results of the evaluation showed that 4 evaluators said it was good, and 1 evaluator said it was very good, meaning that the physics blended learning device planning program is appropriate and very suitable to be applied. And all the evaluators gave the same conclusion that the blended learning program for physics of mechanical wave material is valid, meaning that the design of synchronous and asynchronous learning devices and platforms is feasible to apply. Blended learning consists of online learning components and face-to-face learning components so that the application of these two learning components needs to be considered in these two components in accordance with the learning competency targets [12].

Program Implementation components are carried out in: 1) Synchronous through face-to-face in class, Phet simulation and through the zoom meeting application; 2) Asingkronus via WhatsApp and google classroom. The wave motion can be learned from the autonomous work done by the student because it is a process in which he can appreciate and understand how the energy is transmitted from one part to another without the transfer of matter [20]. Evaluation standards are determined by teachers because it is difficult to have uniform and specific specifications [10]. Evaluation of measurement aspects, namely: 1) Students' basic abilities in the physics subject of wave mechanics through a pretest; 2) Participation in student learning through attendance, training and task fulfillment;

3) Representation of Students' theoretical/mathematical abilities through posttest. The results of student learning measurements refer to the Minimum Completeness Criteria (KKM) for Physics at SMA Negeri 1 Timpah, which is 70 with the predicate:

$\text{Value} < 70 = \text{Less (D)}, 70 \leq \text{Score} < 80 = \text{Enough (C)}, 80 \leq \text{Score} < 90 = \text{Good (B)}, \text{Score} \geq 90 = \text{Very Good (A)}$. The results of the evaluation of program implementation vary for each student and in the aspect of the assessment, namely quite good, good, and very good. This means that blended learning physics is carried out according to program planning and student learning needs, but must pay attention to implementation constraints. Thus, the conclusion of the validator is valid. This means that the implementation of the program has been able to describe the process of learning activities. This illustrates that teachers can combine face-to-face learning strategies with e-learning-based learning strategies [24].

Program Improvement component, processing the collected data and evaluated by the same validator. The evaluation results showed that all 5 evaluators said they were good. At the conclusion, 4 validators stated that the improvement program was valid, and 1 validator stated that it was valid with improvements to the parts that became obstacles. This means that the physics blended learning program can be applied to measure student learning abilities. It is necessary to record the entire student learning process dynamically [14]. Thus, the overall data projection of each component and its aspect is valid. This means that all the data collected can describe the ongoing program evaluation. The blended learning model is a solution to the limited hours of lessons a teacher has by placing student involvement in a learning agenda that focuses on students or Student Centered Learning [12].

Certification Program Component, by a validator consisting of the principal, physics subject teacher, and the head of the Physics MGMP. In the aspect of data correctness, program relevance, consistency, objectivity on student involvement, documented, measurable, interrelated data results are obtained for projecting research results on the function of the program evaluation process. The results of the evaluation of the certification program components are valid, meaning that all data used can be projected and meet all aspects of data validity. In general, evaluation can be interpreted as an activity of collecting, analyzing, and presenting related information about a certain object / program / policy / service based on measurements using an appropriate evaluation model and appropriate and valid evaluation instruments so that the results can be used as recommendations in making decisions [7]. So, in general, the preparation, planning, process, results, and benefits are valid. This means that all instrument data used can be projected systematically and interrelated. The implementation of blended learning is a systematic project, which needs to start from determining teaching goals, designing teaching links, building online platforms, teaching guides in the classroom, and evaluating [14]. So it can be concluded that the physics blended learning program at SMA Negeri 1 Timpah supports the policy of learning from home.

3.2. Supporting factors and obstacles to the implementation of program evaluation

Evaluation is useful in identifying the supporting factors or obstacles found in the physics blended learning program for wave mechanics at SMA Negeri 1 Timpah. This evaluation uses the CSE-UCLA evaluation model, as an effort to maximize the potential and minimize the obstacles to present a feasible solution.

The System Assessment component, the physics teacher explained that the supporting factor was smooth teacher and student communication, while the obstacles were that not all students had laptops, lack of understanding of laptop use, and unstable internet networks, so that they became obstacles in implementing virtual applications. The principal explained that the supporting factors were the attention of the government, school committees and parents, while the obstacles were learning materials and physics books that still needed to be completed. Instructional approaches and strategies that enable students to communicate, collaborate, and explore beyond the knowledge transferable by books and teachers, which are not strictly traditional teaching where the only available resources are teachers, books, and whiteboards, is an application of blended learning [13]. So, responding to it is to maintain smooth communication by constantly coordinating and sharing information related to education or learning from and to all related parties, both the Central Kalimantan provincial education office government, schools, teachers, school committees, parents, and students, and with identify material that can be delivered synchronously or asynchronously so that learning can be carried out. Education-related parties should develop a new performance evaluation system suitable for mixed learning to identify and measure the actual level of student learning and promote overall student development [10].

The program planning component, the physics teacher explained that the supporting factor was that the blended learning program had been running in schools, while the obstacle was the lack of student discipline in collecting assignments. Through social networks can improve interpersonal relationships and facilitate communication [24]. The principal explained that the supporting factor is good and smooth communication between schools, teachers, students, parents and school committees, while the problem is that not all learning materials are available. So, responding to it is to design a blended learning device that simultaneously fosters student discipline and contains practical learning materials such as embedding a short video about a material for students to analyze according to learning objectives. It is necessary to record the entire student learning process dynamically [14].

The program implementation component, the physics teacher explained that the supporting factors were that they had provided additional material from links or other references, and had used synchronous and asynchronous learning applications, while the obstacle was learning applications that could not be applied optimally. Science process skills as basic competencies in understanding physics [22]. The principal explained that the supporting factor was adequate classrooms according to the size of the national standard, while the problem was the need to add physics books. Thus, responding to it is to arrange synchronous and asynchronous meetings by paying attention to learning needs and problems. Asynchronous is carried out in a virtual laboratory or embedding a link on how to use the tool. In sync, recorded and embedded in learning applications so that students who do not understand or miss can repeat again. If the mixture is too simple then the meaning of mixed learning is lost and the ideal teaching effect cannot be achieved and if the mixture is too complex then it is difficult to promote teaching activities, so integrated learning should focus on teaching objectives, design teaching, continue to deepen, and advance step by step [14].

The improvement program component, the physics teacher explained that the supporting factor was to anticipate the use of simple practical tools, such as Figure 2(a). Students' comments on Google Classroom are as shown in Figure 2(b), it is known that students' physics learning problems are difficult to focus, difficult to remember theory, unstable network and difficulty understanding concepts and formulas, such as the relationship between period and wave frequency expressed as $T = \frac{1}{f}$ or $f = \frac{1}{T}$ [25]. The principal explained that the supporting factor was the presence of 2 physics teachers, while the obstacle was the standard student's physics ability.

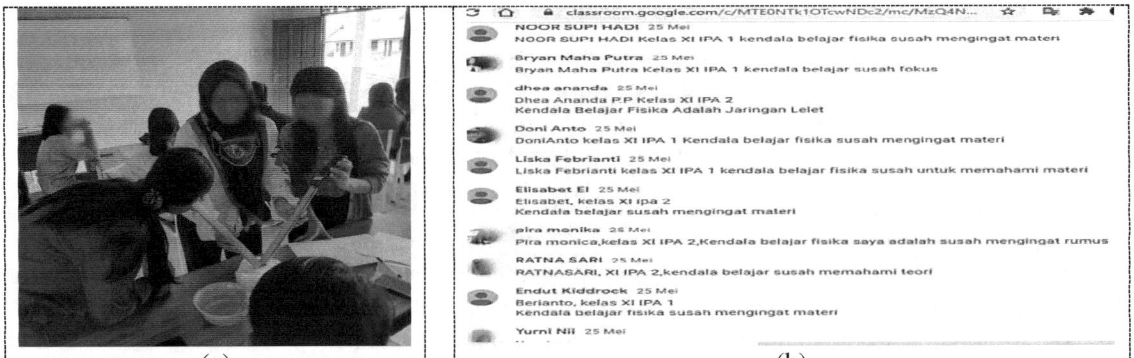

(a) (b)

Figure 2. (a) Wave reflection practicum with simple tools, (b) Students state the obstacles to learning physics.

Students cannot do physics learning well without mastering good process skills [22]. So, responding to it is by adapting learning so that students can follow it without reducing the content of the material. Meaningful learning, namely learning is an experiential process because it is related to the natural context and the ability to adapt to the surrounding environment [18].

The component of the certification program, the physics teacher was found to explain the supporting factor was the evaluation of the program to get a picture of the success of blended learning physics. Compared to the simple assessment model of traditional teaching-testing, mixed learning evaluation is more comprehensive, systematic and scientific [14]. The principal explained that the supporting factors for the program are that every year students are entered in science competitions and there is an evaluation of grade promotion. So, responding to it is by evaluating the program to get a picture of students' physics abilities and the success of blended learning physics. This takes into account the academic culture of the school, where the use of computers and other devices is sought to be used massively and extensively in the teaching and learning process [13].

So the evaluation of the program obtained can be stated that blended learning is relevant to the supporting factors and existing obstacles to support learning from home and as a solution to the policies implemented during the current covid-19 pandemic. Advantages of blended learning [14]: 1) Integrated learning is conducive to increasing students' autonomous learning; 2) Mixed learning content is more targeted; 3) Evaluation of mixed learning is more scientific. So that the benefits that can be obtained from the benefits of blended learning for educational or training institutions are in the form of: expanding the range of learning/training, ease of implementation, efficiency, optimal results, adapting various student needs, and increasing the attractiveness of learning.

4. Conclusion

Based on the results of the evaluation of the CSE-UCLA model in general, each aspect is declared good, which describes the implementation of the program in each of its components saying that the component system assessment is valid, program planning is valid, program implementation is valid, program improvement is valid, and program certification is valid. This means that the physics blended learning program at SMA Negeri 1 Timpah has supported the Central Kalimantan Education Office Government Policy regarding learning from home for the COVID-19 outbreak and its recovery period. The blended learning program also cannot be separated from the supporting factors and constraints that are relevant as its implementation, namely smooth communication, fostering student discipline, students can repeat their learning, adaptation of learning media and programs. This study offers blended learning in physics subjects, which still have problems, only learning from home or conventional learning.

References

[1] Feng, G., & et al. 2020. *Chin. Phys. B, 29 (04).*

[2] Anderson, R. M., Heesterbeek, H., Klinkenberg, D., & Hollingsworth, T. 2020. *Lancet, 931-943.*

[3] Ray, S., & Srivastava, S. 2020. *Journal of Proteins and Proteomics, 77-80.*

[4] Maulana, I., Sumarto, Nurafiati, P., & Puspita, R. H. 2018. *IOP Conf. Series: Materials Science and Engineering 306 012059.*

[5] Sudana, I. M., Apriyani, D., & Suryanto, A. 2019. *International Conference on Education, Science and Technology 2019 Journal of Physics: Conference Series 1387 012075.*

[6] Henderi, & et al. 2020. *ICComSET 2019 Journal of Physics: Conference Series 1477 022027.*

[7] Divayana, D. G., Ardiarta, A., & Abadi, I. B. 2018. *IOP Conf. Ser.: Mater. Sci. Eng. 296 012003.*

[8] Divayana, D. G., Suyasa, P. W., & Abadi, I. B. 2019. *J. Phys.: Conf. Ser. 1402 022075.*

[9] Atmaja, I. M., Divayana, D. G., & Setemen, K. 2021. *IConVET 2020. Journal of Physics: Conference Series 1810 012036.*

[10] Gong, X. 2020. *IWAACE 2020 Journal of Physics: Conference series 1550 032063.*

[11] Dewantara, D., Misbah, M., & Wati, M. 2020. *1st IC-MSCEdu Journal of Physics: Conference Series 1422 012002.*

[12] Budiharti, R., & Waras, N. S. 2018. *ICoSMEE IOP Publishing: Jounal of Physics: Conf. Series 1022 012024.*

[13] Hipol, A. I., Cabahug, R., & Bongon, R. 2020. *The 7th South East Asia Desian Research International Conference (SEADRIC 2019) IOP conf. Series: Journal of Physics: Conf. Series 1470 012052.*

[14] Chuanzen, S. 2020. *2020 International Conference on Machine Learning and Computer Application Journal of Physics: Conference Series 1682 012086.*

[15] Setyaningrum, W. 2019. *ISIMMED 2018 IOP Conf. Series: Journal of Physics: Conf. Series 1320 012089.*

[16] Halliday, D., Resnick, R., & Walker, J. 2011. *Fundamentals of Physics 9th Edition.* America: Jhon Wiley & Sons, Inc.

[17] Tumanggor, A. M., & et al. 2020. *The 5th International Seminar on Science Education Journal of Physics: Conference Series 1440 012059.*

[18] Afrizon, R., Dewi, W. S., & Dwirida, L. 2020. *J. Phys.: Conf. Ser. 1481 012109.*

[19] Jasron, J. U., & et al. 2019. *Conference on Mechanical Engineering 494 012019.*

[20] Perez, H. J., Ortega, M. V., & Suarez, J. P. 2020. *IV International Meeting on Mathematics Education, Journal of Physics: Conference series 1672 012017.*

[21] Ardnil, P. P., & et al. 2020. *IOP Conf. Ser.: Earth Environ. Sci 589 012028.*

[22] Bohori, M., & et al. 2020. *International Conference on Mathematics and Science Education 2019 (ICMScE 2019) Journal of Physics: Conference Series 1521 022050. IOP Publishing.*

[23] Indarti, Nugroho, A. P., & Syifa, N. H. 2016. *Fisika untuk SMA/MA Peminatan Kelas XI.* Surakarta: CV Mediatama.

[24] Santosa, A. B., & et al. 2019. *The 1st Workshop on Environmental Science, Society, and Technology, Journal of Physics: Conference Series 1363 012064.*

[25] Bueche, F. J., & Hecht, E. 1997. *Schaum's Outline of Theory and Problems of College Physics Ninth Edition.* America: McGraw-Hill.

Identifying High School Students' Misconceptions Using Digital Four-Tier Diagnostic Tests in Distance Learning

M E Munggarani, Y Supriyati and I M Astra
Physics Education Master Program, Universitas Negeri Jakarta
Jl. Rawamangun Muka, Jakarta 13220, Indonesia
Email: meutiaelsafira@gmail.com; yetti.supriyati@unj.ac.id; imadeastra@unj.ac.id

Abstract. The pandemic of the spread of a new virus, namely the corona virus disease (Covid-19) in Indonesia, has forced the government to use an education virtualization system which has been expanded into online learning or distance learning. This makes it difficult for teachers to identify how big the misconceptions experienced by students in Physics subjects with harmonious vibration material during learning take place, so teachers and students need an online assessment instrument that is able to identify the student's misconceptions. This study aims to analyze the misconceptions of high school students using a digital four-tier diagnostic test in distance learning. The method used is a single case study. The research analysis used the Rasch model. The instrument used is a conceptual understanding test using a digital four-tier diagnostic test instrument which consists of 10 questions on the Harmonic Vibration material to identify students' misconceptions. This research was conducted in three stages; (1) the planning stage, (2) the implementation stage, and (3) the final stage. Based on the results of data analysis, there is a percentage of students who understand the concept of 54%, 23% less understanding of the concept, 21% of misconceptions, and 2% non-code (error). The average value of misconceptions experienced by students shows a low category for Harmonic Vibration material.

1. Introduction

In the 21st century, students are required to have four competencies in themselves, namely conceptual understanding, critical thinking, creative thinking and collaboration and communication [1]. Physics is a part of science that emphasizes mastery of concepts. In learning physics, it is directed to find out so that it can help students to gain in-depth mastery of concepts [2]. Mastery of concepts is one of the objectives of physics subjects in high school. These objectives are stated in the Education Unit Level Curriculum (KTSP), namely so that students have the ability to master the knowledge, concepts and principles of physics [3]. Mastery of concepts becomes very important for students because it is included in the indicators where students have fully understood what has been taught, not just memorizing. So that later mastery of this concept can help students in solving problems, not only in the learning process but also in everyday life. Therefore, teachers need to direct students' understanding of the concepts of the material being taught.

However, the World Health Organization (WHO) stated that in 2019, the world is being faced with a pandemic of the spread of a new virus, namely the corona virus disease (Covid-19). The nature of this virus is highly contagious, social distancing and lockdowns have become unavoidable to prevent its rapid spread in society [4]. One of the areas most affected by this pandemic is education. To reduce the number of transmission of the virus, teaching and learning activities in schools must be temporarily suspended. Then, the government made a policy that the implementation of education needed to use an

Journal of Physics: Conference Series **2019** (2021) 012016 doi:10.1088/1742-6596/2019/1/012016

educational virtualization system which was expanded into Distance Learning [5]. Online learning is an effective way as a new avenue for science and technology education and has accelerated educational globally especially during the pandemic [6]. So the Indonesian government decided to organize online learning or Distance Learning (PJJ) through the use of various communication media in accordance with the policies set by the government.

Based on the explanation above, this situation has an impact on students' misconceptions or misconceptions in receiving the material. Misconceptions that occur in students vary with different causes [7]. Conceptual errors or misconceptions are errors in interpreting concepts or not in accordance with scientific concepts that have been believed and applied by experts [8]. The causes of student misconceptions, mostly come from the students themselves, namely students' inadequate prior knowledge and lack of motivation to learn [9]. Misconceptions can be a barrier to organizing knowledge so it needs to be addressed properly [10]. Misconceptions occur if students have a high level of confidence in a concept that is considered wrong [11] where the wrong concept is not in accordance with the scientific understanding or understanding accepted by scientists in the field concerned, especially physics [8]. Misconception can also be interpreted as a contradiction or incompatibility of the concept understood by someone with the concept used by the expert in science [12] or a preconception that is not in accordance with the scientific conception [13]. Thus, the outline of misconceptions is a form of inaccurate understanding of a concept which is characterized by (1) students being inaccurate in accepting concepts, (2) using wrong concepts, (3) classifying examples of incorrect questions, (4) confusion of different concepts. , and (5) the relationship of the concepts does not match.

Based on previous research conducted by Kharisma Fenditasari, the results showed that the value of misconceptions on one of the physics topics was 61.5% [14]. Subsequent research was conducted by Islami et al. The results showed that 47% of students had misconceptions, 7% of the students understood the concept, 36% of students did not understand and 10% of students had errors [15]. Then the research conducted by Zulfikar et al, the results showed that the level of comprehensive understanding was 20%, the level of partial understanding was 27%, misconceptions were 40% and the level of understanding did not understand was 3% [16]. Efforts to identify students' misconceptions must distinguish between students who experience misconceptions and students who do not know the concept. One way to find out misconceptions in students is with a diagnostic test. The four-tier diagnostic test is an extension of the three-tier multiple-choice diagnostic test [17]. The first level is a multiple choice question with three distractors and one answer key, the second level is the level of students' confidence in choosing the answer, the third level consists of three choices of reasons that have been provided and one open reason, the fourth level is the level of students' confidence in choosing reasons [18] . From these results, it is believed that the four-tier diagnostic test item can be implemented as an effective learning strategy to identify students' misconceptions.

In general, the steps that can be used to help students overcome misconceptions are looking for the form of the misconception they have, looking for the cause and determining the appropriate way [19]. Misconceptions cannot be generalized directly because the forms of misconceptions that occur can be different or the same [20] especially with the current distance learning. The misconceptions experienced by students must be understood and discovered by teachers in order to help students improve the misconceptions they experience effectively. However, there are still few teachers who pay attention to how to identify and resolve students' misconceptions. Not only students, teachers are also required to create new innovations in learning and assessment. The creativity of teachers in conducting assessments is the main element in evaluating the ongoing online learning process. One of the new innovations in assessment is the media. Google Forms is a useful digital tool to help plan events, send surveys, give quizzes or collect easy information in an efficient way [21]. Therefore, we need a digital test instrument that can help teachers to identify misconceptions innovatively and effectively. Thus, the purpose of this study was to analyze the misconceptions of high school students using a digital four-tier diagnostic test in distance learning.

2. Method

This research is a single case study. A case study is a model that focuses on exploring a "bounded system" on one case or in some cases in detail with in-depth data mining [22]. The research analysis was carried out using the Rasch model through a conceptual understanding test using a digital four-tier test instrument to identify the misconceptions of high school students where the digital media used were Google Form. This preliminary case study was attended by 126 students of class X SMA in Tangerang City. Student research subjects were selected purposefully, namely subjects who have competencies based on data needs that can support the achievement of research objectives [20]. The data were collected by conducting a concept understanding test in the form of a four-tier diagnostic test of 10 questions on the material of harmonic vibrations.

This case study was conducted in three stages; (1) planning/preparation stage, conducting visits to schools for research permits, contacting physics teachers to determine research time, determining sampling, preparing preliminary case study instruments, (2) implementation stage, giving students' concept understanding tests at home to identify students' misconceptions, and (3) the final stage, processing research data, discussing and analyzing data, providing conclusions and recommendations [22]. The data processing technique was carried out quantitatively in the form of students' conceptual understanding test scores using a digital four-tier diagnostic test to identify students' misconceptions which were analyzed by statistical tests to describe the percentage of students' misconceptions. The analysis technique uses the Rasch model, where the principle of this model uses the principle of probability for each available option in accordance with classical test theory which prioritizes the total value of the test results. In this study, Winstep software was used to process the data obtained from the results of the students' concept understanding test. Calculation of the percentage of students' misconception scores by [23]:

$$average\ score\ percentage = \frac{score\ obtained}{maximum\ score} x100\% \quad (1)$$

Then, the percentage is interpreted according to Table 1. below:

Table 1. Interpretation of Students' Misconceptions

Percentage (%)	Category
61-100	High
31-60	Medium
0-30	Low

In addition to indicators of interpretation of misconceptions, other data used to identify students' misconceptions are; (1) the reliability of the instrument, in Table 2. There are criteria for the reliability of the instrument using the Cronbach alpha value in the Rasch Model. The analysis was applied to the data obtained from tests and student response questionnaires.

Table 2. Instrument Reliability Criteria

Interval	**Criteria**
RC-20 < 0.67	Weak
0.67 < RC-20 < 0.8	Enough
0.8 < RC-20 < 0.9	Good
0.91 < RC-20 < 0.94	Very Good
RC-20 > 0.94	Excellent

(2) the level of difficulty, the index of difficulty is calculated by the formula [23]:

$$P = \frac{B}{J_s} \quad (2)$$

With description; P is the level of difficulty, B is the number of students who answered correctly, J_s is the total number of students. Table 3. shows the classification of the level of difficulty of the questions.

Table 3. Classification of Problem Difficulty Levels

Clasification	**Criteria**
$P < 0.30$	Hard
$0.30 \leq P \leq 0.70$	Medium
$P > 0.70$	Easy

(3) differentiating power, the ability of questions to distinguish students with high abilities from those with low abilities [24]. The index used in distinguishing them is the index of discriminating power. The formula used to determine the discrimination index is as follows:

$$D = \frac{B_a}{J_a} - \frac{B_b}{J_b} \quad (3)$$

With description; D is the value of distinguishing power, B_a is the number of the upper group who answered correctly, B_b is the number of the lower group who answered correctly, J_a is the number of participants in the upper group, J_b is the number of the lower group participants. In Table 4. the following shows the classification of the difference power index of the questions.

Table 4. Classification of Differentiating Power Index Questions

Differing Power Index	**Criteria**
0.00 – 0.20	Bad
0.20 – 0.40	Enough
0.40 – 0.70	Good
0.70 – 1.00	Very Good

3. Result and Discussion

This research was applied to 126 students of class X SMA in Tangerang City. The test consists of 10 multiple choice questions at four levels using Google Form. This test instrument is equipped with reasons and the level of student confidence in doing the test. The analysis technique of the test instrument used the Rasch model, which obtained several categories including the reliability value of the questions, the level of difficulty and the discriminating power of the questions.

Through this analysis, the results of personal reliability are 0.16 and item reliability is 0.89. From these results it was concluded that the consistency of the answers from students was weak, but the quality of the items in the instrument's reliability aspect had good criteria. Cronbach's alpha value to measure the reliability of the interaction between the person and the items is 0.08. The alpha value indicates that the reliability of the test in general is still not satisfactory and is in the weak category. In the analysis of the Rasch model, the level of difficulty of the test instrument can be seen based on the measure column in the Winstep software.

ENTRY NUMBER	TOTAL SCORE	TOTAL COUNT	MEASURE	MODEL S.E.
1	0	37	3.95	1.83
4	0	37	3.95	1.83
9	0	37	3.95	1.83
6	1	37	2.71	1.02
10	1	37	2.71	1.02
7	4	37	1.13	.56
2	11	37	-.35	.40
5	20	37	-1.66	.38
8	23	37	-2.11	.39
3	25	37	-2.43	.41
MEAN	8.5	37.0	1.19	.97
P.SD	9.9	.0	2.49	.61

Figure 1. Column Measure in Rasch Model by Winsteps Program

In Figure 1. above shows the logit value for each item sorted from highest to lowest by Winsteps Program. For item 1, which is 3.95 logit, it shows the most difficult item, while item 3 is the easiest item at -2.43 logit.Likewise, the discriminatory power of questions can be seen from the distribution of questions. The order of distribution of questions according to the level of difficulty. Students with low abilities consist of 21 people with person codes 07, 15, 22, 24, 27, 30, 39, 42, 43, 48, 52, 65, 73, 97, 107, 111, 116, 118, 121, 124 and 125. Students with high abilities consist of 9 people with person codes of 01, 03, 11, 25, 36, 67, 78, 82 and 93. The data on the results of the diagnostic test of misconceptions using digital four-tier diagnostic test questions are grouped and counted. Students who are included in the category of understanding concepts, do not understand concepts and misconceptions so that the results are obtained as shown in Table 5.

Table 5. Number of Students by Category

No. Item	Total students			
	UC	NUC	M	E
1	107	19	0	0
2	115	11	0	0
3	87	13	24	2
4	102	20	2	2
5	35	9	82	0
6	82	17	23	4
7	97	18	11	0
8	22	35	63	6
9	31	70	18	7
10	2	78	45	1
Total	680	290	268	22
Percentage	54%	23%	21%	2%

UC is conceptual understanding, NUC is not understanding the concept, M is misconception and E is the error.

From Table 5. above, students are grouped into four groups for each sub-concept. The percentage data from these are shown in Table 6.

Table 6. Percentage of Results of UC, NUC, M and E data analysis

Sub Concept	Percentage				Interpretation
	UC	NUC	M	E	Misconception
Harmonic Motion on the Pendulum	85	15	0	0	Low
Simple Pendulum System Period and Frequency	91	9	0	0	Low
Spring Period and Frequency	69	10	19	2	Low
Harmonic Vibration Deviation	81	15	2	2	Low
Harmonic Vibration Speed	28	7	65	0	High
Harmonic Vibration Acceleration	65	14	18	3	Low
Kinetic Energy of Harmonic Motion	77	14	9	0	Low
Harmonic Motion Potential Energy	17	28	50	5	Medium
Harmonic Motion Mechanical Energy	25	56	14	5	Low
The Speed of Harmonic Vibrating Objects	2	61	36	1	Medium

From the data that has been obtained, Table 6. shows the percentage of analysis results of students who understand concepts, do not understand concepts, mis conceptions and errors in the concept of harmonic vibration using a digital four-tier diagnostic test with Google Form media. While in Table 6.

shows the percentage of the results of data analysis of students' misconceptions and their interpretation. Based on these data, it could be seen that in the sub-concept of harmonic motion of a pendulum (item number 1), students experience 0% misconceptions with low interpretations. In the sub-concept of the period and frequency of the simple pendulum system (item number 2), students experience 0% misconceptions with low interpretations. In the period and frequency sub-concepts on the spring (item number 3), students experienced 19% misconceptions with low interpretations. In the sub-concept of harmonic vibration deviation (item number 4), students have misconceptions of 2% with low interpretation. In the sub-concept of the speed of harmonic vibrations (item number 5), students experience a misconception of 65% with a high interpretation. In the sub-concept of harmonic vibration acceleration (item number 6), students have 18% misconceptions with low interpretations. In the sub-concept of harmonic motion kinetic energy (item number 7), students have 9% misconceptions with low interpretations. In the sub-concept of potential energy of harmonic motion (item number 8), students experience a misconception of 50% with a moderate interpretation. In the sub-concept of harmonic motion mechanical energy (item number 9), students experience misconceptions by 14% with low interpretation. In the sub-concept of the speed of objects that vibrate harmonics (item number 10), students experience misconceptions by 36% with low interpretation.

Based on the tests carried out using a digital four-tier diagnostic test to identify students' misconceptions, it can be seen in Table 7. which shows the highest value, namely in item number 5, namely the sub-concept of harmonic vibration speed and the lowest value in item number 1, namely harmonic motion on a pendulum. Item number 2, namely the period and frequency on a simple pendulum. After the implementation of the digital four-tier diagnostic test instrument, the highest results were obtained in question number 5 with the sub-concept of harmonic vibration speed of 65%, while the lowest results were in question number 1 with sub-concept of harmonic motion on a pendulum of 0% and question number 2 with sub-concept period and frequency on a simple pendulum is 0%. In Table 5. the average value of misconceptions experienced by 126 students is 21%. This shows that the misconceptions about the concept of harmonious vibration for class X SMA in Tangerang City are included in the low category. Based on the presentation of the results of data analysis using the Rasch model that has been carried out, it states that the digital four-tier diagnostic test instrument is very effective in identifying the misconceptions of class X high school students in the Tangerang City on the material of harmonic vibrations.

The results of data analysis and discussion, it can be concluded that students are divided into four groups, namely understanding concepts, not understanding concepts, misconceptions and non-codes (errors). The category of students who understand the concept is moderate, lack of conceptual understanding is low, misconception is low and non-code (error) is low. The category of the highest percentage value of misconceptions is in item number 5 with the sub-concept of harmonic vibration speed while for the category of the lowest percentage value of misconceptions is found in item number 1 with sub-concept of harmonic motion on the pendulum and item number 2 with the sub-concept of period and frequency on the pendulum. The category of the average percentage of misconceptions in class X high school students in Tangerang City is low.

4. Conclusion

The pandemic of Covid-19 makes it difficult for teachers to identify how big the misconceptions experienced by students in Physics subjects with Harmonic Vibration material during learning take place, so teachers and students need an online assessment instrument that is able to identify the student's misconceptions. This study aims to analyze the misconceptions of high school students using a digital four-tier diagnostic test in distance learning. The method used is a single case study. The research analysis used the Rasch model by Winsteps Program. The instrument used is a conceptual understanding test using a digital four-tier diagnostic test instrument which consists of 10 questions on the Harmonic Vibration material to identify students' misconceptions. This research was conducted in three stages; (1) the planning stage, (2) the implementation stage, and (3) the final stage. Based on the results of data analysis, there is a percentage of students who understand the concept of 54%, 23% less understanding of the concept, 21% of misconceptions, and 2% non-code (error). The average value of misconceptions experienced by students shows a low category for Harmonic Vibration material.

Suggestions that can be put forward after the case study is carried out to improve students' conceptual understanding by conducting practicum or changing methods, models or approaches that emphasize students' ability to understand concepts in depth. Thus, students have a sense of responsibility to learn the concept of Harmonic Vibration material well.

References

[1] Astuti A P Aziz A Sumarti S S & Bharati D A 2021 Preparing 21st Century Teachers: Implementation of 4C Character's Pre-Service Teacher through Teaching Practice *IOP Publishing* **1233** 1-2
[2] Dahar R W 2011 *Teori Belajar dan Pembelajaran* Jakarta: Penerbit Erlangga
[3] Depdiknas 2003 *Kurikulum Tingkat Satuan Pendidikan* Jakarta: Depdiknas
[4] World Health Organization 2020 March 29 Naming the Corona Virus Disease (Covid-19) and The Virus that Causes It
[5] Ray S & Srivastava S 2020 Virtualization of Science Education: A Lesson from The Covid-19 Pandemic *Nature Publishing Group* **11** 8-10
[6] Kitao S K & Kitao K 1998 Selecting and Developing Teaching/Learning Materials *the Internet Journal* **4** 7-10
[7] Widya dkk S B 2018 Identifikasi Miskonsepsi Siswa Menggunakan Four Tier Diagnostic Test pada Materi Optik Geometri *Jurnal Pembelajaran Fisika* **7** 147
[8] Suparno 2005 *Miskonsepsi & Perubahan Konsep dalam Pendidikan Fisika* Jakarta: Grasindo
[9] Enawaty E & Sartika R 2015 Description of Students' Misconception in Chemical Bonding *Proceeding of International Conference on Research, Implementation and Education of Mathematics and Sciences* **8** 17
[10] Mahmudah A 2019 Perbandingan Instrumen Tes Diagnostik Two-Tier Multiple Choice, Three-Tier Multiple Choice dan Four-Tier Multiple Choice dalam Mengidentifikasi Miskonsepsi Siswa pada Materi Ikatan Kimia *Jurnal Pendidikan Sains* **2** 6
[11] Ainiyah M Ibrahim M & Hidayat M 2018 The Profile of Student Misconceptions on The Human and Plant Transport Systems *IOP Publishing* **947** 2-3
[12] Berg E d 1991 *Physics of Misconceptions and Remediation* Amsterdam: Springer
[13] Novak J & Gowin D 1985 *Learning How to Learn* New York: Cornell University
[14] Fenditasari K et al 2020 Identification of Misconceptions on Heat and Temperature Among Physics Education Students Using Four-Tier Diagnostic Test *IOP Publishing* **1470** 3
[15] Islami D Munawaroh F Hadi W P & Wulandari A Y 2018 Analisis Miskonsepsi Siswa pada Mata Pelajaran IPA Materi Listrik Statis Menggunakan Four-Tier Test *Proceedings SENCO* **3** 4-5
[16] Zulfikar A Samsudin A & Saepuzaman D 2017 Pengembangan Terbatas Tes Diagnostik Force Concept Inentory Berformat Four-Tier Test *Jurnal WaPFI (E-Journal)* **10** 3-5
[17] Afif, N Nugraha M & Samsudin A 2017 Developing Energy and Momentum Conceptual Survey (EMCS) with Four-Tier Diagnostic Test Items *SIP Publishing* **1848** 2
[18] Hermita N Suhandi a Syaodih E Samsudin a Johan H Rosa F & Safitri D 2017 Constructing and Implementating a Four-Tier Test About Static Electricity to Diagnose Pre-Service Elementary School Teacher' Misconceptions *IOP Publishing* **895** 2
[19] Jubaedah D Kaniyawati I Suyana I & Suhendi E 2017 Pengembangan Tes Diagnostik Berformat Four-Tier untuk Mengidentifikasi Miskonsepsi Siswa pada Topik Usaha dan Energi *Prosiding Seminar Nasional Fisika (E-Journal) SNF2017* **6** 4
[20] Krebs R E 1999 *Scientific Deelopment and Misconceptions Through the Age: a reference guide* USA: Greenwood Press
[21] Priyanda E Fadhelina N Ariska R N 2021 Analysis of The Effectiveness of Students' Worksheets Assisted by Google Form in Junior High School in Langsa Kota as an Alternative of Independent Learning during Covid-19 Pandemic *IOP Publishin* **1806** 2-4

[22] Sugiyono 2017 *Metode Penelitian Kuantitatif, Kualitatif dan R&D* Bandung: Alfabeta CV
[23] Smith R & Plackner C 2009 The Family Approach to Assessing Fit in Rasch Measurement *J Appl Meas* **7** 5-7

Design of Virtual Physics Laboratory (VPL) on Collision Topic

S Arianti[1], I M Astra[2] and E Budi[2]
[1]Physics Education Master Program, FMIPA, State University of Jakarta
Jl. Rawamagun Muka, Rawamangun, 13220, Indonesia
[2]Department of Physics, FMIPA, State University of Jakarta
Jl. Rawamagun Muka, Rawamangun, 13220, Indonesia
Email: silfiaarianti14@gmail.com, imadeastra@unj.ac.id, esmarbudi@unj.ac.id

Abstract. Collision phenomena are very difficult for teachers to visualize in distance learning. Facts in the field show that there are still many students who have difficulty understanding the concept of Collision. This is what causes the Collision concept to be one of the physics materials that are difficult for students to reach. This study aims to produce an interactive multimedia design in the form of a Virtual Physics Laboratory (VPL) in the Collision sub-material. The research method used is Research and Development (R&D) using the ADDIE development model (Analyze, Design, Develop, Implement, and Evaluate). However, this research was only carried out until the design stage. The results of the analysis of the distribution of the questionnaire data analysis of the needs of teachers and students stated that students needed to meet the needs of new learning media used to assist them in visualizing the Collision phenomenon in Distance Learning. The media used to visualize the collision is the Virtual Physics Laboratory (VPL) media. With the Virtual Physics Laboratory (VPL) design, it is hoped that it will produce interactive multimedia that is effective in helping students understand Collision material.

1. Introduction

Currently the world has entered the era of the industrial revolution generation 4.0 which is marked by increased connectivity, interaction and development of digital systems, artificial and virtual intelligence. With the increasingly convergent boundaries between humans, machines and other resources, information and communication technology certainly have an impact on various sectors of life. One of them is the impact on the education system in Indonesia [1]. The rapid development of Information and Communication Technology (ICT) gave birth to a trend of using computer-based multimedia as a medium of learning in schools. The skills that become the focus of learning competencies in the 21st Century are skills in mastering Information and Communication Technology (ICT) media. The 21st century learning paradigm emphasizes the ability of students to find out from various sources, formulate problems, think analytically and collaborate and collaborate in solving problems [2]

Physics is a part of science that studies natural phenomena and phenomena empirically, logically, systematically and rationally which involves scientific processes and attitudes [3]. Physics is one of the basic sciences that plays an important role in the development of science and technology [4]. Physics learning emphasizes on providing direct experience to develop student competencies. Part of teaching that aims to give students the opportunity to test and implement in real situations what is obtained in theory is to carry out practical activities.

One of the physics materials that still contains abstract concepts, namely momentum and impulse. For example, the law of conservation of kinetic energy does not necessarily apply to collisions. The difference between the types of collisions and their characteristics [5]. Collisions are one of the sub-materials of physics course whose application is widely encountered in everyday life. Every human day is not far from the collision. For example, when playing billiards, when a ball that is styled by a stick moves towards another ball and hits it, besides playing soccer, softball, tennis and many others. In physics, material is taught about collisions, so these events can be analyzed in two ways, namely first by theoretical calculations and assisted by the second way, namely using practicum which functions as direct observation activities carried out by students on physical events that occur.

The world is currently shocked by the outbreak of a virus, the spread of this virus almost throughout the world very quickly. The World Health Organization (WHO) has declared this outbreak a global pandemic. Indonesia is one of the countries affected by the Covid-19 virus, which has had a major impact on people's lives. With the stipulation of the covid-19 outbreak as a pandemic, the Indonesian government began to be fast and responsive to encourage its people to establish social distancing [6]. Online learning is a learning process mechanism that is far from the center of education and is independent. Technological developments allow online learning to be carried out well. The emergence of the COVID-19 pandemic has pushed the demands of the use of digital technology so that learning will be easier for students [7]. The current learning situation, teachers are required to present contextual, creative, efficient and fun learning, so teachers must try to update their abilities according to the demands of the times [8].

The implementation of online learning is considered close to face to face, namely using video conferencing applications, including the Zoom webinar. Webinar Zoom is a face-to-face conference platform where educators and students can interact directly like meeting face-to-face [9]. Research on the use of webinars in the learning process compared to face-to-face meetings, it was found that student participation was much higher in Zoom webinars when the teacher encouraged students to speak [10]. Physics learning and laboratories are inseparable parts. The application of the practicum process with an online learning system is one of the new things that must be done by educators.

A virtual laboratory is a system that can be used to support a conventional practicum system. This virtual laboratory is commonly referred to as a Virtual Laboratory or V-Lab. Virtual laboratories as supporting factors of real laboratories enrich students' learning experiences and offer students to conduct experiments, control materials and equipment, collect data, conduct experiments interactively, and prepare experimental reports and develop experimenting skills [11]. The laboratory approach is based on the idea that experiments to prove basic information about science should be carried out in the laboratory by students. Individual differences will be eliminated in laboratory studies in some way. This is because all the equipment and methods used in conducting experiments in the study laboratory are also elements of individual training [12]. In addition, the use of laboratories as a teaching method develops reasoning, critical thinking, scientific perspectives and problem solving abilities of students [13]. With this skill, students are encouraged to think, learn and conduct experiments like a scientist [14]. Laboratory activities can make students able to solve problems according to the physics theory learned in class [15].

There are a number of studies that were previously developed in generating and utilizing virtual labs in Physics subjects. Among them is this research aims to produce interactive virtual laboratory learning media that is effective and practical for Optical Physics subjects [16]. Other studies [17] This study aims to provide students with access via the internet to various experiments in control techniques, which are located in control laboratories in several universities. Other research [18] The research study referred to here is to produce an application virtual physics laboratory called ViPhyLab using an Android-based smartphone. This research can also determine the suitability and quality of the virtual physics laboratory application that has been developed and to describe the improvement of learning. Other research [19] This study aims to describe the feasibility of Algodoo VPL-based physics practicum software on Newton's Law of Gravity. Other studies [20]. This study aims to examine the application of discovery learning models assisted by virtual laboratory media to influence the mastery of concepts in students.

Based on the results of the analysis of student needs, as many as 77.7% of students stated that the teacher had never presented virtual practicum activities needed during distance learning. As many as 71.2% of students have difficulty in distinguishing the criteria for collision in working on the description test questions. As many as 83.5% of students are interested in learning Physics using the Virtual Physisc Laboratory. As many as 94.2% of students agree that learning media in the form of Virtual Physic Laboratory (VPL) is developed for Collision topic.

Based on the description above, this can provide an opportunity to conduct research by designing a Virtual Physics Laboratory on the Collision topic. This study will discuss the design of the Virtual Physics Laboratory (VPL) media which contains an overview of the comparison between collisions.

2. Method

The method used in this research is the Research & Development method. Media design development is carried out by referring to the ADDIE development model. ADDIE stands for Analyze, Design, Develop, Implement, and Evaluate [21]

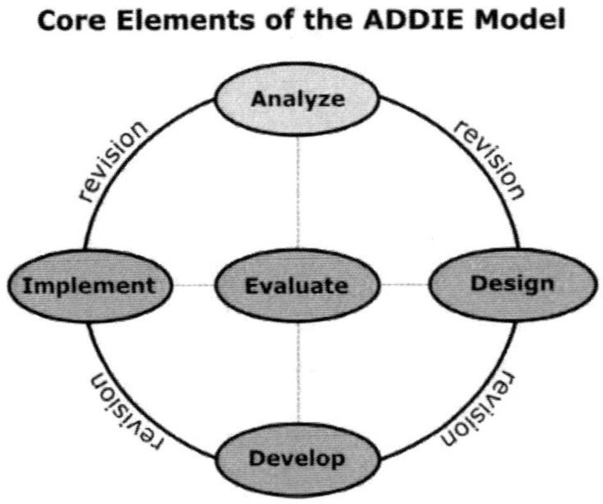

Source: (Branch, 2010)
Figure 2.1 Core elements of ADDIE

However, the ADDIE development model used is only up to the design stage. The analysis stage is the stage to determine the needs of physics learning media in schools. The analysis carried out includes literature studies in the form of journal analysis and the availability of simulation media, analysis of student needs, and analysis of teacher needs. While the design stage is the stage that determines how a learning program is designed according to the needs and objectives that have been determined. The design carried out includes a design (storyboard) from the Virtual Physics Laboratory (VPL).

3. Results and Discussions

3.1. Literature Study

The literature study was carried out by analyzing and synthesizing various international and national journals available from various sources of database providers. The stages carried out in the literature study are reviewing abstracts from each study, collecting library data, and comparing other literature to be processed and produce conclusions. From these stages, an overview of the problems in physics learning can be obtained so that they can be solved in research. The articles analyzed contain the results of studies on the development and use of computer simulations in learning physics. The scientific articles used were obtained from Computers and Education, Asia-Pacific Forum on Science Learning and Teaching, Science Education, IOP Conf. Series, Journal of Physics, Procedia Technology, Physical

Journal of Physics: Conference Series **2019** (2021) 012017 doi:10.1088/1742-6596/2019/1/012017

Review Physics Education Research, Science Direct, American Scientific Publishers, Instructional Science, Research in Higher Educational Journal, American Journal of Physics, American Physical Society and several other sources. The search process is limited to journal articles published in the range of 2000 to 2020. In addition to journal analysis, an analysis of the availability of simulation media is also carried out on several platforms. The availability of the collision simulation platform can be seen in Table 3.1.

Table 3.1 Availability of Collision simulation across multiple platforms

No	*Developed virtual lab*	Source	Information
1.	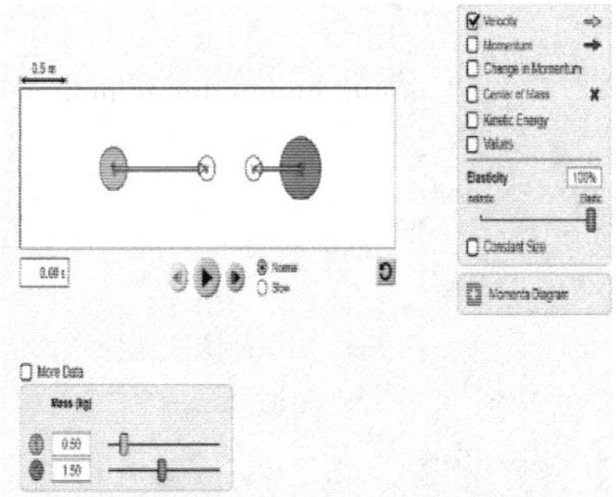	https://phet.colorado.edu/sims/html/collision-lab/latest/collision-lab_en.html	This simulation shows 2 balls colliding with a flat plane with a mass that can be varied, but the results of observations is not placed in the table directly. So students have to transfer the results of the observation data to a notebook.
2.	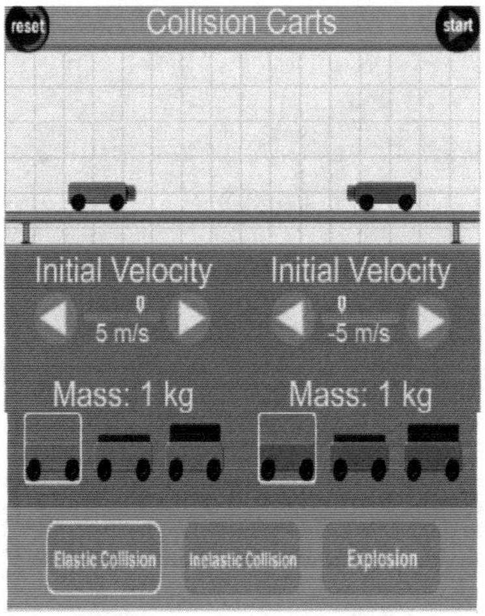	https://www.physicsclassroom.com/Physics-Interactives/Momentum-and-Collisions/Collision-Carts/Collision-Carts-Interactive	This simulation displays 3 collision processes in the same experiment. Existing experiments include: 1. Perfectly Elastic Collision, 2. Collision is not elastic at all and 3. Explosion Process There have been no experiments on the partially elastic collision process.
3.	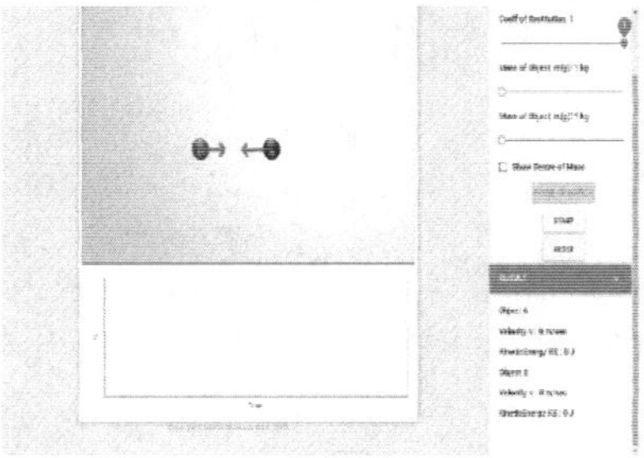	https://vlab.amrita.edu/?sub=1&brch=68&sim=197&cnt=4	This simulation shows the process of 3 collisions, namely: perfectly elastic collisions, partially elastic collisions and completely elastic collisions. However, the user must make arrows on the ball so that the ball moves according to the direction of the arrow, and this is not in the instructions for use.

Journal of Physics: Conference Series **2019** (2021) 012017 doi:10.1088/1742-6596/2019/1/012017

No	*Developed virtual lab*	Source	Information
4.	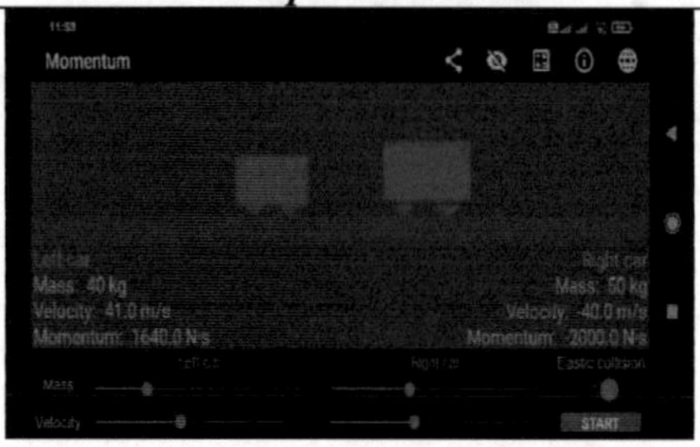	https://towardbrightspark.info/physicsapp	This simulation only shows perfectly inelastic collisions and completely inelastic collisions. There is no simulation of a partially elastic collision yet.

Table 3.1 shows the simulations of Collision available on several platforms. Based on the table above, it can be synthesized that the available simulations only show Collision simulations for 2 Dimensions. The results of the collision process simulation shown have not directly shown the difference between the three collisions.

3.2 Student Needs Analysis and Teacher Needs Analysis
Analysis of student needs was carried out in several high schools in the city of Jakarta, as many as 132 students contributed in filling out the questionnaire. Analysis of teacher needs was carried out in several high schools, as many as 24 physics teachers contributed in filling out the questionnaire.

3.3. *Virtual Physics Laboratory (VPL)*

Virtual Physics Laboratory (VPL) is an interactive multimedia that provide facilities and infrastructure for virtual practicum activities with the help of computers. The virtual laboratory design is made to resemble a real laboratory. Users can prepare practical materials and simulate the media to obtain observational data. The design of the VPL in the Collision sub-material can be seen in Figure 1.

Figure 1. Front view

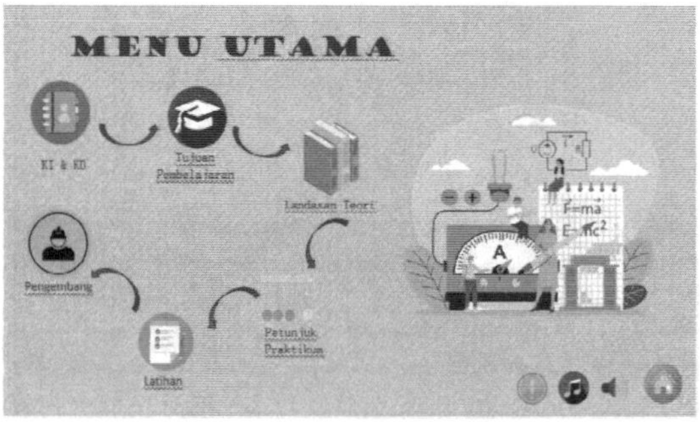

Figure 2. Main Menu

Figure 1 shows the main menu page on the VPL which contains a login page for students who will do a practicum. Figure 2 shows the main menu consisting of KI & KD, Learning Objectives, Theoretical Foundations, Practical Instructions, exercises and developer profiles. As for each sub-material, there is a virtual simulation and a description of the tools and materials that can be used by users during the practicum activity

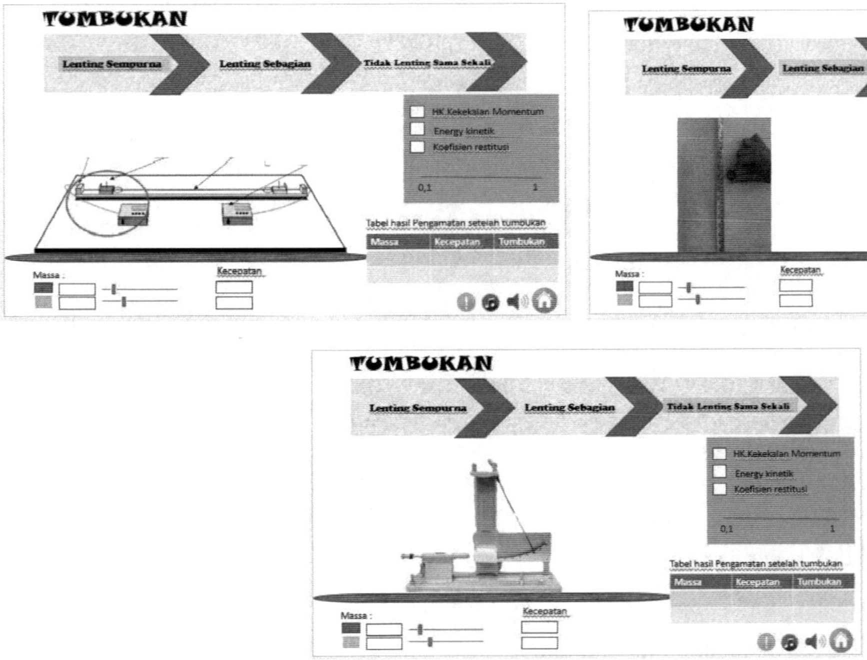

Figure 3. Perfectly Elastic Collisions, Partially Elastic Collisions and Totally Inelastic Collisions

Figure 3 shows an atmosphere that is made similar to a real laboratory, students can prepare the tools and materials that will be used during virtual practicum. The button used provides a description of the instructions for use so that it can reduce the confusion faced by students when carrying out practical activities. The available columns are also equipped with recommended instructions for filling in. The table of observations will be automatically filled if the user has made 3 experiments on each material.

4. Conclusion

Based on the literature study, the analysis of the needs of teachers and students can be concluded that the available learning media have not helped students to visualize, describe, and describe the Collision sub-material correctly. Therefore, it is necessary to develop interactive multimedia that can help students understand the Collision material correctly. The media used to visualize the collision is the Virtual Physics Laboratory (VPL). Virtual Physics Laboratory (VPL) is an interactive multimedia that provide facilities and infrastructure for virtual practicum activities with the help of computers. The virtual laboratory design is made to resemble a real laboratory. Users can prepare practical materials and simulate the media to obtain observational data. With the development of VPL, it is expected that learning will be more effective and help students to understand the sub-material Collisions in Distance Learning. This VPL design will be refined and further developed to meet the research stages of the ADDIE model.

References

[1] Lase D 2019 Pendidikan di Era Revolusi Industri 4.0 *SUNDERMANN J. Ilm. Teol. Pendidikan Sains Hum dan Kebud* vol 12 no 2 pp 28–43
[2] Kemdikbud L 2013 Kurikulum 2013 Pergeseran Paradigma Belajar Abad-21
[3] Muldiani R F *et al* 2014 The Development Of Computer-Assisted Applied Physics Module To Improve The Efficiency And Effectiveness Of Learning Applied Physics At Bandung State Of Polytechnic Tabel 1 Pelaksanaan Praktikum Fi *Sigma-Mu* vol 06 no 1 pp 117

[4] Rofiqoh F I, Mahardika and Yushardi Y 2015 Pengaruh Model Pembelajaran Kooperatif Tipe Numbered Heads Together (Nht) Disertai Media Monopoli Games Terintegrasi Pendekatan Problem Solving Pada Pembelajaran Fisika Di Sma *Journal Pembelajaran Fis* vol 4 no 3 pp 198-203 [Online] Available: http://jurnal.unej.ac.id/index.php/JPF/article/view/2637

[5] Mulhayatiah D, Ningsih R F, Suhendi H Y, Nasrudin D and Yuningsih E K 2019 TUMPULS teaching aids as an alternative media for physics learning *J Phys Conf Ser* vol 1402 no 4

[6] Napsawati N Analisis 2020 Situasi Pembelajaran Ipa Fisika Dengan Metode Daring Di Tengah Wabah Covid-19 *Karst J Pendidik Fis DAN Ter* vol 3 no 1 pp 96–102

[7] Schröder-Turk G E and Kane D M 2020 How will COVID-19 change how we teach physics post pandemic? *Phys Eng Sci Med* vol 43 no 3 pp 731–733

[8] Dewa O E, Mukin M U J and Pandango 2020 Persepsi Kesesuaian Penempatan Pegawai Dalam Menunjang Peningkatan Kinerja BBPK Ciloto *JARTIKA J Ris Teknol dan Inov Pendidik* vol 3 no 2 pp 312–321

[9] Kusuma J W and Hamidah H 2020 Perbandingan Hasil Belajar Matematika Dengan Penggunaan Platform Whatsapp Group Dan Webinar Zoom Dalam Pembelajaran Jarak Jauh Pada Masa Pandemik Covid 19 *JIPMat* vol 5 no 1

[10] Weiser Y, Blau O and Eshet 2016 The Role of Pedagogy Media and Students'personality in Synchronous Learning Comparing Face to Face and Videoconferencing Participation *INTED2016 Proc* p 5005

[11] Tatli Z and Ayas A 2013 International Forum of Educational Technology & Society Effect of a Virtual Chemistry Laboratory on Students Achievement Technologies for the Seamless Integration of Formal and Informal Learning *J Educ Technol Soc* vol 16 no 1 pp. 159–170

[12] Tatli Z and Ayas A 2010 Virtual laboratory applications in chemistry education *Procedia Soc Behav Sci* vol 9 pp 938–942

[13] Odubunmi O and Balogun T A 1991 The effect of laboratory and lecture teaching methods on cognitive achievement in integrated science *J Res Sci Teach* vol 28 no 3 pp 213–224

[14] Bozdogan N and Yalcin 2004 The rate of experiments being carrıed out in science lessons at elementary education and the problems encountered during the physics experiments *J Kirsehir Educ Fac* pp 5 59-70

[15] Darman D R. *et al* 2019 Virtual media simulation technology on mathematical representation of sound waves *J Phys Conf Ser* vol 1188 no 1

[16] Billah A and Widyatmoko A 2018 The Development of Virtual Laboratory Learning Media for The Physical Optics Subject *J Ilm Pendidik Fis Al-Biruni* vol 7 no 2 p 153

[17] Roehrig C and Jochheim 1999 A Virtual lab for controlling real experiments via internet *Proc IEEE Int Symp Comput Control Syst Des* pp 279–284

[18] Arista F S and Kuswanto H 2018 Virtual physics laboratory application based on the android smartphone to improve learning independence and conceptual understanding *Int J Instr* vol 11 no 1 pp 1–16

[19] Rohman S and Admoko 2017 Pengembangan Software Praktikum Fisika Berbasis Vpl Algodoo Untuk Membelajarkan Konsep Hukum Newton Tentang Gravitasi Melalui Penyelidikan *Inov Pendidik Fis* vol 6 no 3 pp 323–328

[20] Sari P I, Gunawan G and Harjono A 2017 Penggunaan Discovery Learning Berbantuan Laboratorium Virtual pada Penguasaan Konsep Fisika Siswa *J Pendidik. Fis dan Teknol* vol 2 no 4

Meaningful Learning of Work and Energy Materials Through Implementation of Basic Locomotor Movements in Basketball Games for High School Students

E F N Sari[1,a)], M Asmawi[1,b)], F Dlis[1,c)], D Susanti[2,d)] and N M Siregar[1,e)]
[1] Physics Education Program, Universitas Negeri Jakarta
Jl. Rawamangun Muka, Jakarta 13220, Indonesia
[2] Physics Education Program, Universitas Negeri Jakarta
Jl. Rawamangun Muka, Jakarta 13220, Indonesia
Email: [a)]efnovita@unj.ac.id, [b)]moch.asmawi@unj.ac.id, [c)]firmansyahdlis@unj.ac.id, [d)]dwisusanti@unj.ac.id, [e)]nfsiregar@unj.ac.id

Abstract. It is necessary to bring up meaningful learning, especially on work and energy material. Meaningful learning can be achieved through direct experience. Physics learning can be integrated through sports and health physical education learning, so it is possible to bring up meaningful learning. So this study aims to determine the effect of meaningful learning on the material of work and energy through direct experience in basic locomotor movement activities in basketball games on Physical Education, Sports and Health subjects. This research method is a quasi-experimental design with a Post-test only Control Group Design. The average post-test scores for the experimental and control classes were 84.34 and 77.33, which means there is a difference between classes that learn through direct experience and conventional ones. The independent sample t-test results give the value of Sig. of 0.017 that inform there is a significant difference in students' understanding of the concept of work and energy between classes that learn through direct experience with conventional ones. The interviews showed an average score of 89.06% of students indicated that they had experienced a meaningful learning process. So it can be concluded that meaningful learning in the learning process becomes the main focus that teachers should implement. Meaningful learning of the material of work and energy through direct experience in basic locomotor movement activities in basketball games in Physical Education, Sports and Health is proven to be better for student learning outcomes. In addition, through meaningful learning that has been done, it also informs that students become easier to understand the concept of the material through authentic experiences in everyday life.

1. Introduction

Basic locomotor movement is one of the basic materials in Physical Education, Sports and Health at the elementary level. The locomotor motion itself is a basic motion [1] that moves a body to change position [2]. Learning basic locomotor movements at the elementary level aims to improve the quality of the student movement to move optimally, such as walking, running, jumping, and so on [3]. This movement is also helpful in improving body functions to achieve proficiency in performing basic movements that ultimately improve body health [4]. At the high school level, basic locomotor movements such as walking, running, and jumping are directly integrated into some sports games. One of the sports games that integrate basic locomotor movements is basketball.

Physics is one of the basic sciences that discusses all phenomena that exist in life. Physics is a science to find answers to "why" and "how" to phenomena that occur in life [5]. Physics is a science that can be integrated with various other sciences, one of which is in sports. This means that Physics learning can be carried out with students' direct experience in learning sports at school.

In the high school curriculum, one of the materials studied in physics is work and energy. Work and energy material is one of the essential materials in high school physics learning. This material is abstract, so often, the learning process that is carried out does not lead to meaningful learning. In this material, many students have not been able to connect, work and energy [6].

Meaningful learning is an essential factor in the learning process [7] and becomes the ideals of the desired learning process. Meaningful learning can be achieved through direct experience. Meaningful learning occurs when the learning process is active, constructive, intentional, authentic, and cooperative [8,9]. Meaningful learning is realized when students can connect the knowledge they have learned with everyday life [10], thereby increasing the benefits of learning [11].

Based on these reasons, it is necessary to bring up meaningful learning, especially work and energy. Physics learning can be integrated through sports and health physical education learning, so it is possible to bring up meaningful learning. So based on these things, this study aims to determine the effect of meaningful learning on the material of work and energy through direct experience in basic locomotor movement activities in basketball games on Physical Education, Sports and Health subjects.

2. Method

This research is quantitative. This research method is a quasi-experimental design with a Post-test only Control Group Design. The quasi-experimental research method was chosen because the class groups used were not random [12,13].

This study used two classes, namely the experimental and control classes. The two classes used will be tested for normality and homogeneity first. After the two classes were declared normal and homogeneous, the Physics learning process in the experimental class was carried out by integration through basic locomotor movement activities in the basketball game. At the same time, the control class of the Physics learning process is carried out conventionally in the classroom without direct experience. After the learning process is complete, both classes will take a post-test.

The post-test instrument used was questions from the question bank. The results of the post-test will then be analyzed to determine the difference between learning outcomes in the Business and Energy material between the learning process through actual experience and the conventional learning process with the following hypothesis:

- H0 = There is no difference in the ability to understand the concept of work and energy material between the experimental class and the control class
- Ha = There is a difference in the ability to understand the concept of work and energy material between the experimental class and the control class

Testing is conducted with an independent sample t-test with the following criteria:

- If the value of Sig. (2 tailed) > 0.05, then H0 is accepted, and Ha is rejected
- If the value of Sig. (2 tailed) < 0.05, then H0 is rejected, and Ha is accepted

In addition, interviews were also conducted with students from the experimental class to determine whether students received meaningful learning in the learning process carried out.

3. Results and Discussion

3.1. Normality Test

The normality test was carried out to find out the distribution of students in each class was normal. The normality test is based on the Shapiro-Wilk test due to the limited number of students. The normality test results give the value of Sig. of 0.61 for the experimental class and 0.74 for the control class. Both show a value greater than 0.05, which means that the distribution of students in each class is normal.

Tests of Normality

	Kolmogorov-Smirnov[a]			Shapiro-Wilk		
	Statistic	df	Sig.	Statistic	df	Sig.
Experimental_Class	.640	32	.052	.941	32	.061

a. Lilliefors Significance Correction

Figure 1. Normality test of Experiment Class.

Tests of Normality

	Kolmogorov-Smirnov[a]			Shapiro-Wilk		
	Statistic	df	Sig.	Statistic	df	Sig.
Control_Class	.672	33	.055	.926	33	.074

a. Lilliefors Significance Correction

Figure 2. Normality test of Control Class.

3.2. Homogeneity Test

A homogeneity test was carried out to find out that students in each class were homogeneous. The results of the homogeneity test give the value of Sig. of 0.110. A valuc was grcater than 0.05 means that the students in the experimental and control classes are homogeneous.

Test of Homogeneity of Variances

Posttest

Levene Statistic	df1	df2	Sig.
1.139	1	63	.110

Figure 3. Homogeneity test.

3.3. Post-test

The post-test was carried out to determine students' understanding of the concept of work and energy. The average post-test score for the experimental class was 84.34, and the control class was 77.33. This significant difference in numbers descriptively shows a difference between classes that learn through direct experience and conventional ones.

Group Statistics

	Class	N	Mean	Std. Deviation	Std. Error Mean
Posttest	Experimental Class	32	84.3438	5.45722	.96471
	Control Class	33	77.3333	8.26766	1.43921

Figure 4. Posttest mean score.

3.4. Independent Sample T-Test

An Independent sample t-test was conducted to determine the significant differences in students' understanding of concepts on the work and energy materials. The test results give the value of Sig. of 0.017. This value smaller than 0.05 informs that the first hypothesis (H0) is rejected and the second hypothesis (Ha) are accepted. This explains the significant difference in students' understanding of concepts in work and energy between classes that learn through direct experience and conventional ones.

Independent Samples Test

		Levene's Test for Equality of Variances		t-test for Equality of Means						
									95% Confidence Interval of the Difference	
		F	Sig.	t	df	Sig. (2-tailed)	Mean Difference	Std. Error Difference	Lower	Upper
Posttest	Equal variances assumed	1.139	.110	2.021	63	.017	7.01042	1.74330	3.52670	10.49413
	Equal variances not assumed			2.046	55.624	.017	7.01042	1.73263	3.53903	10.48181

Figure 5. Independent sample t-test.

3.5. Meaningful Learning

Students in the experimental group carried out the learning process of work and energy physics through real experience by performing basic locomotor movements in basketball games. After the learning process, students were interviewed to find out whether students received meaningful learning or not. The interviews showed an average score of 89.06% of students indicated that they had experienced a meaningful learning process. The questions in the interview are as follows.

Table 1. Meaningful Learning Questions.

No	Question	Score
1	Is the learning process with real experience through basic locomotor movements in basketball games very closely related to the concepts of work and energy?	90.62%
2	Can the learning process with real experience through basic locomotor movements in basketball games facilitate understanding the concept of work and energy?	93.75%
3	Can the learning process with real experience through basic locomotor movements in basketball games connect the basic concepts of work and energy that they already have?	84.38%
4	Can the learning process with real experience through basic locomotor movements in basketball games connect new concepts about work and energy?	87.50%

From the interviews, it was found that 90.62% of students stated that the learning process with real experience through basic locomotor movements in basketball games was closely related to the concept of work and energy. Then, 93.75% of students agreed that the learning process with real experience through basic locomotor movements in basketball games could facilitate understanding the concept of work and energy. In addition, 84.38% of students stated that the learning process with real experience through basic locomotor movements in basketball games was able to connect the basic concepts of business and energy they already had, and 87.50% were able to connect new concepts about work and energy.

3.6. Discussion

This study aims to determine the effect of meaningful learning on the material of work and energy through direct experience on basic locomotor movement activities in basketball games in Physical Education, Sports and Health. This study used two classes, namely the experimental and control classes. Both classes were initially in normal and homogeneous conditions.

The normality test was carried out to find out the distribution of students in each class was normal. The normality test is based on the Shapiro-Wilk test due to the limited number of students. The normality test results give the value of Sig. of 0.61 for the experimental class and 0.74 for the control class. Both show a value greater than 0.05, which means that the distribution of students in each class is normal. Normality was a vital part of the research. The normal class can show actual analysis.

A homogeneity test was carried out to find out that students in each class were homogeneous. The results of the homogeneity test give the value of Sig. of 0.110. The value was greater than 0.05 means that the students in the experimental and control classes are homogeneous. Like the normality test, the homogeneity test is also crucial in this study.

The post-test was carried out to determine students' understanding of the concept of work and energy. The average post-test scores for the experimental and control classes were 84.34 and 77.33, respectively. This significant difference in numbers descriptively shows a difference between classes that learn through direct experience and conventional ones. Descriptively, it is clear that there are differences in learning outcomes between the two classes.

An Independent sample t-test was carried out to find out the significant differences in students' understanding of concepts in work and energy. The test results give the value of Sig. of 0.017. This value smaller than 0.05 informs that the first hypothesis (H0) is rejected and the second hypothesis (Ha) are accepted. This explains the significant difference in students' understanding of work and energy between classes that learn through direct experience with conventional ones.

Students in the experimental group carried out the learning process of work and energy physics through real experience by performing basic locomotor movements in basketball games. After the learning process, students were interviewed to find out whether students received meaningful learning or not. The interviews showed an average score of 89.06% of students indicated that they had experienced a meaningful learning process.

From the interviews, it was found that 90.62% of students stated that the learning process with real experience through basic locomotor movements in basketball games was closely related to the concept of work and energy. Then, 93.75% of students agreed that the learning process with real experience through basic locomotor movements in basketball games could facilitate understanding the concept of work and energy. In addition, 84.38% of students stated that the learning process with real experience through basic locomotor movements in basketball games was able to connect the basic concepts of work and energy that they already had, and 87.50% could connect new concepts about work and energy.

Physics learning that is integrated with sports turns out to be interesting for students [14,15]. This is in line with the statement that learning effectively increases student interest and learning outcomes [16]. Both statements are following other statements that support that learning Physics should focus on meaningful learning [17].

This is in line with several studies related to meaningful learning. The learning process with meaningful learning and less dialogue gives better results [18]. Meaningful learning gives the importance of new knowledge gained with its very close application in everyday life [19,20]. So that meaningful learning in the learning process becomes the main focus that teachers should implement.

4. Conclusion

Meaningful learning is the main focus of the actual learning process. Meaningful learning of the material of work and energy through direct experience in basic locomotor movement activities in basketball games in Physical Education, Sports and Health is proven to be better for student learning outcomes. In addition, meaningful learning that has been conducted also informs that students become easier to understand the concept of the material through authentic experiences in everyday life.

References

[1] Khajooei M, Wochatz M, Baritello O and Mayer F 2020 Effects of shoes on children's fundamental motor skills performance *Footwear Science*

[2] Asnaldi A, Yelis R, Zulman, Atradinal, Putri L P and Bakhtiar S 2020 The Differences of Ability in the Level of Development on Control Objects of Early Childhood Education Students in Pariaman City and Padang Panjang City

[3] Hernawan H, Sukarya Y and Solahuddin S 2019 Locomotor basic motion learning model based on traditional game for basic school students *Journal of Physics: Conference Series* vol 1318

[4] Sari E F N, Julianti R R, Siregar N M and Sukiri 2020 Locomotor basic movement levels in improving the health of elementary school students *International Journal of Human Movement and Sports Sciences*

[5] Saleh S 2021 Malaysian students' motivation towards Physics learning *European Journal of Science and Mathematics Education*

[6] Gutierrez-Berraondo J, Zuza K, Zavala G and Guisasola J 2018 University students' ideas on the relations between work and energy in Mechanics in introductory Physics courses *Revista Brasileira de Ensino de Fisica*

[7] Hung B P 2019 Meaningful learning and its implications for language education in Vietnam *Journal of Language and Education*

[8] Mystakidis S, Berki E and Valtanen J 2019 The patras blended strategy model for deep and

meaningful learning in quality life-long distance education *Electronic Journal of e-Learning*
[9] Ghazali N, Nordin M S, Abdullah A and Mohd Ayub A F 2020 The relationship between students' MOOC-efficacy and meaningful learning *Asian Journal of University Education*
[10] Polman J, Hornstra L and Volman M 2020 The meaning of meaningful learning in mathematics in upper-primary education *Learning Environments Research*
[11] Fan K K, Xiao P wei and Su C H 2015 The effects of learning styles and meaningful learning on the learning achievement of gamification health education curriculum *Eurasia Journal of Mathematics, Science and Technology Education*
[12] Miller C J, Smith S N and Pugatch M 2020 Experimental and quasi-experimental designs in implementation research *Psychiatry Research*
[13] Hastjarjo T D 2019 Rancangan eksperimen-kuasi *Buletin Psikologi*
[14] Mathelitsch L 2017 Sport and physics *Scientia in educatione*
[15] Martínez-Muñoz M, Muñoz-Jiménez J, García-Gómez M J, Valiente-Blanco I and Díez-Jiménez E 2019 Making physics attractive to high-school students through sports gymnasium
[16] Budi S 2019 Effectiveness of the implementation of meaningful instructional design learning model in improving interest and student learning outcomes of physics in class x senior high school *Jurnal Geliga Sains: Jurnal Pendidikan Fisika*
[17] Moreira M A 2021 Desafios no ensino da física *Revista Brasileira de Ensino de Física*
[18] López de Aguileta G and Soler Gallart M 2021 Ausubel's meaningful learning and educational segregation *Multidisciplinary Journal of Educational Research*
[19] Agra G, Formiga N S, Oliveira P S de, Costa M M L, Fernandes M das G M and Nóbrega M M L da 2019 Analysis of the concept of meaningful learning in light of the ausubel's theory *Revista brasileira de enfermagem*
[20] Kurniawati H and Maulana S 2019 The implementation of chordophone enrichment book as an essential factor to create meaningful learning *Journal of Physics: Conference Series*

Journal of Physics: Conference Series **2019** (2021) 012019 doi:10.1088/1742-6596/2019/1/012019

The Video-Based STEM Experiment: An Observation of the Momentum of a Bouncing Ball

I Sugihartono[1,a], Sunaryo[2], C Rustana[3], J Luhukay[4], A A Yaqin[5], K Jalassuad[6], A Apriyansa[6], A Nugraha[7]
[1,2]Department of Physics, Faculty of Mathematics and Sciences, Universitas Negeri Jakarta
Jl. Rawamangun Muka, Jakarta Timur 13220
[3]Physics Department, Faculty of Science and Technology, State Islamic University of Maulana Malik Ibrahim Malang, Jawa Timur
[4]CV Jyotis Cemerlang Bekasi, Indonesia
[5] Faculty of Mathematics and Sciences, Universitas Negeri Jakarta
Jl. Rawamangun Muka, Jakarta Timur 13220
[6]Lembaga Penelitian dan Pengabdian kepada Masyarakat (LPPM), Universitas Negeri Jakarta
Jl. Rawamangun Muka, Jakarta Timur 13220
[7]SMA Labschool Jakarta
Jl. Pemuda Komplek Univeritas Negeri Jakarta, Jakarta Timur 13220
Email: [a]iwan-sugihartono@unj.ac.id

Abstract. We have performed a video-based STEM experiment of a bouncing ball using a force plate logger sensor and Neulog application. Our objective was to observe the students understanding the relationship between impulse and momentum, examine this relation by bouncing a ball on a force plate logger sensor, and calculate impulse and change in momentum. All the data, such as raw data of force and time, video recorded, and the questions given to the 10th-grade senior high school students. Then, their answer will be analyzed to confirm the understanding of momentum during the learning process by video-based STEM experiments.

1. Introduction

As a fundamental concept in Physics, momentum is one of the topics in Physics Subject for senior high school students and undergraduate students. However, the basic knowledge of momentum is probably already introduced by the teacher for the students in the primary. In Indonesia, according to the national curriculum 2013 (revision), momentum is provided for senior high school students 10th grade[1]. There are some studies on momentum. Kaniawati et al reported that the student has difficulty understanding the conservation of energy and momentum[2]. Xu et al report that the connections between net force, time, and change in momentum become the main difficulties for the students[3]. Furthermore, the efforts to improve the understanding of momentum have been implemented by the teachers using teaching aids[4], learning media[1], and learning approach[5], [6].

Among these efforts, the learning approach believed can be effectively improved student understanding during the learning process[7]. As well known, by using teaching and learning approach-based science, technology, engineering, and mathematic (STEM) believed that teaching in physics laboratory or experiment could be more effectively improved physics understanding[8]. Others believed that the integration of science, technology, engineering, and mathematics stimulates the students to have a scientific behavior include engagement in inquiry, logical reasoning, collaboration, and investigation[7], [8]. Hence, by the STEM, systematical a set of thinking, reasoning, teamwork, investigative, and creative skills of the students could be developed.

The covid-19 outbreak causes all the activities to include the schools were to be closed. Online learning is becoming the best alternative to keep the learning process keep running. It is a novel condition for the students, video-based STEM experiments combined with e-learning activities are supposed to be a good choice for the students to obtain experience during the learning process[9]. Others reported that of the video-based problem, the students able to correlate theoretical objects and procedures with experimental observations and measurements[10].

In this paper, the video-based STEM experiments for senior high school student 10th grade to observe the momentum have been reported. The physics understanding of the student will be explored based on the observation by a video-based STEM experiment to determine the value of velocity, impulse, and momentum of the ball motion.

2. Method

To facilitate the video-based STEM experiments, we have prepared a force plate logger sensor, laptop, camera, Neulog application, USB 200 module, and basketball (0.49kg). We then dropped a basketball on a force plate with the height of 100 cm. The ball will bounce as many times as possible on the force plate. The bouncing ball which represented by the graph (force versus time) recorded by the Neulog application. All the experiment activities were presented by video streaming. Then, the STEM experiment presented by the video was observed and analyzed by all the participants who are 10th-grade senior high school students (MAN 1 Bekasi, West Java, Indonesia). There were 40 participants during the activity. Furthermore, the recorded video, and experiment result were given to the students to be analyzed and answered. To analyze the experimental experiment, we provided the instruction as follows: 1) determine the velocity before hitting the plate, 2) determine the velocity after hitting the plate, 3) determine the change in momentum, and 4) analyze the ratio between impulse and change in momentum.

3. Results and discussion

As revealed in the feedback from the students, video-based STEM experiments can be used to replace the role of experiments in classroom activity. However, by using video-based, there is a limited learning experience as students obtained. Most of the students (>90%) agreed that they cannot have experience in direct investigative and creative skills to operate the experiment. Meanwhile, by using the video-based, about 80% agreed that thinking, reasoning, and teamwork can be performed virtually by the online discussion.

Furthermore, to elaborate the understanding of momentum of the bouncing a ball, we analyze the student skill to plot the data as the graph from the data. The graph represents the relationship between force and time. The basketball dropped on a force plate, then will bounce a few times. Furthermore, the force applied will be measured as a function of the time of the experiment and can be recorded by Neulog application. Figure 1 shows the graph pattern of the bouncing ball which represented by force versus time.

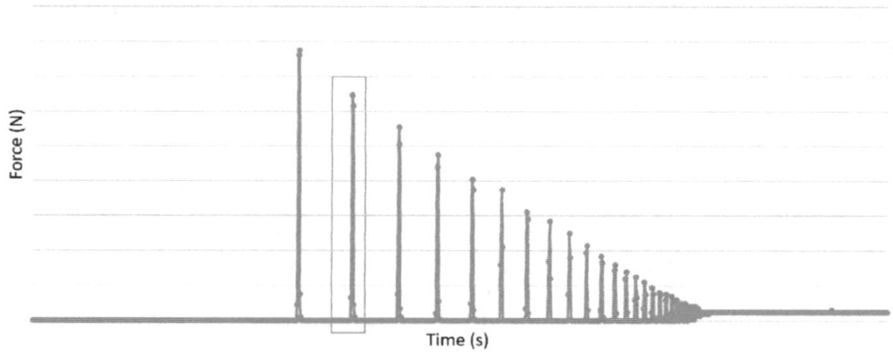

Figure 1. The bouncing ball pattern.

Figure 1 shows the graph pattern of the bouncing ball. Based on the sheets answer, most of the students able to plot the graph as seen in the figure. Meanwhile, to observe the impulse, all the students have difficulties determining it from the graph. They just mentioned that impulse is the value of force versus time. Most recently, Amin et al reported that the factors causing these difficulties are the ability to read the graph and understanding the connection between the equation and the graph[11]. Hence, we assumed that the students do not understand the mathematical principle to interpret the graph as physics phenomena. In this case, an impulse as the value of multiplication force and time can be calculated from the graph by observing the area of the spike. Let us consider the second pike of the graph (figure 1). There are six areas under the spike (figure 2), hence, we obtained the value of the impulse is 2.77 Ns.

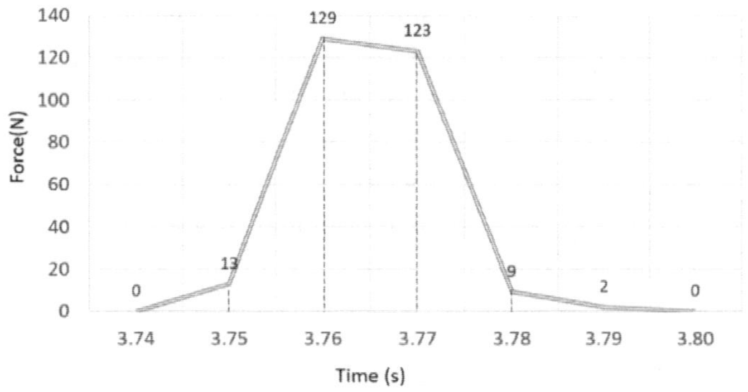

Figure 2. Second pike of the graph of force applied to a force plate during the time of the hit.

In the bouncing ball experiment, the student observed the ball with a height of about 100 cm falls without any external force and hits the surface of a force plate. By consider the second spike, to calculate the velocity before hitting the force plate, we can use the equation of $v_1 = -\frac{gt}{2}$. As well known, g is the gravitational acceleration. A minus sign indicates the direction of the ball in the same axis with a negative y-coordinate. Meanwhile, the time (t) is the time between first hitting and before the second hitting a plate. Further, the velocity after hitting the plate can be observed in figure 1. As seen from figure 1, the velocity can be calculated by considering the time after the second hitting and before the third hitting a plate. Hence, the change in momentum could be determined (2.94 Ns). If we compared to the impulse, we obtained the ratio between impulse and change in momentum of 94%.

In contrast, most of the students used the momentum equation to determine the velocity. It is a bottleneck for the student to find out the change in momentum. We predicted, most of the students focus on the equation of momentum without considering the graph. It's meant the student has difficulties correlate the physics concepts (such as the law of momentum conservation and impulse) with the experimental observation and measurement as presented through the video. Nevertheless, other reported that the use of video able to effectively attain and improve student knowledge[10].

4. Conclusions

In summary, we conclude that video-based STEM experiments can be used as a learning process to understand the Physics of momentum. Even though, there is limited experience in direct investigative and creative skills to operate the experiment. Further investigation, we found that the students have difficulties to determine the Physics variable such as the velocity of the bouncing ball, change in momentum, and the ratio between impulse and momentum. We predict that the student has difficulties correlate the physics concepts with the experimental observation and measurement.

Acknowledgement
We thank Zainul Ali for his support and help during the preparation and designed of the video-based STEM experiment. Financial support from Hibah Pengabdian kepada Masyarakat FMIPA Universitas Negeri Jakarta 2020 is gratefully acknowledged.

References
[1] J V D Wirjawan, D Pratama, E Pratidhina, A Wijaya, B Untung, and Herwinarso 2020 Development of smartphone app as media to learn impulse-momentum topics for high school students *Int. J. Instr.* **13** no. 3, pp. 17–30, doi: 10.29333/iji.2020.1332a.
[2] I Kaniawati, A Samsudin, Y Hasopa, A D Sutrisno, and E Suhendi 2016 The influence of using momentum and impulse computer simulation to senior high school students' concept mastery *J. Phys. Conf. Ser.* **739** no. 1, pp. 4–8, doi: 10.1088/1742-6596/739/1/012060.
[3] W Xu *et al.* 2020 Assessment of knowledge integration in student learning of momentum *Phys. Rev. Phys. Educ. Res.* **16** no. 1, p. 10130, doi:10.1103/PHYSREVPHYSEDUCRES.16.010130.
[4] H Al Asy'ari and R B Sitepu 2019 Development of props atmega 16 based momentum with sensor system *J. Curric. Indones.* **2** no. 2, pp. 36–40.
[5] E N Rahmawati, Jumadi, and D P Astuti 2020 Development of e-handout assisted by PhET simulation with problem based learning (PBL) model about momentum conservation law and collision to train students' conceptual understanding *J. Phys. Conf. Ser.* **1440** no. 1, doi: 10.1088/1742-6596/1440/1/012048.
[6] M Sahin 2010 The impact of problem-based learning on engineering students' beliefs about physics and conceptual understanding of energy and momentum *Eur. J. Eng. Educ.* **35**, no. 5, pp. 519–537, doi: 10.1080/03043797.2010.487149.
[7] A Thahir, C Anwar, A Saregar, L Choiriah, F Susanti, and A Pricilia 2020 The effectiveness of STEM learning: scientific attitudes and students' conceptual understanding *J. Phys. Conf. Ser.* **1467** no. 1, doi: 10.1088/1742-6596/1467/1/012008.
[8] C Chanthala, T Santiboon, and K Ponkham 2018 Instructional designing the STEM education model for fostering creative thinking abilities in physics laboratory environment classes *AIP Conf. Proc.* **1923** doi: 10.1063/1.5019501.
[9] T H Lee, P Di Shen, and C W Tsai 2008 Enhancing computing skills of low-achieving students via e-learning: A design experiment of web-based, problem-based learning and self-regulated learning *Cyberpsychology Behav.* **11** no. 4, pp. 431–436, doi: 10.1089/cpb.2007.0080.
[10] S Gröber, P Klein, and J Kuhn 2014 Video-based problems in introductory mechanics physics courses *Eur. J. Phys.* **35** no. 5, doi: 10.1088/0143-0807/35/5/055019.
[11] B D Amin, E P Sahib, Y I Harianto, A J Patandean, Herman, and E H Sujiono 2020 The interpreting ability on science kinematics graphs of senior high school students in South Sulawesi, Indonesia *J. Pendidik. IPA Indones.* **9** no. 2, pp. 179–186.

Journal of Physics: Conference Series **2019** (2021) 012020 doi:10.1088/1742-6596/2019/1/012020

Design of Massive Online Simulation in The Learning Physics of Thermodynamics Process

M A Lestari[1], F C Wibowo and M Delina
Physics Education Master Program, State University of Jakarta, Jakarta, Indonesia
Jl. Rawamangun Muka, Jakarta 13220, Indonesia
Email: megaaulialestari_1310819009@mhs.unj.ac.id, fcwibowo@unj.ac.id, mutia_delina@unj.ac.id

Abstract. The thermodynamic process is an important material in learning physics. The needs analysis stated that 96.8% of students still experienced difficulties in understanding the thermodynamic process material. One of the media that can be used to help students understand the phenomenon of thermodynamic processes is simulation. This study aims to design a physics learning media in the form of a Massive Online Simulation (MOS) that can be accessed via an internet connection. The development model used in this study is ADDIE (analysis, design, develop, implement, evaluate), but in this study it was only carried out until the design stage. Analysis stage is obtained through literature study and student and teacher needs analysis. At the design stage, it focuses on making Massive Online Simulation (MOS) media designs on thermodynamic processes. This research is expected to produce a physics learning media design in the form of a Massive Online Simulation (MOS) on thermodynamic processes which can be an alternative to online-based interactive learning media that can be accessed anywhere and anytime.

1. Introduction

Thermodynamics is one of the materials in the learning of physics. In the material of thermodynamics there is a sub-material that discusses the processes of thermodynamics. Based on the results of needs analysis has been done by using the google form, the majority of students expressed are still having difficulty in understanding the thermodynamic process with percentage of 96.8%. Research conducted stated that the difficulties experienced by students in the process of thermodynamics, the first difficulty that is not consistent in the use of the concept of law ideal gas so that the difficulties in explaining a phenomenon, the second difficulty that students could not understand the phenomenon of macroscopic relationship between the pressure, temperature and volume [1]. Another study conducted stated that there are still many students who do not understand about the ideal gas and the first law of thermodynamics, as well as experiencing difficulties that are not able to connect the magnitude of the macroscopic such as pressure, temperature and volume in an ideal gas [2]. Another difficulty experienced by students in learning that still have the misconception in understanding the material process thermodynamics, among others, as follows: (1) a wrong connection between the pressure with temperature, (2) the work done by the system, and (3) students have difficulties in applying the first law of thermodynamics and define the diagram of the PV [3].

Physics consists of concepts that are macroscopic and microscopic [4]. To be able to understand the thermodynamics so required an understanding of the magnitudes of the macroscopic and the microscopic process of thermodynamics [1]. If the phenomenon of microscopically can be understood, so the phenomenon of macroscopic is also being easy to understand [4]. Simulation of laboratory can be used as learning media for students [5]. With the use of laboratory simulation can visualize the properties of molecules and the physical properties of the gas [6]. As well as a media of simulation can help visualize physical phenomena microscopically.

One of the online learning media that is currently a trend and widely used is the Massive Open Online Course (MOOC). MOOC is a new idea for online education [7]. Massive Online is an online learning media that can meet global learning needs with the main advantage of providing free education and virtual learning [8]. Massive Online has a bigger impact compared with the Learning Management System (LMS), where students prefer the use of Massive Online than LMS, because of the Massive Online offer uses as the technology is a new learning for students [9]. MOOC can increase the enthusiasm and efficiency of student learning with online learning [10]. With the advantages of Massive Online, so need to develop the design of media learning Massive Online Simulation (MOS) that can provide free simulation. Massive Online Simulation (MOS) is a simulation that can be done online on a large scale. Massive Online Simulation (MOS) is an online-based open learning media that can be followed by students anywhere and anytime using an internet connection. This Massive Online Simulation (MOS) media design was developed through a website that can be accessed via the internet. Massive Online Simulation (MOS) is also useful to clarify students in understanding learning. Simulation of the virtual laboratory can be a solution for distance learning via online [11]. Media-based learning simulation is effective to introduce the students to the concept of physics, and to provide the opportunity for students to engage in higher order thinking process [12]. The simulation has a positive impact in developing the conceptual ideas of the scientific students [13]. The use of simulation in physics can overcome the difficulties of students' conceptual and facilitate the conceptual understanding of students [14,15].

However, some of the simulation platform cannot display the chart of the relationship of pressure, temperature and volume as well as not displaying the thermodynamic process microscopically. Therefore, the need for the development of the design of learning media Massive Online Simulation (MOS) processes in thermodynamics is the process of isobaric, isochoric, isothermal and adiabatic displaying a simulation of the interactive, graphic displays the relationship of pressure, temperature and volume as well as the state of the particles are microscopic. This research aims to design physics learning media in the form of a web-based Massive Online Simulation (MOS) that can be used anytime, anywhere and can be accessed via an internet connection.

2. Method

This research method using Research and Development (R&D). The design model of learning media using the stages of the ADDIE model. The stages of the ADDIE model consist of several stages, namely: (1) analysis, (2) design, (3) development, (4) implementation, and (5) evaluation [16][17]. The following scheme of the stages of the ADDIE model in Figure 1.

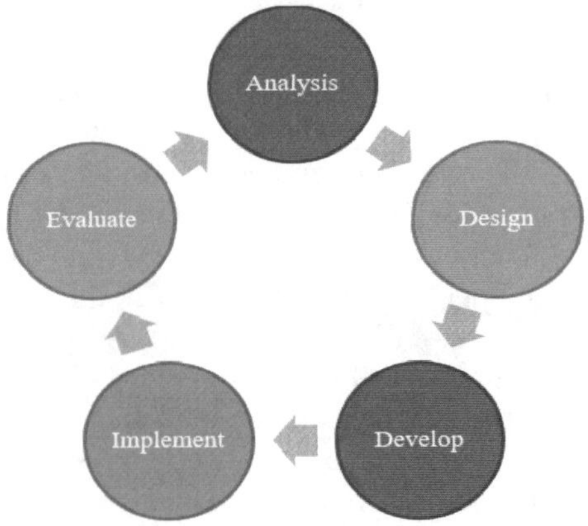

Figure 1. The stages of ADDIE model

Based on Figure 1. The stages of the ADDIE model consist of five stages, but in this research it was only carried out at the design stage. At the stage of analysis, the analysis of the availability of a media platform that has been developed and conduct a needs analysis to determine the design of learning media of physics that will be developed. At the design stage is done making the design of learning media of physics that will be developed.

3. Results and Discussion

3.1. Main Display

At the start of the simulation will appear in the main display before entering into the main menu of the application of learning media Massive Online Simulation (MOS). In the main view, there is the sign in button and the button to sign up. For users who don't have an account to sign in to the app, the user can select the sign up button and fill in the data themselves, among others, an email, username and password. After registering, the user can enter the application by completing the username and password that was previously created. The main display as shown in Figure 2.

Figure 2. Main Display

3.2. Start Menu

The start menu is the menu to start and sound settings in the app. On this start menu there is a play button so that users can enter in media Massive Online Simulation (MOS) and the sound button to enable or disable the sound. The start menu as shown in Figure 3.

Figure 3. Start Menu

Journal of Physics: Conference Series **2019** (2021) 012020 doi:10.1088/1742-6596/2019/1/012020

3.3. Home

The home menu is the main view from media Massive Online Simulation (MOS). Inside the home there is a menu: basic competencies, objectives, materials & simulation, as well as a quiz. As shown in Figure 4.

Figure 4. Home Menu

Based on Figure 4. There are basic competencies that have been adjusted and the goals to be achieved in learning activities.

3.4. Material and Simulation

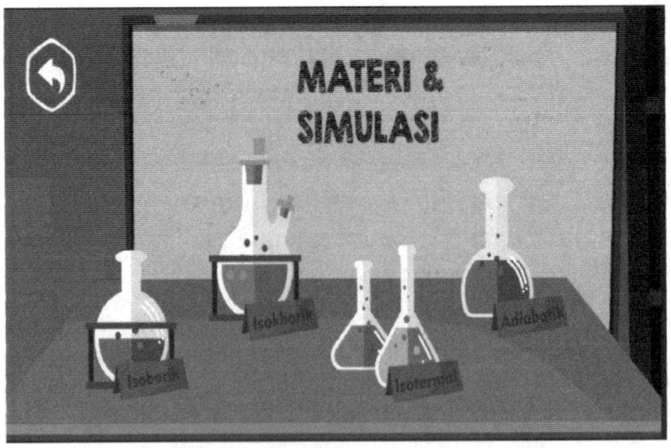

Figure 5. Material & Simulation Menu

Based on Figure 5. On the menu material & simulation there is some selection of material, among others, isobaric, isochoric, isothermal and adiabatic. If the user selects one of the materials so there is a material explanation of the process thermodynamics therein, after the user read the material, there is a button simulation in the corner top right which serves to display the page of the simulation. And there is an exit button that functions to return to the home menu.

3.5. Isobaric Simulation

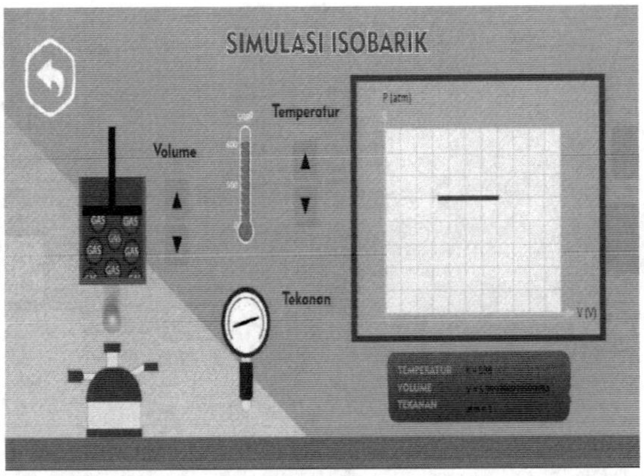

Figure 6. Simulation on the Process of Isobaric

Based on Figure 6. On the simulation page there are macroscopic quantities, namely pressure, volume and temperature. In addition, there are down and up buttons that function to adjust the increase in pressure, volume and temperature. Also, there are pressures, volume and temperature drop buttons. When the user adjusts the pressure, volume, and temperature, the gas particles in the piston will move and automatically the PV graph will display the relationship between the thermodynamic processes that occur. In Figure 6 is a simulation design for the isobaric process. In the isobaric process, the pressure button is not activated because the process changes the state of the gas at constant pressure. After the simulation is done, there is an exit button that functions to return to the material & simulation menu, then the user can try several other thermodynamic process simulations.

3.6. Quiz

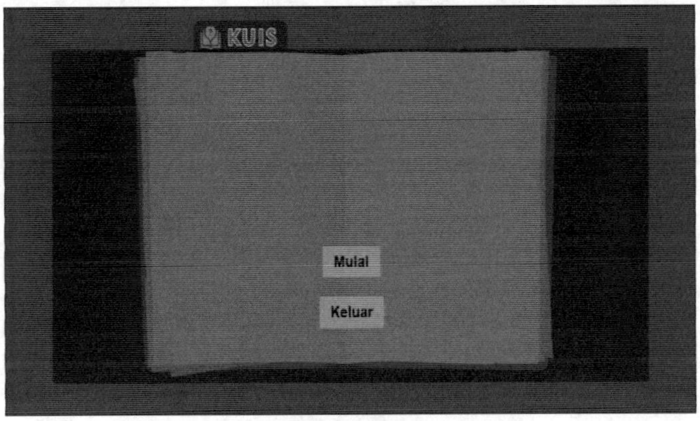

Figure 7. Start Quiz Page

Figure 7. Is the display to start the quiz, there is a start button that serves to continue to work on the quiz and an exit button that functions to return to the start menu. If you click the start button on the Figure 7. So the display will appear in Figure 8.

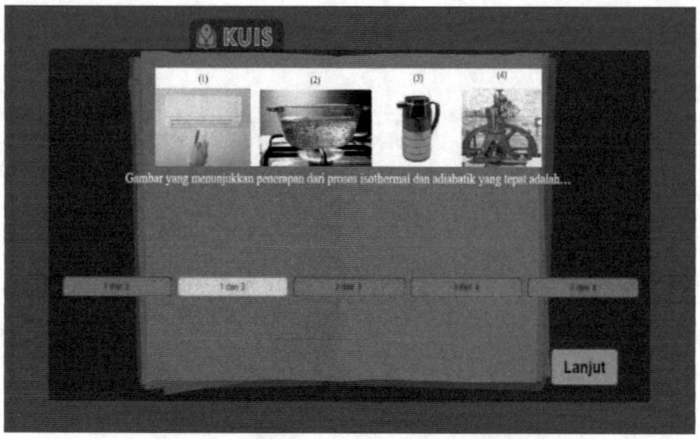

Figure 8. Question

Figure 8 is a quiz page display, on this page there are several multiple choice questions in it. The user can choose one of the options the correct answer from the four options provided. After the user selects an answer, it would appear the answer key directly. The correct answer is shown on the green color, while the red color is the wrong answer. To continue to the next questions, users can click the next button colored blue in the bottom corner. After all the questions have been completed, then the last question will appear a finish button which functions to finish the quiz.

Figure 9. Quiz Score

After the quiz is completed, it will automatically display the score of the correct answer. The score page is as shown in Figure 9. The score page displays the score obtained from the correct answer choices. On the score page there is a start button that serves to repeat the work on the question and an exit button that functions to exit the simulation. If the user does not want to repeat the task, the user can click the exit button on this page which will redirect the user back to the start menu.

4. Conclusions

This research resulted in the media design of web-based a Massive Online Simulation (MOS), where this simulation can be accessed anytime and anywhere via an internet connection. The design of this Massive Online Simulation (MOS) media presents four thermodynamic processes, namely isobaric, isochoric, isothermal, and adiabatic processes. This media design also displays a graph of the PV relationship and displays the phenomenon of microscopic gas particles. In addition, the Massive Online Simulation (MOS) media design presents quizzes in it. The design of Massive Online Simulation (MOS) media that has been developed can be an alternative to online-based interactive media that can be accessed anywhere and anytime with an internet connection. This research is expected to be continued in further research so that the designs that have been made can be developed to the final stage.

References

[1] Loverude M E, Kautz C H, and Heron P R L 2002 *Journal of Physics* 70(2) 137–148
[2] Kautz C H, Heron P R L, Loverude M E, and McDermott L C 2005 *Journal of Physics* 73(11) 1055–1063
[3] Wattanakasiwich P, Taleab P, Sharma M D, and Johnston I D 2013 *Journal of Innovation in Science and Mathematics Education* 21(1) 29–53
[4] Astra I M, Wibowo F C, Susanti D, and Darman D R 2021 *Journal of Physics: Conference Series* 1869(1)
[5] Astra I Made, Nasbey H, and Nugraha A 2015 *Journal of Mathematics Science and Technology Education* 11(5) 1081–1088
[6] Chao J, Chiu J L, DeJaegher C J, and Pan E A 2016 *Journal of Science Education and Technology* 25(1) 16–33
[7] Zhang Y, Chen J, Miao D, and Zhang C 2018 *Journal of Emerging Technologies in Learning* 13(7) 111–123
[8] Sneddon J, Barlow G, Bradley S, Brink A, Chandy S J, and Nathwani D 2018 *Journal of Antimicrobial Chemotherapy* 73(4) 1091–1097
[9] Tarmuji N H, Nassir A A, Ahmad S, Abdullah N M, and Idris A S 2018 *Students' acceptance of e-learning in mathematics: Comparison between LMS and MOOC using SEM PLS approach AIP Conference Proceedings* 1974
[10] Ji Z 2016 *Journal of Emerging Technologies in Learning* 11(5) 62–67
[11] Khoroshko L L, Ukhov P A, and Keyno P P 2019 *Journal of Engineering Pedagogy* 9(4) 4–15
[12] Falloon G 2019 *Using simulations to teach young students science concepts: An Experiential Learning theoretical analysis Computers and Education* 135 138–159
[13] Park M 2019 *Journal of Mathematics Science and Technology Education* 15(7)
[14] Stephens A L and Clement J J 2015 *Use of physics simulations in whole class and small group settings: Comparative case studies Computers and Education* 86 137–156
[15] Meli K and Koliopoulos D 2019 *Journal of Physics: Conference Series* 1287(1)
[16] Henukh A, Reski A, Nikat R F, Waremra R S, and Bahri S 2020 *Journal of Physics: Conference Series* 1569(2)
[17] Serevina V, Raihanati, Sunaryo, and Andriana W 2019 *Journal of Physics: Conference Series* 1280(5).

Design of Massive Online Simulation (MOS) on kinetic theory of gases

W T Priane[1], F C Wibowo[2] and M Delina[2]
[1] Student Master of Physics Education, State University of Jakarta, Indonesia
[2] Lecturers of Master of Physical Education, State University of Jakarta, Indonesia
Email: widiatripriane95@gmail.com

Abstract. The purpose of this study was to design a Massive Online Simulation (MOS) to describe the behavior of gas particles to make it easier for students to understand the microscopic kinetic theory of gases. This research is a type of development research (Research and Development) with the ADDIE model consisting of analysis (analysis), design (design), development (develop), implementation (implement) and evaluation (evaluate). But, in this study, researchers only use the initial stage is the stage of an analysis and d design. The result of this research is the design of Massive Online Simulation (MOS) learning media on a gas kinetic theory which discusses monatomic, diatomic and polyatomic sub-materials, in which there is a main display menu, start menu, material menu, simulation menu and quiz menu that can be used. Students for learning the kinetic theory of gases.

1. Introduction

At the current level of digitization software and technology are playing an increasing role in almost all areas of society and every aspect of life [1]. Advances in communication technology also indirectly change the social order of society in their social interactions. This phenomenon provides many opportunities for humans as potential beings to actualize their potential. However, its actualization of course requires tools as a condition to build a logical line of thought [2]. In physics learning that is studied in SMA Class XI there is material on the kinetic theory of gases, this material is material that is abstract. The kinetic theory of gases is the concept that gases consist of atoms that move randomly continuously. Even if we are able to carry out a class experiment using a pressure gauge and a syringe, the results of this experiment do little to explain the kinetic theory of gases [3]. Basically, the concepts of gas pressure and gas kinetic energy are microscopic concepts and cannot be observed directly. It also relates to Newton's laws of motion; in terms of microscopic aspects, namely speed, collision, gas pressure, the principle of equipartition of energy [4]. In grade 8 students and prospective science teachers showed that eighth grade students used examples from everyday life when talking about gases, teachers tended to emphasize the properties of gases in their explanations [5]. It is recommended to overcome this conceptual understanding with an interactive teaching model and supplemented with some multi-visualization-based teaching materials because the kinetic theory of gases is a microscopic concept [4]. Therefore, teachers are required to innovate in order to make learning interesting and students are actively involved in it. In the teaching and learning process, there needs to be a two-way relationship between students and teachers. It is intended that there is good cooperation during the teaching and learning process [6]. Based on the results of a preliminary study conducted in several senior high schools (SMA) in the Banten area using google forms which aims to determine the level of understanding possessed by students, it can be seen in table 1.1.

Table 1. Results of Students' Level Understanding

Score	Criteria	Many participants Educate	Percentage amount participants learners
80 <X≤ 100	Very High	8	10 %
60 <X≤ 80	High	6	7 %
40 <X≤ 60	Medium	10	12 %
20 <X≤ 40	Low	30	36 %
0 <X≤ 20	Very Low	28	34 %
Total		82	100 %

Based on table 1.1, the students' test results have a very low level of understanding, with a percentage of 34%. Meanwhile, students who have a very high level of understanding are only 10%. Students who have low criteria are 36%. Based on the preliminary study, it was concluded that the level of understanding, skills possessed by most of the students was still in the very low category. A preliminary study was also carried out by distributing a needs analysis questionnaire through the google form which had been carried out in several high schools in Banten. A total of 82 students contributed in filling out the questionnaire. Based on the results of the analysis of student needs, 92.7% of teachers have used various learning media. The physics learning media that are often used by teachers are shown in Figure 1.

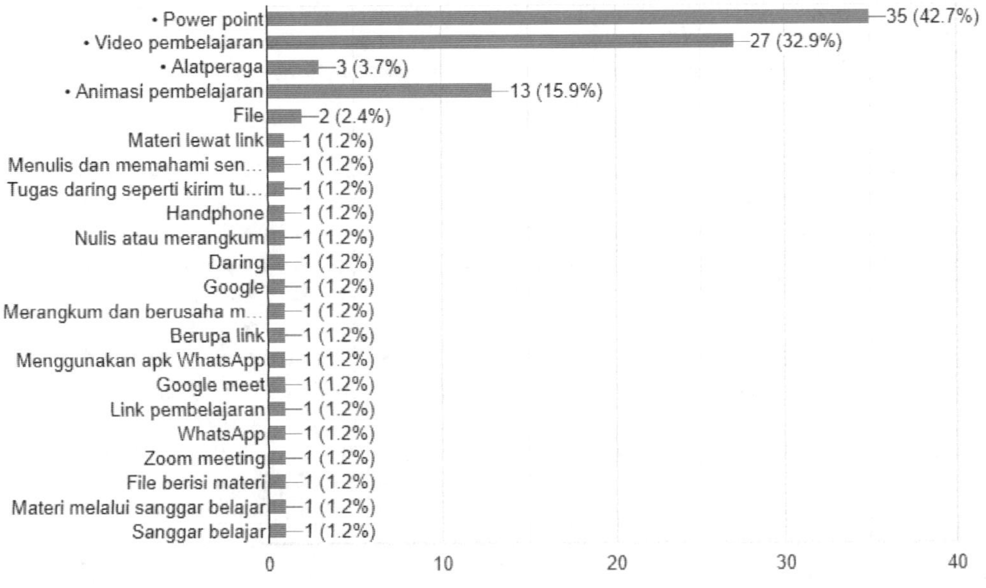

Figure 1. Media used by teachers

In the picture the percentage of learning media used by students given by the teacher, the most widely used learning media by teachers is using power point with a percentage of 42.7%, while learning videos are 32.9% teaching aids 3.7% learning animation 15.9%, 2.4% file and other media used by teachers to students as much as 1.2%. Based on the results of the questionnaire that has been distributed to students, it is known that the teacher has provided learning media to students with various kinds of media, but as many as 41.5% of students stated that they were not focused and did not understand the learning of gas kinetic theory at the time of learning, and 47.6 % of students feel bored while learning. The results of the analysis of teacher needs were also carried out by distributing questionnaires via a google form. A total of 24 teachers from several high schools in Banten have filled out the questionnaire. From the results of the questionnaires that have been distributed, it is known that as many as 79.2% of students still do not have a good level of understanding of the kinetic theory of gases, as many as 45.8% of the learning media that teachers use have not been able to provide a complete picture of the movement of gas particles in space. Closed. So that 100% of teachers stated that they needed new learning media that could make students more active, creative and interactive in learning, and 100% of teachers stated that they agreed that it was necessary to develop a Kinetic Gas Theory learning medium using Massive Online Simulation (MOS). One of the current trending online learning media that is widely used by students and teachers is the Massive Open Online Course or often known by the abbreviation MOOC. Massive Online is an online learning media that can meet global learning needs with the main advantage of providing free education and virtual learning [7]. In addition, Massive Online learning media have another advantage, namely it can make education more accessible, because it tends to be interesting for students, the courses are easily accessible by anyone around the world with an internet connection [8]. This finding sheds new light on the influence of MOOC which shows that in terms of exploring students' knowledge, teachers also have a strong tendency to be able to see MOOC participation as a way to make participation able to explore knowledge on their own [9]. Simulation learning media, according to [10] has proven to be a useful tool for teaching and learning physics. With the open learning resources contained in MOOCs, accessed easily and flexibly, parents and teachers will have the opportunity to gain broader knowledge [2]. In previous research on the developed media, it was concluded that most of the teachers interviewed had explained that there were several challenges in producing Massive Open Online Course (MOOC) files. One teacher revealed that his lectures on campus had not worked out in a format, and he was motivated by the opportunity to learn new aspects of teaching. Experience with Massive Open Online Course (MOOC) projects will allow the possibility to use new teaching materials and activities on campus [11]. Learning physics will be more interesting if there is a simulation. Based on the problems described above, the authors are interested in conducting development research to design, simulation learning media on the kinetic theory of gas material called Massive Online Simulation (MOS) Design on Gas Kinetic Theory.

In an interesting learning process, you should use learning media that can make it easier for students to understand difficult material. Research [3] entitled Discovering the gas laws and understanding the kinetic theory of gases with an iPad app. This study aims to visualize the behavior of particles and to relate the behavior of gas properties such as pressure and temperature using the iPad application. The results showed that the iPad application called Atoms in Motion can demonstrate Boyle's Law and Gay Lussac's Law, to explain data to students about the kinetic theory of gases, namely the relationship between motion between particles and the nature of pressure and temperature. Research [4] entitled Analyzing Students' Level of Understanding on Kinetic Theory of Gases. This study aims to analyze the level of students' understanding of the kinetic theory of gases. The results showed that in the ideal gas law, 33% of the students had fully understood it. However, 14% of them have many misconceptions about the sub-material pressure and kinetic equipartition energy of gases. It is recommended to overcome this conceptual understanding with an interactive teaching model and supplemented with some multi-visualization-based teaching materials because the kinetic theory of gases is a microscopic concept. Research [12] entitled Designing MOOCS with Virtual Microscopic Simulation (VMS) for increasing students' levels of understanding. This study aims to determine the effect of MOOCS with Virtual Microscopic Simulation (VMS) on students' level of understanding. The results showed that students fully understood 76.67%, partially understood 15.56%, and misunderstood 5.56%, and did not understand at 2.22% and the lowest level of understanding was at the 0% level, namely students did not answer. It was concluded that the average level of understanding experienced a significant increase with learning through MOOCS with VMS. Research conducted by Carolyn Penstein Rosé & Oliver Ferschke [13] with the research title Technology Support for Discussion Based Learning: From Computer Supported Collaborative Learning for the Future of Massive Open Online Courses aims to determine the effect of Massive Open Online Courses (MOOCs) through a learning approach problem-based, team project-based learning, and collaborative reflection. This research method applies a problem-based learning approach, team project-based learning, and collaborative reflection. The results show that Massive Open Online Courses (MOOCs) for scale-based learning can improve problem-based learning, team project-based learning, and facilitate collaborative reflection on students. The researcher concludes that the use of Massive Open Online Courses (MOOCs) can increase collaborative reflection on students. MOOCS (Massive Open Online Courses) the current trend in online learning experiences significant changes towards a student-centered approach with the aim of being a new challenge for autonomous and independent learning.

Research conducted by Filippo Sciarrone and Marco Temper in [14] with the research title K-Open Answer: a simulation environment to analyze the dynamics of massive open online courses in smart cities, aims to provide a teacher platform with an environment capable of simulating MOOC (RQ1)to give the teacher some didactic insight about the classroom to set the right teaching strategy in real MOOC (RQ 2). This research method applies a method for modeling users that involve cross-analysis of student representations in the system, based on dependencies, and information flows between teachers and students. The results showed that the use of MOOC assisted by K-Open Answer for simulated learning for teachers could improve the quality of automatic assessment provided by the system. The researcher concludes that the use of simulation-based MOOC can improve the quality of automated assessments.

2. Method

Type Research This is a research development with models ADDIE (Analysis, Design, Development, Implementation, and Evaluation). The procedure of research that is used in the study is referring to the model of R & D that is the model of ADDIE because the steps are used more systematically and clearly. The stages that must be taken in the development model [15] consist of five stages, namely: analysis (analysis), design (design), development (development), implementation (implementation), and evaluation (evaluation). However, this study performed only at the design stage (design). The stages of the ADDIE model can be seen in Figure 2.

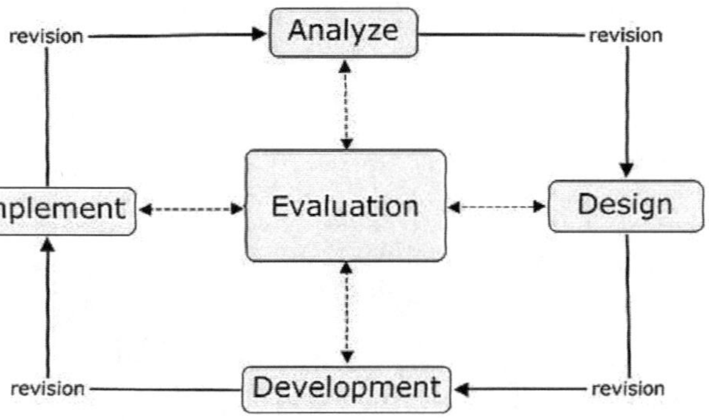

Figure 2. ADDIE Model

The initial step is to analyze the needs of the media teacher and participating students about the device learning online, students on the material difficulties the kinetic theory of gases and media required for online learning. The needs analysis was carried out by distributing questionnaires to several high school physics teachers and distributing questionnaires to several high school students in Banten Province. Data obtained from the teacher and participant students through a questionnaire that was distributed through a google form. Furthermore, learning media Massive Online Simulation (MOS) in design to facilitate teachers convey instructional materials that are abstract and microscopic as well as help students to understand the material the kinetic theory of gases.

3. Result and Discussion

The result of the design of this form of media learning Massive Online Simulation (MOS) which contains the main page to log in and then after their login menu basic competence (KD) and the objectives, content learning, simulation and q is in it. Phase manufacture medium of learning is divided into two parts the first stage of preparation and that the second stage of the manufacture of media design that use methods ADDIE which consists of five stages, namely: analysis, design, development, implementation and evaluation. However, in this study only arrived at the design stage. This analysis stage was carried out by researchers to collect the data needed in making Massive Online Simulation (MOS). At the stage of analyzing a few things such as problems / solutions that experienced by the students during the teaching and learning activities (KBM), the availability of resources to learn at school, information regarding the material subjects, and the theory that needed to put into Massive Online Simulation (MOS). Stage design of the study this is done the design of the initial applications based on the results of the analysis that has been done, such as determining the contents of the content Massive Online Simulation (MOS) which in accordance with the matter in the eyes of subjects kinetic theory of gases, making the layout of the initial application, determine the content of Massive Online Simulation (MOS), make a description of each part, makes the contents of the content that consists of menu basic competence (KD) and the objectives, content learning, simulation and quiz. Design early application Massive Online Simulation (MOS) can be on view in the figure below this:

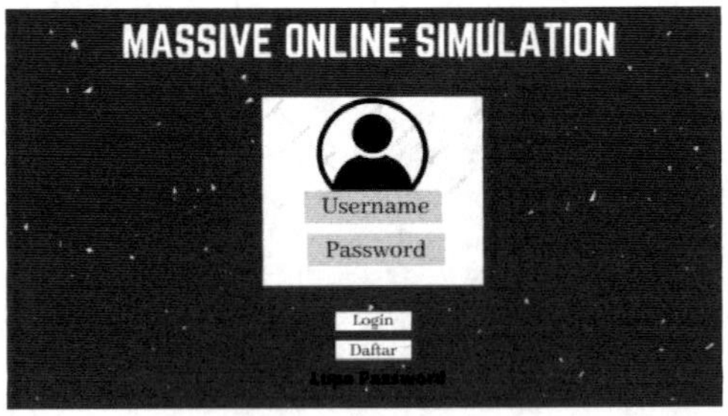

Figure 3. Main View

The main screen is the initial screen that appears when the user opens the Massive Online Simulation media. On this main display screen, there is a login for students who already have an account, and a list column for students who don't have an account.

Figure 4. MOS Start Menu

The start menu is a menu to enter Massive Online Simulation. On this start menu there is a start button for users to enter Massive Online Simulation media. Button of sound to enable or disable sound. I am a media maker info, and book pictures are for reference used.

Figure 5. Home Menu

The Home Menu displays program menus, including: Basic Competencies, Objectives, Materials, Simulations and Quizzes.

Figure 6. KD Menu and Destinations

Menu displays to view basic competencies and objectives. The box can be clicked and go directly to the section you want to see.

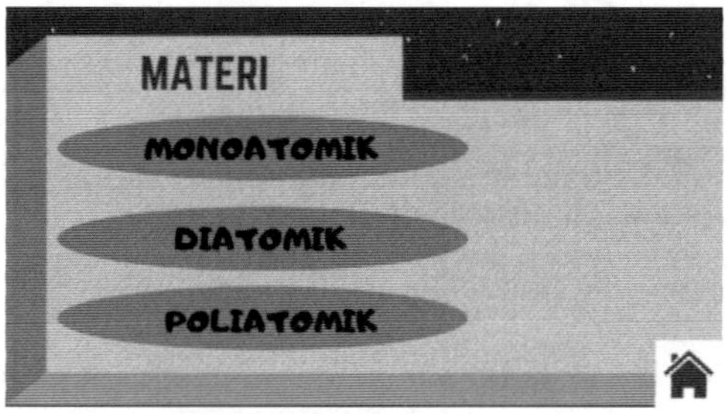

Figure 7. Material Menu

Display menu to select the material you want to learn.

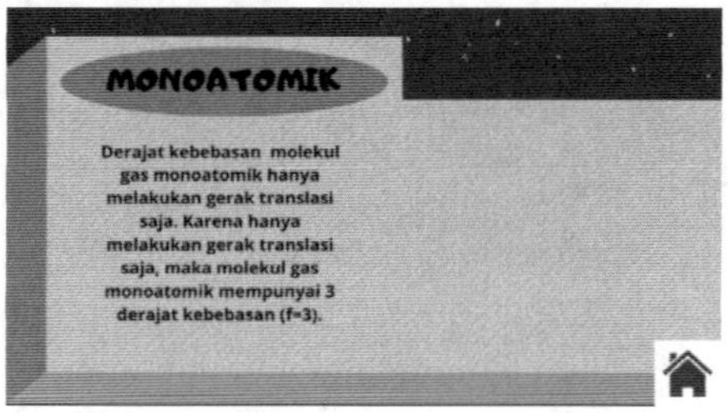

Figure 8. Monatomic Matter

Monatomic material that can be studied by students.

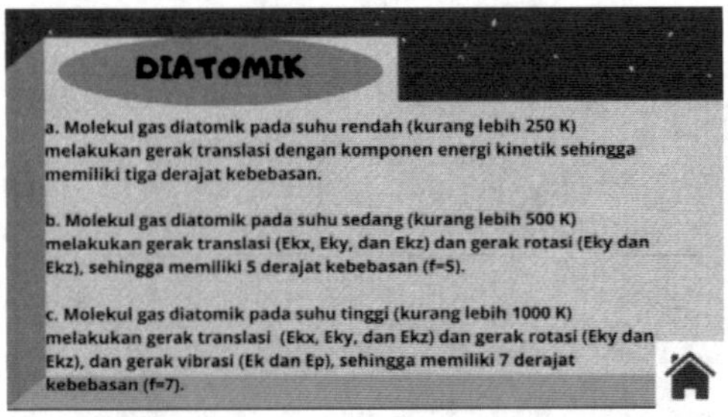

Figure 9. Diatomic Material

The material in atomizers that can be studied by students.

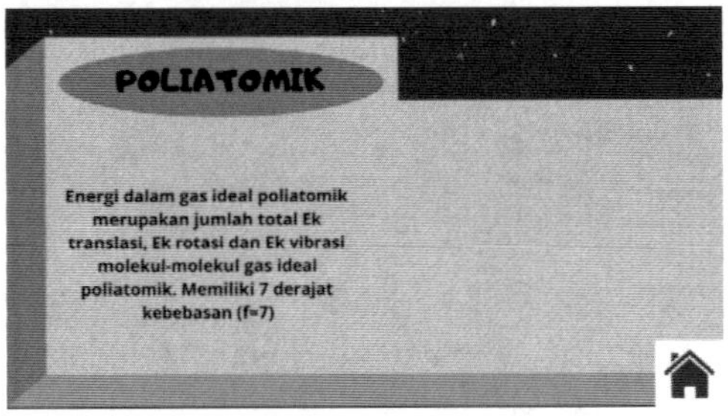

Figure 10. Polyatomic Material

Polyatomic material that can be studied by students.

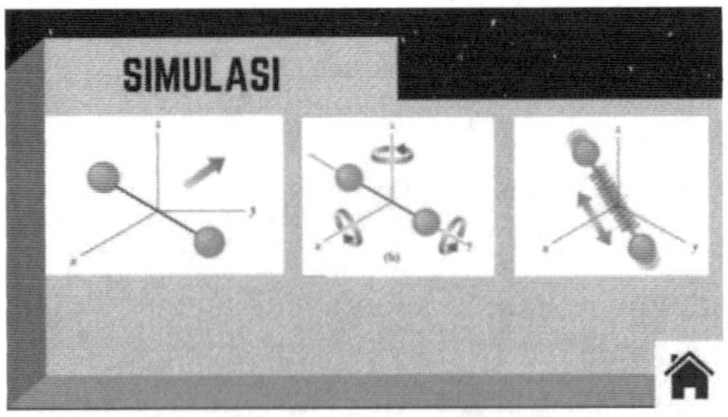

Figure 11. Simulation

Simulation displays that will display monatomic, diatomic and polyatomic simulations.

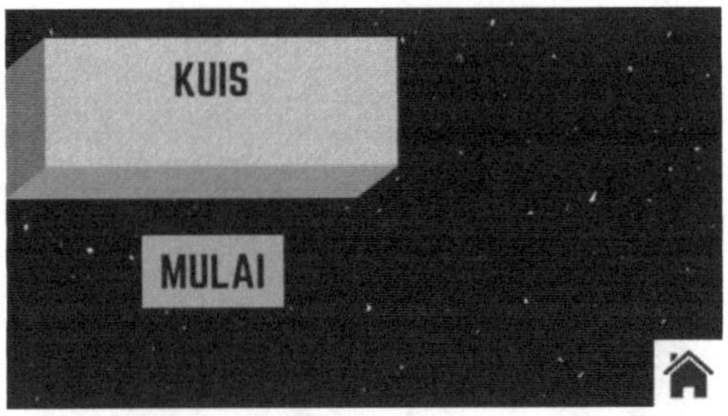

Figure 12. Quiz Start Display Menu

Menu display to start the quiz.

Figure 13. Quiz Display Menu

The menu display is for displaying the quiz, while the arrow is showing the remaining time in working on the questions. Next is used to continue the quiz when the quiz is finished. Based on the results of the Massive Online Simulation learning media design, the researchers hope that this media can be used by teachers and students to easily study the kinetic theory of gases in monatomic, diatomic and polyatomic sub-materials.

4. Conclusion

Massive Online Simulation online learning media on gas kinetic theory material can make it easier for students and teachers in the online teaching and learning process or distance learning. The result of this study is the design of Massive Online Simulation (MOS) learning media on a gas kinetic theory which discusses monatomic, diatomic and polyatomic sub-materials, which includes the main display menu, start menu, material menu, simulation menu and quiz menu that can be used. Students for learning the kinetic theory of gases. So that in this journal a Massive Online Simulation (MOS) learning media design has been produced on the kinetic theory of gases that can assist teachers in conveying the material of the kinetic theory of gases and is expected to make it easier for students to understand the material of the kinetic theory of gases.

Reference

[1] J. Nouri, L. Zhang, L. Mannila, and E. Norén. *Educ. Inq.*, vol. 11, no. 1, pp. 1–17, 2020, doi: 10.1080/20004508.2019.1627844.
[2] S. S. Dewi, S. Suherman, F. C. Wibowo, D. R. Darman, A. P. A. Rino, and I. A. Darmawan. *J. Panelist. Pengemb. Pendidik. Fis.*, vol. 6, no. 1, pp. 17–24, 2020, doi: 10.21009/1.06103.
[3] G. B. Davies. *Phys. Educ.*, vol. 52, no. 4, 2017, doi: 10.1088/1361-6552/aa6d9f.
[4] S. N. Kane, A. Mishra, and A. K. Dutta. *J. Phys. Conf. Ser.*, vol. 755, no. 1, 2016, doi: 10.1088/1742-6596/755/1/011001.
[5] M. Çalik and A. Ayas . *J. Res. SCI. Teach.*, vol. 42, no. 6, pp. 638–667, 2005, doi: 10.1002/tea. 20076.
[6] W. Katrina, H. J. Damanik, F. Parhusip, D. Hartama, A. P. Windarto, and A. Want. *J. Phys. Conf. Ser.*, vol. 1255, no. 1, 2019, doi: 10.1088/1742-6596/1255/1/012005.
[7] Y. Li, J. Sun, and M. Y. Sun. *Kuram ve Uygulamada Egit. Bilim.*, vol. 18, no. 6, pp. 3442–3448, 2018, doi: 10.12738/estp. 2018. 6. 251.
[8] K. F. Hew. *Br. J. Educ. Technol.*, vol. 47, no. 2, pp. 320–341, 2016, doi: 10.1111/bjet. 12235.
[9] S. Ardavani, "How MOOCs Can Develop Teacher Cognition: The Case of in-service English Language Teachers," *Eur. J. Open, Distance E-Learning*, vol. 22, no. 2, pp. 56–71, 2020, doi: 10.2478/eurodl-2019-0010.
[10] S. Y. Lye, L. K. Wee, Y. C. Kwek, S. Abbas, and L. Y. Tay. *Educ. Technol. Soc.*, vol. 17, no. 3, pp. 121–137, 2014.
[11] K. Alsaadat. European Journal of Open Education and E-learning Studies pp. 46–53, 2019, doi:

10.5281/zenodo. 3706329.

[12] F. C. Wibowo and B. H. Iswanto, *J. Phys. Conf. Ser.*, vol. 1402, no. 6, 2019, doi: 10.1088/1742-6596/1402/6/066094.

[13] C. P. Rosé and O. Ferschke, *Int. J. Artif. Intell. Educ.*, vol. 26, no. 2, pp. 660–678, 2016, doi: 10.1007/s40593-016-0107-y.

[14] F. Sciarrone and M. Temperini, *Soft Comput.*, vol. 24, no. 15, pp. 11121–11134, 2020, doi: 10.1007/s00500-020-04696-z.

[15] B. A. Jones, "Instructional Design in a Business English Context," no. 1, pp. 683–696, 2014, [Online]. Available: http://www.umich.edu/~ed626/define.html%0Ahttp://www.brentjones.com/presentations/2008/id_business_english_handout.pdf.

Journal of Physics: Conference Series **2019** (2021) 012022 doi:10.1088/1742-6596/2019/1/012022

The Effect of The Use of Harmonic Movement PhET Interactive Simulation in Online Learning Process on Mastering the Concept of High School Students

Sunaryo[1], C E Rustana[2], Raihanati[1], S N Khalifah[1] and I Sugihartono[1]
[1] Physics Departement, Faculty of Mathematics and Science State University of Jakarta East Jakarta, Indonesia (13220)
[2] Physics Department, Faculty of Science and Technology, State Islamic University of Maulana Malik Ibrahim Malang-East Java, Indonesia (65144), Indonesia
E-mail: naryounj@yahoo.com, CE.Rustana@yahoo.com

Abstract. This research aimed to determine the effect of using the PhET Interactive Simulation in the online learning to the conceptual mastery ability of high school students. The method used in this research was an experimental method with the one group pretest posttest research design. The research sample was selected using purposive sampling technique from a population of the Science and Mathematics Group Students of Public High School (SMAN) 1, Jakarta. The research instrument used was an objective test of students' mastery of concepts that had been tested for validity, reliability, difficulty level, and distinguishing power. Based on the results of data processing, the average of pretest and posttest scores were 47.42 and 73.55, respectively, and the N-gain of 0.51 with the medium category. The analysis result shown that count (10.909) was larger than table (2.042) which indicated that the null hypothesis (H0) was rejected and the alternative hypothesis (HA) was accepted. Meanwhile, the strength of the effect of using the PhET simulation in the Online Learning Process to Mastery the Concept of Harmonic Motion for High School Students on a numeric scale was categorized high with an effect size value of 1.79. Thus, it can be concluded that the PhET simulation can be applied to mastery of the physics concept of Harmonic Motion in the online learning process of high school students.

1. Introduction

The outbreak of coronavirus diseases 2019 (covid-19) in Indonesia provide a restriction on various public activities including teaching and learning process at school. As an alternative, an online learning process is applied by utilizing internet technology and other supporting learning facilities. The online learning is a learning process that is carried out when students and teachers do not physically meet at school [1]. However, this online learning rises up a problem that hampered, especially to the learning activities for example, likes physics that use experimental approach through practicum in the laboratory.

Meanwhile, based on a preliminary survey to the students conducted in the SMAN 1 Jakarta, as many as 80% of respondents said that they found difficult to understand the concept of physics during the online learning process. This condition will certainly have an impact on the lack of students' mastery of physics concepts. Whereas, the students are required to master physics concepts in order to be able to develop their knowledge and abilities in solving a problem.

Therefore, effective learning strategies need to be applied in helping students to master the concepts of physics. One of them, this can be done by using an alternative learning medium which can practically be utilized by students in mastering the physics concept they are learning. One of the learning media that can be used to meet the needs and interests of students for mastering those physics concepts is the PhET (Physics Education Technology) Interactive Simulation that is containing virtual experiment developed by the University of Colorado, Boulder, USA [2]. As known that PhET Interactive Simulations are designed to be flexible to improve the way of teaching and learning physics; so it can be used as a teaching demonstration, laboratory, or homework activity. Thus, the students can gain significant learning experiences that are synthesized into three big themes: learning physics is fun, learning physics is real, and learning physics is simple and easy [3].

Many relevant research results for example by [4, 5, 6, 7] generally, they agree that the use of PhET Interactive Simulation has a positive impact on learning achievements or outcomes of the students in physics. Unfortunately, the use of PhET Interaction Simulation has not been integrated with online learning physics at SMA N 1 Jakarta. About 80% of students stated that their physics teacher had never used the PhET Interactive Simulation in the online learning process at their school, Based on this description as described above, this research was conducted to analyze the effect of the PhET Interactive Simulation when was applied in the online learning process to the students' concept mastery ability, especially, in the Harmonic Motion. The results of this study are expected to benefit physics teachers, especially at SMA N 1 with an idea how to select an alternative learning media PhET Interactive Simulation in order to support their online teaching and learning process for improving students' conceptual mastery abilities. Furthermore, by using this PhET Interactive Simulation, it is expected that can improve understanding of physics, material that is conceptually difficult for students.

2. Concept Mastery

Mastery of concepts is defined as the ability to understand, comprehend, apply, classify, generalize, synthesize, and conclude objects so that they are not only a simple understanding about objects, but can also to describe it [8]. According to Dahar quoted by [8], mastery of concepts is the result of a learning process which is indicated by the ability of students to analyze and explain concepts in their own language. One of the indicators of students in understanding what they have learned can help students in solving problems, not only in learning at school, but also in everyday life [10]. Whereas, Arends quoted by [9] stated that mastery of concepts can help students in constructing an understanding of previously owned concepts in achieving mastery of the concepts being studied.

It is very important and prioritized in the science learning process compared to memorize and in learning it meaningful, students must carry out cognitive processes that build relationships between information on concepts [12]. Mastery of concepts is very necessary to achieve student success in learning, with good mastery of concept, students are expected to be able to manage cognitive skills so that improvements can be made in further learning [13]. Students' mastery of concepts has a positive correlation with learning outcomes where if students' mastery of concepts is high, their cognitive learning outcomes will also be higher [4]. Based these definitions, it can be synthesized that concept mastery is a person's ability to understand the concept and meaning of learning and be able to apply it in her/his life.

3. PhET Interactive Simulation

Learning media is a set of tools used to help teachers convey information to students in the learning process. Along with the times and science and technology development, learning media are exist now not only media in their actual form, but there is already a medium called virtual laboratory. An example of such media is PhET (Physics Education Technology). PhET was created and developed by the University of Colorado, USA [6]. According to Mubarak quoted by Anggriani, PhET is an interactive simulation media in the form of software and can be used to clarify concepts or phenomena that have been practiced to replace the practicum activities at laboratory [4]. PhET can be downloaded for free for the benefit of learning both in the classroom (interest group) and outside the classroom for individual or personal interests and or group.

According to Fatik quoted in Hadiyanti, the advantage of the PhET simulation media is that it can ideally carry out experiments, which cannot be done using real tools. PhET simulation media is suitable for use as a physics learning medium to support the achievement of learning objectives [14].

Based on this explanation, it can be concluded that PhET is an interactive simulation in the form of a virtual laboratory that functions as a learning medium to clarify and prove the phenomena and concepts that have been studied. Learning to use that is done using a laptop or mobile phone as the medium, without having to practice directly using real tools or practicum kits. For this reason, that can be applied through distance learning activities.

As known that distance learning is a learning method that was originally developed in the United States, Germany, France, and England in 1800 using the media of letters. Then in 1980, International Correspondence Schools (ICS) built a system or lecture method at home, which became the forerunner of the distance learning system [15]. Distance learning is a learning method that is rapidly developing along with the development of technology, especially the internet. This method is not new to some schools, because previously they have implemented e-learning in their learning process. Based on various definitions by [16, 17, 18], it can be synthesized that distance learning is learning that is carried out at a distance without direct meetings between teachers and students at school by utilizing technology and the internet as a liaison.

4. Method

As a quantitative research, the research used the experiment method by providing treatment to research subjects to find the effect of such treatments on research subjects. The research design used was pre-experimental with a one group pretest-posttest design model. The following is a one group pretest posttest design research:

Table 1. One Group Pretest – Posttest Research Design.

Class	*Pretest*	Treatment	*Posttest*
Eksperiment	P1	O	P2

P1 : the pre-test scores before any treatment was given.
O : online learning process using the PhET Interactive Simulation.
P2 : the post-test scores after the treatment is given.

The sample of this research was selected using purposive sampling technique from all populations of class X of the science and mathematics group students of SMAN 1 Jakarta of the 2019/2020 academic year. The research data was collected by using an objective test in form multiple choices to measure the physics concept mastery ability, especially the harmonic motion. Prior to data collection, the research instrument was tested through various kinds of tests such as validity tests, reliability tests, and difficulty level and distinguishing tests of questions. The normality test as a prerequisite test is also carried out before testing the hypothesis using the paired sample t test and the N-gain test. Furthermore, an effect size test was also conducted to determine the strength of the effect of using the PhET interactive simulation in the online learning process to the high school students' mastery of the Harmonic Motion concepts.

5. Results and discussion

In this study, the treatment given to the experimental class was carried out in two lessons with a time allocation of 1x60 minutes per meeting. The allocation of time used adjusts to the online learning schedule for SMA Negeri 1 Jakarta. Online learning is carried out by utilizing the WhatsApp application as a medium for discussion and a liaison between teachers and students. The WhatsApp media were chosen because this application is the most popular application by the community as a means of communication and almost all students use it in their daily life. Virtual experimental assignments are given individually. Prior to the PhET assisted learning process, students are first given a pretest to measure the initial ability of students' mastery concepts.

After the PhET assisted learning process, students are given a posttest to determine the ability of students' mastery of concepts after learning with the PhET simulation. During the practicum activities of the harmonic vibration learning material by using PhET Simulation, discussions between teachers and students were also intensively carried out well. The students were asked and was identified be able to answer properly and correctly the questions on the practicum sheet given. Furthermore, students should collect the results of the virtual experiment by sending them in the form of photos.

The research instrument was tested through various kinds of tests such as validity tests, reliability tests, and difficulty level and distinguishing tests of questions and declared to meet the requirements of the various forms of these tests for further used in collecting research data through pretest and posttest. Table 2 and figure 2 below shows the average scores of pretest and posttest.

Tabel 2. The results of Pretest and Posttest of Studet' Concept Mastery.

Data	Number of Students	Minimum Score	Maximum Score	Average Score	Deviation Standar
Pretest	31	20	50	47,42	13,89
Posttest	31	70	100	73,55	15,18

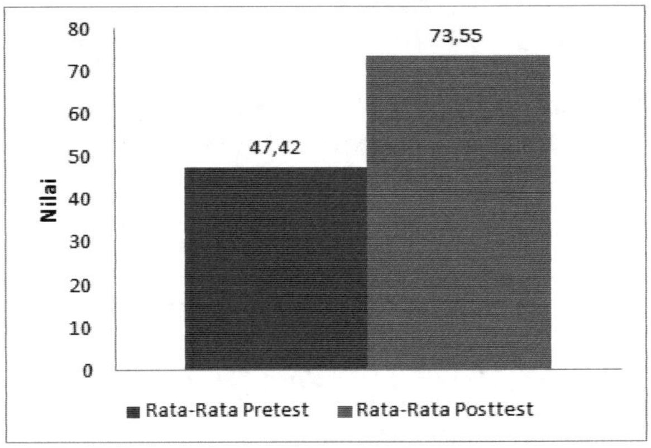

Figure 1. Graph of the average Scores of pretest and posttest

Based on the graph above, it can be seen that there is an improvement of students' concept mastery as shown by the posttest results with an average of 73.55 after online learning using the PhET interactive simulation. This result was higher than pretest results of only 47.42 in average that indicated students' concept mastery before online learning using PhET interactive simulation. However, prior to the hypothesis testing, the normality test was done to determine whether the data obtained is normally distributed or not. This normality test was conducted using the SPSS application with the Shapiro Wilk normality test since the number of samples is less than 50. The results of normality tests indicated that the collected data were normally distributed as shown with a significance value of higher than 0.05. The resulting data of normality test of students' concept mastery abilities are stated in Table 3 below.

Table 3. The result of Normality test.

Data	Sig.	Note
Pretest	0,084	Normal
Posttest	0,058	Normal

Furthermore, the hypothesis test was carried out using the Paired Sample T-test by applying SPSS 23 to find out whether there is a to determine whether or not there is an effect of using PhET on mastery of the concept of harmonic vibration. The analysis result stated that the tcount (10,909) was higher than table (2.042). It meant that Ho in this study was rejected and Ha was accepted. This can be interpreted that there is an effect of use that simulation of online learning physics on students' mastery of concepts, especially related Harmonic Vibration. The improvement of students' concept mastery was also measured by applying the normalized gain test. N-gain of 0.51 was achieved to describe the improvement of students' concept mastery with medium category. This improvement can be described based on the major indicators of concept mastery as shown in figure 2.

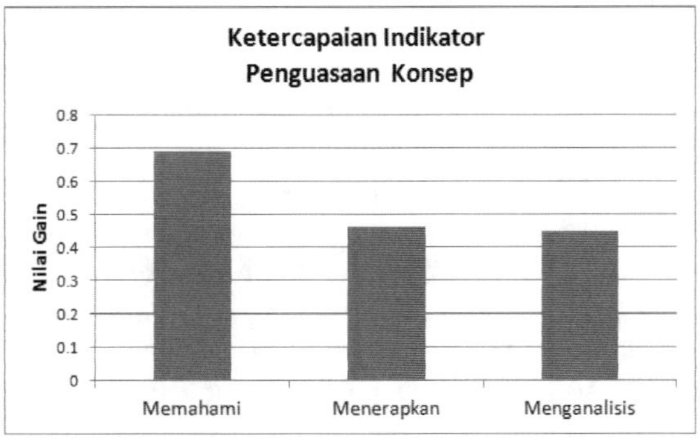

Figure 2. Achievement of Concept Mastery Indicators.

Based on the research results as described above, it can be stated that the use of PhET Interactive Simulation in online learning physics can improve students' ability to master the concept of harmonic vibration. PhET Interactive Simulation helps students to be able to understand (memahami), apply (menerapkan), and analyze (menganalisis) concepts of physics, especially Harmonic Vibration. The results of this study are in line or relevant to the results of previous studies that have been conducted by several researchers. Concluded that learning with the experimental method in the classroom that was applying the PhET Interactive Simulation was more effective compared to use laboratory kit for learning the Gas Kinetic Theory material [19]. In addition, Susanto also concluded in his research that that can have a significant positive effect on students' problem-solving abilities in the subject of Hooke's Law [7]. Finally, also conducted a similar study and obtained the result that the use of that software as a learning medium has a positive effect on student physics learning achievement [6]. Generally, it is proven that the use of PhET Interactive Simulation in online learning physics has a positive effect on the high school students' concept mastery.

6. Conclusion

The Result of result can be concluded that the use of PhET Interactive Simulation in online learning physics, has a positive effect on increasing the ability to master lonsep. That simulation can be used as a medium in online learning physics, especially for Harmonic Vibration

Acknowledge
Thank you to the Physics Teacher and Principal of SMAN 1 Jakarta for giving permission to research and helping during the research.

References
[1] Setiawan, D. F. 2018 *Prosedur Evaluasi dalam Pembelajaran* (Sleman: Deepublish).
[2] Fithriani, S. L., Halim, A., & Khaldun, I. 2016 *Jurnal Pendidikan Sains Indonesia* Vol. 04, No.02, 45-52.
[3] Batuyong, C. T., & Antonio, V. V. 2018 *Asia Pacific Journal of Multidisciplinary Research* Vol. 6, No. 2, 121-131.
[4] Anggriani, S. 2019 *Jurnal Ikatan Alumni Fisika Universitas Negeri Medan* Vol. 5 No. 1, 1-4.
[5] Saputra, B. 2020 *Jurnal Pendidikan dan Pembelajaran Kimia* 9(1), 42-53.
[6] Sunni, M. A. 2019 *Jurnal Explore STMIK Mataram* Vol. 9 No. 1, 54-55.
[7] Susanto, I. 2019 *Jurnal Penelitian Fisikawan* Vol 2 No. 2, 1, 1-7.
[8] Asmawati, E. Y. 2015 *Jurnal Pendidikan Fisiska* Vol. III. No. 1.
[9] Jannah, A. N., Yuliati, L., & Parno 2016 *Jurnal Pendidikan* Vol. 1 No. 3, 409 – 419.
[10] Sukma, F. B., Handayanto, S. K., & Kusairi, S. 2017 *Jurnal Pendidikan: Teori, Penelitian, dan Pengembangan* 2(6), 826-832.
[11] Rizaldi, D. R., Jufri, W., & Jamal 2020 *Jurnal Ilmiah Profesi Pendidikan* Vol. 5 No. 1, 10-14.
[12] Cakir, M. 2008 *International journal of environmental and science education* 3(4), 193-206.
[13] Ramdani, A., Jufri, A., & Setiadi, D. 2020 *Jurnal Penelitian Pendidikan IPA* 6(1), 119-124.
[14] Hadiyanti, D. R., & dkk. 2018 *Jurnal Seminar Nasional Fisika 2018 Univ. Jember* Vol. 3, No. 2, 119-124.
[15] Taufik, A. 2019 *Jurnal Pendidikan&Konseptual* 3(2), 88-98.
[16] Setiawan, A. R. 2020 *Jurnal Edukatif: Jurnal Ilmu Pendidikan* Vol. 2, 28-37.
[17] UPI, T. P. 2007 *Ilmu dan Aplikasi Pendidikan* (Bandung: PT Imperial Bhakti Utama).
[18] Tanduklangi, A., & Amri, C. 2019 *Manajemen Sumber Daya Pembelajaran Bahasa Berbantuan Komputer* (Sleman: Deepublish).
[19] Rustana, CE, W Andriana, V Serevina, D Junia 2020 *J. Phys.: Conf* Ser. 1567 022011

Journal of Physics: Conference Series **2019** (2021) 012023 doi:10.1088/1742-6596/2019/1/012023

Visualization Lorentz Force with Tea Leaves for Studying Magnetic Field in Senior High School

A A Suryani, B H Iswanto and W Indrasari
Physics Education Master Program, Universitas Negeri Jakarta
Jl. Rawamangun Muka, Jakarta 13220, Indonesia
Email: agnesiaastrisuryani_1310818001@mhs.unj.ac.id, bhi@unj.ac.id, cwidyaningrum-indrasari@unj.ac.id

Abstract. The purpose of this study is to visualize the motion of the Lorentz force using tea leaves for learning magnetic fields in high school. Based on assessment needs results of students who attend school in Jakarta, 56.7% of students have difficulty determining the direction of the Lorentz force, 95% of students have a better understanding of the direction of the Lorentz force if it can be visualized. This research uses the R&D method, which refers to the ADDIE model. The Lorentz force visualization (Vigalo) consists of a power generator, tea leaves, two copper wires, two neodymium magnets, and an adjustable voltage source of 1.5 volts - 6.0 volts. The two permanent magnets are placed separately at a distance of 2 cm and able to produce a magnetic field of 88.6634 Gauss. The voltage source is connected to the experiment box and can see the movement of tea leaves and visualize Lorentz force direction lines that match the right-hand rule. This experiment shows that visualization Lorentz force with tea leaves can be used in learning magnetic field material in senior high school.

1. Introduction

Physics is one of the branches of science that is developing in science and technology (IPTEK) which can educate students to have intellectual and religious attitudes in life and study daily activity phenomena, so that learning physics requires activities in the form of observations or experiments [1]. In high school, students need to improve multi-representation skills include verbal skills; visual skills such as pictures, charts, diagrams, tables, graphs; symbolic skills are symbols [2], codes, symbols [3], and mathematical skills are equations or formulas [4]. Therefore, students are required to be able to face changes in all fields, act based on logical thinking, think critically, creatively, and innovatively [5].

The level of understanding of students in basic physics learning that fully understood can be categorized as low [6]. One of the physics learning materials that are abstract and difficult for students to understand is electricity and magnetism [7]. Some students have misconceptions in presenting magnetic field lines and the direction of the Lorentz force [8]. There are misconceptions in understanding the Lorentz force and magnetic field include determining the direction of the magnetic force, magnetic field, and determining the direction of electric current using the right-hand rule [9]. In addition, the difficulties faced by students in studying magnetic electricity are that students often experience errors in distinguishing between magnetic and electric forces, students have difficulty in determining the direction of the magnetic field on electric charges, students have difficulty in applying the concepts of electric fields and magnetic fields and students have difficulty knowing changes in the magnetic field. Magnetic flux and implement Faraday's law and Lorentz force [10]. Therefore, learning media is needed that can be understood by students regarding the concept of the Lorentz force.

The first step in this research is to analyze the needs of teachers and students in schools. Based on the results needs analysis on physics teachers from Jakarta and Bekasi areas, it starts learning media, one of which was teaching aids, helped deliver subject matter and help students understand the concepts. But there is not enough teaching aids in schools to be used in physics learning. In the analysis of learning media needs with the number of respondents 60 students randomly in schools in Jakarta and Bekasi that 61.7% have difficulty understanding the Lorentz force material, 56.7% of students have difficulty determining the direction of the Lorentz force, 65% didn't understand if only explained by teacher, and only 10% of learning in schools that used teaching aids as learning media. As many as 95% of students think they understand more about the direction of the Lorentz force if it can be visualized and seen direct so that students agree on the development of teaching aids that can visualize the direction of the Lorentz force for magnetic field learning. The following is the data from the needs analysis by respondents:

Table 1. Result of student needs analysis

Aspects	Percentage
Difficult to understand the Lorentz force	61.7 %
Difficult determining the direction of the Lorentz force	56.7 %
Not understand Lorenz force if only explained by teachers	65.0 %
Using media for learning	10,0 %
Need visualization to learn Lorentz force	95.0 %

Respondents agreed that visualize the motion of the Lorentz force can help students understand the magnetic field to motivate student learning in schools. The development of Lorentz forces the name is Vigalo can visualize the phenomena of the movement of charged objects due to the interaction between electric and magnetic fields that can be applied to learning magnetic fields in the class. The purpose of this research is to visualize the motion of Lorentz force with tea leaves for learning physics in senior high school, especially on magnetic field material.

2. Method

The method used is Research and Development (R&D) using the ADDIE research model. The stages in the ADDIE model are Analyze, Design, Develop, Implement and Evaluate [11]. The ADDIE model has a simple step [12] and there is an evaluation at each step [13]. The first step is preliminary research, including analysis of student needs and analysis of teacher needs regarding learning media, teaching aid facilities in schools, and difficulties in studying the Lorentz force. Needs analysis using a questionnaire. The second step is to design 2D regarding the development of learning media need in schools. The third step is to develop Lorentz force props or called Vigalo (visualization of Lorentz force). The design of the experiments shown in figure 1.

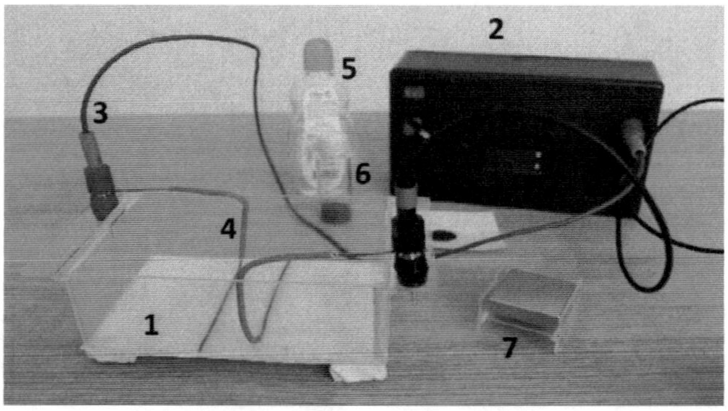

Vigalo model:
1. Box experiment
2. Power generator
3. Red and black wires
4. Copper wires
5. Baby oil
6. Tea leaves
7. Neodymium magnets

Figure 1. Set experiment Vigalo (visualization Lorentz force)

Figure 1 shows that set experiment Vigalo (visualization of Lorentz Force). Developments in the Vigalo props include tea leaves to visualize the movement of particles when there is an interaction between electric and magnetic fields, power generation using batteries, baby oil liquid, copper wire, and neodymium magnets. The power generator used a high voltage inverter to increase the voltage source. The voltage source used batteries 2 x 1,5 volt plus 9 volts. The fourth step is testing the Lorentz force Vigalo. The props developed can show the interaction of charged objects in electric and magnetic fields. Tea leaves as a charged object when interacting with electric and magnetic fields will move in a direction that is by the right-hand rule, so this teaching aid can visualize the concept of the Lorentz force. The fifth step is an evaluation of the teaching aids that have been developed.

3. Result and Discussion

The Lorentz force teaching aid was developed using a 10 cm x 5 cm x 5 cm experimental box with acrylic materials, a power generator, red and black wires, two copper wires, baby oil, tea leaves and two neodymium magnets assembled as shown in the figure 2.

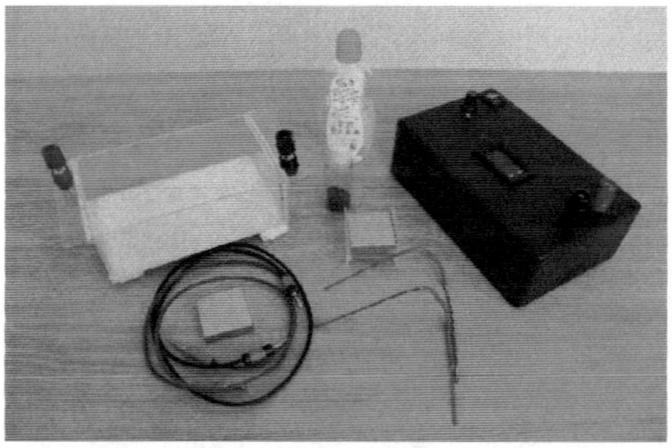

Figure 2. Kit experiment before installation

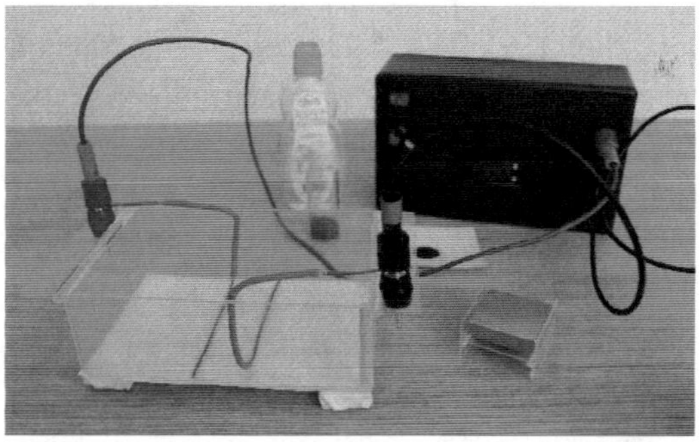

Figure 3. Kit experiment after installation.

The power supply has components, including a battery, a motor speed controller to adjust the desired voltage between 1.5 volts to 6.0 volts, an on/off switch, red and black binding posts, and a digital voltmeter to see the voltage and current. Figure 3 showed that the kit experiment after installation. The Vigalo teaching aid experiment was divided into two activities: before being given a neodymium magnet and after being given a neodymium magnet. The following are the results of the experiment before being given a neodymium magnet.

Journal of Physics: Conference Series **2019** (2021) 012023 doi:10.1088/1742-6596/2019/1/012023

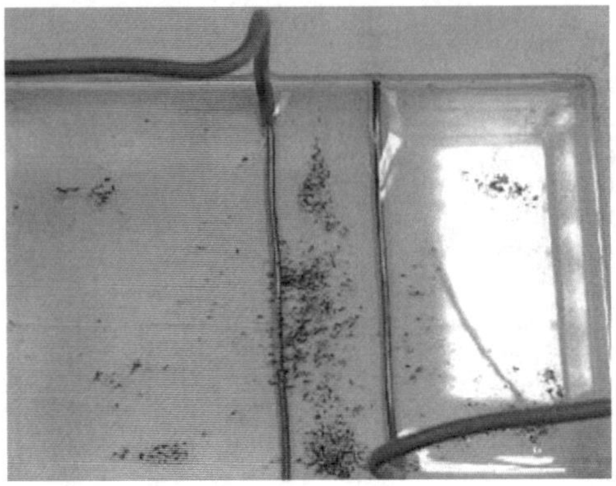

Figure 4. First experiment visualization electric field before applying voltage

Figure 5. First experiment visualization electric field after applying voltage

In figure 4, when the tea leaves are sprinkled into the box experiment without an electric current, the tea leaves will remain motionless, which means that the tea leaves are at no charge. After being given a voltage, the tea leaves will move like charged particles. The motion of the tea leaves form the lines of the electric field because the electric field moved from high potential (+) to a low potential (-) and the magnetic field lines are form see in figure 5.

The second experiment used two neodymium magnets placed at the top and bottom in the box experiment. The results of measurements using a gauss meter, the magnitude of the magnetic field at a distance of 2 cm is 88.6634 gauss. Put a north magnet under the experimental box and a south magnet above the wires with a distance of 2 cm. Current flows from positive (left) to negative (right). The magnetic field is north to south (upward). The voltage is given from 1.5 volts to 6 volts. Tea leaves will move faster if given a large electric voltage.

Figure 6. Second experiment visualization Lorentz force south pole before applying voltage

Figure 7. Second experiment visualization Lorentz force south pole after applying voltage

In Figure 6, after being given a magnet and before an electric current is applied, the tea leaves remain stationary. The tea leaves do not go up to stick to the magnet because the nature of the tea leaves is diamagnetic, it cannot be attracted by a magnet. In Figure 7, when the copper wires are an electric field, the tea leaves that were previously silent will move from high potential (+) to a low potential (-) because it is affected by a magnetic field from the north (bottom) to the south magnetic field (top). Lorentz's force is directed towards the observer in figure 8. So that the tea leaves will move towards the observer. The direction is following the right-hand rule. The thumb as the direction of the electric current goes to the right, the other fingers as the magnetic field from north to south point up, then the Lorentz force will go to the observer. Therefore the tea leaves will move obliquely towards the observer see figure 9

Journal of Physics: Conference Series **2019** (2021) 012023 doi:10.1088/1742-6596/2019/1/012023

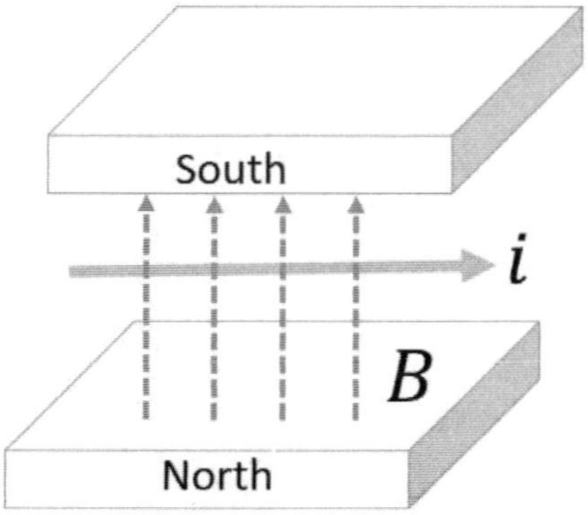

Figure 8. Direction of current (i) and magnetic field (B)

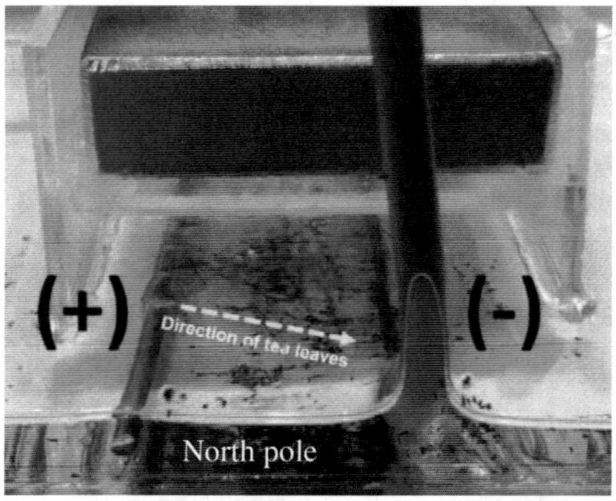

Figure 9. Direction of tea leaves

In the third experiment, the south pole is placed under the box experiment and the north pole is placed above the box experiment. The current flows from positive (left) to negative (right) and the magnetic field is north to south (downward). The voltage given is from 1.5 volts to 6 volts, which can be adjusted by turning the motor speed controller and can be seen on the voltage display. The greater the electric current is given, the faster the movement of the tea leaves.

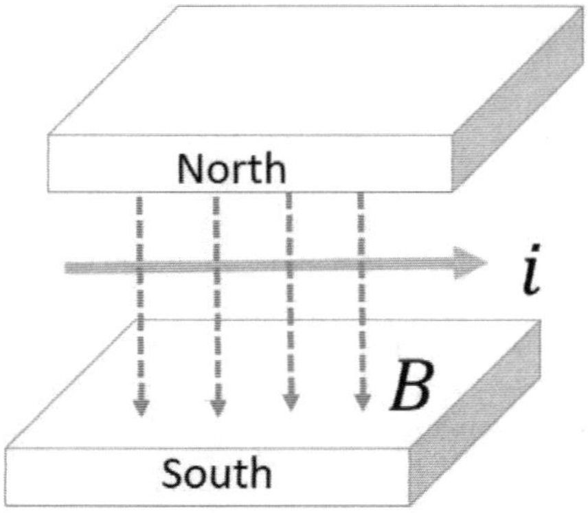

Figure 10. Direction of current (i) and magnetic field (B)

Figure 11. Direction of tea leaves

When the tea leaves are magnetized and before an electric current is applied, the tea leaves remain stationary. When a copper wire is electrified, the previously stationary tea leaves will move from high potential (+) to a low potential (-) because it is affected by a magnetic field from the south (bottom) to the north magnetic field (top) so there will be a Lorentz force whose direction is away from the observer. Illustration in figure 10. So the tea leaves will move away from the observer. This direction is following the right-hand rule. The thumb as the direction of the electric current goes to the right, the other fingers as the magnetic field from north to south point down, then the Lorentz force will move away from the observer. Therefore the tea leaves will move obliquely away from the observer see in figure 11.

This experiment is to visualize the direction of the electric field and the Lorentz force using tea leaves. In the first experiment, the direction of motion of the tea leaves can visualize the direction of the electric field. In the second and third experiments, the direction of motion of the tea leaves can be turned because it is impacted by electric and magnetic fields. So that the Vigalo props can visualize the direction of the Lorentz force using tea leaves according to the right-hand rule. Vigalo can help students determine the direction of the Lorentz force. Students can be motivated in learning physics at school with learning media [14].

4. Conclusion

Based on the experiments, visualization of Lorentz force (Vigalo) is able to visualize the direction of the Lorentz force with tea leaves, used a voltage variation of 1.5 - 6.0 volts and two variations in the direction of the magnetic field. The deflection of the direction of the tea leaves is visualized according to the right hand rule. So that the Vigalo can be used in learning magnetic fields in high school.

References

[1] Hikmawati and Gunanda 2013 *Buku Ajar Kajian Fisika SMA* (Mataram: FKIP Press)
[2] Heron P R L and Meltzer D E 2005 *Journal of Physics* 73(5)
[3] Kohl P B and Finkelstein N D 2008 *Pattern of Multiple Representation use by Expert and Novices during Physics Problem Solving Physical Review Special Topics-Physics Education Research* 4(1) 1-13
[4] Heuvelen V and Zou X L 2001 *Journal of Physics* 69(2)
[5] Yance R D, Ramli E and Mufit F 2013 *Pillar of Physics Education* 48-54
[6] Wibowo F C and Iswanto B H 2019 *Designing MOOCS with Virtual Microscopic Simulation (VMS) for increasing of students' levels of understanding 4th Annual Applied Science and Engineering Conference*
[7] Guisasola J, Almudi M J and Zubimendi L 2004 *Difficulties in Learning the Introductory Magnetic Field in the First Years of University Sciences Educations* 443-464
[8] Scaife T M and Heckler A F 2010 *Journal Of Physics* 868-878
[9] Pateda A B, Kendek Y and Saehana S 2015 *Jurnal Pendidikan Fisika Tadulako*
[10] Munfarikha N, Kusairi S and Zulaikhah S 2017 *Pros. Seminar Pend. IPA Pascasarjana UM* 226-231
[11] Branch R M 2009 *Instructional Design : The ADDIE Approach* (Boston: Springer)
[12] Mahoney D J 2018 *Learning Using the ADDIE Model* (Alexandria: ATD)
[13] Hess and Greer 2016 *Designing for Engagement: Using the ADDIE Model to Integrate High-Impact Practices into an Online Information Literacy Course Communications In Information Literacy* 264-279
[14] Puspita, Suyatna and Rosidin 2020 *Optimizing the use of Smartphone for M-Learning as a Supplement for Magnetic Learning with a Scientific Approach International Conference on Theoretical and Applied Physics (ICTAP)* 1-8

Journal of Physics: Conference Series **2019** (2021) 012024 doi:10.1088/1742-6596/2019/1/012024

Faraday's Law Teaching Aids Using Magnetometers on Smartphone and Infrared Sensors for Electromagnetic Induction Learning

B Andrianto, B H Iswanto and W Indrasari
Physics Education Department, Universitas Negeri Jakarta, Jl. Rawamangun Muka No.1, Jakarta 13220, Indonesia
Email: bagusandrianto_1310818002@mhs.unj.ac.id, bhi@unj.ac.id, widyaningrumindrasari@unj.ac.id

Abstract. Faraday's law is one of the abstract physics materials. This research aims to develop Faraday's Law teaching aids by combining smartphones and infrared sensors. This study uses the Research and Development method, which refers to the ADDIE model. The teaching aids consist of a 45 cm coil whose area is , a 4000 μT neodymium permanent magnet, an Arduino integrated infrared sensor, and a smartphone's magnetometer sensor using the Phyphox application. The research trials were carried out by varying the number of coils and the height of the falling magnet. The test results obtained data in a magnetic field, the magnetic time passing through the coils, and voltage. Based on the induced emf obtained from observations and calculation results, there is an increase in the number of coils and the height of the falling magnet. This study concludes that Faraday's law teaching aids can show the effect of the number of coils and the distance of the magnet falling on the magnitude of the induced emf. The developed Faraday's law teaching aids can be used in the learning process of Faraday's law.

1. Introduction

Education is a planned-effort process of mentoring and learning for individuals to develop and grow into independent, responsible, creative, knowledgeable, healthy, and noble human beings, both seen from the physical and spiritual aspects. One of the abstract physics materials is Faraday's Law material. Students have difficulty because there is no clear explanation of abstract concepts [1]. Students find physics difficult because they have to deal with different representations such as experiments, formulas and calculations, graphs, and conceptual explanations at the same time [2]. Various supporting facilities and infrastructure are needed to realize effective physics learning. One of the supporting facilities and infrastructure is learning media.

Learning media can overcome student difficulties such as physics subjects with abstract concepts [3]. One of the learning media that can be used is teaching aids. However, some teachers prefer to explain orally and provide material to students without supporting teaching aids. The lack of teaching aids in schools is caused by several things: the high price of teaching aids and the lack of funds to buy these teaching aids [4]. One of the efforts to maximize the learning process on Faraday's law material is to use visual aids combined with a smartphone.

Currently, most students have grown up accompanied by technology such as computers, mobile phones, and video consoles in almost every activity, such as studying, working, or just entertainment [5]. Smartphones are active 24 hours on average, but not fully used for learning [6]. The proper use of smartphones helps students in obtaining information anywhere and anytime. One of the uses of

Journal of Physics: Conference Series **2019** (2021) 012024 doi:10.1088/1742-6596/2019/1/012024

smartphones in education is sensors. The use of sensors can guide students' learning interactions with learning content to easily track students' learning interactions [7].

This article describes the development of Faraday's Law teaching aids by combining smartphones and infrared sensors. This teaching aid is expected to overcome students' difficulties in understanding Faraday's Law material. In addition, these teaching aids is expected to make it easier for teachers to carry out an interactive, economical, safe, and effective learning process.

2. Method

This study uses a research and development model. The research and development carried out referring to the stages of the ADDIE development model. The ADDIE model was chosen because of its simple steps [8]. In addition, the ADDIE model is iterative, involving review and revision during the design process [9]. The steps for developing the ADDIE model include Analyze (Analysis), Design (Planning), Development (Development), Implementation (Implementation), and Evaluation (Evaluation). Faraday's law teaching aids are in the form of physical tools. This tool uses the help of a magnetometer to measure the field strength of the magnet used in the experiment. This tool also uses an infrared sensor to measure the time the magnet is in the coil. The developed Faraday's law props is a physics tool form that uses a magnetometer to measure the magnetic-field intensity used in the experiment. This tool is also equipped with an infrared sensor to measure the time the bar magnet is in the coil.

The analysis phase was conducted to determine the effectiveness of the Faraday law learning process and the methods used in several schools, both from the point of view of teachers and students. The analysis carried out includes several stages: analysis of student needs, analysis of teacher needs, and literature study. Based on the needs analysis results, around 70% of schools in Jakarta and Bekasi do not yet have Faraday's legal aids. The next step is to create a media design. The tools and materials to be used to develop Faraday's law teaching aids are also determined. The next step is the development stage, which includes the making of Faraday's law teaching aids, at this stage. The developed teaching aids consist of jumper cables, voltmeter, LCD screen, magnetic drop case, 300 and 600 turn coils, 4000 μT neodymium permanent magnets, magnetometer on smartphones Arduino-based infrared sensors.

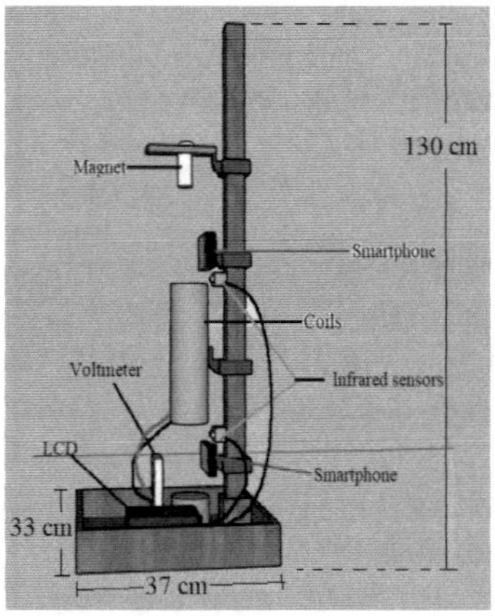

Figure 1. Faraday's law teaching aids design from the side

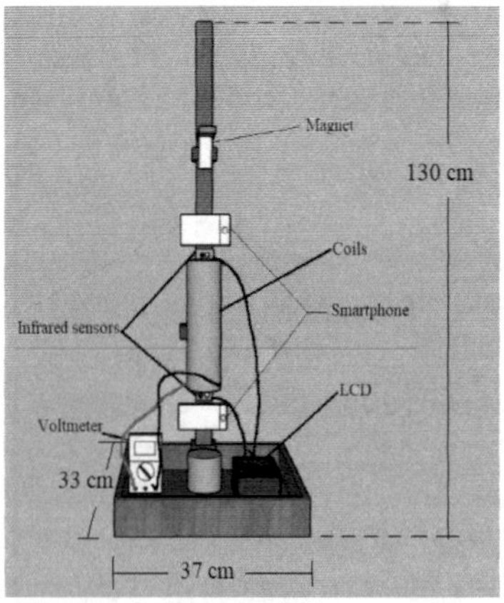

Figure 2. Faraday's law teaching aids design from the front

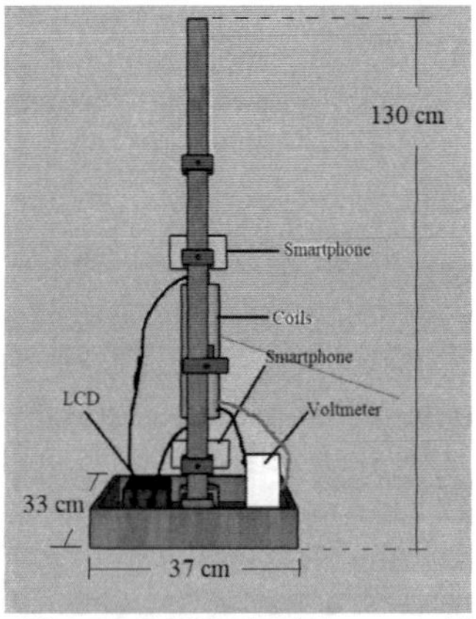

Figure 3. Faraday's law teaching aids design from behind

The next step is implementation. In this step, the teaching aids are tested to obtain results in a magnetic field, time and voltage. At the data collection stage, variations are made to the number of coils and the distance the magnet falls to the coil. After the implementation phase is complete, the teaching aids are evaluated regarding the effectiveness of the output results based on observations with calculations.

Journal of Physics: Conference Series **2019** (2021) 012024 doi:10.1088/1742-6596/2019/1/012024

3. Result and Discussion

The teaching aids developed using coils with 300 and 600 turns, infrared sensors used to measure the time the magnet was in the coil, voltmeters, 4000 T neodymium magnets, prop for teaching aids, jumper cables, voltmeters, LCD screens, magnetic drop containers, as well as a magnetometer sensor on a smartphone using the phyphox application.

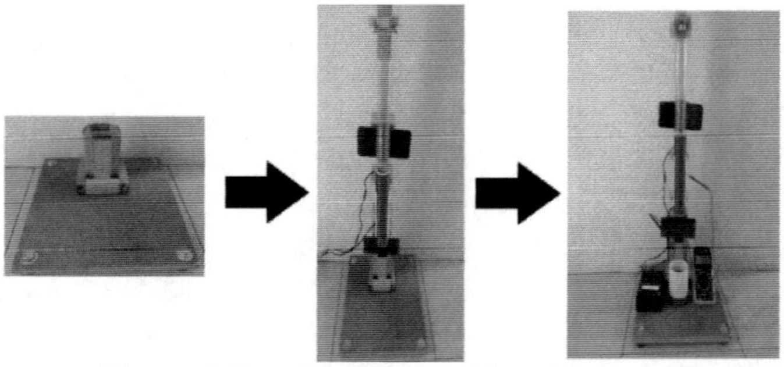

Figure 4. Faraday's law teaching aids

The data obtained using Faraday's law teaching aids include magnetic field strength, time, and voltage. Data retrieval is done by repetition ten times. Data were collected from varying the height (25 cm and 40 cm), a magnetic field of 4000 μT, and the number of turns of the coil (300 turns and 600 turns). The output results obtained from the voltmeter are then compared with the calculation results using the following equation [10].

$$\varepsilon = -N\,A\,\frac{\Delta B}{\Delta t}$$

with :

ε	: EMF induction (Volt)
N	: Coils
A	: Area (m^2)
ΔB	: Magnetic field change (T)
Δt	: Time interval (s)

3.1. Coil 300 turns at the height of 25 cm

In the experiment using a coil of 300 turns, it was carried out using a 4000 μT neodymium magnet and by dropping the magnet at the height of 25 cm. The experimental results are in the following table.

Table 1. The experiment's result used coils of 300 turns and a height of 25 cm

No	Coils	Area (m^2)	Time (s)	Magnetic Fields (T) Start	Magnetic Fields (T) Finish	dB (T)	Experiment Voltage (V)	Calculation Voltage (V)
1	300	0,001386	0,115	0,004815	0,000311	-0,0045	0,019	0,016
2	300	0,001386	0,105	0,004822	0,000265	-0,00456	0,013	0,018
3	300	0,001386	0,116	0,004729	0,000193	-0,00454	0,015	0,016
4	300	0,001386	0,113	0,004604	0,000308	-0,0043	0,017	0,016
5	300	0,001386	0,114	0,004738	0,000332	-0,00441	0,019	0,016
6	300	0,001386	0,106	0,004751	0,000270	-0,00448	0,014	0,018
7	300	0,001386	0,110	0,004787	0,000271	-0,00452	0,015	0,017
8	300	0,001386	0,105	0,004864	0,000225	-0,00464	0,014	0,018
9	300	0,001386	0,113	0,004792	0,000227	-0,00457	0,018	0,017
10	300	0,001386	0,105	0,004808	0,000240	-0,00457	0,014	0,018
				Average			0,0158	0,0170

Based on the table, the average voltage obtained from the observations is 0.0158. While the average voltage from the calculation results is 0.0170. So that obtained a relative error of 16% in the experiment.

3.2. Coil 300 turns at the height of 40 cm

In the experiment using a coil of 300 turns, it was carried out using a 4000 μT neodymium magnet and by dropping the magnet at the height of 40 cm. The experimental results are in the following table.

Table 2. The experiment's result used coils 300 turns and a height of 40 cm

No	Coils	Area (m^2)	Time (s)	Magnetic Fields (T) Start	Magnetic Fields (T) Finish	dB (T)	Experiment Voltage (V)	Calculation Voltage (V)
1	300	0,001386	0,086	0,005141	0,00019	-0,00495	0,027	0,024
2	300	0,001386	0,087	0,004865	0,000171	-0,00469	0,024	0,022
3	300	0,001386	0,087	0,004822	0,000192	-0,00463	0,028	0,022
4	300	0,001386	0,088	0,004823	0,000151	-0,00467	0,026	0,022
5	300	0,001386	0,085	0,005011	0,000172	-0,00484	0,024	0,024
6	300	0,001386	0,088	0,005025	0,00019	-0,00484	0,023	0,023
7	300	0,001386	0,091	0,005026	0,000185	-0,00484	0,024	0,022
8	300	0,001386	0,090	0,004608	0,000212	-0,0044	0,024	0,020
9	300	0,001386	0,086	0,004796	0,000136	-0,00466	0,025	0,023
10	300	0,001386	0,087	0,004962	0,000151	-0,00481	0,026	0,023
				Average			0,0251	0,022

Based on the table, the average voltage obtained from the observations is 0.0251. While the average voltage from the calculation results is 0.0225. So that obtained a relative error of 12% in the experiment.

Journal of Physics: Conference Series **2019** (2021) 012024 doi:10.1088/1742-6596/2019/1/012024

3.3. Coil 600 turns at the height of 25 cm

In the experiment using a coil of 600 turns, it was carried out using a 4000 μT neodymium magnet and by dropping the magnet at the height of 25 cm. The experimental results are in the following table.

Table 3. The experiment's result used coils 600 turns and a height of 25 cm

No	Coils	Area (m^2)	Time (s)	Magnetic Fields (T)		dB (T)	Experiment Voltage (V)	Calculation Voltage (V)
				Start	Finish			
1	300	0,001386	0,105	0,004587	0,00033	-0,00426	0,039	0,034
2	300	0,001386	0,105	0,004766	0,000336	-0,00443	0,038	0,035
3	300	0,001386	0,107	0,004882	0,000234	-0,00465	0,035	0,036
4	300	0,001386	0,104	0,004813	0,000423	-0,00439	0,033	0,035
5	300	0,001386	0,100	0,005025	0,00045	-0,00458	0,035	0,038
6	300	0,001386	0,104	0,004884	0,000466	-0,00442	0,027	0,035
7	300	0,001386	0,103	0,004947	0,000251	-0,0047	0,046	0,038
8	300	0,001386	0,104	0,004891	0,000266	-0,00463	0,032	0,037
9	300	0,001386	0,107	0,004802	0,000331	-0,00447	0,034	0,035
10	300	0,001386	0,106	0,003821	0,000413	-0,00341	0,028	0,027
				Average			0,0347	0,0350

Based on the table, the average voltage obtained from the observations is 0.0348. While the average voltage from the calculation results is 0.0350. So that obtained a relative error of 11% in the experiment.

3.4. Coil 600 turns at the height of 40 cm

In the experiment using a coil of 600 turns, it was carried out using a 4000 μT neodymium magnet and by dropping the magnet at the height of 40 cm. The experimental results are in the following table.

Table 4. The experiment's result used coils 600 turns and a height of 40 cm

No	Coils	Area (m^2)	Time (s)	Magnetic Fields (T)		dB T)	Experiment Voltage (V)	Calculation Voltage (V)
				Start	Finish			
1	300	0,001386	0,086	0,00497	0,000604	-0,00437	0,044	0,043
2	300	0,001386	0,087	0,004916	0,000693	-0,00422	0,043	0,039
3	300	0,001386	0,087	0,004907	0,000375	-0,00453	0,032	0,043
4	300	0,001386	0,088	0,004958	0,000611	-0,00435	0,046	0,043
5	300	0,001386	0,085	0,004925	0,000573	-0,00435	0,058	0,042
6	300	0,001386	0,088	0,004859	0,00037	-0,00449	0,057	0,041
7	300	0,001386	0,091	0,004893	0,000682	-0,00421	0,06	0,040
8	300	0,001386	0,090	0,004967	0,00043	-0,00454	0,05	0,043
9	300	0,001386	0,086	0,004886	0,000511	-0,00438	0,039	0,042
10	300	0,001386	0,087	0,004874	0,000662	-0,00421	0,055	0,040
				Average			0,0484	0,0417

Based on the table, the average voltage obtained from the observations is 0.0484. While the average Voltage from the calculation results is 0.0417. So that obtained a relative error of 19% in the experiment.

The research trials were carried out by varying the number of coils and the height of the falling magnet. The test results obtained data in a magnetic field, the magnetic time passing through the coil and voltage. Data retrieval using Faraday's law teaching aids was carried out four times with ten repetitions. Data collection is also done by varying the number of coils and the height of the falling magnet.

The first experiment was carried out by dropping a magnet from a height of 25 cm into a coil of 300 turns with an area of. In the first experiment, data were obtained in the form of changes in the magnetic field (dB) of -0.005, time-lapse (dt) of 0.11 s , and the induced emf of 0.0158 V. Meanwhile, according to calculations, the results of the induced emf were 0.0170 V. The second experiment is done by dropping a magnet from a height of 40 cm into a coil of 300 turns with an area of . In the second experiment, data were obtained in the form of changes in the magnetic field (dB) of -0.005, time-lapse (dt) of 0.088 s, and the induced emf of 0.0251 V. Meanwhile, according to calculations, the results of the induction emf of 0.0225 V. The third experiment was carried out by dropping a magnet from a height of 25 cm into a coil of 600 turns with an area of . In the third experiment, data obtained in the form of changes in the magnetic field (dB) of -0.004, time interval (dt) of 0.105 s, and the induction emf of 0.0347 V. Meanwhile, according to calculations, the results of the induction emf of 0.0349 V. The fourth experiment was carried out by dropping a magnet from a height of 40 cm into a coil of 600 turns with an area of . In the fourth experiment, the data obtained in the form of changes in the magnetic field (dB) of -0.004, the time interval (dt) of 0.087 s, and the induction emf of 0.0484 V. Meanwhile, according to the calculation, the results of the induction emf were 0.04169 V. Based on the results of the induced emf The results obtained from observations and calculation results increase with the increase in the number of coils and also the height of the falling magnet with a relative error of 14.5%.

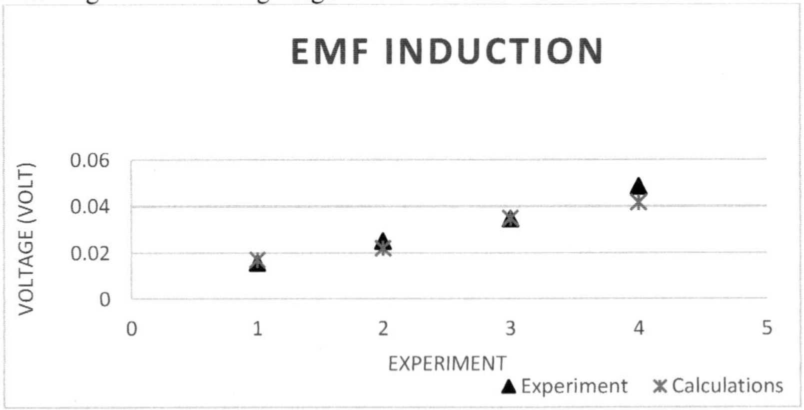

Figure 5. Graph of experiment voltage and calculation voltage

Based on the research results, Faraday's law teaching aids can show the effect of the number of coils and the distance of the magnet falling on the magnitude of the induced emf. The height of the falling magnet can affect the time the magnet passes through the coil measured using an infrared sensor. The use of infrared sensors was chosen to increase the accuracy of the resulting time data. Magnetic field data retrieval is done using a magnetometer sensor available in the phyphox application. At the same time, the voltage data retrieval is obtained by using a voltmeter. After calculating using the induced emf equation, there are differences in the induced emf obtained. This is due to several things, such as the lack of accurate magnetic field data in the phyphox application. This is because the phyphox application is very dependent on the specifications of the smartphone used. In addition, the inaccuracy of the voltmeter measuring instrument also causes differences in the results of the induced emf obtained. Collecting data using electronic devices will increase accuracy in reading data so that students can manipulate and analyze the data that has been obtained [11].

Based on the results obtained from the experiment, the developed Faraday's law teaching aids can be used in the physics learning process on Faraday's Law. The use of Faraday's law teaching aids will further assist students in overcoming misconceptions about Faraday's law material [12]. In addition, the use of cheap and easy-to-use Faraday law teaching aids can, of course, be readily accepted by students and teachers [13]. Faraday's law teaching aids that are developed will undoubtedly help students understand the material of Faraday's law. This is because teaching aids can increase the effectiveness of physics learning on Faraday's law [14]. Interactive learning in this pandemic period is necessary so that students are motivated to participate in physics learning [15]. The use of faraday law teaching aids can create activity in students. In addition, students can also analyse and compare the induced emf read by the voltmeter with the results of calculations using the induced emf equation.

4. Conclusion
Based on the research conducted, Faraday's law teaching aids can show the effect of the number of coils and the distance of the magnet falling on the amount of induced emf produced, although the results obtained are the difference between the voltage in the experiment and the voltage from the calculation results so that the developed Faraday's law teaching aids can be used for the learning process of Faraday's law.

Acknowledgments
In this section, thanks are given to several parties who helped in carrying out this research. Thank you to Physics Lecturers of the Universitas Negeri Jakarta for providing convenience in carrying out this research.

References
[1] K. Zuza, J. Guisasola, M. Michelini and L. Santi 2012 *European Journal of Physics* 33 pp. 397-406

[2] F. Oernek, W. Robinson and M. Vaughan 2008 *International Journal of Environmental and Science Education* 3 pp. 30-34

[3] F. Ariyani, T. Nayana, A. Saregar, Y. Yuberti and A. Pricilia 2018 *Jurnal Ilmiah Pendidikan Fisika Al Biruni* 7 pp. 227-237

[4] S. An'nur, M. Sari, M. Wati, M. Misbah and D. Dewantara 2020 *Journal of Physics* pp. 1-6

[5] M. Minovic 2012 *Simposio Internacional de Informática Educativa* pp. 315-320

[6] S. Irna 2019 *International Conference on Mathematics and Science Education* pp. 1-5

[7] C. K. Sik 2017 *FutureTech 2017: Advanced Multimedia and Ubiquitous Engineering* pp. 459-465

[8] D. J. Mahoney 2018 Lean Learning Using the ADDIE Model (Alexandria: ATD)

[9] A. K. N. Hess and K. Greer 2016 *Communications In Information Literacy* 10 pp. 264-279

[10] J. D. Cutnell and K. W. Johnson 2012 Physics 9th Edition (Hoboken: John Wiley & Sons Inc)

[11] R. Trumpery and M. Gelbman 2014 *IOP Publishing Physics Education* pp. 90-95

[12] S. Arabasi and H. Al-Taani 2017 *Journal of Physics* pp. 1-8

[13] I. E. C. S. A. N. A. B. Enrique Arribas 2015 *European Journal of Physics* 36 pp. 4-7

[14] N. M. Puspita, A. Suyatna and a. U. Rosidin 2020 *International Conference on Theoretical and Applied Physics (ICTAP)* 9 pp. 1-8

[15] E. K. Yuningsih, M. M. Chusni and R. Sidik 2018 *Annual Applied Science and Engineering Conference* pp. 1-6

Enhancing Students' Learning Activities Using Problem-Based Learning Model on Temperature and Heat Concept

I M Astra[1], D Aminudin[2] and A Henukh[3]
[1]Department of Physics Education, Faculty of Mathematics and Natural Sciences Universitas Negeri Jakarta, Jalan Rawamangun Muka, Jakarta Timur 13220, Indonesia
[2] Post-SM-3T Teacher Professional Education, Physics Education Department, Universitas Negeri Jakarta, Jalan Rawamangun Muka, Jakarta Timur 13220, Indonesia
[3]Department of Physics Education Faculty of Teacher Training and Education Universitas Musamus, Jl. Kamizaun Mopah Lama, Merauke, 99611
Email: imadeastra@unj.ac.id

Abstract. This study aimed to improve the learning activities of SMA Labschool Jakarta students in grade XI of MIPA 1 by applying the Problem-Based Learning model (PBL). The study was undertaken by employing a classroom action research method with 34 participating students on the topic of temperature and heat. The study was conducted in three cycles. The cycles are extended to; planning, action, observation, and reflection. The success indicator of the study was that students' learning activities could reach > 75%. The data collection was through the use of an observation sheet and the implementation of the learning model. Based on the data analysis, the results indicated that students' learning activities in the first cycle reached 41,7%. In cycle II, the score reached 61,76%. Finally, the score increased to 85,27% in cycle III. Therefore, it can be concluded that student learning activity increased. In other words, the results verified that Problem-Based Learning can significantly enhance students' learning activities.

1. Introduction

Act Number 33 of 2003, concerning the National Education System, requires learning that can facilitate students to take an active role in the learning process. In education, the teaching and learning process is considered the most important since it determines the success of achieving the goals of education. There are two major subjects in a learning process; students as active learners and teachers who guide the students[1].

The success of a learning process highly depends on the interaction between teachers and students. Teachers have been known to teach a lot of subjects. One of which is Physics. Physics is one of the branches of the natural science learning process. Essentially, Physics is a science that requires an understanding of concepts and scientific models that can make physics interesting. Hence, the physics learning process emphasizes providing direct experience through laboratory activities and experiences in daily life. Such learning will challenge students to develop analytical, critical and creative thinking skills in solving problems[2].

In the first observation of the class, there had been no interactive class atmosphere. Also, the researcher encountered a lack of students' direct involvement. Some students were hesitant in asking questions, and the courage of students to deliver opinions and ask questions was also lacking. Often students participated in learning by experimenting. However, the process was still a teacher-centered, so it was not optimal. In addition, students' results of daily tests were below the minimum completeness criteria.

Therefore, if the teacher explains the material without relating it to the experience of students, students will not understand the purpose of the learning and get bored. In the teaching and learning process, the lack of student activity can be seen from the attention and involvement of students during the learning process. Students are more likely to talk with friends, daydream, hold cell phones, and leave class for various reasons. When given assignments, students prefer to copy their friends' assignments without trying to do them on their own. Then, students will not ask or answer when asked by the teacher. They prefer to be silent[3].

The importance of student activities in the teaching and learning process should be of particular concern to teachers since they, as educators, are obliged to guide students' learning activities so that students want to play an active role in the learning process. Furthermore, students' activities are crucial to optimize the learning outcomes as activities greatly determine learning outcomes. A student with high learning activity will get satisfactory learning outcomes. In contrast, a student with low learning activity will obtain low learning outcomes[4][5].

Problem-Based Learning is a student-centered learning model that is based on problems that are often encountered by students in the surrounding environment pertinent to the learning process. Thus, there will be a two-way interaction between students and their environment. With this regard, students are involved in further analysis so that the problem can be solved properly and become a learning experience for students. In other words, when teaching in class, the teacher must optimally create a learning environment that can educate students by picturing a problem that can encourage students to learn to solve it. Besides, the teacher should allow students to play an active role in constructing the concepts of the material they are learning gradually[6][7][8].

Based on the aforementioned, it is expected that the Problem-Based Learning model can increase student activity in learning physics in class, especially in grade XI of science in the senior high school. For this reason, the author conducted a study on students' activities through the use of Problem-Based Learning in SMA Labschool of Jakarta regarding the topic of Heat and Temperature. The study was entitled: "Enhancing Students' Learning Activities Using Problem-Based Learning Model on Temperature and Heat Concept".

Based on the background, the formulation of the problem is: "Has there been an increase in students' learning activities after using the Problem-Based Learning model on Temperature and Heat Concept?". The goal of the study is to determine if Problem-Based Learning can increase students' learning activities on temperature and heat concept.

2. Method

The study was carried out in Labschool high school in Jakarta. The subjects of the study were students of class XI MIPA 1 which consisted of 34 students, 15 male students, and 19 female students. The study employed Action Research with a qualitative approach. The type of study was Classroom Action Research (CAR). In general, CAR is open action research that is reflective and collaborative. It is carried out to improve the quality of learning practice in the classroom[8][9][10].

A classroom action research consists of 4 stages, namely (1) planning, (2) implementation, (3) observation, and (4) reflection. After conducting the reflections which include analysis, synthesis, and assessment of the process and the results of the actions taken, problems or thoughts about the need for improvement usually arise, so it is necessary to re-plan, re-act, re-observe, and re-reflect. These stages of activity continue until a problem is considered resolved.

The implementation of the learning model is considered successful if the percentage of success reaches 100%. About students' learning activities, they are analyzed descriptively based on the description that appears on the student learning activity observation sheet. Students' learning activities are said to be complete if >75% of students in the class are in the active category. Percentage analysis was employed to analyze the data of students' responses after participating in learning using the Problem-Based Learning model. As for each appearance of the descriptor using a sign (√), the appearance of 1 checkmark is proportional to the acquisition score of 1. The appearing score of each indicator is summed, and the result is called the total score. While the assessment of student learning activities is given 1 (one) if it is done and given 0 (zero) if it is not done. Regarding the aspects of Visual

Journal of Physics: Conference Series **2019** (2021) 012025 doi:10.1088/1742-6596/2019/1/012025

activities, Oral activities, Listening Activities, and Writing activities, each of them is respectively developed into two aspects. In addition, the calculation encompasses 0,5 points for activity and 1 point for two activities of each aspect.

3. Result and discussion

Before the implementation of the PBL model, a pre-cycle was first carried out with a direct learning model. Then, it was ended with a test. During the research, each cycle used a problem-based learning model with steps extended to (1) orienting students to the problem; (2) organizing students to learn; (3) guiding investigations independently or in groups; (4) developing and presenting work results; and (5) analyzing and evaluating the results of problem-solving [11][12]. Meanwhile, according to previous research the learning activities are classified into: the visual activities (reading, viewing pictures, watching experiments, conducting demonstration, exhibiting, and watching other people work or play), oral activities (suggesting a fact or principle, connecting an event, asking questions, giving advice or opinions, interviewing, and doing discussion and interruptions), listening activities (listening to the presentation of the material, and listening to the conversation or discussion groups), writing activities (writing reports and summaries, doing tests, and completing questionnaires), drawing activities (drawing graphics and diagrams), motor activities (conducting the experiments, choosing tools and equipment, and creating models), mental activities (memorizing, troubleshooting, analyzing the factors, analyzing relationships, and making decisions), and emotional activities (having interest, differentiating, being couraged, being calm, and feeling bored and nervous)[13][14][15].

The results of cycles I, II, and III indicated the existence of students' learning activities. This can be seen in table 1 and table 2.

Table 1. The percentage of students who completed (active) each cycle

Cycle	Activities (%)	Categories
I	41.17	moderate
II	61.76	moderate
III	85.29	Active

Table 2. Students' activities in each aspect per cycle

No	Activity aspects	Cycle I	% Total students	Cycle II	% Total students	Cycle III	% Total students
1	Visual activities	30	88	34	100	34	100
2	Oral activities	16.5	49	17	50	20	59
3	Listening activities	21	62	27	79	29	85
4	Writing activities	19.5	57	22.5	66	25.5	75
5	Drawing activities	15	44	20.5	60	27	79
6	Motor activities	23	68	28	82	31	91
7	Mental activities	13	38	19	56	23	68
8	Emotional activities	25	74	29	85	34	100

Students' learning activities, for the study, were divided into 8 aspects. The aspects include Visual activities, Oral activities, Listening activities, Writing activities, Drawing activities, Motor activities, Mental activities, and Emotional activities.

Based on table 2, the encountered students' learning activities in the cycle I were as follows; Visual activities by 88%, Oral activities by 16.5%, Listening activities by 21%, Writing activities by 19.5%, Drawing activities by 15 %, Motor activities by 23%, Mental activities by 13%, and Emotional activities by 25% with 41.17% completeness of active students (as seen in table 1). As the percentage of completion of active students is less than 75%, this indicates that students' learning activities in cycle I was still not complete. However, the learning process has reached 94%. When re-analyzing the results of observations, it turned out that students were active at the beginning of learning in which 30 students were involved in the visual activities aspect. At the same time, regarding 7 other aspects, 26 students were still below or more than 75% of the expected results. Furthermore, many students still rely on their friends in making a group decision during a discussion. It can be seen from the aspect of mental activities that have the lowest percentage within 38% or 13 students.

In cycle II, as seen in Table 2, the encountered students' learning activities were extended to; Visual activities by 100%, Oral activities by 50%, Listening activities by 27%, Writing activities by 22.5%, Drawing activities for 20.5%, Motor activities by 28%, Mental activities by 19%, and Emotional activities by 29%. Additionally, the percentage of completion of active students is 62.76% (seen in table 1). In cycle II, there is an increase in all aspects. However, the percentage of completion of active students is less than 75%. This means that regarding students' learning activities, cycle II is still considered not incomplete as in the cycle I. However, the learning process has reached 100%. When re-analyzing the results of observations, it turned out that students were active at the beginning of learning in which 30 students were involved in the visual activities aspect. Altogether, regarding 7 other aspects, 26 students were still below or more than 75% of the expected results. Furthermore, many students still rely on their friends in making a group decision during a discussion. It can be seen from the aspect of mental activities that have the lowest percentage within 38% or 13 students[16][17].

Based on table 1 cycle III, it appears to have increased as in cycle II with the percentage of Visual activities by 100%, Oral activities by 59%, Listening activities by 85%, Writing activities by 75%, Drawing activities by 79%, Motor activities are 91%, Mental activities are 68%, and Emotional activities are 100%. Meanwhile, the percentage of completion of active students was 85.29% (as seen in table 1). With this regard, it can be assumed that the student's learning activities in this study have reached the success indicator, that is 75%.

The comparison of aspects of student learning activities in each cycle can be seen in Figure 1 below

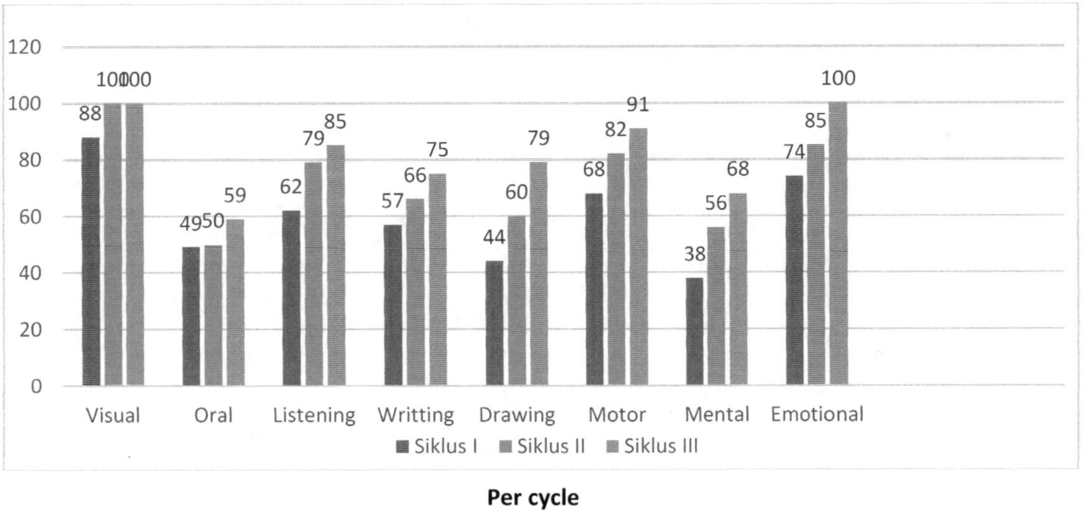

Figure 1. Graph of the comparison of each aspect of student learning activities in each cycle

The data distribution of students' learning activities for each cycle can be represented graphically, as shown in the following Figure 2.

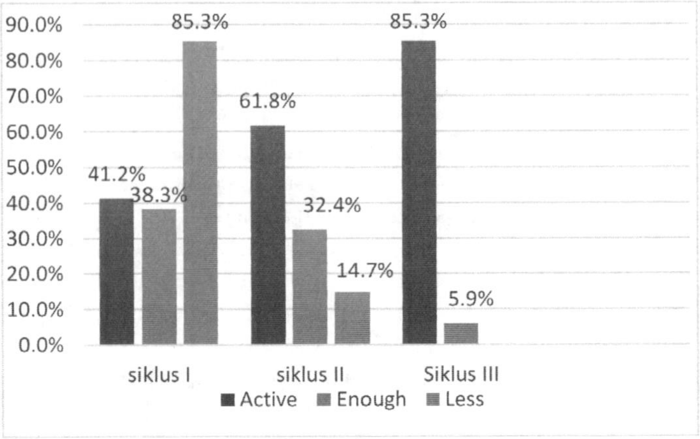

Per cycle

Figure 2. Graph of the percentage of student learning activities for each cycle

It can be seen in Figure 1 the increase in every aspect of the cycle I, cycle II, and cycle III. This can be said because learning in cycle III is the result of reflection in cycle II, and cycle II is the result of reflection from cycle I. Thus, the shortcomings or weaknesses that once occurred, do not happen again. When each aspect has increased, students' learning activities have increased on each cycle (seen in Figure 2). These findings were in accordance with previous research that there is an increase in students' learning activities when using the problem-based learning model in classroom learning [18].

4. Conclusion

Based on the description of the research implementation during the three cycles of this study, it can be concluded that the application of Problem-Based Learning can increase students' learning activities. This is proven by the increase in the percentage of achievement of students' learning activities in each cycle. Cycle I reached1.17%, and Cycle II increased to 70%. At last, Cycle III reached 85,27%. As previously mentioned, the success indicator of this action research is if students' learning activities are > 75%. Regarding this indicator, cycle I and cycle II of this research were considered incomplete. However, in cycle III, the research is said to be complete as the percentage of the students' learning activities reached 75%.

References

[1] G. Resbiantoro and D. *J. Inkuiri*, vol. 4, no. I, pp. 2252–7893, 2015, [Online]. Available: http://jurnal.fkip.uns.ac.id/index.php/sains.

[2] A. Henukh, R. F. Nikat, M. Simbolon, C. Nuryadin, and Y. S. *IOP Conf. Ser. Earth Environ. Sci.*, vol. 343, no. 1, 2019, doi: 10.1088/1755-1315/343/1/012160.

[3] C. Umamah, N. Norhasan, and J. Rofi'ah, *LENSA (Lentera Sains) J. Pendidik. IPA*, vol. 8, no. 2, pp. 67–74, 2018, doi: 10.24929/lensa.v8i2.35.

[4] I. M. Astra, E. I. Rosita, and R. Raihanati, *AIP Conf. Proc.*, vol. 2169, no. November, 2019, doi: 10.1063/1.5132637.

[5] I. M. Astra, A. Henukh, and Algiranto, *J. Phys. Conf. Ser.*, vol. 1876, no. 1, p. 012064, 2021, doi: 10.1088/1742-6596/1876/1/012064.

[6] D. D. Lestari, I. Ansori, and B. Karyadi, *Diklabio J. Pendidik. dan Pembelajaran Biol.*, vol. 1, no. 1, pp. 45–53, 2017, doi: 10.33369/diklabio.1.1.45-53.

[7] F. Herliana, I. M. Astra, Y. Supriyati, H. Mazlina, and Musdar, *J. Phys. Conf. Ser.*, vol. 1460, no. 1, pp. 0–5, 2020, doi: 10.1088/1742-6596/1460/1/012125.

[8] N. W. Parwati, N. K. Suarni, I. W. Suastra, and P. B. Adnyana, *J. Phys. Conf. Ser.*, vol. 1318,

no. 1, 2019, doi: 10.1088/1742-6596/1318/1/012096.
[9] R. Pujiastuti, *Lembaran Ilmu Kependidikan*, vol. 42, no. 2, pp. 93–100, 2013.
[10] Nurul Iskandar, Mustaji, Miftakhul Jannah, and Soetam Rizky Wicaksono, *IJORER Int. J. Recent Educ. Res.*, vol. 2, no. 2, pp. 237–249, 2021, doi: 10.46245/ijorer.v2i2.93.
[11] Albert Efendi Pohan, "PROBLEM BASED LEARNING; AN EFFICIENT STRATEGY ON ENHANCING STUDENTS' GRAMMAR ABILITY," vol. 10, no. 1, pp. 56–66, 2019.
[12] P. Chung, R. C. Yeh, and Y. C. Chen, *Int. J. Technol. Des. Educ.*, vol. 26, no. 2, pp. 285–307, 2016, doi: 10.1007/s10798-015-9306-3.
[13] I. P. Suarbawa, *Indones. J. Educ. Res. Rev.*, vol. 2, no. 2, p. 162, 2019, doi: 10.23887/ijerr.v2i2.17624.
[14] L. Akhsani, Ahmad, and F. Eko Subekti, *J. Phys. Conf. Ser.*, vol. 1778, no. 1, 2021, doi: 10.1088/1742-6596/1778/1/012025.
[15] R. Amini, B. Setiawan, Y. Fitria, and Y. Ningsih, *J. Phys. Conf. Ser.*, vol. 1387, no. 1, 2019, doi: 10.1088/1742-6596/1387/1/012082.
[16] B. Maharani and Yohandri, *J. Phys. Conf. Ser.*, vol. 1481, no. 1, 2020, doi: 10.1088/1742-6596/1481/1/012061.
[17] A. K. A. Faozi, Hobri, M. Fatekurohman, K. Aini, and D. Yuniar, *J. Phys. Conf. Ser.*, vol. 1538, no. 1, 2020, doi: 10.1088/1742-6596/1538/1/012070.
[18] M. Widyasmah, Abdurrahman, and K. Herlina, *J. Phys. Conf. Ser.*, vol. 1467, no. 1, 2020, doi: 10.1088/1742-6596/1467/1/012072.

Journal of Physics: Conference Series **2019** (2021) 012026 doi:10.1088/1742-6596/2019/1/012026

Analysing Students' Problem Solving Skills on The Topics of Modern Physics

I E Santosa
Department of Physics Education, Sanata Dharma University
Paingan, Maguwohardjo, Sleman, Yogyakarta 55282, Indonesia
Email: edi@usd.ac.id

Abstract. The study of problem solving skill has been performed on the subject of Modern Physics. The main objective of this study was to investigate the student's ability in solving physics problems. Here 45 students were involved in this study. The students worked on two essay tests consisting of two numerical problems and two explanation questions within each test. The score distributions were then analysed qualitatively. It was found that the student's achievement in the numerical problem is higher than that of the explanation question. The students possess a low skill in selecting and using the data.

1. Introduction

Physics concepts are often taught through explanation and assignment. The students learn to understand these concepts by recognizing, analysing and constructing conceptual connections on their own. They work on many exercises and problems to achieve the goal of teaching and learning i.e. Understanding the physics concept. Problem solving skills are important in studying physics.

The conception of quantitative problem-solving in physics requires one to integrate conceptual reasoning i.e. A qualitative analysis to select relevant equations and checking the reasonable solution [1]. There are often gaps between the physics concepts and the mathematical expressions. Niss [2] found that physics students are not able to solve real-world problems that involve mathematization. It seems that mathematics and physics concepts are unrelated. Many students following physics courses, learn to solve different types of quantitative exercises, but are often unable to explain the meaning of their own numerical solutions to the problems. They focus on manipulating equations without a physical understanding of the problem. Students often skip the qualitative step and jump to the equation [1-4].

It has been realized that to avoid having students focusing on calculation, qualitative type of problem solving based on the physics concept is required [5-6]. This test in the sense no numerical value is used to strengthen students' understanding of fundamental concepts and processes of physics. The importance of considering the argumentation and explanations has been recognized in the science teaching as a central scientific activity that students must learn. The analysis of students' explanations quality can be used to identify students' misconception [7]. Hernández and Tecpan [8] determined the relationship between the level of conceptual knowledge and the quality of the explanations. Their studies incorporated a multiple-choice answer in the Force Conceptual Inventory test and writing explanation. However, there are relatively few studies on this subject. Here we present a study where students are tested in solving both numerical and explanation types of problem. The purpose of this study is to investigate the students' ability in solving different types of physics problem.

2. Method

This research utilized a descriptive qualitative approach to explore the students' problem solving skills. Participants of this study were 45 students taking part in the Modern Physics course. This course covers the black body radiation, photoelectric effect, Compton effect, and special theory of relativity. Students took two essay tests as midterm test 1 and midterm test 2 after teaches on those subjects. Those two tests had different topics. Each test consisted of two types i.e. Numerical and explanation test. Students were required to solve the problems to determine the unknown using the proper equation in the numerical test. The explanation test asked the student to interpret, classify, interfere, compare and explain physics phenomenon e.g. The difference between photoelectric effect and Compton effect. Students worked on individual written test for 90 minutes.

In the first test students were presented with 4 problems consisting of 2 numeric problems and 2 explanation questions. A maximum score of 25 for each problem was available. The average of the cumulative score was determined for each type of question. In total each student had two average scores, i.e. For the numerical and the explanatory problems. Those maximum average scores were also 25. The distribution of student scores was analysed to obtain the profile of the student's ability in problem solving.

The second test was done on a different topic than the first test. In this second test, 3 numerical questions and 3 explanation questions were asked. Students worked on 4 numbers out of 6 given possibilities. Similar to the first test, students had to work on both numerical and explanation questions. The scoring process in this second test was the same as in the first test.

3. Result and Discussions

The results of student scores for the first and second tests are presented in Figure 1 and Figure 2. Each point in the figure represents the score of one student. These points are a pair of average numerical scores and average explanatory scores. The average score of each type is calculated from two problems or questions, which has a maximum score of 25.

A diagonal line, horizontal and vertical lines are also included in both figures as a guideline for data analysis. The diagonal line in the figure shows the positions of the points with the average numerical score equals to the average explanation score. The horizontal line in the middle is the 50% limit for the student achievement on the explanation. Points that are above the horizontal line indicate the average explanation score as being greater than 50% (12.5 out of 25). In the same sense the vertical line in the middle shows a limit for 50% student achievement in the numerical score. The points to the right of the vertical line show the average numerical score as being greater than 50%.

The horizontal and vertical lines divide the data distribution into 4 quadrants. Quadrants I and II show the average score of explanatory questions above 50%, in contrast to quadrants III and IV whose score are below 50%. Students who have a numerical score above 50% are represented in quadrants I and IV, while those scored below 50% are represented in quadrants II and III.

Figure 1 shows the random distribution of all data points; there is no tendency to approximate the diagonal line. This fact indicates that in general the achievement on numerical problem is not followed by the achievement on explanations and vice versa. There is no correlation between the explanation understanding and the numerical ability. Indeed, there are some cases in which students achieve the maximum points in both numerical score and the explanation score i.e. 25. One must note that most of the data points in figure 1 are below the diagonal line; which reveals the tendency of the numerical score to be greater than the explanation.

A clear data representation of individual quadrant is shown in figure 1. There are only a few data points in quadrant II, which results in almost no data in a high explanation score with a low numerical score. On the other hand, many data points are represented in quadrant IV. This is a sign that many students achieve a higher score in numerical problems as compared to the explanation ones. In one extreme case, we found that some students whose numerical scores reaching a maximum of 25, would have a very low explanation score, namely 5. These two quadrants confirm the fact that there is a tendency for students to be better at numerical problems than in explanations question. This finding is also supported by the fact that many students achieve maximum score in numerical problems, but only two students scored maximum on the explanatory question.

There are eight students in quadrant III. These students have a low achievement in both problems. Many students are represented in quadrant I; this indicates their achievement in both problems is more than a half of the maximum.

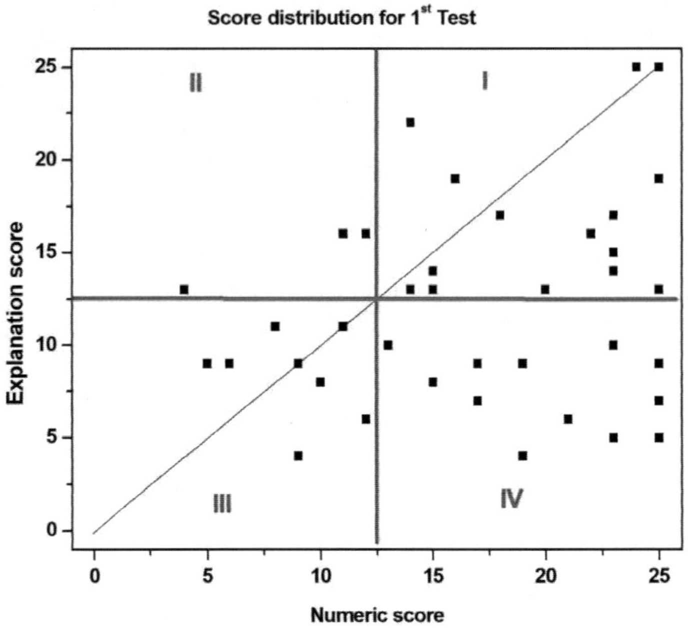

Figure 1. Distribution of average student numerical score and explanation score in the first test.

A similar situation in the first test is observed in the second test as in Figure 2. The same pattern can be observed in this figure, namely the data do not show a relationship between the explanation and the numerical score. The number of students having explanation scores of more than 50% is almost equal to those with explanation scores of less than 50%. Many students achieve a maximum numerical score of 25. In fact, only two students achieved the maximum score for explanation problem. A small number of data points are shown in quadrant III. Many students represented in quadrant IV show an indication of having a higher ability in solving numerical problems than explanation questions. Both test results agree that students tend to score higher on numerical questions than on the descriptions. This situation is summarized in table 1.

Table 1 shows the mean and standard deviation of the 45 students' score. Here the maximum score should be 25. In the first test the mean score of numerical and explanation questions are 17 and 12, respectively. The same behaviour is found in the second test, from a maximum score of 25, the mean score of the numerical and explanation problems are 19 and 13, respectively. Overall, if we compare to the maximum score, the mean numerical score in the two tests is about 70%, while the mean explanation score is 50%.

Journal of Physics: Conference Series **2019** (2021) 012026 doi:10.1088/1742-6596/2019/1/012026

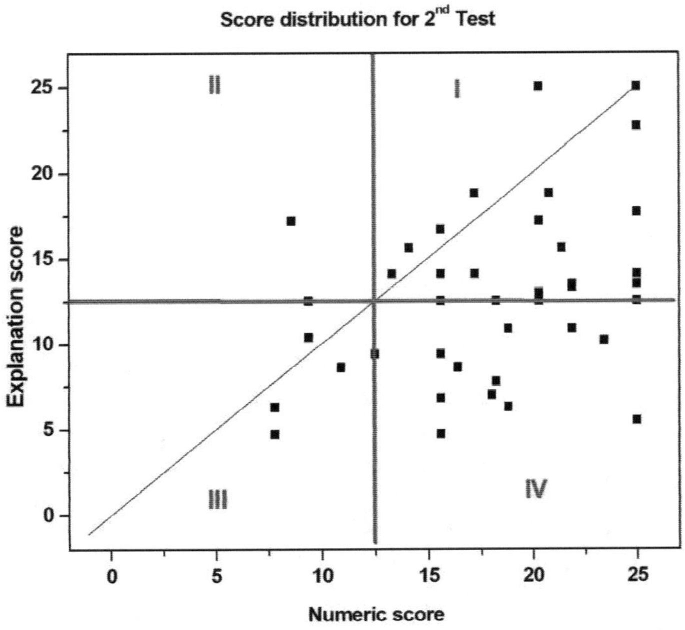

Figure 2. Distribution of average student numerical score and explanation score in the second test.

Table 1. Mean score and standard deviation for numerical and explanation problem. The maximum score is 25.

Test		Numerical	Explanation
Test I	Mean	17	12
	SD	6	5
Test II	Mean	19	13
	SD	3	5

The number of students who obtain scores greater than 12.5 from the maximum of 25 is shown in table 2. A positive result is observed in the numerical problems. In the first test 38 out of 45 students, achievement score greater than 12.5 of the maximum 25. Overall for the two numerical tests, about 70% of students obtain scores greater than 12.5. Whereas in the explanation problem, only about half of the students obtain a score greater than 12.5.

The numerical problems used in this study are intended to strengthen the student's understanding of the physic concept rather than the matter of calculation. Here is one example of the numerical problems:

The threshold wavelength for the photoemission of electrons from a metal surface is 2400 Å. Calculate the work function of electron on this metal expressed in a unit of electron volt!! Determine the speed of photoelectron emitted when light of 1200 Å is incident on the same metal!

The student should understand the concept and the process of the photoelectric effect before doing the calculation. First, they learn different quantities and unit e.g. Wavelength, frequency, photon energy, kinetic energy, speed, electron volt, and angstrom. Afterwards, they use the relevant equation that relates the unknown and the known quantity. Furthermore, they must understand the connection between the first illumination (wavelength 2400 Å) and the second illumination (wavelength 1200 Å).

Table 2. Number of students and their score. The total number of students is 45, and the maximum score is 25

		Number of students	
Test	*Score*	*Numerical*	*Explanation*
Test I	≥ 12,5	38	24
	< 12,5	7	21
Test II	≥ 12,5	32	20
	< 12,5	13	25

Figure 1 and 2 shows plenty of students who managed to achieve perfect scores in numerical problem. The obstacles faced by students to solve the numerical problems, are related to students' attitude and motivation [9]. In this study the obstacles are connected to the lack of conceptual understanding and computing skills. As an example in this study, some students do not understand the fact that the calculated work function of the same metal would remain constant. Due to this lack of understanding, they are unable to complete the subsequent question. Problem solving involves conceptual reasoning to facilitate the manipulation of the equation itself [1].

We expect the student who has solved such kind of numerical problems to really understand the concept and process of the photoelectric effect. They should also be able to explain all aspects related to the photoelectric effect in the explanatory question. This assumption can be expanded into a more general situation i.e. The achievement scores in numerical problems related to the score of explanation question. Table 1 shows the mean score of the numerical problem is higher than the explanatory question which shows both tests. This observation is different from the assumption.

The fact that the numerical score is higher than the explanation score is similar to the study of Hernández and Tecpan [8]. Aside from giving a multiple choice test, Hernández and Tecpan also asked the participants to explain their answer. Their result showed a contrast between the numbers of correct answers obtained with the test (quantitative result) and the amount of explanation that was considered correct (qualitative results).

This high score on the numerical test is related, among other things, to students' learning tendencies. Students usually prioritize the calculation problem type. Numerical exercise is important to a desirable extent. There is an assumption that teaching problem solving leads to understanding of the subject), however various studies present the side effect. They often are unable to explain the meaning of their own numerical solutions [3]. Students try to reach the solution as soon as possible by manipulating mathematical equations without having a clear conceptual insight into the problem situation [10]. Those researches show that traditionally formulated numerical exercise, which are mostly used in physics teaching, do not develop students' abilities in higher-order thinking.

About a half of the student achieves a score less than 50% of the maximal score for the explanation test. Students encounter difficulties in understanding the concept as well the way to express it. Similar results were qualitative assignments are not easily completed by students had been reported in the literature [11]. One of the questions of this type is regarding the analysis of the photoelectric effect and Compton effect. Students are able to describe both the photoelectric and the Compton Effect, however unable to explain the difference between them. They do not realize that their description contains the difference as well as the similarity between those effects. In order to explain scientific phenomena one needs skill to collect, select and use the data as evidence to support the statement / answer [8]. Furthermore the evidence should be linked with the scientific principles involved. In our case it seems that students possess a low skill in selecting and using the data.

Byun and Lee [12] reported that high school physics teachers in South Korea mostly used traditional lectures and solving or discussing quantitative or mathematical problems in their instruction. Their study showed that students who had solved many physics problems are not always adept at understanding the physics concepts. Adams and Wieman [13] showed various individual sub-skills that correlate to students' overall problem solving ability. Students' current state should be considered for effective teaching. The teacher can help students to become better problem solver by teaching a problem-solving strategy that can be applied to any situation [11, 14, 15]. Qualitative problem solving strategies can improve students' achievement in explaining and understanding of the underlying physics concepts and principles.

4. Conclusion

Problem solving in physics is closely related to the understanding of its concept. Although the numerical problems seem to be computing problems, it can easily be solved if the concept is understood clearly. The study presented here shows that the students achieved a higher score in the numerical problems rather than those of the explanation questions. Students have difficulties in collecting, selecting and using the data to explain the answer of the problems/questions.

Acknowledgments

The author would like to thank K. Heskett for helpful comments on the manuscript.

References

[1] Kuo E, Hull M M, Gupta A and Elby A 2013 How students blend conceptual and formal mathematical reasoning in solving physics problems *Sci. Educ.* **97** 32–57
[2] Niss M 2017 Obstacles related to structuring for mathematization encountered by students when solving physics problems *Int. J. of Sci. and Math. Educ.* **15** 1441–62
[3] Zuza K, Garmendia M, Barragués J I and Guisasola J 2016 Exercises are problems too: implications for teaching problem-solving in introductory physics courses *Eur. J. Phys.* **37** 055703
[4] Harper K A 2006 Student problem-solving behaviors *Phys. Teach.* **44** 250-1
[5] Tao P K 2001 Developing understanding through confronting varying views: the case of solving qualitative physics problems *Int. J. Sci. Educ.* **23** 1201- 1218
[6] Docktor J L, Strand N E, Mestre J P and Ross B H 2015 Conceptual problem solving in high school physics *Phys. Rev. ST Phys. Educ. Res* **11** 020106
[7] Khwanda M N and Kriek J 2020 An evaluation of student's understanding of DC circuit concepts through students' written explanations *J. Phys.: Conf. Ser.* 1512 012020
[8] Hernández C and Tecpan S 2018 Correct answers with wrong justifications? analysis of explanations in classical mechanics with FCI test *J. Phys.: Conf. Ser.* 1043 012056
[9] Güneş I, Güneş Z O, Derelioğluc Y and Kırbaşlar F G 2015 Relations between operational chemistry and physics problems solving skills and mathematics literacy self-efficacy of engineering faculty students *Procedia Soc. Behav. Sci.* **174** 457–63
[10] Marusic M, Erceg N and Slisko J 2011 Partially specified physics problems: university students' attitudes and performance *Eur. J. Phys.* **32** 711–22
[11] Mualem R and Eylon B S 2007 Physics with a smile—explaining phenomena with a qualitative problem-solving strategy *Phys. Teach.* **45** 158-63
[12] Byun T and Lee G 2014 Why students still can't solve physics problems after solving over 2000 problems *Am. J. Phys.* **82** 906-13
[13] Adams W K and Wieman C E 2015 Analyzing the many skills involved in solving complex physics problems *Am. J. Phys.* **83** 459-67
[14] Hill N B 2016 MAUVE: a new strategy for solving and grading physics problems *Phys. Teach.* **54** 291-4
[15] Mason A J and Singh C 2016 Impact of guided reflection with peers on the development of effective problem solving strategies and physics learning *Phys. Teach.* **54** 295-9

Journal of Physics: Conference Series **2019** (2021) 012027 doi:10.1088/1742-6596/2019/1/012027

Development of Online Learning Videos: An Innovation for Higher Education Students in Electromagnetic Experiment Course During a Pandemic COVID 19

C I Yogihati[1], D S Putri, D A Suaidi, Hartatiek and Yudyanto
Department of Physics Education, Universitas Negeri Malang
Jl. Semarang No. 5, Sumbersari, Kec. Lowokwaru, Kota Malang, Jawa Timur 65145
Email: chusnana.insjaf.fmipa@um.ac.id

Abstract. The ongoing COVID-19 pandemic has led to government decisions regarding the continuation of online lectures. This transformation becomes a problem in the ongoing compulsory course for students, namely electromagnetic experiment. Until finally the researchers chose video media as the answer to these problems. In the process, video development was carried out in two cycles. Through this development, the researcher wanted to know the ability of video in presenting data and its feasibility. The subjects of this study were students of S1 Physics Education, Universitas Negeri Malang in batch 2019 with a total of 100 people. This video was developed using the Four-D method which consists of defining, designing, and developing stages. The first cycle video acts as the preceding video and the second cycle video acts as the development result video. Practical reports and student assessment questionnaires are used as a review of development results. As a result, students obtained an average experiment report score of 93 in the first cycle and 96 in the second cycle. Student assessments related to the first cycle of media showed that 70.57% stated that they were feasible and became 79.86% in the second cycle.

1. Introduction

As it is known that during the COVID-19 pandemic lectures were conducted online [1][2]. Online learning is one manifestation of the application of information technology developments that have brought major changes to education. This technological development is manifested in its use in the teaching and learning process from conventional methods to more modern ways [3]. With advances in technology and communication, online lectures can be well planned. One of the technological media that can be used is video learning [2]. Video is all things related to what is seen, especially moving images which in the process require a recording process and the display requires technological assistance [4].

Electromagnetic experiment is one of the subjects that must be taken by students majoring in physics at FMIPA UM. The learning process consists of experiment activities, report generation, and presentations. However, during the pandemic student activities are carried out online. To know the expected student competencies, it is necessary to strive for learning media that can represent students in reviewing the data obtained related to experiment material. Through these considerations, the video becomes one of the solutions to overcome this problem. The selected electromagnet experiment materials are induction coils, transformers, electromagnets, and AC/DC generators. With this video, it is hoped that students can gain an understanding of the practical material presented. Furthermore, students can practice skills in obtaining data, analyzing, making reports, and presentations.

Based on the literature review that has been carried out, the electromagnetic experiment learning video has not been obtained by the expected competencies. Therefore, an electromagnet experiment learning video was developed to determine the ability of the video to provide an overview of the observational data and its feasibility to be used in learning activities. In the manufacturing process, an android phone is used with the help of the kinemaster video editor application. Broadly speaking, there are three main stages needed in making videos, including pre-production, production, and post-production [5][6][7].

2. Method

This type of research is the development (Research and Development) with the 4D method (four D model) which was developed by S. Thiagarajan, et al [8]. Several stages of the 4D method are defined, designing, developing, and disseminating [9][10]. However, in this study, it was only applied to the development stage. In the early stages, several aspects need to be analyzed to provide the basis for video development, namely the initial analysis to raise and determine the basic problems faced. After that, an analysis of the students was carried out to find out their character to develop media according to their academic abilities. Through these two things, a task analysis is carried out to detail the outline of the teaching material and concept analysis to identify the essential concepts to be taught. Then formulated the learning objectives of the material to be made.

The design stage is the second stage of the 4D method. In this stage, the initial manufacture of the product (prototype) of the media that will be used is carried out and in this case, is represented by a video made in cycle 1 of the electromagnetic experiment. The videos made are based on the formulation of learning objectives from the results of the analysis, topic selection, media, and source selection, learning strategy selection. Development is the last stage of the methodology used. At this stage, Thiagarajan (1976) divided it into two, namely expert appraisal and development testing. An expert appraisal is a technique to assess or validate the feasibility of the media to be developed. In this case, the researcher acts as a validator of video media content because he has become an expert in the field created. As for development testing, it is a product design test activity consisting of individual trials, small groups, and large groups. This was represented by students of S1 Physics Education Offering AC, B, C, and D class of the 2019 Universitas Negeri Malang. The purpose of this stage is to produce learning tools that have been revised based on the input obtained from the validator. At this stage, there are several stages carried out including product validation by the team, design revisions, limited trials, product revisions, and usage trials.

This research is divided into two cycles. The first cycle is used as the first video which is made as a provision for video development in the second cycle. The first cycle raised the topic of induction coils and transformers. Meanwhile, in the second cycle, the topic of AC/DC generators and electromagnetic forces was raised. Because the assessment was carried out for the ability of the video to display observational data and its feasibility, a questionnaire was provided to answer its feasibility and an experiment report was used as an indication of the ability of the video to display observational data.

In the classroom implementation, three stages are carried out in each cycle, namely pre-activity, planning, implementation, and reflection. In the pre-activity stage, an analysis is carried out as a form of defining the method used. Through this stage, it is known that students do experiments independently with the help of a virtual lab or project-based. Followed by the planning stage covering RPS, SAP, CPMK, making videos, and research questionnaires. This second stage represents the design stage of the media. Then all the things that have been prepared are carried out in class and produce some reflections contained in the cycle questionnaire and or in the discussions that take place in class. These results are then compared with the criteria for the assessment score (%) of the implementation of learning presented by Arikunto (2002). The results of this reflection are the basis of the video development for the second cycle.

Table 1. Criteria for assessment scores (%) of learning implementation.

Score (%)	Criteria
92-100	Very good
75-91	Good
50-74	Pretty good
25-49	Not good
00-24	Bad

3. Result and Discussion

In the process of developing this electromagnetic experiment video, students' opinions were reviewed through questionnaires and experiment reports. A review is carried out on the content and appropriateness of the media used.

Table 2. Scores of electromagnetic experiment report.

Offering	Transformer	Induction Coil	AC/DC Generator	Force Electromagnetic
AC	91,4	92,2	97,6	97,8
B	93,2	91,8	97,4	94,6
C	96,2	90,6	95,8	96,6
D	95	91	93,2	95
Average	93,95	91,4	96	96

Through the data on the value of the electromagnetic experiment report, the average value of the first cycle for the topic of transformers and induction coils is 93.95 and 91.4, respectively. This rating is lower than the second cycle with the topic of AC/DC generator and electromagnetic force which has an average value of both 96.

Table 3. Student opinions about video content.

Questions	Cycle 1	Cycle 2
The video can display the symptoms according to the material from the topic	66,7%	73,3%
Students are free to observe the existing independent variables	67,6%	69%
Students are free to observe the existing dependent variables	73,3%	74%
Students are free to explore related material	71,1%	74,4%
Average	69,675%	72,675%

Even the video content assessment questionnaire shows the same symptoms as the experiment report. In assessing the suitability of the symptoms shown in the video with the material, the first cycle of the video, 66.7% of students agreed with the statement and 33.3% disagreed. When compared to the second cycle, there were 73.3% of students agreed with the statement and 26.7% of the other students disagreed. This shows that this media has increased in presenting symptoms for the topics discussed to students.

Not only in presenting symptoms, but the experiment videos made also show developments in the student's discretion in observing existing variables. The first cycle showed that 67.6% of students agreed to the discretion of the independent variable observation and 69% agreed to the discretion of the dependent variable observation. An increasing pattern also appears for the independent and dependent variables in the second cycle with percentages of 73.3% and 69% respectively. The last aspect that was assessed about the content was the students' flexibility in exploring each topic in each cycle. The first cycle has a percentage of 71.1% and thc sccond cycle has a percentage of 74% of students agreeing with the question. Finally, students agree that the media used can provide flexibility for students to explore material related to the percentage of 71.1% for the first cycle and 74.4% for the second cycle. So that through this review the second cycle of media is said to have increased from the videos made in the first cycle. Through a comparison with the assessment criteria for the implementation of learning presented by Arikunto (2002) the second cycle media has a percentage of 72.675% in terms of content or has a pretty good predicate.

Table 4. Student opinion on media quality.

Questions	Cycle 1	Cycle 2
Video image quality	73,3%	83,8%
Video color quality	78,3%	86%
Video lighting quality	66,7%	81,4%
Sound quality in video	73,3%	81,4%
There is no noise in the video	70%	81,4%
References in the video are clear and supports	63,3%	69,9%
Readability of text on video	73,4%	81,4%
The video contains important things related	63,4%	72,1%
Clarity of delivery of material in the video	73,4%	81,3%
Clarity of scale in video	51,7%	73,25%
Average	68,68%	79,195%

The review of the quality of the video media used also showed an increase when compared to the percentage of the first cycle of questionnaires and the second cycle of questionnaires. The first cycle has an assessment of 73.3% of students agree that the media used has good image quality. This increase is indicated by 83.8% of students agreeing that the second cycle media have good image quality. Likewise for color quality, the second cycle has a percentage of 86% of students agree. This is different from the first cycle media which only obtained 78.3% of students agree. The lighting quality of the second cycle media has better quality, with a percentage of 81.4%, when compared to the first cycle media which only has a percentage of 73.3%.

Not only visuals, but the audio is also an assessment that should be taken into account. In terms of clear sound quality and minimal audio interference, the second cycle video has a percentage of 81.4% each. In contrast to the first cycle of video which only has a percentage of 73.3% for clear sound quality and 70% for minimal audio interference. The same thing also happened to the assessment of supporting information contained in the video. The first cycle video has a percentage of 63.3% and the first cycle video has a percentage of 69.9% for the point rating. In addition, the second cycle video has a percentage of 81.4% for the clarity and legibility of the available text, and the first cycle video has a 73.4% percentage for this rating. Even the first cycle video had 63.4% student approval in the availability of important information in data collection. This increase was significant in the second cycle with a percentage of 72.1%. Students assess the clarity of the speaker in delivering the material in the second cycle has a percentage of 81.3% and the first cycle has a percentage of 73.4%. Finally, 73.25% of students agree that the video has a clear scale in the second cycle. This assessment increased when compared to the first cycle with a percentage of 51.7%. Through all assessments in terms of video quality, the second cycle of media has a percentage of 79.195% with a good predicate. So, if the average rating of the second cycle, content and video quality, is 75.935% with a good predicate.

However, several things become the evaluation of the making of this electromagnetic experiment video. The first thing to evaluate is that the video is less stable because the video was taken manually with a cellphone camera without the help of a tripod. Then, it would be better if the detailed scale was taken when shooting, not zooming during the editing process. In addition, students provide suggestions to include what data will be taken and how to collect it so that students are not confused in the data collection process. The emergence of observational data needs to be extended in duration so that students can obtain data and observe phenomena better. According to students' opinions, videos with the topic of transformers and videos of electromagnetic forces for the first and second cycles are videos that need to be improved. This is a reflection for researchers to always make improvements for the next videos.

4. Conclusion

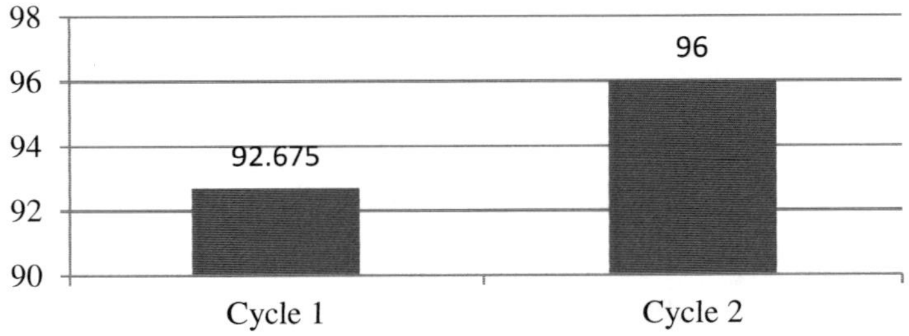

Figure 1. Comparison of practicum report values.

The learning activities that took place in cycles I and II showed that there were two aspects to assess the success of this experimental video, namely the value of the experimental report and media analysis. The results presented in Figure 1 show that the experimental value of students in cycle 2 increased when compared to cycle 1. The students' ability in compiling reports shows that this video can make students take experimental data according to the topic easily. This is because the video contains experimental data which is a provision for students in observing the symptoms that occur according to the topic of discussion.

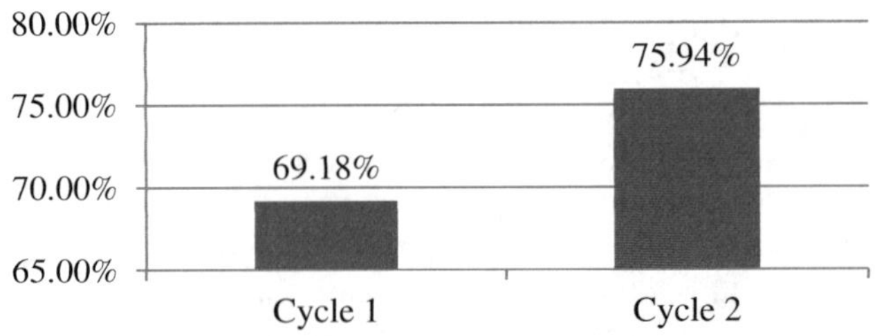

Figure 2. Comparison of media ratings.

The results of the questionnaire given to students, in Figure 2, show that the video has the appropriate quality when used as a learning medium. This is because videos can provide a more comprehensive picture, and by utilizing the replay and pause features students are able to obtain more accurate results and material related to the topic of discussion. In this experiment, the video can represent the position of students to observe the tools used in the data collection process. This is reflected in an assessment questionnaire for students. Through the quality of the content and media obtained through the questionnaire, the second cycle of videos has better quality from both sides as shown in Figure 2. This shows that this video is suitable to be used as a learning medium for electromagnetic experiments. However, there are several evaluations contained in the development of this video. Writing the script, taking pictures, and compiling the material to be delivered needs to be well planned so that students are able to accept it well. There is also a need for tools that support making this video well so that the video has good quality (image and sound).

Acknowledgment

First of all, we thank God almighty for the abundance of grace and strength so that we are able to complete "Development of Online Learning Videos: An Innovation for Higher Education Students in Electromagnetic Experiment Course during a pandemic COVID 19" in facing all the challenges encountered during the process. We also thank all parties involved in this research. Undergraduate students of physics education in the Universitas Negeri Malang class of 2019 have taken the time to evaluate the experiment videos that have been developed. We also express our gratitude to the lecturers of the physics department and education staff who we cannot mention one by one for their willingness to be involved in helping this research as a bridge between the research team and students.

References

[1] C Nuritha and A Tsurayya 2021 Pengembangan video pembelajaran berbantuan geogebra untuk meningkatkan kemandirian belajar siswa *J. Cendekia J. Pendidik. Mat.* **5** no. 1, pp. 48–64.

[2] L Tarida Pemanfaatan google classroom dan video pembelajaran berbasis problem solving sebagai solusi kegiatan belajar mengajar di era pandemi Covid-19 Oct. 2020.

[3] F Martin and A K K Betrus, *Digital media for learning: Theories, processes, and solutions*. 2019.

[4] M Ario 2019 Pengembangan video pembelajaran materi integral pada pembelajaran flipped classroom *AKSIOMA J. Progr. Stud. Pendidik. Mat.* **8** no. 1, pp. 20–31.

[5] Z Apriliani, U Hasanah, and A S Anas 2019 Pembuatan video profil dengan efek vintage kampung wisata adat sengkoah sebagai media informasi *JTIM J. Teknol. Inf. dan Multimed.* **1** no. 1, pp. 57–65.

[6] N Rimayanti, A Yulianti, B Nasution, and E E Lubis 2019 Pembuatan video profil lembaga penelitian dan pengabdian universitas riau sebagai media promosi berbasis multimedia *It J. Res. Dev.* **3** no. 2, pp. 84–95.

[7] C Prihantini and G K Nugroho 2013 Pembuatan video profil sekolah menengah pertama negeri 4 polokarto kabupaten sukoharjo dengan menggunakan komputer multimedia *J. Teknol. dan Sist. Komput.* **5** no. 1, pp. 1–5.

[8] I M Tegeh, A H Simamora, and K Dwipayana 2019 Pengembangan media video pembelajaran dengan model pengembangan 4d pada mata pelajaran agama hindu *Mimb. Ilmu* **24** no. 2, p. 158.

[9] D Lawhon 1976 Instructional development for training teachers of exceptional children: A sourcebook *J. Sch. Psychol.* **14** no. 1, p. 75.

[10] I P P Suryawan and D Permana 2020 Media pembelajaran online berbasis geogebra sebagai upaya meningkatkan pemahaman konsep matematika *Prisma* **9** no. 1, p. 108.

[11] S Arikunto 2002 *Prosedur Penilaian : Suatu Pendekatan Praktik* (Jakarta: PT Adi Mahasdya).

Implementation Management Strategies for Modern Physics Experiment with Online Systems Approach to Improve The Quality of Learning or Physics Students

Sutrisno, A Simaremare, A Suyudi, M R A Taqwa and Parno
Department of Physics Education, Universitas Negeri Malang
Jl. Semarang No.5, Malang 65145, Indonesia
E-mail: sutrisno.fmipa@um.ac.id

Abstract. Since COVID-19 broke out in early 2020, State University of Malang has implemented online lectures. This transformation creates new problems, including in Modern Physics experiment lectures. The fact is that when students was founded with the problem of assembling complex tools they are unable to implement their knowledge. The problems with Modern Physics experiment lectures that have existed for a long time are now exacerbated by the emergence of the COVID-19 pandemic. In overcoming this problem, the researcher implementation, management strategy with a systems approach as a solution with the consideration that this strategy is able to direct students to think critically. This study aims to improve students' practical understanding and learning quality. This research was conducted in 2 cycles. The research instruments were questionnaires and experiment reports. Based on the data analysis of the questionnaire cycles I and II, it was concluded that the quality of the implementation of modern physics experiment learning was good with an average result of 88.38%. In addition, there is an increase in students' practical understanding which can be seen from the average score of the experiment report, where the first cycle is 91.33 and the second cycle reaches 95.

1. Introduction

Currently, the discussion about COVID-19 (Corona Virus Disease 2019) is a hot topic in various countries. COVID-19 first appeared at the end of 2019, precisely in Wuhan, China [1]. COVID-19 is a disease caused by SARS-CoV-2 whose transmission is very fast and it is difficult to know the characteristics of people who have contracted this virus because the incubation period is approximately 14 days [2]. Almost all countries in the world have been affected by COVID-19, including Indonesia. COVID-19 has affected all aspects of life, including education [3]. Since WHO (World Health Organization) declared COVID-19 as a pandemic on March 11, 2020 [4], various policies have been issued by the Indonesian government, one of which is a policy in the field of education through the circular letter of the Minister of Education and Culture Number 15 of 2020 concerning "Guidelines for the Implementation of Learning From Home in an Emergency Period. The spread of COVID-19".

Such conditions make the State University of Malang implement an online lecture policy or online lectures. Online learning is an educational innovation that involves elements of information technology in learning [5]. Therefore, many application facilities are available on telecommunication tools to facilitate interaction between lecturers and students [6]. Lectures that were originally offline must be transformed into online, this transition process requires every educator to apply a new learning model [7], especially in the Modern Physics experiment course. Modern Physics experiment course is a compulsory subject for Physics Education Study Program students. In the experiment, this experiment is separate from the activities of Modern Physics theory. The implementation of the Modern Physics experiment usually begins with the introduction of both passive and active components, then students observe the characteristics of the components by following the experiment instructions.

However, in fact, in the Modern Physics experiment, there are weaknesses, among others, when faced with the problem of assembling complex tools they are not able to implement knowledge about the characteristics of each component in understanding complex circuits. Sometimes experiment activities are only seen as formalities that must be met to be able to take Modern Physics experiment courses. Realizing this reality, the author tries to examine the implementation of the Modern Physics experiment by paying attention to the following 3 things: (1) experiment by studying the characteristics of each component separately does not stimulate students' thinking and creativity; (2) practical students feel that understanding the characteristics of the components of modern physics experiment has not become a necessity for them; (3) the acquisition of experiment scores has not been meaningful for students, it can be seen from the inability to implement their knowledge on other problems.

To find alternative solutions to problems, a series of discussions have been carried out by the supervisor of the modern physics experiment course, especially regarding the low effectiveness of this experiment activity. The conclusion of this discussion recommends an experiment management strategy using a systems approach, where from the beginning students are introduced to a complete system composed of components of online experiment learning implementation. The basic consideration, with this strategy from the beginning, students have been directed to think critically about how the system works and how the contribution of each component of the learning participants in the system. In the end, the students' curiosity about the working of systems will motivate them to know more about the characteristics of the components of modern physics. Student creativity can also be built by questioning the possibility of developing an existing system [8]. Practical learning with a systems approach in question is learning the strategy for managing experiment activities which consist of three stages, namely (1) the preparation stage; (2) the exploration stage; (3) the application stage [9].

Considerations that Another is in assigning modern physics experiment tools and materials cannot be provided outside the modern physics laboratory because the tools and materials are special and expensive. Therefore, the use of the online system needs to be optimized in carrying out quality and effective modern Physics experiment learning, therefore this research takes the title "Implementation Management Strategies for Modern Physics Experiment with Online Systems Approach to Improve the Quality of Learning for Physics Students" which aims to apply Modern Physics experiment management strategies with a systems approach to improve understanding of concepts for students participating in Modern Physics experiment courses through online learning.

2. Method

This type of research is classroom action research called classroom action research, which is a systematic study of classroom learning experiments to improve or improve the quality of the learning process and results carried out by certain actions [10]. Suharsimi Arikunto (2002) stated that classroom action research is practical research, because classroom action research is carried out to solve factual problems that are faced by teachers as managers of learning activities [11]. This research departs from the problems that exist in the field and then is reflected and analyzed based on the theory that supports then implemented action.

This classroom action research was conducted by the Department of Physics, Faculty of Mathematics and Natural Sciences, State University of Malang Jl. Semarang No. 5 Malang. This research involves students of the physics education study program, a lecturer in charge of experiment courses, and a student who is taking a Modern Physics experiment course in the even semester of 2020-2021. This research was carried out in 2 cycles. Based on the initial evaluation, discussion, and lecturer's reflection on the performance of previous lectures, the action given in the first cycle is the approach of enforcing the implementation strategy of the Modern Physics Experiment. With a systems approach to students participating in the Modern Physics Experiment course.

In more detail, the procedure for this classroom action research is as follows: (1) planning the first cycle, at this stage preparing the 1st and 2nd experiment videos, uploading the 1st and 2nd experiment videos to the student via the HTTPS link, carrying out the experiment by downloading the video on the sipejar 1st and 2nd experiments. Within one week, students explore by making a report that is uploaded to the student. To evaluate the students conducted a discussion of the results of the experiment report via zoom. Reflecting on the results of the implementation of cycle 1, to plan cycle 2; (2) action planning cycle 2, at this stage preparing the 3rd and 4th experiment videos, uploading the 3rd and 4th experiment videos to the student via the HTTPS link, carrying out the experiment by downloading the 3rd and 4th experiment videos on the student. Within one week, students explore by making a report that is uploaded to sipejar. To evaluate the students conducted a discussion of the results of the experiment report via zoom. Reflect and analyze the results of the implementation of cycle 2 to obtain conclusions. In this study, the instrument was used in the form of questionnaire results in cycle 1 and cycle 2 and an experiment report. In analyzing the data, RPS is needed for the Modern Physics experiment course. In the Modern Physics experiment, the implementation is based on the description of the Modern Physics experiment course. For learning devices, they were analyzed qualitatively, while for research data, such as experiment reports in cycle 1 and cycle 2, the average class score was then compared whether there was an increase in learning scores and to measure the quality of learning referring to the Arikunto classification (2002) used the average questionnaire in the cycle. 1 and 2.

Table 1. Criteria for assessment score (%) implementation of learning.

Score (%)	Criteria
92-100	Very good
75-91	Good
50-74	Pretty good
25-49	Ugly
00-24	Bad

3. Results and Discussion

Based on the questionnaire data analysis conducted by researchers in the implementation of the first cycle activities are as follows:

(1) The use of experiment videos as learning media for modern physics experiment courses has produced symptoms that are following the theory in the practical determination of electron-specific charge and Michelson interferometer experiment. This is known from the results of the first cycle of the questionnaire regarding the suitability of the experiment video with symptoms in theory.

Table 2. The results of the questionnaire regarding the suitability of the video with theory.

Experiment	Agree	Disagree
Determination of the specific charge of electrons	93.1%	6.9%
Michelson interferometer experiment	95.6%	4.4%

(2) The use of experiment videos as a learning medium for modern physics experiment, subjects makes students free to take data in the experiment of determining the specific charge of electrons and Michelson interferometer experiment. This is known from the results of the first cycle of the questionnaire regarding the freedom of students in obtaining data through experiment videos.

Table 3. The results of the questionnaire regarding the freedom to obtain data.

Experiment	Agree	Disagree
Determination of the specific charge of electrons	86.3%	13.7%
Michelson interferometer experiment	90%	10%

(3) The use of experiment videos as a learning medium for modern physics experiment, subjects makes students take data on the dependent variable according to the measured symptoms in the experiment of determining the specific charge of electrons and Michelson interferometer experiment. This is known from the results of the first cycle of questionnaires regarding the suitability of data collection for the dependent variable with the measured symptoms through experiment videos.

Table 4. The results of the questionnaire regarding the suitability of the data and symptoms.

Experiment	Agree	Disagree
Determination of the specific charge of electrons	90.8%	9.2%
Michelson interferometer experiment	90%	10%

(4) The use of experiment videos as a learning medium for modern physics experiment courses presents measurements with good video quality where the scale is clear to the smallest scale in the experiment of determining the specific charge of electrons and Michelson interferometer experiment. This is known from the results of the first cycle of the questionnaire regarding the clarity of the measurement scale to the smallest scale through experiment videos.

Table 5. Questionnaire results regarding the clarity of the measurement scale.

Experiment	Agree	Disagree
Determination of the specific charge of electrons	77%	23%
Michelson interferometer experiment	73.3%	26.7%

(5) The use of experiment videos as a learning medium for modern physics experiment courses can increase student creations in developing material in the experiment of determining the specific charge of electrons and Michelson interferometer experiment. This is known from the results of the first cycle of questionnaires regarding the effect of experiment videos on the level of student creation in concluding and developing material.

Table 6. Questionnaire results regarding human creations in concluding material.

Experiment	Agree	Disagree
Determination of the specific charge of electrons	92%	8%
Michelson interferometer experiment	88.9%	11.1%

From the average result obtained in the first cycle, the researcher carried out the second cycle to correct the deficiencies in the first cycle. The following are the results of the second cycle of questionnaire data analysis:

(1) The use of experiment videos as a medium of learning for modern physics experiment courses has produced symptoms that match the theory in Franck Hertz's experiment and nuclear radiation count. This is known from the results of the second cycle of the questionnaire regarding the suitability of the experiment video with the symptoms in theory.

Table 7. Questionnaire results regarding the suitability of the video with theory

Experiment	Agree	Disagree
Franck Hertz's Experiment	94.3%	5.7%
Nuclear radiation count	92.2%	7.8%

(2) The use of experiment videos as a learning medium for Franck Hertz's experiment and nuclear radiation count make students free to take data in the experiment. This is known from the results of the second cycle of questionnaires regarding the freedom of students in obtaining data through experiment videos.

Table 8. The results of the questionnaire regarding the freedom to obtain data.

Experiment	Agree	Disagree
Franck Hertz's Experiment	87.4%	12.6%
Nuclear radiation count	83.4%	16.6%

(3) The use of experiment videos as a learning medium for modern physics experiment subjects makes students take dependent variable data according to the symptoms measured in the Franck Hertz's experiment and nuclear radiation count. This is known from the results of the second cycle of questionnaires regarding the suitability of data collection for the dependent variable with the measured symptoms through experiment videos.

Table 9. The results of the questionnaire regarding the clarity of the measurement scale.

Experiment	Agree	Disagree
Franck Hertz's Experiment	89.7%	10.3%
Nuclear radiation count	94.5%	5.5%

(4) The use of experiment video as a learning medium for modern physics experiment courses presents measurements with good video quality where the scale is clear to the smallest scale in the Franck Hertz's experiment and nuclear radiation count. This is known from the results of the second cycle of the questionnaire regarding the clarity of the measurement scale to the smallest scale through experiment videos.

Table 10. The results of the questionnaire regarding the suitability of the data and symptoms.

Experiment	Agree	Disagree
Franck Hertz's Experiment	81.6%	18.4%
Nuclear radiation count	92.2%	7.8%

(5) The use of experiment videos as a medium of learning for modern physics experiment courses can increase students' creativity in developing material in Franck Hertz's experiment. This is known from the results of the second cycle of questionnaires regarding the effect of experiment videos on the level of student creation in concluding and developing material.

Table 11. Questionnaire results regarding human creations in concluding material.

Experiment	Agree	Disagree
Franck Hertz's Experiment	92%	8%
Nuclear radiation count	93.3%	6.7%

Based on the results in a cycle I and cycle II, the average of each cycle is presented in the following diagram:

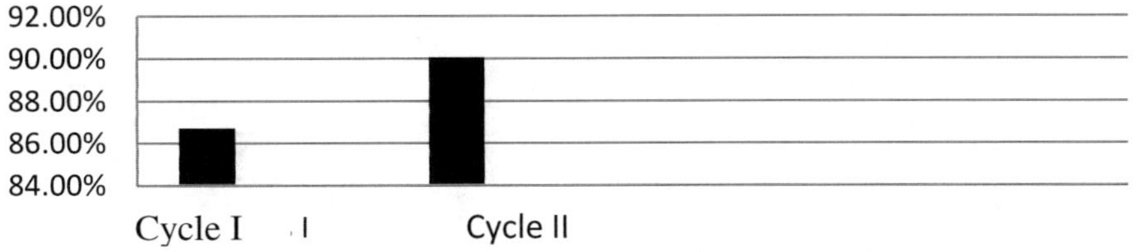

Figure 1. The average results of the first cycle of the questionnaire and the second cycle.

To find out the understanding of the concepts of students participating in the Modern Physics experiment course through online learning, the researchers reviewed the results of the student experiment reports. The following is the data from the student experiment report:

Table 12. The average value of student experiment reports.

Offering	Load determination electron specific	Michelson interferometer experiment	Franck Hertz's Experiment	Nuclear radiation count
AC	90.38	89.38	93.75	95.13
B	91.79	93.75	94.6	95.4

Based on the average value of student experiment reports in the cycle I and cycle II, the following diagram is obtained:

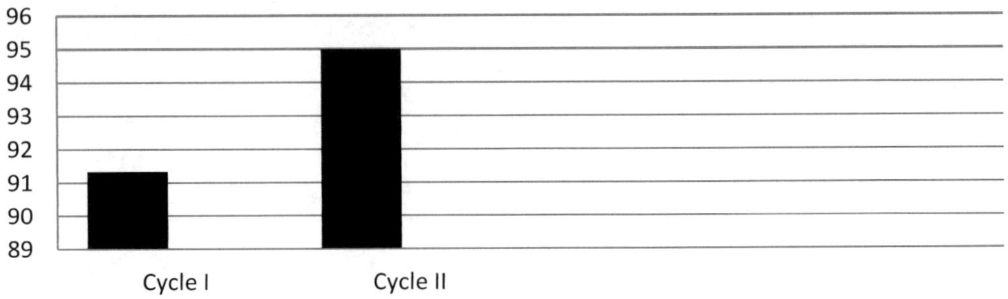

Figure 2. The average value of the first cycle and second cycle practicum reports.

4. Conclusion

The quality of the implementation of modern physics experiment lessons using experiment videos can be seen from the results of the distribution of research instruments, namely questionnaires. In the first cycle, the quality of the implementation of learning that is included in the good criteria and has reached the expected criteria, it can be seen from the average quality of the implementation of learning through experiment videos in the first cycle reaching 86.7%. In the implementation of the second cycle of learning reached 90.06%. So the average cycle I and cycle II are 88.38% and is in the good category. So it can be concluded that the quality of the learning implementation of modern physics experiment courses using experiment videos is good.

Improved understanding of the concept of practicing modern physics can be seen from the value of student experiment reports. A good experiment report score was obtained in both cycles. In the first cycle, the students' understanding of practical concepts was included in good criteria and had reached the expected criteria, it was seen from the average student experiment report score of 91.3. Understanding the concept of the experiment in cycle II obtained an increase in achievement to 95 and is in very good criteria. This increase occurs because students have been able to take data and their errors well, are able to explain phenomena that occur in the experiment, and develop material through independent observations made by the student.

This online learning method is recommended to use modern physics experiment videos as an alternative to other physics experiment.

Acknowledgment

We would like to thank for the Department of Physics at the State University of Malang in 2021 who participated in this research.

References

[1] M Suryaman *et al.* 2020 COVID-19 pandemic and home online learning system: does it affect the quality of pharmacy school learning? *Syst. Rev. Pharm.* **11** no. 8, pp. 524–530.

[2] A Mohamad, Akhmad, M., S., Habib, Z. 2020 *Covid-19 (Corona Virus Disease 2019)*, (Malang: PT. Cita Intrans Selaras).

[3] Z Ilmi, D C Darma, and M Azis 2020 Independence in learning, education management, and industry 4.0: habitat Indonesia during COVID-19 *J. Anthropol. Sport Phys. Educ.* **4** no. 4, pp. 63–66.

[4] K Prem *et al.* 2020 The effect of control strategies to reduce social mixing on outcomes of the COVID-19 epidemic in Wuhan, China: a modelling study *Lancet Public Heal.* **5** no. 5, pp. e261–e270.

[5] O I Handarini and S S Wulandari 2020 Pembelajaran daring sebagai upaya study from home (SFH) selama pandemi COVID-19 **8** no. 1, pp. 496–503.

[6] P Sari 2015 Memotivasi belajar dengan menggunakan E-learning **VI**, no. 2, pp. 20–35.

[7] Meda, Yuliani, et al., (ed). 2020 *Pembelajaran Daring Untuk Pendidikan; Teori Penerapan* (Medan; Yayasan Kita Menulis).

[8] A Kiray 2018 *Highlights in Stem* (U.S.A: Iowa State University).
[9] S K Handayanto 2001 Konstruktivisme dalam pembejalaran fisika *Makalah Seminar, FMIPA UM*.
[10] S K Alkarhami 2000 Sikap ilmiah sebagai wahana pengembangan unsur budi pekerti: kajian melalui sudut pandang pengajaran IPA *Jurnal Pendidikan dan Kebudayaan* **6** 27.
[11] S Arikunto 2002 *Prosedur Penelitian Suatu Pendekatan Praktik* (Jakarta: PT Adi Mahasdya).

The Improvement of Critical Thinking Skills Through Problem Based Learning Models Assisted by Trackers Video on Parabolic Movements

E Supriyanto[1]and Mustika[2]
Senior High School 3 Pandeglang
Jl. Perintis Kemerdekaan Km. 02 Labuan, Kec. Labuan, Kab. Pandeglang 42264
Email: [1]edifis.tpm71@yahoo.com and [2]mustikafisika16@gmail.com

Abstracts. This study aims to improve critical thinking skills using a problem-based learning model assisted by video tracker on parabolic movements. This research was conducted in class X Senior High School Pandeglang. This study uses one group pretest-posttest design with paired sample t-test analysis technique. Learning is done by conducting demonstration activities of parabolic movement based on problem based learning. The results of the video footage of the parabolic motion demonstration were then analyzed using a video tracker application. Based on the results of the study obtained the average score of critical thinking skills increased by 53.936% with a moderate N-Gain average (0.536) at a confidence level of 95%. These results show that learning using a tracker can improve critical thinking skills well. Based on the test results obtained signification of 0.00, it can be concluded that there are significant differences in critical thinking skills before and after learning using the tracker.

1. Introduction

Physics is often thought of as a subject that contains problem solving work by remembering formulas and performing some mathematical calculations. This phenomenon often occurs causing physics to be considered abstract and a difficult subject [1]. Physics learning must be equipped with experimental activities and fundamental in building understanding of the concepts associated with its application in everyday life. However, the availability of appropriate and adequate laboratory equipment in every high school is sometimes considered a luxury, let alone to present a good science experiment. In terms of the availability of suitable laboratory equipment, the experimental data produced is meaningless without proper processing and analysis. So, it is very important that high school students have good skills in processing and analyzing experimental data. Experimental data can be presented in various ways (multiple representations), such as by tables, graphs, or diagrams [2]. Students must be proficient in making cross-representations in order have a basis for scientific reasoning to solve physics problems [3] [4] and conduct physics experiments. Ironically, based on the case study [5], it was found that the problem solving skills of students with a multi-representational approach were still low.

Visual information, such as the chart is an easy way to display the information in a large amount to a form that is easily understood [6]. Graphs are commonly used in many gateway subjects (mathematics and science) to convey important information [7]. Students' understanding of graphs is discussed in many Physics Education Research (PER) studies [8][9]. The first is how the PER studies student difficulties in connecting the graph with the physical concept of a [9]. Graphs in the topic of kinematics help students explain the comparison between position, velocity, and acceleration versus time [9][10]. But in fact, students have difficulty in understanding the graph [10]. Some students experience problems with motion graphs even when they understand mathematical concepts [11].

In this study, in the use of video-based learning analytics as an alternative tool for students to practice the skills of critical thinking and multi-representation. Technological advances have provided many better ways to teach physics, one of which is that physical quantities can be presented interactively in a dynamic format [12]. Open Source Physics (OSP) is one of the computational modeling software that is suitable to be applied in high school to improve students' exploration abilities at the application level in the learning cycle [13]. The tracker is one of the video analysis and modeling software built with the OSP Java Framework, which is designed as a supporter to meet various needs in physics learning [14]. Tracker video analysis is free software and can be downloaded from the open source physics website. The benefits of using video analytics tracker are inferred value of video analysis is consistent with the data real-world acceleration due to gravity at Earth's surface [15].

The tracker provides data analysis tools such as graphs and tables, curve fittings, where we can understand examples of particle motion during the experiment. Behavior in interpreting is a way for students to identify and understand the main ideas contained in the information presented, and understand the relationship between ideas or ideas. Interpretations in relation to physics learning include: a) the ability to interpret verbal statements; b) the ability to interpret pictures, graphs, diagrams, and mathematical equations; c) ability to interpret different types of data; d) ability to make appropriate qualifications in interpreting data; and e) ability to distinguish contradictory conclusions from data arrays [16].

Previous studies have reported the results of video analysis Tracker to solve the problems of education Physics related to everyday life like a pendulum [17], projectiles [18], and the free fall motion [19]. Results from previous studies [18] [19] in the report that, when teachers deliver lessons using video analysis Tracker Physics and is linked to the concept of physics, the teacher got a lot of positive response from students.

Learning physics using tracking video analysis can be combined with learning models. One model of learning that can improve the ability of students to find the answer to a question is m model problem based learning. The main principle of the problem based learning model is the existence of a problem, student-centered learning, and learning in small groups.

Based on the things mentioned above, a research was conducted that aims to determine the improvement of students' critical thinking skills through the use of Problem Based Learning-based video trackers on Parabolic Motion material.

2. Method

The population of this study, namely all students of class X Senior High School Pandeglang who were distributed into 8 study groups in the odd semester of the 2020/2021 academic year. The sample in this study is class X MIPA totaling 35 students. The design used in this study is the One Group Pretest Posttest Design. The class will be given a pretest to measure the students initial abilities. Then, given the treatment by using a video tracker based on problem based learning on the parabolic motion material. Then the class was given a posttest in the form of the same questions as the initial test. To measure the difference in the results of the critical thinking skills assessment, students conducted pretest and posttest. The learning tools, namely syllabus, lesson plan, worksheet, were validated by constructs by experts. The question instrument consists of 10 questions with reasoned descriptions and tests the validity and reliability of the students of Class X Senior High School Pandeglang.

Research data in the form of the use of a video tracker of parabolic motion learning of critical thinking skills obtained by calculating the N-Gain between pretest and posttest through the equation:

$$g = \frac{S_{post} - S_{pre}}{S_{maks} - S_{pre}}$$

Table 1. Criteria for Interpretation of N-Gain

N-Gain (g)	Interpretation Criteria
0.7 < g	High
0.3 ≤ g ≤ 0.7	Medium
g < 0.3	Low

Normality test and hypothesis testing were carried out using statistical tests. Hypothesis testing using Paired Sample T-test.

3. Results and Discussion

3.1. Instrument Trial

This research was conducted from September to October 2019 at Senior Hight School Pandeglang in the class X MIPA as many as 35 students. Prior to use in the implementation phase of the research instrument pretest and posttest tested to determine the feasibility of it. The question instrument was tested on 35 respondents. Based on the results of the validity and reliability test, of the 10 questions that were tested for validity, only 8 questions had a Pearson correlation > 0.361 and in the reliability test, 8 valid questions were obtained with a reliability value of Cronbach's Alpha 0.674, meaning it has a high degree of reliability. So that the question instrument is feasible to use.

3.2. Application of Video Tracker Analysis in Learning

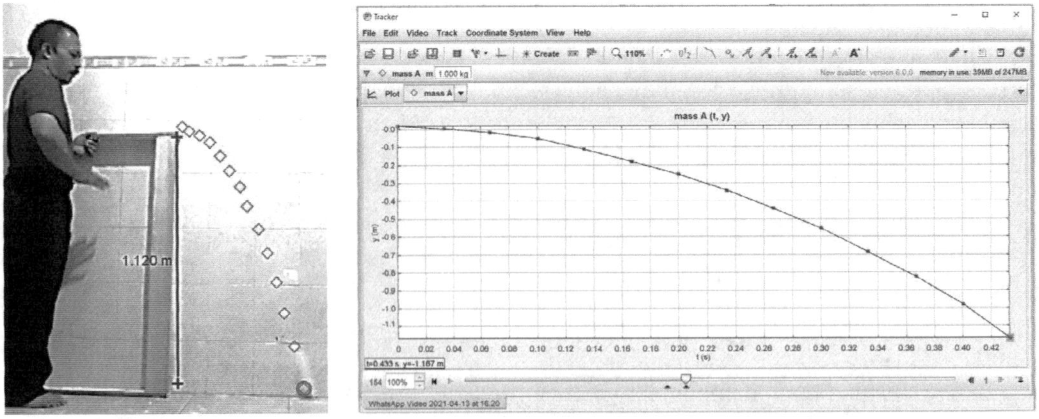

(a) (b)

Figure 1 (a) Demonstration of parabolic motion, (b) Parabolic motion graph.

In Figure 1a, the teacher demonstrates a parabolic motion by throwing a ball from a certain height in a horizontal direction. The activity was recorded using a camera. The recorded video was analyzed using a video tracker. The results of the ball motion analysis using a video tracker are shown in Figure 1b. The shape of the trajectory of the ball is a parabolic graph.

Journal of Physics: Conference Series **2019** (2021) 012029 doi:10.1088/1742-6596/2019/1/012029

Table 2 shows the summary data for mass A, namely time (s), position on the *x*-axis (m), position on the *y*-axis (m), and velocity components in the x-axis (v_x m/s), and velocity components in the direction of the y-axis. (v_y m/s)). Based on Table 2 it is known that the magnitude of the velocity component in the direction of the x axis (v_x) from 0.100 second to 0.400 second has the same relative value, which is 1.2 m/s. This means that the velocity component in the direction of the x-axis of parabolic motion is constant and the trajectory is straight or can be said to be moving in a straight line. The velocity component in the sumby y direction (v_y) from 0.100 second to 0.400 second has a greater value with the increase in speed every time interval of almost the same magnitude or constant acceleration. This means that the velocity component in the direction of the y-axis of the parabolic motion is uniformly changing or can be said to be moving in a straight line and changing uniforms.

Table 2. Summary of Mass Data A

t (s)	x (m)	y (m)	v_x (m/s)	v_y (m/s)
0.100	0.102	-5.09	1.337	-1.400
0.133	0.144	-0.110	1.274	-1.910
0.167	0.187	-0.178	1.337	-2.101
0.200	0.233	-0.250	1.209	-2.419
0.233	0.267	-0.339	1.209	-2.863
0.267	0.314	-0.441	1.337	-3.183
0.300	0.356	-0.552	1.274	-3.631
0.333	0.399	-0.683	1.210	-4.075
0.367	0.437	-0.823	1.273	-4.392
0.400	0.484	-0.976	1.273	-5.157

3. 3. Research Data Analysis

Quantitative data on the results of the pretest carried out at the beginning of the lesson, while the results of the posttest carried out at the end of the lesson can be seen in Table 3. The results of the N-Gain value is shown in Table 3. Based on Table 4, it is known that there is an increase in the average value of the pretest and posttest of 23.97, from 56.89 to 80.86.

Table 3. Results pretest and posttest Critical thinking skills

Parameter	Pretest	Posttest
Total students	35	35
Lowest value	35.00	70.00
The highest score	76.00	93.00
Maximum value	100.00	100.00
Average score	56.89	80.86

Table 4. Data *N*-Gain Percent Critical Thinking Skills

Parameter	Information
Highest *Gain*	58,652
Lowest *Gain*	48,576
Average *Gain*	53.614
Increase in average scores	53.936%
Average *N-Gain*	0, 536
Interpretation Criteria	Medium

Based on Table 4, the value of N-Gain obtain an increase of 53.936% with an average N-Gain 0,536 which shows in the medium category. Based on the category of effectiveness interpretation N-Gain is known that the average value of n-Gain amounting to 53.614 are at intervals of 56 to 76 with less affective category.

Based on the average critical thinking skills data in Table 5, it is known that the highest N-Gain on the indicator determines the name of the variable at coordinates (x,y) of 0.80 and the lowest N-Gain on the indicator determines the independent variable and the dependent variable into the graph. of 0.43.

Table 5. Average Data Critical thinking skills

Indicator	*Pretest*	*Posttest*	*N-Gain*
Identifying graphs from data	39.43	86.29	0.77
Determine the independent variable and the dependent variable in the graph	10.29	49.14	0.43
Specifies the name of the variable at the coordinates (x, y)	5.14	81.14	0.80
Determining the value of a physical variable/quantity from the graph	4.00	69.71	0.67
Determine the relationship between variables	12.00	77.71	0.75

Table 6. Tests of Normality

	Kolmogorov-Smirnova			Shapiro-WI k		
	Statistics	Df	Sig.	Statistics	df	Sig.
Pretest Score	0.129	35	0.154	0.970	35	0.431
Posttest Score	0.122	35	0.200	0.973	35	0.531

Based on Table 6 in the data normality test is known that the significance of pretest sig. (0.154)> α (0.05) and posttest sig. (0.20 0) > α (0.05), so that both the data normally distributed.

Table 7. Paired Samples Test

		mean	Std. Deviation	Std. Error Mean	95% Confidence Interval of the Difference Lower	Upper	t	df	Sig. (2tailed)
Pairs 1	Pretest Score - Posttest Score	-14.257	9.992	1.689	-17.690	-10.825	-8.441	34	0.000

Hypothesis testing in this study used the paired sample t-test technique. Test paired sample t-test aims to determine differences in average-average results of critical thinking skills before and after being treated. After being given treatment, there was an increase in students' critical thinking skills. This is based on the *N-Gain value* with moderate criteria, meaning that learning using a problem based learning based video tracker can improve critical thinking skills well. Based on the results of hypothesis testing using a paired sample T-test to see the difference in average critical thinking skills before and after the learning using video tracker in Table 7 obtained value *sig.* (0.00) < α 0.05. So it can be said that there is a difference average-average critical thinking skills before and after the learning using video tracker.

This is in accordance with the research conducted by Fitriyanto and Sucahyo which stated that the analysis of the pretest and posttest question sheets, learning using a video tracker improved skills with a medium gain category [19]. So it is clear that learning using tracker can improve students' critical thinking skills. This is because by using a video tracker, you can display results in the form of images, tables, graphs and find the equations of motion of objects with great precision. The tracker has the ability to track the motion of an object, so that various information needed in the analysis of a motion event can be obtained [20]. After going through the activity of recording a real motion phenomenon using a video recorder, the recording results can be processed on the tracker application. One use of video tracker contributes to practicing the ability to interpret or interpret a representation of data. The interpretation of the data that can be seen on the video tracker is the interpretation of images, tables, graphs and equations of motion [19].

In physics learning, interpretation includes the ability to interpret pictures, graphs, diagrams, mathematical equations, and the ability to interpret various types of data [19]. Based on the test results, students' critical thinking skills have increased. The result is consistent with the understanding of the link between verbal and visual capabilities to improve knowledge of the concept and the process of science. The following are the results of the critical thinking skills test on the pretest and posttest scores. The very significant difference in the average pretest and posttest indicators is supported by student activities in the learning process in class. Activities carried out by students ranging from observing a problem, conducting live demonstrations, recording demonstrations, analyzing demonstration data using a tracker application, for communicating demonstration results. Based on the results of the attitude assessment, which is 82.5%, it shows that student activities during the learning process are active. This is because the tracker application is relatively new to be applied in physics learning, so students feel very interested in knowing the tracker application in various physics demonstrations [19].

Learning activities emphasize the process of thinking critically and analytically to find the answer to a problem for yourself. This learning leads to a problem based learning model that focuses more on real life problems and is meaningful for students. Students must be able to use and integrate the knowledge that they already have or seek the knowledge needed in order to overcome the problem.

Problem based learning is able to provide opportunities for students to be active in interacting, reflecting, and taking the initiative in discussion activities. (1) Students better understand the concepts being taught because they themselves discovered the concept; (2) involves actively solving problems and demands students' higher thinking skills; (3) embedded knowledge based on students' schemas so that learning is more meaningful; (4) students can feel the benefits of learning because the problems that are solved are directly related to real life, this can increase students' motivation and interest in the material being studied; (5) make students more independent and mature, able to give aspirations and accept the opinions of others, instill positive social attitudes among students; and (6) student conditioning in group learning that interacts with students and their friends so that the achievement of student learning mastery can be expected.

Students conduct critical thinking at the time of air discussions by analyzing data from the results of the demonstration. Students are guided by the teacher to read the table and compare it with the graph formed on the tracker. Based on this stage, students can know that the graph is a representation of the table data. Skills of critical thinking that exist on items divided into five indicators, namely (1) identify the graph of the data, (2) determine the independent variables and the dependent variable in the graph, (3) determine the variable name at coordinates (x, y), (4) determine the value of the physical variable/quantity on the graph, and (5) determine the relationship between variables. The average critical thinking skills on each indicator after learning can be observed in Table 5.

Based on Table 5. It is shown that the highest critical thinking skill indicator is indicating (1) with an average value of 86.29. Critical thinking skills are high due to the students have been able to identify the shape of the graph parabolic motion there are currently analyzing the data demonstration using the video tracker, explaining the amount indicated, and understand the relationship between variables so that students can take information from a chart very well. Critical thinking skills are defined as the ability to translate from graphs to verbal expressions. Students can extract the information from graphs they need to solve problems and make different types of interpretations or focus on different aspects of graphs [19].

Critical thinking skills in indicator (2) show the lowest critical thinking skill indicator, namely determining the independent variable data and the dependent variable in the graph, obtaining an average value of 49.14. This is because students have difficulty understanding between the independent variables and the dependent variable contained in the critical thinking skills test questions.

Meanwhile, the results of Subekti's research show that there is a significant increase in learning outcomes of cognitive aspects of physics and science process skills in terms of initial physics abilities in the problem based learning model through the experimental method [19]. The science process skills studied are conducting research/experiments such as carrying out experiments related to the subject of parabolic motion in accordance with the independent variables, controlled variables and dependent variables that have been determined and can test the hypotheses that have been made. This process skill is not understood by students so they cannot solve the problem if the variable is changed.

Critical thinking skills in indicator (3) determine the name of the variable in the coordinates (x,y) to obtain an average value of 81.14. This indicator is in the second rank of the five critical thinking indicators observed. This happens because students can determine the name of the variable in the coordinates (x,y) which is presented on the parabolic motion graph in the video tracker analysis. The graph shown in the video tracker analysis is a vertical line that shows the deviation axis (y) and the horizontal axis which is the time axis (t). The use of video tracker analysis also shows the ability to interpret the equations of motion so that students can find out the mathematical interpretation.

Critical thinking skills in indicator (4) determine the data value of the variable obtaining an average value of 68.71. This value is the second lowest value because students have not been able to determine the value of physics quantities on the graph variables presented to be able to find the required data information. If the concepts and interpretations are not good, then students will find it difficult to solve parabolic motion graph problems. Prasetya argues that it is impossible to draw and think critically without understanding the concept of the quantity in question. Therefore, to determine the variables of a graph, an understanding of the concept is needed and finding the required information [19].

Critical thinking skills in indicator (5) determine the relationship between variables obtaining an average value of 77.71. This is because students already understand the relationship between variables so that students can take the information from the graph very well. However, students have not been able to describe it in graphical form. Therefore, this indicator is in third place.

4. Conclusions

The conclusions of this study are learning to use a video tracker -based problem-based learning in a parabolic motion effectiveness, improve critical thinking skills of students with a significance of 0.00. Achievement of critical thinking skills of students learning parabolic motion on problem based learning to identify the graph of the data reaches N-Gain 0.77; the ability to determine the independent variable data and the dependent variable into the graph reached N-Gain 0.43; the ability to determine variable names in (x, y) coordinates reaches N-Gain 0.80; the ability to determine the data value of the variables reached an N-Gain of 0.67 and the ability to determine the relationship between variables reached an N-Gain of 0.75.

Video-based learning using the Tracker video analysis is very good to be developed in learning Physics. This method could be an alternative solution though experiments do direct activity in the absence of a complete laboratory equipment and adequate. The tracking software provides facilities for the representation of some experimental data. Video tracking analysis using Tracker software can train students' multiple representation skills in the context of Physics.

Based on the research that has been done, it is suggested that a sovereign, when teachers or researchers else wants to train critical thinking skills, then can use the video analytics tracker as an alternative to learning physics. A video tracker analysis is used as a way to process data to get graphs, tables, and equations of motion. The video tracker analysis has very high accuracy. Video recording in analysis uses trackers to practice graphic interpretation skills. Students must gain experience in processing their respective data even though the implementation is carried out in groups.

References

[1] Funda O, William R dan Mark P 2008 *Int J Environ Sci Educ* jilid 3 (1):30–4.
[2] Alan V 2001 *Am J Phys* jilid 69 pp 1139–46
[3] David R, Alan. V dan Eugenia E 2009 *Phys. Spesifikasi Rev. Puncak-Fisik. Pendidikan Res* jilid 5 pp 1–13.
[4] Alan V 1991 *Am. J.Phys* Vol. 59 (10) pp 891–7.
[5] Almira A, Fourier D dan Parlindungan S 2016 Seminar Kontribusi Fisika
[6] Stephen B 2011 *Penulisan akademis untuk studi internasional bisnis* (New York: Taylor & Francis)
[7] Zucker A, Kay R dan Staudt C 2014 *J. Sci. Educ.Technol* pp 23 441
[8] Leelakrishna R. 2016 *Int. Conf. Matematika. Sci. Technol. Educ.* 377
[9] Susac A, Bubic A, Kattoji E, Planinic M, dan Palmovic M. 2018 *Phys. Pdt. Phys. Educ. Res.* 14
[10] Zavala G, Tejeda S, Barniol, dan Beichner RJ. 2017 *Phys. Pdt. Phys. Educ. Res.*
[11] Cil E. 2015 *Procedia - Soc. Berperilaku. Sci.* 191 1818
[12] Zacharias Z dan O. Roger A 2003 *Am. J. Fisik* jilid 71 p 618.
[13] Wolfgang C, Francisco E dan Lyle. B 2011 *Sumber Terbuka Fisika Sains* vol. 334 pp 1077–1078
[14] Alat *Analisis dan Pemodelan Video Pelacak 2012 untuk Pendidikan Fisika* http://physlets.org/tracker/.
[15] Wee LK, Tan KK, Leong TK dan Tan C. 2015 *Phys. Educ.* 50 436
[16] Vinka Raflesiana at. all. 2019 http://jurnal.untirta.ac.id/index.php/Gravity Gravity Vol. 5 No. 1
[17] Douglas B 2008 *Pertemuan Musim Panas AAPT*
[18] Loo K, Charles C, Giam H, Samuel T dan Tat L 2012 *Phys. Pendidikan* jilid 47 pp 448–455
[19] Loo K, Kim K, Tze K dan Ching T 2015 *Phys. Pendidikan* jilid 50 pp 436–442.
[20] Nurohman, S. 2017 *Analisis Gerak Benda menggunakan Program Tracker*

The Development of Virtual Laboratory Assisted by Flash and PhET to Support Distance Learning

V Serevina and D Kirana
Department of Physics Education, Universitas Negeri Jakarta
Jl. Rawamangun Muka, Jakarta 13220, Indonesia
Email: vserevina@unj.ac.id, datukiranaaa@gmail.com

Abstract. This research aims to develop and produce a virtual laboratory, assisted by Adobe Flash and Pet simulation that is suitable for use as a learning resource and can help support physics practicum during the distance learning process on the fluid material for students of class XI in high school. The research method used is research and development method to ADDIE model (Analysis, Design, Development, Implementation, Evaluation). This virtual laboratory is made with the help of Adobe Flash applications and PhET simulations. Based on the results of needs analysis in several high schools in Jakarta and Bogor with total of 135 respondents, 69.6% of respondents stated that they did not do physics practicum during distance learning, 71.1% of respondents said they could better understand physics with virtual laboratory, 68.1% of respondents said they needed a virtual laboratory for fluid material, and 88.1% stated that virtual laboratory on fluid material practicum helped understand the material during distance learning. A feasibility test of virtual laboratory on fluid material practicum assisted by Adobe Flash and PhET simulation was validated by material and media experts using a validation test questionnaire. Then, trials were carried out by students and teachers of class XI in high school.

1. Introduction

At the end of 2019, coronavirus disease 2019 or COVID-19 appeared in Wuhan, China [1]. The emergence of COVID-19 has a major impact on human activities in all fields of life, including education. The change from face-to-face learning to distance learning is one of the impacts. In education, the Indonesian government issues a policy to limit the spread of COVID-19 by changing the teaching and learning process where is usually done face-to-face into distance learning [2].

Distance learning is learning in which teacher and student are separate and the learning uses various learning resources through the application of the principles of educational technology [3]. The absence of direct interaction and the difference in location between teacher and student are characteristics of distance learning [4]. However, during this distance learning, many schools are unable to do practicum on physics learning. Based on the results of a needs analysis survey that was carried out. 69.6% of 135 respondents stated that they did not do practicum on physics learning during the distance learning process. From Serevina and Khaerunisa's research, the need for distance learning devices with Newton's Law material on the analysis of student needs at 89.2% [13], but in this research students need for distance learning devices with fluid material at 71.1%. Learning devices are needed to help students learning physics in distance learning. Therefore, virtual practicum using a virtual laboratory can be a solution for practicum during the distance learning process.

A virtual laboratory (virtual lab) is one of the ICT-based learning processes that can be used as an alternative solution for learning with practical methods [5]. A virtual laboratory contains a simulation program for activities carried out in a computer-assisted laboratory [6]. A virtual laboratory can simulate practical activities in a laboratory-like being in a real laboratory. The existence of a virtual laboratory also makes it easier for schools that do not have a laboratory. Based on the results of the needs analysis survey that has been carried out, 71.1% of 135 respondents felt they can better understand physics learning with the virtual laboratory. The materials that are used to develop the virtual laboratory are static fluid and dynamic fluid. Fluids materials are closely related to everyday life and there are rarely practical tools for fluid materials available [7]. Based on the results of the needs analysis survey, 68.1% of 135 respondents felt they needed a virtual laboratory for fluid practicum. Then, 88.1% of 135 respondents felt that the existence of a virtual laboratory for practicum helped to understand fluid learning materials.

The applications used to help develop the virtual laboratory in this research are Adobe Flash and PhET Simulation. One of advantages of Adobe Flash CS Professional is that it is the most flexible program in creating animations, such as interactive game animations, company profiles, presentations, movies, e-cards, and animations used in websites [8]. Peat contains abstract physics animations or cannot be seen by an open eye [9]. From the analysis survey, 25.6% of 135 respondents felt that having a virtual laboratory for practicum made it easier to understand learning materials during the distance learning process, and 73.3% felt that having a virtual laboratory for practicum was able to make learning fun during the distance learning process, and 65.9% felt that having a virtual laboratory for practicum made them more interested in learning during this distance learning process. Based on that reason, this research was carried out with the title "Development of a Virtual Laboratory in Adobe Flash-Assisted Fluid Material Practicum and PhET Simulation to Support Distance Learning".

2. Research Method

The research method used in this research is research and development methods with the ADDIE model. ADDIE is an acronym for Analyze, Design, Develop, Implement, and Evaluate.

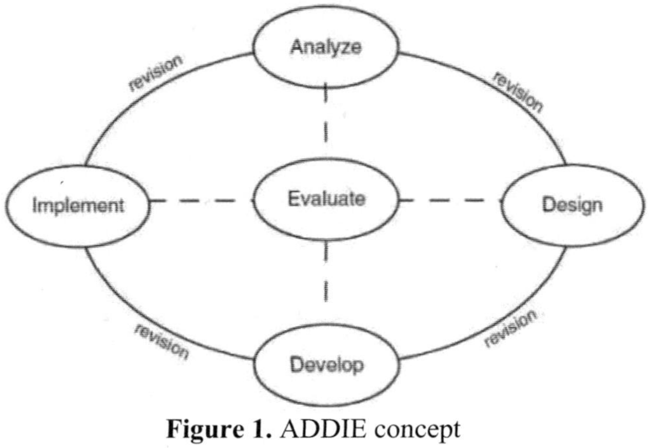

Figure 1. ADDIE concept

In the ADDIE model, there are five stages which include analyzing, design, develop, implement and evaluate (Figure 1) [10]. At the analyze stage, a needs analysis will be carried to identify the problems and needs of students. At the design stage, a design for the product will be made and created. Then at the development stage, the product will be developed and will be validated by material and media experts. At the implementation stage, the product will be tested on student and teacher in high school. After the product has been tested, student and teacher will be given a questionnaire regarding responses, criticisms, and suggestions related to the product that has been tested. At the evaluation stage, a formative evaluation will be carried out to determine the advantages and disadvantages of the product. The results of this evaluation determine whether the product has revisions or not.

3. Results and Discussion

Based on the steps of the ADDIE model, the results of the following research procedures are as follows:

3.1 Analyze

A needs analysis to identify the problems of students in learning physics during distance learning has been carried out. Based on the results of needs analysis in several high schools in Jakarta and Bogor with a total of 135 respondents, 69.6% stated that they did not do physics practicum during distance learning, 71.1% of respondents said they could better understand physics with virtual laboratory, 68.1% of respondents said they needed a virtual laboratory for fluid material, and 88.1% stated that virtual laboratory on fluid material practicum helped understand the material during distance learning.

3.2 Design

At this stage, the design of the virtual laboratory was carried out by determining the subject matter, designing the storyboard for the virtual laboratory to be developed. In this research, the subject matter that will be used fluids material.

3.3 Develop

At this stage, after the virtual laboratory was designed. Then, the virtual laboratory developed using Adobe Flash CS6 and PhET simulation. After the virtual laboratory was done. The virtual laboratory validated by material and media experts. If the virtual laboratory has flaws after being validated. Then, it will be revised. If the virtual laboratory is valid then is ready to be tested at the next stage. The following is the virtual laboratory that has been developed:

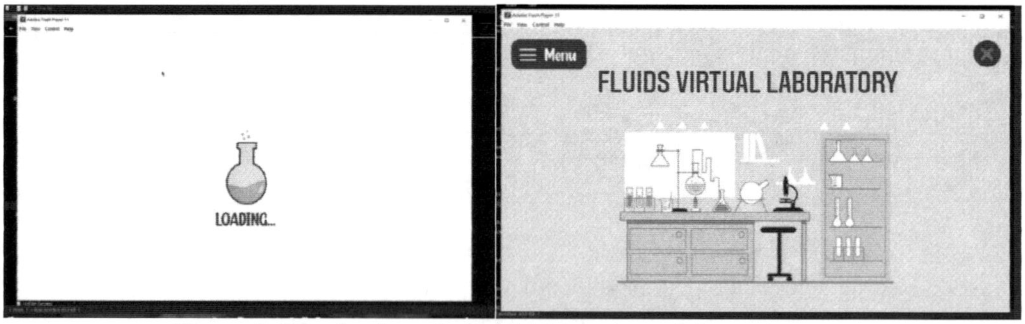

Figure 2. Introduction of virtual laboratory and the main menu

In the top left corner of the main menu, there is a tab menu. In the top right corner of the main menu, there is the exit.

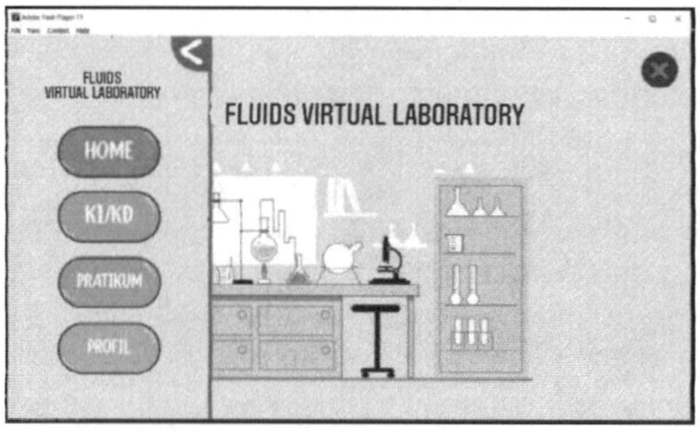

Figure 3. The tab menu

Journal of Physics: Conference Series **2019** (2021) 012030 doi:10.1088/1742-6596/2019/1/012030

The tab menu consists of home, core competencies and basic competencies (KI and KD), fluids practicum list, and profile. The following is part of the list of practicums:

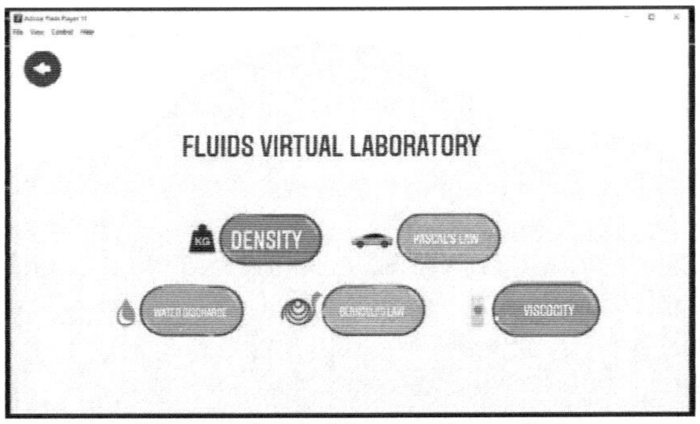

Figure 4. Fluids practicum list

The fluids practicum list consists of five practicum. One of them comes from pet simulation, shown in Figure 5. For pascal's law, water discharge, Bernoulli's law and viscosity made using Adobe Flash CS6. The animation shown is a complex animation in which the object, background and movement of the display from adobe flash can be created by the user [12]

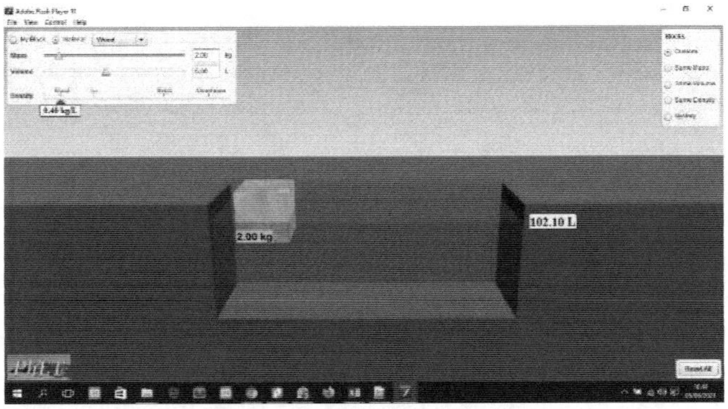

Figure 5. PhET simulation

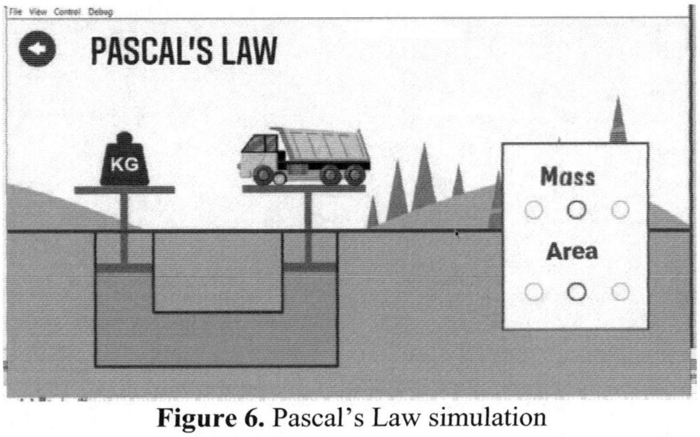

Figure 6. Pascal's Law simulation

In figure 6 is a simulation of Pascal's law on a hydraulic lift in a virtual laboratory made using Adobe Flash CS6. A virtual laboratory is a laboratory used by students in conducting experiments without the need for real laboratory equipment [11]. In this pandemic situation, students can do a practicum using virtual laboratory.

3.4 Implementation

At this stage, the virtual laboratory validated by material and media experts using a validation test questionnaire. Then, trials were carried out by students and teachers of class XI in high school tested in Jakarta.

3.5 Evaluation

At this stage, determine whether the product is suitable for fluids practicum learning media or not.

4. Conclusion

This research produced a virtual laboratory in fluid practicum assisted by adobe flash and PhET simulation to help support distance learning of class XI in high school. A virtual laboratory helps students to understand physics and do physics practicum during distance learning. This research is still in the development stage. Hope researchers can develop more perfect virtual laboratories for various other physical materials.

Acknowledgement

The researchers would like to extend their heartfelt thanks to the grade XI science students and teachers to support throughout the study. To all physics lectures in Universitas Negeri Jakarta for their continuous support.

References

[1] WHO 2020 [Online] Available: https://www.who.int/news/item/13-04-2020-public-statement-for-collaboration-on-covid-19-vaccine-development
[2] Mendikbud RI 2020 [Online] Available: https://www.kemdikbud.go.id/main/blog/2020/03/se-mendikbud-pelaksanaan-kebijakan-pendidikan-dalam-masa-darurat-penyebaran-covid19.
[3] Mendikbud RI 2014 Peraturan Mendikbud Republik Indonesia Nomor 119 Tahun 2014 Tentang Penyelenggaraan Pendidikan Jarak Jauh Jenjang Pendidikan Dasar dan Menengah, no. 1650, pp. 1–12 [Online] Available: http://luk.tsipil.ugm.ac.id/atur/bsnp/Permendikbud119-2014PJJ-Dikdasmen.pdf.
[4] Munir 2009 Pembelajaran Jarak Jauh Berbasis Teknologi Informasi Dan Kominukasi Bandung: Penerbit Alfabeta
[5] M. Masril, H. Hidayati, and Y. Darvina 2018 IOP Conf. Ser. Mater. SCI. Eng., vol. 335, no. 1, doi: 10.1088/1757-899X/335/1/012069.
[6] M. R. Hakim, Epinur, and Fuldiaratman 2017 vol. 6, pp. 5–9
[7] E. D. Jannati, A. Setiawan, P. Siahaan, C. Rochman, D. Susanti, and Y. Samantha 2019 J. Phys. Conf. Ser., vol. 1280, no. 5 doi: 10.1088/1742-6596/1280/5/052025.
[8] F. Yuliawati 2017 Trihayu J. Pendidik. Ki-SD-an, vol. 3, no. 3, pp. 129–138 [Online]. Available: http://jurnal.ustjogja.ac.id/index.php/trihayu/article/view/1874/1043.
[9] M. R. Nurulhidayah, P. H. M. Lubis, and M. Ali 2020 pp. 95–103
[10] R. M. Branch 2009 Instructional Design The ADDIE Approach
[11] P. I. Sari, G. Gunawan, and A. Harjono 2017 J. Pendidik. Fish. dan Teknol., vol. 2, no. 4, p. 176, doi: 10.29303/jpft. v2i4. 310.
[12] M. B. Darari 2017 J. Handayani, vol. 7, no. 2, pp. 33–41
[13] Serevina V, Khaerunisa N A 2021 https://iopscience.iop.org/article/10.1088/1742-6596/1876/1/012075/

Journal of Physics: Conference Series **2019** (2021) 012031 doi:10.1088/1742-6596/2019/1/012031

Review of Trends Project Based Learning (PjBL) Integrated STEM in Physics Learning

A Solihin[1], F C Wibowo and I M Astra
Department of Physics Education, Universitas Negeri Jakarta
Jl. Rawamangun Muka, Jakarta 13220, Indonesia
Email: sepsholi7@gmail.com

Abstract. For physics teachers, how physics should be taught is an important question. This study, examines project based learning (PjBL) integrated in Science, Technology, Engineering and Mathematics (STEM). The aims of this study is to identify trends in PjBL integrated STEM in physics education. The method used is a bibliometric analysis method. 80 articles were collected from the literature of scholarly journals, both nationally and internationally through the Publish or Perish (PoP) application. The articles collected were analyzed using the VOSviewer application to obtain information. After analysis, information on the trend of PjBL integrated STEM was obtained by further analysis and by synthesizing it. The result of this study is that in recent years, Project Based Learning (PjBL) integrated STEM has become one of the trends and has become an option in implementing physics learning in schools. This can be seen from the many research, studies and discussions about PjBL integrated STEM in physics education. Based on the results of this literature study, several recommendations for future research were also obtained, namely so that PjBL integrated STEM can be better, it needs to be tried by adding various learning media in it, especially on-line learning media.

1. Introduction

For teachers of physics subjects, the question of how physics should be taught is an important issue. Various studies were conducted to present the current situation, advantages, challenges, limitations and trends in the use of models and approaches in learning. Until now, in the field of science education, research studies that look at trends are still often carried out [1]. Trends in physics education can be seen by revealing what topics researchers have studied from the past to the present, what topics have received more or less attention, and how changes in interest have occurred over time. Recent physics education, research shows a sharp increasing trend in the topics of learning models, PCK, assessment, and gender [2].

The learning model used in schools, of course, must be in accordance with the demands of the applicable curriculum. Implementation of the 2013 curriculum in learning is to use a scientific approach [3]. The strategy of the scientific approach is discovery learning, project based learning and problem based learning. Project Based Learning (PjBL) is understood as a promising approach to improve student learning. Empirical studies on project based learning have been reviewed with a focus on student outcomes. Affective results, namely the perception of the benefits of PjBL and the perception of the experience of PjBL were the most widely applied, as measured by questionnaires, interviews, observations, and self-reflection journals [4]. Project Based Learning (PjBL) is an innovative learning that encourages students to conduct investigations, work collaboratively in researching and creating projects that apply their knowledge of discovering new things, proficient in the use of technology and able to solve problems. Projects should challenge students to do and complete them. In implementing project based learning, the learning environment must be designed in such a way that students can conduct investigations into real problems, including deepening a material from a subject topic and carrying out other meaningful tasks [5]. The learning steps in project based learning developed by The George Lucas Educational Foundation in [5] consist of: (1) Starting with essential questions; (2) Planning of project work rules; (3) Make a schedule of activities; (4) Monitoring student project progress; (5) Assessment of student work; (6) Evaluation of students' learning experiences.

In its development PjBL can be integrated with other approaches, such as the STEM (Science Technology Engineering and Mathematics) approach. The word STEM was launched by the US National Science Foundation in the 1990s as the theme of the education reform movement in these four disciplines to grow the workforce in STEM fields, develop STEM literate citizens, and increase US global competitiveness in science and technology innovation. STEM education provides an opportunity for teachers to show students how concepts, principles, and techniques from science, technology, engineering, and mathematics are used in an integrated manner in the development of products, processes, and systems that are used in their daily lives [6].

One pattern of integration that might be implemented without restructuring the primary and secondary education curriculum in Indonesia is to incorporate engineering, technology, and mathematics content in science learning based STEM [6]. From research [7], it is known that both male and female students showed positive and happy responses to the implementation of PjBL STEM. According to students, learning is interesting and motivating, can help understand teaching materials, form creative attitudes, and students are increasingly aware of the importance of protecting the environment. Students feel happy to work in groups so they want PjBL STEM learning to be re-applied to other materials. Even an experiment showed that after carrying out learning, the use of the project based learning integrated STEM model (PjBL-STEM) could improve student learning outcomes (cognitive and psychomotor aspects) and students' creativity in solving problems compared to the use of the PjBL model [8].

Based on the explanation that has been presented, the author tries to do a literature study. This study examines research trends in the field of physics education, particularly PjBL Integrated STEM with bibliometric analysis. Organizing and examining research trends can help us identify interesting trends from the past as well as become milestones for future research directions [1]. First, researchers can examine important issues in the field of physics education by identifying the extent to which these issues are addressed in pedagogical research. Second, information about trends in physics education research can be obtained by studying how interest in a topic has changed over time. Third, topics can be explored that have been relatively marginalized or neglected and can then be used in determining future research topics.

The review of trends PjBL integrated STEM was carried out by collecting the results of the scholarly journals literature. Representative articles were collected and data were taken using the Publish or Perish (PoP) application. The data are in the form of a collection of titles, abstracts and other data that is searched based on keywords and restrictions on the time span of the search. Once collected, the data is then processed using the VOSviwer application. VOSviewer is software for creating maps based on network data and for visualizing and exploring these maps. VOSviewer can be used to build a network of scientific publications, scientific journals, researchers, research organizations, countries, keywords, or terms. The items in this network can be linked by co-authoring, co-occurrence, citations, bibliographic merging, or co-citing links. To build a network, data from Web of Science, Scopus, PubMed, RIS, or Crossref JSON files can be used [9]. VOSviewer is used to see the relationship between documents. This trend review significantly contributes to the literature, listing articles individually and categorizing them by features. This review also discusses articles with an inductive approach by focusing on the current situation, advantages and challenges associated with STEM integrated PjBL in physics education.

In short, this study examines PjBL integrated STEM with the aim of identifying trends in PjBL integrated STEM in physics education. The method used is a bibliometric analysis method. Bibliometric analysis is applied to quantitatively measure and analyze certain indicators in the published literature in a particular domain and to generate knowledge maps based on large databases

[9]. This allows researchers to summarize published information regarding the distribution of articles by year, author, institution, journal, and discipline, collaboration between authors and institutions, and co-word analysis. The results of this study are expected to provide various information to physics teacher about the trend of PjBL integrated STEM, so that educators can find new ideas in improving the quality of learning in schools.

2. Method

This type of research is library research using bibliometric analysis methods. Bibliometric analysis is carried out by looking at the distribution of publications to evaluate the contribution of articles to the advancement of knowledge of various literatures using a statistical approach and can provide a broader understanding of the entire discipline [10]. The data collection technique of this research used secondary data. The research data is in the form of research articles related to Project Based Learning (PjBL) integrated STEM in Physics Learning. Data collection was carried out using the Publish or Perish (PoP) application with the keywords "Project Based Learning (PjBL) integrated STEM in physics learning" and the limitation of the search time span for the year of publication of the last 5 years, namely between 2017 and 2021. Then carried out data collection and obtained 719 articles. The data is then selected with a limit of articles in the form of Article, Journal, so that 80 articles are obtained to be analyzed. The data for these articles is exported in RIS format and will be processed using the VOSviwer application. VOSviwer application is a software that has the ability to visualize and explore the results of bibliometric studies.

The data analysis technique in this study refers to the results of bibliometric mapping from the VOSviewer application that applies Co-Occurrence calculations. Co-occurrence analysis reveals the research topic statistically, with the condition that the more frequent pairs between two keywords, the closer the relationship between these keywords [10]. In this analysis, VOSviewer extracts the titles and abstracts of articles obtained by the binary counting method, meaning that each word found related to either one or more topics will be considered one. The minimum number of occurrences of a term is 5, meaning that the words that will be displayed are words that have 5 relatedness. Therefore, of the 674 words found, only 31 words meet the threshold. The following is the flow of data collection and processing methods that have been carried out:

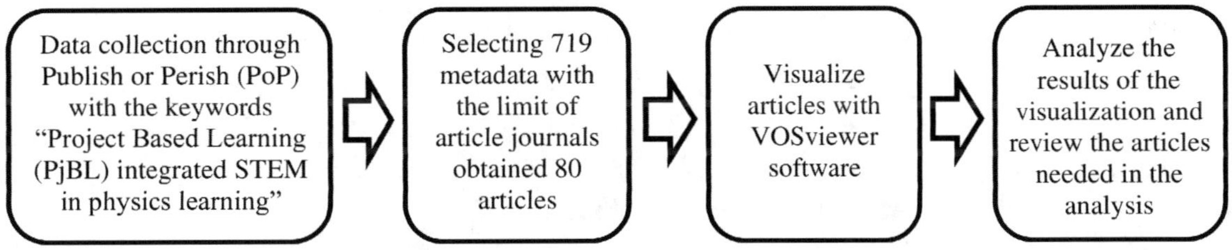

Figure 1. Data collection and processing methods

3. Results and Discussion

The results of the study obtained at the beginning of the gradual selection using the Publish or Perish (PoP) application obtained 80 articles on Project Based Learning (PjBL) integrated STEM in physics Learning. Figure 2 shows the publication of this research on physics learning in high school throughout 2017 to 2021. The number of Project Based Learning (PjBL) integrated STEM articles on physics learning throughout the year has increased. In 2021 the data seems to be declining because the data was only obtained for half a year (6 months). Articles on Project Based Learning (PjBL) integrated STEM in physics learning were most published in 2020, while the last was in 2017. This explanation shows that STEM integrated Project Based Learning (PjBL) in physics learning still has the potential to be investigated further.

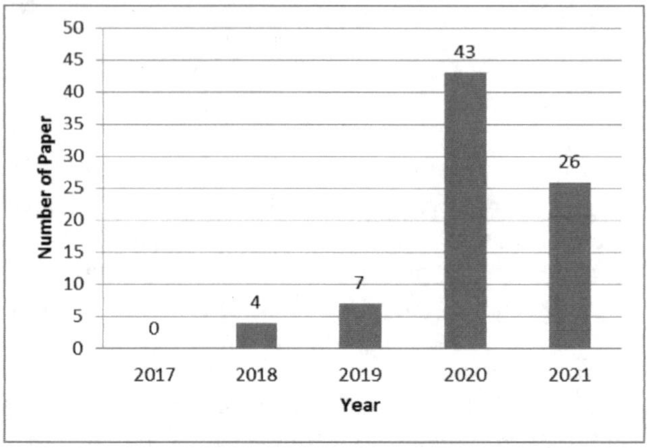

Figure 2. Publications of Project Based Learning (PjBL) integrated STEM in physics Learning research for the period 2017-2021

The results of the research on Project Based Learning (PjBL) integrated STEM in physics Learning were obtained from keyword analysis of 80 research articles using VOSviewer software. This is done to find research variables regarding Project Based Learning (PjBL) integrated STEM in physics learning. Based on the VOSviewer mapping, it can be found several parameters of the relationship between variables, which are related to learning, including design, environment, development, approach, model, implementation, effectiveness and learning outcomes. It is interesting to note, related to learning media, it is only represented by the Student Worksheet (LKPD) variable. Figure 3 shows the mapping of the results of the similarity of article keywords, obtained 4 groups (clusters) according to each color with 31 related keyword terms regarding Project Based Learning (PjBL) integrated STEM in physics Learning. Selection of keywords is then carried out according to the research objective, namely to identify trends in STEM Integrated PjBL using bibliometric analysis methods.

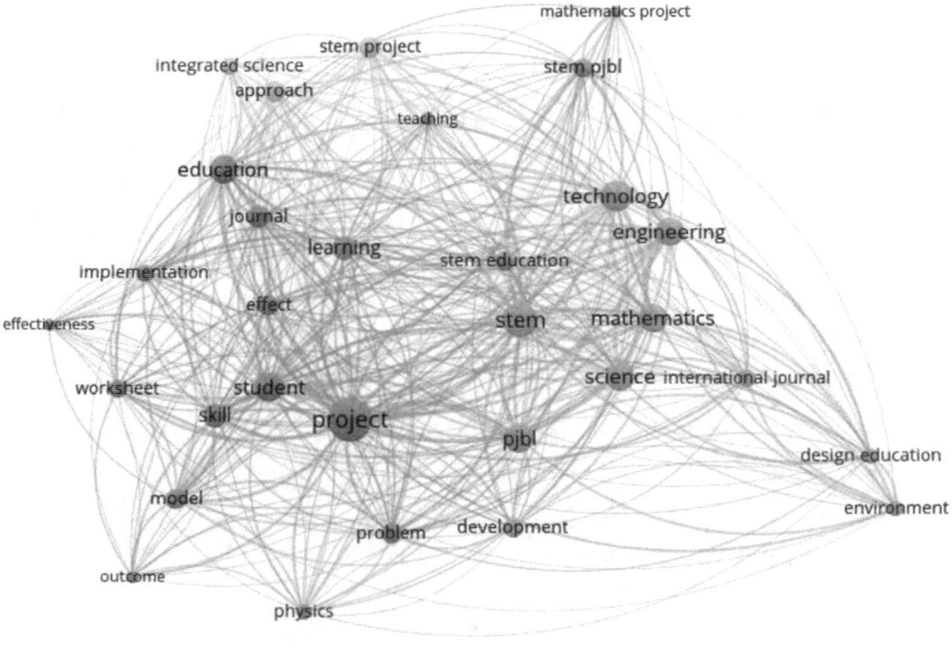

Figure 3. Network visualization map of keywords

From the results obtained, the larger the circle on a keyword indicates that the keyword is widely used by article writers and has a strong relationship with other keywords. Figure 3 has the 4 largest circles for each cluster, namely project, STEM, STEM project and learning, which shows that there is a relationship between Project Based Learning (PjBL) and the STEM approach. Even the integration of PjBL and STEM approaches or STEM based PjBL have been widely applied or it can be said that Project Based Learning (PjBL) integrated STEM is one of the preferred learning models that can be implemented among existing learning models.

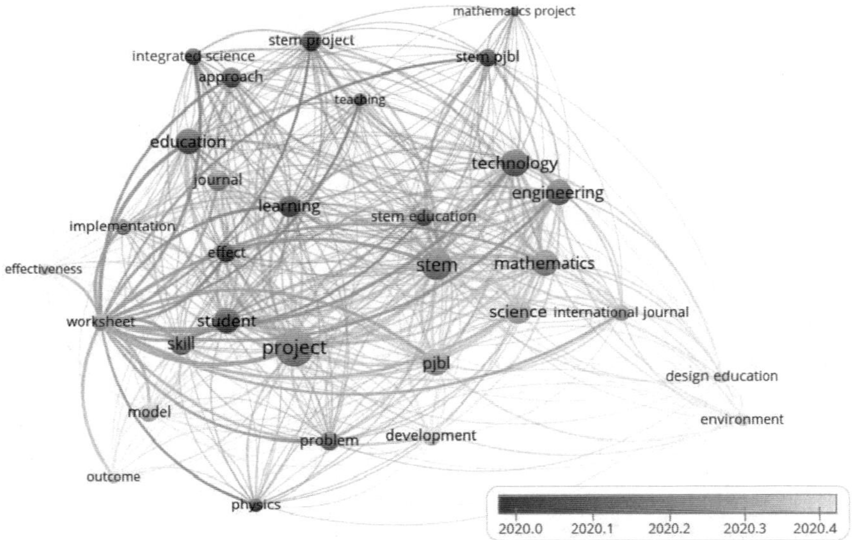

Figure 4. The relationship of learning media (worksheets) in Project Based Learning (PjBL) integrated STEM

Figure 4 shows one keyword, namely worksheets as learning media that can be used in Project Based Learning (PjBL) integrated STEM. Figure 4 also shows the use and the relationship of worksheet keywords as one of the media in learning. It can also be seen that the word worksheet was used around 2020. This worksheet as a learning medium is the only keyword found in the VOSviewer mapping. This shows that there are not too many studies in Project Based Learning (PjBL) integrated STEM that observe the influence of media in it. This is also an opportunity for further research to be able to try and find alternative learning media that can be used in Project Based Learning (PjBL) integrated STEM. Thus, the learning process and the expected results will be even better.

Learning media is a tool that can transfer messages or information from an educator to students which aims to facilitate the learning process in the classroom (Articles on media). By using learning media during the learning process in the classroom, it is hoped that the thoughts, feelings, concerns and interests of students can be raised and students can receive and understand the subject matter from the teacher well. In today's digital era, especially during the COVID-19 pandemic, an educator must be an expert in creating creative and innovative learning media. Learning media that need to be developed mainly is online learning media for the distance learning process. The development of the times will make students creative in all things so that the educational process will be balanced.

4. Conclusions

Based on the results and discussions that have been carried out, identifying trends in PjBL integrated STEM using bibliometric analysis, it can be concluded that in recent years, Project Based Learning (PjBL) integrated STEM has become one of the trends and has become an option in implementing physics learning in schools. This can be seen from the many research, studies and discussions about PjBL integrated STEM in physics education.

Based on the results of the literature study, several recommendations for future research were obtained, namely so that Project Based Learning (PjBL) integrated STEM could be better, it needs to be tried by adding various learning media in it, especially on-line learning media.

References

[1] B. Cavas 2015 *Sci. Educ. Int.*, vol. 26, no. 4, pp. 573–588
[2] E. Yun 2020 *J. Balt. Sci. Educ.*, vol. 19, no. 3, pp. 388–400
[3] N. Nurdyansyah and E. F. Fahyuni 2016 *Inovasi model pembelajaran sesuai kurikulum 2013* (Nizamia Learning Center)
[4] P. Guo, N. Saab, L. S. Post, and W. Admiraal 2020 *Int. J. Educ. Res.*, vol. 102, p. 101586
[5] N. M. Y. Suranti, G. Gunawan, and H. Sahidu 2017 *J. Pendidik. Fis. dan Teknol.*, vol. 2, no. 2, pp. 73–79
[6] H. Firman 2015 "Pendidikan sains berbasis STEM: Konsep, pengembangan, dan peranan riset pascasarjana," in *Disampaikan pada Seminar Nasional Pendidikan IPA dan PLKH Universitas Pakuan, Agustus*
[7] J. Afriana, A. Permanasari, and A. Fitriani 2016 *J. Inov. Pendidik. IPA*, vol. 2, no. 2, pp. 202–212
[8] L. M. I. Furi, S. Handayani, and S. Maharani 2018 *J. Penelit. Pendidik.*, vol. 35, no. 1, pp. 49–60
[9] S. Kurnia 2021 *Science, Technology, Engineering, Art And Mathematics (STEAM) di Pendidikan Sains: Analisis Bibliometrik dan Pemetaan Literatur Penelitian Menggunakan Perangkat Lunak Vosviewer* (UIN Raden Intan Lampung)
[10] C. F. H. Ayudha and W. Setyarsih 2021 *J. Pendidik. Fis. Undiksha*, vol. 11, no. 1

Perception of Novice Learners Using Blended Learning Approach During the Covid-19 Pandemic

F Herliana[1], A Farhan[1], Elisa[1], M Syukri[1], E Mahzum[1]
[1]Department of Physics Education, Universitas Syiah Kuala, Jl. Teuku Nyak Arief, Banda Aceh, 23111, Indonesia
Email: fitriaherliana@unsyiah.ac.id

Abstract. This study aims to determine the perceptions of students in physics education department, faculty of teacher training and education - Universitas Syiah Kuala who have just learned (novice learners) using the blended learning approach during the Covid-19 pandemic. This research is quantitative research with data collection using the form of questionnaires and interviews. The questionnaire to measure novice learners' perceptions used a semantic differential scale. The indicators measured consist of 5 aspects, including: technology, classroom management, instructors, interaction, and instruction. The population of this study was all students of the Physics Education Department, faculty of teacher training and education - Universitas Syiah Kuala. Purposive sampling technique was used to select sample and obtained 250 novice learners of blended learning approach. The data analysis technique used is descriptive statistical analysis through the process of data collection, data reduction, data presentation, and conclusions. The results of this study indicate that the perception of novice learners of the blended learning approach is positive at 87.92%. However, there is still a negative perception on each indicator with a different percentage value. The Indicator of technological at 14.52%, class management at 11.32%, instructors at 10,99%, interaction at 11,70%, and instruction at 12.60%. The results of the research are expected to be of concern for lecturers in the physics education department to find out the obstacles of students in distance learning during the pandemic period and can be used as a reference for improving and developing distance learning using a blended learning approach to make it better.

1. Introduction

The educational paradigm in this era has been forced to shift from direct learning in the classroom to online/distance learning. The COVID-19 pandemic is the issue that causes the sudden shift in the Education paradigm [1]. However, technology that continues to develop from time to time does not make this a barrier for education to change according to the current COVID-19 pandemic conditions [2]. Technological advances help all aspects of life to be able to continue even though they cannot interact directly in the same place [3] including aspects of education. The learning platform continues to be developed so that it can become a medium for student learning and communication, both with teachers and peers [4] in order to keep the learning process going. Although the learning process cannot be carried out directly in schools, with current technological advances, learning can still be carried out in different places using a learning platform and requires an internet network [5].

This change happened very suddenly where people who usually interact directly with each other, can no longer gather in large numbers. Public places, which may be a gathering place for many people, are closed for fear that they can become places with a high risk of virus transmission [6]. Schools are one of the public places that are prohibited from opening [7]. However, education must continue because education is the most important aspect of life [8] so that an appeal from the government was issued to shift learning to a distance/online learning system [9]. Almost all Indonesian people learn to use the conventional learning system, namely face-to-face learning activities in schools with a number of students from 20-40 students per class. Students and teachers are accustomed to interacting directly to meet their daily learning needs. Currently, they cannot interact directly anymore, but learning needs must still be met so that they must be able to adapt to a new learning system using a new learning approach [10].

This condition is a new thing for both teachers and students. Currently, learning can only take place online, but if it is not properly supervised and the process is carefully planned, learning will be less effective [11]. The appropriate learning approach to be applied in this condition is the blended learning approach. This approach combines face-to-face learning and independent learning assisted by digital/online media [12]. Face-to-face learning is carried out online using video conferencing applications such as zoom meeting/ google meet (synchronous) [13] and independent learning assisted by digital/online media is carried out using learning platforms such as e-learning/ google classroom (asynchronous) [14]. The combination of these two learning environments is a learning process that can be applied by teachers and students for the sake of continuity of education in this era of the covid-19 pandemic. By applying this approach, students get new experiences in the learning process because they have never experienced this condition before. New learners using this approach because of the paradigm shift in education in the era of the COVID-19 pandemic can be called novice learners using a blended learning approach [15].

The success of a learning process can be seen from the perceptions of students who experience the learning process and their learning outcomes [16]. Likewise with this blended learning approach, because this approach is a solution offered to overcome the problem of the education paradigm shift in the era of the covid pandemic, the perception of students who are just learning to use this approach needs to be measured in order to see the effectiveness of the application of this approach.

2. Method

This research is quantitative research with data collection techniques using questionnaires and interviews. The population of this study was all students of the Department of Physics Education, FKIP - USK, and the samples used in this study were taken using a purposive sampling technique that only focused on students who had just experienced distance learning based on blended learning (novice) totaling 250 students. The questionnaire used as an instrument to measure student perceptions uses a differential semantic scale with points 1 to 7. The indicators measured consist of 5 aspects, including technology, classroom management, instructors, interaction, and instruction [17] with a total of 43 items after being validated and measuring the reliability using Cronbach's alpha. The reliability test results show the instrument has high reliability of 0.96. The grid of the questionnaire instrument to measure the perceptions of novice learners using a blended learning approach can be seen in Table 1 below:

Table 1. The grid of novice learners perception using blended learning approach

No	Aspect	Indicator	Description	No Item	Total Item
1	Technology	ICT Skills	The lecturers skills to use ICT in Blended Learning	1,2	2
		Blended Learning Facility	Availability of facilities that support the implementation of blended learning	3,4,5	3
2	Manajemen Kelas	Lesson Plan	Preparatory activities carried out by lecturers and students before the start of the learning process	6,7,8	3
		Managing Class	Activities carried out by lecturers by involving students in the learning process	9,10,11,12,13	5
		Learning Evaluation	Evaluation activities for the results of the learning process carried out by lecturers for students in the form of assignments/exams	14,15,16,17,18,19,20	7
3	Instructor	Frequency of using blended learning applications	How often do lecturers use learning platforms for activities from preparation, teaching, to responding to student questions and opinions in chat forums)	21,22,23,24,25	5
		Give Motivation	The behavior shown by the lecturer in motivating students	26,27	2
		Discipline	Behavior that shows the discipline of the lecturer in the learning process	28,29,30,31,32	5
		Material Mastery	The ability of the lecturer to explain the material with various media, answer questions, and relate the material to developing issues	33,34	2

No	Aspect	Indicator	Description	No Item	Total Item
4	Interaction	Skills in discussion	The ability of lecturers to manage discussion rooms between lecturers and students, or among fellow students	35,36	2
		Application use for online discussion	Student satisfaction with the devices used by lecturers for online discussions	37	1
5	Instruction	Instructions in the Blended Learning	How lecturers involve students in the Blended Learning Process	38,39,40,41,42	5
		Satisfaction with learning achievement	Student satisfaction with the blended learning process that has been carried out	43	1
			Total Item		43

The data analysis technique used is descriptive statistical analysis in the form of percentages using the equation:

$$P = \frac{Total\ score\ each\ item}{\text{Maximum score}} \text{ x } 100\ \%$$

Then the percentage results are compared with the criteria for interpretation of novice learning perceptions using a blended learning approach to determine the measurable percentage result categories by referring to the following criteria table:

Table 2. Interprate of novice learners' perception using blended learning approach

No.	Range Score	Criteria
1	86-100	Very Good
2	76-85	Good
3	60-75	Fair
4	55-59	Poor
5	≤54	Very Poor

In addition, an analysis of the interview results was also carried out. The final results of the analysis were obtained through the process of data collection, data reduction, data presentation, and drawing conclusions.

3. Result and Discussion

Novice learners in this study were students who did not have experience learning using a blended learning approach. The application of this approach was experienced by novice learners for the first time and the results obtained by novice learners after using the blended learning approach as shown in the graph below:

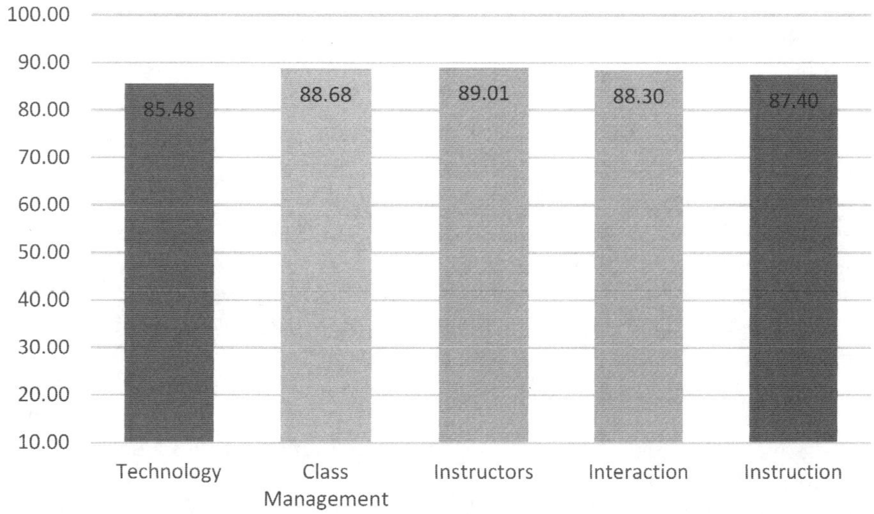

Figure 1. Novice Learners' Perceptions Using Blended Learning Approach

Of the five aspects that measure students' perceptions after using the blended learning approach, it shows that novice learners have positive perceptions of 87.92%. The highest perception is found in the aspect of instructors at 89.01%. This is caused by several factors, including the frequency of lecturers using LMS in the application of the blended learning approach, the way lecturers provide motivation to learn, the suitability between the instructions given and the discipline of the lecturers in the learning process, and the mastery of lecturer materials in giving instructions using the blended learning approach. In addition, the lowest perception is on the technology aspect at 85.48%. This is caused by several factors, including the skills of lecturers and students in using IT and facilities owned by lecturers and students to learn using a blended learning approach. The percentage of novice learning perceptions using a blended learning approach in detail can be seen in the following table:

Table 3. Percentage of novice learning perceptions using a blended learning approach per indicator

No	Aspect	Indicator	Percentage of Novice Learners Perceptions (%)
1	Technology	ICT Skills	87.00
		Blended Learning Facility	83.96
2	Class Management	Lesson Plan	88.51
		Managing Class	89.63
		Learning Evaluation	87.88
3	Instructors	Frequency of using blended learning applications	90.00
		Give Motivation	88.54
		Discpline	87.43
		Material Mastery	90.06
4	Interaction	Skills in discussion	90.43
		Application use for online discussion	86.17
5	Instruction	Instruction in the Blended Learning	88.29
		Satisfaction with learning achievement	86.51

3.1. Technology

The data above shows that 87% of students and lecturers have excellent IT skills. This is indicated by the implementation of teaching and learning activities without being present directly in the campus environment. All students and lecturers carry out lectures in their respective homes in accordance with government recommendations which prohibit direct learning activities in the university environment during the COVID-19 pandemic [18]. Learning activities are carried out using the help of information and communication technology entirely in this era [19]. In addition, novice learners' perception of this distance learning facility is 83.96%. This figure shows the perception of novice learners using blended learning is quite good for the facilities used in learning. However, this figure is the lowest percentage of all indicators measuring novice learners' perceptions using blended learning. Most students consider the learning facilities to support the implementation of this blended learning approach, such as complete lecture materials that they can access in e-learning/ google classroom, the availability of a zoom meeting/ google meet platform without a time limit for face-to-face lectures, and a chat application for discussions when only through WhatsApp-group/ discussion forum in e-learning. Novice learners' are quite helped by the learning facilities prepared by the lecturers because these are the main elements needed in learning [20].

In addition, in distance learning using this mixed approach, the strength of the internet network is a major factor in the effective running of learning activities [4] because the entire process of learning activities are in a network, both virtual face-to-face learning using applications video conference (synchronous) [13] as well as independent learning assisted by learning platforms or learning management systems (asynchronous) such as e-learning, google classroom, or Microsoft teams [14]. Mixed learning between these two learning environments (synchronous and asynchronous) requires lecturers and students to have the skills to use IT and good learning facilities for effective learning [21]. Based on the data above, the perception of novice learners' using blended learning shows that they have very good IT skills and good learning facilities, so it can be concluded that learning using the blended learning approach is effective for novice learners who are just learning to use this approach from the technology aspect.

3.2. Class Management

Novice learners' in this study are students who are just learning or experiencing the learning process using a blended learning approach. Previously, they generally studied using a conventional approach with a direct instruction model in the classroom [22]. In this era, students are required to learn in 2 environments, namely synchronous and asynchronous to optimize the results achieved from the learning process [23]. The application of blended learning also requires class management like other approaches in general, from planning to evaluation. The perception of novice learners using blended learning on class management carried out on the application of the blended learning approach is classified as very good. This is illustrated by the high percentage of each indicator, such as planning, process, and evaluation of lectures. Students admitted that almost all lecturers had very good planning in their lectures because they showed lecture contracts at the first meeting of lectures, provided references to books and media that would be used during lectures, and explained the lecture system that would be used during lectures. This shows that lecturers prepare/plan their lectures very well because the initial planning determines the process and results achieved in the lecture [24]. Students will be better prepared to attend lectures if they begin with a clear plan [25].

The learning process using the blended learning approach is also considered very good by students with the ability of lecturers to manage the class by 89.63%. Novice learners using blended learning argue that they experience a structured lecture process. Lecturers not only give assignments but carry out face-to-face lectures virtually and provide complete lecture materials on the learning platform. The learning process runs very dynamically using various strategies, such as face-to-face using video conferencing, sending text messages via chat forums, providing lecture materials (PPT, PDF, animation) through e-learning, and students are given the opportunity to ask questions and discuss anytime. through the platform provided by the lecturer. Merging all these activities is a strategy used in the blended learning approach [26]. Varied learning media can increase students' learning motivation [27] so that students take learning seriously and achieve maximum results [28]

At the end of the lesson, novice learners' using blended learning said they got assignments to be completed independently/in groups and uploaded them into the learning management system (e-learning/ google classroom). Mid-semester exams and end-of-semester exams are also given by lecturers through LMS with a transparent scoring system. Evaluation is the last stage in the learning process that is indispensable in classroom management [29]. Without evaluation, learning cannot be measured validly [30]. Novice learners using blended learning perceptions of the planning, process, and evaluation of learning are very good, so it can be concluded that the class management approach of blended learning is very good for novice learners using blended learning.

3.3. Instructors
This aspect is the aspect that has the highest percentage of perception compared to other aspects, which is 89.01%, meaning that novice learners using blended learning are very satisfied with the instructors/lecturers who guide them in learning. The role of the instructor in learning is one of the factors that greatly influence the success of a lesson [31]. Especially in this blended learning approach, where the instructor is tasked with fully facilitating learning [32]. If the instructor is not good at preparing, managing, and evaluating learning, then this approach will not work well [33].

Novice learners argue that lecturers have a high frequency of accessing learning platforms to teach or just responding to student questions on chat forums. Although not all lecturers do the same thing, most do this. Lecturers attend and end lectures on time, just like lectures in class. In addition, lecturers are also able to deliver material using various digital media that make it easier for students to receive lecture material. The role of this instructor is equally important in both conventional and blended learning approaches [34] there is no difference of both.

3.4. Interaction
In the learning process using a blended learning approach, learning should not only run in one direction (from teacher to student). Learning must run in two directions, namely from teacher to student and from student to teacher [35] so that interaction is needed in learning in this approach [36]. The perception of novice learners' using blended learning on the interaction aspect is 88.30% with discussion proficiency of 90.3% and the use of applications for discussion is 86.17%. This figure shows the perception of novice learners using blended learning is very good in this aspect. Novice learners admit that lecturers provide time for students to discuss during virtual face-to-face lectures and outside class hours using chat forums in the learning management system so that students can interact with lecturers and their peers through this media [37]. The existence of platforms such as zoom meetings, WhatsApp, chat forums, etc. really helps lecturers and students to keep interacting even though they don't meet face to face in class. This does not reduce the essence of the interaction needed in the learning process because the interaction can continue because of the media that connects it.

3.5. Instruction
Instruction in the learning process is another important factor that affects the outcome of a learning process [38]. The clarity of the instructions given determines the direction of the learning process [36]. If the instructions given are not clear enough, students will experience confusion and reduce their motivation in learning [39] so that the results achieved by students at the end of learning will not be optimal [28]. After using the blended learning approach, students felt that the instructions/instructions given by the lecturers during lectures were very clear and well received by students. This is indicated by the perception of novice learners using blended learning in the instruction aspect of 87.40% which is classified as very good. Therefore, it can be concluded that students can follow lectures very well using a blended learning approach during this covid-19 pandemic.

4. Conclusion
Novice learners have a positive perception of learning using a blended learning approach. This positive perception shows that the blended learning approach is quite effective during the COVID-19 pandemic in the Department of Physics Education, FKIP USK. However, there are still negative

perceptions in every aspect that is measured. The largest percentage of negative perceptions is found in aspects of technology and instruction. This shows that there are still obstacles faced by novice learners in terms of the technology used and the instructions given during the learning process. This obstacle must be a special concern for the lecturer of the Department of Physics Education, FKIP USK to be able to find a solution so that learning can run effectively and efficiently.

References

[1] Mondol, M. S., & Mohiuddin, M. G. (2020) *International Journal of Social, Political and Economic Research*, *7*(2), 231-247.
[2] He, W., Zhang, Z. J., & Li, W. (2021) *International journal of information management*, 57, 102287.
[3] Huang, R. H., Liu, D. J., Tlili, A., Yang, J. F., & Wang, H. H. (2020). *Handbook on facilitating flexible learning during educational disruption: The Chinese experience in maintaining undisrupted learning in COVID-19 Outbreak* (Beijing: Smart Learning Institute of Beijing Normal University)
[4] Herliana, F., Astra, I. M., Supriyati, Y., & Mazlina, H. (2020) *Journal of Physics: Conference Series* (Vol. 1460, No. 1, p. 012125). IOP Publishing.
[5] Schneider, S. L., & Council, M. L. (2020). *Archives of dermatological research*, 1-2.
[6] Nagaj, R., & Žuromskaitė, B. (2021) *Energies*, 14(7), 2000.
[7] Onyema, E. M., Eucheria, N. C., Obafemi, F. A., Sen, S., Atonye, F. G., Sharma, A., & Alsayed, A. O. (2020) *Journal of Education and Practice*, 11(13), 108-121.
[8] National Research Council. (2012) *Education for life and work: Developing transferable knowledge and skills in the 21st century* (National Academies Press)
[9] Rahiem, M. D. (2020) *International Journal of Learning, Teaching and Educational Research*, 19(6), 1-26.
[10] Lockee, B. B. (2021) *Educational Technology Research and Development*, 69(1), 17-20
[11] Bacow, L. S., Bowen, W. G., Guthrie, K. M., Long, M. P., & Lack, K. A 2012 *Barriers to adoption of online learning systems in US higher education* (pp. 39-51) (New York, NY: Ithaka)
[12] Graham, C., Bonk, C. J., & Graham, C. R. (2006*). Handbook of blended learning: Global Perspectives, local designs* (San Francisco: Pfeiffer)
[13] Bower, M., Dalgarno, B., Kennedy, G. E., Lee, M. J., & Kenney, J 2015 *Computers & Education*, 86, 1-17.
[14] Supriyanto, A., Hartini, S., Irdasari, W. N., Miftahul, A., Oktapiana, S., & Mumpuni, S. D. (2020) *Counsellia: Jurnal Bimbingan dan Konseling*, 10(2), 176-189.
[15] Wen, M. H., Chen, C. M., & Hsu, I. T. (2016) *Eurasia journal of mathematics, science and technology education*, 12(5), 1347-1363.
[16] Mathew, J. S. (2017) *Indian Journal of Science and Technology*, 14(14), 1-6.
[17] Abou Naaj, M., Nachouki, M., & Ankit, A 2012 *Journal of Information Technology Education: Research*, 11(1), 185-200.
[18] Crawford, J., Butler-Henderson, K., Rudolph, J., Malkawi, B., Glowatz, M., Burton, R., ... & Lam, S 2020 *Journal of Applied Learning & Teaching*, 3(1), 1-20.
[19] García-Alberti, M., Suárez, F., Chiyón, I., & Mosquera Feijoo, J. C 2021 *Education Sciences*, 11(2), 59
[20] Elisa, E., Farhan, A., Herliana, F., Wahyuni, A., & Susanna, S 2021 *Journal of Physics: Conference Series* (Vol. 1882, No. 1, p. 012030). IOP Publishing.
[21] Osgerby, J. (2013) *Accounting Education*, 22(1), 85-99.
[22] Herliana, F., Supriyati, Y., & Astra, I. M. (2015). *Prosiding Seminar Nasional Fisika (E-Journal)* (Vol. 4, pp. SNF2015-II).
[23] Sabah, N. M. (2020) *Behaviour & Information Technology*, 39(8), 875-898.
[24] Syukri, M., Herliana, F., Soewarno, Rizal, S., & Halim, L. (2021) *AIP Conference Proceedings* Vol. 2320, No. 1, p. 020036 AIP Publishing LLC.
[25] Abdullah, H 2015 *2015 IEEE 3rd International Conference on MOOCs, Innovation and Technology in Education (MITE)* (pp. 290-295). IEEE.

[26] Herliana, F., Halim, A., Farhan, A., & Kasli, E. (2020) *Asian Journal of Science Education*, 2(2), 106-113.
[27] Lin, C. H., Zhang, Y., & Zheng, B. (2017) *Computers & Education*, 113, 75-85.
[28] Santrock, J. W. 2009 *Psikologi Pendidikan Edisi 3 Buku 2* (Jakarta: Salemba Humanika)
[29] Akgün, E., Yarar, M., & Dinçer, Ç 2011 *Pegem Journal of Education and Instruction*, 1(3), 01-09.
[30] Marlinsa, E., Saputri, F., & Walid, A. (2021) *Asian Journal of Science Education*, 3(1), 37-43.
[31] Soewarno, S., Herliana, F., & Musdar, M. (2020) *Asian Journal of Science Education*, 2(1), 71-77
[32] Benson, V., & Kolsaker, A 2015 *The International Journal of Management Education*, 13(3), 316-325.
[33] Hung, M. L., & Chou, C. (2015) *Computers & Education*, 81, 315-325.
[34] Alutu, A. N 2006 *Journal of Instructional Psychology*, 33(1), 44.
[35] Istiningsih, S., & Hasbullah, H. (2015) *Jurnal Elemen*, 1(1), 49-56.
[36] Huang, Q. (2016) *Ortesol Journal*, 33, 14-33.
[37] Wang, M. J. (2010) *Australasian Journal of Educational Technology*, 26(6).
[38] Shand, K., & Farrelly, S. G. (2017) *Journal of Online Learning Research*, 3(1), 5-30..
[39] Schober, A., & Keller, L. (2012). *International Journal of Emerging Technologies in Learning (iJET)*, 7

Journal of Physics: Conference Series **2019** (2021) 012033 doi:10.1088/1742-6596/2019/1/012033

Flipped Learning Models and Students' Scientific Literacy on Physics Achievement Test

W Widjaja[1], I M Astra[2] and F C Wibowo[2]
[1] Physics Education Master Program, FMIPA, Universitas Negeri Jakarta Jl. Rawamangun Muka, Jakarta 13220, Indonesia
[2] Department of Physics, FMIPA, Universitas Negeri Jakarta Jl. Rawamangun Muka, Jakarta 13220, Indonesia
Email: wellywidjaja_1310819004@mhs.unj.ac.id; imadeastra@unj.ac.id; fcwibowo@unj.ac.id

Abstract. This study aimed to determine the influence of flipped learning models and scientific literacy on students' achievement test. The study used a quasi-experimental research method with two-way analysis of variance (ANOVA). The subjects of this study were Grade 10 high school physics students. The samples were two experimental groups where one group used inquiry flipped learning approach and the other group used mastery flipped learning. Both groups were given a scientific literacy test to separate students into two categories, which were students who had high scientific literacy and students who had low scientific literacy. The research showed that physics achievement test for students who learned by using inquiry flipped learning was similar than those who learned by using mastery flipped learning. Also, the achievement test for students who had high scientific literacy was greater than students who had low scientific literacy. The results of the research test indicated that there is a significant difference in achievement test between students who had high scientific literacy and students who had low scientific literacy.

1. Introduction

Science education in Indonesia has been developed since starting to participate in the Program for International Student Assessment (PISA) in 2000. In 2015, Indonesia is ranked fifth in the development of the education system out of 72 countries who participate in PISA. If Indonesia have continuous development, then in ten years, it is estimated that Indonesian students can compete internationally. Recently, Indonesia got an average PISA 2015 test score below the average score of the Organisation for Economic Co-operation and Development (OECD) countries on the science literacy basis [1]. The PISA assessment framework uses a four-factor approach, namely complexity, familiarity, forming a response and cognitive operation. PISA categorizes the cognitive processes required for the competencies that form the basis of scientific literacy together with consideration of the depth of knowledge required in the assessment of cognitive learning outcomes in the field of sciences including physics topic. PISA defines three scientific literacy competencies, namely scientifically explaining natural phenomena, evaluating and designing scientific investigations, and scientifically interpreting data and evidence. The assessment of students reading texts to understand and answer written questions about scientific literacy raises the problem of the level of reading literacy required. [2] Scientific literacy is that someone can ask, seek, or determine answers to questions that come from curiosity about everyday experiences. This means that a person has the ability to describe, explain, and predict natural phenomena. Scientific literacy requires the ability to read by understanding articles about science to engage in social conversations about the validity of conclusions [3].

Another fact obtained from preliminary studies by distributing a questionnaire based on needs analysis in one of a private high school in Banten, Indonesia. It was found that students were still low in applying science concepts in everyday life. Questionnaires were distributed to 83 students in class XI senior high school. There are 50.6% of students have not solved problems in everyday life with science concepts, as many as 48.2% of students have not been able to analyze science articles using science concepts. 43.4% of students have not yet demonstrated the understanding of science concepts.

In the analysis of a study related to the results of the Indonesian PISA test, it is suggested that further research is needed in order to improve science learning in Indonesia, one of which is pedagogy using active learning [4]. In recent years, a popular learning model that uses active learning is the flipped classroom learning approach [5] [6]. When using the flipped classroom, students performed significantly on the assessment and were allowed to do 50% more hands-on activities in the classroom [7]. Through flipped classroom, students gain cognitive skills such as the acquisition of knowledge before lessons and focus on application, analysis, synthesis and evaluation [8]. Flipped Classroom has an advantage over more traditional forms of teaching and can be taken to the next level, thus shifting to deeper learning strategies, such as flipped-mastery, or a more project-based model. Flipped Learning is the second stage of the flipped classroom, where teachers can use classroom time to become more content-rich, inquiry-based, and project-based [9].

The need for active learning is also supported by preliminary studies that have been carried out by a researcher. A total of 23 teachers from private schools in Banten filled out a questionnaire on the implementation of Physics learning. 95.65% agree that the subject matter of Physics is easier for students to understand through learning videos than the lecture method. 73.9% of teachers strongly agree that active learning such as discussions, experiments or projects can further enhance 21st century skills than delivering subject matter through teacher lectures. 82.60% of teachers strongly agree that students can better understand learning in the classroom if they have prior knowledge about the material to be taught compared to students who attend without preparation. 91.30% of teachers agree that by watching learning videos before attending class, teachers can carry out active learning. The use of active learning is supported by a research on the effect flipped classroom model on physics lesson had been compared with a traditional teaching approach. This study showed that flipped classroom model can improve physics learning outcome [10]. Research on the implementation of the development of the science literacy with a critical thinking framework aimed to improve the quality of science education [11]. Flipped classroom can be said to be a learning model that assigns homework before coming to class. Class time is spent solving problems, learning concepts and engaging in collaborative learning [12]. These activities will need a cognitive process that can predict academic performance [13].

The Cognitive Level was first presented by Bloom et al. where only one dimension that was ordered in such a way: Knowledge, Understanding, Application, Analysis, Synthesis and Evaluation [14]. Anderson et al. has revised Bloom's Taxonomy. Bloom's Revised Cognitive Taxonomy consists of two dimensions, namely: cognitive process dimensions (remembering, understanding, applying, analyzing, evaluating, making) with sub-categories and knowledge dimensions (factual, conceptual, procedural and metacognitive) [15]. Revised Bloom Taxonomy can be used to plan learning outcomes in physics lesson. Learning outcomes are statements that describe the knowledge or skills that students must acquire at the end of a particular assignment, class, course, or program, and help students understand why the knowledge and skills will be useful to them. They focus on context and the ability to apply knowledge and skills, help students relate education in various contexts, and help guide evaluation and assessment. Good learning outcomes emphasize the implementation and integration of knowledge. Not only focusing on the scope of the material, educational learning outcomes articulate how students will be able to use the material, both in the classroom context and more broadly [16].

Based on the previous description, the authors are interested in conducting research on the influence of flipped learning and science literacy on the physics achievement test. In this study, researchers used

Journal of Physics: Conference Series **2019** (2021) 012033 doi:10.1088/1742-6596/2019/1/012033

two models of flipped learning, namely inquiry flipped learning model and mastery flipped learning model.

2. Method

The study was conducted at one of the private high schools in Banten, Indonesia from May – June 2021. This study used a quasi-experimental method and the design used was 2 x 2 levels of treatment. The method aims to predict conditions that can be achieved in actual experiments, but there is no control or manipulation of all variables. In this research design, there are two independent variables which are learning model and scientific literacy and one dependent variable which is achievement test.

Table 1. Research design

Scientific literacy	Flipped Learning Models	
	Inquiry	Mastery
High	A_1B_1	A_2B_2
Low	A_1B_2	A_2B_2

In this study, one group learned a physics lesson by using the Inquiry Flipped Learning model, and the other group learned by using Mastery Flipped Learning model. Both groups were given scientific literacy test.

The score of the scientific literacy test were then sorted from the highest to the lowest. Based on this value, 27% of the higher score was taken as the high scientific literacy group, and 27% of the lowest score was taken as the low scientific literacy group. After the treatment, a physics achievement test on the Newton's Laws of Motion topic was given at the end of the meetings. Both instruments had been tested for validity and reliability. Data obtained from these two instruments were tested for analysis prerequisites which are normality and homogeneity test, followed by two-way ANOVA hypothesis test.

3. Results and Discussions

The mean scores for physics achievement test on Newton's Laws of Motion topic in each group are presented in Table 2.

Table 2. Mean Scores for Each Group

Scientific literacy	Flipped Learning Models	
	Inquiry	Mastery
High	79.38	77.75
Low	67.50	66.75

Before testing the hypothesis, a normality test and a variance homogeneity test were carried out using IBM SPSS software. The normality test was carried out using the Shapiro-Wilk test, which resulted in a significant value of $p > 0.05$ which means that students' achievement test data were normally distributed. The homogeneity of the variance by the Levene's test also resulted in a significant value of $p > 0.05$. This shows that students' achievement test from two groups of learning models were homogeneous. Then, hypothesis testing was carried out using two-way ANOVA. On Table 3 presents dependent variable ANOVA using IBM SPSS.

Table 3. Two-way ANOVA Output using IBM SPSS

Dependent Variable: Achievement Test					
Source	Type III Sum of Squares	df	Mean Square	F	Sig
Corrected Model	1059.344[a]	3	353.115	2.199	.110
Intercept	169798.781	1	169798.781	1057.260	<,001
Learning Models (A)	11.281	1	11.281	.070	.793
Scientific literacy (B)	1046.531	1	1046.531	6.516	.016
A * B	1.531	1	1.531	.010	.923
Error	4496.875	28	160.603		
Total	175355.000	32			
Corrected Total	5556.219	31			

[a] R Squared = .191 (Adjusted R Squared = .104).

The testing outputs show that there was no significant difference in achievement test between the learning models ($F = 0.070$ and $p = 0.793$), while there was a significant difference in achievement test between students who had high scientific literacy and students who had low scientific literacy ($F = 6.516$ and $p = 0.016$). Meanwhile, there was no significant interaction between the learning models and scientific literacy. The interaction between learning models and scientific literacy on students' achievement test is shown in Figure 1.

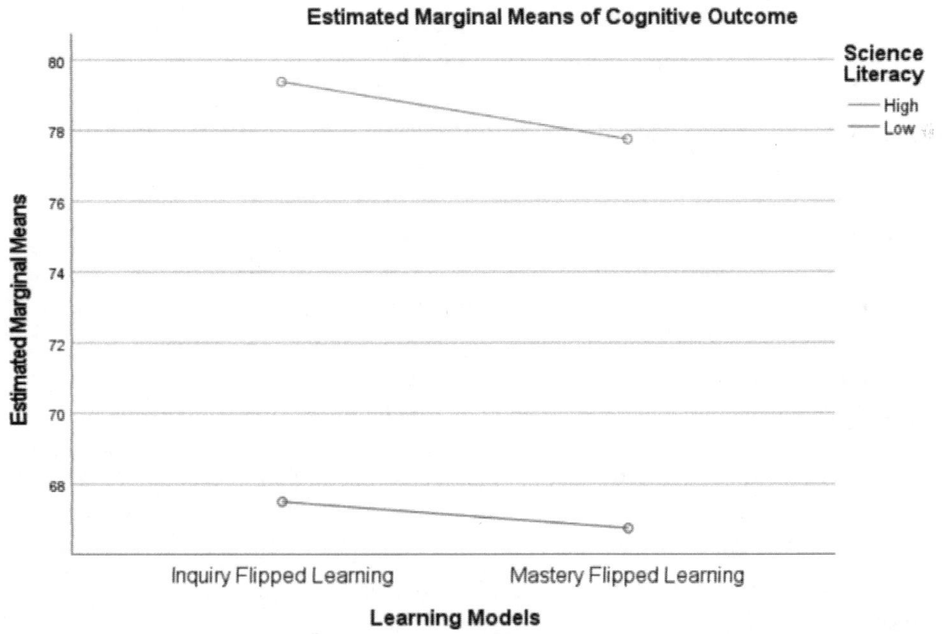

Figure 1. Interaction between learning models and scientific literacy on students' achievement test

The estimated marginal means of inquiry flipped learning model and mastery flipped learning model shows no significant difference is no surprise since both flipped learning models are active learning model. These results are in line with research on flipped classroom and non-flipped classroom but using a constructivist approach. The learning gains in either condition are most likely a result of active learning style of instruction [17].

The difference in students' scientific literacy shows a significant difference in the achievement test score. The result obtained are in line with DeBoer which stated that scientific literacy requires the ability to read by understanding articles about science to engage in social conversations about the validity of conclusions [3]. This is because, when students with higher science literacy will more likely to understand physics learning materials better than the students with lower science literacy.

4. Conclusion

Based on the research results obtained from data analysis, it can be concluded that the achievement test for students who learned by using inquiry flipped learning model was similar to students who learned through mastery flipped learning model. Also, there is no significant interaction between the learning models and scientific literacy on students' achievement test. However, the achievement test for students who had high scientific literacy is greater than students who had low scientific literacy. Therefore, based on hypothesis testing, there is a significant difference in achievement test between students who had high scientific literacy and students who had low scientific literacy.

Acknowledgments

The research was funded by Yayasan Tunas Manunggal.

References

[1] OECD 2016 *Programme for International Student Assessment (PISA) Results* (Paris: OECD Publishing)

[2] OECD 2017 *PISA 2015 Assessment and Analytical Framework: Science, Reading, Mathematic, Financial Literacy and Collaborative Problem Solving* (Paris: OECD Publishing)

[3] G. E. DeBoer. Scientific Literacy: Another Look at Its Historical and Contemporary Meanings, *Journal of Research on Science Teaching,* p. 590. 2000.

[4] N. Suprapto. What should educational reform in Indonesia look like? Learning from the PISA science scores of East-Asian countries and,*Asia-Pacific Forum on Science Learning and Teaching,* vol. 17(2). 2016.

[5] Y. Chen, Y. K. Wang and N. S. & Chen. Is FLIP enough? Or should we use, *Computers and Education,* vol. 79, p. 16–27. 2014.

[6] J. Bergmann and A. Sams 2015 *Flipped Learning for Elementary Instruction* (United States of America: International Society for Technology in Education)

[7] J. Bergmann and A. Sams 2012 *Flip Your Classroom: Reach Every Student in Every Class Every Day* (Alexandria: International Society for Technology in Education)

[8] G. Aşıksoy and F. Özdamlı 2016 *Flipped Classroom adapted to the ARCS Model of Motivation and applied to a Physics Course* (Turkey: Eurasia Journal of Mathematics, Science & Technology Education)

[9] J. Bergmann and A. Sams 2015 *Flipped Learning for Science Instruction* (United States of America: International Society for Technology in Education)

[10] I. M. Astra and S. I. Khumaeroh. The effect of flipped classroom model on student's physics, *IOP Conf. Series: Journal of Physics.* 2019.

[11] R. M. Vieira. Fostering Scientific Literacy and Critical Thinking, *International Journal of Science and Math Education.* 2014.

[12] S. J. Delozier. Flipped Classrooms: a Review of Key Ideas and Recommendations for Practice, *Education Psychol Rev.* 2016.

Journal of Physics: Conference Series **2019** (2021) 012033 doi:10.1088/1742-6596/2019/1/012033

[13] A. S. e. a. Finn. Cognitive Skills, Student Achievement Tests, and,*Psychological Science,* vol. XX(X), p. 1–9. 2014.

[14] B. S. Bloom 1956 *Taxonomy of Educational Objectives Book 1: Cognitive Domain* (London: Longmans Publishing)

[15] L. W. Anderson and K. D. R. 2001 *A taxonomy for learning teaching and assessing: a revision of Bloom`s taxonomy of educational objetives* (New York: Longman)

[16] M. Battersby. *So, What's a Learning Outcome Anyway?*. 1999.

[17] J. L. Jensen, T. A. Kummer and P. D. d. M. Godoy. Improvements from a Flipped Classroom May Simply Be the Fruits of Active Learning, *CBE—Life Sciences Education,* vol. 14, no. 1, 2017.

Journal of Physics: Conference Series **2019** (2021) 012034 doi:10.1088/1742-6596/2019/1/012034

Application of Online Science Practicum by *Using Microsoft Teams* and Learning Management System (LMS) During the Covid 19 Pandemic

W A Pertiwi[1], L A Sanjaya[2], Nadiyah[1], Hasirah[1], M Haramen[3], A S Budi[2], E Zarvianti[4], R W Puspa D[5], L K Sianipar[6], W Qoiriyah[7] and H Sitompul[6]

[1] Islamic Education Management Study Program, Faculty Islamic of Education and Teacher Training, Muhammad Azim Institute, Jambi, Indonesia

[2] Department of Physics, Faculty of Mathematics and Natural Science, Universitas Negeri Jakarta, Indonesia.

[3] Sharia Economics Study Program, Faculty of Sharia and Law, Muhammad Azim Islamic Institute, Jambi, Indonesia

[4] Physics Education Study Program, STKIP Muhammadiyah Sungai Penuh, Indonesia

[5] Kartika VIII-1 Senior High School, Jakarta, Indonesia

[6] Department of Physics Education, Universitas Pelita Harapan, Tangerang 15811, Indonesia

[7] Informatics Engineering study program, Faculty of Engineering, Sheikh Yusuf Islamic University, Tangerang, Indonesia

Email: wulananna1892@gmail.com ; lari@unj.ac.id ; nadiyahspog@gmail.com ; hasirahmpdi@gmail.com ; mohdharamen@gmail.com ; ennyzarvianti@gmail.com

Abstract. The research objective of this is to see how to apply science practicum online using Microsoft Teams and Learning Management System (LMS). The form of this research is classroom action research which aims to apply the online science practicum during the Covid 19 pandemic. Therefore, it is carried out using two cycles in the application of online science practicum. The subjects of this study were students of the seventh semester in the Natural Science practicum course. Through observation and evaluation, it is shown that in the cycle I the percentage of practicum implementation which includes aspects of preparation, process and data and practicum results is 87.5%, in cycle II it has increased to 92.5%. In the assessment aspect of the practicum report, the average practicum assessment obtained in the first cycle was 3.09 from the highest score of 4 and in the second cycle was 3.7, while the aspect of student knowledge about the use of the Microsoft team which was taken using google form was 93%. In the aspect of student knowledge about the use of Learning Management System (LMS) by 75% and aspects of student understanding in the implementation of online science practicum using Microsoft Teams by 75%.

1. Introduction

The COVID-19 pandemic has changed the way of teaching and learning in Indonesia, lecturers and teachers are competing to find alternatives and learning innovations during the COVID-19 pandemic [1]. The alternative used is of course IT-based where during the COVID-19 pandemic, face-to-face learning is not allowed because it is to avoid the transmission of the corona virus. In conducting online learning, lecturers can combine two approaches, namely synchronous and asynchronous. Synchronous learning is a learning interaction between lecturers and students carried out at the same time using video conference or chat technology [2]. Asynchronous learning is online learning with the lecturer preparing the material first and learning interactions are carried out flexibly and do not have to be at the same time, for example using discussion forums or independent study/student assignments [3].

Practical learning is learning that has a role in the development of science process skills [4]. Where these science process skills students can apply the scientific method in understanding, developing and discovering knowledge [5]. In science process skills, there are three components that need to be developed, namely 1) the ability to use the mind (intellectual), 2) the ability to reason, 3) Efficient and effective actions to achieve certain results, including creativity, therefore science process skills are typical skills used by science students and can be applied to understand phenomena [6]. The components in the science process skills are very much needed during practical learning, because the application of science process skills as well as the development of scientific attitudes that support the knowledge process in students is very possible in practical activities so that in this science lesson practicum has a very important position [7].

Practicum is one of the learning strategies that can attract students' interest in developing scientific concepts, because practicum can provide students with direct knowledge and experience to simultaneously develop scientific attitudes that support the process of acquiring knowledge (scientific products) within students [8]. In practicum activities, students can observe, interpret data, predict, use tools and materials, plan practicum, communicate practicum results and ask questions [9]. Practicum is a good learning tool for students/students in developing science process skills with practicum students/students are invited to directly discover new knowledge and are trained to develop all their five senses [10]. Practicum and science process skills cannot be separated because the development of science process skills in students is to use the practicum method because in the practicum activities psychomotor, cognitive and affective skills can be developed [11].

In March 2020 until the beginning of 2021 and even now all learning carried out both in schools and universities which are usually done face-to-face has turned into distance learning (PJJ) due to the covid 19 pandemic which requires not to meet in person to prevent the spread of the COVID-19 virus. Therefore, the atmosphere of the teaching and learning process changes and the teaching and learning process must immediately adapt to new habits, namely learning from home. From this case, blended learning is applied by combining or combining synchronous and asynchronous learning. The application of synchronous and asynchronous combinations here is to combine learning using the Web Conference from Microsoft Teams and the Learning Management System (LMS) in the form of power point teaching materials, videos and teaching materials in pdf format. Microsoft Teams is an example of a synchrony that functions as a Web Conference with Microsoft Teams of students and lecturers being able to meet or meet face-to-face online by activating the video feature. In addition to face-to-face online, there is also a chat feature with this feature students can submit messages or questions in writing, this feature can be used to accommodate student questions during the online learning process. Learning Management System (LMS) is an example of asynchronous which has functions as attendance, delivery of material in pdf and ppt formats, assignment and collection of assignments and management of assessments, not only that in LMS there is also a discussion forum feature where you can carry out discussions and lecturers can provide an assessment in each of these discussions. Therefore, the combination of synchronous and asynchronous aims to support the learning process that will be taught.

Before implementing synchronous and asynchronous, students and lecturers are provided with first how to use the platform. The debriefing is carried out online and using video tutorials on the website. In this case, lecturers and education staff work together and learn to use this application quickly, so that learning services for students can be optimized. As a result of the briefing, students and lecturers can make good use of the Web Conference from Microsoft Teams and the Learning Management System (LMS) platform. This is due to the enthusiasm between lecturers and students to be able to adapt in online learning for practicum courses.

Based on the initial observations that the author made by interviewing science practicum lecturers via WhatsApp, the use of synchronous and asynchronous has advantages and disadvantages. The advantage is that the learning process can still be carried out during the pandemic with the synchronous and asynchronous methods. The drawback in using synchronous and asynchronous uses is that students and lecturers must have a strong and stable signal network, and students and lecturers must learn more about the use of Microsoft teams and the Learning Management System (LMS) for the implementation of distance learning in science practicum courses. Students and lecturers must be able to adapt in the implementation of online science practicums, the things that students must prepare are internet networks, knowledge of the use of Microsoft teams and learning management systems (LMS), as well as preparing tools and materials for the implementation of science practicums. Meanwhile, the things that must be prepared by lecturers are the same as students, namely knowledge about the use of Microsoft teams and learning management systems (LMS), then preparing Tutorial Activity Plans (RAT) and Tutorial Program Units (SAT), power points of teaching materials and video tutorials.

Therefore, based on the explanation of the background above, the author is interested in conducting a research entitled "*Implementation of Online Science Practicum by Using Microsoft Teams and Learning Management System (LMS) During the Covid 19 Pandemic Period at UPBJJ-UT Jambi.*"

2. Method

The research method used in this research is Classroom Action Research (CAR). Classroom Action Research (CAR) is a research that is carried out systematically reflective of various actions taken by teachers/lecturers as researchers, from the beginning of the preparation of a plan to the end of the assessment of real actions in the classroom in the form of teaching and learning activities, in order to improve conditions learning [12].

Broadly speaking, there are four stages in Classroom Action Research (CAR) which are carried out repeatedly or in cycles, namely (1) Planning, (2) Implementation, (3) Observation and (4) Reflection [13].

a. Planning

Before carrying out the action, it is necessary to have a plan in advance and the forms of activities included in the plan are:

1) Create tutorial activity units
2) Prepare a practical guide
3) Prepare the observation sheet
4) Setting up the Microsoft Teams link
5) Prepare the Learning Management System (LMS)

b. Action Implementation

After the preparation has been carried out, the next stage is the implementation of the action. The action plan consists of: 1) lesson plan, 2) activity plan, 3) what plan will be observed, and 4) data analysis plan [14]. The forms of activities in this research are as follows:

1) Checking the availability of practical tools and materials

The tutor checks the availability of practical tools and materials by activating the camera so that the tutor can see the availability of these tools and materials.

2) Practicum Implementation

a. The tutor conveys the title, purpose and workings of the practicum through Microsoft Teams.
b. Tutors guide students in the implementation of practicum.
c. Students follow the tutor's directions.
d. Students present the results of the practicum that has been done.
e. The tutor provides feedback on the presentation of student practicum results.
f. The tutor asked the obstacles in the implementation of the practicum.
g. Students upload practicum reports in the Learning Management System (LMS)

c. Observation

After planning and implementing actions, the next stage is observation and evaluation. The forms of activities from observation and evaluation are as follows:

Journal of Physics: Conference Series **2019** (2021) 012034 doi:10.1088/1742-6596/2019/1/012034

1. Fill out the student practicum process assessment sheet
2. Provide assessment of student practicum reports

d. Reflection

The reflection in this research is on the application of science practicum and the obstacles found during the practicum. In this reflection, researchers can review the obstacles and problems that occur so that the results of the reflection can be used to determine further steps to achieve the goals achieved [15].

3. Result and Discussion

Table 1. Percentage of Science Practicum Implementation in Each Cycle

Aspect	Percentage %	
	Cycle 1	Cycle 2
Practical Preparation	88	91
Practical Process	87.84	93.8
Practical Data and Results	86.74	92.7
Average	87.5	92.5

Table 1 and table 2 explain that the percentage of practicum implementation in each cycle is 87.5% in cycle 1 and in cycle 2 it increases to 92.5%, and the average practicum assessment in cycle 1 is 3.09 and in cycle 2 it is 3.7 with an average The highest average 4. It can be said that the application of science practicum using Microsoft teams and the Learning Management System during the covid 19 pandemic can be applied in the implementation of science practicum during the covid 19 pandemic, besides being an alternative to the implementation of science practicum during the covid 19 pandemic.

Table 2. Average IPA Practicum Assessment

Aspect	Assessment Component	Cycle 1	Cycle 2
Practicum Report Assessment	Practicum Title and Objectives	4	4
	Tools and materials	4	4
	Theoretical basis	3	3
	Trial Procedure	3	4
	Observation result	3	4
	Questions	3	4
	Discussion	2	3
	Conclusion	2	3
	Bibliography	3	4
	Confusion experienced/suggestions and input	4	4
	Practicum Photo/Video	3	4
	Average	**3.09**	**3.7**

In addition to the percentage of science practicum implementation and the average science practicum assessment, this study also looks at students' understanding of the knowledge of using Microsoft Teams as measured by using a google form questionnaire. The following is the result of a questionnaire on student understanding of Microsoft Teams knowledge.

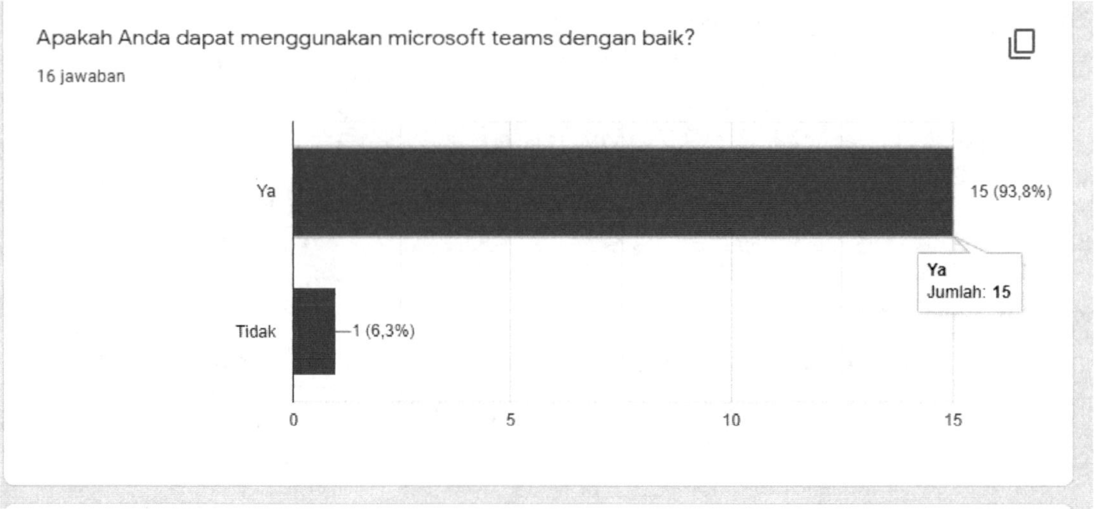

Figure 1. Questionnaire results of students' understanding of the knowledge of using Microsoft teams

From Figure 1 above, which is about the results of the questionnaire on students' understanding of the knowledge of using Microsoft Teams, that 93.8% of students understand the use of Microsoft Teams and in Figure 2, the results of the questionnaire on students' understanding of using the Learning Management System (LMS) that 75% of students can use LMS to science practicum report collection. The following are the results of the questionnaire on students' understanding of the use of the Learning Management System (LMS):

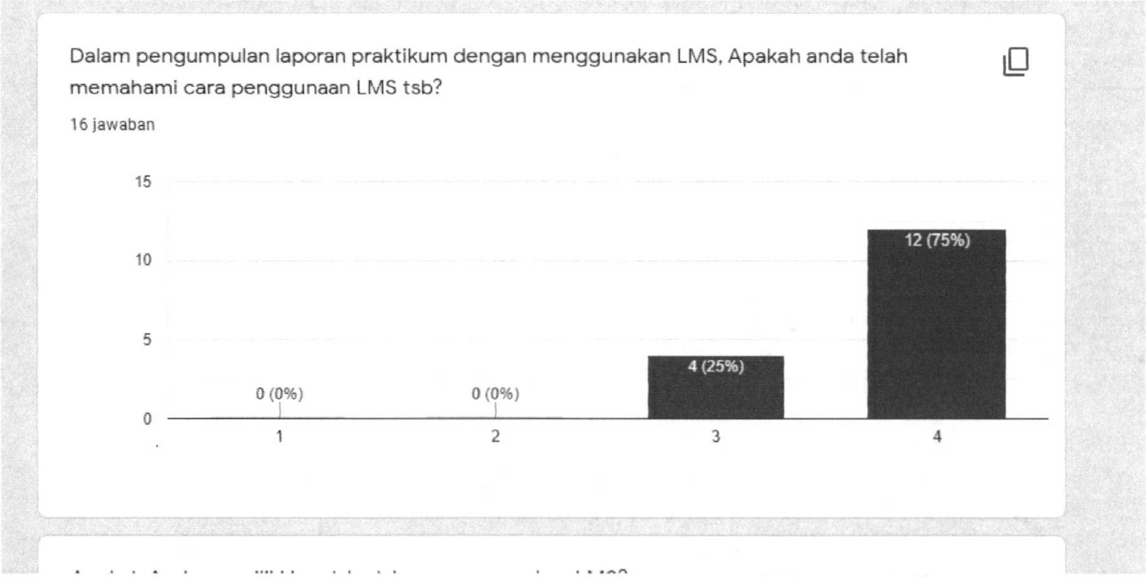

Figure 2. The results of the questionnaire on students' understanding of the use of the Learning Management System (LMS)

In addition to a questionnaire on understanding the use of Microsoft Teams and the Learning Management System (LMS) in this study, this study also looked at student understanding when tutors provided direction and guidance during the implementation of science practicum using Microsoft Teams. The results obtained are that 75% of students understand the direction and guidance of science practicum which is carried out using Microsoft Teams. Attached are the results of the questionnaire to

Journal of Physics: Conference Series **2019** (2021) 012034 doi:10.1088/1742-6596/2019/1/012034

see student understanding when tutors provide direction and guidance in the implementation of science practicum using Microsoft Teams:

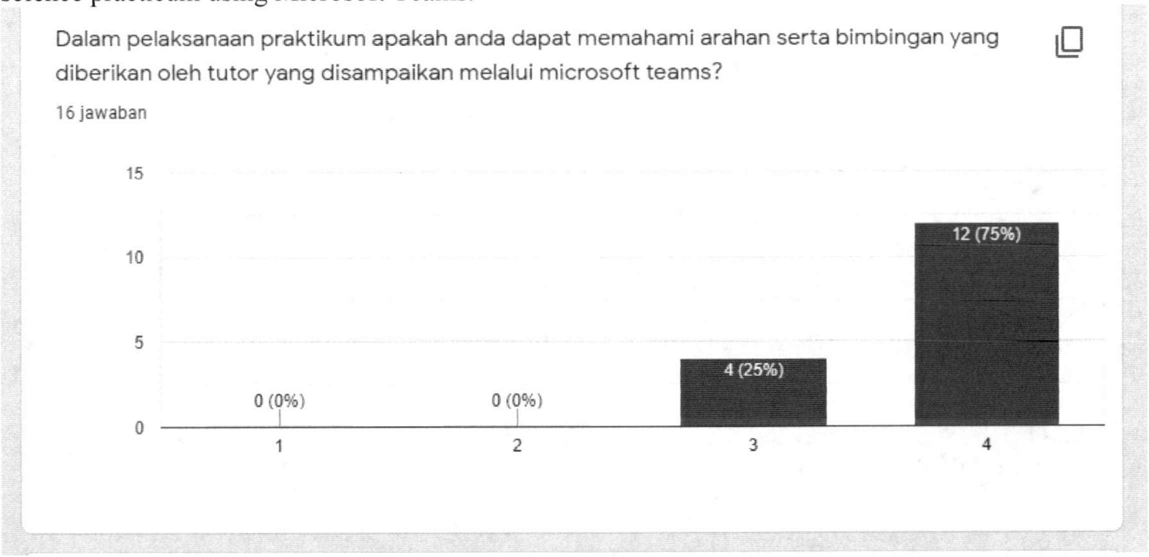

Figure 3. The results of the questionnaire see students' understanding when the tutor provides direction and guidance in the implementation of science practicum using Microsoft Teams

Figure 4. Documentation of the Implementation of Science Practicum using Microsoft Teams

4. Conclusion

The results obtained in applying the online science practicum using Microsoft teams and the Learning Management System (LMS) during the covid 19 pandemic, namely in the process of implementing the science practicum where there are aspects of preparation, process, data and practicum results in cycle 1 the average percentage is obtained by 87.5% and in cycle 2 it increased to 92.5%, while the average science practicum assessment obtained in cycle 1 was 3.09 and in cycle 2 it increased to 3.7 with the highest average of 4. In the aspect of student knowledge about the use of Microsoft Teams taken using Google Form are 93.8%, while aspects of student knowledge about the use of the Learning Management System (LMS) are 75% and aspects of student understanding in the implementation of online science practicum using Microsoft Teams are 75%. Based on the results of the research obtained, it can be concluded that the application of online science practicum using Microsoft teams and the Learning Management System (LMS) during the covid 19 pandemic can be carried out well and can be an alternative for science practicum during the covid 19 pandemic where learning must be carried out from home without face to face. The obstacle found in the implementation of this online science practicum is the signal network that is less supportive in each student's area. This research is still limited to the application of online science practicum, it is hoped that further research on other aspects can be carried out.

References

[1] SÃ¡, Maria JosÃ©; Serpa, Sandro 2020 *Sustainability* 12(20) p. 8525
[2] Samson, Perry J 2020 *Journal of Geoscience Education 1 p. 10*
[3] https://pjj.ui.ac.id/ufaqs/sinkronus-atau-asinkronus/
[4] Johannesson, Peter 2020 *Educational Action Research,* p. 1-16
[5] Matthews, Michael R 2018 *[Science: Philosophy, History and Education] History, Philosophy and Science Teaching || Epistemic Practices and Science Education. , 10.1007/978-3-319-62616-1(Chapter 5), 139–165.* doi:10.1007/978-3-319-62616-1_5
[6] Purwandono, E 2000 Penerapan Pertanyaan Produktif dalam Mengembangkan Keteramilan Proses Sains Siswa pada Pembelajaran Konsep Pemencaran organisme (Bandung : UPI Press).
[7] Turrini T, Dörler D, Richter A, Heigl F, Bonn A 2018 *Biological Conservation* 225 *p.*176–186
[8] Daniah, D 2020 *PIONIR: JURNAL PENDIDIKAN* 9(1)
[9] Gerde H K, Pierce S J, Lee K, Van Egeren L A 2017 *Early Education and Development* p. 1–21
[10] Huguet C, Pearse J, Noè L F, Valencia, Diego M, Ruiz N C, Heredia A J, Avedaño M 2019 *Journal of Geoscience Education* p. 1–13
[11] Sholikah T 2020 Indonesia Journal of Science Learning
[12] Darmadi H 2015 *Desain dan Implementasi Penelitian Tindakan Kelas (PTK)* (Bandung: Alfabeta)
[13] Markley C T, Miller H, Kneeshaw T, Herbert B E 2009 *Journal of Geoscience Education* 57(4) p.264–274
[14] Slameto S 2015 *Scholaria : Jurnal Pendidikan Dan Kebudayaan.*
[15] Niemi R 2018 *Five approaches to pedagogical action research. Educational Action Research p. 1–16*

Effect of Feedback on Learning Motivation of Primary Teacher Education Students in Primary School Physics Courses

L K Sianipar[1], H Sitompul[1], L A Sanjaya[2], R W Puspa D[3], W A Pertiwi[4] and W Qoiriyah[5]

[1] Department of Physics Education, Universitas Pelita Harapan, Tangerang 15811, Indonesia

[2] Department of Physics, Faculty of Mathematics and Natural Science, Universitas Negeri Jakarta, Indonesia.

[3] Kartika VIII-1 Senior High School, Jakarta, Indonesia

[4] Islamic Education Management Study Program, Faculty Islamic of Education and Teacher Training, Muhammad Azim Institute, Jambi, Indonesia

[5] Informatics Engineering study program, Faculty of Engineering, Sheikh Yusuf Islamic University, Tangerang, Indonesia

Email: henni.sitompul@uph.edu; lia.sianipar@uph.edu; lari@unj.ac.id

Abstract. Based on the results of the questionnaire distributed to Primary Teacher Education Students, 96% of students said that Physics is difficult to understand because the material is very complex, many formulas, especially the learning process is done online. This study aims to describe the influence of providing feedback on the motivation of Primary Teacher Education Students in primary school physics courses. The method used in this research is descriptive qualitative. Meanwhile, the instrument used was a questionnaire which was distributed to 115 elementary school education students. The result of the research shows that giving feedback consistently is able to motivate students. This can be seen from the results of the questionnaires distributed, each indicator is of very good value.

1. Introduction

The covid-19 pandemic is still sweeping countries around the world including Indonesia. The covid-19 pandemic affects many areas, including education. The learning process that was originally done face-to-face turned into online learning. This change overwhelmed teachers, parents, students, and students because they were not in a ready condition [1]. Online learning is a solution provided by the government to inhibit the spread of the Covid-19 virus. Online learning is applied at all levels of education at both school and college levels. This change is certainly a challenge for students, students, parents, especially educators. For the learning process to be effective even though it is done online, lecturers must be able to plan and design a learning process that can teach students. An effective online learning process results from careful planning, instructional design and is designed and developed using a systematic model [2]. This instructional design is expected to be able to encourage students' interest and motivation to learn.

In the learning process, it is important for a lecturer to build student motivation, especially in online learning situations like today. Lack of motivation is one of the factors causing dropout in online learning [3]. In general, motivation is an impulse within a person to do something [4]. Learning motivation can arise from within a person and due to the influence of others [5]. Indicators of motivation are classified into several parts, namely: 1) the desire and desire to succeed; 2) there is encouragement and need in learning; 3) there are hopes or aspirations for the future; 4) there is an appreciation in learning; 5) there are interesting activities in learning; 6) conducive learning environment [6]. There are several factors that affect the low motivation to learn in a person, one of which is the difficulty in understanding the subject matter. Someone who has difficulty in understanding the lesson, it will affect his learning behaviour [7].

Physics is a subject that is considered difficult to understand [8] [9]. One of the factors that causes Physics lessons to be considered difficult is because Physics material is very complex and abstract to understand [10]. The results of the questionnaire distributed to Elementary School Education (PGSD) students taking Elementary School Physics courses show that 94% percent of PGSD students consider Physics to be a difficult subject because it is very complex. In addition to complex physics material factors, online learning is also one of the factors that causes physics to be considered difficult. This perception of difficult physics can affect students' initial motivation in studying physics. To address this, one of the solutions that can be given by lecturers is to provide feedback in the learning process. Giving feedback is very important because it will affect students' learning motivation [11].

In the learning process feedback is interpreted as information conveyed to learners with the aim of changing behaviour and thinking to improve the teaching and learning process [12]. Providing feedback plays an important role for lecturers and students. By providing feedback to students, lecturers can know the extent to which students understand the materials that have been taught. As for students giving feedback this will provide information on the extent to which students master the materials studied, increase knowledge while motivating students [13]. Feedback can be done during synchronized and asynchronous learning. In synchronous learning, feedback can be done directly during discussion activities, Q&A or when the teacher feels it is necessary. In asynchronous classes, feedback can be done when correcting student assignments with the aim of sharpening student understanding. So, with feedback, it can help students ensure their understanding.

Based on the above explanation, this study aims to describe the influence of providing feedback on the motivation of Primary Teacher Education Students in primary school physics courses.

2. Method

The type of research used in this study is descriptive qualitative. The population sampled in this study was elementary school education students (PGSD) who took physics courses in elementary school. The instrument used to measure student motivation is a questionnaire consisting of 13 questions. This questionnaire was distributed at the end of the semester to 115 elementary school students. The type of questionnaire used is a closed questionnaire using a Likert scale [14].

Table 1. Likert scale

Statement	Score
Strongly agree	5
Agree	4
Undecided	3
Disagree	2
Strongly Disagree	1

Data obtained from research respondents were analyzed using qualitative descriptive analysis using the formula: [15]

$$p = \frac{f}{n}$$

where: p = percentage of each score

f = respondent's answer on each score

n = number of maximum scores

After the data is presented, the mean score for each indicator of learning motivation will then be calculated using the formula:

$$\text{Mean (R)} = \frac{\Sigma f.x}{N}$$

Based on the results of the calculation of the mean, the criteria for each aspect are determined by a range of values as follows:

Table 2. Criteria for student learning motivation

Score Mean(R)	Criteria
$4 \leq R \leq 5$	Excellent
$3 \leq R < 4$	Good
$2 \leq R < 3$	Fair
$1 \leq R < 2$	Poor
$0 \leq R < 1$	Very Poor

3. Result and Discussion

Primary Physics lectures conducted online using Microsoft application teams. Lectures are carried out synchronously and asynchronously, which are arranged in such a way on each meeting topic. Asynchronous lecture activities are explaining material using video recordings, doing physics exploration using physics classroom, exercises, and doing practicum. Synchronous sessions are held to discuss and provide feedback on student work in general. Feedback is also given personally to student accounts for assignments, quizzes, and practicum reports that have been collected. The feedback given is not only limited to corrections for what has been done by students but also encourages positive words.

The effect of feedback on student learning motivation is measured through questionnaires and student's reflections. Questionnaire compiled based on indicators of motivation to learn and distributed to students to be filled out online. The questionnaire was filled out by 115 students from 122 students who took primary physics courses, while student reflection is done personally in the middle of the semester. The indicators used in this study are 1) the desire to succeed, 2) the encouragement and need for learning, 3) the appreciation for learning, and 4) the expectation or aspiration of the future.

The desire to succeed is the first indicator that shows students' learning motivation. There are four statements in the questionnaire that measure this indicator. The results of the questionnaire for this indicator are shown in Figure 1. Based on the questionnaire results showed that 54.78% of the students Strongly Agree (SA) and 44.35% Agree (A) that the feedback given by lecturers encourages them to do their best in the learning process. Similar results were also found in the student reflection. From these results, it can be concluded that the feedback given by the lecturers encourages students to be enthusiastic about doing their best in the learning process. It is also confirmed by the second statement in which students feel encouraged to follow this course (25.22% SA and 68.7% A). Lecturer feedback also helps students to learn from the mistakes found (50.43% SA and 46.96% A), and were challenged to explore material that had not been mastered or felt difficult during learning (34.78% SA and 59.13% A). The mean calculation results on this indicator is 4.37, meaning that this indicator is included in the excellent category.

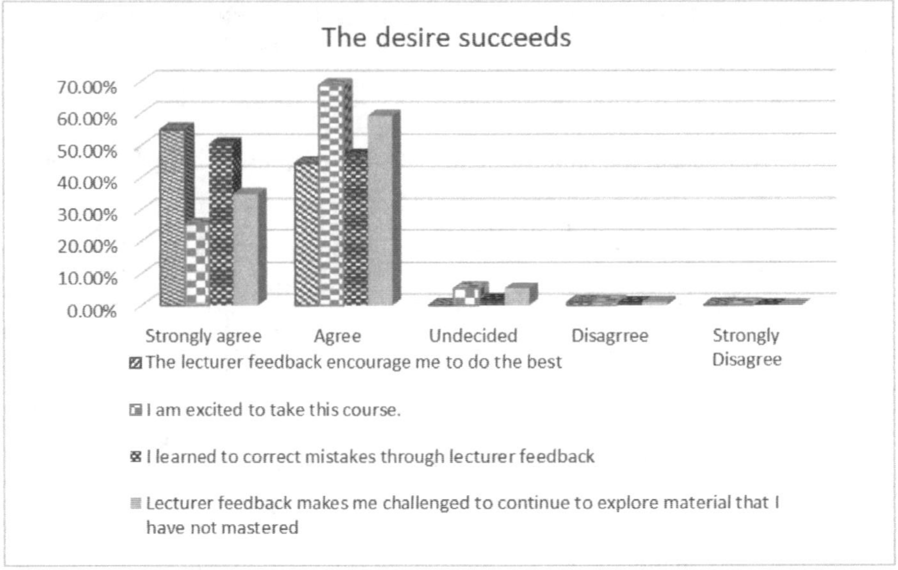

Figure 1. The questionnaire results in an indicator 1: The desire to succeed.

The second indicator used in this study is the existence of encouragement and need in learning. This indicator was identified through five statements in the questionnaire distributed to students. The results of the questionnaire for this indicator are presented in Figure 2. Based on Figure 2, it can be seen that lecturer feedback can help students to measure understanding of the learning material carried out. The same statement was also expressed by students in their reflections. Student responses to the administration of a compliment to the work duties are quite varied. Sixty percent of students want appreciation for the work they have done, but the rest don't want it. This indicator has a mean of 4.20, which means this indicator is in the category excellent.

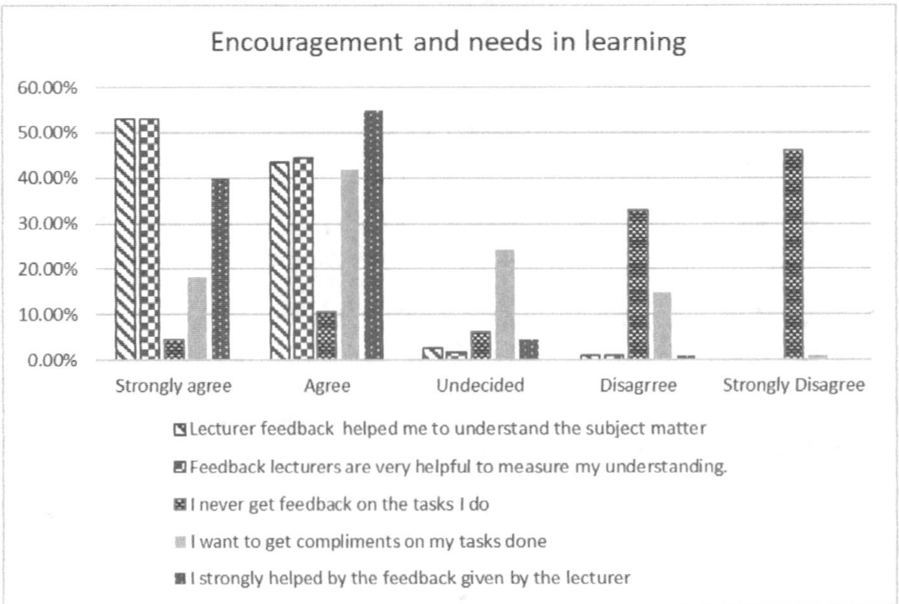

Figure 2. The questionnaire results in an indicator 2: The encouragement and need in learning

The next indicator used in this research is the appreciation of learning. This indicator measures through three statements contained in the questionnaire, two of which are negative statements. The statements in this indicator further confirm other indicators. The survey results on this indicator show that students feel more enthusiastic and appreciated when lecturers provide feedback on every task they do. The mean value of this indicator is 4.19. This indicator is also included in the excellent category. The results of the questionnaire for this indicator are presented in Figure 3.

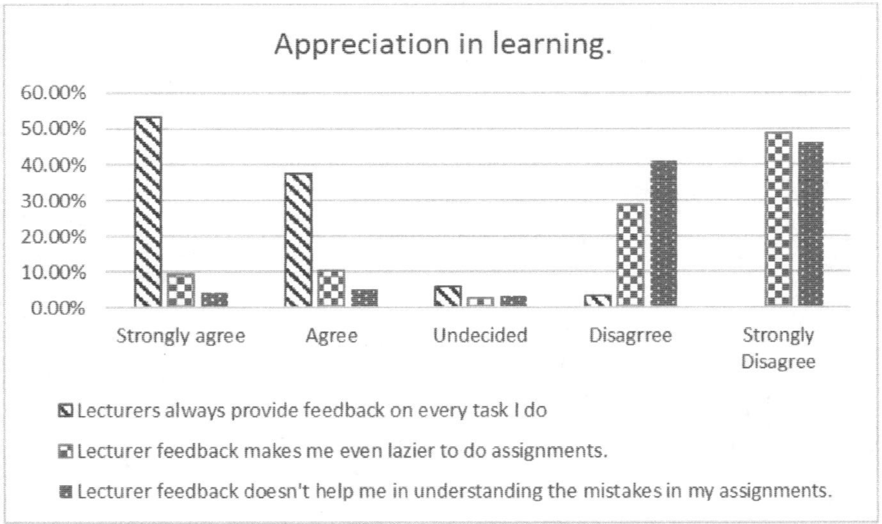

Figure 3. The questionnaire results in an indicator 3: The Appreciation in learning

The fourth indicator used in this study is the expectation or aspiration of the future. This indicator is seen from the results of the survey and student reflection. Based on the results of the questionnaire, it is known that 88.7% of students (24.35% SA & 64.35% A) believe that they will get good grades in this course. Student reflection also gave the same result where most of the students believed that they were able to complete this course with good grades even though it was done online. The mean value for this category is 4.11 and is included in the excellent category. The results of the student survey on this point are shown in Figure 4.

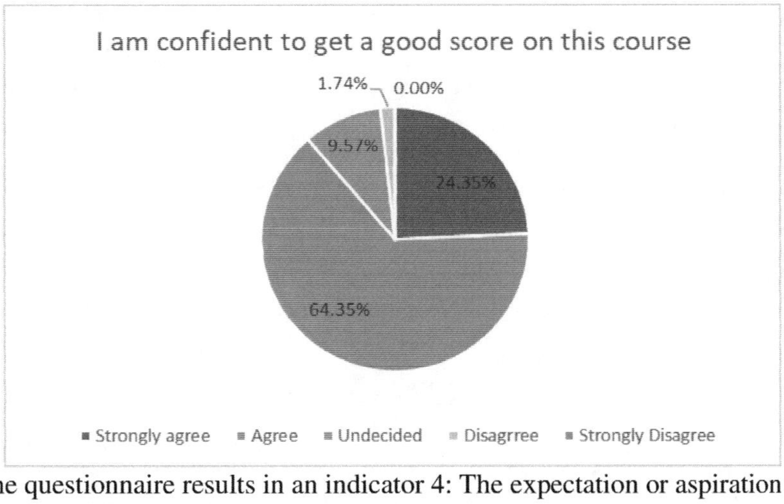

Figure 4. The questionnaire results in an indicator 4: The expectation or aspiration of the future

The percentage of students for each indicator based on achievement category is shown in Figure 5. Based on these data, we can see that more than 74% of the students in the excellent category in learning motivation. These four indicators are interrelated with each other. The feedback given consistently in both synchronous and asynchronous sessions makes students have the desire to succeed in this course. It is clear from the attitude strive to do better during the lectures. Lecturer feedback lets students know the level of understanding that has been achieved and helps them learn from mistakes that still exist. This encourages them to explore and study material that is difficult to understand. The appreciation of the lecturers in their feedback makes students excited and confident that they can take this course well. Students' fears and worries at the beginning of the lecture turned into a passion for learning during the lecture. Some students in their reflection admitted that the lecturer's feedback made them more motivated to study physics and even started to like it. It is also confirmed from student achievement at the end of the semester in which all the students graduating from this course with 90% of them passed with good grades.

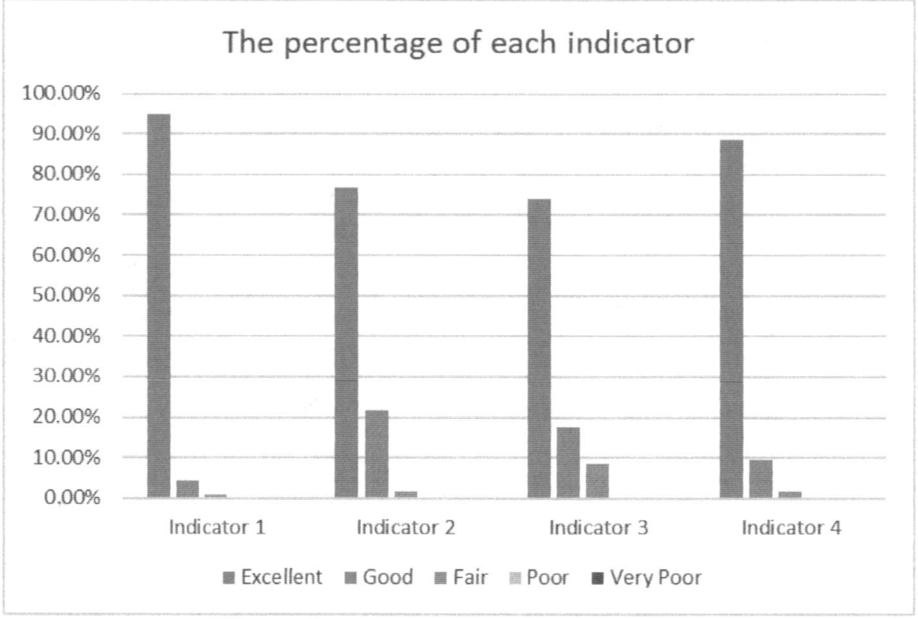

Figure 5. The percentage of each indicator

4. Conclusion

Lecturer feedback given consistently in primary physics lectures during online learning succeeded in making students learn to take lessons. Although most of the students have a fear of starting this course, all students successfully pass this course and 90% of them have good grades. Learning motivation measured using a questionnaire shows that all indicators used in the study are in the excellent category. Student reflections also show that some students have done their best in the learning process and have even started to like learning physics.

Acknowledgments

The authors would like to thank the Pelita Harapan University Research and Community Service Institute (LPPM) which funded this research.

References

[1] Suni Astini N K 2020 *Cetta: Jurnal Ilmu Pendidikan 3*(2) p. 241-255 https://doi.org/10.37329/cetta.v3i2.452

[2] Robert M. Branch and Tonia A. Dousay 2015 "Survey of Instrucsional Design Model" Association for Educational Communications and Technology (AECT)

[3] Hartnett M, St. George A & Dron J 2011 International Review of Research in Open and Distributed Learning 12(6) p. 20–38 https://doi.org/10.19173/irrodl.v12i6.1030

[4] Uno H B 2014 Teori Motivasi & Pengukurannya Analisis di bidang pendidikan (Jakarta: PT. Bumi Aksara)

[5] Cook David A, Artino, Anthony R 2016 Medical Education 50(10) p. 997–1014 doi:10.1111/medu.13074
[6] Uno H B 2011 Teori Motivasi & Pengukurannya Analisis di bidang pendidikan (Jakarta: PT. Bumi Aksara)
[7] Huitt W 2011 Motivation to learn: An overview *Educational Psychology Interactive* (Valdosta, GA: Valdosta State University) Retrieved from http://www.edpsycinteractive.0rg/topics/motivation/motivate.html
[8] Prastowo S A B, Jatmiko B, Supardi Z A I 2019 *Jurnal Penelitian Pendidikan Sains* 9 (1) p. 1745-1749 https://journal.unesa.ac.id/index.php/jpps/index
[9] Erinosho S Y 2020 *International Journal for Cross-Disciplinary Subjects in Education* (IJCDSE) 3(3) p. 1510 -1515
[10] Pramudia R P, Agustin A A 2018 *Jurnal Riset Pendidikan Fisika* 3 (1) p. 8-14 http://journal2.um.ac.id/index.php/jrpf/
[11] Erhel S and Jamet E 2013 *Computers & Education* 67 p. 156–167 doi:10.1016/j.compedu.2013.02.019
[12] Sumarno 2016 *Jurnal Ilmiah Pendidikan Pancasila dan Kewarganegaraan* 1(2) p. 115-125
[13] Geister S, Konradt U, Hertel G 2006 *Small Group Research* 37(5) p. 459-489 doi:10.1177/1046496406292337
[14] Djaali 2008 Skala Likert (Jakarta: Pustaka Utama)
[15] Sugiyono 2012 Metode Penelitian Pendidikan Pendekatan Kuantitatif, Kualitatif, dan R&D (Bandung: Alfabeta)

Journal of Physics: Conference Series **2019** (2021) 012036 doi:10.1088/1742-6596/2019/1/012036

The Influence of Discovery Learning Model using PhET and Scientific Skills of Class XI High School Students

L A Ramadhan[1], A S Budi[1] and M Delina[1]
[1]Physics Education Master Program, Universitas Negeri Jakarta
Jl. Rawamangun Muka, Jakarta 13220, Indonesia
Email: lidaniarifah@gmail.com , agussb@unj.ac.id , mutia_delina@unj.ac.id

Abstract. This study aims to determine the effect of interaction between discovery learning models and scientific attitudes on students metacognitive skills. The method used in this study is a quasi experimental design of treatment by level 2x2. The number of samples in this study was 144 high school grade XI students. The sample was divided into two, namely the control class and the experimental class. The control class was given learning using the free discovery model while the experimental class uses the guided discovery model by having high and low student scientific attitudes. After treatment was given to both groups, a test was conducted to find out the students metacognitive skills. Then compare the two results to see the interaction between learning models and scientific attitudes towards students metacognitive skills.

1. Introduction

One of the keys to a person's success in the world of reform is the mastery of metacognitive abilities [1]. Metacognition is one of the main issues in science education. Metacognitive skills are one of the skills that students must possess in the curriculum. are important skills that can equip students in the life of the century [2].

Educational experts view the exact sciences as not only consisting of facts, concepts, and theories that can be memorized, but also consisting of active activities or processes using scientific thoughts and attitudes in studying problems or natural phenomena that cannot be explained. The ability of students to process solving problems is one of the benchmarks in achieving physics learning [3].

There is one interactive simulation learning media called Physics Education Technology (PhET). PhET is an interactive simulation of physical phenomena, based on research that is provided free of charge. There are several physics simulations in PhET, including simulations of waves. By using PhET students can encourage students to investigate various kinds of real and abstract material.

The use of the Guided Discovery learning model helps to outline the early stages of 21st-century skills, learning developed to encourage educators, service providers, learners, administrators, and policymakers to achieve learning goals. In the 21st century, educators must help build an understanding of knowledge and skills in the learning environment. Learning in the 21st century emphasizes 4c, namely communication, collaboration, critical thinking, and creativity. Several studies have shown that the Guided Discovery learning model has a positive impact on the scientific and metacognitive attitudes of students. The research conducted shows that the practical guide based on the discovery learning model can improve metacognition in every student [4].

Based on the description of the results of observations and preliminary studies, it was found that learning activities have not raised problem-solving problems for students, but only invite students to verify the facts of the concepts that have been conveyed by the teacher. Likewise, in practicum-based learning, students are not faced with problems that must be solved, so that students do not find facts, concepts, theories as a result of their findings. This is due to the limited time for effective learning and the lack of teacher ability in understanding the appropriate model in the delivery of each physics material which has different characters.

This study aims to determine the comparison between the Guided Discovery learning model and the Free Discovery learning model on the metacognitive abilities of students in terms of scientific attitudes in class XI MIPA students.

2. Method

The method used in this study is a quantitative research method with an experimental design, namely quasi-experimental or Quasi-Experimental Design with a two-way Anova research design or treatment by level 2x2. In this research design, there are two independent variables, namely the learning model as a treatment variable and scientific attitude as an attribute variable, and one dependent variable, namely metacognitive skills. Each independent variable is classified into two types, namely the guided discovery learning model and the free discovery model, while the scientific attitude consists of a high scientific attitude and a low scientific attitude.

Table 1. Research design.

Scientific Skills	Learning Models	
	Free Discovery	Guided Discovery
High	A_1B_1	A_2B_1
Low	A_1B_2	A_2B_2

The quasi-experimental form has a control group, but cannot function fully to control external variables that affect the implementation of the experiment. In this design, neither the experimental group, nor the control group was selected randomly[5]. Instead, before being given treatment to both groups, a pretest was carried out to determine the extent to which the students' basic abilities were in the concepts to be studied. Furthermore, the two groups were given different treatments, namely, the experimental group was given learning treatment using the Guided Discovery model and had high and low scientific attitudes. While the control group was given learning treatment using the Free Discovery model and had high and low scientific attitudes. After being given treatment to both groups, a posttest will be conducted to determine the extent of the students' metacognitive skills [6].

3. Result and Discussion

The results of the analysis of the average metacognitive score in each data group are presented in Table 2. Based on the data in Table 2 explains that: 1) the metacognitive of students who learn by applying the free discovery learning model is lower than students who learn by the guided discovery learning model, 2) the metacognitive of students who have high scientific attitudes is greater than students who have low scientific attitudes, 3) students who have high scientific attitudes, metacognitive students who are taught by applying the free discovery model are lower than students who are taught by guided discovery learning models, and 4) in students who have low scientific attitudes, metacognitive students who are taught by applying the free discovery model are lower than students who are taught by the guided discovery learning model.

Journal of Physics: Conference Series **2019** (2021) 012036 doi:10.1088/1742-6596/2019/1/012036

Table 2. Average Metacognition Scores for Each Group.

Scientific Skills	Learning Models		Total
	Free Discovery	Guided Discovery	
High	42.65	42.85	42.75
Low	38.80	46.30	42.55
Total	40.28	43.88	

The results obtained to state that in achieving learning outcomes, the guided discovery learning model is better than the free discovery learning model. This is because, in the free discovery learning model, students determine the goals and desired to learn experiences, the teacher only gives students' problems and learning situations. Students examine the facts or relationships contained in the problem and conclude (generalizations) from what students find. While in the guided discovery learning model, the teacher provides examples of specific topics and guides students to understand the topic.

Before testing the hypothesis, the normality test and homogeneity test were carried out using IBM SPSS software. Normality test was performed using the Shapiro-Wilk test. From the results of the calculation of the normality test, it was found that the students' metacognitive data were normally distributed. Homogeneity variations were tested using Levene's test. The results obtained from the Levene test stated that the students' metacognition between groups of learning models was homogeneous. After the normality test and homogeneity test were carried out, the hypothesis was tested using ANOVA [7].

The results of hypothesis testing indicate that the effect of the learning model on students' metacognition is significant so that students who learn through the free discovery learning model get lower scores than students who are taught by applying the guided discovery learning model, while the main influence of scientific attitudes on metacognition is not significant. There is also a significant interaction between learning models and scientific attitudes. For students who have a low scientific attitude, the guided discovery learning model produces higher metacognition than the free discovery learning model. For students who have a high scientific attitude, the learning model has no significant effect. The interaction between the discovery learning model and scientific attitude states that the discovery learning model has a significant impact on increasing students' metacognition. The Discovery learning model is one of the methods in teaching cognitive theory by prioritizing the teacher's role in creating learning situations that involve students in active and independent learning. The Discovery learning method is a teaching method that regulates teaching in such a way that children acquire knowledge that they previously did not know, not through notification, partially or wholly discovered by themselves [8].

During the learning process, students who have a low scientific attitude learn by applying the free discovery model have difficulty adapting to the stages in the free discovery learning model. Meanwhile, students who have a high scientific attitude seem easy to adapt to the stages in the free discovery and guided discovery learning models. Students show a more positive attitude towards guided discovery learning than free discovery learning. Students prefer to study with research procedures provided by the teacher. Whereas in free discovery learning, learning is student-centered and not teacher-centered, so that students determine the goals and desired to learn experiences, the teacher only gives problems and learning situations for students. Students examine the facts or relationships contained in the problem and conclude what students find.

4. Conclusion

Based on the research results obtained by analyzing the data, it can be concluded that the metacognitive of students who learn by applying the free discovery learning model is lower than students who learn through the guided discovery learning model. Based on hypothesis testing, there is a significant difference in metacognition between students who are learned to apply the free discovery learning model and students who learn to apply the guided discovery learning model, and there is a significant interaction effect between the learning model and students' scientific attitudes towards students' metacognition learning outcomes.

Acknowledgments
The authors would like to thank the principal and physics teacher of SMAN 10 Bogor for granting permission and providing assistance in conducting this research.

References

[1] Baskoro A P and Bowo S 2015 *Jurnal Penelitian Sosial Keagaaman* p. 305-324
[2] Mansour N 2016 *International Conference on Education by theme Education in the 21 th Century: Responding to Current Issues*
[3] Rahayu S 2018 *Momentum: Physics Education Journal* 2(2) p. 64-71
[4] Endang A and Ernawati S K 2016 *Jurnal Penelitian Pendidikan INSANI* 19(2) p. 74-86
[5] Sugiyono 2017 Metode Penelitian Pendidikan (Pendekatan Kuantitatif, Kualitatif, dan R&D) (Bandung: ALFABETA)
[6] Kipnis, M., & Hofstein, A. (2007). The Inquiry Laboratory as a Source for Development of Metacognitive Skills. International Journal of Science and Mathematics Education, 6(3), 601–627. doi:10.1007/s10763-007-9066-y
[7] Schultz, B. B. (1985). Levene's Test for Relative Variation. Systematic Zoology, 34(4), 449. doi:10.2307/2413207
[8] Thillmann, H., Gößling, J., Marschner, J., Wirth, J., & Leutner, D. (2013). Metacognitive Knowledge About and Metacognitive Regulation of Strategy Use in Self-Regulated Scientific Discovery Learning: New Methods of Assessment in Computer-Based Learning Environments. Springer International Handbooks of Education, 575–588. doi:10.1007/978-1-4419-5546-3_37

Digital Storytelling of Physics (DiS-Phy): Learning Physics from Home Through Stories

L A Sanjaya[1], E A Putri[1], F C Wibowo[1], D K Robby[2], R W Puspa D[3], W A Pertiwi[4], L K Sianipar[5], W Qoiriyah[6] and H Sitompul[5]

[1] Department of Physics Education, Universitas Negeri Jakarta, Jakarta 13220, Indonesia

[2] Department of Management Education, Universitas Negeri Jakarta, Jakarta 13220, Indonesia

[3] Kartika VIII-1 Senior High School, Jakarta, Indonesia

[4] Management of Islamic Education, STAI AHSANTA, Jelutung Kota Jambi 36125, Jambi, Indonesia

[5] Department of Physics Education, Universitas Pelita Harapan, Tangerang 15811, Indonesia

[6] Informatics Engineering study program, Faculty of Engineering, Sheikh Yusuf Islamic University, Tangerang, Indonesia

Email: lari@unj.ac.id; fcwibowo@unj.ac.id; ratnawidayanti@eduphysics-unj.ac.id

Abstract. This study aims to develop a product such as "Digital storytelling of Physics (DiS-Phy)" that can be used as a physics' educational media for distance learning (PJJ) on magnetic field material. DiS-Phy developed as a digital storytelling media which is able to combine several learning media such as writing, pictures and learning videos in one medium. The developed DiS-Phy also equipped with scientific literacy questions. DiS-Phy can be accessed on all types of distance learning devices such as smartphones, laptops and tablets that are connected to the internet. DiS-Phy including the type of research and development (Research & Development) which refers to the ASSURE model. The result of validation test shows the percentages of 91,07%, according to subject-matter expert, 93,16%, according to media expert and 87,50% according to learning expert. DiS-Phy also tested by 4 physics teachers at senior high school in Jakarta that achieved 92,57% points that make this product has been qualified as a worthy physics' distance learning device. Also of 35 as respondents in this study shows the percentages of 87,48% points that make this product worthy to use as an educational media. Based on the results, it can be concluded that the DiS-Phy has been qualified as a worthy physics' distance learning device for teachers and students.

1. Introduction

The emergence of the Covid-19 virus has an impact not only in the health sector, but on various sectors of life, one of which is education [1, 2]. In Indonesia, the emergence of this pandemic has changed the face-to-face learning process to distance learning [3, 4]. The role of learning media during the distance learning process is very important because through the use of learning media educators can connect with students so that the learning process can take place [5]. However, many educators have not been able to use technology properly to make variations in learning media [6]. As a result, the use of learning media during distance learning (PJJ) tends to be less varied and has an impact on the difficulty of students understanding the material and the emergence of boredom during the PJJ. Digital storytelling is an art of storytelling using digital technology [7]. The use of digital storytelling in learning can increase students' motivation during learning and students' interest in physics [8, 9]. Digital storytelling can unite various kinds of learning media such as writing, pictures and learning videos into one medium [10, 11]. To access digital storytelling, students only need an internet connection on their distance learning device [12].

By utilizing the internet, digital storytelling can be accessed anytime, anywhere by students easily [13]. Therefore, digital storytelling can be a practical and easy-to-use learning medium. Educators can also modify the questions that will be included in digital storytelling with a specific purpose, such as adding scientific literacy questions to familiarize students with scientific literacy. Based on this background, this research was conducted to develop digital storytelling of physics (DiS-Phy) which is equipped with scientific literacy questions on the magnetic field material. The purpose of this research is to produce a digital storytelling of physics (DiS-Phy) equipped with scientific literacy questions on magnetic field material that is suitable for use as a physics distance learning medium.

2. Method

This study used a research and development (R&D) methods with the ASSURE model. This research model was chosen because this research model contains steps to integrate technology development into learning [14]. The research steps based on the ASSURE model are:

1) Analyze learner. At this stage, a questionnaire is distributed to analyze students from aspects of general characteristics and initial competencies that students have.
2) State standards and objective. The learning objectives in DiS-Phy are formulated based on the 2020 emergency curriculum.
3) Select strategies, technology, media and materials. The learning strategy used is student centered-learning. The technology that will be integrated into the learning process is a multimodal technology. The learning material in DiS-Phy is a magnetic field.
4) Utilize technology, media and materials. At this stage, researchers used an application program from Microsoft 365, namely Microsoft sway as the main application for making DiS-Phy, while Photoshop and movie maker were chosen as supporting applications for making DiS-Phy.
5) Require learner participation. In the DiS-Phy, student participation is obtained by making quizzes and exercises using Microsoft forms that students can fill out as feedback during the learning process.
6) Evaluate and revise. At this stage DiS-Phy is evaluated to identify deficiencies and improvements that need to be made. The results of the DiS-Phy evaluation are used to produce DiS-Phy which is feasible as a learning medium that can be used in distance learning.

3. Result

The development of digital storytelling of physics named "DiS-Phy". DiS-Phy was created using the Microsoft sway application as a digital storytelling application provided by Microsoft 365. In DiS-Phy physics material is presented through writing, pictures and learning videos accompanied by facts about everyday phenomena that are explained in physics. At the end of learning using DiS-Phy there is an evaluation in the form of scientific literacy questions.

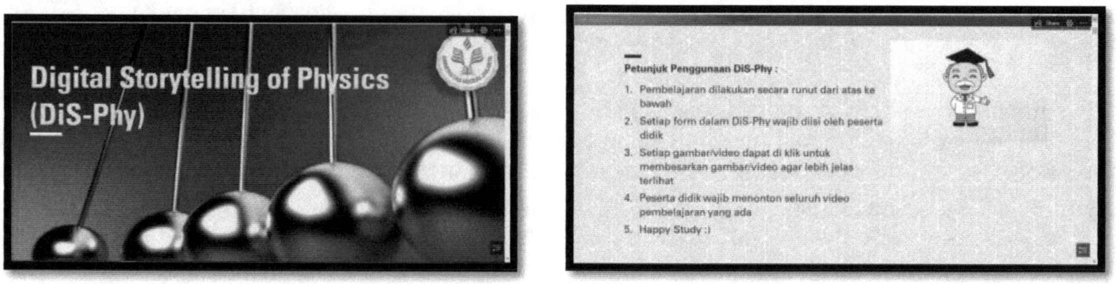

(a) (b)

Figure 1. (a) DiS-Phy initial display and (b) Instructions for using DiS-Phy in DiS-Phy

The initial display of DiS-Phy contains the name of the product, namely "Digital Storytelling of Physics (DiS-Phy)" and there is the logo of the State University of Jakarta on the top right. On DiS-Phy there are instructions for use. This provides information to students on how to use DiS-Phy during learning.

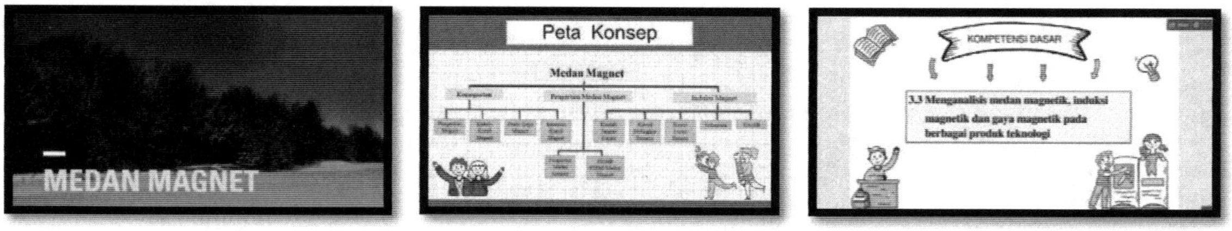

(a) (b) (c)

Figure 2. (a) Title of material, KD and indicators in DiS-Phy (b) Concept Map (c) Learning activities for basic competencies (KD) and indicators contained in DiS-Phy are adjusted to the applicable curriculum.

There is a title of the material to be studied using DiS-Phy (Magnetic Field). The background of the material is the aurora phenomenon which symbolizes the magnetic field phenomenon in real life. The concept map serves to briefly show the material that will be studied by students. The Basic Competencies used are adapted to the emergency curriculum on magnetic fields.DiS-Phy also has a concept map for the material to be studied. The concept map serves to show briefly the material that students will learn. Learning activities show the sequence of material that students will learn in sequence. Learning materials in DiS-Phy are presented in various formats, such as writing, pictures, and learning videos.

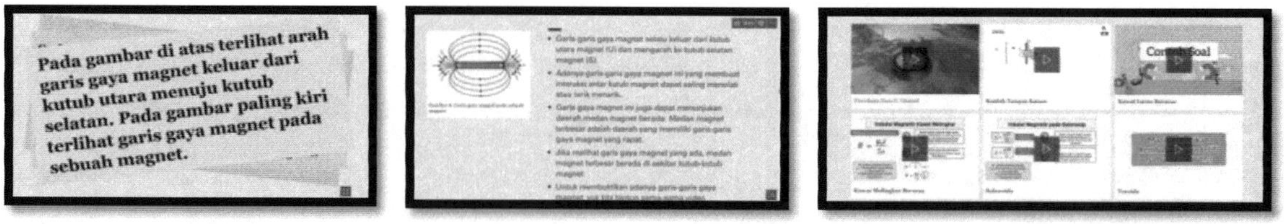

(a) (b) (c)

Figure 3. (a) Learning material in text format in DiS-Phy (b) Learning material in text and picture formats in DiS-Phy (c) Learning material in video format in DiS-Phy.

The existence of variations in the delivery of learning materials aims so that students do not feel bored during learning using DiS-Phy. The magnetic field material in DiS-Phy is also equipped with explanations of real-life phenomena related to magnetic fields. The explanation is presented in the form of fun facts. To measure students' understanding of the learning material, questions are given for each sub-discussion of the material. The questions given are in the form of scientific literacy questions that are adapted to the material being studied.

(a)

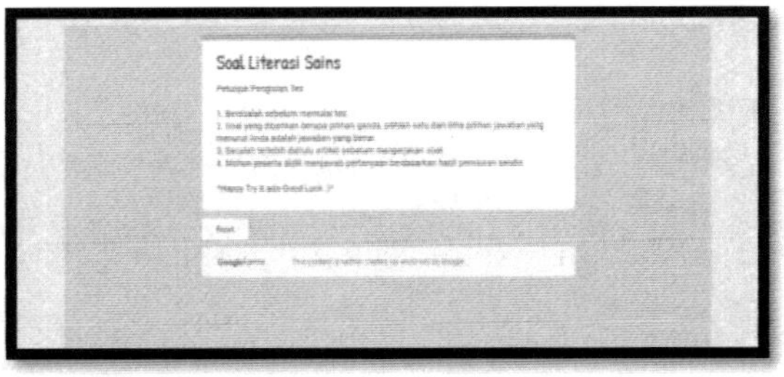

(b)

Figure 4. (a) Funfact in DiS-Phy relates to everyday phenomena (b) Evaluation of learning materials.

Table 1. Assessment of media expert.

No.	Aspect	Interpretation
1	Learning function and effectiveness	Very feasible
2	Advantage	Very feasible
3	Usability and interactivity	Very feasible
4	Efficiency	Very feasible
5	Probability	Very feasible
Average Aspect interpretation		Very feasible

Table 2. Assessment of material expert

No.	Aspect	Interpretation
1	Suitability of material content	Very feasible
2	Language	Very feasible
3.	Presentation technique	Very feasible
Average Aspect interpretation		Very feasible

Table 3. Assessment of learning expert

No.	Aspect	Interpretation
1	Use of media in learning	Very feasible
2	Science literacy	Very feasible
Average Aspect interpretation		Very feasible

When compared with previous studies [9, 15] both have not been equipped with scientific literacy questions. Therefore, the researchers developed a digital storytelling of physics (DiS-Phy) on magnetic field material equipped with scientific literacy questions to be used as distance learning media. The development of digital storytelling of physics (DiS-Phy) should be helping students understood the physics material. The material contained in DiS-Phy is presented in various formats such as text, pictures, and learning videos. The final DiS-Phy results created using the Microsoft Sway application can be easily shared with students. To access DiS-Phy students only need to be connected to the internet, so students have the convenience to access it anywhere and anytime [16].

4. Conclusion
Based on the results of the feasibility test of materials, media and learning from experts, it can be concluded that the digital storytelling of physics (DiS-Phy) magnetic field material equipped with scientific literacy questions for distance learning that was developed is suitable for use as a distance learning medium.

Acknowledgements
The research was funded by the faculty of mathematics and natural sciences, the state university of Jakarta, the Ministry of Education and Culture, through a Community Service grants 2021, with contract number: 26 /SPK PENGABDIAN MASYARAKAT/5.FMIPA/2021.

References
[1] Crawford, J., Butler-Henderson, K., Rudolph, J., & Glowatz, M. (2020). COVID-19: 20 Countries' Higher Education Intra-Period Digital Pedagogy Responses. *Journal of Applied Teaching and Learning (JALT), 3(1).*

[2] Nicola, M., Alsafi, Z., Sohrabi, C., Kerwan, A., Al-Jabir, A., Iosifidis, C., ... & Agha, R. (2020). The socio-economic implications of the coronavirus pandemic (COVID-19): A review. *International journal of surgery (London, England)*, *78*, 185.

[3] Ichsan, I. Z., Rahmayanti, H., Purwanto, A., Sigit, D. V., Irwandani, I., Ali, A., ... & Rahman, M. M. (2020). COVID-19 Outbreak on Environment: Profile of Islamic University Students in HOTS-AEP-COVID-19 and PEB-COVID-19. *Tadris: Jurnal Keguruan dan Ilmu Tarbiyah*, *5*(1), 167-178.

[4] Pakpahan, R., & Fitriani, Y. (2020). Analisa Pemanfaatan Teknologi Informasi dalam Pembelajaran Jarak Jauh di Tengah Pandemi Virus Corona Covid-19. *Journal of Information System, Applied, Management, Accounting and Research*, *4*(2), 30-36.

[5] Maphosa, C., & Bhebhe, S. (2019). Digital literacy: A must for open distance and e-learning (ODEL) students. *European Journal of Education Studies*.

[6] Putra, C. A. (2017). Pemanfaatan Teknologi Gadget Sebagai Media Pembelajaran. *Bitnet: Jurnal Pendidikan Teknologi Informasi*, *2*(2), 1-10.

[7] Dewi, N. R., Kannapiran, S., & Wibowo, S. W. A. (2018). Development of Digital Storytelling-based science teaching materials to improve students' metacognitive ability. *Jurnal Pendidikan IPA Indonesia*, *7*(1), 16-24.

[8] Kevser Hava. (2019). Exploring the role of *Digital Storytelling* in student motivation and satisfaction in EFL education, *Computer Assisted Language Learning, DOI: 10.1080/09588221.2019.1650071*

[9] Kotluk, N. & Kocakaya, S. (2017). The effect of creating digital storytelling on secondary school students' academic achievement, self-efficacy perceptions and attitudes toward physics. *International Journal of Research in Education and Science (IJRES), 3(1), 218-227*

[10] Prins, E. (2017). *Digital Storytelling* in adult education and family literacy: A case study from rural Ireland. *Learning, Media and Technology*, *42*(3), 308-323.

[11] Wihartanti, L. V., & Wibawa, R. P. (2017). Development of e-Learning Microsoft Sway as Innovation of Local Culture-Based Learning Media. *Dinamika Pendidikan*, *12*(1), 53-60

[12] Sudarmoyo, S. (2018). Pemanfaatan Aplikasi Sway Untuk Media Pembelajaran. *Edudikara: Jurnal Pendidikan dan Pembelajaran*, *3*(4), 346-352.

[13] Prananta, Y. R., Setyosari, P., & Santoso, A. (2016). Pemanfaatan Digital storytelling Sebagai Media Pembelajaran Tematik di SD. Seminar Inovasi Pendidikan di Era Big Data dan aspek psikologinya. digilib. mercubuana. ac. id.(online).

[14] Bajracharya, J. R. (2019). Instructional Design and Models: ASSURE and Kemp. *Journal of Education and Research*, *9*(2), 1-9.

[15] Saritepeci, M. (2020). Students' and Parents' Opinions on the Use of Digital Storytelling in Science Education. *Technology, Knowledge and Learning*, 1-21.

[16] Usodo, B., Kurniawati, I., & Kuswardi, Y. (2016). Training on Application of Several Applications from Microsoft: Office Mix, Onenote, Sway in Learning for Mathematics Teachers. Journal of Electronic Mathematics Learning, 4(9), 743–752.

Journal of Physics: Conference Series **2019** (2021) 012038 doi:10.1088/1742-6596/2019/1/012038

Website of Physics Instructional (WoPI): Learning Physics from Home During COVID-19

L A Sanjaya[1], I Nurkholis[1], B H Iswanto[1], F C Wibowo[1], N F Fidhyallah[2], R W Puspa D[3], N Ilmi[4], W A Pertiwi[5], L K Sianipar[6] and D P Rasmi[7]
[1] Department of Physics Education, Universitas Negeri Jakarta, Jakarta 13220, Indonesia
[2] Faculty of Economics, Universitas Negeri Jakarta, Jakarta 13220, Indonesia
[3] Kartika VIII-1 Senior High School, Jakarta, Indonesia
[4] 11th Public Senior High School Pandeglang, Indonesia
[5] Islamic Education Management Study Program, Faculty Islamic of Education and Teacher Training, Muhammad Azim Institute, Jambi, Indonesia
[6] Department of Physics Education, Universitas Pelita Harapan, Tangerang 15811, Indonesia
[7] Pendidikan Matematika dan Ilmu Pengetahuan Alam, Fakultas Keguruan dan Ilmu Pendidikan, Universitas Jambi, Jambi, Indonesia
Email: lari@unj.ac.id; iqbal132nurkholis@gmail.com

Abstract. In recent years, the education system, especially in Indonesia, has begun to adapt to distance learning. Technology in the field of education has emerged. Therefore, this research has the aim of producing an E-module based on a STEM with website-assisted in learning Physics on rotational dynamics. This research was conducted by applying the research method of FODEM (Formative Development Method) and was limited to the development phase. The results of this study are E-module learning media assisted by a website that can be used through various electronic devices, with learning material based on the STEM (science, technology, engineering, and mathematics) approach that can help students understand physics subjects in rotational dynamics. Further research is needed regarding the validation of this STEM-based E-modules assisted with website.

1. Introduction

Technology continues to develop and progress quite rapidly, then accurate information becomes very important for everyday life. Technological advances are able to change the norms that already exist in society. The field of education itself can be assessed as a form of development of information technology as a medium of information during the teaching and learning process [1]. Information technology is used by humans to help humans solve existing problems. Then, when the world was hit by a disaster that caused learning not to be done directly, then education was carried out remotely with the help of the internet. Technology successfully solves problems regarding learning that should not be done directly, but there are other problems that need to be considered, there are three classifications of problems that are obtained when conducting distance learning, namely; The teaching and learning activities carried out still need development, the use of appropriate technology, and in the personal or residential environment [2].

Physics is the science that underlies the natural sciences obtained from the results of experiments and theory development [3]. Physics subjects become a problem for students. Most students have problems with physics. In addition, teaching materials in physics subjects are also less developed [4]. A study was conducted related to understanding the material in physics subjects, then it was found that as many as 56.45% of students stated that the rotational dynamic material was the least easy material to understand. While the percentage of understanding of other materials is 1.61% for elasticity, 6.45% for static fluid, 9.68% for dynamic fluid, 6.45% for temperature and heat, and 19.35% for gas kinetic theory [5]. Then another research was conducted on the students' ability to understand the topic of rotational dynamics to 29 high school students. To measure students' understanding ability with the pretest device, it was found that one question did not answer correctly, and three other questions students answered incorrectly (<30% answered correctly) of the 9 questions given. This is because many students still have difficulty in understanding the basic concepts of rotational dynamics [6].

The thing that causes rotational dynamic material is difficult to understand because the material is presented pragmatically. Teaching materials are only composed of definitions, display formulas, and practice questions. The concept of the material being studied is only on the surface. Then students also find it difficult to solve various problems related to rotational dynamic material [7]. To be able to solve problems related to appropriate learning media, one of them is E-module with the help of a website. This website-based E-Module is very well used as a learning medium [8].

Problems related to learning activities and understanding of learning can be overcome with a STEM learning approach (Science, Technology, Engineering, and Mathematics). STEM is a learning approach that is popular at the world level, this approach is said to be effective in the application of learning in integrative thematic because it integrates four areas of knowledge: science, technology, mathematics and engineering [9]. A study related to the application of the STEM approach to vocational students for small group trials obtained a score of 79.7%, and trials in large groups obtained an average score of 68%, so it can be said to be effective and valid [10].

Referring to the explanation above, it is necessary to develop learning media in the form of an E-module based on a website-assisted STEM approach to learning physics on rotational dynamic material and it is hoped that this media can have a positive impact on students and can help students understand the learning material. Therefore, the author plans to develop a website-assisted STEM-based E-module for learning physics on rotational dynamic material. The e-module is intended for students who are in the upper secondary education level. In addition, this E-module will be used as a choice of learning media that can be applied in the learning process by teachers.

2. Method

This research was conducted by applying the FODEM (Formative Development Method) research method. FODEM is a product design approach that aims to support the design and development of various kinds of technological innovations in education, such as learning tools and online learning programs. This method has three important components, namely Need Analysis (NA), Implementation (I), and Formative Evaluation (FE) [11]. The number of threads of research can be more than one and the thread of research can be carried out in parallel or can go hand in hand without waiting for another thread to finish first in carrying out the development. In simple terms the flow of FODEM as shown below [12].

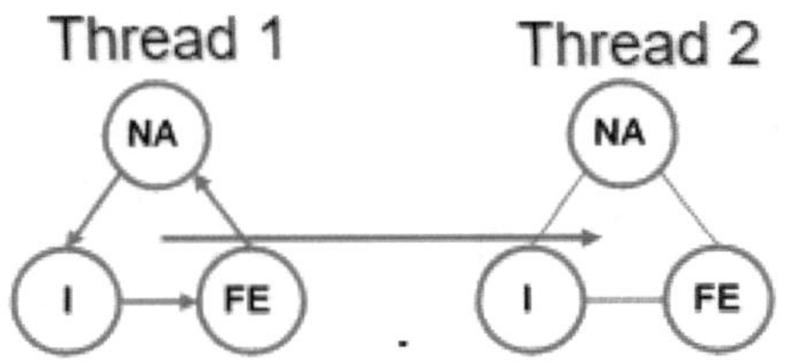

Figure 1. Simple flow of Formative Development Method (FODEM).

From this simple flow, it is applied to the research stage by limiting research only to the development stage. The research steps to be carried out are depicted in the following figure.

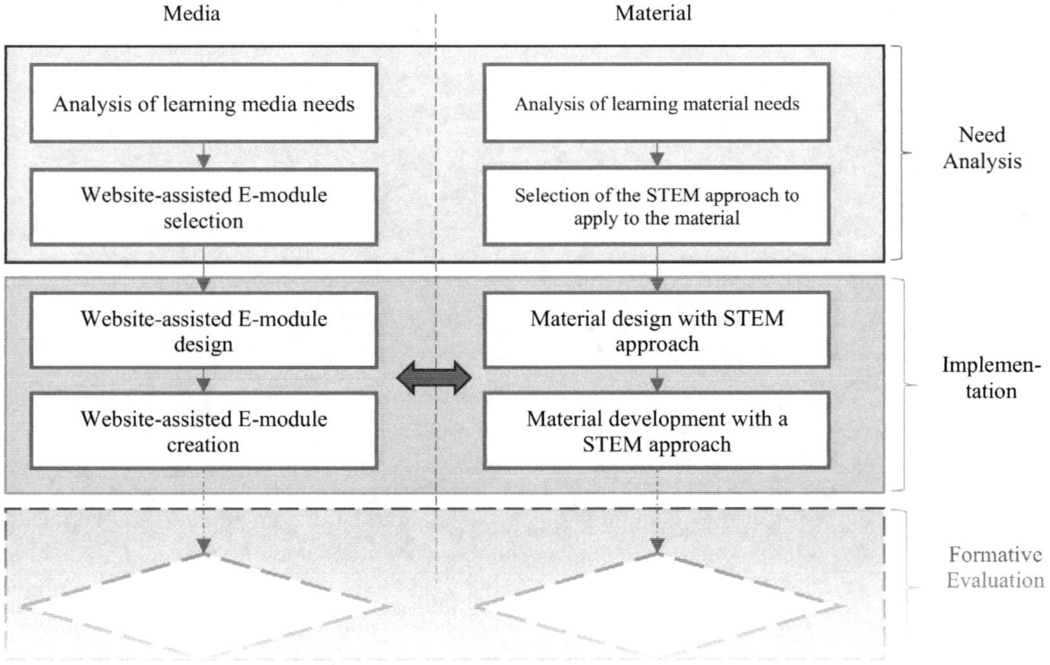

Figure 2. Website Assisted STEM-Based E-module development research flow

3. Result

The e-module resulting from this development is named WoPI (Website of Physics Instructional). This e-module discusses Rotational Dynamics material by integrating the fields of STEM (science, technology, engineering, and mathematics) to make these teaching materials more up-to-date and this teaching material prepares students in the future by practicing critical and creative thinking skills, systematically and logically. So that they are able to meet 21st century human resource standards and are able to face increasingly complex global challenges [13]. The form of this E-module is a website, so that it can be accessed from all devices, both mobile phones, laptops, and even smart televisions can access it.

Figure 3. Login page view

Before learning, students are required to login using an account that has been registered first. This account registration function is useful for storing student learning progress and other profiles.

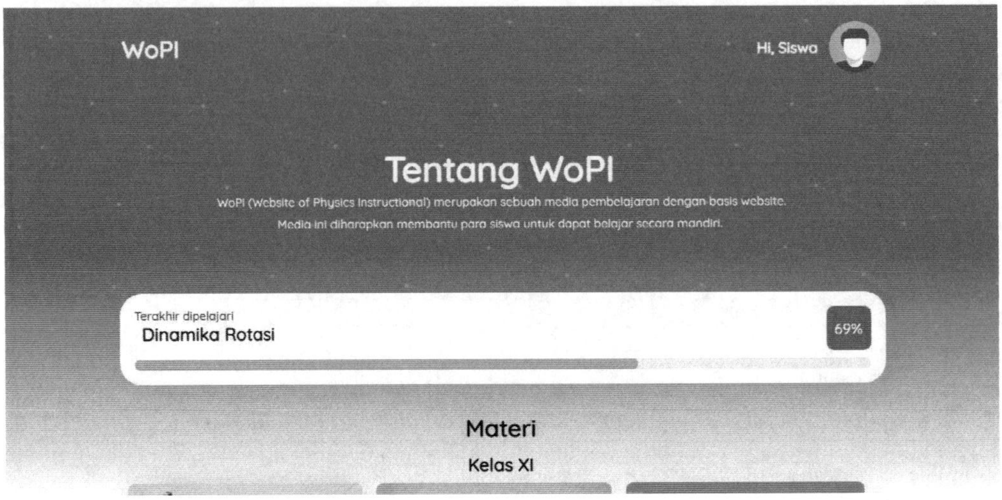

Figure 4. Home page view

After logging in with an account that has been registered, students will go to the start page, where on this page a student account is presented, then displays student learning progress on the material that was last worked on, and displays a list of materials that can be studied..

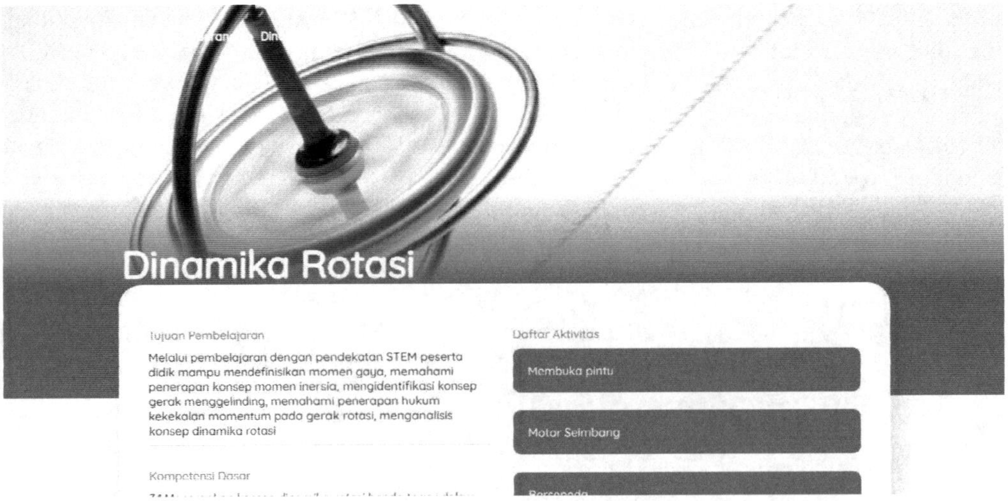

Figure 5. Material information page view

After students select the material to be studied, students will enter a page that displays information on the material such as learning objectives, basic learning competencies and also a list of activities to be followed in carrying out learning.

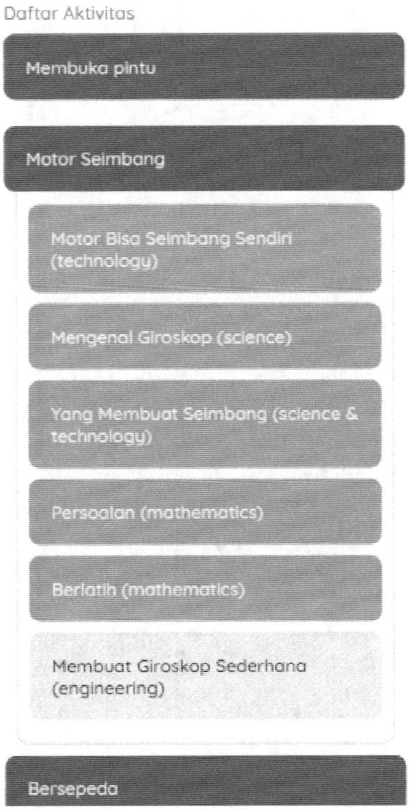

Figure 6. View of the activity list section

In the list of activities on the information page, when the activity is opened, other activities will appear. There are activities that are green and gray. The green color on the activity indicates that the activity has been studied, while the gray color indicates that the activity has not been studied and cannot be opened, this system is made so that students learn the material in sequence without missing a single learning activity.

Because the learning materials are made based on the STEM approach, the material presented integrates four fields of science which include natural sciences, technology, mathematics, and engineering. The following is how the material presented is applied based on the STEM approach:

Table 1. Student learning activities on rotational dynamics

Aspect	Activity
	Open the door
Science	- Students are given the questions related to opening doors - After that, the material is presented to verify the students' answers whether they are in accordance with the existing theory.
Technology	- Connecting materials with technology that applies the concept of torque, one of the door closer technologies, a tool that functions to close the door automatically.
Mathematics	- Students are given practice questions by applying the mathematical formulas contained in the concept of torque.
Engineering	- Students are asked to make a simple project assignment from the door closer using the principle of

Aspect	Activity
	- In working on project assignments, students are asked to start from planning, designing the tool to be made, until the tool is finished.
	Self-balance motorcycle
Science	- Students are asked how a motor can balance itself - After that, the moment of inertia and angular momentum were presented to verify the student's answer whether it was in accordance with the existing theory
Technology	- Connecting materials with gyroscope technology, which is the main factor in making the motor self-balanced
Mathematics	- Students are given practice questions by applying the mathematical formulas contained in the concepts of moment of inertia and angular momentum.
Engineering	- Students are asked to make a simple gyroscope with a cassette that can be found at home - In working on project assignments, students are asked to start from planning, designing the tool to be made, until the tool is finished.
	Cycling
Science	- Students are asked how the kinetic energy is experienced by a moving bicycle - After that, the kinetic energy material on the rolling object is presented to verify the student's answer whether it is in accordance with the existing theory
Technology	- Connecting materials with existing technology on a modern bicycle to improve performance
Mathematics	- Students are given practice questions by applying the mathematical formulas contained in the concepts of moment of inertia and angular momentum.
Engineering	- Students are asked to design a bicycle design by considering the kinetic energy of the bicycle

4. Conclusion

The website-assisted STEM-based E-module learning media for physics learning on rotational dynamic material was developed using the FODEM (Formative Development Method) research method. This e-module is presented in the form of a website so that it can be accessed from all devices, whether mobile phones, laptops, even smart televisions can access it. This e-module is designed and structured based on the STEM approach, so that the material presented integrates science, technology, engineering, and mathematics. This research still requires further research related to the validation of STEM-based E-modules assisted by websites until finally the media is suitable for use by teachers in the learning process.

Acknowledgements

The research was funded by the faculty of mathematics and natural sciences, the state university of Jakarta, the Ministry of Education and Culture, through a Community Service grants 2021, with contract number: 10/SPK PENELITIAN/5.FMIPA/2021.

References

[1] M. Yusup, Q. Aini, D. Apriani and P. Nursaputri, "Pemanfaatan Teknologi Blockchain pada Program Sertifikasi Dosen," *SENSITIF (Seminar Nasional Sistem Informasi & Teknik Informatika),* vol. 2019, 2019.

[2] S. Yudhistira and D. Murdiani, "Pembelajaran Jarak Jauh: Kendala dalam Belajar dan Kelelahan Akademik," *MAARIF,* vol. 15 No.2, no. Pendidikan Masa Pandemi Covid-19: Strategi, Adaptasi dan Transformasi, 2020.

[3] B. Z. S. a. H. P. Fauzi Bakri, "Rancangan Website Pembelajaran Terintegrasi dengan Modul Digital Fisika Menggunakan 3D PageFlip Professional," *Jurnal Penelitian & Pengembangan Pendidikan Fisika,* vol. 2(2), pp. 113 - 118, 2016.

[4] R. Rizaldi, "Analisis Kebutuhan Peserta Didik dalam Pengembangan Perangkat Pembelajaran Fisika SMA Berbasis Model Pembelajaran Inkuiri Terbimbing untuk menstimulus Keterampilan Proses Sains," *Jurnal Sintaksis,* p. 1, 2020.

[5] Y. Apriyani, S. Siswoyo and V. Serevina, "Pengembangan Media Pembelajaran Fisika Berupa Permainan Monopoli pada Pokok Bahasan Dinamika Rotasi dan Kesetimbangan Benda Tegar Kelas XI SMA," *WAPFI (Wahana Pendidikan Fisika),* pp. 42-48, 2019.

[6] D. Saputri and A. Suyudi, "Pembelajaran Interactive Demonstration dengan Diagram Gaya melalui Real dan Virtual Lab untuk Meningkatkan Pemahaman Konsep Dinamika Rotasi," *JRPF (Jurnal Riset Pendidikan Fisika),* p. 4, 2020.

[7] F. Nikmah, A. Yulianto and S. Wahyuni, "Bahan Ajar Kesetimbangan dan Dinamika Rotasi dengan Pendekatan Filosofis," *EKSAKTA : Jurnal Penelitian dan Pembelajaran MIPA,* pp. 122-132, 2020.

[8] T. Solihudin, "Pengembangan E-Modul Berbasis Web untuk Meningkatkan Pencapaian Kompetensi Pengetahuan Fisika pada Materi Listrik Statis dan Dinamis SMA," *Jurnal Wahana Pendidikan Fisika,* pp. 51-61, 2018.

[9] R. W. Sukmana, "Pendekatan Science, Technology, Engineering And Mathematics (STEM) sebagai Alternatif dalam Mengembangkan Minat Belajar Peserta Didik Sekolah Dasar," *Pendas: Jurnal Ilmiah Pendidikan Dasar,* p. 2, 2017.

[10] H. Hasanah and dkk, "Pengembangan Bahan Ajar Matematika Berbasis STEM Pada Materi Bangun Ruang," *Indonesian Journal of Learning Education and Counseling,* 2020.

[11] J. Suhonen, M. R. d. Villiers and E. Sutinen, "FODEM: a multi-threaded research and development method for educational technology," *Education Tech Research Development, vol.60,* p. 287–305, 2012.

[12] J. Suhonen and E. Sutinen, "FODEM: A formative method for developing digital learning environments in sparse learning communities," *Proceedings of the Fifth IEEE International Conference on Advanced Learning Technologies (ICALT'05),* 2005.

[13] R. Z. P. Angin, "Penerapan STEM Pada Pembelajaran IPA Materi Bioteknologi," *Prosiding Seminar Nasional Pendidikan Biologi UMM,* pp. 300-307, 2019.

Interactive Book Augmented Reality (IBAR) for Lesson Physics on STEM

F C Wibowo[1], H Nasbey[1], L A Sanjaya[1], D R Darman[2] and N J Ahmad[3]
[1] Department of Physics Education, Universitas Negeri Jakarta, Jakarta 13220, Indonesia
[2] Department of Physics Education, Universitas Sultan Ageng Tirtayasa, Serang City, Banten 42117, Indonesia
[3] School of Educational Studies, Universiti Sains Malaysia, Penang, Malaysia
Email: fcwibowo@unj.ac.id; hadinasbey@unj.ac.id; lari@unj.ac.id; dina_rd@untirta.ac.id; jahan@usm.my

Abstract. The scientific development process can be carried out with STEM (Science Technology Engineering Mathematics) education where this interdisciplinary approach trains students to learn where the concept of science is combined with the context in applying science, technology, engineering, and mathematics. The purpose of the study is the development of Interactive Book Augmented Reality (IBAR) for Lesson Physics on STEM. The research method is the ADDIE Model which consists of the stages of Analysis, Design, Development, Implementation, and Evaluation. The sample for the implementation of IBAR used in the study amounted to 90 students aged 19-22 years from a province in Indonesia. The results showed that IBAR contributed to students' concepts in physics. In the questionnaire given to students, information was obtained that the IBAR application was more useful, realistic, and interesting for their learning; help them understand and analyze problems and STEM scenarios. It has been suggested that IBAR technology could be a potent and effective tool to enable Lesson Physics students in the STEM process. In addition, the implications of using IBAR for physics education and recommendations for further studies are also discussed in this study.

1. Introduction

Physics is a science that examines the interaction between energy and matter, is the basis of natural science obtained from the results of experiments and theory development [1]. Physics is one of the branches of science education held to develop the ability to think analytically to solve problems related to the surrounding environment both qualitatively and quantitatively [2]. In short, physics is a branch of natural science that can explain all phenomena that occur in life in this universe, including one in the marine field [3]. The process of inquiry, which is a special feature of physics, has three main integrated components, namely aspects of the process, products, and scientific attitudes. The scientific development process can be carried out with STEM (Science Technology Engineering Mathematics) education where this interdisciplinary approach trains students to learn where the concept of science is combined with the context in applying science, technology, engineering, and mathematics [4]. In line with technological advances, currently emerging technology that can combine real objects and virtual objects with a real outside environment in real-time, which is currently known as Augmented Reality (AR) [5]. AR is a technology that combines two-dimensional and or three-dimensional virtual objects into a real three-dimensional environment and then projects these virtual objects in real time [6]. AR is currently experiencing rapid development and has touched various lives, one of which is in the world of education. The world of education is currently required to innovate and be creative with the aim of increasing effectiveness in learning and the quality of education, namely by using Augmented Reality [7]. AR applications in the education of these studies have positive results related to increased collaboration, deeper understanding, and student motivation when using Augmented Reality-based learning media [8] [9].

The implementation of 21st Century education in this context creates a link between school, community, and work, which allows the development of STEM literacy with the aim of enabling the ability to compete in the economy [10] [11]. STEM is a step taken by scientists to carry out investigations to find explanations for natural phenomena. These steps include formulating problems, formulating hypotheses, designing experiments, collecting, and analyzing and finally concluding. Therefore, the process of discovering concepts has been done in a cookbook, not through contextualism and constructivism. In addition, based on the results of preliminary research, information was obtained that student really like playing with learning, so this is an opportunity to try to transform STEM for learning [12]. Through STEM learning activities, students can learn the principles and concepts of science, according to an understanding model to develop higher-order thinking skills, one of which is 21st-Century Skills.

Meanwhile, Interactive Book Augmented Reality has indeed been developed, but not many have been able to demonstrate or practice both microscopic and microscopic physics concepts. So, referring to the findings and results of the study above, the purpose of research is development of Interactive Book Augmented Reality (IBAR) for Lesson Physics on STEM. In addition, IBAR with the developed STEM Model must be able to provide solutions to the limitations of virtual learning equipment and be able to implement independent learning according to the demands of the government.

2. Method

The research method is the ADDIE Model which consists of the stages of Analysis, Design, Development, Implementation, and Evaluation [13]. The sample for the implementation of IBAR used in the study amounted to 90 students aged 19-22 years from a province in Indonesia. The comparison of the number of Boy and Girl student's implementation of IBAR is shown in Figure 1.

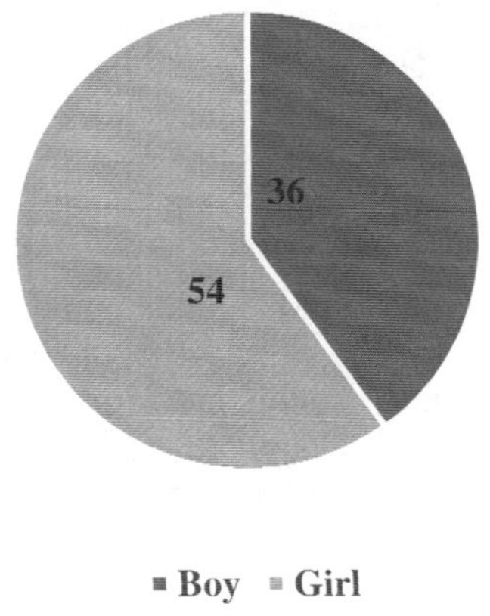

Figure 1. Comparison of Boy and Girl student's implementation of IBAR (%)

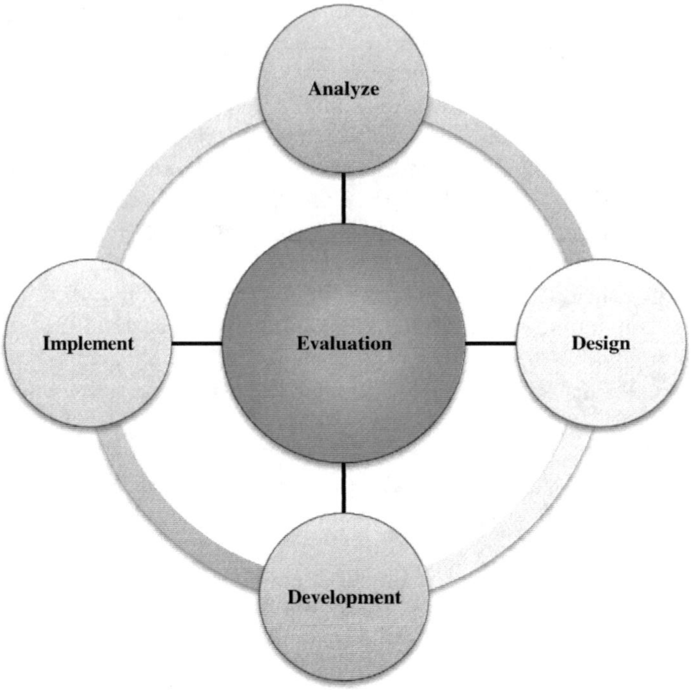

Figure 2. IBAR Development with ADDIE

Figure 2 shows the method of research in this study, the development of IBAR uses the ADDIE development model which consists of 5 stages, namely the analysis stage (Analyze), the planning stage (Design), the development stage (Development), the implementation stage (Implement), and the evaluation stage (Evaluate). This analysis was carried out through interviews and observations of prospective IBAR users. The purpose of holding interviews and observations is to find out the difference between real conditions and ideal conditions. The second stage is Design in the form of design planning, this activity aims to determine the design of the Augmented Reality application interface. The third stage, Design Development, this stage performs objective analysis, capability analysis, and design planning, then the IBAR book design is developed in the form of storyboards. The next stage is implementation, namely with validation tests, product validation is carried out by experienced experts to assess, identify weaknesses and strengths, and propose improvements to the IBAR. The result is in the form of inputs that can be used as a basis for revising the developed product and as a basis for product testing on students. The next implementation, with Trial to Users by conducting IBAR trials on users who are students, by providing links to users who contain IBAR with the application. The evaluation stage is always carried out at every stage of ADDIE.

3. Result

3.1. Characteristic IBAR

The implementation of IBAR in the experimental group was carried out at two weeks (8 hours). IBAR was taught to the experimental group through an interactive book based on AR, while the control group used traditional methods using physics textbooks on the structure of the atomic nucleus. In the experimental group, the concept was taught by the teacher with AR-based IBAR, where students actively participate in the class and each student can use the IBAR in every activity. During classroom learning, the virtual objects of the AR-based IBAR use the students' own smartphones, thus creating a learning environment that facilitates the construction of the atomic nucleus structure concept [14]. The procedures applied during the implementation phase and visuals for the implementation of IBAR are shown in Figure 3 and Figure 4.

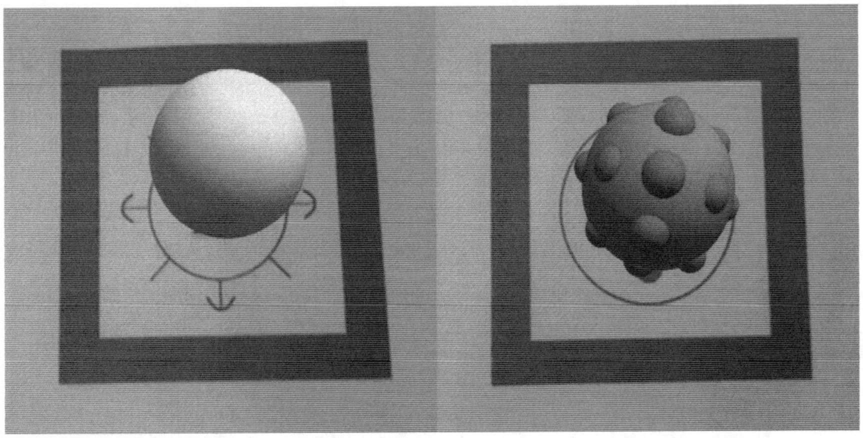

(a) The result of moving Image Dalton's Atom

(b) The result of moving Image Atom Thompson

Figure 3. Results of moving images of Dalton and Thompson's Atoms

In the first week of the meeting, students were given lessons about introduction to the concept of atomic nucleus structure using IBAR on the concept of Atom Dalton and Atom Thompson. Students obtain information about learning materials prepared with IBAR technology.

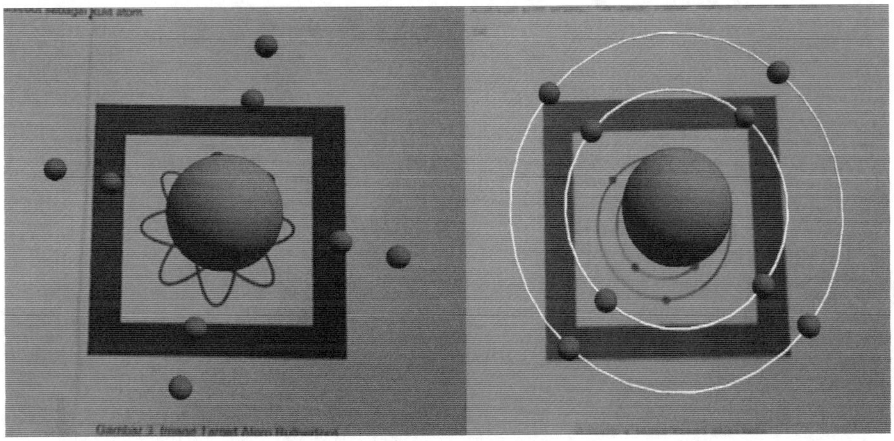

(a) The result of moving Image Rutherford's Atom

(b) The result of moving Image Atom Bohr

Figure 4. Results of moving images of Dalton and Thompson's Atoms

The second week, Characteristics of the Rutherford Atom, and the Bohr Atom, teaches students how the Rutherford Atom and the Bohr Atom work and how their nuclei move. The results showed that IBAR contributed to students' concepts in physics. In the questionnaire given to students, information was obtained that the IBAR application was more useful, realistic, and interesting for their learning; help them understand and analyze problems and STEM scenarios [15]. It has been suggested that IBAR technology could be a potent and effective tool to enable Lesson Physics students in the STEM process. In addition, the implications of using IBAR for physics education and recommendations for further studies are also discussed in this study.

3.2. IBAR Questionary to Students'

The results showed that IBAR contributed to students' concepts in physics with total sample 90 students. In the questionnaire given to students, information was obtained that the IBAR application was more useful, realistic, and interesting for their learning [16] and them understand and analyze problems and STEM scenarios. It has been suggested that IBAR technology could be a potent and effective tool to enable Lesson Physics students in the STEM process. The full results are shown in Figure 5.

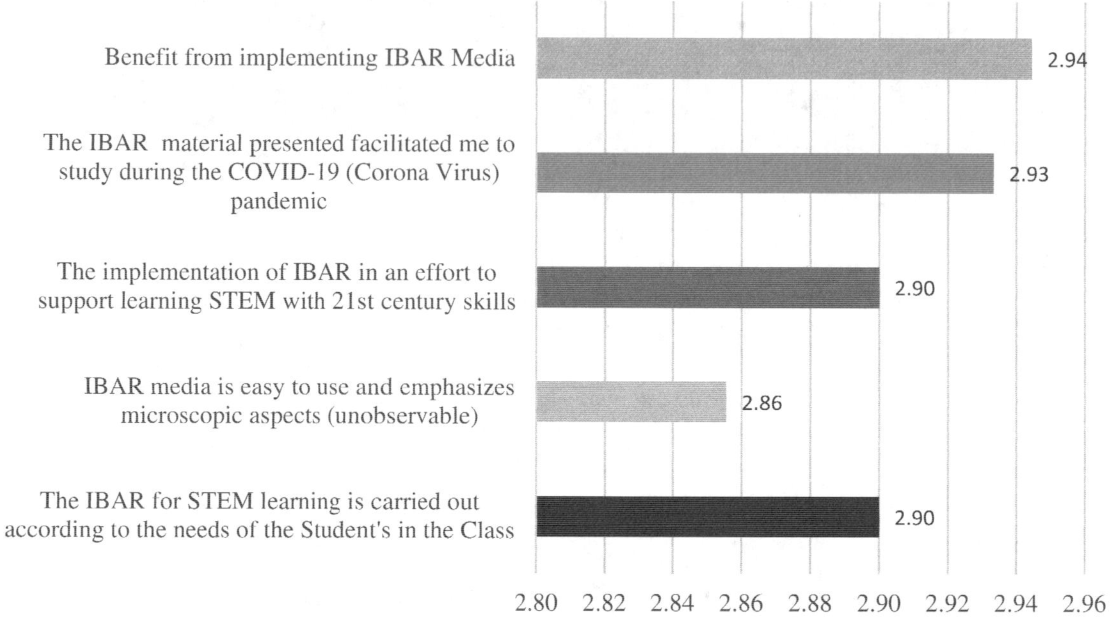

Figure 5. Result of IBAR Questionary to Students'

Based on Figure 5, information is obtained that The IBAR for STEM learning is carried out according to the needs of the Student's in the Class of 2.90 with a very good category from a scale of 1-3. Meanwhile, IBAR media are easy to use and emphasizes microscopic aspects (unobservable) of 2.86 with a very good category on a scale of 1-3. And The implementation of IBAR to support learning STEM with 21st century skills is 2.90 with a very good category from a scale of 1-3. For The IBAR material presented facilitated to study during the COVID-19 (Corona Virus) pandemic of 2.93 with very good category on a scale of 1-3. Students feel that the benefits from implementing IBAR Media are 2.94 with a very good category from a scale of 1-3.

4. Conclusion

The conclusion of the study is Interactive Book Augmented Reality (IBAR) for Lesson Physics on STEM The results showed that IBAR contributed to students' concepts in physics. In the questionnaire given to students, information was obtained that the IBAR application was more useful, realistic, and interesting for their learning; help them understand and analyze problems and STEM scenarios. It has been suggested that IBAR technology could be a potent and effective tool to enable Lesson Physics students in the STEM process. In addition, the implications of using IBAR for physics education and recommendations for further studies are also discussed in this study.

Acknowledgments

The research was funded by research grant research at Research and Community Service Institutions (LPPM), Universitas Negeri Jakarta, Ministry of Education and Culture 2021.

References

[1] Cavagna, A. I. Giardina, and T. S. Grigera. (2018). The Physics of Flocking: Correlation as a Compass from Experiments toTheory, *Phys. Rep.* 728, 1.

[2] Pont, D., Rocle, M., Valentini, A. et al. Environmental DNA reveals quantitative patterns of fish biodiversity in large rivers despite its downstream transportation. *Sci Rep*, 8, 10361 (2018). https://doi.org/10.1038/s41598-018-28424-8

[3] Hwang, WY., Purba, S.W.D. (2021). Effects of Ubiquitous-Physics App on Students' Inquiry Behaviors and Learning Achievements. *Asia-Pacific Edu Res* https://doi.org/10.1007/s40299-021-00585-7.

[4] Brown, R., Ernst, J., DeLuca, B., & Kelly, D. (2017). STEM curricula. *The Technology and Engineering Teacher*, 77(1), 26–29.

[5] Radu, I. (2014). Augmented reality in education: A meta-review and cross-media analysis. *Personal and Ubiquitous Computing*, 18(6), 1533–1543

[6] Abdusselam, M. S., & Karal, H. (2020). The effect of using augmented reality and sensing technology to teach magnetism in high school physics. *Technology, Pedagogy, and Education*. 29:4, 407-424.

[7] Akcayir, M., Akcayir, G., Pektas, H. M., & Ocak, M. A. (2016). Augmented reality in science laboratories: The effects of augmented reality on university students' laboratory skills and attitudes toward science laboratories. *Computers in Human Behavior,* 57, 334-342.

[8] Chang, H. Y., Hsu, Y. S., Wu, H. K., & Tsai, C. C. (2018). Students' development of socio-scientific reasoning in a mobile augmented reality learning environment. *International Journal of Science Education*, 40(12), 1410–1431.

[9] Kapp, S., Thees, M., Strzys, M. P., Beil, F., Kuhn, J., Amiraslanov, O., ... Wehn, N. (2019). Augmenting Kirchhoff's laws: Using augmented reality and smartglasses to enhance conceptual electrical experiments for high school students. *The Physics Teacher*, 57(1), 52–53.

[10] Li, Y., Wang, K., Xiao, Y. et al. Research and trends in STEM education: a systematic analysis of publicly funded projects. IJ STEM Ed 7, 17 (2020). https://doi.org/10.1186/s40594-020-00213-8

[11] Ibáñez, M.-B., & Delgado-Kloos, C. (2018). Augmented reality for STEM learning: A systematic review. *Computers & Education,* 123, 109–123.

[12] Holmlund, T. D., Lesseig, K., & Slavit, D. (2018). Making sense of "STEM education" in K-12 contexts. *International Journal of STEM Education*, 5(1). doi:10.1186/s40594-018-0127-2

[13] Branch, R. M. (2009). Instructional *design: The ADDIE* approach. Chicago

[14] Fidan, M., & Tuncel, M. (2019). Integrating augmented reality into problem based learning: The effects on learning achievement and attitude in physics education. *Computers & Education*. 142.

[15] Sırakaya, M. & Sırakaya, D A., (2020): Augmentedreality in STEM education: a systematic review, *Interactive Learning Environments*, DOI:10.1080/10494820.2020.1722713

[16] Altmeyer, K, Kapp, S. Thees, M., Malone, S. , Kuhn, J., Brünken, R. (2020). The use of augmented reality to foster conceptual knowledge acquisition in STEM laboratory courses—Theoretical background and empirical results. *British Journal of Educational Technology,* 51 (3), 611-628. bjet.12900–. doi:10.1111/bjet.12900

Dissemination of GOOPI (Game Open Online Physics Instructional) to Sparking Innovation in Education

F C Wibowo[1], L A Sanjaya[1], D K Robby[2], N J Ahmad[3], D R Darman[4], U Alizkan[1], C Fathurohman[1], B K Prahani[5] and B R Kurniawan[6]
[1]Department of Physics Education, Universitas Negeri Jakarta, Jakarta 13220, Indonesia
[2]Department of Education Management, Universitas Negeri Jakarta, Jakarta 13220, Indonesia
[3]School of Educational Studies, Universiti Sains Malaysia, Penang, Malaysia
[4]Department of Physics Education, Universitas Sultan Ageng Tirtayasa, Serang City, Banten 42117, Indonesia
[5]Department of Physics Education, Universitas Negeri Surabaya, Surabaya City, East Java 60213, Indonesia
[6]Department of Physics Education, Universitas Negeri Malang, Malang City, East Java 65145, Indonesia
Email: fcwibowo@unj.ac.id; lari@unj.ac.id; jahan@usm.my; dimaskurniarobby@unj.ac.id; dina_rd@untirta.ac.id, binarprahani@unesa.ac.id, bakhrul.rizky.fmipa@um.ac.id

Abstract. This study of dissemination of GOOPI (Game Open Online Physics Instructional) on page https://goopi.id/ (use UC browser) to sparking innovation in education. The method of the study is the Interactive Lecture Demonstration (ILD) approach, the ILD stages are carried out with three stages of activities, namely preparation, implementation, and evaluation. Dissemination with training for physics teachers who are members of the Physics Subject Teacher Meeting (PSTM) from a province in the west of Indonesia and Malaysian Teacher Association. The participants of this dissemination were 120 physics teachers for the first meeting and 86 teachers for a second meeting with GOOPI. The results of the study in the form of this dissemination were obtained information that physics teachers experienced in accessing and developing physics learning media. The results of the questionnaire responses to GOOPI with average 2,90 very good category of dissemination obtained information that GOOPI is effective for media for learning physics during the Covid-19 epidemic. The suggestion in this study is that dissemination training should be done frequently in order can provide skills or expertise that are not too complicated but are widely applied and needed by physics teachers.

1. Introduction

Higher education should make Independent Learning in the face of social change, the world of work and technological advances, in accordance with the needs of the industrial revolution era 4.0 [1] [2]; a link and match between researchers and the community or the world of work is necessary. University is required to innovate and develop innovative learning at the same age as DUDI demands. This demand is in line with the target of achieving the learning achievement of the Higher education merdeka campus curriculum [3] and problem-solving abilities [4]. Merdeka Belajar or Independent Learning (IL) is a tangible form of interesting learning challenges to create an innovative learning culture, and according to student needs [5].

Learning through the listening process without doing other things such as taking notes, it only contributes to the mastery of the material by 5%. If followed by reading, it contributes 10% mastery of concepts and when learning with audio-visual assistance, it will contribute to mastering the material by 30% [6]. The results of these studies indicate that if students can see the physical material visually, students will find it easier to understand and will avoid the construction of wrong conceptions. Evidence that not many games have been developed for learning Physics is obtained based on a survey conducted that at Bekasi district, the Information obtained that physics teachers who actively develop physics learning media are 27%, while physics teachers who are not actively developing learning media are 73%. So, it can be concluded that there are still many physics teachers in Physics Subject Teacher Meeting (PSTM) from a province in the west of Indonesia has not developed much physics learning media.

Based on information obtained from partners from the School of Education Studies, Universiti Sains Malaysia, that the School of Education Studies collaborates with the Malaysian Teacher Association for assistance. However, the science learning that has been trained has not yet combined learning science with online games. So that this becomes an opportunity to collaborate in community service in terms of improving teacher quality in the use of digital learning media.

Efforts that have been made in empowering PSTM have been carried out by the government through the relevant agencies, namely the District Education Office. Bekasi through relevant programs, but still the impact is not too significant in increasing the ability of physics teachers in developing learning media, especially learning media. The effort to empower PSTM teachers is not only the responsibility of the government alone, but also the responsibility of all components of the nation. UNJ as a tertiary institution that has the role of Tri Dharma, one of which is the task of Community Service, should have contributed to community empowerment programs in accordance with the scientific fields developed in each unit at UNJ. The Physics Education Study Program as a unit under UNJ must also strive to achieve the vision and mission of UNJ, especially those related to the third Dharma, namely Community Service by increasing studies on the application of research results in the field of Physics to help overcome various problems that occur in society. Physics education in meeting the demands of the 4.0 industrial revolution, one of which is the Development of Media, is very relevant to issues related to the development of learning media.

In the last two years, the expertise group of lecturers in the Department of Physics Education, Faculty of Mathematics and Natural Sciences, UNJ who are members of Physics Education Research (PER) has conducted a series of studies to study the working principles of GOOPI (Open Online Physics Instructional Game) as a medium for learning physics and can be accessed for free on the following page: https://goopi.id/ (with UC Browser) developed by researchers [8] [9]. In addition, GOOPI's research results have also obtained Intellectual Property Rights (IPR) on the following page: https://s.id/IPR_GOOPI. Several researchers have published the results of their investigations related to use, including: enabling students to apply their previous strategies and knowledge to solve problems through online games [10]; Online games can involve student activeness, motivation and learning outcomes [11] [12]; educational games increase student interest in continuous motivation [13]; training certain behaviors that might be a critical strategy [14]. In addition, online games for learning can solve problems [15] and Instructional Games for Challenges & Motivation [12].

The show that on the one hand physics teachers are low in the ability to develop media and the number of teachers who are not actively developing media is also high, and it is suspected that it causes many negative excesses, on the other hand learning around the area is still lecturing and using media. Because the conditions are in accordance with the objectives and targets of this ICCS activity, this area was chosen as the location for the ICCS activities to be carried out. Although the area is limited in PSTM Physics in the City of East Jakarta, it is hoped that it will be able to provide simulation and animation training for physics learning in other areas around it. The ICCS activity is entitled "International Collaborative Community Services (ICCS): “Dissemination of GOOPI (Open Online Physics Instructional Game) to Sparking Innovation in Education in Universitas Negeri Jakarta and Universiti Sains Malaysia”.

Journal of Physics: Conference Series **2019** (2021) 012040 doi:10.1088/1742-6596/2019/1/012040

2. Method

The audience in the Community Service Program for the ICCS Faculty Community Partnership Program is Physics teachers who are members of the Physics Subject Teacher Meeting (PSTM) from a province in the west of Indonesia. As for the number of teachers of participant is 120 for the first meeting and 86 teachers for a second meeting and implemented in 2 forms face to face and face to face of online meetings. For face-to-face plans are planned for 2 times and for virtual meetings it is done through zoom meetings and the result meeting at the YouTube at URL: http://www.youtube.com/watch?v=wUkaon9MZM4. The community service system owned by the physics education study program on the page: https://p2mfisika.smart-unj.id/. The method used to introduce work principles and important parts of GOOPI (Open Online Physics Instructional Game) Media Training is a combination of informative and using the Interactive Lecture Demonstration (ILD) method. The informative method is used to provide a brief explanation of the working principles of the Online Game and its main parts. The demonstration method is used to demonstrate modelling how virtual simulation can be applied to training [16]. Other things that can be done to solve partner problems in the GOOPI training, as well as improve the skills they already have through. Training on designing GOOPI for learning in other forms Mentoring as GOOPI for independent learning. The Figure 1. ICCS activity program implementation of GOOPI.

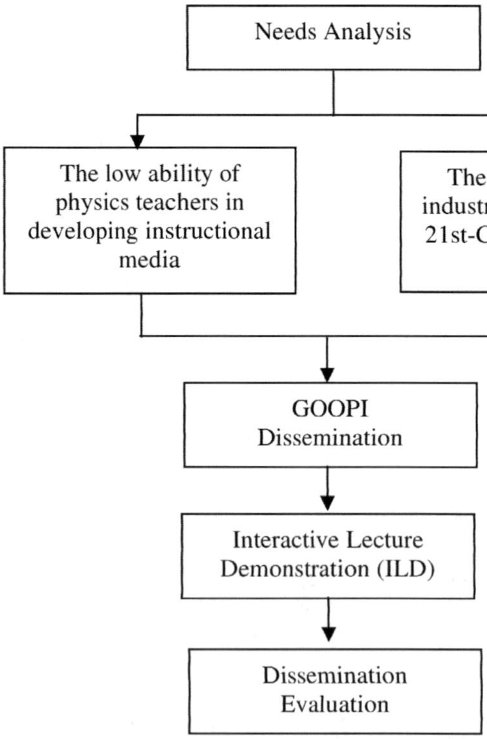

Figure 1. ICCS Activity Program Implementation of GOOPI

3. Result

The GOOPI media used in this study is the result of the development by researchers for approximately 2 years in the multimedia laboratory of the State University of Jakarta, Indonesia. Researchers who are members of Physics Education Research (PER) have conducted a series of studies to examine the working principles of GOOPI (Open Online Physics Instructional Game) as a physics learning medium and can be accessed for free on the following page: https://goopi.id/ (UC Browser) developed by researchers [8] and has been published [9]. The characteristics of GOOPI are shown in Figure 2.

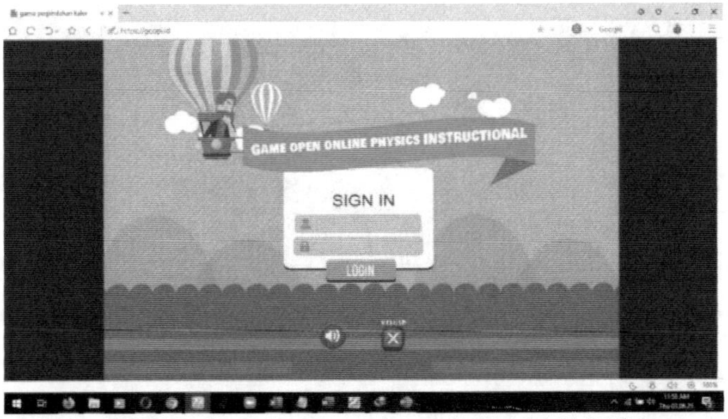

Figure 2. GOOPI Opening Page https://goopi.id/ (UC Browser)

Figure 2 shows the start page of GOOPI, this page can be accessed on the https://goopi.id/ page to make it more optimal to use UC Browser. To be able to Sing In you can enter a user and password, the user and password are "physics" and the user is "123". The GOOPI user and password are intentionally given so that students and teachers can optimize it for learning, especially during the Covid-19 pandemic. After entering the user and password, click login to enter. On this home page there is also a sound button if you want to mute the sound. If students can answer, the finish button will appear, and they can continue to the Mission 3 menu. Mission 3 This GOOPI menu is the last menu, and this menu contains an online game on the topic of radiation. For more details, menu 1, menu 2 and menu 3 are shown in Figure 3.

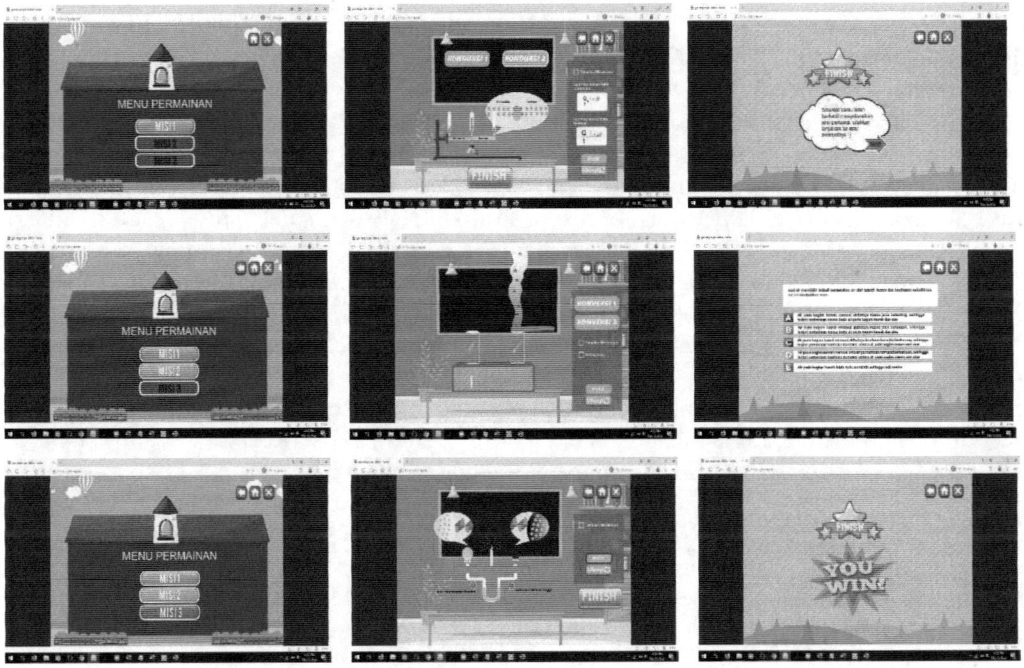

Figure 3. Mission 1, Mission 2, and Mission 3 of GOOPI

Based on Figure 3. Information obtained that the GOOPI process in facilitating physics learning the concept of heat transfer. The advantage of GOOPI is that it can visualize microscopic physical material or phenomena. Characteristics of physics learning that is oriented to conception construction is physics learning in which there is a process of building a physics conception from the initial state of not understanding the concept to understanding the concept. Therefore, real steps are needed in the process that can construct knowledge and correct (reconstruct) concepts if they are not in accordance with the scientific context [11]. The teacher's response to the implementation of this ICCS is as shown in Figure 4.

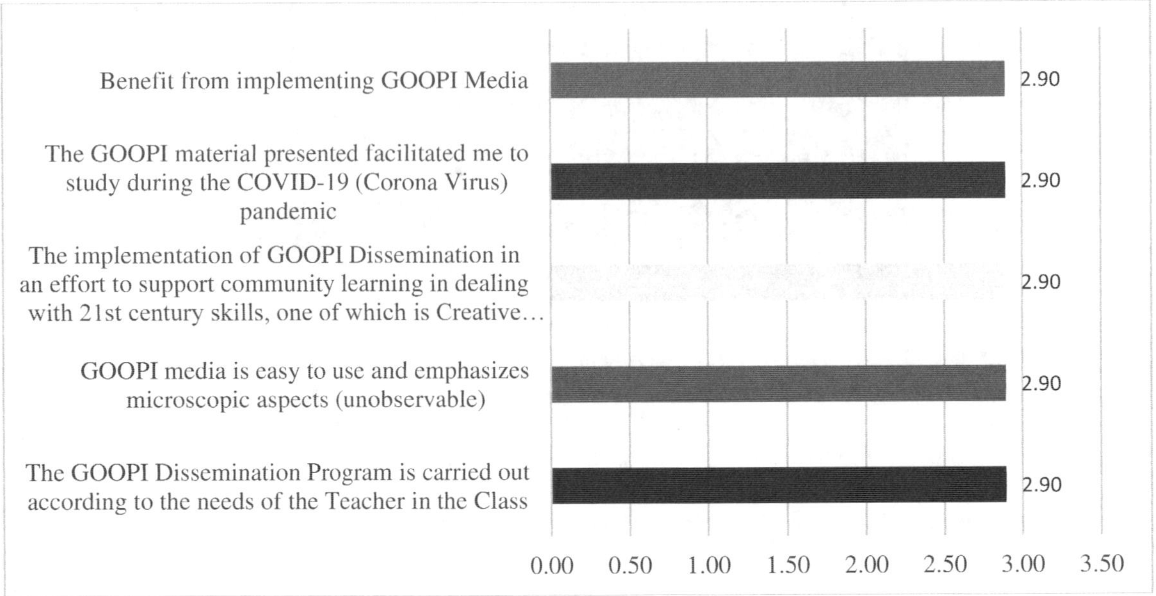

Figure 4. Mission 1, Mission 2, and Mission 3 of GOOPI

Based on Figure 4, information is obtained that the GOOPI Dissemination Program is carried out according to the needs of the Teacher in the Class, the results of the teacher's responses to GOOPI are obtained with an average result of 2.90, this is a very good category. GOOPI media is easy to use and emphasizes microscopic aspects (unobservable) the results of the teacher's responses to GOOPI are obtained with an average result of 2.86, this is a very good category. Meanwhile, the implementation of GOOPI Dissemination to support community learning in dealing with 21st century skills, one of which is Creative Thinking Skills, obtained teacher responses to GOOPI with an average result of 2.93, this is a very good category. Meanwhile, the GOOPI material presented, facilitated me to study during the COVID-19 (Corona Virus) pandemic, the results of the teacher's response to GOOPI with an average result of 2.93, the category is very good. Another response, Benefit from implementing GOOPI Media, the results obtained from teachers' responses to GOOPI with an average result of 2.94, the category is very good.

The results of the study show that learning with media can reduce the quantity of students who have misconceptions on optical material; The results of the research [17] stated that learning physics using magnetic electricity media can change the concept of students who are not scientific to a scientific one; The results of the study [18] show that the use of light media and colour dispersion is effective in helping students construct their conceptions so that their understanding ability can be improved; The results of the study [7] show that the use of virtual simulation media on the concept of light waves can improve understanding of the concept of waves and the process of changing the conception. If in the past lecture learning was the prima donna, now it has shifted to learning using media. This is because the media has the potential to improve conceptual understanding and more effective learning experiences. Students can give virtual responses and will get feedback. Opportunities as a form of real devotion are quite encouraging if physics teachers have the skills to develop media for learning. If the media produced is effectively in use and easy to manufacture, it will be a special attraction for PSTM teachers. The physics PSTM teacher community will tend to use the media for learning as a demand as professional educators who have skills in the Development of Media. With the aim that teachers are confident in facing the 21st century skills so that they can get along in industry 4.0.

4. Conclusion

Having the skills to develop learning media is a teacher's demands and needs in teaching. The conclusion of this dissemination is the results of the questionnaire responses to GOOPI with the very good category of dissemination obtained information that GOOPI is effective for media for learning physics during the Covid-19 epidemic. The suggestion in this study is that dissemination of training should be done frequently in order can provide skills or expertise that are not too complicated but are widely applied and needed by physics teachers. GOOPI is a medium that can show microscopic or unobservable physical concepts. Of course, teachers must be carefully trained in this training so that the skills and knowledge are obtained completely. Suggestions in this ICCS activity were based on input and during the question-and-answer session, information was obtained that the teacher was still weak in developing learning media. So, the recommended suggestion is that this kind of activity should be carried out more often with training or dissemination so that teachers can improve their skills.

Acknowledgments

The research was funded by research grant research at Research and Community Service Institutions (LPPM), Universitas Negeri Jakarta, Ministry of Education and Culture 2021.

References

[1] Ahmad T 2019 Scenario based approach to re-imagining future of higher education which prepares students for the future of work *Higher Education, Skills and Work-Based Learning* 10(1) 217–238 doi:10.1108/heswbl-12-2018-0136
[2] Jackson N C 2019 Managing for competency with innovation change in higher education: Examining the pitfalls and pivots of digital transformation *Business Horizons* doi:10.1016/j.bushor.2019.08.002
[3] Landry M H 2019 The Efficacy of Teaching Independent Study Skills Within English for Academic Purposes Programs *BC TEAL Journal* 4 (1): 1-12.
[4] Balta N and Asikainen M 2019 Introductory students' attitudes and approaches to Physics problem solving: Major, achievement level and gender differences *Journal of Technology and Science Education* 9(3) 378 doi:10.3926/jotse.666
[5] Abidah A, Hidaayatullaah N H, Simamora R M, Fehabutar D, Mutakinati, L 2020 The Impact of Covid-19 to Indonesian Education and Its Relation to the Philosophy of "Merdeka Belajar" *Studies in Philosophy of Science and Education (SiPoSE)* 1 (1) 38-49
[6] Kåre Letrud and Sigbjørn Hernes 2018 Excavating the origins of the learning pyramid myths *Cogent Education* 5:1 1518638
[7] Wibowo Firmanul Catur, Andi Suhandi, Nahadi Nahadi, Bayram Coştu 2017 Virtual Microscopic Simulation (VMS) to promote students' conceptual change: A case study of heat transfer *Asia-Pacific Forum on Science Learning and Teaching* 18(2) pp 1-32
[8] Wibowo F C, Nasbey H, Sanjaya L A, Darman D R. 2020 Development of Game Open Online Physics Instructional (GOOPI) for Improving 21st-Century Careers: Creativity Skill (21-CC:CS) *International Journal of Advanced Trends in Computer Science and Engineering* 9 (3) 2625-2631 https://doi.org/10.30534/ijatcse/2020/20932020
[9] Wibowo F C et al 2021 Game Open Online Physics Instructional (GOOPI) for technology improving 21st-century careers *IOP Conf Ser: Mater Sci Eng* 1098 032079
[10] Yildirim I 2017 The effects of gamification-based teaching practices on student achievement and students'attitudes toward lessons *The Internet and Higher Education* 33 86–92.
[11] Srisawasdi N and Kroothkeaw S 2014 Supporting Students Conceptual Development of Light Refraction by Simulation-Based Open Inquiry with Dual-Situated Learning Model *Journal Computer & Education* 1 (1) 49-79
[12] Jong M S Y 2019 Does online game-based learning work in formal education at school? A case study of VISOLE *Curriculum Journal* 26(2) 249-267
[13] Huizenga J, Admiraal W, Dam G ten and Voogt J 2019 Mobile game-based learning in secondary education: Students' immersion, game activities, team performance and learning outcomes *Computers in Human Behavior* 1-13 doi:10.1016/j.chb.2019.05.020

Journal of Physics: Conference Series **2019** (2021) 012040 doi:10.1088/1742-6596/2019/1/012040

[14] Hamalainen R H, Niilo-Rama M, Lainema T and Oksanen K 2018 How to raise different game collaboration activities: The association between game mechanics, players' roles and collaboration processes *Simulation & Gaming* 49, 50–71 https://doi.org/10.1177/1046878117752470.
[15] Cicchino M I 2015 Using game-based learning to foster critical thinking in student discourse *Interdisciplinary Journal of Problem-based Learning* 9(2):1-13
[16] Cresswell J 2016 *Research Design: Qualitative, Quantitative and Mixed methods Approach* (London: Sage publication lnc)
[17] Eber J M, M Wagner and Hajo G Boomgaarden 2016 Are Perceptions of Candidate TraitsShaped by the Media? The Effects of Three Types of Media Bias *The InternationalJournal of Press/Politics22* (1) 111–132
[18] Katz O, Ramaz F, Gigan S and Fink M 2019 Controlling light in complex media beyond the acoustic diffraction-limit using the acousto-optic transmission matrix Nature Communications 10(1) doi:10.1038/s41467-019-08583-6

Journal of Physics: Conference Series **2019** (2021) 012041 doi:10.1088/1742-6596/2019/1/012041

TPACK and Augmented Reality in Kinematics Practicum Module: Forming HOTS Physics Education Students

F Bakri, R Kusuma and AH Permana
Department of Physics Education, Universitas Negeri Jakarta
Jl. Rawamangun Muka, Jakarta 13220, Indonesia
Corresponding author's email: fauzi-bakri@unj.ac.id

Abstract. High Order Thinking Skills (HOTS) are needed to support 21st-century skills. Students are required to think critically in overcoming a problem. HOTS can be trained through the lab activities. The method used in this research is research and development (R&D) with a Dick & Carey approach. This article described the implementation of TPACK and augmented reality in the kinematics practicum module to forming HOTS for physics education students. This practicum module has implemented integrated technology, pedagogy, content, and knowledge. The addition of augmented reality media makes it easier for students to understand practicum steps and forms of phenomena related to practicum. The module was declared feasible in terms of material, media, and pedagogical concepts in forming the HOTS abilities of physics education students. The stages of HOTS formation in the module present a video of real-world phenomena as a stimulus, and students are asked to answer the pre-test related to the stimulus video, students are asked to create observation tables according to the steps in the practicum, students are asked to process data, and analyze concepts regarding practical material which is related.

1. Introduction

The development of education in the 21st century requires logical, analytical, critical, and innovative thinking [1]. The 21st century competence requires higher-order thinking skills known as 4C skills, including communication, collaboration, critical thinking, and problem-solving, as well as being creative and innovative [2]. Physics education students must be able to develop HOTS to understand physics concepts [3]. HOTS can be trained through lab activities [4].

The model for implementation of the physics practicum that has been carried out is the cookbook model, which is all things related to the practicum, from practicum instructions to tools and materials that have been provided by module practicum. This model has a weakness; the enthusiasm for exploring student knowledge is low [5]. Problem-based learning is a model that can help students forming HOTS to solve problems and bridge the gap between theory and practice [6]. Problem-oriented students ensure that they find relevant solutions, this is because they practice finding solutions independently without any help from lecturer or lab assistant [7].

The practicum module should support the developments and demands of the 21st century, namely the integration of technology in pedagogy and content [8]. The practicum module supporting the development of the 21st century has three major components, including content, technology, and pedagogy, as well as the relationship between the components represented as PCK (Pedagogical Content Knowledge), TCK (Technological Content Knowledge), TPK (Technological Pedagogic) and TPCK (Technological Pedagogical and Content Knowledge) [9]. Material-oriented learning media and learning objectives that utilize technology must be designed within the TPACK framework [10].

The application of AR technology in education has the potential to increase the effectiveness of education and academic training by delivering real-time information enriched with 3D media [11]. In practicum activities, AR is packaged in the form of a 3-5 minute stimulus video related to practicum topics so that students can get an overview of the practicum that will be performed and can improve students' ability to work in the laboratory [12] [13]. The addition of AR media to the practicum module can be used as a learning medium with good quality and feasible as teaching material in physics learning. The use of AR media in the form of videos in experimental laboratory activities can improve the development of students' laboratory skills. In addition, AR videos can help students complete practical activities in a shorter time, thus providing more time to discuss experimental results [14]. AR videos are placed on the initial stage in the form of physical phenomena related to the concepts to be practiced, at the practical stage in the form of how to assemble and use tools. In addition, the AR video can help students complete lab activities in a shorter amount of time, giving them more time to discuss the results of the experiment [15].

Based on the above description, this article presented the results of research into the development of TPACK and augmented reality in the kinematics practicum module to forming HOTS physics education students.

2. Methods

2.1. Research design

This article is the result of research and development of TPACK and augmented reality in the kinematics practicum module using the Dick & Carey approach. In order that this book can forming HOTS physics education students, the R&D process is carried out in the following steps:

2.1.1. Identify learning goals

Identifying learning objectives that refer to the Course Learning Outcomes of the fundamental physics course module Jakarta State University.

2.1.2. Conduct a learning analysis

Identify skills, competencies, and concepts that must be formed in fundamental physics practicum. Learning analysis is carried out by formulating learning objectives in the affective, cognitive, and psychomotor fields.

2.1.3. Analyze learners and context

Analyzing the prerequisite abilities and affective, cognitive, and psychomotor abilities that will be formed in the fundamental physics practicum. Analyze the form of fundamental physics practicum based on the equipment available in the laboratory.

2.1.4. Write performance objective

Formulate learning objectives for each fundamental physics practicum topic based on the equipment available in the fundamental physics laboratory. The formulation of learning objectives describes the implementation of TPACK in practicum to form HOTS for physics education students.

2.1.5. Develop assessment instrument

Develop an instrument for measuring learning objectives for each developed practicum module.

2.1.6. Develop instructional strategy

Develop a practicum strategy that implements TPACK in forming HOTS for physics education students.

2.1.7. Develop and select instructional materials

Develop practical modules that are relevant to the learning objectives of fundamental physics and the equipment available in the laboratory. The module is prepared by implementing TPACK and the stages to form HOTS for Physics Education students.

2.1.8. Design and conduct formative evaluation

Develop a formative evaluation instrument for fundamental physics module products that implement TPACK with the stages of forming HOTS for physics education students. Formative evaluation was carried out for the feasibility of the material, the feasibility of the media, and the feasibility of learning. Formative evaluation instruments are arranged on a continuum scale of 4.

2.1.9. Revise module
Make improvements to the practicum module which, according to the expert is not feasible. Revisions are carried out until the expert states that the practicum module is suitable to be used to form HOTS for physics education students.
2.2. Instrumentation research
The instrument used was a questionnaire to test the feasibility of the research results on the development of the fundamental physics practicum module. The questionnaire used is a questionnaire for the feasibility test of the material, a questionnaire for the media feasibility test, a questionnaire for the pedagogical feasibility test.
2.2.1. Physics concept feasibility test
2.2.1.1. The suitability of the practicum material with the fundamental physics course syllabus, consisting of (a) the suitability of the practicum material with fundamental physics concepts; (b) the suitability of the practicum material with learning outcomes for fundamental physics courses; (c) the suitability of practicum activities with the description of fundamental physics courses.
2.2.1.2. The suitability of practicum activities with the learning outcomes of fundamental physics courses, consisting of (a) the suitability of the practicum objectives with the fundamental physics syllabus; (b) the suitability of the practicum objectives with the learning outcomes of fundamental physics courses; (c) the suitability of practicum activities with the established HOTS.
2.2.1.3. Feasibility of the preparation of the concept of physics, consisting of (a) the accuracy of the stages in the preparation of the concept of physics; (b) the relevance of the stages in the practicum module with the physics concepts learned by students; (c) the relevance of the practicum stages with the formation of student scientific literacy; (d) the relevance of the practicum stages with the established HOTS.
2.2.1.4. The feasibility of the language in the practicum module, consists of (a) the accuracy of the language used; (b) ease of understanding the language used; (c) the effectiveness of sentences in supporting scientific literacy.

2.2.2. Media feasibility test
2.2.2.1. Display of the practicum module, consisting of (a) proportional cover layout (text and image layout); (b) clarity of module title; (c) suitability of font type selection and cover font size (font type and numbers); (d) proportional layout of module content (layout of text and images); (e) the suitability of the selection of the font type and the font size of the content of the module (font type and numbers); (f) the relevance of the instructions for use with the contents of the practicum module; (g) the effectiveness of user-friendly instructions for use.
2.2.2.2. The practicum module media, consists of (a) the relevance of augmented reality-based media to the stages in the practicum module; (b) the quality of augmented reality-based media is clearly presented; (c) duration of augmented reality-based media; (d) interactive augmented reality-based media; (e) the completeness of augmented reality-based media to support independent study students.
2.2.2.3. Augmented reality, consisting of (a) easy installation of augmented reality applications; (b) easy access to augmented reality applications; (c) ease of user interaction with augmented reality applications

2.2.3. Pedagogical feasibility test
2.2.3.1. The practicum stages, consist of (a) orientation students to problems; (b) organize students; (c) guide individual or group investigations; (c) develop and present the work; (d) analyze and evaluate the problem solving process.
2.2.3.2. Implementation of TPACK, consisting of (a) implementation of technological knowledge from the material practiced; (b) implementation of pedagogical knowledge for the practicum stage; (c) implementation of content knowledge for the practicum stage; (d) implementation of technological knowledge and pedagogy for the presented augmented reality media; (e) implementation of technology knowledge and content for stimulus videos and applications in each practicum module; (f) implementation of pedagogical knowledge and content for the formation of student HOTS at each stage of the practicum; (g) implementation of TPACK in the formation of HOTS in each practicum module;

(h) implementation of TPACK in the formation of scientific literacy in activities in the practicum module.

2.3. Data is analyzed

The expert assessment questionnaire on the practicum module produces research data. Experts provide an assessment using a continuum scale of 4 on each indicator of each aspect of the assessment.

Table 1. Continuous Scale

No	Alternative Answer	Score
1.	Very Good	4
2.	Good	3
3.	Bad	2
4.	Very Bad	1

The interpretation of the assessment results is calculated using the technique of determining the score range and score presentation.

$$\text{IS} = \frac{\text{maximum scale} - \text{minimum scale}}{\text{number of scale}}$$

$$\text{Scor percentage} = \frac{\sum \text{earning score}}{\sum \text{maximum score}} \times 100\%$$

The result of obtaining the percentage score is then measured using the interpretation of the score and the percentage score for the continuum scale, as shown in Table 2.

Table 2. Score Interpretation

Scale	Score Percentage	Interpretation
$1 \leq \text{IS} < 1{,}75$	$25\% \leq \text{SP} < 43{,}8\%$	Very Not Feasible
$1{,}75 \leq \text{IS} < 2{,}50$	$43{,}8\% \leq \text{SP} < 62{,}5\%$	Not Feasible
$2{,}50 \leq \text{IS} < 3{,}25.$	$62{,}5\% \leq \text{SP} < 81{,}3\%$	Feasible
4. $3{,}25 \leq \text{IS} < 4{,}00$	$81{,}3\% \leq \text{SP} < 100\%$	Very Feasible

3. Result and discussion

3.1. Product

3.1.1. The pre-stage of the practicum module

This section consists of cover, preface, laboratory rules, list of practicum activities, and initial questions in physical phenomena videos related to practical materials integrated with AR applications.

3.1.2. The main stage of the practicum module

This stage consists of practicum objectives that students must achieve, tools and materials needed for practicum activities, brief theoretical material related to practicum material, work steps equipped with video assembling tools and materials, observation tables, data processing, and concept analysis. Video assembly tools integrated with AR applications.

3.1.3. The post-stage of the practicum module

In this stage, there are post practicum assignments that students must do to measure student knowledge after doing a practicum.

3.1.4. AR application development

In the practicum module, several photos with unique markers are integrated with videos to support practicum activities, such as videos of physical phenomena at the beginning and videos of assembling tools and materials in the main section. The videos contained in this practicum module can be displayed using an augmented reality application. This augmented reality application was developed with the Unity software and installed on android with a minimum android version of 4.1 Jelly Bean. The kinematics practicum module has several menu sections in the augmented reality application, namely Scan, Hint, About, and Exit.

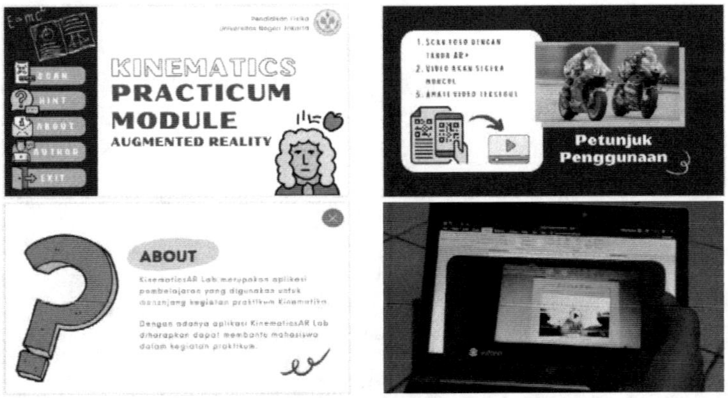

Figure 1. Interface of Kinematics AR Lab

3.1.5. HOTS indicators in the practicum module

This kinematics practicum module develops HOTS, which consists of 3 categories, namely analyzing (C4), evaluating (C5), and creating (C6). These three categories can be seen in the following table.

Table 3. HOTS indicators in the practicum module

No.	HOTS indicators	Part of the module	Information
1.	Analyze (C4)	**Main stage** Observation table	-□ Students can analyze the observation table into a meaningful graph.
		Data processing	-□ Students can interpret graphs following physics concepts.
2.	Evaluate (C5)	**Pre-stage** Pre-lab assignment	-□ Students can criticize AR videos related to physics concepts in the initial question section.
		Main stage Data processing	-□ Through questions in data processing and concept analysis, students can evaluate a situation based on the graphs that have been obtained.
3.	Create (C6)	**Main stage** Data processing	-□ Students can develop concepts based on data observations and graphs.
			-□ Students can make mathematical equations from the concepts and graphs they get from practical activities.
		Post-stage Post-lab assignment	-□ Students create problem-solving by combining concepts obtained through the practicum.

3.2. Description of validation results

The development of this practicum module has passed the validation test by media experts, material experts, and pedagogical experts.

3.2.1. Physics concept feasibility test

As shown in Table 4, the results of material expert validation for this research product obtained an average value of 97%. These results indicate that the kinematics practicum module equipped with augmented reality technology to forming HOTS for physics students is considered very good and feasible to use.

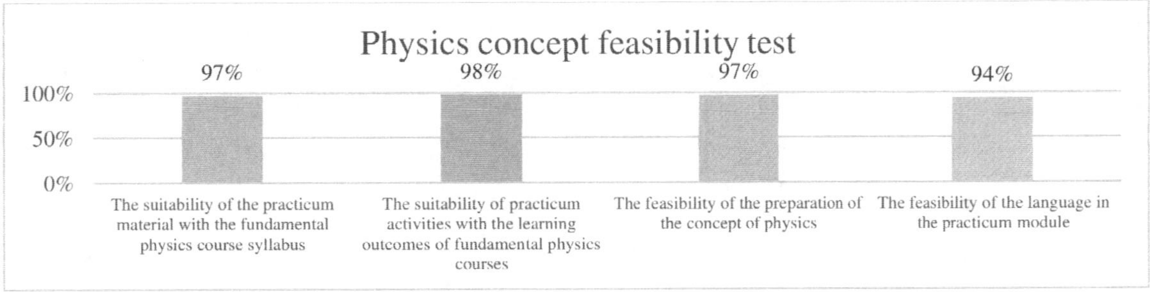

Figure 2. Graphics of formative evaluation results for material feasibility

3.2.2. Media feasibility test

As shown in Figure 3, the results of media expert validation for this research product obtained an average value of 94%. These results indicate that the kinematics practicum module equipped with augmented reality technology to forming HOTS for physics students is considered very good and feasible to use.

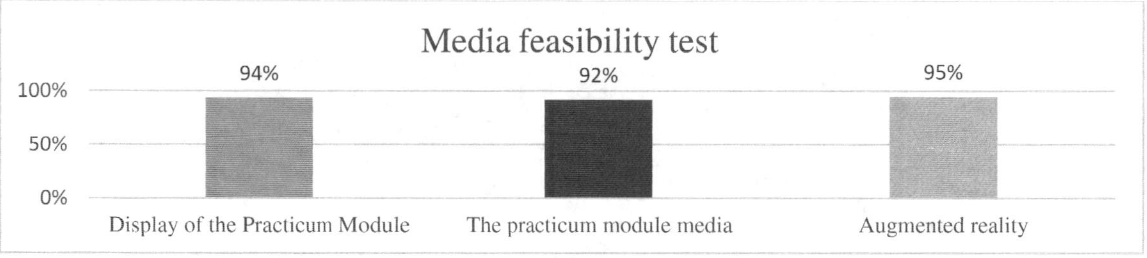

Figure 3. Graphics of formative evaluation results for media feasibility

3.2.3. Pedagogical feasibility test

As shown in Figure 4, the results of pedagogical expert validation for this research product obtained an average value of 94%. These results indicate that the kinematics practicum module equipped with augmented reality technology to forming HOTS for physics students is considered very good and feasible to use.

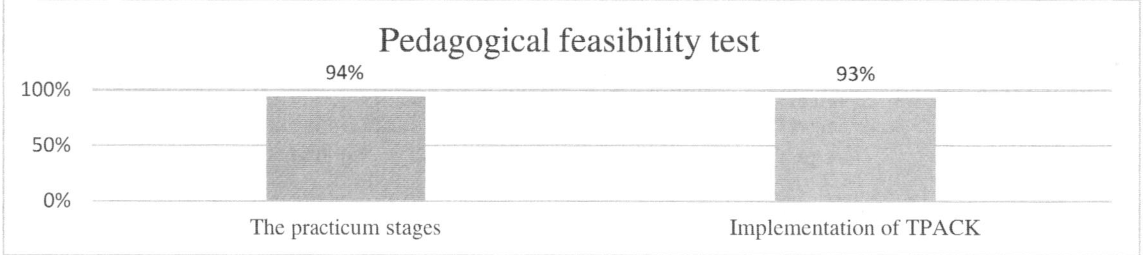

Figure 4. Graphics of formative evaluation results for pedagogical feasibility

3.3. Discussion

The development of the 21st century has brought education to be more advanced, as seen from the many learning media equipped with technology [1]. A kinematics practicum module has been produced by utilizing augmented reality to form HOTS for physics education students. This kinematics practicum module was developed using a problem-based learning model. The implementation of TPACK in collaborating between problem-based learning models and media-based augmented reality technology can form the stages of training HOTS through practical activities [9]. With the addition of videos about real-world phenomena, students identify and solve physics problems in everyday life [12]. The addition of videos on instructions for assembling tools and materials, helps students practicum more quickly and independently [15].

4. Conclusion

The implementation of TPACK and augmented reality-based media in the kinematics practicum module is declared feasible and can be used in learning to form HOTS for physics education students. The use of augmented reality technology-based media is very supportive of student independence in practicum activities.

Acknowledgments
We gratefully acknowledge funding support from the *Hibah Penelitian Terapan Fakultas* by the decision of the Dean Faculty of Math and Science Universitas Negeri Jakarta No 0597/UN39.5/FMIPA/DT/2021.

Reference
[1] Hilton M 2010 *Exploring the Intersection of Science Education and 21st Century Skills A Workshop Summary* (Washington DC: The National Academies Press)

[2] Prihadi E 2018 Pengembangan Keterampilan 4C Melalui Metode Poster Comment pada Mata Pelajaran PAI aan Budi Pekerti (Penelitian di SMA Negeri 26 Bandung) *Jurnal Pendidikan Islam Rabbani*

[3] Dosinaeng W B N Leton S I and Lakapu M 2019 Kemampuan Mahasiswa dalam Menyelesaikan Masalah Matematis Berorientasi HOTS *JNPM (Jurnal Nasional Pendidikan Matematika)* pp 250-264

[4] Malik A 2018 Pengembangan Higher Order Thinking Laboratory (Hot-Lab) Untuk Meningkatkan Transferable Skills Mahasiswa Calon Guru Fisika *Doctoral dissertation Universitas Pendidikan Indonesia*

[5] Subali B 2010 Penerapan Model Praktikum Problem Solving Laboratory Sebagai Upaya Untuk Memperbaiki Kualitas Pelaksanaan Praktikum Fisika Dasar *Jurnal Pendidikan Fisika Indonesia 6,* pp 90-97

[6] Choi S Y 2014 Effects of Problem-Based Learning vs. Traditional Lecture on Korean Nursing Students' Critical Thinking, Problem-Solving, and Self-Directed Learning *Nurse Education Today* p. 52–56

[7] Sumarmi 2012 *Model-model Pembelajaran Geografi* (Yogyakarta: Aditya Media)

[8] Sholihah M and Yuliati L 2016 Peranan TPACK Terhadap Kemampuan Menyusun Perangkat Pembelajaran Calon Guru Fisika Dalam Pembelajaran Post-Pack *Jurnal Pendidikan: Teori, Penelitian, dan Pengembangan* pp 144-153

[9] Koehler M J and Mishra P 2014 What Is Technological Pedagogical Content Knowledge? *Contemporary Issues in Technology and Teacher Education*

[10] Nevrita N, Nasikin N and Amelia T 2020 Analisis Kompetensi TPACK Guru Melalui Media Pembelajaran Biologi SMA *JurnalPendidikanSainsIndonesia (Indonesian Journal of Science Education)* pp 203-217

[11] Kangon L 2012 Augmented reality in education and training *TechTrends* pp 13-20

[12] Bakri F, Wulandari S and Muliyati D 2020 Students Worksheet With Augmented Reality Media: Scaffolding Higher Order Thinking Skills Of High School Students On Uniform Accelerated Motion Topic *Journal of Physics: Conference Series*

[13] Nadelson L S, Scaggs J, Sheffield C and McDougal O M 2015 Integration of video-based demonstrations to prepare students for the organic chemistry laboratory *Journal of Science Education and Technology 24(4)* pp 476-483

[14] Bakri F, Pratiwi S and Muliyati D 2019 Video-enriched Worksheet Based on Augmented Reality Technology: The Heat Experiment is Easier *AIP Conference Proceedings 2169* p 020010

[15] Bakri F, Ervina E and Muliyati D 2019 Practice The Higher-Order Thinking Skillsin Optic Topic Through Physics Worksheetequipped With Augmented Reality *AIP Conference Proceedings*

Journal of Physics: Conference Series **2019** (2021) 012042 doi:10.1088/1742-6596/2019/1/012042

E-learning in Sains Learning: A-Review of Literature

R Mulyanengsih and F C Wibowo
Physical Education Study Program, Faculty of Mathematics and Natural Sciences, State University of Jakarta, Jl. Rawamangun Face No. 1, Rawamangun, Pulogadung, East Jakarta 13220, Indonesia.
Email: nengrat34@gmail.com, firmanul27@gmail.com

Abstract. Developments in information technology encourage changes in various sectors of human life, not least in the world of education. Information and communication technology is a tool to improve the quality of education. This study discusses e-learning in physics learning in terms of the characteristics, and opportunities of e-learning to support 21st century skills. This article examines 20 articles on physics learning using e-learning which is presented in references in the 2017 to 2019 period and analyzed using this method content analysis. It was found that this study was conducted in both high school and university and the results of learning by implementing e-learning were able to meet the diverse needs of students, very effective in helping students improve motivation, learning experience and learning outcomes better than conventional learning methods. And provide opportunities to improve 21st century skills.

1. Introduction

One of the real challenges of education in the era of globalization is to be able to produce human resources who have complete competence, known as 21st century competence. The main competencies that students must have in order to be able to take part in real life in the 21st century. Important skills in the 21st century include learning to know, learning to do, learning to be and learning to live together. Each of these four principles contains specific skills that need to be empowered in learning activities, such as critical thinking skills, problem solving, scientific literacy, and communication skills. Achieving 21st century skills can only be done by updating the quality of learning, helping students develop participation, and personalizing learning. To prepare it all, of course, a new paradigm is needed in the learning process in the classroom. Along with the development of information technology, a new trend in the world of education is to use computers in the learning process. Computers allow learning to be more flexible and able to serve individual students. One of the uses of computer and internet technology is a learning system through electronic learning or better known as e-learning. E-learning is a popular information technology that has been the focus of educational research over the last decade (Sucheta V. Kolekar1 & Radhika M. Pai1 & Manohara Pai MM1, 2018, Mohammed Simko Abdullah 1, Mehmet Toucan 1*, 2017, Lyubov Alekseevna Krasnova and Viktor Yurjevich Shurygin, 2020, I Kadek Suartama, (2020), e-learning is used at all levels of education from elementary to university levels, Sucheta V. Kolekar, Radhika M. Pai*, Manohara Pai MM(2020), Lyubov Alekseevna Krasnova and Viktor Yurjevich Shurygin*(2020), D Mulhayatiah1*, A Kindi1, Y Dirgantara(2018). So this study focuses on the use of e-learning in learning, and looks for answers to what are the characteristics of e-learning, and what opportunities can be taken. In e-learning studies?

2. Literature Review

The development of technology, especially in the field of information and communication in the 21st century, has had a significant influence on educational services. Currently we are in an increasingly competitive era, it takes careful preparation to form superior human resources. An important component in turning challenges into opportunities is to continue to practice developing yourself, learning new skills that suit the needs of today's students. The skills most needed today are cultural competence skills and mastery of technology.

According to Horton in Ramadhani (2012) e-learning is any utilization or use of the internet and web technology to create a learning experience. E-learning can be seen as an innovative approach to design a good delivery medium, user-centered, interactive and as an open, flexible and distributed learning environment.

E-learning according to Oetomo in Darmawan (2014) is a concept in an effort to integrate traditional learning processes (traditional learning), distance learning (distance learning), and learning that combines various learning models (blended learning), which combines various learning models that aimed at optimizing learning processes and services both remote, traditional, mediated, and even computer-based. Students who study can take advantage of online teaching materials, facilities, then print and download and study classical both in class and outside of class, after that they discuss with the help of print, electronic, and online media.

3. Method

The type of research used is the content analysis method. Content analysis is a research that is an in-depth discussion of the content of a pre-existing article. Research data collection is done by searching for articles contained in online journals, using Google Cedekia. The keyword used by the researcher in searching the article is "e-learning", in science learning.

From a search using the keywords "e-learning" and "science learning" several articles were obtained and then selected articles that met the criteria for e-learning and science learning. Then a descriptive qualitative data analysis was carried out to see the characteristics of e-learning used in science learning.

4. Results and Discussion

The research results obtained 20 relevant articles, then analyzed and synthesized because the data from the research reports were still very broad, so only 10 relevant articles were taken. The data of the article were processed by summarizing and determining the essence of the research results. Then the data is reported back by means of qualitative and quantitative descriptions. The data from the analysis of the characteristics of e-learning can be seen as follows:

Table 1. Data analysis of characteristic E-Learning

NO	Author	Year	Study	Subject	Result
1	Sucheta V. Kolekar, Radhika M. Pai*, Manohara Pai M.M.	2018	e-learning-learning style	College student	Effectively improve learning style and learning experience
2	Mohammed Simko Abdullah 1, Mehmet Toycan 1*	2017	e-learning – learning outcome	College student	Efektif meningkatkan hasil belajar

NO	Author	Year	Study	Subject	Result
3	Lyubov Alekseevna Krasnova and Viktor Yurjevich Shurygin*	2020	e-learning- physics teacher professionalism	physics teacher	Effectively improve the competence of physics teachers in terms of teaching methods
4	I Kadek Suartama	2020	E-learning development	High School	e-learning is very good for use in the learning process
5	D Mulhayatiah1*, A Kindi1, Y Dirgantara	2018	e-learning- problem solving skills	High School	Effectively improve problem solving skills
6	Garry Falloon	2019	e-learning- learning, achievement, and group collaboration	College student	Effectively improve cooperative learning
7	Abdullah AlhabibSebuah, Jennifer Rowleyb,*	2019	e-learning- learning effectiveness	High School	Effectively improve learning activities
8	Ahmad Alshehri, MalcolmRutter, Sally Smith	2019	E-learning design – motivation to learn	High School	E-learning design increases students' learning motivation
9	Ching-Huei Chen	2014	e-learning- learning motivation	High School	Effectively increase learning motivation
10	Che Ahmad Azlan, PhD1 , Jeannie Hsiu Ding Wong, PhD1*, Li Kuo Tan, PhD1, Muhammad Shahrun Nizam AD Huri, MMedPhys1 , Ngie Min Ung, PhD2 , Vinod	2020	e-learning – learning activities	College student	Effectively improve learning activities

NO	Author	Year	Study	Subject	Result
	Pallath, PhD3 , Christina Phoay Lay Tan, PhD4 , Chai Hong Yeong, PhD5 , Kwan Hoong Ng, PhD1,6*				

Based on the results of the study, it was found that e-learning is very effective in helping the learning process in the classroom, e-learning is not a substitute for teachers or tutors, but as a medium that can improve the shortcomings of the classical learning process, e-learning is quite effective in increasing learning, motivation, adapting style diverse student learning, and make the learning process student-oriented, so it is hoped that the development of e-learning can be used as a vehicle for developing 21st century skills for students. The results showed that there were no research results linking e-learning with 21st century skill variables, especially scientific literacy skills. This is a gap and the basis for further research on applied physics courses.

5. Recommendations

Based on the literature review, there are recommendations regarding the use of e-learning in classroom learning.

- E-learning is able to meet the diverse needs of students, so that the nature of student-centered science learning can run well.
- the ease of using e-learning can increase students' motivation and learning activities
- Based on the literature review, research is still rare to measure the effect of e-learning on 21st century skills, especially regarding scientific literacy. Therefore, further research is expected to be able to test the effectiveness and influence of the application of e-learning on students' scientific literacy skills

6. Conclusion and Suggestion

Based on the literature review above, it can be concluded that the use of e-learning in learning has a positive effect on learning in the classroom. E-learning is able to meet the diverse needs of students, increasing activities and learning outcomes.

References

[1] Keenan A Pituch and Yao-Kuei Lee 2006 The Influence of System Characteristics on E-Learning Use pp 222–244

[2] Abdullah Alhabeeb and Jennifer Rowley 2018 E-Learning Critical Success Factors: Comparing Perspectives from Academic Staff and Students pp 1-12

[3] Ahmed Alshehri, Malcolm Rutter and Sally Smith 2019 Assessing the Relative Importance of an E-learning system's Usability Design Characteristics Based on Students' Preferences 3 (8) pp 839-855

[4] Mitja Decˇman 2015 Modeling the acceptance of e-learning in mandating environments of higher education: The influence of previous education and gender 49 pp 272-281

[5] Ezzat A, José H and Carmen Penades 2017 on The Way To Learning Style Models Integration: A Learner's Characteristics Ontology

[6] Yinan Liu, Bong Gee Jang and Zaline Roy-Campbell 2017 Optimum Input Mode In The Modality And Redundancy Principles For University ESL Students' Multimedia Learning *Computers & Education* do: 10.1016/j.compedu.2018.08.025

[7] Sucheta V Kolekar1, Radhika M Pai1 and Manohara Pai M 2018 Rule Based Adaptive User Interface For Adaptive E-Learning System Springer *Science+Business Media part of Springer Nature*

[8] Kyle C and Huff 2015 The Comparison Of Mobile Devices To Computers For Web-Based Assessments, *Computer In Human Behavior* 49 pp 208-212

[9] Weirui Wang, Sigal Segev and Yu Liu 2015 Use of Social Networking Sites for Product Communication: Computer and Human Behavior 49 pp 86-93

[10] Che Ahmad Azlan at all 2020 Pengajaran Dan Pembelajaran Fisika Kedokteran Pascasarjana

Menggunakan E-Learning Berbasis Internet Selama Pandemi COVID-19-studi kasus dari Malaysia S1120-1797(20)30245-3

[11] Weirui Wang, Sigal Segev and Yu Liu 2015 Use Of Social Networking Sites for Product Communication: A Comparative Study of Hispanics and Non-Hispanic Whites, *Computers in Human Behavior* 49 pp 86-93

[12] Sucheta V Kolekar, Radhika M and Manohara 2020 Adaptive User Interface for Moodle based E-learning System using Learning Styles *Procedia Computer Science* 135 pp 606–615

[13] Mohammed Simko Abdullah and Mehmet Tycoon 2020 Analysis of the Factors for the Successful E-Learning Services Adoption from Education Providers' and Students' Perspectives: A case study of Private Universities in Northern Iraq. ISSN: 1305-8223 (online) 1305-8215 (print) 2018 14(3):1097-1109

[14] Lyubov Alekseevna Krasnova and Viktor Yurjevich Shurygin 2020 Blended learning of physics in the context of the professional development of teachers *Int. J. Technology Enhanced Learning V*ol 12 No 1

[15] Gabriel Badea, Elvira Popescu, Andrea Sterbini and Marco Temperini 2020 Integrating Enhanced Peer Assessment Features in Moodle Learning Management System *Lecture Notes in Educational Technology* https://doi.org/10.1007/978-981-13-6908-7_19

[16] Mulhayatiah D, Kindi A and Dirgantara Y 2018 Moodle-Blended Problem Solving on Student Skills in Learning Optical Devices *Journal of Physics: Conf. Series* do: 10.1088/1742-6596/1155/1/012073

[17] Suleiman Alhaji Ahmad, Umar BawaChinade and Abdu Muhammad Gambaki 2012 The Need for Moodle as A Learning Management System In Nigerian Universities: Digesting University Utara Malaysia Learning Zone As A Case Study Vol 2(3)

[18] Garry Falloon 2018 Using Simulations to Teach Young Students Science Concepts: An Experiential Learning Theoretical Analysis *Computers & Education* pp 138-159

[19] Winfred Arthur, Dennis Doverspike Gonzalo, Jason E Taylor and Alison E Carr 2014 *International Journal of Selection and Assessment* Vol 22(2)

[20] Ezra Açikgül Firata and Mustafa Serdar Köksal 2019 Effects of Instruction Supported by Web 2.0 Tools, On Prospective Teachers' Biotechnology Literacy *Computers & Education* pp 61-74

Journal of Physics: Conference Series **2019** (2021) 012043 doi:10.1088/1742-6596/2019/1/012043

Project Based Learning (PjBL) Learning Model in Science Learning: Literature Review

I J Nurhidayah, F C Wibowo and I M Astra
Physical Education Study Program, Faculty of Mathematics and Natural Sciences, State University of Jakarta, Jl. Rawamangun Face No. 1, Rawamangun, Pulogadung, East Jakarta 13220, Indonesia.
Email: isnajuitanh@gmail.com

Abstract. The world of education today not only requires students to be experts in the cognitive realm, but is required to be able to achieve 21st century skills. Based on the analysis and synthesis of journals, the appropriate learning model to face the 21st century is the Project Based Learning (PjBL) learning model. The purpose of this article is to identify PjBL from the characteristics, effectiveness and implementation aspects of science learning. This structured review reviewed 20 articles on PjBL for science learning based on the available Scopus database reference from 2017 to 2021. The data obtained were analyzed using content analysis methods. The results showed that on average PjBL can be categorized as a learning model that can improve student learning outcomes in science learning and train students in problemving (critical thinking). The review reveals that PjBL has an influence on student learning, especially in science learning. From this article, it can be concluded and can be recommended three recommendations related to the essential success of PjBL in schools.

1. Introduction

Project-Based Learning (PjBL) is a form of learning that focuses on students. Students are actively involved in the learning process. Through this PjBL learning process, it will train students' thinking in dealing with problems. In PBL, students work collaboratively with others and reflect on what they have learned. In addition, students can become active in the search, and decision-making process by improving their practical thinking skills (Harris, 2002; McGrath, 2002; Solomon, 2003) [1][2][3]. In addition, PBL develops students' scientific process skills. Therefore, students who develop scientific process skills produce solutions to their scientific problems by asking questions, discussing ideas, making observations and predictions, conducting experiments, collecting and analyzing data (Westwood, 2006). PjBL aims to involve students in the learning process. So that with project-based learning students are more active and learning will be more effective and efficient.

Finally, PjBL contributes to students' academic achievement by creating a more equal learning environment (Solomon, 2003) [3]. PjBL Project work is assigned to individuals or groups of students. Then, the project work begins with the selection of a particular topic by the learners with the teacher's scaffolding. Topics are generally problems that students can solve by experiment or observation (Bell, 2010) [5]. Project work provides opportunities for students to promote their achievements. In addition, these works allow students to work together in a real-world environment by collaborating on an assignment (Baş, 2011) [6]. PjBL according to Baran et al (2018) [7] which shows a positive impact on student learning outcomes in physics lessons on Newton's law concepts by showing there is a significant increase in the average test

Journal of Physics: Conference Series	**2019** (2021) 012043	doi:10.1088/1742-6596/2019/1/012043

scores obtained by participants in the experimental group and the control group ($p < 0,05$). In this case, it can be said that the PjBL learning model has an influence on student learning. Moreover, education is currently being faced with 21st century skills, so this PjBL learning model is considered capable of being used to cultivate 21st century skills.

Project-based learning is learning that is more student-centered. Project-Based Learning (PjBL) is one of the teaching strategies in which students are guided to solve problems. In addition, PjBL develops students' scientific process skills. Therefore, students who develop scientific process skills produce solutions to their scientific problems by asking questions, discussing ideas, making observations and predictions, conducting experiments, collecting and analyzing data (Kizapan & Bektas, 2016) [8]. In fact, in PjBL, students collaborate with each other and take on their responsibilities as team members. In addition, students recognize the similarities between what they learn and what happens outside of school. In addition, PjBL improves students' metacognitive skills, thereby; they make successful plans, and evaluate their solutions (Blumenfeld et al., 1991)[9]. Finally, PBL contributes to student academic achievement by creating a more equal learning environment. In PjBL learning, students are given a project in groups by the teacher.

2. Method

This research is a research with content analysis method and collects data through literature study, which is searched by researchers through Google Cendikia and publish or perish with the keywords project based learning (PjBL). Then the results of the search on PjBL obtained several articles relevant to PjBL. Then the articles were analyzed and synthesized to determine the role of PjBL in learning. The problems studied in this study are about the characteristics of PjBL, the effectiveness of PjBL, and the implementation of PjBL in learning.

3. Results and Discussion

3.1 Identifying the Characteristics of Project-Based Learning (PjBL)

Project work begins with selecting a topic that will lead students to a problem. Which of these problems can be solved by experiment or observation. Each individual or group begins to investigate information about a particular topic (Kizapan & Bektas, 2016) [8]. Thus, students offer suggestions for solving problems; these suggestions are called "Project Plans" in schools. Then, each individual or group collects the information and documents needed to prove or support the topic. They conduct experiments, make observations, collect and interpret data and record the results. At the end of the session, each group made a report and presented the results of their study in class. The presented project is discussed and evaluated by all students and teachers (Frank & Barzilai, 2004; Herron, Magomo & Gossard, 2008) [10,11]. In essence, it can be concluded that Project-Based Learning is a learning model designed to develop desired student characteristics such as research skills, self-confidence, responsibility and cooperation, through activities where students work individually or in groups to design plans. And program within the specified time. The PjBL syntax is:

Table 1. Project Based Learning Syntax

No	Syntax	Activities
1	Reflection	Reflection Students are led into the context of the material and relate what is known to what needs to be learned. Then students are given a problem around.
2	Research	Research Students collect relevant sources of information to solve problems. Each group is directed to joint discussion
3	Discovery	Discovery Each group designs an experiment
4	Application	Application Each group will test the results of the experiments that have been designed.
5	Communication	Communication Each group reports the results of their research, and draws conclusions together

In Table 1 it is explained that there are six learning syntaxes regarding PjBL. Based on this syntax, students can be trained to combine the concepts they have learned with the problems around them. In addition, students are also trained to collect information related to the material. Then students are trained to design experiments and work with their groups. In PjBl syntax, students are also trained to communicate the results of their work in front of the class. Based on the literature review regarding PjBL, there are several characteristics that can be stated, including those listed in table 2.

Table 2 Characteristics of PjBL in Learning

No	Researcher (Year)	Country	Study	Key Findings
1	Pengyue Guo, Nadira Saab, Lysanne S. Post, Wilfried Admiraal (2020)	Netherlands	The PjBL -The Learning	PjBL improves studen t learning outcomes in higher education
2	Mohd Ali Samsudin , Seyedh Mahboobeh Jamali, Ahmad Nurulazam Md Zain, Nader Ale Ebrahim (2020)	Turkey	The PjBL -Self Efficacy Students	PjBL increases student s' self-efficacy in solving phy sics problems
3	I Wayan Santyasa, Ni Ketut Rapi, I Wayan Windu Sara. (2020)	Indonesia	PjBL - Academic Value	PjBL improves studen ts' academic scores
4	Muchsin and Mariati (2020)	Indonesia	PjBL- Creative Thinking	PjBL improves studen ts' creative thinking
5	Mohd Ali Samsudin , Ahmad Nurulazam Md Zain, Seyedh Mahboobeh Jamali, and Nader Ale Ebrahim. (2018)	Malaysia	PjBL - Problem solving s kills	The PjBL improve Enhanc ing the ability of solving the problem of students

Based on the findings in Table 2, it can be seen that PjBL learning plays an important role in the student learning process at school. A review of the literature listed in Table 2, PjBL does not only affect student learning outcomes, as revealed by Pengyue Guo, Nadira Saab, Lysanne S. Post, Wilfried Admiraal (2020) [12]. Other studies listed in the table as research by Mohd Ali Samsudin, Seyedh Mahboobeh Jamali, Ahmad Nurulazam Med Zain, Nader Ale Ebrahim (2020) [13] stated that PjBl after being applied to students there was an increase in student self-efficacy, especially in problems in the field of physics. . Whereas according to my Wayan Santyasa, Ni Ketut Rapi, I Wayan Windu Sara (2020) [14] stated that PjBL can increase students' academic scores. Muchsin and Mariati (2020) [15] stated that PjBL can improve students' creative thinking in applying learning materials. So it can be concluded that project-based learning can train students' ability to solve problems. Mohd Ali Samsudin, Ahmad Nurulazam Md Zain, Seyedh Mahboobeh Jamali, and Nader Ale Ebrahim. (2018) [16] PjBL improves students' problem solving abilities. In addition, project-based learning can improve students' creative thinking. So students are trained to be creative in working on projects and products. So that it can be concluded that PjBL learning greatly affects students' academics, through PjBL students can be trained to be independent in solving problems, and making projects according to the direction of the teacher.

3.2 Identifying Evidences of the Effectiveness of Project-Based Learning (PjBL)

Most of the literature reviews on Project-Based Learning (PjBL) involved students, both students in the control group using the conventional learning model and students in the experimental group using the PjBL model. It was found that PjBL had a positive influence on student learning. Good effectiveness of learning outcomes, student motivation, students' creative level, and increasing scientific literacy. As for the literature review that discusses the evidence of the effectiveness of PjBL in learning, it can be seen in Table 3.

Table 3 Effectiveness of Project- Based Learning in Learning

No	Author (Year)	Study	Key Findings
1	Ashley M. Craft, Robert M. Capraro (2017)	PjBL- Student academic enga gement	Effective in increasing student academic engagement
2	Horse Admawati & J umadi (2018)	PjBL - Rasa want to know the students, the attitude of science, cooperation with others	Effective in improving the attitude of science students, develop a sense of wanting to know, and cooperation with an other
3	Fikriye Kirbağ Zengi n, and Afak Yucasu (20 17)	PjBL - Results Postes students	Effective in improving student posttest results
4	Derya Basera , c, M. Yasar Ozdenb and Hasan Karaarslan (20 17)	PjBL - student advanced tech nology and collaboration	Effective and efficient in impr oving students' advanced techn ology and collaboration
5	Kristin Huysken, Harold Olivey, Kevin McElmurry, Ming Gao, and Peter Avis (2019)	PjBL- Student understanding	Effective in improving student s' understanding
6	Medine Baran, Abdulkadir Maska n , and eyma Yaşar (2018)	The PjBL -The Learning	Effective in improving student physics learning outcomes
7	Muchsin and Mari ati (2020)	PjBL-Thinking creatively	Effective in improving student s' creative thinking
8	Adwan Mohammad Hasan Bani-Hamad, Abdul Hakim Abdullah (2019)	The PjBL-Skills Century 21	Effective in improving skills of Century 21

Based on Table 3 it can be seen that the results of the study show that academic rigor and relevance provided through STEM PBL lessons increase students' academic engagement (Ashley et al., 2017) [17]. Then the results of the study also stated that STEM PjBL had a significant effect on students' scientific attitudes. This shows that PjBL STEM can develop students' curiosity, open-mindedness, and cooperation with others (Horse Admawati Jumadi, 2018) [18]. The results of the study revealed that project-based learning had a significant effect on improving student posttest outcomes (Fikriye Kirbağ Zengin and Afak Yucasu, 2017) [19]. Interdisciplinary, collaborative PjBL studies are efficient in improving students' advanced collaborative and technological skills (Derya Basera, c, M. Yasar Ozdenb and Hasan Karaarslan, 2017) [20]. Then the results of research from articles that have been explored show that PjBL improves student performance and student understanding of the subject matter (Kristin, 2019) [21]. The results show that PjBL has a positive impact on student learning outcomes in physics lessons on Newton's law concepts by showing a significant increase in the average test scores obtained by participants in the experimental group and the control group ($p < 0.05$) (Medine Baran, Abdulkadir Maskan, and email Yaşar, 2018) [7].

The results show that PjBL according to can improve students' creative thinking with an average n-gain acquisition in the experimental class I of 0.75 belonging to the high category and in the experimental class II of 0.77 and belonging to the high category, with an average acquisition the value of n-gain for both classes is 0.76 in the high category (Muchsin and Mariati, 2020) [16]. The results showed that there was a statistically significant difference in the level ($\alpha = 0.05$) of students' 21st century skills test (4Cs) between the scores before and after the test, indicating better performance in the post-test. In other words, the project-based learning approach has a significant effect on improving 21st century skills (4C) (Adwan Mohammad Hasan Bani-Hamad, Abdul Hakim Abdullah, 2019) [22]. So it can be concluded that PjBL has a positive influence on learning. So the literature review can illustrate that students are more honed in their learning.

3.3 Implementation of Project-Based Learning (PjBL)
Implementation of the use of the project-based learning model (PjBL) will require quite a lot of implementation time. Careful preparation of subject teachers before implementing PjBL is an important step. To prepare for project-based learning, a teacher is not only required to prepare what projects are in accordance with the learning material. However, teachers are also required to be mature in preparing student worksheets (LKPD), tools and materials used during the project-based learning process.

The successful implementation of PjBL in the classroom lies in the ability of teachers to effectively assist student learning, motivate, and guide students during the learning process. In the PjBL process, collaboration between students and students, then students and teachers are very important. The project-based learning model is a learning model that provides opportunities for teachers to manage learning in the classroom by involving project work carried out by students both in groups (Nasir et al., 2019) [23]. In addition, project-based learning is an instructional approach that offers the potential to help students develop flexible understanding and lifelong learning skills (Hmelo-Silver, 2004) [24].

The use of project-based models has been suggested in the literature review as the first effective learning model in assisting students in developing the required knowledge and skills. To then design and manufacture products independently or in groups.

Recommendations in Learning Based on the literature review, there are recommendations that can make the implementation of the project-based learning model (PjBL) more effective in the school environment. Then also recommendations regarding the achievements that can be obtained by students using PjBL, especially the achievements of the 21st century.

1. Student support: Students need to be mentored and supported effectively while implementing PjBL. Make the best use of the time possible to work on projects that have been guided by the teacher. In addition, students must be taught to be able to work with groups. and directed
2. Teacher support: Support for peer teachers is essential for the successful implementation of project-based learning models. Exchanging experiences regarding the implementation of PjBL will make the implementation of the model more focused.
3. Application of PjBL to measure 21st century skills: Based on a review of the literature, it is still rare for research to measure the effect of PjBL on 21st century skills, particularly regarding critical thinking. Therefore, further research is expected to examine the effectiveness and influence of a PjBL implementation on students' critical thinking skills.

4. Conclusion
Based on the literature review above, it can be concluded that PjBL learning has a positive influence on student learning in schools. And it is known that PjBL can improve students' thinking power in dealing with a problem, then improve learning outcomes as well. In addition, PjBL learning can make students more active when participating in classroom learning.

References
[1] Harris, J. (2002). Learning and Leading with Technology, 27(7), 42-50
[2] McGrath, D. (2002). Learning and Leading with Technology, 30(3), 42-50
[3] Solomon, G. (2003). *Project-Based Learning: a primer, Technology and Learning*, 23(6), 20-30.
[4] Westwood, P. (2006). *Teaching and Learning Difficulties: Cross-Curricular Perspectives,*

Camberwell, Vic: ACER. Press
[5] Bell, S. (2010). The Clearing House, 83(2), 39-43
[6] Baş, G. (2011). TOJNED: The Online Journal of New Horizons in Education, 1(4), 1-15.
[7] Baran, Medine., Maskan, Abdulkadir.,& Yaşar, Şeyma. (2018). International Journal of Instruction April 2018, Vol.11, No.2 e-ISSN: 1308-1470, p-ISSN: 1694-609X pp. 221-234.
[8] Kızkapan, Oktay & Bektaş, Oktay. (2017). International Journal of Instruction January 2017.Vol.10, No.1 e-ISSN: 1308-1470. www.e-iji.net p-ISSN: 1694-609X pp. 37-54.
[9] Blumenfeld, P. C., Soloway, E., Marx, R. W., Krajcik, J. S., Guzdial, M., & Palincsar, A. (1991). Educational psychologist, 26 (3-4), 369-398.
[10] Frank, M. & Barzilai, A. (2004*)*. Assessment & Evaluation in Higher Education, 29(1), 41-61.
[11] Herron, S., Magomo, D., & Gossard, P. (2008). International Journal of Social Sciences, 3(1), 44-53.
[12] Guo, Pengyue., Saab, Nadira., S. Post, Lysanne., & Admiraal, Wilfried. (2020). International Journal of Educational Research 102 (2020) 101586. https://doi.org/10.1016/j.ijer.2020.101586.
[13] Samsudin, Mohd Ali., Jamali, , Seyedh Mahboobeh., Md Zain, Ahmad Nurulazam ., & Ebrahim, Nader Ale. (2020). Journal of Turkish Science Education, *17* (1), 94-108.
[14] Santyasa, I Wayan., Rapi, Ni Ketut., & Sara, I Wayan Windu. (2020). International Journal of Instruction January 2020 Vol.13, No.1 e-ISSN: 1308-1470. www.e-iji.net p-ISSN: 1694-609X pp. 489-508.
[15] Muchsin & Mariati. (2020*)*. Budapest International Research and Critics Institute-Journal (BIRCI-Journal) Volume 3, No 2, May 2020, Page: 1453-1458 e-ISSN: 2615-3076 (Online), p-ISSN: 2615-1715 (Print). DOI: https://doi.org/10.33258/birci.v3i2.1008.
[16] Samsudin, Mohd Ali., Md Zain, Ahmad Nurulazam., Jamali, Seyedh Mahboobeh., and Ebrahim, Nader Ale. (2018). The Asia Pacific Journal of Educators and Education, 32. Pp 21-28.
[17] Craft, Ashley M., & Capraro, Robert M. (2017*)*. Electronic International Journal of Education, Arts, and Science, Volume: 3, Issue: 6, 140-158, Ohio, USA.
[18] Admawati, Harsi & Jumadi. (2018). Advances in Intelligent Systems Research (AISR), Mathematics, Informatics, Science, and Education International Conference (MISEIC 2018), volume 157.
[19] Zengin, Fikriye Kirbağ & Yucasu, Afak. (2017). International Congress Of Eurasian Social Sciences (ICOESS) Özel Sayısı, Vol:8, Issue: 28.
20] Baser, Derya., Ozden, M. Yasar., & Karaarslan, Hasan. (2017). Research in Science & Technological Education, 35:2, 131-148, DOI: 10.1080/02635143.2016.1274723.
[21] Huysken, Kristin., Olivey, Harold., McElmurry, Kevin., Gao, Ming., & Avis, Peter. (2019). Journal of the Scholarship of Teaching and Learning, Vol. 19, No. 1, February 2019, pp. 6-28. Do: 10.14434/josotl. v19i1. 26777.
[22] Bani-Hamad, A. M. H., & Abdullah, A. H. (2019). International Journal of Academic Research in Business and Social Sciences, 9(12), 560–573.
[23] Nasir, Muhammad., Fakhrunnisa, Rini., Nastiti, Luvia Range. (2019). International Journal of Environmental & Science Education, Vol. 14, No. 5, 229-238.
[24] Hmelo-Silver, C. E. (2004). 16(3), 235-266. https://doi.org/10.1023/B:EDPR.0000034022.16470.f3.

Trends of Flipped Classroom Studies for Physics Learning: A Systematic Review

S S Amanah, F C Wibowo and I M Astra
Department of Physics Education, Universitas Negeri Jakarta
Jl. Rawamangun Muka, Jakarta 13220, Indonesia
Email: ssasulis@gmail.com

Abstract. This study focused to review the trends of the Flipped Classroom (FC) studies for physics learning. This review seeks to address the overall trends of the existing studies, contribution of integrated model or strategy and media into flipped classroom approach, the advantages of flipping classroom for physics courses, and the challenges of these studies. This structured review examined 33 article journal about flipped classroom for physics learning. The obtained data were analyzed using content analysis method. It was found that the studies in this field were generally carried out at undergraduate students, which pure flipped model, used discussion method and institution website for online media. The implementation of flipping classroom approach in a physics course yielded positive impact on student achievement when compared to traditional classroom. On the other hand, some challenges were identified, such as students' challenges, teacher challenges, and technical challenges.

1. Introduction

The development of information and communication technology in the 21st century is increasing in all fields, including in the field of education. Flipped Classroom (FC) is a new and popular learning model, in which learning activities that are usually carried out in the classroom are turned into home activities, and activities that are usually homework are turned into classroom activities [1]. Interactive activities involving higher order thinking skills are the main activities in the classroom [2]. Abeysekera and Dawson [3] define the flipped classroom as a learning approach that has the following characteristics:

1) move most of the teaching/delivery of subject matter outside the classroom
2) use class time for active and interactive learning activities
3) requires students to complete pre- and/or post-class activities in order to fully benefit from in-class activities

In other words, the more specific characteristics of the flipped classroom approach include:

1) Changes in the use of class time
2) Changes in the use of time outside of class
3) Doing activities that are traditionally considered 'homework' in class
4) Doing activities are traditionally considered as 'work in class' outside of class
5) Classroom activities that emphasize active learning, peer learning, and problem solving
6) There are pre-class activities
7) There are post-class activities
8) Use of technology, especially video.

Broadly speaking, the flipped classroom has two stages in its learning, namely pre-class learning and classroom learning. Pre-class self-learning is carried out before class starts which refers to the delivery of knowledge in students' private spaces, usually at this stage teaching materials are given in video format and exercises related to the material being studied. In the next stage, activities are carried out in class in groups or in pairs using discussion, exercise or project methods [2][4]. Flipped classroom is a technology-based learning approach that emphasizes problem solving in peer-assisted learning in the classroom [5][6]. Flipped classroom is seen as a learner-centered learning approach [4]. So that the teacher only acts as a facilitator to motivate, control and provide feedback on the results that students provide[7]. Along with the trend of flipping classroom research, the application of the flipped classroom in various fields of study has been carried out [7], one of which is in learning physics. The view that physics is a difficult subject is always attached to students [8]. Therefore, studying the subject matter before learning in class is one way to increase student understanding. In addition, the use of interesting media can change students' bad perspective on physics. The application of flipping classroom is quite effective in improving student learning outcomes in class [7]. This is also expected in physics learning research. From the results of the identification, several research questions were formed including:
1) What is the trend of flipping classroom research on physics learning from 2016 to 2021?
2) What learning models and methods are usually integrated with the flipped classroom approach?
3) What media and platforms are usually integrated with the flipped classroom approach?
4) What are the benefits of applying flipped the classroom to physics learning?
5) What are the challenges of applying flipped the classroom to physics learning?

2. Method

This study analyzed and synthesized articles about flipped classrooms in physics learning from 2016 to April 2021. Articles were collected using Publish or Perish software using the keywords "flipped" or "flipped physics" in the Google Scholar database. There are several criteria used to select the articles to be analyzed. The criteria are as follows.

Table 1. Article included and excluded.

Included	Excluded
English	Non-English
Journal articles, book chapter	Non-journal article (e.g. proceedings, conference series, book review, web article, theses, etc.)
Empirical research	Literature review, meta analysis
Physics learning	Non-physics learning (e.g. EFL, math, etc.)
Full document	Abstract only
2016 – April 2021	May 2021 and later

The first search found 108 articles. Then the articles were selected by reading the abstract and adjusted to the criteria. So that at the final stage, 33 articles were obtained that met the criteria.

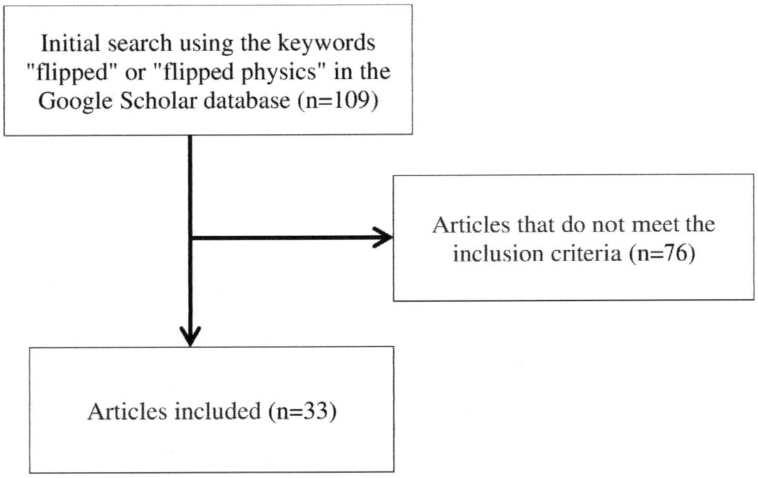

Figure 1. Method for selection article.

3. Results and Discussion

3.1 Flipped Classroom research trends in physics learning

This study analyzes articles on the flipped classroom approach to physics learning from 2016 to mid-2021. Based on the 33 articles collected, flipped classroom research for physics learning was most published in 2018 (n=8) and the last in 2019 (n= 4) and in 2021 (n=4). However, because the data was collected only until mid-2021, which indicates that flipped classroom research on physics learning may still increase over time. So it can be observed that there was a significant decline from 2018 to 2019 which issued the fewest publications.

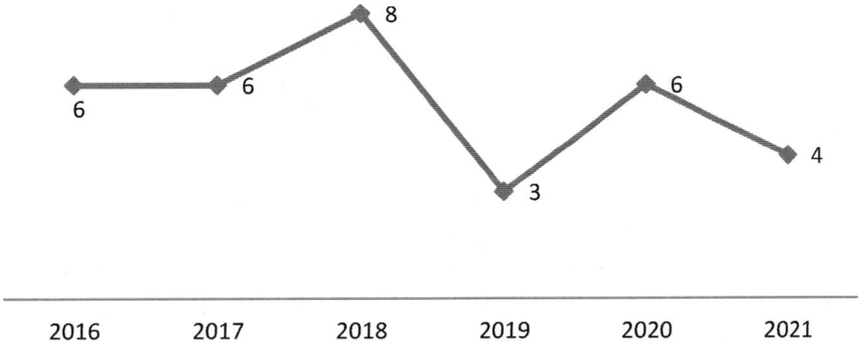

Figure 2. The number of journal articles published by year.

Research in flipped classrooms will mostly be applied in universities. This is because the learning independence factor for university students is quite good compared to high school students.

Table 2. Learner type.

Learner type	Number	Article
Secondary school	14	[9][10][11][12][13][14][15] [5][16][17][18][19][20][21]
Undergraduate	17	[22][23][24][25][26][27][28] [29][30][31][32][33][34][35] [36][37][38]
Postgraduate	1	[39]
Diploma Program	1	[40]

3.2 Learning models and methods

In several studies it was found that the flipped classroom approach can be integrated with various supportive learning models. However, 73% of the articles analyzed showed that the application of the flipped classroom approach was pure without being integrated with other learning models. However, several studies have integrated the flipped classroom approach with other learning models that support active learning.

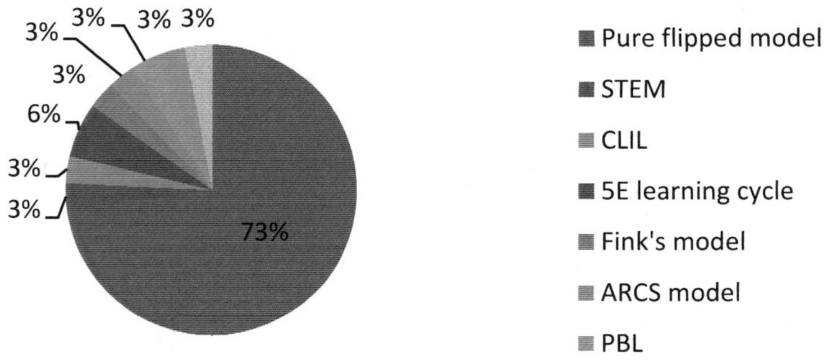

Figure 3. Integrated model.

In its application, a learning approach usually applies several learning methods in it. The learning methods that are more integrated in the flipped classroom are discussion methods (31%), quizzes (17%) and experiments (11%). The methods of discussion, experimentation, question and answer, demonstration, problem solving, and peer instruction support active learning in the classroom at the in-class stage. Gamification, quiz and simulation methods are usually applied with the help of online platforms and support pre-class activities.

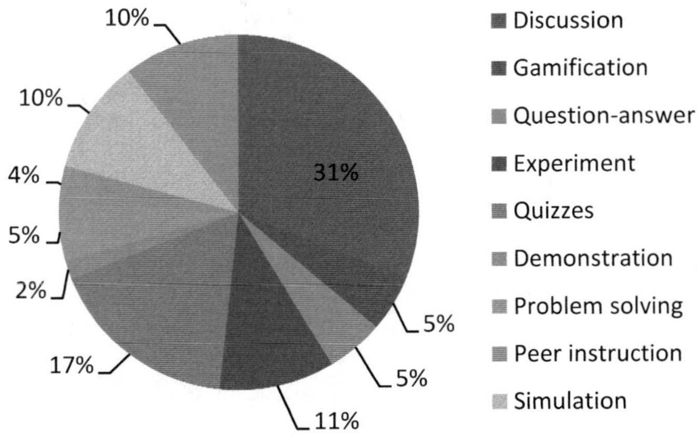

Figure 4. Learning method.

3.3 Learning media and platforms

In flipped classroom -based learning, of course, requires online-based learning media, especially the Learnings Management System (LMS) which is used to support pre-class activities. The LMS platforms found in this study include Moodle, Edmodo, Google Classroom, and institutional websites designed specifically for online learning at these institutions. A total of 13 studies show that the use of special websites designed by institutions or specially designed by researchers is able to support the effectiveness of the application of the flipped classroom approach. Several websites that have been specially developed by institutions/institutions include LAMS (Learning Activity Managenet System), online annotation system and M-learning media. In addition to using the LMS, several studies also use social media such as WhatsApp and blogs to support pre-class learning activities. At the stage of in-class, integrated multiple platforms learning that support active learning, quiz games like Edpuzzle, Kahoot, quizzes, Flipquiz. The platforms are used with the hope to make the learning process more active and certainly fun to motivate students to more contribute during the learning process of physics class.

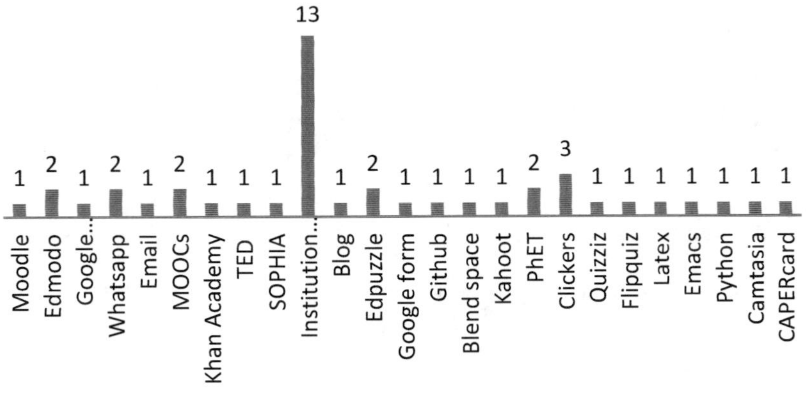

Figure 5. Learning platform.

3.4 Advances by applying flipped the classroom to physics learning.

The application of the flipped classroom has many benefits for learning physics. The findings of this study indicate that the most common benefits of applying flipped classrooms are to improve learning outcomes in learning physics which is often considered difficult.

Table 3. Advances by applying flipped the classroom to physics learning.

Advances	Number	Sample Article
Improve physics learning achievement	16	[34]
Improve students' motivation	5	[9]
Increase students' self-direction in learning	5	[31]
Positive responses from students	5	[28]
Increase students' understanding	4	[27]
Increased student participation	4	[13]
Good time management	4	[23]
Develop interpersonal interaction	3	[21]
Increase students' interest	2	[15]
Increase positive students' attitudes	2	[22]
Increase student engagement	2	[37]
Develop specific skills	1	[30]
Increase students' awareness	1	[14]
Decrease students' anxiety	1	[28]
Active learning	1	[28]

3.5 The challenge of applying the flipped classroom to physics learning

Some studies do not write about the difficulties in implementing the flipped classroom and only focus on the benefits of its implementation. This study analyzes the difficulties in implementing the flipped classroom in physics learning which are grouped into three categories, namely student difficulties, teacher difficulties and technical difficulties (table 4). The results of the analysis show that the most difficult student difficulties or during pre-class activities. Some students do not do pre-class activities well, for example, students do not watch the video material that has been provided by the teacher [10]. Some studies also show that students find it difficult to participate in dynamic and independent learning [17]. The most common problem found from the teacher's point of view is that the flipped classroom approach requires quite a lot of preparation, such as preparing learning videos, homework assignments, and the use of the LMS as a support for pre-class activities [8]. Technology for teachers who will apply this learning approach.

Table 4. The challenge of applying the flipped classroom to physics learning.

	Challenges	Number	Sample Article
Students	Students need more motivation to do pre-class activities	4	[39]
	Students have difficulty in dynamic learning	4	[28]
	Increase students' study hours	1	[25]
	Students unable to ask questions while pre-class	1	[16]
	Students do not prefer it	1	[32]
Teachers	Need more teachers' preparation	7	[17]
	Teachers should be more trained	2	[12]
	Teachers have difficulty controlling students' pre-class activities	2	[26]
	Time consuming	2	[27]
Technical	Problem of technology and internet access	4	[13]

4. Conclusion

The study is analyzing the study of literature about the flipped classroom approach to learning physics from 2016 to mid 2021. The publication of research on the flipped classroom on learning physics at most that in 2018 but experienced a drastic decline in subsequent years. This approach is applied more at the university level than at the secondary school level. The independence of students at the university level is more supportive in the application of the flipped classroom in physics lessons. In its application, the flipped classroom approach is mostly applied without being combined with other learning models, but several studies explain that the integration of certain learning models can support the implementation of the flipped classroom. The discussion method is the most widely integrated learning method with the flipped classroom, especially in in-class activities. The application of the flipped classroom must be supported by adequate technology and platforms. A total of 13 studies use and implement their respective institution's website as an online platform to support pre-class activities in the flipped classroom.

Basically, the application of flipping classroom in physics learning has a very positive impact, especially in improving physics learning achievement. However, in its implementation, there are several obstacles, especially during pre-class activities. The study shows that there are still many students who are not motivated to participate in pre-class learning activities and follow the subject matter with dynamic methods. The difficulty experienced by the teacher is that it takes more time to prepare before learning activities. The next challenge that is often found in this study is technology and internet access.

The flipped classroom approach has indeed supported the active learning process. However, integrating certain learning models and methods that also support active learning can be a good choice. There are not many studies on integrating the flipped classroom approach with learning models and methods. Therefore, more studies are needed on this matter.

In addition, students who are less motivated in pre-class activities are always a recurring problem and are often found in the application of flipping classrooms. Therefore, it takes a fun way and method of learning in pre-class activities that can also control student participation.

The determination of the articles analyzed in this study is less specific because it is only searched using the Google Scholar database. In addition, the article selection process that only reads the abstract manually can produce different results if analyzed by different researchers. Therefore, an efficient selection method, or tool is needed to select articles.

References

[1] J. Bergmann and A. SAMs, 2012.

[2] D. Zou, *J. Comput. Educ.*, vol. 7, no. 2, pp. 213–228, 2020, doi: 10.1007/s40692-020-00153-w.

[3] L. Abeysekera and P. Dawson," *High. Educ. Res. Dev.*, vol. 34, no. 1, pp. 1–14, 2015, doi: 10.1080/07294360.2014.934336.

[4] H. T. Hung, *ELT J.*, vol. 72, no. 3, pp. 296–308, 2018, doi: 10.1093/it/ccx055.

[5] C. K. Lo, C. W. Li, and K. F. Hew, *Comput. Educ.*, vol. 118, pp. 150–165, Mar. 2018, doi: 10.1016/j.compedu.2017.12.003.

[6] H. C. Lin and G. J. Hwang, *Interact. Learn. Environ.*, vol. 27, no. 8, pp. 1011–1027, 2019, doi: 10.1080/10494820.2018.1467462.

[7] Z. Zainuddin and S. H. Halili, *Int. Rev. Res. Open Distance Learn.*, vol. 17, no. 3, pp. 313–340, 2016, doi: 10.19173/irrodl. v17i3. 2274.

[8] I. S. Colon, Y. S. M. Tan, and Y. H. Chou, *Res. SCI. Educ.*, vol. 48, no. 1, pp. 117–149, 2018, doi: 10.1007/s11165-016-9562-6.

[9] K. H. Li and H. K. Fun, in *Shaping the Future of Education, Communication and Technology*, 2019, pp. 91–106.

[10] N. Hadriani, *Sang Pencerah J. Elm. Univ. Muhammadiyah But.*, vol. 7, no. 1, pp. 33–39, Jan. 2021, doi: 10.35326/pencerah. v7i1. 771.

[11] I. A. D. Astuti, Y. B. Bhakti, R. A. Sumarni, D. Sulisworo, and M. Two-four, *Int. J. Educ. Managua. Innov.*, vol. 2, no. 2, p. 175, May 2021, doi: 10.12928/ijemi. v2i2. 3491.

[12] C. S. Ugwuanyi, C. C. Nduji, U. C. Elejere, and N. E. Amok, *Int. J. SCI. Basic Appl. Res.*, vol. 52, no. 2, pp. 136–148, 2020, [Online]. Available:

http://gssrr.org/index.php?Journal=JournalOfBasicAndApplied.
[13] F. S. Say and F. S. Yıldırım, 2020. [Online]. Available: http://iojet.org/index.php/IOJET/article/view/759.
[14] E. F. N. Putri and E. Purwaningsih, *Rev. Mex. Fish. E*, vol. 18, no. 1, pp. 131–135, Jan. 2021, doi: 10.31349/REVMEXFISE.18.131.
[15] R. Herawati, D. Sulisworo, and S. Fayanto, *IOSR J. Res. Method Educ.*, vol. 9, no. 4, pp. 51–58, 2019, doi: 10.9790/1959-0904015158.
[16] A. Koray and B. E. Üniversitesi, 2018. [Online]. Available: https://www.researchgate.net/publication/330384810.
[17] N. Diamantopoulos, A. Brami, and I. Spanos, 2018. [Online]. Available: http://lamsfoundation.org.
[18] J. Wang, M. Jou, Y. Love, and C. C. Huang, *Comput. Human Behav.*, vol. 84, pp. 36–48, Jul. 2018, doi: 10.1016/j. chub. 2018. 02. 018.
[19] R. Capone, M. R. Del Sorbo, and O. Fire, *Eurasia J. Math. SCI. Technol. Educ.*, vol. 13, no. 10, pp. 6579–6582, 2017, doi: 10.12973/ejmste/77044.
[20] K. Y. Al-Derbashi, *J. Educ. Pract.*, vol. 8, no. 29, pp. 102–105, 2017, [Online]. Available: www.iiste.org.
[21] Z. Atwa, R. Din, and M. Hussin, 2016.
[22] A. G. Ç. ŞEKERCİOĞLU and E. YÜNKÜL, 2021.
[23] H.-C. Hung, S. Shwu, and C. Young, 2021.
[24] A. Pattiserlihun and S. J. J. Setiadi, *J. Inov. Pendidik. IPA*, vol. 6, no. 1, Apr. 2020, doi: 10.21831/jipi. v6i1. 28109.
[25] H. Torío, *J. Technol. SCI. Educ.*, vol. 9, no. 3, pp. 404–419, 2019, doi: 10.3926/JOTSE.554.
[26] K. Miller, B. Lukoff, G. King, and E. Mazur, *Front. Educ.*, vol. 3, no. 8, Mar. 2018, doi: 10.3389/feduc. 2018. 00008.
[27] J. L. Cagande and R. R. Jaguar, 2018.
[28] G. Aşıksoy and Y. Sorakin, 2018. [Online]. Available: http://iojet.org/index.php/IOJET/article/view/389/238.
[29] D. Addenda, *J. Pendidik. Teknol. dan Kejuru.*, vol. 24, no. 1, pp. 13–22, 2018, doi: 10.21831/jptk. v24i1. 18319.
[30] S. L. D. Martínez, K. Lombaerts, and C. L. Celaya, *Am. J. Phys. Educ*, vol. 11, no. 2, 2017, [Online]. Available: http://www.lajpe.org.
[31] J.-C. Jao, *World Trans. Eng. Technol. Educ.*, vol. 15, no. 1, pp. 28–33, 2017.
[32] G. Aşıksoy and F. Ozdamli, *Croat. J. Educ.*, vol. 19, no. 4, pp. 1131–1166, 2017, doi: 10.15516/cage. v19i4. 2564.
[33] G. Aşıksoy, *Qual. Quant.*, vol. 52, pp. 129–145, Dec. 2018, doi: 10.1007/s11135-017-0597-1.
[34] G. Asiksoy and F. Özdamli, *Eurasia J. Math. SCI. Technol. Educ.*, vol. 12, no. 6, pp. 1589–1603, Jun. 2016, doi: 10.12973/Eurasia. 2016. 1251a.
[35] A. K. Wood, R. K. Galloway, R. Donnelly, and J. Hardy, *Phys. Rev. Phys. Educ. Res.*, vol. 12, no. 1, Jun. 2016, doi: 10.1103/PhysRevPhysEducRes.12.010140.
[36] J. C. Y. Sun, Y. T. Wu, and W. I. Lee, *Br. J. Educ. Technol.*, vol. 48, no. 3, pp. 713–729, May 2017, doi: 10.1111/bjet. 12444.
[37] K. Miller, S. Zyto, D. Karger, J. Yoo, and E. Mazur, *Phys. Rev. Phys. Educ. Res.*, vol. 12, no. 2, 2016, doi: 10.1103/PhysRevPhysEducRes.12.020143.
[38] J. Lazendic-Galloway, M. Fitzgerald, and D. H. Mckinnon, 2016. [Online]. Available: http://www.caperteam.com/classroom-resources.
[39] C. Stöhr, C. Demazière, and T. Adawi, *Comput. Educ.*, vol. 147, pp. 1–12, Apr. 2020, doi: 10.1016/j.compedu.2019.103789.
[40] S. F. Widyastuti and E. D. Wuryanto, *Sci. Educ.*, vol. 9, no. 2, p. 153, Dec. 2020, doi: 10.24235/sc.educatia.v9i2.7254.

Journal of Physics: Conference Series **2019** (2021) 012045 doi:10.1088/1742-6596/2019/1/012045

Unveil Problem Based Learning on Physics Learning: A Literature Review

Mustika and F C Wibowo
Department of Physics Education, Jakarta State University
Jl. Rawamangun Muka, Jakarta 13220, Indonesia
Email: mustikafisika16@gmail.com, Firmanul27@gmail.com

Abstracts. Scientific literacy is one of the benchmarks for the ability of words in the divinity of daily life. This study is about *problem-based learning* (PBL) on learning physics from aspects of the word, not, and its implementation. The writing of this article uses literature reviews. First of all, the study problem is formulated, then while it is being experienced by existing and relevant researchers to be analyzed. Data collection techniques with electronic journals through google scholar, Scopus, Microsoft academies, Harzing's Publish or Parish, and crossref. From the results of 20 articles from journals and 3 e-books. Based on the results of the analysis concluded that the Problem Based Learning model is a skill skill, thinking analysis skills, and scientific attitudes. The results of the journal synthesis show that there have been no research results linking the learning-based problem model to the variety of science literacy skills. This becomes a loophole and basis for advanced research and development of e-Module-based learning problems. The thing behind the varied skills of science skills is one of the problems one sign of learning there is a problem, student-centered learning, and learning in a small group.

1. Introduction

The education has a very important role for the development of a country. The quality of education a n Agra influenced by many factors. The quality of education is determined by several factors such as curriculum, teachers or teaching staff, facilities, and learning resources. Teachers have an important role in improving the quality of learning in the classroom. To improve the quality of learning, teachers can conduct innovative learning in the classroom. Innovative learning prioritizes students at the center of learning. One model of student- centered learning is Problem Based Learning (PBL).

It's ringing in a good way from the person who's doing it. Retention, learning attitudes, and active use of learning only by learning where pupils are about, and sucking with, what they learn [1].

1.1 Problem Based Learning

Good learning in the classroom is able to foster an understanding of students' concepts and ways of thinking. Many models are able to foster understanding of the concepts and ways of thinking of students, one of which is Problem Based Learning (PBL) model. PBL was introduced by Barrow and Tamblyn (1980) and is suitable for use because it can strengthen students' ability to learn a principle and concepts and to gain knowledge from many different situations. PBL is a constructivist learning model that helps students solve problems and thinking [2]. Model problem based learning is a set of teaching models that use problems as a focus to develop problem solving skills, materials, and self-regulation [3]. PBL is one of the learning models that is primarily designed to develop students' critical thinking skills, problem solving skills, intellectual skills, learn the role of adults with experience through real and simulated situations, and become independent learners [4].

Ali mason suggested that Problem Based Learning, is a learning model that uses problems as a first step in collecting and integrating new knowledge [5]. In addition, Syahroni Ejin stated that problem based learning is learning model in which students are faced with real-life (contextual) problems from the environment so as to improve students' conceptual understanding and critical thinking skills [6]. According to Rahmadani and Anugraheni problem based learning emphasizes problem solving activities in learning. Through the problem based learning model, students learn through problem solving activities that can hone students' thinking skills. Problem based learning is learning that uses real world problems as a context for students to learn about critical thinking and problem solving skills, as well as to acquire essential knowledge and concepts from the subject matter [7].

According to Paul Eggen and Don Kauchak, the phases of the problem based learning model include: (1) reviewing and presenting problems, (2) developing strategies, (3) implementing strategies, (4) discussing and evaluating results. In the phase of reviewing and presenting problems, the teacher is able to review the knowledge needed to solve problems and give students specific and concrete problems to solve. In the strategy formulation phase, students can formulate strategies to solve problems and the teacher provides feedback to students. In the strategy implementation phase, students are able to apply strategies in solving problems and the teacher carefully monitors and provides feedback to students. In the phase of discussing and evaluating the results, the teacher guides the discussion about the students' efforts and the results they get [1]. Heller created problem-solving steps in science through five stages: (1) visualizing the problem, (2) describing the problem in a physics description, (3) planning a solution, (4) implementing the plan, (5) examining and evaluating [4].

Hmelo-Silver & Barrows stated that the problems that arise in problem based learning, learning are the questions given do not have a single answer, meaning that students must be involved in exploration with several answer solutions. Student involvement in problem based learning activities can help develop students' critical thinking skills, because in problem based learning activities, students are fully involved in learning process activities through problem solving. In problem based learning activities, students can develop critical thinking skills as a step in solving problems and can draw conclusions based on what they understand [8].

So the problem based learning Model or problem-based learning model is a learning model that involves students in learning activities and prioritizes real problems in the school, home, or community environment as a basis for acquiring knowledge and concepts through critical thinking and problem solving skills.

1.2 Critical Thinking

According to Eggen and Kauchak critical thinking is the ability and tendency of a person to make and make assessments of conclusions based on evidence. Eggen and Kauchak stated that with critical thinking, a person can adjust, adjust, change, or improve his thoughts, so that he can make decisions to act more appropriately [1]. Critical thinking skills in learning mathematics can be developed in schools to colleges. B erpikir critically focused on systems, structures, concepts, principles, and a tight connection between one element and the other element.

Based on the opinion above, it is necessary to have a learning process that involves many students actively, especially in the learning process in the classroom. The success of the learning carried out is determined by the applied learning process. Therefore, the researchers tried to analyze the improvement of students' thinking processes.

2. Method

The type of research used is article review. Article review is done by summarizing and analyzing research data from several previous research results. Research data collection was carried out by browsing articles in Online Journals, Google Scholar, Scopus, Microsoft Academics, Harzing's Publish or Parish, and Crossref. The keywords used by the researcher in searching the articles are problem based learning, and critical thinking.

From a search using the keyword problem based learning and critical thinking by a few articles then have articles that meet the criteria of Problem Based Learning to improve skills be r thought critical is the availability of the data before the procedure and after the action in the form of scores and N-gain. The analysis technique is carried out using a comparative method to determine the impact of the application of the problem based learning model. The analysis used in this study is to compare the difference in scores before and after the action using the problem based learning model.

3. Results and Discussion

The results of the study obtained 20 articles related to the problem based learning model in improving students' critical thinking skills. Data on article titles and authors are contained in the discussion table.

Data from the research report is still very wide, but only taken 18 articles that are relevant. The data of the article are processed by summarizing and determining the results of research with Problem Based Learning. Then the data is reported back by means of qualitative and quantitative descriptions. The data from the analysis of the Problem Based Learning model can be seen as follows:

Table 1. Percentage Critical Thinking Skills Students

No	Description	Initial Earning Score	Final Score
1	Students' Critical Thinking Skills	499	583
2	Maximum scores	725	
3	Percentage students' critical thinking skills	68.82%	80.41%

Based on T Abel 1 above shows that the learning problem based learning model can improve the ability of critical thinking students. The average percentage increase in students' critical thinking before using Problem Based Learning was 68, 82 % increased to 80.41 %. The average value before problem based learning and after problem based learning had a significant increase of 11.59 %. This is also shown from Table 2 regarding the following summary of critical thinking skills data.

From Table 2 above, it is known that the Problem Based Learning Model can improve critical thinking skills, analytical thinking skills, scientific attitudes, and thinking skills. Of the nine articles analyzed, the average increase in critical thinking skills of students was 17.29. Most of the nine articles analyzed the relationship between problem based learning models and critical thinking skills, analytical thinking, and scientific attitudes.

Based on the results of the analysis of the journal, a synthesis was obtained that provides an overview of research that is able to summarize and follow up on the analysis of the journal, namely by conducting research to determine the relationship between problem based learning (PBL) models and scientific literacy skills. The results of the journal synthesis show that there are no research results linking PBL-based e-Modules with science literacy skill variables. This is a gap and the basis for research on the development of PBL-based e-Modules. The background of the scientific literacy skill variable is that one of the characteristics of PBL is the existence of a problem, student centered learning or student-centered learning, and students learning in small groups.

Table 2. Critical Thinking Skills Data Summary

No	Research Title	Code	Average Value		The Improved
			Pretest	posttest	
1	Project Based Learning and Problem Based Learning: Are They Effective To Improve Student's Thinking Skills?	[9]	46,85	59,77	12,92

No	Research Title	Code	Average Value		The Improved
			Pretest	posttest	
2	The Influence of Problem Based Learning on Critical Thinking Skills and Competence	[10]	54,17	93,75	39,58
3	Developing Critical-Thinking Skills through the Collaboration of Jigsaw Model with Problem-Based Learning Model	[11]	48,44	69,23	20,79
4	Development of *PBL* Model Science Comic Media to Improve Analytical Thinking Ability	[12]	43,10	73,65	30,55
5	Development of *PBL* Model Science Comic Media to Improve Scientific Attitude	[13]	23,60	28,53	4,93
6	Development of Devices Problem Based Learning Models Assisted by Virtual Experiment to Improve Mastery of Students Concept on Material Optical Instruments	[14]	29,9	62,8	32,9
7	The Effect of *Problem Based Learning on* Students' Critical Thinking Skills	[15]	126,43	140,91	14,48
8	The Improving Critical Thinking Skills using Problem Based Learning	[16]	9,52	19,69	10,17
9	The Effectiveness of Problem-Based Learning (PBL)	[17]	17,95	48,66	30,71
	Average	RAT	44,58	61,87	17,29

Table 3 shows that: (1) the results of the study on the effectiveness of problem-based learning on improving students' science literacy skills with code [17] have a n-gain score of 0.55 which means the PBL model effectively improves students' science literacy skills. (2) The results of the study on the effectiveness of problem-based learning on improving students' scientific attitudes with code [17] have a n-gain score of 0.57 which means the PBL model effectively improves students' science literacy skills. (3) The development of problem-based learning model devices assisted by virtual experiments to improve the mastery of student concepts on material optical instruments with code [13] having a n-gain score of 0.47 means the development of pbl models effectively improves mastery of student concepts. (4) The effect of problem based learning on students' critical thinking skills with code [14] has a n-gain score of 0.25 which means the PBL model affects students' critical thinking skills. (5) The development of PBL model IPA comic media to improve analytical thinking skills with code [12] has a n-gain value of 0.51, meaning pbl-based comic media effectively improves students' critical thinking skills. (6) The results of research on the development of comic media IPA PBL model to improve analytical thinking skills with code [12] has a n-gain score of 0.43 which means the development of comic media IPA PBL model improves the analytical thinking skills of students. (7) Problem based learning approach with support interactive multimedia in physics course: its effects on critical thinking disposition with code [18] has a value of n-gain of 0.53, which means that pbl interactive multimedia-assisted approach affects students' critical thinking skills.

Based on the results of the research above, it can be seen that each research conducted to obtain different results of increasing the percentage and N-gain scores for critical thinking, analytical thinking, and scientific attitudes. The differences in the results of the research that have been analyzed are caused by internal factors and external factors. Internal factors are factors that exist within the students themselves (such as health, interests, talents, intelligence, body condition), while external factors are factors that come from outside of students, namely family factors, school environment, and community environment.

Table 3. N-gain value

No	Research focus	Code	N-gin	Criteria	Conclusion
1	The Effectiveness of Problem-Based Learning in Improving Students Scientific Literacy Skills	[17]	0.55	medium	Increased
2	The Effectiveness of Problem-Based Learning in Improving Students Scientific Attitudes	[17]	0.57	medium	Increased
3	Development of Devices Problem Based Learning Models Assisted by Virtual Experiment to Improve Mastery of Students Concept o and Material Optical Instruments	[13]	0.47	medium	Increased
4	The Effect of *Problem Based Learning on* Students' Critical Thinking Skills	[14]	0.25	medium	Increased
5	Development of *PBL* Model Science Comic Media to Improve Analytical Thinking Ability	[12]	0.51	medium	Increased
6	Development of *PBL* Model Science Comic Media to Improve Scientific Attitude	[12]	0.43	medium	Increased
7	Problem - Based Learning Approach with Supported Interactive Multimedia in Physics Course: Its Effects on Critical Thinking Disposition	[18]	0.53	medium	Increased

The background of the research place also affects the research results obtained because they come from different regions. The level of student ability is also different so that it also affects the learning outcomes obtained. The health condition of students is also an influence, because when researchers take research data, students' health conditions are sick so they are not optimal in doing the tasks given by the teacher. The application of different curricula can also have an impact on learning outcomes. In addition, the learning process carried out by the teacher even though using the same type of model, there is a possibility that the application will experience differences.

4. Conclusions and Suggestions

Based on the results of the analysis, it is concluded that the Problem Based Learning model can improve critical thinking skills, analytical thinking skills, and scientific attitudes. The results of the journal synthesis show that there is no research linking the problem based learning model with the variable science literacy skills. This is a gap and a basis for further research and development, e of problem based learning based. Things underlying science literacy skill variables that one of the characteristics of problem based learning is the existence of a problem, learning centered on students and learning in small groups.

References

[1] Eggen, P., & Kauchak, D. (2012). *Strategi dan model pembelajaran: Mengajarkan konten dan keterampilan berpikir.* Jakarta, Indonesia: Indeks.

[2] Nurul Syazwani Ismail at all. 2018. DOI: https://doi.org/10.1016/j.tsc.2018.04.002
[3] Hmelo-Silver, C. E., & Barrows, H. S. (2006). *The Interdisciplinary Journal of Problem-Based Learning, 1(1),* 21-39. DOI: https://doi.org/10.7771/1541-5015.1004. diunduh pada tanggal, 4 Juni 2021, 21.15.
[4] Erta Sri Wahyu*, Sahyar, Eva Marlina Ginting. 2017. *American Journal of Educational Research, 2017, Vol. 5, No. 6, 633-638* Available online at http://pubs.sciepub.com/education/5/6/7 ©Science and Education Publishing DOI: 10.12691/education-5-6-7
[5] Indri Anugraheni. 2018. A Journal of Language, Literature, Culture, and Education POLYGLOT Vol.14 No.1 January 2018
[6] Again, Syahroni. (2016). *Jurnal Pendidikan, 1(1), 65 – 71.* DOI: https://doi.org/10.26740/jp.v1n1.p66-72.
[7] Mafia, Y. N., & Suyanto, W. (2014). *Jurnal Pendidikan Vokasi, 4(1)*, 125 – 143. DOI: http://dx.doi.org/10.21831/jpv.v4i1.2540
[8] Susanti, A. E, & Suwu, S. E (2016). *Journal POLYGLOT, 12(1)*, 66-81. DOI: https://doi.org/10.19166/pji.v12i1.383
[9] R. D. Anazifa*1, Djukri2. 2017. Http://Journal.Unnes.Ac.Id/Index.Php/Jpii. Do: 10.15294/Jpii.V6i2.11100. Diunggah Pada Tanggal 29 Mei 2021. 20.30.
[10] D M Aswan1*, L Lufri2, R Sumarmin. 2018. IOP Conf. Series: Materials Science and Engineering 335 (2018) 012128 doi: 10.1088/1757-899X/335/1/012128.
[11] Maskhur Dwi Saputra, dkk. 2018. *International Journal of Instruction*, 12(1), 1077-1094. *e-ISSN: 1308-1470 ● www.e-iji.net*
[12] Davy Indah Lestari 1 *, Anti Kolonial Projosantoso 2. 2016. http://journal.uny.ac.id/index.php/jipi.
[13] Baiq Ewik Jiniarti1, Ahmad Harjono2, Muh Makhrus2. 2019. J. Pijar MIPA, Vol. 14 No. 2. 2019: 25 – 30 ISSN 1907-1744 (Cetak) DOI: 10.29303/ jpm. v14i2. 1233
[14] Pricilla Anindyta, Suwarjo. 2014. Volume 2 - Nomor 2, 2014, diunduh pada tanggal 7 June 2021, 10.15.
[15] Syahroni again. 2016. Jurnal Pendidikan Volume 1 Nomor 1 Tahun 2016. ISSN: 2527-6891
[16] Miranti Diah Prastika, Mustika Wati & S. Suyidno. 2019. DOI: 10.20527/bipf. v7i3. 7027, diunduh pada tanggal, 8 Juni 2021: 11.30.
[17] Jungyoun Ha. 2018. ISSN (Print) 2288-8675 ISSN (Online) 2508-9145, Vol.5, No.1, April, 2018. Diunduh pada tanggal, 5 Juni 2021, 10.30.
[18] Gunawan GUNAWAN1, Ahmad HARJONO2, Lovy HERAYANTI3 and Sadam HUSEIN4. 2019. Journal for the Education of Gifted Young. Scientists, 7(4), 1075-1089, December 2019. e-ISSN: 2149- 360X. http://jegys.org

TPACK in Blended Learning Media: Practice 4C Skills for Rotational Dynamics in Senior High School

A B Susila, I Indiyahni and F Bakri
Department of Physics Education, Universitas Negeri Jakarta
Jl. Rawamangun Muka, Jakarta 13220, Indonesia
Email: anggarabs@unj.ac.id

Abstract. In the 21st century, the fields of ICT and education are like two sides of a coin that cannot be separated. The skills of students must support the rapid development of science and technology. Workers in the 21st century have at least 4C competencies, namely: critical thinking, communication, collaboration, and creativity. The emergence of the Covid-19 in 2020 demands a blended learning model as an ideal learning process. This article describes research on website development as a blended learning medium for rotational dynamics in high school. The research method used is Research and Development with a Dick & Carey approach. Media has been produced in a website that combines technological knowledge, pedagogical knowledge, and physics content knowledge (TPACK) in multiple representations with stages to train students' 4C skills. The stages of 4C skills on the website as a blended learning medium that developed were: (1) criticizing initial information; (2) developing and compiling arguments, reasons, assumptions, and evidence; (3) communicating arguments, reasons, assumptions, and evidence; (4) criticizing the quality of arguments, reasons, assumptions, and evidence; (5) open-minded; (6) seek new information; (7) conclude. Media also facilitates communication between users to build networks and improve collaboration capabilities. The results of this study have been tested for eligibility by material physicists, learning experts, and media experts. The eligibility test results of the website as a blended learning medium for rotational dynamic material in high school were declared eligible to practice 4C skills.

1. Introduction

The 21st-century learning process stimulates 4C skills: critical thinking, communication, collaboration, and creativity in problem-solving [1] [2]. Where the learning process, which still contains many misconceptions, interferes with the formation of 4C skills [3] [4] [5]. Learning physics that is not contextual will lead to misconceptions in students [6] [7] [8]. The use of technology in learning can minimize misconceptions in students [9] [10] [11] [12].

Blended learning is technology-based learning method [13] [14]. The blended learning model combines face-to-face learning with independent learning based on e-learning [15] [16] [17]. Blended learning materials utilize audio, visual, video, and animation to support online learning process activities [18]. Blended learning can be done in computer-based learning, web-based learning, virtual classrooms, and digital collaboration [19].

Learning during the Covid-19 pandemic separates students and teachers in different spaces [20] [21]. The blended learning model requires learning media to learn independently under teacher direction and supervision [22]. Physics blended learning media must be designed by utilizing technology for the delivery of physical concepts in contextual learning stages and multiple representations [23] [24] [25].

Exposure to contextual material will shape scientific literacy in students [26]. Website-based blended learning can display more scientific representations that support contextual learning [27].

The implementation of TPACK in physics blended learning media will make it easier for students to understand physics concepts independently [28]. TPACK in blended learning will integrate technology and pedagogical concepts at the stage of exposure to physics concepts in blended learning website media [29]. The basic steps in the Blended Learning model refer to TPACK-based learning, one of which is the Acquisition of Information model used in forming 4C skills [30].

Based on the description above, this article presents research on the development of a blended learning media website based on TPACK to train 4C skills in class XI SMA students with the rotational dynamic material.

2. Method

2.1. Research and Design

This article describes a website-based blended learning media resulting from research and development using the Dick and Carey approach. It can use this blended learning website media to practice 4C skills of class XI high school students on rotational dynamic material. An R&D process has been carried out according to Dick and Carey, which has staged:

1. Identify the objectives of learning physics in high school, referring to the Kementerian Pendidikan Dan Kebudayaan No. 21 of 2016 concerning the standard of content for primary and secondary education.
2. Conduct an analysis of learning physics in high school, which refers to content standards to produce basic competency indicators.
3. Analyze the learner and the context. Prepare material maps by the indicator formulation and determine prerequisite materials and materials relevant to basic competencies that will form according to learning objectives.
4. Formulate specific learning objectives. Formulate specific learning objectives according to learning indicators relevant to basic competencies and physics learning objectives in high school. Students' specific learning objectives are learning objectives achieved by students using blended learning website media.
5. Develop assessment instruments for physics learning in a blended learning website media. Develop test items for measuring indicators of achievement of basic competencies in physics teaching in high school.
6. Develop instructional strategy, physics learning in blended learning website media. The physics learning strategy in the blended learning website media is designed to train 4C skills. The learning component of the blended learning website media consists of an introduction, material description, illustrations, sample questions, conclusions, and self-evaluation.
7. Develop and select instructional materials. The introductory part consists of perception which displays animations, linking students' prior knowledge with the material to be studied. Discussion activities to train students' critical thinking, creative and communication skills about the perception displayed. In the material description section, the essential materials of rotational dynamics are presented contextually and with multiple representations. Illustrations and sample questions support the description of the material. In the end, there is a conclusion of essential concepts and self-evaluation.
8. Design and conduct a formative evaluation of instructions. Conduct a formative assessment to determine the appropriate blended learning website media used in physics learning to train students' 4C skills. Evaluation of material feasibility, media feasibility, and pedagogical feasibility was measured using instruments on a continuous scale of 1 – 4. A formative assessment was carried out by material physicists, blended learning media experts, and physics learning experts.
9. Revise blended learning physics website media. The physics blended learning website media that is not feasible according to the informative expert evaluation is revised, and re-formative assessment is carried out by the expert declares it possible.

2.2. Instrumentation research

Research and development of blended learning website media test for eligibility by using an instrument in the form of a questionnaire. There are three questionnaires developed, namely a questionnaire to test the eligibility of the material, a questionnaire to test the eligibility of the media, and a questionnaire to test the feasibility of pedagogy.

2.2.1. The Physics Concepts Eligibility on the Blended Learning Website Media

The physics concepts eligibility on the blended learning website media is assessed in three parts, which has staged:

2.2.1.1 Part of presentation of physics concepts on the blended learning website media, with assessment indicators, including: (a) The sequence of physics, material presented; (b) Accuracy of scientific representation in formulating physics concepts; (c) Video illustrations on blended learning media websites are relevant to the physics concepts discussed; (d) Animated examples on blended learning media websites are relevant to the physics concepts discussed; (e) The illustrations on the blended learning website media are relevant to the physics concepts discussed; (f) The graphic picture on the blended learning website media is relevant to the physics concept being discussed; (g) The table illustrations on the blended learning website media are accurate and relevant to the physics concepts discussed; (h) The mathematical equations presented are relevant to the physics concepts discussed; (i) Ease of understanding the presentation of physics material exposure; (j) The integrity of the meaning of the physics material is shown; Dan (k) Exposure to physics material is free from SARA, gender bias, pornography, and bullying.

2.2.1.2 Part of science representation on blended learning website media, with assessment indicators, including: (a) Effective Indonesian sentences support verbal representation; (b) The symbolic representation of mathematical equations is presented in a precise, easy, and sequential manner; (c) The animate representation of science supports the understanding of the physics concepts presented; (d) The visual representation of the images and graphs shown is relevant to the physics concepts presented; (e) The use of language in the preparation of physics, material is precise, straightforward, and clear; (f) Non-verbal illustrations by the concepts of the physics material shown; (g) The order of presentation of communicative and easy-to-understand physics concepts; (h) The order in which facts, data, and illustrations are presented contextually; Dan (i) The suitability of the preparation of physics material with language rules.

2.2.1.3 Part of blended learning website media, with assessment indicators, including: (a) Illustrations from videos on blended learning media websites relevant to the physics concepts presented; (b) Illustrations of animations on blended learning media websites relevant to the physics concepts presented; (c) Placement of blended learning website media supports the process of understanding the physics concepts being studied; and (d) The blended learning website media that is presented is appropriate and not excessive.

2.2.2. The Eligibility of the Blended Learning Website Media

The eligibility of the blended learning website media is assessed in two parts, which has staged:

2.2.2.1. Part of the performance of blended learning website media, with assessment indicators including: (a) Domain name; (b) Login to website system; (c) Speed of page access; (d) The process of downloading materials; (e) Task file upload process; (f) Video upload process; (g) Functioning of discussion forums; (h) The functioning of the message forum among users; and (i) Self-evaluation.

2.2.2.2. Part of the media design of the blended learning website, with assessment indicators including: (a) Quality of media banners; (b) The quality of the background display; (c) Quality of graphic display; (d) Header font size; (e) Material font size; (f) Header font color; (g) Material font color; (h) Font type alignment; (i) Functionality of the navigation buttons; (j) Image quality; and (k) Video quality.

2.2.3. The Pedagogic Eligibility on the Blended Learning Website Media

The pedagogic eligibility on the blended learning website media is assessed in five parts, which has staged:

2.2.3.1. Part of presentation of physics concepts on blended learning website media, with assessment indicators, including: (a) The suitability of the material with the basic competencies of physics subjects in high school; (b) Coherence of presentation of physics, material; (c) The simplicity of presentation of

Journal of Physics: Conference Series **2019** (2021) 012046 doi:10.1088/1742-6596/2019/1/012046

physics, material; (d) Contextual presentation of the material; (e) Ease of understanding physics, material; (f) Continuity in the presentation of physics, material; (g) Relevance of sample questions to indicators of achievement of basic competencies; (h) Relevance of sample questions to the material; and (i) the Sufficient number of questions for each indicator.

2.2.3.2. Part of use of science illustrations on blended learning website media, with assessment indicators, including: (a) Scientific illustrations are presented in multiple representations (verbal, symbolic, digital, and visual); (b) Illustrations on perception foster curiosity; (c) Illustration of apperception stimulates critical thinking ability; (d) A series of illustrations of physics concepts on blended learning media websites support the formation of scientific literacy skills; (e) Examples of problem-solving stimulate creativity; and (f) All illustrations are presented contextually.

2.2.3.3. Part of the components of blended learning website media, with assessment indicators, including: (a) The description of the material is presented contextually and multiple representations as independent study material; (b) Discussion forums as an indirect (asynchronous) component of learning media; (c) Virtual synchronous teachers in discussion forums; (d) discussion groups on live chat as a virtual synchronous learning medium; and (e) Virtual synchronous teachers in discussion groups on live chat.

2.2.3.4. Part of the 4C skills on blended learning website media, with assessment indicators, including: (a) perception as a stage of critiquing initial information to train critical thinking skills; (b) Discussion forum as a stage of developing and compiling arguments, reasons, assumptions, and evidence on a problem to exercise creativity; (c) Discussion forum as a stage for communicating arguments, reasons, assumptions, and evidence to practice communication skills; (d) Comment column in the discussion forum as a stage to criticize the quality of arguments, reasons, assumptions, and evidence to practice critical thinking skills; (e) Activities in discussion forums as an open-minded stage to train critical thinking skills and creativity; (f) There are stages of learning in finding new information to train creativity; (g) There are stages of learning in making conclusions to practice the ability to conclude; (h) Group creation facilities on discussion forums and live chats to practice networking skills in small groups; and (i) Formation of small groups in problem solving in discussion forums to practice collaboration skills.

2.2.3.5. Part of TPACK implementation on blended learning website media, with assessment indicators, including: (a) The material presentation fulfills the physics content knowledge aspect; (b) The material presentation fulfills the pedagogical knowledge aspect of physics; (c) The presentation of the material is supported by aspects of technological knowledge (technological knowledge); (d) The material presentation meets the technological knowledge aspect of the physics content being discussed; (e) The presentation of the material fulfills the aspect of material exposure that is by the students' thinking processes (pedagogical content knowledge); (f) Material presentation using technological knowledge in the presentation (technological pedagogical knowledge) of the physics concepts discussed; and (g) The presentation of the material as a whole fulfills the relationship between knowledge of technology, pedagogy, and physics content (TPACK).

2.3. Analyze Data

The data processing of the eligibility test results in the form of a questionnaire by material experts, media experts, and pedagogical experts uses a continuous scale of 1 - 4 to get the results of the eligibility of blended learning website media which there are four categories of assessment based on interpretation criteria, which:

Table 1. Continuum Scale Interpretation Criteria

No	Criteria	Score
1.	Very bad	1
2.	Bad	2
3.	Good	3
4.	Very good	4

The interpretation of the score is calculated based on the acquisition score of each item, with the following equation.

$$\text{Score Interpretation} = \frac{\sum \text{Acquisition Score}}{\sum \text{Maximum Score}}$$

This study uses a continuum scale range of 1 to 4 so that the maximum score is 4, the minimum score is 1, and the total capacity of scores is 4. So, it can be calculated:

$$\text{Score Range} = \frac{\text{Maximum Score} - \text{Minimum Score}}{\text{Number of Score Categories}}$$

$$\text{SR} = \frac{4-1}{4} = 0{,}75$$

Then the eligibility criteria were obtained with a difference of 0.75 with a very low interpretation starting from 1.00 – 4,00, as shown in the following table.

Table 2. Eligibility Interpretation Criteria

Amount of Value	Percentage of Score Interpretation	Eligibility Category
$1{,}00 \le \text{Is} < 1{,}75$	$25\% \le \text{IS} < 43{,}75\%$	Not feasible
$1{,}75 \le \text{Is} < 2{,}50$	$43{,}75\% \le \text{IS} < 62{,}50\%$	Quite decent
$2{,}50 \le \text{Is} < 3{,}25$	$62{,}50\% \le \text{IS} < 81{,}25\%$	Worth it
$3{,}25 \le \text{Is} \le 4{,}00$	$81{,}25\% \le \text{IS} \le 100\%$	So worth it

3. Result and Discussions

3.1. Product

The research and development carried out have resulted in learning media in the form of a blended learning website media based on TPACK with the advantages of training 4C skills. The stages of the 4C skills contained in the media are as follows.

1. The perception section is the stage of critiquing the initial information to link the knowledge to be learned by the knowledge and experience possessed by students (Figure 1).

Figure 1. Apperception Section

2. Showing the discussion forum feature as a stage for developing and compiling students' arguments against a problem. The discussion forum also acts as a stage for communicating the student's arguments against the perceptions that have been displayed. Finally, there is a comment column on the discussion forum to criticize ideas between students to practice critical thinking skills.

(a) (b) (c)

Figure 2. Softskill training activities (a) discussion forum; (b) comment on discussion forum; and (c) reply the comment.

3. The description of the material consisting of essential materials, the remembering section as an answer to the problem in perception, the CATING section, and sample questions.

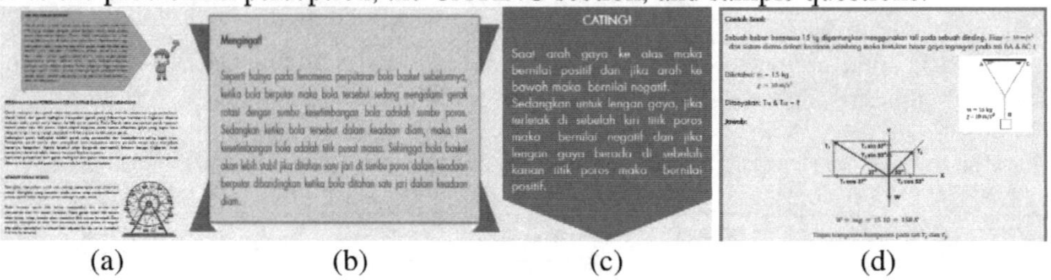

Figure 3. Stages of material presentation (a) essential concepts; (b) review phenomena; (c) postscript; and (d) sample questions.

4. The existence of group facilities in live chat to train the ability to create networks in small groups and group collaboration features between students.
5. The final section consists of conclusions and self-evaluation.

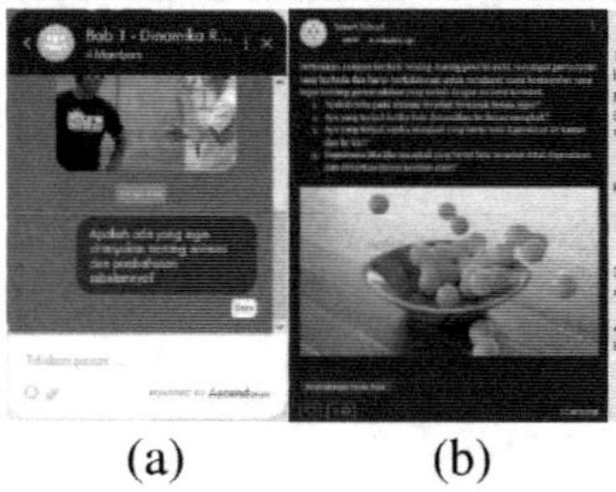

Figure 4. Collaboration skill (a) live chat; and (b) collaboration group

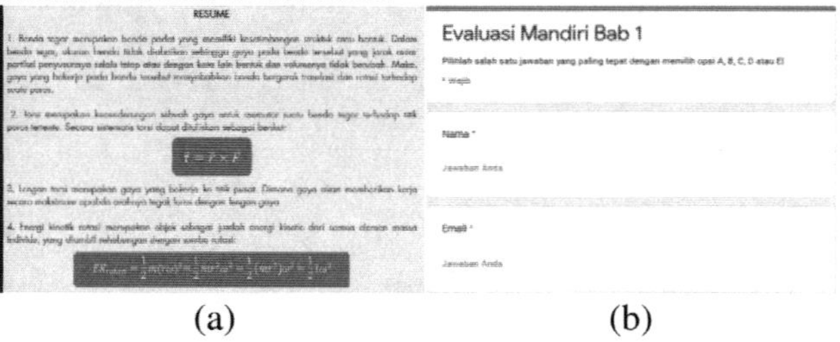

Figure 5. The final section (a) resume of material; and (b) self-evaluation

3.2. *Description of Eligibility Test Results*

The eligibility test results for the concept of physics of the blended learning website media by a physicist material expert received a very decent assessment in the three aspects assessed. The evaluation results are shown in Figure 6.

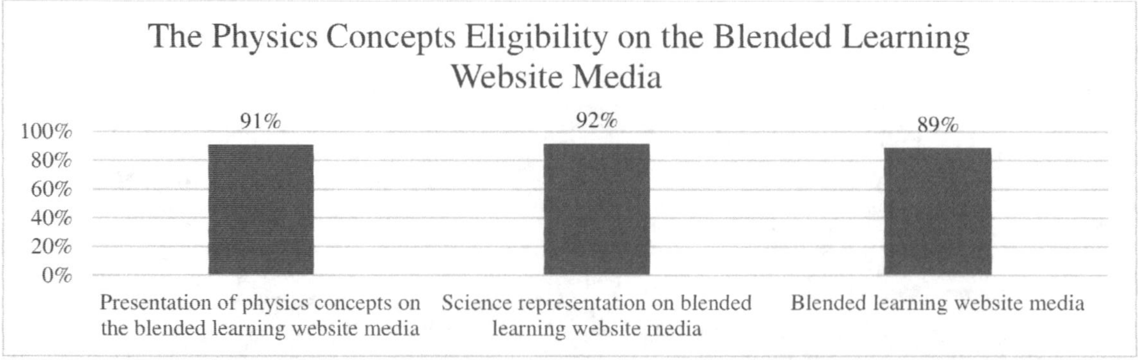

Figure 6. The Chart of Material Eligibility Test Results

The results of the blended learning website media eligibility test by media experts received a very decent assessment in the two aspects assessed. The evaluation results are shown in Figure 7.

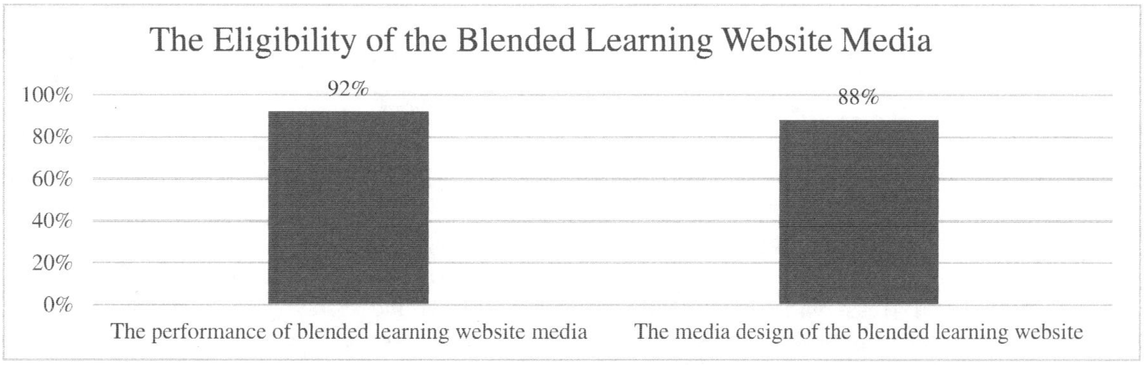

Figure 7. The Chart of Madia Eligibility Test Results

The eligibility test results for learning physics about the blended learning website media by pedagogical experts get a very decent assessment in the five aspects assessed. The evaluation results are shown in Figure 8.

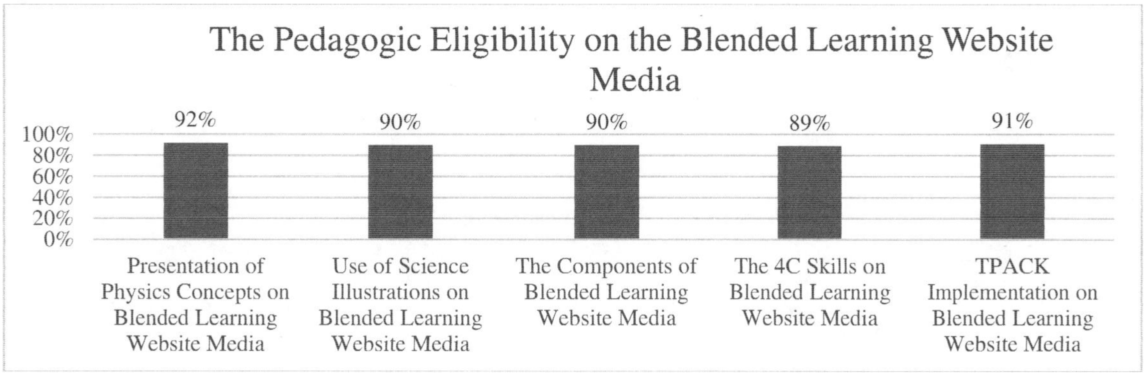

Gambar 8. The Chart of Pedagogical Eligibility Test Results

3.3. Discussions

The resulting TPACK-based physics blended learning website media is considered feasible in material, media and learning. Descriptions of essential material are presented contextually, and multiple representations are supported by tables, mathematical equations, sample questions, and self-evaluations [25]. Blended learning website media utilizes audio, visual, video, and animation to support synchronous and asynchronous online learning process activities [18]. Pedagogically, learning physics on the TPACK-based blended learning website media are declared feasible to train 4C skills [30]. The media displays perception in the form of scientific illustrations that can stimulate students' critical thinking skills. In the perception section, students are encouraged to develop and communicate arguments related to the problems presented. There is a discussion forum as a supporting facility so that students can practice 4C skills. Discussion forum activities remain under the direction and supervision of the teacher. This media can also train students' communication and collaboration skills displayed on discussion forums, live chat, and "HOPES" group [19].

4. Conclusions

TPACK-based physics blended learning website media has been produced on rotational dynamics. The stages of TPACK-based learning about the media are considered capable of training students' 4C skills. Based on the results of formative evaluations by material experts, media experts, and pedagogical experts, it can be concluded that the TPACK-based blended learning website media on the rotational dynamics produced is considered suitable to be used as a learning medium to train the 4C skills of high school students.

References

[1] S. Handajani, H. Pratiwi and Mardiyana, Journal of Physics: Conf. Series, pp. 1-7, 2018.
[2] A. Khoiri, E. Evalina, N. Komariah, R. T. Utami, V. Parameter, S. Siswandi, J. Janudin and D. Sunrise, Journal of Physics: Conference Series, pp. 1-10, 2020.
[3] E. Prihadi, JPI_Rabbani, pp. 464-479, 2018.
[4] F. A. Zulfia, A. K. Sudrajat, H. Susilo and F. T. Fajar, Advances in Social Science, Education and Humanities Research, volume 446, pp. 1-14, 2019.
[5] S. Soeharto, B. Csapó, E. Sarimanah, F. I. Dewi and T. Sabri, JPII, 8 (2), pp. 1-20, 2019.
[6] T. Neidorf, A. Arora, E. Erberber, Y. Tsokodayi and T. Mai, Student Misconceptions and Errors in Physics and Mathematics: Exploring Data from TIMSS and TIMSS Advanced, Washington: International Association for the Evaluation of Educational Achievement (IEA), 2020.
[7] A. Akmam, R. Anshari, H. Amir, N. Galenas and A. Amran, IOP Conference Series: Materials Science and Engineering, pp. 1-8, 2018.
[8] K. Hamzah, W. Liliawati and A. Samsudin, Journal of Physics: Conference Series, pp. 1-7, 2019.
[9] T. N. Diyana, Sutopo and Sunaryono, Jurnal Pendidikan IPA Indonesia, pp. 219-230, 2020.
[10] A. Almubarak and R. Rusmansyah, Journal of Physics: Conference Series, pp. 1-12, 2020.
[11] W. D. Aryani, E. Suhendi, I. Suyana, A. Samsudin and I. Kaniawati, Journal of Physics: Conference Series, pp. 1-8, 2019.
[12] Y. Kurniawan, R. Muliyani and S. NassimJurnal Ilmiah Pendidikan Fisika Al-BiRuNi 08 (2), pp. 1-10, 2019.
[13] R. Resien, H. Sitompul and J. Situmorang, BirLE Journal, pp. 1-15, 2020.
[14] A. R. Lorenzo, The Turkish Online Journal of Educational Technology, pp. 1-11, 2017.
[15] A. Kristanto, M. Mustaji and A. Marino, International Education Studies; Vol. 10, No. 7, pp. 1-8, 2017.
[16] A. M. Nortvig, A. K. Petersen and S. H. Balle, he Electronic Journal of e-learning Volume 16, pp. 1-10, 2018.
[17] I. K. Suartama, iJET Vol. 14, pp. 1-19, 2019.
[18] C. J. Bonk and C. R. Graham, The Handbook of Blended Learning: Global Perspectives, Local Designs, San Francisco: Pfeiffer, 2012.
[19] T. A. Urdan and C. C. Weggen, Corporate e-learning: Exploring a New Frontier, New York: WR Hambrecht + Co, 2000.
[20] O. B. Adedoyin and E. Soykan, Interactive Learning Environments, pp. 1-14, 2020.
[21] A. Bahasoan, W. Ayuandiani, M. Mukhram and A. Rahmat, International Journal Of Science, Technology & Management, pp. 1-7, 2020.
[22] J. Simarmata, A. Djohar, J. Purba and E. A. Juanda, Journal of Physics: Conf. Series, pp. 1-6, 2018.
[23] A. Aisyah, N. Bukit and D. Derlina, Journal of Physics: Conference Series, pp. 1-9, 2020.
[24] L. A. Krasnova and V. Y. Surgeon, Int. J. Technology Enhanced Learning, pp. 1-15, 2020.
[25] W. Suana, N. Maharta, I. D. P. Nyeneng and S. Wahyuni, JPII, 6 (1), pp. 1-9, 2017.
[26] M. Ma'ruf, A. Setiawan, A. Suhandi and P. Siahaan, Journal of Physics: Conference Series, pp. 1-7, 2019.
[27] V. Chandra, The Impact of a Blended Web-based Learning Environment on the Perceptions, Attitudes, and Performance of Boys and Girls in Junior Science and Senior Physics, Bentley: Curtin University of Technology, 2004.
[28] A. Doering, G. Veletsianos, C. Scharber and C. Miller, Journal of Educational Computing Research, 2009.
[29] A. A. A. Qasem, Contemporary Educational Technology, pp. 1-13, 2016.
[30] G. Ramsay, Success Through a Whole School, 2011.

Development of Electronic Physics Learning Module With Phet (Physical Education and Technology) Simulation on Elasticity and Hooke Law in Class Xi Senior High School

V N Rahma, C E Rustana and Umiatin
Department of Physics Education, Universitas Negeri Jakarta
Jl. Rawamangun Muka No. 1, Jakarta 13220, Indonesia
Email: rahmavellia@gmail.com; ce.rustana59@gmail.com; umiatin@unj.ac.id;

Abstract. One way to use technology in learning is by using Simulation Physics Education and Technology (PhET). This article describes the results of the development of an electronic module for learning physics about elasticity and Hooke's law in class XI SMA. This research method is Research and Development with ADDIE model. The steps in this development are making needs assessments, analyzing materials, designing electronic modules, developing assessment tools, realizing electronic module designs, product validation by experts, testing electronic modules by teachers and students and performing evaluations. Based on the needs analysis conducted in class XI MIPA SMAN 107 Jakarta on 58 students, the results were obtained: 98.23% of students have difficulty understanding the physics material during PJJ, 94.83% of students are interested in the use of electronic modules as a companion for self-study at home and 87.72% of the students chose a module with instructions for independent practicum. The electronic module supported by that simulation is an alternative teaching material that can be used by students for self-study and can be used to conduct virtual experiments. The results of this research have been validated by physicists, learning experts and media experts.

1. Introduction

Physics is a science that studies nature and natural phenomena or natural phenomena and all the interactions in it that can be observed by humans. Learning physics can benefit from experimentation to help students understand physics lessons. The problem that students often face when learning physics is the difficulty of understanding the material [1]. This can happen because teachers only teach abstract physics through classroom learning and not all of them are equipped with lab experiments. Experiments in labs often face obstacles, including concepts that are difficult to observe through experiments, are dangerous to perform through experiments, incomplete tools, and others [2]. In addition, the fact that Covid-19 has disrupted all segments of human life on earth, without exception of education. The impact of this pandemic is that all students have to undertake the online learning process or popularly referred to as a Distance Learning (PJJ). The impact of this pandemic is that all students have to undertake the online learning process or popularly referred to as a Distance Learning (PJJ) [3]. Therefore, we need teaching materials that are practical, interesting and interactive and can be used by students to study independently at home, and can replace the role of conducting experiments in school labs. An alternative is through the use of other teaching materials, such as using technology-based teaching materials that are not limited by place and time. Based on these problems, this research will develop an electronic module, supported by PhET (Physical Education and Technology) simulation on Elasticity and Hooke's Law. Get interactive simulation is an example of multimedia applications in the education world that are expected to support students' understanding of physics concepts assessed through comprehension tests. Therefore, this research focuses on developing an electronic module, supported by that simulation on Elasticity and Hooke's Law, as an independent learning material to help increase students' understanding.

2. Method

The method that is used in this research is the research and development method to ADDIE (Analysis, Design, Development, Implementation, and Evaluation) model.

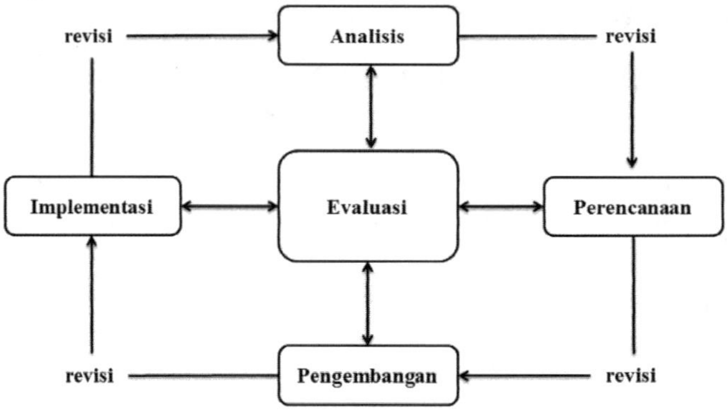

Figure 1. Addie Model Flow

2.1. Analysis

At this step, questionnaires are distributed to students to identify learning objectives, determine appropriate media, approaches, and materials for students, and find out what processes and facilities are available in learning. Based on the needs analysis, 94.83% of students are interested in using electronic modules as a medium for self-study at home and 87.72% of students choose modules that contain steps for using physical materials simulation with PhET. In addition, it was found that students had difficulty understanding the material because there were no experiments in the laboratory in distance learning.

2.2. Design

The goal in this step is to design an electronic module. At this step, the researcher collects information about the product to be developed. In addition, at this step, the researcher makes a research schedule flow to make the development of electronic modules on time according to the schedule and planned research activities.

2.3. Development

At this step, the realization of the development of an electronic module, supported by that simulation on Elasticity and Hooke's Law. On the first page of the module there is a cover, then the next sheet contains core competencies, basic competencies, concept maps, introduction to the material, material, sample questions, practice questions, steps in conducting experiments with PhET, a summary of the material, formative tests and answers on formative tests.

2.4. Implementation

In the implementation phase, validation tests were carried out by material experts, media experts, and learning experts. This validation test is intended to analyze the feasibility of the electronic module developed as a form of teaching materials/materials for students. In this case, the validation test by the expert was carried out through the distribution of questionnaires using a Likert scale. Improvements will be made based on the input given to the validation process by experts. After that, the product readability test was carried out by high school physics teachers and high school students in class XI.

2.5. Evaluation

At this step, an evaluation of the electronic module that has been tested on teachers and students is carried out. Are there still deficiencies in the developed electronic module. If there are still shortcomings, it is necessary to carry out an evaluation step to perfect the developed electronic module, namely the evaluation of unmet needs in the develop electronic module.

Journal of Physics: Conference Series **2019** (2021) 012047 doi:10.1088/1742-6596/2019/1/012047

3. Result and Discussion

The result of this research and development is an electronic module for physics learning supported by that simulation on the elasticity and material of Hooke's law for class XI SMA as a learning medium that can facilitate students in independent learning that can be accessed anywhere, anytime. These are the components of the developed electronic module:

Table 1. Components of Electronic Module Assisted with PhET Simulation

Early Section	Contents Section	Final Section
Cover	Material introduction	Evaluation
Foreword	Learning material	Bibliography
Table of contents	Problem's example	Lesson plan
Basic competencies	Exercises	Author's identity
	Instruction for use that	
	Summary	

Below is the initial design of the Phet simulation-assisted physics learning electronic module on elasticity and Hooke's law material for class XI developed:

Figure 2. Display of electronic module cover

Figure 3. Display of Basic Competemcies and Achievement Indicators

Figure 4. Display of material introduction

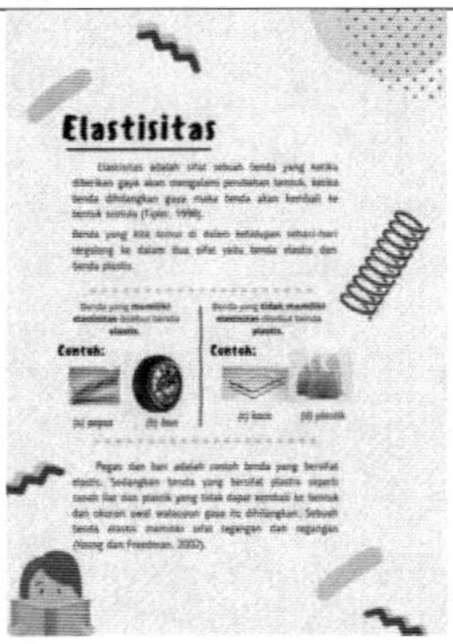

Figure 5. Display of the material

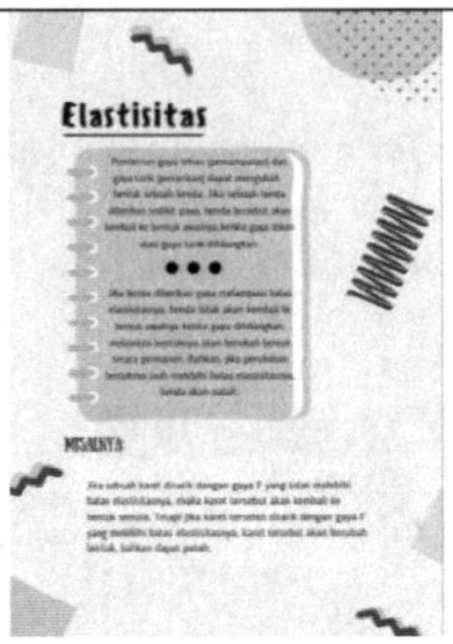

Figure 6. Display of the material

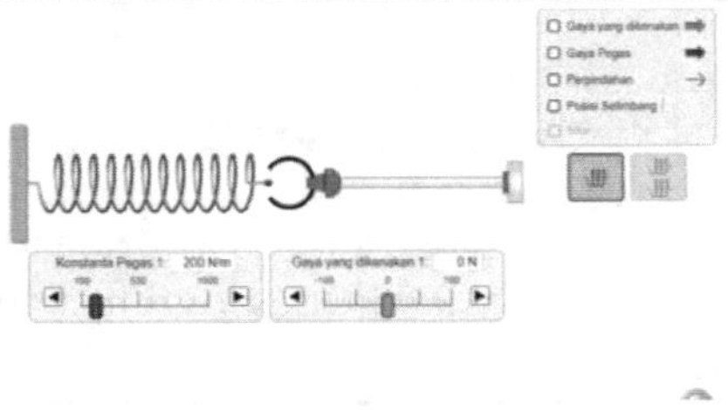

Figure 7. Display of PhET Simulation

The Phet Simulation Assisted Electronic Module is a digital module systematically arranged to cover all components of a module and equipped with a Phet simulation so that it can be used as independent teaching material to help students understand the topic of Elasticity and Hooke's law. The electronic module supported by that simulation is an alternative teaching material that can be used by students for self-study and can be used to conduct virtual experiments. Phet simulation makes students more active and effective as a teaching material [4]. Moreover, through the use of Phet teaching materials, students better understand the concept as they are directly involved in conducting experiments [5].

Research by Siti Nur Khalifah [6] concluded that Phet simulation of distance learning physics has a positive effect on increasing the concept mastery ability of high school students. The electronic module was chosen because it is flexible, accessible from anywhere and only requires supporting electronic devices. The Phet Simulation Assisted Physics Module is also a teaching material that can increase students' enthusiasm for independent learning, as the Phet Simulation Assisted Physics Electronic Module contains materials, sample questions, practice questions, and formative tests and answers so that students can ask can practice. Independent [7]. In addition, the PhET Simulation-Assisted Physics module is equipped with guidelines and experimental steps for accessing and using PhET simulations, allowing students unfamiliar with that simulations to experiment. Based on research conducted by Adams, et. Al. [8] Showed that the Phet media can visualize the concept of the material well. Phet simulation experiments are easily accessible to students as they can be accessed on their own electronic devices. That simulation can be done repeatedly so that it helps the students to understand the learned physics concepts. Research by C E Rustana et al [9]. Showed that students who taught using interactive PhET simulations had better learning outcomes than those taught using the experimental method using laboratory kits, especially for the kinetic theory of gases.

Based on the formulation and research objectives, and various related theoretical studies, by with the ADDIE research and development model, this Phet simulation-assisted electronic module was created. Starting with creating a needs analysis completed by students to use as initial data, then continuing with a literature search related to electronic modules and PhET simulations. Then create product designs, develop products, validate with materials experts, learning experts, and media experts, after which it continues to review the validated products. After the product has been declared valid, field trials are carried out. If the results of the student product trial are successful, they can be used without reviewing the product. But if it fails, the product must be repaired and revalidated from a material perspective, then tested to show the desired results. Finally, the results of the research in the form of a Phet-supported electronic module on elasticity and Hooke's law are expected to be useful and become one of the alternative teaching materials for students to use when conducting virtual experiments.

4. Conclusion

In this research, a product is an electronic module for learning physics with the aid of Phet simulation of elasticity and Hooke's law material for class XI SMA, which was used as an independent learning teaching material to help increase students' understanding. This study uses research and development methods using the ADDIE model (Analysis, Design, Development, Implementation, and Evaluation). Researchers hope that the developed electronic module can be used as an alternative for independent teaching materials for students and can help in the learning process.

References

[1] Han, K., Yi, M. Y., Gweon, G., & Lee, J.-G. (2013). Understanding the Difficulty Factors for Learning Materials: A Qualitative Study. Artificial Intelligence in Education, 615–618. doi:10.1007/978-3-642-39112-5_72

[2] Samudra G B, Suastra I W, Suma K. (2014). e-Journal Program Pascasarjana Universitas Pendidikan Ganesha. 04

[3] N R H S. (2020). Dampak Covid-19 pada Pendidikan di Indonesia: Sekolah, Keterampilan, dan Proses Pembelajaran. SALAM; Journal Sosial & Budaya Syar-I. 07(5) 395-402

[4] Rahayu N.P & Dwikoranto. (2020). *Efektivitas Model Pembelajaran Simulasi Berbantuan PhET untuk Pembelajaran Fisika Ditinjau dari Peningkatan Hasil Belajar Peserta Didik SMA*. 09 (2) 104 – 110

[5] Clark, R C & Mayer, R E. 2003. E-learning and the science of instruction: San Francisco. CA: Jossey-Bass.

[6] Khalifah S N. (2020). JPPPF (Jurnal Penelitian dan Pengembangan Pendidikan Fisika

[7] Ganasen, S., & Shamuganathan, S. (2017). The Effectiveness of Physics Education Technology (PhET) Interactive Simulations in Enhancing Matriculation Students' Understanding of Chemical Equilibrium and Remediating Their Misconceptions. Overcoming Students' Misconceptions in Science, 157–178. doi:10.1007/978-981-10-3437-4_9

[8] Adams W K et. Al. (2008). Journal of Interactive Learning Research, 19(4) 551-577

[9] Rustana C. E, et.al. 2020. 6th International Conference on Mathematics, Science, and Education (ICMSE 2019).

Problem Solving Ability of Distance Education Student on Electrostatic Topic

H Safitri[1, 2], I Hamidah, W Setiawan and I Kaniawati
[1]Graduate School, Universitas Pendidikan Indonesia, Jl. Dr. Setia Budhi No. 229, Bandung 40154, Indonesia
[2]Departement of Physics Education, Universitas Terbuka, Jl. Cabe Raya, Tangerang Selatan 15437, Indonesia
Email: henis@ecampus.ut.ac.id

Abstract. In this paper, we present the result of a study on the identifying problem-solving ability on distance education students' physics. Descriptive-qualitative is used for method of research. The subjects of the study were in-service students' physics of distance education. Sample were 19 students from all of student taking electricity and magnetism course in term 2019/2020. Essay test of electrostatics problem solving is used for collecting data technique. Test of problem solving is given at the beginning of the course in asynchronous tutoring of electric and magnetic cores. The finding of the study indicated problem solving ability of in-service students' physics of distance education are categorized low (44,56%) in solving the problems. Therefore, it is necessary to make efforts to improve students' ability to solve problems, through asynchronous tutorials by applying problem-solving strategies in interactive discussion and collaborative work in groups.

1. Introduction

Era of globalization that is growing so fast and demands of imposing curriculum have made educational institutions carry out the 21st century skill innovations that are beneficial to education world [1]. The abilities are in line with the demands of the 21st century include problem solving skills. Learning and assessment of problem-solving skills is very necessary and driven by the needs of students at the institution and career level to be able to work in groups and apply problem solving skills in real social situations [2-4]. Problem solving skills are the main skills that must be developed in 21st century learning practice, in addition to critical thinking skills, communication and collaboration skills [5]. Problem solving skills are defined as the ability to identify, analyze, design or plan, implement or practice, evaluate, and find solutions in formulating new answers to solve a problem [6]. In addition, the problem-solving process includes the process of identifying problems, formulating problems, and formulating hypotheses, proving hypotheses through experimental activities or discussion processes, and choosing the most appropriate alternative solutions [7].

Problem solving skills are important in physics [8, 9]. One of the goals in learning physics is to teach problem-solving skills. Physics really needs this ability because the problems in physics are complex problems and are related to everyday life [10]. Physics, especially mechanics and electricity and magnetism topics, often emphasize quantitative aspects of problem solving such as mathematical equations and procedures rather than qualitative analysis to select appropriate concepts and principles [8]. Learning mechanics and electricity and magnetism requires students' ability to understand formulations, graphs, drawings, read tables, and express relationships between variables, which are very important skills that students must possess. The concepts of physics, especially electricity and magnetism include electrostatic topic, have many complex concepts when explained verbal [11, 12], but they will be simpler if they are formulated into mathematical formulas, graphs, or sketch [13, 14]. Physics problem solving can be trained by enabling strategies to develop rational solutions to these problems that must be adapted to various kinds of problems faced in life so that it leads to analytical thinking [15, 16]. Mastery of these abilities can help students simplify problems and organize knowledge so that it is easier to communicate and understand physics [17, 18]. Integrating conceptual knowledge with problem solving is a desired goal in learning physics, but students must be able to integrate conceptual knowledge and problem solving to develop knowledge of physics concepts and help students solve problems that occur by designing authentic problems [8, 19, 20].

Various research related to problem solving have been carried out, for example the development of problem-solving skills in science learning has been widely studied starting from primary education [21, 22], secondary education [23-26], higher education [27, 28], to research for science teachers [29, 30]. These various studies were carried out in different scientific contexts (biology, chemistry, physics). Research that tries to develop various instruments to measure problem-solving skills and problem-solving processes with various models [31-34]. The use of technology in developing problem-solving skills has also been widely developed, for example the use of mobile learning [35, 36], interactive multimedia [37], Virtual laboratories [38-40] and online learning [41, 42].

Universitas Terbuka as a distance education uses online tutorials as a learning aid for students. Online tutorial using Modular objective oriented dynamics, learning (Moodle) as learning management system [43]. This research is intended to determine the problem-solving ability of students in electrical and magnetic courses with learning assistance through online tutorials and to increase the use of facilities in Moodle for learning, which in the end is designed to improve problem solving skills.

2. Method

This research is a descriptive research, which is a method of examining the status of a group of people, an object, a set of conditions, a system of thought or a class of events in the present. The goal of descriptive research is to describe a phenomenon and its characteristics. This research is more concerned with what rather than how or why something has happened. Therefore, observation and survey tools are often used to gather data. In such research, the data may be collected qualitatively, but it is often analyzed quantitatively, using frequencies, percentages, averages, or other statistical analyses to determine relationships [44]. Research subjects are students of physics education program of Faculty of teaching and education of Universitas Terbuka who had taken electric and magnetism courses in the 2019/2020 academic year. The number of students who became research subjects was 19 people. The data studied were answer sheets for student tutorial assignments on electrostatic topics. The instrument used in this study was a test structured so that students answered based on five problem solving indicators. The five indicators of problem-solving ability are (1) the ability to understand problems, (2) the ability to describe problems, (3) the ability to plan solutions through choosing the right problem-solving strategy, (4) the ability to solve problems through correct answer and appropriate calculation process, and (5) the ability to check to ensure the results obtained are correct. The type of test used in this study is a subjective test, which is generally in the form of a description test in which students express their thoughts in answering questions. In this way students can explore their knowledge as much as possible.

3. Result and Discussion

The results of the research carried out contain the demographic characteristics of students and students' problem-solving abilities

3.1. Demographic characteristics of students

Based on studies, the respondents consist of 58% female and 42% male. Respondents' ages ranged from 21-45 years, with predominantly under 30 years of age (47%) for those who have both taught high school. Respondents are scattered throughout Indonesia, but domicile is still dominated by the island of Java (47%) then Sumatra (37%), Kalimantan (11%) and Maluku (5%).

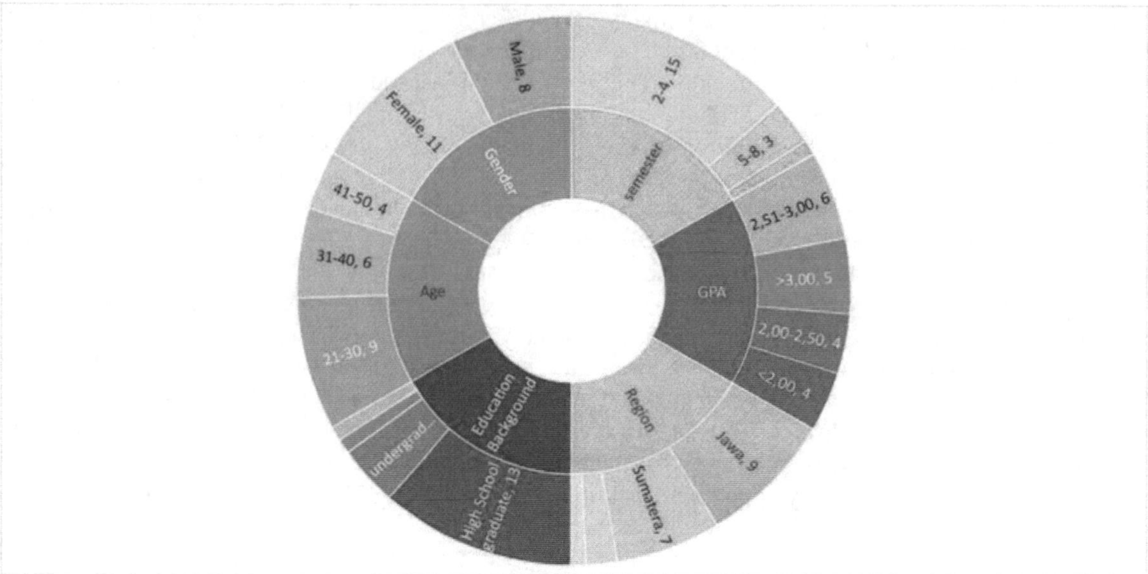

Figure 1. Student Demographic

Figure 1 shows that the educational background of respondents is dominated by high school graduates (68%) then diploma graduates (5%), undergraduate and graduate (27%) with a specialty come from engineering who will study physical education as a provision for teaching in secondary schools. When the research was conducted, most of the respondents were in semesters 2-4 as many as 79% were in accordance with the curriculum that the electric and magnetic course was an advanced course after taking basic physics courses. For students who are in high semesters taking this course due to various reasons, namely course exams clash with other courses so that they have not been taken in the previous semester or due to repeating to get better grades. Meanwhile, for the academic level, the respondents were represented balanced for all levels from low to high, for students who had a high academic GPA> 3.00, as much as 26%, followed a good GPA from 2.50 to 3.00 (22%), Moderate GPA 2.00-2.50 (21%) and low GPA <2.00 (21%).

3.2. Level of Problem-Solving Abilities

The scoring results obtained from the students' answers are as follows.

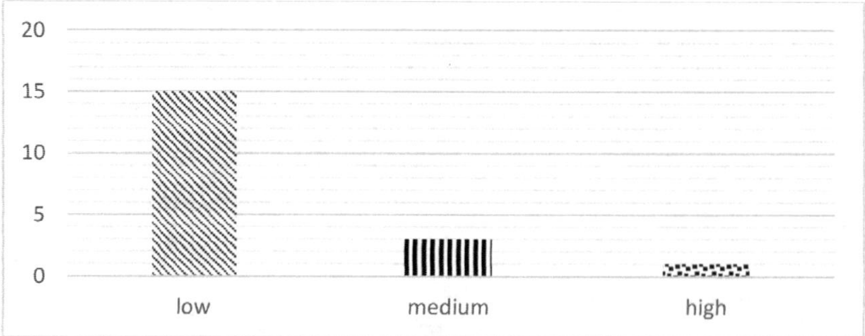

Figure 2. Frequency of Student Problem solving abilities

Based on the 19 students who became respondents, 15 students (78%) got low scores, so it can be said that most students have not been able to solve problems in electricity and magnetism course, especially on the topic of electrostatics. Meanwhile, the student's ability in each indicator can be shown in Figure 3.

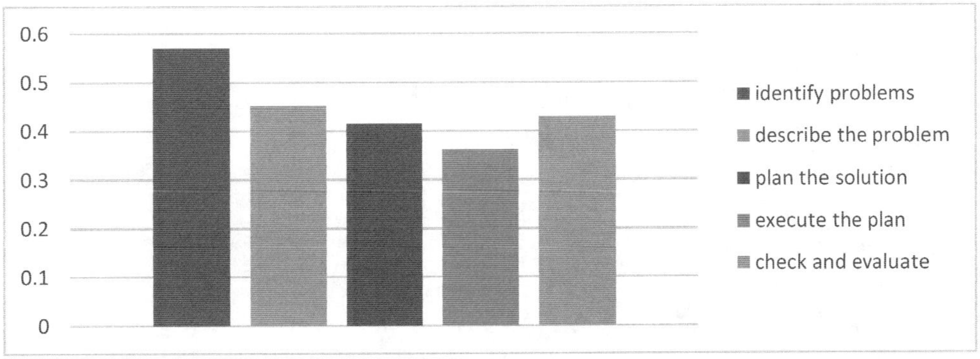

Figure 3. Student ability in every indicator of Problem Solving

The result shows the lowest indicator is in executing the plan (36.29%), this shows that most students have not been trained in doing calculations that involve mathematical language, so they often make fundamental mistakes in doing calculations. That most students experience difficulties in answering calculation or proof questions. The following is an example of a student's answer to the electric fields topic. The selection of respondent student answers is based on similarities in student demographics so that only problem-solving abilities will be seen. Both students were chosen because they have the same educational background, region, age, so the assessment is only based on the ability of students to solve problems.

Figure 4 explained student low ability when student answered between there no connect between concept to another item when solving problems. More picture explained, figure 5, student with high abilities. There are connected between concept when solving problems. Even though there are mistakes made by student experts, especially at the execute solution to solve problems stage, from the way they answer, there is a relation between one answer and the answer to another question.

In the field of physics, solving physics problems regarding the concepts of physics. Factors that influence solving physics problems are the knowledge structure of students who solve problems and the character of the problems [45]. The difference between students who have low (novice) and high (expert) abilities in solving physics problems is how students organize and use knowledge and connect one concept to another when solving problems [46-50]. Students who have high ability in solving physics problems tend to use qualitative arguments based on the physics concept that underlies the problem (deep features), evaluate solutions, and tend to use representational tools. On the contrary, students who have low ability in solving physics problems tend to recognize problems based on the problem presentation, do not evaluate, and tend to use formulas in solving problems [47, 50].

The results of the research conducted indicate that the ability of students in planning and implementing plans in solving problems is still low, it is necessary to enrich learning supports for students, especially in online tutorials with examples of problem solving in the presentation of material and collaborative discussions.

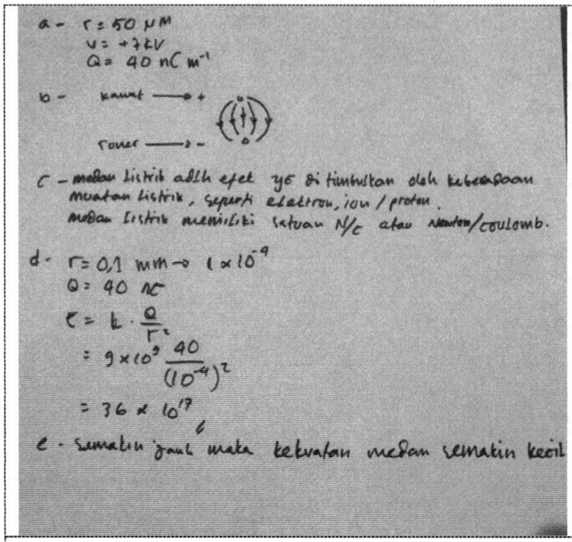

Figure 4. Native Student Answer

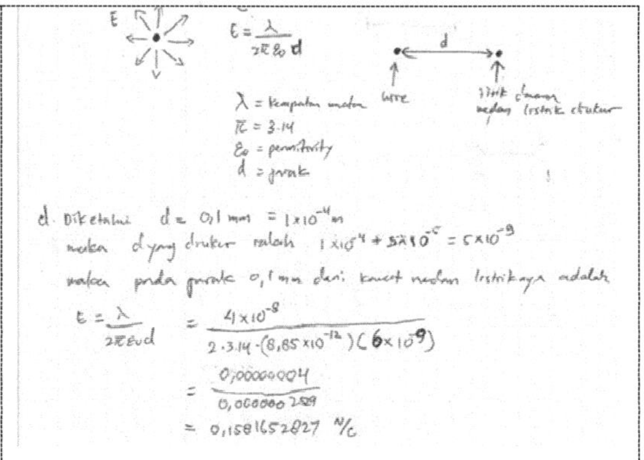

Figure 5. Expert Student Answer

4. Conclusion

From the results of research and discussion, it was concluded that the ability student problem solving in electricity and magnetism courses are classified in the low category with indicator problem solving. The lowest indicator is executing a solution plan for problem solving. With this condition of students, weaknesses, and strengths of students in problem solving abilities. So that this can underlie the steps next thing to do in order student weaknesses can be fixed, and the advantages can be increased. Through learning support, especially online tutorials, it can be suggested that there are many examples of how to solve problems and collaborate discussions between students.

Acknowledgments

Authors wishing to acknowledge assistance or encouragement from colleagues, the special work of technical staff or financial support from BUDIDN-LPDP scholarship, Ministry of Finance Republic of Indonesia

References

[1] Becker SA, Cummins M., Davis A, Freeman A., Hall Giesinger, C., and Ananthanarayanan, V. 2017 *NMC horizon report: 2017 higher education edition*. Austin, Texas: The New Media Consortium.
[2] Jonassen, D. H. 2011 *Learning to solve problems*. New York: Simultaneously
[3] Malan, S. B., Ndlovu, M., & Engelbrecht, P 2014 *South African of Education*, 34(1), p1-16.
[4] Morze, N. V., Smyrnova-Trybulska, E., & Glazunova, O 2017 *Design of a University Learning Environment for SMART Education Smart Technology Applications in Business Environm* p 221-248 IGI Global.
[5] Boo and Koenig Disciplinary and Interdisciplinary Science Education Research 2019 1:2
[6] Erozkan A 2013 *Educational Sciences: Theory & Practice*.
[7] Moreno R 2010 *Educational Psychology*. John Wiley and Sons.
[8] Docktor, J. L, Strand, N. E., Mestre, J. P., & Ross, B. H. 2015 *Physical Review Special Topics-Physics Education Research*, 11(2), 020106
[9] Gok T and Silay I 2010 *Latin American Journal of Physics Education*, 4(1), p 7-21.
[10] Alif D, Berlian A., Fatah K, Rania N., all. A, Parno. P 2020 Jurnal Pendidikan Fisika dan Keilmuan (JPFK), 6(1), p 55-64.
[11] Cert Çıbık, A. 2017 *Educational Sciences: Theory & Practice*, 17, 1061–1090

[12] Kalman, C S 2017 Successful Science and Engineering Teaching in Colleges and Universities: IAP Article I
[13] Caballero M D, Doughty, L, Turnbull A M, Pepper R, E, & Pollock S J 2017 *Phys. Rev. Phys. Educ. Res.* ***13**, 010118*
[14] Pratt-Smith, S 2017 *Transformations of Electricity in Nineteenth-Century Literature and Science*: Taylor & Francis
[15] Ross B H 2007 Cognitive Science: Problem Solving and Learning for Physics Education. *Physics Education Research*, 951, p11-18
[16] Adams, W K 2012 Problem Solving Assessment. *Journal Acoust. Soc. Am., 132*, 1923-1931
[17] Hall N and Webb, D 2014 Instructors' support of student autonomy in an introductory physics course. *Physical Review Special Topics-Physics Education Research, 10(2)*, p 1-22
[18] Zwickl B, M, Hu D, Finkelstein, N and Lewandowski H J 2015 *Physical Review Special Topics – Physics Education Research, 11(2*), 1-12
[19] Kuo E, Hull M, M, Gupta A, and Elby A 2013 *Science Education, 97(1*), 32-57
[20] Kelley T, R, Knowles J G 2016 *International Journal of STEM Education*, 3(1), 11
[21] Watson, J. R. 1994 *International Journal of Science Education, 16:1*, 27-43
[22] Steven R. Rose 1987 *Social Work with Groups, 10:1*, 85-95
[23] Bascones J., Venezuela, and Novak, J D 1985 *European Journal of Science*, 7(3), 253–261.
[24] Fortus D, Krajcik J, Dershimer RC, Marx R W and Naaman RM. 2014 *International Journal of Science*, (October), 37–41.
[25] Gaigher E, Rogan J M, and Braun, M. W. H 2007 *International Journal of Science Education*, *29*(9), 1089–1110.
[26] Kim M, and Tan H T 2013 *International Journal of Science Education*, *35*(3), 357–387
[27] Kelly R, McLoughlin E, and Finlayson O. E. 2016 *International Journal of Science Education*, *38*(11), 1766–1784
[28] Taylor P., Ranade S, M, & Corrales, A. 2012 *European Journal of Engineering Teaching Problem Solving: Don't Forget the Problem Solvers*. *38* (May 2013), 7–41
[29] Freitas IM, Jimenez R, and Mellado V, 2004 *Research in Science Education*, *30*(1), 110–130
[30] Park J, and Lee L 2004 *International Journal of Science Analyzing Cognitive or Non-Cognitive Factors Involved in the Process of Physics Problem - Solving in an Everyday Context*. 37–41
[31] Doctor J L, Dornfeld J, Frodermann E, Heller K, Hsu L, Jackson K A, Yang J 2016 *Physical Review Physics Education Research*, *12*(1), 1–18
[32] Dossey JA, and Funke, J 2016 *Canadian Journal of Science, Mathematics and Technology Education*, *16*(1)
[33] Grigg S, Van Dyken J, Morkos B, and Benson L 2013 *120th ASEE Annual Conference & Exposition*, 1–12. United State of America: Atlanta
[34] Scherer R, Meßinger-Koppelt J, and Tiemann R 2014 *International Journal of STEM Education*, *1*(1), 2.
[35] Al-Khateeb M, 2018 *The Effect of Teaching Mathematical Problems Solving Through Using Mobile Learning on the Seventh Grade Students' Ability to Solve them in Jordan* International Association of Online Engineering
[36] Bai, H. 2019 Pedagogical Practices of Mobile Learning in K-12 and Higher Education Settings. *Tech Trends* ***63***, 611–620
[37] Ng K.H, Chiu T.K.F. 2017 Emotional Multimedia Design for Developing Mathematics Problem Solving Skills. In: Ma W., Chan CK., Tong K., Fung H., Fong C. (Eds) *New Ecology of Education* — Communication X Learning. Springer, Singapore.
[38] G. Gunawan, A. Harjono, H. Sahidu, L. Herayanti 2017 *Jurnal Pendidikan IPA Indonesia*, vol. 6, no. 2 pp. 257-264
[39] Naukkarinen J and Sainio T, 2018 *Education for Chemical Engineers, 22*, 61-68.

[40] Yuliati L, Riantoni C, Mufti N 2018 *International Journal of Instruction, v11 n4* p123-138
[41] Klein P, Ivanjek L, Dahlkemper M N, Jeličić K, Geyer M A, Küchemann, A and Susac A 2021 *Phys. Rev. Phys. Educ. Res. 17*, 010117
[42] Koehler A A, Cheng Z, Fiock H, Janakiraman S, and Wang H. 2020 Asynchronous online discussions during case-based learning: A problem-solving process. *Online Learning, 24(4),* 64-92
[43] Susilo A, and Suhardi D. 2018 *Edcomtech Jurnal Kajian Teknologi Pendidikan, 3(2)*, 53-62.
[44] Nassaji H 2015 Qualitative and descriptive research: Data type versus data analysis *Language Teaching Research, Vol. 19(2*) 129–132
[45] Reddy, M V B and Panacharoensawad B, 2017 *Journal of Education and Practice Vol.8, No.14, p59-62*
[46] Chi, M. T. H., & Glaser, R. 1985 Problem solving ability. In R. J. Sternberg (Ed.), *Human abilities: An information-processing approach* (pp. 227-257). San Francisco, CA: W. H. Freeman & Co.
[47] Chi MTH, Feltovich J, Glaser R, 1981 Categorization and representation of physics problems by experts and novices *Cognitive Science Vol 5 (2),* p 121-152
[48] Singh, C 2008 *Phys. Rev. ST Phys. Educ. Res.* 4, 010105
[49] Mason A and Singh, C 2011 *Phys. Rev. ST Phys. Educ. Res. 7*, 020110
[50] Lin, S Y and Singh, C 2013 *Phys. Rev. ST Phys. Educ. Res. 9*, 020114

Journal of Physics: Conference Series **2019** (2021) 012049 doi:10.1088/1742-6596/2019/1/012049

Development of Online Learning Resource for Accelerated Linear Motion Material with the Problem Based Learning (PBL) Model During the COVID-19 Pandemic

Sudarwanto[1], V Serevina[1], D A Nugroho[1] and H F Lipikuni[2]
[1] Faculty of Mathematics and Natural Sciences, Universitas Negeri Jakarta Jl. Rawamangun Muka, Jakarta 13220, Indonesia
[2] Faculty of Mathematics and Natural Sciences, Universitas San Pedro Jl. Veteran, Nusa Tenggara Timur 85111, Indonesia
Email: sudarwanto@unj.ac.id; vina.serevina77@gmail.com; drajat.agung@gmail.com fridolinary@gmail.com

Abstract. Coronavirus Disease or COVID-19 causes a serious affect in different divisions of life, including in education. At the present time, exercises at school have been incidentally suspended, but learning must proceed to be carried out to give what students need. The solution offered in dealing with the current pandemic is by doing online learning. However, the conditions in the field of teachers were not ready for online learning, it was found that the percentage of availability of online learning resources at Public Vocational School 53 Jakarta was only 44%. Therefore, this study aims to develop Edmodo application and Zoom-based online learning resource with problem-based learning (PBL) models on accelerated linear motion material. This type of research is an educational research and development with the ADDIE model (analysis, design, development, implementation, and evaluation). Assessment of online learning resource was carried out by physics, material experts, learning experts and media experts. Testing was carried out at Public Vocational School 53 Jakarta by 35 class X students and filling out the teacher response questionnaire. The result of this research is an online learning resources using the PBL model on accelerated linear motion material (GLBB). The online learning resource which developed include the Learning Implementation Plan (RPP), Student Worksheets (LKS), and Handout. The results of the media feasibility validation test were obtained 98%, for the feasibility of physics, material obtained 81%, and the feasibility of learning physics was obtained 86%. From the results of the feasibility test with the teacher, it was obtained 94%. Based on the interpretation of the Likert scale, online learning resource is categorized as "very good". So, it can be concluded that the online learning resources that has been developed is suitable for use in learning physics at Public Vocational School.

1. Introduction

The Corona Virus (COVID-19) disease caused a serious disturbance in different socio-economic areas. The education sector also experienced critical disturbances. Schools and colleges have been closed, either broadly or locally in a few nations influenced by COVID-19 [1].

Online learning must be supported by fitting classroom plan and conveyance strategies so that online learning can empower students to reflect on their convictions; give a secure environment for talking about various viewpoints; guide them to explore, validate and broaden new points of view; and support them to develop new roles [2]. In addition, access is the key word for optimal online learning. Lembani, et al (2019) said that open remote learning increases opportunities for access to education because of the availability of online material. However, this condition is greatly influenced by various variables such as age, access to computers and the internet, as well as the digital divide. Facilitating adequate access, for example, is one of the most basic things in supporting the smooth running of open distance learning [3].

In responding to the COVID-19 circumstance, the teacher must be able to encourage students to provide material and indeed the evaluation system. Online learning utilizes technology, where students attempt to manage with different assignments and make choices at a time. However, in reality, within the field, instructors are not prepared to do online learning. Based on the results of the needs analysis conducted at Public Vocational School 53 Jakarta, it was found that the availability of online learning resource was only 44%. Therefore, we need an online learning resource designed to respond to the COVID-19 situation. Learning resources are a set of tools that support the success of a lesson that has been arranged already to be used within the learning process. The learning resources comprise a syllabus, lesson plans, teaching materials, and assessment instruments. The usage of high school material science learning within the 2013 curriculum requires learning gadgets in understanding the characteristics of scientific/scientific learning [4]. During the online learning process, various platforms can be utilized effectively such as learning applications, learning websites, or learning management systems (LMS). Edmodo can be used to share learning materials, give assignments, and make assessments. The Zoom application can also be used to conduct video conferences to monitor learning activities legitimately.

Edmodo is a learning platform which is free and secure. Edmodo is accessible at www.edmodo.com [5]. Edmodo's interface is just like social media that permits teachers to make a class, and anyone with the class code can join the class. At the class, the teacher can give some materials, assignments, etc. Students can also view the materials that the teacher gave, and submit their assignments and view the teacher's feedback, notes and advice. [6] [7]. Edmodo is also one of the top 25 website in "Social Networking" Category by the American Affiliation of School Custodians in 2011 [8].

The development of physics tools certainly cannot be separated from the learning model used. However, the reality shows that so far most educators have used conventional learning models and many are dominated by educators. Circumstances like this tend to be boring and will ultimately affect the learning outcomes of students. One strategy that is suitable for increasing the motivation of students to learn more effectively is to carry out variations in learning, namely by implementing physics learning with a problem-based learning model (PBL). One of the learning strategies that are seen as in line with the standards of the scientific approach is Problem Based Learning (PBL). PBL is an approach that gives the students any problems or issues that need to be solved and supposes them to find solutions to the real-world problem by applying their knowledge and skills as the learning process. [9].

Learning resources are all things needed to be used by the teacher to carry out the learning process. Teaching materials are all types of material that teachers use to assist them in carrying out learning activities. The learning resource developed include lesson plans, basic physics textbooks, student worksheets, and student learning outcomes tests.

The problem-based learning (PBL) model is based on many issues that require true examination, namely investigations that require solutions for real problems. PBL is a learning method that the relevant problem is presented early within the instruction cycle and is utilized to provide setting and motivation for the learning that takes after [10]. The PBL model has a syntax consisting of five stages of learning: 1) organizing students into problems, 2) organizing students to learn, 3) helping groups and independent investigations, 4) developing and presenting works and exhibitions, 5) analyzing and evaluating the problem solving process [11].

One of the sub-concepts discussed in this basic competency are accelerated linear motion. Accelerated linear motion (GLBB) is a motion whose path is straight and its speed changes in an orderly/patterned manner.

2. Method

The purpose of this study was to develop edmodo and zoom-based online learning resource with problem-based learning models on accelerated linear motion material suitable for use in physics learning. The learning resources that have been developed in this study are Learning Implementation Plan (RPP) that integrates Edmodo and Zoom, handouts, Student worksheet (LKS). This type of research is an educational research and development with the ADDIE model (analysis, design, development, implementation, and evaluation). A lot of teaching designers have adopted the ADDIE model [12] as a standard model for education with technology based. ADDIE is a plan that presents an arrangement of five stages that construct effective education and training [13].

3. Results and Discussion

Based on the steps of the ADDIE model, the results of the following research procedures are as follows:

3.1 *Analysis*

The analysis stage in this research includes needs analysis, curriculum analysis, and analysis of the characteristics of students. The analysis of conditions in the field is needed, to find out whether online learning can be applied. A needs analysis has been done at Public Vocational School 53 Jakarta to determine the quantity of students who have used the internet or online learning to study. It was found that as many as 74.3% of students had used the internet or online learning to independently study the physics subject matter. Some of the platforms that are often used by students include the Ruang Guru, Zenius, YouTube, Google Classroom, and Brainly. In line with the results of the needs analysis, online learning is one of the right solutions for learning during the COVID-19 period.

Figure 1 Percentage of Online Learning Resource Availability.

The material of accelerated linear motion is one of the materials that students find difficult to understand, if the teacher does not use the right learning resource. This can be seen from the results of the student's conceptual achievement needs analysis of linear motion material only 49%. The low achievement of students' concepts is caused by inadequate linear motion learning resource. It was found that the availability of linear motion learning devices in Public Vocational School 53 Jakarta was 59%. Based on the description that has been given, this study aims to develop edmodo and zoom-based online learning resource with problem-based learning models on linear motion material. This is supported by needs analysis, in which 95% agree to need online learning resources, accelerated linear motion material.

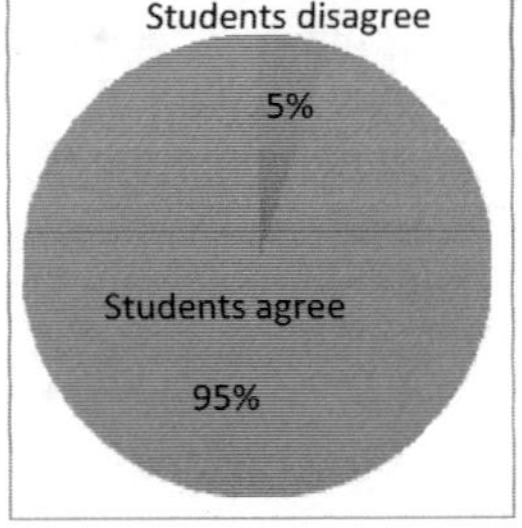

Figure 2 Percentage of Needs for Online Learning for Linear Motion Materials content.

Journal of Physics: Conference Series **2019** (2021) 012049 doi:10.1088/1742-6596/2019/1/012049

3.2 *Design*

At this stage, the preparation of online learning resource, reference collection, and instrument preparation is carried out.

3.3 *Development*

The development of online learning resource is carried out based on the design in the previous stage. The following is an explanation of the results obtained at the development stage.

a. Development of a Learning Implementation Plan (RPP)
b. Development of Student Worksheets (LKS)
c. Handout Development
d. Validation
   Furthermore, the learning device is validated by 6 validators, 2 each for each learning expert, media, and material.
e. Revision
   Then the revisions were made according to the validator's suggestions and input.

This research produces an online learning resource using the PBL model on accelerated linear motion material. The online learning resource developed include lesson plans, worksheets, and handouts. The Learning Implementation Plan (RPP) that integrates Edmodo and Zoom, handouts, Student worksheet (LKS), and Edmodo that have been developed can be seen at:
https://drive.google.com/file/d/1R_gEEr3JFUjDluvQjvIsk1049RTT8Km6/view?usp=sharing

3.4 *implementations*

At this stage, several activities were carried out, namely testing the learning resource and filling out the teacher and student response questionnaires. Learning resource that has been validated and revised are then tested. The trial of learning devices was carried out at Public Vocational School 53 Jakarta by 35 Automotive Light Vehicle Engineering (TKRO) class X students on June 30, 2020.

3.5 *Evaluation*

The last stage in the development of learning resource is evaluated. During the trial process, there were several suggestions from students and teachers that were used as suggestions for improvement or revision of stage II.

Instruments for collecting data include: (1) a questionnaire to analyze the needs of teachers and students, (2) a validated questionnaire for material, media, and learning experts, (3) a field trial questionnaire for teachers and students. The data in this study were obtained by non-test. Quantitative data were obtained from the results of the assessment of lesson plans, worksheets, and handouts by the material, media, and learning validators, as well as the results of the teacher and student field trial questionnaire assessments.

Furthermore, the learning resource that have been developed are validated by 6 validators, 2 each for each learning, media, and material expert. The media expert validation test aims to determine the feasibility of the media on based online learning resource edmodo and zoom with a learning model problem based learning for accelerated linear motion material that has been developed. The assessment is given through the media expert validation test sheet. This validation test sheet contains 23 questions from four aspects, namely (1) Components of Learning Activities in Edmodo Web which consists of 5 question items, (2) Components of worksheets consisting of 7 questions, (3) Handout components consisting of 8 questions, and (4) Video/Animation Components and Phet which consists of 3 questions. The following table of data is obtained from the results of the validation test by media experts:

Table 1. Validation Test Results by Media Experts

No	Measured Aspects	Percentage Achievements	Interpretation
1.	Components of Learning Activities in Edmodo Web	98%	Very good
2.	Components of Worksheets	97,14%	Very good
3.	Handout Components	96,25%	Very good
4.	Component Video / Animation and Phet	100%	Very good
Average of all Measured Aspects		**98%**	**Very good**

The following is a picture of the percentage of achievement of validation results by media experts:

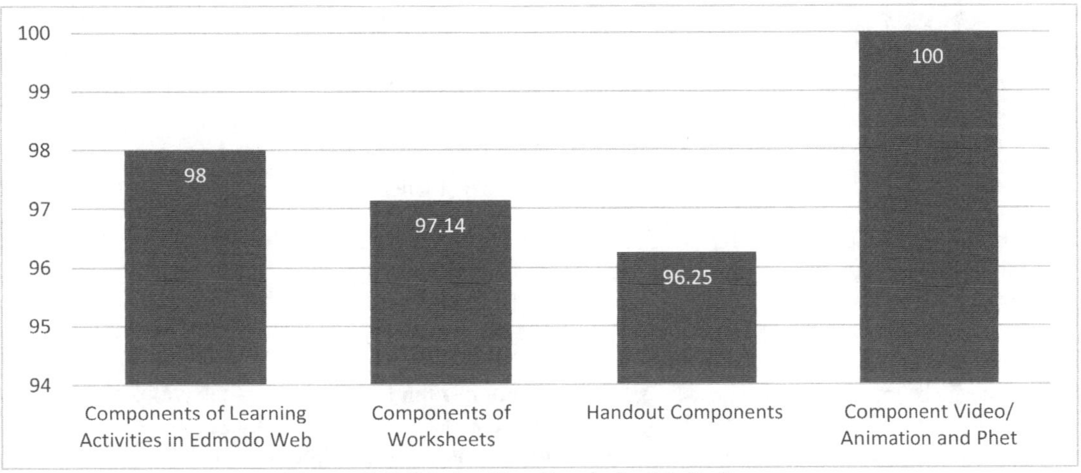

Figure 3 Validation Test Results by Media Experts.

Based on the results of the media feasibility validation test, the average percentage of achievement in all aspects was 98%. Based on scale interpretation Likert, These results indicate that the online learning resource created is very good for use in learning physics.

The physics, material expert validation test aims to determine the feasibility of the material on the based online learning resource edmodo and zoom with a learning model problem based learning for accelerated linear motion material that has been developed. The assessment is given through the physics material expert validation test sheet. This validation test sheet contains 21 questions from four aspects, namely (1) Suitability of the Handout Component with the Physics, Material which consists of 9 questions, (2) The suitability of the material in the Handout which consists of 4 questions, (3) the suitability of the material on the video / Animation which consists of 5 questions, and (4) Language of Writing Material which consists of 3 questions. The following table of data is obtained from the results of the validation test by material physics experts:

Table 2. Validation Test Results by Material Experts

No.	Measured Aspects	Percentage Achievements	Interpretation
1.	Suitability of the Handout Components with the Physics	81,11%	Very Good
2.	Material in the Handout	77,5%	Good
3.	Suitability of the material on the video / animation	82%	Very Good
4.	Use of language in writing	83,33%	Very Good
Average of all Measured Aspects		**81%**	**Very Good**

The following is a histogram, image of the percentage of achievement of validation results of material physics experts:

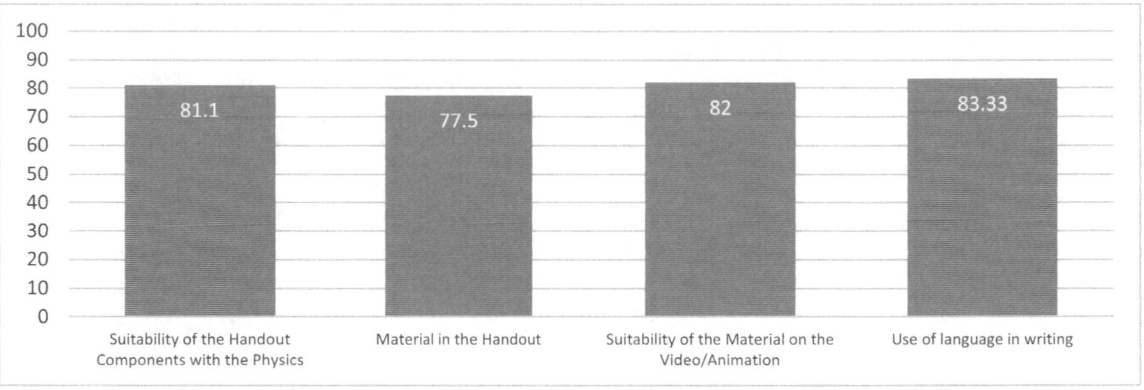

Figure 4 Validation Test Results by Material Experts.

Based on the results of the feasibility validation test of the physical material, the average percentage of the overall aspects of the achievement is 81%. Based on scale interpretation Likert, these results indicate that the online learning resource developed is very good for use in learning physics.

The learning expert validation test aims to determine the feasibility of learning on based online learning resource edmodo and zoom with a learning model problem based learning for accelerated linear motion material that has been developed. The assessment is given through the learning expert validation test sheet. This validation test sheet contains 20 questions from five aspects, namely (1) Characteristics of the RPP, which consists of 5 questions, (2) Stages of PBL, which consists of 5 questions, (3) Characteristics of the Handout which consists of 3 questions, (4) Learning Assessment which consists of 4 questions and (5) Characteristics of worksheets consisting of 3 questions. The following table of data is obtained from the results of the validation test by learning experts.

Table 3. Validation Test Results by Learning Experts

No.	Measured Aspects	Percentage Achievements	Interpretation
1.	RPP Characteristics	90%	Very Good
2.	PBL Phases	84%	Very Good
3.	Handout Characteristics	90%	Very Good
4.	Learning Assessment	77,5%	Good
5.	Characteristics of worksheets	90%	Very Good
Average of all Measured Aspects		**86%**	**Very Good**

The following is a histogram, image of the percentage of achievement of validation results by learning experts:

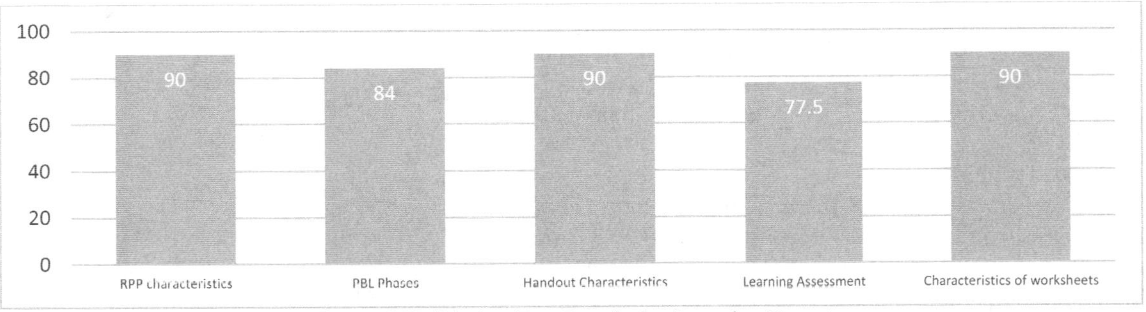

Figure 5 Validation Test Results by Learning Experts.

Based on the results of the feasibility validation test for learning physics, the average percentage of achievement in all aspects is 86%. Based on scale interpretation Likert, these results indicate that the online learning resource developed is very good for use in learning physics.

Learning devices that have been validated by experts and revised are then tested for learning devices and filling out teacher response questionnaires. This test sheet for teachers contains 17 questions from four aspects, namely (1) Characteristics of the lesson plan which consists of 5 questions, (2) Components of Learning Activities in Edmodo Web which consists of 5 questions, (3) Language of Writing Material which consists of from 3 questions, and (4) Learning Assessment which consists of 4 questions. The following table of data is obtained from the results of the validation test by the teacher:

Table 4. Validation Test Results by the Teacher

No.	Measured Aspects	Percentage Achievements	Interpretation
1.	RPP characteristics	92%	Very good
2.	Components of Learning Activities in Edmodo Web	96%	Very good
3.	Use of language in writing	93,33%	Very good
4.	Learning Assessment	93,33%	Very good
Average of all Measured Aspects		**94%**	**Very Good**

The following is a histogram, image of the percentage of teacher validation results:

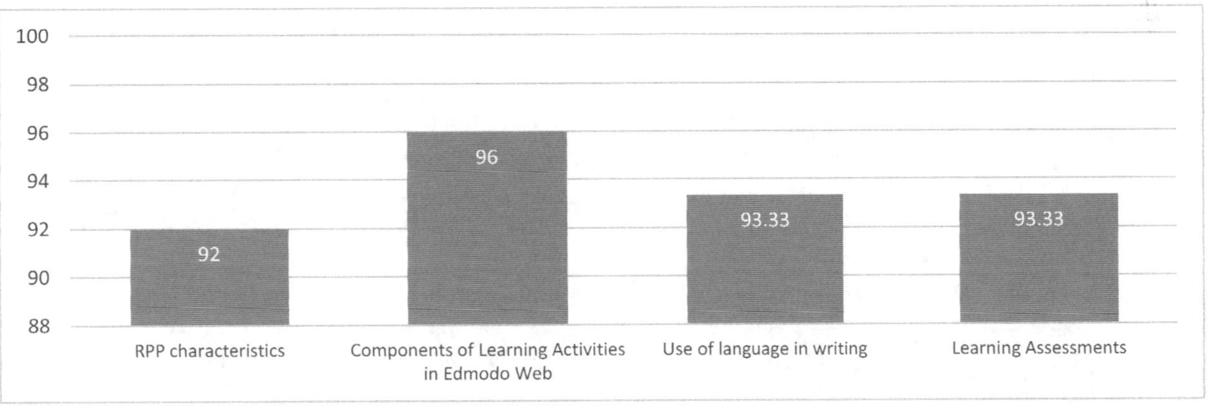

Figure 6 Validation Test Results by the Teacher.

Based on the results of the feasibility validation test with the teacher, an average percentage of the achievement of all aspects was 94%. Based on scale interpretation Likert, these results indicate that the online learning resource developed is very good for use in learning physics.

4. Conclusion

The conclusion of this research is the Edmodo application and Zoom-based online learning resource with problem-based learning (PBL) models on the material of accelerated linear motion developed has met the criteria very well and is suitable for use in learning physics.

Acknowledgement

The researchers would like to thank to the Automotive Light Vehicle Engineering (TKRO) class X students and teachers from the Public Vocational School 53 Jakarta for support through this study.

References

[1] Gunawan, Dec. Variations of Models and Learning Platforms for Prospective Teachers During the COVID-19 Pandemic Period *Journal of Teacher Education, Vol. 1 No. 2, 2020: 61-70.* 2020

[2] Barbara J. Hoskins. Is Distance Learning Transformational?, *The Journal of Continuing Higher Education*, 61:1,62-63, DOI: 10.1080/07377363.2013.759488. 2013

[3] Reuben Lembani, Ashley Gunter, Markus Breines & Mwazvita Tapiwa Beatrice Dalu. The same course, different access: the digital divide between urban and rural distance education students in South Africa, *Journal of Geography in Higher Education*, 44:1,70-84, DOI: 10.1080/03098265.2019.1694876. 2020

[4] Mutia, M., bode, A. S., & Serevina, V. Pengembangan Perangkat Pembelajaran Fisika Sma Berbasis Problem Based Learning Sebagai Implementasi Scientific Approach Dan Penilaian Authentic, *Prosiding Seminar Nasional Fisika (E-Journal)*, *3*, 169 – 184. 2014

[5] Chada Kongchan. 2013 *How Edmodo and Google Docs can change traditional classrooms* (Brighton, United Kingdom: The European Conference on Language Learning)

[6] Jarc, J. Edmodo-a free, web 2.0 classroom management tool, Http://trendingeducation.com/?P=190, August 24, 2010.

[7] Witherspoon, A. Edmodo A Learning Management System, Http://www.plugintotechnology.com/2011/01/edmodoa -learningmanagement.html. January 05,2011. 2010

[8] Habley, J. AASL announces 2011 best websites for teaching and learning, http://americanlibrariesmagazine.or/news.ala/aasl-announces2011-best-websites-teaching-and-learning, June28, 2011.

[9] Bligh, J. Problem-based learning in medicine: an introduction, *Post-Graduate Medial Journal*, 71 (8), 323-326. 1995

[10] Michael, P. Does Active Learning Work? A Review of the Research, *Journal of Engineering Education, 93*(3), 223-231. 2004

[11] Agustina, K., Kristiyanto, W. & Noviandini, D. Learning Design of Problem Based Learning Model Based on Recommendations of Syntax Study and Contents Issues on Physics Impulse Materials with Experimental Activities, *International Journal of Active Learning, 2*(2), 68-81. Retrieved July 20, 2020 from https://www.learntechlib.org/p/208679/. 2017

[12] Morrison, G.R. Designing Effective Instruction, *6th Edition, John Wiley & Sons, Hoboken, New Jersey.* 2010

[13] Almomen, R. , Kaufman, D. , Alotaibi, H. , Al-Rowais, N. , Albeik, M. And Albattal, S. Applying the ADDIE—Analysis, Design, Development, Implementation and Evaluation—Instructional Design Model for Continuing Professional Development for Primary Care Physicians in Saudi Arabia, *International Journal of Clinical Medicine*, **7**, 538-546. Do: 10.4236/ijcm. 2016. 78059. 2016

Journal of Physics: Conference Series **2019** (2021) 012050 doi:10.1088/1742-6596/2019/1/012050

Analysis of Students' Misconception and Strategic Thinking Skill on Parabolic Motion Material Using Two-Tier Multiple Choice Items Instrument

D Susanti[1,a], Y Supriyati[1,b], S Maulana[2,c], A R Hakim[1,d]
[1] Physics Education Program, Universitas Negeri Jakarta
Jl. Rawamangun Muka, Jakarta 13220
[2] Labschool Cibubur Senior High School
Jl. Raya Hankam Kampus Labschool No.15-20, Kota Bekasi, Jawa Barat 17432
Email: [a]dwisusanti@unj.ac.id, [b]yeti.supriyati@unj.ac.id, [c]slametmaulana1996@gmail.com, [d]hakimabdulrahman10@gmail.com

Abstract. This study aims to develop an instrument test in the form of two-tier multiple-choice items that can be used to analyze students' misconceptions and measure students' strategic thinking skills. This research uses the research and development method with the ADDIE model. The resulting test instrument has ten questions that refer to indicators of strategic thinking skills. The results of the validation of the average score of 89.20% with very good criteria. The results of the Rasch Model analysis report that the test instrument that has been developed is declared feasible with acceptable criteria, all items that have been developed have reliability in the good category, and all items that have been developed are valid. The results of the Rasch Model analysis are also informed that in the implementation of the test, it is implausible that cheating, carelessness, and lucky phenomena will occur when doing the test. The results of the Rasch Model analysis also group students into three, namely high, medium, and low abilities, which convey information on students' strategic thinking skills. The analysis results showed that 13.64% of students understood the overall concept of parabolic motion, 69.32% of students had misconceptions about the parabolic motion material, and 11.36% of students did not understand the parabolic motion material at all. The remaining 5.68% cannot be predicted with certainty whether the student understands, has misconceptions, or does not understand at all. This study concludes that the two-tier multiple choice items instrument developed is valid, reliable, and feasible to be used in analyzing students' misconceptions and measuring students' strategic thinking skills on parabolic motion material.

1. Introduction

Today, education in Indonesia is growing. Even during this pandemic, education in Indonesia must continue to be developed. Many studies have contributed to the world of education. Many educational products are able to analyze problems in the learning process, such as students' misconceptions. In addition, many educational products can foster important abilities that students should possess, one of which is strategic thinking skills.

Student misconceptions are one of the urgent problems to be discussed. Misconceptions become a critical issue to be discussed because the concepts that have been obtained previously can affect the new concepts that will be obtained [1]. There are still many students who often experience misconceptions [2]. Misconceptions have a significant impact on student performance in learning [3]. Misconceptions in students contribute to causing other errors that students will face [4].

Misconceptions in students can also make students frustrated [5]. This, of course, clearly states that misconceptions hinder students in the learning process [6,7].

Strategic thinking skill can sort out problems and find the right solution to solve these problems. Strategic thinking skills usually consist of correct data, analyze problems, and structured ways to solve these problems [8]. Strategic thinking skills are the ability to think creatively, critically, and analytically, which triggers solving a problem [9,10].

Parabolic motion is one of the most challenging physics materials. The preliminary study results showed that 137 out of 162 students stated that the parabolic motion material was problematic. The previous study results also showed that the average score on the Daily Test material for the parabolic motion was low, namely 53.

In order to solve these two problem topics, a new research product is needed that can analyze students' misconceptions and students' strategic thinking skills. One product that can be developed is a test instrument in the form of two-tier multiple-choice items.

Based on the descriptions above, this study aims to develop a test instrument in the form of two-tier multiple-choice items that can be used to analyze students' misconceptions and measure students' strategic thinking skills.

2. Method

This research uses the research and development method with the ADDIE model. The ADDIE model was chosen because it is easy to implement [11]. The ADDIE model consists of five steps: Analyze, Design, Development, Implementation, and Evaluation.

In the analyze stage, this research begins by analyzing the needs and problems that are often encountered in the world of education. At this stage, it is known that the problem of misconceptions and strategic thinking skills is one of the problems that must be solved. The solution offered is the development of a test instrument in the form of two-tier multiple-choice items. At the design stage, this study develops a blueprint of the test instrument in the form of two-tier multiple-choice items that will be developed. At the development stage, the test instrument in two-tier multiple-choice items began to be developed. After being developed, the test instrument will be theoretically validated by several experts to determine its feasibility. At the implementation stage, the theoretically validated test instrument will be tested on many students. The test results will then be analyzed using the Rasch Model measurement model, which focuses on determining students' validity, reliability, and strategic thinking skills. The results of the trial will also be analyzed to determine the possibility of student misconceptions. At the evaluation stage, an evaluation of each step has been carried out to conclude at each stage.

3. Results and Discussion

3.1. Two-Tier Multiple Choice Items Instrument

This research produces educational products in the form of two-tier multiple-choice items instrument. The resulting test instrument has ten questions. Each item developed refers to indicators of strategic thinking skills. Each item has two levels of answers. The first level requires students to provide answers to the problems given. The second level requires students to give reasons in choosing the answers given at the first level.

Journal of Physics: Conference Series **2019** (2021) 012050 doi:10.1088/1742-6596/2019/1/012050

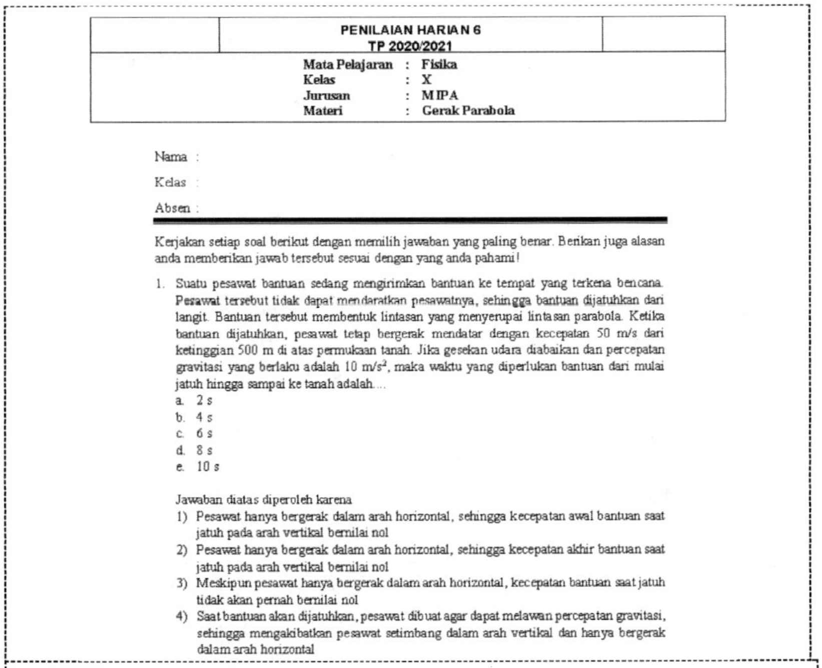

PENILAIAN HARIAN 6
TP 2020/2021

Mata Pelajaran : Fisika
Kelas : X
Jurusan : MIPA
Materi : Gerak Parabola

Nama :

Kelas :

Absen :

Kerjakan setiap soal berikut dengan memilih jawaban yang paling benar. Berikan juga alasan anda memberikan jawab tersebut sesuai dengan yang anda pahami!

1. Suatu pesawat bantuan sedang mengirimkan bantuan ke tempat yang terkena bencana. Pesawat tersebut tidak dapat mendaratkan pesawatnya, sehingga bantuan dijatuhkan dari langit. Bantuan tersebut membentuk lintasan yang menyerupai lintasan parabola. Ketika bantuan dijatuhkan, pesawat tetap bergerak mendatar dengan kecepatan 50 m/s dari ketinggian 500 m di atas permukaan tanah. Jika gesekan udara diabaikan dan percepatan gravitasi yang berlaku adalah 10 m/s^2, maka waktu yang diperlukan bantuan dari mulai jatuh hingga sampai ke tanah adalah....
   a. 2 s
   b. 4 s
   c. 6 s
   d. 8 s
   e. 10 s

Jawaban diatas diperoleh karena
1) Pesawat hanya bergerak dalam arah horizontal, sehingga kecepatan awal bantuan saat jatuh pada arah vertikal bernilai nol
2) Pesawat hanya bergerak dalam arah horizontal, sehingga kecepatan akhir bantuan saat jatuh pada arah vertikal bernilai nol
3) Meskipun pesawat hanya bergerak dalam arah horizontal, kecepatan bantuan saat jatuh tidak akan pernah bernilai nol
4) Saat bantuan akan dijatuhkan, pesawat dibuat agar dapat melawan percepatan gravitasi, sehingga mengakibatkan pesawat setimbang dalam arah vertikal dan hanya bergerak dalam arah horizontal

Figure 1. Two-Tier Multiple Choice Item.

3.2. Theoretical Validation

The test instrument that has been developed will then be validated theoretically by several experts. The validation results show an average score of 89.20%, with very good criteria. The score is divided into material aspects of Physics by 89.58%, media aspects by 86.36%, and assessment aspects by 91.67%, with all three having very good criteria.

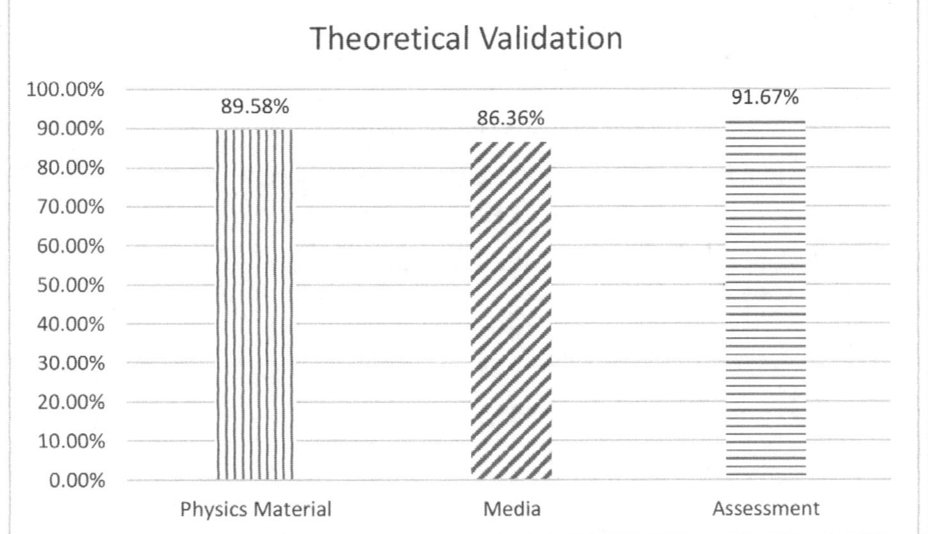

Figure 2. Theoretical Validation.

3.3. Rasch Model Analysis

The test instrument that has been declared valid is then tested on many students. The answers given by students will be analyzed using the Rasch Model measurement model.

3.3.1. Unidimensionality. The Rasch model provides information about whether or not a test instrument that has been developed is valid. The results of the Rasch Model analysis give a unidimensionality value of 21.07%. This value informs that the test instrument that has been developed is declared feasible with acceptable criteria.

3.3.2. Item Reliability. The Rasch model provides information about the reliability of the items that have been developed. The results of the analysis of the Rasch Model give an item reliability value of 0.83. This value informs that all the items that have been developed have good reliability.

3.3.3. Item-Fit. The Rasch model provides information about whether or not each item developed is valid. The results of the Rasch Model analysis show that each item developed meets at least one of the item-fit criteria. This informs that all the questions that have been developed are valid.

3.3.4. Person-Fit. The Rasch model provides information about whether or not students taking the test are valid in working on the test instrument that has been developed. The results of the Rasch Model analysis show that each student taking the test meets at least one of the person-fit criteria. This informs that every student taking the test has a fit ability. It also informs that in implementing the test, it is implausible that cheating, carelessness, and lucky phenomena will occur when taking the test.

3.3.5. Person Measure. The Rasch model provides information about the abilities of each student taking the test. The results of the Rasch Model analysis classify test-taking students into three categories. The categories are categories of students with high, medium, and low abilities. The results of the Rasch Model analysis regarding student abilities also inform students' strategic thinking skills.

3.4. Students' Misconception

The answers obtained from the test results are also used to obtain how big the student's concept is. The analysis results showed that 13.64% of students understood the whole concept of parabolic motion, 69.32% of students had misconceptions about the parabolic motion material, and 11.36% of students did not understand the parabolic motion material at all. The remaining 5.68% cannot be predicted with certainty whether the student understands the concept, has misconceptions, or does not understand at all. The results of the analysis for each item are as follows.

Table 1. Students' Misconception for Each Item.

Item Number	Fully Understanding	Misconception	No Understanding	Unpredictable
1	25.00%	56.82%	11.36%	6.82%
2	6.82%	78.41%	13.64%	1.14%
3	18.18%	53.41%	23.86%	4.55%
4	4.55%	88.64%	4.55%	2.27%
5	7.95%	81.82%	5.68%	4.55%
6	20.45%	35.23%	29.55%	14.77%
7	1.14%	84.09%	10.23%	4.55%
8	12.50%	72.73%	5.68%	9.09%
9	12.50%	78.41%	7.95%	1.14%
10	18.18%	63.64%	4.55%	13.64%

Journal of Physics: Conference Series **2019** (2021) 012050 doi:10.1088/1742-6596/2019/1/012050

3.5. Discussion

This research produces educational products in the form of two-tier multiple-choice items instrument. The resulting test instrument has ten questions. Each item developed refers to indicators of strategic thinking skills. Each item has two levels of answers. The first level requires students to provide answers to the problems given. The second level requires students to give reasons in choosing the answers given at the first level.

Based on some previous studies, the development of a two-tier multiple-choice items instrument can measure High Order Thinking Skills (which requires strategic thinking skills) [12–14] and also student scientific literacy [15]. This is in line with a study that states that developing a two-tier multiple choice item instrument can help identify students' misconceptions [16]. This certainly supports that with developing a two-tier multiple-choice item instrument, measurement of strategic thinking skills and analysis of student misconceptions can be carried out.

The test instrument that has been developed will then be validated theoretically by several experts. The validation results show an average score of 89.20%, with very good criteria. The score is divided into material aspects of Physics by 89.58%, media aspects by 86.36%, and assessment aspects by 91.67%, with all three having very good criteria. These results inform that when viewed from the aspect of physics material, media aspect, and assessment aspect, the two-tier multiple-choice items instrument that has been developed can be declared feasible with very good criteria,

The results of the Rasch Model analysis give a unidimensionality value of 21.07%. This value informs that the test instrument that has been developed is declared feasible with acceptable criteria. The results of the analysis of the Rasch Model also provide an item reliability value of 0.83. This value informs that all the items that have been developed have good reliability. In addition, the results of the Rasch Model analysis show that each item developed meets at least one of the item-fit criteria. This informs that all the questions that have been developed are valid. Based on the study results, the two-tier multiple-choice items instrument can be declared valid, reliable, and can be used in the learning process.

The results of the Rasch Model analysis are also able to show that each student taking the test meets at least one of the person-fit criteria. This informs that every student taking the test has a fit ability. It also informs that in implementing the test, it is implausible that cheating, carelessness, and lucky phenomena will occur when taking the test. The results of the Rasch Model analysis classify test-taking students into three categories. The categories are categories of students with high, medium, and low abilities. The results of the analysis of the Rasch Model also inform students' strategic thinking skills. Based on the results of this study, the two-tier multiple-choice items instrument that has been developed can measure students' strategic thinking skills.

Strategic thinking skills have proven to be one of the skills needed to face challenges in this era of globalization [17]. Students' strategic thinking skills are proven to change student behaviour in digging for information [18]. Strategic thinking skills are currently needed; given that evaluation questions are developed with the HOTS category, students must think strategically in solving a problem. [19]. Several opinions above indeed become the basis that the need for a tool to measure students' strategic thinking skills is critical. So that with the development of the two-tier multiple-choice items instrument has answered the problem of measuring students' strategic thinking skills.

The answers obtained from the test results are also used to obtain how big the student's concept is. The analysis results showed that 13.64% of students understood the whole concept of parabolic motion, 69.32% of students had misconceptions about the parabolic motion material, and 11.36% of students did not understand the parabolic motion material at all. The remaining 5.68% cannot be predicted with certainty whether the student understands the concept, has misconceptions, or does not understand at all. Based on the study results, the two-tier multiple choice items instrument that has been developed can also analyze the students' misconceptions.

Students' misconceptions have become a common phenomenon in the learning process [20]. Misconceptions of a concept have been shown to affect student performance [21] directly. Identifying misconceptions in students can help teachers determine gaps in the curriculum [22]. If the misconception is maintained, then the concept to be studied next will be wrong and create difficulties in the learning process [23]. This is supported by recommendations from a previous study that asked to pay attention to the learning model to minimize misconceptions in students [24]. So with the development of a two-tier multiple-choice item instrument, it can help teachers analyse students who have misconceptions.

4. Conclusion

Based on the results of the research that has been obtained and the discussions that have been described, it can be concluded that the two-tier multiple-choice items instrument that has been developed is valid and reliable. The two-tier multiple-choice items instrument that has been developed is also suitable for analyzing student misconceptions and measuring students' strategic thinking skills on parabolic motion material.

References

[1] Andariana A, Zubaidah S, Mahanal S and Suarsini E 2020 *Journal for the Education of Gifted Young Scientists* 8
[2] Pakpahan T R, Hernawati D and Ardiansyah R 2020 *BIOEDUSCIENCE: Jurnal Pendidikan Biologi dan Sains* 4
[3] Chen C, Sonnert G, Sadler P M, Sasselov D and Fredericks C 2020 *Journal of Research in Science Teaching* 57
[4] Qian Y and Lehman J 2017 *ACM Transactions on Computing Education* 18
[5] Chen C, Sonnert G, Sadler P M and Sunbury S 2020 *CBE Life Sciences Education* 19
[6] Soeharto, Csapó B, Sarimanah E, Dewi F I and Sabri T 2019 *Jurnal Pendidikan IPA Indonesia* 8
[7] Vitharana P 2021 *European Journal of Science and Mathematics Education* 3
[8] Al-Qatamin A A and Esam A M 2018 *International Journal of Business and Management* 13
[9] Pisapia J, Pang N S-K, Hee T F, Lin Y and Morris J D 2009 *International Education Studies* 2
[10] Alsaaty F M 2011 *Journal of Business & Economics Research (JBER)* 5
[11] Cahyadi R A H 2019 *Halaqa: Islamic Education Journal* 3
[12] Rintayati P, Lukitasari H and Syawaludin A 2020 *International Journal of Instruction* 14
[13] Laksono P J 2019 *Orbital: Jurnal Pendidikan Kimia* 2
[14] Ramadhan G, Dwijananti P and Wahyuni S 2018 *Unnes Physics Education Journal* 7
[15] Haka N B, Hamid A, Nurhidayah N, Kesumawardhani A D, Rudhini M and Riski R A 2019 *Biosfer: Jurnal Tadris Biologi* 10
[16] Widiyatmoko A and Shimizu K 2018 *Jurnal Pendidikan IPA Indonesia* 7
[17] Julita J, Darhim D and Herman T 2020 *Journal of Physics: Conference Series* vol 1477
[18] Ardakan M A, Ghadyani F and Alidadi T 2018 *Journal of Information Technology Management* 10
[19] Firmansyah E, Mubarika M P and Saniah L 2021 *Journal of Physics: Conference Series* vol 1778
[20] Mubarak S and Yahdi 2020 *Jurnal Pendidikan IPA Indonesia* 9
[21] Bordes S J, Gandhi J, Bauer B, Protas M, Solomon N, Bogdan L, Brummund D, Bass B, Clunes M and Murray I V J 2020 *Advances in Physiology Education* 44
[22] Neidorf T, Arora A, Erberber E, Tsokodayi Y and Mai T 2020 *IEA Research for Education* vol 9
[23] Halim A S, Finkenstaedt-Quinn S A, Olsen L J, Gere A R and Shultz G V. 2018 *CBE Life Sciences Education* 17
[24] Kurniawan Y, Muliyani R and Nassim S 2019 *Jurnal Ilmiah Pendidikan Fisika Al-Biruni* 8

Project Based Blended Learning and Independent Learning on Critical Thinking Skill

O Aliftika[1], I M Astra and Y Supriyati
[1] Physics Education Master Program, Universitas Negeri Jakarta, Jl. Rawamangun Muka, Jakarta 13220, Indonesia
[2] Department of Physics, Universitas Negeri Jakarta, Jl. Rawamangun Muka, Jakarta 13220, Indonesia
Email: [1]oliviaaliftika_1310819015@mhs.unj.ac.id, [2]imadeastra@unj.ac.id, [3]yetti.supriyati@unj.ac.id

Abstract. This study aimed to determine the effect of the project based blended learning and independent learning of critical thinking skill. The study used the quasi-experimental research method in the 2x2 treatment by level model. The subjects of this study were vocational high school students in Grade 10. The samples were divided into two groups; one group learned through the project based blended learning model with 73 students while the other group learned through the blended learning model with 74 students. Both groups were then given an independent learning test to divide students into categories of students who had high independent learning and students who had low independent learning. The research showed that the critical thinking skill of students who learned through the project based blended learning models was higher than those who learned through only blended learning models. Also, the critical thinking skill of students who had high independent learning was greater than students who had low independent learning. The results of the hypothesis test indicated that there is a significant difference in critical thinking skill between students who learned through the project based blended learning model and students who learned through the blended learning model, and there is a significant interaction effect between learning models and independent learning on critical thinking skill.

1. Introduction

Graduates' competencies are based on the Indonesian National Education Goals that must consider 21st century learning and skills such as critical thinking skills, creativity, collaboration, and communication. Therefore, classroom learning must help students in training and improving 21st century skills so that students can be successful in their future. Critical thinking skills are one of the 21st century skills that must be possessed by students. Critical thinking skills are mental processes in the form of higher-order thinking skills, including knowledge and skills to analyze, evaluate, draw conclusions, make decisions and solve problems [1]. Critical thinking skills can be trained and developed by adjusting appropriate educational strategies or models [2]. According to Duron, the right strategy or learning model to improve critical thinking skills is learning that make students as the center so that students can practice their skills in critical thinking [3].

Journal of Physics: Conference Series 2019 (2021) 012051 doi:10.1088/1742-6596/2019/1/012051

Research conducted by Suharno, Pambudi & Harjanto (2020) states that learning approaches still tends to be theoretical and less relevant. In addition, 74% of students feel bored while studying [4]. Some of these things are also in line with the results of a preliminary study using a survey method via Google Form which was distributed to class X students and physics teachers at the vocational education level. With the number of responses from 167 students and 15 physics teachers, it shows that 152 of 167 or 91% of students feel that classroom learning has not led to the reconstruction of knowledge, analyzing a phenomenon and solving problems. Classroom learning occurs only in one direction between the teacher and the students. This has an effect on not being able to develop and train the skills needed. A total of 149 out of 167 or 89% of students also stated that the learning process currently being carried out still prioritizes theoretical studies and has not trained in processing skills in depth.

Learning must be able to integrate knowledge, skills, attitudes and abilities of ICT. Higher education institutions also continue to familiarize themselves with online and blended learning [5]. In addition, considering the current conditions that make many educational institutions adjust to blended learning as a result of the COVID-19 pandemic. One of the learning models of the development of blended learning that can improve critical thinking skills is Project Based Blended Learning (PjB2L). Project Based Blended Learning is a combination of the Project Based Learning (PjBL) model with the Blended Learning model. This is because of the characteristics of PjBL learning models that can improve students' knowledge and skills of working within a period of time to investigate and respond to complex questions, problems or challenges [6]. By combining the blended learning model, students can be more active, communicative and interactive in learning and increase their knowledge. Because the advantages of blended learning are providing individual learning experiences, supporting and encouraging independent and collaborative learning, increasing teacher pre-service involvement in education, and being flexible in learning, meaning anytime and anywhere [7]. Through the concept of PjB2L learning, students are trained to express facts, ideas, and conclusions through giving project assignments. Not only that, the skills to search and dig up data from various internet-based sources are also trained [8].

In distance learning due to the COVID-19 pandemic and the application of blended learning, students must be able to learn independently. In addition, according to the Victoria State Government (2018) learning independence is one of the important principles in learning that is recommended to be in accordance with classroom settings in the 21st Century [9]. Independent learning will be realized if students have control over personal awareness, motivation, competence, and skills that they will achieve [10]. Students who are able to study independently have greater potential to achieve high academic achievement because they have the ability to identify and choose their own problems, plan activities, and submit results at the end of the activity [11] [12]. Independent learning means developing active and participatory learning methods for students to improve skills and abilities in the learning process without being bound by the teacher or classmates, the teacher only acts as a facilitator. So that the independence of student learning has an important influence. Independent learning that can be applied properly by students will certainly affect the intellect of these students [13]. Which means that independent learning has an influence on critical thinking skills.

Based on the previous description, the authors are interested in conducting research on the influence of project based blended learning and independent learning of critical thinking skill. In this study, researchers used two learning models, which is project based blended learning model and blended learning model.

2. Method

The method used in this study is a quasi-experimental research method using a treatment by level 2 x 2 design. The dependent variable in this study is critical thinking skills while the independent variable is the learning approach with the moderator variable being independent learning. The treatment variables were divided into two, namely the learning model through project-based blended learning and blended learning model. The moderator variable is high independent learning and low independent learning. The method aims to predict conditions that can be achieved in actual experiments, but there is no control or manipulation of all variables. The design that will be used in this experimental research is as follows Table 1.

Table 1. Research Design

Independent Learning (X_2)	Learning Model (X_1)	
	Project Based Blended Learning (X_{1a})	Blended Learning (X_{1b})
High (X_{2a})	($Y_{1a\text{-}2a}$)	($Y_{1b\text{-}2a}$)
Low (X_{2b})	($Y_{1a\text{-}2b}$)	($Y_{1b\text{-}2b}$)

The population in this study was all students of grade 10 in the 2020/2021 semester II academic year at SMK Negeri 2 Sukabumi. While the sample in this study was 74 students of grade 10 SMK Negeri 2 Sukabumi which were divided into 2 classes. The experimental class consists of 37 students who will learn to use a project based blended learning model. While the controller class consisting of 37 students will learn using blended learning. Furthermore, as many as 74 students were given a test to determine the learning independence of students. The values obtained from the results of the questionnaire will be sorted starting from the highest value to the lowest. Based on this value, 27% of the top scores were taken as the upper group, namely students who had the high learning independence and the lowest 27% as the lower group, namely students who had low independent learning. Based on these proportions, students are then given a critical thinking skill test. Data obtained from these two instruments were tested for analysis prerequisites which are normality and homogeneity test, followed by two-way ANOVA hypothesis test.

3. Result and Discussions

The research data is in the form of critical thinking skills test scores. The mean scores for critical thinking skill test in each group are presented in Table 2.

Table 2. Mean Scores for Each Group

Independent Learning	Learning Model (X_1)	
	Project Based Blended Learning	Blended Learning
High	87.21	83.84
Low	74.83	81.83

In table 2, it can be seen that project based blended learning with high independent learning has the highest average score of 87.21. And the lowest average score is project based blended learning with low independent learning. Before testing the hypothesis, the normality test was carried out using the Shapiro-Wilk test, which resulted in a significant value of $p > 0.05$ which means that students' critical thinking skill test data were normally distributed. The homogeneity of the variance of the Levene's test also resulted in a significant value of $p > 0.05$. This shows that students' critical thinking skill test for two groups of learning models were homogeneous. Then, hypothesis testing was carried out using two-way ANOVA. On Table 3 presents dependent variable ANOVA using SPSS.

Table 3. Two-way ANOVA Output using SPSS

Dependent Variable: Critical Thinking Skill

Source	Type III Sum of Squares	Def	Mean Square	F	Sig.
Corrected Model	3215.063a	3	1405,021	31,740	0,334
Intercept	323244,188	1	323244,188	7302,141	<,001
Learning Approach	873,021	1	963,021	21,755	0,461
Independent Learning	4056,879	1	3056,021	69,036	0,243
Learning Approach * Independent Learning	196,333	1	196,021	4,428	0,298
Error	1947,750	44	44,267		
Total	329677,000	48			
Corrected Total	6188,813	47			

R Squared =, 684 (Adjusted R Squared =, 662)

The results of the tests that have been carried out in table 3 show that there are significant differences in the critical thinking skills test between learning model (F = 21.755 and Sig. 0.461). It means that there are differences in students' critical thinking skills with the implementation of project-based blended learning and blended learning. In table 3 also show that there are significant differences in the critical thinking skills test between independent learning (F = 69,036 and Sig. 0,243). It means that there are differences in students' critical thinking skills among students with high independent learning and law students. And there is significant interaction between the learning models and independent learning (F = 4,428 and Sig. 0,298). It means that there is an influence on learning approach and independent learning of students' critical thinking skills. The interaction between learning models and independent learning on students' critical thinking test is shown in Figure 1.

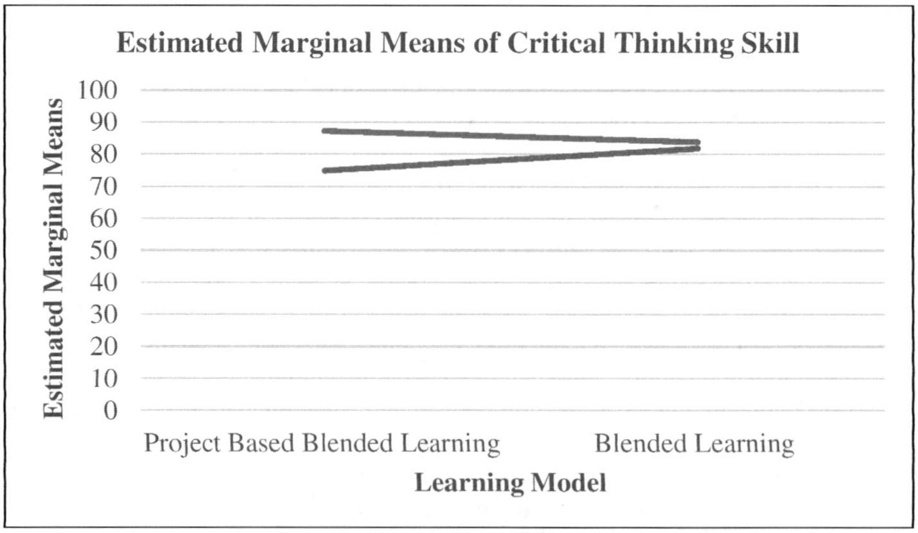

Figure 1. Interaction between learning models and independent learning on students' critical thinking skill test

In Figure 1, it shows that there are differences in the results of students' critical thinking skills among students with project based blended learning model and blended learning models. In addition, there are also differences result between students with high independent learning and low independent learning. Students who study with a project based blended learning model and have higher independent learning get the highest results compared to other students. This means that project based blended learning is effective for improving critical thinking skills [14]. In addition, this is because project based blended learning allows students to learn independently, so students can learn better [15].

4. Conclusion

Based on the research results obtained from data analysis, it shows that the critical thinking skills test for students who learn through using the project based blended learning model is different from the students who learn through the blended learning model. In addition, there are significant differences between students who have high independent learning and low independent learning. Also, there is a significant interaction between the learning model and independent learning to the students' critical thinking skills test.

Acknowledgement

The author is grateful to several parties involved in this research, such as school institutions that have given permission and Jakarta State University

References

[1] F. Fajrianthi, W. Hendriani and B. G. Septarini 2016 Jurnal Penelitian dan Evaluasi Pendidikan, p. 45–55
[2] R. Paul, R. Niewoehner and L. Elder 2019 The thinker's guide to engineering reasoning: Based on critical thinking concepts and tools, Maryland, USA: Rowman & Littlefield
[3] Mustaqim, S. D. Corebima and S. Mahanal 2019 Jurnal Pendidikan Sains, p. 145–150
[4] Suharno, N. A. Pambudi and B. Harjanto 2020 "Vocational education in Indonesia: History, development, opportunities, and challenges," Children and Youth Services Review
[5] M. Elmubarak 2020 IJAEDU- International E-Journal of Advances in Education
[6] S. Boss 2015 PBL for21st Century Success: Teaching Critical Thinking, Collaboration, Communication, and Creativity, Novato: CA: Buck Institute for Education
[7] J. S. Martin, J. E. Kreiger and A. L. Apicerno 2015 Journal of the Scholarship of Teaching and Learning, pp. 72-81
[8] W. Wahyudi and A. Winanto 2018 Journal of Educational Science and Technology (EST), pp. 91-102
[9] Victoria State Government 2018 Victoria State Government - Education and Training
[10] A. B. Bruin, E. M. Kok, J. Lobbestael and A. De Grip 2017 Metacognition and Learning, pp. 21-43
[11] H. Suhandri and T. Mardalena 2013 Jurnal Formatif
[12] F. S. Maltby, N. L. Gage, D. Berliner and C. David 2005 Educational Psychology: an Australia and New Zealand Perspectiv, Brisbane: Jhon Willey & Sons
[13] M. Yamin, 2008 Jakarta: Gaung Persada Press
[14] S. U. Poetry and Y. Hendawati 2018 "Blended Project Based Learning: Strategy for Improving Critical Thinking of Pre-Service Teachers in Science Education," in Proceeding of The 1st UR International Conference on Educational Sciences
[15] F. Silvia, Basori and D. Maryono 2019 "The Influence of Project-Based Blended Learning Toward Outcomes Student Learning," in International Conference on Online and Blended Learning 2019

Blackbody Radiation Experimental Apparatus (Arabeta) to Support Physics Learning for Senior High School Students

I Safari[1], A B Susilo and C E Rustana
Physics Education Master Program, Universitas Negeri Jakarta
Jl. Rawamangun Muka, Jakarta 13220, Indonesia
Email: [1]mnsfr2@gmail.com, anggarabs@unj.ac.id, crustana@unj.ac.id

Abstract. As part of ongoing research to develop a tool for teaching thermal radiation, we have constructed an Experimental Apparatus of Blackbody Radiation (ARABETA). Some previous researches have been conducted to investigate students understanding of Modern Physics, but very little on the topic of Black-body radiation. Apart from being more abstract, the experimental apparatus on this topic are also almost difficult to find in school laboratories. A preliminary study that involved 53 physics teachers in Jakarta region confirmed that black body radiation is a concept that is difficult to visualize to students during learning. Therefore, we developed a simple apparatus that is capable of measuring the temperature and thermal radiation of the samples used. This research uses research and development model by following some steps of Borg and Gall. The results of this measurement are shown in a graph of intensity against the multiplication result of the Stefan-Boltzmann and the fourth power of its absolute temperature, so that the emissivity value of the sample can be obtained. Through this development, we may provide an alternative solution for the experimentation issue in the learning of blackbody radiation at senior high level.

1. Introduction

Physics is an empirical science, which is based on the phenomena that occur, can be observed and measured [1]. Therefore, in the study of physics subjects accompanied by observation of phenomena or experiments in the laboratory in which measurements are also carried out. In Putra's research, et al., 2018 it was obtained that work in the laboratory is rarely done by high school teachers, thus allowing potential problems for students in understanding the subjects studied.

Blackbody radiation is one of the difficult and abstract physics learning materials for senior high school students. This data was obtained from preliminary studies to 53 high school physics teachers in Bekasi and Jakarta. According to 53 respondents, black matter radiation material is rarely observed and conducted experiments in the laboratory, due to the absence of real practicum tools for students.

Blackbody radiation experiment apparatus have actually been widely developed by experts such as experiments for the measurement of coefficients of emission of pine wood biomass pellets, rice husks and cypress wood by the method of multi-wave radiation thermometric [2], coefficient of emission of leaf revitalization conditioned in a greenhouse using an infrared thermometer and a thermocouple wire [3], measurement of blackbody emission in steel threads. Illuminated objects with infrared imaging radiometers and observed with Scanning Electron Microscopy and Energy Dispersive X-ray Spectroscopy [4], Experiments in measurement of black matter cavity radiation designed using monte carlo method [5-9], measurement of exhaust pipe emission by Monte Carlo method [10], measurement of radiation emissions of black material made of ceramic pipes with rolled heating wire [11], measurement of emissions of nickel and aluminum particle powder using Bragg fiber regeneration/RFBG lattice [12], development of normal spectral emissivity measuring instrument of solid material at high temperatures from 1073K to 1873K and wavelength from 2 μm to 25 μm [13], development of data logger-based thermal radiation practical device using the DHT11 Arduino temperature sensor[14] and the development of SNR V-1.4SL thermal radiation devices made of easily accessible and environmentally friendly materials and equipment [15].

Some experiments and practical device designed in addition to using expensive materials and tools, are also still too complicated compared to the achievement of competency standards formulated by indonesia's national education curriculum in 2013 on quantum phenomena learning. This study aims to develop a blackbody radiation experiment apparatus (ARABETA) that is adapted to the achievement of competency standards of Indonesia's national education curriculum in 2013, namely to qualitatively analyze quantum phenomena that include the radiation properties of black matter, photoelectric effects, Compton effects, and X-rays in daily. This study developed a blackbody radiation experiment apparatus that can show quantum phenomena and use materials that are easily obtained in everyday life.

2. Method

The radiation intensity of black matter varies with the temperature indicated by the equation:

$$I = \frac{P}{A} = \frac{e\,\sigma A\,T^4}{A} \; and \; I = e\,\sigma\,T^4 \qquad (1)$$

With P is the power in watts that radiations at all wavelengths from the surface of an object.

$$\sigma = 5{,}670 \times 10^{-8}\ W/m^2K^4 \qquad (2)$$

is a Stefan–Boltzmann constant, A is the cross-sectional area of the object in m2, e is the surface emissions of the object, T is the surface temperature in kelvin[16] This research uses research and development model by following some steps of Borg and Gall, they are; research and information collecting, planning, develop preliminary form of product, preliminary field testing, main product revision and final product revision. The previous design, sample was not covered by box but only covered by quartz glass as shown as figure below:

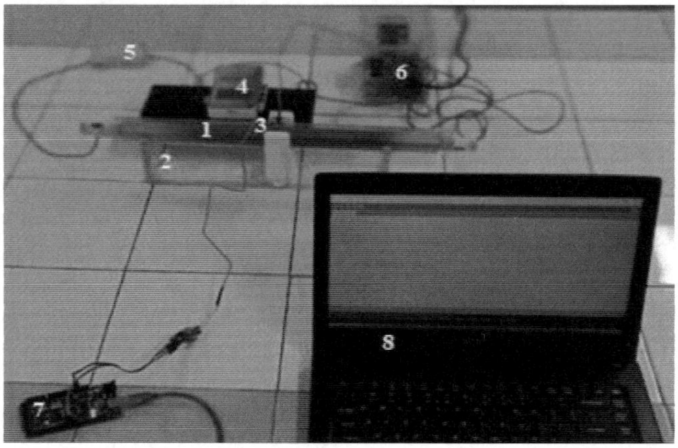

Figure 1. The previous design of blackbody radiation experiment apparatus. 1-Alumunium as sample (covered by quartz glass); 2-Supporting wall; 3-Termocouple; 4-Linshang Infrared Power meter; 5-Fuse; 6-Power supply; 7-Arduino board; 8-Laptop

The result of this design shows that the temperature of quartz glass as cover and the noise intensity of radiation from environment are too high. The next design using heater filament connected to power supply with temperature controller. The heater was covered by sample (alumunium, copper and stainless Steel rod with 3cm of inner diameter and 2 mm of thickness). The temperature of sample was measured by termocouple Arduino which are connected to laptop. The experimental apparatus was covered by wooden box and the intensity of radiation emmited by sample was measured by using Linshang infrared intensity. The recorded intensity and the fourth power of absolute temperature of sample were plotted in a graph to find the gradient of graph as the coefficient emmisivity of sample.

The experiment was applied with the similar duration, length and thickness of sample. Each sample was measured for 5 minutes every 50 kelvin of increasing temperature which started from 323 K and be limited to 823 K of heater temperature and recorded every second by using arduino platform. The intensity of radiation emmits by sample was measured with Linshang infrared power meter and recorded by using smartphone. Both data (temperature of sample and intensity of radiation) were plotted into a graph. The design of experiment is shown by picture below;

Figure 2. Blackbody radiation experiment apparatus model. 1- Heater Wire 1000 watt, 220 Volt; 2- cable conector; 3- sample rod; 4-termocouple; 5- supporting wall; 6-cover box

Figure 3. Set experiment apparatus of blackbody radiation. 6-cover box; 7- Linshang infrared intensity meter; 8- cable connector to power; 9-Fuse; 10- electric power and temperature controller of heater; 11- Arduino board; 12-laptop

3. Result and Discussion

The result of ARABETA using three sample rod (stainless steel, copper and alumunium) is presented by graph below;

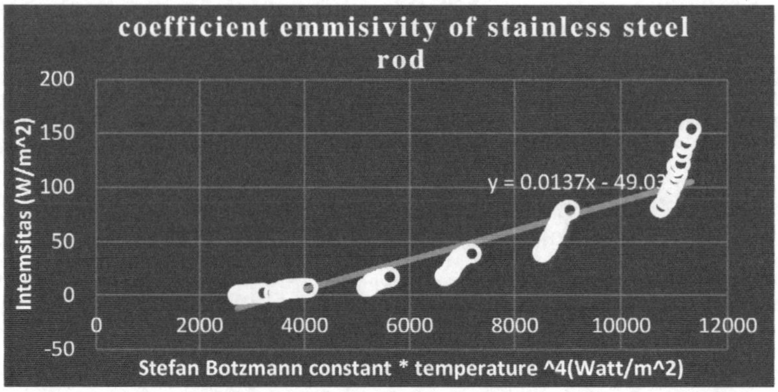

Graph a. The experiment result using stainless steel rod as sample

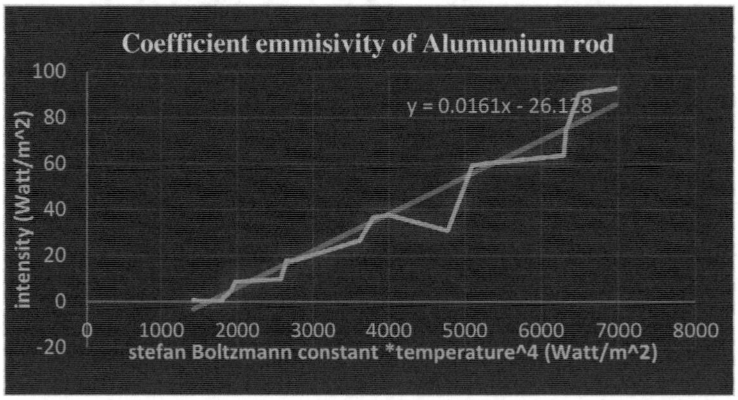

Graph b. The experiment result using alumunium rod as sample

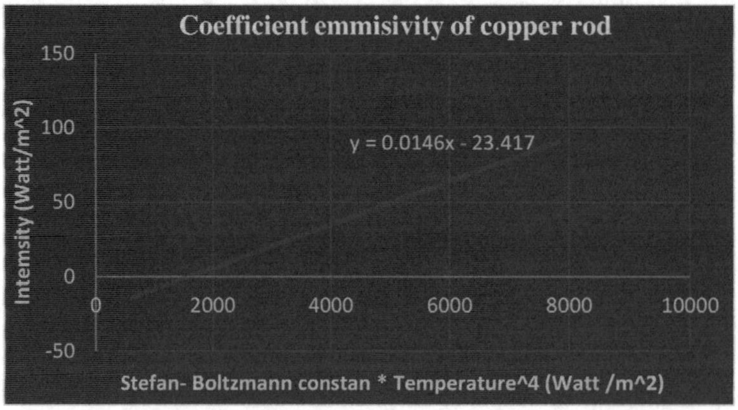

Graph c. The experiment result using copper rod as sample

All samples measurement starts from 300 K to 773 K of sample's temperature (when the temperature of stainless steel was 668,4 K, Alumunium was 591,1 K and Copper was 609.4 K). Intensity or radiation emmits by stainless steel start from 467.5 K of sample's temperature, 398.15 K of alumunium rod's temperature and 325.5 K of copper rod's temperature.

According to data from three graphs above in the same temperature every sample emmits different intensity of radiation. From the gradient of graph a known the coefficient emmisivity of stainless steel rod is 0.0137, the gradient of graph b known the coefficient emmisivity of alumunium rod is 0.0161 and the gradient of graph c known the coefficient emmisivity of copper rod is 0.0146. According to literature, the coefficient emmisivity of stainless steel polished is 0.075, coefficient emmisivity of alumunium polished is 0.039-0057 and the coefficient emmisivity of copper polished is 0.023-0052 [17]. For experiment from 0 to 2500 C, the coefficient emmisivity of stainless steel polished is 0.1-0.15, coefficient emmisivity of alumunium polished is 0.05-0.1 and the coefficient emmisivity of copper polished is 0.1 [18].

The different result may be effect of radiation background of environment such as wood as the material of cover box, alumina and teflon material as supporting wall and also ceramic as connecting cover. The significancy of infrared power meter may be also effects to the result.

The implication of this research are to use more significant infrared power meter and which is also connected to data logger in order to get more accurate data and to use slimmer designed cover box in order to reduce noise intensity of blackbody radiation by environment.

4. Conclusion

The developed experiment apparatus for blackbody radiation may show the blackbody radiation as simple quantum phenomena, eventhough the result of experiment is still farther than literature.

References

[1] I. Y. Putra, D. Dasmo, D. L. Saraswati, I. A. D. Astuti, N. Nurullaeli, Y. B. Bhakti and I. B. Rangka 2018 *Developing of physics practical module based on scientific method for students* (Jakarta: MSCEIS)

[2] S. Zheng, Y. Yang, X. Li, H. Liu, W. Yan, R. Sui and Q. Lu. Temperature and emissivity measurements from combustion of pine wood, rice husk and fir wood using flame emission spectrum, *Fuel Processing Technology* 204, pp. 1-7. 2020.

[3] Chen and Chiachung. Determining the Leaf Emissivity of Three Crops by Infrared Thermometry, *Sensors*, 15, 11387-11401; doi:10.3390/s150511387, pp. 11388-11401. 2015.

[4] M. Larciprete, Y. Gloy, R. L. Voti, G. Cesarini, G. Leahu, M. Bertolotti and C. Sibilia. Temperature dependent emissivity of different stainless steel textiles, *International Journal of Thermal Sciences*, 113 , pp. 130-135. 2017.

[5] J. G. C. B. Yikun Zhao. Comparative Study on Radiation Properties of Blackbody Cavity Model Based on Monte Carlo Method, *International Journal of Thermophysics 2020*, pp. 1-11. 2020.

[6] J. D. Lucas and J. J. Segovia. Measurement and Analysis of the Temperature Gradient of Blackbody Cavities, for Use in Radiation Thermometry Int J Thermophys, pp. 39-57. 2018.

[7] P. Saunders. Optimising Blackobody Cavity Shape for Specially Uniform Integrated Emissivity, *International Journal of Thermophysics*, pp. 1-9. 2017.

[8] C. J. Donlon, W. Wimmer, I. Robinson, G. FIisher, M. Ferlet, T. NIightingale and B. Brass. A Second-Generation Blackbody System for the Calibration and Verification of Seagoing Infrared Radiometers, *Journal of Atmospheric and Oceanic Technology*, pp. 1104-11027. 2014.

[9] V. A. Arkhipov, V. D. Gol'din, I. K. Zharova and K. G. Perfilieva 2018 *Technique of measuring the emissivity coefficient* (Tomsk Rusia: in Thermophysics 2018)

[10] G. Mei, J. Zhang, S. Zhao and Z. Xie. Design and analysis on fume exhaust system of blackbody cavity sensor for continuously measuring molten steel temperature, *AIP ADVANCES,* 7, pp. 1-6. 2017.

[11] W. G. Z. R. E. B. P. R. Poprawski. Investigation of black body radiation with the aid of a self-made pyroelectric infrared detecto, *European Journal of Physics*, pp. 1-9. 2015.

[12] M. L. G. L. J. R. M. F. Z. S Grohmann 2019 *Characterisation of the activation and reaction behaviour and determination of the emissivity of reactive nickel-aluminium particles with regenerated fibre Bragg gratings* (Munich: in Materials Science and Engineering)

[13] D. Ren, H. Tan, Y. Xuan, Y. Han and Q. Li. Apparatus for Measuring Spectral Emissivity of Solid Materials at Elevated Temperatures, *Int J Thermophys*, pp. 1-20. 2016.

[14] B. H. I. H Permana 2018 *Development of Thermal Radiation Experiments Kit Based on Data Logger for Physics Learning Media* (Jakarta: in Materials Science and Engineering)

[15] A. Malik, R. Zakwandi, S. Nurfalah, N. Nurhayati, C. Rochman and D. Nasrudin 2018 *Measuring the coefficient of emissivity using thermal radiation equipment SNR V-1.4 SL* (Bandung: in 3rd Annual Applied Science and Engineering Conference (AASEC 2018))

[16] R. D. Knight 2007 *The emission and absorbtion of Light," in Physics for scientis and engineer* (San Fransisco: Pearson) pp. 1231-1233

[17] E. T. Box, "*Emissivity Coefficient Materials,*" [Online]. Available: https://www.engineeringtoolbox.com/emissivity-coefficients-d_447.html. [Accessed 26 5 2021].

[18] Optotherm, "*Emmisivity in the infrared*" 2018 Optotherm, Inc., [Online]. Available: https://www.optotherm.com/emiss-table.htm. [Accessed 26 5 2021].

Journal of Physics: Conference Series **2019** (2021) 012053 doi:10.1088/1742-6596/2019/1/012053

Determination of Springs Constant by Hooke's Law and Simple Harmonic Motion Experiment

S Wulandari[1], B H Iswanto and I Sugihartono
Physics Education Master Program, Universitas Negeri Jakarta
Jl. Rawamangun Muka, Jakarta 13220, Indonesia
Email: [1]algieba892@gmail.com, bhi@unj.ac.id, iwan-sugihartono@unj.ac.id

Abstract. This study aims to calculate the spring constants of two types of stainless using Hooke's Law principle and simple harmonic motion methods. Two types of springs (spring I and II) with different sizes in diameter have spring constant k values of 24 N/m and 5.4N/m, respectively. Based on the experiment and calculation, the spring constant of spring 1 calculated by Hooke's Law and the simple harmonic motion experiment has an average value of 23.22 N/m with an error of 3.24% and 23.06 N/m with an error of 3.90%. On the other hand, the spring constant of spring 2 calculated by Hooke's Law and the simple harmonic motion experiment has an average value of 5.47 N/m with an error of 1.44% and 5.32 N/m with an error of 1.48 %. Hence, we concluded that the determination of spring constant using Hooke's Law has better accuracy.

1. Introduction

Spring is a component used for experiment in physics lessons both in junior high school and senior high school. In learning at school, springs are commonly used in experiment to calculate the value of the spring constant. The spring includes an elastic material, which if stretched will return to its original state, due to the presence of an elastic reaction force over the stretching tension force [1]. The spring restorer force is opposite to the deviation so that it can return to the equilibrium position. The restorer force evoked by a spring is determined by Hooke's law [2]. In addition, the spring can also dampen vibrations so that in the selection of the spring needs to be adjusted to the working force [3].

Each spring has a constant value usually called a spring constant. The value of each spring constant differs depending on the given force and the length of the spring [4]. Some factors that affect the length of the spring that can affect the value of spring constants include the type of spring, the diameter of the spring material, the diameter of the spring, and also the coil of the number of spring coils [5]. The greater the value of the spring constant, the greater the force required to stretch the spring. So the stiffer the spring, the larger the spring constant [6].

In line with the progress of technology now, the development of smartphones is very fast. Generally smartphones are used as a means of communication. Smartphones is mini computers that are easy to carry and contain applications that make it easy for users [7]. Smartphones are usually equipped with several sensors, including accelerometers, magnetometers, gyroscopes, light sensors, and GPS [8] [9]. The use of smartphones can be used in physics learning by utilizing built-in sensors and downloading existing open-source applications. Then the load will be used in the form of smartphones and case holders.

The value of spring constants can be determined in both static and dynamic ways [2]. The method of determining spring constants using Hooke's Law can be done by measuring the length of the initial spring then hanging the smartphone as a load on the spring and measuring the length of the spring after

the smartphone is hung. From here we can obtain the increase in the length of the spring. The method of determining spring constants with simple harmonic motion can be done by giving deviation to the suspended smartphone so that the spring will produce vibrations with a certain period. Therefore, it can be done by comparing the value of the spring constant of the two methods.
Based on the description above will be conducted experiments using both methods so that we will see which method has good accuracy in determining the value of spring constants.

2. Experiment Setup
This experiment aims to calculate spring constants of two types of stainless steel spring sizes using hooke's law principle and simple harmonic motion method. First, calculate the value k of the literature using the help of Spring Calculator with the following data.

Table 1. Specifications of the spring used

	Pegas 1 ***(Extended Hooks)*** **(mm)**	**Pegas 2** ***(Machine Hooks)*** **(mm)**
	Hook Length 1 : 30 Hook Length 2 : 22	-
Wire diameter	1	0.77
Outer diameter	20	15
Length inside hooks	105	170

From the results obtained the value of k literature for spring 1 is 24 N /m and for spring 2 is 5.4 N / m. The experiment set used consists of a stative pole made of iron, two springs, ruler, and a smartphone and case holder. At first, the load in the form of a smartphone and case holder was weighed, so that a mass of 0.218 kg was obtained. From here, it is known which F value will work.

Figure 1. Weigh the smartphones and case holders.

Figure 2. Prepared tools for experiment

Figure 3. Do the experiment

If based on the theory of the value of k obtained using both methods should be the same. In this case, we'll compare the two methods and we can see which method is more accurately used to determine the value of spring constants based on the experiments performed.

3. Hooke's Law
Experiment is done by measuring the length of the initial spring, the spring is hung on the stative then hangs the smartphone and case holdervon the spring and measures the length of the spring after the smartphone is hung so that the length of the spring is obtained.
Using the Hooke's Law method is combined with Newton's Law II. Where we obtain the increase in the length of the spring by using the following hooke's law.

$$F = -k \,.\, y \tag{1}$$

Where is Newton's II Law,

$$F = m \,.\, g \tag{2}$$

So that it can be substituted:

$$m \,.\, g = -k \,.\, y \tag{3}$$

$$-k = \frac{m \,.g}{y} \tag{4}$$

From the results of experiments for spring1 calculated using the Hooke's Law method obtained an average length increase of 0.092 m. As for spring 2 calculated using the Hooke's Law method obtained an average length increase of 0.39 m.

Table 2. Comparison of Hooke's Law method with k value

Springs	Methods 1 (N/m)	k Value (N/m)	Diff%
1	23.22	24	3,24%
2	5.47	5.4	1,44%

4. Simple Harmonic Motion

For methods with simple harmonic motion periods, the load is given a deviation of 2-3 cm and turn on the application to display the period by utilizing sensors from the smartphone. Here, make sure the spring moves harmoniously. So that the value of the spring constant uses the equation of harmonious motion asfollows.

$$F = m \,.\, a_y \tag{5}$$

Where a_y the acceleration of harmonic motion,

$$a_y = -\omega^2 y \tag{6}$$

So obtained,

$$-k \,.\, y = m \,. -\omega^2 y \tag{7}$$

$$k = m \,.\, \omega^2 \tag{8}$$

$$\omega^2 = \frac{k}{m} \tag{9}$$

$$\omega = \sqrt{\frac{k}{m}} \tag{10}$$

With angular velocity $\omega = \frac{2\pi}{T}$, so that

$$\frac{2\pi}{T} = \sqrt{\frac{k}{m}} \tag{11}$$

$$\frac{2\pi^2}{T^2} = \frac{k}{m} \tag{12}$$

Then obtained the value of k,

$$k = \frac{4\pi^2 . m}{T^2} \tag{13}$$

From the results of experiments for spring 1 average period using the simple harmonic motion method of 0.611. For spring 2 average period using the the simple harmonic motion method of 0.272.

Table 3. Comparison of Simple Harmonic Motion method with k value

Springs	Methods 2 (N/m)	k Value (N/m)	Diff%
1	23.06	24	3,90%
2	5.32	5.4	1,48%.

5. Conclusion

Based on the results of this experiment it can be concluded that the method of calculating spring constants by measuring the increase in the length of the spring whenweighed, using the concept of Hooke's Law has better accuracy. This can be seen from the percented different for the method. For spring 2 has a smaller k value than spring 1. When using the simple harmonic motion method, it is necessary to ensure that the spring oscillates properly. The magnitude of the spring constant greatly affects the given force. For future research it is recommended to varies the type of spring.

References

[1] E. L. Ikhtiardi, R. D. Handayani and A. D. Lesmono. Analisis Pengaruh Suhu terhadap Konstanta Pegas dengan Variasi Jumlah Lilitan dan Diameter Pegas Baja, *Jurnal Pembelajaran Fisika FKIP Universitas Jember*, vol. Vol. 3 No. 4, pp. 349 - 354. 2015.

[2] Elisa and Y. Claudya. Penentuan Konstanta Pegas degan Cara Statis dan Dinamis, *Jurnal Fisika Edukasi*, vol. 3, pp. 46 - 50. 2016.

[3] A. Pratama and M. Fitri. Rancang Bangun Alat Uji onstanta Pegas Ulir Tekan Untuk Kapasitas 50 N/mm menggunakan Metode VDI 2221, *Jurnal Ilmiah Teknik Mesin*, vol. 6 No. 2, pp. 41 - 49. 2020.

[4] M. Z. Pratama and M. Fitri. Desain Komponen Utama Alat Uji Konstanta Pegas Untuk Kapasitas 50 N/mm, *Jurnal Teknik Mesin*, vol. 10, pp. 15 - 21. 2021.

[5] I. Setiawan and D. Sutarno. Pembuktian Eksperimental Pengaruh Jumlah Lilitan Pegas dan Diameter Pegas terhadap Konstanta Pegas, *Prosiding Simposium Nasional Inovasi Pembelajaran dan Sains.* 2011.

[6] D. C. Giancoli 2001 *Fisika Edisi Kelima Jilid 1* (Jakarta: Erlangga)

[7] X. Gao and Nianqiang. Smartphone-Based Sensors, *The Electrochemical Society*, pp. 79 - 81. 2016.

[8] P. Vogt and J. Kuhn. Analyzing free fall with a smartphone acceleration sensor, *Department of Physics University of Kaiserslautem*, vol. 50, pp. 182 - 183. 2012.

[9] M. Opera and C. Miron. Mobile Phones in the Modern Teaching of Physics, *Romanian Report in Physics*, vol. 66 No. 4, pp. 1236 - 1252. 2013.

MIX Reality Based Media Prototype For Learning Physics of Gravity And Kepler's Law

L W Munawaroh[1], V Serevina and B H Iswanto
Department of Physics Education, Universitas Negeri Jakarta
Jl. Rawamangun Muka, Jakarta 13220, Indonesia
Email: [1]lindaaawm@gmail.com , vserevina@unj.ac.id , dan bhi@unj.ac.id

Abstract. Gravity and Kepler's Laws are materials that study natural phenomena, events or phenomena that reveal all the secrets of the universe. Based on the observations of several schools, there are difficulties for students understanding gravity and Kepler's laws because they can not directly present 3D learning media in interactive classes. This study aims to develop a *mixed reality*-based media prototype for learning gravity physics and Kepler's laws using special *VR box glasses* and *markers* to display 3D objects equipped with audio explanations, 3D animations, and evaluation questions. The research was conducted using the ADDIE development model *(Analyze, design, develop, implement, and evaluate).* To make it easier for students to use mix reality applications, the mix reality software design employs blender version 2.78, unity 5.1, Photoshop, the vuforia AR extension for unity, and the c # programming language.The trial of the feasibility of mix reality media was conducted by averaging the results of the validator's assessment of each aspect of the assessment. Mixed *reality* media prototypes can help students understand gravity and Kepler's laws.

1. Introduction

Industry 4.0 has become the main topic of the industrial and manufacturing economy with the fourth industrial revolution driven by the advancement of the internet with the main component of *cyber* physical system [1]. *Mix Reality* is one such technological advancement.*Mix Reality* (MR) is a combination of real and *virtual* that requires special hardware such as *glasses* (spectacles) [2]. Use of Reality Mix media are limited as the eclipse phenomenon [2]. One of them is the emergence of *Mix Reality* technology, namely the combination of *Augmented Reality* (AR) and *Virtual Reality* (VR) technology [1]. The results of research conducted by [3] Show that *Mix Reality* (MR) is a digital technology that can provide a more realistic learning sensation. *Augmented Reality* (AR) is a combination of *virtual reality* and *world reality* [4]. The use of *Augmented Reality* (AR) applications can be used easily by students and teachers anywhere with the help of *smartphones* by displaying 3D models on markers [5]. *Virtual Reality* (VR) is a technology that allows users to enter and interact in a virtual *(virtual)* environment. The use of this VR system can perform an activity many times without fear of damaging the object because it is only a visual illusion [6]. VR technology used in physics is limited to the topic of quantum mechanics [7]. *Mix Reality* (MR) is a combination of *Augmented Reality* (AR) and *Virtual Reality* (VR) where AR is closer to the real world while AV is closer to the virtual world, as shown below: [8]

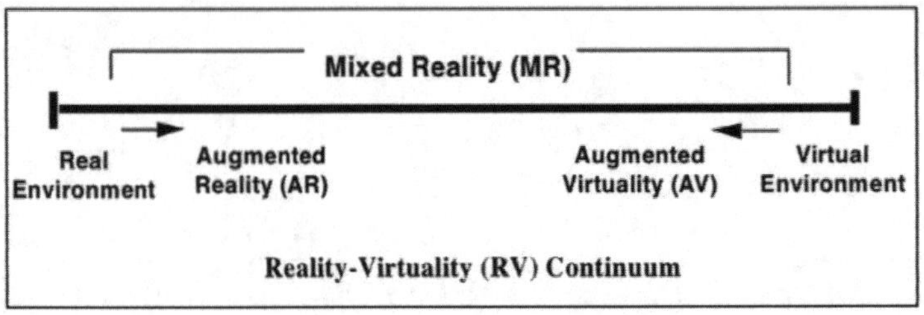

Figure 1. *Reality Virtuality Continuum*

In this study, a single entity that is becoming Mix Reality (MR) was created using marker color and special tools. and *pointers* are already available on the display medium of learning *Mix Reality* (MR) that allows users to learn the material gravity and Kepler's laws.

2. Method

The purpose of this study is to develop a *mix of reality* media on gravity and Kepler's laws that are appropriate as interactive learning media. The Systematic ADDIE model with a clear framework produces effective, creative, and efficient products (Learning, 2012). ADDIE Model Development Flow [9]. Phase analysis can be subdivided into a needs analysis (*needs analysis)*, analysis of the task *(task analysis)*, analyzing the students (*learner analysis*) and analysis of the performance (*performance analysis*) (Cheung, 2016). The development stage is the embodiment of the *blue-print* or design in the previous stage into reality (Aldoobie, 2015). The implementation phase is a real step to implementing the learning system that we are making. That is, at this stage, everything that has been developed is *installed* or set in such a way according to its role or function so that it can be implemented. The evaluation stage is a process to see whether the learning system being built is successful, in accordance with initial expectations or not. The procedure in this development research refers to the ADDIE model development research flow. The product development procedure in this study can be seen in the following figure.

Table 1 Steps to develop Mix Reality media

No	ADDIE Stages	Development Steps
1.	*Analyze*	- Needs analysis questionnaire for students and teachers - Study of literature on learning media
2.	*Design*	- Designing the display of *Mix Reality* learning media - Create components of mixed reality komponen - Develop expert validation tools
3.	*Development*	- Creating media that fits the framework - Programming in 3D in Blender, Unity 3D, Vuforia for AR and VR, and C# programming language - Testing *mixed reality* media before being validated by experts
4.	*Implementation*	- 2nd SMAN field trial - Filling of response questionnaires by students
5.	*Evaluation*	- Conduct a feasibility test for media, material and learning experts - Product revision based on validator's suggestion - Field test (Gain Test) - Media enhancements

Final Product : *Mix Reality* (MR) Based Media Prototype for Learning Physics of Gravity and Kepler's Law

3. Results and Discussion

Media *Mix Reality* shows that MR is a digital technology that can give the sensation of learning to be more realistic in a classroom or learning environment. The developed MR media is a *mixed reality*-based media prototype for learning gravity physics and Kepler's laws. The MR media design is divided into two, namely displays and *markers.* The first part is the display consisting of 3D objects with audio explanations, 3D animations with audio explanations, and using VR Box *Glasses* to see the MR display. The second part is a *marker* that is packaged in *HVS* size with a standard A4 size made *colorfully,* equipped with instructions for use, core competencies and basic competencies. The media design that will be developed can be seen as shown below:

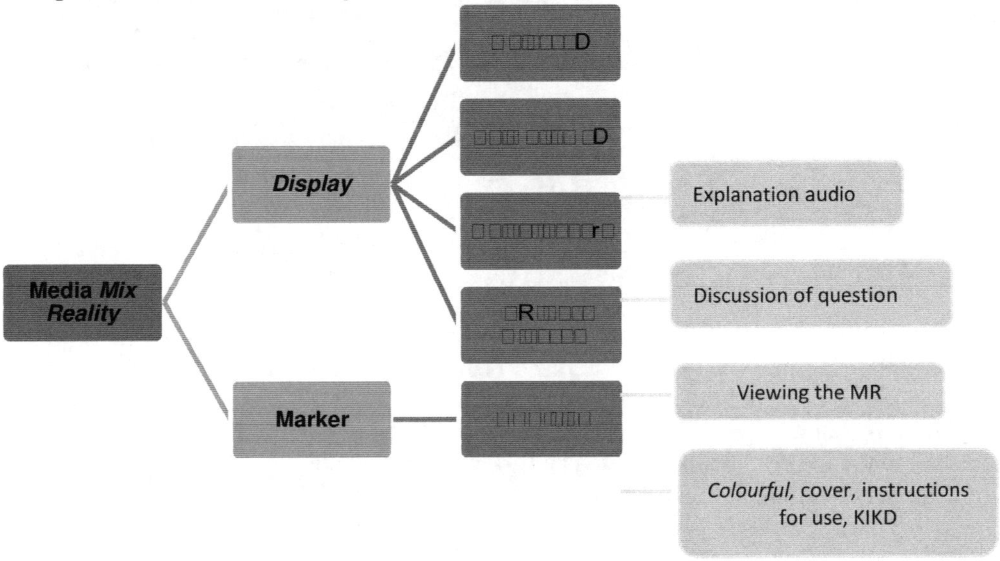

Figure 2. MR media design

Software selected to create applications The software Blender version 2.78, Unity 5.1, Photoshop, Vuforia AR extensions for Unity, and the programming language C # are used in M ix R eality.Making MR media software begins with creating 3D objects in Blender, then exporting 3D objects into.fbx format, then importing 3D objects into Unity 3D, then importing vuforia developer markers, then inputting markers into Unity 3D, then using the C # programming language, then building the application using the android SDK, then connecting to the camera on the smartphone equipped with MR media software.

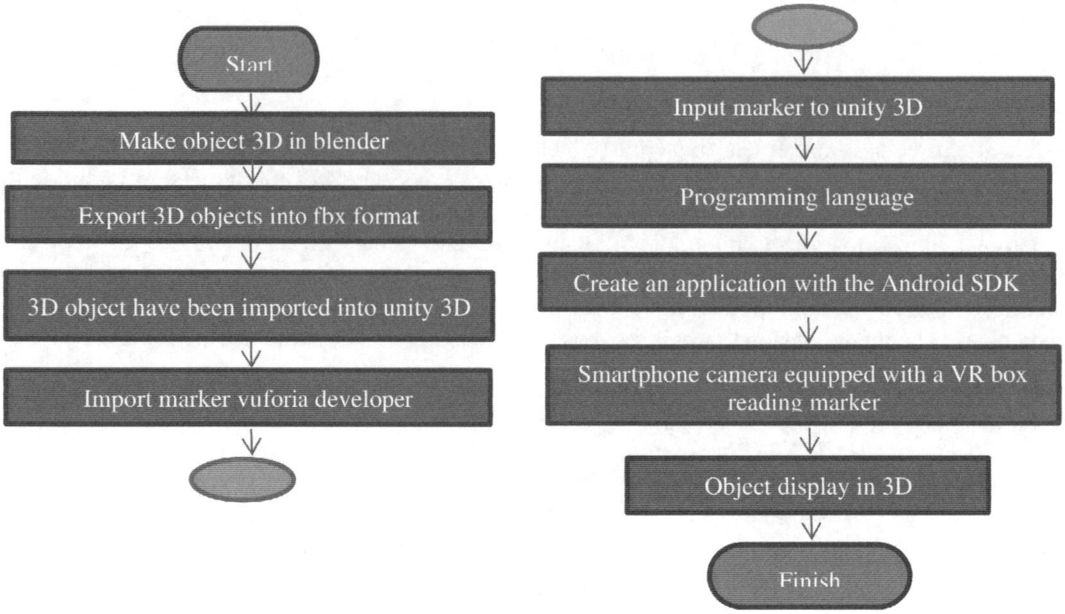

Figure 3. The process of making Mix Reality software

The process of presenting the MR that will be used begins by providing a *cellphone* that already has the MR application, then attaching the *cellphone* to the VR Box *glasses* (glasses), then attaching the VR Box to the eye directly, then focusing the VR Box glasses to the available *marker*, Then a 3D image will appear according to the material being taught. The following presentation of MR media for gravity and Kepler's laws can be seen in the following figure.

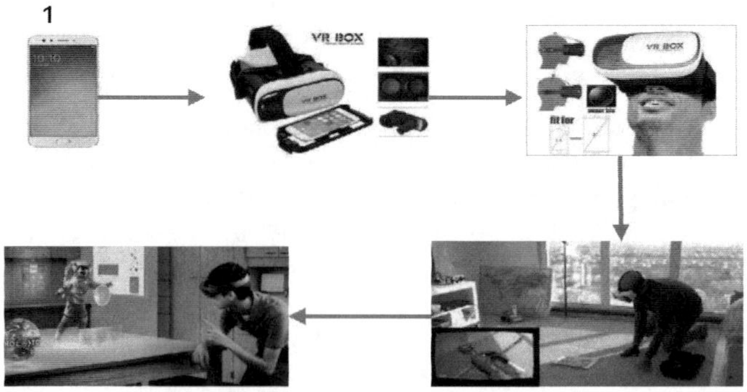

Figure 4. *Mix Reality* media presentation process

The display of the *mixed reality* media prototype is as follows: The first, second and third pages that will appear the first time this application is run are the initial view of the creator of the topic of gravity and Kepler's laws that will be discussed. Then the name of the student and the origin of the school that must be filled in by the student appear. There are instructions for calibrating the *gyro*. Look at the image below!

(a) (b)

Figure 5. Application Display (a).Initial display, (b).Gyro Calibration

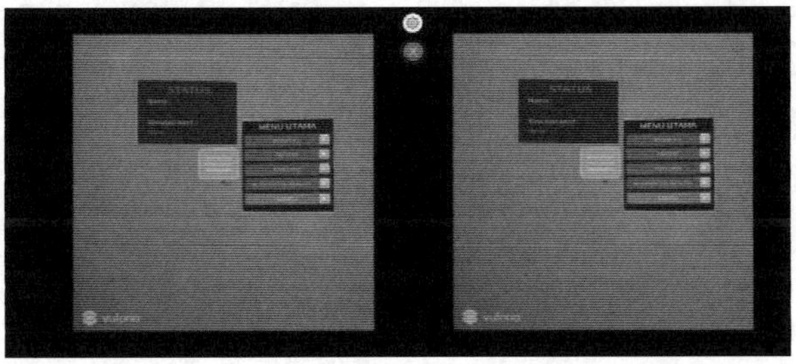

Figure 6. Display of the main menu of Gravity and Kepler's Law

The main menu consists of a status containing the name and the active simulation. Active simulation is a sign when the *augmented reality* sub menu is clicked. In addition, on the main menu there are:

1. Core competencies and basic competencies (KI and KD)
2. Material
3. Evaluation
4. *Augmented Reality* (AR)
5. Lights. The lamp here is used to add light.

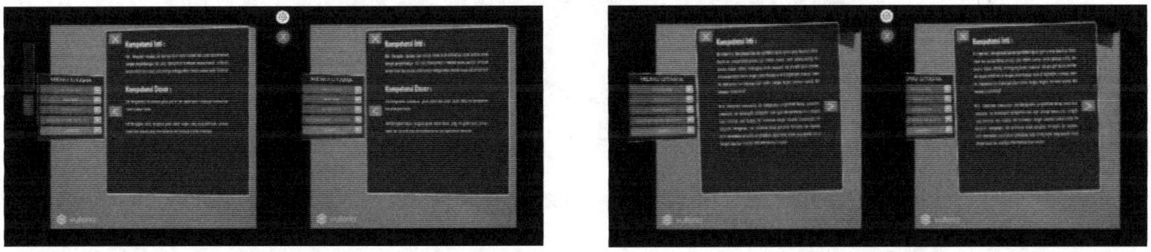

(a) (b)

Figure 7.(a). Core competencies, (b). basic competencies

Submenu content includes: 1) Newton's law of gravity, 2) the general gravitational constant, and 3) the resultant gravitational force.4). Calculating the mass of the earth, 5). Calculating the mass of the sun, 6). Gravitational field and gravitational field strength, 7). Satellite movement, 8). Planets and Kepler's laws

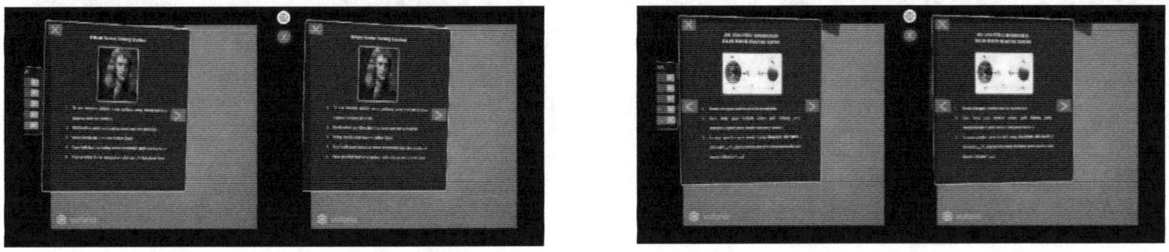

Journal of Physics: Conference Series **2019** (2021) 012054 doi:10.1088/1742-6596/2019/1/012054

(a) (b)

Figure 8.(a).(b) Newton's law of gravity

In the general gravitational constant, it is shown how the Cavendish balance experiment was carried out by *Henry Cavendish.*

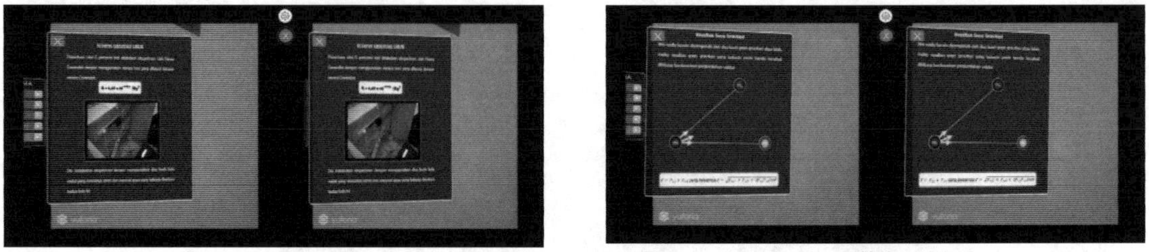

(a) (b)

Figure 9.(a) General gravitational constant (G), (b). Resultant gravitational force

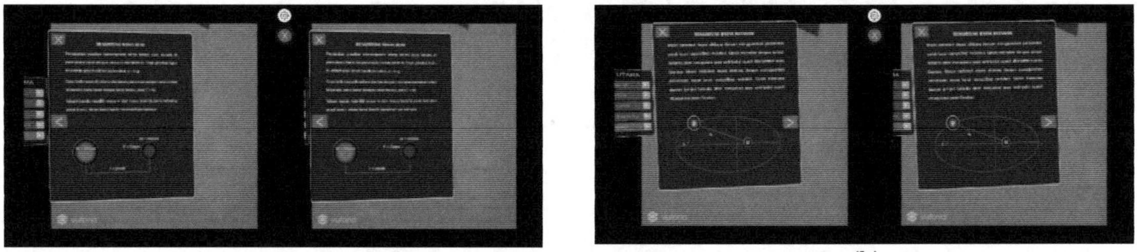

(a) (b)

Figure 10.(a) Calculating the mass of the earth, (b). Calculating the mass of the sun

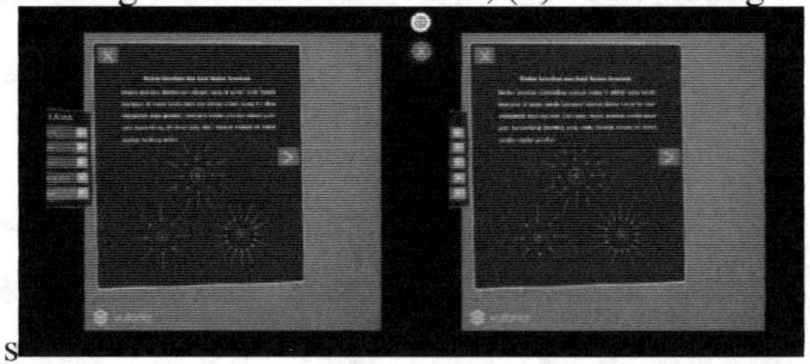

s

Figure 11. Gravitational field and gravitational field strength

In the image below, the motion of the satellite is accompanied by a visual explanation or recorded.

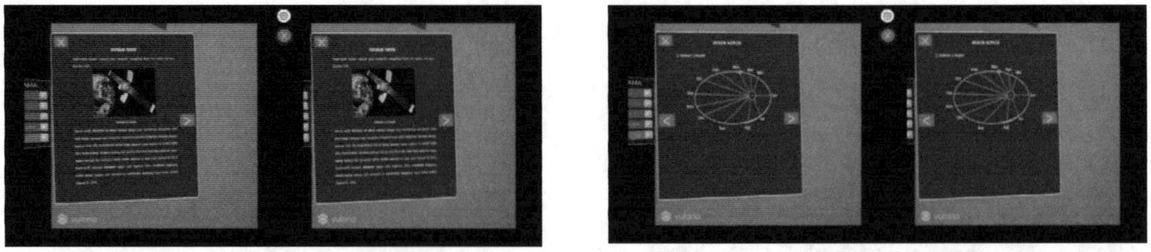

(a) (b)

Figure. 12.(a) Motion of the satellite, (b). Planets and Kepler's laws (complete with sound recordings when explaining material).

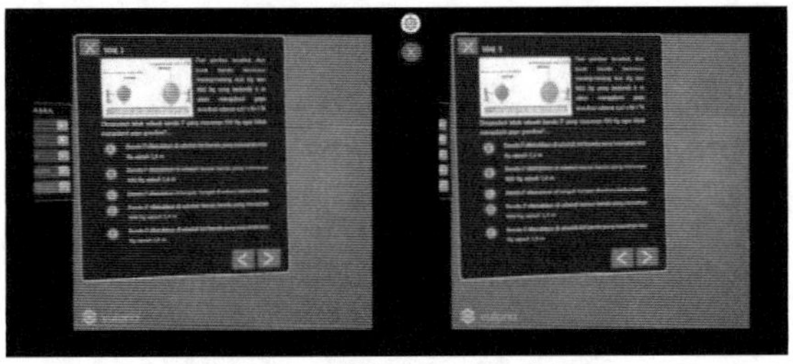

Figure 13. Evaluation

When using augmented reality, there is a submenu that includes: AR gravity of two bodies, AR resultant two gravity, A R mass of the earth, AR earth around the sun, AR satellite movement, AR geocentric model, and AR heliocentric model.

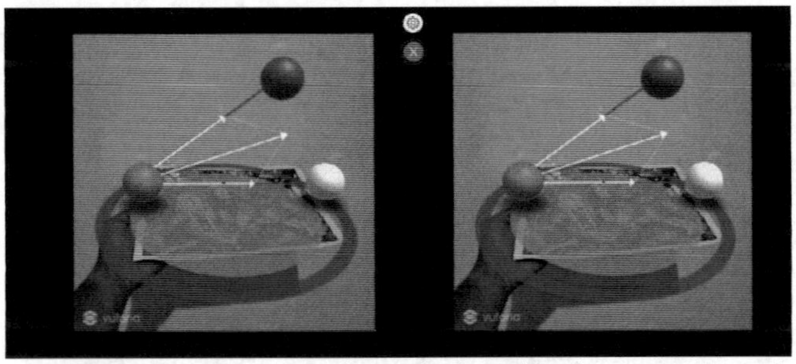

Figure 14. Augmented reality (AR) by pointing the glasses at the printed marker.

4. Conclusion

Based on the results of the discussion on the development of the *Mix Reality* (MR) media prototype for gravity and Kepler's laws, the application of *mixed reality* media can be used by students optimally because it can display 3D material equipped with audio and practice questions, so that it can make it easier for students to understand the material. This media also uses a point center as a convenience for students to apply it.

References

[1] Herausgeber, B. Zinn and R. Tenberg. Learning in a Mixed Reality System in the Context of Industrie 4.0, Journal of Technical Education (JOTED), p. 6. 2015.

[2] V. S. Pratiwi, A. Mayub and D. Hamdani, Pengembangan Media Pembelajaran Planetarium Gerhana sebagai Alat Bantu dalam Pembelajaran Ilmu Pengetahuan Bumi Antariksa (IPBA) pada Materi Gerhana, Jurnal Kumparan Fisika , p. 2. 2018.

[3] J. Birt and M. Cowling. Toward Future 'Mixed Reality' Learning Spaces for STEAM Education, International Journal of Innovation in Science and Mathematics Education, p. 2. 2017.

[4] Mustika, C. G. Rampengan, R. Sanjaya and Sofyan. Implementasi Augmented Reality Sebagai Media Pembelajaran Interaktif, Citec Jurnal, p. 1. 2015.

[5] M. Figueiredo, J. Gomes, C. Gomes and J. Lopes. Augmented Reality Tools for Teaching and Learning, International Journal on Advances in Education Research, p. 22. 2014.

[6] P. M. Herlambang and L. Aryoseto. Potensi Virtual Reality Berbasis Smartphone sebagai Media Belajar Mahasiswa Kedokteran, Jurnal Kedokteran, p. 2. 2017.

[7] C. Savage , D. McGrath, T. McIntryre, M. Wegener and M. Williamson. Teaching Physics Using Virtual Reality, Research Gate, p. 2 2014.

[8] J. Carmigniani and B. Furht 2011 Handbook of Augmneted Reality (Florida : Springer-Verlag New York)

[9] R. Branch 2009 Instructional Design: The ADDIE Approach (London: Springer New York Dordrecht Heidelberg London)

The Role of Digital Modules on Cognitive Ability of High School Students in Newton's Law Material

Wirdiyatusyifa[1], W Sunarno and A Supriyanto
[123]Physics Education of Post Graduate Program, Universitas Sebelas Maret, Surakarta, Indonesia
[2]Physics Education, Universitas Sebelas Maret, Surakarta, Indonesia
[3]Physics Department, Universitas Sebelas Maret, Surakarta, Indonesia
Email: [1]wirdiyatusyifa@student.uns.ac.id

Abstract. Students perceive physics as a difficult subject to understand because the material contains a lot of formulas, mathematical calculations, and abstract concepts. One of the physics materials for high school students in Newton's law. Newton's law is one of the materials considered difficult to understand by students because the basic concepts in Newton's law material, namely the concept of force have not been fully mastered by students, resulting in students' cognitive physics abilities that are still low and have not reached the minimum completeness criteria (KKM). This study aims to examine the role of the digital module in improving students' cognitive abilities in Newton's Law material. The research subjects were 108 students of class X MIA in 3 high schools in the Surakarta residency. The study used a quantitative descriptive method with a test technique using 20 multiple choice questions about Newton's Law. Data were analyzed using N-gain. The results showed that the average scores before using the digital module at 3 SMA in Surakarta residency were 53, 58, and 60 respectively, while the average scores after using the digital module were 73, 74, and 75. Based on the N gain value, the increase in ability students' cognitive belonged to the medium category, namely > 0.3. The digital module has a significant effect on improving students' cognitive abilities in Newton's Law material.

1. Introduction

Physics is one of the important subjects to be mastered by students. However, most students still perceive physics as a difficult subject to understand because it contains many formulas, mathematical calculations, and abstract concepts [1]. This perception causes students to have difficulty in understanding and mastering the concept of physics. One of the physics concepts that are difficult for students to understand and master is the concept of force and its application in Newton's Laws [2]. This is evidenced by the results of mastery of the 2019 High School National Examination material in physics subjects reporting only 55.09% of students who answered correctly on the concept of Newton's Law. This result is still relatively low when compared to students' mastery of physics concepts in other materials. The low understanding of the concept of Newton's Law material affects the cognitive abilities of students in physics. The results of the study [3] that the concept of Newton's Law which is difficult for students to understand causes students' cognitive abilities to be low. This is evidenced by the results of the cognitive ability test of Newton's Law material showing that most students are still low and have not reached the minimum completeness criteria (KKM).

Efforts to improve the understanding of students' cognitive abilities on Newton's Law material have so far been carried out using student-centered learning models [4] [5] and the application of modules as supporting teaching materials in learning Physics [6][7]. This study using a digital module as support in learning physics. The digital module was chosen as supporting teaching material in physics learning because it utilizes technology and is practical in its use. Digital modules provide various features that help students get learning materials quickly and easily. Students can easily access learning through digital modules anytime and anywhere via smartphones, tablets, and computers [8]. Several studies have been carried out, namely [7] the application of digital modules to Newton's Law material in the form of an android application. However, the digital module in this study only contains Newton's Law material which is equipped with interesting videos and pictures, evaluations, and student worksheets. This digital module is not yet equipped with physics animations and virtual laboratories that can help improve the understanding of physics concepts. The novelty of this research is adding animated physics content and virtual laboratory activities to Newton's Law digital module. The addition of this content is done so that students can better understand the concept so that students' cognitive abilities on Newton's Law material increase. The purpose of this study is to examine the role of digital modules to improve cognitive abilities.

2. Method

The study was conducted at 3 high schools in the residency of Surakarta in January 2021. The subjects of the study were 108 students of class X MIA in 3 high schools in the residency of Surakarta. The study used a quantitative descriptive method with a test technique using 20 multiple choice questions on Newton's Law. Improving students' cognitive abilities on Newton's Law material is done by comparing the results of the pre-test and post-test of Newton's Law cognitive test. A cognitive ability pre-test was conducted before applying the digital module in physics learning. A post-test of cognitive abilities was carried out after applying the digital module in learning physics. Data were analyzed using N-gain. The normalized N-gain score <g> was used to obtain the criteria for increasing the value of students' cognitive abilities. The normalized N-gain score formula is shown in the equation below [9]:

$$\langle g \rangle = \frac{(S_{post} - S_{pre})}{(S_{max} - S_{pre})} \quad (1)$$

Table 1. Interpretation of Mean of N-Gain Value.

Value <g>	Classification
$\langle g \rangle \geq 0{,}7$	High
$0{,}7 < \langle g \rangle \leq 0{,}3$	Medium
$\langle g \rangle < 0{,}3$	Low

3. Result and Discussion

The results showed that the average value of the pre-test of cognitive physics abilities in each high school was relatively low when compared to the post-test average (Table 1 & Table 2). The average pre-test before using the digital module at 3 SMA in Surakarta Residency is 53 at the first school, 58 at the second school, and 60 at the third school. The post-test average of students' physical cognitive abilities in 3 SMA in Surakarta Residency namely in the first school of 76, the second school of 74, and the third school of 75.

Table 2. The Results of Student's Pre-Test

School	Mean	Std. Deviation	Minimum	Maximum	Range
School 1	53	14,04	33	81	48
School 2	58	10,89	38	85	47
School 3	60	11,65	34	82	48

Table 3. The Results of Student's Post-Test

School	Mean	Std. Deviation	Minimum	Maximum	Range
School 1	75,6	11,57	55	95	40
School 2	73,6	10,93	50	90	40
School 3	74,9	11,43	55	95	40

The results showed that the average value of the pre-test of cognitive physics abilities in each high school was relatively low when compared to the post-test average (Table 1 & Table 2). The average pre-test before using the digital module at 3 SMA in Surakarta Residency is 53 at the first school, 58 at the second school, and 60 at the third school. The post-test average of students' physical cognitive abilities in 3 SMA in Surakarta Residency namely in the first school of 76, the second school of 74, and the third school of 75.

The increase in students' cognitive abilities was seen from the average N-gain score from the pre-test and post-test results of students' cognitive abilities (Table 4). The results of the average N-Gain in the first school is 0.46, at the second school is 0.46 and at the third school is 0.37. Learning physics using digital modules can improve students' cognitive abilities on Newton's Law material. This can be seen from the increase in the average pre-test and post-test scores of students' cognitive physics abilities, namely the post-test average value is higher than the pre-test average value (Table 2 & Table 3). Improved cognitive abilities of students at 3 high schools in Surakarta residency with an average N-Gain score greater than 0.3 (> 0.3) in the medium category[9]. This proves that the process of learning physics on Newton's Law material using digital modules can improve the cognitive abilities of 3 high school students in Surakarta residency.

Table 4. Data on N-Gain of Students' Cognitive Abilities

School	Mean	Std. Deviation	Minimum	Maximum	Range
School 1	0,46	0,24	0,04	0,92	0,88
School 2	0,46	0,33	0,01	1,26	1,25
School 3	0,37	0,23	0,03	0,37	0,34

The use of Newton's Law digital module in physics learning is proven to improve students' cognitive abilities. This is because Newton's Law material in the digital module is presented contextually, namely connecting Newton's Law material with phenomena that occur in real life. This phenomenon is presented in the form of video shows, images, animations, and simple virtual experiments. Contextual presentation is done to make it easier for students to learn Newton's Law material. This study uses a digital module that begins with watching a short video or observing the images available in the digital module. This video contains natural phenomena, interesting events, or situations that illustrate real-life problems in Newton's Law material such as the phenomenon of a bus speeding and braking, rowing a boat, and the traditional game of marbles. The digital module also presents Newton's Law phenomena in the form of pictures of real-life phenomena such as the event of a person pushing a sand cart. The presentation of Newton's Law phenomena in digital modules contextually based on real-life can facilitate students in mastering Newton's Law concepts so that students' cognitive physics abilities increase (Table 4). This supports the research results of [10] that contextual physics learning can improve students' physics learning outcomes, especially learning outcomes in the cognitive domain.

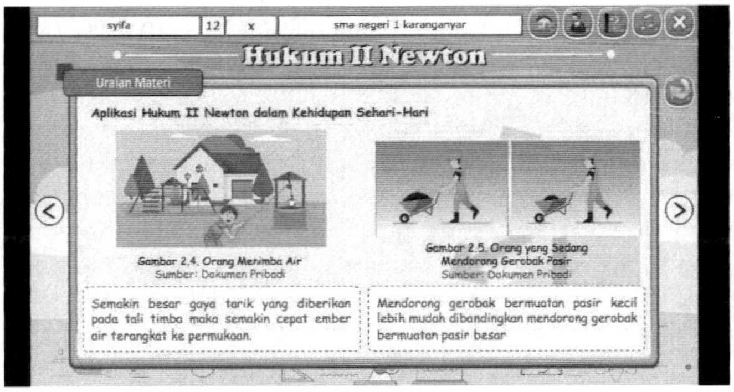

Figure 1. Presentation of Newton's Law Material Contextually

The improvement of students' physical cognitive abilities at 3 SMA Residency Surakarta is also influenced by the presentation of Newton's Law material on digital modules that are supported by visualization of video shows, images, and physics animations. Through the visualization, it can make it easier for students to understand and master Newton's Law concepts. High understanding and mastery of students' physics concepts have a positive effect on increasing students' physical cognitive abilities[11]. The results of research by [12][13] that the use of digital modules as physics teaching materials equipped with multimedia elements such as audio, video, and images can create more interesting learning. Presentation of physical phenomena in the form of visualization of video shows, images, and animations can increase students' interest and curiosity so that students are more motivated in learning physics and easier to understand the concept of learning material.

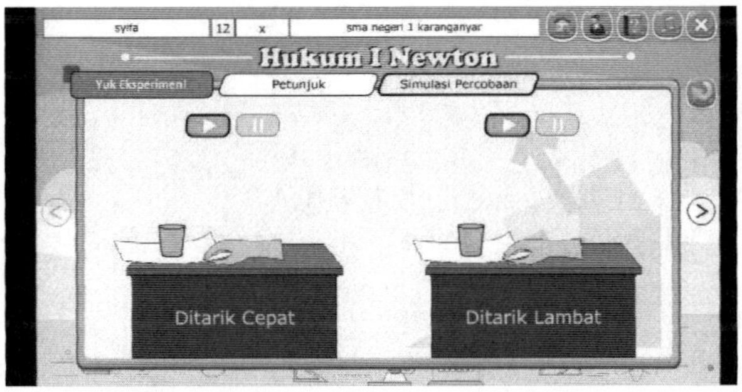

Figure 2. Virtual Laboratory of Newton's Law Digital Module

The digital module in this study is equipped with virtual laboratory activities. Virtual laboratory activities in the form of experimental simulations for Newton's First Law, Second Law, Newton's Third Law, and Frictional Force. For example, in the experimental simulation of Newton's First Law, the phenomenon of pulling paper quickly and slowly under the glass. Students can observe the movement of the paper when it is pulled at different speeds. Through this activity, students have direct experience so that they better understand the concept of inertia is Newton's First Law. This is evidenced by the results of the study that the average post-test of cognitive abilities is higher than the pre-test of cognitive abilities (Table 2 and Table 3). A high understanding of physics concepts has a positive effect on increasing students' cognitive abilities [14]. Through virtual laboratory activities, students have real experiences by actively participating in direct experimental activities. Students can also find and prove a physics concept through the process of taking and analyzing experimental data so that students get a complete physics concept. This discovery activity makes students better understand the concept of physics. This is in line with the research results of [15] & [1] that the implementation of virtual laboratory activities in physics learning can improve students' understanding of physics concepts. Through virtual laboratory activities, students get hands-on experience.

The use of digital modules equipped with media in the form of animations, videos, and virtual laboratories can make it easier for students to master the concepts presented and increase student motivation in learning physics. This is supported by the results of research by [16] that students who study with digital modules have higher learning motivation than students who study traditionally. The increase in student motivation is due to the presentation of material in the module equipped with animation, video shows, and a virtual laboratory. The attractive display of the digital module will increase student interest so that students do not feel bored in learning physics. Students' high interest in learning physics will increase students' learning motivation. High student learning motivation will improve students' cognitive abilities.

4. Conclusions

The average pre-test score at three high school in Surakarta residency is 53 at the first school, 58 at the second school, and 60 at the third school. After using the digital modules the first school is becomes 76, the second school is 74, and the third school is 75. The N-gain scores in three high schools in Surakarta residency are 0.46, 0.46, and 0.37. This proves that is the use of digital modules as supporting teaching materials in learning physics can improve the cognitive abilities of 3 high school students in Surakarta residency.

References

[1] Zakaria N H Phang F A and Pusppanathan J *J. Emerg. Technol Learn. 14* 24 p. 167–183.
[2] Hermanto I., M. Muslim, M. Samsudin A., and Maknun J 2019 *J. Phys. Conf. Ser. 1280* 5.
[3] Meltzer D. E. and Thornton, R. K. 2012 *J. Phys. 80* 6 p. 478–496.
[4] Fratiwi, N. J. et al. 2020 *J. Educ. Res. 9* 2 p. 699–708.
[5] Baran, M. Maskan A., and Yasar S., *J. Instr. 11* 2 p. 221–234.
[6] Zain Z. A., Haris A., and Khaeruddin, 2020 *Icope 2019* p. 76–79.
[7] Ardiyati T. K., Suparno S., Eveline E., and Dasilva B. E. 2020 *Jurnal Pendidik. Indones* 9, 3 p. 368.
[8] Darmaji, et al., 2019 *J. online Biomed. Eng. 15* 15 p. 4–17.
[9] Hake R. R., 1999 *ANALYZING CHANGE/GAIN SCORES* USA : Dept. of Physics, Indiana University.
[10] Juniwati Yusrizal, and Khaldun I, 2020 *J. Phys. Conf. Ser. 1460* 1.
[11] Manurung S. R. and Panggabean D. D. 2020 *Cakrawala Pendidik. 39* 2 p. 460–470.
[12] Istuningsih W. Baedhowi, and Sangka K. B. 2018 *Ijere 03* 03 p. 75–85.
[13] Putri M. D., Rusdiana D., and Rochintaniawati D. 2019 *J. Phys. Conf. Ser. 1157* 2.
[14] Gunawan G., Nisrina N., Suranti N. M. Y., Herayanti L., and Rahmatiah R. 2018 *J. Phys. Conf. Ser. 1108* 1.
[15] Arista F. S. and Kuswanto H. 2018 *J. Instr. 11* 1 p. 1–16.
[16] Perdana F. A., Sarwanto S., Sukarmin S., and Sujadi I. 2017 *J. Sci. Appl. Sci. Conf. Ser. 1* 1 p. 45.

Journal of Physics: Conference Series **2019** (2021) 012056 doi:10.1088/1742-6596/2019/1/012056

Transition of Learning Physics Technology Using Virtual Practicum to High School Physics Teachers in Aceh Barat District

A Farhan[1], A Wahyuni[1], F Herliana[1] and A Akhyar[2]
[1]Department of Physics Education, Syiah Kuala University, Banda Aceh, 23111, Indonesia
[2]Department of Mechanical and Industrial Engineering, Faculty of Engineering, Syiah Kuala University Darussalam, Banda Aceh, 23111, Indonesia
E-mail: farhan@unsyiah.ac.id

Abstract. Practicum is one of the essential learning aspects of physics learning that is required in the 2013 curriculum. The problem of High School Physics teachers in West Aceh Regency is the limited facilities and infrastructure for the Physics laboratory. The development of Information Technology (IT) enables the Virtual Laboratory of Physics is implemented using a smartphone device. Technology transfer using IT in learning is carried out for High School Physics teachers in the West Aceh district. The method used is physics training using virtual experiments. The training participants are all High School Physics teachers in West Aceh District. The aim of the activity is to train physics teachers' ability of IT which can be applied in learning activities at High Schools in the West Aceh district. The results of the activity were that 100% of the participants could master IT technology using a Smart Phone for a virtual practicum on Wave material. Therefore, they feel that there is an increase in their pedagogical ability in carrying out their duties and the optimists can apply this in the learning process in their place of duty. 75% of trainee teachers believe that the application of virtual practicum methods in learning can improve students' understanding of physics subject matter; 12.5% of participants were not sure and 6.25% of participants were not sure that learning Physics using the virtual experiment method could improve students' understanding. The benefits of the activity are expected to increase the human resources of High School Physics teachers in West Aceh district in utilizing the Virtual Laboratory so that it supports the physics learning process using practicum methods and online learning.

1. Introduction

Physics is a natural science that studies the interactions between matter and energy. The concept of science is obtained through experiments and empirical conceptual approaches. Understanding the material will be easier if the material is obtained through practicum [1] because the learning process through practicum is proven to improve science process skills [2] which directly affects the improvement of student physics learning outcomes [3]. Through the practicum, students can learn actively by directly practicing the material they are learning so that students can more easily understand the Physics learning material being taught [4].

Learning Physics by applying the scientific approach where the experimental method is one of the learning processes that must be carried out in every school refers to the 2013 curriculum [5]. However, the physics laboratory facilities and infrastructure actually needed for experimental activities are very limited in West Aceh District. There are even high schools that do not have laboratory facilities yet, due to the damage that occurred during the 2004 Tsunami and which have not been repaired [6]. As a result, physics learning is generally carried out using the conventional method (one-way learning). This situation is one of the factors causing the low quality of education in Aceh because several studies have found that the implementation of the conventional method in learning has lower results compared to other learning methods [7,8,9,10].

The virtual laboratory model is a solution that can be applied to overcome the limitations and unpreparedness of physics laboratory facilities in Aceh [11,12]. By utilizing this model, experimental learning methods can be carried out by utilizing technology that is rapidly developing today [13]. In the current industrial era 4.0, many Physics virtual laboratory applications are available on the Google Play Store to support practicum in learning. Therefore, current physics teachers need to have teaching skills using a virtual laboratory model, so that physics learning using experimental methods can be carried out, even though real laboratory facilities and infrastructure are limited.

Based on preliminary information obtained from the head of the Physics Teachers Association (MGMP of Physics) in West Aceh district, the description of the situation for learning Physics in West Aceh Regency includes:

a. Physics teachers in high school are still lacking skills in using physics laboratory equipment.
b. Conventional learning methods are still widely applied in the physics teaching and learning process
c. Lack of equipment in the laboratory.
d. The physical condition of the laboratory space is inadequate.

Therefore, the transfer of learning technology using virtual laboratory learning methods is very appropriate to be carried out immediately in the district.

Transfer of learning technology using the virtual laboratory method for high school physics teachers in West Aceh district is carried out with the aim of training mastery of IT which can be applied in learning activities at high schools in West Aceh district. The benefits of the activity are expected to increase the human resources of Physics High School teachers in West Aceh district in the use of virtual laboratories, thus supporting the Physics learning process using practicum and learning methods. In the long term, it is expected that it can improve the quality of school graduates who can compete with other regions.

2. Method

This activity used a training method in which physics teachers at high schools in the West Aceh district are trained directly with the supporting devices (learning tools) that have been prepared (Figure 1). The number of participants was 16 people, consisting of physics teachers from public and private high schools in the West Aceh district (Table 2). The location of the activity was carried out at the MGMP office of West Aceh district, Meulaboh in 2019.

The subject chosen in this activity is the Physics Learning of Wave material using virtual laboratory experimental methods. This material is one of the subjects that are abstract in nature and there are no practical equipment facilities in Senior High School in West Aceh district, so it needs to be visualized in the form of a virtual laboratory [13]. After mastering this subject, it is expected that teachers can train independently on other subjects. Activities are carried out using the inquiry learning model. Teachers are directed to be able to use the application in learning. The implementation of activities is equipped with modules, worksheets, and evaluation instruments. The activity stages are carried out as follows:

1. Introducing the application that available on Smart Phone devices and downloading it on the google play store
2. After the teachers can run the application, the activity continues to use it in the physics learning process by using an experimental method using a virtual laboratory.
3. Carry out the experiment according to the instructions on the worksheets.
4. Analyze data and make conclusions.
5. Evaluation of activities

As illustrated in the flow chart below:

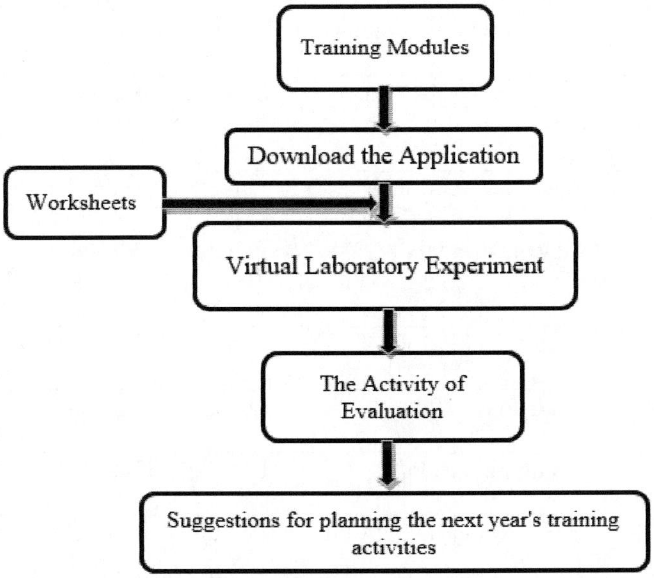

Figure 1. The Stages of Training Activity

The evaluation of training is carried out to determine the ability to absorb training material from the participants. The data were obtained by a questionnaire. The calculation of the participant's answer score is done using equation [14]:

$$P = \frac{\text{the total score of each question}}{\text{maximum score}} \times 100\ \%$$

The results of the questionnaire analysis were interpreted qualitatively using the criteria in Table 1.

Table 1. Criteria of Training Activity Result

No.	Percentage (%)	Criteria
1	86-100	Very good
2	76-85	Good
3	60-75	Enough
4	55-59	Less
5	≤54	Very Less

3. Result and Discussion

The transfer of technology using Smartphone devices in physics learning with the virtual laboratory experimental method was attended by 16 Physics teachers in high school, West Aceh district (Table 2). The results of the activity will be presented in the following description:

Table 2. Senior High School Physics teachers in West Aceh district and their professional qualifications who participated in the training activities.

No	School Name	Teacher's Name	Gender	Degree	Graduated
1.	SMA Negeri 1 Meulaboh	Suhartatik	Female	Bachelor	UNY
2.	SMA Negeri 2 Meulaboh	Srimisbah, S. Pd	Female	Bachelor	FKIP Fisika Unsyiah
3.	SMA Negeri 3 Meulaboh	Dra. Zuraidah, M. Pd	Female	Master	FKIP Fisika Unsyiah/ PPs Unsyiah
4.	SMA Negeri 3 Meulaboh	Lismawati, S. Pd	Female	Bachelor	FKIP Fisika Unsyiah
5.	SMA Negeri 3 Meulaboh	Safrida, S. Pdi	Female	Bachelor	Tarbiyah PMIPA UIN
6.	SMA Negeri 1 Meureubo	Yulianty, S. Pd	Female	Bachelor	FKIP Fisika Serambi Mekkah
7.	SMA Negeri 1 Meureubo	Eka Ratna Jaya, S. Pd	Female	Bachelor	FKIP Fisika Unsyiah
8.	SMA Negeri 1 Meureubo	Rosmanidar, S. Pd	Female	Bachelor	FKIP Fisika Unsyiah
9.	SMA Muhammaddiah 6 Meulaboh	Nurmalahayati, S. Pd	Female	Bachelor	FKIP Fisika Unimed
10.	SMA Negeri 1 Arongan Lambalek	Roudiatur Rahmi, S. Pd	Female	Bachelor	FKIP Fisika Unsyiah
11.	SMA Negeri 1 Samatiga	Dedek Juliana, S. Pd	Female	Bachelor	FKIP Fisika Unsyiah
12.	SMK 1 Setia Bakti	Maisarah, S. Pd	Female	Bachelor	FKIP Fisika Unsyiah
13.	SMK Negeri 2 Meulaboh, Lapang	Cut Fonny Nazira, S. Pd	Female	Bachelor	FKIP Fisika Unsyiah
14.	SMK Negeri 2 Meulaboh, Lapang	Erlita	Female	Bachelor	FKIP Fisika Unsyiah
15.	SMK Negeri 4 Meulaboh; Suak Sigadeng	Nasriyetie N	Female	Bachelor	MIPA Fisika
16.	SMA Negeri 2 Kaway XVI; Pasie Kumbang	Agustia Warni, S. Pd. I	Female	Bachelor	Tarbiyah PMIPA UIN

3.1 Responses of the Training Participant to the Instructors' abilities and services.

The participants' responses to the services provided by the instructor in the implementation of the training are summarized in Table 3. All the participants (100%) felt the services and guidance provided by the instructors were very good. The ability of instructors to provide good explanations is due to their relatively long experience in learning. They are senior lecturers in the Department of Physics Education,

FKIP Unsyiah. Preparation in the form of modules and worksheets are also arranged in a simple, so as to help understanding of the material for participants. Participants' questions during the discussion were also responded to well. This situation caused the training participants to feel satisfied with the instructor's guidance and services in carrying out activities.

Table 3. Responses of the Training Participant to the Instructors' abilities and services.

No	Statements	Yes	Uncertain	No	Criteria
1	The guidance provided by the instructor is easy to understand	16	0	0	Excellent
2	Tutors provide sufficient time for discussion and question and answer	16	0	0	Excellent
3	The material provided by the tutor during the training is easy to understand	16	0	0	Excellent
4	The way the tutor listens, responds, and explains answers so that it can facilitate and understand the training content	16	0	0	Excellent

3.2 The Effect of Community Service on the Improvement of Teachers' Pedagogical Ability and understanding of the subject

The implementation of the training provided is expected to provide additional knowledge and skills of physics subject teachers at Senior High School in West Aceh district in mastering IT technology, especially the application that available on smartphone devices. The results of the evaluation using a questionnaire are summarized in Table 4.

Table 4. The absorption of learning skills using the virtual experimental method by Physics teachers at Senior High School in West Aceh District

No	Statement	Yes	Uncertain	No	Criteria
1	In your opinion, were the preparations for this training activity well prepared by the implementer?	16	0	0	Excellent
2	Can you participate in this training smoothly?	15	1	0	Very Smoothly
3	Was this training activity useful and added to your skills?	16	0	0	Very Useful
4	Can you apply physics lessons using virtual experimental methods obtained from this training in the implementation of tasks and improve your pedagogical abilities?	16	0	0	Very Confident
5	Are you sure that learning Physics material that you will implement with the virtual laboratory experimental method can facilitate understanding of Physics material?	12	2	1	Very Confident

Based on the questionnaire answers from the participants, it was found that all participants (100%) stated that the training activities carried out were well prepared. Therefore, almost all participants (93.75%) were able to attend the training smoothly. The remaining 1 participant or 6.25% of the participants experienced problems. Based on observations during the implementation of activities, the teacher is not familiar with using a smartphone device with the Android operating system. Communication activities carried out so far use old mobile devices with the Symbian operating system. The instructor pays special attention to the teacher. In addition, other participating teachers who nearby also gave directions. At the end of the implementation of activities, the participant teachers who experience problems with the initial activities can follow the instructions given by the instructor.

The materials obtained during the training add to their skills stated by 100% of the participants. Teachers gain additional knowledge and skills, especially related to the use of smartphones in learning physics. They have not previously obtained this information, so they feel a significant additional knowledge. They believe that the skills acquired in this training can be applied in the classroom where they work, so as to improve their pedagogical skills. The use of Smartphone devices as learning media is not an obstacle. In general, students already have and are used to using it [15]. The positive effect that may be obtained is to increase the usefulness and function of the tool in catch-up activities.

The perceptions of the participants on the effect of teaching physics using virtual laboratory experimental methods in making it easier for students to understand physics subject. 75% of participating teachers are optimistic that using the virtual experimental method has an effect on facilitating students' understanding of the material. The reason given by the optimistic participating teachers was that virtual experiments can clarify/detail the stages of natural phenomena being taught so as to improve students' critical thinking skills [16]. The appearance of virtual laboratory animations is quite interesting so that can increase students' interest in learning and creative thinking skills [17]. In addition, this application in virtual experimental learning is something new for senior high school students in the West Aceh district. Thus the learning process becomes more interactive and interesting [18].

12.5% of participants stated that only part of the students could improve their understanding. One training participant teacher (6.25%) had another opinion; the use of virtual laboratories in physics learning is not able to improve students' understanding of the subject matter. The teachers argued that students' understanding of physics subject matter apart from being influenced by external factors (in this case learning methods and media) was also influenced by internal factors, namely students' intellectual abilities, and the way students processed or digest physics material in their minds [19].

3.3 Perception of the Training Participants' Physics Teachers on Other Physics Subjects.

All teachers participating in the training (100%) stated that learning with the practicum method using a virtual laboratory could be developed for other physics subject matter (Table 5). They argue that abstract matter exists in almost all subjects of physics. Based on the information obtained in the discussion activities, the training participants argued that the animations shown in the virtual video help to detail the processes of natural phenomena that are difficult to explain using the lecture method. The visuals displayed to motivate students to keep learning so that it helps students develop their imagination to understand concepts [20]. Thus, students will absorb more material.

Table 5. Perception of the Training Participants' Physics Teachers on Other Physics Subjects.

No	Statements	Yes	Uncertain	No	Criteria
Which physics materials can be developed using virtual laboratory methods?					
1	Basic concepts of mechanics, both related to kinematics and dynamics	16	0	0	Really can
2	Magnetic electricity concepts (related to static, dynamic, electric circuits, etc)	16	0	0	Really can
3	Advanced physics concepts (relativity, photoelectric effect, atomic spectrum, atomic model, etc)	16	0	0	Really can
4	Physics materials that require practicum activities in the laboratory	16	0	0	Really can

Based on the data above, all participants believed that many other physics materials could be developed using this virtual experimental method. After this training activity, it is hoped that the participants can develop learning using this virtual laboratory independently and be implemented in their respective schools.

4. Conclusion

The transfer of physics learning technology using the experimental virtual laboratory method for physics teachers at senior high school in West Aceh district can be controlled by 100% of the participants. This activity gave the effect of increasing pedagogic abilities stated by 100% of participants. 100% of participants believe they are able to apply the skills gained from training in learning in the schools where they work. 75% of the training participants believed that the application of the virtual practicum method in learning Physics was able to improve students' understanding of physics subject matter. 12.5% of the participants were not completely convinced and 6.25% stated that they were not sure that the application of teaching physics using the virtual laboratory method could improve students' understanding of the material. They assume that the understanding of learning material is not only influenced by method factors but also by other external and external factors. 100% of participants are optimistic that the learning method obtained from the training can be developed for all other physics subject matter. All the participants are satisfied with the services provided by the service team and expect to carry out similar activities for other physics subjects.

Acknowledgement

Thank you to the head of the Unsyiah Research and Community Service Institute (LPPM) and his staff, Dean of FKIP Unsyiah and their staff, the head of Physics Education Department and his staff, and the head of the West Aceh District Physics Association (MGMP) and Senior High School Physics teachers in West Aceh for the support of the cooperation given.

References

[1] Anies E 2017 *Jurnal Pembelajaran Fisika* **6(1)** 75-82.
[2] Wiwin E and Kustijono R 2018 *Journal of Physics: Conference Series* **997(1)** 1-8
[3] Nasir M, Fakhrunnisa R and Nastiti L R 2019 *International Journal of Environmental and Science Education* **14(5)** 229-238.
[4] Krisnawan I K, Candiasa I M and Sunu G K A 2013 *Jurnal Administrasi Pendidikan Indonesia* **4(1)**
[5] Sani R A 2014 Pembelajaran saintifik untuk implementasi kurikulum 2013.
[6] Yuliana S, Marwan A R and Wahyuni A 2017 *Jurnal Ilmiah Mahasiswa Pendidikan Fisika* **2(3)** 303-306.
[7] Halim A and Hamid A 2020 *Asian Journal of Science Education* **2(1)** 33-40.
[8] Merisa N S, Halim A and Mahzum E 2020 *Asian Journal of Science Education* **2(1)** 24-32.
[9] Tarmizi T and Hamid A 2019 *Asian Journal of Science Education* **1(1)** 16-21.
[10] Herliana F, Supriyati Y and Astra I M 2015 *Prosiding Seminar Nasional Fisika (E-Journal)* **4**
[11] Kusumaningsih R Y R, Iswahyudi C and Susanti E 2014 *Prosiding Seminar Nasional Aplikasi Sains & Teknologi (SNAST)* **15**
[12] Sutrisno 2011 Pengantar pembelajaran inovatif (Jakarta: Gaung Persada Press)
[13] Gunawan G, Harjono A and Sahidu H 2017 *Jurnal Pendidikan Fisika dan Teknologi* **1(2)** 140-145.
[14] Ngalim P 2011 *Evaluasi hasil belajar* (Yogyakarta: Pustaka Pelajar)
[15] Yu F and Conway AR 2012 *Proceedings of the 2012 Southwest Decision Sciences Institute* 831-839.
[16] Koretsky M, Kelly C and Gummer E 2011 *Journal of Engineering Education* **100(3)** 540-573.
[17] Pujiati A 2015 *Formatif: Jurnal Ilmiah Pendidikan MIPA*, **2(3)**
[18] Konak A, Clark T K and Nasereddin M 2014 *Computers & Education* **72** 11-22.
[19] Juliartini N M, Hatibe A and Darsikin D 2020 *Musamus Journal of Science Education* **2(2)** 81-90.
[20] Gunawan G, Harjono A and Sutrio S 2017 *Jurnal Pendidikan Fisika dan Teknologi* **1(1)** 9-14.

TPACK Implementation in Physics Textbook: Practice Problem-Solving Skill in Newton's Law of Motion for Senior High School Students

F Bakri, KF Kusuma and AH Permana
Department of Physics Education, Universitas Negeri Jakarta
Jl. Rawamangun Muka, Jakarta 13220, Indonesia
Email: fauzi-bakri@unj.ac.id

Abstract. One of the skills needed in the 21st century is a problem-solving skill. This article describes the result of research on the implementation of TPACK in the physics textbook. The implementation of TPACK in the physics textbook is aimed to practice the problem-solving skill of senior high school students. The research method used is Research and Development with Dick & Carey model. Has been produced a physics textbook that integrates technological knowledge, pedagogical knowledge, and physics content knowledge in multiple representations with the stage to practice a problem-solving skill in Newton's Law of Motion concept. The stages of problem-solving in the developed physics textbook are: (a) presents the problem at the beginning of the chapter; (b) asking several questions to focus the problem; (c) formulating hypotheses on the problem; (d) present the material in multiple representations based on pedagogical and technological studies, and (e) explain phenomena based on the physics concepts of being studied. The results of this study have been validated by experts, are physics material experts, learning experts, and textbook media experts. The validation result of the physics textbook was declared feasible to practice the problem-solving skill for senior high school students.

1. Introduction

The success of education and the learning process is very important in improving students' cognitive, psychomotor, and affective abilities [1]. Therefore, the learning process must be designed actively, innovatively, and relevant to the needs of the 21st century [2].

Physics is very widely applied in human life. Therefore, the process of learning physics must be related to the environment of everyday life [3]. One of the goals of learning physics in high school is to develop analytical thinking skills, both inductively and deductively, using the concepts and principles of physics [4]. Furthermore, in the 4.0 revolution in the 21st century, students are required to have communication skills, collaboration, critical thinking, problem-solving, and being creative and innovative [5].

The demand for skills needed to face the 21st century, namely: 1) learning and innovation skills, which include critical thinking and being able to solve problems, be creative and innovative, and able to communicate and collaborate; 2) skills to use media, technology, information and communication (ICT); 3) the ability to live life and career, including the ability to adapt, flexible, initiative, develop themselves, have social and cultural abilities, productive, trustworthy, have leadership spirit, and responsibility [6].

In 21st century problem-solving skills, students must use their experience, understanding, skills, and expertise in solving existing challenges [7, 8]. Students have high curiosity, creative ideas and use a critical mindset in making rational decisions to solve the problems they face [9, 10]. Innovation in the learning process is needed to improve students' soft skills and hard skills [11].

Textbooks have an important role in achieving mastery of students' concepts and independent learning patterns in solving problems [12, 13]. However, many physics textbooks cause misconceptions, exposure to material that lacks scientific representation and is not contextual [14]. There are many misconceptions in physics textbooks [15, 16] regarding the explanation of concepts, writing formulas, symbols, and units, presenting images, tables, and graphs [17] makes it very difficult for students to develop thinking skills to build their physics concepts [18]. The study of Newton's law has quite a large misconception [19, 20].

The available physics textbooks only describe physics material and have not practiced scientific literacy. Therefore, physics textbooks should be displayed in multiple representations of science to minimize students' learning difficulties [21, 22, 23].

Textbooks must support the development of 21st-century competencies. Therefore, it is necessary to integrate technological and pedagogical knowledge in formulating physics concepts in textbooks needed in 21st-century learning. This integration is built within the TPACK (Technological, Pedagogical, and Content Knowledge) framework [24, 25, 26, 27]. The integration of TPACK in textbooks can optimize learning activities, improve conceptual understanding, and build students' scientific attitudes [28, 29].

The TPACK framework can be applied to textbooks [30]. The TPACK framework will integrate technology and pedagogy into content and content presented pedagogically using technology [31, 32]. In textbooks that implement TPACK, the presentation of the material will be supported by technology, and the application of technology supports pedagogical stages, the concept of the material, the pedagogic stage will make students easier to understand the material supported by technology [33, 28].

There are five criteria for assessing the TPACK framework, namely 1) the material is presented with the support of Augmented Reality technology that can visualize abstract concepts; 2) representation to change difficult and abstract material to be real and easy for students to understand; 3) the steps of presenting the material that is arranged pedagogically will make it easier for students to build the concepts being studied; 4) the selection of appropriate technology and the use of effective pedagogics; and 5) learner-centered learning strategies combined with appropriate technology [34].

Based on the description above, this article will describe the research results on the development of physics textbooks that implement TPACK to practice problem-solving skills for 10th grade high school students on Newton's Laws of Motion.

2. Method

2.1. Research Design

This article is the result of research and development of physics textbooks that implement TPACK using the Dick & Carey approach. This textbook to be able to practice the problem-solving skills of class 10th high school students on Newton's Laws of Motion, an R&D process has been carried out according to Dick & Carey, namely:

2.1.1. Identify learning objectives

The identification of physics learning objectives in the cognitive domain is carried out by Permendikbud No. 21 of 2016 concerning Content Standards for Primary and Secondary Education.

2.1.2. Conduct learning analysis

A learning analysis was carried out referring to the 2013 revised 2018 curriculum on Newton's Law of Motion by identifying learning competencies 3.7, then compiling indicators of competency achievement hierarchically and systematically.

2.1.3. Analyze students and learning context

An analysis of the basic skills that students must master is carried out before studying Newton's Law of Motion and the skills acquired by students at the end of the lesson. The results of the analysis are presented in a concept map and prerequisite materials.

2.1.4. Formulate specific learning objectives

The formulation of learning objectives is based on basic competencies and indicators of competency achievement.

2.1.5. Develop test items using textbook tools

The development of learning measuring tools is made to practice problem-solving skills that refer to learning indicators.

2.1.6. Develop learning strategies

The strategy applied in this textbook uses the TPACK framework and problem-solving learning model.

2.1.7. Developing learning materials

The development of physics textbooks was carried out by determining the scope of the material developed in the textbook, designing textbook using the TPACK framework, compiling and designing textbook, and making learning videos packaged in AR markers.

2.1.8. Develop and execute formative assessment

The product feasibility instrument was developed in the form of a questionnaire to test the physics concept feasibility, a questionnaire to test the pedagogic feasibility, and a questionnaire to test the media feasibility.

2.1.9. Review textbook

A revision was made to the physics textbook based on the criticism and suggestions given after conducting the validation test.

2.2. Instrumentation Research

To test the feasibility of the results of development research on the physics textbooks, an instrument in the form of a questionnaire is used. Thus, there are three questionnaires developed, namely a questionnaire to test the physics concept feasibility, a questionnaire to test the pedagogic feasibility, and a questionnaire to test the media feasibility.

The eligibility of physics concept in text book is assessed in 3 aspects, namely: (a) *Presentation of physics concept in textbook*; (b) *The representation of science in physics textbook*; (c) *The physics concept in augmented reality media.*

The eligibility of pedagogic in phyisics textbook is assessed in four aspects, namely: (a) *Presentation of the physics concept in textbook*; (b) *The use of science illustrations in physics rextbook*; (c) *The stages of problem-solving in physics textbook*; (d) *Implementation of TPACK in physics textbook.*

The eligibility of physics textbook media is assessed in 5 aspects, namely: (a) *The quality of front part of the physics textbook*; (b) *The quality of the content section of the physics textbook*; (c) *The quality of graphics aspects of the physics textbook*; (d) *The quality of back part of the physics textbook;* (e) *The quality of augmented reality media.*

2.3. Data Analysis

The expert assesses the feasibility of the physics concept, pedagogic, and media instruments using four continuum scale for each aspect assessed. The interpretation of the assessment refers to the following table.

Table 1. Interpretation of Four Continuum Scale

No	Response	Score
1	Very good	4
2	Good	3
3	Bad	2
4	Very bad	1

The result of interpretation score calculated based on the acquisition score of each item by the following formula

$$Score\ Percentage\ (\%) = \frac{\sum acquisition\ score}{\sum maximum\ score}\ x\ 100\%$$

The range of scales on the continuum scale used in this research is 1 to 4, with a minimum score of 1 and a maximum score of 4, so the total score range is 4. This can be calculated using the following formula

$$Scale\ Range = \frac{maximum\ score - minumum\ score}{maximum\ score} = \frac{4-1}{4} = 0.75$$

The result of obtaining the score percentage is then measured using the interpretation table as follows. The scale range obtained is 0.75. The lowest interpretation is 1 to 1.75, with each range plus 0.75. Then, change the scale range in percent form. For more details, see the table below.

$$Scale\ Range\ (\%) = \frac{score}{maximum\ score}\ x\ 100\%$$

Table 2. Interpretation Criteria Eligibility of Physics Textbook

Scale Range	**Category of Eligibility**
$25\% \leq SI < 43.75\%$	Not Feasible
$43.75\% \leq SI < 62.5\%$	Quite Decent
$62.5\% \leq SI < 81.25\%$	Feasible
$81.25\% \leq SI \leq 100\%$	Very Feasible

3. Result and Discussion

3.1 Product research

Physics textbooks that implement TPACK have been produced from this development research. The advantages of this book is on TPACK framework components are described below.

3.1.1. Technological knowledge, pedagogic knowledge, and content knowledge

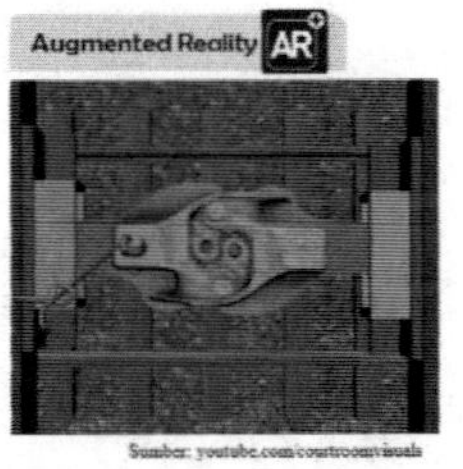

Figure 1. Technology Knowledge

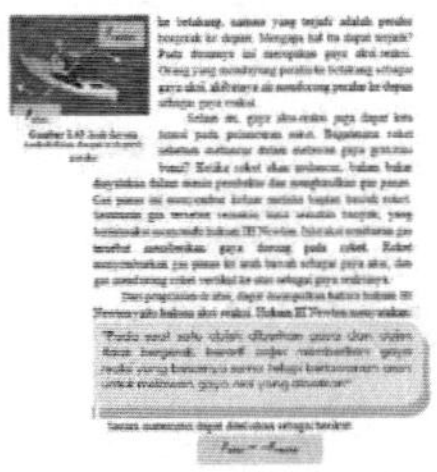

Figure 2. Content Knowledge

The technology applied in the physics textbook is augmented reality media. Newton's law of motion phenomena explained in accordance with the concepts of physics (**Figure 2**).

3.1.2. Technological pedagogical knowledge

Figure 3. Technology for pedagogic knowledge

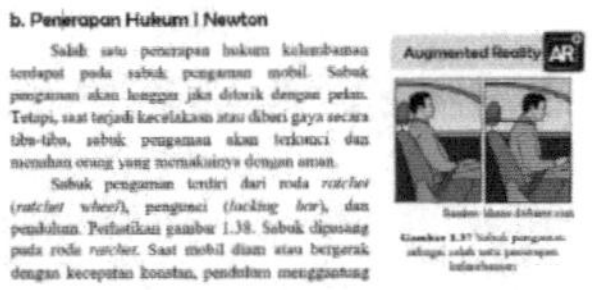

Figure 4. Pedagogic for technology knowledge

To make students easier to visualize abstract material, learning is assisted by technology-based 3-dimensional media. Augmented reality media does not just display it, but learning allows students to analyze the concept being studied. Augmented reality media in the apperception section (**Figure 3**). Application of Newton's law of motion packaged with augmented reality media (**Figure 4**).

3.1.3. Technological content knowledge

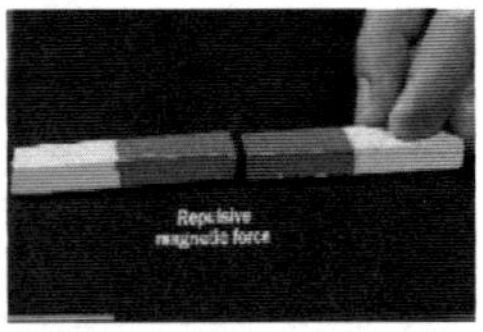

Figure 5. Technology for content knowledge

Figure 6. Application technology from content knowledge

Augmented reality media have a role in conveying the abstract concept of Newton's laws of motion to make it easier for students to understand. Augmented reality media for explain magnetic field as non-touch force (**Figure 5**). Technology that applies Newton's first law concept (**Figure 6**).

3.1.4. Pedagogical Content knowledge

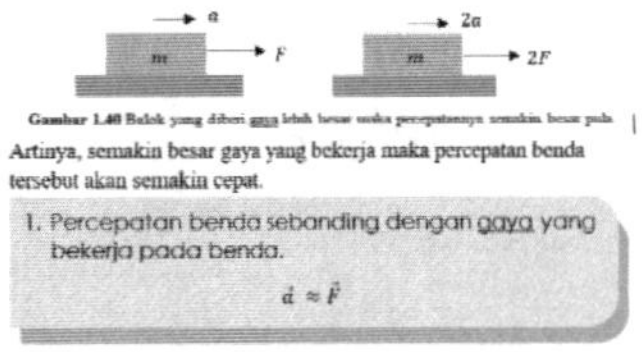

Figure 7. Illustrations and mathematical relationship

Figure 8. Dimensional analysis relates related variables

Newton's law of motion concept is built using the stages of problem solving learning, multiple representations and building science literation with the illustrations provided.

3.1.5. Technological, pedagogical, and content knowledge

The material is presented pedagogically by involving technology so that physics concepts are easy to understand. The physics textbook applied problem-solving stages.

3.1.5.1. Aperception, problem orientation, and hypotheses

.

Figure 9. Students' apperception

Figure 10. Problem orientation of Newton's law of motion

Figure 11. Student hypotheses box

Students watch and listen to the phenomenon of Newton's laws of motion in augmented reality media as students' apperception (**Figure 9**). Students are given problems related to the concept of Newton's laws of motion that have been built from apperception and then analyze it (**Figure 10**). Students make hypotheses from the problems they face (**Figure 11**).

3.1.5.2. Students are presented with material equipped with illustrations and augmented reality media, also questions that stimulate students to think critically.

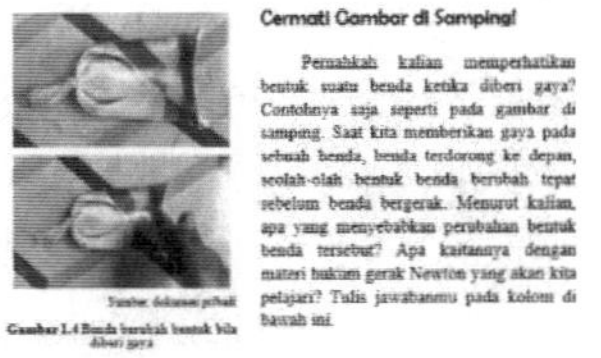

Figure 12. Ilustrasion of Physics Concept

Figure 13. Explanation of concept

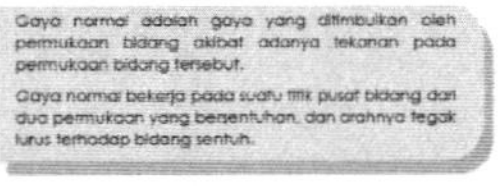

Figure 14. Note

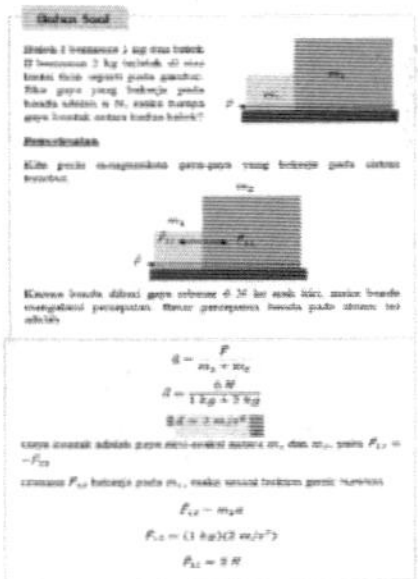

Figure 15. Problem example

3.1.5.3. Students learn the concept of Newton's laws of motion which are equipped with phenomenons review with the help of illustrations and augmented reality media, sample questions, and simple experiment that help students to understand the concepts, summaries and competency test.

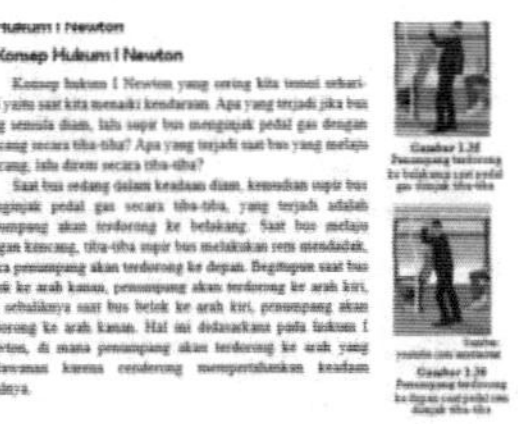

Figure 16. Phenomena review

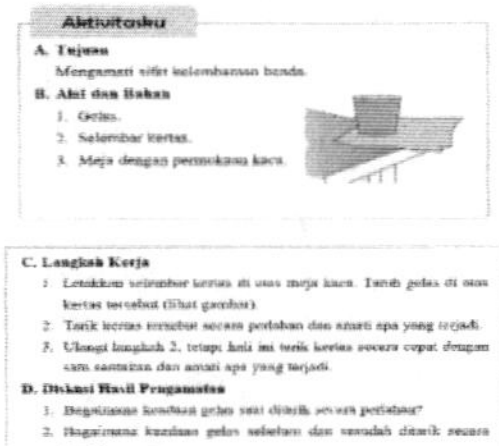

Figure 17. Simple experiment

Figure 18. Summary

Figure 19. Evaluation test

3.2. Description of formative evaluation result

The development of the physics textbook has passed the formative evaluation by material experts, pedagogic experts, and media experts. The eligibility of physics concepts in textbook is validated by material expert covering three aspects with the result as shown in Figure 20.

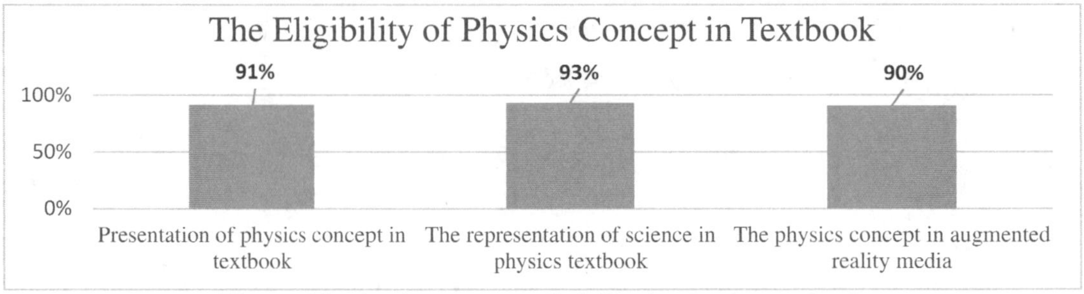

Figure 20. Physics concept eligibility in physics textbook

The eligibility of pedagogic in physics textbook is validated by pedagogic expert covering four aspects with the result as shown in Figure 21.

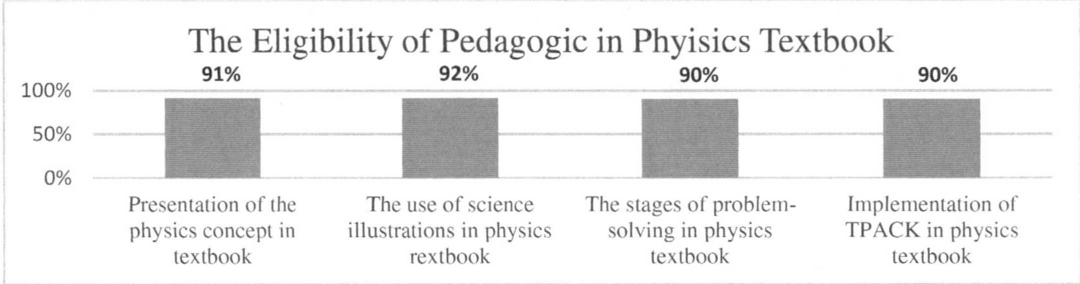

Figure 21. Pedagogic eligibility in physics textbook

The eligibility of physics textbook media is validated by media expert covering five aspects with the result as shown in Figure 22.

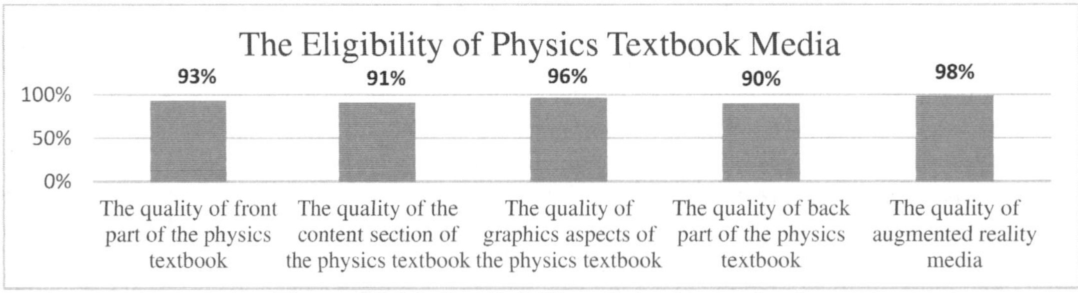

Figure 22. Media eligibility of physics textbook

3.3. Discussion

Implementation of TPACK and media based on augmented reality technology in physics textbooks can practice students to solve problems [28]. The contextual presentation in textbook and multiple representations make students easier to understand physics concepts [21]. The addition of augmented reality technology-based media makes students easier to understand abstract concepts to avoid misconceptions [19, 29].

Problem-solving stages in the presentation of concepts make students easier to understand the studied physics concepts [10]. The presentation of physics concepts supported by augmented reality technology and problem-solving learning stages make students understand Newton's law of motion concept easier and students' problem-solving skills are practiced well with this physics textbook [8, 27].

4. Conclusion
The TPACK framework implemented in the physics textbook is compiled to practice problem-solving skills on Newton's laws of motion concept. Based on formative evaluation, physics textbooks that implement TPACK are very suitable for practicing students' problem-solving skills on Newton's laws of motion concept.

References
[1] W. Syafi'i, E. Suryawati and A. R. Saputra, "Kemampuan Berpikir Kreatif dan Penguasaan Konsep Siswa Melalui Model Problem Based Learning (PBL) dalam Pembelajaran Biologi Kelas XI IPA SMAN 2 Pekanbaru Tahun Ajaran 2010/2011," *Jurnal Biogenesis,* vol. 8, no. 1, pp. 1-7, 2011.

[2] Sutiah, Teori Belajar dan Pembelajaran, Sidoarjo: Nizamia Learning Center, 2016.

[3] A. R. Harefa, "Peran Ilmu Fisika dalam Kehidupan Sehari-hari," *Jurnal Warta,* vol. 60, 2019.

[4] N. D. Permana, "Penerapan Model Pebelajaran Learning Cycle 7E Berbantuan Website Untuk Meningkatkan Keterampilan Berpikir Kritis Siswa Pada Materi Kinematika Gerak Lurus," *Journal of Natural Science and Integration,* vol. 1, no. 1, pp. 11-41, 2018.

[5] S. D. Aji, M. N. Hudha and A. Y. Rismawati, "Pengembangan Modul Pembelajaran Fisika Berbasis Problem Based Learning untuk Meningkatkan Kemampuan Pemecahan Masalah Fisika," *Jurnal Terapan Sains dan Teknologi.,* vol. 1, no. 1, pp. 36-51, 2017.

[6] A. H. Subarjo, "Utilization of QR - Code in Citizenship Education," in *Prosiding Seminar Nasional Teknologi Informasi dan Kedirgantaraan : Transformasi Teknologi untuk Mendukung Ketahanan Nasional*, Yogyakarta, 2018.

[7] J. Carson, "A Problem With Problem Solving : Teaching Thinking Without Teaching Knowledge," *The Mathematics Educator,* vol. 17, no. 2, p. 7–14, 2007.

[8] N. A. P. Yaman and A. Anwar, "Terampil dalam Pemecahan Masalah: Kompetensi Matematika Siswa Abad 21," *Seminar Matematika dan Pendidikan Matematika UNY,* pp. PM 489-PM 496, 2017.

[9] J. A. Koenig and Rapporteur, Assessing 21st Century Skills: Summary of a Workshop, Washington, DC: The National Academies Press., 2011.

[10] T. F. Wijayanti, "Potensi Model Pembelajaran Problem Solving Disertai Argument Mapping untuk Memberdayakan Berpikir Kritis," in *Prosiding Seminar Nasional Pendidikan*, Palembang, 2016.

[11] A. Ayob, A. Hussain and R. A. Majid, "A Review of Research on Creative Teachers in Higher Education," *International Education Studies,* vol. 6, no. 6, pp. 8-14, 2013.

[12] G. Rahmawati, "Buku Teks Pelajaran Sebagai Sumber Belajar Siswa di Perpustakaan Sekolah di SMAN 3 Bandung," *EduLib,* vol. 5, no. 1, pp. 102-113, 2015.

[13] P. Sothayapetch, J. Lavonen and K. Juuti, "A comparative analysis of PISA scientific literacy framework in Finnish and Thai science curricula," *Science Education International,* vol. 24, no. 1, pp. 78-97, 2013.

[14] L. Piranti and D. Muliyati, "Pengembangan Buku Referensi Berbasis Multi Representasi dengan Pendekatan Kontekstual pada Materi Kalor dan Termodinamika," in *PROSIDING SNIPS 2016*, Bandung, 2016.

[15] N. Respatiningrum, Y. Radiyono and E. Wiyono, "Analisis Miskonsepsi Materi Fluida pada Buku Ajar Fisika SMA," in *Prosiding Seminar Nasional Fisika dan Pendidikan Fisika (SNFPF) Ke-6*, Surakarta, 2015.

[16] D. K. Gurel and A. Eryılmaz, "A Review and Comparison of Diagnostic Instruments to Identify Students' Misconceptions in Science," *Eurasia Journal of Mathematics, Science and Technology Education,* vol. 11, no. 5, pp. 989-1008, 2013.

[17] H. Khoiri, A. K. Wijaya and I. Kusumawati, "Identifikasi Miskonsepsi Buku Ajar Fisika SMA Kelas X pada Pokok Bahasan Kinematika Gerak Lurus," *Jurnal Ilmu Pendidikan Fisika,* vol. 2, no. 2, pp. 60-64, 2017.

[18] Matsun, D. F. Saputri and Triyanta, "Analisis Miskonsepsi dan Tingkat Keterbacaan Buku Ajar Fisika SMA Kelas XII pada Materi Listrik Statis," *Jurnal Pendidikan Informatika dan Sains,* vol. 5, no. 2, pp. 227-236, 2016.

[19] S. Wuryanti, Yennita and Fakhruddin, "Analisis Miskonsepsi Siswa pada Materi Dinamika Gerak Menggunakan Tes Diagnostik Pilihan Ganda Tiga Tingkat," *Jurnal Geliga Sains,* vol. 5, no. 2, pp. 110-118, 2017.

[20] F. Fitria and Novitrian, "Diagnostik Miskonsepsi Siswa di Lingkungan Sekolah Menengah Atas di Bandung untuk Topik Hukum Newton tentang Gerak," in *Prosiding Seminar Nasional Quantum*, 2018.

[21] D. Rosengrant, E. Etkina and A. V. Heuvelen, "An Overview of Recent Research on Multiple Representations," in *Physics Education Research Conference*, New York, 2006.

[22] B. Waldrip, "Improving learning through use of representations in science.," in *Proceeding The 2nd International Seminar on Science Education*, Bandung, 2008.

[23] C. Angell, O. Guttersrud and E. K. Henriksen, "Multiple Representations as a Framework for a Modelling Approach to Physics Education," pp. 1-4, 2007.

[24] N. Baya'a and W. Daher, "The Development of College Instructors' Technological Pedagogical and Content Knowledge," *Procedia - Social and Behavioral Sciences,* no. 174, pp. 1-11, 2014.

[25] J. Rosenberg and M. J. Koehler, "Context and Technological Pedagogical Content Knowledge (TPACK): A Systematic Review," *Journal of Research on Technology in Education,* vol. 47, no. 3, pp. 47(3):186-210, 2015.

[26] P. Mishra and M. J. Koehler, "Technological Pedagogical Content Knowledge: A Framework for Teacher Knowledge," *Teachers College Record,* vol. 108, no. 6, p. 1017–1054, 2006.

[27] A. M. Ilmi, Sukarmin and W. Sunarno, "Development of Macro VBA as a TPACK Based-Physics Learning Media to Improve Critical Thinking Skills," in *The 2nd International Conference on Science, Mathematics, Environment, and Education AIP Conference Proceeding*, Surakarta, 2019.

[28] Robby, Sutrisno and W. D. Ernawati, "Pengembangan Perangkat Pembelajaran Kerangka Kerja TPACK untuk Meningkatkan Pemahaman Konsep Siswa pada Materi Kesetimbangan Kimia di Kelas XI IPA5 SMA Negeri 1 Kota Jambi," *Artikel Ilmiah Pendidikan Kimia FKIP Universitas Jambi,* 2014.

[29] S. A. Garba, T. K. Singh and N. M. Yusuf, "Integrating Technology in Teacher Education Curriculum and Pedagogical Practices: the Effects of Web-based Technology Resources on Pre-service Teachers' Achievement in Teacher Education Training," in *International Conference on Information Science and Technology Application (ICISTA)*, Paris, 2013.

[30] A. Y. A. Putranti, "Analisis TPACK Buku Guru Kelas 5 Tema 1 Edisi Revisi 2017 pada Muatan IPA," FKIP Universitas Muhammadiyah Surakarta, Surakarta, 2020.

[31] N. Brouwer, P. J. Dekker and J. V. D. Pol, E-Learning Cookbook. TPACK in Professional Development in Higher Education, Amsterdam: Amsterdam University Press, 2013.

[32] M. J. Koehler, P. Mishra and W. Cain, "What Is Technological Pedagogical Content Knowledge (TPACK)?," *Journal of Education,* vol. 193, no. 3, pp. 13-19, 2013.

[33] S. Cox and C. R. Graham, "Diagramming TPACK in Practice: Using An Elaborated Model of The TPACK Framework to Analyze and Depict Teacher Knowledge," *Tech Trends,* vol. 53, no. 5, pp. 60-69, 2009.

[34] C. Angeli and N. Valanides, "Epistemological and Methodological Issues for the Conceptualization, Development, and Assessment of ICT-TPCK: Advances in Technological Pedagogical Content Knowledge (TPCK)," *Computers & Education,* vol. 52, no. 1, pp. 154-168, 2009.

A Review of Research on The Use of Augmented Reality in Physics Learning

S Wulandari[1], F C Wibowo and I M Astra
Department of Physics Education, Universitas Negeri Jakarta
Jl. Rawamangun Muka, Jakarta 13220, Indonesia
Email: suciwulandari@mahasiswa.unj.ac.id

Abstract. The aim of this study is to reveal augmented reality on the effect of models in learning physics. This article is a qualitative synthesis of quantitative and qualitative research that investigates the design of learning with augmented reality which is applied in physics learning and its impact on learning. Articles published between 2015 and 2021 were reviewed, and 30 articles were included for detailed analysis and synthesis. The results of the study indicate that integrating augmented reality with teaching materials and learning models is capable to support the process and student learning outcomes. It was also found that problem-based augmented reality made learning more meaningful. Studies related to AR that measure students' critical thinking skills are still minimal. Future research is urgently needed to look in detail at the effect of problem-based augmented reality media on students' critical thinking skills.

1. Introduction

Physics is a part of science that is very influential in the development of technology and natural sciences. One of the lessons that are difficult to reach by most high school students is physics learning [1]. Students with visual thinking have difficulty understanding and mastering the material because they cannot understand and study a phenomenon without visualizing it [2]. Therefore, learning strategies include learning media used by educators to deliver material that greatly influences the process of forming students' concepts [3].

The development of IT technology opens up opportunities for the development of learning media tools that are an attraction and understanding that can be more accepted by students [4]. Digital skills are the main capital for students in facing the progress of all aspects of life, including learning now on digitalization [5]. Digital learning can encourage the formation of other skills for students, therefore a suitable digital media is needed [6]. Augmented reality is one of the media that creates a digital learning environment [6]. Augmented reality (AR) is a digital technology that provides content in the form of virtual images generated by a computer or mobile device to be spread out in the real world in real time. The generated content can be in the form of three-dimensional (3D) models, videos, images, sound and text [7]. Currently, AR is popularly used in the learning process.

In this case, research related to AR in several learning topics has been carried out including those related to human anatomy, language learning, history, chemistry, and mathematics [8]–[12]. Research related to AR in physics learning has also been carried out, this research was made as a form of information through a literature study on the implementation of AR in physics learning [13]–[16]. The main purpose of this study is to reveal augmented reality on the influence model in learning physics.

This study presents a list of literature with the type of AR used, the integrated learning model, the chosen physics concept, and the results of the use of AR on students in physics learning. The formulation of the problem that can be made from this study are:

a. What is the role of AR on the model and its influence in learning physics?
b. How is AR presented in learning?
c. On what physics topic is AR integrated?

2. Method

The type of research used is literature study, by collecting data and information from various related scientific sources [17]. The data collection technique that will be used is secondary data. The secondary data sources used in this study are a number of journals with international publications that can be accounted for, related to the use of Augmented Reality media in physics learning. This study uses a qualitative approach by generating descriptive data from several journals obtained. The stages of data analysis in qualitative research by Miles and Huberman [18] are as follows:

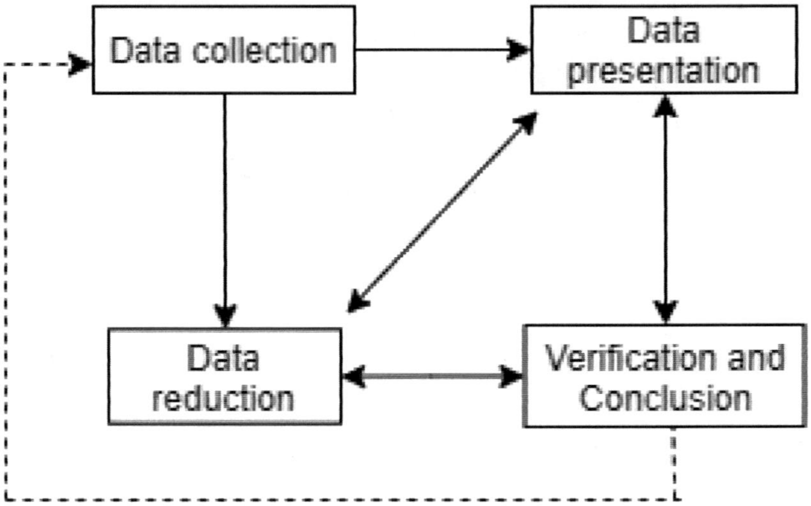

Figure 1. Data Analysis Process from Miles and Huberman

The first stage in qualitative data analysis is carried out by reviewing journals to obtain data and information needed in research. Second, data reduction is by summarizing, selecting points, and focusing the discussion so that the data obtained has a clearer picture and makes it easier for researchers. Then perform the presentation of the data, namely by presenting a narrative text in the form of a brief description that displays the relationship between subjects and the like. The last stage in qualitative data analysis is drawing conclusions and verification.

At the stage of data collection and reduction, the PRISMA model was adopted with 4 steps; Identification, screening, eligibility, and included. First, a search was conducted through Google Scholar with the keyword "Augmented Reality in Science Education" in the last 10 years. From the 86 articles obtained, then the articles were re-selected based on the subject of "physics", after which they were re-screened by category: research articles and international publications. Figure 2 shows the process of selecting articles in stages by following the PRISMA model [19].

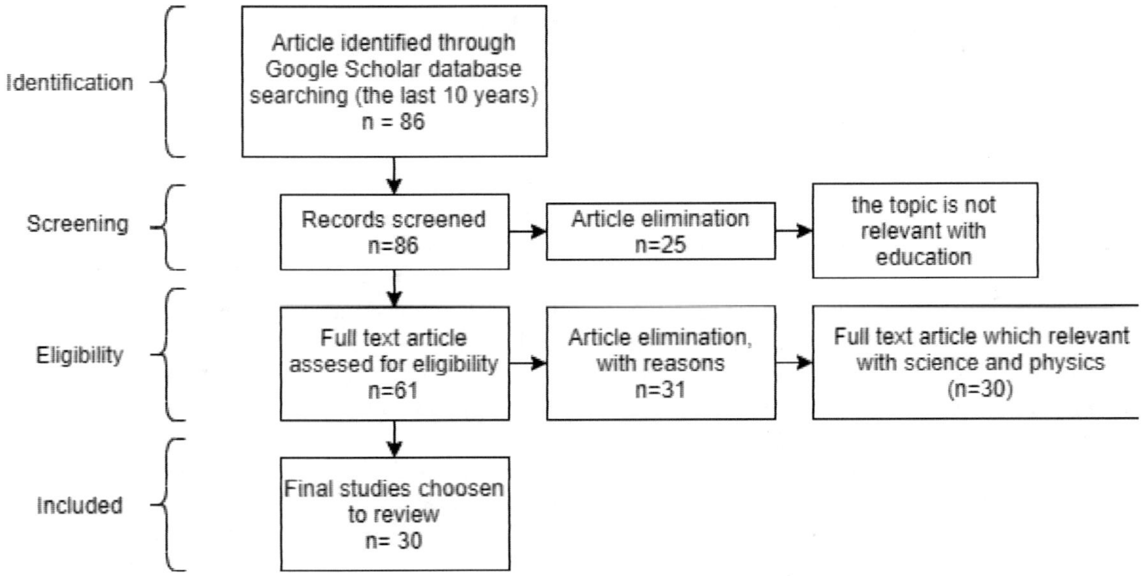

Figure 2. PRISMA Model for reduction articles

3. Results

Augmented Reality (AR) is one of the emerging technologies in various fields of education such as mathematics and science, because it can be an efficient technological tool to improve learning techniques [20] and increase student activity by bringing them to new learning environments [21]. AR is derived from two words, namely "augmented" which means addition and "reality" which means reality, or if combined it is "addition of reality" (Language Development and Development Agency, 2019). The addition of this reality refers to technology that enhances the sense of reality or "sense of reality" that allows the emergence of digital information in the form of objects in the form of 2D, 3D, animation, or simulation that can be accompanied by sound into the real environment [16]. A study review was conducted to find the relationship between physics learning activities carried out with AR and the results obtained as shown in Table 1.

Table 1. Learning activity and result of AR implementation

No.	Article	Learning activity	Result
1	[22]	Discovery learning	AR with discovery learning creates motivation, confidence and improves learning outcomes because of a good learning environment.
2	[23]	Case-based learning	The use of AR integrated worksheets can train and improve critical thinking skills.
3	[24]	Problem based learning	The use of the PBL learning model with AR media can improve learning outcomes, learning motivation, and student attitudes during learning.
4	[25]	Traditional lectures	The use of AR increases the desire to learn and has an impact on students' positive attitudes during the learning process.
5	[26]	Collaborative Inquiry Learning	AR simulation supports collaborative inquiry learning more than 2D simulation

No.	Article	Learning activity	Result
6	[2]	Laboratory activities	AR learning environment supports abstract visualization of physics concepts, has a positive impact on critical thinking skills

Based on table 1, it is found that learning activities with the Discovery Learning model can create motivation, self-confidence and learning outcomes because of the formation of a good learning environment. The same results were obtained when AR was implemented with a Problem Based Learning model, even AR had an impact on changing student attitudes towards a better direction in learning. The same principle is that problem-based learning can train students' critical thinking skills by integrating worksheets and AR. Critical thinking skills can also be improved by integrating AR in laboratory activities. As for other results, it can be seen that the use of AR with traditional learning only has an impact on the desire to learn and positive attitudes of students. While the use of the model can measure more variables. The type of AR that is developed will certainly affect the research variables you want to measure. Table 2 shows the materials and types of AR and AR-based that have been developed in physics learning development research over the last 10 years with the research category having international publications.

Table 2. Topic, kind of AR and AR-based in physics development

No.	Article	Topic	Kind of AR	AR-based
1	[27]	Optical Devices	3D object	Image-based
2	[28]	Light and Optic	3D object and video	Image-based
3	[25]	Convex Lens	3D object	Image-based
4	[29]	Mechanics	Simulation	Image-based
5	[24]	Force and motion	3D Object, simulation	Image-based
6	[30]	Mechanics	Simulation	Image-based
7	[23]	Physics harmonic	3D animation and videos	Image-based
8	[26]	Elastic collision	Simulation	Image-based
9	[24]	Fluids	3D object and simulation	Image-based
10	[31]	Heat transfer	3D animation	Natural interaction
11	[32]	Electromagnetism	3D animation	Image-based
12	[33]	Electromagnetism	3D Object	Image-based
13	[22]	Electricity	Simulation	Image-based
14	[34]	Magnetic Field	3D Object	Natural interaction
15	[35]	Model atom	3D Object	Image-based
16	[36]	Electricity	Videos	Image-based
17	[37]	Double Slit	Simulation	Image-based
18	[2]	Magnetic Field	Simulation	Image-based
19	[38]	Free fall	Simulation	Real time
20	[39]	Electromagnetism	Simulation	Image-based

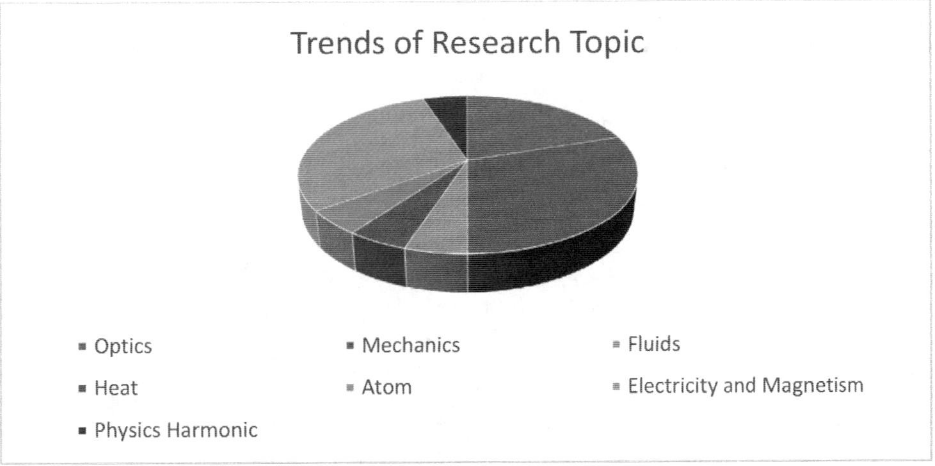

Figure 3. Trends of research topic in physics

Table 2 shows that there are 4 studies on the topic "Optics". 6 studies on the topic "Mechanics", 1 study on the topic "Fluids", 1 study on the topic "Temperature and Heat", 1 study on "Atoms", and 6 studies on the topic "Electricity and Magnetism", and 1 study on "Motion". Simple Harmonics". Based on the data, it can be concluded that there is no research on the topic "Kinetic Gas Theory (KGT)", while the KGT topic is considered to contain abstract material and is difficult to understand. As research which states that from 0-4 levels of understanding, students who have partial understanding at the third level are 19%, and students who have partial understanding at the second level are 81% with clear misconceptions [40]. So it is necessary to develop AR media for the topic of KGT as a recommendation from this research. The types of AR used vary, consisting of 3D objects, simulations, videos, animations. The process of displaying AR tends to use image-based.

4. Conclusion

Based on the results of the study review, several conclusions were obtained in response to the formulation of the problem proposed, including:

1. Augmented reality can be used as a support for learning carried out with certain models. Augmented reality can increase student motivation, student involvement in learning, student learning outcomes, and critical thinking skills. AR applied with certain learning models has a more significant impact than AR without learning models.
2. Augmented reality in learning is presented in the form of 3D objects, simulations, videos and animations. AR appearance can be image-oriented, direct or real-time interaction.
3. Physics topics that have been developed in integrating AR include: optics, mechanics, fluids, temperature and heat, atoms, electricity and magnetism, and simple harmonic motion.

Acknowledgement

Further research on AR on the topic of gas kinetic theory needs to be done, with certain learning models to measure critical thinking skills as one of the 4C skills that needed in this time.This study review was conducted under the guidance of Mr. Dr. Firmanul Catur Wibowo, M.Pd and Prof. Dr. I Made Astra, M.Si.

References

[1] E. Ekici, "'Why Do I Slog Through the Physics?': Understanding high school students' difficulties in learning physics," *J. Educ. Pract.*, vol. 7, no. 7, pp. 95–107, 2016.

[2] H. Faridi, N. Tuli, A. Mantri, G. Singh, and S. Gargrish, "A framework utilizing augmented reality to improve critical thinking ability and learning gain of the students in Physics," *Comput. Appl. Eng. Educ.*, vol. 29, no. 1, pp. 258–273, 2021.

[3] K. C. Stephen, S. N. Mailu, and P. K. Koech, "European Journal of Social Sciences Studies

Relationship Between Learning Strategies And Student Performance In Physics In Public Secondary Schools In Nakuru East Sub-County, Kenya," *Phys. Public Second. Sch. Nakuru East Sub-County*, vol. 3, p. 238, 2018.

[4] K. H. Cheng and C. C. Tsai, "Affordances of Augmented Reality in Science Learning: Suggestions for Future Research," *J. Sci. Educ. Technol.*, vol. 22, no. 4, pp. 449–462, 2013.

[5] E. van Laar, A. J. A. M. van Deursen, J. A. G. M. van Dijk, and J. de Haan, "The relation between 21st-century skills and digital skills: A systematic literature review," *Comput. Human Behav.*, vol. 72, pp. 577–588, 2017.

[6] M. J. Sousa and Á. Rocha, "Digital learning: Developing skills for digital transformation of organizations," *Futur. Gener. Comput. Syst.*, vol. 91, pp. 327–334, 2019.

[7] K. Lee, "Augmented Reality in Education and Training," *TechTrends*, vol. 56, no. 2012, pp. 13–21, 2012.

[8] S. S. Jamali, M. F. Shiratuddin, K. W. Wong, and C. L. Oskam, "Utilising Mobile-Augmented Reality for Learning Human Anatomy," *Procedia - Soc. Behav. Sci.*, vol. 197, no. February, pp. 659–668, 2015.

[9] Salin Boonbrahm, C. Kaewrat, and P. Boonbrahm, "Using Augmented Reality Technology in Assisting English Learning for Primary School Students," *Springer Int. Publ.*, vol. 9192, p. 24, 2015.

[10] Y. Blanco-Fernández, M. López-Nores, J. J. Pazos-Arias, A. Gil-Solla, M. Ramos-Cabrer, and J. García-Duque, "REENACT: A step forward in immersive learning about Human History by augmented reality, role playing and social networking," *Expert Syst. Appl.*, vol. 41, no. 10, pp. 4811–4828, 2014.

[11] S. Cai, X. Wang, and F. K. Chiang, "A case study of Augmented Reality simulation system application in a chemistry course," *Comput. Human Behav.*, vol. 37, pp. 31–40, 2014.

[12] H. C. K. Lin, M. C. Chen, and C. K. Chang, "Assessing the effectiveness of learning solid geometry by using an augmented reality-assisted learning system," *Interact. Learn. Environ.*, vol. 23, no. 6, pp. 799–810, 2015.

[13] M. Akçayır and G. Akçayır, "Advantages and challenges associated with augmented reality for education: A systematic review of the literature," *Educ. Res. Rev.*, vol. 20, pp. 1–11, 2017.

[14] P. Chen, X. Liu, W. Cheng, and R. Huang, "A review of using augmented reality in education from 2011 to 2016," *Lect. Notes Educ. Technol.*, no. 9789811024184, pp. 13–18, 2017.

[15] M. B. Ibáñez and C. Delgado-Kloos, "Augmented reality for STEM learning: A systematic review," *Comput. Educ.*, vol. 123, pp. 109–123, 2018.

[16] F. Saltan and Ö. Arslan, "The use of augmented reality in formal education: A scoping review," *Eurasia J. Math. Sci. Technol. Educ.*, vol. 13, no. 2, pp. 503–520, 2017.

[17] H. Snyder, "Literature review as a research methodology: An overview and guidelines," *J. Bus. Res.*, vol. 104, no. July, pp. 333–339, 2019.

[18] J. R. Fraenkel, N. E. Wallen, and H. H. Hyun, *How To Design and Evaluate Research in Education*, vol. 4, no. 1. 2016.

[19] D. Moher, A. Liberati, J. Tetzlaff, and D. G. Altman, "Preferred reporting items for systematic reviews and meta-analyses: The PRISMA statement," *BMJ*, vol. 339, no. 7716, pp. 332–336, 2009.

[20] A. Martin-Gonzalez, A. Chi-Poot, and V. Uc-Cetina, "Usability evaluation of an augmented reality system for teaching Euclidean vectors," *Innov. Educ. Teach. Int.*, vol. 53, no. 6, pp. 627–636, 2016.

[21] F. Arici, P. Yildirim, Ş. Caliklar, and R. M. Yilmaz, "Research trends in the use of augmented reality in science education: Content and bibliometric mapping analysis," *Comput. Educ.*, vol. 142, no. August, p. 103647, 2019.

[22] M. B. Ibanez, A. Di-Serio, D. Villaran-Molina, and C. Delgado-Kloos, "Augmented Reality-Based Simulators as Discovery Learning Tools: An Empirical Study," *IEEE Trans. Educ.*, vol. 58, no. 3, pp. 208–213, 2015.

[23] L. D. Herliandry, H. Kuswanto, and W. Hidayatulloh, "Improve Critical Thinking Ability Through Augmented Reality Assisted Worksheets," *Proc. 6th Int. Semin. Sci. Educ. (ISSE 2020)*,

vol. 541, no. Isse 2020, pp. 470–475, 2021.
[24] M. Fidan and M. Tuncel, "Integrating augmented reality into problem based learning: The effects on learning achievement and attitude in physics education," *Comput. Educ.*, vol. 142, no. May, p. 103635, 2019.
[25] S. . b Cai, F.-K. . Chiang, and X. . Wang, "Using the augmented reality 3D technique for a convex imaging experiment in a physics course," *Int. J. Eng. Educ.*, vol. 29, no. 4, pp. 856–865, 2013.
[26] H. Y. Wang, H. B. L. Duh, N. Li, T. J. Lin, and C. C. Tsai, "An investigation of university students' collaborative inquiry learning behaviors in an augmented reality simulation and a traditional simulation," *J. Sci. Educ. Technol.*, vol. 23, no. 5, pp. 682–691, 2014.
[27] A. Kristanto, C. A. P. Rahayu, and S. C. Wibawa, "The Development of Augmented Reality Media for Physics Subject in Learning Optical Devices Material at SMK Multimedia," *Adv. Soc. Sci. Educ. Humanit. Res.*, vol. 372, no. ICoET, pp. 198–206, 2019.
[28] P. Wozniak, O. Vauderwange, D. Curticapean, N. Javahiraly, and K. Israel, "Perform light and optic experiments in Augmented Reality," *Educ. Train. Opt. Photonics ETOP 2015*, vol. 9793, no. 0, p. 97930H, 2015.
[29] Y. Daineko, M. Ipalakova, D. Tsoy, A. Shaipiten, Z. Bolatov, and T. Chinibayeva, "Development of practical tasks in physics with elements of augmented reality for secondary educational institutions," *Lect. Notes Comput. Sci. (including Subser. Lect. Notes Artif. Intell. Lect. Notes Bioinformatics)*, vol. 10850 LNCS, pp. 404–412, 2018.
[30] A. D. Morales, S. A. Sanchez, C. M. Pineda, and H. J. Romero, "Use of Augmented Reality for the Simulation of Basic Mechanical Physics Phenomena," *IOP Conf. Ser. Mater. Sci. Eng.*, vol. 519, no. 1, 2019.
[31] M. Thees, S. Kapp, M. P. Strzys, F. Beil, P. Lukowicz, and J. Kuhn, "Effects of augmented reality on learning and cognitive load in university physics laboratory courses," *Comput. Human Behav.*, vol. 108, p. 106316, 2020.
[32] S. Techakosit and P. Nilsook, "Using Augmented Reality for Teaching Physics," *Int. E-Learning Conf. 2015, Thail. Cyber Univ.*, no. July, pp. 282–287, 2015.
[33] M. B. Ibáñez, Á. Di Serio, D. Villarán, and C. Delgado Kloos, "Experimenting with electromagnetism using augmented reality: Impact on flow student experience and educational effectiveness," *Comput. Educ.*, vol. 71, pp. 1–13, 2014.
[34] S. Cai, F. K. Chiang, Y. Sun, C. Lin, and J. J. Lee, "Applications of augmented reality-based natural interactive learning in magnetic field instruction," *Interact. Learn. Environ.*, vol. 25, no. 6, pp. 778–791, 2017.
[35] N. Suprapto, W. Nandyansah, and H. Mubarok, "An evaluation of the 'PicsAR' research project: An augmented reality in physics learning," *Int. J. Emerg. Technol. Learn.*, vol. 15, no. 10, pp. 113–125, 2020.
[36] M. Akçayır, G. Akçayır, H. M. Pektas, and M. A. Ocak, "Computers in Human Behavior Augmented reality in science laboratories : The effects of augmented reality on university students ' laboratory skills and attitudes toward science laboratories," vol. 57, pp. 334–342, 2016.
[37] T. Wang, H. Zhang, X. Xue, and S. Cai, "Augmented Reality-Based Interactive Simulation Application in Double-Slit Experiment," vol. 2, 2018.
[38] N. J. Sung, J. Ma, Y. J. Choi, and M. Hong, "Real-time augmented reality physics simulator for education," *Appl. Sci.*, vol. 9, no. 19, pp. 1–12, 2019.
[39] Harun, N. Tuli, and A. Mantri, "Experience Fleming's rule in electromagnetism using augmented reality: Analyzing impact on students learning," *Procedia Comput. Sci.*, vol. 172, no. 2019, pp. 660–668, 2020.
[40] T. Nurhuda, D. Rusdiana, and W. Setiawan, "Analyzing Students' Level of Understanding on Kinetic Theory of Gases," *J. Phys. Conf. Ser.*, vol. 755, no. 1, 2017.

Development Of Android Physics Applications (APA) As Learning Media On Dynamic Fluid Concepts

C Fathurohman[1,a], F C Wibowo[1,b], and B H Iswanto[1,c]
[1]Physics Education Master Program, Universitas Negeri Jakarta
Jl. Rawamangun Muka, Jakarta 13220, Indonesia
Email: [1,a]cecepfathurohman_1310819005@mhs.unj.ac.id, [1,b]fcwibowo@unj.ac.id, [1,c]bhi@unj.ac.id

Abstract. Android is one of the learning media based on 21st-century learning styles. Android is open source and provides opportunities for users to develop other application features as needed. This study aims to create Android-based physics learning media using the Smart App Creator on Dynamic Fluid material. Media is developed using the methods of research and development R&D (Research and Development) Model FODEM (Formative Development Method) consisting of Need Analysis (NE), Implementation (I), and Formative Evaluation (FE). The results of the validity of the Android Physics Applications (APA) media obtained an average score of 83% in the very feasible category, while the results of the limited trial of students obtained an average score of 82.1% with the very good category. Based on these results, it can be concluded that the Android Physics Applications (APA) media can be used as a learning medium in the classroom. The development of Android-based Physics Applications is expected to be an alternative for educational institutions or educators to develop learning media that use multimedia.

1. Introduction

Learners in the 21st century should have the opportunity to reflect on their ideas, hone skills anali s is to them, strengthen the capacity to think critically and creatively and show initiative in solving problems [1,2]. In solving problems it is necessary to have a learning media. The selection of technology-based learning media has a very important role [3,4]. The Android operating system is one of the 21st-century learning media [5], android media is an open-source media and provides an opportunity for users to develop other application features as needed [6,7], and can be accessed easily anywhere. anytime and anywhere [4,6,8].

Physics is one of the basic sciences that plays an important role in the development of science and technology [9,10]. One of the topics in physics is Fluid Dynamics [11,12]. Abstract fluid dynamics concepts such as Bernoulli's principle and ideal fluid concepts need to be visualized so that students have a complete concept [13,14] The next step is that teachers need to change their role from content deliverers to content creators, creating content and applications to turn technology into a tool for learning [1,15].

Android media is a complex mixture of OS, middleware, and applications that can increase the interest and attractiveness of students with a multi-representative presentation [16,17]. Mobility is one of the main factors that attract distance learners to use mobile learning [18]. Learning media using the android system is more attractive and flexible [19-21]. the device may smartphone also minimalist so that students can bring the material easily [21].

Technology-based learning innovation is assessed as a medium that has strong potential in improving the quality of education [22]. The selection of technology-based learning media has a very important role [14,24,24], therefore, physics learning innovation is needed as a solution to problems that arise in every school [8,25]. Researchers late this developing learning media in the form of Android Physics Application (APA) on material Dynamic fluid in high school.

2. Method

In this study, the method used is R & D [26], using the FODEM (Formative Development Method) model [27]] which consists of three components, as follows:

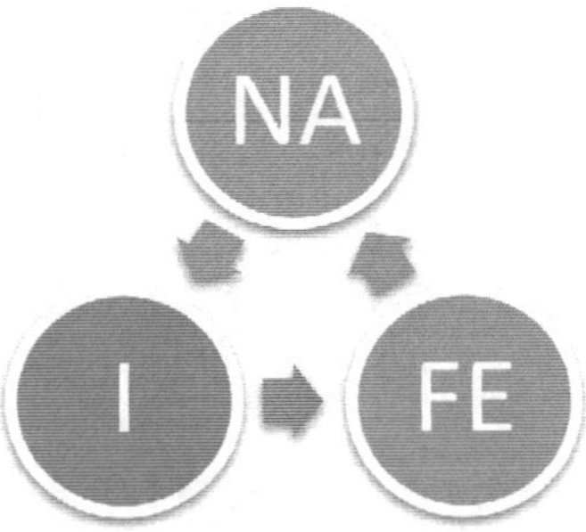

Figure 1. FODEM circuit [27]

In the first stage of Needs Analysis (NA), at this stage, the needs and shortcomings of android media design and development are analyzed. In the early stages of development. The definition of pedagogical objectives can be based on both theory and practice. Design context requirements can be based on previous development experience in similar situations. A needs analysis can also include the effects of other sets.

In the second stage of implementation (I), at this stage, a trial and evaluation of learning are carried out so that the android media that has been made can be used optimally. The emphasis in the implementation component is on rapid prototyping of the design ideas specified in the requirements analysis. Its main purpose is to test design ideas for information about feasible and failed solutions.

The last stage is Formative Evaluation (FE), at this stage as a revision between stages starting from the Need Analysis stage to Implementation and this stage also measures the extent to which the android application can function.

3. Results and Discussion

The product produced in this research is a learning application, hereinafter referred to as APA (Android Physics Application). This application was developed based on the FODEM development model, with the Needs Analysis (NA) stage carried out by analyzing problems, needs, media materials, development specifications, and media creation tools.

The Implementation (I) phase begins by integrating the APA (Android Physics Application) concept into the dynamic fluid concept. by making learning media according to the needs of students with specific media development specifications. This development is carried out by utilizing the Smart App Creator application as software for developing this learning media. Then carried out by conducting a small-scale test. The Formative Evaluation (FE) stage was carried out by analyzing the validation of material data and media expertise, as well as the results of the implementation of small-scale media tests.

3.1. APA Development

The display will be visible after clicking the application tab on the mobile screen. The screen contains sub-menus, guides, materials, videos, simulations, practice questions. APA development flowchart on dynamic fluid materials can be seen in Figure 2.

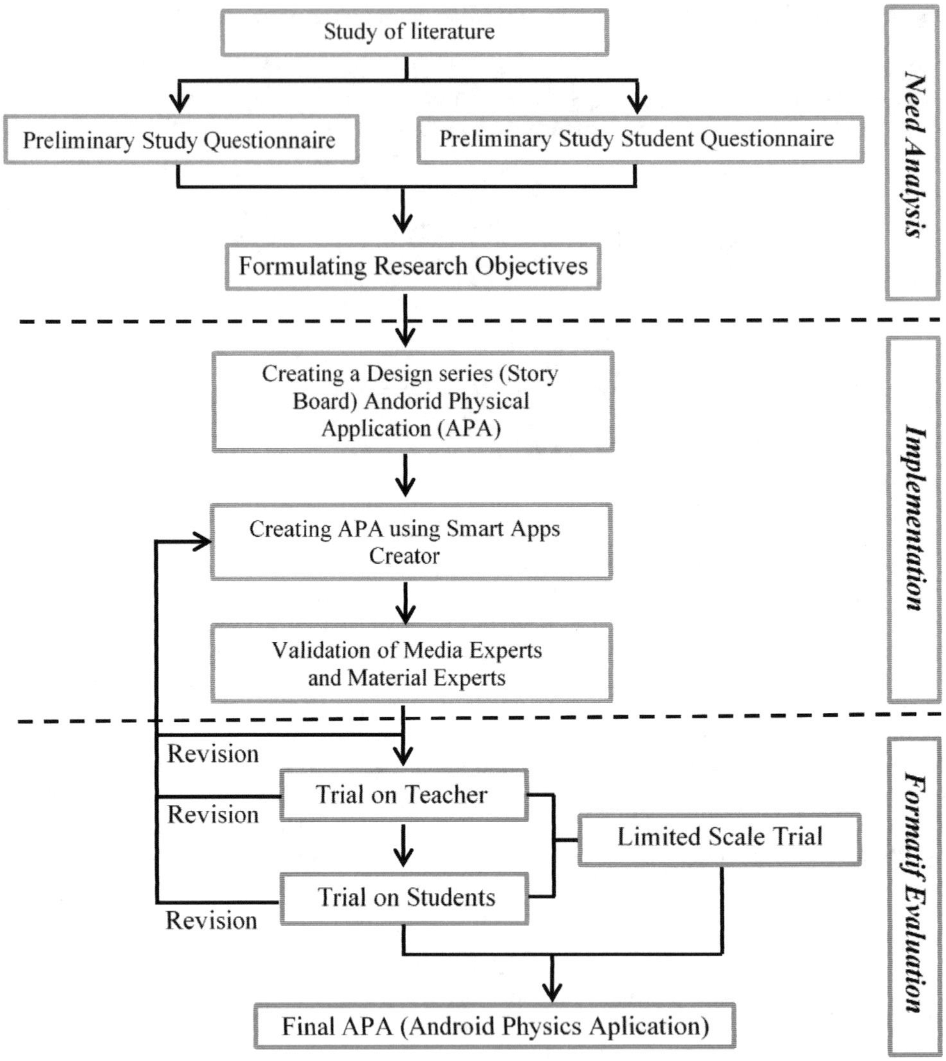

Figure 2. APA Development Flowchart

3.2. Final Product Review

Android Physics Application (APA) is a medium of learning physics on the concept of dynamic fluid. This application contains guides, materials, videos, simulations, and practice questions. This application is developed, analyzed, and improved based on comments and suggestions at the validation and implementation stages. Overall the media developed is feasible to use.

Journal of Physics: Conference Series **2019** (2021) 012059 doi:10.1088/1742-6596/2019/1/012059

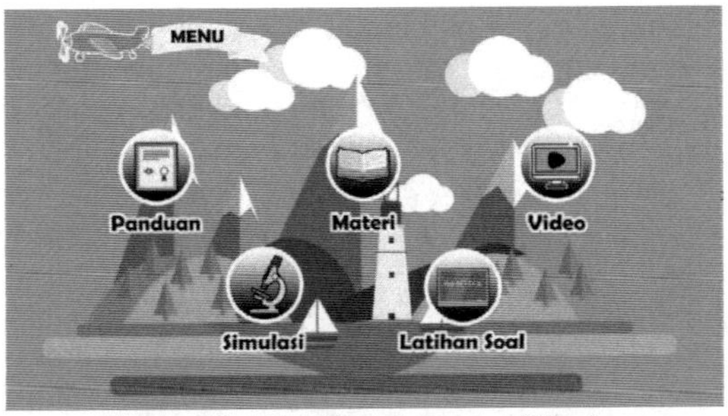

Figure 3. APA Development Display

Figure 4. Guide icon display

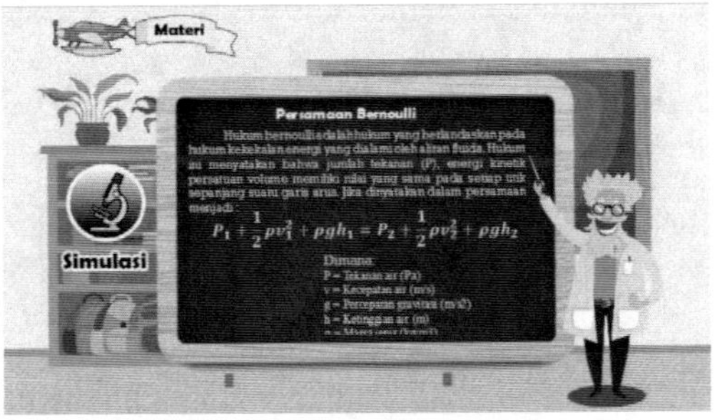

Figure 5. Material icon display

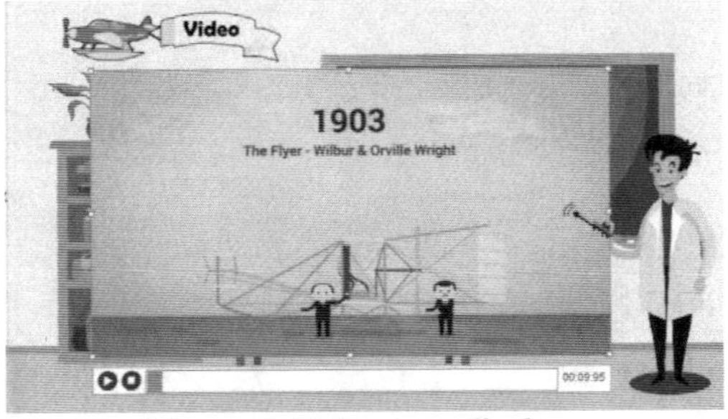

Figure 6. Video icon display

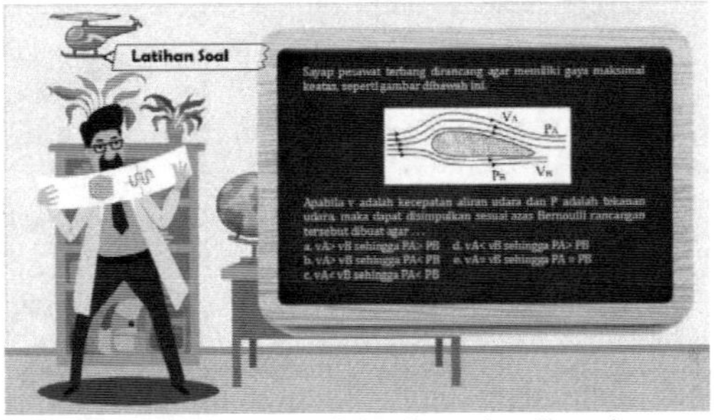

Figure 7. Display of the Question Practice Menu

Based on Figure 3. there are several main menus in APA development including 1) guide icon, 2) material icon, 3) video icon, 4) simulation icon and 5) practice questions icon. students can select each menu by clicking each icon. Based on Figure 4. there is a description of each icon in the main menu of the developed APA application. There is a next button to change the explanation of the icon on the main menu, and there is a back button. Based on Figure 5 . contains material about dynamic fluid equations and concepts in full, and there is a simulation icon to go to the simulation menu to make it easier for students to understand the material with the simulation. Based on Figure 6 . contains learning videos about dynamic fluid concepts. Figure 7 contains practice questions as student evaluations in learning.

3.3. Discussion

The results of the research conducted by the researchers were then compared with the results of previous studies. Based on research conducted by Aji, et al [25] Astra, et al [6] Saputra & Kuswanto [5] and Rosyid, et al [7].

The results obtained that the three studies are categorized as suitable for use in the learning process. The results of the study are not much different from the results obtained by the researchers, the results of the validity of the Android Physics Applications (APA) media obtained an average score of 83% with a very feasible category, while the results of the trial were limited to students obtaining an average score of 82.1%. with very good category. Android Physics Applications (APA) is an innovative learning media that can adapt to technological developments. This application can be used offline on all Android smartphone resolutions.

4. Conclusion

The results obtained that the three studies are categorized as suitable for use in the learning process. The results of the study are not much different from the results obtained by the researchers, the results of the validity of the Android Physics Applications (APA) media obtained an average score of 83% with a very feasible category, while the results of the trial were limited to students obtaining an average score of 82.1% with very good category. Physics Education Game Application is an innovative learning media that can adapt to technological developments. This application can be used offline on all Android smartphone resolutions.

References

[1] C. L. Scott. THE FUTURES OF LEARNING 3: WHAT KIND OF PEDAGOGIES FOR THE 21st CENTURY?, *EDUCATION RESEARCH AND FORESIGHT WORKING PAPERS*. 2015.

[2] K. Ananiadou dan M. Claro. 21st Century Skills and Competences for New Millennium Learners in OECD Countries, *OECD Education Working,* pp. No. 4*1*. 2009.

[3] N. Mardiana dan H. Kuswanto. Android-Assisted Physics Mobile Learning to Improve Senior High School Students' Divergent Thinking Skills and Physics HOTS, *AIP Conference Proceedings.* 2017.

[4] N. Liliarti dan H. Kuswanto. Improving the Competence of Diagrammatic and Argumentative Representation in Physics through Android-based Mobile Learning Application, *International Journal of Instruction.* 2018.

[5] M. R. D. Saputra dan H. Kuswanto. Development of Physics Mobile (Android) Learning Themed Indonesian Culture Hombo Batu on the Topic of Newton's Law and Parabolic Motion for Class X SMA/MA, *Journal of Physics: Conference Series.* 2018.

[6] I. M. Astra, D. Susanti dan A. Novriansyah. Development of e-Handout materials physics-based android for improvement learning outcomes senior high school student, *IOP Publishing.* 2020.

[7] F. A. Rosyid, S. Arjo, N. Suminten dan D. P. Sandjaj. The Development of Android-Based Physical Learning Media Brain Quiz Game Assisted on Momentum and Impulse Materials, *Journal of Physics: Conference Series.* 2020.

[8] F. P. Sari, R. L dan I. Wiluje. Development of Android Comics media on Thermodynamic Experiment to Map the Science Process Skill for Senior High School, *Journal of Physics: Conference Series.* 2019.

[9] F. Rofiqoh. Pengaruh Model Pembelajaran Kooperatif Tipe Numbered Heads Together (NHT) Disertai Media Monopoli Games Terintegrasi Pendekatan Problem Solving Pada Pembelajaran Fisika di SMA, *Jurnal Pembelajaran Fisika, Vol.4 No.3: 198-203.* 2015.

[10] J. Docktor dan J. Mestre. Physical Review Special Topics, *Physics Education Research.* 2014.

[11] P. A. Tipler dan G. Mosca 2008 *Physics for Scientists and Engineers with Modern Physics, 6 penyunt* (New York: W. H. Freeman and Company)

[12] D. C. Giancoli. Physics for Scientists & Engineers with Modern Physics, 4 penyunt., *Pearson.* 2014.

[13] G. D. Putra, A. Samsudin dan D. Saepuzaman. Computer simulation-assisted conceptual change text (CS-CCT): a FODEM study on fluid dynamics," *Journal of Physics: Conference Series.* 2019.

[14] C. Fathurohman, Y. Ruhiyat dan R. F. Septianto. Penerapan Media Simulasi PhET Untuk Meningkatkan Pemahaman Konsep Siswa SMA Pada Materi Fluida, *Gravity Journal Research and Learning Physics.* 2018.

[15] C. Yung-Ting. SSCLS: A Smartphone-Supported Collaborative Learning System, *Telematics and Informatics.* 2014.

[16] N. Safaat 2012 *Android: Pemrograman Aplikasi Mobile Smartphone dan Tablet PC Berbasis Android* (Bandung: Penerbit Informatika)

[17] T. M. Miangah dan N. Amin. Mobile-Assisted Language Learning, *International Journal of Distributed and Parallel Systems,* vol. 3, pp. 1-11. 2012.

[18] K. Yin Yin. Pocket learning: a new mobile learning approach for distance learners, *J. Mobile Learning and Organisation.* 2015.

[19] W. Setya dan R. Zakwandi. Development of Android-base media on the point of glass and lens, *Journal of Physics: Conference Series.* 2019.

[20] B. E. Dasilva dan S. Development of The Android-Based Interactive Physics Mobile Learning Media (IPMLM) to Improve Higher Order Thinking Skills (HOTS) of Senior High School Students, *IOP Publishing.* 2019.

[21] L. Jianye dan Y. Jiankun. Research on Development of Android Applications, *Fourth International Conference on Intelligent Networks and Intelligent Systems.* 2011.

[22] Y. J. Joo, K. Y. Lim dan E. Lim. Investigating the structural relationship among perceived innovation attributes, intention to use and actual use of mobile learning in an online university in South Korea. Australasian Journal of Educational Technology, *Australasian Journal of Educational Technology,* vol. 30, pp. 427-439. 2014.

[23] F. C. Wibowo dan B. H. Iswanto. Designing MOOCS with Virtual Microscopic Simulation (VMS) for increasing of student's levels of understanding, *Journal of Physics: Conference Series 1402 066094.* 2019.

[24] F. C. Wibowo, D. R. Darman, H. Abizar, S. S. M. Leksono, S. R. N. Hodijah, L. Nulhakim dan A. Istiandaru. Virtual simulation instructional training for students' drop out of mathematical science digital entrepreneurs, *Journal of Physics: Conference Series.* 2019.

[25] S. H. Aji, Jumadi, A. T. Saputra dan R. N. Tuada. Development of physics mobile learning media in optical instruments for senior high school student using android studio, *Journal of Physics: Conference Series.* 2020.

[26] Sugiyono 2012 *Metode Penelitian Pendekatan Kuantitatif, Kualitatif, dan R&D* (Bandung: Alfhabeta)

[27] J. Suhonen, M. R. de Villiers dan E. Sutinen. FODEM: a multi-threaded research and development method for educational technology, *Education Tech Research Dev.* 2012.

Trends of Augmented Reality in Science Learning: A Review of the Literature

U Alizkan[1], F C Wibowo[1], L Sanjaya[1], B R Kurniawan[2] and B K Prahani[3]
[1]Department of Physics Education, Universitas Negeri Jakarta
Jl. Rawamangun Muka, Jakarta 13220, Indonesia
[2]Department of Physics Education, Universitas Negeri Malang
Malang City, East Java 65145, Indonesia
3 [2]Department of Physics Education, Universitas Negeri Surabaya
Surabaya City, East Java 60213, Indonesia
Email: ubedalizkan@gmail.com, fcwibowo@unj.ac.id, lari@unj.ac.id, binarprahani@unesa.ac.id

Abstract. This study provides a literature review on the trend of using Augmented Reality (AR) in science learning. A relative literature review was carried out by collecting 30 journals from the Science Direct and Google Scholar databases in the period 2012 to 2021. This paper examines the research that has been done on the use of AR in science learning and classifies it into 11 topics, namely about inquiry skills, critical thinking, conceptual knowledge, misconception, conceptual understanding, knowledge construction, learning outcomes, attention, attitude, achievement, motivation, collaborative and autonomous learning. A review of the research results shows that the overall use of AR technology in recent years has increased, especially on the topic of attitude and achievement which has a positive impact on science learning. While on the topic of misconceptions, inquiry skills, attention, collaborative and autonomous learning has not been widely studied by researchers. In addition, the use of AR in science learning has challenged such as students, teachers, and technical. This is interesting for further research.

1. Introduction

Science is a science that contains systematic, rational, and objective theories about the universe Science can be claimed as scientific knowledge, namely science that has been tested for truth through systematics or scientific methods [1]. Science contains abstract concepts so that it requires the use of learning media to visualize abstract concepts. The integration of technology in developing appropriate learning media can make it easier for students to understand abstract concepts in their entirety. One of the uses of technology in education is the use of Augmented Reality technology in learning science.

Augmented reality is a technology that visually augments the real world environment by projecting computer-generated information into the eye [2]. Augmented Reality is a rapidly growing field of research that aims to fully integrate the virtual with the real environment [3]. This AR can be interpreted that real objects in real time are added to virtual objects that appear when using a tool or device on the real object. So that there is a relationship between the virtual world and the real world with the help of the camera. The resulting content can be in the form of three-dimensional (3D) models, videos, images, sound and text [4].

Research on augmented reality has been widely carried out, based on published journals reporting the advantages, limitations, effectiveness challenges, etc. of the use of augmented reality in education. However, since augmented reality is an emerging technology, it is important to get an idea of the progress and real impact of its use in education especially on science learning.

Previous studies on augmented reality in science learning can be used to guide future studies. Therefore, a systematic review is very important to present the current situation and to shed light on future studies. Reviewing previous studies helps researchers to make decisions about issues such as topics, methods and sampling. It is possible to find many systematic reviews in the literature on the use of technology in education [5, 6]. However, there have only been a few systematic reviews examining augmented reality studies. This augmented reality study, therefore, aims to bridge the existing gap in the literature by analyzing all educational studies found in various databases (ScienceDirect and Google Scholar). From the identification results,

1) What journals are used in this literature study?
2) How will augmented reality research trends in science learning be from 2012 to 2021?
   What are the challenges of implementing augmented reality in science learning?

2. Method

This study analyzed and synthesized articles about the use of Augmented Realty media in science learning from 2012 to May 2021. Articles were collected using Publish or Perish software using the keyword "Augmented Realty Science Education" in the Science Direct and Google Scholar databases. There are several criteria used to select the articles to be analyzed which are shown in table 1.

Table 1. Article Criteria Used

Criteria	Inclusion	Exclusion
Article type	Indexed journal	Journal not indexed
Types of research	Empirical research	Meta analysis, literature review
Research field	Science Learning	Non-science learning
language	English	Non-English
Timeline	2012- May 2021	Less than 2012

3. Result and Discussion

3.1 Number of Journal Publications

Among the 30 studies augmented reality were analyzed from 15 journal and 3 conference, consisting of dairy journal Interactive Learning Environments (2), Advances in Human-Computer Interaction (1), British Journal of Educational Technology (2), Computers & Education (4), Computers in Human Behavior (6), Computer Applications in Engineering Education (1), Contemporary Educational Technology (1), Education and Information Technologies, European Journal of Physics, International Journal of Computer Supported Collaborative Learning (1), Journal of Computer Assisted Learning (1), Journal of Research in Education and Teaching (1), Journal of Science Education and Technology (1), Science Educational International (1), Human Interaction Conference (1), International Conference on Advanced Learning Technologies (1), International Symposium on Mixed and Augmented Reality Adjunct (1). The distribution of articles from 2012- Mei 2021 is shown in table 2

Table 2. Publication Type and Number of Articles Selected in Each

Publication Type	Journals and Conference Selected	Number of Articles Selected	% of Articles Selected
Conference	Human Interaction Conference	1	3,3%
Conference	International Conference on Advanced Learning Technologies	1	3,3%
Conference	International Symposium on Mixed and Augmented Reality Adjunc	1	3,3%
Journal	Interactive Learning Environments	2	6,6%
Journal	Advances in Human-Computer Interaction	1	3,3%
Journal	British Journal of Educational Technology	2	6,6%

Publication Type	Journals and Conference Selected	Number of Articles Selected	% of Articles Selected
Journal	Computers & Education	4	16,7%
Journal	Computers in Human Behavior	6	20,0%
Journal	Computer Applications in Engineering Education	1	3,3%
Journal	Contemporary Educational Technology	1	3,3%
Journal	Education and Information Technologies	1	3,3%
Journal	Educational Technology & Society	1	3,3%
Journal	European Journal of Physics	1	3,3%
Journal	International Journal of Computer Supported Collaborative Learning	1	3,3%
Journal	Journal of Computer Assisted Learning	1	3,3%
Journal	Journal of Research in Education and Teaching	1	3,3%
Journal	Journal of Science Education and Technology	3	10,0%
Journal	Science Educational International	1	3,3%
Total		30	100 %

3.2 Augmented Reality trends in Science Learning

The use of AR in the educational environment is a research topic that has recently become popular. In this study, 30 articles were used regarding research on the use of augmented reality in science learning which were grouped into 11 topics, namely about inquiry skills, critical thinking, conceptual knowledge, misconceptions, conceptual understanding, construction knowledge, learning outcomes, attention, attitude, achievement, motivation, collaborative and autonomous learning. In addition, the use of augmented reality is mostly applied in universities. This is because the student's learning independence factor is quite good compared to students at middle and elementary school levels.

Table 3. The Study on the use AR in Science Learning

Study	Subject	School level	Topic
[7]	Physics (Electromagnetic)	University	Critical Thinking
[8]	Biology (Anatomy)	Secondary School	Attitude, Achievement
[9]	Chemistry (Molecular Geometry)	University	Motivation
[10]	Physics (Electric circuit)	University	Conceptual Understanding
[11]	Chemistry (Chemical Elements)	Junior high school	Conceptual knowledge
[12]	Physics (Heat Transfer)	University	Knowledge Construction
[13]	Physics (Electric circuit)	Middle school	Attitude
[14]	Biology (Cell)	University	Motivation, Achievement
[15]	Physics (Force and energy)	Junior high school	Achievement, Attitude
[16]	Physics (Electric circuit)	Junior high school	Conceptual knowledge, Inquiry Skill
[17]	Biology (Anatomy)	University	Motivation
[18]	Natural Science	Elementary School	Achievement, Motivation, Crticial Thinking
[19]	Physics (Heat Transfer)	Senior High School	Conceptual Understanding
[20]	Physics (Solar system)	Junior high school	Achievement, Misconception
[21]	Physics (Heat Transfer)	University	Conceptual Understanding
[22]	Physics (Magnetic Field)	University	Motivation, Learning Outcomes
[23]	Physics (Electricity)	University	Attitude

Study	Subject	School level	Topic
[24]	Physics (Magnetic Field)	University	Attitude, Learning Outcomes
[25]	Biology (Ecology)	Elementary School	Achievement, Attitude
[26]	Biology	Secondary School	Motivation
[27]	Physics (Electrical Machines)	University	Collaborative, Autonomous Learning
[28]	Chemistry (Atoms)	Junior high school	Attitude
[29]	Natural Science	Elementary School	Achievement, Motivation
[30]	Physics (Electromagnetics)	High school	Attitude
[31]	Physics (Elastic collision)	University	Learning outcomes
[32]	Astronomical	Elementary School	Conceptual knowledge
[33]	Physics (Lens)	High school	Attitude, motivation, attention
[34]	Physics (Space)	University	Knowledge Construction
[35]	Physics (Electromagnetics)	High school	Achievement, motivation
[36]	Physics (Newtonian force and motion)	Primary school	Learning outcomes

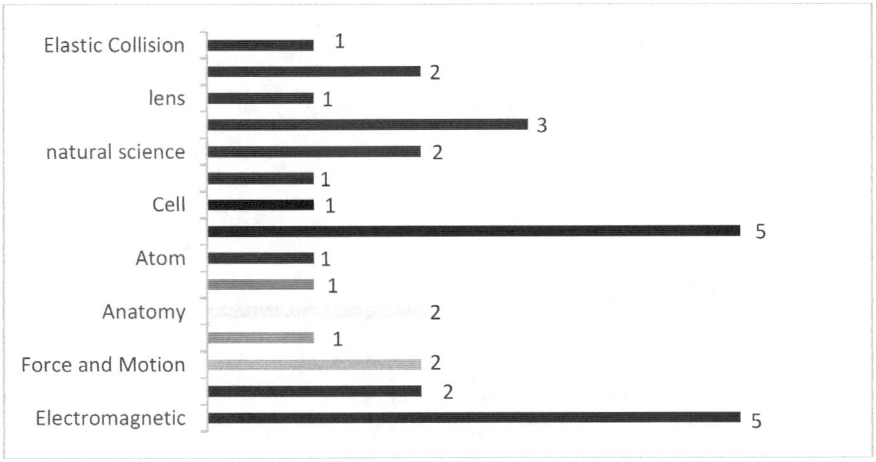

Figure 1. Trends of Research Subject in Science Learning

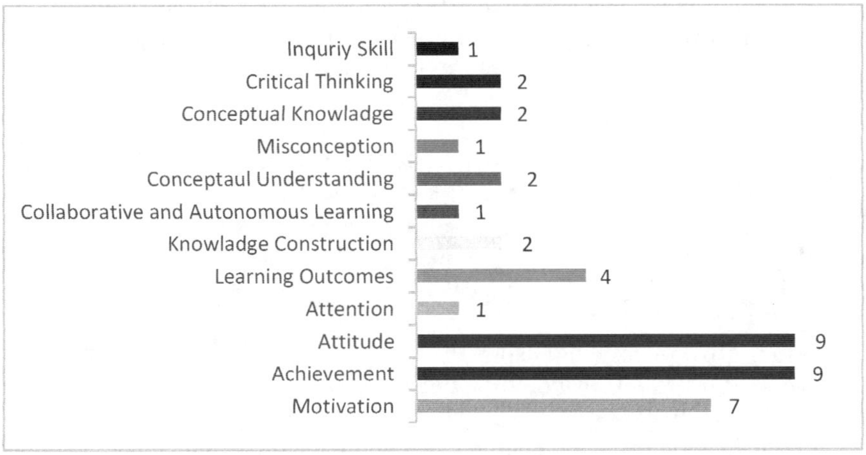

Figure 2. Tren of Research Articles on the use of AR for Several Variables in Science Learning from 2012 – 2021

Based on Figure 1 and 2, it can be concluded that the research trend of using AR in science learning from 2012-2019 is on the subject of electricity and electromagnetic. Wheres for the topic of attitude and achievement which can have a positive impact on science learning. While on the topic of inquiry skills, misconceptions, and attention have not been studied by many researchers. This is interesting for further research.

3.3 Challenges of Using Augmented Reality in Science Learning

Based on the results of the analysis that has been done, there are several studies that do not write about the difficulties in using augmented reality in science learning and only focus on the benefits of its implementation. However, there are also some limitations to using AR for educational purposes. The technical problems experienced when using AR is the main and most important limitation [39] . The use of location-based AR is generally limited by the technical problems experienced with GPS [38] The use of marker-based AR is limited by the technical problems experienced with perceiving the markers [31]. Other limitations include the lack of adequate information from teachers to develop AR materials [37]. In addition, there are challenges faced by students regarding students who are not accustomed to using augmented reality media in learning [32, 34]. Based on the results of the analysis that has been made the use of augmented reality in science learning is grouped into three categories, namely student difficulties, teacher difficulties, and technical difficulties as shown in table 4.

Table 4. Challenges of Application of Augmented Reality in Science Learning

Challenge	Challenge	Article Example
Student	Unaccustomed to using and difficulty in using it	[32, 34, 37]
Teacher	Lack of information in developing AR	[37]
Technical	GPS and markers	[37-39]

4. Conclusion

This study examines 30 articles on augmented reality on science learning indexed in the ScienceDirect and Google Scholar databases using content analysis methods. Data analysis set out to establish trends in augmented reality studies of science learning. The number of educational AR studies has increased over the years. The importance of using Augmented reality for educational purposes and the increasing number of AR studies will continue in the years to come. It can be said that the results of this systematic review are important to guide future studies.

The results show that the research trend of using AR in science learning from 2012-2019 is on the topic of attitude and achievement that can have a positive impact on science learning. Apart from the advantages of using augmented reality in science learning, there are challenges for students, teachers, and technical challenges.

References

[1] Bromme, R., & Beelmann, A. (2016). Transfer Entails Communication: The Public Understanding of (Social) Science as a Stage and a Play for Implementing Evidence-Based Prevention Knowledge and Programs. Prevention Science, 19(3), 347–357.

[2] Azuma, R., Baillot, Y., Behringer, R., Feiner, S., Julier, S., & MacIntyre, B. (2001). Recent Advances in Augmented Reality. IEEE Comput. Graph. Appl., 21(6), 34–47. http://doi.org/10.1109/38.963459

[3] Pai, Y. S., et al (2016). Armswing: using arm swings for accessible and immersive navigation in AR/VR spaces. *Proceedings of the 16th International Conference on Mobile and Ubiquitous Multimedia.* ACM (189-198) doi: org/10.1145/3152832.3152864

[4] K. Lee, "Augmented Reality in Education and Training," TechTrends, vol. 56, no. 2012, pp. 13–21, 2012

[5] Kucuk, S., Aydemir, M., Yildirim, G., Arpacik, O., & Goktas, Y. (2013). Educational technology research trends in Turkey from 1990 to 2011. Computers & Education, 68, 42–50.

http://doi.org/10.1016/j.compedu.2013.04.016

[6] Pimmer, C., Mateescu, M., & Gröhbiel, U. (2016). Mobile and ubiquitous learning in higher education settings. A systematic review of empirical studies. Computers in Human Behavior, 63, 490–501. http://doi.org/10.1016/j.chb.2016.05.057

[7] H. Faridi, N. Tuli, A. Mantri, G. Singh, and S. Gargrish, "A framework utilizing augmented reality to improve critical thinking ability and learning gain of the students in Physics," *Comput. Appl. Eng. Educ.*, vol. 29, no. 1, pp. 258–273, 2021, doi: 10.1002/cae.22342.

[8] O. Article, "Determining the Effect of Science Teaching Using Mobile," vol. 32, no. 2, pp. 137–148, 2013.

[9] M. Abdinejad, B. Talaie, H. S. Qorbani, and S. Dalili, "Student Perceptions Using Augmented Reality and 3D Visualization Technologies in Chemistry Education," *J. Sci. Educ. Technol.*, vol. 30, no. 1, pp. 87–96, 2021, doi: 10.1007/s10956-020-09880-2.

[10] K. Altmeyer, S. Kapp, M. Thees, S. Malone, J. Kuhn, and R. Brünken, "The use of augmented reality to foster conceptual knowledge acquisition in STEM laboratory courses—Theoretical background and empirical results," *Br. J. Educ. Technol.*, vol. 51, no. 3, pp. 611–628, 2020, doi: 10.1111/bjet.12900.

[11] S. Y. Chen and S. Y. Liu, "Using augmented reality to experiment with elements in a chemistry course," *Comput. Human Behav.*, vol. 111, p. 106418, 2020, doi: 10.1016/j.chb.2020.106418.

[12] M. Thees, S. Kapp, M. P. Strzys, F. Beil, P. Lukowicz, and J. Kuhn, "Effects of augmented reality on learning and cognitive load in university physics laboratory courses," *Comput. Human Behav.*, vol. 108, p. 106316, 2020, doi: 10.1016/j.chb.2020.106316.

[13] B. Baran, E. Yecan, B. Kaptan, and O. Paşayiğit, "Using augmented reality to teach fifth grade students about electrical circuits," *Educ. Inf. Technol.*, vol. 25, no. 2, pp. 1371–1385, 2020, doi: 10.1007/s10639-019-10001-9.

[14] C. Erbas and V. Demirer, "The effects of augmented reality on students' academic achievement and motivation in a biology course," *J. Comput. Assist. Learn.*, vol. 35, no. 3, pp. 450–458, 2019, doi: 10.1111/jcal.12350.

[15] M. Fidan and M. Tuncel, *Integrating augmented reality into problem based learning: The effects on learning achievement and attitude in physics education*, vol. 142. Elsevier Ltd, 2019.

[16] H. O. Kapici, H. Akcay, and T. de Jong, "Using hands-on and virtual laboratories alone or together—which works better for acquiring knowledge and skills?," *J. Sci. Educ. ...*, 2019, [Online]. Available: https://link.springer.com/article/10.1007/s10956-018-9762-0.

[17] T. Khan, K. Johnston, and J. Ophoff, "The Impact of an Augmented Reality Application on Learning Motivation of Students," *Adv. Human-Computer Interact.*, vol. 2019, 2019, doi: 10.1155/2019/7208494.

[18] S. C. Chang and G. J. Hwang, "Impacts of an augmented reality-based flipped learning guiding approach on students' scientific project performance and perceptions," *Comput. Educ.*, vol. 125, pp. 226–239, 2018, doi: 10.1016/j.compedu.2018.06.007.

[19] P. Knierim, F. Kiss, and A. Schmidt, "Look Inside: Understanding Thermal Flux through Augmented Reality," *Adjun. Proc. - 2018 IEEE Int. Symp. Mix. Augment. Reality, ISMAR-Adjunct 2018*, pp. 170–171, 2018, doi: 10.1109/ISMAR-Adjunct.2018.00059.

[20] M. Sirakaya and E. K. Cakmak, "The effect of augmented reality use on achievement, misconception and course engagement," *Contemp. Educ. Technol.*, vol. 9, no. 3, pp. 297–314, 2018, doi: 10.30935/cet.444119.

[21] H. Search *et al.*, "Ac ce d M us," 2017.

[22] M. B. Ibanez, A. J. De Castro, and C. D. Kloos, "An Empirical Study of the Use of an Augmented Reality Simulator in a Face-to-Face Physics Course," *Proc. - IEEE 17th Int. Conf. Adv. Learn. Technol. ICALT 2017*, pp. 469–471, 2017, doi: 10.1109/ICALT.2017.105.

[23] M. Akçayir, G. Akçayir, H. M. Pektaş, and M. A. Ocak, "Augmented reality in science laboratories: The effects of augmented reality on university students' laboratory skills and attitudes toward science laboratories," *Comput. Human Behav.*, vol. 57, pp. 334–342, 2016, doi: 10.1016/j.chb.2015.12.054.

[24] G. J. Hwang, P. H. Wu, C. C. Chen, and N. T. Tu, "Effects of an augmented reality-based

educational game on students' learning achievements and attitudes in real-world observations," *Interact. Learn. Environ.*, vol. 24, no. 8, pp. 1895–1906, 2016, doi: 10.1080/10494820.2015.1057747.

[25] S Cai, F K Chiang, Y Sun, C Lin & J J Lee. Applications of augmented reality-based natural interactive learning in magnetic field instruction. *Interactive Learning Environments*, 1744-5191 http://dx.doi.org/10.1080/10494820.2016.1181094

[26] G. W. Ng, Y. B. Oon, H. Y. Lee, and E. H. Teoh, "An Augmented Reality System for Biology Science Education in Malaysia," *Int. J. Innov. Comput.*, vol. 6, no. 2, pp. 8–13, 2016.

[27] J. Martín-Gutiérrez, P. Fabiani, W. Benesova, M. D. Meneses, and C. E. Mora, "Augmented reality to promote collaborative and autonomous learning in higher education," *Comput. Human Behav.*, vol. 51, pp. 752–761, 2015, doi: 10.1016/j.chb.2014.11.093.

[28] S. Cai, X. Wang, and F. K. Chiang, "A case study of Augmented Reality simulation system application in a chemistry course," *Comput. Human Behav.*, vol. 37, pp. 31–40, 2014, doi: 10.1016/j.chb.2014.04.018.

[29] T. H. C. Chiang, S. J. H. Yang, and G. J. Hwang, "An augmented reality-based mobile learning system to improve students' learning achievements and motivations in natural science inquiry activities," *Educ. Technol. Soc.*, vol. 17, no. 4, pp. 352–365, 2014.

[30] M. B. Ibáñez, Á. Di Serio, D. Villarán, and C. Delgado Kloos, "Experimenting with electromagnetism using augmented reality: Impact on flow student experience and educational effectiveness," *Comput. Educ.*, vol. 71, pp. 1–13, 2014, doi: 10.1016/j.compedu.2013.09.004.

[31] H. Y. Wang, H. B. L. Duh, N. Li, T. J. Lin, and C. C. Tsai, "An investigation of university students' collaborative inquiry learning behaviors in an augmented reality simulation and a traditional simulation," *J. Sci. Educ. Technol.*, vol. 23, no. 5, pp. 682–691, 2014, doi: 10.1007/s10956-014-9494-8.

[32] J. Zhang, Y. T. Sung, H. T. Hou, and K. E. Chang, "The development and evaluation of an augmented reality-based armillary sphere for astronomical observation instruction," *Comput. Educ.*, vol. 73, pp. 178–188, 2014, doi: 10.1016/j.compedu.2014.01.003.

[33] S. Cai, F. K. Chiang, and X. Wang, "Using the augmented reality 3D technique for a convex imaging experiment in a physics course," *Int. J. Eng. Educ.*, vol. 29, no. 4, pp. 856–865, 2013.

[34] T. J. Lin, H. B. L. Duh, N. Li, H. Y. Wang, and C. C. Tsai, "An investigation of learners' collaborative knowledge construction performances and behavior patterns in an augmented reality simulation system," *Comput. Educ.*, vol. 68, pp. 314–321, 2013, doi: 10.1016/j.compedu.2013.05.011.

[35] A. Dünser, L. Walker, H. Horner, and D. Bentall, "Creating interactive physics education books with augmented reality," *Proc. 24th Aust. Comput. Interact. Conf. OzCHI 2012*, pp. 107–114, 2012, doi: 10.1145/2414536.2414554.

[36] N. Enyedy, J. A. Danish, G. Delacruz, and M. Kumar, *Learning physics through play in an augmented reality environment*, vol. 7, no. 3. 2012.

[37] Chang, K.-E., Chang, C.-T., Hou, H.-T., Sung, Y.-T., Chao, H.-L., & Lee, C.-M. (2014). Development and behavioral pattern analysis of a mobile guide system with augmented reality for painting appreciation instruction in an art museum. Computers & Education, 71, 185–197.

[38] Cheng, K.-H., & Tsai, C.-C. (2013). Affordances of augmented reality in science learning: Suggestions for future research. Journal of Science Education and Technology, 22(4), 449–462.

[39] Lu, S.-J., & Liu, Y.-C. (2015). Integrating augmented reality technology to enhance childrens learning in marine education. Environmental Education Research, 21(4), 525–541.

Development of Video Education using Problem-Based Learning (PBL) to Support M-Learning on the Kinetic Gas

Raihanati and H Nasbey
Department of Physics Education, Universitas Negeri Jakarta
Jl. Rawamangun Muka, Jakarta 13220, Indonesia
Email: raihanati57@gmail.com, hadinasbey@unj.ac.id

Abstract. The teaching materials in the form of video education using problem-based learning models (PBL) accessible via smartphones to support mobile learning (M-Learning) is developed. the kinetic theory of gases is chosen after analysing the needs of the students in the development of this teaching material. The Research and Development (R&D) and refers to the ADDIE development model (Analysis-Design-Development-Implementation-Evaluation) are used as the method of research. The learning videos that are applied in the smartphone apps conform to the stages of the problem-based learning model. The sequences are basic concept, problem definition, self-learning, knowledge exchange and evaluation. In the activities of the learning process, the steps carried out which are orienting problems, organizing students to learn, guiding individual / group experiences, develop and presenting work, analysing and evaluating problem-solving processes. The teaching material in the form of videos that has been developed is accessible via smart devices (smartphones) in the hope that they can be a source of learning in the process of mobile learning (M-Learning). The videos produced are validated by material experts, media experts and learning experts. The Materials Expert gave a score of 95.83 out of a maximum score of 100 with a very good interpretation. Media experts gave a score of 83.6 out of a maximum score of 100 with a very good interpretation. The learning expert gave a score of 89.4 out of a maximum score of 100 with very good interpretation. In addition to being validated, the resulting video was tested for the feasibility of its use by teachers and students. The teacher's feasibility test scored 90.5 out of a maximum score of 100 with very good interpretation. The students gave a usability score of 87 out of a maximum score of 100 with very good interpretation. Based on the validation and feasibility tests, it can be said that the videos developed as learning material can be used for classroom learning.

1. Introduction

The development of information technology is very rapid in today's world. Its development accelerated from the lower strata to the upper strata of society. The use of digital technology has become commonplace in everyday life, from waking up in the morning to sleeping in the evening [1]. It also has an impact on the learning process of students in school. The use of various digital media in learning in schools provides more meaningful learning. Among the media used are learning videos.

Video education are materials that provide audio and visual elements of good learning messages containing concepts, principles, procedures and knowledge translation theories to help understand learning material. Video is visible and heard (audio-visual) learning material that can be used to convey messages / topics [2]. It is said to be seen to hear because both hearing (audio) and visual / video (visible) elements can be presented simultaneously. Video is learning material that is packaged on videotape and can be viewed through a video / VCD player connected to a television monitor [3]. The use of learning videos has a positive impact on understanding and preventing the spread of disease [4]. A significant positive effect of learning videos is also demonstrated on health knowledge literacy. The use of learning videos increases knowledge about health [5].

The right learning method is also important in addition to the right learning materials. The problem-based learning method is one of the many methods which shows a good influence on the learning process of students. The use of problem-based learning in learning improves students' intra-personal attitudes and teamwork [6]. Problem-based cooperative learning applied in the learning process improves the cooperative skills of students in an online learning environment [7]. A positive impact is also shown in the use of problem-based learning when used to teach physics [8]. In learning, students are encouraged to find themselves and transform complex information, to verify new information with what is already in memory and to develop it into information or capacities appropriate to the environment and time. This context conforms to one of the learning models which uses a scientific approach, namely problem-based learning (PBL).

Therefore, the authors are interested in conducting research on the development of problem-based learning physics learning videos for the high school student, in particular subject of kinetic theory of gases.

2. Method

The research method used in this study is a research and development method with a model developed by Dick and Carry, namely the development model consists of five steps; Analysis (analysis), design (planning), development (development), implementation (implementation) and evaluation (evaluation).

The learning video making mechanism is realized to produce physics learning videos using the steps of the ADDIE model as follows; Literature review and needs analysis, Program review, Selection of material to be presented, Objectives of the learning video, Analysis and preparation of required material, Research and collection of references related, Creation of a production schedule, Realization of scenarios, Realization of storyboards, Taking of photos, recording of stories, selection of background sounds, Evaluation validations, School test, Evaluation of learning videos.

3. Result and Discussion

The medium produced by this development research is a physics video learning the kinetic theory of gases in high school. The material in this learning video includes material on the ideal gas law, ideal gas equations, and equipartition equations. This learning video contains material for the formation of concepts by students, the application of concepts in everyday life and technology, as well as simple experiments that can encourage and motivate students to be able to perform simple experiments. This learning video that has been developed is 15 to 20 minutes long. Most of the videos shown in this learning video are taken from YouTube), but some videos were taken and made by themselves, as one of the hands-on activities.

This video uses a problem-based learning model that has 5 steps, namely orienting students to problems, organizing students to learn, orienting group experiences, presenting work and l 'assessment of the problem-solving process. In the first step, students are faced with a problem about a flying hot air balloon, what factors make the hot air balloon fly. In the next step, the students are divided into several groups to answer the problems that were presented in the video. Then, after the students were divided into several groups, each group observed and analyzed the video containing the experiments in the video. After the students analyzed the video, representatives from several groups presented reports related to the experiences of the video. In the last step, the teacher evaluates the learning by showing an explanatory video related to the experiments that were carried out in the video.

The feasibility of using the developed learning video can be determined by the validation of several experts (materials experts, media experts and learning experts) and high school physics teachers. The validation results obtained serve as a benchmark for reviewing the developed product both in terms of material, video display presentation, language used in storytelling and use of video in the learning process. After that, a test was conducted for the high school students who had studied the material on the kinetic theory of gases to find out whether the resulting product could increase the knowledge of the students by studying the concepts of physics in the material of the kinetic theory of gases. Figure 1 shows the screen capture of the developed learning video. The screen capture shows an experiment with Gay Lussac's law.

Figure 1. The screen capture shows an experiment with Gay Lussac's law.

The results of the validation of the learning video by materials experts showed a percentage of 95.83%. with a very good interpretation of the value on several aspects. The highest percentage of material expert data is presented in the aspects of material accuracy, learning and skill development, and motivation. This shows that the physics learning videos that have been developed already meet the requirements as learning aids. Here are some suggestions given by material experts: The choice of words and phrases in the story should be improved, for example by using too many words now in the narration of the video; Addition of problems to the kinetic theory of gases; Subtract the liquid sample and add the gas sample.

The results of the validation of the learning videos by media experts showed a percentage of 83.6% with very good and good value interpretations. The highest percentage of data on media experts is presented in the experience aspect. This shows that the physics learning video that was developed met the requirements as a good learning aid in terms of appearance. videos. The advice given by media experts is that more videos are made by themselves.

The results of the validation of the learning videos by learning experts showed a percentage of 89.4% with very good value interpretations in all aspects. The highest percentage of data on learning experts is shown in the motivation aspect. In this aspect, there are 2 indicators, namely to help students to conduct experiments and to prove abstract concepts in problem solving. This shows that the physics learning videos that have been developed already meet the requirements as learning aids. Suggestions given by learning experts are as follows: simple experiences of all tools and materials are attempted to be measurable; Dubbing and video must be in sync.

The results of the validation of the learning videos by the high school/vocational physics educators show a percentage of 90.5% with very good value interpretations in all aspects. The highest percentage of data by secondary / vocational physics teachers is shown in the material coverage aspect. In this aspect, there are indicators to encourage and motivate students in learning. This shows that the physics learning videos that have been developed already meet the requirements as learning aids. The suggestions given by high school physics educators / MA are, when a video on energy equipment can combine water and oil in order to stimulate students' curiosity; The service life must be distinguished between the formula and the image (illustration); Formula display must exist when read aloud.

4. Conclusion

The evaluation by materials experts, media experts and learning experts on the kinetic gas theory physics learning videos is very good, with average scores of 95.83%, 83 , 6% and 89.4%, respectively. And supported by data on the average test score of 70% of students with good interpretation. Based on the results of the development carried out, it can be concluded that this physics learning video using the problem-based learning model meets the criteria and can be used as learning material for high school students.

Acknowledgments
This research financially was supported by Ministry of Research, Technology and Higher Education of Indonesia, "Dana BLU Pok Fakultas Matematika dan Ilmu Pengetahuan Alam Universitas Negeri Jakarta, Nomor : 30 /SPK PENELITIAN/5.FMIPA/2021, 11 May 2021".

References

[1] Spiteri, Marthese; Rundgren, Chang; Shu-Nu 2018 *Literature Review on the Factors Affecting Primary Teachers' Use of Digital Technology* (Technology, Knowledge and Learning) (), –. doi:10.1007/s10758-018-9376-x
[2] Reid, Kimone R.Y.; Reid, Kathryn; Esquivel, Jill Howie; Thomas, S. Craig; Rovnyak, Virginia; Hinton, Ivora; Campbell, Cathy 2019 Using video education to improve outcomes in heart failure. Heart & Lung, (), S0147956319300378–. doi:10.1016/j.hrtlng.2019.05.004
[3] Dalvi, Tejaswini; Wendell, Kristen 2016 Using Student Video Cases to Assess Pre-service Elementary Teachers' Engineering Teaching Responsiveness, *Research in Science Education*, (), –. doi:10.1007/s11165-016-9547-5
[4] Lenczowski E, Tung-Hahn E, Higareda J 2018 Video education to improve recognition of common benign and malignant cutaneous lesions and skin cancer prevention in the public International Journal of Women's Dermatology 4(2) 80 – 82.
[5] Laurie A F 2012 Implementing a Video Education Program to Improve Health Literacy The Journal for Nurse Practitioners 8(8) e17-e22.
[6] Dawn M , Bos 2020 Use of problem based learning activity to improve attitudes regarding inter-professional collaborative teamwork in mental health nursing students: An evaluation study Nurse Education in Practice 49 102908.
[7] Kawita P, Panita W 2015 Design of Cooperative Problem-based Learning Activities to Enhance Cooperation Skill in Online Environment Procedia - Social and Behavioral Sciences 174 2184-2190.
[8] Pinar C, Fatih O, Ilhan S 2011 The effects of problem-based learning on the students' success in physics course Procedia - Social and Behavioral Sciences 28 656-660.

Development of Educational Adventure Game on Fluid Physics Material

D Muliyati[1], N Wahdaniyah and F Bakri
Department of Physics Education, Universitas Negeri Jakarta
Jl. Rawamangun Muka, Jakarta 13220, Indonesia
Email: dmuliyati@unj.ac.id

Abstract. An educational game is a form of technological progress that can be implemented in the field of education. This article describes the results of research on the development of educational games that can be used as an alternative mentoring for students. The method used in this research is the Research and Development (R&D) method with the ADDIE approach, which includes five stages, namely Analysis, Design, Development, Implementation, and Evaluation. An HTML5-based educational game has been created using Articulate Storyline 3. This educational game contains a person's journey with the background of applying static and dynamic fluids and has several levels containing fluid for the next round. Material experts and media experts have validated the educational games produced. Experts provide a comprehensive opinion on all aspects assessed.

1. Introduction

An educational game is one form of technological progress that can be implemented in the field of education. Initially, games were often considered to give negative stigma to children, but in fact, games as a means of education are not wrong because games are entertaining [1]. Educational games can increase students' learning, motivation [2]. Educational games can make learning materials enjoyable because they are packaged so that students can feel the sensation of playing while learning [3]. Educational games are effective as a companion for learning in the classroom. The research conducted shows the results if learning using educational games shows better learning outcomes [4].

Characteristics of educational games, namely, fantasy, rules and goals, sensory stimuli, challenges, mystery, and control [5]. Therefore, one of the main advantages of educational games is the visualization of real problems [6]. Using educational games can make the fact that the subject matter is blurred with the characteristics of the game and the interaction between students and the game [7]. From the game, players, consciously or not, are also trained to complete every mission in the game. It requires several skills such as literacy skills, counting, coordinating, setting strategies, and many more [8].

One of the current interactive media innovations is Articulate Storyline software because it is easier to use, but can compete with Adobe Flash media [9]. Global Incorporation is the maker of Articulate Storyline software that can be used to create interactive learning media. The resulting output is a format that can be used for web-based media or HTML5, tablet users, Mac iOS users, Android users, and PC users [10]. Its features make it easier for users to use it. The complete features of Articulate Storyline so that it can compete with Adobe Flash and an easy interface, such as using PowerPoint makes Articulate Storyline an option for creating interactive multimedia [11].

Based on the description above, this article describes the results of research on the development of educational adventure games on static fluid and dynamic fluid physics material.

2. Method

2.1. Research and Design

This research is qualitative research that is descriptive and tends to use analysis. The research method used in this research is the research and development (R&D) method, namely by developing a product and testing the effectiveness [12]. The method used in this research is the ADDIE research and development method. Based on Instructional design: The ADDIE approach, the ADDIE stage consists of five main stages, namely Analyze, Design, Development, Implementation, and Evaluation [13].

2.1.1. Analysis

The stage to analyze the need for product development and analyze the requirements of development. At this stage, a needs analysis and literature study is carried out.

2.1.2. Design

At this stage, a design will be carried out in the development of educational games with HTML5 output that can be accessed via a browser.

2.1.3. Development

At this stage, begins to carry out product realization.

2.1.4. Implementation

At this stage, will be carried out in a limited manner at the designated school at the research site. To produce a decent product, of course, it is necessary to conduct a trial of the product.

2.1.5. Evaluation

This stage is the researcher's final revision stage of educational games to improve product quality. The final result of this research is an educational game on the fluid physics material.

2.2. Instrumentation Research

To test the feasibility of the research results on the development of educational games, an instrument in the form of a questionnaire was used. Questionnaires that have been developed include a questionnaire to test the feasibility of the media and a questionnaire to test the feasibility of the material.

2.2.1. Media Eligibility Instrument

Media feasibility is assessed on five aspects: the aspect of the storyline presented, the element of interactivity, the design aspect of the game, the aspect of the appearance of the game and the part of the ease of operation of the game.

2.2.2. Material Eligibility Instrument

The feasibility of the material is assessed on three aspects, namely the aspect of the presentation of the material, the element of the content of the material and the element of the quality of the questions provided

3. Results and Discussion

3.1. Product Research

The development research carried out has produced a product in the form of an educational game. This educational game contains a journey related to the fluid physics material. In this game, there are several stages as follows:

3.1.1. Main Menu

The main menu contains a start page, and players must fill in their names first to proceed to the next stage.

(a) (b)

Figure 1. Main menu, (a) start page, (b) login page

3.1.2. Map Page

This educational game consists of five levels related to static and dynamic fluid physics material.

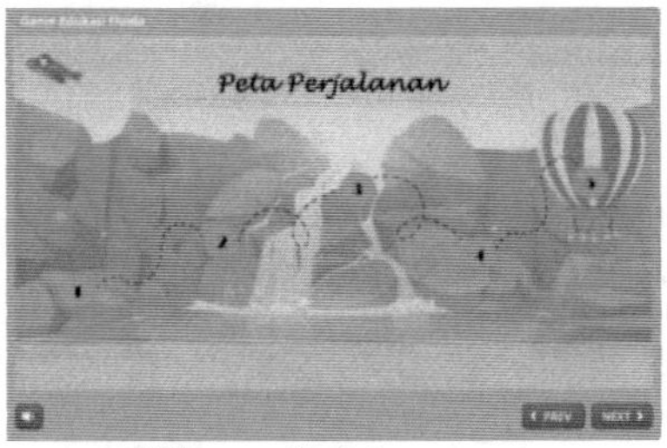

Figure 2. Map page

3.1.3. Journey

At each level, there will be a journey. The player must pass this journey to get to the next level.

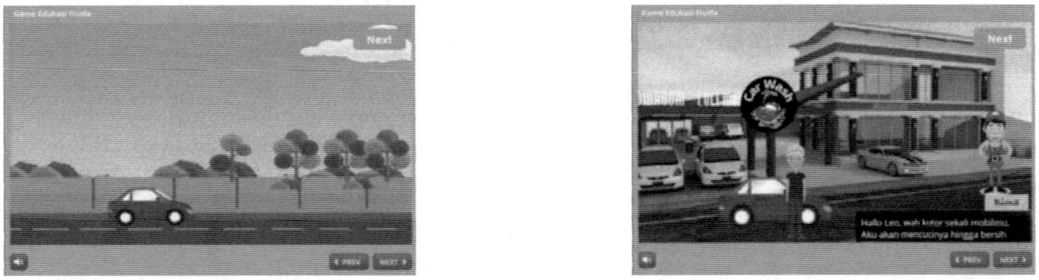

(a) (a)

Figure 3. Journey, (a) journey process, (b) arrives at destination

3.1.4. Question

At the end of the journey, there will be several questions related to static and dynamic fluid physics material. Players must answer these questions.

(a) (b)

Figure 4. The question

3.1.5. Feedback
In this educational game, there is feedback on player answers. It consists of input on correct answers, feedback on wrong answers, and scores obtained while playing educational games.

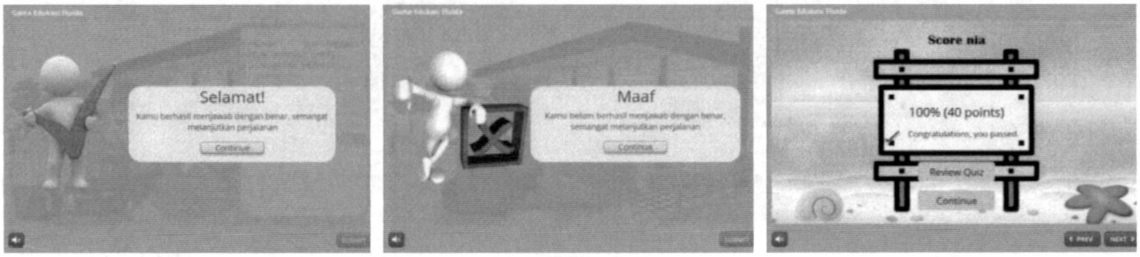

(a) (b) (c)

Figure 5. (a) feedback when answering correctly, (b) feedback when answering incorrectly, (c) the score received

3.2. Description of Validation
The development of adventure-type educational games on fluid physics, material using articulate storylines has passed the feasibility test stage by media experts and material experts. Interpretation of validation data made qualitative level intervals which are presented in Table 1. The assessment indicators and validation results by experts are shown in Table 2.

Table 1. Interpretation Score

Percentage	Interpretation
$20\% \leq score < 40\%$	Bad
$40\% \leq score < 60\%$	Quite Good
$60\% \leq score < 80\%$	Good
$score \geq 80\%$	Very Good

Table 2. Validation Results by Media Experts

No.	Aspect	Score Average	Interpretation
1.	Storyline	73,33%	Good
2.	Interactivity	100%	Very Good
3.	Design	100%	Very Good
4.	Display	100%	Very Good
5.	Convenience	73,33%	Good
Average of All Aspects		89,33%	Very Good

Table 3. Validation Results by Material Experts

No.	Aspect	Score Average	Interpretation
1.	Material Presentation	86,67%	Very Good
2.	Theory	100%	Very Good
3.	Questionable Quality	100%	Very Good
Average of All Aspects		95,33%	Very Good

3.3. Discussion
This research succeeded in developing an educational game with a fluid physics material, which was considered very feasible—making this game using Articulate Storyline software, which can be accessed on the browser.

4. Conclusion
This research has produced a physics educational game on the fluid physics material. Based on the formative evaluation results, the resulting educational game was considered very feasible to be used as a learning companion for students in fluid physics material.

References
[1] N. Whitton and M. Maclure, *Discourse*, vol. 38, no. 4, pp. 561–572, 2017.
[2] S. M. Alessi and S. R. Trollip, *Eur. J. Educ. Stud.*, 2001.
[3] Y. Siswanto and B. E. Putra, *Ranc. Bangun APL. Mob. Game Edukasi Ilmu Pengetah. Alam Untuk Anak Kelas VI Sekol. Dasher*, vol. 5, no. 4, p. 6, 2013.
[4] S. Hidayatulloh, H. Praherdhiono, and A. Wadi, *JKTP J. Kaji. Teknol. Pendidik.*, vol. 3, no. 2, pp. 199–206, 2020.
[5] M. Rothschild, *US Dep. Educ.*, pp. 61–70, 2008.
[6] M. Prensky, in *From Digital Natives to Digital Wisdom: Hopeful Essays for 21st Century Learning*, 2014.
[7] M. Pivec, O. Dziabenko, and I. Schinnerl, *Proc. I-KNOW '03*, 2003.
[8] D. Clark, "Games and Learning Games and Learning," 2006.
[9] S. Purnama and I. G. P. Asto B, *J. Pendidik. Tek. Elektro*, 2014.
[10] S. Harnett, *Learning Articulate Storyline*. 2013.
[11] Darnawati, L. Batia, Irawaty, and Salim, *J. Pengabdi. Kpd. Masy.*, 2019.
[12] Sugiyono, in *METODE PENELITIAN ILMIAH*, 2014.
[13] R. M. Branch, *Instructional design: The ADDIE approach*. 2010.

Journal of Physics: Conference Series **2019** (2021) 012063 doi:10.1088/1742-6596/2019/1/012063

Analysis of Phase Change of $BaTiO_3$ Ferroelectric Material from a Tetragonal Crystal Structure ($BaTiO_3$) to Orthorhombic ($Ba_4Ti_{13}O_{30}$) to Monoclinic ($Ba_6Ti_{17}O_{40}$)

Irzaman[1]
[1] Department of Physics, IPB University, Campus IPB Dramaga, Bogor 16680, Indonesia
Corresponding author: irzaman@apps.ipb.ac.id

Abstract. $BaTiO_3$ ferroelectric material is very interesting to study, because $BaTiO_3$ ferroelectric material has spontaneous polarization properties when given an electric field from outside the system. This spontaneous polarization property of $BaTiO_3$ ferroelectric material can be applied as a light sensor (ferroelectric properties), temperature sensors (pyroelectric properties), pressure sensors (piezoelectric properties). Has successfully analyzed the phase change of $BaTiO_3$ ferroelectric material from a tetragonal crystal structure ($BaTiO_3$) to orthorhombic ($Ba_4Ti_{13}O_{30}$) to monoclinic ($Ba_6Ti_{17}O_{40}$) using the Cramer-Cohen method. This shows that the Cramer-Cohen method is very accurate in calculating the lattice constants of the crystal structure of ferroelectric materials and other materials.

1. Introduction

$BaTiO_3$ ferroelectric material is very interesting to study, because $BaTiO_3$ ferroelectric material has spontaneous polarization properties when given an electric field from outside the system. This spontaneous polarization property of $BaTiO_3$ ferroelectric material can be applied as a light sensor (ferroelectric properties), temperature sensors (pyroelectric properties), pressure sensors (piezoelectric properties)[1 - 6].

Previous researchers have succeeded in developing analysis of lattice parameters and phase change in crystal structures using the Cohen method, including: Cohen's method revisited [7], Determination of lattice parameters with the aid of a computer [8], Crystallographically engineered $BaTiO_3$ single crystals for high-performance piezoelectrics [9], Confirmation of MB-type monoclinic phase in PMNTO: A powder neutron diffraction study [10], Precision Lattice Constant Determination [11], The Crystal Structure of the Monoclinic Form of n-Hexadecanol [12], A simple method for the determination of lattice parameters from powder X-ray diffraction data [13], A Modification of the Cohen Procedure for Computing Precision Lattice Constants from Powder Data [14], Android based XRD data analysis software design for cube crystal structure with analytic and cohen methods [15].
The purpose of this study was to Analysis of Phase Change of $BaTiO_3$ Ferroelectric Material from a Tetragonal Crystal Structure ($BaTiO_3$) to Orthorhombic ($Ba_4Ti_{13}O_{30}$) to Monoclinic ($Ba_6Ti_{17}O_{40}$) using Cramer-Cohen Method.

2. Method

There are two steps in this program, including: (i) the first step is to implement the Cramer-Cohen method as an algorithm in the excel program, (ii) The second step is to test the performance of the excel program in performing calculations by comparing the results with the actual lattice parameter

Journal of Physics: Conference Series **2019** (2021) 012063 doi:10.1088/1742-6596/2019/1/012063

data. The International Center for Diffraction Data (ICDD) was referred to obtain sample lattice parameter data.

The algorithm used in this excel *program* is the Cramer-Cohen method. When the program runs, the program will ask the user to type in the code of crystal structure. After that, the program will ask the user to type in the number of rows of x-ray diffraction data they have along with the diffraction angles and miller indexes one by one.

The "MDETERM" command from the Excel program facility can solve the value of the Determinant of the Cramer-Cohan Method of any matrix of order n x n and its application to obtain the lattice parameter values of the crystal structure: (i) Tetragonal (order 3 x 3) for $BaTiO_3$, (ii) Orthorhombic (order 4 x 4) for $Ba_4Ti_{13}O_{30}$, (iii) Monoclinic (order 5 x 5) for $Ba_6Ti_{17}O_{40}$.

3. Result and Discussions

3.1. Bragg's law

Bragg's law is the main principle used in the x-ray diffraction method. Two x-rays hitting some parallel lattice planes must interfere each other. Constructive interference happens when the difference path length of two rays is equal to multiple of positive integers of the x-ray wavelength used. Bragg's law is mathematically formulated in equation (1)(1),

$$2d\sin\theta = m\lambda \tag{1}$$

where d is the distance between the parallel lattice planes [16]. For analysing two rays in two neighboured parallel planes, substitute m = 1, so the equation (2)(2) is obtained.

$$\frac{1}{d} = \frac{2\sin\theta}{\lambda} \tag{2}$$

3.2. XRD data analysis for tetragonal crystal structure

A tetragonal crystal structure is a crystal structure with the unit cell dimensions are equal on two axis only, but all the axis of the unit cell remain perpendicular to each other. Therefore, a tetragonal crystal has 2 output lattice parameter which is usually denoted as '*a*' and '*b*'. The distance between the lattice planes in the tetragonal crystal is shown in equation (3)**Kesalahan! Sumber referensi tidak ditemukan.**,

$$\frac{1}{d^2} = \frac{h^2+k^2}{a^2} + \frac{l^2}{c^2} \tag{3}$$

with h, k, and l are unit cell dimensions on x, y and z axis, respectively [17]. Equation **Kesalahan! Sumber referensi tidak ditemukan.** is then compared with equation (2), so equation (4)**Kesalahan! Sumber referensi tidak ditemukan.**, (5) are obtained.

$$\frac{h^2+k^2}{a^2} + \frac{l^2}{c^2} = \frac{4\sin^2\theta}{\lambda^2} \tag{4}$$

$$\sin^2\theta - \frac{\lambda^2}{4}\left(\frac{h^2+k^2}{a^2} + \frac{l^2}{c^2}\right) = 0 \tag{5}$$

By replacing zero in the equation above with error during measurement that is expressed as $F\sin^2 2\theta$, equation (6), (7), (8) are obtained,

$$\sin^2\theta - \frac{\lambda^2}{4a^2}\left(h^2+k^2\right) - \frac{\lambda^2}{4c^2}\left(l^2\right) = F\sin^2 2\theta \tag{6}$$

$$\sin^2\theta = \frac{\lambda^2}{4a^2}\left(h^2+k^2\right) + \frac{\lambda^2}{4c^2}\left(l^2\right) + F\sin^2 2\theta \tag{7}$$

$$\sin^2\theta = C\alpha + B\gamma + A\delta \tag{8}$$

with $C=\dfrac{\lambda^2}{4a^2}, B=\dfrac{\lambda^2}{4c^2}, A=\dfrac{F}{10}, \alpha=h^2+k^2, \gamma=l^2$, and $\delta=10\sin^2 2\theta$. Transform equation (6), (7), (8) **Kesalahan! Sumber referensi tidak ditemukan.** into a linear matrix equation (order 3 x 3) shown in equation (9)**Kesalahan! Sumber referensi tidak ditemukan.**.

$$\begin{pmatrix}\Sigma\alpha\sin^2\theta\\ \Sigma\gamma\sin^2\theta\\ \Sigma\delta\sin^2\theta\end{pmatrix}=\begin{pmatrix}\Sigma\alpha^2 & \Sigma\alpha\gamma & \Sigma\alpha\delta\\ \Sigma\alpha\gamma & \Sigma\gamma^2 & \Sigma\gamma\delta\\ \Sigma\alpha\delta & \Sigma\gamma\delta & \Sigma\delta^2\end{pmatrix}\begin{pmatrix}C\\ B\\ A\end{pmatrix} \tag{9}$$

Solution of the equation (9) can be found using the Cramer method for tetragonal crystal structure.

3.3. XRD data analysis for orthorhombic and monoclinic crystal structure

The same steps as the Transform equation (6), (7), (8) into a linear matrix equation (order 4 x 4) for orthorhombic shown in equation (10):

$$\begin{pmatrix}\Sigma\tau\sin^2\theta\\ \Sigma\phi\sin^2\theta\\ \Sigma\chi\sin^2\theta\\ \Sigma\psi\sin^2\theta\end{pmatrix}=\begin{pmatrix}\Sigma\tau^2 & \Sigma\tau\phi & \Sigma\tau\chi & \Sigma\tau\psi\\ \Sigma\tau\phi & \Sigma\phi^2 & \Sigma\phi\chi & \Sigma\phi\psi\\ \Sigma\tau\chi & \Sigma\phi\chi & \Sigma\chi^2 & \Sigma\chi\psi\\ \Sigma\tau\psi & \Sigma\phi\psi & \Sigma\chi\psi & \Sigma\psi^2\end{pmatrix}\begin{pmatrix}Q\\ R\\ S\\ T\end{pmatrix} \tag{10}$$

and into a linear matrix equation (order 5 x 5) for monoclinic shown in equation (11)**Kesalahan! Sumber referensi tidak ditemukan.**:

$$\begin{pmatrix}\Sigma\sigma\sin^2\theta\\ \Sigma\tau\sin^2\theta\\ \Sigma\phi\sin^2\theta\\ \Sigma\chi\sin^2\theta\\ \Sigma\psi\sin^2\theta\end{pmatrix}=\begin{pmatrix}\Sigma\sigma^2 & \Sigma\sigma\tau & \Sigma\sigma\phi & \Sigma\sigma\chi & \Sigma\sigma\psi\\ \Sigma\sigma\tau & \Sigma\tau^2 & \Sigma\tau\phi & \Sigma\tau\chi & \Sigma\tau\psi\\ \Sigma\sigma\phi & \Sigma\tau\phi & \Sigma\phi^2 & \Sigma\phi\chi & \Sigma\phi\psi\\ \Sigma\sigma\chi & \Sigma\tau\chi & \Sigma\phi\chi & \Sigma\chi^2 & \Sigma\chi\psi\\ \Sigma\sigma\psi & \Sigma\tau\psi & \Sigma\phi\psi & \Sigma\chi\psi & \Sigma\psi^2\end{pmatrix}\begin{pmatrix}P\\ Q\\ R\\ S\\ T\end{pmatrix} \tag{11}$$

Solution of the equation (10) can be found using the Cramer method for orthorhombic crystal structure, and solution of the equation (11) can be found using the Cramer method for monoclinic crystal structure.

3.4. Algorithm of the excel program

The algorithm used in this excel *program* is the Cramer-Cohen method. When the program runs, the program will ask the user to type in the code of crystal structure. After that, the program will ask the user to type in the number of rows of x-ray diffraction data they have along with the diffraction angles and miller indexes one by one.

The "MDETERM" command from the Excel program facility can solve the value of the Determinant of the Cramer-Cohan Method of any matrix of order n x n and its application to obtain the lattice parameter values of the crystal structure: (i) Tetragonal (order 3 x 3) for $BaTiO_3$, (ii) Orthorhombic (order 4 x 4) for $Ba_4Ti_{13}O_{30}$, (iii) Monoclinic (order 5 x 5) for $Ba_6Ti_{17}O_{40}$.

3.5. Evaluation of the program's performance

The program is tested by analysing the diffraction angle and the miller index to determine the lattice parameter of the sample. The literature data of lattice parameter of these samples were obtained from the International Center of Diffraction Data (ICDD) [18, 19, 20].

Table 1. Results of lattice parameter measured by Cramer Cohen methods.

Molecule	Structure	Lattice constants					
		Lietrature (ICDD) [18]			Cramer-Cohen Analyze		
		a (Å)	b (Å)	c (Å)	a (Å)	b (Å)	c (Å)
$BaTiO_3$	Tetragonal	3.994	3.994	4.038	3.986	3.986	4.029
		Lietrature (ICDD) [19]			Cramer-Cohen Analyze		
$Ba_4Ti_{13}O_{30}$	Orthorhombic	14.059	17.065	9.867	13.939	16.913	9.782
		Lietrature (ICDD) [20]			Cramer-Cohen Analyze		
$Ba_6Ti_{17}O_{40}$	Monoclinic	16.930	17.103	9.891	19.245	17.385	10.065

Based on table 1 shows that the result of calculating the lattice constants using the Cramer-Cohen method is close to the literature data, this means that he Cramer-Cohen method is very accurate in calculating the lattice constants of the crystal structure of ferroelectric materials and other materials.

4. Conclusion

Has successfully analyzed the phase change of $BaTiO_3$ ferroelectric material from a tetragonal crystal structure ($BaTiO_3$) to orthorhombic ($Ba_4Ti_{13}O_{30}$) to monoclinic ($Ba_6Ti_{17}O_{40}$) using the Cramer-Cohen method. This shows that the Cramer-Cohen method is very accurate in calculating the lattice constants of the crystal structure of ferroelectric materials and other materials.

Acknowledgments

This research was funded by Hibah Penelitian Terapan Unggulan Perguruan Tinggi (PTUPT) from Deputy of Research and Development, Ministry of Research and Technology/National Research and Innovation Agency of Republic Indonesia with contract number: 1/E1/KP.PTNBH/2021 dated March 8th, 2021.

References

[1] Irzaman, Ridwan S, Aminullah, Irmansyah, Husin A 2016 *Integrated Ferroelectrics* 168 (1) p 130
[2] Ade K, Irzaman, Brian Y, Mochammad Z F and Ferdiansjah 2020 *Ferroelectrics* 544 p 160
[3] Johan I, Renan P J, Ulfah J S, Brian Y and Irzaman 2020 *Ferroelectrics* 544 p 134
[4] Irzaman, Ridwan S, Brian Y, Mochammad Z F and Ferdiansjah 2020 Application of $Ba_{0.5}Sr_{0.5}TiO_3$ (BST) Film Doped with 0%, 2%, 4% and 6% Concentrations of RuO_2 as an Arduino Nano-Based Bad Breath Sensor Chemosensors 8 (3) p 1
[5] Ulfie A, Mochammad N I and Irzaman 2021 *Ferroelectrics* 570 p 162
[6] Jinghui G, Dezhen X, Wenfeng L, Chao Z and Xiaobing R 2017 *Actuators* 6 (24) p 1
[7] Giorgio S and Filippo M 1999 *Powder Diffr.,* 14 (3) p 208
[8] Melvin Mr, Leroy H and Miller K 1960 *Acta Cryst.* 13 p 828
[9] Seung-E P, Satoshi W, Cross L and Thomas S. 1999 *Journal of Applied Physics* 86 (5) p 2746
[10] Akhilesh K S and Dhananjai P 2003 *Physics Review* B 80 p 172103
[11] Bond W 1960 *Acta Cryst.* 13 p 814
[12] Sixten A, Gunnar L and Erik S 1960 *Acta Cryst.* 13 p 770
[13] Tsai, Chin, Hsu and Hung 1989 *Materilas Transaction, JIM.* 30 (7) p 474
[14] Hess J 1951 *Acta Cryst.* 4 p 209
[15] Ade K and Irzaman 2021 *AIP Conference Proceedings* 2320 p 030001

[16] Nguyen T, Sierra E, Eguiraun H and Lizundia E 2018 *European Journal of Physics* 39 (4) p 045803.
[17] Zul J, Idris M S, Irzaman and Mochammad B 2007 *Journal of Nuclear and Related Technology* 4 p 137
[18] Swanson and Fuyat 1954 *Natl. Bur. Stand (US). Circ.* 539 (3)p 45
[19] Tillmms E 1971 *Inorg. Nucl. Chem. Lett.* 7 p 1169
[20] Tillmanns E and Baur W.H 1970 *Acta Crystallog. Sect.* B B26 p 1645

Synthesis of Silicon Dioxide (SiO_2) from Reeds Biomass (Imperata cylindrica) and The Analysis of Electrical Properties (Study Cases: Impedance, Resistance, Capacitive Reactance, Inductive Reactance)

D A P Wardani[1], B Hariyanto[2], N Kurniawati[2], N P Har[3], N Darmawan[4], and Irzaman[3]
[1]Department of Chemistry, FMIPA, University of Palangka Raya, Palangka Raya,Central Kalimantan, Indonesia, 73111
[2]Department of Physics, FMIPA, University of Palangka Raya, Palangka Raya, Central Kalimantan, Indonesia, 73111
[3]Department of Physics, FMIPA, IPB, Campus of IPB Dramaga Bogor, East Java, Indonesia, 16680
[4]Department of Chemistry, FMIPA, IPB, Campus of IPB Dramaga Bogor, East Java, Indonesia, 16680
Email: dayupwardani@mipa.upr.ac.id

Abstract. Silicon dioxide (SiO2) has successfully synthesized from reeds biomass (imperata cylindrical) by annealing temperature variation at 500, 600, 700 oC. The result of the synthesis has analyzed using LCR meters to calculate the electrical properties including resistance (R), reactance (X) (delta inductive reactance (XL) and capacitive reactance (XC)), impedance (Z) in the frequency range at 50 Hz to 5 MHz. The resistance (R), reactance (X) (delta inductive reactance (XL) and capacitive reactance (XC)), impedance (Z) value according to reversible electric current rules.

1. Introduction

Reeds (Imperata cylindrica) are natural vegetation that is very abundant in tropical and subtropical areas. In Indonesia, It has a community of 8.5 million hectares, especially in extended rainy areas. Imperata plants are wild plants and easy to reproduce [1]. Reed is also known as one of the most harmful and difficult to eradicate pest plants. It is caused by controlling the scope of the soil or rhizomes. People use reed for making roofs, briquettes, and livestock, but their use is limited [2]. People also usually use the roots of the reed as medicine, and the leaves dispose of as agro-waste. The ash from the leaves of the reed can use as a material for making silica.

Naturally, silica is found in quartz sand (quartz), opal rock, flint, and gemstones. Silica has the properties of colorless crystals, specific gravity 2.2 -2.6 insoluble in water and acids except for HF, soluble in molten alkali, and has a melting point between 1600 °C-1750 °C and will sublimate at a temperature of 1750 °C [3].

This study aims to study and analyze the ignition temperatures on the silica content of reeds and measure its electrical properties, including resistance (R), reactance (X) (delta inductive reactance (XL) and capacitive reactance (XC)), impedance (Z) in the frequency range at 50 Hz to 5 MHz.

2. Method

The method was carried out in 6 stages: (i) 1 g of reed leaves soaked in 12 ml of 3% technical HCl at 200 °C for 3 hours; (ii) rinsed with aquabides until neutral (pH = 7).); (iii) reed leaves sample was dried in an open space; (iv) reed charcoal calcinated at annealing temperatures of 500 °C, 600 °C, and 700 °C; (v) extraction of SiO2 from reed ash samples dried in the furnace at a temperature of 800 °C with a velocity temperature of 100 °C/hour (calcination process); and (vi) the analysis of electrical properties includes resistance (R), the difference between capacitive reactance and inductive reactance (X), and impedance (Z) using a LCR meter.

3. Result and Discussion

Silicon dioxide from Reed biomass was successfully synthesized and analyzed using XRD at 500 °C, 600 °C, and 700 °C. The XRD results show that the synthesis of silicon dioxide formed is amorphous, as shown in Figure 1 [4].

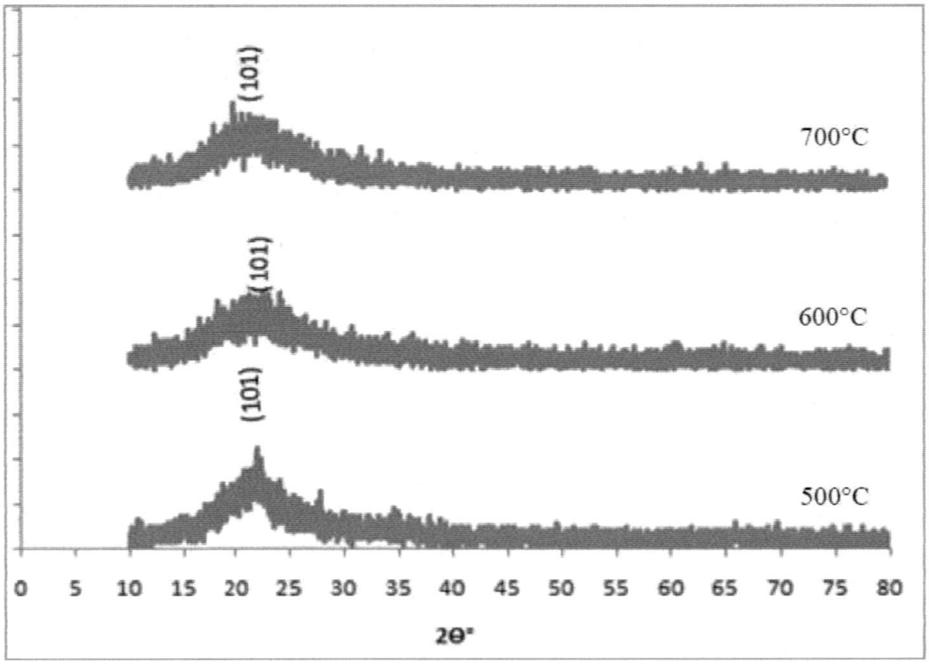

Figure 1. XRD graph of silicon dioxide from cogon grass at ashing temperature of 500 °C, 600 °C, and 700 °C

The result of the synthesis have analyzed using LCR meters to calculated the electrical properties including resistance (R), reactance (X) (delta inductive reactance (XL) and capacitive reactance (XC)), impedance (Z) in the frequency range at 50 Hz to 5 MHz.

3.1. Resistor circuits (R)

A resistor (R) carries alternating current when connected to an alternating voltage source. The resistor circuit alternating current reduces the electric potential in the circuit or limits the incoming electric current. The current and voltage in the resistor circuit have the same phase when connected to an alternating voltage source.

Figure 2 shows the resistance data (R) at frequency (a) 50 Hz to 100.000 Hz (low frequency); (b) 100.000 Hz to 3.000.000 Hz; (c) 3.000.000 Hz to 5.000.000 Hz of SiO_2 from reeds at an annealing temperature of 500 °C (black line), 600 °C (blue line), and 700 °C (red line). It shows the similar patterns, as the increasing frequency causes the resistance value decreases. It probably occurs because the increasing frequency causes the lower current and voltage produced by the resistor of SiO_2 from reeds at annealing temperatures of 500 °C, 600 °C, and 700 °C.

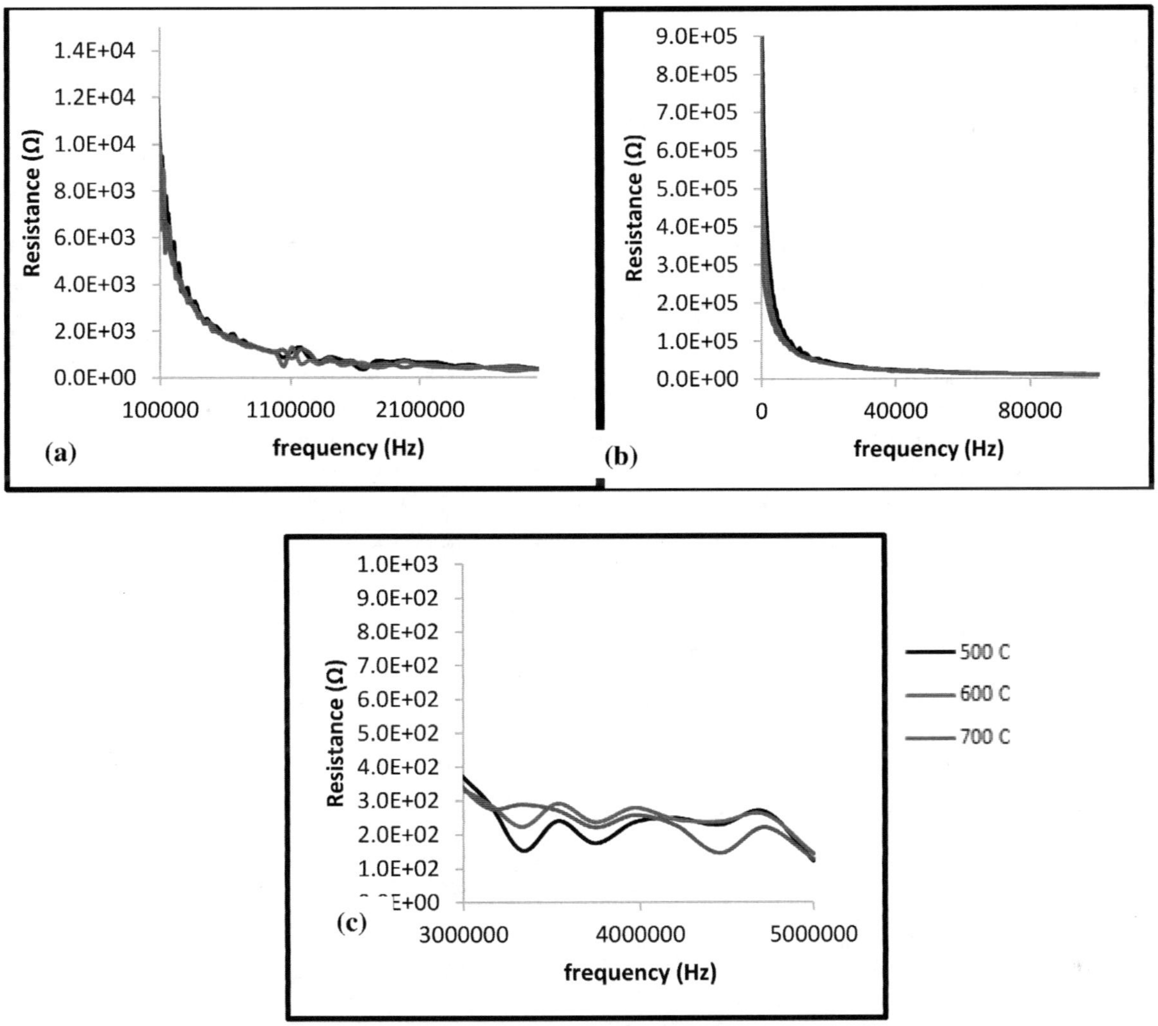

Figure 2. The resistance data at frequency (a) 50 Hz to 100.000 Hz (low frequency); (b) 100.000 Hz to 3.000.000 Hz; (c) 3.000.000 Hz to 5.000.000 Hz of SiO_2 from reeds at an annealing temperature of 500 °C (black line), 600 °C (blue line), and 700 °C (red line).

Silicon has the properties of (a) high mobility, (b) small dielectric constant, (c) large thermal conductivity, (d) good heat dissipation, and (e) low impurity ionization energy. These properties make silicon widely used as a semiconductor material, such as a rectifier diode, thyristor (SCR), and others. Silicon compound, SiO_2 (quartz) is often used in optical instruments with a refractive index of 1.54. Semiconductors have resistance with a negative temperature coefficient, meaning that the resistance of the semiconductor decreases with increasing temperature [5].

3.2. Reactance circuit (X) (difference between inductor reactance (XL) and capacitive reactance (XC)

3.2.1. Inductor reactance (XL)

An inductor has a resistance called inductive reactance (XL) connected to an alternating voltage source. The voltage across the inductor reaches its maximum value a quarter of a period faster than the current. The inductive reactance (XL) depends on the angular frequency of the current and the self-inductance of the inductor and written in equation (1):

$$X_L = \omega L \tag{1}$$

3.2.2. Capacitive reactance (XC)

A capacitor has a resistance called capacitive reactance (XC) connected to an alternating voltage source. The electric current in the inductor reaches its maximum value four times faster than its voltage. The capacitive reactance (XC) depends on the magnitude of the capacitance value of the capacitor and the angular frequency of the current and written in equation (2):

$$X_c = \frac{1}{\omega C} \tag{2}$$

Reactance X is the difference between inductor reactance (XL) and capacitive reactance (XC) when connected to an alternating voltage source.

Figure 3 shows the reactance data (X) (difference between inductor reactance (XL) and capacitive reactance (XC)) at the frequency of (a) 50 Hz to 100.000 Hz (low frequency); (b) 100.000 Hz to 3.000.000 Hz (medium frequency); (c) 3.000.000 Hz to 5.000.000 Hz (high frequency) of SiO_2 from reeds at an annealing temperature of 500 °C (black line), 600 °C (blue line), and 700 °C (red line). It shows the similar patterns, as the increasing frequency causes the reactance (X) value to decrease. It probably occurs because the increasing frequency causes the lower current and voltage produced by the reactance (X) of SiO_2 from reeds at annealing temperatures of 500 °C, 600 °C, and 700 °C.

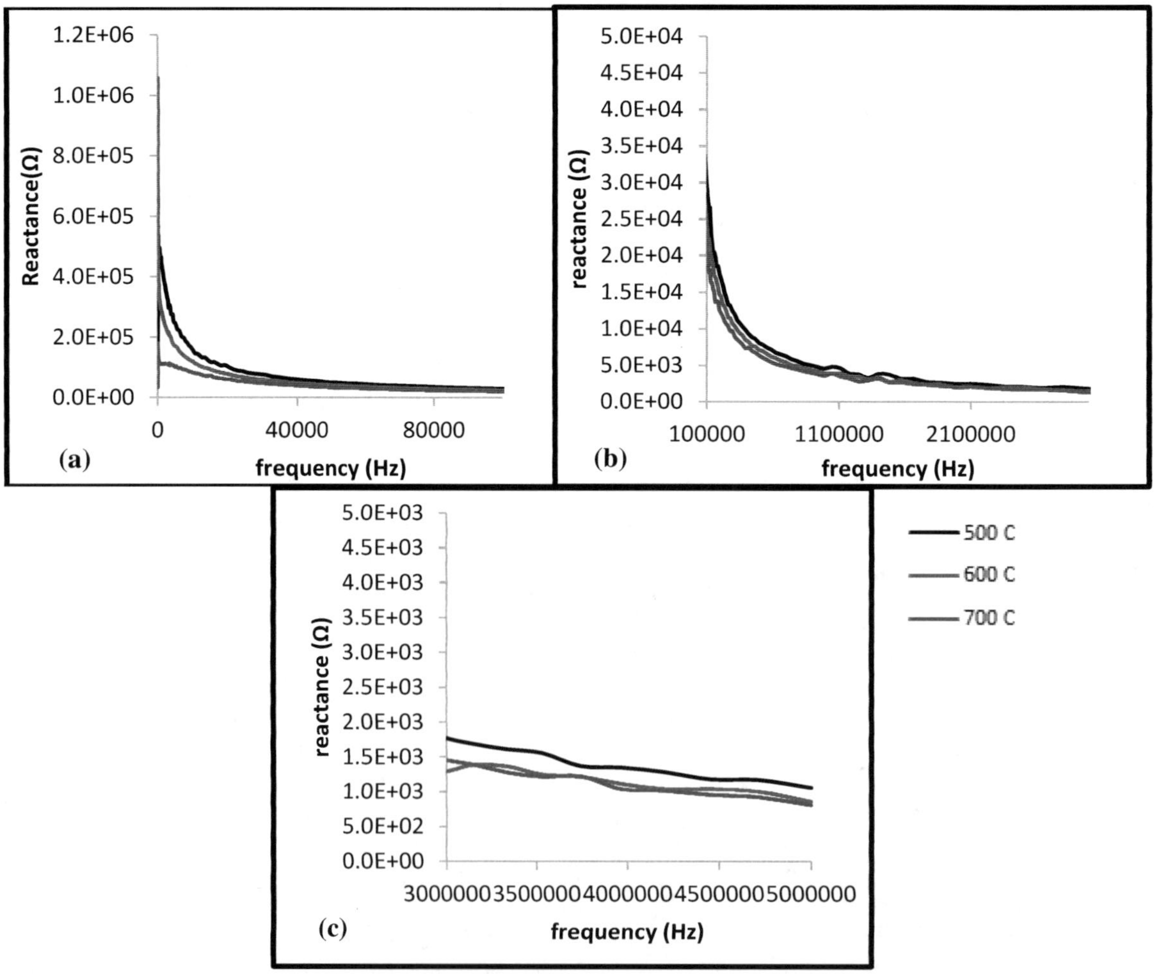

Figure 3. The reactance data (X) (difference between inductor reactance (XL) and capacitive reactance (XC)) at frequency of (a) 50 Hz to 100.000 Hz (low frequency); (b) 100.000 Hz to 3.000.000 Hz (medium frequency); (c) 3.000.000 Hz to 5.000.000 Hz (high frequency) of SiO_2 from reeds at an annealing temperature of 500 °C (black line), 600 °C (blue line), and 700 °C (red line).

The equation or formula for inductive reactance above shows that if either the Frequency or Inductance increases, the Inductive Reactance value will also increase as a whole. If the Frequency is close to infinity, then the value of inductive reactance will also increase to infinity as in an open circuit.

However, if the frequency is close to 0 or as in DC voltage, the inductive reactance will decrease until it reaches 0 in a short circuit. It means the value of Inductive Reactance is proportional to Frequency. In other words, the Inductive Reactance will increase as the frequency increases.

The capacitive reactance is inversely proportional to the Frequency. If the Frequency increases, then the capacitive reactance will decrease and vice versa. Capacitors are equal, and this applies only to ideal capacitors [5].

3.3. Impedance circuits (Z)

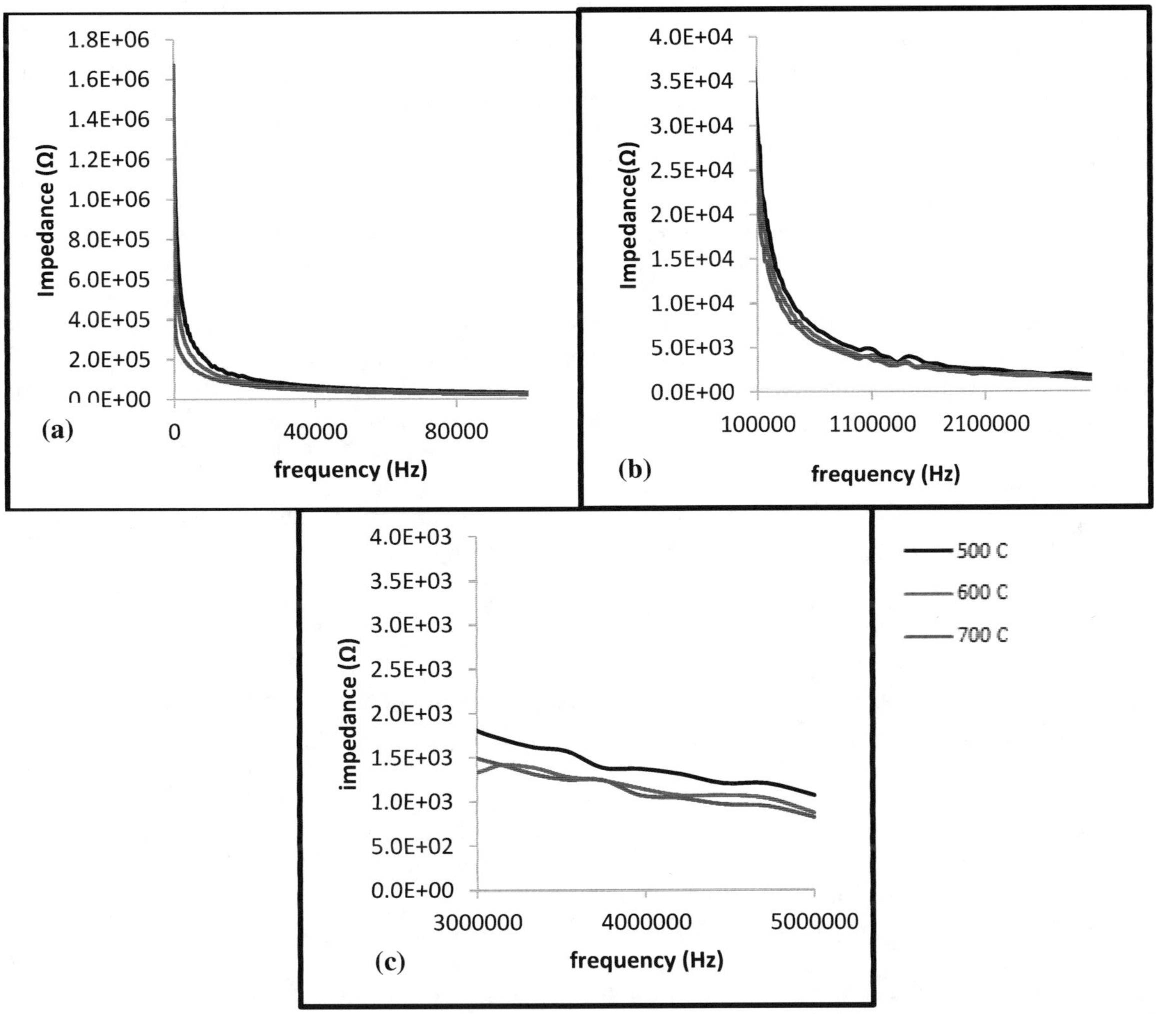

Figure 4. The impedance data (Z) at frequency of (a) 50 Hz to 100.000 Hz (low frequency); (b) 100.000 Hz to 3.000.000 Hz (medium frequency); (c) 3.000.000 Hz to 5.000.000 Hz (high frequency) of SiO_2 from reeds at an annealing temperature of 500 °C (black line), 600 °C (blue line), and 700 °C (red line).

Electrical impedance is the electrical resistance of an electronic component to the current flow in a circuit at a particular frequency. Impedance is fundamentally more complex than resistance (R) due to the effects of capacitive reactance (XL) and inductive reactance (XC), which vary with the frequency of the current passing through the circuit. Electrical Impedance Formula according to equations (3) and (4):

$$Z = \sqrt{}R^2 + (XL - XC)^2 \quad (3)$$

$$Z = \sqrt{}R^2 + (X)^2 \quad (4)$$

The impedance data analysis (Z) in equation (4) combines data from resistance (R) with reactance data (X) in equations (1) and (2), which produces impedance data (Z) according to Figures 3(a), 3(b), and 3(c).

Figure 4 shows the impedance data (Z) at the frequency of (a) 50 Hz to 100.000 Hz (low frequency); (b) 100.000 Hz to 3.000.000 Hz (medium frequency); (c) 3.000.000 Hz to 5.000.000 Hz (high frequency) of SiO2 from reeds at an annealing temperature of 500 °C (black line), 600 °C (blue line), and 700 °C (red line). It shows the similar patterns, as the increasing frequency causes the reactance (X) value to decrease. It probably occurs because the increasing frequency causes the lower current and voltage produced by the reactance (X) of SiO2 from reeds at annealing temperatures of 500 °C, 600 °C, and 700 °C.

The magnitude of the impedance Z of a circuit is equal to the maximum value of the potential difference, or voltage, V (volts) across the circuit, divided by the maximum value of the current I (amperes) through the circuit, or simply Z = V/I. The unit of impedance, like that of resistance, is the ohm. Depending on the nature of the reactance component of the impedance (whether predominantly inductive or capacitive), the alternating current either lags or leads the voltage. It is consistent with the silicon dioxide impedance data from reed biomass. The difference in potential value decreases with increasing frequency perpendicular to the decrease in voltage and electric current [6].

4. Conclusion

The synthesis results have analyzed using LCR meters to calculate the resistance, capacitive reactance, and inductive reactance, the impedance value according to reversible electric current rules. The increasing frequency causes the value decrease of the resistance (R), reactance (X) (difference between inductor reactance (XL) and capacitive reactance (XC)), and impedance (Z).

Acknowledgement

This research was funded by the PKPT DRPM Grant of the Ministry of Education and Culture, Research and Technology of the Republic of Indonesia in 2021.

References

[1] A. F. Ramdja, R. A. Silalahi, and N. Sihombing 2010 *J. Tek. Kim* 7
[2] Z. S. Osvaldo, P. Putra, and M. Faizal 2012 *J. Tek. Kim.* 18
[3] E. Kurniati, 2009 (Surabaya: Penerbit UPN Press)
[4] B. Irzaman, I., Yustaeni, D., Aminullah, A., Irmansyah, I., Yuliarto 2021 *Egypt. J. Chem.*, p 1–10
[5] J. Allison 1990 *Electronics Engineering Semiconductor and Devices*. McGraw-Hill College
[6] T. Britannica 2008 *Encyclopedia Britannica*

Performance Testing of Motorcycle Centrifugal Clutch Lining Made from Composite Wood Powder, Coconut Fibre, and Green Mussel Shell

A Kholil[1], S T Dwiyati, Riyadi, and Randika H P
Department of Mechanical Engineering, Engineering Faculty, Universitas Negeri Jakarta, Kampus A Jl. Rawamangun Muka, Jakarta, 13220, Indonesia.
Email: [1]ahmadkholil@unj.ac.id

Abstract. This study aims to determine the performance of a motorcycle fitted with a CVT centrifugal clutch lining made from a composite of wood powder, coconut fiber and green clam shell powder with epoxy resin as a constituent matrix. The process of making clutch lining is carried out by mixing the composite constituent materials in the form of powder. The specimens were made in two variations where clutch B was 20% wood powder, 20% coconut fibre, 10% green shell powder. Clutch C was 10% wood powder, 30% coconut fiber, and 10% green shell powder. Genuine clutch A was tested by the same method as the comparison. Meanwhile, the performance testing of automatic motorcycle was carried out by a chassis dynamometer to determine power and torque on wheels. Clutch lining with natural composite materials from a mixture of wood powder, coconut fiber, and shell powder have a greater power and torque than the genuine clutch lining. The performance of clutch B is better than C because the composition of B has 10% more wood powder which cellulose that affects the characteristics of natural fiber composites.

1. Introduction

Automatic motorcycles use a centrifugal clutch as part of the power transfer transmission. Centrifugal clutch works by centrifugal force which pushes the clutch to the drum so that the clutch is connected to the pulley. The centrifugal clutch on the automatic motor will transfer kinetic energy from the rotating crankshaft to the CVT transmission and then transmitted to the wheels. The process of merging into a power source in the clutch utilizes the frictional force on the surface of the friction material with a certain coefficient of friction. The frictional force is used to start the driven shaft from rest and gradually bring it to the proper speed without excessive slipping on the friction surface [1].

The clutch serves to disconnect and connect the engine speed to the transmission quickly and gently. The clutch must be resistant to friction and high rotation so that there is no damage when transferring power from engine speed to transmission. Some of the requirements that must be met on the clutch, among others: can continue the engine speed to the transmission; can disconnect and connect the engine speed to the transmission; and can continue the rotation of the engine to the transmission gradually, evenly, and without pounding [2].

The centrifugal clutch is usually incorporated into the motor pulley. Consists of a number of shoes on the inside edge of the pulley, as shown in Figure 1. The outer surface of the shoe is coated with a friction material. These shoes, which can move radially in guides, are held against the boss (or spider) on the drive shaft by means of a spring. The spring exerts a radial inward force which is assumed to be constant. The weight of the shoe, as it rotates, causes it to exert a radial outward force (i.e. Centrifugal force). The magnitude of this centrifugal force depends on the speed of rotation of the shoe. A little consideration will show that when the centrifugal force is less than the spring force, the shoe remains in the same position as when the drive shaft is at rest, but when the centrifugal force equals the spring force, the shoe simply floats. When the centrifugal force exceeds the spring force, the shoe moves outward and comes into contact with the moving parts and compresses it. The increase in speed causes the shoe to press harder and allows more torque to be delivered [3]. Figure 1 shows the parts of the centrifugal clutch.

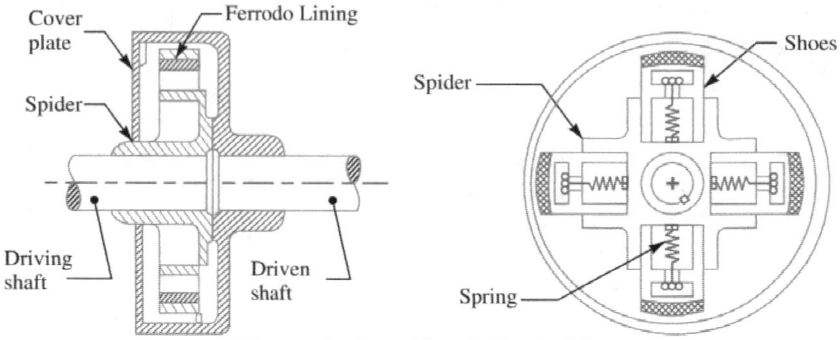

Figure 1. Centrifugal clutch [3].

In general, clutch use asbestos materials which are harmful to human health and the environment [1]. Non-asbestos based friction material is a multi-material system to achieve the desired mixture of performance properties [4]. Natural composite friction material for clutches and brakes can be classified in the following groups: fibers, binders, friction modifiers, and fillers [5]. Composites are to produce better material compared to conventional materials. Composite fibers and particles are embedded in a matrix of other material. The wood powder is a waste produced by wood processing such as sawdust[6]. Wood powder is used because it has a high cellulose content [7,8]. An investigation was carried out on the use of sawdust to develop brake pads [8,9]. The composites prepared from the seashells are tested for various mechanical properties like toughness, tensile strength, impact strength, flexural strength[10]. Green mussel shell wastes consist of calcium carbonate and organic matrices, with the former accounting for 95-99% by weight. Being the richest source of biogenic $CaCO_3$, shell wastes are suitable to prepare high purity $CaCO_3$ powders [11]. The shell waste could be further processed to be the filler of polymer composites [12]. But there are limited studies in mussel shells used as filler in epoxy-based composites. In previous studies, characterization of the alloy clutch material has been carried out, especially the characteristics of hardness, coefficient of friction, wear and microstructure [1,13].

The purpose of this study is to test the performance of an automatic motorcycle using a sample of the centrifugal clutch lining made from natural fiber composites from wood powder, coconut fiber and green mussel powder. The test results are compared with genuine parts centrifugal clutch pads to see the characteristics of power and torque produced against engine speed. This will give an idea of the ability of the clutch to transmit power and torque to the wheels. This research will provide an overview for the application of the clutch lining and also for further research.

2. Materials and Methods

The material used in this study is a natural fiber composite whose ingredients are wood powder, coconut fiber, and green mussel powder which are used as a mixture with a percent by weight. Phenol-formaldehyde resin was used as a binder. The composition of the test sample can be seen in Table 1.

Journal of Physics: Conference Series **2019** (2021) 012065 doi:10.1088/1742-6596/2019/1/012065

Table 1. The composition of sample testing.

No	Sample for testing	Compositions
1	Clutch A	Genuine clutch
2	Clutch B	20% wood powder, 20% coconut fiber, 10% green shell
3	Clutch C	10% wood powder, 30% coconut fiber, 10% green shell

The process of making clutch lining samples refers to a previous study [1]. The pressed material is then prepared for the manufacture of clutch lining samples. The process of preparation of the clutch lining sample can be seen in Figure 2. The cut of the lining material is adjusted to the size of the genuine parts, centrifugal clutch shoe to facilitate assembly. The lining material that is the right size is then attached to the centrifugal clutch shoe using destine glue. Each set of automatic motorcycle centrifugal clutches for this test has three clutch shoes. The clutch that is ready for testing can be seen in Figure 3.

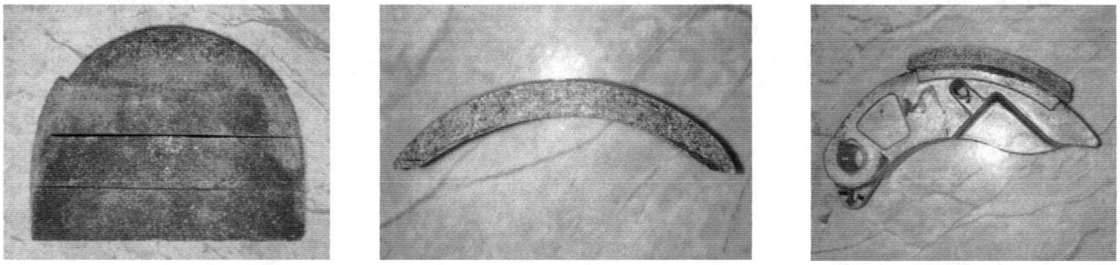

Figure 2. The process of preparing the test sample into a clutch.

(a) Clutch B (b) Clutch C

Figure 3. Specimen of centrifugal clutch.

Performance testing uses test specimens that are applied to a 150 cc motorcycle engine with CVT transmission. Motorcycle performance testing is done using chassis dynamometer [14]. The testing process can be seen in Figure 4. The motorcycle performance test is carried out by varying the clutch lining material and comparing it with genuine parts, centrifugal clutch lining so that the performance characteristics of the power and torque of the motorcycle against engine speed can be seen in the form of a graph. In the performance test with the chassis dynamometer, the rear wheel of the motorcycle is raised above the dynamometer roller and the motorcycle is tied using a safety rope according to the standard operating procedure of the test. Next, the dynamometer is set up by determining the parameters to be displayed in the graph and the data to be generated by the dynamometer. The motorcycle is started and the throttle is opened to the full position until the engine reaches maximum capacity. Torque, power, rpm, and time data from the motorcycle display in the monitor so that they can be viewed and analyzed. The testing process is the same for all clutch lining samples.

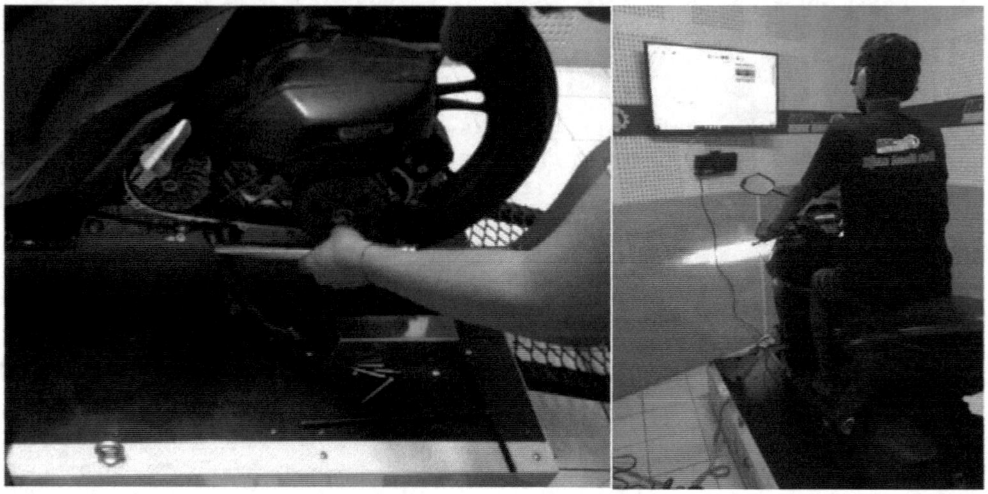

Figure 4. Testing with chassis dynamometer.

3. Results and Discussion

The performance of the dynamometer on the clutch lining sample produces power and torque data. From the test data, the maximum torque on the clutch pad A which is a genuine part has a value of 13.17 Nm at 4698 rpm in 1.98 seconds. While the maximum power on this clutch is 10.3 hp at 6278 rpm in 4.18 seconds. Table 2 shows the overall data obtained from the dynamometer using genuine parts centrifugal clutch pads.

Table 2. Results of genuine parts clutch lining.

Engine speed (rpm)	Power (hp)	Torque (Nm)	Time (s)
4000	4.6	7.65	1.34
4500	7.9	12.56	1.7
4698	8.7	13.17	1.98
5000	9	12.84	2.38
5500	9.7	12.53	3.04
6000	9.8	11.63	3.76
6278	10.3	11.6	4.18
6500	10.2	11.16	4.5
7000	9.6	9.74	5.34
7500	9.1	8.62	6.3
8000	8.7	7.77	7.36
8500	8.8	7.34	8.52
9000	8.5	6.71	9.74

Table 3 shows the overall data using the sample clutch B. The maximum torque on the clutch B is 17.06 Nm at 4034 rpm in 1.76 seconds. While the maximum power on the B clutch is 10.6 hp at 6206 rpm in 4.42 seconds.

Table 3. Results of the clutch lining B.

Engine speed (rpm)	Power (hp)	Torque (Nm)	Time (s)
4000	9.6	17.05	1.72
4034	9.7	17.06	1.76
4500	10.1	15.97	2.24
5000	10.1	14.39	2.8
5500	10.2	13.19	3.44
6000	10.5	12.45	4.12
6206	10.6	12.11	4.42
6500	10.3	11.27	4.82
7000	10.1	10.25	5.64
7500	9.6	9.12	6.52
8000	9.4	8.35	7.5
8500	9.5	7.93	8.56
9000	9.1	7.2	9.7

Table 4 shows the overall data using the sample clutch C. The maximum torque on the clutch C lining is 15.81 Nm at 4391 rpm in 1.86 seconds. While the maximum power on the clutch C is 10.5 hp at 6233 rpm in 4.38 seconds.

Table 4. Results of the clutch lining C.

Engine speed (rpm)	Power (hp)	Torque (Nm)	Time (s)
4000	6.4	11.47	1.4
4391	9.7	15.81	1.86
4500	9.8	15.6	1.96
5000	10.1	14.4	2.54
5500	9.8	12.68	3.18
6000	10.1	12	3.88
6333	10.5	11.69	4.38
6500	10.1	11.08	4.62
7000	9.9	10.11	5.44
7500	9.6	9.15	6.36
8000	9.4	8.38	7.34
8500	9.4	7.89	8.42
9000	9	7.12	9.56

Figure 5 is a graph of power and torque against engine speed in the engine speed range of 4000-9000 rpm. This graph is a plotting of the three results of the clutch lining test using the dynamometer.

The power and torque generated from the test are the power and torque of the motorcycle wheel as the dynamometer test.

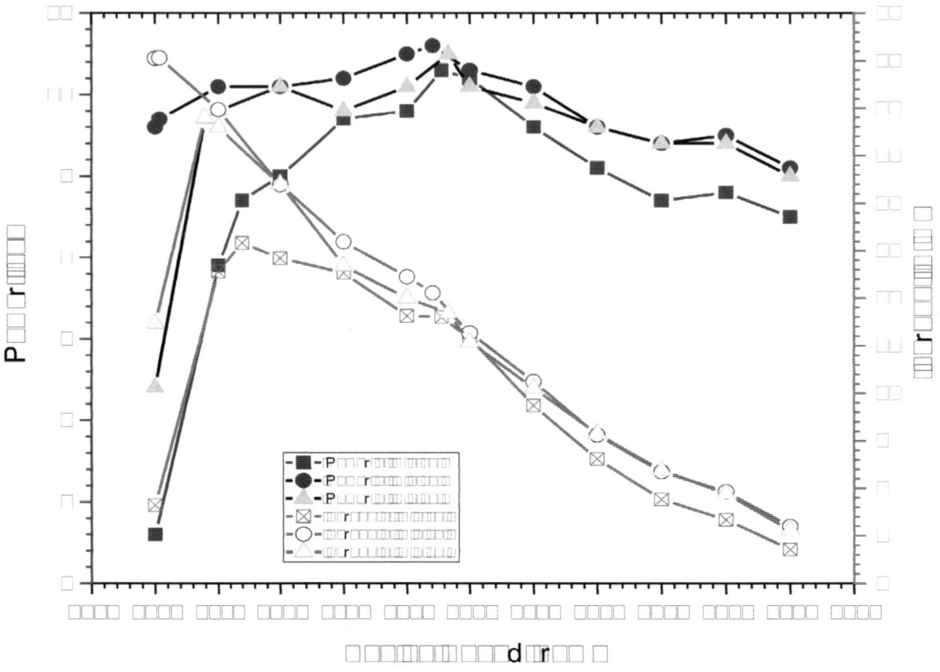

Figure 5. Power and torque of sample test result.

Clutch pads B and C with natural composite materials from a mixture of wood powder, coconut fiber and green mussel shells have higher power and torque values than clutch pads A, which are genuine clutch parts. Clutch B has a mixture of 20% wood powder, 20% coconut fiber, and 10% green mussel shells, while Clutch C has a mixture of 10% wood powder, 30% coconut fiber, and 10% green mussel shells. The power value produced in Clutch B is greater than Clutch C. Both have a different mixture of wood powder and coconut fiber. Differences in the mixture will affect the properties of the coefficient of friction, hardness, and wear. In previous studies, tests have been carried out which indicate that the friction coefficient value of the previous test results is between 0.3 - 0. 4 [1] and the value is in the range of the conventional clutch bearing friction coefficient of the resin mold usually in the range of 0.3 - 0.5 [3]. The composition of the larger wood powder affects the hardness of the composite because it contains a lot of cellulose [7,13]. As research by kholil et al, the hardness value for the composite type of material used in clutch B has a higher hardness value than the type of material used in clutch C [1,13]. In addition, the coefficient of friction of clutch B is also higher than that of Clutch C, thus affecting the dynamometer results. High clutch torque at low rpm is very good for giving the motorcycle thrust so that it can increase speed. The results of this study can provide initial information for further research related to the reliability of the clutch material.

4. Conclusions

Clutch pads B and C with natural composite materials from a mixture of wood powder, coconut fiber and shells have higher power and torque values than clutch pads A, which are genuine clutch parts. The performance of the B clutch lining is better than C because the composition of the material in the B clutch has 10% more wood powder containing cellulose. Cellulose content and other material compositions will have an influence on the characteristics of hardness, coefficient of friction and wear on natural fiber composites to become friction materials for centrifugal clutches.

Acknowledgment
The author expresses his gratitude and highest appreciation to the Directorate of Research and Community Service of the Ministry of Education, Culture, Research and Technology for providing PTUPT grants. And also thanks to the National Research and Innovation Agency.

References

[1] A Kholil, S T Dwiyati, R Riyadi, J P Siregar, N G Yoga, and A I Aji 2021 Characteristics of wood powder, coconut fiber and green mussel shell composite for motorcycle centrifugal clutch pads *IOP Conf. Ser. Mater. Sci. Eng.* **1098**, no. 6, p. 062034.

[2] A Kholil and R Wirawan, Pengaruh pattern pada sepatu kopling sentrifugal terhadap daya dan torsi sepeda motor matic, no. Snttm Xiii, pp. 15–16, 2014.

[3] R S Khurmi and J K Gupta, A textbook of machine design, no. I. Eurasia Publishing House (PVT.) LTD., 2005.

[4] J Bijwe 1997 Composites as friction materials : recent developments in non-asbestos fiber reinforced friction materials-a review, *Polym. Compos.* **18**, no. 3.

[5] P V Gurunath and J Bijwe 2007 Friction and wear studies on brake-pad materials based on newly developed resin, *Wear* **263**, pp. 1212–1219.

[6] O L Rominiyi, B A Adaramola, O M Ikumapayi, O T Oginni, and S A Akinola 2017 Potential utilization of sawdust in energy, manufacturing and agricultural industry; waste to wealth, *World J. Eng. Technol.* **05** no. 03, pp. 526–539.

[7] R C Pettersen 1984 The chemical composition of wood, pp. 57–126.

[8] A Kholil, S T Dwiyati, and J P Siregar 2020 Development brake pad from composites of coconut fiber, wood powder and cow bone for electric motorcycle *Int. J. Sci. Technol. Res.* **9**, no. 02, pp. 2938–2942.

[9] S S Lawal, K C Bala, and A T Alegbede, Development and production of brake pad from sawdust composite, no. 30, pp. 47–56, 2017.

[10] B V Ramnath, J Jeykrishnan, G Ramakrishnan, and B Barath 2018 Sea shells and natural fibres composites : a review, *Mater. Today Proc.* **5**, no. 1, pp. 1846–1851.

[11] S Kocaman, G Ahmetli, A Cerit, A Yucel, and M Gozukucuk 2016 Characterization of biocomposites based on mussel, *World Acad. Sci. Eng. Technol. Int. J. Mater. Metall. Eng.* **10**, no. 4, pp. 438–441.

[12] J Abutu, S A Lawal, M B Ndaliman, R A Lafia-araga, O Adedipe, and I A Choudhury 2018 Engineering science and technology, an international journal effects of process parameters on the properties of brake pad developed from seashell as reinforcement material using grey relational analysis, *Eng. Sci. Technol. an Int. J.*

[13] A Kholil, S T Dwiyati, A Sugiharto, and I W Sugita 2019 Characteristics composite of wood powder, coconut fiber and green mussel shell for electric motorcycle brake pads, *J. Phys. Conf. Ser.* **1402**, no. 5.

[14] I Wayan Sugita, D Rio Budi Syaka, and A Irianto Wahyudi 2019 Effect of pertalite – methanol blends on performance and exhaust emission of a four-stroke 125 cc motorcycle engine, *KnE Soc. Sci.* **3**, no. 12, p. 384.

Journal of Physics: Conference Series **2019** (2021) 012066 doi:10.1088/1742-6596/2019/1/012066

Analyze The Strength of Ceramics Made from Clay, Sinabung Volcanic Ash and Sea Water in The Term of The Structure

E Sembiring[1], D Bonardo[1], K Sembiring[1] and Z Sitorus[1,2]
[1]Department of Physics, FMIPA, Universitas Sumatera Utara
Jl. Bioteknologi 1 Kampus USU,Medan 20155, Indonesia
Email: zuriahsitorus@usu.ac.id

Abstract. Ceramics made from clay, Mount Sinabung ash and sea water have been made using the die pressing method. Sinabung clay and ash were sieved through a 200 mesh sieve. Ceramics are printed with a mixture of clay, Sinabung mountain ash and sea water in a ratio of 10:0:0 gr; 10:9:1 gr; 10:8:2 gr; 10:7:3 gr; 10:6:4 grams; 10:5:5 g, etc. Ceramics are sintered with a sintering temperature of 1000 OC with a holding time of 3 hours. Ceramics are characterized by determining physical properties (shrinkage, porosity and density), mechanical properties (compressive strength), surface morphology and pore size (SEM) and XRD. The characterization results show that the optimum value occurs in a mixture of clay, Mount Sinabung ash and seawater at a mixed variation of 10:5:5 gr with a shrinkage value of 31.2%; porosity = 27.16%; density = 2. 09×103 kg/m3 %; compressive strength = 86.66 MPa. Surface morphology (SEM) showed that the optimum variation experienced agglomeration and significant unification of grains occurred and the structure formed was hexagonal with a crystal size of 45.44 nm.

1. Introduction

With the increasing growth of the construction sector in Indonesia. For this reason, in order to accelerate the construction of structures, it is necessary to encourage development in strengthening the domestic construction industry, so that it is expected to reduce our dependence on imported raw materials (1). In science and technology that continues to advance, many scientists quickly, precisely and practically in creating something new but useful for life in the material field, one of which is ceramic technology that can support the construction industry in terms of development.

In 2010 to 2021, Mount Sinabung, which is located in the highlands of Karo Regency, North Sumatra, is still experiencing prolonged eruptions so that Sinabung emits thick volcanic ash. In a previous study conducted by Nain Felix Sinuhaji (2011), volcanic ash from Mount Sinabung contains silica (SiO_2) as much as 59.92%. Therefore, volcanic ash from Mount Sinabung can be used for ceramic raw materials as a substitute for quartz.

The use of seawater with a percentage of 15% reaches a concrete compressive strength of 17.71 MPa which has increased from normal concrete of 13.33 MPa. At 30% seawater mixture, the concrete compressive strength is 18.59 MPa. So that the percentage of sea water use of 30% is the maximum increase from mixing sea water in making concrete (2)

In this study, a ceramic material was made as a raw material for clay by mixing Mount Sinabung ash and sea water with conventional compression molding techniques. So that the mixture is expected to produce a ceramic material that has good physical and mechanical properties, characterization test for construction ceramics.

2. Method

2.1 Fabrication Of Clay Based Ceramic

Clay and mountain ash as the main ingredients for making samples were obtained from Mount Sinabung, while sea water was obtained from the coast of the Mirror Beach, Deli Serdang. In this study, the clay was dried in the sun for 3 days to remove its moisture content. The results of the drying will produce clay powder. After that, the clay and ash of Mount Sinabung were each sieved using a 200 mesh sieve. 10 grams of clay mixed with Mount Sinabung ash and sea water using a mixer. The mass between Mount Sinabung ash and sea water is determined by 3 variations i.e. 9:1, 8:2, 7:3, 6:4. 5:5, 4:6, 3:7, 2:8 and 1:9 in grams. All of the samples were denoted as L10, L20, L30, L40, L50, L60, L70, L80 and L90, respectively. As a control variable, samples without the addition of Mount Sinabung ash were also prepared and denotes as L0. All of the above samples were printed in a cylindrical shape with a diameter of 5 cm and a height of 1 cm using the Die Pressing method. The printed sample was dried for 24 hours in the sun to remove the moisture content, then the sample was sintered at the 1000 ^{O}C for 3 hours and dried at room temperature.

2.2 Characterizatons

The structural properties of ceramic-based clay samples were studied by using X-ray diffraction (XRD) patterns recorded using an X-ray diffractometer (Miniflex series Benchtop powder X-ray diffraction Instruments, Rigaku Corporation, Japan). The scanning rate of 5/min was applied to record the patterns in the range of 7° – 90°. The mean crystallite size of these ceramic based clay samples was calculated using a modification of Debye Scherer's formula,

$$D = \frac{\kappa\lambda}{e^{\text{int}\,ersept}} \tag{1}$$

Where k is the shape factor and has a typical value of 0.9, is the wavelength (Cu Kα = 1.5405 am), is the full width at half maximum of the several peaks (in radians), and θ is the Bragg angle of the peak. For calculation, the several peaks were taken for calculation. The surface morphologies of ceramic based clay samples were characterized by scanning electron microscopy (Hitachi SU-3500) operated at 10 kV with a thin layer of gold sputter coated prior to analysis. And the grain size in each sample was measured through quantitative analysis using the ImageJ application. Physical tests such as porosity, density and volume loss were also carried out manually and for mechanical testing in the form of compressive strength using a universal testing machine (UTM).

3. Results and Discussions

3.1 Volume Loss

The volume loss graph in figure 1 shows that the burn loss value for each variant of the construction ceramic mixture has increased and decreased. The higher the sintering temperature, the ceramic will undergo a high compaction process so that it will produce a relatively high volume loss value. The lowest burn loss value was owned by sample L20 of 6.02%, while the highest value of burn loss was owned by sample L50 of 31.21%.

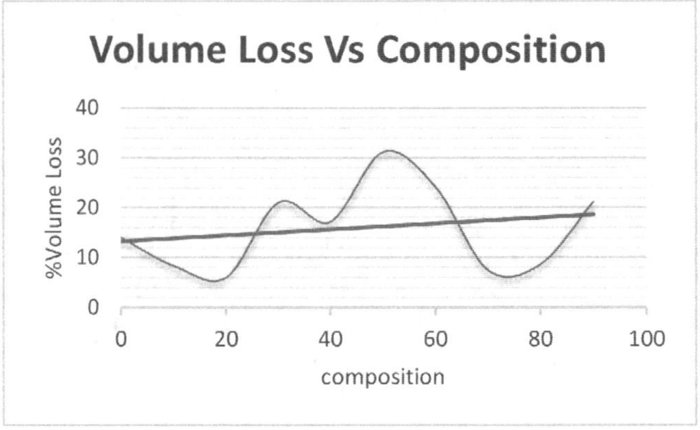

Figure 1. Volume loss graph with different compositions on clay-based ceramic samples.

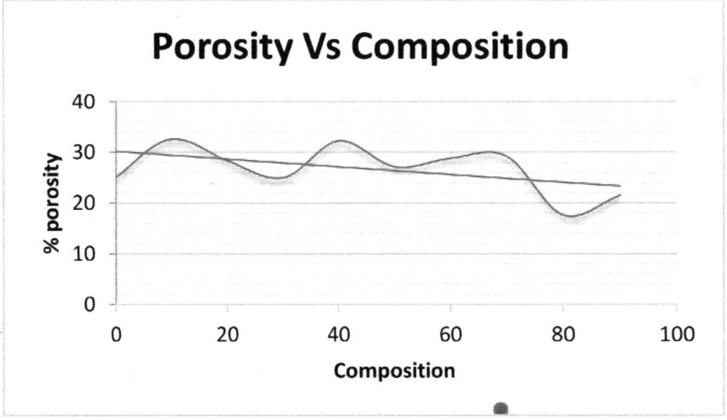

Figure 2. Graph of the relationship between porosity and compositional variations in clay-based ceramic samples.

Figure 2 shows a graph of the porosity in each sample. Low porosity indicates that the material has a solid alloy because the pores in the material are relatively lower, the L80 sample has the most optimum level of porosity with the value of 17,68%. The low porosity in sample L80 when compared to other samples indicates that the cavities formed in sample L80 are fewer and smaller so as to form a material with an optimum level of homogeneity and interparticle occupancy.

Density test data on samples using seawater solvents show that the L50 sample has the highest density as shown in figure 3. An increase in density indicates that the mass density of the sample is getting higher in each volume, or in other words the L50 sample has the strongest bond between particles because the mass density reaches its optimum point.

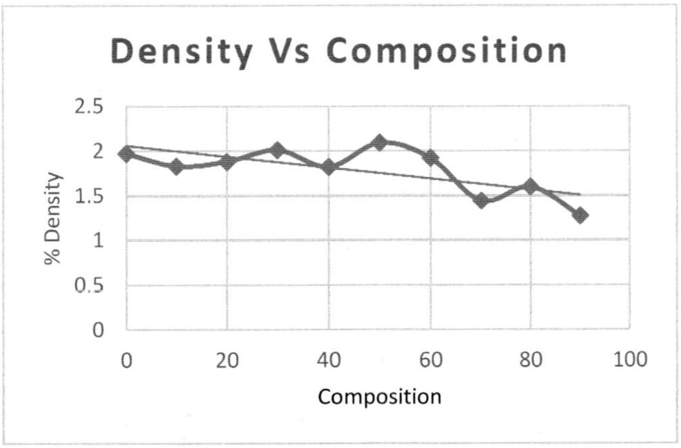

Figure 3. Plot of density values in each seawater composition in clay-based ceramic samples.

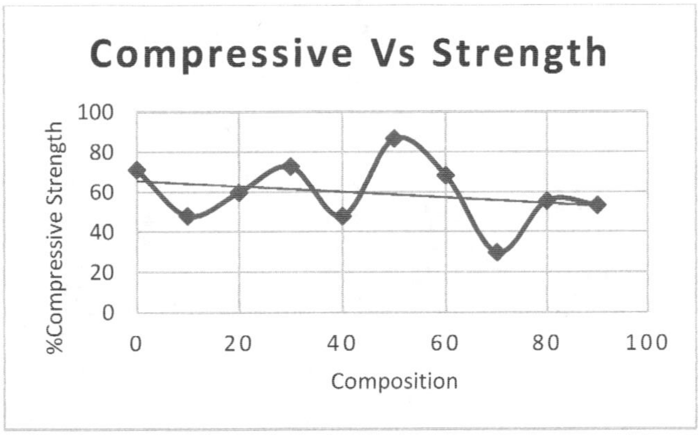

Figure 4. Plot of density values in each seawater composition in clay-based ceramic samples.

From the test data on the compressive strength of the sample on the figure 4, the lowest compressive strength was found in the L90 sample with a value of 47.96 MPa, while the highest compressive strength was found in the L50 sample with a value of 86.66 MPa. This value is even higher when compared to the previous study of 15.2086 MPa (Hendra L. Nababan).

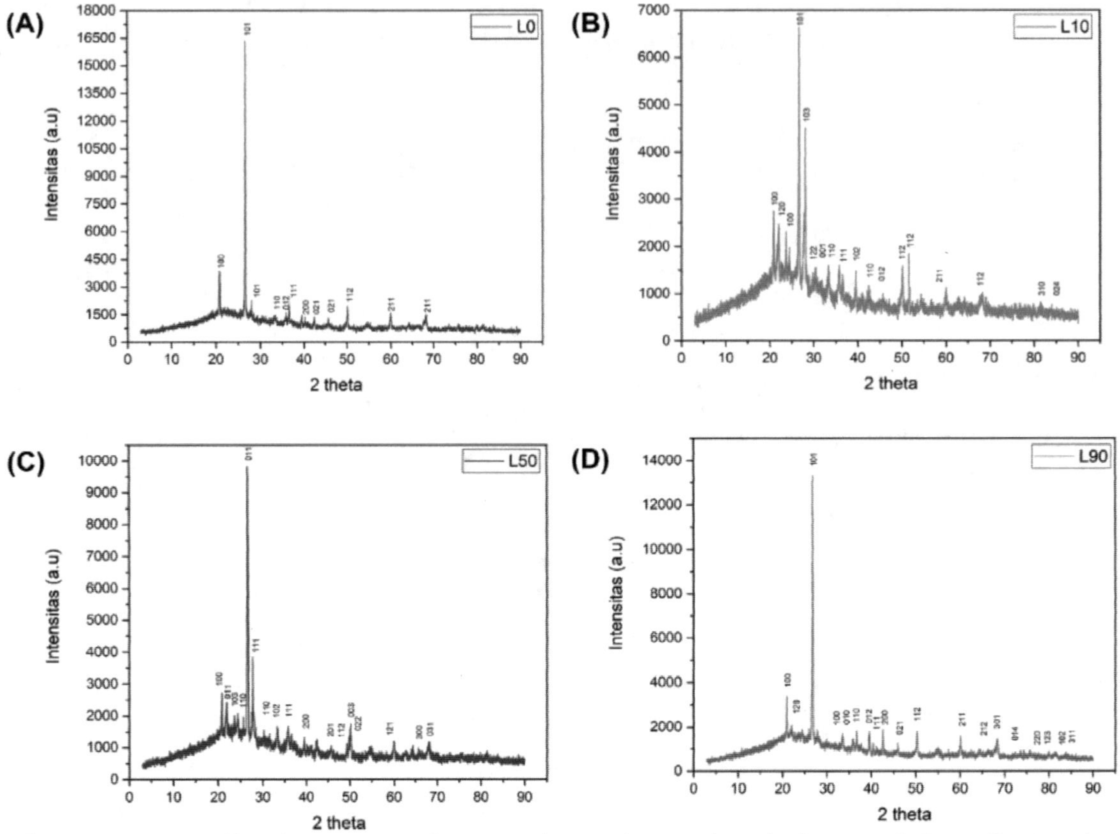

Figure 5. X-ray diffraction patterns of clay-ased ceramic samples of L0 (a), L10 (b), L50 (c) and L90 (d).

In the picture above, which displays the XRD diffraction pattern in the sample using seawater, it shows the same crystalline phase, namely SiO2, according to the peaks formed. However, there are some differences between the diffraction patterns formed, where the L10 sample produces several new peaks when compared to the L0 sample. Based on the data sheet obtained, the new peaks that appear still showing the SiO2 phase. As the concentration of Mount Sinabung ash decreases to 50% at L50 to 90% in the L90 sample, the peaks formed have decreased in intensity, this indicates that a reduction in the ash concentration of Mount Sinabung by more than 10% will reduce the level of crystallinity in the sample.

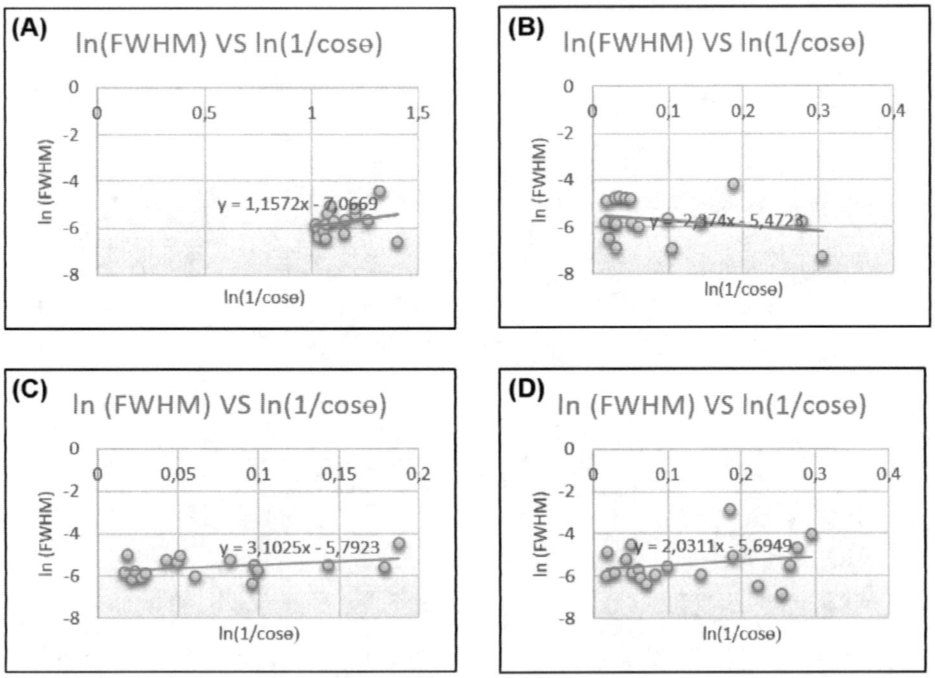

Figure 6. Graph of the relationship of the regression equation of clay-based ceramic samples of L0 (a), L10 (b), L50 (c) and L90 (d)

As well as the crystal size can be determined using the Debye Scherrer equation. The largest crystal size was found in the sample L0 which was 162.57 nm, while the L10 sample had a significant decrease in crystal size to 33 nm. In sample L50, the crystal size increased again with the decrease in the concentration of Mount Sinabung ash and along with the increase in the concentration of seawater. The crystal size again decreased in sample L90 with a relatively low decrease. From the results of the crystal size analysis above, it was found that the concentration between Mount Sinabung ash and sea water sufficiently affected the size of the crystals formed, resulting in a certain crystal size. According to Beknalkar, the concentration of Mount Sinabung ash as an additive and seawater as a concentration solvent in the sample has a major role in changing the morphology of the sample surface and the size of the crystals formed in the sample can only be influenced by microscopy by these two variables (Beknalkar et al., 2020).

Table 1. Crystal size in ceramic samples using seawater

Sample	Crystal size (nm)
L0	162,57
L10	33
L50	45,44
L90	41,23

Journal of Physics: Conference Series **2019** (2021) 012066 doi:10.1088/1742-6596/2019/1/012066

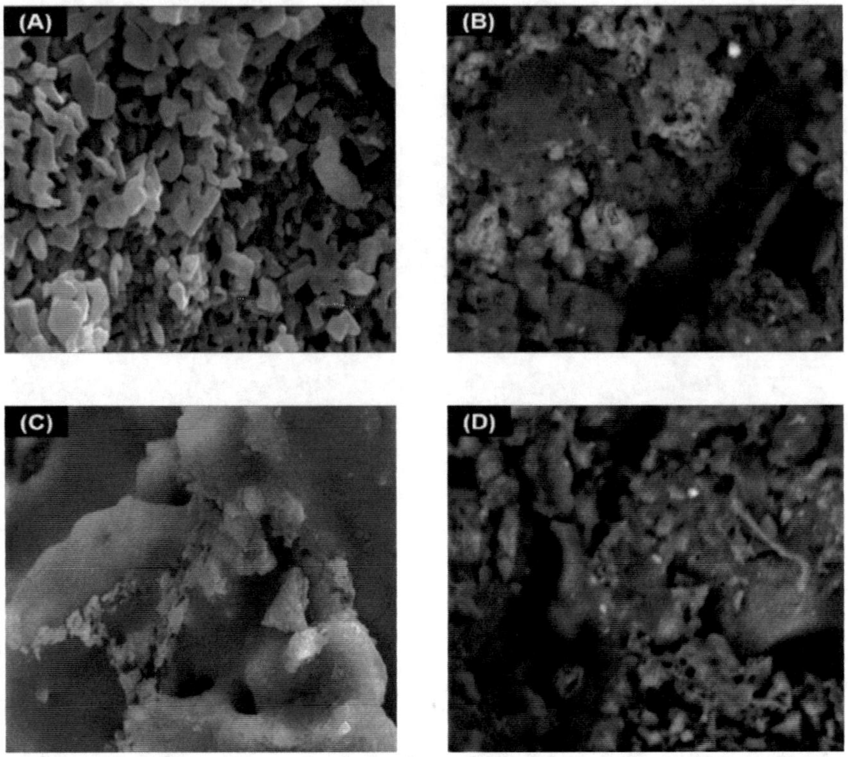

Figure 7. SEM images of of clay-based ceramic samples of L0 (a), L10 (b), L50 (c) and L90 (d)

In figure 7.a, which shows sample L0 shows a much more even distribution of granules when compared to what happened in the sample with Aquadest solvent. The shape of the grains also looks more regular and the grain boundaries are clearer, after further analysis the grains are flat and rectangular. Figure 7. B shows that the L10 sample has agglomerated, causing the distance between the grains to be closer so that it has an impact on decreasing the number of grain boundaries formed. In sample L50, which is shown in Figure 7.c, the phenomenon of agglomeration is getting stronger in the sample, this can be seen from the occurrence of a significant unification process between grains so that they lose grain boundaries. In Figure 7.d, the L90 sample has decreased grain density, the level of agglomeration in this sample is also quite high, but much lower than that of the L50 sample. This also proves that the agglomeration phenomenon will reach its maximum condition when the concentration between Mount Sinabung ash and sea water has the same ratio.

4. Conclusion

The size of the crystal atoms in clay-based ceramics have been investigated with variations in the composition of Mount Sinabung ash with seawater or aquadest of L0: 162.57 nm; L10: 33 nm; L50: 45,44 nm and L90: 41,23 nm. The composition of Mount Sinabung ash with seawater or aquadest in the sample can affect the physical and mechanical properties of ceramics. The optimum physical properties are indicated by the lowest porosity value, namely 17.68% in sample L80, while the optimum density is obtained at 2.09 x 103 Kg/m3 which is the highest density possessed by sample L50. The mechanical properties of the L50 sample also showed an optimum value of 86.66 Mpa. The crystal structure and morphology of the sample have a significant influence on the mechanical strength of the ceramic. The crystal structure with hexagonal shape has relatively more optimal mechanical strength when compared

Journal of Physics: Conference Series **2019** (2021) 012066 doi:10.1088/1742-6596/2019/1/012066

to the monoclinic structure, this is because the crystal bonds in the hexagonal structure are stronger so that it has an impact on increasing the morphological structure formed.

References

[1] Adha, Rahmat Fajri 2019 *Pengaruh Pengunaan Air Laut Terhadap Kuat Tekan Beton* (Univesitas Andalas: padang)

[2] Ariswan 2016 *Hand Out Struktur Kristal* (Yogyakarta: Universitas Negeri Yogyakarta)

[3] Ahmad, Syamsul Bahri 2018 *Investigasi Pengaruh Air Laut Sebagai Air Pencampuran Dan Perawatan Terhadap Sifat Beton* vol 5 (1) (Makassar : Universitas Politeknik Negeri Ujung Pandang/Journal INTEK) Hal 48-52

[4] Beiser, Arthur 1992 *Konsep Fisika Moderen Edisi ke Empat. (alih bahasa: DR.The Houw Liong* (Jakarta : Penerbit Erlangga)

[5] Bonardo. D., Siberian. R. 2021 *Analisis Struktur Kristal Silika Dari Abu Ampas Tebu Menggunakan Metode XRD* (Einstein Journal)

[6] Nababan, Hendra L 2019 *Pembuatan Keramik Konstruksi dari Debu Gunung Sinabung dan Abu Sekam Padi Dengan Polyvinyl Alchohol Sebagai Perekat* (Medan: Universitas Sumatra Utara)

[7] Sidabutar, Tri Exaudi 2017 *Pembuatan Dan Karakterisasi Keramik Magnesium Alumina Silika Dari Abu Vulkanik Gunung Sinabung* (Jakarta : Universitas Mercu Buana/Jurnal Teknik Mes)

[8] Shen. M, Qin. W. *Mechanical properties of 3D printed ceramic cellular materials with triply periodic minimal surface architectures*, 41, 2. 1481-1489. 2021.

[9] Subriyer, N 2013 *Aplikasi Filter Keramik Berbasis Tanah Liat Alam Dan Zeolit Pada Pegolahan Air Limbah Hasil Proses Laundry* vol 13 (1) (Palembang:Universitas Sriwijaya/Jurnal Bumi Lestari) Hal 45-51

[10] Sinuhaji, P, Maghrifah, A Dan Nababan, Sahat M 2019 *Characterization of the Volcanic Rocks of Mount Sinabung, Simacem Village, Karo Regency, Conducted With XRD, SEM-EDX* (Medan: Universitas Sumatra Utara)

[11] Smallman, R. E & Bishop, R. J 2000 *Metalurgi Fisika Modern & Rekayasa Material Edisi Keenam. (Alih Bahasa: Ir. Sriati Djaprie, M. Met*) (Jakarta: Erlangga)

[12] Sinuhaji, N.F 2011 *Analisis Logam Berat dan Unsur Hara Debu Vulkanik Gunung Sinabung Kabupaten Karo-Sumatra Utara* (Medan : Universitas Sumatra Utara)

[13] Van Vlack 2004 *Elemen-elemen & Ilmu Rekayasa Material* (Jakarta: Erlangga)

[14] Wang M Di, Li YY, Yao BH, Zhai K, Li ZJ, Yao HC. *Sensors, Actuators, B Chem*, 656–66, 288. 2019

[15] Wei S, Han L, Wang M, Zhang H, Du W, Zhou M. Mater Lett. 259–62, 186 (2017).

Journal of Physics: Conference Series **2019** (2021) 012067 doi:10.1088/1742-6596/2019/1/012067

Preliminary Study on High-Temperature Oxidation of Ni-AlN-TiN/Si3N4 Electrodeposition Composite Coatings

E Budi[1,2], C Tandri[1], N Darsono[3], A W Andiani[1], M Fajrin[1], W Indrasari[1], I Sugihartono[1], and T B Prayitno[1]

[1]Department of Physics, Faculty of Mathematics and Science, Universitas Negeri Jakarta
Jl. Rawamangun Muka, Jakarta 13220, Indonesia

[2]Department of Physics Education, Faculty of Mathematics and Science, Universitas Negeri Jakarta
Jl. Rawamangun Muka, Jakarta 13220, Indonesia

[3]Research Center for Metallurgy and Material, Indonesian Institute of Science (LIPI), Building 470, Kawasan Puspiptek Serpong, Tangerang Selatan, Indonesia

Email: r d d, dr D

Abstract. High-temperature oxidation is a serious problem for several materials during their application in aggressive environments, such as high temperatures, high-speed machining, and corrosive environments. This research was conducted to analyze the effect of high-temperature oxidation on the surface morphology of the Ni-AlN-TiN/Si3N4 composite coating. The coatings were electrodeposited at a current of 5 me for 15 minutes and various electrodeposition temperatures of 35°C, 40°C, and 45°C. This composite coating was formed on tungsten carbide substrates with the electrolyte solution consisting of 6 g/L AlN, 6 g/L TiN, 40 g/L H3BO3, 0.6 g/L Si3N4, 0.17 M NiCl2.6H2O, 0.38 M NiSO4.6H2O, and 0.6 g/L Sodium Dodecyl Sulfate. The high-temperature oxidation process was carried out at a temperature of 700°C for 2 hours in the furnace. The surface morphology was characterized by using Scanning Electron Microscopy (SEM). The results showed that the surface morphology of the Ni-AlN-TiN/Si3N4 composite coating deposited at a temperature of 40°C had a scaled growth, reduced cracks, and reduced pore. It was due to the effect of high-temperature treatment.

1. Introduction

High-temperature oxidation is a type of oxidation process at high temperatures that can extent be a corrosion site. High-temperature oxidation resistance played an essential role in selecting materials used for construction, from gas turbines to heating furnaces. The types of high-temperature corrosion that usually caused problems in industrial equipments, including oxidation, sulfidation, carburizing, nitriding, halogen attack corrosion, and salt melt corrosion [1]. The high-temperature oxidation process in metals is a diffusion process where diffusion will depend on the diffusion coefficient and increase due to high temperatures. High-temperature oxidation was a severe problem for several materials during their application in aggressive environments, such as high temperature, high-speed machining, and corrosive environments. Oxidative damage of metals during high-temperature used in aggressive environments caused severe losses every year [2]. This is an especially case that material at temperatures up to 800°C and in various oxidizing atmospheres, such as conditions containing H_2O^-, O_2^- or CO_2 [3].

High-temperature oxidation resistance is a crucial parameter for optimizing coating performance. In Tribology, high-temperature conditions are associated with high-speed dry machining operations. In these conditions, oxidation may occur due to diffuse out of the coating element and reacts with oxygen (O_2) to form an oxide coating on the surface coating [4]. Surface modification with coatings has become an important step to improve surface properties such as wear, corrosion, and oxidation. Various conventional techniques have been used to deposit the desired materials onto the substrate to achieve surface modification [5]. The coatings could protect the substrate by providing a barrier between the metal and the environment or through the presence of corrosion-inhibiting chemicals within it [6].

Ni electrodeposited coatings have been widely used in various applications due to their unique combination of properties such as wear resistance, corrosion resistance, nonmagnetism, and uniformity of coating thickness. Electrodeposition is a low-temperature process used to fabricate nanocomposite coatings in a single step without secondary treatment. The nickel matrix prepared by electrodeposition has a unique high density, minimum porosity and has been extensively studied [7]. The nickel matrix has also been widely used in the chemical, mechanical and electronic industries because of its wear, corrosion, and oxidation resistance [8]. Nickel metal (Ni) with its alloys is widely used as a matrix metal in composite coatings because of its good ductility and corrosion resistance. The crystal structure of nickel is FCC with a melting point of 1455°C provided its high thermal stability [9]. Nickel corrosion resistance occurs due to forming a passive oxide coating on its surface which acts as a protector against the corrosive environment.

The high-temperature oxidation of Ni and Ni-based alloys have become an essential part of the literatures on high-temperature oxidation for the last 50 years. One reason is the number of applications for Ni-based alloys at high temperatures due to neo formation. The NiO system was also of considerable scientific interest, with only one oxide stable at high temperatures [10]. Pure nickel metal does not protect against corrosion through the formation of nanocrystals but the formation of a NiO oxide coating. However, at high temperatures in the range of 700°C - 900°C, the formation of Ni was reduced due to the growth of NiO grains, which caused the reduction of Ni to diffuse out through the NiO coating and also the soft nature of nickel at high temperatures can potentially reduce its wear and corrosion resistance. Therefore, adding a complex compound of nitride particles into the nickel coating could maintain wear resistance and corrosion resistance, especially at high temperatures. Several complex particle nitride compounds that can be used were AlN-TiN [11] [12], Si3N4 [13], TiN [14], AlN [15], and ZrN [16].

On-TiN compounds have the ability to form very hard and dense layers with a hardness that can be maintained at high temperatures [17]. The oxidizing behavior of these hard coatings plays a vital role in tool applications as they are frequently exposed to oxidative atmospheres at high temperatures during use as wear and corrosion-resistant coatings [18]. The AlN-TiN coating will start to oxidize at a temperature range of 700°C - 900°C [19][20]. Most ceramic compounds have poor resistance to thermal shock; however, Si_3N_4 has good resistance to thermal stress, excellent high-temperature strength, and resistance [21][22]. The formation of a super hard coating of AlN-TiN/Si_3N_4 compound using deposition rules produces a level of diamond hardness with thermal stability at high temperatures. This occurs through a thermodynamic phase segregation mechanism that leads to the spontaneous formation of nanostructures [23]. The nanocomposite formed is stable to grain growth at high temperatures [24].

In this study, a Ni-based system of Ni-AlN-TiN/Si_3N_4 composite coating was deposited by electrodeposition process. A preliminary study was conducted to investigate the effect of high temperature on the surface coating morphology. The correlation between the electrodeposition process parameters and coating surface morphology after high temperature oxidation treatment was discussed.

2. Method

Ni-AlN-TiN/Si_3N_4 composite coating was electrodeposited onto Tungsten Carbide (WC) at current 5 MA for 15 minutes and various electrodeposition temperatures of 35°C, 40°C, and 45°C. Before electrodeposition, the substrates were sanded first and washed with soap to remove dirt attached to the substrate. The substrates, then were rinsed with distilled water and alcohol using an ultrasonic cleaner with an electrolyte solution composed of 6 g/L AlN, 6 g/L TiN, 40 g/L H_3BO_3, 0.6 g/L Si_3N_4, 0.17 M $NiCl_2.6H_2O$, 0.38 M $NiSO_4$. $6H_2O$, and 0.6 g/L Sodium Dodecyl Sulfate. All ingredients were mixed in 5 ml of distilled water and stirred for 24 hours using a magnetic stirrer. In this study, three electrodes were used, consisting of AgCl as the reference electrode, put as the supporting electrode, and Tungsten Carbide (WC) as the working electrode. The high-temperature oxidation process was carried by heating the samples at temperatures of 700°C for 2 hours in the furnace. After the high-temperature oxidation test, the surface morphology of Ni-AlN-TiN/Si_3N_4 composite coatings was characterized by using Scanning Electron Microscopy (SEM) JEOL-JED 2300 (voltage 15 kV, 1 now) with a magnification of 500x.

3. Results and discussion

Figure 1. Shows the surface morphology of the Ni- AlN-TiN/Si3N4 electrodeposition composite coating before being given high-temperature oxidation behavior. Based on the results of the characterization, when the temperature is 35°C, the surface of the coating tends to appear with small cracks on the entire surface of the coating. When the temperature increases from 35°C to 40°C, the surface morphology of the coating occurs with large cracks, pores, and agglomeration. At 45°C, the crack disappears, but becomes more porous on the surface of the coating [25].

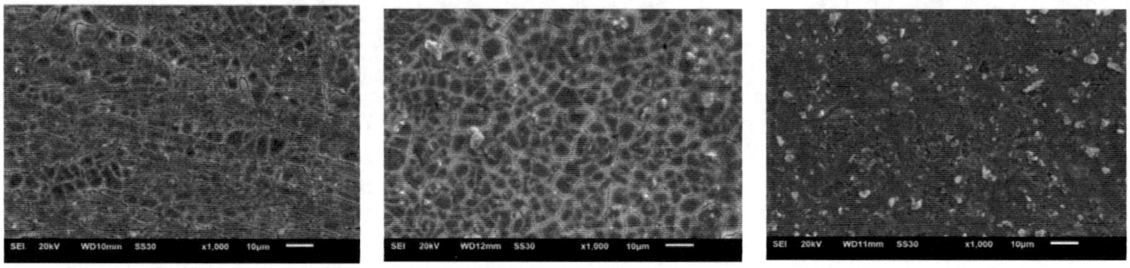

Figure 1. SEM morphology of Ni-AlN-TiN/Si_3N_4 composite coating electrodeposited at temperature of (a) 35 °C (b) 40 °C and (c) 45 °C [25].

As the electrodeposition temperature increases, the cracks on the surface of the coating tend to disappear and become smoother, this is in accordance with the previous research that the higher the temperature used in the coating, the smoother the coating [26]. This happens because when the temperature increases to 45°C, the crystal size gets smaller and makes the coating smoother [27]. However, as the temperature increases, the formation of hydrogen gas increases, so that more hydrogen gas diffuses into the coating, causing the coating to be more porous [28]. The addition of temperature at the time of coating will help reduce the morphological cracks of the coating [29]. This happens because the temperature reduces the absorption of hydrogen in the coating, reduces stress, and reduces the tendency towards cracking [30].

Meanwhile, Figure 2. Shows the surface morphology of the Ni-AlN-TiN/Si3N4 electrodeposition composite coating after being given high-temperature oxidation behavior. Based on the characterization results, the morphology of the coating showed the presence of scale growth on the surface of the coating, reduced cracks, and reduced pore. The growth of scale on the coating is a characteristic of oxidation [31] and indicates that a protective oxide coating has deposited the above surface coating after high-temperature oxidation [32]. The pores on the coating surface are reduced because the pores have been oxidized on exposure to high temperatures [33].

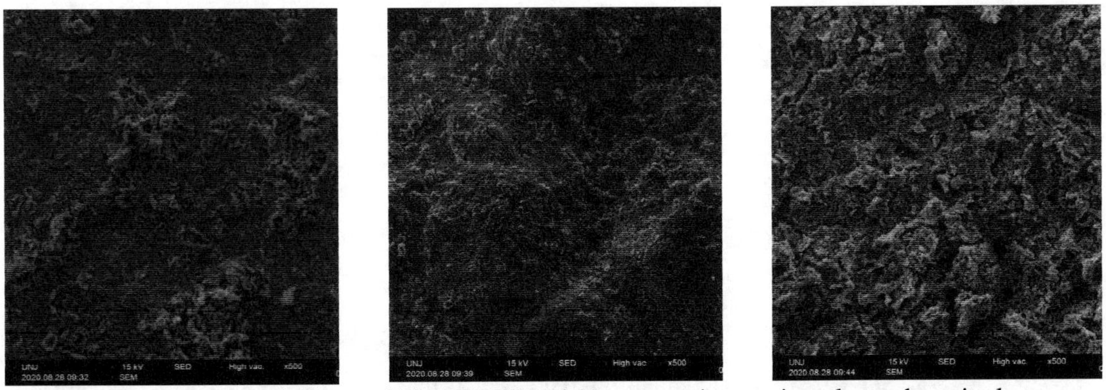

Figure 2. SEM morphology of Ni-AlN-TiN/Si_3N_4 composite coating electrodeposited at a temperature of (a) 35°C (b) 40°C and (c) 45°C. All sample were exposed at temperature of 700°C for 2 hours within air condition.

The nickel surface is morphologically porous due to the kinetics of electrochemical deposition, nucleation, and crystal growth [33], and the pores will oxidize on exposure to high temperatures [34]. Higher resistance to high-temperature oxidation is obtained with the Ni composite coating embedded by TiN or AlN particles. However, its properties depend on the fraction of embedded particles. The good corrosion resistance of the Ni-TiN composite coating is overcome by the presence of TiN particles in the coating, which eliminates intergranular dots [35]. Since the Ni coating with a rough surface becomes potentially susceptible to reacting with corrosive elements, the co-deposition of AlN from the nickel composite coating causes a fine morphology of the nanostructured layer, which efficiently prevents corrosive elements from entering the coating [36]. Si_3N_4 particles also play a role in refining the grains of the Ni composite layer, which causes lower corrosion rates [13].

4. Conclusion

In summary, the morphology of the Ni-AlN-TiN/Si_3N_4 coating showed the presence of scale growth on the surface of the coating, reduced cracks, and reduced pore after the high temperature oxidation process treatment at 700°C.

Acknowledgments

The authors gratefully acknowledge research facilities support from the Laboratory of Material Physics, Physics and Physics Education Programs, Faculty of Mathematics and Science, Universitas Negeri Jakarta and Laboratory of Research Center for Metallurgy and Material, Indonesian Institute of Science (LIPI). This research was financially supported by Kemenristek with contract No. 1/E4.1/DSD/LPPM/2021.

References

[1] Fontana MG, Greene ND. Corrosion engineering Singapore. *McGraw-Hill Book Company*. 1987.

[2] Gunthner M, Kraus T, Dierdorf A, Decker D, Krenkel W, Motz G. Advanced Coatings on the Basis of Si(C) N Precursors for Protection of Steel Against Oxidation, *Journal of the European Ceramic Society*, **29**: p. 2061-68. 2009.

[3] Pond RB, Shifler DA. *High Temperature Corrosion Related Failures: ASM Handbook*. 2002.

[4] Joshi A, Hu HS. Oxidation Behavior of Titanium-Aluminium Nitrides, *Surface and Coatings Technology,* p. 499-507. 1995.

[5] Jagielski J, Khanna AS, Kucinski J, Mishra DS, Racolta P, Sioshansi P, et al. Effect of chromium nitride coating on the corrosion and wear resistance of stainless steel, *Applied Surface Science*, **156**: p. 47–64. 2000.

[6] Gray JE, Luan B. Protective coatings on magnesium and its alloy: a critical review, *J. Alloys comped*, **336**: p. 88. 2002.

[7] Aruna ST, Bindu CN, Selvi VE, Grips VK, Rajam KS. Synthesis and properties of electrodeposited Ni/Cora nanocomposite coatings, *Surf. Coat. Technol*, **200**: p. 6871. 2006.

[8] Harvested A, Janssen LJ. Electrochemical codeposition of inert particles in a metallic matrix, *J. Apple. Electrochem*, **40**: p. 519-27. 1995.

[9] Xi FF, Liu C, Wang F, Wu MH, Wang JD, Fua HL, et al. Preparation and characterization of Nano Ni–TiN coatings deposited by ultrasonic electrodeposition, J*ournal of Alloys and Compounds*, **490**: p. 431-435. 2010.

[10] Haugsrud R. *High Temperatur Oxidation of Ni-20 wt.% Cu from 700 to 1100°C,* p. 55. 2001.

[11] Khrais SK, Lin YJ. Wear mechanisms and tool performance of TiAlN PVD coated inserts during machining of AISI 4140 steel, Wear*,* **262**: p. 64-9. 2007.

[12] Yoon SY, Kim JK, Kim KH. A comparative study of the tribological behavior of TiN and TiAlN coatings prepared by arc ion plating technique, *Surface and Coatings Technology*, **161**: p. 237-42. 2002.

[13] Khazrayie MA, Aghdam AR. Si3N4/Ni nanocomposite formed by electroplating: Effect of average size of nanoparticulates, *Trans. Nonferrous Met. Soc*, **201**: p. 1017-23. 2010.

[14] Veprek , Reiprich S. *A Concept for The Design of Novel Superhard Coatings*, **268**: p. 64-71. 1995.

[15] Aal AA, Bahgat M, Radwan M. Nanostructured Ni-AlN composite coatings, *Surf. Coatings Technol*, **201**: p. 2910-918. 2006.

[16] Erdemir A, Eryilmaz OL, Urgen M, Kazmanli K, Mehta N, Prorok B. *Tribology of Nanostructured and Composite Coatings*. 2006.

[17] Veprek S, Haussmann M, Reiprich S, Shizhi L, Dian J. Novel Thermodynamically Stable and Oxidation Resistant Superhard Coating Material*s*, *Surface and Coating Technology*, **86-87**: p. 394-401. 1996.

[18] Derflinger VH, Schutze A, Ante M. Mechanical and structural properties of various alloyed TiAlN-based hard coatings, *Surf. Coat Technol*, **200**: p. 4693. 2006

[19] Joshi A, Hu HS. *Surf. Coat. Technol*, **76**: p. 499-507. 1995.

[20] Chen L, Paulitsch J, Du Y, Mayrhofer PH. Thermal stability and oxidation resistance of Ti–Al–N coating*s, Surf. Coat. Technol*, **206**: p. 2954-960. 2012.

[21] Pehlke RD, Elliott JF. High-Temperature Thermodynamics of the Silicon, Nitrogen, Silicon Nitride System*, Trans. AIME*, **215**: p. 781-85. 1959.

[22] Croft WJ, Cutler IB. Review of Silicon Nitride, *ONR Report*. 1973.

[23] Holubar P, Jilek M, Sima M. Present and possible future applications of superhard nanocomposite coatings, *Surface and Coatings Technology*, **133-134**: p. 145-51. 2000.

[24] Veprek S, Jilek M. Superhard nanocomposite coatings. From basic science toward industrialization, *Pure Appl. Chem,* **74**: p. 475-81. 2002.

[25] Andiani AW, Budi E, Sugihartono I. Pembentukan Lapisan Komposit Ni-TiN-AlN/Si3N4 Menggunakan Metode Elektrodeposisi dengan Variasi Temperatur, *Prosiding Seminar Nasional Fisika*, p. 145-48. 2019.

[26] Sadiku A, Sadiku ER, Ojo OI, Akanji OL, Biotidara OF. Influence of Operation Parameters on Metal Deposition in Bright Nickel-plating Process, *Port. Electrochim. Acta*, **29**: p. 91-100. 2011.

[27] Sen R, Das S, Das K. The effect of bath temperature on the crystallite size and microstructure of Ni–CeO2 nanocomposite coating, *Material Characterization*, p. 257-262. 2011.

[28] Lazic MS, Simovic K, Miskovic S, Kicevic D. The Influence of the Deposition Parameter on the Porosity of Thin Alumina Films on Steel, *Serbian Chemistry Society*, p. 239-249. 2004.

[29] Lu W, Guang P, Li K, Yan P, Yan B. Effect of Bath Temperature on the Microstructural Properties of Electrodeposited Nanocrystalline FeCo Films, *International Journal of Electrochemical Science,* p. 2354 - 364.2013.

[30] Kumar S, Pande S, Verma P. Factor Effecting Electro-Deposition Process, *International Journal of Current Engineering and Technology*, p. 2347 – 5161. 2015.

[31] Talbot D. Corrosion Science and Technology New York, *CRC Press*. 1998.

[32] Sugiarti E, Zaini KA, Sundawa R, Wang Y, Ohnuki S, Hayashi S. Influence of Oxidation Temperature on The Oxide Scale Formation of NiCoCrAl Coatings, *Journal of Physics*, p. 817. 2017.

[33] Kamaraj AB, Shresta H, Speck E, Sundaram M. Experimental Study on the Porosity of Electrochemical Nickel Deposits, *Procedia Manuf*, p. 478-85. 2017.

[34] Bei ZX, Chao C, Qu ZG, Zhao Z, Feng LJ. Electrodeposition and corrosion behavior of nanostructured Ni-TiN composite films, *Transactions of Nonferrous Metals Society of China*, p. 2216-24. 2011.

[35] Xia FF, Liu C, Wang F, Wu MH, Wang JD, Fu HL, et al. Preparation and characterization of Nano Ni–TiN coatings deposited by ultrasonic electrodeposition, *Journal of Alloys and Compounds*, p. 431-35. 2010.

[36] Ma C, Yu W, Jiang M, Cui W, Xia F. Jet pulse electrodeposition and characterization of Ni–AlN nanocoatings in presence of ultrasound, *Ceramics International*, **44**: p. 5163-170. 2018.

Journal of Physics: Conference Series **2019** (2021) 012068 doi:10.1088/1742-6596/2019/1/012068

Brake Pad Characteristics of Natural Fiber Composites from Coconut Fibre and Wood Powder

A Kholil[1*], S T Dwiyati[1], R Wirawan[2] and Elvin M[1]
[1]Department of Mechanical Engineering, Engineering Faculty, Universitas Negeri Jakarta Campus A Jl. Rawamangun Muka, Jakarta, 13220, Indonesia.
[2]Faculty of Mechanical and Aerospace Engineering, Institut Teknologi Bandung Jl. Ganesha No.10 Bandung 40132, Indonesia.
e-mail: ahmadkholil@unj.ac.id

Abstract. The aim of the research is to develop brake pad of natural fibre composite from coconut fiber waste and wood powder. The characteristics of hardness, braking time and wear compared to the commercial brake pad had been studied. Test samples were produced by varying content of coconut fiber from 0 - 40% weight, wood powder content from 40% - 0 weight with 10% intervals. The percentage of polyester resin was made constant. The process of mixing ingredients started from filtering coconut fiber and wood powder using a filter with mesh size 40 μm. Then, the materials combined with certain composition into a container and stirred until evenly mixed. The composite poured into the mold was design appropriate the motorcycle disc brake pad dimensions. The test results of natural fiber composites on brake pads showed different characteristic values on hardness, wear, and braking time tests.

1. Introduction

Brakes are a vital component of a vehicle that serve to help reduce speed, stop the vehicle and maintain the speed of the vehicle when passing on a downhill road [1]. Brakes are a very important regulator of motion and vehicle safety. Brakes that do not function properly can cause accidents because they crash into a vehicle that stops suddenly. The vehicle can't be controlled and finally crashes. Crash into a vehicle when passing a downhill road or falling into a ravine. This is due to the ability of the braking force which depends on the characteristics of the brake pad material. Brake pads are steel backing plates with friction material bound to the surface facing the brake disc [2]. The test equipment used was a brake pad test rig for determining the brake pad wear, brake disk temperature rise, and braking time under different braking conditions [3].

The brake pads generally consist of asbestos fibers embedded in polymeric matrix along with several other ingredients. The use of asbestos fiber is being avoided due to its carcinogenic nature. Therefore a new asbestos free friction material and brake pads has been developed [2].

Coconut fiber shows a good stiffness and is used in products such as floor mats, doormats, brushes, mattresses, coarse filling material, and upholstery. Coconut fiber, obtained from unripe coconut, is a natural fiber extracted from the husk of coconut. The coconut is steeped in hot seawater, and subsequently, the fibers are removed from the shell by combing and crushing, the same process as jute fiber. The individual fiber cells are narrow and hollow with thick walls made of cellulose, and each cell is about 1 mm long and 10–20 μm in diameter. The raw coconut fibers show length varying from 15 to 35 cm and diameter from 50 to 300 μm. When they are immature and then become hardened and yellowed because a layer of linen is deposited on their walls [4].

One coconut fruit produces 0.4 kg of husk containing 30% fiber. Coconut fiber consists of fiber and cork which connects one fiber with other fibers. Coconut fiber has the following physical properties: rough, colored, and stiff. The composition of the content of coconut husk includes hemicellulose (8.50%), cellulose (21.07%), lignin (29.23%), pectin (14.25%) and water (26.0%). Cellulose a compound like clay, insoluble in water and found in the cell walls of plant protectors [5]. Fiber an essential part of coir. Each coconut contains 525 grams of fiber (75% of coir) and 175 grams of cork (25% of coir) [6]. Figure 1a shows coconut fiber.

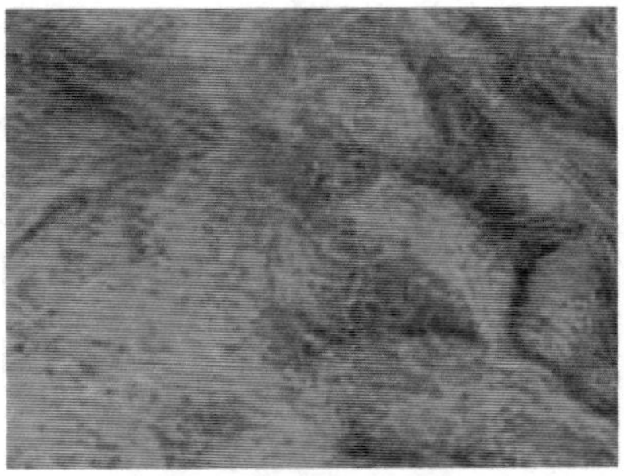

a. Coconut fiber

b. Wood powder

Figure 1. Natural material

The wood powder a waste produced by wood processing such as sawdust [7]. Sawdust possesses characteristics similar to wood but due to the fact that it is in particles, some structural properties have been altered. However, in general, the wood powder has chemical compositions such as hollocellulose (70.52%), cellulose (40.99%), lignin (27.88%), pentosan (16.89%), ash (1.38%), and water (5.64%) [8]. Figure 1b shows wood powder. An investigation was carried out on the use of sawdust to develop brake pads [9].

Palm kernel fibers (PKFs) were used to produce asbestos-free automobile brake pads [10]. Similarly, Development of asbestos-free brake pad using barges was investigated with a view to replace the use of asbestos whose dust is carcinogenic. The sieve bags were used in production of brake pad in a ratio of 70% bags - 30% resin using compression moulding [2]. A research was carried out on the use of banana peels [11] to replace asbestos in brake pad with phenolic resin (phenol formaldehyde) as a binder. The resin varied from 5 to 30% weight in an interval of 5% weight. This research therefore aimed to develop brake pad from agro waste material composite of coconut fiber and wood powder with a view to investigate the characteristics of hardness, braking time and wear as compared with the commercial brake pad.

2. Method

The materials for this test were taken from the waste. Coconut fiber was taken from the peel of coconut fruit and wood powder from sawdust. As a matrix, is used polyester resin with a fixed composition of 60% for all samples. This test aims to see the effect of changes in the composition of coconut fiber and wood powder so that the polyester resin matrix is fixed. The process of mixing ingredients started with filtering coconut fiber and wood powder using a mesh filter of size 40 μm to get fine material. Then mixing done in the container so that the mixing of coconut fiber and the wood powder was evenly mixed. Mixing ingredients according to Table 1, as follows:

Table 1. The composition of brake pad specimens

Specimen	Composition (% weight)		
	Coconut fiber	Wood powder	Polyester resin
A	0	40	60
B	10	30	60
C	20	20	60
D	30	10	60
E	40	0	60

Test samples were produced by varying coconut fiber from 0 - 40% by weight, wood powder from 40% - 0 weight at 10% intervals and fixed percentage polyester resin. The formulation was mixed in a container, stirred to reach a homogeneous mixture and transferred to the mold plate with dimensions of 100 mm x 100 mm x 10 mm at a pressure of 2 tons for 60 minutes. The results were dried naturally by sunlight for three days. The material is cut partially with a size of 25 mm x 25 mm x 10 mm as a hardness testing sample for each sample. The remaining part of the canvass material, then cut with a saw following the size of the commercial brake pads. The lining material is glued then affixed to the bearing. The shape of the brake pads can be seen in Figure 2.

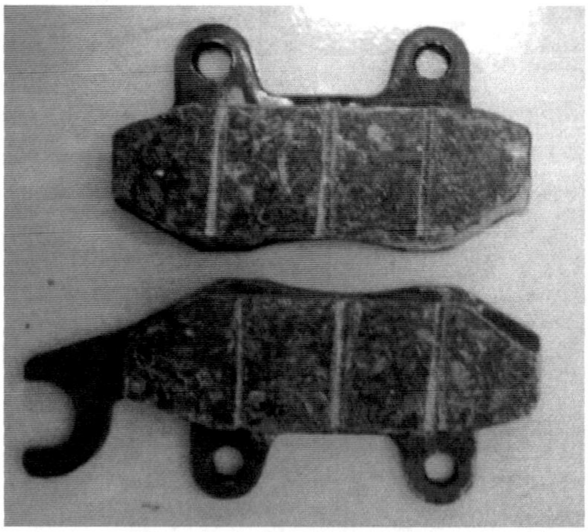

Figure 2. Test specimens

Hardness testing based on ASTM E 92 using the Vickers FV-300e test equipment. Each sample was tested three times and taken data-average. Testing hardness by giving a load of 10 kg.

The test rig was used to determine the performance of the brake pads produced with the different composition. The pads were tested for wear and braking time. A set of commercial brake pads was subjected to similar test for comparison. Figure 3 shows the schematic diagram of the brake pad test rig. It has a 1.5 kW motor with a provision for speed variation by using an inverter. The motor provides the energy required to set the wheel weights and the brake disc in angular motion. When a set of brake pad was fixed into the brake caliper assembly of the test rig, the system is switched and the drive shaft begins to rotate, then it allowed to attain a desired speed. Thereafter, a manual force was applied to the brake pedal that was similar to that of a motorcycle. Subsequently the braking time and brake pad material lost are recorded [3]. Tests carried out at speeds of 20 km/h at temperatures of 25°C, 50°C, 75°C, 100°C at disc brake. Every speed that has determined the test carry out three times so that three data obtained. Then the average value of the braking time obtained. The maximum load was given to the brake lever for 4 kg refers to the average grip of the human hand. This test to determine the effect of the composition of coconut fiber and wood powder at the time of braking with changes in temperature. This test was carried out on disc brake pads samples.

Wear testing was carried out using a test rig as shown in Figure 3. This test aims to obtain wear data after braking testing with a speed of 20 km/hr. Sample testing using disc brake pads. The test begins by weighing the brake pads using a digital scale to measure the initial weight of the brake pad. After braking testing was carried outweighing again to obtain the mass of disc brake pads after testing. The mass difference between the brake pads after testing with the mass of the brake pad before testing becomes the mass reduction data of the disc brake pads. For this test the data were taken using Equation 1.

$$w = m_0 - m_1 \quad (1)$$

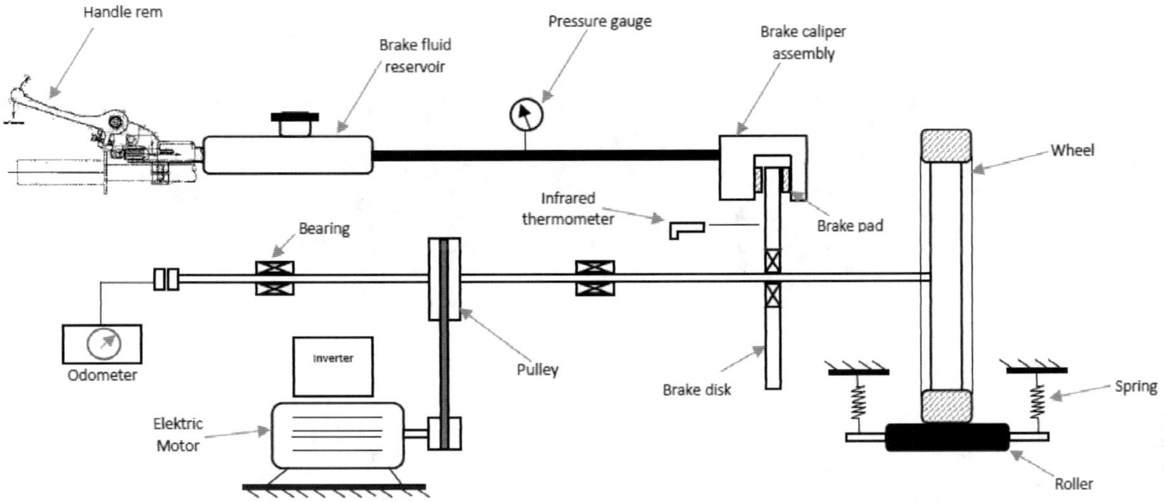

Figure 3. Schematic of brake pad test rig.

3. Results and Discussion

The result of the hardness values is shown in Figure 4, shows the variation of coconut fiber and wood powder composition to the hardness value. Specimen A with a composition of 0% coconut fiber and 40% wood powder has the highest hardness value of 59 VHN. The hardness decreases along with increasing of the composition coconut fiber or reduction of wood powder. The hardness value of the specimen with composition 40% coconut fiber and 0% wood powder is 44.3 VHN. The hardness value influenced by lower cellulose content in coconut fiber, it is about 21.09% [5] compared with wood powder which has 40.99% of cellulose [8]. The lower cellulose content in coconut fiber brings on the mechanical properties of coconut fiber is softer than wood powder. Consequently, specimen D and E with more coconut fiber composition have lower hardness compared to specimens A, B, and C which contain less coconut fiber composition. When compared with commercial brake pad hardness, the D and E specimen have, the closer violence to commercial brake pads violence.

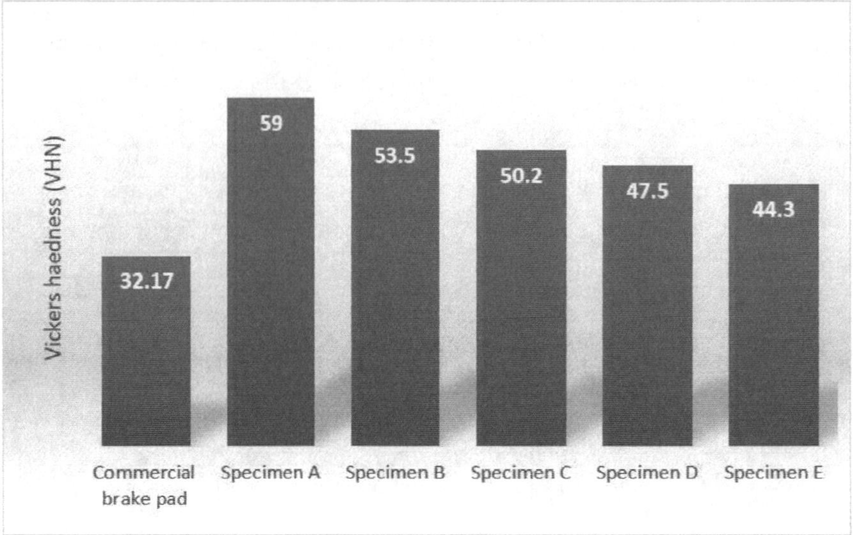

Figure 4. Vickers hardness test result of specimens and commercial brake pad.

The braking time was obtained from the braking test measurement. Figure 5 shows the variation of material composition and temperature of braking to the breaking time. Specimen E results a value of 20°C in 0.2 seconds, 50°C at 0.25 seconds, 70°C at 0.31 seconds and 100°C at 0.37 seconds and specimen D with a value of 20°C at 0.31 seconds, 50°C at 0.34 seconds, 70°C at 0.42 seconds and 100°C at 0.51 seconds. Braking time of specimen D and E are closer to braking time of commercial breaks. In addition, the braking time of specimens D and E are faster than specimens A, B, and C. It is happening because specimen E contains 40% coconut fiber and specimen D contain 30% coconut fiber and 10% wood powder, have a rough surface so their braking ability are better than the others. This graph shows higher temperature, give slower braking time. It causes by polyester matrix used in the brake pad will be soft at higher temperature, thus reduced braking ability. In addition, at high temperatures the specimen can releases of molecules such as H_2O and other volatile molecules from decomposition process that can affect the braking ability of brake pads.

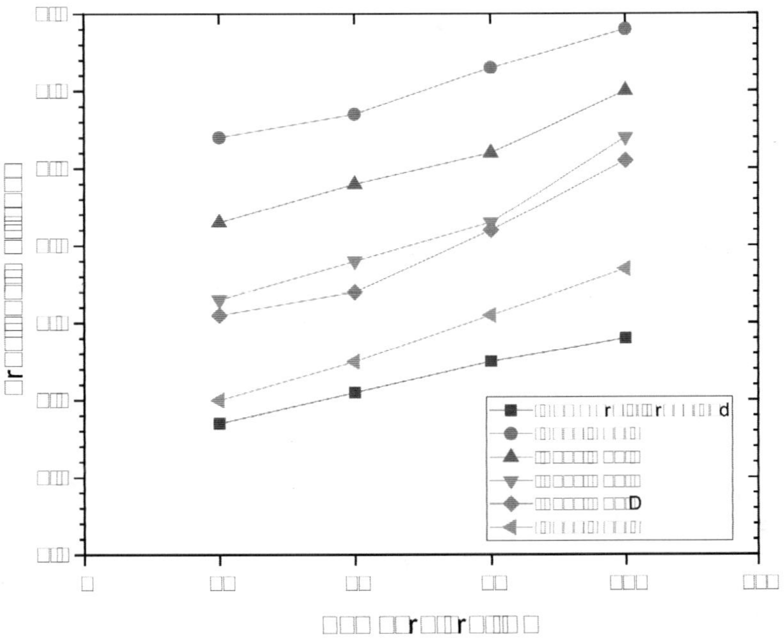

Figure 5. Disc braking time test result of specimens and commercial brake pad.

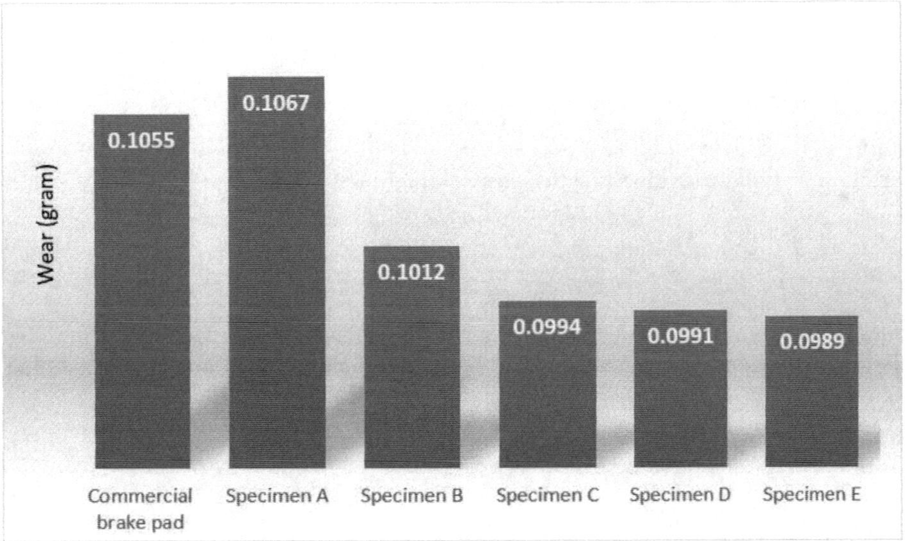

Figure 6. Wear test result of specimens and commercial brake pad.

Figure 6 shown relation between the wear of the brake pad to composition of wood powder and coconut fiber after braking test. The wear decreases as a composition of coconut fiber increasing. This graph as similar trend in the hardness value (see figure 5). The increasing of composition coconut fiber contributed to increasing hardness value which has affected to decrease wear of brake. From figure 6, it can be seen that specimens D and E have lower wear than specimen A, B, C and commercial brake pad, so that they have higher wear resistant than others. As the result, increasing coconut fiber content shown good wear resistance in braking.

Journal of Physics: Conference Series **2019** (2021) 012068 doi:10.1088/1742-6596/2019/1/012068

4. Conclusions

The test results of natural fiber composites from coconut fiber and wood powder on brake pads showed different characteristic values on hardness, wear, and braking time tests. The hardness value decreased along with the decrease in the composition of wood powder containing cellulose. The braking time test shows that the decrease in wood powder content affects the decrease in braking time due to the changed hardness properties. The decrease in the hardness value in the test results is also followed by the wear value in the variation of the specimen for each test. Natural composite material brake pads with excess coconut fiber composition looks close to the test results of commercial brake pads. The results of this study prove that coconut fiber waste and wood powder can be used for motorcycle brake pads through appropriate composition comparisons. It is still too early to conclude which natural fiber composition is better, further testing is needed to obtain a suitable composition to meet the brake pads criteria.

Acknowledgment

Thanks to the Head of LPPM UNJ which has provided an opportunity to complete this research through PTUPT scheme 2019, as well as to the Director of Research and Community Service of the Ministry of Research Technology and Higher Education who has provided this research grant.

References

[1] T. S. Document. *Light Vehicle Brake Systems*, Hal. 135. 2015.

[2] S. B. H. V. S. Aigbodion., U. Akadike and J. O. A. F. Asuke. Development of Asbestos - Free Brake Pad Using Bagasse, *Tribol. Ind.*, vol. 32, pp. 12–18, 2010.

[3] A. O. A. Ibhadode and I. M. Dagwa. *Development of Asbestos-Free Friction Lining Material from Palm Kernel Shell,* vol. XXX, no. 2, pp. 166–173. 2008.

[4] Albrecht W, Fuchs H;, and W. Kittelmann 2003 *non-woven: fabrics, raw materials, manufacture, applications, characteristics, test processing* (Weinheim: Wiley-Vch Verlag GmbH & Co. KCaA)

[5] S. Kalia et al. Cellulose-based bio- and nanocomposites: A review, *Int. J. Polym. SCI.*. 2011.

[6] E. Pelita, T. R. Hidayani, and A. Akbar. Analysis physical properties of composite polymer of coco fiber and polypropylene plastic waste with Maleic anhydrate as crosslinking agent, *IOP Conf. Ser. Mater. Sci. Eng*., vol. 223, no. 1. 2017.

[7] O. L. Rominiyi, B. A. Adaramola, O. M. Ikumapayi, O. T. Oginni, and S. A. Akinola. Potential Utilization of Sawdust in Energy, Manufacturing and Agricultural Industry; Waste to Wealth, *World J. Eng. Technol.*, vol. 05, no. 03, pp. 526–539. 2017.

[8] R. C. Pettersen. *The Chemical Composition of Wood*, pp. 57–126. 1984.

[9] S. S. Lawal, K. C. Bala, and A. T. Alegbede. *Development and production of brake pad from sawdust composite*, no. 30, pp. 47–56. 2017.

[10] K. K. Ikpambese, D. T. Gundu, and L. T. Tuleun. Evaluation of palm kernel fibers (PKFs) for production of asbestos-free automotive brake pads, *J. KING SAUD Univ. - Eng. Sci.*, pp. 1–9. 2014.

[11] U. D. Idris, V. S. Aigbodion, I. J. Abubakar, and C. I. Nwoye. Eco-friendly asbestos free brake-pad: Using banana peels, *J. King Saud Univ. - Eng. Sci.*, vol. 27, no. 2, pp. 185–192. 2015.

Effect of Milling Time on The Desorption Temperature of Mgh_2-10wt% Fe_3O_4 for Hydrogen Storage Materials Application

A Ulandari[1], A Rahwanto[1], E Handoko[2] and Z Jalil[1,*]
[1] Physics Department, FMIPA Universitas Syiah Kuala (USK), Banda Aceh, Indonesia
[2] Physics Department, FMIPA Universitas Negeri Jakarta (UNJ), Jakarta, Indonesia
Email: [1,*]zjalil@unsyiah.ac.id

Abstract. The effect of milling time on the desorption temperature of hydrogen storage material MgH2-Fe3O4 has been carried out. This study aims to determine the effect of milling time variation and the effect of adding Fe3O4 catalyst on the characterization of MgH2. Milling time variation using mechanical alloying technique, with milling time given for 10 hours, 20 hours, 30 hours with a ball ratio of 10: 1 and a speed of 350 rpm. MgH2–10wt% Fe3O4 was characterized using XRD and DSC. The results showed the phases of MgH2, MgO, and Fe3O4. The crystal grain size of MgH2 decreased from 11.81 nm to 8.35 nm. Variation of ownership time and the addition of Fe3O4 catalyst can reduce the desorption time from 377oC to 369oC.

1. Introduction

Hydrogen is an alternative energy that can be used as a substitute for fossil fuels whose availability is dwindling. Hydrogen was chosen because it has the advantages of high abundance, reproducibility, and zero-pollutant or environmentally friendly emissions during combustion [1] The advantages of hydrogen have prompted researchers to use hydrogen as a substitute fuel for transportation. Hydrogen can be used as a substitute for fuel (fuel cell) by converting energy into electricity, a fuel cell is an electrochemical device that converts chemical energy into electrical energy [2].

Fuel cell-based fuels are often used in the automotive industry in motor vehicles. However, when applied, it still inhibits the hydrogen storage system (hydrogen storage). Hydrogen storage consists of three types of storage, namely in the form of gas, liquid, and solid. In this study, hydrogen storage is used in solid form. This is because hydrogen in solid form is better able to store or absorb hydrogen in large quantities, is lighter and at a more economical price. Light metal elements that have the ability to absorb hydrogen in large quantities are magnesium (Mg). Magnesium was chosen because it has the ability to absorb hydrogen in large amounts (7.6 weight%). However, magnesium has a weakness, namely the temperature required for desorption is very high (> 300 ^{Oz}) and the time required is quite long. The World Energy Agency (IEA) is able to absorb 5wt % hydrogen at a temperature of 100 ^{0}C and the desorption time is less than 1 hour.

The weakness of MgH_2 can be minimized by adding a catalyst, the catalyst that is commonly used is Fe_3O_4. The addition of Fe_3O_4 catalyst can reduce the desorption temperature from 310°C to 200°C [3]. The addition of Fe_3O_4 catalyst can keep the desorption temperature lower [4]. Jalil *et al*. [5] Tried to add Ni catalyst, the results obtained changed the crystalline nature of MgH_2, namely a widening of the diffraction peak and it could reduce the desorption temperature peak. Variations in milling time can also affect the crystal grain size and widen the diffraction peaks of MgH_2 [6-9]. In this study, we present the results of adding Fe_3O_4 catalyst and produced by variations in milling time using mechanical alloying method.

Journal of Physics: Conference Series **2019** (2021) 012069 doi:10.1088/1742-6596/2019/1/012069

2. Method

The materials used were magnesium hydride (Sigma Aldrich, Germany 99.9%) and Fe_3O_4 powder (Sigma Aldrich, Germany 99.9%). The mechanical alloying process was then carried out with a planetary ball mill machine (Fritsch, P6). The ratio of the weight of the ball and the powder used is 10:1. The composition of the weight used is 2 grams with a composition of 10wt% Fe_3O_4. Then the milling time was varied from 10 hours, 20 hours, and 30 hours with a milling speed of 350 rpm. Identification of the phase composition present in the material was carried out using X-Ray diffraction (XRD) Shimadzu D6000, Cu-Kα with = 1.54060. Furthermore, the thermal properties of hydrogen desorption were tested using the Shimadzu D50 Differential Scanning Calorimetry (DSC). The data obtained from the XRD test results were calculated by calculating the crystal size of the identified material, using the Scherrer method.

3. Results and Discussion

The X-ray diffraction pattern with variations in milling time is shown in Figure 1, where the detected phases are MgH_2, Fe_3O_4, and a small peak MgO. The MgH_2 phase appears at angles of $2\theta = 38.05°$, 37.94°, 37.93°, 52.46°, and 44.65°, while the Fe_3O_4 phase appears at angles of $2\theta = 16.7018$ and 35.46°. The MgO phase appears in samples that have been milled, this is influenced by outside air contamination during sample preparation. The MgO phase is considered as an impurity because it can affect the desorption properties of hydrogen [8,10].

Figure 1. Shows the X-ray diffraction pattern after milling time variations of 10 hours, 20 hours, and 30 hours, it is seen that there is an increase in the intensity value and peak widening. The widening of the peaks indicates that the crystal size reduction process occurs, resulting in an increase in the material surface of grain volume ratio. Therefore, hydrogen will easily interact and absorb in the MgH_2 material [5,7,11]. The Fe_3O_4 phase becomes a minor phase due to the peak of the pattern that appears at a very low intensity. This proves that Fe_3O_4 which acts as a catalyst, but does not participate in the reaction in MgH_2. So that Fe_3O_4 is expected to accelerate the hydrogen desorption process as needed.

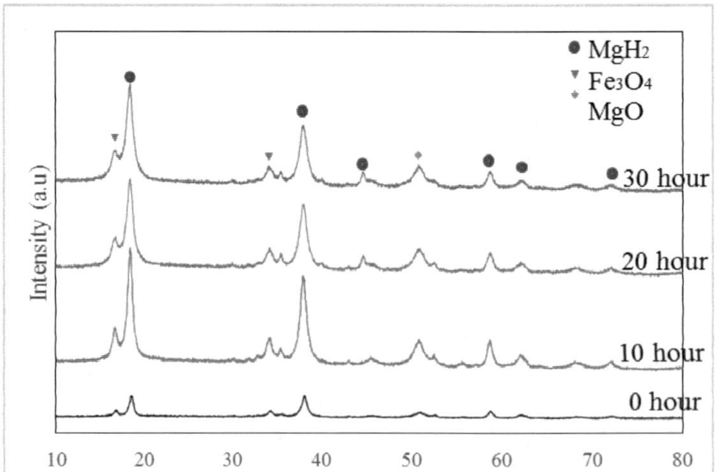

Figure 1. XRD pattern of MgH_2-10wt%Fe3O4 material with variations in milling time.

The results of the calculation of the crystal size of MgH_2–10wt% Fe_3O_4 are shown in Table 1, where the calculation results for each MgH_2 phase field with milling time variations since 10, 20, and 30 hours. From the calculation results show that the longer the milling time, the smaller the crystal grain size [12]. This shows that the mechanical alloying process is quite effective in reducing the grain size of the MgH_2-10wt% Fe_3O_4 material. The smaller crystal grain size is influenced by the presence of a balanced collision between the milling balls with the sample and the vial wall in the milling process [13]. The smaller crystal size is also influenced by the occurrence of peak widening which is influenced by the X-ray reflected field [14].

Table 1. Crystal Size of the MgH_2 phase with variations in milling time

Nr.	MgH_2–10wt% Fe_3O_4	θ (°)	FWHM	Crystal size (nm)
1	0 hour	19,0257	0.75050	11.81
2	10 hourss	18,9664	0.86640	10.22
3	20 hours	18,9735	1.02960	8.61
4	30 hours	18,9720	1.06150	8.35

In Figure 2, information is obtained that the peak desorption temperature of the MgH_2 material before milling is at a temperature of 377 ^{O}C. After milling for 10 hours, 20 hours, and 30 hours there was a significant decrease in temperature as shown in Table 2. The decrease in desorption temperature was still very large, namely 369 with the release of hydrogen of 1.463 mg in 426 seconds. This is due to the presence of impurities during the sample preparation process. The decrease in desorption temperature is influenced by the role of the Fe_3O_4 catalyst [15]. Based on XRD testing, the crystal size of MgH_2 has shown nanocrystals size, so that the hydrogen diffusion process was occurring and can be released [16,18]. The addition of Fe_3O_4 proves that according to the desorption temperature, the crystal size of MgH_2 is getting smaller. Thus, improving thermal properties and maintaining hydrogen storage capacity [19,20].

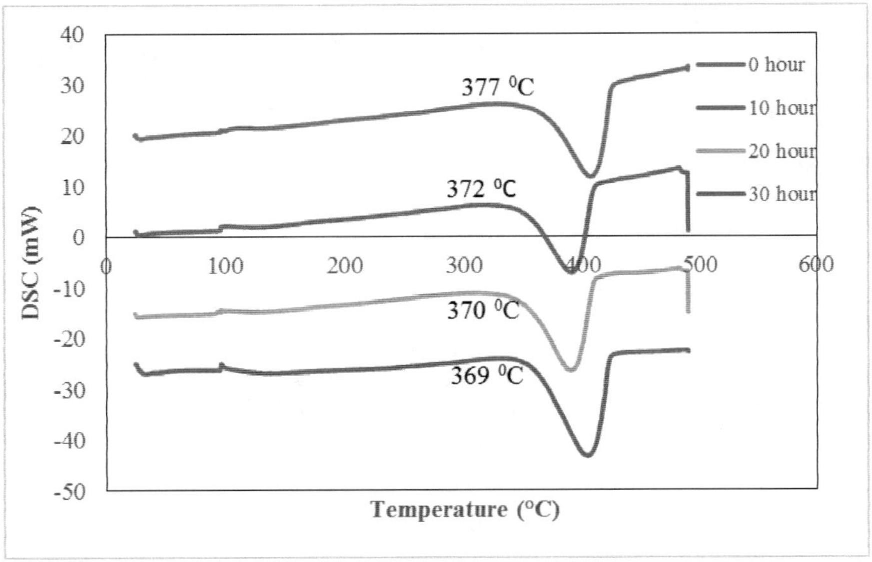

Figure 2. DSC curve of MgH_2-10wt% Fe_3O_4 with variations in milling time.

Table 2. DSC results on MgH_2–10wt% Fe_3O_4 with variations in milling time

Nr.	Milling Time	Temperature (°C)
1	0 hour	377
2	10 hourss	372
3	20 hours	370
4	30 hours	369

4. Conclusion
Based on the results of the study, it was found that variations in milling time can affect the crystalline of MgH_2 and an increase in intensity and peak widening occurs. So that the reduction of the crystal grain size resulted in a decrease in the crystal size from 11.81 nm to 8.35 nm with the longer milling time. The addition of Fe_3O_4 catalyst and variations in milling time can also reduce the desorption temperature from 377 ^{O}C to 369 ^{O}C.

Acknowledgement
After completing this research, the author would like to thank the Material and Energy Laboratory FMIPA, the Chemical Engineering Laboratory of Syiah Kuala University, Banda Aceh and the Geology Laboratory in Bandung in supporting the structural and thermal properties measurements.

References
[1] Wang Y and Wang Y 2017 *Recent Advances in Additive-Enhanced Magnesium Hydrides for Hydrogen Storage Progress in Natural Science: Materials International* 27, pp. 41-49.
[2] Hanada N, Ichikawa T, Fujii H 2005 *Catalytic effect of nanoparticle 3d-transition metals in hydrogen storage properties in magnesium hydride MgH2 prepared by mechanical milling, J Phys Chem B*, 109 (15) 7188−7194.
[3] Wang H, Zhang J, Liu J W, Ouyang, L Z and Zhu, M 2013 *Improving Hydrogen Storage Properties of MgH2 by addition of Alkali Hydroxide, International Journal of Hydrogen Energy,* pp. 38 10932–10938.
[4] Sabitu S T and Goudy A J 2012 *Dehidrogenation Kinetics and Modelling Studies of MgH2 Enhanced by Transition Metal Oxide Catalysts Using Constant Pressure Thermodynamic Driving Force, Open Access Metal*, ISSN 2075-4701
[5] Jalil Z, Rahwanto A, Handoko E 2018 *Effect of nanostructure Ni on the sorption properties of mechanical milled MgH_2 MATEC, Web of Conferences,* pp. 197 02014
[6] Jalil Z, Rahwanto A, Ismail I, Sofyan H, Handoko E 2018 *The use of nano-silicon carbide and nickel as catalyst in magnesium hydrides (MgH_2) for hydrogen storage material application*, *Mater. Res. Express,* pp. 5 064002
[7] Malahayati, Ismail, Mursal, Jalil Z 2018 *The use of silicon oxide extracted from rice husk ash as catalyst in magnesium hydrides (MgH_2) prepared by mechanical alloying method*, *J. Phys.: Conf. Ser.*, pp. 1120 012061
[8] Rahwanto A, Jalil Z, Akhyar, Handoko E 2020 *IOP Conf. Series: Materials Science and Engineering*, pp. 931 012012
[9] Sharbati M and Kashani-Bozorg S F 2012 *Acta Physica Polonica A,* pp. 121 211-213
[10] Zhang Q, Zang L, Yike H, Gao P, Jiao L, Yuan H and Wang Y 2017 *Improve Hydrogen Storage Properties of MgH2 with Ni-based Compound, International Journal of Hydrogen Energy,* pp. 42 24247-24255

[11] Jalil Z, Rahwanto A, Sofyan H, Usman M and Handoko E 2018 *The Use of Silica from beach sand as catalyst in Magnesium Based Hydrides for Hydrogen Storage Materials*, *IOP Conf. Series: Earth and Environmental Science*, pp. 105 012093

[12] Ranjbar A, Guo Z P, Yu X, B, Attard D, Calka A and Liu H K 2009 *Effects of SiC nanoparticles with and without Ni on the hydrogen storage properties of MgH_2*, *International of Hydrogen Energy*, pp. 34 7263-7268.

[13] Jalil Z, Rahwanto A, Akhyar H, Razali R and Handoko E 2018 *MgH_2-SiC Based Hydrogen Storage Material Prepared by Reactive mechanical alloying method*, *IOP Conf. Series: Earth and Environmental Science,* pp. 105 012098

[14] Kwak YJ, Lee SH, Mumm DR and Song MY 2015 Development of a Mg-based Hydrogen Storage Material by Addition of Ni and NBF5 Via Milling Under Hydrogen. International Journal of Hydrogen Energy 40 11908-11916

[15] Li-Shuai X, Jin-Shan L, Tie-bang Z, Hong-Chao, K 2017 Role of milling time and Ni content on dehydrogenation behavior of MgH2/Ni composite Trans. Nonferrous Met. Soc. China 27 569−577

[16] Wang H, Zhang S F, Liu J W, Ouyang, L Z, Zhu M 2012 The MgH2 dehydrogenation property improvement was attributed to the nanosized effect of mesoporous silica. Materials Chemistry and Physics 136 146-150

[17] Jalil Z, Rahwanto A, Mustanir, Akhyar, Handoko E 2017 AIP Conference Proceedings 1862 030023

[18] Ismail I, Aini Q, Zulfalina, Jalil Z, Fadzullah S H S M 2018 Journal of Physics: Conf. Series 1120 012014

[19] Malahayati M, Yufita Z, Ismail I, Mursal M, Idroes R, Jalil Z 2021 *Hydrogen Desorption Properties of MgH_2 + 10 weight% SiO2 + 5 weight% Ni Prepared by Planetary Ball Mill*, *Bulletin of Chemical Reaction Engineering & Catalysis,* pp. 16 (2) 280-285

[20] Rahwanto A, Ismail I, Nurmalita N, Mustanir, Jalil Z 2021 *Nanoscale Ni as a catalyst in MgH_2 for hydrogen storage material, Journal of Physics: Conference Series,* 1882(1)

Python-based Program for Analysing Lattice Parameter of Cubic and Tetragonal Crystal Structure

A Rizky[1], A Fadhillah[1], Irzaman[1] and Irmansyah[1]
[1]Department of Physics, IPB University Campus Dramaga, Bogor, Indonesia
Email: ahmad_rizky@apps.ipb.ac.id

Abstract. A program for analysing lattice parameter of a crystal structure has been successfully created based on Python programming language using the Google Collaboratory service, so it can be accessed through a PC or smartphone as long as internet access is available. This program can be used to calculate lattice parameter of a crystal structure with x-ray diffraction data as the input. Crystal structures that can be calculated for its lattice parameter are cubic and tetragonal. The program will ask for the type of crystal structure of the data, along with diffraction angles and miller indices. The input will be processed using the Cramer-Cohen method according to the previously entered crystal structure. By also entering the wavelength used, the output of this program is the lattice parameter in the angstrom unit. The percentage of error of this program's output is extremely low.

1. Introduction

Solid substances are often found in various areas of human daily life. Utilization of solids is carried out by analysing the characteristics of these solids. Various characteristics of the solid material can be determined by analysing the structure and phase of the crystal, also its lattice parameter [1]. Research on analysing the crystal structure of a solid material is very interesting, as did previous researchers. Previous researchers have succeeded in analysing and determining the crystal structure of various materials, including lithium aluminum silicate glass-ceramics [2] and selenium-doped hydroxyapatite [3].

One of the common experimental methods used in determining the structure, phase and lattice parameter of a crystal is x-ray diffraction (XRD). X-ray diffraction uses the phase difference that occurs between x-rays hitting the parallel lattice planes to calculate the crystal lattice parameter based on Bragg's Law [4]. The output resulting from x-ray diffraction is a graph imaging the relationship between the diffraction angle for each miller index and the intensity of the x-rays received back by the detector.

There are seven types of crystal structures, i.e. Cube, Tetragonal, Orthorhombic, Hexagonal, Rhombohedral, Monoclinic and Triclinic [5]. Each structure has a different equation for distance between parallel lattice planes (d-spacing). The simplest method of lattice parameter's analysis is the Cramer-Cohen method. The Cramer method is a method to solve linear matrix equations by counting the determinant of the matrix involved in the equation [6].

Over time, mathematical equations that previously had to be solved by hand can be done using applications or programs. One programming language that has been widely used to solve mathematical equations is Python. Research on the Python programming language is very interesting, as did previous researchers. Previous researchers have successfully applied the python programming language to simplify various types of mathematical and physical cases, including Python to learn programming [7], PDB File Parser and Structure Class Implemented in Python [8], Emcee V3: A

Python Ensemble Sampling Toolkit for Affine-Invariant MCMC [9], pyhf: Pure-python Implementation of HistFactory Statistical Models [10].

Most people use Microsoft Excel to analyse lattice parameter. Research on making application to analysing lattice parameter has been carried out previously as an android-based application [11]. Apart from using the application, lattice parameter analysis can also be done with code-based program using Python as the programming language. This research was conducted with the aim of providing an analysis tool for crystal structure lattice parameter that is easier and more efficient to use without additional occupied memory on the device and can be accessed anywhere as long as the internet service is provided.

2. Method

The first step taken in this research is the literature review. The next step is building the code-based program with Cramer-Cohen method as an algorithm and Python as the programming language. The final step in this research is testing the performance of the program in performing calculations.

2.1. Literature Review

The literature review is the first step taken in this research. Topics that being reviewed from the literature are about Bragg's law, its applications for XRD, crystal structure and Cramer-Cohen method for analysing lattice parameter in cubic and tetragonal crystal structure.

2.1.1. Bragg's law.

Bragg's law is the main principle used in the x-ray diffraction method. Two x-rays hitting some parallel lattice planes must interfere each other. Constructive interference happens when the difference path length of two rays is equal to multiple of positive integers of the x-ray wavelength used. Bragg's law is mathematically formulated in equation (1),

$$2d\sin\theta = n\lambda \tag{1}$$

which d represents path's length between two lattice planes with same *hkl* value [12]. For analysing two rays in two neighboured parallel planes, substitute n = 1, so equation (2) is obtained.

$$\frac{1}{d} = \frac{2\sin\theta}{\lambda} \tag{2}$$

2.1.2. Lattice parameter.

Those seven types of crystal structure has their unique characteristic of lattice parameter compare to the others [5]. Lattice parameter is important to be determined, because it describes the size of the unit cell of a material. This knowledge could lead to many different potential applications of the material [13]. Lattice parameter contains a set of constants denoted as "a", "b", "c", "α", "β" and "γ". "a", "b" and "c" describe distance between atoms on unit cell's x, y and z axis, respectively. Meanwhile, "α", "β" and "γ" describe angle between unit cells's y and z axis, x and z axis, also x and y axis, respectively.

2.1.3. Cramer-Cohen method for cubic crystal structure.

A cube crystal structure is a crystal structure that has unit cells with the lattice parameter a = b = c and α = β = γ = 90° [5]. Therefore, a cube crystal needs 1 lattice parameter only to describe its characteristics, which is usually denoted as 'a'. The d-spacing in the cube crystal is shown in equation (3),

$$d = \frac{a}{\sqrt{h^2 + k^2 + l^2}} \tag{3}$$

with *h, k* and *l* are plane's dimensions on *x, y* and *z* axis, respectively [14]. Equation (3) is then compared with equation (2), so equation (4) is obtained.

$$\frac{h^2+k^2+l^2}{a^2}=\frac{4\sin^2\theta}{\lambda^2} \tag{4}$$

$$\sin^2\theta-\frac{\lambda^2}{4a^2}\left(h^2+k^2+l^2\right)=0 \tag{5}$$

By replacing zero in the equation above with error during measurement, according to what Cohen has derived, that is expressed as $D\sin^2 2\theta$ [15], equation (6) is obtained,

$$\sin^2\theta-\frac{\lambda^2}{4a^2}\left(h^2+k^2+l^2\right)=D\sin^2 2\theta \tag{6}$$

$$\sin^2\theta=\frac{\lambda^2}{4a^2}\left(h^2+k^2+l^2\right)+D\sin^2 2\theta \tag{7}$$

$$\sin^2\theta=S\chi+T\psi \tag{8}$$

with $S=\frac{\lambda^2}{4a^2}, T=\frac{D}{10}, \chi=h^2+k^2+l^2$ and $\psi=10\sin^2 2\theta$. Transform equation (8) into a linear matrix equation shown in equation (9).

$$\begin{pmatrix}\Sigma\chi\sin^2\theta\\ \Sigma\psi\sin^2\theta\end{pmatrix}=\begin{pmatrix}\Sigma\chi^2 & \Sigma\chi\psi\\ \Sigma\chi\psi & \Sigma\psi^2\end{pmatrix}\begin{pmatrix}S\\ T\end{pmatrix} \tag{9}$$

Solution of the equation above can be found using the Cramer method. This is Cramer-Cohen method for cubic crystal structure [11].

2.1.4. Cramer-Cohen method for tetragonal crystal structure.
A tetragonal crystal structure is a crystal structure that has unit cells with the lattice parameter a = b ≠ c, but α = β = γ = 90° [5]. Therefore, a tetragonal crystal need 2 lattice parameters describe its characteristics, which are usually denoted as 'a' and 'c'. The d-spacing in the tetragonal crystal is shown in equation (10),

$$\frac{1}{d^2}=\frac{h^2+k^2}{a^2}+\frac{l^2}{c^2} \tag{10}$$

with *h, k* and *l* are plane's dimensions on *x, y* and *z* axis, respectively [16]. Equation (10) is then compared with equation (2), so equation (11) is obtained.

$$\frac{h^2+k^2}{a^2}+\frac{l^2}{c^2}=\frac{4\sin^2\theta}{\lambda^2} \tag{11}$$

$$\sin^2\theta-\frac{\lambda^2}{4}\left(\frac{h^2+k^2}{a^2}+\frac{l^2}{c^2}\right)=0 \tag{12}$$

By replacing zero in the equation above with error during measurement, according to what Cohen has derived, that is expressed as $D\sin^2 2\theta$[15], equation (13) is obtained,

$$\sin^2\theta-\frac{\lambda^2}{4a^2}\left(h^2+k^2\right)-\frac{\lambda^2}{4c^2}\left(l^2\right)=D\sin^2 2\theta \tag{13}$$

$$\sin^2\theta=\frac{\lambda^2}{4a^2}\left(h^2+k^2\right)+\frac{\lambda^2}{4c^2}\left(l^2\right)+D\sin^2 2\theta \tag{14}$$

$$\sin^2\theta=R\phi+S\chi+T\psi \tag{15}$$

with $R=\dfrac{\lambda^2}{4a^2}, S=\dfrac{\lambda^2}{4c^2}, T=\dfrac{D}{10}, \phi=h^2+k^2, \chi=l^2$ and $\psi=10\sin^2 2\theta$. Transform equation (15) into a linear matrix equation shown in equation (16).

$$\begin{pmatrix}\Sigma\phi\sin^2\theta\\ \Sigma\chi\sin^2\theta\\ \Sigma\psi\sin^2\theta\end{pmatrix}=\begin{pmatrix}\Sigma\phi^2 & \Sigma\phi\chi & \Sigma\phi\psi\\ \Sigma\phi\chi & \Sigma\chi^2 & \Sigma\chi\psi\\ \Sigma\phi\psi & \Sigma\chi\psi & \Sigma\psi^2\end{pmatrix}\begin{pmatrix}R\\ S\\ T\end{pmatrix} \tag{16}$$

Solution of the equation above can be found using the Cramer method. This is Cramer-Cohen method for tetragonal crystal structure.

2.2. Algorithm of the program

The algorithm used in this program is the Cramer-Cohen method. When the program runs, the program will ask the user to type in the code of crystal structure (1 for cubic and 2 for tetragonal). After that, user to type in the number of rows of the x-ray diffraction data. Then, the diffraction angles and miller indices have to be input by the user one by one. The program will process the inputs with Cramer-Cohen method according to the structure. As the output, the program will display the result of the lattice parameter calculation also its crystal structure's characteristics.

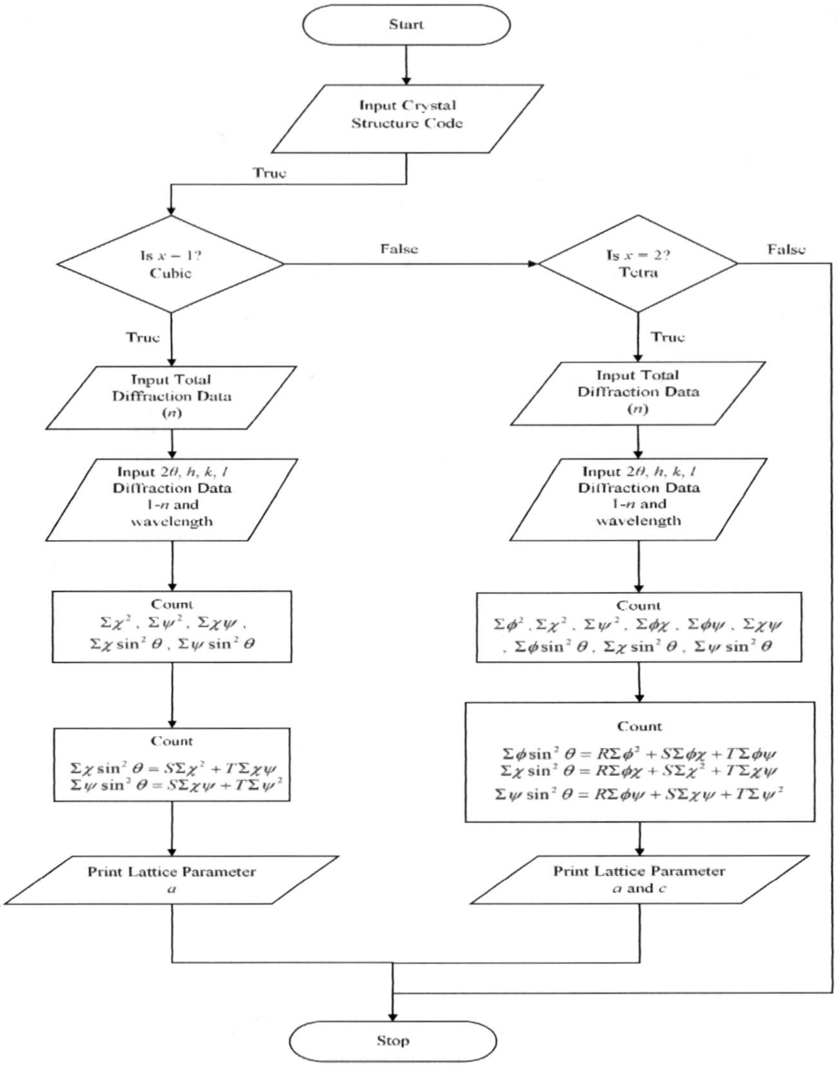

Figure 1. Algorithm of the program using Cramer-Cohen method

3. Result and Discussions

The final output of this research is the code-based program. Using Python as the programming language and Cramer-Cohen method as the algorithm, the program runs accurately to analysing lattice parameter of XRD data.

3.1. User interface of the program

This program is created through Google Collaboratory service, a cloud service based on Jupyter Notebook provided by Google. This service can be used for programming in various languages, including Python. Therefore, the program that has been created can be accessed on a PC or smartphone as long as the program files are accessible and the internet network is available. Usage of different version of browsers does not affect the calculation results obtained, but may provide a different user experience and affect the length of time it takes to run the program. User interface in Microsoft Edge 91.0.864.48 version is shown in Figure 2.

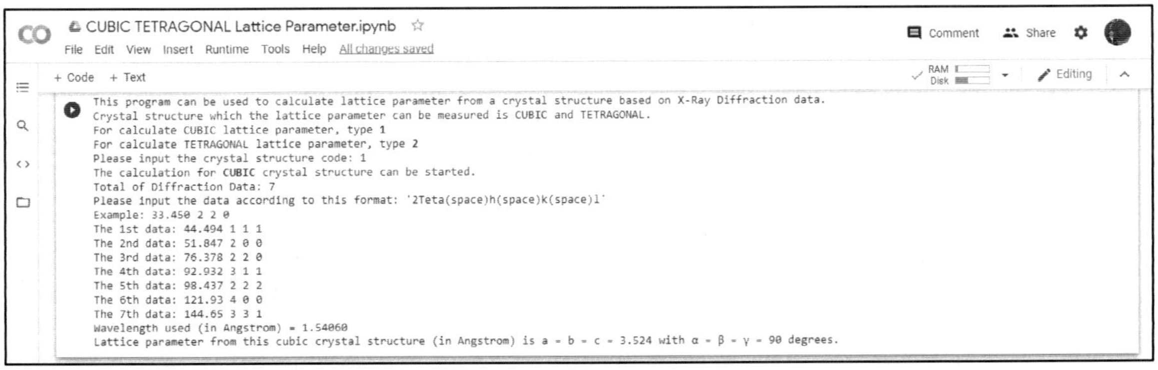

Figure 2. The user interface of the program for calculating Nickel (Ni) lattice parameter in Microsoft Edge 91.0.864.48 version.

The user has to be careful while inputting *hkl* and diffraction angle value from the XRD data. Data should be inputted according to the format shown in the initial program instruction. If any inputted data is wrong or not as same as the format, the program will stop at the time and the user should reinput the data from the beginning. It is well-tolerated if this program compared to Microsoft Excel, because this program does not need multiple file according to the structure. It only needs a file and the code structure, then the program will analyse the XRD data smoothly.

3.2. Evaluation test for the program's performance

The program is tested for analysing the diffraction angle and the miller index to determine the lattice parameter of the sample. The literature data of lattice parameter of these samples were taken from the International Centre for Diffraction Data (ICDD) [17].

Table 1. Results of lattice parameter measured by the program.

Crystal Structure	Sample	Lattice Parameter (Å)					
		a			*c*		
		Lit[a]	Mea[b]	%Error	Lit[a]	Mea[b]	%Error
Cubic (a = b = c)	Cesium Chloride (CsCl)	4.123	4.123	0	4.123	4.123	0
	Barium Strontium Titanium Oxide ($Ba_{0.5}Sr_{0.5}TiO_3$)	3.947	3.947	0	3.947	3.947	0
	Cadmium Telluride (CdTe)	6.481	6.482	0.015	6.481	6.482	0.015
	Nickel (Ni)	3.524	3.524	0	3.524	3.524	0
	NiO (Nickel Oxide)	4.194	4.194	0	4.194	4.194	0
	CaO (Calcium Oxide)	4.810	4.811	0.021	4.810	4.811	0.021
	CdS (Cadmium Sulfide)	5.818	5.818	0	5.818	5.818	0
Tetragonal (a = b ≠ c)	Zirconium Oxide (ZrO_2)	3.598	3.599	0.028	5.152	5.152	0
	Copper Selenium Telluride (Cu_2TeSe_4)	5.449	5.459	0.184	5.578	5.587	0.161
	Silver Indium ($AgIn_2$)	6.869	6.867	0.029	5.604	5.606	0.036
	Lithium Iron Oxide ($Li_2Fe_2O_4$)	4.049	4.049	0	8.742	8.742	0
	Chromium Tungsten Oxide (Cr_2WO_6)	4.580	4.580	0	8.867	8.867	0

[a] Literature data from ICDD [17].
[b] Result provided by the program.

As seen from the results above, the percentage of error of the calculation result that program provide is extremely low. The percentage of error is less than 0.2 percent. It means that this program is very accurate for analysing lattice parameter of XRD data and also can be an alternative tool that is easier to use compared to the existing tools, such as Microsoft Excel.

4. Conclusion

A python-based program for calculating lattice parameter of cubic and tetragonal crystal structures has been successfully created using the Cramer-Cohen method as the algorithm. This program is created using the Google Collaboratory service, so it can be accessed through a PC or smartphone as long as the program's file is accessible and the internet access is available. This program is very accurate and can be an efficient tool for analysing lattice parameter.

Acknowledgments

This research was funded by Hibah Penelitian Terapan Unggulan Perguruan Tinggi (PTUPT) from Deputy of Research and Development, Ministry of Research and Technology/National Research and Innovation Agency of Republic Indonesia with contract number: 1/E1/KP.PTNBH/2021 dated March 8th, 2021.

References

[1] Oponowicz A, Marciszko-Wiackowska M, Baczmanski A, Klaus M, Genzel C, Wronski S, Kollbek K and Wrobel M 2020 *Metallurgical and Materials Transactions A* 51(11) 5945-57
[2] Xia L, Yang Y, Zhang X, Zhang J, Zhong B, Zhang T, Wang H and Wen G 2018 *Ceramics International* 44(12) 14896-900
[3] Wei L, Pang D, He L and Deng C 2017 *Ceramics International* 43(18) 16141-8
[4] Pozo C, Rodriguez-Llamazares S, Bouza R, Barral L, Castano J, Muller N and Restrepo I 2018 *Journal of Polymer Research* 25(12) 266
[5] Kittel C 2005 *Introduction to Solid State Physics: Eighth Edition* (New Jersey: John Wiley & Sons, Inc) p 9
[6] Kyrchei I 2018 *Advances in Applied Clifford Algebras* 28(1) 1-16

Journal of Physics: Conference Series **2019** (2021) 012070 doi:10.1088/1742-6596/2019/1/012070

[7] Bogdanchikov A, Zhaparov M and Suliyev R 2013 *Journal of Physics: Conference Series* 423 012027
[8] Hamelryck T and Manderick B 2003 *Bioinformatics Applications Note* 19(17) 2308-10
[9] Foreman-Mackey D *et al.* 2019 *The Journal of Open Source Software* 43(4) 1864
[10] Heinrich L, Feickert M, Stark G and Cranmer K 2021 *The Journal of Open Source Software* 6(58) 2823
[11] Kurniawan A and Irzaman 2021 *AIP Conference Proceedings* 2320 030001
[12] Nguyen T, Sierra E, Eguiraun H and Lizundia E 2018 *European Journal of Physics* 39(4) 045803
[13] Maskery I, Aremu A O, Simonelli M, Tuck C, Wildman R D, Ashcroft I A and Hague R J M 2015 *Experimental Mechanics* 55(7) 1261-72
[14] Akl A A S and Elhadi M 2020 *Journal of Ovonic Research* 16(5) 323-35
[15] Suryanarayana C and Norton M G 2013 *X-ray Diffraction: A Practical Approach* (Berlin: Springer Science & Business Media) p 160
[16] Jamal Z, Idris M S, Irzaman and Barmawi M 2007 *Journal of Nuclear and Related Technology* 4 137-42
[17] PDF-ICDD International Centre for Diffraction data 1998 *PCPDFWIN* 2 (1)

Journal of Physics: Conference Series **2019** (2021) 012071 doi:10.1088/1742-6596/2019/1/012071

Analysis of Lattice Constants and Error for The Hexagonal Crystal Structure of Silicon Dioxide Using The Cramer-Cohen Method

N Kurniawati[1*] D A P Wardani[2], B Hariyanto[1], N P Har[3], N Darmawan[4] and Irzaman[3]
[1]Program Studi Fisika, FMIPA, Universitas Palangka Raya, Palangka Raya, Kalimantan Tengah, Indonesia, 73111
[2]Program Studi Kimia, FMIPA, Universitas Palangka Raya, Palangka Raya, Kalimantan Tengah, Indonesia, 73111
[3]Departemen Fisika, FMIPA, IPB, Kampus IPB Dramaga Bogor, Jawa Barat, Indonesia, 16680
[4]Departemen Kimia, FMIPA, IPB, Kampus IPB Dramaga Bogor, Jawa Barat, Indonesia, 16680
Email: *nenykurniawati@mipa.upr.ac.id

Abstract. Has successfully analyzed the lattice constants and the error of the hexagonal crystal structure of silicon dioxide (SiO2) using the Cramer-Cohen method. The peak data for the silicon dioxide material used are secondary data from the ICDD. The Cramer-Cohen method calculations show that the lattice constants are relatively the same as the secondary data from the ICDD, with an average error analysis value of 0,001531805%. This shows that the analysis of the lattice constant and the error of the hexagonal crystal structure of silicon dioxide (SiO_2) using the Cramer-Cohen method is very accurate.

1. Introduction

XRD is beneficial for studying the crystal structure, chemical composition, and physical properties of nanomaterials [1]. X-ray diffraction occurs in the elastic scattering of X-ray photons by atoms in a periodic lattice. The monochromatic scattering of X-rays in this phase provides constructive interference [2]. The data obtained in this characterization are the intensity value to form a diffraction pattern.

The purpose of this study was to analyze the lattice constants of the hexagonal SiO2 structure processed using the Cramer-Cohen method based on secondary data from ICDD hexagonal SiO2 and its error analysis.

2. Methods

Analyzed the lattice constants and the error of the hexagonal crystal structure of silicon dioxide (SiO2) using the Cramer-Cohen method. The peak data for the silicon dioxide material used are secondary data from the ICDD. The Cramer-Cohen method calculations show that the lattice constants are relatively the same as the secondary data from the ICDD. The lattice constants of the SiO2 hexagonal structure were processed using the Cramer-Cohen method based on the first 15 peaks of secondary ICDD data according to Table 1.

Journal of Physics: Conference Series **2019** (2021) 012071 doi:10.1088/1742-6596/2019/1/012071

Table 1. SiO_2 Powder ICDD Secondary Data

46-1045 ★

SiO_2

Silicon Oxide

Quartz, syn

Rad. CuKα_1 **λ** 1.540598 **Filter** Ge Mono. **d-sp** Diff.
Cut off **Int.** Diffractometer **I/I$_{cor.}$** 3.41
Ref. Kern, A., Eysel, W., Mineralogisch-Petrograph. Inst., Univ. Heidelberg, Germany, *ICDD Grant-in-Aid*, (1993)

Sys. Hexagonal **S.G.** P3$_2$21 (154)
a 4.91344(4) b c 5.40524(8) A C 1.1001
α β γ Z 3 mp
Ref. Ibid.

D_x 2.65 D_m 2.66 **SS/FOM** F_{30}=539(.002,31)

εα nωβ 1.544 εγ 1.553 **Sign** + **2V**
Ref. Swanson, Fuyat, *Natl. Bur. Stand. (U.S.), Circ. 539*, **3** 24 (1954)

Color White
Integrated intensities. Pattern taken at 23(1) C. Low temperature quartz. 2θ determination based on profile fit method. O_2Si type. Quartz group. Silicon used as internal standard. PSC: hP9. To replace 33-1161. Structure reference: Z. *Kristallogr.*, **198** 177 (1992).

See following card.

dÅ	Int	hkℓ	dÅ	Int	hkℓ
4.2550	16	100	1.1530	1	311
3.3435	100	101	1.1407	<1	204
2.4569	9	110	1.1145	<1	303
2.2815	8	102	1.0816	2	312
2.2361	4	111	1.0638	<1	400
2.1277	6	200	1.0477	1	105
1.9799	4	201	1.0438	<1	401
1.8180	13	112	1.0346	1	214
1.8017	<1	003	1.0149	1	223
1.6717	4	202	0.9896	<1	115
1.6592	2	103	0.9872	<1	313
1.6083	<1	210	0.9783	<1	304
1.5415	9	211	0.9762	<1	320
1.4529	2	113	0.9608	<1	321
1.4184	<1	300	0.9285	<1	410
1.3821	6	212	0.9182	<1	322
1.3750	7	203	0.9161	2	403
1.3719	5	301	0.9152	2	411
1.2879	2	104	0.9089	<1	224
1.2559	3	302	0.9009	<1	006
1.2283	1	220	0.8972	<1	215
1.1998	2	213	0.8889	1	314
1.1978	<1	221	0.8814	<1	106
1.1840	2	114	0.8782	<1	412
1.1802	2	310	0.8598	<1	305

[3]

X-ray beams that come into the crystal plane will produce peaks of diffraction intensity if the scattering angle and wavelength of the X-rays satisfy Bragg's law in equation (1) [4]:

$$d\sin\theta = n\lambda \tag{1}$$

The index of the diffraction pattern of the hexagonal crystal system must satisfy the equation (2): [4]

$$\sin^2\theta = A(h^2 + hk + k^2) \tag{2}$$

where,

$$A = \frac{\lambda^2}{3a^2} \quad \text{and} \quad C = \frac{\lambda^2}{4c^2} \tag{3}$$

Analysis of lattice constants based on the first 15 peaks of secondary data of ICDD SiO_2 with a hexagonal crystal structure (Figure 1) using a matrix of order 3 x 3 in the equation (4) of Cramer and Cohan: [4]

$$\begin{aligned} \textstyle\sum \alpha \sin^2\theta &= C\textstyle\sum \alpha^2 + B\textstyle\sum \alpha + \alpha\delta \\ \textstyle\sum \gamma \sin^2\theta &= C\textstyle\sum \alpha\gamma + B\textstyle\sum \gamma^2 + A\textstyle\sum \gamma\delta \\ \textstyle\sum \delta \sin^2\theta &= C\textstyle\sum \alpha\delta + B\textstyle\sum \gamma\delta + A\textstyle\sum \delta^2 \end{aligned} \tag{4}$$

Analysis of the error according to equation (5)

$$A\ \delta \tag{5}$$

Information: A is a constant in the Cramer Cohen matrix
$\delta = 10\ \sin^2 2\theta$

3. Result and Discussions

We were assisted by data processing with Excel software, changing the distance between atoms d(Å) into an angle of 2θ based on the first 15 peaks of ICDD SiO_2 secondary data with a hexagonal crystal structure according to Table 2. The 2θ angle is essential for further data analysis

Table 2. Modifier d(Å) to 2θ

d(Å)	**2θ**
4,255	20,85997
3,3435	26,63972
2,4569	36,54356
2,2815	39,46489
2,2361	40,30051
2,1277	42,45028
1,9799	45,79199
1,818	50,13767
1,8017	50,6231
1,6717	54,87591
1,6592	55,32459
1,6083	57,234
1,5415	59,96129
1,4529	64,03528
1,4184	65,78656

With the aid of data processing using Excel software, the analysis of lattice constants was based on the first 15 peaks of secondary data of ICDD SiO2 with a hexagonal crystal structure using the Cramer and Cohan method and equations (1), (2), (3), (4), (5) according to Table 3.

Table 3.a. Analysis of lattice constants secondary data of ICDD SiO_2 with hexagonal crystal structure

Peak	h	k	l	2θ (deg.)	θ (deg.)
1	1	0	0	20,85997	10,42998
2	1	0	1	26,63972	13,31986
3	1	1	0	36,54356	18,27178
4	1	0	2	39,46489	19,73244
5	1	1	1	40,30051	20,15025
6	2	0	0	42,45028	21,22514
7	2	0	1	45,79199	22,89599
8	1	1	2	50,13767	25,06884
9	0	0	3	50,6231	25,31155
10	2	0	2	54,87591	27,43796
11	1	0	3	55,32459	27,66229
12	2	1	0	57,234	28,617
13	2	1	1	59,96129	29,98065
14	1	1	3	64,03528	32,01764
15	3	0	0	65,78656	32,89328

Table 3.b. Analysis of lattice constants secondary data of ICDD SiO_2 with hexagonal crystal structure (continued)

α	α^2	γ	γ^2	αγ	$\sin^2 2\theta$	$\sin^2\theta$
1	1	0	0	0	0,126797	0,032773
1	1	1	1	1	0,201044	0,053078
3	9	0	0	0	0,354541	0,098298
1	1	4	16	4	0,403994	0,113993
3	9	1	1	3	0,418346	0,118669
4	16	0	0	0	0,455558	0,131068
4	16	1	1	4	0,513821	0,151367
3	9	4	16	12	0,589189	0,179527
0	0	9	81	0	0,597513	0,182791
4	16	4	16	16	0,668973	0,212325
1	1	9	81	9	0,676322	0,215537
7	49	0	0	0	0,707092	0,229395
7	49	1	1	7	0,749415	0,249708
3	9	9	81	27	0,808316	0,281091
9	81	0	0	0	0,831788	0,294932
	$\Sigma\alpha^2$		$\Sigma\gamma^2$	$\Sigma\alpha\gamma$		
	267		295	83		

Table 3.c. Analysis of lattice constants secondary data of ICDD SiO_2 with hexagonal crystal structure (continued)

δ	δ^2	$\gamma\delta$	$\alpha\delta$	$\alpha\sin^2\theta$	$\gamma\sin^2\theta$	$\delta\sin^2\theta$
1,267967	1,607739	0	1,267967	0,032773	0	0,041555
2,010435	4,041851	2,010435	2,010435	0,053078	0,053078	0,10671
3,545413	12,56995	0	10,63624	0,294893	0	0,348506
4,03994	16,32112	16,15976	4,03994	0,113993	0,455972	0,460524
4,183458	17,50132	4,183458	12,55037	0,356006	0,118669	0,496446
4,555578	20,75329	0	18,22231	0,524273	0	0,597092
5,138211	26,40121	5,138211	20,55284	0,605469	0,151367	0,777757
5,891894	34,71442	23,56758	17,67568	0,538582	0,71811	1,057757
5,975126	35,70213	53,77614	0	0	1,645115	1,092197
6,689733	44,75253	26,75893	26,75893	0,849302	0,849302	1,4204
6,763225	45,74121	60,86902	6,763225	0,215537	1,93983	1,457723
7,070924	49,99797	0	49,49647	1,605767	0	1,622037
7,494147	56,16224	7,494147	52,45903	1,747953	0,249708	1,871345
8,083157	65,33743	72,74841	24,24947	0,843274	2,529821	2,272104
8,317877	69,18708	0	74,86089	2,654384	0	2,453204
	$\Sigma\delta^2$	$\Sigma\gamma\delta$	$\Sigma\alpha\delta$	$\Sigma\alpha\sin^2\theta$	$\Sigma\gamma\sin^2\theta$	$\Sigma\delta\sin^2\theta$
	500,7915	272,7061	321,5438	10,43528	8,71097	16,07536

In accordance with equation (4), the Cramer and Cohan methods in Table 3 are obtained:

$$\begin{array}{cccc} \Sigma\alpha\sin^2\theta & C\Sigma\alpha^2 & B\Sigma\alpha\gamma & A\Sigma\alpha\delta \\ \Sigma\gamma\sin^2\theta & C\Sigma\alpha\gamma & B\Sigma\gamma^2 & A\Sigma\gamma\delta \\ \Sigma\delta\sin^2\theta & C\Sigma\alpha\delta & B\Sigma\gamma\delta & A\Sigma\delta^2 \end{array}$$

Then the values in Table 3 are arranged in Table 4.

Table 4. The values are compiled from Table 3.

10,43528	267	C+	83	B+	321,5438	A
8,71097	83	C+	295	B+	272,7061	A
16,07536	321,5438	C+	272,7061	B+	500,7915	A

Then Cramer Cohen's analysis is obtained in the 3 x 3 order matrix

$$\begin{array}{ccccc} 267 & 83 & 321{,}5438 & C & 10{,}43528 \\ 83 & 295 & 272{,}7061 & B & 8{,}71097 \\ 321{,}5438 & 272{,}7061 & 500{,}7915 & A & 16{,}07536 \end{array}$$

Finding the Determinant of Matrix A using the Cramer method [5]

267	83	321,5438	
83	295	272,7061	194328,9
321,5438	272,7061	500,7915	

Finding the Determinant of the A1 Matrix using the Cramer method [5]

10,43528	83	321,5438	
8,71097	295	272,7061	6368,77
16,07536	272,7061	500,7915	

Finding the Determinant of the A2 Matrix using the Cramer method method [5]

267	10,43528	321,5438	
83	8,71097	272,7061	3946,9
321,5438	16,07536	500,7915	

Finding the Determinant of the A3 Matrix using the Cramer method [5]

267	83	10,43528	
83	295	8,71097	0,551064
321,5438	272,7061	16,07536	

Then the values of the constants C, B, A from Cramer Cohen's analysis in the 3 x 3 order matrix are obtained: [4]

$\lambda =$ 1,540598 Å (from literature)

C = Det (A1/A) = 0,032773154

B = Det (A2/A) = 0,020310413

A = Det (A3/A) = 2,83573E-06

Lattice constant analysis based on equation (3) is obtained:

Analysis Lattice constants		**Literatur**
a (Å)	4,913258166	4,91344
c (Å)	5,405052947	5,40524

The analysis of Lattice constants compared with the literature is very close, meaning that the analysis of Lattice constants using the Cramer Cohen method is very thorough.

Journal of Physics: Conference Series **2019** (2021) 012071 doi:10.1088/1742-6596/2019/1/012071

Error Analysis (Aδ) according to Table 5.

Table 5. Error Analysis

A	D	Aδ
2,83573E-06	2,83573E-05	3,59561E-06
2,83573E-06	2,83573E-05	5,70105E-06
2,83573E-06	2,83573E-05	1,00538E-05
2,83573E-06	2,83573E-05	1,14562E-05
2,83573E-06	2,83573E-05	1,18631E-05
2,83573E-06	2,83573E-05	1,29184E-05
2,83573E-06	2,83573E-05	1,45706E-05
2,83573E-06	2,83573E-05	1,67078E-05
2,83573E-06	2,83573E-05	1,69438E-05
2,83573E-06	2,83573E-05	1,89703E-05
2,83573E-06	2,83573E-05	1,91787E-05
2,83573E-06	2,83573E-05	2,00512E-05
2,83573E-06	2,83573E-05	2,12514E-05
2,83573E-06	2,83573E-05	2,29216E-05
2,83573E-06	2,83573E-05	2,35872E-05
		ΣAδ
		0,000229771
Average Error	In %	
1,5318E-05	0,001531805	%
=	0,015318048	permil

Then an average error analysis (Aδ) is obtained of 0.001531805%, meaning that the average error analysis based on the Cramer-Cohen method is comprehensive because the mean value of error is less than 0.1%.

4. Conclusion

Has successfully analyzed the lattice constants and the error of silicon dioxide (SiO2) hexagonal crystal structure using the Cramer-Cohen method. The result of the program calculation shows that the error is less than or equal to 0.1 percent.

Acknowledgement

This research was funded by the PKPT DRPM Grant of the Ministry of Education and Culture, Research and Technology Republic of Indonesia in 2021.

References

[1] Sharma R, Bisen D P, Shukla U and Sharma B G 2012 X-Ray Diffraction: A Powerful Method of Characterizing Nanomaterials *RRST*

[2] Mursal I L P 2018 Karakterisasi XRD Dan SEM Pada Material Nanopartikel Serta Peran Material Nanopartikel Dalam Drug Delivery System *Pharma Xplore : Jurnal Sains dan Ilmu Farmasi* **3**

[3] Anon ICDD Database Search *ICDD*

[4] Cullity B D 1978 *Elements of X-Ray Diffraction* (Massachusetts: Addison Wesley Publishing Company)

[5] Song G-J, Wang Q-W and Chang H-X 2011 Cramer Rule for The Unique Solution of Restricted Matrix Equations Over the Quaternion Skew Field *Computers & Mathematics with Applications* **61** 1576–89

Critical Temperature in Zigzag Graphene Nanoribbon: a First-principles Study

T B Prayitno and E Budi

Department of Physics, Faculty of Mathematics and Natural Science, Universitas Negeri Jakarta, Kampus A Jl. Rawamangun Muka, Jakarta Timur 13220, Indonesia

E-mail: `teguh-budi@unj.ac.id`

Abstract. The critical (Néel) temperature in the zigzag graphene nanoribbon was calculated using the mean-field approximation within the generalized Bloch theorem. This calculation was carried out over the Brillouin zone of the magnon spectrum. We found a nearly flat magnon dispersion at the high energy in one-third of the Brillouin zone. Our calculation showed the critical temperature below room temperature, in good agreement with the prediction in the previous works. Our last work (Prayitno 2021 *Physica* E **129** 114641) revealed that the critical temperature may be enhanced by increasing the ribbon width. In this brief report, we justified that the critical temperature becomes almost constant up to a certain ribbon width. This result indicates that the critical temperature in the graphene nanoribbon will never reach room temperature for any ribbon widths, thus it is likely difficult to apply pristine graphene nanoribbon in any practical devices working near room temperature.

Keywords: graphene nanoribbon, spin stiffness, critical temperature

1. Introduction

Recent studies on the zigzag graphene nanoribbon (ZGNR) and its area give rise to the interesting studies in magnetism due to localized edge states [1], which are responsible for the spintronic devices. Generally, the magnetism in materials can be found in the elements having either the d- or the f- electron shells. Compared to the d- or the f- elements, ZGNR has many benefits such as low density, high conductivity, and gap semiconductor. Son *et al.* predicted the half metallicity in the ZGNR under the electric field for the application of spin transport [2], see also Ref. [3] for similar cases. On the other side, controlling the magnetic order in ZGNR can also be realized by taking the carrier doping [4].

To utilize the ZGNR for the spintronic devices, the critical temperature T_C should be higher or close to the room temperature. For the d- shell elements such as Fe, Co, and Ni, the T_C is higher than the room temperature, thus indicating the d-elements-based practical devices can operate properly. However, the T_C of the ZGNR is still unclear until now. Previously, Yazyev and Katsnelson predicted the critical temperature that reaches the room temperature in the ZGNR is possible when the spin correlation length in the order of few nanometers was taken into account [5]. In practical devices, such order of magnitudes is very subtle to produce at present, thus the ZGNR-based spintronic devices are only reliable at the low temperature. This prediction is also supported by Kunstmann *et al.* who justified the instability of the magnetic order in the ZGNR at room temperature [6]. They also claimed that the magnetic order can only be stabilized at a temperature below 10 K.

An optimistic hope is addressed to the spin stiffness D in the ZGNR, which is much higher than those in the d- elements. This high value is believed to give information how to improve T_C by controlling some parameters in the ZGNR [5]. The recent finding, which compared the experimental result with the Hubbard approach, showed the possibility of the stable magnetic order in the ZGNR even at room temperature, thus ZGNR-based devices should operate well [7]. They claimed that the possibility may occur when the magnetic order is controlled by the orientation of the edge. Therefore, it is still disputed whether if ZGNR-based devices can work or not at room temperature.

A suitable choice to investigate the T_C is addressed to the spin-wave (magnon) excitations. Although previous studies explored the spin-wave excitations using the Hubbard model, there is no sufficient information, except Ref. [5], if the T_C can reach the room temperature or not. You *et al.* focused on the feature of the relativistic magnon for ferromagnets and antiferromagnets [8] while Culchac *et al.* investigated the lifetime of the spin-waves [9]. At the same time, Rhim and Moon concentrated on the response of D under external electric field [10]. The purpose of the present work is to investigate the T_C and its relationship with the D by using the generalized Bloch theorem (GBT) to calculate the magnon energy in the Brillouin zone. We suppose that this approach is very efficient since it only uses the primitive unit cell.

While Yazyev and Katsnelson [5] used the spin correlation as a parameter, and Kunstmann *et al.* [6] used the energy gap as a parameter to explain the low T_C in the ZGNR, we use the magnon spectrum to explain why the T_C in the ZGNR never reaches the room temperature. In our last work [11], increasing the ribbon width N can increase the T_C, thus making a possibility to increase the T_C through increasing N. Here, N means the number of stripes along the ribbon width, as shown in Fig. 1. In this brief report, we demonstrate that increasing N as large as possible leads to a constant T_C below room temperature. This means that the T_C of pristine ZGNR will never surpass the room temperature. We also show the dependence of D and T_C on N. We observe the same trend not only for D but also for T_C in all N.

2. Computational Details

We carried out the self-consistent calculation using the GBT as implemented in the OpenMX code [12], a package exploiting the linear combination of pseudo-atomic orbital (LCPAO) as basis sets [13] as well as utilizing the norm-conserving pseudopotential [14]. The implementation of the spin spiral in this code was done by formulating the Kohn-Sham wavefunction as an LCPAO [11]

$$\begin{aligned}\psi_{\nu\mathbf{k}}(\mathbf{r}) &= \frac{1}{\sqrt{N}}\left[\sum_{n}^{N} e^{i\left(\mathbf{k}-\frac{\mathbf{q}}{2}\right)\cdot\mathbf{R}_n}\sum_{i\alpha} C^{\uparrow}_{\nu\mathbf{k},i\alpha}\phi_{i\alpha}(\mathbf{r}-\tau_{\mathrm{i}}-\mathbf{R}_{\mathrm{n}})\begin{pmatrix}1\\0\end{pmatrix}\right.\\ &\left.+\sum_{n}^{N} e^{i\left(\mathbf{k}+\frac{\mathbf{q}}{2}\right)\cdot\mathbf{R}_n}\sum_{i\alpha} C^{\downarrow}_{\nu\mathbf{k},i\alpha}\phi_{i\alpha}(\mathbf{r}-\tau_{\mathrm{i}}-\mathbf{R}_{\mathrm{n}})\begin{pmatrix}0\\1\end{pmatrix}\right], \end{aligned} \tag{1}$$

where the localized basis function $\phi_{i\alpha}$ is produced via a confinement method. Meanwhile, the magnetic moments of magnetic carbon atoms at the edges will be continuously rotated with respect to the azimuthal angle φ, which is given by

$$\mathbf{M}_i = M_i(\cos[\varphi_0+\mathbf{q}\cdot\mathbf{R}_i]\sin\theta_i + \sin[\varphi_0+\mathbf{q}\cdot\mathbf{R}_i]\sin\theta_i + \cos\theta_i) \tag{2}$$

Then, the magnon energy for each wavevector $\mathbf{q}$ was given by [15]

$$\hbar\omega_{\mathbf{q}} = \frac{\mu_B}{M}\frac{E(\mathbf{q},\theta)-E(\mathbf{0},\theta)}{\sin^2\theta}. \tag{3}$$

We used the conical spiral configuration ($\theta = 10°$) instead of the flat spiral configuration ($\theta = 90°$) for the magnetic carbon atoms at the edges in the antiferromagnetic order, which gives the most stable state, as shown in Fig. 1. Note that the antiferromagnetic order as a ground state also holds in the bilayer ZGNR [16]. Even though the calculation using the flat

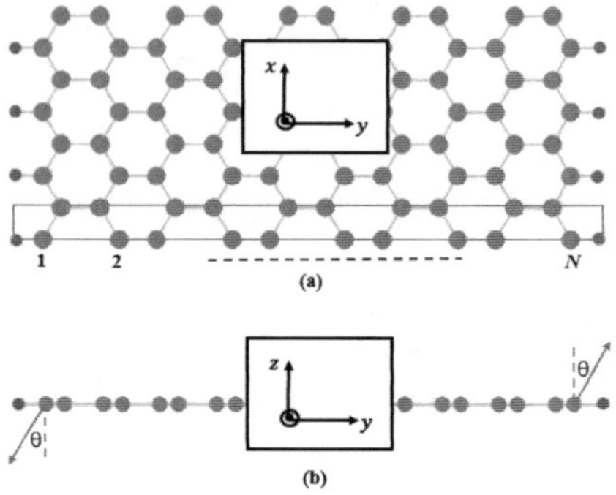

Figure 1. (Color online) Crystal structure of one-dimensional N-ZGNR from top view (a) and side view (b). The carbon and hydrogen atoms are represented by the large and small filled circles while the primitive unit cell is shown by the solid-lined rectangle. The magnetic moments represented by the arrows of the carbon atoms at the edges are deviated antiferromagnetically with the small angle θ.

spiral configuration is more stable, in this case, it only fixes the magnetic moment close to Γ point [15]. Therefore, it cannot be used to evaluate the critical temperature since the calculation should be carried out over the Brillouin zone. The instability of the conical spiral configuration is due to the tendency of the system to go back to the ground state. For keeping the direction of magnetic moment, the penalty functional within the constraint method therefore was applied [17].

We used the lattice constant of 2.46 Å as the experimental lattice constant for graphite for the unit cell (x-axis) and took the vacuum length of 50 Å to create a non-periodic cell in y- and z-axes. To get the converged and reliable results, two s- and two p- primitive orbitals were assigned for the carbon atoms within the cutoff radius of 4.0 Bohr. At the same time, two s- and one p- orbitals were set to the hydrogen atoms within the cutoff radius of 6.0 Bohr. For the electron-electron interaction, the generalized gradient approximation (GGA) [18] was used for the self-consistent calculation with a k-point mesh of $90 \times 1 \times 1$ and a cutoff energy of 150 Ryd.

3. Results and Discussions

Figure 2 shows the magnon spectra of ZGNR with $N = 6, 10, 14, 18$ over the Brillouin zone. It shows the nearly flat dispersion in one-third of the Brillouin zone, which gives the most contribution for the high T_C. While the D is obtained by fitting the magnon energy $\hbar\omega_{\mathbf{q}} = D\mathbf{q}^2(1 - \beta\mathbf{q}^2)$ near Γ point as shown in Fig. 3, the T_C is calculated by using the mean-field approximation (MFA) over the Brillouin zone [19]

$$k_B T_C^{\mathrm{MFA}} = \frac{M}{6\mu_B}\frac{1}{N_{\mathbf{q}}}\sum_{\mathbf{q}} \hbar\omega_{\mathbf{q}}, \tag{4}$$

where $N_{\mathbf{q}}$ is the number of $\mathbf{q}$. Therefore, the D and T_C originate from the contribution of the low magnon energy near the Γ point and all the magnon energies in the Brillouin zone, respectively.

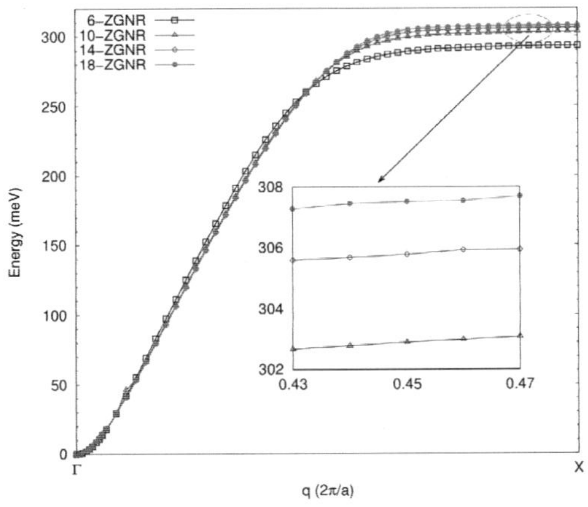

Figure 2. (Color online) Spectra of magnon of ZGNR in the Brillouin zone.

Table 1. Evaluated D and T_C for several N.

N	D (meVÅ^2)	T_C (K)
6	3345	102
10	3595	107
14	3801	108
18	3976	109

Note that the T_C here refers to the Néel temperature since the ground state of the ZGNR system is antiferromagnetic.

We notice that the change of D will be followed by the change of T_C, thus indicating a relationship between the D and the T_C. It can also be found that the D and T_C increase as N increases, as shown in table 1. For the large N, we predict that both the D and T_C will go to the constant values, exhibiting the critical N. This means that the T_C in a pristine ZGNR will never gain room temperature although the N increases. As a consequence, the magnetism in the ZGNR will be lost at room temperature. These results are consistent with those of previous works [5, 6].

From Eq. (4), we deduce that the T_C is determined by the average of magnon energies $\hbar\omega_{\mathbf{q}}$ while the high T_C contribution is given only in one-third of the Brillouin zone. According to Ref. [19], the Green function random phase approximation (GF-RPA) gives a better result as matched with the MFA. However, it needs much more $\mathbf{q}$ in the Brillouin zone due to the harmonic average of magnon energies. As shown in Fig. 2, no strong fluctuations in the magnon spectra, so we claim that there is no large discrepancy of T_C using those two approaches. The large discrepancy may happen in the case of strong fluctuations in the magnon spectra, as reported in Ref. [19].

Compared to the $3d$ ferromagnetic metals [19] or the oxide systems [20], the scale of magnon energy in the ZGNR is difficult to give the high T_C. From our point of view, this feature may be influenced by the dimensionality of a system when the spin-orbit interaction is very weak. When the dimensionality of systems is lowered, the T_C tends to never reach the room temperature, for example in the case of two-dimensional metal dihalides [21]. This leads to the absence of magnetism at room temperature in the lower-dimensional materials, which has been stated by the Mermin-Wagner theorem [22]. One of the effective ways to preserve the magnetism in the lower-dimensional systems is to include the magnetic anisotropy. However, this way cannot be

Journal of Physics: Conference Series **2019** (2021) 012072 doi:10.1088/1742-6596/2019/1/012072

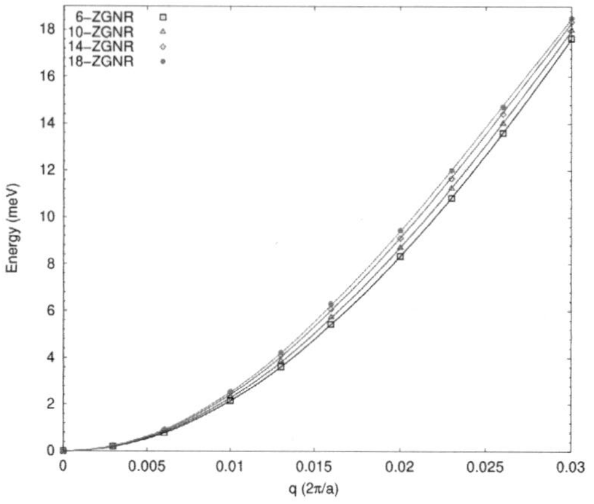

Figure 3. (Color online) Spectra of magnon of ZGNR near Γ point for obtaining spin stiffness. Here, the boxes, triangles, diamond, and filled circles denote the calculated magnon energies while the solid line represents the fitting curve $\hbar\omega_{\mathbf{q}} = D\mathbf{q}^2(1 - \beta\mathbf{q}^2)$.

applied in the ZGNR due to the negligible spin-orbit interaction.

To discuss the change of D as well as T_C when N increases, we consider the electron hopping from one edge magnetic carbon atom to the other one as an exchange magnetic interaction J_{ij}. Indeed, the strength of J_{ij} is determined by the easiness of electron hopping. Based on the distance of separated edge magnetic carbon atoms, the largest J_{ij} will be possessed by the smallest N and vice versa. Therefore, the large D is the implication of small J_{ij} (large N), namely, the large D is caused by the difficulty of an electron to hop. At the same time, when N is so large, J_{ij} remains unchanged, indicating that there is a critical N due to the electron hopping. Therefore, the enhancement of D and T_C can only be performed up to a critical N. This means that T_C in the ZGNR will never surpass the room temperature although N increases as large as possible, as shown in table 1.

Here, we comment on our calculated D, which is quite higher than that in Ref. [5], which obtained D=2100 meVÅ^2. This large discrepancy may be caused by the different approaches. Although both of us employed the frozen magnon method to calculate the D, we use the GBT while the authors [5] used the supercell calculation. For the very small $\mathbf{q}$ close to the Γ point, it is very difficult to calculate the D in the supercell approach since the large number of cells with the orientation of magnetic moments of carbon atoms at the edges should be appropriately arranged. So, we think that they used the number of $\mathbf{q}$ sufficiently far from the Γ point. Since the Stoner excitations may not be neglected in the *sp*- elements such as ZGNR [23], some magnon energies for the large $\mathbf{q}$ should decay into the Stoner excitations. Therefore, we think that our approach should be more suitable.

Our findings show that the T_C in the pristine ZGNR will never surpass the room temperature although the D is quite high, in the opposite way with the opinion by Yazyev [5]. From the previous studies in Refs. [10, 24] who investigated the influence of D in the ZGNR under external electric field E along N (y-axis), they found that the D never surpasses that for the non-E case, thus the T_C has the same feature. Although some computational results show that this applied E along N is very useful to generate such as half-metallicity, this will not become feasible in this situation. Even for the hole-electron doping case, which can be realized by chemical doping, the D never exceeds that for the non-doped case [15].

From table 1, we see that the increase of D is followed by the increase of T_C for all N. The situation can be different, in our case, when the electric field E is taken into account along x-axis (periodic direction). As shown in Fig. 4, we observe the small enhancements of the D as well as the T_C up to the critical E for all N with the different tendency. Figure 4(a) shows the small increase of D up to $E = 6$ V/nm for $N > 6$ and $E = 3$ V/nm for $N = 6$, a different tendency of D for the small N. Meanwhile, the same tendency of T_C occurs for all N, as shown in Fig. 4(b). For all N, the T_C initially increases up to $E = 9$ V/nm and then tends to constant, thus never reaching the room temperature too. This tendency never occurs when the E is employed along y-axis, i.e., both the D and T_C reduce as the E increases for all N as shown in our previous work [11].

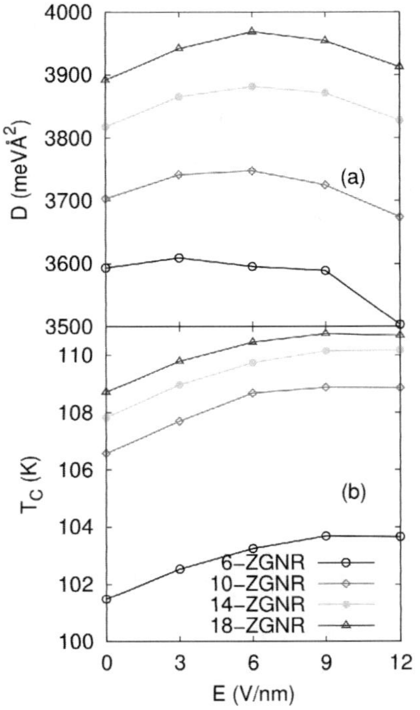

Figure 4. (Color online) Dependence of D and T_C on applied E along periodic direction (x-axis).

The different tendency between the D and T_C originates from the different contributions of the magnon energy, as previously mentioned. This means that the applied E gives instability to the low magnon energy for the small N, i.e., the D reduces with a different critical E. This reduction is addressed to the reduction of the magnon energy close to Γ point. It seems that this anomaly may be affected by the weaker J_{ij} in the small N. On the contrary, this applied E will not affect the magnon energy far from the Γ point, i.e., the tendency of T_C is the same for all N.

In the previous report [6], Kunstmann *et al.* stated that the destruction of magnetism in the ZGNR at room temperature is caused by each small magnetic moment of two edge carbon atoms, which is close to 0.3 μ_B. We also obtain the magnetic moment close to 0.3 μ_B for all N both in the non-electric-field and in the electric field cases. This may be caused by the lost dangling bond when introducing the hydrogen atom to passivate the edges, as shown previously by Song *et al.* [27]. When the metal atom is included in the ZGNR [25, 26], the energy scale drastically increases due to the charge transfer from the metal atom to the magnetic carbon atoms, creating a strong bonding. This report is to support the previous work of Magda *et al.* [7] who included the Au atoms. This means that the stability of magnetism at room temperature in the ZGNR is due to the strong bonding between the metal atom and the magnetic carbon atom, which enhances the magnetic moments of two edge carbon atoms.

4. Conclusions

By using the mean-field approximation (MFA), we show that the critical (Néel) temperature T_C in the ZGNR never reaches room temperature even though the ribbon width N, which can increase the T_C, increases. This is due to a critical N that gives a constant T_C as well as the spin stiffness D. The main reason for this finding is the small magnetic moments of carbon atoms at the two separated edges. This implies that the magnetic feature can only maintain at low temperature.

Although introducing the electric field can also increase the T_C for all N, it is still not possible to reach the room temperature. This is also due to a flat energy dispersion in one-third of the Brillouin zone around 300 meV. It is also supported by the small magnetic moments of two edge carbon atoms, thus destroying the magnetism at room temperature. Therefore, the ZGNR-based devices cannot be possible to operate well at room temperature.

Acknowledgments

We acknowledge Faculty of Mathematics and Science Universitas Negeri Jakarta to support this research with the contract number: 57 /SPK PENELITIAN/5.FMIPA/2021.

References

[1] Fujita M, Wakabayashi K, Nakada K and Kusakabe K 1996 Peculiar localized state at zigzag graphite edge *J. Phys. Soc. Jpn.* **65** 1920.
[2] Son Y -W, Cohen M L and Louie S G 2006 Half-metallic graphene nanoribbons *Nature* **444** 347.
[3] Rudberg E, Sałek P and Luo Y 2007 Nonlocal exchange interaction removes half-metallicity in graphene nanoribbons *Nano Lett.* **7** 2211.
[4] Sawada K, Ishii F, Saito M, Okada S and Kawai T 2009 Phase control of graphene nanoribbon by carrier doping: appearance of noncollinear magnetism *Nano Lett.* **9** 269.
[5] Yazyev O V and Katsnelson M I 2008 Magnetic correlations at graphene edges: basis for novel spintronics devices *Phys. Rev. Lett.* **100** 047209.
[6] Kunstmann J, Özdoğan C, Quandt A and Fehske H 2011 Stability of edge states and edge magnetism in graphene nanoribbons *Phys. Rev.* B **83** 045414.
[7] Magda G Z, Jin X, Hagymási I, Vancsó P, Osváth Z, Nemes-Incze P, Hwang C, Biró L P and Tapasztó L 2014 Room-temperature magnetic order on zigzag edges of narrow graphene nanoribbons *Nature* **514** 608.
[8] You J -S , Huang W -M and Lin H -H 2008 Relativistic ferromagnetic magnon at the zigzag edge of graphene *Phys. Rev.* B **78** 161404(R).
[9] Culchac F J, Latgé A and Costa A T 2011 Spin waves in zigzag graphene nanoribbons and the stability of edge ferromagnetism *New J. Phys.* **13** 033028.
[10] Rhim J and Moon K 2009 Spin stiffness of graphene and zigzag graphene nanoribbons *Phys. Rev.* B **80** 155441.
[11] Prayitno T B 2021 Impossibility of increasing Néel temperature in zigzag graphene nanoribbon by electric field and carrier doping *Physica* E **129** 114641.
[12] Ozaki T *et al.*, Open source package for Material eXplorer (http://www.openmx-square.org).
[13] Ozaki T and Kino H 2004 Numerical atomic basis orbitals from H to Kr *Phys. Rev.* B **69** 195113.
[14] Troullier N and Martins J L 1991 Efficient pseudopotentials for plane-wave calculations *Phys. Rev.* B **43** 1993.
[15] Prayitno T B and Ishii F 2019 Carrier-induced antisymmetric–symmetric tendencies of spin stiffness in zigzag graphene nanoribbons *J. Phys.: Condens. Matter* **31** 365801.
[16] Prayitno T B 2020 Spin stiffness of bilayer zigzag graphene nanoribbon for several configurations *Physica* E **118** 113916.
[17] Gebauer R and Baroni S 2000 Magnons in real materials from density-functional theory *Phys. Rev.* B **61** 6459(R).
[18] Perdew J P, Burke K and Ernzerhof M 1996 Generalized gradient approximation made simple *Phys. Rev. Lett.* **77** 3865.

[19] Pajda M, Kudrnovský J, Turek I, Drchal V and Bruno P 2001 Ab initio calculations of exchange interactions *Phys. Rev.* B **64** 174402.

[20] Essenberger F, Sharma S, Dewhurst J K, Bersier C, Cricchio F, Nordström L and Gross E K U 2011 Magnon spectrum of transition-metal oxides: Calculations including long-range magnetic interactions using the LSDA+U method *Phys. Rev.* B **84** 174425.

[21] Kulish V V and Huang W 2017 Single-layer metal halides MX_2 (X = Cl, Br, I): stability and tunable magnetism from first principles and Monte Carlo simulations *J. Mater. Chem.* **5** 8734.

[22] Mermin N D and Wagner H 1966 Absence of ferromagnetism or antiferromagnetism in one- or two-dimensional isotropic Heisenberg models *Phys. Rev. Lett.* **17** 1133.

[23] Edwards D M and Katsnelson M I 2006 High-temperature ferromagnetism of sp electrons in narrow impurity bands: application to CaB_6 *J. Phys.: Condens. Matter* **18** 7209.

[24] Prayitno T B and Budi E 2020 Applied electric field on zigzag graphene nanoribbons: reduction of spin stiffness and appearance of spiral spin density waves *J. Phys.: Condens. Matter* **32** 105802.

[25] Huang C, Wu H, Deng K and Kan E 2017 Edge-modified graphene nanoribbons: appearance of robust spiral magnetism *J. Phys. Chem.* C **121** 1371.

[26] Lü X -L, Zhang C -X, Wang W -J, Cheng X and Xie H 2019 Excitation and phase transitions of spin *J. Phys.: Condens. Matter* **31** 455501.

[27] Song L L, Zheng X H, Wang R L and Zeng Z 2010 Dangling bond states, edge magnetism, and edge reconstruction in pristine and B/N-terminated zigzag graphene nanoribbons *J. Phys. Chem.* C **114** 12145.

Journal of Physics: Conference Series **2019** (2021) 012073 doi:10.1088/1742-6596/2019/1/012073

Permittivity Measurement Using the Resonance Circuits

I E Santosa
Department of Physics Education, Sanata Dharma University
Paingan, Maguwohardjo, Sleman, Yogyakarta 55282, Indonesia
Email: edi@usd.ac.id

Abstract. A pair of LC circuits has been used to demonstrate the resonance condition of electromagnetic waves. The first circuit comprises of an inductor, a capacitor and a power supply; while the second pair employs an inductor, a variable capacitor and a neon bulb. The same inductance is applied in both circuits. The variable capacitor is a parallel plate capacitor. By varying the capacitance in the second circuit, we achieve the resonance condition as indicated by the lit neon bulb. By doing so, we observe the equal oscillation frequency in both, the first and second circuit. The condition is then used to obtain the permittivity of air to be (6.1±0.2) x10-11 F/m.

1. Introduction

Permittivity of free space is widely known as one of the universal physical constants. In general the permittivity relies on the medium, which is indicated by their dielectric constants (relative permittivity). Students encounter the permittivity in various parts of electromagnetism courses e.g. The Coulomb law, capacitance, and the speed of electromagnetic wave. Thus it is important to understand all aspects related to the permittivity or dielectric constant by employing different methods of measurements.

Many experiments used the parallel plate capacitor to obtain the dielectric constant of the medium located in between the capacitor plates. In this case the capacitance should be measured and subsequently its dielectric constant. A direct measurement of the capacitance can be done using an LCR meter. Riaz and Kanwal used an LCR meter to determine the capacitance of an acrylic filled plate capacitor, and Munguía and Madonaldo used a professional capacitance meter to measure the dielectric constant of the overheated edible oil [1, 2]. The low cost digital multimeters are also capable of measuring capacitance [3]. Although this method of measurement is relatively simple, due to its low resolution, the typical digital multimeter possesses its own limitation.

One can incorporate the flat plate capacitor in a series RLC circuit. The capacitance is determined from the measured resonance frequency [1]. Mohsen-Nia et al measured both input voltage and the voltage on the capacitor of a RC circuit [4]. By comparing these voltages they were able to determine the dielectric constant of water, ethanol, methanol, butanol and acetone. Similar to this, Nogi et al measured the voltage across a resistor in the RC circuit to obtain the dielectric constant of the material inserted in a parallel plate capacitor, e.g. Soda-lime glass, Bakelite, acrylic glass, and Teflon [5]. Another setup used an LC circuit to determine the dielectric constant of gases by measuring its resonance frequency [6].

Mak used a pair of LC circuits with identical resonance frequencies in both the transmitter and receiver. He demonstrated the generation and transmission of electromagnetic wave by adjusting the frequency of the signal generator [7]. Similarly, the generation of electromagnetic wave has also been

Journal of Physics: Conference Series **2019** (2021) 012073 doi:10.1088/1742-6596/2019/1/012073

demonstrated using a simple RLC circuit. A digital oscilloscope is used for a quantitative analysis of the electromagnetic emission [8].

In this paper we show the measurement of capacitance of the parallel plate capacitor based on the transmission of electromagnetic wave. Unlike Mak's experiment, here we tune the capacitance of the receiver to achieve a resonance condition. By graphical analysis we then determine the permittivity of the air which is the medium in between the capacitor plates. For teaching purposes this experiment also demonstrates the principle of a radio receiver.

Theory

Figure 1 shows a pair of LC circuit used in the experiment. The first LC circuit comprises of an inductor L1, a capacitor C1 and power supply. Similarly in the second circuit employs an inductor L2, a variable capacitor C2 and a neon bulb as an indicator.

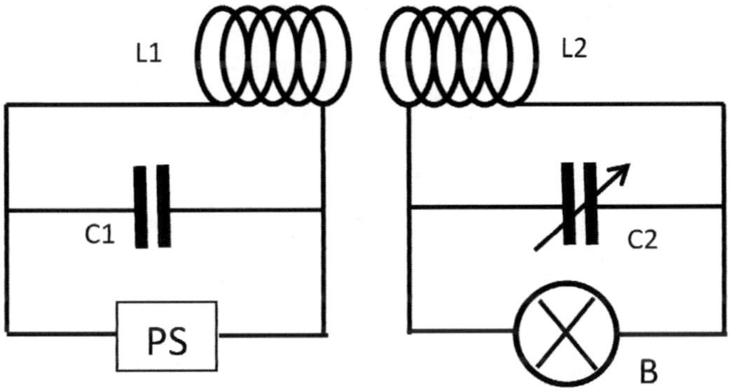

Figure 1. The experiment setup, PS is a power supply and B is a neon bulb.

The oscillation in the first circuit occurs with the frequency f_1 that is given by

$$f_1 = \frac{1}{2\pi}\sqrt{L_1 C_1} \quad (1)$$

Where L_1 and C_1 are the inductance and capacitance.

Due to the oscillation, this circuit generates an electromagnetic wave and behaves as the transmitter. This wave is then to be captured by the second circuit which has an inductance of L_2 and a capacitance of C_2. The frequency oscillation f_2 of the second circuit (receiver) follows

$$f_2 = \frac{1}{2\pi}\sqrt{L_2 C_2}\,. \quad (2)$$

By tuning the variable capacitor C2 the resonance condition is achieved as indicated by the lit neon bulb. In this condition the frequency oscillation is the same for both the first and the second circuit, and hence

$$L_1 C_1 = L_2 C_2. \quad (3)$$

In the experiment we use the same inductor in both circuits, therefore equation (3) becomes

$$C_1 = C_2. \quad (4)$$

Our variable capacitor is a parallel plate capacitor, its capacitance C_2 is given by

$$C_2 = \varepsilon \frac{A}{D} \tag{5}$$

where ε: permittivity
A: surface area of the plate
d: the distance between the plates.
Using equation (4) and (5), in the resonance condition the equation applies

$$C_1 = \varepsilon \frac{A}{D} \tag{6}$$

Our variable capacitor consists of two sets of semicircular metal plates, see figure 2. The first set is a stationary plate or stator, while the second set (rotor) is placed in between the plates of the first set that can be rotated against its axis. The area of overlapping can be varied by rotating the shaft.

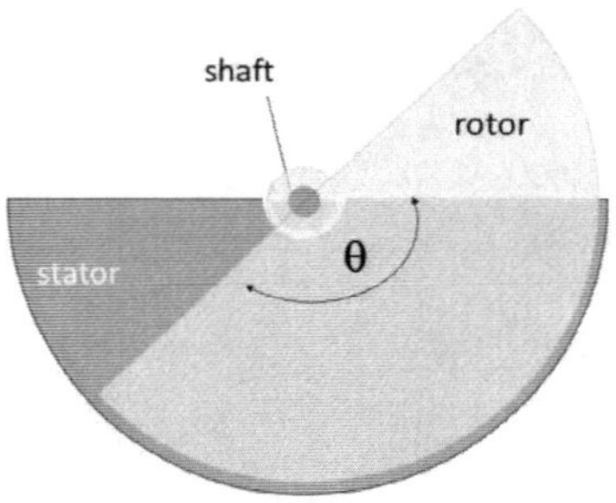

Figure 2. The variable capacitor used in the experiment. The rotor is rotated against the axis to vary the overlapping area.

By measuring the rotated angle θ using a protractor we can determine the overlapping area *A*:

$$A = \frac{\theta}{180} A_{\text{max}} \tag{7}$$

Where A_{max} is the maximum area of the semicircular plate.

2. Method

Experimental setup is depicted in figure 3. It consists of two LC circuits i.e. Transmitter and receiver. A transmitter circuit comprises of a capacitor, an inductor and power supply. Ceramic capacitors were used in the transmitter circuit. In the receiver circuit we used a variable capacitor, an inductor and a small neon lamp as an indicator. The variable capacitor as seen figure 4 is equipped with a protractor for measuring the rotor angle.

In this experiment the inductor used in the transmitter was the same as that in the receiver. Both inductors were mounted close each other. The oscillation frequency of the transmitter was set of mounting a ceramic capacitor. During the experiment the variable capacitor was tuned to achieve a resonance condition. As a result, this process will light the neon indicator lamp. The angle of the rotor position was then measured by observing the scale in the protractor.

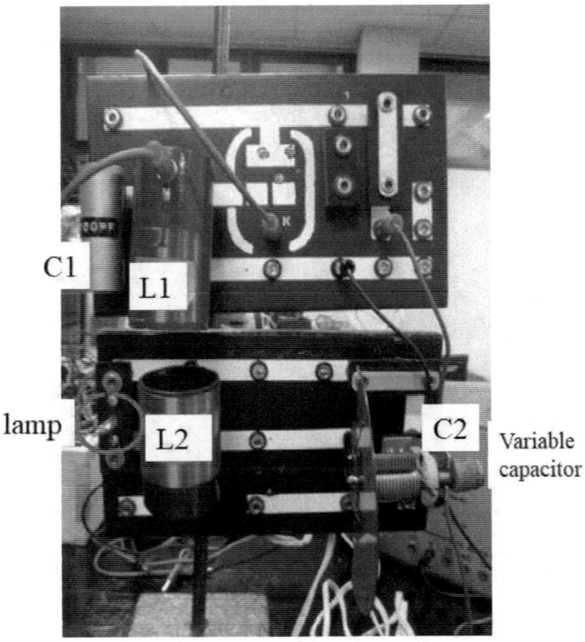

Figure 3. Experimental setup

Figure 4. The variable capacitor used in the experiment.

3. Result and Discussion

The experiment is based on the resonance conditions that incorporate the transmitter and receiver circuits. In the receiver circuit we use a variable capacitor with the maximum area of the semi-circular plate A_{max} = 5443 mm^2. The average distance between plates is 8 $x10^{-4}$ m. By carefully rotating the variable capacitor of the second circuit the neon bulb becomes lit when the stator and rotor from a certain angle. It has been observed that the neon bulb is lit only within a short range of angles. This event reveals that the oscillation occurring in the second circuit is generated by the oscillation in the first circuit. Overall, here we have demonstrated a particular behaviour known as a resonance condition, i.e. The oscillation frequency of the second circuit (receiver) is equal to the oscillation frequency of the first circuit (transmitter). The rotated angle for several transmitters' capacitances is tabulated in Table 1. Equation (7) is used to obtain the overlapping area A from the measured rotated angle θ. The complete result is depicted in figure 5.

Table 1. Experimental results in the resonance condition.

C_1 (pF)	θ (⁰)	A (mm^2)
50	14	423
100	46	1391
150	66	1996
200	86	2600
250	109	3296
300	127	3840
350	149	4506
400	166	5019

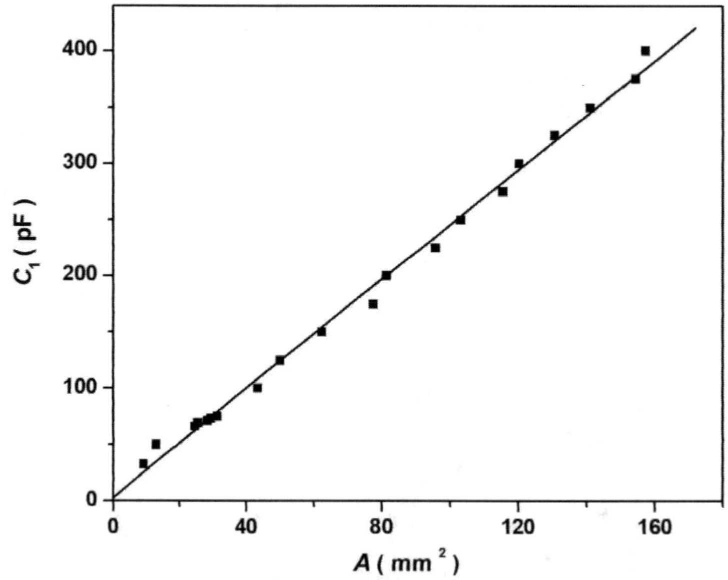

Figure 5. The relation between the capacitance of the transmitter and area of the receiver capacitor. Solid line is the best fit.

Figure 5 shows a clear linear relation between the capacitance in the transmitter and the area of the overlapping area *A*. The overlapping area *A* determines the capacitance of the receiver capacitor, thus we can deduce that figure 5 also shows a linear relation of the transmitter's capacitance and the receiver's capacitance. This behaviour is based on the resonance condition that is formulated in equation (6). The graph gives a gradient of (7.6 ± 0.1) 10^{-8} F/m and an offset of (2 ± 4) pF. In accordance with equation (6) the gradient is (ε /d). Thus the gradient yields the permittivity of medium ε to be (6.1 ± 0.2) x10^{-11} F/m.

This result is quite high compared to the reported air permittivity i.e. Ranging from 4 to 9 x10^{-12} F/m [9]. In the experiment, for the transmitter we used a common ceramic capacitor available in our lab. This capacitor has a tolerance of about 20% of the nominal value. The most critical quantity is the distance between plates in the variable capacitor. The irregularities of the capacitor plates caused the nonlinearity in the experiment result [3]. The distance between plates should be the same for all [1].When we look into the variable capacitor in detail (figure 4), it shows a set of 30 plate capacitors.

We assume the distance is the same for all the plates; we do not account for the individual capacitor. For the calculation we take the average distances. As seen in figure 4 some plates closer than others and the spacing becomes irregular. Another factor that contributes to the result is the fact that the measured capacitance is in order of Pico Farad. In this very small order of capacitance, a stray capacitance will give a significant contribution.

In general our experiment is intended for pedagogical purposes, which is suitable for teaching the topic of electromagnetic wave. Although the students are familiar with various wireless communications devices, they do not understand the underlying physics [10]. Here students learn the basic of electromagnetic wave and the resonance frequency. During this experiment students are amazed and excited when observing the neon bulb lit up albeit not being directly connected to the power supply. A parallel situation happens when students operate a radio as a common device. Students have a basic understanding that when selecting a radio station, they dial the capacitor within the radio transmitter located in the radio; which is acting like a black box.

4. Conclusion

This simple experiment demonstrates the resonance condition for electromagnetic waves; it incorporates the transmitter and receiver circuits. A pair of LC circuits is in resonance when the oscillation frequency arises from the first circuit is equal to the second. A variable capacitor is used to tune the resonance. The variable capacitor's capacitance depends on the overlapping area; it is determined by measuring the rotated angle of two sets of semicircular metal plates. In this experiment this resonance condition is indicated by the lit neon bulb. For the same inductance the condition is presented by a linear relation of the transmitter's capacitance in the area of the receiver's capacitor. The permittivity of medium ε is obtained to be (6.1 ± 0.2) x10^{-11} F/m. The uncertainty of this measurement is contributed mainly by the irregularities of the capacitor plates.

Acknowledgments

The author would like to thank K. Heskett for helpful comments on the manuscript

References

[1] Riaz M and Kanwal N 2019 *Eur. J. Phys.* 40 (2) 025502
[2] Aguilar H M and Maldonado R F 2015 *Phys. Educ.* 50 662-6
[3] Grove T T Masters M F and Miers R E 2005 *Am. J. Phys.* 73 52-6
[4] Mohsen-Nia M Amiri H and Jazi B 2010 *J Solution Chem.* 39 701–8
[5] Nogi Y Watanabe M Suzuki K and Ohkuma Y 2015 *Am. J. Phys.* 83 782-6
[6] Kraftmakher Y 1996 *Am. J. Phys.* 64, 1209-10
[7] Mak S 1999 *Phys. Teach.* 37 179-80
[8] Faccioa D and Clerici M 2006 *Am. J. Phys.* 74(11) 992-4
[9] Chakarvarti S K 1981 *Phys. Teach.* 19 120-1
[10] Davids M Forrest R and Pata D 2010 *Phys. Teach.* 48 217-21.

Influence of LDA+*U* on band dispersion in monolayer MCl_2 (M: Fe, Co)

T B Prayitno[1], E Budi and R Fahdiran
Department of Physics, Faculty of Mathematics and Natural Science, Universitas Negeri Jakarta, Kampus A Jl. Rawamangun Muka, Jakarta Timur 13220, Indonesia
Email: [1]Teguh-budi@unj.ac.id

Abstract. We investigated the effect of LDA+*U* on the band dispersions in 1T monolayer $FeCl_2$ and $CoCl_2$ within the self-consistent noncollinear calculation. As shown in the band dispersed, $FeCl_2$ is metallic while $CoCl_2$ is insulating. When the effective Coulomb energy takes into account, $FeCl_2$ is metallic without band gap while the band gap in $CoCl_2$ increases as the Coulomb energy increases. Thus, the band gap in $CoCl_2$ can be effectively controlled by the Coulomb energy, thus it is feasible for spontaneous development.

1. Introduction

Nowadays, spintronic devices are built from semiconductor materials. Two famous semiconductors refer to the germanium and silicon having the band gap, which can be passed through by electrons from the valence band to the conduction band. The simple property if the materials can be potential for the semiconductor is the band gap. If any materials have the band gap, even narrow, they can be candidates for the spintronic developments as long as the band gap can be controlled up to the specific value.

Considered from the dimensionality, two-dimensional materials are expected to the prominent candidates for the spintronic devices. Starting from the invention of graphene [1-3], a two-dimensional hexagonal material, many scientists attempt to explore the physical properties of other two-dimensional materials. The other two-dimensional materials, which can also be employed for the spontaneous development are metal dihalides [4-6] or tungsten dichalcogenides [7-11] by observing the band gap in the band structure within the density functional theory (DFT).

This paper investigates the properties of two-dimensional metal dihalides MCl2 (M: Fe and Co). Following the previous study [12], we employ the LDA+*U* method to consider the band gap. This method is very useful when the Coulomb repulsion between electrons becomes essential, especially for the 3*d* transition metals. The other method for the same purpose refers to the GW method [13-15]. When the Coulomb repulsion *U* increases, we find two significantly different trends for $FeCl_2$ and $CoCl_2$. While $FeCl_2$ maintains the metallic property (no band gap is available), $CoCl_2$ insulating feature for is observed with the increasing band gap.

2. Method and Material

A crystal structure of two-dimensional 1T metal dihalides for monolayer $FeCl_2$ and $CoCl_2$ was shown in Fig. 1, in which the large and small balls represent the metal cations and anions, respectively. We used the lattice constant of the bulk structure of 3.57 Å for $FeCl_2$ and 3.54 Å for $CoCl_2$ [16]. Meanwhile, the vacuum region was taken larger than 17 Å for the non-periodic cell. The detailed computation was performed by using the OpenMX package [17] with a 20 × 20 × 1 k-point mesh without including the spin-interaction. The electron-electron interaction was expressed by the generalized gradient approximation (GGA) [18] within the self-consistent calculations.

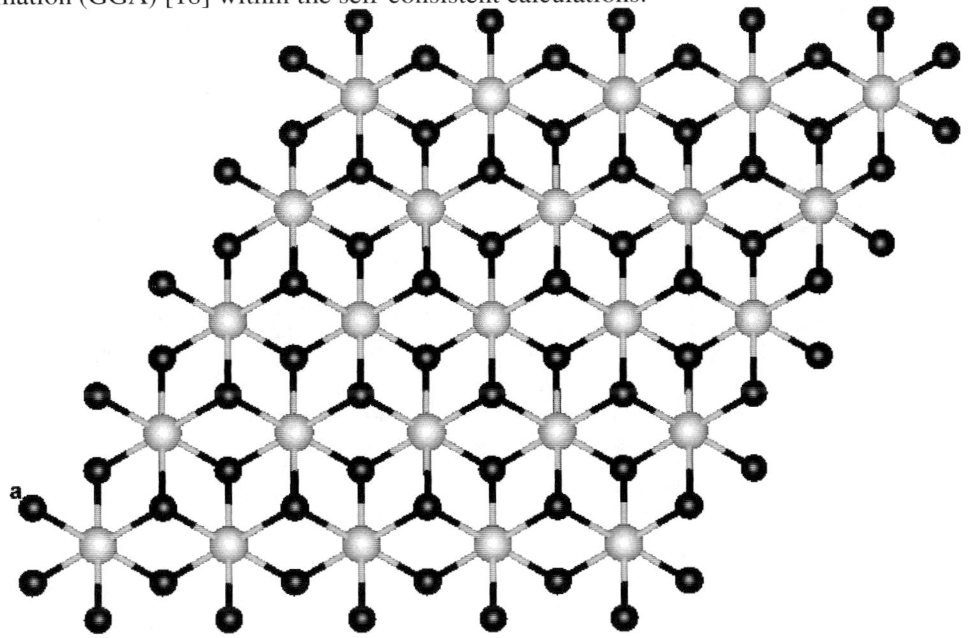

Figure 1. Crystal structures of monolayer MCl_2 (M: Fe and Co) from top view. The large and small spheres are addressed to the magnetic atom M and Cl atom, respectively.

We specified the basis orbits with the abbreviation M4.0-s3p3d3f2, which means that Fe and Co have the cutoff radius of 4.0 Bohr and 41 abilities that consist of three s primitive orbits, three p primitive orbits, three d primitive orbits, and two f primitive orbits, as selected in Ref. [19]. At the same time, we also specified Cl7.0-s2p2d1 for Cl atoms, which means that Cl has a cutoff radius of 7.0 Bohr and 13 abilities that consist of two s primitive orbits, two p primitive orbits, and one d primitive orbital. To obtain the convergent results in the numerical integration, we apply the cutoff energy of 200 Ryd. The crystal structures have been optimized by the non-magnetic case where the force acts on the atom is less than 0.0001 Hartree/Bohr. We then increased U from 0 eV to 4 eV to investigate the trends of the band structures of $FeCl_2$ and $CoCl_2$.

3. Results and Discussions

Figures 2(a) -2(d) show the robustness of the metallic property in the monolayer $FeCl_2$. Even though there are band displacements both in the conduction and valence bands as U increases, this system is a metal. We can see the bands crossing the Fermi energy from the valence band to the conduction band, and vice versa, thus indicating the metallic properties. This property is a good agreement with the collinear version [5] where the band structure in a spin up shows the metallic property while the other shows the insulator. In the nonlinear version, indeed, we cannot separate the band structures for the spin up and spin down since they are mixed. Note that the metallic property will be destroyed when the spin-orbit interaction is taken into account as reported by Botana and Norman [20].

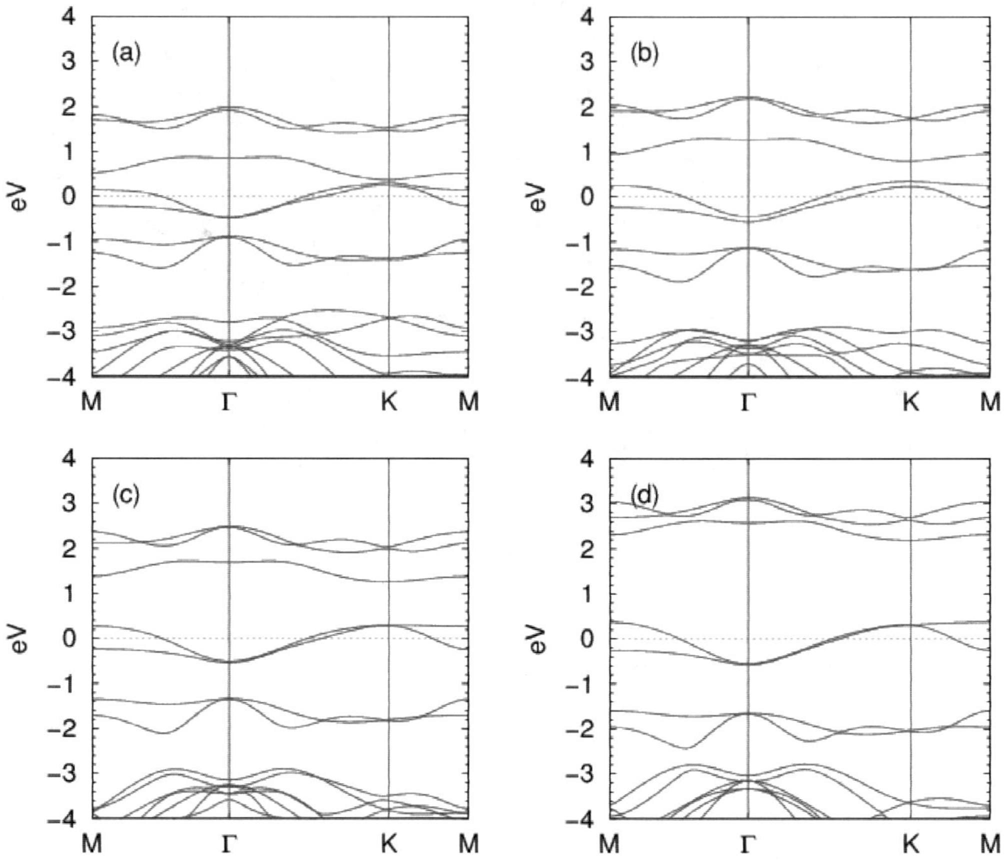

Figure 2. Band structures of monolayer $FeCl_2$ for $U = 0$ eV (a), $U = 1$ eV (b), $U = 2$ eV (c), and $U = 4$ eV (d).

The significant effect of increasing U arises when we consider the monolayer $CoCl_2$, as shown in Figs. 3(a)-3(d). We can see the increase of the band gap in the monolayer $CoCl_2$ when U increases, which means that the monolayer $CoCl_2$ is an insulator. This feature indicates that the band gap in this material can be controlled by *us*. This means that the spontaneous devices can be constructed of this material as long as the controlled band gap is suitable.

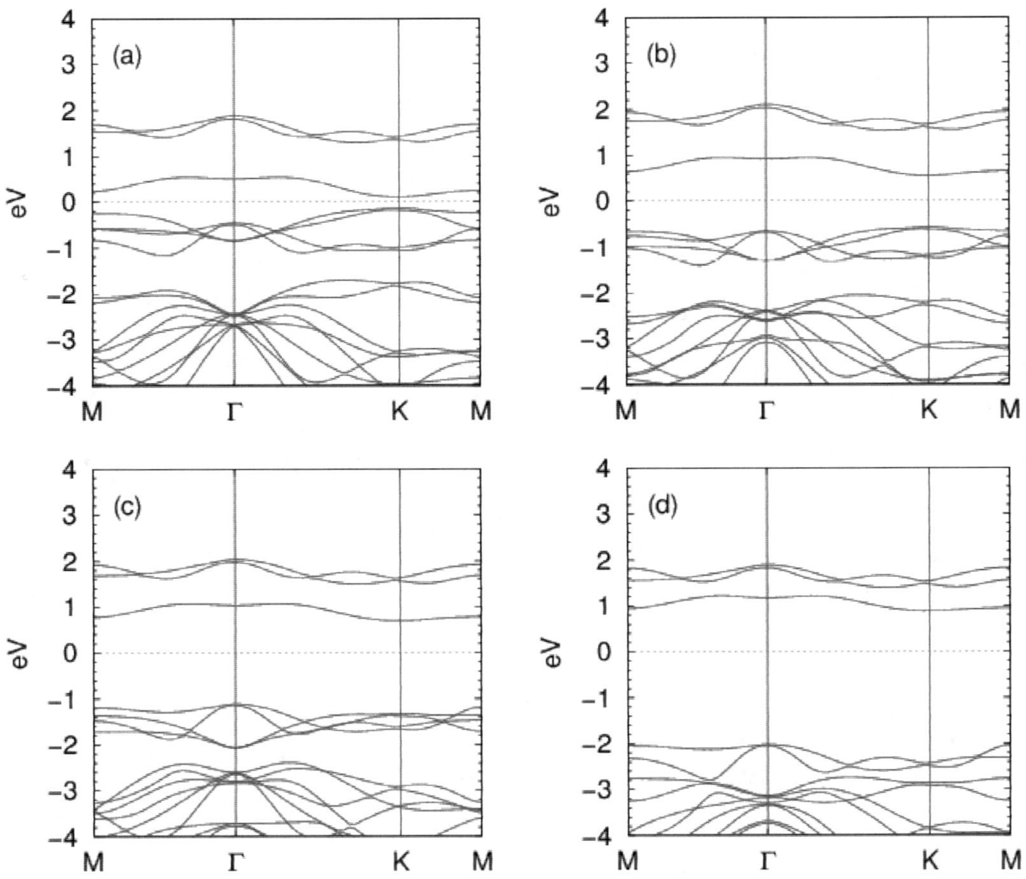

Figure 3. Band structures of monolayer $CoCl_2$ for $U = 0$ eV (a), $U = 1$ eV (b), $U = 2$ eV (c), and $U = 4$ eV (d).

If we compare to the monolayer $MnCl_2$ [12], which also shows an insulating property with a spiral ground state [21], there is no critical U at which the band gap decreases as U increases. Also, nor does the magnetic moment decreases as U increases, as shown in table 1. This means that there is a tendency that increasing U leads to the insulating property of monolayer $CoCl_2$. In addition, while the spin state in monolayer $MnCl_2$ splits into two parts, i.e., the high and low spin states as U increases, we never find the split state in the monolayer $CoCl_2$. This means that the spin state in the monolayer $CoCl_2$ cannot be controlled by U.

Table 1. Calculated band gap and moment of Co atom for the monolayer $CoCl_2$ as U increases.

U (eV)	0	1	2	3	4
Band gap (eV)	0.22	1.13	1.97	2.5	2.92
Co moment (μ_B)	2.55	2.6	2.64	2.68	2.72

For the last discussion, we calculate and compare our Curie temperature of $FeCl_2$ by using the mean-field approximation (MFA) $\boldsymbol{k_B T_C = \frac{3}{2} J}$ as proposed by Torun *et al.* [5]. Here, the exchange parameter is given by $\boldsymbol{J = \left(\frac{1}{12}\right)(E_{AFM} - E_{FM})/24m^2}$ [5]. We then obtain $\boldsymbol{T_C = 6.3}$ K, which is much smaller than that of Torun *et al.* [5] who achieve $\boldsymbol{T_C = 17}$ K. Note that we use the experimental lattice constant while Torun *et al*. Use the optimized one. So, we think the discrepancy comes from the different lattice constants. Furthermore, we cannot use this approach when U is applied. This is due to a transition from the ferromagnetic ground state to the antiferromagnetic ground state when U is employed, as shown in table 2.

Table 2. Calculated total energy difference ΔE between the antiferromagnetic state and the ferromagnetic state of the monolayer $FeCl_2$ as U increases.

U (eV)	0	1	2	3	4
ΔE (MeV)	0.36	-0.47	-3.58	-5.9	-8.22

4. Conclusions

We show that the strength of U cannot create the band gap in the monolayer $FeCl_2$, but can increase the band gap in monolayer $CoCl_2$. This result shows that the monolayer $FeCl_2$ tends to a metallic property while tends to an insulating property. Even though the monolayer $FeCl_2$, does not show the band gap, it can still be considered as a semiconductor since the band structure of collinear calculation shows the half-metallic property [5].

We also find there is a transition of the ground state from the ferromagnetic state in the antiferromagnetic state in $FeCl_2$ as U increases. As a consequence, we cannot predict the critical temperature for the non-zero U as proposed by Torun *et al*. [5] Which only holds for the ferromagnetic ground state.

Acknowledgements

This research is funded by Faculty of Mathematics and Natural Science Universitas Negeri Jakarta under number of contract 57 /SPK PENELITIAN/5.FMIPA/2021.

References

[1] Novoselov K S, Geim A K, Morozov S V, Jiang D, Zhang Y, Dubonos S V, Grigorieva I V and Firsov A A 2004 *Science* 306 666
[2] Novoselov K S, Geim A K, Morozov S V, Jiang D, Katsnelson M I, Grigorieva I V, Dubonos S V and Firsov A A 2005 *Nature* 438 197
[3] Geim A K and Novoselov K S 2007 T *Nature Mater*. 6 183
[4] Tokunaga Y, Okuyama D, Kurumaji T, Arima T, Nakao H, Murakami Y, Taguchi Y and Tokura Y 2011 *Phys. Rev.* B 84 060406
[5] Torun E, Sahin H, Singh S K and Peeters F M 2015 *Appl. Phys. Let.* 106 192404
[6] McGuire, M A 2017 *Crystals* 7 121
[7] Ramasubramaniam A 2012 *Phys. Rev.* B 86 115409
[8] Jiang H 2012 *J. Phys. Chem.* C 116 7664
[9] Zeng H, Liu G B, Dai J, Yan Y, Zhu B, He R, Xie L, Xu S, Chen X, Yao W and Cui X 2013 *Sci. Rep.* 3 1608
[10] Lin L, Xu Y, Zhang S, Ross I M, Ong A C M and Allwood D A 2013 *ACS Nano* 7 8214
[11] Eftekhari A 2017 *J. Mater. Chem.* A 5 18299
[12] Prayitno T B and Fahdiran R 2019 *J. Phys.: Conf. Ser.* 1402 044081
[13] Aryasetiawan F, Gunnarsson O 1998 The GW method *Rep. Prog. Phys.* 61 237
[14] Kotani T, van Schilfgaarde M and Faleev S V 2007 *Phys. Rev.* B 76 165106
[15] Friedrich C, Blügel S V and Schindlmay A 2010 *Phys. Rev.* B 81 125102
[16] Kulish V V and Huang W 2017 *J. Mater. Chem.* C 5 8734.
[17] Ozaki T *et al.* OPENMX website (http://www.openmx-square.org)
[18] Perdew J P, Burke K and Ernzerhof M 1996 *Phys. Rev. Lett.* 77 3865
[19] Prayitno T B 2021 *J. Magn. Magn. Mater.* 517 167386.

[20] Botana A S and Norman M R 2019 *Phys. Rev. Mater.* 3 044001
[21] Prayitno T B and Ishii F 2019 *J. Phys. Soc. Jpn.* 88 104705.

Journal of Physics: Conference Series **2019** (2021) 012075 doi:10.1088/1742-6596/2019/1/012075

Development of a temperature detector and room capacity system to mitigate the spread of COVID-19

W Indrasari[1, a)], H Firmansyah[2)], F Subekti[1)], N A Rahadian[1)] and T A Lita[1)]
Department of Physics, Faculty of Mathematics and Sciences, Universitas Negeri Jakarta, Jakarta 13220, Indonesia [1)]
Department of Electronic Engineering, Faculty of Engineering, Universitas Negeri Jakarta, Jakarta 13220, Indonesia [2)]
Email: [a)]widyaningrum-indrasari@unj.ac.id

Abstract. Several efforts to mitigate the spread of COVID-19 and use a minimum of three layers of masks, it is also necessary to apply physical distancing in all work activities both indoors and outdoors. This study describes one of the efforts that researchers can do to prevent the spread of COVID-19, namely, designing a room capacity detection system equipped with a temperature detector. The research method is an experiment focusing on a temperature detector and room capacity system design. This system uses ultrasonic sensor HCSR04, temperature sensor MLX90614, PIR sensor (Passive Infra-Red) HCSR501, and infrared barrier sensor based on Arduino UNO to measure temperature and count the number of visitors in a room. The system designed based on Arduino UNO has worked well, where the system can detect the body temperature of building visitors and detect the number of visitors entering and leaving the building. The test results show that the temperature measurement on the forehead has a higher accuracy with an error of 1.17%. The results of this study will be used to design automatic door designs to mitigate the spread of Covid-19 in offices.

1. Introduction

The Covid-19 pandemic is still engulfing the world with the number of positive cases continuing to grow to more than 198 million and claiming more than 4.2 million lives worldwide [1]. This pandemic is caused by a new type of coronavirus with the name Severe Acute Respiratory Syndrome Corona Virus (SARS-CoV-2) [2]. This virus is one of the main pathogens that attack the human respiratory system [3]. COVID-19 spreads very quickly, especially in public places because this virus can be transmitted from human to human through respiratory droplets produced during coughing [4]. Respiratory droplets can spread up to 2 meters and settle on various surfaces, where they can persist for hours or even days [5]. The infection is transmitted either by inhaling droplets or touching a contaminated surface and then touching the nose, mouth, or eyes.

There are four types of COVID-19 patients: asymptomatic, mild, moderate, and severe [6]. Fever, dry cough, diarrhea, difficulty breathing (dyspnea), headache, and pneumonia are the most common symptoms of COVID-19. Monitoring body temperature is critical, especially for the early detection of suspected COVID-19. SARS-CoV-2 can be stable in infected patients for up to 4 days [7]. The stability of this virus at conditions around average body temperature implies that temperature can play a crucial role in the transmission and severity of COVID-19. Various body temperature measuring instruments (thermometers), both contact and non-contact, are available and easy to obtain. However, because this virus can be transmitted through fluids released by the patient's body, it is recommended to use a non-contact clinical electronic thermometer to measure body temperature. In this COVID-19 pandemic, non-contact handheld infrared thermometers (thermogun) are often used to check fever. This thermometer has the advantage of being a rapid, non-contact, and easy-to-use fever screening [8].

As an effort to mitigate the spread of COVID-19, apart from using a minimum of three layers of masks, it is also necessary to implement physical distancing in all work activities both indoors and outdoors. This of course has an impact on the decrease in the capacity of a room. Based on the Instruction of the Minister of Home Affairs of the Republic of Indonesia Number 3 of 2021, it has been determined that during PPKM (Enforcement of Restrictions on Community Activities) a workspace can only be filled with a maximum of 50% of normal capacity. So, it is necessary to have a room user capacity detector to ensure the number of users does not exceed the established rules. To calculate the number of users entering and leaving a room, ultrasonic sensors [9], infrared sensors [10] -[11], Quick Response Code technology [12].

This article describes a room capacity detection system equipped with a temperature detector. The results of this study will be used in the design of automatic door designs as an effort to mitigate the transmission of Covid-19 in offices.

2. Method

The research method is an experiment focusing on designing a temperature detector and room capacity system design. This system uses ultrasonic sensor HCSR04, temperature sensor MLX90614, PIR sensor (Passive Infra-Red) HCSR501, and infrared barrier sensor based on Arduino UNO to measure temperature and count the number of visitors in a room. The system consists of 2 separate parts that are connected by cables. Part 1 device consists of Arduino UNO R3 microcontroller as a data management center, temperature sensor MLX90614, ultrasonic sensor HCSR04, and PIR sensor HCSR501. The Part 1 device is placed at the front door and functions as a temperature gauge for visitors, and automatically counts the number of visitors entering the room. When a visitor is detected, the MLX 90614 temperature sensor as a body temperature gauge will measure the visitor's body temperature, which is in the ultrasonic sensor range. In comparison, the second part of the device uses PIR and infrared sensors to detect and count the number of visitors leaving the room. All sensors used in this device are characterized to determine the measuring range of the sensor. The sensor characterization was conducted by comparing the output sensor and the standard laboratory equipment.

3. Result and Discussion

The temperature sensor and proximity sensors are previously characterized to obtain more accurate measurement results of the device. The following presents the results of a temperature sensor, PIR sensor, ultrasonic sensor, and infrared sensor characterization, as well as the results of the device measuring on some parts of the human body.

Table 1. Optimization results of the MLX90614 temperature sensor measurement distance

Distance (cm)	MLX90614 Temperature Sensor Output (°C)										Average of output Sensor (°C)
0.2	35.07	35.27	35.51	35.45	35.47	35.51	35.15	35.19	35.21	35.37	35.320
0.4	35.37	35.31	35.19	35.55	35.29	35.43	35.29	35.29	35.43	35.15	35.330
0.6	35.23	35.01	35.07	35.29	35.09	35.21	35.31	35.31	35.09	35.15	35.176
0.8	35.07	35.15	35.37	35.29	35.31	35.29	35.31	35.39	35.43	35.37	35.298
1	35.15	35.27	35.35	35.23	35.31	35.15	35.13	35.13	35.01	35.09	35.182
1.2	34.05	34.13	34.15	34.21	35.09	35.07	34.15	34.13	35.15	35.15	34.528
1.4	34.23	34.19	34.21	34.13	34.19	34.21	34.21	34.13	34.21	34.15	34.186
1.6	34.09	34.39	34.15	34.19	34.27	34.23	34.01	34.09	34.05	34.19	34.166
1.8	34.01	34.19	34.21	34.01	34.07	34.05	34.09	34.21	34.35	34.07	34.126
2	34.05	34.01	34.21	34.21	34.09	34.15	34.21	34.07	34.09	34.05	34.114

3.1. Characterization of MLX90614 temperature sensor

The MLX90614 sensor is a sensor that is used to measure temperature by utilizing infrared radiation. The MLX90614 sensor is designed to detect infrared radiation energy and calibrate infrared radiation energy into a temperature scale. In this study, the sensor characterization was carried out to optimize the sensor measurement distance range and determine the accuracy of the sensor measurement results against a standard measuring instrument, namely the forehead thermogun of the Omron Mc 720 types. The measurement results are shown in Table 1. The data in Table 1 show that the optimal distance from the MLX90614 temperature sensor reading is 1 cm from the object. Based on these results, the sensor calibration was then carried out by repeatedly measuring the temperature of five different things at 1 cm using a sensor and a forehead thermometer. Graph plotting was carried out to find out the relationship between the sensor output and the forehead thermogun (Figure 1).

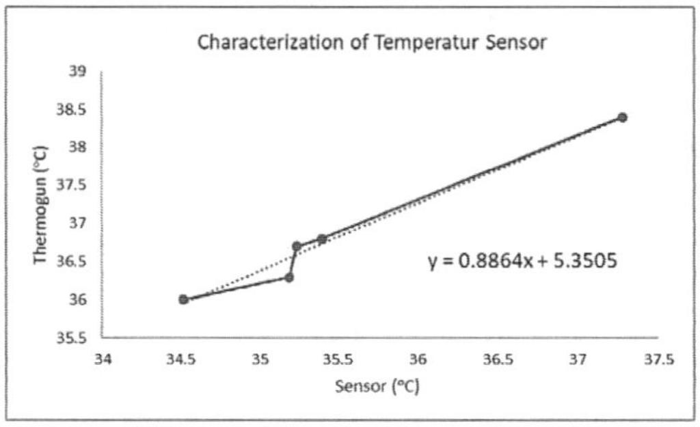

Figure 1. Curve of temperature sensor's characterization

Based on the results of plotting the curve in Figure 1, there are differences in the measurement results between the temperature sensor and the thermogun. So that a correction is made in the readings at the temperature sensor output on the microcontroller using equation 1, namely:

$$y = 0.8864x + 5.3505 \quad (1)$$

Where x is the result of measuring body temperature using a temperature sensor (0C), and why is the result of the conversion of body temperature measurement (0C) to determine the accuracy of temperature measurement, the measurement error is calculated using equation (1). The results are shown in Table 2. There is an average deviation of 0.265%.

Table 2. The error of the MLX90614 temperature sensor measurement

Calibration			
Sensor (°C)	**Termogun (°C)**	**Result(°C)**	**Error (%)**
34.518	36	35.947	0.147
35.192	36.3	36.545	0.674
35.242	36.7	36.589	0.302
35.404	36.8	36.733	0.183
37.278	38.4	38.394	0.016
	Average		0.265

3.2. Characterization of PIR sensor

The HCSR501 PIR sensor is used to detect infrared radiation emitted by an object. The PIR-sensing element is responsive to mid-and far-infrared radiation within a spectral range from approximately 4 to 20 m [13]. This Passive Infrared sensor technology is critical for detecting human presence [14] because humans produce infrared radiation with wavelengths between 9-10 m. The characterization of the HCSR501 PIR sensor with a length measuring instrument aims to find the farthest range of sensor readings. The data obtained are shown in Table 3. From Table 3, the PIR sensor used can work optimally to detect objects in the distance range of 5 - 30 cm.

Table 3. Characterization of the HCSR501 PIR Sensor

Distance (cm)	Object
5	Detected
10	Detected
15	Detected
20	Detected
25	Detected
30	Detected
35	Undetected
40	Undetected

3.3. Characterization of Infrared barrier Sensor

The infrared sensor is composed of infrared LEDs as infrared transmitters and phototransistors as infrared light receivers. An infrared sensor has a function to detect objects when objects block infrared light. Distance characterization is carried out to determine the farthest distance from the sensor reading. The infrared sensor output is affected by the colour barrier that interferes with the radiation from the infrared LED. So that two types of barriers colours are used, namely white screens and red screens. Observations were made repeatedly, each at 1 - 20 cm with an interval of 1 cm. The data (Table 4) show that for white barriers, the sensor can detect obstructions up to 14 cm, while for red barriers, the sensor can work to detect obstructions up to 12 cm.

Table 4. Characterization of the HCSR501 PIR Sensor

Distance (cm)	White Object	Coloured Object
1	Detected	Detected
2	Detected	Detected
12	Detected	Detected
13	Detected	Undetected
14	Detected	Undetected
15	Undetected	Undetected
16	Undetected	Undetected

3.4. Characterization of Ultrasonic Sensor

This characterization was conducted to determine the range of the sensors and identify the ultrasonic sensor's resolution [15]. Measurements were only carried out from the closest to the furthest distance and the farthest distance to the most relative distance with variations in lengths of 5, 10, 15, 20, 25, and 30 cm, with three repetitions. The measurement results are shown in Figure 2. From the results of the characterization of the ultrasonic sensor HCSR04, the average error of the sensor reading is 1.094%. This value has increased from the results of the ultrasonic sensor calibration, which has a relative error value of 2.02% [16].

Journal of Physics: Conference Series **2019** (2021) 012075 doi:10.1088/1742-6596/2019/1/012075

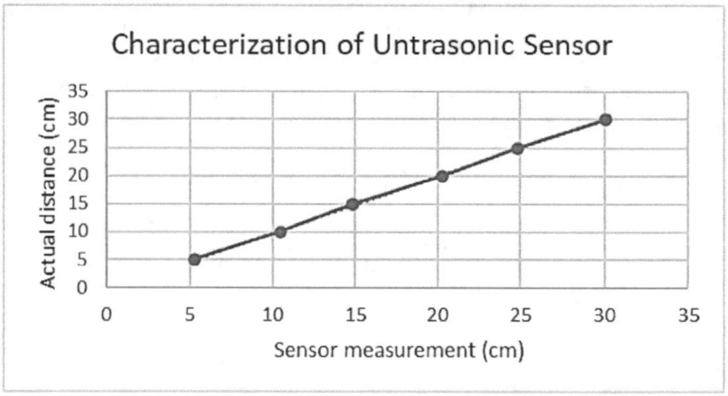

Figure 2. The result of ultrasonic sensor characterization

3.5. Device Trial

The system testing phase is carried out to determine the operational capability of the device. The test results show that the part 1 device can detect the body temperature of visitors and count the number of visitors who enter the building automatically. In comparison, the part 2 device can automatically count outgoing visitors. Indoor capacity is limited to a maximum of 50 visitors. If the power exceeds, there will be a warning in the form of a buzzing sound, and the sentence "Capacity is Full". The LCD will display the results of measuring the visitor's body temperature and the remaining room capacity, as shown in Figure 3. To determine the accuracy of the measurement tools, measurements are made in several different parts of the body surface, namely the forehead, right hand, left hand, and cheeks. The test results show that the forehead temperature measurement has a higher accuracy with an error of 1.17%.

Figure 3. Results of room temperature and capacity measurements

As for measuring the distance range of the infrared and PIR sensors on the system, it shows that the infrared sensor can read the presence of obstacles at the furthest distance of 10 cm with a reading delay of ±3 seconds. Meanwhile, the PIR sensor can detect objects at the farthest distance, which is 30 cm. The weakness of this device is that it can only count incoming or outgoing visitors one by one. So, if visitors enter or leave simultaneously (directly two or more people), then only one visitor will be counted whose position is closest to the sensor.

4. Conclusions

A Temperature Detector and Room Capacity system has been successfully built using an ultrasonic sensor HCSR04, temperature sensor MLX90614, PIR sensor HCSR501, and an Arduino UNO-based infrared barrier sensor that works well in detecting the body temperature of building visitors and noticing the number of visitors entering and leaving the building.

Acknowledgments
The authors wish to thank the colleague in the Department of Physics for very helpful comments on an earlier draft of this paper. Also, the authors acknowledge the valuable supports of Universitas Negeri Jakarta in this research.

References
[1] Worldometer 2021 Covid-19 Coronavirus Pandemic https://www.worldometers.info/coronavirus/.
[2] Coronaviridae Study Group of the International Committee on Taxonomy of Viruses 2020 *Nat Microbiol* 5 536
[3] Bogoch I I, Watts A, Thomas B, A, Huber C, Kraemer M U, G and Khan, K 2020 *Journal of Travel Medicine*, 27 1.
[4] Wang C, Horby P W, Hayden F G, and Gao G F 2020 *The Lancet* 395 470
[5] Agarwa K M, Mohapatra S, Sharma P, Sharma S, and Mishra D B N 2020 *Sensor International* 1 100037.
[6] Chang, M C, Park Y K, Kim B, O, and Park D 2020 *BMC Infect. Dis.* 20 445.
[7] Pawar, Stanam A, Chaudhari M, and Rayudu D 2020 *Medrxiv* 03.
[8] Ray R, Low Y, L, Zhao X, Cook A R, and Lee V J 2015 *Public Health* 129 1471.
[9] Ardiansyah E, Fitriyah H, and Syauqy D 2018 *Jurnal Pengembangan Teknologi Informasi dan Ilmu Komputer* 3 673.
[10] Paramananda R G, Fitriyah H, and Praset B H 2018 *Jurnal Pengembangan Teknologi Informasi dan Ilmu Komputer* 2 921.
[11] Argon R and Susanto I M I 2013 *Majalah Ilmiah Teknologi Elektro* 11.
[12] Sinuhaji N 2018 *Majalah Ilmiah Politeknik Mandiri Bina Prestasi* 7.
[13] Fraden J 2016 *Handbook of Modern Sensor: Physics, Designs, and Applications* (Springer) p. 302.
[14] Akhter F, KhadivizandS, Siddiquei H R, Alahi M E E, and Mukhopadhyay S 2019 *Sensors* 19 3374.
[15] Indrasari W, Iswanto B, H, and Andayani M 2018 *IOP Conf. Ser.: mature. SCI. Eng.* 335 012005.
[16] Indrasari W, Rustana C E, and Zulfikar 2021 *J. Phys.: Conf. Ser.* 1816 012109

Design and Development a Wind Tunnel using the Honeycomb as a Wind Turbine Characterization Instrument

H Nasbey[1] and E Budi
Department of Physics Education, Universitas Negeri Jakarta
Jl. Rawamangun Muka, Jakarta 13220, Indonesia
Email: [1]hadinasbey@unj.ac.id

Abstract. In this research, an open circuit wind tunnel system was developed with variations in the use of the honeycomb and not in the settling chamber. This system will be used as an instrument for testing the characteristics of wind turbines. Settling chamber, cylindrical with a diameter of 56 4 mm with a length of 150 mm. Contraction cone, a truncated cone with a large diameter of 564 mm and a small diameter of 460 mm and a height of 300 mm. The test section is cylindrical in shape with a diameter of 460 mm and a length of 500 mm. Diffuser, in the form of a truncated cone with a large diameter of 560 mm and a small diameter of 460 mm and a height of 400 mm. The use of honeycomb is aimed at reducing the value of the intensity of turbulence. The honeycomb consists of a PVC cylinder with a diameter of 8.5 mm and a length of 100 mm, placed in the settling chamber. The intensity of turbulence is defined as the ratio of the standard deviation to the mean value of the wind speed. The wind speed in the test section was measured using an anemometer at nine different measuring points around the center of a cross section of the test section. The nine-point wind speed was used to calculate the intensity of the turbulence. The turbulence intensity of the wind tunnel without honeycomb and with honeycomb in the settling chamber is approximately 0.864 and 0.178, respectively.

1. Introduction

The wind tunnel is an important measuring instrument for aerodynamic research. Wind tunnels are used to characterize the flow of air or gas during the passage of certain objects such as the design or modernization of airplanes, vehicles, bridges, high-rise buildings and wind turbines [1]. Open type wind tunnels can be used as a means of developing the field of aerodynamics, in particular wind turbine testing. The open line type of wind tunnel is used with several considerations, including lower costs and smaller building area than the closed type [2]. In the wind tunnel simulation, the model is assumed to be at rest and the wind is moving at a certain speed. Therefore, the air flow in the wind tunnel test section must meet certain requirements, including low intensity of turbulence. Various methods have been described to describe the behavior of wind speed characteristics in a certain area. Stochastic modeling uses the parameters of mean wind speed and turbulence intensity to configure each resulting model. The best parameterization of the Ornstein-Uhlenbeck (OU) process was used by comparing the set of wind speeds in seconds and their characteristics and determined by turbulence intensity analysis [3, 4]. In another study, the distribution of flexible families was used which was compared to the Weibull distribution to analyze the distribution of wind speed [5]. In an effort to minimize the estimated error, the Johnson distribution with the multi - parameter distribution is also used in addition to the Weibull distribution to determine the flow characteristics of wind speed [6].

The use of ten two-component mixed models including gamma, Weibull, Gumbel and truncated normal mixtures was also used to characterize the wind speed distribution [7 – 9].

Wind tunnel dengan desain rentang kecepatan angin 30 sampai 100 m/s dengan rentang tekanan pengunaan 0.15 sampai 1.0 atm berpotensial digunakan pada pengujian-pengujian micro-aerodynamic [10]. In another study, a two-dimensional incompressible flow breaker simulation based on a finite volume scheme was used, successfully predicting flow along the core and near the wall from three preselected wall shape contractions [11]. The use of three-dimensional geometry in open circuit wind tunnel results in the recommended contraction ratio, maximum uniformity values in the central plane of the part, prevention of separation, absence of Gortler vortices contraction and minimization of the thickness of the boundary layer at the entrance to the workpiece [12].

The low intensity turbulence of wind tunnel is a major requirement. An aluminum honeycomb tested in a wind tunnel at the Central Aerohydrodynamic Institute (TsAGI) exhibits good free flow conditions. [13]. The use of honeycombs and sails, in an open-circuit wind tunnel, installed at the Indian Institute of Technology in Guwahati (IITG) has demonstrated a significant reduction in turbulence, a key component of subsonic wind tunnels [14].

In this research, an open circuit wind tunnel is designed and developed. The intensity of the turbulence in the test chamber was reduced using honeycomb made of PVC tubes. The turbulence intensity reduction test was performed using the main method. The reduction in intensity that occurs in the test chamber is clearly shown in the experiment.

2. Method

The open-circuit wind tunnel is made using an iron plate formed into a cylinder consisting of four main parts. At the very front is a settling chamber in which there is a removable honeycomb. Air enters through the cone of contraction to the test section, finally exiting the diffuser. Figures 1 (a) and (b) show cinematic and photographic diagrams of the developed wind tunnel.

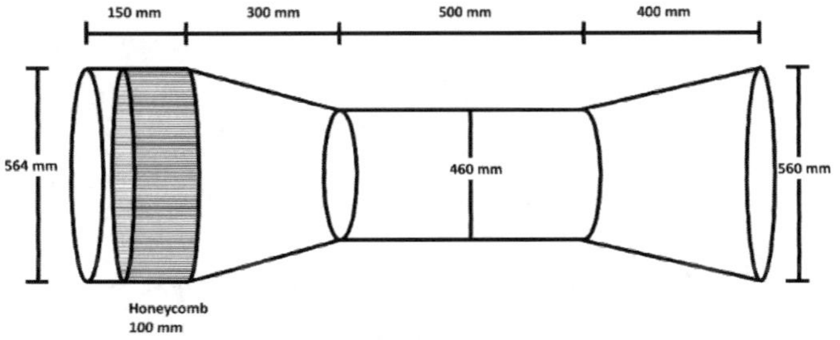

(a)

(b)

Figure 1. (a) Schematic diagram in a wind tunnel, (b) photographic drawing developed in a wind tunnel.

Settling chamber, cylindrical with a diameter of 56 4 mm with a length of 150 mm. Contraction cone, a truncated cone with a large diameter of 564 mm and a small diameter of 460 mm and a height of 300 mm. The test section is cylindrical in shape with a diameter of 460 mm and a length of 500 mm. Diffuser, in the form of a truncated cone with a large diameter of 560 mm and a small diameter of 460 mm and a height of 400 mm. A digital anemometer is used to measure the wind speed in the cross section of the test chamber. Figure 2 (a) shows a schematic diagram of nine wind speed test points in the test section, the midpoint of the cross section of the test section is positioned at the point of origin. Figure 2 (b) shows a photographic image of the cross section of the test section in the process of collecting wind speed data.

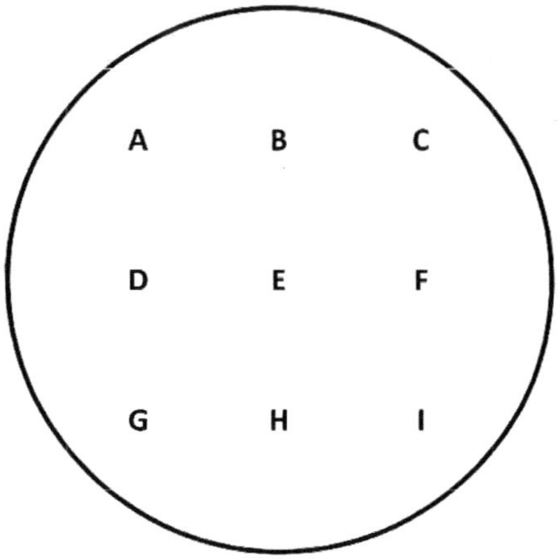

(a)

(b)

Figure 2. (a) Nine wind speed test points in the test section, (b) photographic image of the cross section of the test section.

Figure 3 shows the honeycomb which is attached to the settling chamber. The honeycomb is made of 8.5 mm diameter PCV tubes that are arranged tightly. The space that forms between the two tubes is sealed with an adhesive so that the air only circulates through the holes in the pipe.

Figure 3. Honeycomb with length 100 mm and diameter 8.5 mm.

the intensity of the turbulence and is defined as follows [15];

$$I = \frac{\sigma}{U}$$

with

$$\sigma = \sqrt{\frac{1}{N}\left\{\sum_{i=1}^{N} {x_i}^2 - \frac{\left(\sum_{i=1}^{N} x_i\right)^2}{N}\right\}}$$

$$\overline{U} = \frac{\left(\sum_{i=1}^{N} x_i\right)}{N}$$

Where $\boldsymbol{\sigma}$ is the standard deviation of the variation in wind speed and $\boldsymbol{U}$ is the mean wind speed.

3. Results and Discussion

Figure 4 shows the distribution of wind speed in the cross section of the test section when the wind tunnel is not attached to the honeycomb in the settling chamber.

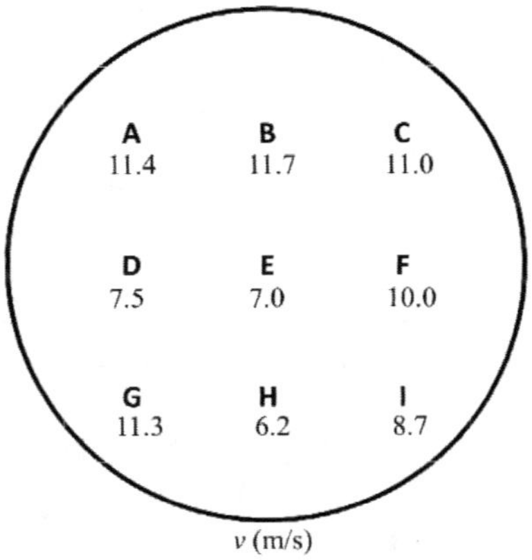

Figure 4. The distribution of wind speeds in the cross section of the test section without honeycomb.

In the calculation, we get $\boldsymbol{\sigma}$ approximately 2.63 and $\boldsymbol{U}$ approximately 3.05. The turbulence intensity (I) of the first model is approximately 0.864.

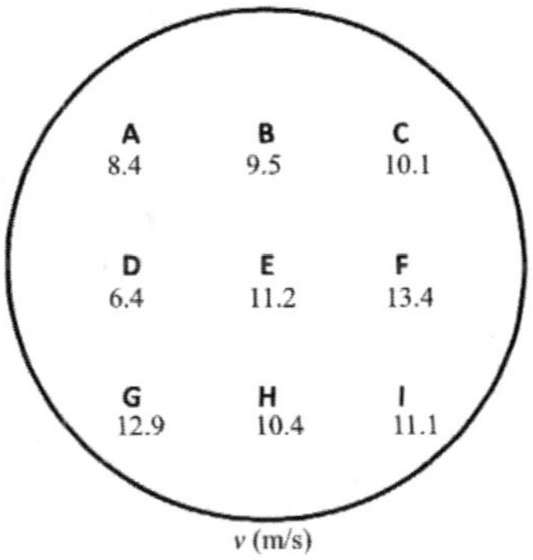

Figure 5. The distribution of wind speeds in the cross section of the test section with honeycomb.

Figure 5 shows the distribution of wind speed in the cross section of the test section when the wind tunnel is attached to the honeycomb in the settling chamber. In the calculation, we get $\boldsymbol{\sigma}$ approximately

1.88 and ***U*** approximately 10.99. The turbulence intensity (I) of the first model is approximately 0.171. Based on the measurement results, the use of honeycomb in the wind tunnel reduces the intensity of wind turbulence in the test section by 0.69 points, or about 80%.

4. Conclusion
A wind tunnel was designed and developed using an open circuit design. The wind tunnel is made by forming a steel plate into a column consisting of four main parts, namely the settling chamber, the contraction cone, the test section and the truncated cone. The intensity of the turbulence in the defined section of the wind tunnel was measured using a mean method. The turbulence intensity in the section sets for the honeycomb-free blower in the settling chamber shows a relatively large value compared to the use of honeycomb in the settling chamber. In further research, the appropriate honeycomb design variation can be used to achieve a smaller turbulence intensity value.

Acknowledgments
This research financially was supported by the Ministry of Research, Technology and Higher Education of Indonesia, "Dana BLU Pok Fakultas Matematika dan Ilmu Pengetahuan Alam Universitas Negeri Jakarta, Nomor : 30 /SPK PENELITIAN/5.FMIPA/2021, 11 May 2021".

References
[1] Jiang Z, Hu Z, Wang Y, Han G 2020 Advances in critical technologies for hypersonic and high-enthalpy wind tunnel *Chinese Journal of Aeronautics* **33**(12) 3027-3038.
[2] Chen F, Peng H, Chan P, Zeng X 2019 Low-level wind effects on the glide paths of the North Runway of HKIA: A wind tunnel study *Building and Environment* **164** 106337.
[3] Arenas-López J P, Badaoui M 2020 Stochastic modelling of wind speeds based on turbulence intensity *Renewable Energy* **1481**(20) 30436-5.
[4] Calif R 2012 PDF models and synthetic model for the wind speed fluctuations based on the resolution of Langevin equation *Applied energy* **99** 173-182.
[5] Usta I, Kantar Y M 2012 Analysis of some flexible families of distributions for estimation of wind speed distributions *Applied Energy* **89** 355-367.
[6] Soukissian T 2013 Use of multi-parameter distributions for offshore wind speed modeling: The Johnson SB distribution *Applied energy* **111** 982-1000.
[7] Ouarda T B M J, Charron C 2018 On the mixture of wind speed distribution in a Nordic region *Energy Conversion and Management* **174** 33-44.
[8] Jaramillo O A, Borja M A 2004 Wind speed analysis in La Ventosa, Mexico: a bimodal probability distribution case *Renewable Energy* **29** 1613-1630.
[9] Carta J A, Ramirez P, Velazquez S 2009 A review of wind speed probability distributions used in wind energy analysis: Case studies in the Canary Islands *Renewable and sustainable energy* **13** 933-955.
[10] Martin M J, Scavazze K J, Boyd I D, Bernal L P 2006 Design of a low-turbulence, low-pressure wind-tunnel for micro-aerodynamics *Journal of Fluids* Engineering **128** 1045-1052.
[11] Zanoun E Z 2018 Flow characteristics in low-speed wind tunnel contractions: Simulation and testing *Alexandria engineering journal* **57**(4) 2265-2277.
[12] Abdelhamed A S, Yassen Y El-S, ElSakka M M 2015 Design optimization of three dimensional geometry of wind tunnel contraction *Ain Shams Engineering* **6** 281-288.
[13] Annaloro J, Ledermann V, Constant E, Spel M, Vasse C, Rivola V, Drozdov S M, Omaly P 2020 Rebuilding with PAMPERO of destructive hypersonic tests on honeycomb sandwich panels in the T-117 wind tunnel *Journal of Space Safety Engineering* **7**(2) 113-124.
[14] Kulkarni V, Sahoo N, Chavan S D 2011 Simulation of honeycomb–screen combinations for turbulence management in a subsonic wind tunnel *Journal of Wind Engineering and Industrial Aerodynamics* **99** 37-45.
[15] Manwell J, McGowan J, Rogers A 2009 Wind energy explained: theory, design and application. 2nd ed *John Wiley & Sons* p 40.

Journal of Physics: Conference Series **2019** (2021) 012077 doi:10.1088/1742-6596/2019/1/012077

Automatic Detection of Breast Calcification in Ultrasound Imaging with Convolutional Neural Network

P D Karunia[1], P Prajitno and D S Soejoko
Department of Physics, Universitas Indonesia
Kampus UI, Depok 16424, West Java, Indonesia
Email: [1]petrananda.dea@ui.ac.id

Abstract. Breast cancer is a common type of cancer that leading death causes of female in the worldwide. Breast calcification can be one of indicator that can be used to detect the breast cancer early. One of the preferred methods used by radiologist to detect breast cancer is ultrasound imaging. Ultrasound imaging is much safer than mammography that followed by radiological effect. However, ultrasound imaging contaminated with speckle noise that looks similar to breast calcification. It can be the cause of the long time diagnosis process. It encourages so many methods of computed aided diagnosis (CADx) that can detect abnormalities automatically. One of them is Convolutional Neural Network (CNN). CNN can be used to classify the normal breast and breast with abnormalities. In this paper, CNN has been proposed for the classification of the ultrasound images into normal breasts and breasts with calcification. Experimental results classification accuracy was 76 % and a sensitivity of 84.61%.

1. Introduction

Breast cancer is a common cancer that cause of death for women in the world. However, with early detection, a small period will provide a greater chance of survival. Based on data from the American Cancer Society, within 5 years after diagnosis, patients with early stage cancer diagnosis have a 100% chance of survival. Whereas in stage two has a chance of survival up to 80-90%, stage three has a chance of survival up to 60% -70%, and for the final stage the chance of survival is only about 15% - 20%. Breast cancer detection can be done by mammography or ultrasound imaging [1].

Ultrasound imaging is one of the most frequently used diagnostic tools to detect and classify abnormalities of the breast, because of its low cost, nonionizing radiation, and real-time capability. However, ultrasound image formation has a weakness in spatial resolution and the presence of speckle noise [2]. Breast abnormalities usually begin with microcalcification. Microcalcification is an early indicator of breast cancer that is not palpable. With breast screening, microcalcification can be categorized by looking at its density, morphological appearance, and the pattern or shape of its distribution [3].

Breast calcifications are a calcium deposits that can be majorly categorized as two types which differ in chemical composition. The first is composed of calcium oxalate, which is mainly found in benign breast lesions. The other is hydroxyapatite, a kind of calcium phosphate, which can be found in both benign and malignant [4,5]. The number, size, morphology and distribution of breast calcification gives important information about the malignancy or benignity of breast lesions. Amorphous calcifications with round, thick rod, diffuse and non-catheter distribution are indicate low risk of malignancy calcifications (Figures 1a, b). Large and heterogeneous calcifications, and amorphous or isolated calcifications are more associated with indeterminate calcifications (Figure 1c, d). While, calcifications that indicates highly suspicious malignant show shows with fine and pleomorphic, or fine linear branching distribution calcifications (Figure 1e, f).

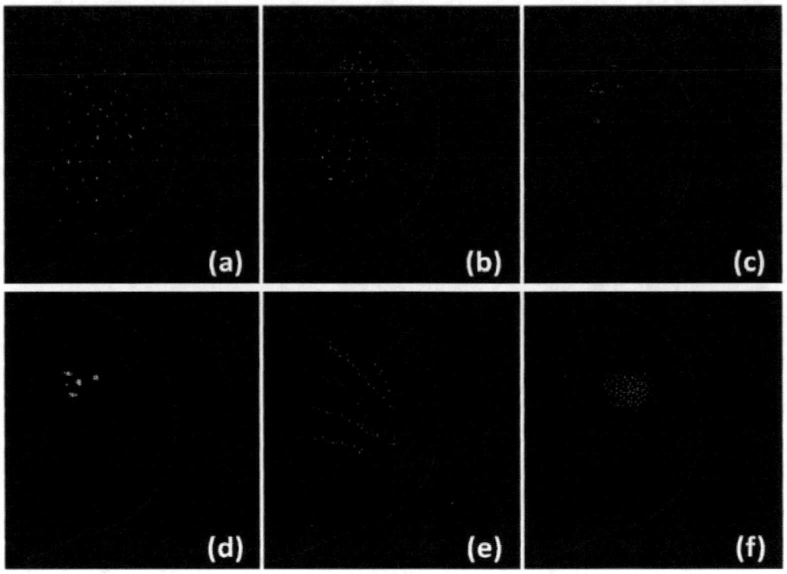

Figure 1. Typical distribution of breast calcification (a) Diffuse, (b) Regional, (c) Isolated, (d) Large and heterogeneous, (e) Linear branching, (f) Clustered [5].

Breast calcification detection is carried out with a human reader, and usually performed by radiologists or oncologists. It takes a long time and may add costs for adding data from other modalities. Another factor that can be a potential error is increasing the length of time for reading and adding information with other modalities is due to noise, so that the reading is not optimal [1]. Noise reduction on ultrasound images still is a serious issue. The error detection can be caused by system that detect the speckle noise as an abnormality. Hence, the filtering of speckles is of prime importance, and in recent years, several studies have been undertaken to address this serious issue [6-7].

Computer-assisted diagnosis (CADx) systems have been developed to reduce costs and increase the radiologist's ability to interpret medical images and differentiate between benign and malignant tissue. One of the method used for CADx is Convolutional Neural Network (CNN). Convolutional Neural Network (CNN) is a deep neural network that is devoted to image recognition. This technique shows how significant the change of the deep layer (node layer) is for image processing. CNN mimics the workings of our brains in processing and recognizing an object [8]. The CNN system consists of three layers of processes, namely the convolution, pooling and fully connected (FC) layers. The convolution and pooling layers are part of the system whose function is to extract image properties, which will then be mapped to the fully connected layer to become output, such as the results of object classification. The output results are evaluated with a loss function of the loss value which is used to update the kernel factor and weight. This process is called training. After the training process, the classification system in the CNN system has been formed and is expected to be able to be used to classify new images accurately [9]. The CNN system includes at least one convolutional layer. Figure 2 shows a 2-dimensional (2D) CNN used to recognize images [10].

2. Literature Review

The conventional approach to analyzing 2D ultrasound images of the breast is likely to be a time-consuming and operator-dependent procedure [11]. Various digital image processing has been developed to obtain a more efficient image analysis method [12-13]. In recent years, several machine learning-based methodologies have been developed, one of which is Convolutional Neural Networks (CNNs). The CNN has also been included in medical image analysis to aid in patient diagnosis [14-15].

This system performs predictions for each pixel value and assign a class label to each pixel in an image [13].

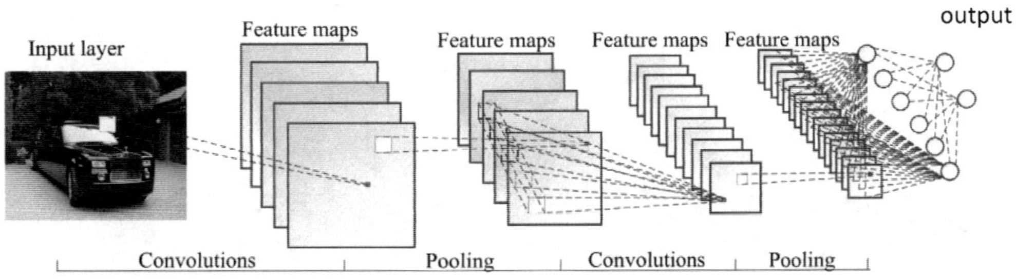

Figure 2. 2D CNN [8]

Recently, CNN has developed due to its high accuracy, great power and flexibility. This feature has been applied to the classification of microcalcification clusters or breast lesions. Becker et al. [16] Demonstrated that the cutting-edge network for general image analysis could detect cancer in mammography with an accuracy similar to that of radiologists. Breast cancer detection and breast calcification based on mammography images have also been developed in several studies [17, 18]. Chai at all [17] obtained detection results with a precision of 89.32% and a sensitivity of 86.89%. Meanwhile, analysis based on ultrasound images is carried out by several methods, for example, in [3], analysis is carried out by image segmentation. In this paper, we will use CNN to detect breast calcification on ultrasound imaging based on Latif 2020 [19], with 2 convolutional layers and 2 pooling layers with differences in sample size.

Noise reduction in this paper obtained by mean filter. Mean filter is a linear filtering technique that is also known as an averaging filter, which replaces each value of the pixel in an image of the average the gray levels of the neighbourhood. The technique does not eliminate speckle noise as a whole, but it Decreases to some extent. The filter has an effect of blurring and smoothing the image [20].

3. Materials

3.1. Data Set

The data used in this study are ultrasound image data obtained from RSUD Dr. Moewardi. The data taken are breast images between August 2019 and August 2020. The images used are limited to patients aged between 20-60 years, with an average age of 46 years. The ultrasound examination process is carried out by an experienced radiologist. The data used in this experiment are ultrasound images with a diagnosis of breast cancer with calcification components and normal breasts. Calcifications are observed as echogenic dots within a mass or lesion or may be distributed as we can see in Figure 1. In this study, the breast images used included 97 images which contained about 48 normal breast images (negative sample), and 49 calcified breast images (positive sample). The data will be used as sample data for the data training, data validation and data testing. The positive and negative sample showed at Figure 3.

For testing purposes, 25 sample data were taken, which included 15 positive samples and 10 negative samples. And other data will be used as sample data in the CNN training and validation process. Due to the limited amount of data, for the training and validation process, the data is multiplied by data augmentation. 38 negative samples and 34 positive samples were multiplied by applying 4 kinds of image transformations, which is rotated (90° and 180°) and flipped the image up down and left right. The total data obtained from the CNN training process is 360 images. Furthermore, for the training and validation processes, the data will be split, 80% in the training process and 20% for the validation.

Journal of Physics: Conference Series **2019** (2021) 012077 doi:10.1088/1742-6596/2019/1/012077

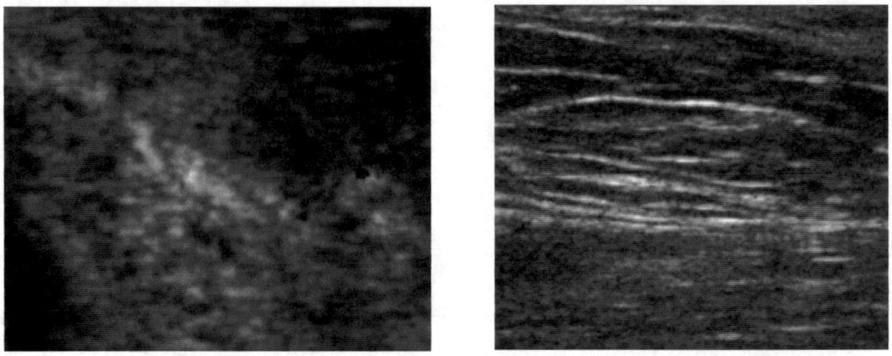

(a) (b)
Figure 3. Sample of (a) Breast with calcification, (b) Normal breast

3.2. Preprocessing
The first step in this research is the preprocessing. This step has been done before the image was processed in the training process. The data obtained are intact data captured on the ultrasound examination. In one chapter, an image can consist of 2 to 4 Region of interest (ROI), so each image has a different size. Therefore, it is necessary to equalize the size of the data and crop it to focusing on positive samples. For positive sample data, cropping was carried out which focused on the calcification area. Furthermore, it will be a positive sample in the CNN training process.

The same treatment was carried out for the negative samples. Negative samples were taken randomly from the image with a normal breast diagnosis. The same size was also applied to the negative sample. This size uniformity was done because the CNN system can only accept samples of the same size. The initial image which is in Dicom format, after the cropping and resizing process was saved in thief to keep a good quality. Next, the image was corrected by normalizing and filtering with mean filter. The purpose of this stage is to reduce speckle noise that can reduce the performance of the CNN. Mean filter used has a 3x3 matrix size. And the result of this process will be evaluated with MSE and PSNR parameters.

$$MSE = \frac{1}{MN}\sum_{j=0}^{m}\cdot\sum_{j=0}^{m}[I(i,j) - I'(i,j)]^2 \quad (1)$$

With MN is total pixel, matrix size mxn, I (I, j) am the original image, and I'(I, j) is filtered image. And the PSNR was described in Equation (2).

$$PSNR = 20log10\left(\frac{255}{\sqrt{MSE}}\right) \quad (2)$$

3.3. Training CNN
At this stage, the CNN model is developed to classify the breast ultrasound image automatically, to classify the image into normal or calcification. The CNN used in this study refers to Latif, 2020, with differences in the size of the input image. The CNN developed is composed of an input layer, two convolutional layers, two pooling layers and a fully connected layer. The input layer has a size of 64x64, the two convolutional layers have a window size of 3x3, and the pooling layer was 2x2. The convolutional layers have several map features and makes use of variable weights. They are responsible for the extraction of various features of each part of the input image. To perform the same feature detection in all possible locations of the input image, all units in one feature map must have the same weight and bias. As a result, each feature map attempts to perform a different local feature detection.

The role of pooling layers is to present information on convolutional layer outputs in a concise and simplified form. With such a pooling layer setting (2x2), the output data size will be half times the input size. The connected output layer is entirely responsible for generating the results of the classification procedure. Training scheme (optimizer) used to update the kernel weight values is Stochastic Gradient

Descent (SGD). SGD is a scheme that is commonly used in the CNN training process. The training process using the SGD scheme is faster than the batch scheme because in the SGD scheme the kernel weight values are updated every time one training data is evaluated. The weights are updated by using an optimization scheme based on stochastic gradient descent, during the training phase [22].

$$W_k = W_{k-1} - \eta \frac{\partial E^m(w)}{\partial w} \tag{3}$$

Where E (W) is the cost function, m is the sample size and η is the step size. In order to deal with the issue of overfitting, a dropping out scheme has been used [28] which tends to ignore the randomly selected neurons while training. This technique leads to a generalization of CNNs, by aiding in learning multiple internal representations. Furthermore, to counter the issue relating to overflow, a minor weight penalty is incorporated during training.

3.4. Evaluation

After the training process, the output of the system has been tested with a new image. The CNN system that has been compiled is used to classify new images with a normal or calcified diagnosis. The data used at this stage consisted of 25 images consist of 10 normal data and 15 images of calcified breasts. The system will automatically classify images that are included in the category of normal or calcified images, according to the CNN system that has been formed. We evaluate the number of true positives, false negatives, false positives and true negatives. True positive (TP) is positive data that are clarified as true by the system, in this case is a calcification sample that is classified correctly by the system. False negative (FN) is the number of negative data, but incorrectly clarified by the system. FN in this study is the amount of positive data (calcification) which is classified as a normal image. False positive (FP) is negative data that are correctly detected, meaning that the normal image data is classified as classified. While true negative (TN) is negative data that is correctly detected by the system, normal images are classified correctly.

Based on the obtained TP, FN, FP and TN values, the CNN system accuracy and sensitivity were calculated. Accurately represents the percentage of normal and abnormal structures that can be classified correctly. While the sensitivity is the percentage of abnormalities that can be classified by the system. The amount of sensitivity, accuracy that can be described in Eq. (4) And (5).

$$Accuracy = \frac{TP+TN}{TP+TP+FP+FN} \times 100\% \tag{4}$$

$$Sensitivity = \frac{TP}{TP+FN} \times 100\% \tag{5}$$

4. Experimental Result

Positive and negative samples obtained at the preprocessing stage of the image were uniformed with a size of 64x64. This size was chosen because it is the optimum size based on observations that have been made. After the image sizes are equalized, the images are stored in two different folders, according to the classification based on the doctor's diagnosis. The naming of this folder will later become the image label in the CNN training process. The CNN system will classify each image according to the label. The CNN system hierarchy formed includes an input layer (64x64 size), convolutional layer I (32x64x64 size), pooling layer-I (32x32x32 size), convolutional layer II (64x32x32 size), pooling layer II (64x16x16 size) and the output layer respectively.

The research was conducted using unfiltered image data (original data) and filtered data. The averaging filter was chosen because this filter is the simplest function and ethics applied in 2D grayscale images can provide an angle effect so that a clearer image is obtained. With better sample image quality, it is hoped that the performance of the CNN system that will be formed will also be better.

In the filtration process, the parameters used to analyze the similarity of the image are Mean Square Error (MSE) and Peak-Signal to Noise Ratio (PSNR). The processed image can be said to be similar to the original image if the MSE value is close to 0, and the PSNR value is more than 30 dB. In this study, the smallest MSE was 0.65 and the highest PSNR was 39.5 dB. The results of this filtration are then used as a CNN training sample. Table 1 shows the results of CNN training for each datum, both filtered and original data. The CNN system classifies images into normal breast images and calcified breast images. For both data, experiments were carried out with epochs of 25, 50, 75, 100, 125 and 150. The epochs will be stopped if the error is constant or increases with increasing epochs. Based on the result, it can be seen that the smallest loss value is achieved at epoch 125 and starts to be constant for epoch 125 and 150. Therefore, the model at these epochs will be stored and used in the testing process.

For unfiltered data, the test validation accuracy achieved for the epoch 125 is 76.04, and 91.67 for the filtered data. The CNN will be tested with new data, to testing the performance of the CNN for classifying ultrasound breast images into normal and calcification. For the unfiltered data sample, classification testing accuracy was 68% and a sensitivity of 73.33%. While, for the filtered data sample, classification testing accuracy was 76 % and 84.61% for the sensitivity.

Table 1. Experimental Result for CNN classification

Data	Test Validation Result		
Result	Epoch	Loss	Accuracy
Unfiltered Sample	25	0.7722	33.44
	50	0.5174	46.79
	75	0.5097	76.04
	100	0.5095	76.04
	125	0.5094	76.04
	150	0.5094	76.04
Filtered Sample	25	0.6724	46.88
	50	0.4856	75.79
	75	0.4688	91.67
	100	0.4682	91.67
	125	0.4681	91.67
	150	0.4681	91.67

From the experimental result, for the filtered data sample using a mean filter, accuracy increased to 76% for accuracy, and to 84.61% for sensitivity, which is increased for about 9% of the unfiltered data sample. This shows that the filtering process provides better results and effective to improve the performance of the CNN classification system. The better image sample quality, the better performance of CNN classification system.

5. Conclusion

In this research, A CNN models have been proposed for classifying of the Breast Ultrasound Images into either normal or calcification breast. For the filtered and unfiltered data samples, a 76% classification accuracy and 84.61% classification, sensitivity has been achieved with the proposed model with filtered data samples using a mean filter. In addition, the segmentation of breast calcification will be investigated in future studies.

References

[1] Yousef, W.A., Abouelkahire, A.A., Almahallawi,D., Marzouk,S.O., Mohamed, S.K.,Mustafa, W,A., Osama,O,M., Saleh, A.A., Abdelrazek,N,M. arXiv:1908.10661v1 [eess. IV] 26 Aug 2019. Do: //arxiv.org/abs/1908.10661.

[2] Hiremath, P. S., Prema T. Akkasaligar, and Sharan Badiger. In Advancements and breakthroughs in ultrasound imaging. Intechopen, 2013.

[3] Chang RF, Hou YL, Huang CS, Chen JH, Chang JM, Moon WK. Automatic detection of microcalcifications in breast ultrasound. Med Phys. 2013 Oct;40(10):102901.

[4] Liu, Shengfeng, Yi Wang, Xin Yang, Baiying Lei, Li Liu, Shawn Xiang Li, Dong Ni, and Tianfu Wang. Engineering (2019).

[5] Ouyang, Y. L., Zhou, Z. H., Wu, W. W., Tian, J., Xu, F., Wu, S. C. et al. A review of ultrasound detection methods for breast microcalcification. Math. Biosci. Eng. 16, 1761–1785 (2019)
[6] B. Peng, Y. Xian and J. Jiang, 2018 IEEE International Ultrasonics Symposium (IUS), Kobe, 2018, pp. 206-212
[7] Jeng, Geng-Shi, Maria Zontak, Nripesh Parajuli, Allen Lu, Kevinminh Ta, Albert J. Sinusas, James S. Duncan, and Matthew O'Donnell. IEEE Access 6 (2018): 17415-17428
[8] Chiang, Tsung-Chen, Yao-Sian Huang, Rong-Tai Chen, Chiun-Sheng Huang, and Ruey-Feng Chang. IEEE transactions on medical imaging 38, no. 1 (2018): 240-249.
[9] Yamashita, R., Nishio, M., Kinh, R., Do, G., dan Togashi, K. (2018): Convolutional neural networks : an overview and application in radiology, *Insights into Imaging*, 9, 611–629
[10] Kim, P. (2017). *MATLAB Deep Learning: With Machine Learning, Neural Networks and Artificial Intelligence.*
[11] T. M. Kolb, J. Lichy, J. H. Newhouse, Radiology, vol. 225, no. 1, pp. 165-175, Oct. 2002.
[12] Anderson ME, Soo MS, Bentley RC, Trahey GE. The detection of breast microcalcifications with medical ultrasound. J Acoust Soc Am. 1997 Jan;101(1):29-39. doi: 10.1121/1.417973. PMID: 9000730.
[13] Brem RF, Lenihan MJ, Lieberman J, Torrente J. Screening breast ultrasound: past, present, and future. AJR Am J Roentgenol. 2015 Feb;204(2):234-40. doi: 10.2214/AJR.13.12072. PMID: 25615743
[14] Latif, G., Iskandar, D. A., Alghazo, J., Butt, M., & Khan, A. H. (2018). Deep CNN based MR image denoising for tumor segmentation using watershed transform. International Journal of Engineering & Technology, 7(2.3), 37-42.
[15] Pinheiro, P. and Collobert, R., (2014), In International Conference on Machine Learning, pp. 82-90.
[16] A. S. Becker, M. Marcon, S. Ghafoor, M. C. Wurnig, T. Frauenfelder, and A. Boss, Investigative Radiology, vol. 52, no. 7, pp. 434–440, 2017
[17] Hongmin Cai, Qinjian Huang, Wentao Rong, Yan Song, Jiao Li, Jinhua Wang, Jiazhou Chen, Li Li, Computational and Mathematical Methods in Medicine, vol. 2019, Article ID 2717454, 10 pages, 2019. https://doi.org/10.1155/2019/2717454
[18] A. Papadopoulos, D.I. Fotiadis, A. Likas,An automatic microcalcification detection system based on a hybrid neural network classifier,Artificial Intelligence in Medicine, Volume 25, Issue 2, 2002, Pages 149-167,ISSN 0933-3657, https://doi.org/10.1016/S0933-3657(02)00013-1.
[19] G. Latif, M.O. Butt, F.Yousif Al Anezi and J.Alghazo. 2020. *RIVF International Conference on Computing and Comunication Tecnologies (RIVF),* 2020, pp.1-5, doi:10.1109/RIVF48685.2020.9140767.
[20] Ahmed S.Bafaraj. 2019. International Journal of Applied Engineering Research ISSN 0973-4562 Volume 14, Number 6 (2019) pp. 1340-1351 © Research India Publications. http://www.ripublication.com
[21] X`, Ghazanfar, Jaafar Alghazo, Loay Alzubaidi, M. Muzzamal Naseer, and Yazan Alghazo. In 2018 IEEE 2nd International Workshop on Arabic and Derived Script Analysis and Recognition (ASAR), pp. 90-95. IEEE, 2018.
[22] Srivastava, N., Hinton, G. E., Krizhevsky, A., Sutskever, I., & Salakhutdinov, R. (2014). Dropout: a simple way to prevent neural networks from overfitting. Journal of Machine Learning Research, 15(1), 1929-1958

Quantification of Pericardial and Epicardial Fat using ImageJ

N Z Yasmin[1], R S Tohir[2], P Prajitno[1] and D S Soejoko[1],
[1] Department of Physics, Faculty of Mathematics and Natural Science, Universitas Indonesia, Depok, West Java 16424, Indonesia
[2] Mayapada Hospital Tangerang, Jl. Honoris Raya Kav. 6, Modernland, Kelapa Indah, Tangerang, Banten 15117, Indonesia
Email: nadia.zakyyah@ui.ac.id

Abstract. Research towards finding new Coronary Artery Disease (CAD) predictors have developed in recent years. One potential CAD predictors are epicardial fat due to its location and function. Another adipose layer of interest is the pericardial fat that is the fat anterior to the epicardial fat. The single slice area method has been developed to estimate volume using a single slice area. In this preliminary study, the aim is to quantify epicardial and pericardial fat using ImageJ and to evaluate the relationship between the fat area and the volume measured. The relationship between the fat parameters, age and gender were evaluated. The samples were cardiac CT images of 50 patients who underwent non-contrast cardiac examination using a 64 CT scanner. ImageJ was used to quantify epicardial and the pericardial fat area and the volume was obtained by the sum of four slices multiplied by the slice thickness. The Pericardial fat area showed a high correlation with the pericardial fat volume ($r=0.92$, $p<0.01$) using the single slice area method. Epicardial fat area and volume showed a moderate significant relationship with age. Epicardial volume in female patients was significantly higher than in male patients. Quantification of epicardial and pericardial fat was highly reproducible using ImageJ.

1. Introduction

Coronary artery disease (CAD) is the leading cause of death in the world [1]. Research towards finding new CAD predictors have developed in recent years. One potential CAD predictor is the epicardial fat. Epicardial fat is referred as the layer of adipose tissue that is located between the visceral pericardium and myocardium of the heart [2]. This particular fat has been an interest in many studies due to its location and function that is linked to coronary artery disease (CAD). Epicardial fat is located close to coronary arteries where plaque can occur. Epicardial fat also functions as an endocrine organ that secretes hormones and cytokines that are believed to play a part in the atherosclerosis process [3]. Other cardiac fats have also been evaluated. These adipose tissue layers consist of the paracardial fat, which is the fat outside of the parental pericardium and the pericardial fat, which is anatomically defined as the fat anterior to the epicardial fat between visceral and parietal pericardium. The pericardial fat is also considered as the total sum of epicardial fat and paracardial fat [4].

Quantification of cardiac fat can be performed using different imaging modalities. Echocardiography is a noninvasive method that produces good reproducibility, but is limited only to thickness measurement. Magnetic Resonance Imaging (MRI) is considered as the golden standard for epicardial fat quantification but the images can be prone to artifacts caused by chemical shift [5]. Computed Tomography (CT) is a preferred method for volumetric quantification due to its spatial resolution and high reproducibility [6].

ImageJ as an open source software with user friendly interface makes it a potential software for CT image processing. ImageJ has been used for many medical cases such as thigh skeletal muscle measurement [7], abdominal circumference measurement [8], and lung adenocarcinoma analysis [9]. Quantification of epicardial fat using ImageJ was studied for patients with HIV [10], atrial fibrillation [11], and peritoneal dialysis [12]. However, the quantification of pericardial fat using ImageJ has not been explored.

One area of cardiac fat quantification research is regarding the single slice area method. This study estimated the volume of epicardial fat using an area of a single slice CT found to be correlated best at the Left Coronary Main Artery (LMCA) landmark [13]. This estimation could potentially decrease the time in volumetric CT quantity of fat. The single area method was used in predicting the presence of obstructive coronary artery disease and it resulted in an independent association in respect to the area and total volume [14].

In this preliminary study, the aim is to quantify epicardial and the pericardial fat area and volume using ImageJ and to evaluate the relationship between the fat area using single slice area method and the volume measured. The quantified fat parameters and risk factors of coronary artery disease such as age and gender were evaluated.

2. Material and Method

2.1. Sample preparation

This study consisted of 50 patients who underwent non-contrast cardiac CT scan examination in Mayapada Hospital Tangerang. Cardiac CT scans were performed using a 64-detector row CT scanner, Revolution Evo (General Electronics Healthcare, Milwaukee, WI). The CT images were reconstructed using the following parameters: slice thickness 2.5 mm; tube voltage 120 kV and matrix 512 x512. The determination of the anatomical landmark of the CT images was consulted beforehand with a radiologist.

2.2. Fat quantification

The quantification of epicardial fat and pericardial fat area and volume were performed using ImageJ software (Ver.1.53, National Institutes of Health, Maryland, USA). The flowchart of fat is a quantification can be seen in Figure 1. The first step is to open the non-contrast CT image. The next step is to delineate the cardiac fat from the heart structure using the ImageJ option Clear Outside in the Edit menu. After obtaining the Region of Interest (ROI), Gaussian filter is applied to the image by using the Gaussian Blur Filter in the Process menu. Gaussian filter is used to reduce the noise of the image [6]. The next step is to segment the fat from the heart structure by thresholding segmentation located in the Image menu and Adjust submenu. The CT attenuation range of fat is between -190 and -30 Hounsfield Unit [3, 6]. In the threshold menu, adjust the lower threshold value to -190 and the upper value to -30 to obtain the fat region. After thresholding, the area of the ROI is selected using Create Selection in the Edit Menu. The last step is to measure the area using the Measure option in the Analyze menu.

Journal of Physics: Conference Series **2019** (2021) 012078 doi:10.1088/1742-6596/2019/1/012078

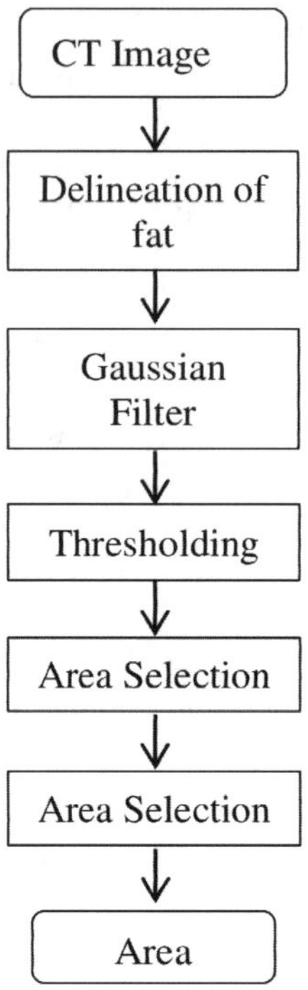

Figure 1. Quantification of Fat Area using ImageJ

2.2.1 Epicardial fat

Epicardial fat area was obtained from a single slice at the LMCA level [10]. It is the adipose tissue within the pericardium. The epicardial fat area quantification was depicted in Figure 2.

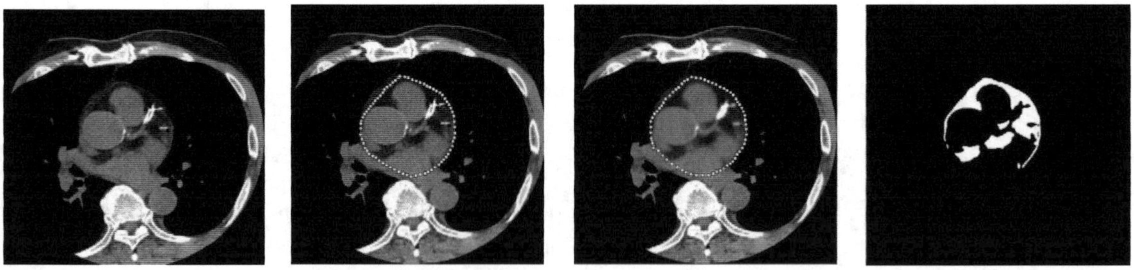

(a) (b) (c) (d)

Figure 2. Epicardial fat area quantification: (a) CT image, (b) delineation, (c) gaussian filter, (d) thresholding

Epicardial fat volume was obtained from the sum of four slices starting from the upper slice limit of pulmonary artery bifurcation to the apex of the heart [10]. The summed area was then multiplied by the slice thickness [3,15].

2.2.2 Pericardial fat

Pericardial fat area is the overall total fat between the epicardial fat and paracardial fat. The Pericardial fat area was obtained from a single slice at the LMCA level depicted in Figure 3. Pericardial fat volume was obtained from the sum of the 4 area slices multiplied by the slice thickness [16]. The pericardial fat was differentiated using the thresholding segmentation of the fat range from -190 H.U to -30 H.U.

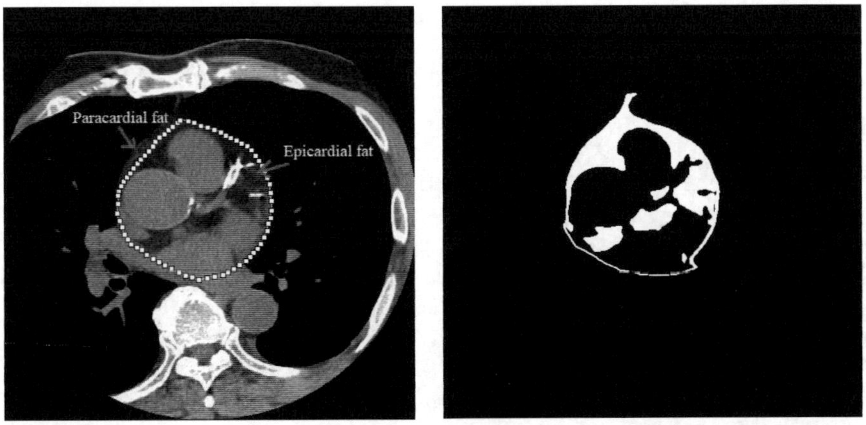

(a) (b)

Figure 3. Pericardial fat area quantification, (a) The sum of paracardial and epicardial fat, (b) Pericardial area

2.3. Statistical analysis

The statistical analyses were carried out using SPSS version 22 (IBM Corp., Armonk, NY, USA). Pearson's correlation coefficient was calculated to evaluate the relationship between the fat areas in the LMCA level with the total fat volume and the quantified parameters regarding to age. The Kolmogorov-Smirnov test was conducted to determine the normality of the data. The area and volume of the epicardial and pericardial fat were normally distributed. The T-test was performed to analyze the mean difference of the quantified parameters with respect to gender. The p value <0.05 indicates significance. Continuous variables are presented as mean values with standard deviations and categorical variables are presented as frequencies with percentages. Inter-observer reproducibility was examined based on the image processing of two observers who are experienced with CT cardiac images.

3. Results & Discussion

3.1 Patient characteristics

In this study, 50 patients consisted of 27 male patients and 23 female patients with the mean age of 55 ± 9.70 years. The mean values of the fat parameters have also been measured and can be seen in Table 1. The area of epicardial and pericardial fat were converted to cm^2 while the volume of both fat is in cm^3.

Table 1. Patient Characteristics

Characteristics	Data
Age (years)	55 ± 9.70
Male gender n (%)	27 (54%)
Epicardial fat area (cm^2)	17.11 ± 7.30
Pericardial fat area (cm^2)	22.05 ± 9.42
Epicardial fat volume (cm^3)	113.20 ± 47.69
Pericardial fat volume (cm^3)	191.95 ± 76.90

3.2 Correlation between epicardial fat and pericardial fat area and volume

The correlation of both epicardial fat and pericardial fat areas at the LMCA level and the total volume was evaluated using Pearson's correlation coefficient. The correlation between the fat measurements are shown in Table 2. The epicardial fat area at LMCA level showed a high correlation ($r= 0.89$, $p<0.001$) with the total fat volume of epicardial fat. This result is in agreement with prior studies [13, 14].

Table 2. Correlation between epicardial and pericardial fat measurement

	Epicardial Fat Area (cm^2)		Pericardial Fat Area (cm^2)	
	Pearson rho	p-value	Pearson rho	p-value
Epicardial fat volume (cm^3)	0.89	<0.001	0.83	<0.001
Pericardial fat volume (cm^3)	0.88	<0.001	0.92	<0.001

The pericardial fat area at LMCA level showed a higher correlation to the pericardial total fat volume ($r=0.92$, $p<0.001$) than epicardial fat. Although pericardial fat area at this level showed a high correlation with the total pericardial fat volume, further studies regarding the best landmark in estimating the volume should be investigated.

3.3 Relationship between epicardial fat and pericardial fat measurement with age

The relationship between epicardial fat and pericardial fat measurements with age was evaluated. Table 3 showed the Pearson's correlation coefficient between these variables.

Table 3. Relationship between fat measurements and age

	Age	
	Person rho	p-value
Epicardial fat area (cm^2)	0.30	0.034
Pericardial fat area (cm^2)	0.24	0.101
Epicardial fat volume (cm^3)	0.33	0.020
Pericardial fat volume (cm^3)	0.24	0.091

The epicardial fat area ($r=0.30$, $p=0.034$) and epicardial fat volume ($r=0.33$, $p=0.02$) showed a moderate and statistically significant relationship with age. This result is in accordance to previously published studies [2, 15]. This is due to the fact that the redistribution of fat tissue occurs during the aging process and is prominent in older women [17]. The pericardial fat area and volume showed no significant value statistically.

3.4 Epicardial and pericardial fat, according to gender

The epicardial fat and pericardial fat measurements were differentiated according to gender, which can be seen in Table 4. Female patients had a higher fat area and volume overall than male patients. There was a significant difference in the epicardial fat volume of female patients (mean volume of 130 ± 53.94 cm^3) two male patients (mean volume of 98 ± 36.89 cm^3). Previous studies, however reported the opposite in which epicaridal fat volume in men tends to be higher than women [16, 17]. Further studies are needed to conclude the impact of gender with regards to epicardial fat.

Table 4. Fat measurements according to gender

	Gender		Mean Difference	p-value
	Female	Male		
Epicardial Fat area (cm^2)	19.02 ± 7.83	15.50 ± 6.53	3.54	0.08
Pericardial Fat area (cm^2)	22.97 ± 9.43	21.27 ± 9.51	1.69	0.53
Epicardial Fat volume (cm^3)	130.00 ± 53.94	98.89 ± 36.89	31.10	0.02
Pericardial Fat volume (cm^3)	196.97 ± 76.95	187.67 ± 78.07	9.30	0.67

3.5. Inter-observer Reproducibility

Two observers measured the pericardial fat area of 15 axial slices of CT images located in the LMCA level and manually traced the pericardium of each slice using ImageJ software. The inter-observer correlation for the pericardial fat area resulted in a high correlation ($r = 0.964$, $p < 0.01$) and can be seen in Figure 4.

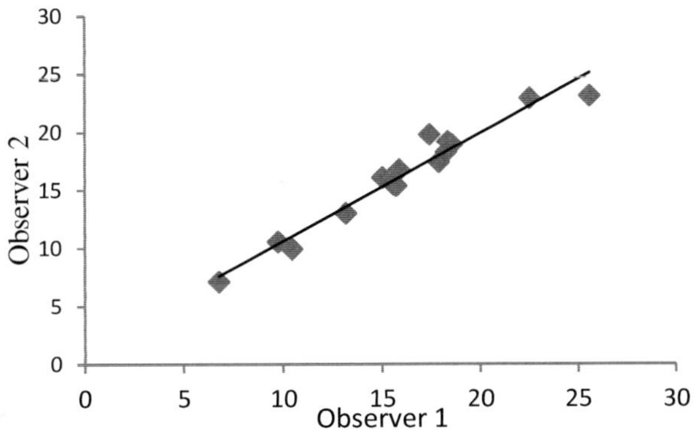

Figure 4. Inter-observer Reproducibility

4. Conclusion

The Pericardial fat area showed a high correlation with the total pericardial fat volume ($r=0.92$, $p<0.01$) using the single slice area method. However, further studies are needed to determine the best landmark of pericardial fat for volume estimation. Epicardial fat area and volume showed a moderate significant relationship with age. Epicardial volume in female patients was significantly higher than that of male patients. Quantification of epicardial and pericardial fat using ImageJ was highly reproducible (or = 0.964, $p < 0.01$).

References

[1] M. A. Khan *et al.*, *curious*, vol. 12, no. 7, 2020, doi: 10.7759/cure us. 9349.
[2] R. Khurana, A. Yadav, T. B. S. Buxi, J. P. S. Sawhney, K. S. Rawat, and S. S. Ghuman, *Indian Heart J.*, vol. 70, pp. S140–S145, 2018, doi: 10.1016/j.ihj.2018.08.009.
[3] P. M. Gorter *et al.*, vol. 197, pp. 896–903, 2008, doi: 10.1016/j.atherosclerosis.2007.08.016.
[4] G. Milanese *et al.*, *Diagnostic Interv. Radiol.*, vol. 25, no. 1, pp. 35–41, 2019, doi: 10.5152/dir.2018.18037.
[5] R. D. Stoffey, *Am. J. Roentgenol.*, 2010, doi: 10.2214/ajr.10.4249.
[6] G. Coppini, R. Favilla, P. Marraccini, D. Moroni, and G. Pieri, no. July, 2010, doi: 10.2174/1874431101004010126.
[7] D. E. Long *et al.*, *PLoS One*, vol. 14, no. 2, pp. 1–11, 2019, doi: 10.1371/journal.pone.0211629.
[8] S. L. Gomez-Perez *et al.*, *J. Parenter. Enter. Nutr.*, vol. 40, no. 3, pp. 308–318, 2016, doi: 10.1177/0148607115604149.
[9] H. J. Koo *et al.*, *PLoS One*, vol. 12, no. 1, pp. 1–14, 2017, doi: 10.1371/journal.pone.0167955.
[10] M. Sadouni *et al.*, *Eur. J. Radiol. Open*, vol. 8, 2021, doi: 10.1016/j.ejro.2020.100317.
[11] M. Vroomen *et al.*, *Eur. J. Cardio-thoracic Surg.*, vol. 56, no. 1, pp. 79–86, 2019, doi: 10.1093/ejcts/ezy472.
[12] H. H. Lin, J. K. Lee, C. Y. Yang, Y. C. Lien, J. W. Huang, and C. K. Wu, *Cardiovasc. Diabetol.*, vol. 12, no. 1, pp. 1–9, 2013, doi: 10.1186/1475-2840-12-127.
[13] N. Oyama *et al.*, *Jpn. J. Radiol.*, vol. 29, no. 2, pp. 104–109, 2011, doi: 10.1007/s11604-010-0524-z.
[14] T. Tran, G. Small, M. Cocker, Y. Yam, and B. J. W. Chow, *Eur. Heart J. Cardiovasc. Imaging*,

vol. 15, no. 4, pp. 423–430, 2014, doi: 10.1093/ehjci/jet175.
[15] A. Razanata, P. Prajitno, and D. S. Soejoko, vol. 2, no. 1, pp. 61–66, 2019.
[16] N. Aslanabadi *et al.*, *J. Cardiovasc. Thorac. Res.*, vol. 6, no. 4, pp. 235–239, 2014, doi: 10.15171/jcvtr.2014.018.
[17] A. G. Bertaso, D. Bertol, B. B. Duncan, and M. Foppa, *Arq. Bras. Cardiol.*, pp. 18–28, 2013, doi: 10.5935/abc.20130138.
[18] M. Dagvasumberel *et al.*, *Cardiovasc. Diabetol.*, vol. 11, no. 1, p. 1, 2012, doi: 10.1186/1475-2840-11-106.
[19] J. Mancio *et al.*, *Int. J. Cardiol.*, vol. 249, pp. 419–425, 2017, doi: 10.1016/j.ijcard.2017.09.178.

Journal of Physics: Conference Series **2019** (2021) 012079 doi:10.1088/1742-6596/2019/1/012079

Model Selection in Peptide-receptor Radionuclide Therapy for an Accurate Determination of Time Integrated Activity Coefficients

N Atikah[1], A Riana[1], A Dwi, Z Anwari[1], Misrawati[1] and D Hardiansyah[1]
[1] Department of Physics, Universitas Indonesia
Jl. Margonda Raya, Jawa Barat 16424, Indonesia
Email: atikahnur369@gmail.com

Abstract. Calculation of accurate time-integrated activity coefficients (TIACs) is desirable in nuclear medicine dosimetry. The accuracy of the calculated TIACs is highly dependent on the fit function. However, systematic studies of determining a good function for peptide-receptor radionuclide therapy (PRRT) in different patients have not been reported in the literature. The aim of this study was to individually determine the best function for the calculation of TIACs in tumor and kidneys using a model selection based on the goodness of fit criteria and Corrected Akaike Information Criterion (AICc). The data used in this study was pharmacokinetic data of ^{111}In-DOTATATE in tumor and kidneys obtained from 4 PRRT patients. Eleven functions with various parameterizations were formulated to describe the biokinetic data of ^{111}In-DOTATATE in tumor and kidneys. The model selection was performed by choosing the best function from the function with sufficient goodness of fit based on the smallest AICc. Based on the results of model selection, function $A_1 e^{-(\lambda_1+\lambda_{phys})t}$ was selected as the best function for all tumor and kidneys in patients with meningioma tumors. By using this function, the calculated of TIACs could be more accurate for future patients with meningioma tumor.

1. Introduction

Molecular Radiotherapy (MRT) is defined as internal radiation therapy which delivers radiopharmaceuticals into the body. Most hospitals apply internal radiation therapy with fixed dose method, which delivers a predetermined fixed dose to certain patients [1]. This fixed dose method does not individualize the patient. Therefore it may cause overdose or underdose which lead to ineffective treatments [1]. Development of dosimetry in internal radiation therapy has enabled each patient to receive a specific dose which suitable with their respective physiological conditions. Individual dosimetry plays an important role in internal therapy because it aims to maximize radiation doses to the tumor site while minimizing radiation dose to organs at risk (OAR) [2]; [3]. Therefore, the real challenge of therapeutic planning in nuclear medicine is how to determine an accurate dosimetry.

In nuclear medicine, the general method which is used to calculate and optimize the injection dose is the method formulated by the Medical Internal Radiation Dose (MIRD) [4]. The internal dosimetry is calculated by the MIRD method, which consists of two parameters that can be optimized, namely the Time-Integrated Activity Coefficients (TIACs) which is related to the Area Under the Curve (AUC) of the activity curve against time and the S-value [5]. Recent studies have indicated parameters that have the highest influence on the inter-individual variability in calculating the absorbed dose in the target organ is the TIACs value [6]. For example, in the kidney, based on the main effect value derived from the global sensitivity analysis (GSA) method, the TIACs value of the kidney has a greater effect, about

90% of the output variance, while the S-value of the kidney only affects about 10 % of the output variance [6]. Therefore, the dosimetry requires an accurate good prediction of TIACs value in order to produce an accurate calculation of absorbed dose.

The sum of Exponential (SOE) functions model is a widely used method to analyze TIACs [7]. SOE function method is a method which is used to analyze the distribution of radiopharmaceuticals and describe the curve of radiopharmaceutical activity in each tissue or organ of the body. SOE modeling was generated based on the assumption that radiopharmaceutical pharmacokinetics can be formulated by an exponential function equation with physical and biological decay rates [8]. The SOE method also requires model selection to determine the best model function because a data may be fitted with many function models. Model selection is needed to provide an assessment of the curve fitting quality or goodness of fit and determine the best model with AICc (Corrected Akaike Information Criterion) parameters to obtain the best final model [9]; [10].

However, systematic studies to individually determine the best function for peptide receptor radionuclide therapy (PRRT) in different patients have not been reported in the literature. Therefore, the purpose of this study was to individually determine the best function for calculating TIACs in tumors and kidneys using a model selection based on goodness of fit and Corrected Akaike Information Criterion (AICc) criteria.

2. Materials and Methods

2.1. Patients/Data

This study used biokinetic data on peptide-receptor radionuclide therapy (PRRT) as secondary data [11]. PRRT biokinetic data were obtained by digitizing the data with Web Plot Digitizer software version 4.4 (https://apps.automeris.io/wpd/). The processed data was biokinetic data which were obtained from 4 patients which were diagnosed with a meningioma tumor after pre-therapy injection of ^{111}In-DOTATATE which was used for model fitting [11]. Pre-therapy dosimetry was performed for 3 and 2 PRRT cycles towards 1st and 2nd patient, respectively. In the pre-therapy measurement process, ^{111}In-DOTATATE with an activity of 140±14 MBq (total peptide amount 75±10 nmol) was injected intravenously as an infusion. In full-body planar scintigraphy, measurements were taken at 2 hours, 4 hours, 1st, 2nd, and 3rd days after injection. The software ULMDOS [12] and NUKDOS [13] were used to determine the activity-time data [11]. There were totally 7 biokinetic data from tumor sites and kidneys which were used in this study.

2.2. Sum of exponentials (SOE) Method

SOE modeling is generated based on the assumption that radiopharmaceutical pharmacokinetics can be formulated as an exponential equation with physical and biological decay rates [8]. There are eleven functions with various parameterizations of mono- and bi-exponential functions (Eq. (1)-(11))

$$f_4(t) = A_1 e^{-(\lambda_1+\lambda_{phys})t} + A_2 e^{-(\lambda_2+\lambda_{phys})t} \tag{1}$$

$$f_{3a}(t) = A_1 e^{-(\lambda_1+\lambda_{phys})t} + A_2 e^{-(\lambda_{phys})t} \tag{2}$$

$$f_{3ap1}(t) = A_1 e^{-(\lambda_1+\lambda_{phys})t} + A_1\, a\, e^{-(\lambda_{phys})t} \tag{3}$$

$$f_{3ap2}(t) = A_1\, b\, e^{-(\lambda_1+\lambda_{phys})t} + A_1\, (1-b)\, e^{-(\lambda_{phys})t} \tag{4}$$

$$f_{3b}(t) = A_1 e^{-(\lambda_1+\lambda_{phys})t} - A_1 e^{-(\lambda_2+\lambda_{phys})t} \tag{5}$$

$$f_{3c}(t) = A_1 e^{-(\lambda_1+\lambda_{phys})t} + (100 - A_1) e^{-(\lambda_2+\lambda_{phys})t} \tag{6}$$

$$f_{2a}(t) = 100\, e^{-(\lambda_1+\lambda_{phys})t} - 100\, e^{-(\lambda_2+\lambda_{phys})t} \tag{7}$$

$$f_{2b}(t) = A_1 e^{-(\lambda_1+\lambda_{phys})t} \quad (8)$$

$$f_{2c}(t) = A_1 e^{-(\lambda_1+\lambda_{phys})t} - A_1 e^{-(\lambda_{phys})t} \quad (9)$$

$$f_{2d}(t) = -A_1 e^{-(\lambda_2+\lambda_{phys})t} + A_1 e^{-(\lambda_{phys})t} \quad (10)$$

$$f_{2e}(t) = A_1 e^{-(\lambda_1+\lambda_{phys})t} + (100 - A_1)\, e^{-(\lambda_{phys})t} \quad (11)$$

Where λ_{phys} is the radioactive decay constant of the radionuclide, A_k and λ_k are the fractional absorption in % activity and the biological decay constant, respectively. These functions are designed to be attached to the data with fraction of the injected dose (%ID) units. For simulation purposes, the numerical software SAAM II version 2.3 (https://tegvirginia.com/software/saam-ii/) was used in the SOE modelling.

2.3. Model Selection

The model selection provides an assessment of the curve fitting results (Goodness of fit) and determines the best model with AICc (Corrected Akaike Information Criterion) parameters to get the best final model. Goodness of fit was determined by graphical visualization, Coefficient of Variation (CV) and Correlation Matrix ($CM_{i \times j}$) [8]. Graphical visualization was conducted by direct observations of the fitted curve. The Coefficient of Variation (CV) describes the level of precision of the fitting parameters and was determined by dividing the standard deviation value with the average value [14]. The requirement for the maximum value of CV of an acceptable fit is 50% [8]. In addition, the off-diagonal values of the Correlation Matrix ($CM_{i \times j}$) show linearity between the parameters that composed a function, acceptable fits are defined as the maximum value of off-diagonal elements equal to $-0.8 \leq CM_{i \times j} \leq 0.8$ [8] [15].

Corrected Akaike Information Criterion (AICc) is modified from Akaike Information Criterion (AIC), where the AICc includes the parameter number of data points (N) [10]. AICc (Eq. (12)) is used to select the best equation in the model that has a small ratio between the number of data points (N) and the number of parameters (K) (N/K < 40) [9] [10]. The weighting of AICc (ω_{AIC_i}) is the value of the weight of AICc which is shown in the following equation (Eq. (13)-(14)).

$$AICc = OF + 2K + \frac{2K(K+1)}{N-K-1} \quad (12)$$

AICc weighting:

$$\Delta i = AICc_i - AICc_{min} \quad (13)$$

$$\omega_{AIC_i} = \frac{e^{-\Delta i/2}}{\sum_{i=1}^{F} e^{-\Delta i/2}} \quad (14)$$

Where N and K are the number of data points and adjustable parameters, respectively. F is the size of the model set, $AICc_{min}$ is the lowest AICc of all fitted functions, Δi is the difference of the AICc values to $AICc_{min}$, and wAICi is the corresponding weight, and −2ln(P) the estimated value of the OF at the minimum. With absolute weighting, the condition of the fit parameter boundary (K) becomes less than and equal to the number of data points (N) minus 2 or K N-2 [8].

2.4. Determination of TIACs
The TIACs value was obtained by integrating the best model that had been gained in the previous process. The selection of integration limits with a range time from 0 to 10 times the half-life ($t1/2$) of radionuclides which is in accordance with Anwari's research [16]. This means that the calculation of TIACs is carried out until the decay activity of radionuclides in the body is very low. The TIACs value is obtained from the following equation:

$$TIACs = \frac{1}{A_0}\int_0^{10t_{1/2}} A(t)dt \tag{15}$$

Where A (t) is the accumulated activity at time t and A_0 is the initial activity.

3. Result and Discussion
It was highly recommended to conduct the selection model in order to obtain the best function on individual dosimetry for Peptide Receptor Radionuclide Therapy (PRRT) treatment. However, this method was not carried out in hospitals because it consumes high workload to perform model selection on each patient's biokinetic data. In this study, the SOE method was applied to tumor and kidney biokinetic data to obtain accurate TIAC values. Based on the results of model selection, function 2b ($A_1e^{-(\lambda_1+\lambda_{phys})t}$) is the best function for all tumor organs and kidneys in patients who suffered meningioma tumor. This function was selected as the best function based on the goodness of fit test and has the smallest AICc value and the largest AICc weighting compared to other functions. Figure 1 shows the results of the curve fitting for the function f2b to the data.

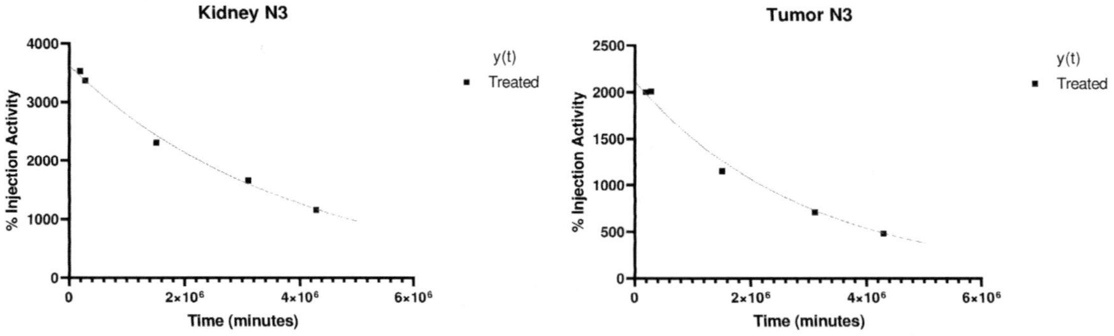

Figure 1. Example of the fitted curves of kidneys and tumor (patient 1, N3)

Based on the graphical visualization, the curve fitting results have shown that the f2b function describes the data well. Physically, f2b has a greater compatibility to be the best function than other functions. The f2b function shows a decrease in the injection dose received by tumor and kidneys organ based on the decay of its physical and biological lambda. After observing the graphic visualization, the goodness of fit test was also checked for the CV and CM values. After that, the AICc value has been calculated to determine the best function. Table 1 shows the results of the model selection, namely the CV value, CM value, AICc, and the AICc weighting of the f2b function.

Table 1. Model Selection of Tumor Biokinetic Data

Measurement	Maximum CV value (%)	Maximum CM value (Abs)	AICc	AICc weighting (%)
N1	31,25	0,75	-39,52	75,58
N2	26,7	0,76	-38,3	97,86
N3	30,84	0,76	-41,47	94,22
N4	39,16	0,74	-43,72	87,94
N5	40,07	0,74	-41,1	89,29
N6	42,92	0,72	-38,42	88,73
N7	36,57	0,73	-34	98,26

Table 2. Model Selection of Kidneys Biokinetic Data

Measurement	Maximum CV value (%)	Maximum CM value (Abs)	AICc	AICc weighting (%)
N1	22,22	0,76	-46,32	99,87
N2	23,38	0,78	-44,3	99,22
N3	23,13	0,77	-45,13	99,63
N4	22,52	0,74	-34,42	99,93
N5	22,5	0,74	-32,08	99,93
N6	22,3	0,78	-76,72	73,87
N7	14,77	0,78	-65,07	99,96

The table above is for the model selection results from the f2b function. When compared to other functions, the f2b function meets the criteria of goodness of fit as seen from the CV and CM values. So, based on the model selection table above, the CV value is below 50% and the CM value is between -0.8 and 0.8. The AICc value of the f2b function also has the smallest value from other functions which are the best candidate functions, while the AICc weighting also has the largest value. By using this function, TIACs calculation is more accurate for patients with meningioma tumors.

In this study, researcher determined the best function in the internal dosimetry of PRRT, especially for cases of meningiomas in tumors and kidneys. In the future, the f2b function can be used for meningioma patients, therefore it can assist clinical activities to determine accurate TIAC. However, reproducibility study to the best function derived using model selection in this study is needed.

4. Conclusion

In this study, we individually determine the best function in the internal dosimetry of PRRT, especially for meningioma cases in tumors and kidneys. Function f2b was found as the best function for most of the data. Therefore, this function might be used for the future meningioma patients.

References

[1] A. R. Haug, "PRRT of neuroendocrine tumors: individualized dosimetry or fixed dose scheme?," *EJNMMI Research,* 2020.

[2] G. Glatting, M. Bardies and M. Lassmann, "Treatment planning in molecular radiotherapy," *Z Med Phys, 23(4),* pp. 262-269, 2013.

[3] D. Hardiansyah, W. Guo and et al., "Time-integrated activity coefficient estimation for radionuclide therapy using PET and a pharmacokinetic model : A simulation study on the effect of sampling schedule and noise," *Med Phys, 43(9),* p. 5145, 2016.

Journal of Physics: Conference Series **2019** (2021) 012079 doi:10.1088/1742-6596/2019/1/012079

[4] J. A. e. a. Siegel, "MIRD pamphlet no 16: Techniques for quantitative radiopharmaceutical biodistribution data acquisition and analysis for use in human radiation dose estimates," *J Nucl Med, 40(2),* pp. 37S-61S, 1999.

[5] W. E. Bolch and et al., "MIRD pamphlet No.21: a generalized schema for radiopharmaceutical dosimetry--standardization of nomenclature," *J Nucl Med, 50(3),* pp. 477-484, 2009.

[6] A. e. a. Zvereva, "Impact of interpatient variability on organ dose estimates according to MIRD schema: Uncertainty and variance-based sensitivity analysis," *Medical Physics,* 2018.

[7] P. Kletting and et al., "Optimal preloading in radioimmunotherapy with anti-CD45 antibody," *Med Phys, 38(5),* pp. 2572-2578, 2011.

[8] P. Kletting and et al., "Molecular radiotherapy: the NUKFIT software for calculating the time-integrated activity coefficient," *Med Phys, 40(10),* p. 102504, 2013.

[9] P. Beier, K. Burnham and D. Anderson, "Model Selection and Inference: A Practical Information-Theoretic Approach," *In The Journal of Wildlife Management,* pp. (Vol. 65, Issue 3), 2001.

[10] G. Glatting, P. Kletting, S. N. Reske, K. Hohl and C. Ring, "Choosing the optimal fit function: Comparison of the Akaike information criterion and the F-test," *Medical Physics, 34(11),* pp. 4285-4292, 2007.

[11] P. Kletting and et al., "Optimized Peptide Amount and Activity for 90Y-Labeled DOTATATE Therapy," *J Nucl Med, 57(4),* pp. 503-508, 2016.

[12] G. e. a. Glatting, "Internal radionuclide therapy: The ULMDOS software for treatment planning," *Medical Physics,* 2005.

[13] P. Kletting and et al., "Physiologically Based Pharmacokinetic Modeling Is Essential in 90Y-Labeled Anti-CD66 Radioimmunotherapy," *PLoS One, 10(5),* p. e0127934, 2015.

[14] W. Panichkitkosolkul, "Confidence Intervals for the Coefficient of Variation in. 2013," 2013.

[15] G. J. Szekely, M. L. Rizzo and Bakirov, N.K., "Measuring and testing dependence by correlation of distances," *The Annals of Statistics, 35(6),* pp. 2769-2794, 2007.

[16] M. Z. Anwari, "Analisa metode ekstrapolasi dan biodistribusi 177Lu-DotaTrastuzumab dari mencit untuk memprediksi dosis radiasi internal di manusia," *Universitas Indonesia,* 2020.

Development of Early Real-Time Disaster Mitigation Warning System Landslide with Gyroscope ADXL345 Sensor

A Apriyansa[1], J Bintoro[2], E Sandi[3]
Department of Electronics Education, Universitas Negeri Jakarta.
Jln. Rawamangun Muka, Jakarta Timur, Jakarta – Indonesia
Email: [1]anak.rangkas@gmail.com, [2]jusufbintorounj@gmail.com, [3]efri.sandi@unj.ac.id

Abstract. The purpose of the research is to develop a Real-Time Landslide Early Disaster Mitigation Warning System with the ADXL345 Gyroscope Sensor and to know the exact system designed in a prototype. The method used is research and development with a prototyping model which is divided into several stages, namely stage (1) problems, (2) data collection, (3) prototype design, (4) prototype testing, and (5) production. The resulting system is a prototype based on an Arduino microcontroller called "ERT-Ladister" with an ADXL345 Gyroscope sensor. Other features include SMS-Gateway, buzzer, and Monitoring via Monitor. Based on the feasibility test and the effectiveness test on a lab scale and field scale, it can be concluded that the system is feasible and effective. The value of the effectiveness test shows that there is a significant difference between before the system is installed and after the system is installed in field trials. The Real-Time Landslide Early Disaster Mitigation Warning System can increase community awareness around landslide-prone areas.

1. Introduction

Indonesia is one of the countries that are prone to natural disasters in the world. This was also stated by the United Nations Agency for the International Strategy for Disaster Risk Reduction (UN-ISDR) based on the data it had released [1].

Geologically, Indonesia is included in the Ring of Fire environment, which has a fairly high potential for natural disasters, because it is located between the trajectory of two mountain paths, namely the Pacific Circum and Mediterranean Circums where there are many volcanoes, and their activities can cause volcanic earthquakes. Indonesia's geological position is at the confluence of three active plates, namely the Indo-Australian plate in the south, the Euro-Asia plate in the north and the Pacific plate to the east. Thus, Indonesia's position is very vulnerable to disasters [2], both from volcanic and tectonic activity. When there is volcanic or tectonic activity, there will be a shift, causing landslides. In addition, landslides can also occur on the banks of the river because it is influenced by the flow of water flowing in the river [3].

Landslide is the movement of slope-forming material in the form of rock, debris, soil, or material [4]. The mixed material usually moves in the highlands to the lowlands. The process of landslides can be explained as follows: water that seeps into the soil will increase the weight of the soil. If the water penetrates to the impermeable soil which acts as a slip plane, the soil becomes slippery and the weathered soil above it will move from the highlands to the lowlands [5].

The landslide disaster mitigation system that is currently being developed by the Center for Physics Research at the Indonesian Institute of Sciences (LIPI) Serpong is a system that uses an inclinometer sensor. This sensor is combined with several other devices to form a system that can monitor a landslide-prone area so that when a landslide occurs, the disaster can be handled quickly by the surrounding community [6]. The landslide disaster mitigation system can be the right choice of tool for landslide-prone areas as an early warning tool to increase the preparedness of citizens against landslides. This is also supported by previous research which shows that in addition to the aspects of knowledge and attitude preparedness, a warning system is also needed to minimize disaster victims[7]. One other sensor that is considered good enough to be used in a landslide disaster mitigation system is the accelerometer sensor. This sensor will activate and change its angle value when the ground starts to shift or move [8]. This sensor will be combined with a microcontroller device, so that if there is a change in angle then the data will be processed by the microcontroller and will be conveyed to the user or the surrounding community [9].

However, the level of public awareness of the disaster-prone archipelago is still very low [10]. This can be seen from the level of awareness or attitude of community disaster response, which is still sluggish. If a disaster occurs in Indonesia, usually it will take quite a lot of victims due to the public's lack of awareness of potential disasters that will occur and their lack of responsiveness in the disaster management process [11]. In this case, we will develop a disaster mitigation system that uses an accelerometer sensor to detect the occurrence of landslides and uses Arduino Mega 2560 to process the information received and SMS gateway as a processing system so that it can be easily monitored via mobile devices and browser monitors. It is hoped that this mitigation system can monitor landslide-prone areas and provide information quickly via mobile devices or cell phones so that people can immediately anticipate the disaster.

2. Method

The research method used to complete this research is to use research and development methods (Research and Development) with a prototype model that includes problem identification, data collection, Prototype Design, Prototype Testing, and Production [12]. Small field trials were carried out at LIPI Puspitek Serpong and field trials were carried out at 5 landslide-prone points in the Subang area, Lebak Regency for a 3 month trial period.

3. Result and Discussion

3.1. Research result

The results of the research on Landslide Disaster Mitigation System Development Using the ADXL345 Accelerometer Based on Web Localhost and SMS Gateway were obtained after conducting a series of trials. This stage is carried out in order to find out how big the value of success in making prototypes is in accordance with the program or system that has been created. The stages are testing the ADXL345 Accelerometer sensor, Ethernet Shield, SMS Gateway, Webcam, and power supply (Power Supply).

a. back view

b. top view

Figure 1. Prototype of Landslide Disaster Mitigation System

3.1.1. Component test results

In the component section there are several tests carried out including Accelerometer sensor testing, Ethernet Shield testing, SMS Gateway delivery testing, webcam testing, and power supply circuit testing.

3.1.2. Tilt test results ADXL345 accelerometer sensor

The test was carried out by changing the tilt position of the ADXL 345 accelerometer sensor using the Arduino IDE 1.0.6 application. The data changes will be displayed through the serial monitor menu in the Arduino IDE application. In addition, to produce a degree value, the data must be calibrated first using a comparison. The following formula is used in the prototype of this system:

$$\text{Sensor tilt angle} = \left((y-5).\frac{90}{(277-5)}\right) \quad (1)$$

Information:

y = Initial slope value
90 = Maximum slope angle
277 = Maximum limit of initial tilt angle
5 = Calibration value

After testing the sensor, the results are obtained as shown in table 1 below.

Table 1. ADXL 345 Test Results on Serial Monitor

No	Degree	Description
1	0	The value taken is the value that often appears, namely 5 with an intensity of 7 times appearing in a row
2	45	The value taken is the value that often appears, namely 141 with an intensity of 5 times appearing in a row
3	90	The value taken is the highest value, which is 277

So it can be concluded that the value of 0 degrees, starts from 5 bits and the maximum limit is 90 degrees, which is 277 bits.

3.1.3. Application test results

At this stage, the results obtained from testing entering data into the localhost database, testing the database calling to the localhost web, and testing the status / conditions displayed on the Opera browser with the Windows 7 operating system.

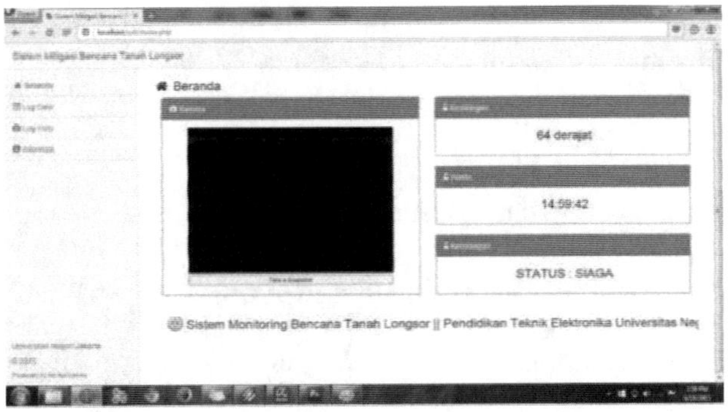

Figure 2. Homepage display on localhost

Figure 2 shows the web homepage on localhost, which contains the slope, time, and webcam display conditions.

3.1.4. Status test result / condition

Table 2 shows the status of testing sensor information at 5 received landslide-prone points. If the sensor is in a state of more than 75 degrees, then the command will be carried out such as sending a short message via SMS Gateway.

Table 2. Status Test Result/ Condition

No.	Condition	Testing Criteria	Description
1	Slope Status "Safe" ≤ 30^0	Status "Safe" appears in Log_data.php	Condition 10 degrees with status "Safe"
2	Slope Status "Standby" $30^0 \leq 75^0$	Status "Standby" appears in Log_data.php	Condition 58 degrees with status "Standby"
`3.	Slope Status "Landslide" $75^0 \leq 90^0$	The "Landslide" status appears in Log_data.php "Landslide" status information received from mobile/HP	Condition 83 degrees with "Landslide" status

3.2. Discussion

Based on the overall test results that have been carried out, it is known that the Development of a Landslide Disaster Mitigation System Using the ADXL 345 Accelerometer Based on Localhost Web and SMS Gateway functions according to the plan. However, there are still some limitations that exist in this system, namely:

The first discussion is that in component testing there are still some shortcomings, such as the IP Address setting on the Ethernet Shield sometimes changes itself when uploading programs on the Arduino IDE so to check it, you must use the help of the console command (CMD). The time frame for sending data from Arduino Mega 2560 to the database using the Ethernet Shield is sometimes interrupted even though the sending time has been set. For sending information via SMS, sometimes it doesn't reach its destination because the provider's signal is experiencing interference, so it is necessary to look for a provider card with a strong signal, reaching all corners, and rarely experiencing interference. When using the webcam, you still have to click the button to take a picture because the button is part of the .swf file so it can't be created automatically.

With the additional interface in the form of a website application at localhost on Landslide Disaster Mitigation System Development Using the ADXL 345 Accelerometer Based on Localhost Web and SMS Gateway, it makes it easier to monitor an area prone to landslides. The researcher believes that the prototype is very good in terms of providing information and can be a new alternative in developing a disaster mitigation system for further research.

4. Conclusions and Suggestions

4.1. Conclusions

After identifying problems, collecting data, designing prototypes, and testing prototypes, the following conclusions can be drawn:

4.1.1 Prototype for Landslide Disaster Mitigation System Development Using the ADXL 345 Accelerometer Based on Localhost Web and SMS Gateway has been designed with a slope limit of 30 degrees and can function as an information system that is quick to respond to landslide hazards through mobile devices.

4.1.2 Mobile / Mobile devices can receive short messages warning of landslide hazards from the prototype of Landslide Disaster Mitigation System Development Using the ADXL 345 Accelerometer Based on Localhost Web and SMS Gateway.

Journal of Physics: Conference Series **2019** (2021) 012080 doi:10.1088/1742-6596/2019/1/012080

4.2. Suggestions

In developing a prototype of a Landslide Disaster Mitigation System Using the ADXL 345 Accelerometer Based on Localhost Web and SMS Gateway, the researcher has several suggestions to overcome and complete the system that will be used in this study.

- For the use of the Accelerometer sensor, use a sensor that is easy to find in the market so that many tutorials can be found in books or the internet.
- First, calibrate the Accelerometer sensor so that the resulting value is as desired.
- Adding a website application on localhost to display the results of the Accelerometer sensor and the data can be stored in the database.
- For database creation, create at least 2 databases to facilitate the data storage process.
- Adjust the delay of the Accelerometer sensor with the delay of sending data to localhost via Ethernet Shield so that the results of the accelerometer can be read and stored in the localhost database.
- When running this prototype open the Arduino IP address and the homepage of its mitigation system in a browser with a different tab.

References

[1] UNDRR, *Country Assessment Report for Indonesia: Strengthening of Hydrometeorological Services in Southeast Asia*. 2017.

[2] C. Setiawan, Muzani, Warnadi, F. M. Akmal, and Ermalia, *IOP Conf. Ser. Earth Environ. Sci.*, vol. 412, no. 1, 2020.

[3] M. Masum and M. Ali Akbar, *IOP Conf. Ser. Earth Environ. Sci.*, vol. 249, no. 1, 2019.

[4] I. M. Patuti, A. Rifa'i, and K. B. Suryolelono, *Int. J. GEOMATE*, vol. 12, no. 29, pp. 1–8, 2017.

[5] S. R. P. Sitorus and A. E. Pravitasari, *Sumatra J. Disaster, Geogr. Geogr. Educ.*, vol. 1, no. 2, pp. 61–71, 2017.

[6] Suryadi, P. Puranto, H. Adinanta, A. Tohari, and P. S. Priambodo, *J. Phys. Conf. Ser.*, vol. 853, no. 1, 2017.

[7] S. Nur *et al.*, *JPIG (Jurnal Pendidik. dan Ilmu Geogr.*, vol. 6, no. 1, pp. 81–90, 2021.

[8] F. Budiman, E. Susanto, D. Perdana, H. Mukhtar, Y. A. Pamungkas, and Y. Y. Kevin, *J. Teknol. dan Sist. Komput.*, vol. 8, no. 4, pp. 255–262, 2020.

[9] Alimuddin, F. Sulaiman, I. H. Parinduri, R. Abdullah, T. Firmansyah, and M. S. Syarif, in *1st Annual Applied Science and Engineering Conference (AASEC)*, 2017, vol. 755, no. 1, pp. 6–12.

[10] M. Martono, S. Satino, N. Nursalam, F. Efendi, and A. Bushy, *Chinese J. Traumatol. - English Ed.*, vol. 22, no. 1, pp. 41–46, 2019.

[11] E. Wahyudianto, *J. Civ. Eng. Forum*, vol. 4, no. 3, p. 265, 2018.

[12] A. Susanto and Meiryani, *Int. J. Sci. Technol. Res.*, vol. 8, no. 7, pp. 141–144, 2019.

Identification 2D Modelling of Subsurface Structure Geothermal Prospect Area by Gravity Method: Case Study in Tanuhi, South Kalimantan

M Nafian[1], N R Permana, A Anjani, B Gunawan and L A Sanjaya[2]
[1] Department of Physics, Syarif Hidayatullah State Islamic University
Jl. Ir. H. Juanda No. 95, Cempaka Putih, Ciputat, South Tangerang City, Banten 15412, Indonesia.
[2] Department of Physics, Faculty of Mathematics and Natural Science, Universitas Negeri Jakarta, Indonesia.
Email: [1]muhammad.nafian@uinjkt.ac.id

Abstract. Geothermal potential in Kalimantan is rare because the island does not have an active volcano. In addition, when viewed from the subsurface structure, Kalimantan Island is in medium-low enthalpy with hot sprung as one of the geothermal potentials, which are generally in the immature water zone. One of the areas is Tanuhi Hot Springs, South Kalimantan. Therefore, this study was conducted to determine the geothermal system of the area using the gravity method. The gravity data processing stage produces the CBA value (Complete Bouguer Anomaly) which is then separated using the Butterworth filter method. Judging from the CBA map, the results of the qualitative interpretation show that the geothermal potential is in a low anomaly with a value of (-19.4 mgal) – (-14. 2 mgal) which is marked in the dark blue. While the results of the quantitative interpretation resulted in the cross-sectional modeling of SW-NE (Southwest – Northeast) and NW-SE (Northwest - Southeast) with a total of 3 layers. The layer consists of granite, clay, and sand mixed with granite. Both models are dominated by sand mixed with granite with the density value at cross-section SW-NE and cross-section NW-SE is the same, namely 2770 kg/m3.

1. Introduction

Plate movements in the territory of Indonesia are very active where Indonesia is located between three plate encounters such as the Eurasian, Indo-Australian, and Pacific Plates. From these plate movement activities, it forms a diverse geographical condition that causes Indonesia to have abundant natural resources. One of these natural resources is the potential of geothermal resources. This potential can be used as environmentally friendly alternative energy to replace non-renewable energy such as coal, oil, and natural gas.

As the economy and population grow, the need for electrical energy is increasing. Of course, this increase in electrical energy does not have to depend on fossil fuels, when viewed from the geographical condition of Indonesia, which is traversed by the ring of fire, the potential for geothermal sources is abundant, this happens because the geothermal system is formed due to ongoing volcanic processes. The utilization of this geothermal resource is more environmentally friendly because it can reduce carbon dioxide emissions from burning fossil fuels. Geothermal is also a clean energy that does not require carbon-intensive fuel to operate power plant. Therefore, geothermal energy can support Indonesia's target in achieving emission reduction of greenhouse gas (GHG) to 29 percent by 2030 [1].

One of the areas in Indonesia that have the potential for geothermal resources is Tanuhi, South Kalimantan, Hulu Sungai Selatan Regency. If viewed from the geological setting, it is included in the Muria-Meratus subduction pathway in the limestone age, which consists of steep, undulating hills dominated by plutonic rock types with limestone composition of granitic and Cretaceous-Miocene sedimentary rocks as well as alluvial deposits in plain areas, geological structure The developing fault is dominated by a normal fault with a southwest-northeast direction known as the Meratus Pattern and a dextral horizontal fault with a northwest-southeast trend called the Andang pattern. [2] The existence of a geothermal manifestation on the surface indicates the presence of geothermal below the surface with a phase change in the hydrothermal geothermal system in the passage of fluids to the earth's surface, causing different types of geothermal manifestations as indicated by the presence of hot springs, geysers, mud pools or hot mud pools, and hot tubs. [3] In this study, geothermal manifestations are in the form of hot springs which are used as hot spring baths in Tanuhi Village, South Kalimantan.

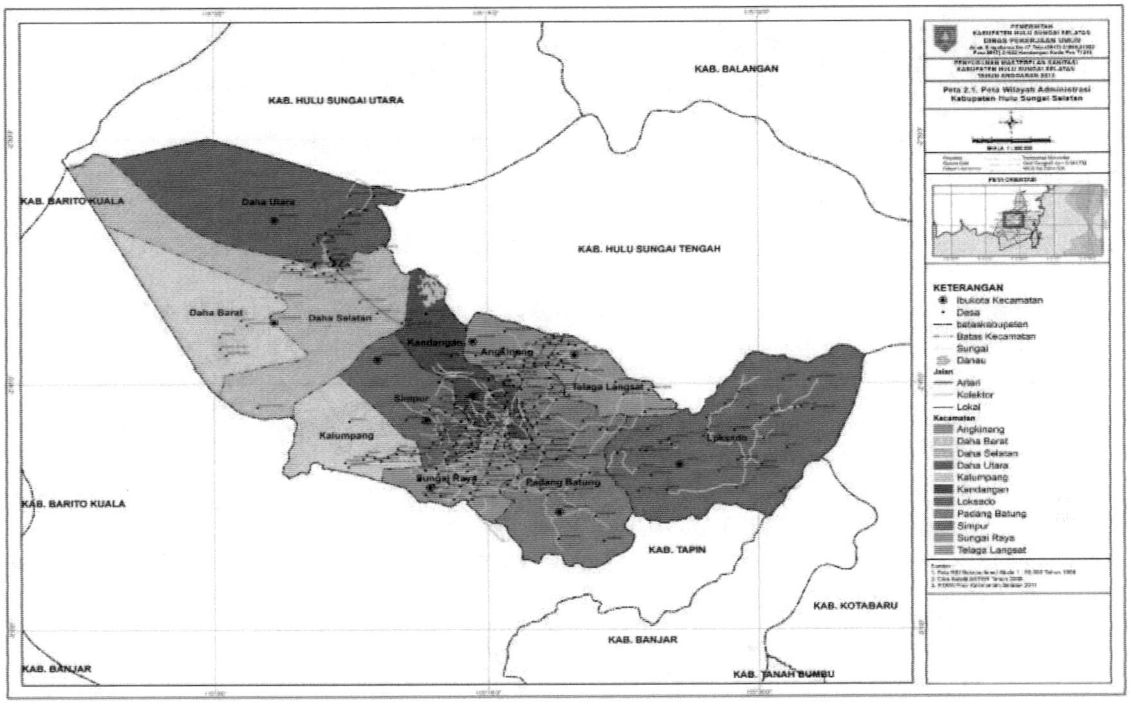

Figure 1. District Boundary of Hulu Sungai Selatan. [4]

To find out the state of the geothermal system, a geophysical method is needed that can provide information about the structure below the earth's surface. In this study, the geophysical method used is the gravity method by looking at the differences in rock density found in Tanuhi Village, South Kalimantan.

1.1 Geothermal System

Geothermal sources can be formed due to plate movements that can cause subduction zones and the formation of volcanic areas. The formation process occurs when rainwater that is absorbed through cracks in the soil will gather in the reservoir rock. The groundwater will go through a healing process that comes from the main heat energy source, namely magma.

Geothermal systems are formed as a result of heat transfer from surrounding heat sources that occur by conduction and convection. Conduction heat transfer occurs through rocks, while convection heat transfer occurs due to contact between water and a heat source that can form hot fluid in the reservoir. The existence of a geological structure in the form of a fault that cuts through the reservoir,

Journal of Physics: Conference Series **2019** (2021) 012081 doi:10.1088/1742-6596/2019/1/012081

causes this hot fluid to come out to the surface in the form of geothermal manifestations, such as hot water. [5]

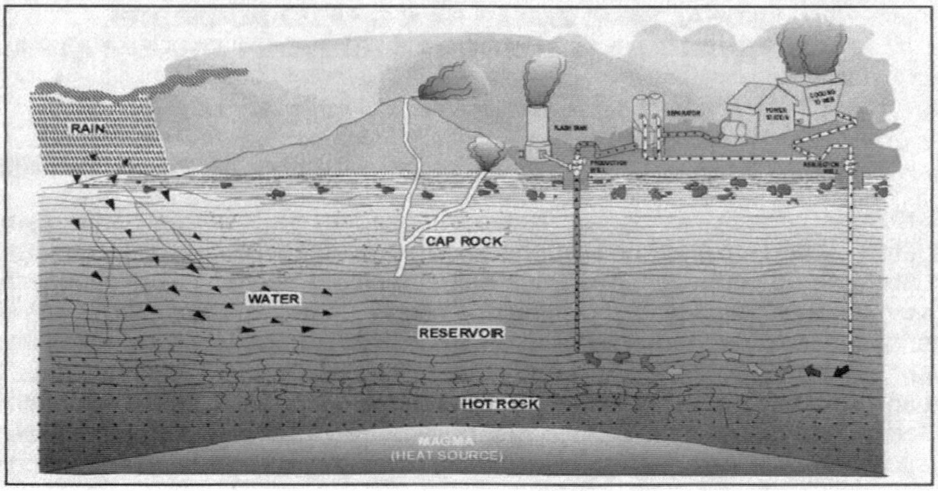

Figure 2. Illustration of Geothermal System. [5]

1.2 Principle of Gravity Method

The basic principle of the gravitational method is to use the basic theory of Newton's law of gravity, where the force between two particles of mass m_1 and m_2 directly proportional to the product of the masses and inversely proportional to the square of the distance between the centers of mass. [7] This is shown in the equation:

$$F = \gamma \frac{m_1 m_2}{r^2} r_1 \quad (1)$$

Where F is the force exerted by a particle of mass m_1 and m_2, r^2 is the distance between the two particles, r_1 is a unit vector, and γ is the gravitational constant which has the value 6.672 x 10^{-11} Nm^2/kg^2 in international units and 6.672 x 10^{-8} dyne cm^2/g^2 cgs units. [7]

In the gravity method, the measured quantity is the acceleration of the earth's gravity by producing various kinds of corrections such as earth tide correction, buoyancy correction, latitude correction, free air correction, Bouguer correction, and field correction. After calculating the corrections in data processing will produce a complete bouguer anomaly value. The acceleration due to Earth's gravity can be shown by the equation:

$$g = \left(\gamma \frac{M_e}{R_{e^2}}\right) r_1 \quad (2)$$

Where M_e and R_e and R_e are the mass of the earth and the radius of the earth.

1.3 Bouguer Correction

Bouguer correction is taken into account because there is a pulling effect of the rock mass of the station and datum plane, assuming it has an infinite radius with a thickness h (meters) and a density ρ (g/cc). [7] This correction is made to eliminate the difference in height h (meters) without ignoring the mass (ρ) below it. The height difference will result in the influence of the mass below the surface which affects the magnitude of the acceleration of gravity at the very point. [9]

$$BC = 0.04191\rho h \quad (3)$$

Figure 3. Bouguer Correction. [6]

1.4 Terrain Correction

Field correction is carried out because at the measurement point there are topographic effects and large elevation differences, such as hills and valleys around the measurement station, so a simple bouguer anomaly correction is carried out, where the topography affects the reading because gravity is conservative and reduces the reading value from the ideal state. [8]

$$\delta g_T(r,\theta) = \gamma\rho\theta\,\{(r_0 - r_i) + \sqrt{(r_i^2 + \Delta z^2} - \sqrt{(r_0^2 - \Delta z^2)}\} \quad (4)$$

δg_T is the field correction (mGal), γ is the universal gravitational constant, Δ_z is the difference in elevation of the compartment (m), r_0 and r_1 is the radius of the outer and inner circle (m), and θ is the angle formed by the compartment (degrees).

1.5 Butterworth Filter

Is a method of smoothing a signal or image. Which can help eliminate noise, because noise is caused by high frequencies. In the function Butterworth lowpass filter (BLPF) with a value of the order to n, and with a cutoff at the D_O distance from the origin. [10] Defined as:

$$H\,(u,v) - \frac{1}{1\,[D\,(u,v)/D_0]^{2\pi}} \quad (5)$$

Where D (u, v) is the distance between the points (u, v) in the frequency domain and the center of the frequency rectangle which is written in the following equation:

$$D\,(u,v) - [\left(u - \frac{\pi}{2)^2}\right] + [v - N/2^2]^{1/2} \quad (6)$$

2. Method

This research was conducted from March 28 to April 30, 2021. The research location for the Tanuhi hot spring is in the Hulu Banyu Village, Loksado District, Hulu Sungai Selatan Regency. The geographical location of Hulu Sungai Selatan Regency is located between 115^0 27' 8.48" East Longitude, and 2 47' 28.12" South Latitude. When viewed based on the topography of Hulu Sungai Selatan Regency, it is located at an altitude of 0-7 meters and a slope of 0-2%, based on an altitude from sea level, 58.3% of the Hulu Sungai Selatan Regency are at an altitude of 0-7 meters and only 0.9% is at an altitude of over 1000 meters. [7]

Journal of Physics: Conference Series **2019** (2021) 012081 doi:10.1088/1742-6596/2019/1/012081

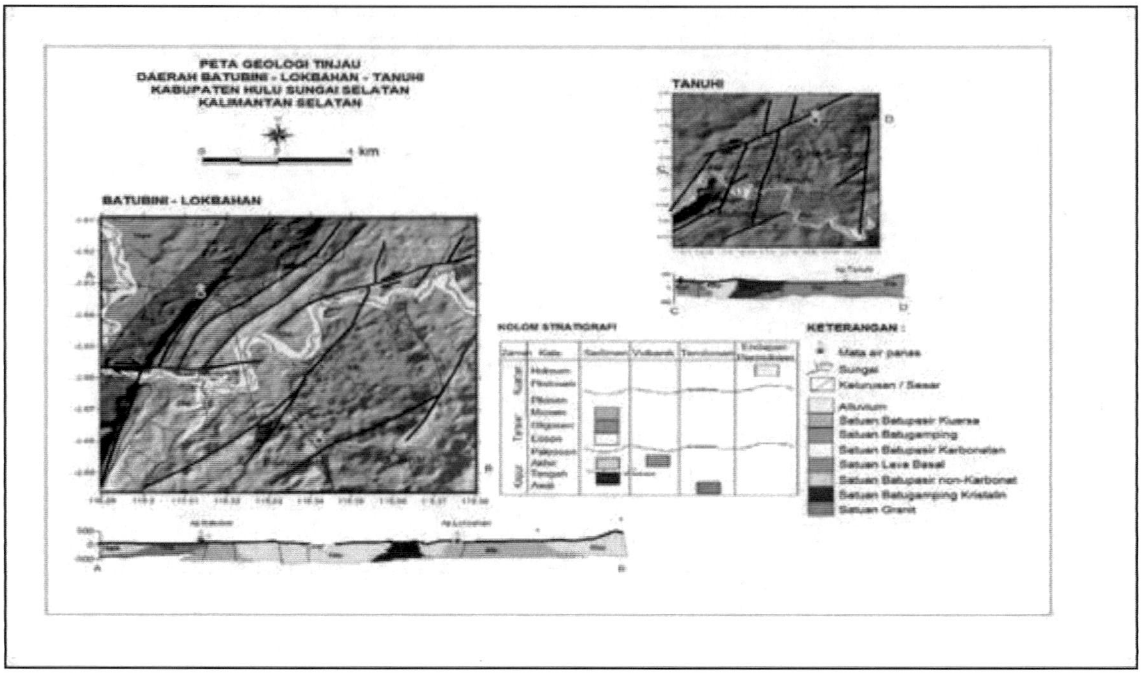

Figure 4. Regional Geological Map of Tanuhi, South Kalimantan. [1]

The processed data are secondary data consisting of gravity and topographic data. Both of these data can be obtained by accessing the TOPEX website, with a total of 3904 data generated. The gravity data only require bouguer correction and field correction because the resulting data has been corrected in the form of FAA (Free Air Anomaly) data. In data processing requires some software, including Microsoft Excel and Oasis Montaj. So to calculate the bouguer and terrain correction values, Microsoft Excel software is needed to get the complete bouguer anomaly (CBA) value. While the oasis montaj software is used for making CBA contour maps, 2D modeling of geothermal systems, determining the residual anomaly slice point and separating regional and residual anomalies using the Butterworth filter method because it is easier to use for clear regional and residual trend selection. In addition, in 2D modeling, forward modeling is used to observe the gravitational response caused by the modeling.

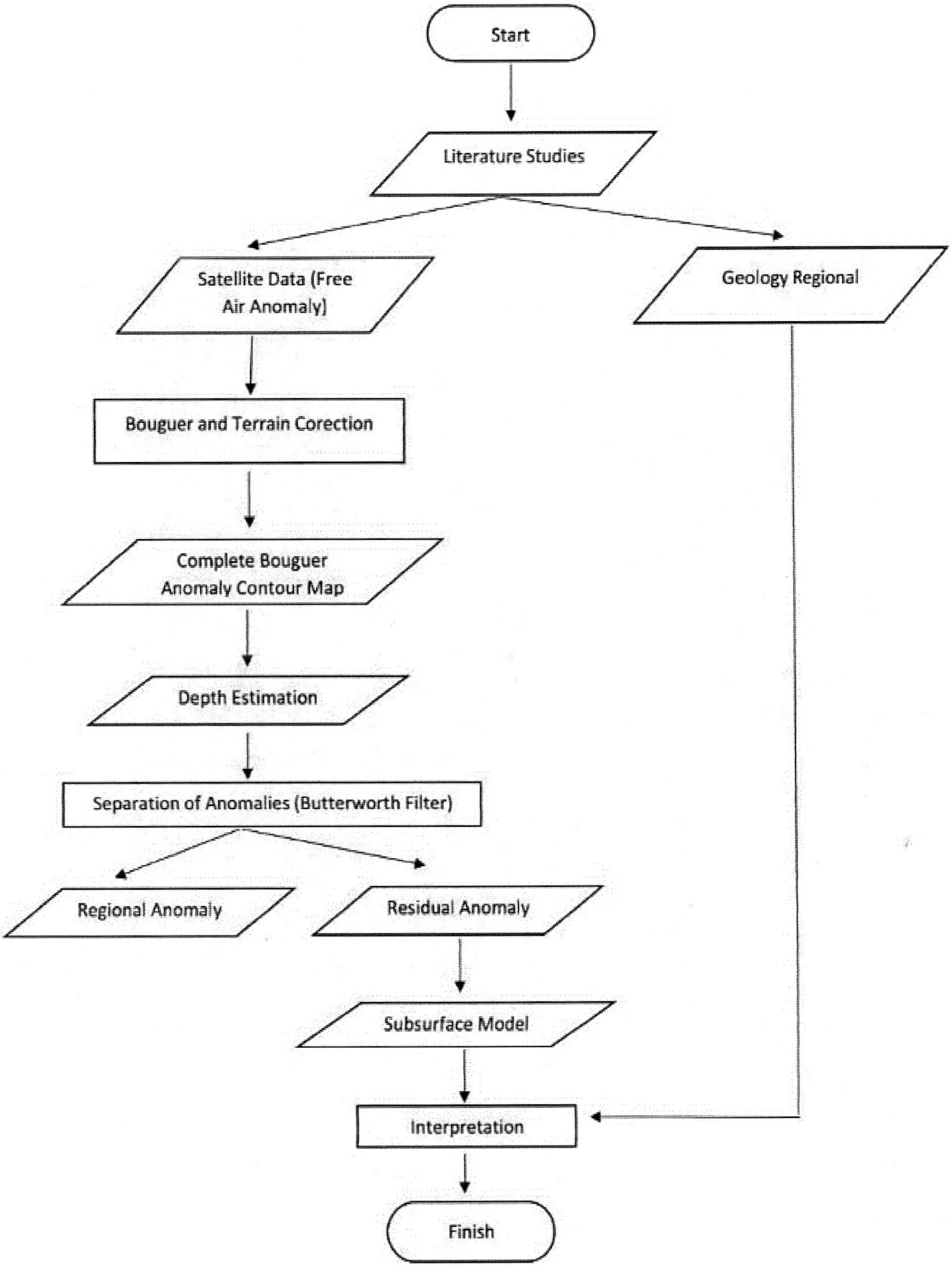

Figure 5. Research Flowchart

Journal of Physics: Conference Series **2019** (2021) 012081 doi:10.1088/1742-6596/2019/1/012081

3. Result and Discussion

3.1 Qualitative Interpretation
In the data processing process, after obtaining the gravity correction value, the CBA (Complete Bouguer Anomaly) value will be obtained. The CBA value is depicted through a CBA contour map which produces an overview of the topographical conditions in the study area.

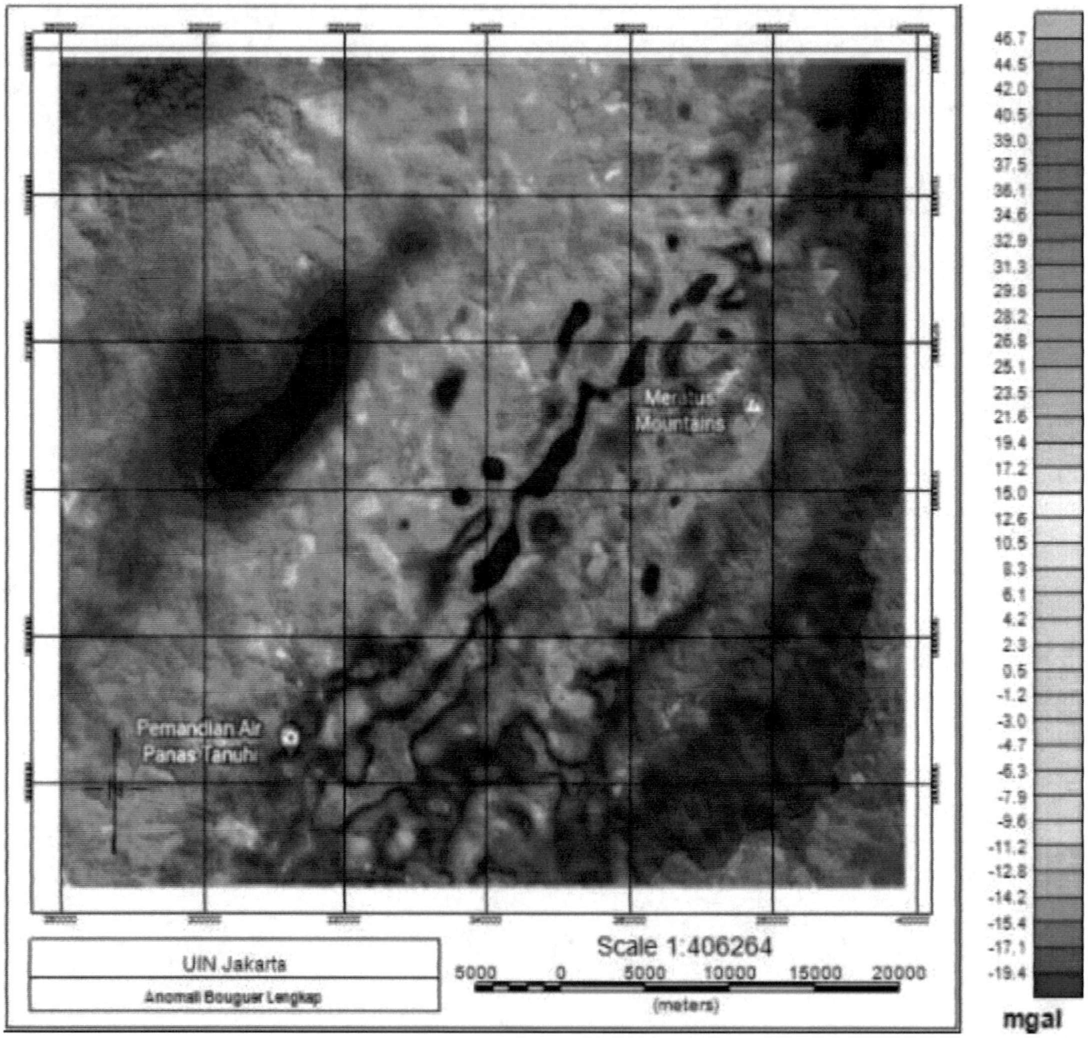

Figure 6. CBA Contour Map (Complete Bouguer Anomaly)

Based on the interpretation of the CBA contour map, there are three anomalous patterns in the research area of Tanuhi Village, South Kalimantan. The three patterns are indicated by gravity anomaly values ranging from low, medium, to high. The first anomalous pattern with low anomaly values ranged from (-194) mgal – (-6.3) mgal. The second anomaly pattern with moderate anomaly values ranged from (-4.7) mgal – (32.9) mgal, and the third anomaly pattern with high anomaly values ranged from 34.6 mgal – 46.7 mgal. This research was conducted with the main focus on areas that have low anomalous values, namely the research area indicated in dark blue with an anomaly value of -194 mgal. This research area is close to Mount Meratus which was formed due to volcanic and tectonic activities.

3.2 Depth Estimation and Separation Anomalies

Gravity data processing will produce a complete Bouguer anomaly value, where the resulting anomaly value will be used to identify the subsurface geological conditions in the research area. The complete Bouguer anomaly value consists of regional and residual anomalies, so it is necessary to separate the two anomalies. In addition to the separation of anomalies, it is necessary to estimate the depth of regional and residual anomalies in the study area, so the first step to be taken is to find the depth value obtained by analyzing the RAPS curve. Based on the RAPS curve, the blue color indicates a regional anomaly with the line equation y = -240.75x + 10.18, so the regional depth obtained is 19 km or deeper. While the red line shows the residual depth.

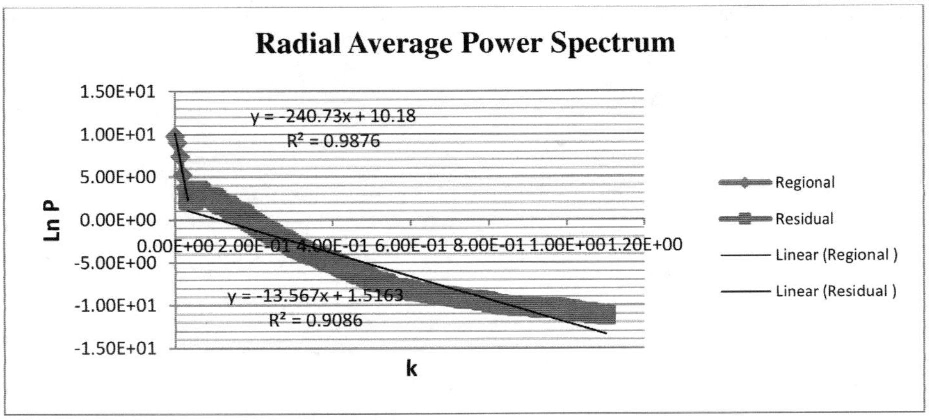

Figure 7. Radial Average Power Spectrum

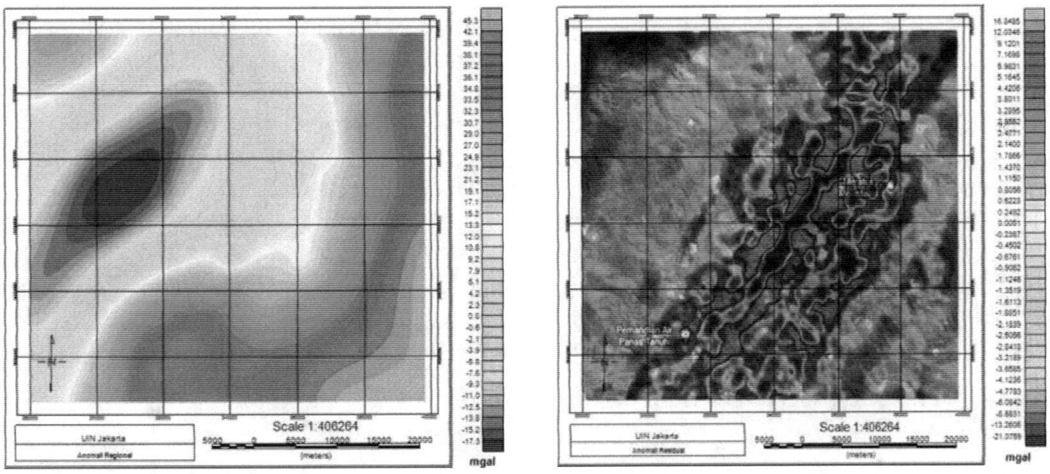

(a) (b)

Figure 8. (a) Regional Anomaly Map, (b) ResidualAnomaly Map

After getting the depth value for the regional and residual anomalies, the next step is to separate the anomalies using the Butterworth filter method which produces a contour map in the form of a regional anomaly map and a residual anomaly map. As seen from Figures 8.a and 8.b, there is a difference between regional and residual, where the resulting regional anomaly map is smoother with values ranging from (-17.3) mgal – (45.3) mgal, while the resulting residual anomaly map is coarser. With values ranging from (-21.0759) mgal – (16.8485) mgal. The difference is because the regional anomaly map has a wider and deeper area when compared to the shallower residual anomaly map.

Journal of Physics: Conference Series **2019** (2021) 012081 doi:10.1088/1742-6596/2019/1/012081

3.3 Quantitative Interpretation

Quantitative interpretation is carried out to see the modeling originating from the gravity anomaly that has been generated to determine a subsurface model, by slicing the residual anomaly to produce 2 cross-sections, namely NW - SE (Northwest - Southeast) and SW - NE (Southwest - Northeast).

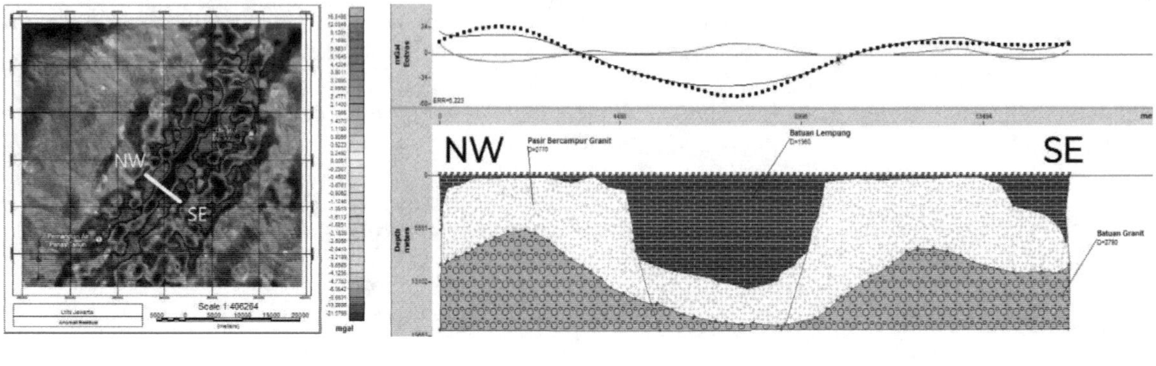

(a) (b)

Figure 9. (a) NW-SE Residual Anomaly Section, (b) NW-SE Section 2D Modeling

Figure 9 is a cross-sectional model of NW-SE which consists of 3 layers below the earth's surface. The first layer is shown in brown, the layer contains clay rock at a depth of about 0 km – 6 km, with a clay rock density value of 1960 kg/m^3. In this layer, clay rock is a rock cover whose role is to protect the geothermal reservoir that is formed, because it can prevent fluid from coming out. The second layer is shown in white, the layer contains sandstone mixed with granite. This layer has a density value of 2770 kg/m^3, which serves as a reservoir of hot water that forms below the earth's surface. And the third layer is shown in gray, the layer contains granite rocks with a density value of 2790 kg/m^3. This third layer helps the formation of a geothermal reservoir because the rock is exposed to heat from magma, so it can heat something above its surface, in this case, a hot water reservoir.

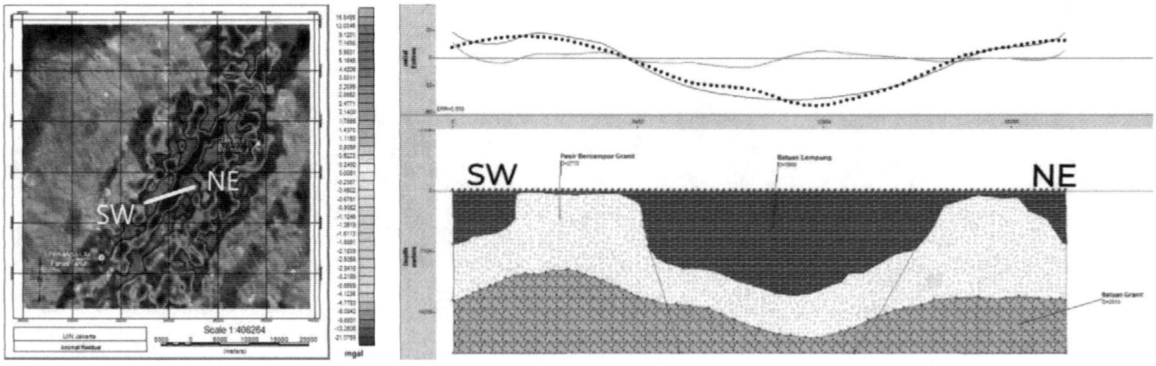

(a) (b)

Figure 9. (a) SW-NE Residual Anomaly Section, (b) SW-NE Section 2D Modeling

Figure 8 is a cross-sectional model of SW-NE which consists of 3 layers below the earth's surface. The first layer is shown in brown, which has a density of 1980 kg/m^3 is estimated that the first layer is at a depth of about 0 km - 8 km, this layer contains clay rock which functions as a rock cover to protect the geothermal reservoir that is formed because it can prevent fluid from escaping. The second layer is shown in white, which has a density 2770 kg/m^3 it is estimated that the second layer is at a depth of about 2 km – 15 km, this layer contains sandstone mixed with granite. In this layer, a hot water reservoir will be formed. And finally, the third layer is shown in gray, which has a density of

2810 kg/m^3 it is estimated that the third layer is at a depth of 12 km - 16 km, this layer contains granite which has a role in the formation of a geothermal system because in this layer the rock will be exposed to heat from magma so that this layer also functions in heating things. Which is above the surface, in this case, is a hot water reservoir.

4. Conclusion

From the results of the qualitative interpretation, it produces a picture through a complete Bouguer anomaly map of the research area which is divided into three anomalous patterns, namely, high, medium, and low anomalies. In this study, it is shown in areas that have a low anomaly with a value of – 19.4 mgal which is shown in dark blue.

Meanwhile, the results of the quantitative interpretation provide a 2D depiction of the rock layers below the ground surface, from slicing residual anomaly resulting in a cross-section of SW-NE (Southwest – Northeast) and NW-SE (Northwest - Southeast) with a total of 3 layers. This layer consists of granite, clay, and sand mixed with granite. Both models are dominated by sand mixed with granite with the density value at cross-section SW-NE and cross-section NW-SE is the same at 2770 kg/m^3.

Reference

[1] U. Siagian, B.B. Yuwono, S. Fujimoro and T. Masui. Energies, 10, 52 (2017); doi:10.3390/en10010052
[2] Geothermal Directorate 2017 *Geothermal Potential* Directorate General of New, Renewable Energy, and Energy Conservation: Ministry of Energy and Natural Resources.
[3] Mass 2018 *Geothermal Energy Exploration* (Darussalam – Banda Aceh: Syiah Kuala University Press).
[4] South Hulu Sungai District Office *Profile of Hulu Sungai Selatan District,* p 1-27.
[5] D. Hermawan, S. Widodo, E. Mulyadi 2012, Vol. 7, p 1-6.
[6] Geothermal Directorate 2017 *Doing Business in Geothermal* Directorate General of New, Renewable Energy, and Energy Conservation: Ministry of Energy and Natural Resources.
[7] W. M. Telford, et al 1990 *Applied Geophysics* (New york: Cambridge University Press)
[8] AD Maulana, DA Prasetyo 2019, Journal of Geoscience, Vol. 5, Central Sulawesi, p 91 -100.
[9] F. Jusmi 2018, Journal of Dynamics, Vol. 9, p 1 – 9.
[10] Hafidz, Ananda, M. Akbar 2015, Journal of Caltex Riau Polytechnic, Vol. 1, p 1-9.

Journal of Physics: Conference Series **2019** (2021) 012082 doi:10.1088/1742-6596/2019/1/012082

Manila Trench Tsunami Source Modeling for West Kalimantan Nuclear Facility Mitigation

S Pribadi[1], Fauzi[1] , T Kurniawan[1], Sunarko[2], H Suntoko[2], A Sudrajat[1], B S Prayitno[1] and N F Riama[1]
[1]Meteorological, Climatological and Geophysical Agency, Kemayoran, Jakarta, Indonesia
[2]National Atomic Energy Agency, Mampang Prapatan, Jakarta, Indonesia
Email: [1]sugengphd52@gmail.com

Abstract. Tsunami modeling with the Manila Trench earthquake source, the Philippines is very necessary as one of the mitigation measures for the prospective nuclear power plant (PLTN) facility in West Kalimantan. In this preliminary study, tsunami modeling using the COMCOT of non linear and the nesting grid method was carried out to obtain an estimate of the height and arrival time of the tsunami at the closest tidal initial points of a nuclear power plant. The Manila Trench is considered as the closed potential sources of earthquake threats to the prospective nuclear power plant site in Gosong Beach, West Kalimantan. The source is divided into 33 multiple segments and accumulated to magnitude 9.1 The results of this study can be used as material for risk assessment for nuclear power plant site development. The results obtained are that the estimated tsunami height at Gosong Beach is around 0.3 – 0.5 meters on the beach with an arrival time of about 550 minutes after the earthquake at its center.

1. Introduction

Kalimantan still lack electricity supply. Even though the infrastructure of a coal-fired Steam Power Plant (PLTU) as a commodity in Kalimantan has been established, the need for electricity as a source of community life is very much needed. Since been proposed as a candidate for the state capital (IKN) in the Paser area, North Penajem Regency, East Kalimantan since early 2020 become the strategic area [1]. Moreover, the existence of this nuclear power plant is estimated to be able to supply 35,000 Mega Watts of national electricity needs.

The Provincial Government of West Kalimantan, in this case the Bappeda of West Kalimantan Province, since 2012 has collaborated with the National Atomic Energy Agency (BATAN) to conduct a study on the selection of potential sites for nuclear power plants in West Kalimantan, in 2015. Then in 2019 BATAN again conducted a Ground Check Plan for the Site Evaluation of Nuclear Power Plants in West Kalimantan, including the Gosong Beach, Sungai Raya Village, Bengkayang Regency.

On March 11, 2011, a massive earthquake (magnitude 9.0) and accompanying tsunami hit the Tohoku region of eastern Japan. Since then, the Fukushima Daiichi Nuclear Power Plants have been facing a crisis due to the loss of all power that resulted from the meltdown accidents [7]. Then the earthquake triggered a tsunami that killed more than 18,000 people and flooded the reactor at the Fukushima nuclear power plant (NPP) caused a huge disaster [8][9],. Therefore, tsunami mitigation is a very important research to do, as the first step in determining the location and safe position of the nuclear power plant site development plan.

2. Far Source of Manila Trench

Even though Indonesia is an active country of the earthquake and tsunami disaster, but the province of Kalimantan, both are very rare based on data of the Tsunami Catalogue of WinITDB for 1559 to 2005.

Agency of Meteorology, Climatology, and Geophysics (BMKG) had listed earthquake inland Kalimantan from East to the West since year 2015 to 2019 under magnitude 6,1 and no caused tsunami.

The author tries to find a distant tsunami source that may be dangerous for the nuclear power plant site. The most likely threat is in the South China Sea (SCS) with a tsunami source in the Manila Trench area, Philippines. In this study, a probabilistic approach is used from the results of the previous author's studies where large tsunamis have been mapped and caused consistent hazards in the past. In this study, 33 multiple segments of Manila Trench, which accumulated as magnitude 9.1 were used with some modification treatments from [3]. See Figure 1, 2 and Table 1.

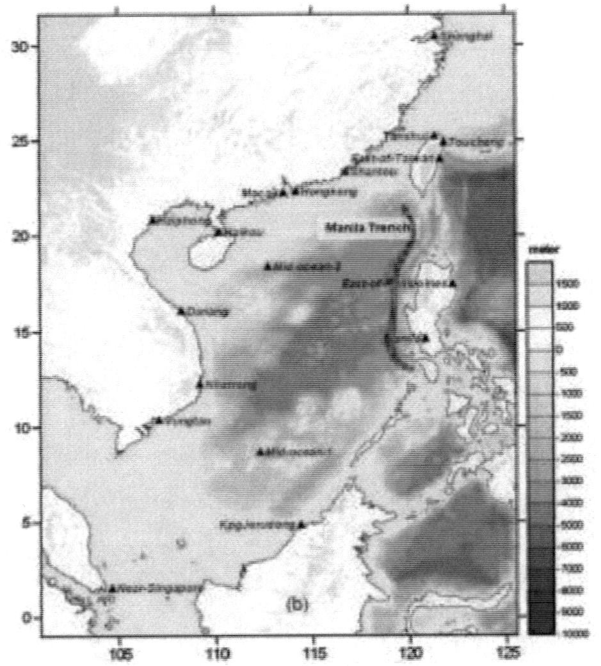

Figure 1. Scenario earthquake at the Manila Trench.

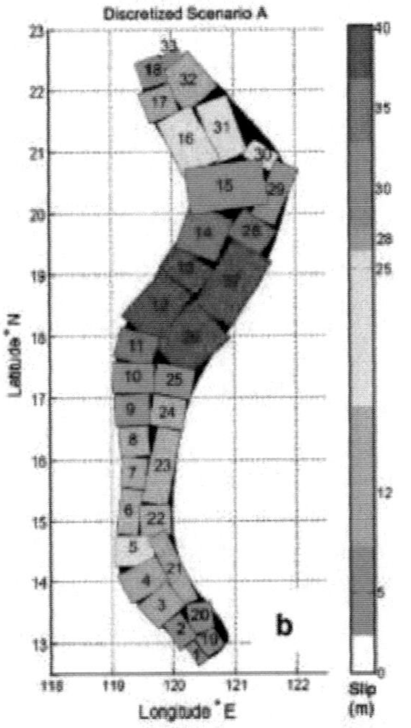

Figure 2. Discrete model of 33 segments (right pane)

Journal of Physics: Conference Series **2019** (2021) 012082 doi:10.1088/1742-6596/2019/1/012082

Table 1. Source of Manila Trech Multiple Segments

No	Long	Lat	Length	Width	Thick	Strike	Dip	Slip	Disloc
1	120,4	12,8	40410	19300	7500	324	22	90	5
2	120,1	12,2	54190	38680	7500	325	11	90	5
3	119,7	13,5	54090	66930	7500	319	7	90	12
4	119,5	14,0	54000	75680	7500	332	6	90	12
5	119,5	14,0	53890	67520	7500	0	6	90	25
6	119,2	15,1	80590	37510	7500	7	12	90	28
7	119,3	15,7	53560	43130	7500	6	10	90	28
8	119,3	16,3	53430	51060	7500	356	8	90	28
9	119,2	16,7	53280	60090	7500	358	7	90	30
10	119,2	17,3	53140	69930	7500	3	6	90	30
11	119,3	17,7	52990	65930	7500	16	7	90	35
12	119,7	18,5	105570	72150	7500	40	6	90	40
13	120,3	19,0	52600	79670	7500	36	5	90	40
14	120,5	19,5	78620	74610	7500	21	6	90	35
15	121,0	20,5	78240	129360	7500	352	3	90	30
16	120,3	21,0	103810	67360	7500	332	6	90	25
17	119,7	21,7	51680	54860	7500	340	8	90	12
18	119,7	22,2	46820	41970	7500	341	10	90	5
19	120,6	13,0	37500	37270	35000	327	29	90	5
20	120,5	13,5	54140	43110	35000	351	25	90	5
21	120,0	14,0	134880	34220	35000	334	31	90	12
22	119,7	15,0	53760	53110	35000	358	21	90	28
23	119,7	16,0	133840	45090	35000	11	24	90	28
24	119,9	16,7	53290	61360	35000	9	18	90	28
25	120,0	17,3	53150	70960	35000	11	16	90	30
26	120,5	16,0	105900	88630	35000	48	13	90	40
27	121,0	18,8	131680	75920	35000	31	15	90	40
28	121,2	119,7	52410	67880	35000	37	16	90	35
29	121,6	20,5	104390	44850	35000	25	24	90	30
30	121,5	21,0	51990	31660	35000	302	32	90	25
31	120,8	21,5	103670	55050	35000	341	20	90	25
32	120,2	22,0	77330	67830	35000	321	16	90	12
33	120,0	22,4	21260	41860	35000	326	25	90	5

3. Data and Method

As a basis for making tsunami modeling, marine and land data are needed. First bathymetric and topographic maps starting from the initial layer or domain with a wide-scale ocean level with The General Bathymetric Chart of the Oceans (GEBCO) aims to provide the most authoritative and publicly available bathymetric data set for the world's oceans within 480 meter resolution. The second layer is National bathymentry (BATNAS) within 160 meter resolution and finally the National Topography (DEMNAS) within 11 meter resolution. Both are taken from the National Bathymetry from the Geospatial Information Agency (BIG). Land cover has been added from the Indonesian

Rupiah Bumi (RBI) by adopting the Manning roughness scale that distinguishes environments such as: settlements, swamps, groves, open land, and so on.

The tsunami modeling method used is COMCOT (Cornell Multi-grid Couple Tsunami) Power from the Institute of Geological & Nuclear Science (GNS Science) of New Zealand [4]. Other supporting applications in this tsunami modeling are Matlab, Arc Map, Global Mapper, Surfer, Octave, QGis and GMT. COMCOT is based on non-linear empirical models and nesting grids so that it can be run in shallow water modeling as described in this formula.

$$\frac{\partial \eta}{\partial t} + \frac{1}{R\cos\varphi}\left\{\frac{\partial P}{\partial \psi} + \frac{\partial}{\partial \varphi}(\cos\varphi Q)\right\} = -\frac{\partial h}{\partial t} \tag{1}$$

Where η stands for the water surface fluctuation; *(P, Q)* denote the volume flux components in the x (Easting) and y (Northing) directions, respectively; (ψ) denote the longitude of the Earth; R is the radius of the Earth; and *the* water depth. The term $-\partial h/\partial t$, reflecting the effect of bathymetry variation, can be used to model tsunami generated by transient seafloor motions such as slow rupturing processes or submarine landslides. f represents the Coriolis force coefficient due to the rotation of the Earth andwhere *(P, Q)* denote the volume fluxes in x direction and y direction, respectively, and both are products of velocity and water depth, i.e., $P = hu$ and $Q = hv$.

4. Results and Discussions

An estimated tsunami source model from the Manila trench, showing areas in the west, so that the tsunami strength is stronger towards the wide-spreading South China Sea. There is a recent paper of **Q** Qiu et al and J P Terry et al, that examines the manila trench, but makes these fault segments separately so that it does not provide an energy accumulation of magnitude 9.1 so the author does not take this paper as a reference. Besides, our original goal was to obtain the worst-case outcome as mentioned by D M Ha et al and K Megawati et al.

The area closest to the tsunami hit is the waters of China and the farthest is the waters around West Kalimantan. The tsunami wave was split by the chain of the Spratly Islands but entered the shallow sea west of Sarawak so that the tsunami height increased again. The strength of the tsunami decreased when it entered the waters of the Riau Islands with a large size so that the tsunami entered from the cracks with the remaining energy. Lack of local bathymetric data around Gosong Beach makes it difficult to analyze tsunami data in this area.

The simulation results for the Gosong Beach area are obtained after 10 hours from the time of the earthquake's rupture from the source. The maximum amplitude of tsunami from the source to the depicted area in Figure 3. The results of the tsunami propagation from the source to the affected area and the arrival time and change in the broad to narrow layer and the initial tsunami gauge distribution of stations 1 to 9 are shown in Figure 4.

Journal of Physics: Conference Series **2019** (2021) 012082 doi:10.1088/1742-6596/2019/1/012082

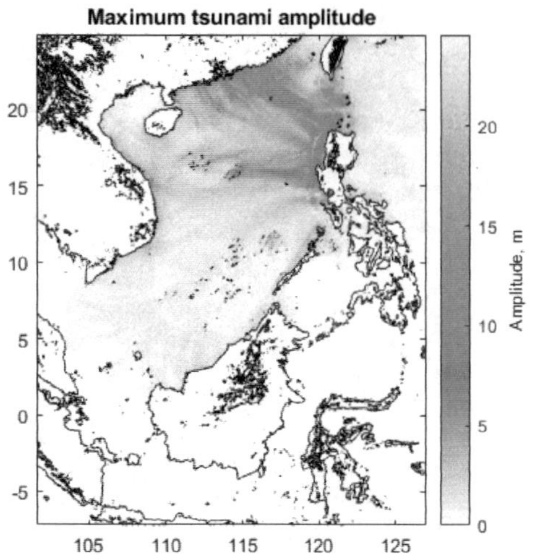

Figure 3. The maximum amplitude of tsunami from the source to the depicted area.

Figure 4. The broad layer to the narrow layer within distribution of initial gauge.

The selection of the arrival time and tsunami height is based on the provisions of the international tsunami warning where the Estimated Time of Arrival (ETA) for the arrival of a tsunami is taken from an indication of a sea level rise of 0.2 meters, the presence of the first tsunami wave at a sea rise of 0.5 meters, and a height of 0.5 meters. Maximum tsunami at the peak measured after the Estimated Wave Height (EWH) of 0.5 meters.

The arrival time and height of a tsunami are closely related to distance. Sequentially from highest to lowest values can be seen from stations 1 to 5. Stations that are closer to the source of the tsunami have a faster tsunami arrival time (1-4 hours) and a higher tsunami height (0.7-1.2 meters). The farthest station number 4 and 5 have a longer arrival time (7-9 hours) and the lowest tsunami height (0.3 meters).

Station 6 is relatively close to the source so it has a fairly fast tsunami arrival time (almost 7 hours) but there is a refractive effect on the islands around Sambas that it passes through so that it gives amplification of the tsunami height (1.2 meters). Stations around Gosong Beach, Bengkayang (stations 7, 8, 9) have low tsunami heights (0.3–0.5 meters) and longer arrival times (7-10 hours) because they are far from the source and blocked by islands near Sambas. See Figure 5, 6 and 7.

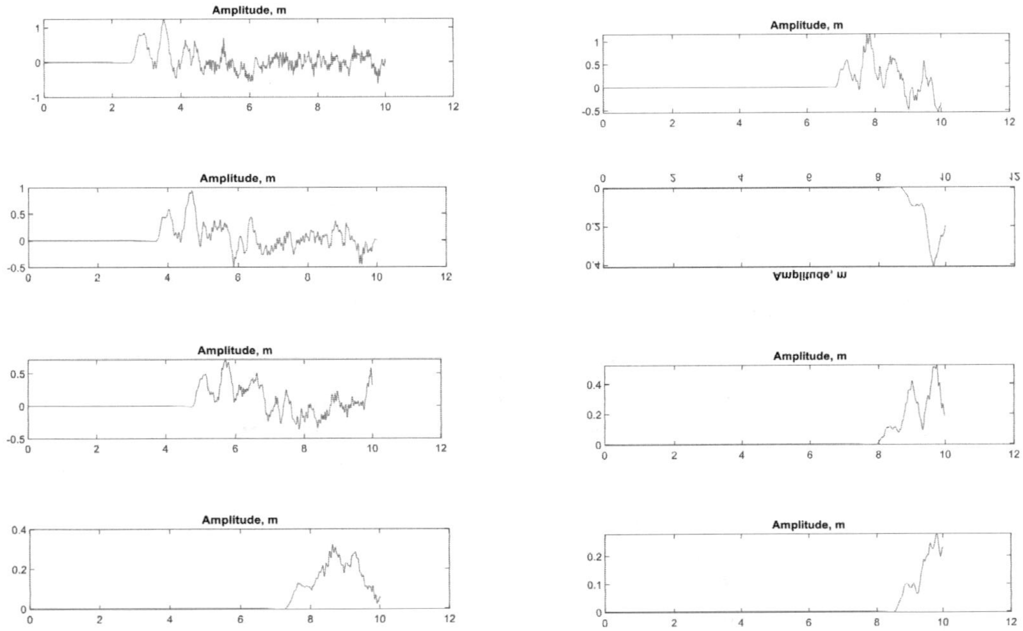

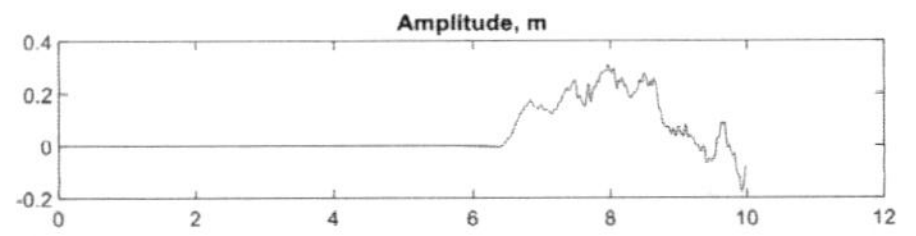

Figure 5. Marigrams from the initial gauge with the lowest to highest serial numbers (1-9) arranged from top to bottom and to the right side.

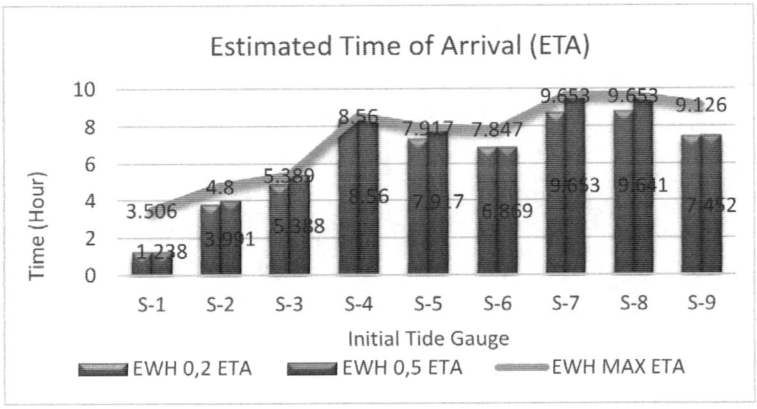

Figure 6. Estimated time of arrival (ETA)

5. Conclusions

The tsunami hazard map in West Kalimantan is based on tsunami modeling with an earthquake scenario in the Manila Trench. The results of tsunami modeling at artificial tidal measurement points around Gosong Beach showed low results because small islands were blocked so that the tsunami arrival time slowed down. Unlike those that are not blocked by islands or off the coast of the South China Sea, the tsunami height is high, and the arrival time is faster.

It is very important to conduct surveys of bathymetric data and field topography spread out, as a reference in conducting topographic data processing to combine elevation contours that have small correction values to satellite data, as well as for bathymetric data, especially the range limit of tens of meters along the coastline.

References

[1] Parliamentary Review 2020 *Transfer of National Capital*, ISSN 2656-923x.
[2] Ha D M, Tkalich P, Soon C E, & Megawati K 2009 *Journal of Asian Earth Sciences*, 36 pp. 67–73
[3] Megawati K, Shaw F, Sieh K, Huang Z, Wu T, Lin Y, Tan S K, and Pan T 2009 *Journal of Asian Earth Sciences*, 36 pp. 13-20
[4] Wang X, and Power W. 2011. GNS Science.
[5] Qiu Q, Li L, Hsu Y, Wang Y, Chan C, Switzer A D 2019 *Nat. Hazards Earth Syst. Sci,* 19 1565–1583
[6] Terry J P, Winspear N, Goff J, Tan, P H H 2017 *Earth-Science Reviews,*167 pp. 47-61
[7] Keiji N, Seiga K, Yoshito O, Kazuki O, and Kazuya Yoshida. 2013. Emergency Response to the Nuclear Accident at the Fukushima Daiichi Nuclear Power Plants using Mobile Rescue Robots. *Journal of Field Robotics* 30(1), 44–63
[8] Costas S, Utku K. 2015. the fukushima accident was preventable, Philosophical Transactions of the Royal Society A 373 (2053), 20140379, 1–23
[9] A. Omoto, Where was the weakness in application of defense-in-depth concept and why? in: J. Ahn, C. Carson, K. Jensen M, Juraku, S. Nagasaki, S. Tanaka (Eds.), Reflections on the Fukushima Daiichi Nuclear Accident, towards Social-Scientific Literacy and Engineering Resilience Springer Open, Springer Cham Heidelberg, Germany, 2015. Ch. P131-164.

Journal of Physics: Conference Series **2019** (2021) 012083 doi:10.1088/1742-6596/2019/1/012083

Bayesian Theorem Application to Model Reservoir Facies Distribution on Deltaic Depositional Environment, Case Study of Browse Basin

M M Adeyosfi[1] and A Harris[2]
[1]Reservoir Geophysics Graduate Program, Physics Dept, Universitas Indonesia
Jl. Salemba Raya no. 4, Jakarta 10430, Indonesia
[2]Physics Department, FMIPA, Universitas Indonesia
Kampus UI Depok, Depok 16424, Indonesia
Email: aharis@sci.ui.ac.id

Abstract. Successful Hydrocarbon exploration and reservoir characterization always related with good understanding of geology and geophysics aspect. One of the fundamental issues is to get reliable parameter distribution and quantify the confidence level of the parameter model in 3D. The case study in this paper will be applied in Browse field that and the result will be a distribution model in 3D of facies and hydrocarbon fluid. The workflow that will be introduced in this paper is the combination between rock physics analysis, simultaneous seismic inversion and Bayesian estimation theorem. Rock physics analysis includes well log conditioning and correlation analysis between reservoir parameter (porosity, saturation, Vshale, etc) with seismic parameter (acoustic impedance, vp/vs, shear impedance, etc) to obtain facies classification in well log scale. The simultaneous seismic inversion method is used to obtain seismic parameter cube to be correlated with rock physics result to drive the facies distribution. Bayesian estimation theorem assembles initial knowledge about a model before observing the inversion attributes. The probability density function later will be used to drive the facies distribution combined with well log data and seismic data; and also estimate the confidence level distribution in 3D. The result can be used to identify new play in the Browse field.

1. Introduction

Browse basin is a northeast-trending depocenter, located at offshore NW of Australia. This basin has high potential of gas hydrocarbon accumulation from Paleozoic, Mesozoic and Cenozoic sediments, but as yet unproduced. One of the reservoir rocks that consist hydrocarbon is sandstone of Plover formation from the early Jurassic (Figure 1). The Plover Formation was deposited during periods of active faulting during Jurassic extensions and has been, therefore, interpreted as a syn-rift succession (Struckmeyer et al., 1998 [1]; Blevin et al., 1998 [2]). The sandstone depositional environment of Plover Formation is mainly fluvial to tidally influenced deltaic system with some mouth bars (Tovaglieri & George., 2012 [3]). Consequently, thickness and lateral continuity of this unit are highly variable throughout the Browse Basin.

Due to variable lateral continuity and depositional environment behaviour, it is more difficult to distribute the facies distribution only with statistical method. In this paper the author used a combination of rock physics analysis, simultaneous inversion, and Bayesian theorem application to characterize the facies distribution in the studied area. The rock physics analysis consists of parameter extraction and sensitivity analysis to define the facies in well based. Continued with simultaneous inversion to characterize the physics parameter distribution in field scale of the studied area. And finally, with Bayesian theorem, multiple physics parameter from simultaneous inversion can be used to extract the facies distribution and probability cube in the studied area.

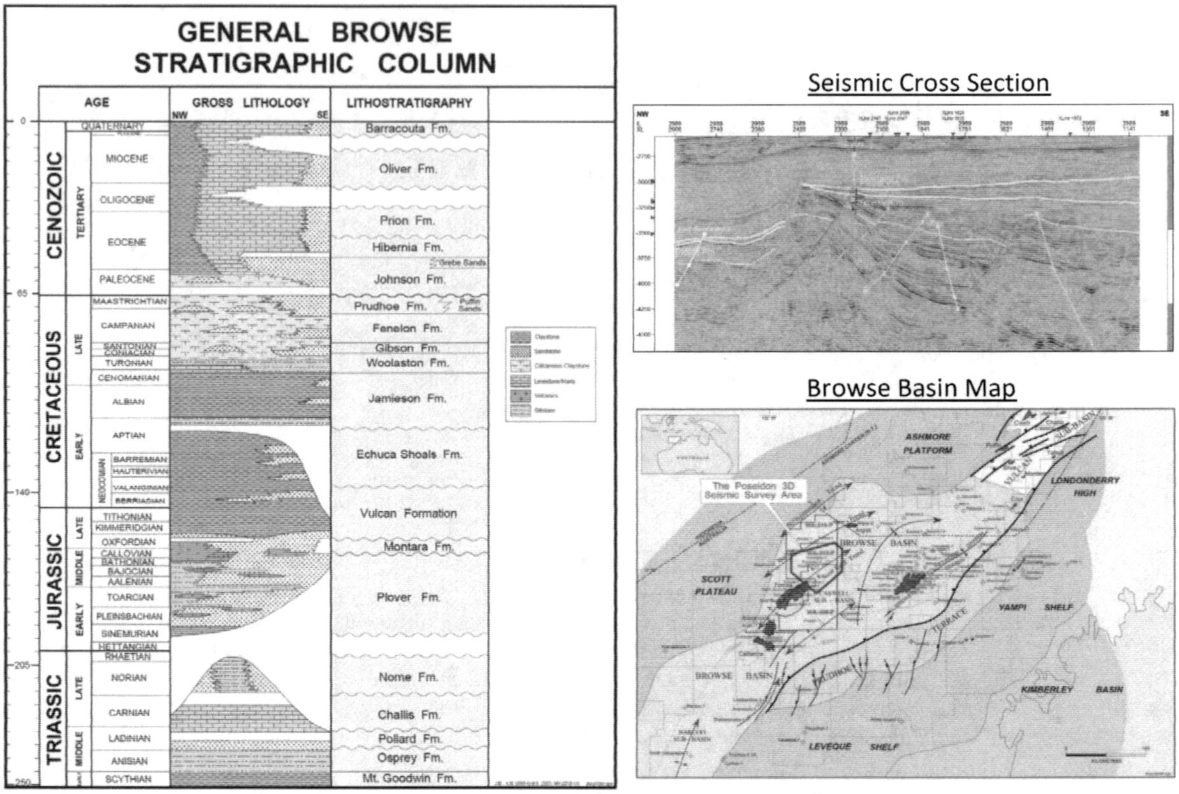

Figure 1. General Browse Basin Stratigraphic Column, seismic cross section and browse basin map [4]

2. Method

This study used angle stack 3D seismic and well log data as primary input. Other input such as velocity cube and borehole image log were used as cross validation and low frequency building supplementary data. The analysis is started with data loading and QC, continued with well tie, wavelet extraction, petrophysical analysis, borehole image processing, seismic interpretation, sensitivity analysis, simultaneous seismic inversion, and finished with the Bayesian theorem application. There will be facies distribution result from this analysis is a hydrocarbon bearing distribution cube.

2.1 Rock Physics

Rock physics method explains the reservoir rock in subsurface with physics properties such as porosity, rigidity, and compressibility. These properties are affecting on how the seismic wave travels trough formation. The objective of rock physics analysis in this paper is to find relationship between rock properties in subsurface with observed seismic response (Figure 2).

Journal of Physics: Conference Series **2019** (2021) 012083 doi:10.1088/1742-6596/2019/1/012083

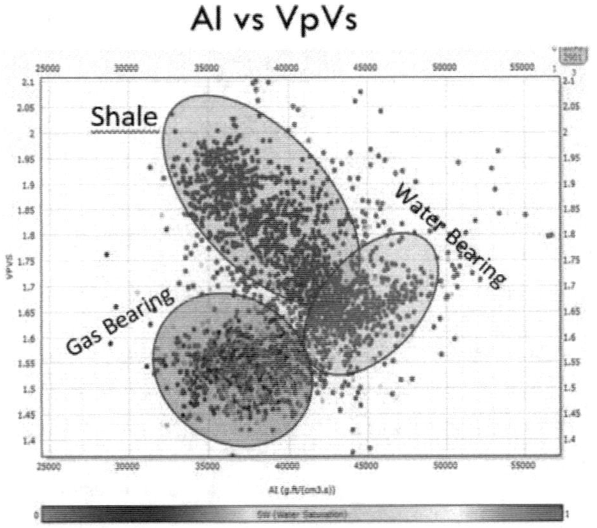

Figure 2. Accoustic Impedance (AI) and VpVs cross plot showing facies classification of gas bearing sand, water bearing sand and shale.

2.2 Simultaneous Inversion

Simultaneous seismic inversion method is an inversion technique with Aki-Richards formula to calculate the reflectivity in different offset in pre-stack seismic data or angle stack. Simultaneous inversion is used to characterize reservoir rock and fluid parameter to inversion result, such as, P-impedance, S-impedance, density, and VP/Vs. Shear wave (S wave) is being used to characterize fluid and compressional wave (P wave) that sensitive to change of fluid in the pore and S wave is related to interaction with rock matrix.

In seismology, seismic wave reflectivity and transmission in bedding boundary is formulated by many scientists, one of them is equation formula from Knott-Zoepprits. This equation is simplified by Aki-Richards, known by Aki-Richards equation (1980) [5]:

$$R(\theta) \approx \frac{1}{2}\left(\frac{\Delta V_p}{V_p} + \frac{\Delta\rho}{\rho}\right) - 2\left(\frac{V_s}{V_p}\right)^2\left(2\frac{\Delta V_s}{V_s} + \frac{\Delta\rho}{\rho}\right) \text{x}\, sin^2\theta + \frac{1}{2}\frac{\Delta V_p}{V_p}\tan^2\theta \tag{1}$$

The assumption that being used is change of property ($\Delta Vp/Vp$, $\Delta Vs/Vs$ and $\Delta\rho/\rho$) relatively small, hence second order can be ignored and Θ with a value less than 90° (Ma, 2002 [6]). Equation (1) can be simplified with S wave impedance and P wave impedance parameter as below:

$$R(\theta) \approx (1 + \tan^2\theta)\frac{\Delta I_p}{2I_p} - 8\left(\frac{V_s}{V_p}\right)^2 \sin^2\theta\,\frac{\Delta I_s}{2I_s} - \left[\tan^2\theta - 4\left(\frac{V_s}{V_p}\right)^2 \sin^2\theta\right]\frac{\Delta\rho}{2\rho} \tag{2}$$

Fatti et al [7] in 1994 simplified the (2) equation by removing ρ parameter where the Vs/Vp ratio is around 0.5 and small angle as below:

$$R(\theta) \approx (1 + \tan^2\theta)\frac{\Delta I_p}{2I_p} - 8\left(\frac{V_s}{V_p}\right)^2 \sin^2\theta\,\frac{\Delta I_s}{2I_s}. \tag{3}$$

During the simultaneous inversion, one of the algorithms that's being used is simulated annealing. A simulated annealing algorithm is a vector solution that optimized with a*n* acoustic impedance value, followed by *n* shear impedance value, where *n* is amount of layers. The algorithm will search for lowest cost function with Monte-Carlo method (Figure 3).

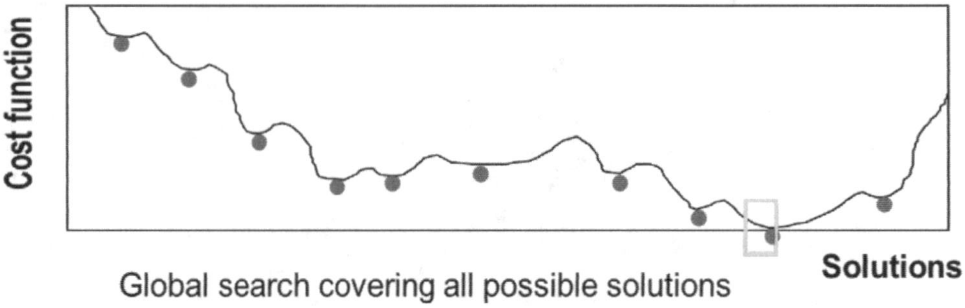

Figure 3. Global search covering possible solutions with Monte Carlo in simulated annealing algorithm [8]

The given initial model is reflection coefficient of zero-offset P and S impedance at interface I can be calculated by the equation below:

$$\frac{\Delta I_p}{2I_p} = \frac{I_p^i - I_p^{i-1}}{I_p^i + I_p^{i-1}}, \qquad \frac{\Delta I_s}{2I_s} = \frac{I_s^i - I_s^{i-1}}{I_s^i + I_s^{i-1}} \tag{4}$$

Hence the ratio of average Is/Ip between layer i-1 and layer I can be calculated by the equation below:

$$\frac{I_s}{I_p} = \frac{I_s^i + I_s^{i-1}}{I_p^i + I_p^{i-1}} \tag{5}$$

By replacing the equation (5) and (4) into the equation (3), reflection coefficient R (Θ) can be calculated in every offset and every layer boundary.

2.3 Bayesian Theorem

A Bayesian theorem in short understanding can be viewed as conditional probabilistic. The theorem itself stated if we have two events x and y, we can design the probability of occurrence of each event by *P(x)* and *P(y)*. If we consider of the probability of these two events, we can design this as joint probability of *(P(x,y))* that is given by:

$$P(x,y) = P(x|y)P(y) = P(y|x)P(x) \tag{6}$$

In This case, *P(x/y)* is the probability of x to occur given that y already occurred, the same situation applied to *P(y/x)* and *P(y/x)* can be defined as conditional probability. From equation (6), if we want to know how the probability of y is, given x already happened, we can re-arrange the equation into:

$$P(y\,|\,x) = \frac{P(x\,|\,y)P(y)}{P(x)} \tag{7}$$

To complete equation (7), according to the total law of probability:

$$P(x) = P(x|y)P(y) + P(x|y^c)P(y^c) \tag{8}$$

In equation (8) above, the *P* (y^c) is defined as the probability of not having *y* to be happening. Therefore the complete Bayesian equation to define the conditional probability of *y* to be happened given *x* that already happened is:

$$P(y\,|\,x) = \frac{P(x\,|\,y)P(y)}{P(x\,|\,y)P(x) + P(x\,|\,y^C)P(y^C)} \tag{9}$$

The product of Bayesian calculation is a posterior PDF from each alternative. Below is a probability density function (PDF) for hydrocarbon bearing facies. Hydrocarbon bearing facies has low AI and low Vp/Vs ratio (Figure 4).

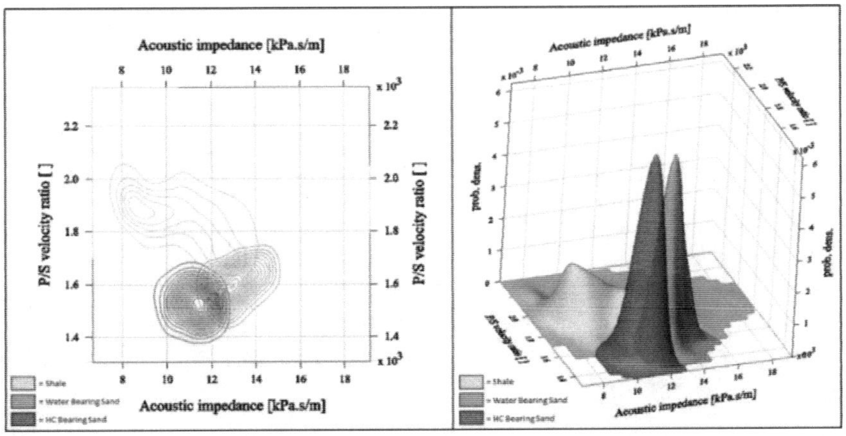

Figure 4. Probability Density Function (PDF) for HC bearing, water bearing and shale facies

3. Result and Discussion

3.1 Seismic Inversion Result

The simultaneous seismic inversion was done in Plover formation only as target formation. The results are acoustic impedance cube (AI), Vp/Vs cube and density cube. The relative misfit of the inversion result is relatively low which is 27% and the residual cube has low amplitude means the simultaneous seismic inversion result is valid to be used for further analysis.

The acoustic impedance in the Plover formation is relatively low compared to Triassic formation and there are some low acoustic impedance reflecting high permeability – gas bearing facies or shale facies. The Montara formation above Plover formation has a mostly lowest acoustic impedance related to shale facies developed in this formation. The cross section with well data in figure 5 below shows good coherency between acoustic impedance from simultaneous seismic inversion with acoustic impedance from well data.

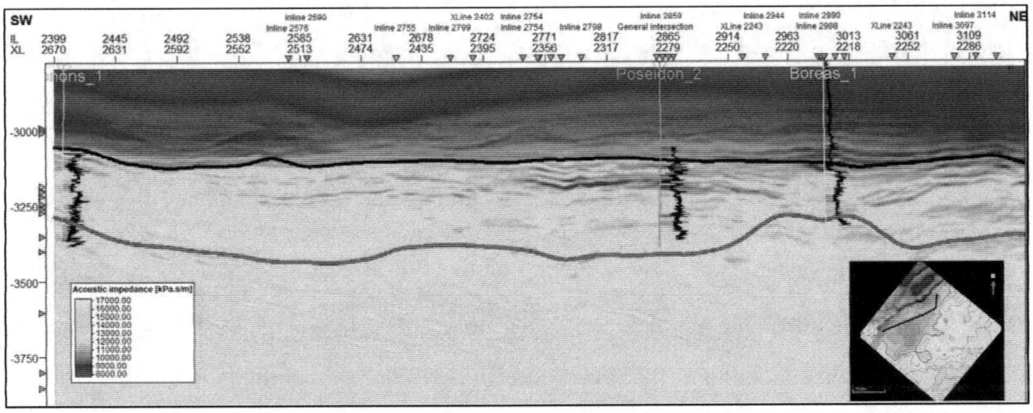

Figure 5. SW-NE cross section of the Acoustic Impedance Cube with Kronos-01, Poseidon-02 and Boreas-01, shows good coherency between inversion result and well data

Another simultaneous inversion result is Vp/Vs cube. NE-SW cross section of Vp/Vs cube with some wells at figure 6 below shows good coherency between the inversion result with Vp/Vs log data from well. The Vp/Vs in Triassic and Montara formation mostly have high value. The Plover formation Vp/Vs is relatively low compared to Triassic formation and Montara formation, with some very low Vp/Vs value at lower Plover formation related to gas bearing sand facies.

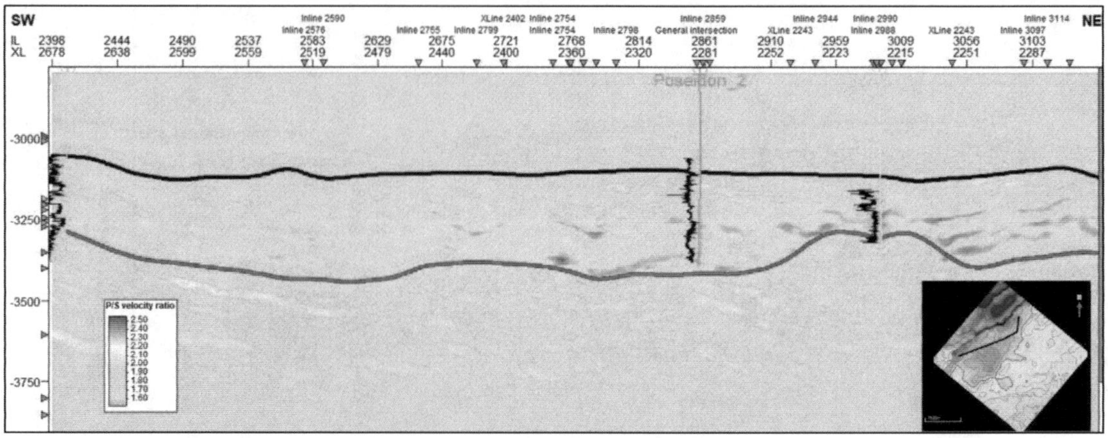

Figure 6. SW-NE cross section of a VP / Vs Cube with Kronos-01, Poseidon-02 and Boreas-01, shows good coherency between inversion result and well data

3.2 Hydrocarbon Bearing Facies Distribution

The probability density function of the gas bearing facies, water bearing facies and shale from acoustic impedance and Vp/Vs cross plot is being used to predict the hydrocarbon bearing facies distribution. Similar to sand facies distribution that act as a reservoir, the hydrocarbon bearing facies are mostly distributed at the lower part of the Plover formation. Based on the NE-SW cross section at figure 7 below, most of the highest hydrocarbon bearing facies probability is at the lower part of the formation. Cross validation with well log data such as water saturation shows good coherency, hence the facies prediction with Bayesian theorem is valid.

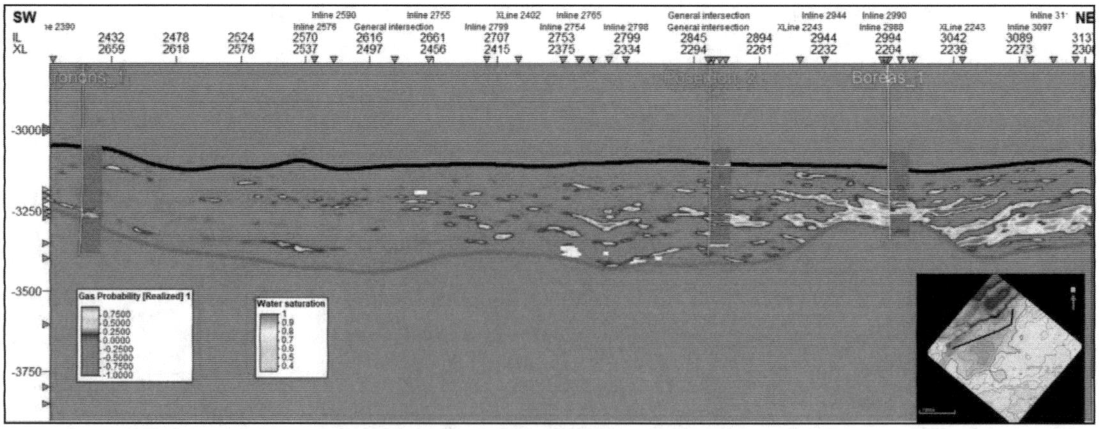

Figure 7. SW-NE cross section of hydrocarbon bearing facies probability Cube with Kronos-01, Poseidon-02 and Boreas-01, shows good coherency between Bayesian theorem and well data (water saturation)

The probability cube can be extracted into horizon probe (geobody) where the facies distribution in field scale can be seen clearly. The hydrocarbon bearing facies distribution shows gas accumulation mostly on the NE part of the field, similar to mud log data and petrophysical analysis result. The geometry of the hydrocarbon bearing facies distribution shows that the sand with gas bearing only at NE part, the sand facies at SW part is mostly filled with water (Figure 8).

Probability Map of Gas Bearing Facies at Plover Fm	Gas Bearing Facies Distribution with accumulation sketch

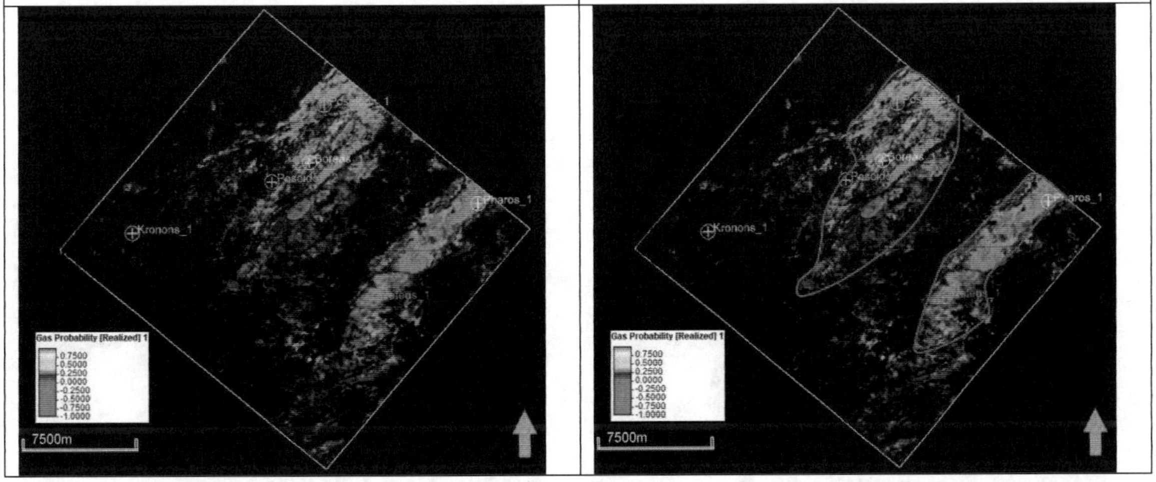

Figure 8. Probability Map of Hydrocarbon Bearing Facies Distribution shows gas mostly accumulated at NE part of the field

4. Conclusions

The application of combination between rock physics, simultaneous seismic inversion and Bayesian theorem application is proven to be a powerful tool to predict facies distribution in tectonically active and poor facies lateral continuity condition compared to geostatistical method. The studied area shows the sand distribution is NE-SW trend similar with browse basin axis with geometry as a channel. The gas accumulation is mainly on the NE part of the field.

Acknowledgments

Authors wishing to acknowledge my colleagues, Abdul Kholiq, Dyna Mariana, Ian Kainama & Levina Aurelia Wirjowedojo, whose given author guidance and input during the analysis process. The author also would like to express gratitude to Australia government (National Offshore Petroleum Information Management System) who provide author this wonderful published data.

References

[1] Struckmeyer 1998 *The Sedimentary Basins of Western Australia 2: Proceedings of the Petroleum Exploration Society of Australia Symposium,* pp. 345-367
[2] Blevin 1998 *The Sedimentary Basins of Western Australia 2: Proceedings of the Petroleum Exploration Society of Australia Symposium,* pp. 369-395
[3] F. Tovaglieri and A. George 2012 *AAPG Annual Convention and Exhibition.*
[4] ConocoPhillips (Browse Basin) Pty Ltd 2009.
[5] K. Aki and P. G. Richards 1980 *Quantitative Seismology: Theory and Methods,* (San Fransisco: W.H. Freeman Company), p. 932pp.
[6] X. Q. Ma 2002 *Geophysics,* 67(6) pp. 1692-2041
[7] Fatti 1994 *Geophysics,* 59 pp. 1362-1376
[8] Schlumberger 2019 *Petrel Quantitative Analysis Training Book,* (Schlumberger)

Identification and Determining The Pseudo Magnitude Of Meissa Star Based on Observation Data From The Eastern Sky In The Mentigen Hills of The Pseudo Magnitude of Meissa Star Based on Observation Data From The Eastern Sky

Sutrisno[1], T J R Fadilla
Department of Physics Education, University State of Malang
Jl. Semarang No.5, Sumbersari, Kec. Lowokwaru, Kota Malang, Jawa Timur 65145
Email: [1]sutrisno.fmipa@um.ac.id

Abstract. Mentigen Hill, Mount Bromo is a tourist area located at an altitude of 2,392 meters above sea level. As a tourist destination in Peak Bromo, the sky at night has almost no light pollution, it will be very good for observing celestial bodies. For this reason, identification and measurements of the pseudo magnitude of star observations are made using a camera. Meissa star is a blue giant star type O which is included in the Lambda Orionis. The rate at which a star is observed from Earth is called its apparent magnitude. Analysis of the identification and measurement of the pseudo magnitude of the stars can be done using the Iris 5.59 software. Based on the results of the analysis in the Iris 5.59 software, it was concluded that the Missal Star was identified and the apparent magnitude measurement was (3.87 ± 1.10) with an error of 2.39%. It is suggested by the results of this study to develop astronomical tourism.

1. Introduction

A clear night sky can show the beauty of a radiating celestial body. Seen from the earth with the naked human eye, there are many celestial bodies with different levels of brightness. Humans see the existence of a variety of material organizations in the form of various celestial bodies in the universe. Examples of celestial bodies such as the earth, solar system, planets, satellites, nebulae, asteroids, meteoroids, comets, space dust, black holes, galaxies and various stars. All celestial bodies appear through light that can be detected with the naked human eye or by using optical binoculars[1]. The light that is detected is not all emitted from the celestial body itself. Some of the celestial bodies receive light reflection from other celestial bodies[2].

Stars are one of the celestial bodies that emit light in the universe [3]. The number of stars is very large, but there are some of the largest stars in this universe such as Aldebaran, Rigel, Betelgeuse, and Antares and so on. The stars do not spread evenly in the universe, but cluster together to form constellations. Meissa star is included in the Orionis lambda which is one of the stars in the Orion constellation. This star is a blue giant type O (blue giant type O) and is part of the cylinder 69 star group. Meissa or Heka comes from Arabic "Al-Maisan" which means "shining one". Heka also comes from Arabic, "Al Hakah", which means "white dot" comes from the additional dim light from the smaller φ1 and yang2 in the background. [4]. The star Meissa has a magnitude of 3.33 and a distance of 1098.18 light years. This star has coordinates + 83°58'43,7 "/ - 13°22'00,7" on Friday, November 1 2019 at 23:16 WIB [5]on the Mentigen Hill, Mount Bromo. Mount Bromo is a good location for stargazing because it is far from cities and settlements so that light pollution disturbances that may occur are very small[6].

The brightness of the stars can be captured using a telescope and an 8 MP digital camera. The brightness level of a star is its magnitude. The magnitude scale is the distance between the earth and the star. The magnitude of a star that can be observed from the earth is called the apparent magnitude[7]. The apparent magnitude of a star can be determined from calculations in the Iris 5.59 software.

2. Method

The method in this observational, descriptive study uses quantitative methods to determine the magnitude of a star using the Iris 5.59 software which refers to the Stellarium software as the reference sky map. The magnitude of the star mesa can be seen from the star Betelgeuse as the reference star. Data collection for this research was carried out using a digital camera which was conducted at Mentigen Hill, Mount Bromo on Friday, November 1, 2019 at 23:16 WIB. The procedures carried out in this study are as follows:

1. Preparing and setting up tools, namely digital cameras.
2. Mounting a tripod along with a Nikon D5500 camera.
3. Check the quality of the sky that will be observed with a sky quality meter.
4. Determine the cardinal directions using a compass.
5. Pointing the camera at the part of the sky that is observed
6. Set the camera to infinity focus, focal length 70.0 mm, diaphragm 5.6, shutter speed 13 "and ISO 16000.
7. Then press the shutter button to take a picture of the stars on Mentigen Hill.
8. Record the information obtained in the camera.
9. Information includes photo, name, exit data, time and part of the sky observed.
10. Analysis with *Stellarium software* and Iris 5.59 software.

To make it easier to understand the research carried out, it is illustrated by the following diagram:

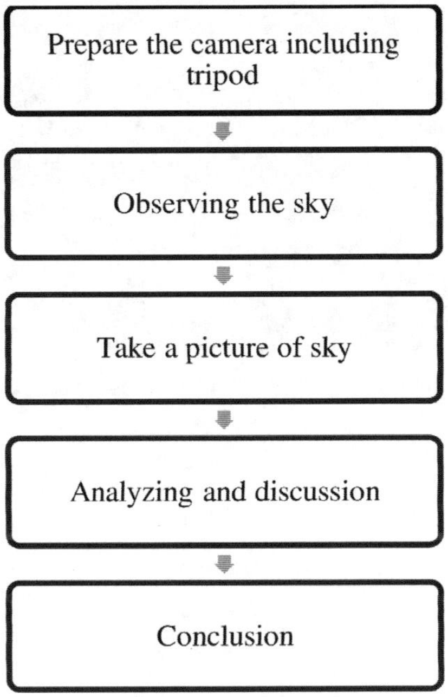

Figure 1. Research Flow Diagram

Journal of Physics: Conference Series **2019** (2021) 012084 doi:10.1088/1742-6596/2019/1/012084

From the research results obtained data, such as the table below:

Table 1. Format Of Observation Data Results

No.	Format	Information
1	File Type	JPEG
2	Sky Quality	20.79 mags / Ac sec2
3	Time	Friday, November 1, 2019, at 23:16 WIB
4	Location	Mentigen Hill, Mount Bromo, East Java, Indonesia
5	Location Coordinates	7 ° 55'39 "S 112 ° 58'09" E
6	Camera	Nikon D5500
7	*Focal Length*	70.00 mm
8	*Diaphragm*	5.6
9	*Shutter Time*	13 "
10	*ISO*	ISO16000

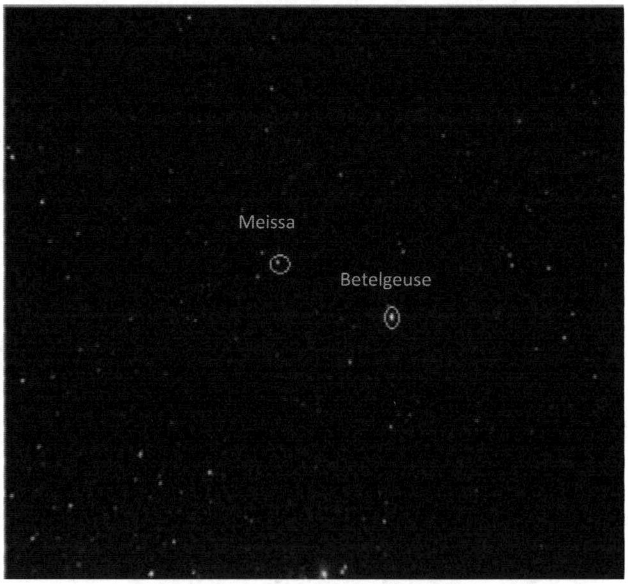

Figure 2. Observation Result Data

3. Discussion

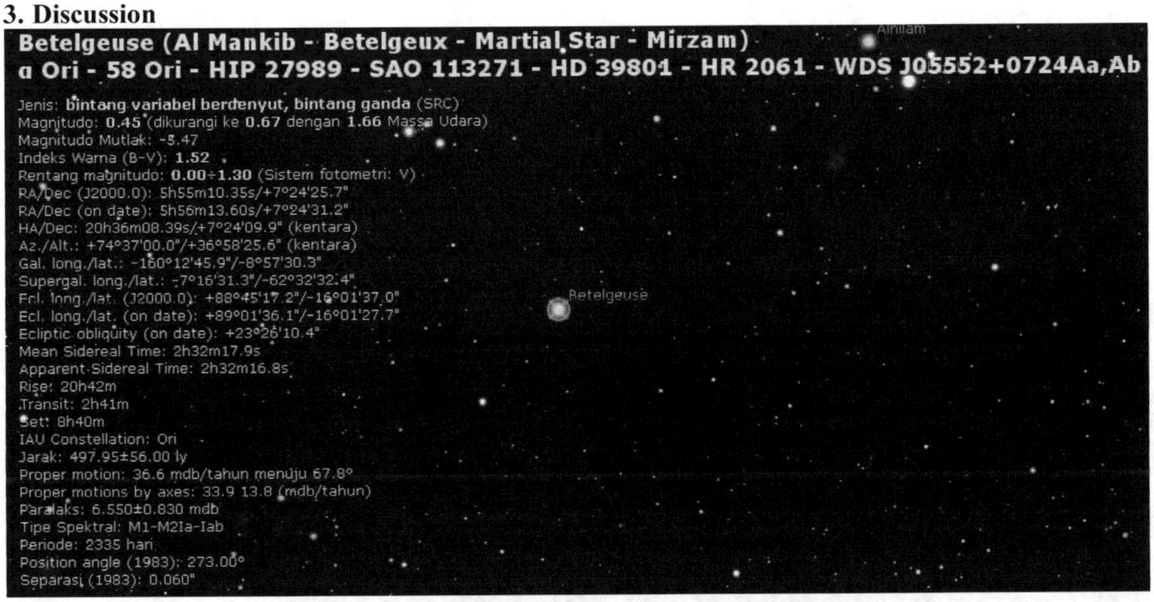

Figure 3. Betelgeuse Star Information From Stellarium

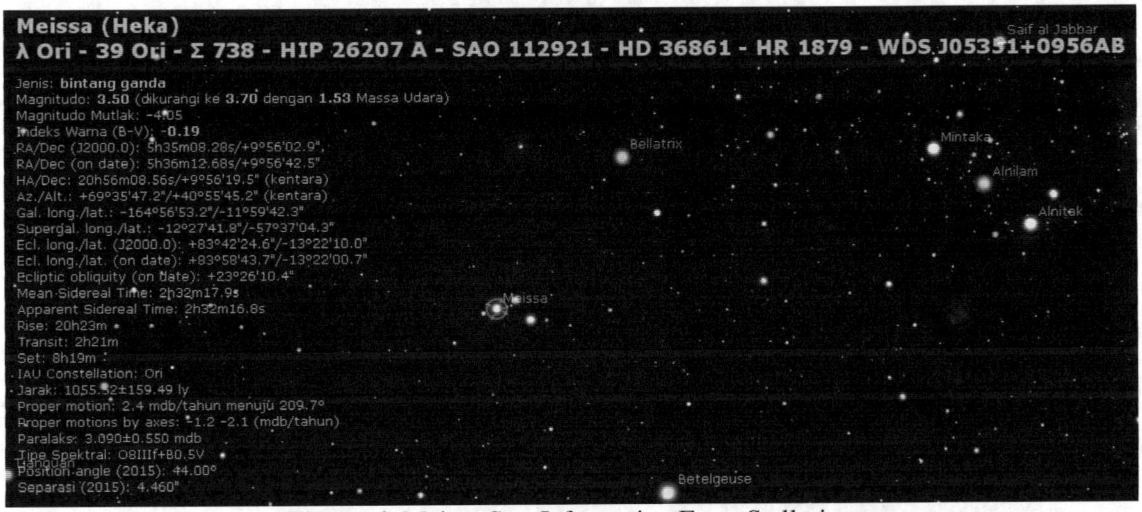

Figure 4. Meissa Star Information From Stellarium

3.1 Determining the Magnitude of the Star Betelgeuse

Based on the information from Stellarium, it is known that the magnitude value of the Betelgeuse star is 0.45. This obtained magnitude is used as a calibration in determining the comparative magnitude constant between the results of the apparent magnitude analysis of the Betelgeuse star on Iris 5.59 software and the information obtained from Stellarium. To determine the magnitude constant of the Betelgeuse star, it is necessary to calibrate the measuring instrument in the Iris 5.59 software by determining the radius of the circle on the photometric aperture setting using two digital rings with a ratio of 1: 6. This radius ratio affects the intensity value of the Meissa star on the inner ring and the intensity of the surrounding sky on the outer ring. Then click on the Betelgeuse star which is at coordinates (359,244). Then set the magnitude constant value to 0 in the Iris 5.59 software (as shown in Figure 4) so that it can be adjusted to the apparent magnitude of the Betelgeuse star on the Stellarium of 0.45 as a reference. The apparent magnitude constant value of the Betelgeuse star in the Iris 5.59 software is 7,997 obtained from calculations using the formula:

$$y = x + c \tag{1}$$

Where

y : magnitude pseudo-star Betelgeuse on the Stellarium

x : Big Betelgeuse star apparent magnitude constant in Iris 5.59 software

c : Big Betelgeuse star apparent magnitude constant in Iris 5.59 software when the magnitude is constant 0

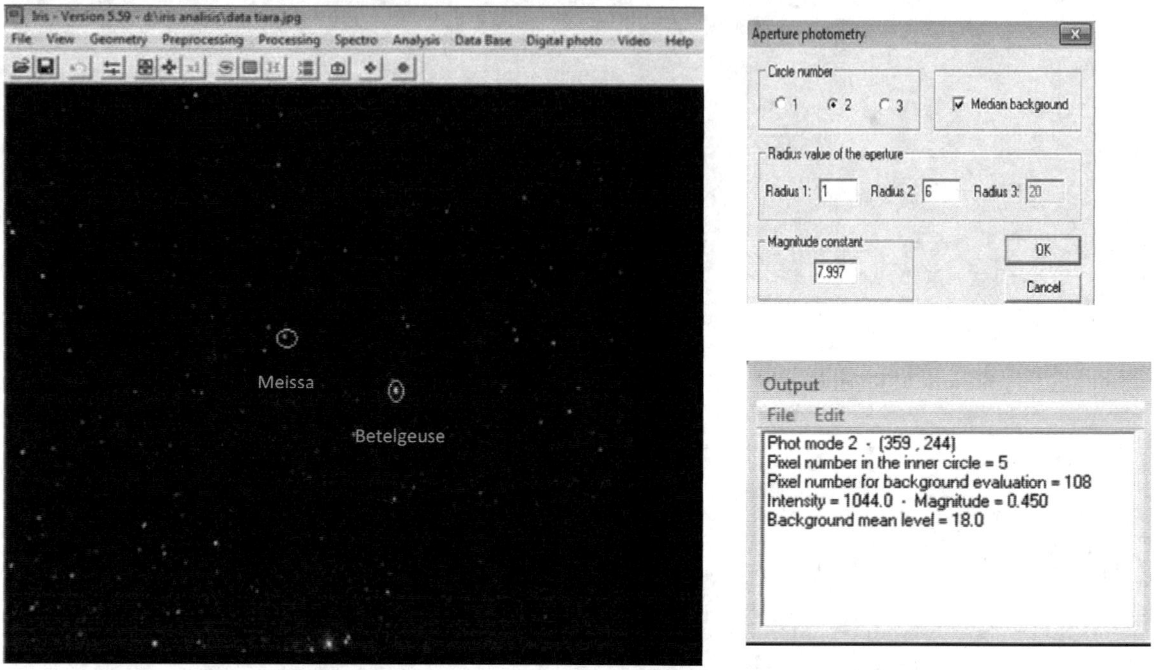

Figure 5. Iris Settings For Determining Betelgeuse Star Magitudo

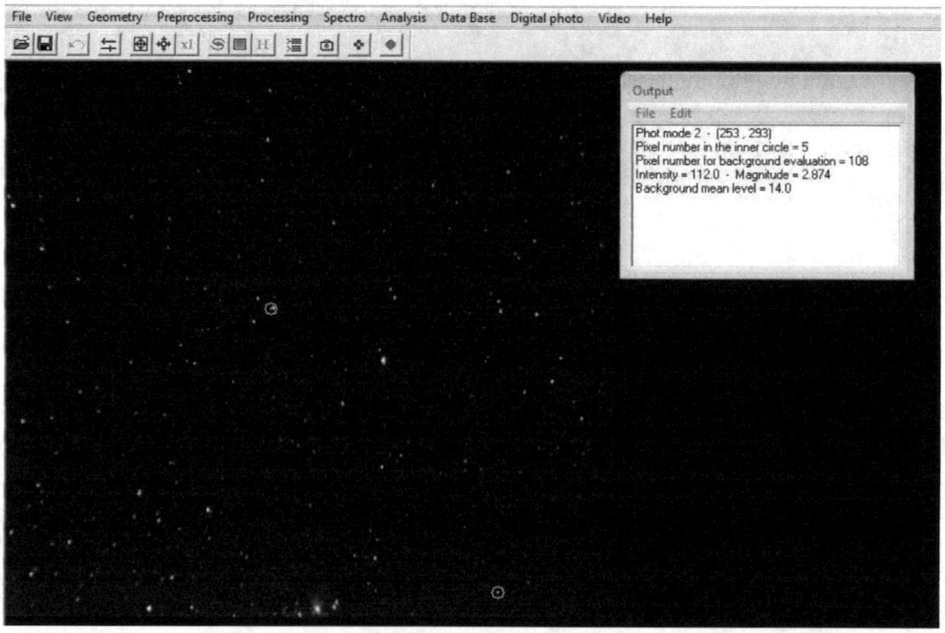

Figure 6. Iris Arrangements To Determine The Magnitudo Star Meissa

Table 2. The Value Of Meissa Star Magnitude

No.	Coordinate	Pseudo magnitude
1	252, 294	3,281
2	255, 290	3,864
3	256, 292	2,579
4	257, 293	3,570
5	254, 297	2,679
6	253, 293	2,874

To determine the apparent magnitude of the Meissa star, an analysis was carried out by determining several points for which the magnitude value would be known by placing the radius of the circle of the photometric aperture. Using the reference to the magnitude constant obtained previously from the standard star. After obtaining several apparent magnitudes from different points on the Meissa star as in the table above, the mean value of the apparent magnitude can be calculated. 3,144167. Then the standard deviation is also searched using equation 8 which gives the result of 0.511972. The apparent magnitude analysis result value of Iris 5.59 software is compared with the apparent magnitude value of *Stellarium* so it can be seen the level of measurement error. So that the apparent magnitude value of the Meissa star is (3.14 ± 0.51) with an error of 2.60% (less than 5%). From the results obtained, it is known that the apparent magnitude value obtained from the calculation is almost close to the magnitude value of Stellarium.

4. Conclusions

Based on research sky, which is conducted On the Mount Bromo Mentigen hill, data is obtained in the form of a picture of the night sky on Friday, November 1, 2019 at 23:16 WIB. The apparent magnitude value of the star Meissa is (3.87 ± 1.10) with an error of 2.39% (less than 5%) with the reference value for the magnitude of Stellarium. The value resulting from the analysis is not that far off. This can prove that determining the apparent magnitude of a star can be done through calibrated using the Iris software.

Reference

[1] Raharto M. Universe, Humans, and the Qur'an. Pulpit. 2004; 20 (1).
[2] Prime. Earth Prayer. J Chem Inf Model. 2018; 53 (9): 1689-1699.
[3] Ghozali MI. Revealing the Interpretation of the Stars in the Qur'an and Astronomy. Published online 2016. http://digilib.uinsby.ac.id/id/eprint/12966
[4] LacusCurtius • Allen's Star Names - Orion. Accessed March 2, 2021. https://penelope.uchicago.edu/Thayer/E/Gazetteer/Topics/astronomy/_Texts/secondary/ALLSTA/Orion*.html
[5] Stellarium-web.org. Stellarium Web Online Star Map. Published 2021. Accessed February 22, 2021. https://stellarium-web.org/
[6] Bramastartya j. Educative and recreative observatory with a critical regionalism approach in gunungkidul, yogyakarta special region. Published online 2017: 1-24. http://e-journal.uajy.ac.id/id/eprint/11371
[7] Tyonila, Cicik C Sophisticated Dwi; Sutrisno; P. NA. Measurement of the Quasi-Magnitude of the Planet Venus in the Quarter Phase Using Iris Software Version 5.59 at the Astronomy Laboratory of the State University of Malang in April 2014. 2014; (April): 5.

Integration of Borehole Image and Sonic to Evaluate Critically-Stressed Fractures to Optimize Production at FORGE Geothermal Field

E Sulistyowati[1, a] and A Haris[2, b]
[1]Reservoir Geophysics Graduate Program, Physics Dept, Universitas Indonesia
Jl. Salemba Raya no. 4, Jakarta 10430, Indonesia
[2]Physics Department, FMIPA, Universitas Indonesia
Kampus UI Depok, Depok 16424, Indonesia
Email: [a)]emiliana.sulistyowati@outlook.com; [b)]aharis@sci.ui.ac.id

Abstract. Geothermal integration solutions as unconventional environment have workflows or methods that must be able to optimize existing data and be adapted to overcome existing challenges. The integration method consists of geology, Petrophysics and Geomechanics domains. Geological analysis to identify fractures, fracture classification, knowing the dip azimuth and the magnitude of the fracture, quantitative analysis (intensity, aperture and porosity) by using the wellbore image log. Petrophysics analysis is to distinguish open / healed fractures that analysed from sonic data. Fracture orientation analysed further for the stress calculation on fracture planes and to determine the effective fractures (most likely to flow) that has a high ratio of shear to normal stress. Geomechanical analysis in geothermal fields, among others, is to determine the dynamic permeability behaviour of fractures during the production / injection phase, determine the orientation and productive fracture interval. The integration is using FORGE well 21-31 in 8.5in section with depth interval 6,226 – 7,920 ft-MD and well 58-32 in 8.5in section with depth interval 7,390-7,527 ft-MD. Well 21-31 shows dominant fracture strike orientation to NNE-SSW, while Well 58-32 shows slightly different set strike orientation to N-S with minor striking to NE-SW and NW-SE. Evaluation of critically stressed fracture intended to have knowledge on production mechanism and based on analysis, fracture/fault that has high potential of shearing is striking to NNE-SSW and dipping to Westerly.

1. Introduction

This paper presents a case study of borehole image and sonic data to understand fracture characterization of geothermal field that needed to evaluate the critically stressed fracture within the study interval. These fractures in many cases are showing enhanced permeability due to shear dilation and lie above the failure line in normal and shear stress spaces. Evaluation of such fractures require an understanding of the stress state, fracture geometry, mineralogy of infill material, and fracture slip properties like coefficient of friction (Anzar et al., 2011 [1]). Fracture characterization is obtained from the analysis of log data, borehole images and Stoneley sonic data. Stress state to obtain three principal stresses (vertical stress, maximum horizontal stress and minimum horizontal stress) and the direction of maximum horizontal stress. Integration of the density log provides the vertical stress in most cases. The minimum and maximum horizontal stress is determined using drilling-induced fracture and breakout information from borehole image as no drilling data, LOT test and mini-freak test are available. The US Department of Energy's (U.S. DOE) Frontier Observatory for Research in Geothermal Energy (FORGE) is a field laboratory built as an ESG (Enhanced Geothermal System) laboratory. This EGS area has natural fractures and very small porosity values that make it possible to regulate water circulation for heat extraction. Utah FORGE is situated in a gently sloping alluvial area, between the peaks of the Mineral Mountains to the east and the central north to west of the Milford valley. The site is located on the southeastern edge of the Great Basin in a broad zone characterized by increased heat flow (Blankenship, D et al., 2018 [2]). The Fallon FORGE, Nevada area located in the Carson Sink is a large half-Graben basin, similar to many locations throughout thc Grcat Basin Region.

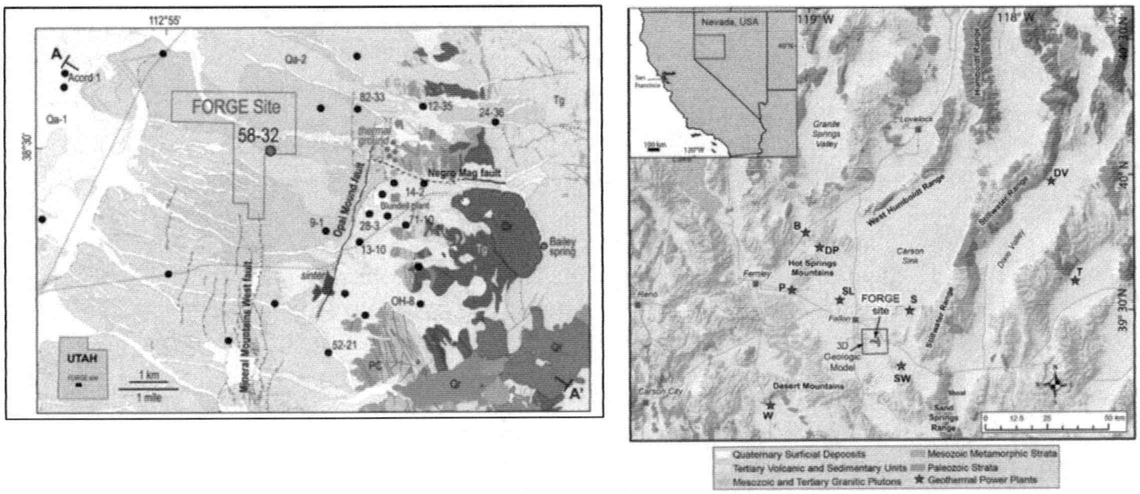

Figure 1. FORGE Location Area: well 58-32 (Left) and well 21-31 (right).

2. Geologic Information about FORGE Field

The regional stratigraphy consists of Paleozoic-Mesozoic folds and carbonate imbrications which are the result of Tertiary-current mafic-felsic magmatic eruptions, including in the Mineral Mountains, and are widespread with extension patterns in basins and mountains. The local lithology is divided into two broadly defined units, consisting of a crystalline bedrock and an overlying basin filled with sedimentary deposits (Moore, J. et al., 2018 [3]).

2.1. Geological Setting of Well 58-32

In 2018, well 58-32 was drilled vertically to a total depth of 2290 m (7515 ft) GL where at 961 m (3154 ft) GL contact with a low permeability crystalline rock and a bottom hole temperature of 199°C (390°F). The well penetrates layered alluvium deposits to 3176 feet (968 m), where it crosses contact with crystalline basement rock to the bottom of the wellbore. The upper interval of layered alluvium (0–3176 ft; 0–968 m) consists of poorly sorted and poorly lithified sediments made of quartz and feldspar eroded from plutonic rocks in the Mineral Mountains. The lower interval (3176–7536 ft; 968–2298 m) consists of granitoids (i.e. Igneous rocks), petrographically identical to those exposed in the Mineral Mountains (Figure 2). These crystalline rocks form strong low porosity reservoir rocks, ranging from granite to monzodiorite to monzodiorite quartz in composition, plus minor dirt. Elite and chlorine are the main clay minerals, but make up <5% of the rock. Other secondary minerals include carbonates and anhydrides, and fractures at intervals that are casted are filled with chlorite or epidote (Moore et al., 2020 [4]).

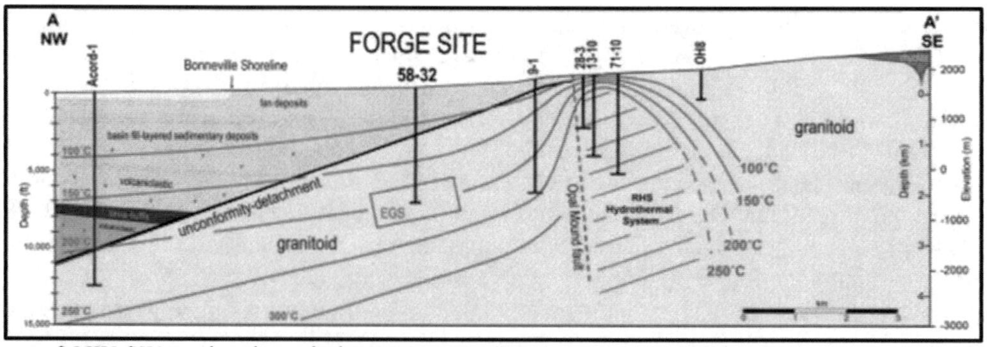

Figure 2 NW-SW section through the UTAH FORGE site showing Well 58-32 location and deposits.

2.2. Geological Setting of Well 21-32

The Fallon FORGE site located in the Carson Sink is a large half-Graben basin, similar to many locations throughout the Great Basin Region. The general stratigraphy of the area is Late Miocene to Quaternary sedimentary basin (<1.5 km thick), overlying Miocene volcanic rocks (0.7-1.3 km thick), and Mesozoic basement consisting of meta-volcanic and meta-volcanic rocks. Sediments breached by a full-size Pluton. This area is located in the western part of the half-graben fold which is widely intersected by normal faults trending north with a displacement of about 200 m (Siler et al., 2018, Figure 3 [5]). The primary rock units that fill the basin found at the Fallon location are: a) Quaternary sediments, b) Quaternary-Tertiary sediments, c) Miocene mafic volcanic rocks, mainly basaltic andesite, but also lithic Tuff, andesite, ryodsite, and volcanic breccias. The infill rock of the basin is located unconformably at the top of the Mesozoic basement, which consists of: a) Mesozoic quartz monzonite, b) Mesozoic meta-basalt and meta-basaltic-andesite, c) Mesozoic quartzite with little pilot, meta-basalt, and marble, d) Mesozoic meta-Rhyolite with lower meta-basalt. The deepest well in the area ending in the Mesozoic basement at depths of 2112-2886 m recorded a maximum downhole temperature of 194 to 214 °C (Blankenship, 2016, Siler et al., 2016 [6]).

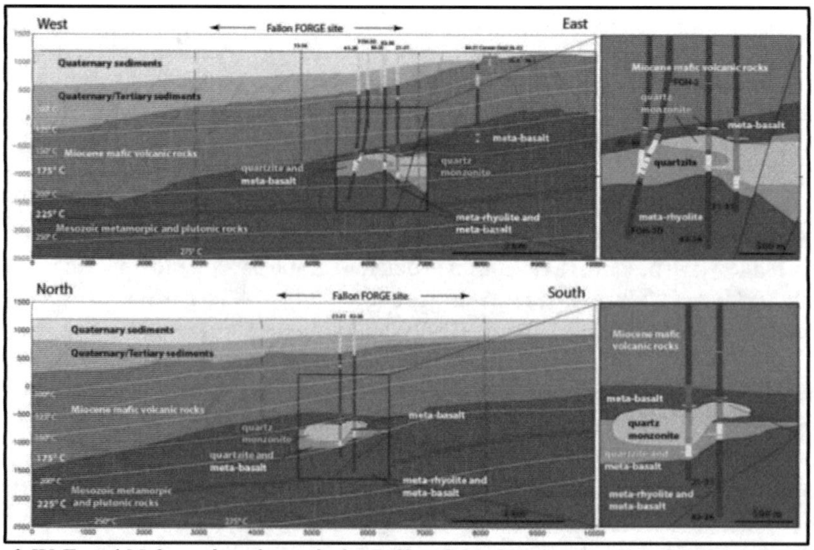

Figure 3 W-E and N-S section through the Fallon FORGE site showing Well 21-31 lithology

3. Method

The flow of research to be carried out is explained as follows: Geological analysis is obtained from borehole image (FMI) data and core data. Petrophysical analysis was obtained from log data (gamma ray, resistivity, neutron porosity, and density) and Stoneley sonic wave. Geomechanical and critically-stressed analysis is the integration of all existing analysis results.

3.1. Borehole Image Interpretation

FMI (Fullbore Formation Microimager) is image data that provide microresistivity formation images for geological, geomechanical and 3D reservoir modeling analysis. FMI imaging is an approach to determine subsurface image data and the direction of layer slope, fracture, fracture. FMI* images are generated with two types of resistivity scales, whereas static and dynamic images. Natural fractures that classified and picked in this study only conductive fracture are which is considered to be open / filled with conductive minerals. A fracture is a planar feature without displacement along the fracture plane. Shear fractures and/or drilling induced fractures are also found throughout the interval and occur in clusters. These fractures are not natural and are related to drilling activities. The orientation of the current geological horizontal stress can be determined by measuring the orientation of the induced fracture as well as the breakout hole in the vertical or near-vertical wall. These features tend to be localized and run vertically down the sides of the borehole and against each other. The orientation of the breakout hole can indicate the orientation of the minimum horizontal stress (σHmin) on the borehole wall, and the orientation of the induced fracture can indicate the direction of the maximum horizontal stress (σHmax). Further quantitative analysis was performed to determine fracture properties: fracture density (corrected), fracture aperture (FVAH) and fracture porosity (P33).

3.2. Sonic Stoneley Wave Analysis

To determine the "openness" of the fracture, two approaches have been used by integrating FMI image and Sonic Scanner data. A reflective analysis of Stoneley was carried out and compared with selected fractures in FMI. When the exposed fracture cuts through the borehole, Stoneley causes well fluid to be pumped in and out of the fracture, dissipating energy. This results in attenuation of the Stoneley wave. The dipole emitter is so strong and directional that it has a strong sensitivity to "open" and "wide" fractures. The fracture zone interval at which significant energy loss was observed was used to identify open and wide fracture intervals. At the same time, a change in acoustic impedance causes a partial reflection of the signal. The greater the energy loss (high attenuation), the wider the propagation of the fracture. Meanwhile, the smaller the energy (low attenuation) lost, the lower the fracture propagation. Fractures with very high magnitude slopes may not be detected by sonic scanners. Thus, the effect of the fracture opening in Stoneley is the reduction of the Stoneley Amplitude (attenuation) and the Stoneley Reflection. A Stoneley reflection is a wave that starts at the transmitter that is received directly by the receiver, but propagates further and can be reflected by the fracture and detected a second time later. Besides being sensitive to fractures (discontinuities), the Stoneley Reflection is also sensitive to changes in borehole size and layer boundaries (lithological changes) in the formation. A similar chevron pattern can be seen in VDL due to the presence of washout and layer boundaries. Due to these limitations, compression and shear lag were used to model the Stoneley reflection response due to non-fracturing effects (borehole enlargement and layer boundary effects).

3.3. 1D Mechanical Earth Model

The development of the Mechanical Earth Model (MEM) (Plumb et al., 2000 [7]) is very important in the formation of field Geomechanics information. MEM is a description of forces, stresses and stresses as a function of depth, referred to a stratigraphic column. MEM consists of a description of the rock material (ie, the mechanical stratigraphy of the formation is considered), rock elastic parameters, including Young's Modulus and Poisson's Ratio; rock strength parameters: Unconfined Compressive Strength (UCS), tensile strength and internal friction angle; Stressor model: vertical stress, minimum and minimum horizontal stress and orientation, and pore pressure.

3.4. Critically stressed Fracture Analysis
Tectonic stress conditions can cause fractures/faults with a directional orientation to slip caused by high shear stresses, potentially resulting in increased permeability due to shear dilatation (Barton et al., 1995 [8]). These fractures are "critically-stressed" and are located above the failure line in the normal and shear stress zones. With increased permeability, these fractures can act as fluid conduits between wells or between different reservoir intervals. Given the stress field and the orientation of the fracture and fault sets, it is possible to determine which fracture and fault sets are susceptible to shear slip, and therefore the likely paths for fluid flow or increased permeability. This is done by determining how stable the fracture or fault is under a current stress field, using the Mohr Coulomb criteria.

4. Results and Discussion
4.1. Fracture Characterization and Critically stressed Fracture of Well 58-32
Well 58-32 in 8.5in section with depth interval 7,390-7,527 ft-MD is comprised entirely by granitoids. The violations were observed in this well, but not picked and analyzed as not contribute in producing (tend to not flow). Three (3) types of dip that was manually picked consists of conductive fracture, drilling induced fracture and breakout. Conductive fracture shows dominant strike direction to N-S with minor striking to NE-SW and NW-SE where dip magnitude range of 20 - 74 deg (Figure 4). No further classification for low and high angle conductive fractures was performed due to all natural conductive fracture dips are needed for further critically-stressed analyses.

Drilling induced fractures mainly observed in the upper interval of study and consistently show NE-SW orientation. The breakout was observed and show perpendicular orientation from drilling induced fractured, show direction to NW-SE. Well 58-32 is vertical well and very small inclination (max 3deg), the strikes of drilling induced fractures and borehole breakouts may align with the trends of maximum and minimum horizontal stress (respectively). As no Stoneley wave sonic is available in this well, thus no sonic fracture analysis was performed. Thus, further classification of probable open fractures cannot be analyzed.

Isotropic was applied to the rock types within the reservoir during 1D construction. Integrating density log was used for Overburden construction where pressure is taken to be hydrostatic pressure at a given depth z from the top of the water table. Key parameters requiring correlation are Unconfined Compressive Strength (UCS) and Tensile Strength also correlated from Young Modulus (static) and Poisson's ratio. The result of MEM parameters was calibrated with rock mechanics test from the lab and caliper data from FMI (breakout / borehole failure indicator). Stress regime in this well is normal regime ($\sigma_v \rangle \sigma_H \rangle \sigma_h$) and using the input (pore pressure, vertical stress, minimum and maximum horizontal), maximum horizontal stress direction 40deg and friction angle 20deg, the result of fracture stability shows that fracture that has high potential of shearing is striking to N-S, dipping to Westerly with dip magnitude range 45-70 deg align with maximum horizontal stress direction (Figure 5).

4.2. Fracture Characterization and Critically stressed Fracture of Well 21-31
Well 21-31 in 8.5in section with depth interval 6226 – 7920 ft-MD is comprised of Rhyolite-Tuff, basalt, quartzite, metabolite, and felsic intrusion rock formations. Foliation in this well is abundant along the study interval. Three (3) types of dip also manually picked consists of conductive fracture, drilling induced fracture and breakout. Conductive fracture shows dominant strike direction to NNE-SSW where dip magnitude range of 25 - 85 deg (Figure 6).

Drilling induced fractures mainly observed in the upper interval of study and consistently show NNE-SSW orientation. The breakout was observed and show perpendicular orientation from drilling induced fractured, show direction to WNW-ESE. Well 21-31 is vertical well and very small inclination (max 4.5), the strikes of drilling induced fractures may align with the trends of the maximum horizontal stress while breakouts align with minimum horizontal stress.

Same workflow as Well 58-32 for 1D MEM construction, parameter and fracture stability were applied in Well 21-31. Fortunately, in Well 21-31 Stoneley wave sonic is available and sonic fracture analysis was performed. All natural conductive fractures picked from FMI that has potential critically stressed fracture was filtered based on potential zone that shows an energy loss (after washout and bed boundary zone reduced). Using the input (pore pressure, vertical stress, minimum and maximum horizontal), maximum horizontal stress direction 30deg and friction angle 20deg, the result of fracture stability shows that fracture that has high potential of shearing is striking to NNE-SSW with dip magnitude range 40-70 deg (Figure 7).

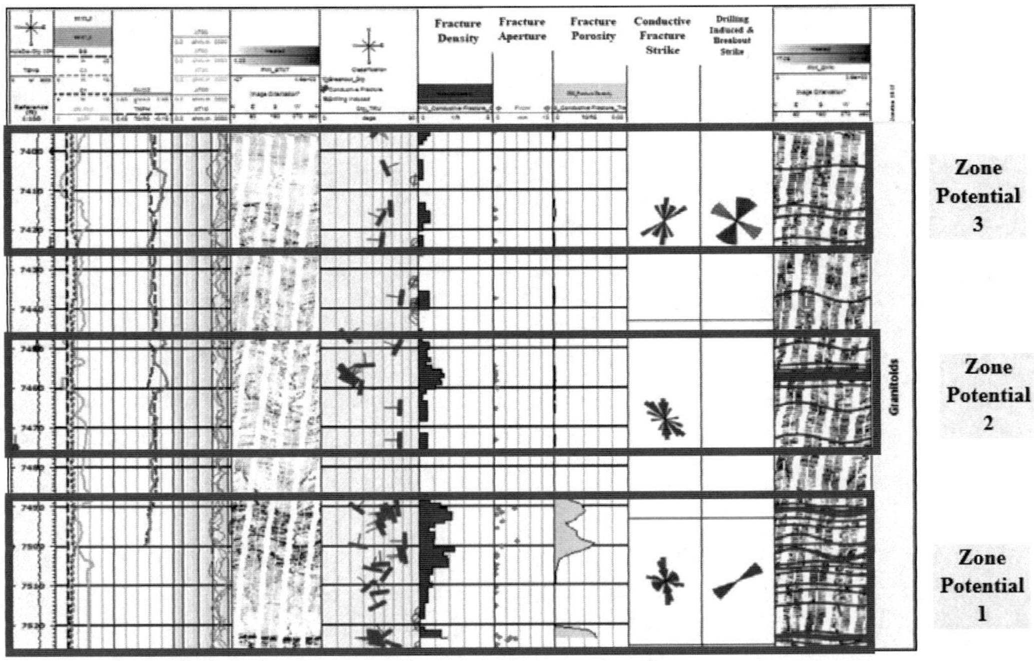

Figure 4. Fracture characterization of Well 58-32 for entire study interval.

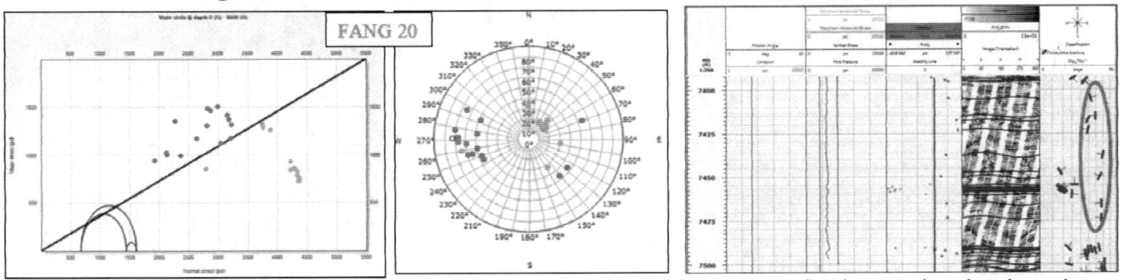

Figure 5. Critically stressed fracture (red points) using FANG 20 dog show dominant azimuth orientation to Westerly, striking to N-S and dip magnitude 45-70 deg..

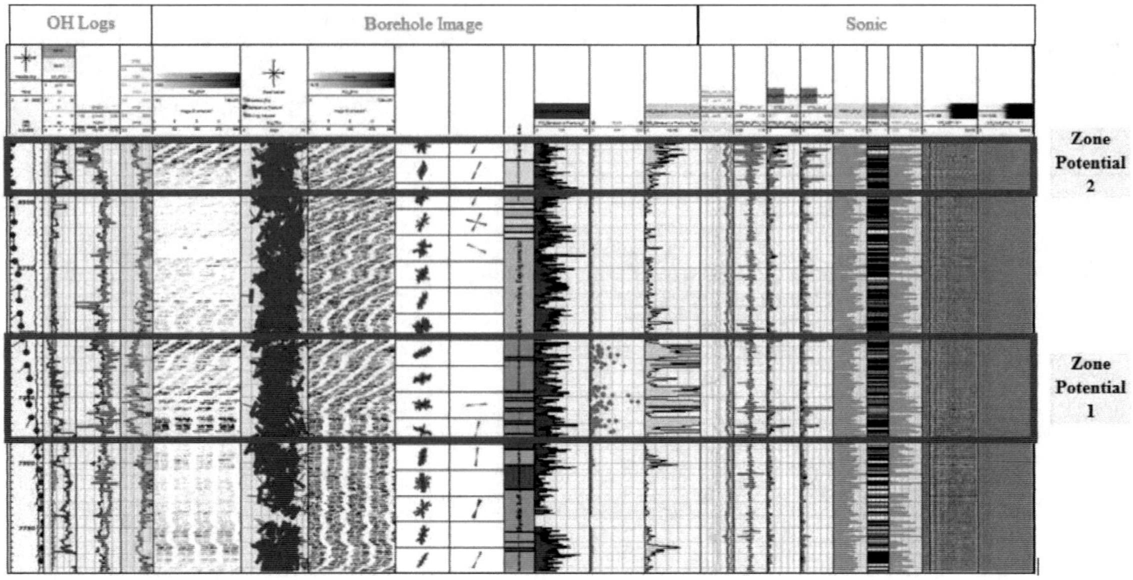

Figure 6. Fracture characterization of Well 21-31 for the entire study interval.

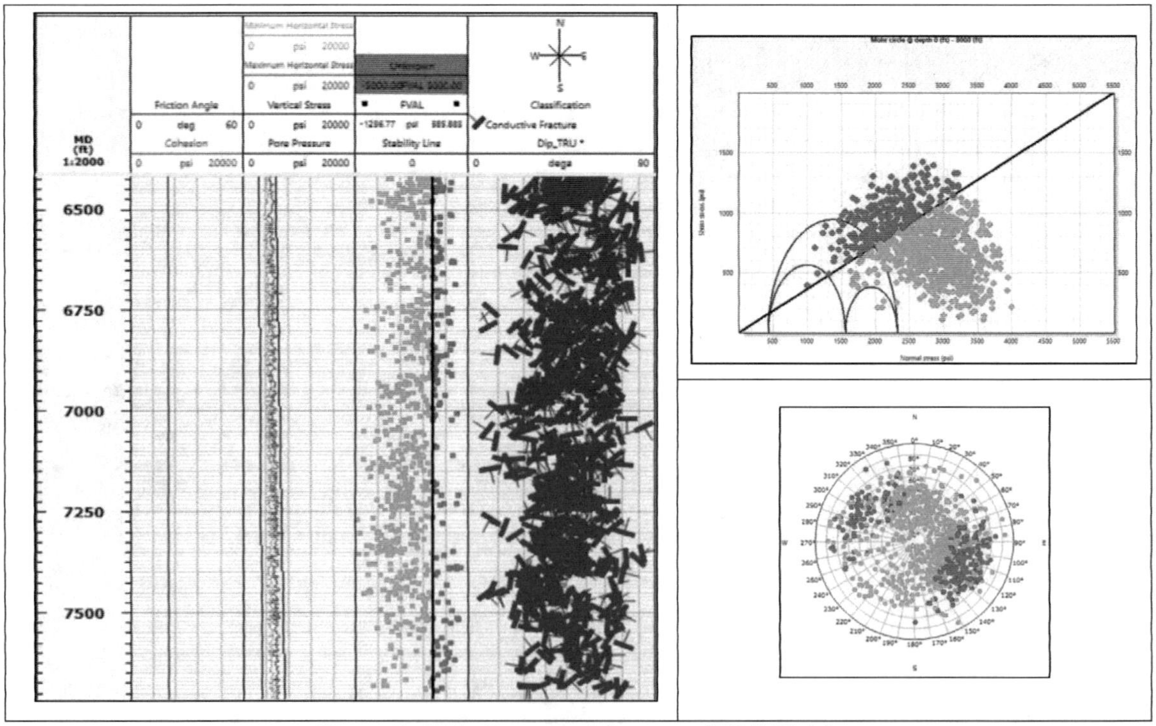

Figure 7. Critically stressed fracture of Well 21-31.

It is also important to analyze the effect of depletion or reduction in pore pressure on critically stressed fractures. With depletion, pore pressure will decrease, and horizontal stress will decrease as well without any change in overburden stress. From analysis in both wells, no change in 10% depletion and an increase in critically stressed fractures at 30% and 50% depleted.

5. Conclusions & Recommendation

To increase production nearby Well 58-32, fractures that have potential to increase production are natural fractures that have a dip azimuth towards Westerly and a dip magnitude > 45deg. While for nearby well 21-31, fractures that have potential to increase production are natural fractures that have dipped azimuth towards the NW and SE and dip magnitude > 32deg. Evaluation of critically stressed fracture intended to have knowledge on production mechanism and based on analysis, fracture that has high potential of shearing is striking to NNE-SSW.

It is important to improve the stress model by improving the understanding of critically stressed fracture. The recommendations are:

- conduct rock mechanical core tests to obtain mechanical properties of the critical formations
- Acquire more stress related data, like Sonic Scanner, extended leak-off tests (XLOT) or Mini-FRAC. Those data very useful to provide stress calibration data
- conduct a 3D Geomechanics analysis over the whole field to identify a sweet spot area based on connection within critically stressed fractures

Acknowledgment

Authors wishing to acknowledge my colleagues, Ricki Marbun, Merza Media Adeyosfi, Tito Perdana and Pasca Siburian whose given author guidance and input during the analysis process. The author also would like to express gratitude to U.S. Department of Energy GDR (Geothermal Data Repository) who provide open/published data.

References

[1] A Syed, C Thomas, and M Akshayeva *Evaluation of critically stressed fractures in a north sea basement reservoir: 9th euroconference on rock physics and Geomechanics*, Trondheim Norway, 2011.

[2] D Blankenship and the Fallon FORGE team 2018 *Conceptual geologic model: Fallon, NV, in Fallon, NV forge phase 2 final report.*

[3] J Moore, et al 2018 *Phase 2B Final Topical Report* (Utah: University of Utah).

[4] J Moore, J McLennan, K Pankow, S Simmons, R Podgorney, P Wannamaker, C Jones, W Rickard, P Xing 2020 *PROCEEDINGS 45th Workshop on Geothermal Reservoir Engineering* (California: Stanford University).

[5] Siler D L, Hinz N H, Faulds J E, Ayling B, Blake K., Tiedeman A, Sabin A, Blankenship B, Kennedy M, Rhodes G, Sophy M J, Glen J M G, Phelps G A, Foruna M, Queen J, and Witter J B 2018 *PROCEEDINGS 43rd Workshop on Geothermal Reservoir Engineering*, (California: Stanford University).

[6] Siler D L, Hinz N H, Faulds J E, Tobin B, Blake K, Tiedeman A, Sabin A, Lazaro M, Blankenship D, Kennedy M, Rhodes G, Dordquist J, Hickman S, Glen J, Williams C, Robertson-Tait A, Calvin W, and Pettitt W 2016 *Geothermal Resources Council Transactions.*

[7] Plumb R, Edwards S, Pidcock G, and Lee D 2000 *The mechanical earth model concept and its application to high-risk well construction projects*; SPE 59128.

[8] Barton C A and Zoback M D 1995 *Fluid flow along potentially active faults in crystalline rock*, Geology, 23, 683-686.

[9] Schlumberger: FMI Borehole Geology, Geomechanics, and 3D Reservoir Modeling, 2002.

[10] Schlumberger: Fracture Analysis, Techlog User Guide, 2019.

Journal of Physics: Conference Series **2019** (2021) 012086 doi:10.1088/1742-6596/2019/1/012086

3D Geomechanics Characterization in V Field

R Marbun[1, a] and A Haris[2, b]
[1]Reservoir Geophysics Graduate Program, Physics Dept, Universitas Indonesia
Jl. Salemba Raya no. 4, Jakarta 10430, Indonesia
[2]Physics Department, FMIPA, Universitas Indonesia
Kampus UI Depok, Depok 16424, Indonesia
Email: [a)]ricki.daniel.marbun@gmail.com, [b)]aharis@sci.ui.ac.id

Abstract. Upstream oil & gas operators faced many challenges in drilling activities during a field development phase where the reservoir pressure depletion might disturb the stresses equilibrium in the formation and triggered stress redistribution in the reservoir and its surrounding rocks. These changes could lead into more problematic drilling experiences such as the narrowing of safe drilling mud weight windows and potential operating losses due to related wellbore instability issues. 3D Geomechanics analysis conducted to identify the stress redistribution and predict the field wellbore instabilities behaviour that honour the geological features in the area. The 1 D Geomechanics model is created for offset wells which consist of rock strength and elastic properties, pore pressure and stresses profiles. Seismic inversion volume is used to populate the properties in the 3D static model and field Geomechanics simulation conducted to obtain the field stresses profile. V field data is used as the data available in this field is complete to support this study (wire-line logs, petrophysical interpretations, VSP, core analysis, production log, drilling reports, and the seismic velocity cube). The results of the integrated reservoir Geomechanics characterization with seismic inversion managed to predict the wellbore instabilities model in the offset wells.

1. Introduction

Upstream oil & gas operators faced many challenges in drilling activities during the development phase of a field. In this phase, the reservoir pressure depletion might disturb the stresses equilibrium in the formation and triggered stress redistribution in the reservoir and its surrounding rocks. This change could lead into more problematic drilling experiences such as the narrowing of safe drilling mud weight windows due to the decreasing of fracture and increasing of breakout gradients in a well scale and even in a field scale, as well as potential operation losses due to related wellbore instability issues.

Reservoir Geomechanics analysis is needed to identify the redistribution of current, pressure, including its magnitude and direction, which is important for modelling borehole stability to simulations of reservoir compaction and surface landslides. In 3-D modelling, to obtain lateral lithological variations, seismic interval velocity data can be used as guidelines and trends in the distribution of good data to field models. Seismic velocity data itself can be obtained from various methods. One example is Amplitude versus Offset (AVO) analysis [Ref. 5: Aki & Richards, 1980] which has been widely used in the oil and gas industry to help detect hydrocarbons, identify lithology, and analyse fluid parameters. From this AVO analysis, seismic amplitude inversion can be performed to extract information about the geological conditions that produce the existing seismic response. Various kinds of seismic inversion methods currently exist, one of the examples is Acoustic Impedance (AI) inversion. Seismic inversion volume will be used as the guideline in distributing geomechanical properties in surrounding formation around the reservoir zone. Meanwhile, reservoir porosity model will be used as the guideline in distributing the geomechanical properties in reservoir.

Data from V Field will be used in this research. This field (Figure 1) is located 200 kilometres west of the town of Stavanger in the far south of the Norwegian area. The availability of data in this field is considered sufficient, such as wire-line logs, petrophysical interpretation, vertical seismic profiling (VSP), rock core analysis, production logs, drilling reports, static and dynamic geological models, and seismic velocity cubes. The hydrocarbons produced from the Hugin formation have good reservoir characteristics. The field was first discovered in 2008 and has produced 2 times the volume of hydrocarbons initially estimated. Seismic acquisitions have been carried out before and after the production period. The field is located west of Stavenger and was decommissioned in 2016 after 8.5 years of production.

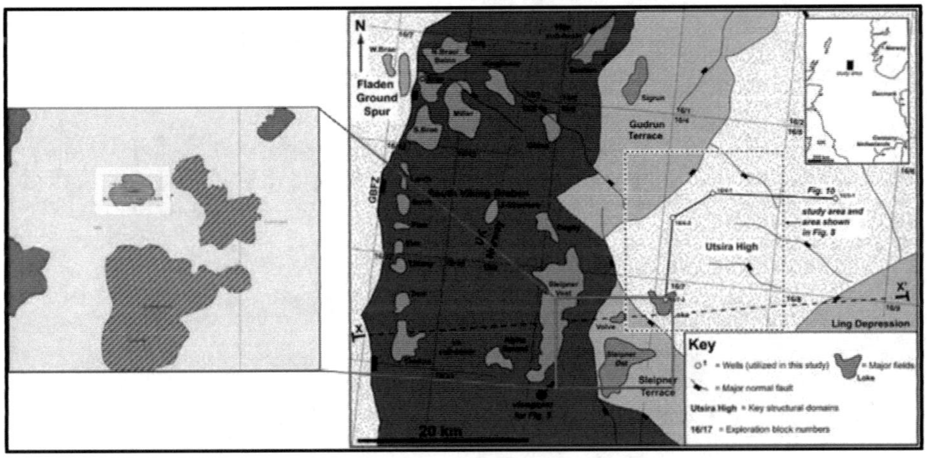

Figure 1. Study Location.

2. Regional Geology

The regional geology of the field is located in the North Sea basin and is structurally part of the South Viking Graben. This sub-chapter will discuss the geology of the research area in the form of physiography, stratigraphy, tectonics and geological structures.

2.1. Physiography

The South Viking Graben is an asymmetrical half-Graben located in the northern area of the triangular rift system in the North Sea [Ref. 1: Jackson et al, 2008]. This Graben has some hydrocarbon accumulation with an estimated recoverable reserve of 340.7 x 106 Sm3 oil equivalent. This half-Graben develops on a north-south axis with the western boundary in the form of GBFZ (Graben Boundary Fault Zone) which is a normal fault with a north-south strike direction. In the east, this Graben is bounded by an elevated area, namely Utsira High.

2.2. Stratigraphy

Sediment deposits in this area start from the oldest Middle Jurassic age of the youngest Quaternary age. The rocks deposited in this area are mostly sedimentary rocks in the form of sandstone, claystone and carbonate. In addition to sedimentary rocks, there are also deposits of classic volcanic rocks in the Balder formation of the Early Eocene age. The depositional environment of rocks in this area is quite varied, ranging from equestrian areas, shallow seas such as deltas and carbonate deposits as well as deep-sea depositional environments (Gradstein, 2010). The reservoir rock candidates in the Volve field are the Hugin formation and the Ty formation, both of which have good porosity [Ref. 2 & 3: Husmo et al. 2003, Isaksen & Tonstad. 1989]. The source rock in this field is the Draupne formation which was deposited in the late Jurassic age. This formation is dominated by organic material in the form of marine algae with type II kerogen. This formation has reached a level of maturity sufficient to produce gas and oil. The cap rock in this field is generally in the form of claystone and shale which was deposited above the reservoir formation and has a fairly low permeability.

2.3. Regional Tectonic and Geology, Structure

Valve Field located in the South Viking Graben area is at the junction between the 3 major elemental structures in the North Sea. These 3 structural elements are the Jurassic fault block at the Utsira High to the west, the Jaren High to the south and the Graben boundary in the form of a normal fault along the Fladen Ground Spur boundary to the west. In the Paleozoic era the stress regime in this area was compressional with right lateral movement. Extensive rifting occurred after the Early Permian to Late Jurassic, with the most intensive rifting occurring in the Late Jurassic. During this rifting the reservoir rock in this field was deposited and at the same time extensional faults were formed in the form of normal faults. Valve Field has a dome-shaped structure formed by the collapse of the adjacent salt deposit ridge in the Middle Jurassic.

3. Method

The outline of the research methodology is shown in Figure 2. The research is divided into two (2) work sections: well scale and field scale. For the well scale, the software that will be used include Microsoft Excel spreadsheet and Techlog™. As for the field scale, the software that will be used is Petrel™ and Visage Geomechanics Simulator™.

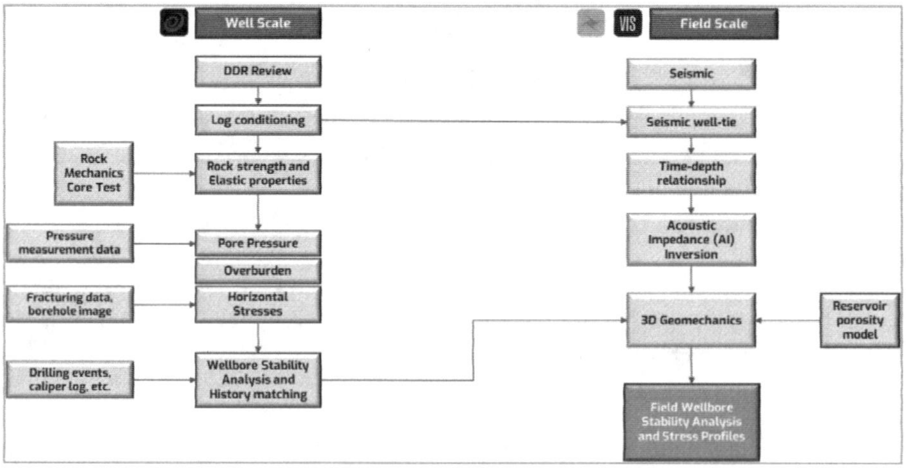

Figure 2. Workflow of the study.

3.1. 1D Geomechanics Model Construction

1D Geomechanics modeling started by understanding the available data and reviewing the offset wells data, for this study are wells 19-A and 19-BT2. Next, mechanical stratigraphy of each wells are built to differentiate the clastic and carbonate formations that will help in the consideration during assigning calculation of geomechanical properties. Elastic properties, Young's modulus and Poisson ratio, are generated using available compressional, shear Sonics; and density log data. Then, rock strength properties, such as: Unconfined Compressive Strength (UCS), Tensile Strength, and Friction Angle are generated using available compressional sonic and gamma ray logs data. Pore pressure modeled using an Eaton method [Ref. 7: Ben A. Eaton, 1975], where the models were calibrated using available reservoir formation testing (RFT) data. Next, stresses profiles are generated for the offset wells using Poroelastic method. Closure pressure data is available in well 19-A and is used to calibrate the minimum horizontal stress magnitude. Moreover, available borehole image data is used to understand the horizontal stresses orientation.

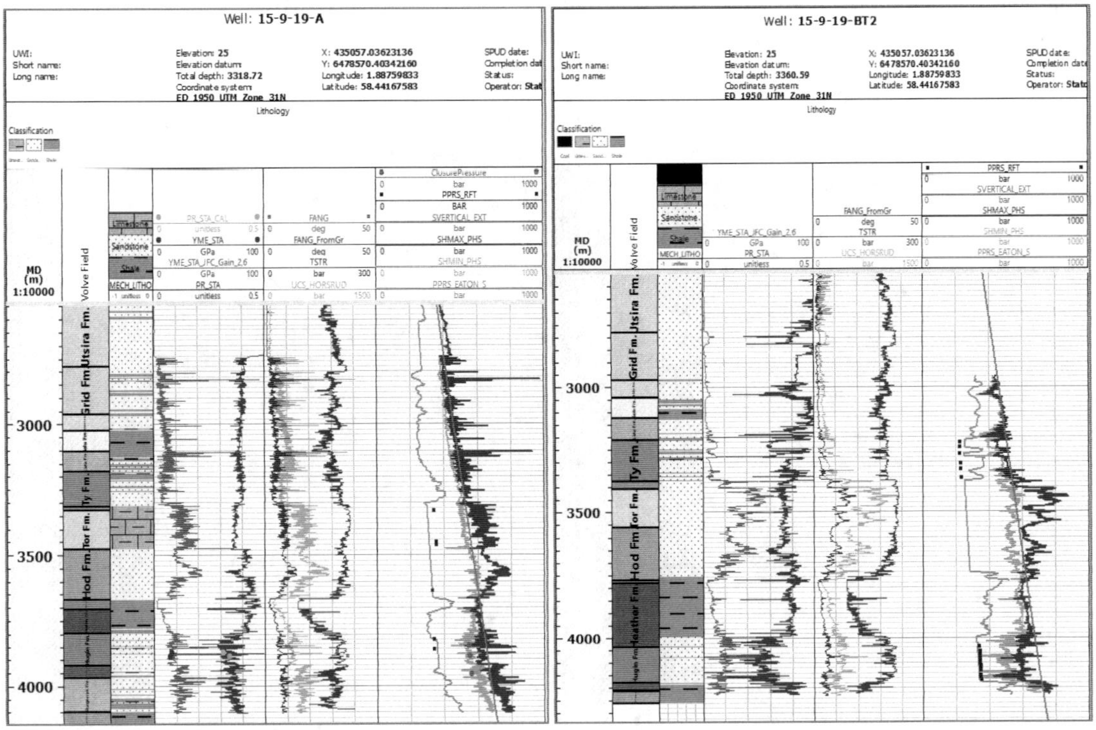

Figure 3. Geomechanics properties in well 19-A and 19-BT2.

3.2. Acoustic Impedance Seismic Inversion

In order to honor the geological variation and features in the area, available seismic data is used to do the Acoustic Impedance inversion. Sonic from wells were calibrated with available VSP Checkshots. Next, wavelets are extracted to perform the seismic to well tie. Extended white method was used to extract the wavelet from the seismic. Low frequency model from available seismic data was created with a high cut filter at 6 Hz and slope of 30 dB/octave. The inversion result has relative misfit of -7.71 dB (16.9% energy), with the relative number of reflectors 25.3%. Figure 4 below shows the result of Acoustic Impedance inversion.

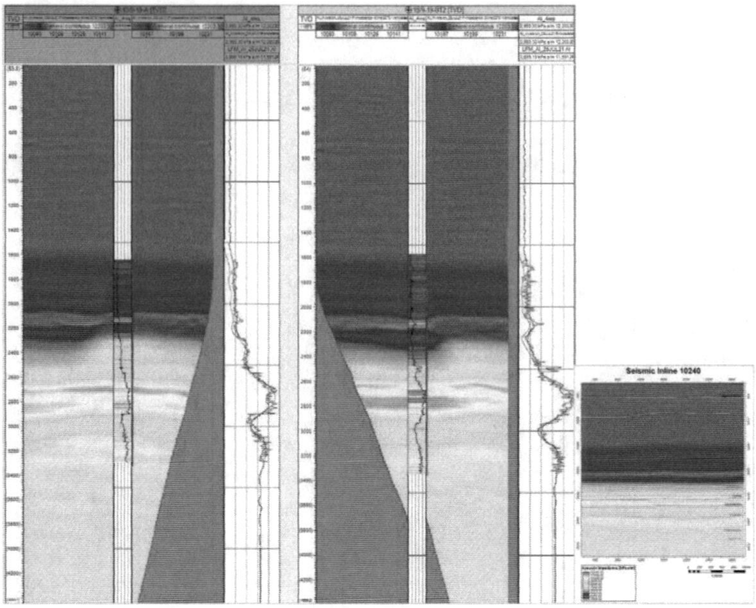

Figure 4. Acoustic Impedance Inversion Result. The QC section is plotted along the wellbore trajectory. The inversion managed to show the impedance contrast of each formations, with range of 3,000 (purple) to 13,000 kPa.s/m (red).

3.3. 3D Geomechanical Modelling

3D Geomechanics modeling started by conducting the model embodiment over the reservoir part. This embodiment consist of overburden, sideburden and underburden zones. The purpose of each zone is to simulate the actual condition of stress propagation from the major tectonic stress event (i.e. Major faults) and tectonic plate movement in below. For pore pressure model, pore pressure prediction conducted by applying Eaton trendline, that obtained during geomechanical analysis in offset well, versus true vertical depth subsea (TVDSS) profile. Figure 5 shows the embedded geomechanical model of the study. Geomechanical properties of offset wells, then populated in the embedded grid of the model. In Reservoir zone, properties are populated based on porosity model. For Overburden zone, properties are populated based on seismic inversion volume. For Sideburden and Underburden zones, properties are populated based on the moving average method. Figure 6 shows the populated properties in the reservoir and overburden zones.

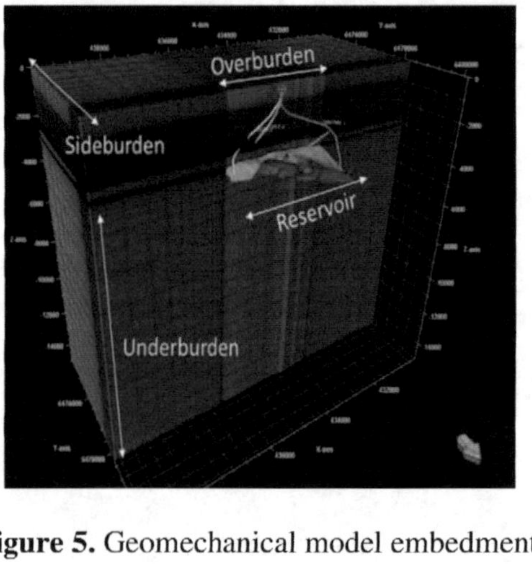

Figure 5. Geomechanical model embedment.

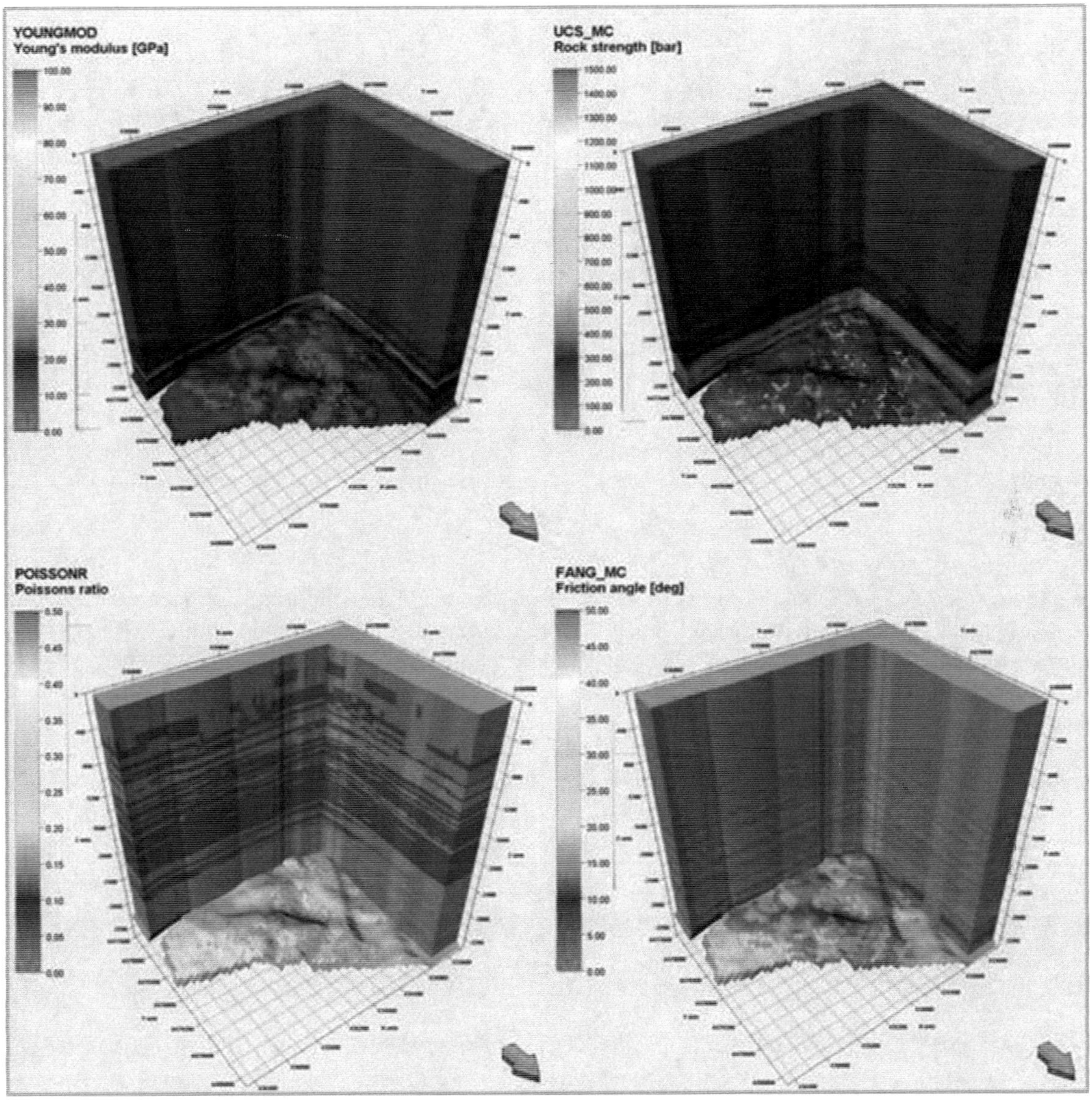

Figure 6. Populated geomechanics properties in Overburden and Reservoir zones.

4. Results and Discussion

4.1. Pre-production Field Stresses Results

Pre-production stress simulation conducted by using the populated and generated Geomechanics parameters above [Ref. 9: Blanton & Olson, 1999]. Equilibrium stresses calculated with the finite element Geomechanics simulator, Visage^{TM}. Geomechanical simulation resulting the stresses profiles across the field as shown in Figure 7. 1D Geomechanics properties, then are compared with 3D Geomechanics simulation results (pore pressure, minimum horizontal stress, maximum horizontal stress, and vertical stress) to ensure the results are inconsistent trend in the well scale data as shown in Figure 8.

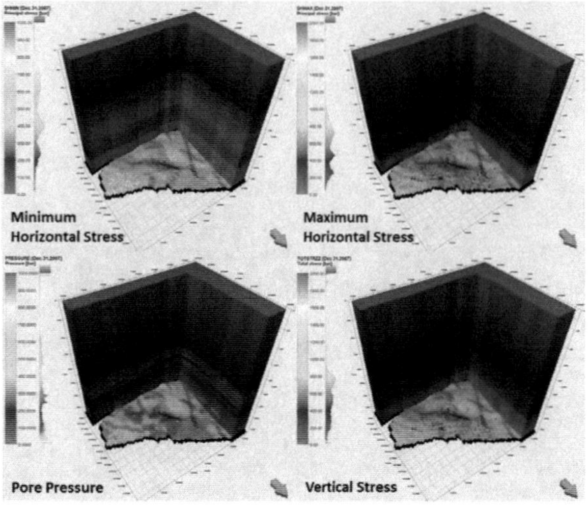

Figure 7. Pre-production stresses simulator results.

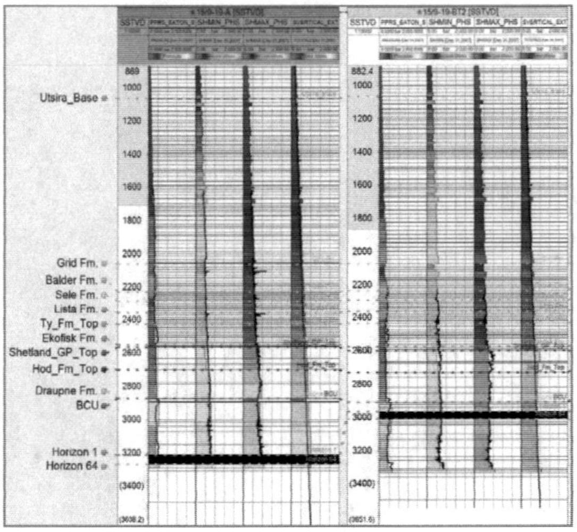

Figure 8. Comparison of offset wells stresses profile and field stresses profile results.

4.2. Wellbore Stability Prediction for Target, Wells

Wellbore stability (WBS) model in the target wells can be calculated and predicted using the resulted stresses profile across the field. The WBS resulted the recommended mud weight window to be used by the drilling team in the drilling operations as shown in Figure 9. The results show a very narrow window in both target wells.

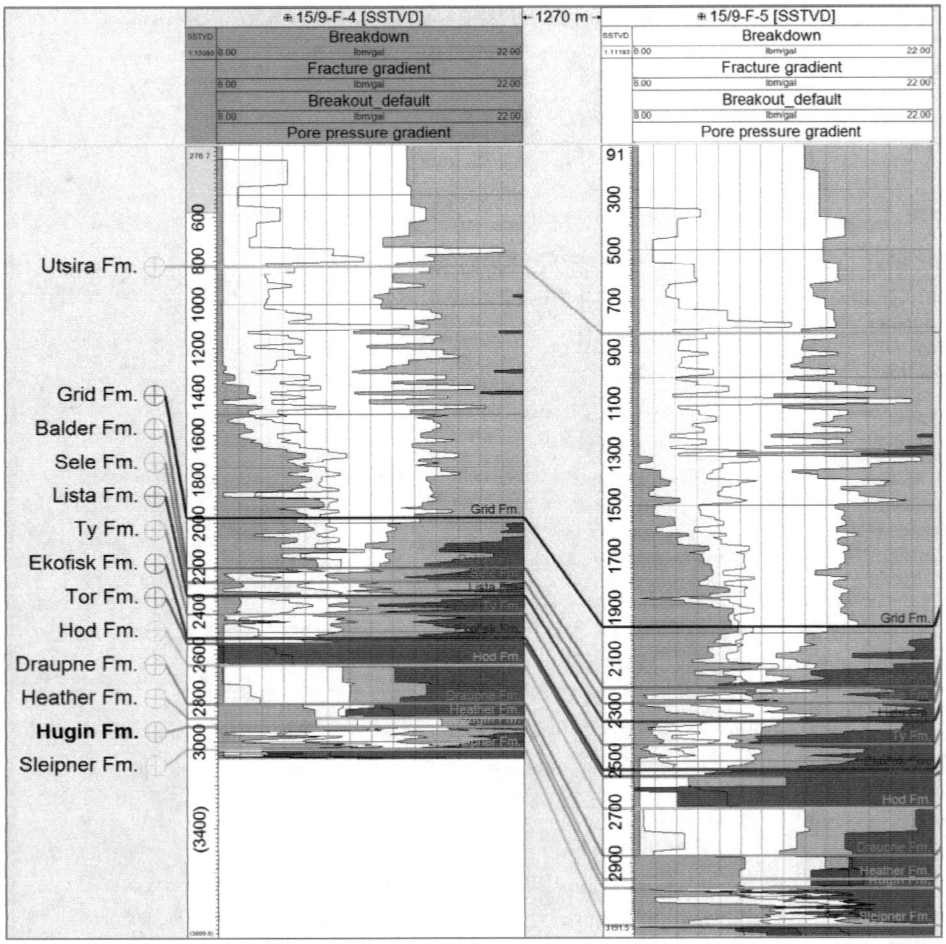

Figure 9. Drilling mud weight window from wellbore stability prediction in F-4 and F-5.

This mud weight window consists of Kick mud weight (MW) range, Breakout MW, Losses MW, and Breakdown MW. Kick MW represents the pore pressure profile along the well trajectory. Breakout MW represents the potential welfare failure due to formation breakout if the MW is in this zone. Losses MW represents the potential reactivation of existing natural fractures in the borehole that can lead to mud weight loss into the formation. Breakdown MW represent the potential of new drilling induced fractures generation due to the high mud weight being used during drilling and is in this zone.
Based on the results of mud weight windows for wells F-4 and F-5, Table 1 and 2 shows the recommended casing points and mud weight.

Table 1. Well F-4 casing points and mud weight recommendation.

Casing Section (inch)	Depth (MMD)	Recommended MW (g/cc)
13-3/8	1,650	1,50
9-5/8	2,525	1.65
7	3,000	1,40

Journal of Physics: Conference Series **2019** (2021) 012086 doi:10.1088/1742-6596/2019/1/012086

Table 2. Well F-5 casing points and mud weight recommendation.

Casing Section (inch)	Depth (MMD)	Recommended MW (g/cc)
13-3/8	2,025	1,50
9-5/8	2,700	1.65
7	3,395	1,20

5. Conclusions

Based on the results of 3D Geomechanical characterization in V Field, following points are concluded:

- Seismic inversion method managed to honour the geological variation and features on the geomechanical propagation.
- Geomechanical simulation resulted the stress profile across the field and this was used to generate the wellbore stability analysis of V Field.
- The wellbore stability analysis gives the recommended casing points and mud weight for the target wells that can help the drilling operation pre-planning hence drilling hazards can be better prevented and mitigated.

Acknowledgment

The authors would like to acknowledge Emiliana Sulistyowati, Merza Media Adeyosfi, Ridho Affandi and Bayu Wisnubroto whose given author guidance and input during the analysis process. Author also would like to express gratitude to Equinor that has made a complete set of data from North Sea oil field available for research, study and development purposes [ref.: https://www.equinor.com/en/what-we-do/digitalisation-in-our-dna/volve-field-data-village-download.html].

References

[1] C A L Jackson, E Larsen 2008 Temporal constraints on basin inversion provided by 3D seismic and well data: a case study of the South Viking Graben *Basin Research* **20** 397-417.

[2] T Husmo, G P Hamar, O Hoiland, E P Johannessen, A Romuld, A M Spencer, Titterton, Rosemary 2003 *Lower and Middle Jurassic. In: The Millennium Atlas: Petroleum Geology of the Central and Northern North Sea.*

[3] D Isaksen and K Tonstad 1989 A revised Cretaceous and Tertiary lithostratigraphic nomenclature for the Norwegian North Sea, *Norwegian Petroleum Directorate Bulletin* **5** 1–79.

[4] K Zoeppritz 1919 Erdbebenwellen VII b: über Reflexion und Durchgang seismischer ellen durch Unstetigkeitsflächen. Nachrichten von der Gesellschaft der Wissenschaften zu Göttingen, Mathematisch-Physikalische Klasse 66-84.

[5] K, Aki and P G Richards 1980 Quantitative seismology: W.H. Freeman and Company.

[6] Plumb R, Edwards S, Pidcock G, and D Lee 2000 The mechanical earth model concept and its application to high-risk well construction projects, SPE 59128.

[7] Ben A Eaton 1975 The equation for geopressure predicted from well logs, SPE 5544.

[8] Plona T J, Kane M, R, Sinha, B, K, Walsh J 2002 Evaluating stress-induced anisotropy and mechanical damage from cross-dipole sonic data using dispersion analysis, SPE 78233.

[9] Blanton T L and Olson, J E 1999 Stress magnitudes from logs – effects of tectonic strains and temperature; SPE 54653.

[10] Addis, M A 1997 The stress-depletion response of reservoirs *SPE Annual Technical Conference and Exhibition* San Antonio, Texas.

Identification of Seawater Intrusion in Kota Lama Semarang and Surrounding Based on Geoelectrical Resistivity Survey

Supriyadi[1], T N Fitrianto[2], Khumaedi[1] dan Sugiyanto[1]
[1]Department of Physics, Universitas Negeri Semarang, Semarang, 50229, Indonesia
[2]Department of Physics, Universitas Gadjah Mada, Yogyakarta, 55281, Indonesia
Email: [1]supriyadi@mail.unnes.ac.id

Abstract. Kota Lama Semarang is a tourism potential that causes a high demand for groundwater, which can cause seawater intrusion. Seawater intrusion can affect groundwater quality and soil conditions. This research aimed to determine the depth and distribution of seawater intrusion in the Kota Lama Semarang and surrounding. The research used electrical resistivity tomography (ERT) and vertical electrical sounding (VES). Geoelectrical resistivity surveyed consist of 4 ERT lines and 3 VES lines with a line length was 100 to 160 m. Lines were scattered north and south of the Kota Lama to determine the distribution of seawater intrusion. Inversion processing used Progress for VES and Res2DInv for ERT. The content of chloride as well in Kota Lama was measured as supporting data. At this location it was estimated that it has 3 layers: top soil, sand and sandy clay. The sand layer at a depth of 3 to 11 m was the layer affected by seawater intrusion. Seawater intrusion on the north side was higher than on the south side. This was indicated by the lower resistivity value of the groundwater layer on the north side. The resistivity value of the groundwater layer on the north side was 0.96 Ωm while the south side was 3.76 Ωm. This result was appropriate to the chloride content which indicates a higher concentration on the north side of the Kota Lama Semarang.

1. Introduction

Seawater intrusion is a global problem affecting groundwater quality and soil conditions along coastlines [1]. Seawater intrusion is the process of entering seawater into freshwater aquifers. So the groundwater, which was originally fresh water becomes brackish water and even saline water. This causes a decrease in groundwater quality in the area [2]. Groundwater is considered as one of the most suitable water sources to meet the needs of most pure water [3][4].

Kota Lama Semarang has good tourism potential to be nominated for UNESCO World Heritage Sites [5]. The area, which is located in the north Semarang district, has a western historic building [6][7]. The growth of Semarang City has increased the demand for water [8]. The step that many people take is to take groundwater from unconfined aquifers. Groundwater exploitation is not only carried out through dug wells, but also drilled wells that can reach deeper aquifers. Groundwater exploitation is mainly for industrial activities such as hotels, hospitals, etc. Overexploitation of groundwater can lead to subsidence and seawater intrusion [9][10]. Coastal aquifers are fragile, because they are generally in direct contact with seawater, which is very vulnerable to seawater intrusion [11].

Seawater can affect electrical properties in the form of higher conductivity than freshwater [12]. The difference of the electrical conductivity of freshwater and saltwater is very significant, and therefore, the geoelectrical resistivity method is suitable for determining seawater intrusion in coastal areas [13].

Based on the description above, the purpose of this study was to determine the depth and distribution of seawater intrusion in the Kota Lama Semarang and its surroundings using the geoelectrical resistivity method with the Schlumberger array and measurement of Chloride content.

2. Method

This research began with studying hydrology and geological condition. Then, measured the chloride content in groundwater as well in Kota Lama Semarang. Next, the geoelectrical resistivity survey used Electrical Resistivity Tomography and Vertical Electrical Sounding with Schlumberger array. Lines were scattered north and south of the Kota Lama to determine the distribution of seawater intrusion. Line 1 was on the south side and lines 2, 3 and 4 on the north side of the Kota Lama Semarang, as can be seen in Figure 1. Line 1 was on the South Parking complex consisting of ERT and VES along 120 m. Line 2 was on Jl. Kom Yos Sudarso consisting of ERT along 150 m. Line 3 was on *Vereenigde Javasche Houthandel Maatschappij* "ex Java wood trading office" complex consisting of ERT along 120 m and VES along 100 m. Line 4 was on Jl. Tawangsari consisting of ERT along 150 m and VES along 160 m.

Figure 1. Map of research location, VES points and ERT lines.

The geoelectrical resistivity survey conducted on March 20 to 21, 2021. This survey used Resistivity Meter OYO Mc Ohm. The ERT survey used a 10 meter electrode spacing unit. The data were measured manually according to a Schlumberger array. The data recorded were current (I_{AB}) and voltage (V_{MN}). After the geoelectrical resistivity acquisition, it was continued by recording the VES point measurement and the position of each ERT electrode using GPS. The basic principle of the geoelectrical resistivity method is to inject electric current into the earth and then measure the electric potential. The purpose of this method is to obtain information about the structure based on the difference in resistivity values [14]. In Schlumberger configuration, two electrodes are positioned symmetrically along a straight line as depicted in Figure 2. The current Electrodes (AB) are outside, whereas the Potential Electrodes (MN) are inside. In order to change the depth range of measurement, the current electrodes are moved outside, while the potential electrodes are fixed [15].

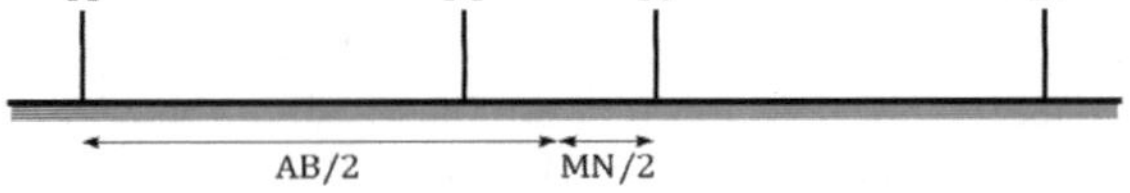

Figure 2. Schlumberger array electrode position.

Journal of Physics: Conference Series **2019** (2021) 012087 doi:10.1088/1742-6596/2019/1/012087

The geometry factor for the Schlumberger array is written in equation (1).

$$K = \pi \frac{\left(\left(\frac{AB}{2}\right)^2 - \left(\frac{MN}{2}\right)^2\right)}{MN} \tag{1}$$

Preliminary data processing used Microsoft Excel software to obtain apparent resistivity value. The apparent resistivity was calculated using equation (2). Inversion processing used software progress for VES and software Res2DInv for ERT. The apparent resistivity value converted using inversion software will be correlated with direct measurements of groundwater samples taken [16]. During the processing stage, the electrode elevation data was entered in the inversion process so that the depth displayed in the resistivity profile refers to the mean sea level [17].

$$\rho_a = K \frac{\Delta V}{I} \tag{2}$$

The difference in resistivity values was then used for identification, with basic knowledge of resistivity aspects such as geological conditions to interpret subsurface conditions [18]. Seawater intrusion will be indicated with a very low resistivity value [19].

3. Result and Discussion

Chloride content measurements were carried out as an early indication of seawater intrusion of the Kota Lama Semarang. Groundwater sampling was taken from 11 wells in the area. The results of the measurement of chloride content in groundwater samples are presented in Figure 3. From the results shown in Figure 3, there were seven groundwater samples with chloride ion content of more than 250mg/l, which is shown in green to red. Seventh groundwater samples had chloride ion levels of more than 250 mg/l that are in KL1, KL2, KL7, KL 8, KL 10, KL 11 and KL 12. Chloride content can indicate seawater intrusion [20][21]. From the image of the distribution of chloride content, it is known that the northern region has a higher content than the southern region. This indicates that the northern region sea water intrusion is higher than the southern region.

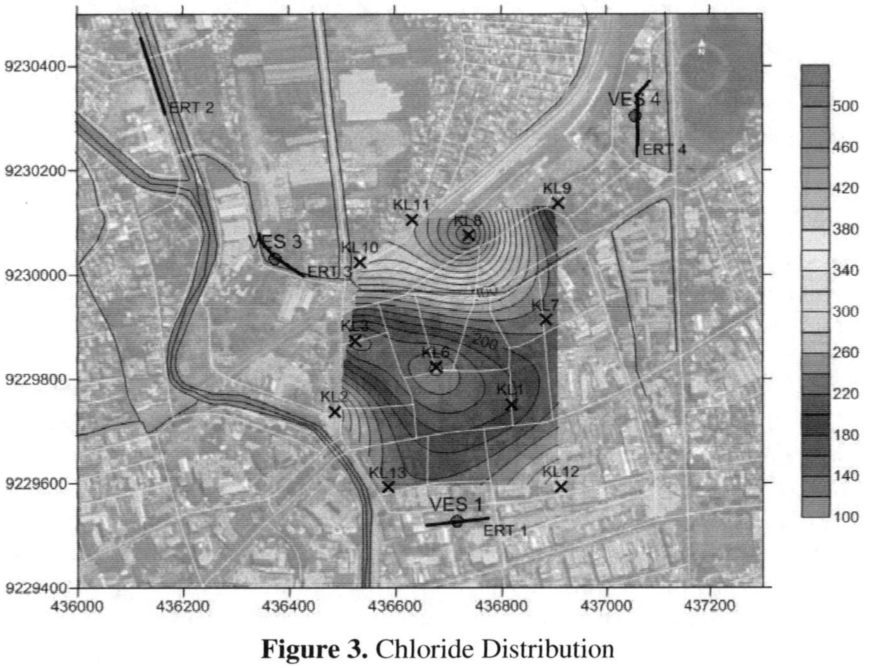

Figure 3. Chloride Distribution

A vertical electrical sounding survey was conducted to determine the depth of seawater intrusion in Kota Lama Semarang. The results of the inversion process can be seen in table 1.

VES 1 was in the south parking lot of the Kota Lama Semarang. At this location it is estimated that it has 3 types of layers: top soil, sand and sandy clay. Top soil layer with a resistivity value of 81.30 Ωm with a thickness of 0.89 m. Sand layer with a resistivity value of 3.76 to 23.96 Ωm. Sand with a resistivity value of 3.76 Ωm is sand saturated with water. This resistivity value indicates the intrusion of sea water with low concentration. This layer was at a depth of 3.24 m. Sandy clay layer with a resistivity value of 19.60 Ωm was at a depth of 11.16 m.

VES 3 was in *Vereenigde Javasche Houthandel Maatschappij* "ex Java wood trading office" complex. At this location it is estimated that it has 3 types of layers: top soil, sand and sandy clay. Top soil layer with a resistivity value of 51.45 Ωm with a thickness of 0.83 m. Sand layer with a resistivity value of 0.96 and 13.88 Ωm. Sand with a resistivity value of 0.96 Ωm is sand saturated with water. This resistivity value indicates the intrusion of sea water with high concentration because it has a resistivity value of less than 1 Ωm. This layer was at a depth of 2.99 m. Sandy clay layer with a resistivity value of 16.79 Ωm was at a depth of 11.84 m.

VES 4 was on Jl. Tawangsari. At this location it is estimated that it has 4 types of layers: top soil, sand and sandy clay and. Top soil layer with a resistivity value of 12.97 Ωm with a thickness of 0.86 m. Sand layer with a resistivity value of 0.96 and 22.14 Ωm. Sand with a resistivity value of 0.96 Ωm is sand saturated with water. This resistivity value indicates the intrusion of sea water with high concentration because it has a resistivity value of less than 1 Ωm. This layer was at a depth of 2.15 m. Sandy clay layer with a resistivity value of 15.40 Ωm was at a depth of 4.42 m. Sand layer saturated by water with a resistivity value of 2.18 Ωm was at a depth of 12.02 m

Table 1. Table type of layer, depth and resistivity value

VES 1		VES 3		VES 4		Lithology
depth	rho	depth	rho	depth	rho	
0	81.30	0	51.45	0	12.97	Top Soil
0.89	23.96	0.83	13.88	0.86	22.14	Sand and Gravel
3.42	3.76	2.99	0.96	2.15	0.96	Groundwater
11.16	19.60	11.84	16.79	4.42	15.40	Sandy Clay
				12.02	2.18	Groundwater

Electrical resistivity tomography survey was conducted to determine the distribution of seawater intrusion in Kota Lama Semarang. An ERT survey in Schlumberger array with an electrode distance of 10 m was used for profiling. Four ERT lines were obtained to a penetration depth of about 21 to 27 m with a total length between 120 to 150 m.

Line 1 was in the south parking lot of the Kota Lama Semarang. The results of the inversion on this line obtained a resistivity value between 2.87 to 12.01 Ωm which can be seen in Figure 4. Line 2 was on Jl. Kom Yos Sudarso. The results of the inversion on this line obtained a resistivity value between 0.28 to 9.03 Ωm which can be seen in Figure 5. Line 3 was in *Vereenigde Javasche Houthandel Maatschappij* "ex Java wood trading office" complex. The results of the inversion on this line obtained a resistivity value between 0.81 to 2.59 Ωm which can be seen in Figure 6. Line 4 was on Jl. Tawangsari. The results of the inversion on this line obtained a resistivity value between 0.27 to 30.29 Ωm which can be seen in Figure 7. Seawater intrusion was detected on Line 2, 3 and 4. Seawater intrusion was indicated by a value less than 1 Ωm shown in blue [22].

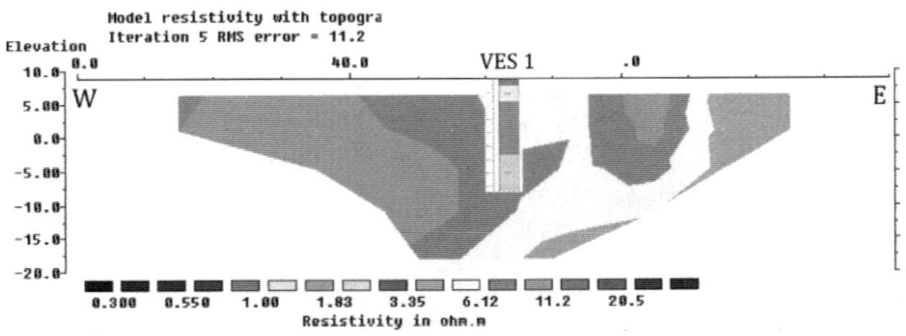

Figure 4. The 2D subsurface resistivity profiling of line 1

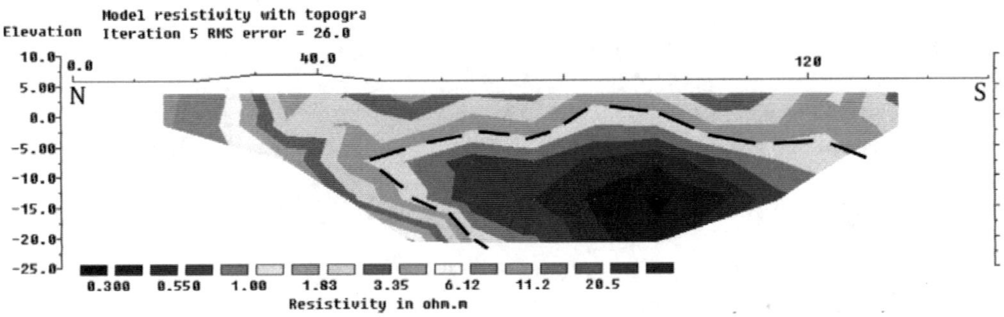

Figure 5. The 2D subsurface resistivity profiling of line 2

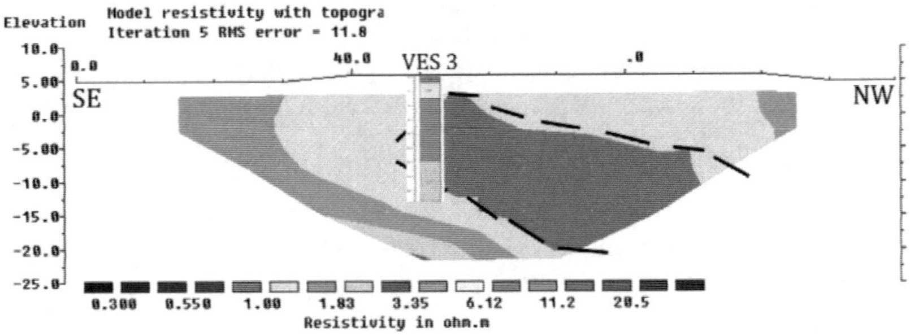

Figure 6. The 2D subsurface resistivity profiling of line 3

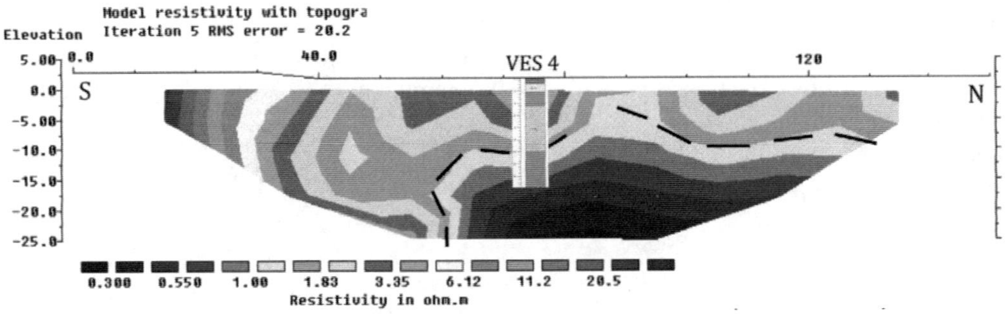

Figure 7. The 2D subsurface resistivity profiling of line 4

Form the resistivity survey and chloride content measurement, seawater intrusion on the north side was higher than on the south side.

4. Conclusion

Seawater intrusion occurred in the sand layer at a depth of 3 to 11 m. The resistivity value of this layer was 0.96 to 3.76 Ωm. Seawater intrusion on the north side was higher than on the south side. This was indicated by the lower resistivity value of the groundwater layer on the north side. The resistivity value of the groundwater layer on the north side was 0.96 Ωm while the south side was 3.76 Ωm. This result was consistent with the chloride content which indicates a higher concentration on the north side of the Kota Lama Semarang.

Acknowledgments

Acknowledgments convey to those who have funded this research through a Basic Research Grant by Kemenristek/BRIN/2021 with contract number: 24.25.3/UN37/PPK.6.8/2021

References

[1] Chen Q, Jia C, Wei J, Dong F, Yang W, Hao D and Ji Y 2020 *Chem. Geol.* 552 119779
[2] Hasan M F R, Fransiska C D, Suaidi D A, Wisodo H, Martina N and Rahmat A 2021 *IOP Conf. Ser. Earth Environ. Sci.* 739 012002
[3] Rezaei A, Hassani H, Hassani S, Jabbari N, Mousavi S B F and Rezaei S 2019 *Groundw. Sustain. Dev.* 9 100245
[4] Alshehri F, Almadani S, El-Sorogy A S, Alwaqdani E, Alfaifi H J and Alharbi T 2021 *Mar. Pollute. Bull.* 165 112094
[5] Yuliati D 2019 *Anuva J. Kaji. Budaya, Perpust. Inf.* 3 157–71
[6] Prabowo B N, Pramesti P U, Ramandhika M and Sukawi S 2020 *IOP Conf. Ser. Earth Environ. SCI.* 402 012020
[7] Amin S, Pramono S E and Atno G F K 2020 *J. Southwest Jiaotong Univ.* 55
[8] Widada S, Rochaddi B, Suryono C A and Irwani I 2018 *J. Kelaut. Troop.* 21 75–80
[9] Rahmawati N, Vuillaume J F and Purnama I L S 2013 *J. Hydrol.* 494 146–59
[10] Wijatna A B, Kayis M, Satrio S and Pujiindiyati E R 2019 *Indones. J. Geosci.* 6 17–28
[11] Kura N U, Ramli M F, Ibrahim S, Sulaiman W N A, Zaudi M A and Aris A Z 2014 *Sci. World J.*
[12] Kazakis N, Pavlou A, Vargemezis G, Voudouris K S, Soulios G, Pliakas F and Tsokas G 2016 *Sci. Total Environ.* 543 373–87
[13] Hasan M, Shang Y, Jin W, Shao P, Yi X and Akhter G 2020 *Water* 12 3408
[14] Yuliatmoko R, S, Kurniawan T, Sunardi B, Hardy T, Martha A A, Rohadi S and Riama N F 2021 *IOP Conf. Ser. Earth Environ. Sci.* 708 012002
[15] Obiajulu O O, Okpoko E I and Mgbemena C O 2016 *ARPN J. Earth Sci.* 5 13–9
[16] Azizah N, Pratiwi N H, Islami A P and Islami N 2019 *J. Phys. Conf. Ser.* 1351 012094
[17] Islami N, Taib S, H, Yusoff I and Ghani A A 2018 *Environ. Earth Sci.* 77 1–18
[18] Ashel H, Taufik A, Pratomo P M, Mawaddah S and Srigutomo W 2021 *AIP Conf. Proc.* 2320 040013
[19] Shang T, Xu Z, Gong X, Li X, Tian S and Guan Y 2021 *J. Hydrol.* 599 126348
[20] Setyawan A, Aribowo Y, Trihadini A, Hastuti D, Ramdhani F, Waskito F and Virgiawan G 2017 a *IOP Conf. Ser. Earth Environ. Sci.* 55 012053
[21] Affaire N and Walraevens K 2018 *Water* 10 143
[22] Putranto T T, Hidajat W K and Alexander K 2018 *Int. J. GEOMATE* 15 173–9

Journal of Physics: Conference Series **2019** (2021) 012088 doi:10.1088/1742-6596/2019/1/012088

Euclidean Action of Electrically Charged Black Holes in the Eddington-inspired Born-Infeld Gravity

A M Kusuma and H S Ramadhan
Departemen Fisika, FMIPA, Universitas Indonesia
Jl. Margonda Raya, Depok 16424, Indonesia.
Email: r d r d d

Abstract. We investigate the Euclidean action of electrically charged black hole in the Eddington-inspired Born-Infeld (EiBI) gravity utilizing a semiclassical approach. The Euclidean action is computed in the canonical ensemble. We found that by imposing an appropriate set of boundary terms, the expression of the black hole's entropy could be found. The obtained form of the entropy is also found to be consistent with previous results.

1. Introduction

The proposal that black holes could be considered as thermal objects started as early as 1973, when Bekenstein conjectured that there should be a generalized second law of black hole mechanics [1]. This conjecture was partly motivated by Hawking's black hole horizon area increase theorem $(\delta A \geq 0)$ Which resembles that of the never decreasing nature of entropy $(\delta S \geq 0)$ [2]. Around the same time this conjecture was being proposed, Bardeen, Carter and Hawking have successfully shown that there is a set of mathematical relations that can be used to explain black hole mechanics, and these relations are analogous to the laws of thermodynamics [3]. Later on, Hawking would give proof that black holes radiate at certain thermal spectrum $(T_H = \tau/2\pi)$, and consequently black hole would have an entropy that is quarter of its area as well $(A_{\mathrm{BH}} = A/4 = \pi r_+{}^2)$ [4]. The fact that these thermodynamics variables could be obtained from the different approach and still yield the same result further cemented the fact that black hole is a thermodynamic object [5].

On the other hand, a modified theory of gravity was proposed in 2010 by Banados and Ferreira, the theory known as the Eddington-inspired Born-Infeld (EiBI) gravity [6]. As the name itself suggest, this particular model of gravity was constructed in an affine and detrimental form, resembling that of Eddington's earlier proposal about affine gravity and Born-Infeld theory of nonlinear electrodynamics [7 - 8]. EiBI gravity is constructed in the Palatini formalism, thus it is able to avoid some of the problems that are usually encountered in models of gravity that are constructed from pure metric formalism, such as the ghost-like instabilities [9].

In this study, we intend to investigate whether it is feasible to obtain the entropy of an asymptotically flat EiBI black hole with Maxwell electrodynamics from a semiclassical approach, namely by evaluating its Euclidean action. The paper is organized as follows. In section 2 we briefly review the solution of electrically charged, static, black holes in EiBI gravity. In section 3 we calculate the Euclidean action on canonical ensemble. Finally, we conclude our work in section 4.

2. Obi Black Holes with Maxwell Electrodynamics

The solution of EiBI black holes with Maxwell electrodynamics has been studied in [10 - 11]. In this section we are going to briefly review the results and highlight certain part of solutions that would be used in the Euclidean action evaluation later. The action of EiBI gravity is given by

$$S = \frac{1}{8\pi\kappa}\int d^4x\left(\sqrt{-|g_{\mu\nu} + \kappa R_{\mu\nu}(\Gamma)} - \lambda\sqrt{-g}\right) + S_M[g, \phi_M]. \tag{1}$$

Employing the Palatini formalism, the gravitational field equations are

$$\sqrt{-q}q^{\mu\nu} = \lambda\sqrt{-g}g^{\mu\nu} - 8\pi\kappa\sqrt{-g}T^{\mu\nu}, \tag{2}$$

and

$$q^{\mu\nu} = g^{\mu\nu} + \kappa T^{\mu\nu}, \tag{3}$$

for $g_{\mu\nu}$ and $q_{\mu\nu}$ Respectively.

Assuming static, spherically symmetric configuration, we choose Ansatz

$$\mathrm{ds}_g{}^2 = -\psi^2(r)f(r)\mathrm{dt}^2 + \frac{1}{f(r)}\mathrm{dr}^2 + r^2\mathrm{d}\Omega_2{}^2, \tag{4}$$

$$\mathrm{ds}_q{}^2 = -G^2(r)F(r)\mathrm{dt}^2 + \frac{1}{F(r)}\mathrm{dr}^2 + H^2(r)\mathrm{d}\Omega_2{}^2, \tag{5}$$

where $\mathrm{ds}_g{}^2$ and $\mathrm{ds}_q{}^2$ Are the physical and auxiliary metrics, respectively.

We are going to couple the EiBI black hole with Maxwell electrodynamics as its source of matter, and the Lagrangian density is given by

$$\mathcal{L}_M = -\frac{1}{16\pi}F_{\mu\nu}F^{\mu\nu}. \tag{6}$$

Based on $\mathcal{L}_M$, we can obtain the energy-momentum tensor as follows

$$T_{\mu\nu} = \frac{1}{4\pi}\left(g^{\alpha\beta}F_{\alpha\beta}F^{\nu\beta} - \frac{1}{4}g_{\mu\nu}F_{\alpha\beta}F^{\alpha\beta}\right). \tag{7}$$

The electromagnetic field strength tensor is defined as

$$F_{\mu\nu} = A_{\nu,\mu} - A_{\mu,\nu}. \tag{8}$$

Considering the Maxwell potential $A^{\mu} = \{\phi(r), 0,0,0\}$ the non-vanishing component of $F_{\mu\nu}$ reads

$$F_{01} = -A_{0,1} = -F_{10}. \tag{9}$$

From the Maxwell's field equation ($\partial_\nu F^{\mu\nu} = 0$), the non-vanishing Maxwell electromagnetic field strength tensor is obtained to be

$$F_{\mathrm{tr}} = \frac{C_0\psi(r)}{r^2}, \tag{10}$$

where C_0 Is an undetermined constant, which later would be found to be equal to Q, the electric charge. Substituting the non-vanishing component of the energy - momentum tensor into Eq. (*3*), we will obtain a set of equations that should be eliminated so that it reduces to a relatively straightforward relation between functions in the physical and auxiliary metric. Solving these relations to the equation of motion would yield the following solutions

$$\psi(\mathrm{r}) = \frac{\sqrt{\lambda}r^2}{\sqrt{\lambda r^4 + \kappa Q^2}}, \tag{11}$$

$$F_{\mathrm{tr}} = \frac{Q\sqrt{\lambda}}{\sqrt{\lambda r^4 + \kappa Q^2}}, \tag{12}$$

$$f(\mathrm{r}) = \frac{r\sqrt{\lambda r^4 + \kappa Q^2}}{\lambda r^4 - \kappa Q^2}\left(\frac{(3r^2 - Q^2 - \Lambda r^4)\sqrt{\lambda r^4 + \kappa Q^2}}{3r^3} + \frac{4}{3}\sqrt{\frac{iQ^3}{\sqrt{\lambda\kappa}}}\,\mathrm{EllipticF}\left[i\mathrm{ArcSinh}\left[r\sqrt{\frac{i\sqrt{\lambda}}{Q\sqrt{\kappa}}}\right], -1\right] - 2\sqrt{\lambda}M + \frac{1}{3}\sqrt{\frac{Q^3}{\pi\sqrt{\lambda\kappa}}}(\mathrm{Gamma}[1/4])^2\right). \tag{13}$$

We can also express the relations between the auxiliary metric and the physical metric functions as follows

$$G(r) = \psi(r)\left(\frac{\lambda r^4 - \kappa Q^2}{r^4}\right), \tag{14}$$

$$\mathrm{F}(r) = f(r)\left(\frac{r^4}{\lambda r^4 - \kappa Q^2}\right), \tag{15}$$

$$H(r) = \frac{1}{r}\sqrt{\lambda r^4 + \kappa Q^2}. \tag{16}$$

In this work we are going to focus on electrically charged EiBI black holes in asymptotically flat space ($\Lambda = 0, \lambda = 1$).

3. Euclidean action calculation of asymptotically flat EiBI Black Holes with Maxwell Electrodynamics

To evaluate the Euclidean action, we start by "Euclideanizing" the initial form of the action [5]. The Euclideanizing is done by employing Wick rotation ($t \to i\tau$) to the previously defined line elements, so that we have

$$\mathrm{d}s_g{}^2 = \psi^2(r)f(r)\mathrm{d}\tau^2 + \frac{1}{f(r)}\mathrm{d}r^2 + r^2\mathrm{d}\Omega_2{}^2, \tag{17}$$

$$\mathrm{d}s_q{}^2 = G^2(r)F(r)\mathrm{d}\tau^2 + \frac{1}{F(r)}\mathrm{d}r^2 + H^2(r)\mathrm{d}\Omega_2{}^2. \tag{18}$$

In this particular model, the Hawking temperature gives [12]

$$T_H = \frac{\kappa}{2\pi} = \frac{1}{4\pi}\left[\frac{\partial_r(g_{00})}{\sqrt{g_{00}g_{11}}}\right], = -\frac{(Q^2 - r^2)}{4\pi r\sqrt{Q^2\kappa + r^4}}. \tag{19}$$

The period of the black hole is then obtained to be

$$\beta = -\frac{4\pi r\sqrt{r^4 + Q^2\kappa}}{Q^2 - r^2}. \tag{20}$$

Note that to obtain T_H, and β consequently, we have utilized the black hole's mass M which can be obtained from $f(r_+) = 0$ and would give

$$M = -\frac{(Q^2 - 3r^2)\sqrt{r^4 + Q^2\kappa}}{6r^3} + \frac{2}{3}\sqrt{\frac{iQ^3}{\sqrt{\kappa}}}\text{EllipticF}\left[i\text{ArcSinh}\left[r\sqrt{\frac{i}{Q\sqrt{\kappa}}}\right], -1\right]$$
$$+\frac{\sqrt{\frac{Q^3}{\sqrt{\kappa}}}\text{Gamma}\left[\frac{1}{4}\right]^2}{6\sqrt{\pi}}. \quad (21)$$

We wish to study the black hole in the situation in the canonical ensemble. Thus, the Euclidean action of electrically charged black hole holes in asymptotically flat EiBI gravity configuration for this case can be written as

$$I = I_b - I_s + I_{\text{GHY}} - I_{\text{ct}}, \quad (22)$$

where I_b is the bulk action, I_s is the surface term, I_{GHY} is the Gibbons-Hawking-York boundary term and I_{ct} Is the counter term. The bulk auction is the Wick rotated form of Eq. (*1*), while I_s Is the surface term that needs to be calculated so that the black hole manifests in the canonical ensemble. The I_{GHY} and I_{ct} Are the boundary terms that need to be evaluated so that we have a well-defined form of Euclidean action [13]. The reason why Gibbons-Hawking-York-like form is used as the boundary term would be explained later. First, we evaluate the bulk action with the line elements given in Eqs. (*17*)-(*18*), along with the Wick-rotated form of the Maxwell electric field strength tensor on the matter's Lagrangian density, which gives

$$I_b = \frac{1}{8\pi\kappa}\int d^4x\left(-\sqrt{q} + \lambda\sqrt{g} - 8\pi\kappa\sqrt{g}\mathcal{L}_M\right),$$

$$= \frac{\beta}{2}\left[-\frac{Q^2\sqrt{r^4 + Q^2\kappa}}{3r^3} - \frac{4}{3}\sqrt{\frac{iQ^3}{\sqrt{\kappa}}}\text{EllipticF}\left[i\text{ArcSinh}\left[r\sqrt{\frac{i}{Q\sqrt{\kappa}}}\right], -1\right]\Big|^{r_+}_{r_b}\right]. \quad (23)$$

Before we continue with the integral evaluation, it is interesting to note that the result above would be finite valued even when $r = r_b$, unlike the usual cases in black hole thermodynamics where some form of divergences would arise in the bulk action. For the first term that appears on the equation above it is evident that the result would be null, whereas the elliptic function that appears on the second term of the result above would yield the following form

$$\frac{4}{3}\sqrt{\frac{iQ^3}{\sqrt{\kappa}}}\text{EllipticF}\left[i\text{ArcSinh}\left[r\sqrt{\frac{i}{Q\sqrt{\kappa}}}\right], -1\right] = -\frac{\sqrt{\frac{Q^3}{\sqrt{\kappa}}}\text{Gamma}\left[\frac{1}{4}\right]^2}{6\sqrt{\pi}}. \quad (24)$$

Thus, the bulk action is obtained as follows

$$I_b = \beta\left(\frac{Q^2\sqrt{r^4 + Q^2\kappa}}{6r^3} - \frac{2}{3}\sqrt{\frac{iQ^3}{\sqrt{\kappa}}}\text{EllipticF}\left[i\text{ArcSinh}\left[r\sqrt{\frac{i}{Q\sqrt{\kappa}}}\right], -1\right]\right.$$
$$\left.-\frac{\sqrt{\frac{Q^3}{\sqrt{\kappa}}}\text{Gamma}\left[\frac{1}{4}\right]^2}{6\sqrt{\pi}}\right). \quad (25)$$

In a fixed ensemble variation of the matter part of the action must be done to ensure that the charge is fixed. The surface term for the matter part of the action is given by

$$I_s = -\frac{1}{4\pi}\int d^3x\sqrt{h}F^{\mu\nu}n_\mu A_\nu. \tag{26}$$

The Wick-rotated form of the Maxwell potential could be written as follows

$$A_\tau = \sqrt{\frac{iQ}{\sqrt{\kappa}}}\text{EllipticF}\left[i\text{ArcSinh}\left[r\sqrt{\frac{i}{Q\sqrt{\kappa}}}\right],-1\right]\Big|^{r_b}_{r_+},$$
$$= -\frac{\sqrt{\frac{Q^3}{\sqrt{\kappa}}}\text{Gamma}\left[\frac{1}{4}\right]^2}{4\sqrt{\pi}} - \sqrt{\frac{iQ}{\sqrt{\kappa}}}\text{EllipticF}\left[i\text{ArcSinh}\left[r\sqrt{\frac{i}{Q\sqrt{\kappa}}}\right],-1\right]. \tag{27}$$

Employing the result above, along with the induced metric from the physical metric, the surface term reads

$$I_s = \beta\left(-\sqrt{\frac{iQ^3}{\sqrt{\kappa}}}\text{EllipticF}\left[i\text{ArcSinh}\left[r\sqrt{\frac{i}{Q\sqrt{\kappa}}}\right],-1\right] - \frac{\sqrt{\frac{Q^3}{\sqrt{\kappa}}}\text{Gamma}\left[\frac{1}{4}\right]^2}{4\sqrt{\pi}}\right). \tag{28}$$

Before we construct the boundary term and counter term action for this particular configuration of black hole in EiBI gravity, first let us note that the action given in Eq. (*1*) could also be written in its alternative form [9]

$$I_b = \frac{1}{16\pi\kappa}\int d^4x\sqrt{-q}\left(\kappa R - 2 + \left(q^{\mu\nu}g_{\mu\nu} - 2\lambda\sqrt{\frac{(-g)}{(-q)}}\right)\right) + I_m. \tag{29}$$

From Eq. (*2*), we have

$$\sqrt{-q}q^{\mu\nu}g_{\mu\nu} = 4\lambda\sqrt{-g} - 8\pi\kappa\sqrt{-g}T. \tag{30}$$

If we were to define $\sqrt{-g}/\sqrt{-q}$ as ϵ, we may express the equation above as follows

$$q^{\mu\nu}g_{\mu\nu} = 4\lambda\epsilon - 8\pi\kappa\epsilon T. \tag{31}$$

From the electromagnetic Ansatz, we know that T is null. Thus, once we substitute the expression above into Eq. (*29*) and evaluate for the case of asymptotically flat space configuration, we will obtain

$$I_b = \frac{1}{16\pi\kappa}\int d^4x\sqrt{-q}(\kappa R - 2 + 2\epsilon) + I_m,$$
$$= \frac{1}{16\pi}\int d^4x\sqrt{-q}\left(R - \frac{2}{\kappa}\right) + \frac{1}{8\pi\kappa}\int d^4x\sqrt{-g} + I_m. \tag{32}$$

We can see that the form above has a somewhat similar structure with the usual Einstein-Hilbert action. This leads us to use a boundary term and a counterterm that follows the usual Gibbons-Hawking-York (GHY) boundary form [5], [13 - 14]. Each of these terms would have the following form,

$$I_{\text{GHY}} = \frac{1}{8\pi}\int d^3x\sqrt{h}K, \tag{33}$$

$$I_{\text{GHY}} = \frac{1}{8\pi}\int d^3x\sqrt{h_o}K_o. \tag{34}$$

To proceed with the GHY boundary term evaluation, we construct the trace of the extrinsic curvature from Eq. (*18*). The result gives

$$I_{\text{GHY}} = \beta\left(\frac{3M}{2} - \frac{\sqrt{\kappa Q^2 + r_b{}^4}}{r_b}\right). \tag{35}$$

The usual procedure to evaluate Eq. (*34*) To match the period of the black hole under consideration in the space-time in which it resides. In our case, we would do the rescaling between the physical and auxiliary metric of the black holes. Consider Eqs (*11*) and (*13*) for $\Lambda = 0, \lambda = 1$ case. As $r \to r_b$, each of these functions would be equal to 1. Thus, the Ricci scalar and tensor shall be constructed from the following boundary metric

$$h_{\text{ij}}{}^o = \{1, r_b{}^2, r_b{}^2 \text{Sin}^2\theta\}. \tag{36}$$

Using the quantities above, the counter term gives

$$I_{\text{ct}} = \beta_o(-r_b). \tag{37}$$

By interpreting that β_o resides in the physical metric, we shall rescale it so that it matches with the auxiliary metric of the black hole,

$$\beta_o\sqrt{\psi^2(r_b)f(r_b)} = \beta\sqrt{G^2(r_b)F(r_b)}, \tag{38}$$

$$\beta_o = \frac{\beta}{\psi(r_b)}\sqrt{G^2(r_b)F(r_b)}\left(\frac{1}{\sqrt{f(r_b)}}\right), \tag{39}$$

and with some approximation we obtain the following form

$$\beta_o \cong \beta\left(-\frac{M}{r_b} + \frac{\sqrt{\kappa Q^2 + r_b{}^4}}{r_b}\right). \tag{40}$$

Substituting the results above into Eq. (*37*) and omitting the null terms (terms that becomes zero as $r \to r_b$), leave us with the following results

$$I_{\text{ct}} = \beta\left(M - \frac{r_b{}^3\sqrt{\kappa Q^2 + r_b{}^4}}{r_b{}^4 - \kappa Q^2}\right). \tag{41}$$

Combining the results that we have obtained so far from Eqs. (*21*), (*23*), (*35*), and (*41*) yields the Euclidean action as follows

$$I = \beta\left(\frac{\sqrt{Q^2\kappa + r^4}(Q^2 + 3r^2)}{12r^3} + \frac{2}{3}\sqrt{\frac{iQ^3}{\sqrt{\kappa}}}\text{EllipticF}\left[i\text{ArcSinh}\left[r\sqrt{\frac{i}{Q\sqrt{\kappa}}}\right], -1\right] + \frac{1}{6}\sqrt{\frac{Q^3}{\pi\sqrt{\kappa}}}(\text{Gamma}[1/4])^2\right). \tag{42}$$

Based on the obtained Euclidean action, we may calculate the black hole's entropy, which would yield the following form

$$S = \beta\left(\frac{\partial I}{\partial \beta}\right)_A - I, \tag{43}$$

$$= \beta\left(\frac{\partial I}{\partial r_+}\right)_A \Big/ \left(\frac{\partial \beta}{\partial r_+}\right)_A - I,$$

$$= \frac{\pi(\kappa Q^2 + r_+{}^4)}{r_+{}^2}. \tag{44}$$

The result above is consistent with the form found previously in [15]. We can see that the entropy for our case is different from that of the ordinary Einstein-Hilbert gravity, due to the existence of κ. Nevertheless, as $\kappa \to 0$ we can see that the entropy would reduce to the usual $\pi {r_+}^2$ Result. Note that the form given in Eq. (*43*) can also be obtained from the black hole's entropy and event horizon area relation calculated from the auxiliary metric

$$S = \frac{A}{4} = \frac{1}{4}\int \sqrt{q_{\theta\theta}q_{\phi\phi}}\,\mathrm{d}\theta\mathrm{d}\phi. \tag{45}$$

4. Conclusion

In this work, we have calculated the Euclidean action for asymptotically flat EiBI black holes with Maxwell electrodynamics. We found that by employing appropriate ensemble in the Euclidean action evaluation, a finite and well-defined expression of the Euclidean action could be obtained. The Euclidean action then is used to obtain the black hole's entropy, which was found to slightly differ from the usual $\pi {r_+}^2$ Expression in general relativity due to the coupling of κ and Q on this model of modified gravity. The form of the entropy is found to be consistent with the previous work on this matter. Further works on the interlinking between the first law of black hole mechanics and the Smarr formula for this particular type of black hole in EiBI gravity would seem to be an interesting inquiry to pursue.

References

[1] Bekenstein J D 1973 *Black Holes and Entropy*, *Phys. Rev D,* pp. **7** 2333–46

[2] Hawking, S W 1971 *Gravitational Radiation from Colliding Black Holes, Phys. Rev Lett.,* pp. 26 1344–6

[3] Bardeen J M, Carter B and Hawking, S W 1973 The Four Laws of Black Hole Mechanics *Commun. Math. Phys,* pp. 31 161–70

[4] Hawking, S W 1975 *Particle Creation by Black Holes*, *Commun. Math. Phys,* pp. 43 199–220

[5] Gibbons, G W and Hawking, S W 1977 *Action integrals and partition functions in quantum, gravity Phys. Rev. D,* pp. 15 2752–6

[6] Bands M and Ferreira P G 2010 *Eddington's theory of gravity and its progeny, Phys. Rev. Lett.,* pp. 105

[7] A. S. Eddington F R S 1923 *The mathematical theory of relativity* Vol 15 (Cambridge University Press)

[8] Born M and Infeld L 1934 *Foundations of the New Field Theory, Proc. Roy. Soc.,* pp. A144

[9] Delsate T and Steinhoff J 2012 *New insights on the matter-gravity coupling paradigm, Phys. Rev. Lett.,* pp. 109 1–5

[10] Sotani H and Miyamoto U 2015 *Properties of an electrically charged black hole in Eddington-inspired Born-Infeld gravity Hajime, Phys. Rev. D - Part. Fields, Gravit. Cosmol.,* pp. 92

[11] Wei S-W, Yang K and Liu Y-X 2015 *Black hole solution and strong gravitational lensing in Eddington-inspired Born-Infeld gravity, Eur. Phys. J. C,* pp. 75

[12] Visser M 1992 *Dirty black holes: Thermodynamics and horizon structure, we Phys. Rev. D,* pp. 46 2445–51

[13] York J W 1986 *Black-hole thermodynamics and the Euclidean instant action, Phys. Rev. D,* pp. 33 2092–9

[14] Arroja F, Chen C Y, Chen P and Yeom D H 2017 *Singular instantaneous in Eddington-inspired-Born-Infeld gravity* Vol 2017

[15] Kusuma A M and Ramadhan H S 2018 *Thermodynamics of black holes in Eddington inspired Born-Infeld theory of gravity, AIP Conf. Proc.,* pp. 2023

Journal of Physics: Conference Series **2019** (2021) 012089 doi:10.1088/1742-6596/2019/1/012089

Mapping of Two Models Peak Ground Acceleration (PGA) of Indonesia

T Zera, M Nafian[1], A Ramadhani, and A R Fauziah
Department of Physics, Universitas Islam Negeri Syarif Hidayatullah
Jl. Ir. H. Juanda No. 95 Ciputat, 15412, Indonesia
Email: [1]tati_zera@uinjkt.ac.id

Abstract. This research was conducted to determine and mapping the value of the PGA in Indonesian territory using two empirical models of attenuation equations from the transition of the second generation to the third generation. The research was conducted by applying the Frisenda's and Luzi's. The calculation is based on a catalog of earthquakes that occurred from 1 January 1970 - 31 December 2020 with a variable earthquake magnitude ≥ 5 SR and hypocentre depth ≤ 60 km (destructive earthquake). The calculation results in the maximum value of PGA respectively 0.6613 g for Frisenda, and 12.257 g in the Luzi. The maximum PGA value in Luzi model is the effects of the Andaman Islands earthquake on December 26th, 2004, with a magnitude of 9,1 SR, and the maximum PGA value in the Frisenda model is the effects of the Flying Fish Cove, Christmas Island earthquake on June 13th, 2013, with the magnitude 6.7 SR. In general, these two maps show different or even opposite patterns.

1. Introduction

Indonesia's territory is located on three main plates, namely the Eurasian plate in the north, the Indo-Australian plate in the south, and the Pacific in the northeast causing high seismic intensity and becoming the main tectonic character of the Indonesian archipelago. Several large earthquakes that have occurred in the last 50 years in Indonesia have caused significant losses and casualties. These include the Banda earthquake (8.5 SR) in 1983, the Sumatra-Andaman Islands earthquake (9.1 SR) in 2004, the North Sumatra/Nias earthquake (8.6 SR) in 2005 the West Coast earthquake of Sumatra (8.6 SR) in 2012 the Lombok (7 SR) in 2018, the Donggala-Palu (7,5 SR) in 2018 and others. Earthquakes as a natural disaster, apart from being unpredictable when they will occur, also have other effects in the form of ground movement. This ground movement causes losses because it can damage buildings and other facilities.

As a preventive effort to overcome the impact of the earthquake, it is necessary to make an earthquake risk map base on the effect it causes. One of the effects of an earthquake that can be used as a reference is a map of the acceleration of ground motion which is known as Peak Ground Acceleration (PGA). These values can then be mapped and an overview of areas that are vulnerable or safe to the risk of ground movement can be seen. As a result of the accumulation of earthquakes, the peak ground acceleration will change every time an earthquake occurs, so this map needs to be updated every certain period [1].

The Researches on earthquakes and Peak Ground Acceleration (PGA) has been carried out by many researchers. Atkinson and Boore have made an empirical relation of ground motion for subduction areas [2], and the Hybrid empirical method in the ground movement was also carried out by Campbell [3]. Research on PGA in several regions of Indonesia, which was conducted separately, Megawati researches ground motion attenuation for major earthquakes in Sumatra [4]. Irsyam it. All proposed a seismic risk map for the islands of Sumatra and Java for the micro-zoning study of the city of Jakarta [5]. Tati Zera et.al a study compared the PGA map based on three different models for Bengkulu [6]. Tati and Nafian also conducted similar research to compare the three PGA models for the West Java region [7]. Research for zoning earthquake hazard areas using the global attenuation function developed by Young et.al has also been carried out by Edy Irwansyah [8] This research mapping the PGA value for all regions of Indonesia as an area with high seismic activity based on two models that are rarely used but are proposed in the BRGM Intermediary Report [9].

2. Peak Ground Acceleration (PGA)

As the effect of an earthquake, the Peak Ground Acceleration (PGA) can be measured using an accelerometer. However, it is difficult to use for large areas. Another alternative to obtain the PGA value is to determine it from the accumulation of earthquakes. Acceleration is a quantity that states the change in velocity from rest to a certain speed. Peak Ground motion acceleration (PGA) is the greatest value of acceleration at a place due to earthquake vibrations during a certain period. One of the effects of an earthquake in a place is the acceleration of the ground on the surface. The magnitude of the acceleration of the ground surface movement indicates the risk of an earthquake that needs to be estimated.

In calculation, the value of ground motion, especially Peak Ground Acceleration (PGA) requires several earthquake parameters, including (1) earthquake magnitude on the surface (Ms). If the Earthquake incident data obtained is not in the form of a Magnitude Surface, it needs to be converted first. (2) Distance to the Site (R), namely the distance from the hypocentre to the recording station. (3) Epicenter (Δ) is the distance from the projection of the hypocenter on the earth's surface of the site or the recording station. Expressed in geographical coordinates, usually accompanied by distance and direction information. Mathematically the value of Δ can be determined from the formulation between two points. (4) Depth of Earthquake (h), is the place/center of the earthquake which is below the earth's surface. Besides these four parameters, local conditions of the soil also affect the PGA value.

Since the acceleration of ground motion decreases with increasing distance, the distribution function of acceleration will be the logarithm of the distance. And the distance is one of the important (main) parameters in the acceleration function of ground motion. In addition, the energy that spreads from the epicenter will decrease due to the attenuation of soil material. Furthermore, Kramer [10] said that the amplitude of the ground motion will decrease exponentially both concerning the distance and magnitude of the earthquake, so the function of peak ground acceleration, in general, expressed as,

$$\log(y) = f(M, R, c) \tag{1}$$

Where y is Peak Ground Acceleration (PGA), M is Earthquake Magnitude at the surface, R is the distance from the hypocenter to the recording station (site) and c is the coefficient.
Research on attenuation of ground acceleration due to earthquakes in various formulations has been carried out for a long time and is divided into two groups, namely Worldwide Attenuation which is the attenuation of ground acceleration using mixed data from various places, various variables, and general attenuation. Next is the specific attenuation where this attenuation takes into account the place, the mechanism of earthquake occurrence, the type of soil, and the activity of certain tectonic movements. This attenuation is then grouped into shallow crustal attenuation in both active and passive areas and then earthquake attenuation in subduction areas [11]. Research on the function of attenuation for the acceleration of ground motion has progressed very well and has produced many models that accommodate conditions in various regions with different geological structures. This research will use a global attenuation function with mixed data will be used.

Journal of Physics: Conference Series **2019** (2021) 012089 doi:10.1088/1742-6596/2019/1/012089

3. Method

In this research, the value of peak ground acceleration was calculated based on a catalog of the earthquake that occurred for 50 years from 1 January 1970 - 31 December 2020 with parameters; magnitude 5 SR and hypocentre < 60 km (destructive earthquake) that occurred within the territory of Indonesia at the limits of 6° N - 11° S and 95° - 141° E. A total of 7488 earthquake data was obtained according to the established criteria [12]. The plot of the data obtained is shown on the Indonesia seismicity map as shown in Figure 1.

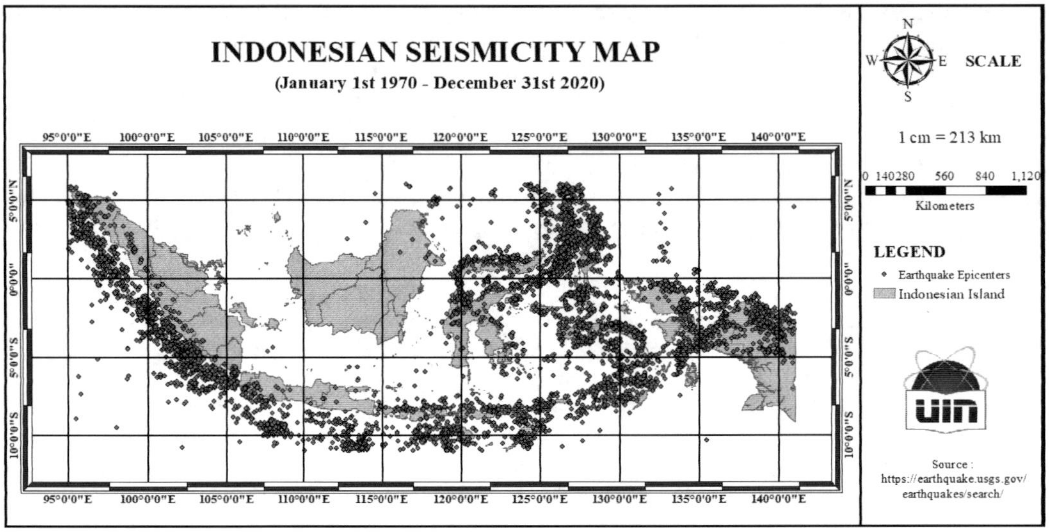

Figure 1. Indonesian Seismicity Map Jan 1970 - December 2021

Calculations were carried out using two equation models proposed by Frisenda and Luzi. These two-equations are included in the 207 ground motion equation models for the prediction of peak ground acceleration listed in the BRGM intermediate report by Douglas [9]. The model developed by Frisenda is based on earthquake data that occurred in northern Italy and data from the Lunigiana – Garfagnana seismic network. Frisenda formulates the empirical method for the PGA as equation (2):

$$\log(PGA) = -3.19 + 0.87M + 0.042M^2 - 1.92\log(R) \tag{2}$$

In this model, there are four Frisenda constants with two of them positive and the other two negative. It is necessary to convert the Magnitude (M) value obtained from the catalog to the surface magnitude value to be used in this Frisenda model. The distance from the epicenter to the recording station or, in this case, is the distance from the hypocentre to all grid corner points specified on the map.

While the model proposed by Luzi is built based on research on 2895 earthquake records from 78 events with a hypocenter depth of < 60 km at 22 stations and several other studies, Luzi (2006) proposes an empirical form of PGA as follows equation (3):

$$log\ (PGA) = -4.417 + 0.77M - 1.097\ log\ R \tag{3}$$

The provisions for a magnitude (M) and distance (R) applied to the Frisenda model are also applied to the Luzi model in this study. PGA results obtained are in units of g while 1 g = 1000 gals. The plot of the calculation results of each model on the base map provides an overview of the PGA values in the form of contours.

4. Results and Analysis

The results of the PGA calculation using the Frisenda model produce a maximum value of 0.6613 g which is at the position 4^0S, 130^0E and is thought to have been caused by an earthquake that occurred at 3.913^0 S – 131.368^0 E Christmas Island on June 13th with the 6.7 SR and the minimum value of PGA is obtained at 0.028 g which is at 14^0S, 95^0E and is thought to have been caused by an earthquake that occurred at 10.004^0 S – 107.236^0 E Seram Sea on January 8th 2008 with the 5,8 SR.

The plot of the calculation results using this model on the PGA contour base map with the maximum value being in the eastern part of Indonesia around the Seram Island and continuing to decline to the west. While the minimum value is around the Sambas area of Kalimantan. This picture asks for compatibility with seismicity data in the eastern region of Indonesia but does not match seismicity data in the western region. Further research is needed to determine the suitability of the Frisenda model with the subduction area which in this study is located on the west coast of Sumatra and continues to the south of Java. The suitability of the Frisenda model with Indonesia seismicity in eastern Indonesia looks quite good, especially the Banda Arc [13]. But overall, for the Indonesian region, this model is not satisfactory. PGA value contour map with Frisenda model can be seen in Figure 2 below.

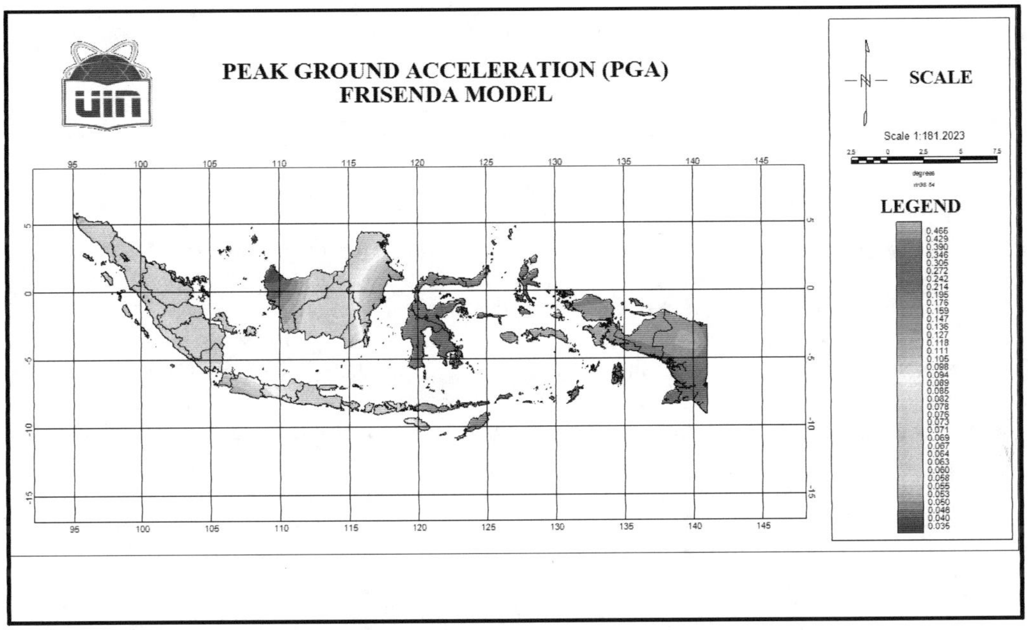

Figure 2. Peak Ground Acceleration (PGA) Map with Frisenda Model.

The results of the PGA calculation using the Luzi model produce a maximum value of 12,257 g which is at the position 6^0S, 95^0E eastern Indonesia (Aceh) and is thought to be caused by an earthquake that occurred at 3.295^0 N – 95.982^0 E Andaman Island on December 26th ,2004 with the magnitude 9.1 SR and the minimum value of PGA is obtained at 5,750 g which is located at 9^0S, 140^0E eastern Indonesia (Papua) and is thought to have been caused by an earthquake that occurred at 3.295^0 N – 95.982^0 E Andaman Island on December 26th ,2004 with the magnitude 9.1 SR.

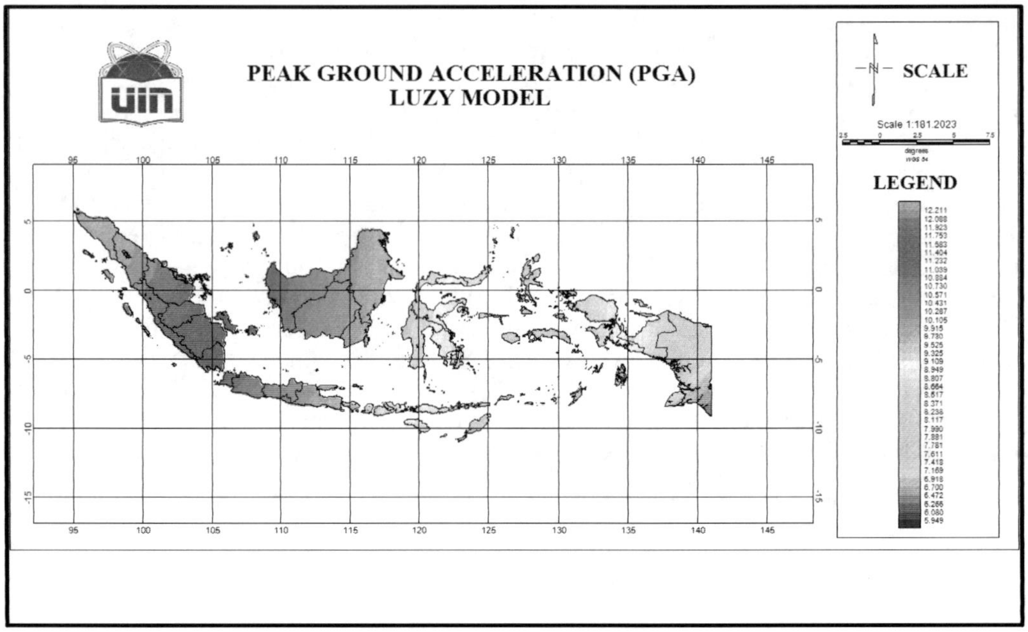

Figure 3. Peak Ground Acceleration (PGA) Map with Luzi Model

Plotting the PGA using the Luzi model on the base map shows the PGA value increasing from Eastern Indonesia to the West. The minimum value is in eastern Indonesia (Papua) and the maximum value is in the north of Aceh which is the western part of Indonesia. This is different from the pattern produced by the Frisenda model where the maximum value of the PGA is around the Seram Islands, the Banda Sea which is the eastern part of Indonesia. This map shows the correspondence with seismicity data in western Indonesia, where the seismicity level is quite high along the subduction zone that stretches along the west coast of Sumatra and south of Java. However, it is not suitable for eastern Indonesia, especially in the Banda Basin area.

The extreme conditions in the form of maximum and minimum PGA values generated from these two models along with the earthquake that caused can be seen in Table 1 below. From the table, it is clear that these two models give different results.

Table 1. The Maximum and minimum value of PGA

Model	PGA Max	PGA Min
Frisenda	0.6613 g 4^0S , 130^0E 2013-06-13 16:47:23 6.7 SR, 3.913^0 S – 131.368^0 E Christmas Island	0.0285 g 14^0S , 95^0E 2008–01–08 19 : 23: 31 5.8 SR, 10.004^0 S – 107.236^0 E Seram Sea
Luzi	12.257 g 6^0S , 95^0E 2004-12-26 T 00:58:53 9.1 SR, 3.295^0 N – 95.982^0 E Andaman Island	5.750 g 9^0S , 140^0E 2004-12-26 T 00:58:53 9.1 SR, 3.295^0 N – 95.982^0 E Andaman Island

5. Conclusion

The calculations and plots of the PGA that have been carried out for the two models tested in this research have given different results. Frisenda gives the maximum value of PGA in Eastern Indonesia (Seram Islands, Banda) and the minimum value is obtained in Eastern Indonesia (Sambas, Kalimantan). On the other hand, the Luzi model gives a maximum value in the west (Aceh) and the minimum in the east (Papua). These two models provide a good fit for different seismicity data groups or regions of Indonesia. The Frisenda model is more suitable for the eastern part of Indonesia and the Luzi model is more suitable for the western part of Indonesia. Further research is needed to obtain an appropriate model for Indonesia's overall seismic conditions.

References

[1] Masyhur I, Wayan S, Fahmi A, Sri W, Wahyu T, Danny H N, Engkon K, Irwan M, Suhardjono, Asrurifak M and Ridwan M 2010. Bandung
[2] Atkinson, G M and Boore D M 2003 *Society of Amerika.* **93(4)** 1703 – 29
[3] Campbell, K W 2003 *Society of Amerika.* **93(3)** 1012 – 33
[4] Megawati K and Pan T C 2010 *Eng. And Struct. Dyn.* **39** 827 – 45
[5] Irsyam M, Dangkua D T, Hendriyawan, Hoedajanto D, Hutapea B M, Kertapati E K, Boen T and Petersen M D 2008 *J. Earth. Sys. And Sci.* **117** 865 – 78
[6] Tati Z, Sutrisno and Agus B 2017 *IOSR J. Apple. Geology and Geophy.* **5** 31 – 36
[7] Tati Z, and Muhammad N 2017 *Int. L Conf. on Sci and Tech* **149** 140 – 42
[8] Eddy I and Edi W 2012 *price. Sem. Nas. If 2012* E14-21
[9] Douglas J 2006 *Errata of addition to Ground Motion Estimation Equation 1964 – 2003* Intermediary Report BRGM 54603-FR
[10] Kramer, S L 1996 *Geotecnical Earthquake Engineering* (New Jersey: Prentice Hall)
[11] Widodo P 2012 *Seismologi Teknik & Rekayasa Kegempaan* (Yogyakarta: Universitas Islam Indonesia) p. 330
[12] http://earthquake.usgs.gov/earthquakes/search/
[13] _Warren H 1979 *Tectonic of Region, Indonesia* (Banner Science: Library US Dept. Of Interior)

Design, Prototype of Arduino and Android-Based Anthropometry System for Toddler's Head and Upper Arm Circumference Measurement

Umiatin[1], R M Mauldy and W Indrasari
Department of Physics, Universitas Negeri Jakarta
Jl. Rawamangun Muka, Jakarta 13220, Indonesia
Email: widyafisikaunj@gmail.com

Abstract. Nutritional status monitoring is a very important in toddler, because in this period children are vulnerable to trauma, infection, or developmental disorders. Nutritional status is one of the health indicators in the successful achievement of MDGs No. 4 related to the reduction of child mortality, which is 45% caused by malnutrition. Head and upper arm circumference are an important anthropometry parameter on toddler growth. The monitoring process for toddler in Posyandu usually by writing and plotting the data in KMS (Kartu Menuju Sehat) manually. Therefore, efforts to modernize Posyandu to facilitate Posyandu officers in taking measurements, as well as become one of the database systems of nutritional status of Indonesian toddler. The purpose of this study was to develop prototype of anthropmometry using ultrasonic sensor HC-SR04 and Hall sensor UGN3503 based Arduino. The system will be integrated with the database through Android applications. The results of this study had an accuracy value of 97.72% for the testing of head circumference measuring instruments and 95.35% of the testing of upper arm circumference measuring instruments. Android applications created to serve to store and provide information about the growth and development and nutritional status of a toddler

1. Introduction

Growth and development of children take place regularly and are interrelated from conception to adulthood. Even though there are variations, each child will pass a certain pattern and experience a very rapid increase, namely the age of 0-5 years, which is commonly referred to as the golden age [1]. The golden age is a very important period to pay attention to the nutritional status of children carefully so that it can be detected in the event of a developmental disorder [2]. Nutritional status is one of the health indicators in the successful achievement of MDGs No. 4 related to reducing child mortality, which 45% is caused by malnutrition [3].

Posyandu is an Indonesian government program to check the nutritional status of toddler using anthropometric measurements. Anthropometry is one of the assessments of nutritional status that measures human individuals to determine human physical variations [4]. The advantages of anthropometry are that the procedure is simple, safe, and can be performed in a large number of samples [5]. Compared to other methods, anthropometric are more practical to assess nutritional status in the public. In collecting data on the growth and development of toddlers, Posyandu provides a data collection book (Kartu Menuju Sehat / KMS) [6]. KMS is used to monitor children's growth, record children's health services and educational tools for parents [7].

Head circumference and upper arm circumference are important parameters in examining the nutritional status of children under five at posyandu. Head circumference describes brain growth from the estimated volume in the head, head circumference is influenced by the nutritional status of children up to 36 months of age [8]. The circumference of the upper arm describes the growth and development of fat tissue under the skin and muscles which is not much affected by the state of body fluids compared to body weight, measuring the circumference of the upper arm aims to determine energy and protein reserves. Measurement of head circumference and upper arm circumference at posyandu still uses an elastic measuring tape and the data is plotted manually in a data collection book (Kartu Menuju Sehat / KMS) [2].

In this study, the development of a toddler growth monitoring system was carried out, namely the Arduino-based and integrated head circumference measurement system for the Android application. This system uses Arduino as a data processor, ultrasonic sensor and Hall effect sensors to detect head circumference and upper arm circumference of a toddler. The Android application plays a role in accessing information on the development of children under five, storing data on the identification of toddler growth and development, and assisting in auditing measurements to monitor the nutritional status of children.

This research was conducted referring to previous research by Kinanthi, et al. Made measurement of weight, body length and head circumference in infants with graphical display using ATMEGA 32 [9]. Further research by Nurfiansyah, et al. Who made an Android-based mobile application to store and provide data about toddler growth, such as body weight, height, and head circumference every month [10]. Recent research by emotion, et al. Who developed a weight and height measurement system for toddler using Arduino with an accuracy of above 98% [11].

2. Method

This system is designed using the Arduino Mega 2560, the HC-SR04 ultrasonic sensor, the UGN3503 Hall sensor and an Android smartphone. In this study, the HC-SR04 ultrasonic sensor and the UGN3503 Hall sensor were characterized as detectors of the head circumference and upper arm circumference of toddler to determine the range and identify the sensitivity of the sensors used.

Arduino Mega 2560 as the main control of this system, Arduino is also a data processor for the head and upper arm circumference data sent by the network, then the data is stored and displayed on the android application.

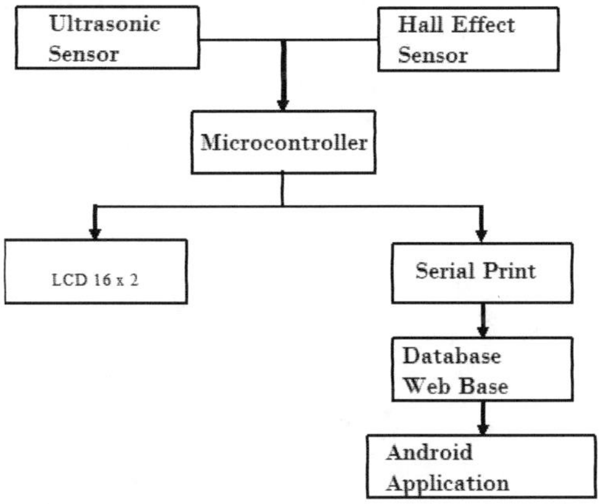

Figure 1. Block diagram of anthropometry prototype

3. Results And Discussion

In this study, 4 HC-SR04 ultrasonic sensors were used to measure the diameter of the head and upper arm circumference of a toddler. Ultrasonic sensor characterization is done by comparing the sensor output value (digital value) with the measurement results of the distance measuring instrument.

Measurement limit starts from 3 cm to 20 cm. The output of the sensor is in centimeters (cm). The results of this characterization can be seen in the graph below.

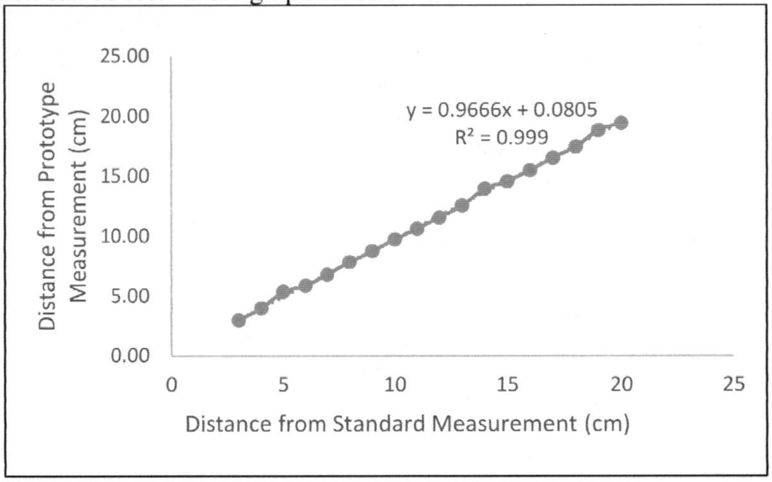

Figure 2. Graph of the ultrasonic sensor's measured distance to the reference distance

Based on these results can also be known relative error values on the sensor by using the following equation:

$$Error(\%) = \frac{|PS-PL|}{PL} \times 100\% \quad (1)$$

PS is the value reading of the sensor used, and PL is the actual value reading or that has been tested using a laboratory measuring instrument. The average relative error value is 1.99%. Thus, with such a small deviation value, it can be interpreted that the ultrasonic sensor can be used as a distance measurement tool to calculate the diameter of the head circumference and the circumference of the toddler's upper arm. The development of the Hall UGN3503 sensor as a proximity sensor is based on its ability to detect changes in the magnetic field. Changes in the magnetic field occur due to the change in distance between the sensor and the object measured in distance. In this case, the output voltage of the sensor is proportional to the change in the magnetic field. The characterization of the UGN3503 sensor is done by measuring the sensor output voltage against the distance from the permanent magnetic source, so that the furthest distance is known from the sensor's increasingly small and stable output voltage value.

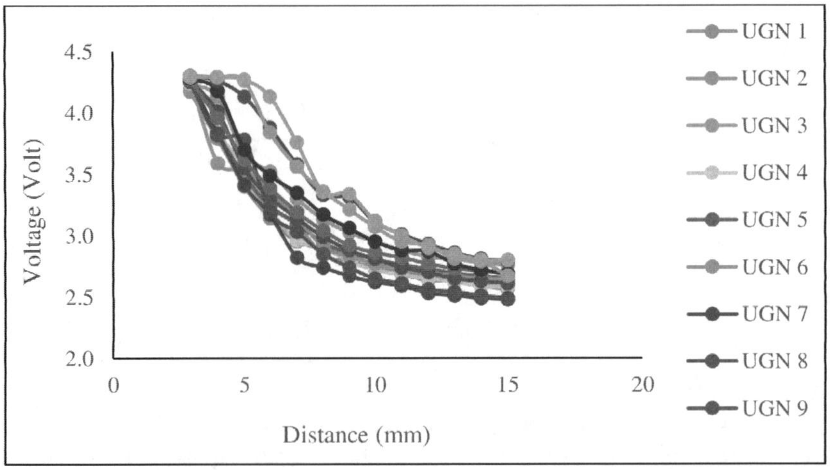

Figure 3. Hall sensor voltage graph versus reference distance

Journal of Physics: Conference Series **2019** (2021) 012090 doi:10.1088/1742-6596/2019/1/012090

The graph explains that the output voltage of the sensor is inversely proportional to the sensor's reference distance, the farther the magnetic field source distance to the sensor, the smaller the sensor output voltage in the range of 3 mm - 15 mm and the farthest detection distance range of 30 mm, this will be used as a reference in the manufacture of prototypes. Testing of prototypes monitoring the circumference of the head and upper arm circumference of toddler in this study was conducted to determine whether or not this circuit system could run properly. Prototype testing was conducted with HC-SR04 ultrasonic sensor readings and Hall UGN3503 sensors simultaneously, as the HC-SR04 ultrasonic sensor has a working range from 3 cm to 400 cm with a sensor resolution of 1 cm, while the Hall UGN3503 sensor has a working range ranging from 0.1 cm to 3 cm with a sensor resolution of 0.1 cm. Therefore the simultaneous use of the HC-SR04 ultrasonic sensor and Hall UGN3503 sensor adds a level of sensory reading accuracy to measure the diameter of the toddler's head circumference and upper arm circumference. Head circumference testing using a combination of 3 ultrasonic sensors and 12 Hall effect sensors, the sensor detection result is the result value of the elliptical circumference equation.

$$\text{Elliptical Circumference} = {}^{1}/_{2}\ \pi\ (R1 + R2) \quad (2)$$

D is the diameter of the circumference of the upper arm. The results of the prototype can be seen in the following image

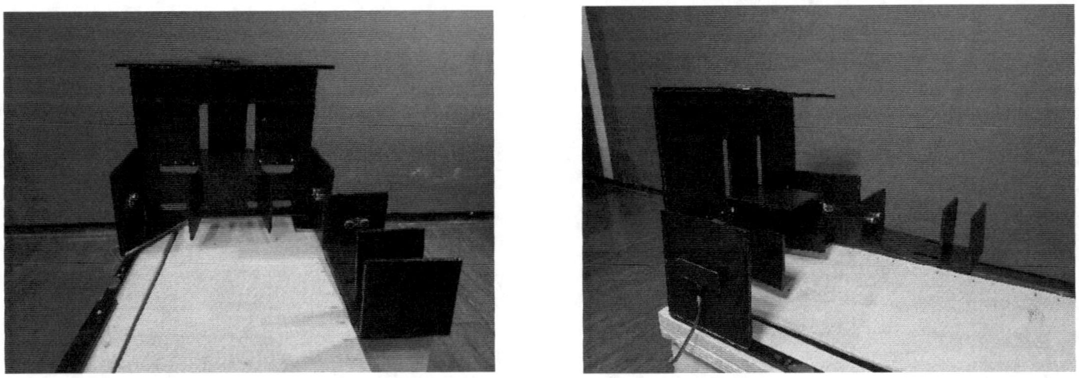

(a) (b)

Figure 4. Prototype head circumference and upper arm circumference of a toddler (a) Front view, (b) Side view

The measurement data of head circumference and upper arm circumference that have been tested are as follows:

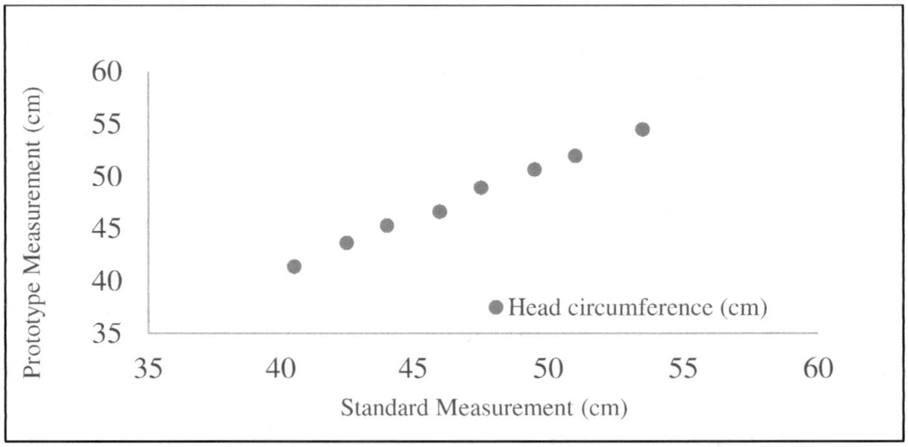

Figure 5. Graph testing prototype head circumference

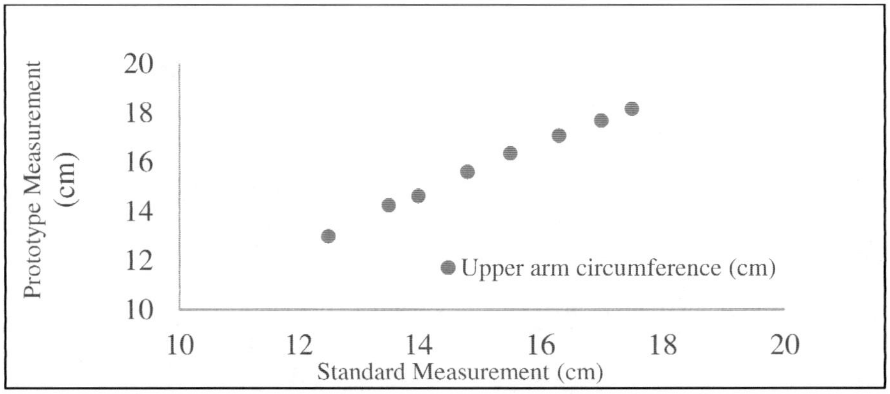

Figure 6. Graph testing prototype upper arm circumference

Based on the test results, the prototype can read the sensor output and read the predetermined algorithm properly. In this prototype, test, the result of head circumference measurement has an average error value of 2.28% and upper arm circumference measurement has an average error value of 4.65%. This average error value is smaller than the error value in previous studies, as in previous studies only used ultrasonic sensors as a gauge, whereas in this prototype it used ultrasonic sensors and hall effects simultaneously. Thus the prototype measurement results become more accurate if the ultrasonic sensor is used simultaneously with the Hall effect sensor because the Hall effect sensor has a very high precision as a distance meter compared to the ultrasonic sensor (can be seen from the resolution of the sensor).

The application of the monitoring system of toddler growth is accompanied by the creation of an account and nutritional status of the toddler. This app called "E-BALITA", this wave-based android app is made to make it easier for parents to monitor the growth of toddler in real-time, fast and efficiently. In this application there is a menu of toddler profiles, parent profiles, data on toddler growth, immunizations, healthy info for toddler and settings.

(a) (b) (c)

Figure 7. Front App Display; (a) Start page; (b) Home page; (c) Toddler Profile

(a) (b) (c)

Figure 8. Inside App Display; a) Measurement page; (b) Growth data of lk toddler; (c) LLA growth data for toddler

4. Conclusion

The design and prototype of an anthropometry system with the head circumference and upper arm circumference of a toddler using the HC-SR04 ultrasonic sensor and Hall sensor UGN3503 have been created. The test result of the prototype head circumference has an accuracy value of 97.72% and the test result of the upper arm circumference prototype has an accuracy value of 95.35%. The prototype integrates with an Android-based toddler growth monitoring app called "E-BALITA". The app features measurements of the toddler's head circumference and upper arm circumference as well as the nutritional status of the toddler in real time, quickly and efficiently.

Acknowledgment
This study was supported by the research grant from Universitas Negeri Jakarta, contract no. 12/ PPI/LPPM/IV/2021.

References

[1] Wicaksono F and Harsanti T. Determinants of stunted children in Indonesia: A multilevel analysis at the individual, household, and community levels, *Kesmas*, 15 48–53. 2020
[2] Kusuma R M and Hasanah R A 2018 *Antropometri Pengukuran Status Gizi Anak USIA 24-60 Bulan di Kelurahan Bener Kota* (Yogyakarta: J. Med. Respati) 13 1970–3887
[3] Nations U. *United Nations Millenium Development Goals*
[4] Fitri M O. Aplikasi Monitoring Perkembangan Status Gizi Anak Dan Balita Secara Digital Dengan Metode Antropometri Berbasis Android, *J. Instek*, 2 140–9. 2017
[5] Ruslianti A I 2014 *Gizi Terapan* (Bandung: PT. Remaja Rosdakarya)
[6] Setyaningsih P H and Kurnia L 2020 *Analisis Hubungan Keaktifan Ibu Dalam Posyandu Dengan Penurunan Jumlah Balita Bawah Garis Merah Dapat Dilihat Berdasarkan Riset Kesehatan,* 4 32–9
[7] Hasyim D I and Sulistyaningsih A. Pemanfaatan Informasi Tentang Balita Usia 12-59 Bulan pada Buku KIA dengan Kelengkapan Pencatatan Status Gizi di Buku KIA *J, Kedokt. Dan Kesehat,*. 15 1. 2019
[8] Ferreira H D S. Anthropometric assessment of children's nutritional status: A new approach based on an adaptation of Waterlow's classification, *BMC Pediatr,* 20 1–11. 2020
[9] Kinanthi R, D, Nugraha P C and Ariswati H G. Alat Pengukur Berat Badan, Panjang Badan dan Lingkar Kepala Bayi dengan Tampilan Grafik (Panjang Badan dan Lingkar Kepala Bayi), *J. Tek. Elektromedik,* 10 1–14. 2017
[10] Nurfiansyah A, Hariyani Y S and Ramadan D N 2018 Desain Dan Implementasi Aplikasi Android Untuk Memantau Status Tumbuh Kembang Balita, 4 2402–12
[11] Umiatin U, Erlandita S, M and Indrasari W. Design baby mass and height monitoring system based on Arduino and Android application *AIP, Conf. Proc.,* 2169. 2019

DC-DC Buck Converter Circuit Simulation on Solar Panel Electricity Storage System

W Indrasari[1], F A I Prasetya[1] and R N Setiadi[2]
[1] Department of Physics, Faculty of Mathematics and Sciences, Universitas Negeri Jakarta, Jakarta 13220, Indonesia
[2] Department of Physics, Faculty of Mathematics and Sciences, Universitas Riau, Riau 28293, Indonesia
Email: widyaningrum-indrasari@unj.ac.id

Abstract. Electrical energy is one of the energies that humans widely use. One of the components used to generate electrical energy is solar cells. By using solar cells, energy can be stored in a battery or used directly. In stores electrical energy into the battery using DC-DC converter is still found to have a loss of power that affects the low efficiency of electricity generated. So it needs a system that can control the process of storing energy to the battery, one of which uses a DC-DC converter with Buck topology that will be used on solar panel electrical power storage system. To optimize the DC-DC Buck converter circuit, we plan a circuit simulation by considering the value of the quality factor Q of the Butterworth filter. The test results showed that the filter with a value of Q = 0.707 was of good quality with a gain of -3.1 dB. If the Q value increase, the gain will be increased, and the time it takes until the voltage stabilizes increases.

1. Introduction

Energy has an essential role in human life to date, one of which is electrical energy. Today's electrical power still uses fossil energy sources, where its availability has been depleted in nature and has an environmentally unfriendly effect [1]. By looking at this condition, new renewable energy that is environmentally friendly is needed. One of the efforts to use renewable energy is using solar cell technology. Solar cells are made of semiconductor materials that can convert solar radiation into electrical energy with the photovoltaic principle [2]. The electrical energy produced by solar cells can be used directly or stored in batteries.

Various kinds of developments in the use of solar cells have been carried out to increase the production of energy produced, ranging from technology by combining solar cells with other energy producers (hybrid) and the use of solar tracker technology [3] -[4]. In addition, further developments are carried out using a DC-DC converter in the energy transmission process. One of these types of converters is the buck converter. Buck converter functions to control the DC voltage by making the output voltage equal to or smaller than the input voltage. This converter works with the switching process on and off using a MOSFET based on the applied duty cycle value. In the working process of this converter, it is still found that there is a loss of power, which affects the resulting efficiency. A previous study [5] found that the power loss in the buck converter applied to the Maximum Power Point Tracker (MPPT) system was 57.69%. Also, another study [6] has a result obtained shows that the buck converter has many power losses when controlled by the four Maximum Power Point Tracker (MPPT) algorithms. To solve the problem, it is necessary to correct and develop the filter design used in the buck converter, a low pass filter with the Butterworth type of order-2. A low pass filter or low pass filter is a filter that will pass frequencies below the cutoff frequency and hold frequencies more significant than the cutoff frequency. The quality of the filtering performance carried out by a filter is influenced by the quality factor Q [7]. In this article, an analysis of the output signal generated by the buck converter is carried out by providing a variety of the Q quality factor, which is simulated with LTSpice software. The data obtained is then used to determine the output quality factor of the filter. The results of this study will be used in the design of the buck converter design for the application of a solar panel electrical power storage system.

2. Method

The developed DC-DC converter circuit uses a buck topology which functions as a voltage reducer. The buck converter uses a Low Pass Filter (LPF), which consists of the main components, namely inductor L, capacitor C, and load resistance R. In addition; a MOSFET acts as an active switch and a diode as a passive switch. In this circuit, the IRF2805 MOSFET is used. The IRF2805 MOSFET is a n-channel MOSFET with fast switching characteristics, with a maximum voltage of 55V and a maximum current of 75A [8]. So that the application of the MOSFET in this study is considered relatively safe because, in the simulation, the buck converter will reduce the supply voltage of 20V. Although the buck converter circuit shown in FIGURE 1 is considered simple, the resulting efficiency is considered relatively high [9].

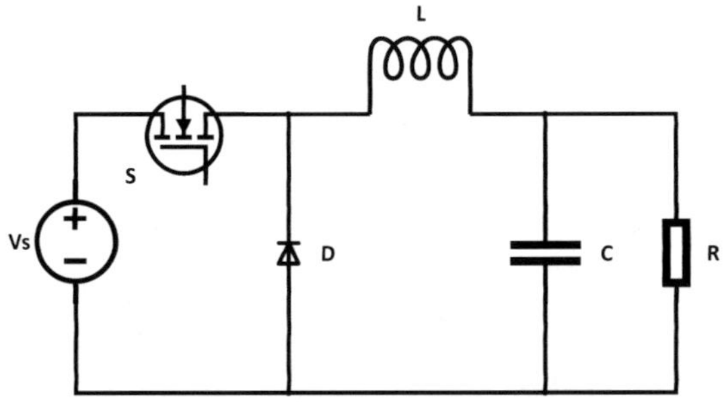

Figure 1. Buck Converter Circuit Schematic [6]

Figure 1 is a schematic of the basic circuit of a buck converter. Inside the buck converter, a switch function to work on and off by adjusting the pulse width based on the duty cycle value applied using Pulse Width Modulation (PWM) [10]. PWM is the basis of a power electronics control system, where the PWM wave formed represents the work cycle performed by a semiconductor device on and off [11]. The equation expresses the duty cycle used by the buck converter:

$$D = \frac{T_{ON}}{T} = \frac{T_{ON}}{T_{ON}+T_{OFF}} = f_s T_{ON} \quad (1)$$

T_{ON} is the period of the switch in the on mode, T_{OFF} is the period of the button in the off mode, and fi_{sh} is the switching frequency (Hz).

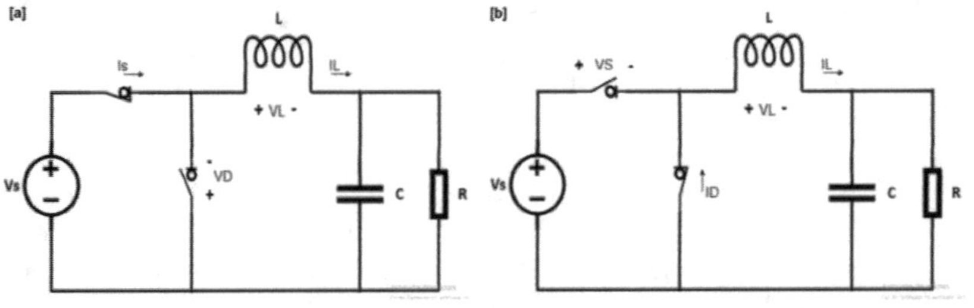

Figure 2. Buck Converter Circuit Schematic (a) When Condition is on, (b) When Condition is off

When the switch is on, as shown in Figure 2a, the diode is in a reverse-biased condition where the current will pass through the inductor, and the inductor will be charged. Then the current is passed to the capacitor for filtering and forwarded to the load, then back to the source. Then when the switch is off like Figure 2b, the diode becomes forward biased where energy discharges from the inductor to the capacitor and load, then passes through the diode back to the inductor until the switch turns on.

$$V_o = D \times V_{in} \tag{2}$$

V_o is the output voltage (V), Vin is the input voltage (V), and D is the duty cycle.
In a low pass filter, the quality of the filter is influenced by the quality factor Q or quality factor. The quality factor is defined as the energy dissipated per radian ratio to the point stored [12]. Then another definition states that the quality factor is the ratio between the resonant frequency and bandwidth [13]. In general, the quality factor of the filter is shown by equation [14]:

$$Q = R_L\sqrt{\frac{C}{L}} \tag{3}$$

Where,

$$\omega_o = 2\pi f_o \tag{4}$$

With, L is the inductor (H), C is the capacitor (F), R is the resistance (Ω), d_o is the cutoff frequency (Hz).

The value of the component inductors, capacitors, and resistors is determined based on the value of the quality factor Q, which in this study uses several variations of the value of Q, which then the value of these components is applied to the circuit in Figure 1. The circuit will be tested that is simulated using LTSpice software.

3. Result and Discussion

Using the MOSFET requires a gate driver as a MOSFET controller to work on and off. MOSFET turn on when the primary side of the transformer energy is the input source. MOSFET works off when the primary coil transfers its energy to the secondary coil, and due to its polarity, the diode becomes forward biased and is now conducting [15]. This simulation circuit using IC4N25 as a gate driver. In planning, a source frequency of 50kHz will be used. However, IC4N25 has a characteristic where the cutoff frequency received is one-tenth of the source frequency, so that a cutoff frequency of 5kHz is obtained. Then the buck converter that is made has a parameter capable of reducing the voltage from an input voltage of 20 V to an output of 12 V. To achieve these parameters, and the duty cycle value is determined using equation (2) so that a duty cycle value of 60% is obtained. In designing the Butterworth low pass filter, an inductor component with a value of 10mH is selected, after which the value of the capacitor and resistor components that will be used in the filter can be determined based on the Q quality factor parameter. In this case, several variations of the Q value are given, namely 0.541, 0.707, and 1.307. The buck converter circuit that has been made is then simulated using LTSpice software to see the output signal generated by the buck converter.

Journal of Physics: Conference Series **2019** (2021) 012091 doi:10.1088/1742-6596/2019/1/012091

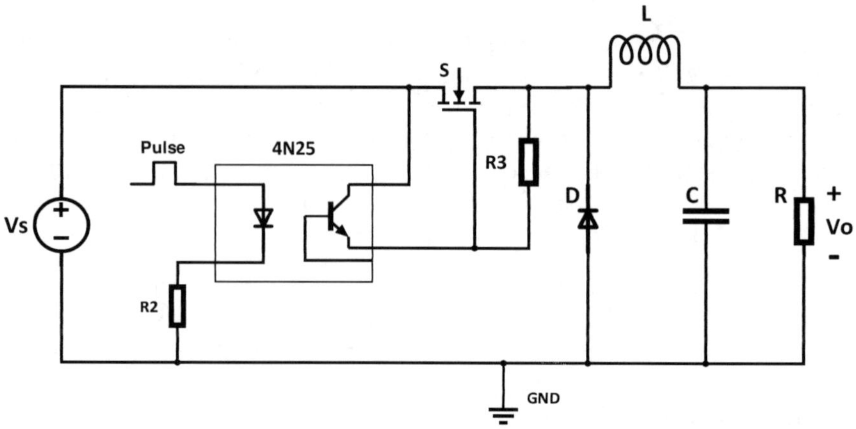

Figure 3. Buck Converter Circuit

Figure 3 shows the schematic of the buck converter circuit used in this study. The Optocoupler uses two resistors, namely R2 and R3, each worth 100 Ω and 410 Ω . In the variation of Q = 0.541, the value of the inductor, capacitor, and load resistor components are 10mH, 100 NF, and 180 Ω , respectively. In the variation of Q = 0.707, the component values of inductors, capacitors, and load resistors are 10mH, 100 NF, and 220 Ω , respectively. In the variation of Q = 1.307, the value of the inductor, capacitor, and load resistor components are obtained, each with a value of 10mH, 100 NF, and 430 Ω . The predetermined component values are then applied to the circuit schematic in FIGURE 3. The circuit is then simulated using a source frequency of 50kHz.

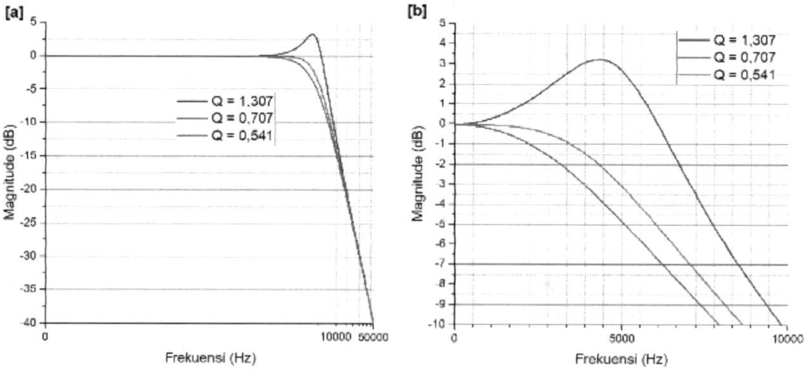

Figure 4. Buck Converter Output Frequency Response Graph (a) Frequency Range 0 - 50kHz, (b) Frequency Range 0 - 10 kHz

Figure 4 is a Bode diagram showing the frequency response at the output of the buck converter. Figure 4a is a graph of the frequency response in the 0-50kHz range, while Figure 4b is in the 0-10kHz range. Before reaching the cutoff frequency, the flat signal is a signal response in the fit band area, which indicates that the filter passes a signal that has a frequency value below the cutoff frequency. In the circuit with Q = 0.541, it was found that at the cutoff frequency, the filter was damped with a gain value ranging from -4.8 dB. Then the circuit with Q = 0.707 has a gain when the cutoff frequency ranges from -3.1 dB. While in the circuit with Q = 1.307, when it reaches the cutoff frequency, the gain increases with a value of around 2.7 DBs. This indicates that the quality of the filter performed by the filter with Q = 1.307 is not good.

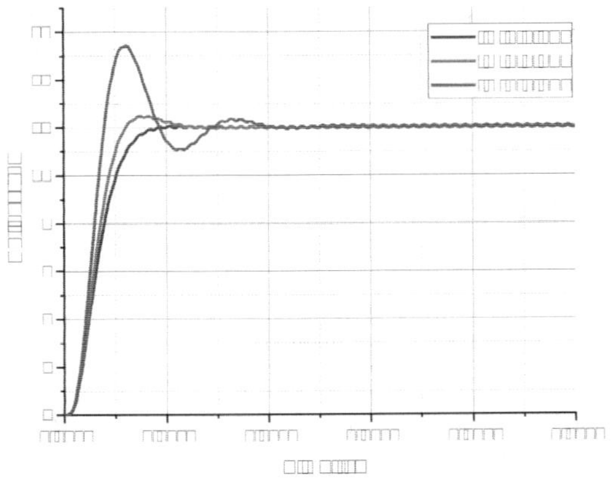

Figure 5. Buck Converter Output Voltage Response Curve

An analysis of the output voltage response is also carried out in the circuit made, as shown in Figure 5. The figure shows that each circuit with a different Q variable still has noise in the resulting output voltage. In addition, it was also found that there was an overshoot voltage before the output voltage produced by the buck converter reached a stable condition. In a circuit with a Q value of 0.541, there is no overshoot voltage found. Then at a variation of Q 0.707, there is an overshoot voltage of 4.8%. In contrast, the most significant overshoot voltage occurs when Q is 1.307, which is 27%.

Based on the data obtained, the filter that has the best quality is a filter with a Q quality factor of 0.707 because the gain value at the cutoff frequency is closest to the gain characteristic of the low pass filter, which is -3dB when it reaches the cutoff frequency, although deviations still occur by 3.3% and still found noise at the output voltage with an overshoot voltage of 4.8%.

4. Conclusion

The design, simulation of the filter used in the Buck converter circuit based on the quality factor value of Q that has been tested shows that the filter design has a good quality with a quality factor value of 0.707. This is because the gain of the simulation circuit when it reaches the cutoff frequency is -3.1 dB, where the value obtained is close to the characteristics of the low pass filter, which has a gain of -3dB when it reaches the cutoff frequency. The greater the value of Q, so the more significant the gain and the time it takes for the Buck converter to achieve a stable output voltage increase.

Acknowledgments

The authors wish to thank the colleague in the Department of Physics for very helpful comments on an earlier draft of this paper. Also, the authors acknowledge the valuable supports of Universitas Negeri Jakarta in this research.

References

[1] Kholiq I 2015 Pemanfaatan Energi Alternatif Sebagai Energi Terbarukan Untuk Mendukung Subtitusi BBM *J. IPTEK* **19** 75

[2] Purwoto B H, Jatmiko M, Alimul A, and Huda I F 2018 Efisiensi Penggunaan Panel Surya Sebagai Sumber Energi Alternatif *Emit. J. Tek. Electra* **18** 1

[3] Indrasari W, Fahdiran F, Umiatin U, and Habiburosid H 2020 The Active Hybrid Solar Panel integrated with Fresnel Lens Concentrator *J. Phys. Conf. Ser.* **1485**

[4] Mehdi G, Ali N, Hussain S, Zaidi A A, Shah A H, and Azeem M M 2019 Design and Fabrication of Automatic Single Axis Solar Tracker for Solar Panel *2019 2nd International Conference on Computing, Mathematics and Engineering Technologies (iCoMET)*

[5] Budi W S 2021 *Skripsi: Rancang Bangun Sistem Penyimpan Energi Listrik Panel Surya Menggunakan Maximum Power Point Tracker* (Jakarta: Universitas Negeri Jakarta)

[6]. Dandoussou A, Kenfack P, Perabi S, and Kamta M 2021 Simulations of the Performance of Maximum Power Point Tracking Algorithms Based on Experimental Data According to the Topologies of DC-DC Converters *Journal of Power and Energy Engineering* **9** 76
[7] Subari A and Manan S 2014 Perencanaan filter pasif untuk meningkatkan kualitas daya listrik di kampus Diploma III fakultas teknik undip *GEMA Teknol* **17** 144
[8] Rectifier I IRF2805 Datasheet *Data Sheet* p. 1
[9] Melfazen O 2018 Stabilisasi Tegangan Keluaran Buck Converter Dengan Metode Fuzzy Logic Controller *ELTEK* **16** 125
[10] Sun J 2012 Pulse-Width Modulation in *Dynamics and Control of Switched Electronic Systems, Advances in Industrial Control* p. 25
[11] Altamimi A and Khan, Z A 2017 A DC-DC buck converter with maximum power point tracking implementation for photovoltaic module application *2017 IEEE Conference on Energy Conversion (CENCON)* p. 305
[12] Hickman I 1999 Passive Circuits *Analog Eectronics* p. 26
[13] Male S B 2016 Simulasi Perancangan Rangkaian Band Pass Filter dengan Frequency Band Variable menggunakan Matlab *ELEKTRA* **1** 25
[14] Texas Instruments 2016 LC Filter Design *Texas Instruments Incorporated* **SLAA701A**.
[15] Sadda A, Keshri J P, Tiwari, H, and Jain V 2022 BLDC Motor Torque Ripple Minimization Technique of Using Isolated Type DC–DC Buck–Boost Converter *Advances in Energy Technology* **766**

Feature Extraction of Tea Leaf Images using Dual-Tree Complex Wavelet Transform and Gray Level Co-occurrence Matrix

B H Iswanto, Alma, I Sugihartono
Department of Physics, Universitas Negeri Jakarta
Jl. Rawamangun Muka, Jakarta 13220, Indonesia
Email: bhi@unj.ac.id

Abstract. Feature extraction is a very important part of machine learning to analyze and find relationships between objects of different categories. This paper aims to analyze the feature vectors of the tea leaf image produced by using a combination of dual-tree complex wavelet transform and gray level co-occurrence matrix techniques. The tea leaf image consists of four different categories, each representing different phases of tea growth and were acquired using a visible camera from eight different orientations. Feature extraction using Principle Component Analysis (PCA) shows that the texture features can identify a different category of the leaf images without being significantly affected by the difference in scale and orientation of the images.

1. Introduction

Nowsday, computer vision technology is being studied intensively and widely used in the field of agricultural automation [1]. In tea farming, this technology is studied among others, for the recognition of tea category and quality [2], detection of leaf diseases [3] leaf classification [4], identification of leaf infections [5], and assessment of leaf lesions and abnormalities [6]. This technology is also a solution to overcome the weaknesses of the human visual system [7] [8].

Computer vision technology uses cameras and computers instead of the human eye to identify, track and measure targets for further image processing [1]. This technology aims to develop an autonomous system that is able to automatically recognize objects based on images with capabilities similar to the human visual system [9]. In its implementation, computer vision tasks include a series of processes, starting from image data acquisition, preprocessing, feature extraction, and image recognition to produce numerical or symbolic information in the form of decisions. So that in this context the image recognition can be seen as a symbolic information transformation of image data with the help of machine learning algorithms, geometry, physics, and statistics [10].

In general, there are two approaches in developing a computer vision-based image recognition system. The first approach is based on the texture and color characteristics of the image, and the second is using threading and clustering techniques. In the first approach, texture, characteristics are generated through feature extraction, then feature data is used to build the machine learning-based prediction models. Some of the popular prediction models are artificial neural networks (ANN) [11], support vector machines (SVM) [2], and k-nearest neighbors (kNN) [12]. To obtain texture characteristics, a number of methods have been proposed in the literatur for different purposes, for example graylevel co-occurrence matrix [13], local binary pattern [14], 2 level wavelet transform [12], convolution neural net (CNN), DSIFT and BOVW [15], generalized eigenvalues proximal SVM [2], Bayesian discriminant principle [16], texture statistical histogram [17], and combination method of dual-tree complex wavelet transform and gray level co-occurence matrix, denoted as DTCWT+GLCM [18].

The combination method of the dual-tree complex wavelet transform and the gray level co-occurrence matrix is interesting to investigate to solve the problem of image recognition, especially images with different scale and object orientation. Therefore, the texture features generated by DTCWT+GLCM will be analyzed. The leaf image samples consist of four categories representing different growth phases. Images were acquired using a digital visible camera from eight different orientation angles and different scale. The feature data will be analyzed using the principal component analysis (PCA).

2. Method

2.1. Data preparation

In this study, the image of fresh tea leaves was taken from a tea plantation using a Canon 4000D camera to identify the growth phase of tea leaves. For this study, 160 tea leaf image samples are used to evaluate the performance of the DTCWT+GLCM method. Leaf images were taken after 3 days of tea growth, then continued every next three days to 12 days of growth. So that the collected tea leaf images consist of four categories (C1, C2, C3, and C4) which represent the tea growth phase after 3, 6, 9, and 12 days. Each category is represented by 40 image samples. Each leaf was taken from eight different camera angles, multiples of 45 degrees.

2.2. DTCWT+GLCM

The dual-tree complex wavelet transform (DTCWT) and the gray level co-occurrence matrix (GLCM), denoted as DTCWT+GLCM proposed by Yang & Yang [18] to overcome pattern recognition problems, especially for scaling and rotated objects. In the method, 1-level two-dimensional dual-tree complex wavelet transform is first implemented to decompose the original image to obtain sub-images at six directions: $\pm\pi/12$, $\pm\pi/4$ and $\pm 5\pi/12$ respectively. Thereafter, a number of texture features may be calculated from the gray level co-occurrence matrix to construct the feature vector. The statistical measures of the gray level co-occurrence matrix are used, e.g. Homogeneity, energy, entropy, correlation, uniformity, contrast, inverse difference, etc. to describe the texture information.

2.3. Feature extraction

A common feature extraction method in machine learning is Principle Component Analysis (PCA). PCA is a technique used to simplify the data, by means of transforming data in a linear manner to form a new coordinate system with variance maximum [19]. PCA also can be used to reduce the number of features by constructing a new variable with a smaller number that captures most of the information found in the original feature. Technically, PCA finds the eigenvector with the highest eigenvalue from the covariance matrix and then uses it to project the data into a new subspace with smaller dimensions. PCA produces principal components, eigenvectors, and eigenvalues. These principal components are the directions in the data that maximize variance when the original data are projected to them; such that the first component explains the most variance in the data. The principal components form the orthogonal basis in the data space. By using these principal components, all high-dimensional feature, data can be projected together into a 2D feature map.

3. Results

Before further processing the original images are converted into a single-channel of 255 grayscale image. Then, the DTCWT+GLCM method is applied to the images. According to the method, the first 1-level two-dimensional DTCWT is implemented to decompose the original image to obtain sub-images at six directions. After that, the gray level co-occurrence matrix of each sub-image is calculated and the corresponding statistical values are used to construct the feature vector. A number of statistical values were calculated for each Metrix as texture features, including parameters as follows: autocorrelation (f1), contrast (f2), correlation (f3), cluster prominence (f4), cluster shade (f5), dissimilarity (f6), energy (f7), entropy (f8), and homogeneity (f9).

The results of principal component analysis are shown in Table 1 and Table 2. From Table 1, the first principal component (PC1) and second principal component (PC2) account for 75.30% and 18.90% of the total variation in the data respectively, or account for about 94.2% of the total variation in the data. Table 2 shows the variables that correlate the most with the first principal component (PC1) are Autocorrelation (38.3%), Contrast (38.2%), Dissimilarity (38.2%), and Entropy (31.6%) . The first principal component is positively correlated with all four of these variables. Therefore, increasing values of Autocorrelation (f1), Contrast (f2), Dissimilarity (f6), and Entropy (f8) increase the value of the first principal component. The first three principal components explain 99.8% of the variation in the data.

Table 1. Eigenanalysis of the feature data

	PC1	PC2	PC3	PC4	PC5	PC6	PC7	PC8	PC9
Eigenvalue	6,7785	1,6998	0,5051	0,0083	0,0061	0,0022	0,0001	0	0
Proportion	0,753	0,189	0,056	0,001	0,001	0	0	0	0
Cumulative	0,753	0,942	0,998	0,999	1	1	1	1	1

Table 2. Eigenvectors

Var	PC1	PC2	PC3	PC4	PC5	PC6	PC7	PC8	PC9
f1	0,383	0,067	-0,016	0,16	-0,053	0,305	-0,176	-0,835	-0,012
f2	0,382	0,024	0,122	0,028	0,281	-0,247	-0,633	0,198	0,508
f3	-0,371	-0,176	0,146	-0,191	-0,110	-0,732	-0,122	-0,459	-0,005
f4	-0,291	-0,369	0,620	-0,113	0,483	0,375	-0,034	-0,083	-0,001
f5	-0,303	0,377	-0,512	-0,271	0,631	0,047	-0,069	-0,158	-0,002
f6	0,382	0,022	0,121	0,02	0,307	-0,263	-0,121	0,100	-0,806
f7	0,031	0,715	0,491	-0,448	-0,207	0,052	0,011	0,006	0
f8	0,316	-0,416	-0,217	-0,806	-0,128	0,118	-0,001	0,012	0,001
f9	-0,382	-0,019	-0,119	-0,009	-0,348	0,285	-0,73	0,108	-0,302

Projection of all feature data to the first two principal components (PC1 and PC2) can be seen in a score plot as in Figure 1. From the figure, it can be seen that there are four clusters of feature data clearly separated by the first principal component. This shows that the nine statistical measures used in DTCWT+GLCM can be used as texture features to identify the growth phase of tea leaves. Figure 1 (b) shows a loading plot that explains the correlation between the principal components and the original variables. The loading plot indicates that the f1, f2, and f6 variables point to the same direction as the first principal component (PC1). The graph also shows that the second principal component (PC2) is primarily in the direction of the f7 variants, with a small shift towards the direction of the f5 and f8. The second principal component is essentially orthogonal to the and f1, f2, f6, and f9 variables. These results provide an indication that there are a number of feature variables that can be reduced, for example, for the set of variables {f1, f2, f6}, as well as {f3, f9}. This result is also in line with its correlation with the eigenvalues of principal components. From this result, dimension reduction can be done without significantly reducing the essential information of the image.

Journal of Physics: Conference Series **2019** (2021) 012092 doi:10.1088/1742-6596/2019/1/012092

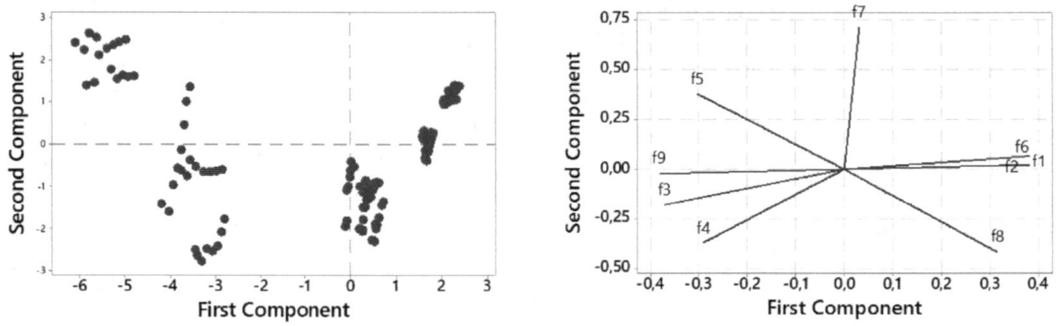

(a) (b)

Figure 1. Score plot (a) and loading plot (b) of the texture features data.

4. Conclusion

In this study, the feature vectors of tea leaf images generated by the combination method of dual-tree complex wavelet transform and gray level co-occurrence matrix (DTCWT-GLCM) were analyzed. From the results of feature extraction using Principal Component Analysis (PCA) it can be explained that the texture feature can identify the growth phase of the tea leaves without being significantly affected by difference scale and orientation of an object in the images. The results of data analysis also found that there were several characteristic features pointing in the same direction as one of the principal components, so that further dimensional reduction is possible.

Reference

[1] Tian, H, Wang, T, Liu, Y, Qiao, X, & Li, Y 2020 Computer vision technology in agricultural automation — A review *Information Processing in Agriculture* 7 pp 1-19
[2] Wang, S, Liu, A, Philips, P, & Du, S 2017 Tea category identification using computer vision and generalized eigenvalue proximal svm *Fundamenta Informaticae* 151(1-4):325-339
[3] Hossain, S, Mou, R M, Hasan, M M, Chakraborty, S, & Razzak, M 2018 Recognition and detection of tea leaf's diseases using support vector machine *IEEE 14th International Colloquium on Signal Processing & Its Applications (CSPA)*
[4] Pydipati R, B T 2006 Identification of citrus disease using color texture features and discriminant analysis *Comput Electron Agric* 52(1) pp 49–59
[5] Molina J F, Gil R, Bojacá C, Gómez F and Franco H 2014 Automatic detection of early blight infection on tomato crops using a color based classification strategy *XIX Symposium on Image, Signal Processing and Artificial Vision* (Armenia) pp 1-5
[6] Tucker C C and Chakraborty S 1997 Quantitative assessment of lesion characteristics and disease severity using digital image processing *J Phytopathology* **145** 7 p 273–278
[7] Park, K 1987 *Human Reliability: Analysis, Prediction and Prevention of Human Error* (Elseivier)
[8] Dhingra G, Kumar V and Joshi H D 2018 Study of digital image processing techniques for leaf disease detection and classification *Multimed Tools Application* **77** 19951–20000
[9] Shapiro, L G, & Stockman, G C 2001 *Computer Vision* (Pearson)
[10] Ponce, F 2011 *Computer Vision a Modern Approach* (Prentice Hall)
[11] Orillo, J, Cruz, J, Agapito, L, Satimbre, P, & Valenzuela 2014 Identification of diseases in rice plant (Oryza Sativa) using back propagation artificial neural network *Int Conf on Humanoid, Nanotechnology, Inf Tech, Comm and Control, Env and Management (HNICEM)* pp 1-6
[12] Wu, X Y 2018 Tea category identification based on optimal wavelet entropy and weighted k-nearest neighbors algorithm *Multimedia Tools and Applications* 77 pp 3745–3759
[13] Chen, Y, & Chen, X 2019 Tea leaves identification based on gray-level Co-occurrence matrix and K-nearest neighbors algorithm *AIP Conference Proceedings* 2073
[14] Tang, Z, Su, Y, Er, M J, Qi, F, Zhang, L, & Zhou, J 2016 A local binary pattern based texture descriptors for classification of tea leaves *Neurocomputing* 168 pp 1011-1023

[15] Chen, J, Liu, Q, & Gao, L 2019 Visual tea leaf disseas recognition using a convolution neural net model *Symmetry* 343
[16] Zhang, S, You, Z, & Wu, X 2019 Plant disease leaf image segmentation based on superpixel clustering and EM algorithm *Neural Computing and Applications* 31 pp1225–1232
[17] Iswanto, B H, & Alma 2021 Texture histogram features for tea leaf identification using visible digital camera *IOP Conf Ser: Mater Sci Eng* 1098 032030
[18] Yang, P, & Yang, G 2016 Feature extraction using dual-tree complex wavelet transform and gray level co-occurrence matrix *Neurocomputing* 197(12) pp 212-220
[19] Miranda, A A, Borgne, Y A, & Bontempi, G 2008 New routes from minimal approximation error to principal components *Neural Processing Letters* 27(3) pp 197–207

Journal of Physics: Conference Series **2019** (2021) 012093 doi:10.1088/1742-6596/2019/1/012093

Synthesis of TiO_2 from $TiCl_4$ Using Morinda Citrifolia Root Extract as a Molecular Template and its Photocatalytic Activity to Methylene Blue Degradation

A Rahman[1], Afrizal[1], B Warden[1] and R Slamet[4]

[1] Department of Chemistry Universitas Negeri Jakarta, Jl. Rawamangun Muka, Jakarta 13220, Indonesia

[2] FMIPA Universitas Indraprasta PGRI, Jakarta, Indonesia

Email: arifrahman@unj.ac.id

Abstract. $TiCl_4$ successfully as precursors for synthesised TiO_2 in nanometer-sized through the sol-gel method on the variation of the concentration of root extract of M. Citrifolia and have shown photocatalyst activity to degrade methylene blue with UV radiation in 365 nm. The characteristics of TiO_2 synthesized such as crystalline, size and morphology using SEM. SEM analysis showed that the morphology of TiO_2 synthesized showed the spherical shape in the nanometer-sized. Examination of the energy bandgap using UV-Vis DRS showed a decrease bandgap for TiO_2 synthesized (TiO_2 Merck = 3.2 eV; TiO_2 10x = 3:16 eV; TiO_2 50x = 3.1 eV; TiO_2 100x = 3.08 eV). The best degradation photocatalytic activity is demonstrated by TiO_2 100x until 84.07%. The reaction rate of TiO_2 photocatalytic for the degradation of methylene blue follow a pseudo-second-order reaction model.

1. Introduction

The textile industry is one of the largest manufacturing industries in Indonesia and the world, which produces large amounts of clothing to meet market demand. This large quantity of textile production requires synthetic dyes as dyes. Not all dyes can be used entirely in the dyeing process, but 15% of the total unused dye is discharged into the environment as waste. Among the reactive dyes available, methylene blue is a dye that is often used in the textile industry. The resulting dye waste can cause damage to the aquatic ecosystem due to the high concentration of organic compounds contained in it. In addition, the textile liquid waste cannot be decomposed naturally, so it is challenging to process it with a simple method. Therefore, an environmentally friendly waste treatment method is needed [1].

Many kinds of research on the treatment of liquid waste originating from the industry have been carried out, one of which is utilized sunlight to produce photocatalytic reactions. Sunlight producing photocatalyst reactions has become a concern because it can fulfil the concept and support innovative technologies from the application of Green Chemistry. Green synthesis offers various methods that can reduce or even eliminate harmful substances to humans and the environment [2]. Furthermore, green synthesis opens up new opportunities in particle synthesis using biomimetic mechanisms such as nanosilver manufacturing using Morinda citrifolia root extract. Among various oxide semiconductor photocatalysts, titanium dioxide is the most exciting semiconductor to be tested. Titanium dioxide (TiO_2) is a stable, relatively inexpensive, non-toxic and environmentally friendly semiconductor. TiO_2 with an energy gap of 3.2 eV in the anatase phase can be activated by ultraviolet light with a maximum wavelength of 385 nm [3].

The synthesis of nanoparticles with plant extracts provides several advantages compared to other methods, such as the process is more straightforward, cheaper, the materials are easy to obtain and tend to produce more stable materials. Green synthesis using plant extracts can be carried out using the sol-gel route with the hydrolysis method. The synthesis of TiO_2 from $TiCl_4$ by a hydrolysis method with the help of none extract (Morinda citrifolia) as a particle shape guide has produced nanometer-sized TiO_2 and shows the stability of the solution, which remains homogeneous after the addition of ammonium. Therefore, testing the ability of the photocatalytic reaction of TiO_2 to decompose methylene blue dye is the idea offered in this study.

2. Method

TiO_2 synthesis was carried out by mixing 20 ml of $TiCl_4$ in demineralized water and dye from Morinda citrifolia root extract with various concentrations. The colloid formed was left at room temperature for 8 hours. After being allowed to stand, the colloid formed was added to 100 ml of 1M NH_4OH slowly into the solution while stirring using a magnetic stirrer at a speed of 300 rpm. After that, it was heated for 2 hours at a temperature of 50°C with a water bath, then allowed to stand again for 42 hours. After the colloid was formed, the solution was centrifuged at 6000 rpm for 15 minutes and then decanted. The TiO_2 obtained after centrifugation was then dried. TiO_2 characterization to analyze the energy value of TiO_2 band gap using UV-Vis DRS (UV-Vis Diffuse Reflectance) and surface morphology analysis by using SEM. TiO_2 photocatalytic activity test to decompose 100 ml methylene blue solution with a concentration of 3 ppm and pH 11. The methylene blue solution to be degraded first measured the initial absorbance value, then added TiO_2 with a dilution of 10x, 50x, 100x as much as 0.1 grams each. In a beaker and then measure the absorbance again.

Furthermore, the glass is irradiated with UV light for 12 hours. During the irradiation process with UV light, the solution is stirred with a magnetic stirrer. Finally, the obtained supernatant was measured its absorbance value by UV-Vis spectrophotometer at the maximum wavelength of methylene blue.

3. Results and Discussion

The synthesis of TiO_2, which was carried out at temperatures below 5°C for exchanging chloride ions with hydroxide ions, aims to convert $TiCl_4$ precursors into $Ti(OH)_4$ so that the TiO_2 particles formed are easily formed directed to nano size. This synthesis must be carried out at low temperatures because if it is carried out at high temperatures, the O-Ti-O condensation process will be fast so that the size of the crystal nucleus of the TiO2 particles formed will be more significant. This is because the condensation is determined by the hydrolysis rate affected by the system's temperature. Therefore, during the ligand exchange process, the temperature must be conditioned to remain stable using an ice bath to change the hydrolysis reaction slowly. The following is the equation for the condensation process that occurs:

$$TiCl_{4(aq)} + 4H_2O_{(l)} \rightarrow Ti(OH)_{4(aq)} + 4HCl_{(aq)}$$

The use of noni root extract in the synthesis process aims as a template to allow the particles that grow to be of uniform size. Stock Solution of many root extracts with water solvent in powder was obtained from a previous study [11]. Making this extract solution is by dissolving the powder into DM water until it is saturated. Then three different concentration variations were made, namely: 10x, 50x and 100x dilutions. This synthesis process was also carried out by adding 100 mmol of ammonium used to control the O-Ti-O condensation process by forming molecular clusters. This cluster can act as a squeezing agent for molecules of various sizes. The cluster formed is influenced by the number of moles of ammonium added to the colloid. The addition of ammonium can also form a salt with HCl resulting from the previous reaction with the following reaction equation:

$$NH_4OH_{(aq)} + HCl_{(aq)} \rightarrow NH_4Cl_{(aq)} + H_2O_{(l)}$$

The next step is heating the colloid formed at 50°C for two hours. Heating is done because energy is needed to direct the atoms to form the desired crystal arrangement during the crystal formation process. The particles that are expected to form during this synthesis are nano-sized. The nanometer-sized particles are colloidal so that the TiO_2 particles dispersed in water will have the characteristics of a colloidal solution [4]. The smaller the particle size formed, the more stable the colloid produced. The muscle structure formed is believed to stabilize the particles in the particle growth or ageing process, so it is expected that the particles can be arranged stably and homogeneously. Furthermore, the muscle structure, a molecular printing agent, will provide the desired morphology and particle size with the addition of dye from non root extract and the right amount of mmol ammonium.

Figure 1 provides information on the morphology and particle size of the synthesized TiO_2 with 10x, 50x and 100x dilutions, showing that the morphology of TiO_2 is elliptical in size in the nanometer range. In Figures 1a and 1b, the morphology of TiO_2 particles is uniformly spherical in certain areas and overlaps each other, while Figure 1c is seen as elongated and close to each other in most certain areas. This indicates that TiO_2 powder has been successfully synthesized. From SEM photography, it can be seen that the TiO_2 particles are close together and agglomerate to form porous spaces. The effect of drying temperature and noni root extract dye on the sample caused the particles to agglomerate.

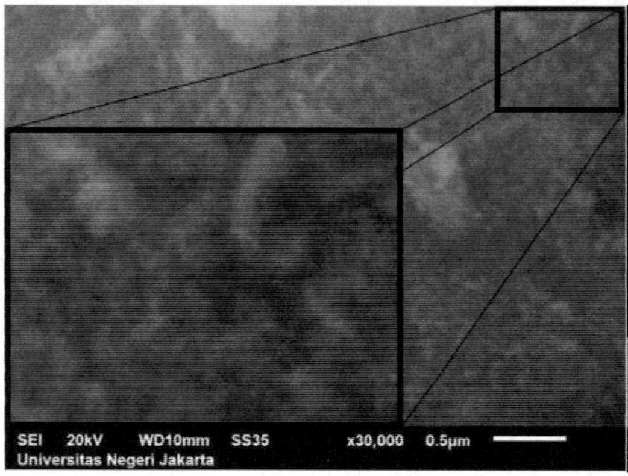

(a)

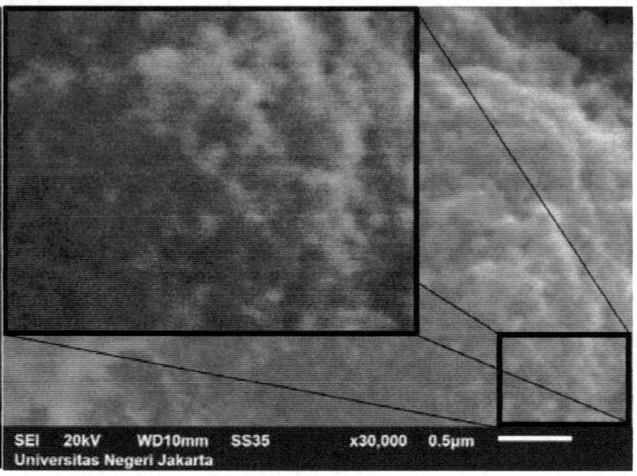

(b)

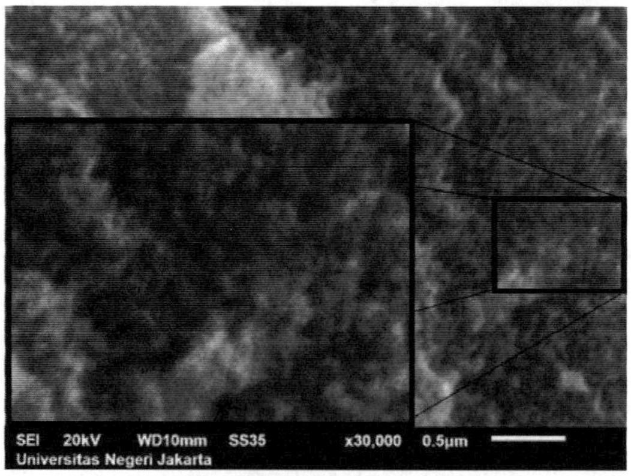

(c)

Figure 1. SEM micrograph of as-synthesized a. TiO_2 10x, b. TiO_2 50x dan c. TiO_2 100x.

Determination of the energy bandgap value was carried out by UV-Vis DRS characterization. The energy band gap value of TiO_2 is between 3.0-3.2 eV. Determination of the energy bandgap value using the Kubelka-Munk equation where Eg is obtained from the graph of the relationship between $[F(R\infty)hv]^2$ to hv (eV) [5]. The energy gap is obtained from the intersection of the straight lines on the curve that intersects the x-axis. Through the UV-Vis DRS characterization, the bandgap value of the synthesized TiO_2 can be obtained. The energy gap value obtained from the sample characterization is the energy of the TiO_2 material required to create electrons and holes that will be used in photocatalytic reactions as photooxidation-reduction initiators. The addition of noni extract with a particular concentration variation affected the energy value of the bandgap on TiO_2. The factor of decreasing the energy gap value is also because the synthesized TiO_2 crystal phase is mixed. Namely, there are anatase and rutile phases.

Table 1. Bandgap energy (Eg) as-synthesized TiO_2

No	Sample	Bandgap energy
1	TiO_2 Merck	3,20 eV
2	TiO_2 10x	3,16 eV
3	TiO_2 50x	3,10 eV
4	TiO_2 100x	3,08 eV

Table 1 shows that the Merck TiO2 band gap is 3.20 eV. This value becomes a reference for comparison with the synthesized TiO_2. The synthesized TiO_2 showed a decrease in the bandgap value compared to Merck's TiO_2, where the bandgap values of TiO_2 10x, TiO_2 50x and TiO_2 100x were 3.16 eV, 3.10 eV and 3.08 eV, respectively. With a smaller bandgap value, it can be assumed that the synthesized TiO_2 with the addition of noni extract will be more responsive towards visible light wavelengths.

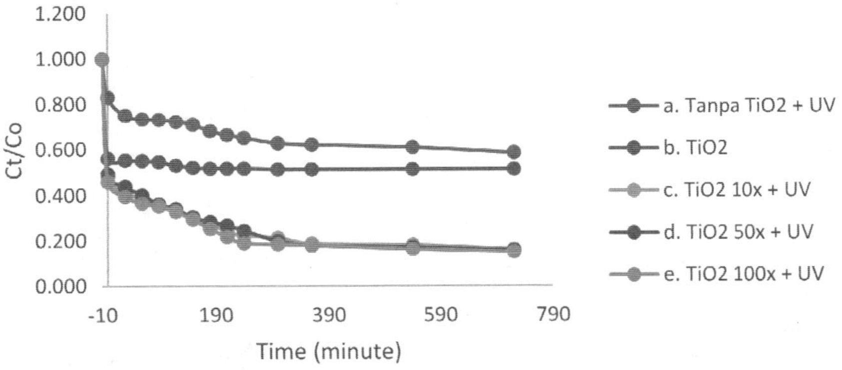

Figure 2. Degradation data of Methylene blue a: without TiO_2+UV, b: TiO_2, c: TiO_2 10x+UV, d: TiO_2 50x+UV dan e: TiO_2 100x+UV.

Figure 2 shows a graph of the decrease in methylene blue concentration with TiO_2 10x, 50x and 100x sharper than without TiO_2 with UV radiation and TiO^2 without UV radiation. In the first 10 minutes, when the methylene blue solution to which TiO_2 was added was homogenized, a very steep decrease in concentration was seen. This indicates that apart from functioning as a catalyst, TiO_2 also has properties as an adsorbent [6]. Titanium dioxide has a surface area that allows methylene blue to be adsorbed on the surface of TiO_2 with weak bonds to be released. Degradation of methylene blue increases with increasing radiation time. Table 2 shows the degradation of methylene blue with TiO_2 photocatalyst with various dilutions as synthesize conditions (10x, 50x and 100x), showed almost the same percentage of degradation, which were 84.07%, 84.42% and 85.10%, respectively. Successively. The graph shows that the optimum time for methylene blue degradation in the three treatments is at the 300 minute, where the graph looks straight from the 300th to the 720th minute. The highest percentage of degradation was obtained from treatment with 100x TiO_2 and UV radiation.

Table 3 shows various calculations for the kinetics of TiO_2 photocatalysis for methylene blue degradation is most suitable following the pseudo-second-order reaction model. They are plotting t/qt as the Y-axis and t as the X-axis produces a straight line with the highest correlation coefficient (R^2), i.e. 0, 9965 at 10x, 0.9971 at 50x and 0.9979 at 100x. The rate constant (k) and adsorption capacity at equilibrium (qe) is generated from the modelling. This is related to the energy band gap value analyzed by the UV-Vis DRS tool that the smaller the bandgap energy value, the higher the correlation coefficient value. Therefore, the high value of R^2 results in this modelling ensures that the binding reaction of methylene blue by TiO_2 proceeds through a second-order reaction.

Table 2. Photocatalyst performance on the degradation of methylene blue.

No	Sample	Percentage degradation	Photocatalyst activity
1	TiO_2	48,70%	-
2	TiO_2 10x + UV	84,07%	35,37%
3	TiO_2 50x + UV	84,42%	35,71%
4	TiO_2 100x + UV	85,10%	36,40%

Table 3. The kinetics model for TiO_2 photocatalytic reaction of methylene blue degradation.

Kinetics Model	Parameter	10x	50x	100x
Orde 1	k (min^{-1})	0,0019	0,002	0,002
	R^2	0,6826	0,7211	0,6636
Orde 2	k ($mM^{-1}min^{-1}$)	0,0019	0,002	0,0021
	R^2	0.8811	0.8965	0.8455
Pseudo Orde 1	k (min^{-1})	0,003	0,0042	0,0052
	R^2	0.9437	0.906	0.8383
Pseudo Orde 2	k ($mM^{-1}min^{-1}$)	0,3265	0,3227	0,3213
	R^2	**0.9965**	**0.9971**	**0.9979**

Based on the rate law, the constant rate values listed in the table can be included in the rate law equation for the reaction. The rate law for the reaction of TiO_2 10x is $v=0,3265\ (q_e-q_t)^2$, TiO_2 50x is $v=0,3227\ (q_e-q_t)^2$ dan TiO_2 100x is $v=0,3213\ (q_e-q_t)^2$. Q_e is the adsorption capacity of methylene blue at equilibrium, and q_t is the adsorption capacity of methylene blue at time t.

4. Conclusions

We successfully synthesised TiO_2 from $TiCl_4$, which have high performance in catalytic activity on methylene blue degradation. TiO_2 from $TiCl_4$ has been successfully synthesized by the sol-gel method using noni root extract (Morinda citrifolia), which produces nanometer-sized TiO_2 crystals. The photocatalytic activity of TiO_2 100x with UV radiation showed the most significant degradation of methylene blue at 85.10%, TiO_2 50x at 84.42% and TiO_2 10x at 84.07%.

References

[1]. Al-Mamun, M.R., et al., *Photocatalytic activity improvement and application of UV-TiO2 photocatalysis in textile wastewater treatment: A review.* Journal of Environmental Chemical Engineering, 2019. **7**(5).

[2]. Li, Y., Y. Fu, and M. Zhu, *Green synthesis of 3D tripyramid TiO2 architectures with assistance of aloe extracts for highly efficient photocatalytic degradation of antibiotic ciprofloxacin.* Applied Catalysis B: Environmental, 2020. **260**.

[3]. Rahman, A., et al., *Enhanced Activity of TiO2/Natural Zeolite Composite for Degradation of Methyl Orange under Visible Light Irradiation&#xD.* International Journal of Technology, 2018. **9**(6).

[4]. Maurya, I.C., et al., *Green synthesis of TiO2 nanoparticles using Bixa orellana seed extract and its application for solar cells.* Solar Energy, 2019. **194**: p. 952-958.

[5]. Saravanan, R., et al., *Crystallinity and lowering band gap induced visible light photocatalytic activity of TiO2/CS (Chitosan) nanocomposites.* Int J Biol Macromol, 2018. **109**: p. 1239-1245.

[6]. Matos, J., et al., *C-doped anatase TiO2: Adsorption kinetics and photocatalytic degradation of methylene blue and phenol, and correlations with DFT estimations.* J Colloid Interface Sci, 2019. **547**: p. 14-29.

Study Surface Area and Pore Size Distribution on Synthetic Zeolite X using BET, BJH and DFT Methods

E Allanas[1], A Rahman[2], E Arlin[3] and E A Prasetyanto[4]
[1,2,3] Department of Chemistry, Universitas Negeri Jakarta
Jl. Rawamangun Muka, Jakarta 13220, Indonesia
[4] Department of Pharmacy, Atma Jaya Catholic University of Indonesia, Jakarta, Indonesia
Email: edithallanas@unj.ac.id

Abstract. The characterization of zeolite X synthesized from Bintan bauxite using nitrogen physisorption technique that employs the BET, BJH and DFT equations was proposed to improve the understanding of the pore characteristics of zeolite X. The BET, BJH and DFT methods were chosen because the characteristics of the synthesized zeolite were characterized using a scanning electron microscope (SEM) showed a typical morphology for zeolite X and zeolite A crystals and the presence of amorphous phase in small abundance. Evaluation is needed to understand the characteristics of micropore, mesoporous and macropore of zeolite X as a synthesis and the best methods. The data obtained show that the DFT method is the most appropriate and superior model to describe the pore size distribution of synthesized zeolite X.

1. Introduction

Porous materials with superior characteristics for various purposes continue to be sought for use in surface area, pore size, surface reactivity or a combination of these properties. For example, catalyst engineering to obtain active catalysts with superior product selectivity can increase the uniformity of pores and surface reactivity of porous materials such as zeolite. Porous materials also offer attractive solutions for storage applications such as carbon capture and sequestration or storage of methane (CH_4) and hydrogen (H_2) for energy use. Nanoporous materials such as sequenced carbon and metal-organic frameworks (MOFs), among others, have been extensively evaluated for the storage of gases such as carbon dioxide (CO_2), CH_4, and H_2 due to their large pore volume and high surface area [1]. The structure of these nanoporous materials must be fully characterized to attribute the favourable structural features to the gas storage performance. Structural characterization includes methods such as X-ray diffraction (XRD), nuclear magnetic resonance (NMR), and scanning electron microscopy (SEM); However, gas absorption may be most appropriate to analyze and evaluate because it can measure the entire range of micropores and mesopores, to macropores in abundant materials [2].

Measurement of pore characteristics in zeolite using the liquid N_2 adsorption technique provides rich information and can be used to study the characteristics of synthetic zeolite in depth. Zeolite is a crystalline material with a porous structure with a size range of nanometers, a three-dimensional framework composed of aluminosilicate tetrahedral with charge balancing ions that can be exchanged on its surface. Having such characteristics makes zeolites can be used as catalysts, ion exchangers, and adsorbents. The application of zeolite as a catalyst is used in petroleum, the adsorbent in natural gas purification, and ion exchange in water purification. Zeolite is also often used in agriculture, namely as a carrier of nutrients in fertilizers [3].

Zeolites can be distinguished according to different standards, with pore openings being the most attention-grabbing due to their applied potential. Zeolites can be classified as (i) small pore zeolites, consisting of channels bounded by eight-membered rings with a pore diameter of 4, (ii) medium pore zeolites, consisting of channels bounded by ten membered rings with a pore diameter of 5– 6, (iii) large pore zeolite, consisting of channels bounded by 12 member rings with a pore diameter of 7, and (iv) extra-large pore zeolite, consisting of channels bounded by more than 12 member rings and pore diameters greater than 7. Channel dimensions represent other zeolite criteria because zeolitic materials can include one-dimensional, two-dimensional, and three-dimensional pore structures, depending on the arrangement of the channels. In addition, zeolitic materials have also been classified as low silica (Si/Al atomic ratio 1), intermediate silica (1.5 $<$Si/Al atomic ratio$<$10), or high silica (Si/Al atomic ratio$>$ 10). Low silica zeolites are hydrophilic due to their high proton content and highly negatively charged framework. On the other hand, high silica zeolitic materials are hydrophobic due to a predominantly covalent Si-O-Si bridge. The NaA zeolite offers excellent CO_2/N_2 selectivity compared to FAU due to the narrow pore window. Many large porous zeolites with channels formed by openings of more than 12 rings are known, but all of them [4].

Gas adsorption is significant for the characterization of various porous materials. Of all the gases and vapours available and can be used as an adsorbent, nitrogen remains universally superior. With the help of easy-to-use commercial equipment and online data processing, it is now possible to use nitrogen adsorption at 77 K for routine quality control and investigation of new materials. Synthetic zeolites from natural materials often produce products with a mixture of crystalline phases such as sodalite, zeolite A and X, and amorphous, so it is essential to evaluate the synthesized product using the N_2 physical adsorption method. Information on the surface area provides information about the surface area formed, the pore distribution provides information about the size distribution of the formed pores. For this purpose, several approaches were used as the most useful evaluation tool for the synthesized zeolite [5].

2. Method

Gas adsorption manometry is a method commonly used to determine nitrogen adsorption isotherms at liquid nitrogen temperature (77 K), volumetric determinations (or as the BET volumetric method). Successive amounts of adsorption are introduced, and at each stage, the system is given sufficient time to reach equilibrium, which is adjusted by a series of single points on the adsorption isotherm.

Adsorption is carried out for measurements at relative external pressures. However, since the filling of the molecular dimension micropores (i.e. ultramicropores) occurs at p/p0 10-4, it is necessary to carry out high-resolution adsorption measurements to assess primary micropore filling. Therefore, sample controlled thermal analysis was used to determine the adsorption isotherm, using Micromeritics ASAP 2020, with liquid nitrogen (N_2) as the working fluid supplemented by helium (He) as filler and cleaning gas. Adsorption-desorption data processing using microactive software.

3. Results and Discussion

The synthesized zeolite X shows a crystalline phase of zeolite X, a crystalline phase of the faujasite type, as shown in Figure 1 shows a typical prismatic morphology of a faujasite zeolite crystal, but with impurities in the form of an amorphous phase and crystal zeolite A. Along with the length of hydrothermal time, zeolite A is formed. Hydrothermal time affects crystal growth. The longer the hydrothermal time, the higher the crystallinity produced, the smaller the size of the crystals formed. The crystallinity formed increases, and the type of phase formed changes with increasing hydrothermal time. Zeolite formed at the beginning of the hydrothermal period, namely zeolite A and faujasite (FAU). These two types of zeolite have a metastable structure, so they are replaced with zeolite P.

Various zeolite phases need to be confirmed using liquid nitrogen physisorption data to picture the structure and pore characteristics accurately. BET can provide surface area information, and further processing using a de-Boer t-plot can provide information on the abundance of internal and external pores. The study using BJH is expected to provide information if the zeolite X synthesized material has meso-sized pores (2-50 nm), while DFT is used to determine the distribution of pore sizes below 2 nm.

Figure 1. SEM of Zeolite X as synthesized from Bintan bauxite shows crystalline and amorphous phase.

Analysis of the SEM images presented in Figure 1 shows the morphology of zeolite X with a prismatic shape dotted with grains and a thin layer of an amorphous phase and a small amount of zeolite A in the form of cubes. The morphology shows the abundance of zeolite X due to the synthesis is the dominant phase and amorphous and zeolite A as mineral impurities. Characterization using gas adsorption-desorption is needed to provide an overview of the pore characteristics of the synthesized zeolite. The characterization results are presented in Figure 2.

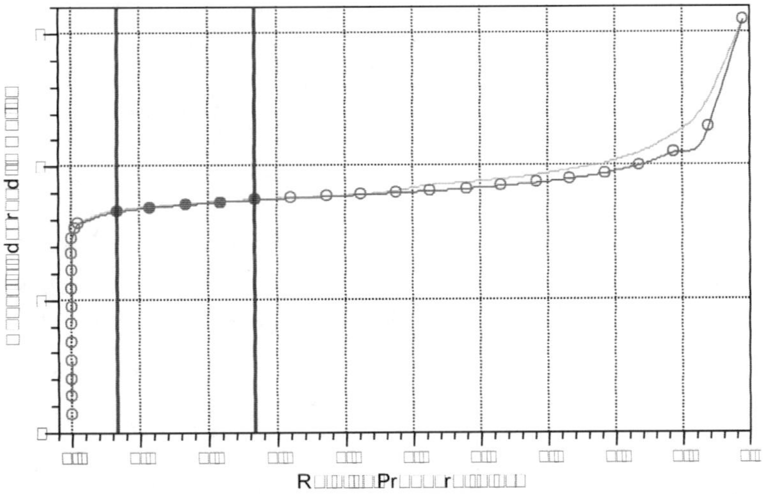

Figure 2. Isotherm adsorption-desorption of N_2 at 77 K into as-synthesized Zeolite X from Bintan bauxite

Figure 2. shows the adsorption-desorption pattern of the as-synthesized zeolite is type I isotherm pattern. The adsorption-desorption pattern of the synthesized zeolite is the type I indicates that the material is microporous. At low P/P_0, the adsorption showed this saturation, indicating that the pores followed the Langmuir adsorption model for micropores. An increase in pressure until P/P0 is close to one does not indicate an increase in the volume of nitrogen gas absorbed. This indicates the absence of mesopore in the zeolite sample. The desorption process shows a small hysteresis, and this may come from the amorphous phase that forms macropores as the external surface. Further data processing using BET confirmed the existence of micropores, and data processing using the BJH and DFT equations showed pore characteristics.

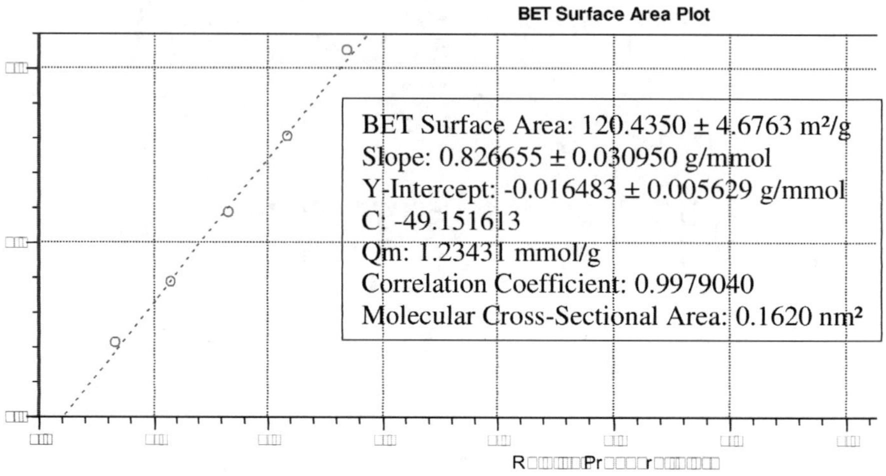

Figure 3. BET plot N_2 physisorption of as-synthesized zeolite

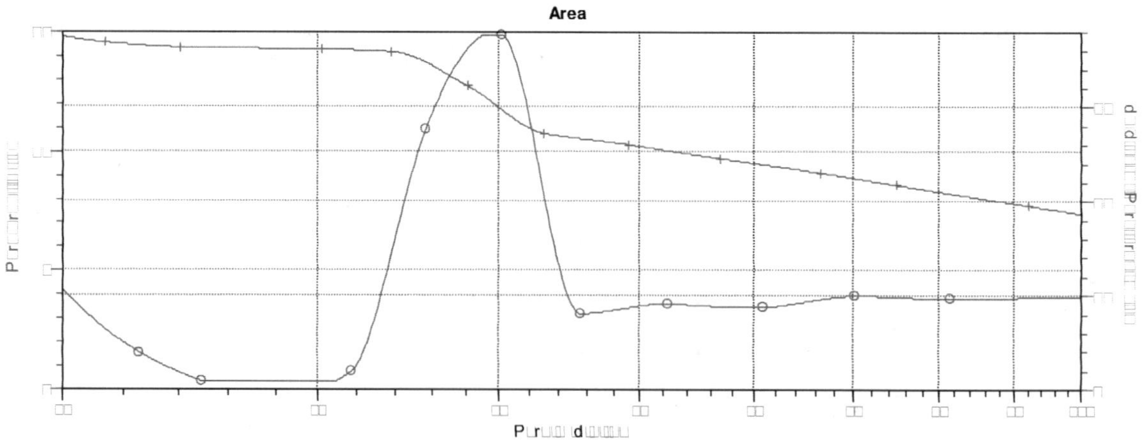

Figure 4. Pore size distribution of as-synthesized zeolite X by BJH plot of desorption N_2 physisorption.

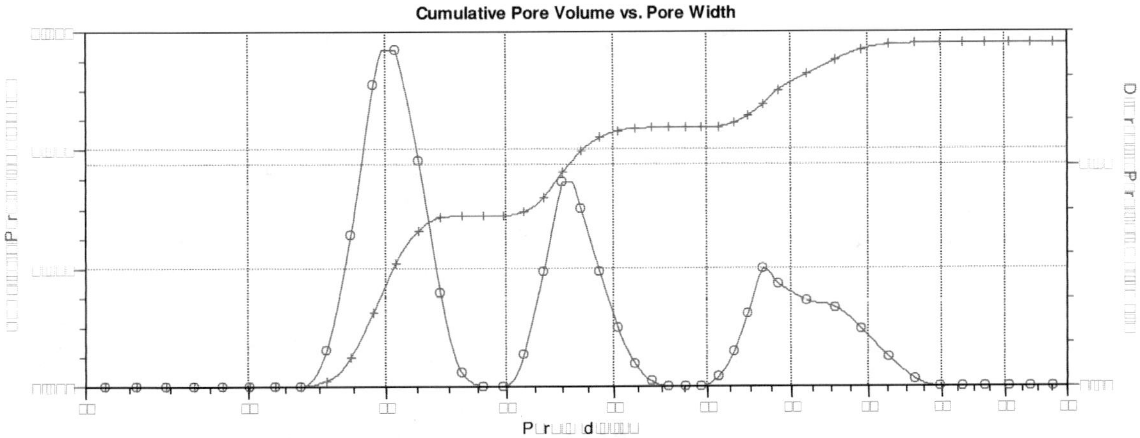

Figure 5. The pore size distribution of as-synthesized zeolite X by DFT plot of N_2 physisorption.

Figure 4 and Figure 5 show a graph of the desorption pattern of BJH and DFT, which shows the pore distribution of the synthesized zeolite sample. The BJH desorption graph shows that the zeolite synthesized in the mesoporous diameter is in the range of 40 Angstroms. This shows that the use of BJH to characterize the pores of the synthesized zeolite is not accurate, considering that the pore of zeolite X is at 13 Angstroms, while that of zeolite A is at 7 Angstroms. This discrepancy is due to the use of BJH, which has high accuracy for porous materials in the mesopore range but is less sensitive for use on microporous materials. Meanwhile, the DFT showed three peaks of the pore distribution at 14, 17 and 22 Angstroms. This shows a very close value to the theoretical, where zeolite X has a pore of 13 Angstroms. These results align with those obtained by other research groups in characterizing shale nanopores using the DFT equation [6].

4. Conclusions

The difference between the calculation results and the simulation of the pore distribution graph using BET, BJH, and DFT is quite significant, so it can be concluded that to analyze the distribution in the micropore range such as the synthesized zeolite, the DFT method is the most appropriate and superior model.

Acknowledgement

We thank the Faculty of Mathematics and Natural Sciences Universitas Negeri Jakarta, who financed this research through Grant Penelitian Dosen Muda, Contract Number: 4/SPK PENELITIAN/5.FMIPA/2021

References

[1] Brea, P., et al. 2019 *Microporous and Mesoporous Materials* 286 pp 187-198
[2] Cychosz, K. A. and M. Thommes 2018 *Engineering* 4(4) pp 559-566
[3] Fan, X. and Y. Jiao 2020 *Sustainable Nanoscale Engineering* pp 115-137
[4] Groen, J. C., et al. 2003 *Microporous and Mesoporous Materials* 60(1-3) pp 1-17
[5] Hammond, K. D. and W. C. Conner 2013 *Advances in Catalysis* pp 1-101
[6] Zhang, L., et al. 2017 *Fuel* 204 pp 1-11.

Preliminary Study on The Effect of Time on Hydrogen Production from Electrolysis of The Seawater

C E Rustana[1], Sunaryo[2], I N Salam[2], I Sugihartono[2], W Sasmitaningsihhiadayah[1], A D R Madjid[1] and F S Hananto[1]
[1] Physics Department, Faculty of Science and Technology, State Islamic University of Maulana Malik Ibrahim, Malang-East Java, Indonesia (65144)
[2] Physics Departement, Faculty of Mathematics and Science State University of Jakarta, East Jakarta, Indonesia (13220)
E-mail: CE.Rustana@yahoo.com; sunaryo@unj.ac.id; ihzanursina@gmail.com; wiwis_hidayah_87@fis.uin-malang.ac.id; armedia@uin-malang.ac.id; faridsamsu@fis.uin-malang.ac.id

Abstract. To know the effect of electrode type on the production of hydrogen gas through the electrolysis of sea water, this research was conducted. At a constant potential difference of 12 Volt, the electrolysis process is carried out by alternately using graphite and copper as electrodes. The electrolysis process time was varied from 10 to 55 minutes with increments every 5 minutes for each electrode. The results showed that the use of copper in the electrolysis of sea water produced a maximum of 82 ml of hydrogen gas better than 76 ml of graphite with a total processing time of 5 hours and 25 minutes. The results also show that the production of hydrogen gas in graphite has the largest hydrogen production rate in the first 10 minutes, but continues to decline, while the copper electrode has the largest hydrogen production rate at 220 minutes and decreases when the electrolysis process reaches 270 minutes when the electric current experiences drop. This can be understood due to the corrosion of the electrode by chlorine, which causes the electrode life to be limited. Meanwhile, the water displacement measurement method is used to determine the volume of hydrogen gas produced from the electrolysis of seawater in this study.

1. Introduction

Seawater as one of the renewable energy sources is abundantly available on our planet, approximately 97% of the total water on our planet. Therefore, seawater is very suitable as a substitute for the fossil energy used at this time [5]. One of the energies that can be produced from seawater is hydrogen gas. This gas can not only be used directly as clean fuel, but also hydrogen energy can be stored because it can be transported over long distances, does not cause pollution, is light, and is sustainable because it can be produced from widely available seawater. Several ways to extract energy from seawater are through the process of a thermal gradient, wind and wave generators, electrolysis, and photolysis. Electrolysis of seawater for the production of hydrogen gas is a promising technology and the simplest.

Departing from the potential of seawater as a renewable energy source that can be extracted to produce hydrogen gas through the electrolysis process, this preliminary research was designed to analyze the effect of the electrolysis process time and the type of electrode used on the production rate and volume of hydrogen gas produced through the electrolysis of seawater. It hopes that through this research it can be seen which types of electrodes are more effective and efficient to produce hydrogen gas at a certain time of gas the electrolysis of seawater. In other words, this research was designed to determine the best type of electrode and processing time to produce hydrogen gas through the electrolysis of seawater. Finally, this applied research can provide basic information about what, why, and how to produce environmentally friendly alternative energy to replace the use of fossil energy in answering the energy crisis problems that may be faced at the future.

1.1 Seawater

Water is the largest part of the earth and seawater alone accounts for 97% of all the water on earth, therefore the use of water is very important for mankind. Water itself is a very useful solvent for dissolving larger amounts of other substances, for example, seawater which can dissolve nearly 48,000 trillion tons of salt consisting of 38,000 trillion tons of sodium chloride, 3000 trillion tons of sulfates, 1600 trillion tons of magnesium, 480 trillion tons of potassium and 83 trillion tons of bromide. Sodium Chloride (NaCl) is a substance that is mostly contained in seawater. NaCl or table salt contains a large percentage of Chloride [4]. Because of the salt content (salinity), seawater has a salty taste. Salinity is the number that shows how many grams of salts dissolved in per kilogram (gr/kg) of seawater is namely.

Other physical property of seawater are temperature, seawater temperature is often referred to as SST (Sea Surface Temperature) or SPL (Sea Surface Temperature) with SI units of Celsius (0°C). When measuring temperature differences (deltas), it must be measured in Kelvin (K) units [1]. The factors that influence sea surface temperature are the location of the altitude from sea level (Altitude), the intensity of sunlight that is received, the season, weather, water depth, air circulation, and cloud cover. Then the physical characteristic of seawater which is no less important is the density. Density is one of the most important parameters in studying ocean dynamics. Density (ρ) is defined as mass per unit volume ($g\ cm^{-3}$). The density of seawater depends on the temperature (t) and salinity (s) of the sample and also the seawater pressure (p) as a result of the compressibility of the water [1].

Seawater has six elements and compounds that make up 99% of seawater, namely 55% chlorine (Cl^-), 31% sodium (Na^+), 8% sulfur ($SO4^{-2}$), 4% magnesium (Mg^{+2}), 1% calcium (Ca^{+2}), and 1% potassium (K^+). There are also atmospheric gases that can be found in seawater, namely: nitrogen, oxygen, carbon dioxide, argon, helium, and neon. Weathering of rocks on land, volcanic gases, and hydrothermal circulation on the seabed are the main sources of these seawater minerals. The chlorine is aggressive in forming compounds that can cause damage or corrosion to the metal (electrode) in the electrolytic process and further it will impact on the length of time of the electrode used [2].

1.2 Electrolysis

Electrolysis is a chemical change or decomposition reaction in an electrolyte by an electric current. Electrolytes dissolve in polar solvents (such as water) by dissociating into positive ions (cations) and negative ions (anions). Negative ions are called anions because during the electrolysis the solution is attracted to the positive charge on the anode; while positive ions are called the cathode because through the solution they move towards the negative charge (cathode). Water as a solvent is a polar. Polar molecules have a charge at the end of the molecule, which is a positive and negative charge. This charge can interact with charges on other polar molecules to dissolve them. Between these molecules, there is a transfer of hydrogen atoms to form hydronium ions.

The addition of electrolytes, such as acids, bases, or salts can increase the conductivity of the water so that the water electrolysis process is faster. In the electrolysis cell, there will be a change in electrical energy into chemical energy. The quantitative relationship between the amount of electric charge used and the amount of substance involved in the reaction was formulated by Faraday. This can occur because it involves a reduction-oxidation reaction which relies on the role of charged particles as conductors of electric charge. Water is a very weak electrolyte, which can ionize to form ions and OH. Therefore, it is possible to electrolyte them into gases and. Gas can be obtained at the cathode due to the ion reduction reaction; meanwhile, gas is obtained at the anode due to the OH^- oxidation reaction. (Isana, 2010). The electrochemical equation of electrolysis can be described as follows

At the cathode: $$4H^+(aq) + 4e^- \rightarrow 2H_2\ (g) \quad (1)$$

At the anode: $$2H_2O\ (l) \rightarrow O_2(g) + 4H^+(aq) + 4e^- \quad (2)$$

Complete reaction: $$2H_2O(l) \rightarrow 2H_2\ (g) + O_2(g) \quad (3)$$

However, the electrolysis of most abundant sea water rather than the purified or fresh water is a more promising way to generate clean hydrogen energy. Hydrogen can be used as an alternative fuel because it can replace fossil energy and has a very minimal negative effect on the environment. One of the factors that influence the electrolysis process of seawater is the type of electrode that has good corrosion resistance due to chlorine, which greatly affects the lifetime of the electrode in the electrolysis process of seawater. Corrosion is defined not only as the destructive oxidation of metal materials, but also the degradation of non-metallic materials and these materials lose their function due to chemical or electrochemical interactions with the environment (Song & Xie, 2018).

Corrosion itself is defined as a process of material degradation due to interaction with the surrounding environment. This interaction causes a corrosion reaction which is generally an electrochemical reaction. Four factors influence and play a role in electrochemical reactions that can lead to corrosion, including Anode, Cathode, Electrolyte solution, and conductors, in the electrolytic process, corrosion will cause the electrode to decrease in effectiveness so that the production rate and volume of hydrogen gas produced will be smaller [3].

1.3 Electrode

An electrode is a conductor through which an electric current passes from one medium to another, usually from a power source to a device or material. The electrodes themselves can take several different forms, including wire, plate, or stick, and are most often made of a metal, such as copper, silver, lead, or zinc, but can also be made of non-metallic electrical conducting materials, such as graphite. Electrodes are used in welding, electricity, batteries, medicine, and industry for processes involving electrolysis.

The process that occurs at the electrode is the occurrence of an oxidation-reduction balance that is formed between the metal and what is in the solution. An electrode consists of an anode and a cathode, The anode receives current or electrons from the electrolyte mixture, so it becomes oxidized, and then when the atom or molecule is close enough to the electrode surface, the mixture in which the electrode is placed donates electrons. This causes the atom to become a positive ion. And, the opposite occurs at the cathode where electrons will be removed from the electrode and the solution around it decreases.

There are various forms of electrodes, namely in the form of wire and plate that usually are made of metal and non-metal with an electrical conducting material such as copper and graphite, respectively. Copper and Graphite were chosen as electrodes in this research. Graphite is formed from carbon, whereas graphite is one of the softest materials and is usually used in pencils. Graphite is a native element mineral with a C (carbon) composition. This mineral has many unique characteristics. Graphite has a hexagonal crystal system appeared as a foliated mass or loose thin sheets, has an opaque structure, and is generally black in color. Graphite is a dimorphism of a diamond but has a very low hardness level (1-2). Graphite has a specific gravity of 2.23, good/clear cleavage, and when you feel it feels oily. This mineral is very resistant to heat and does not dissolve easily in water. Graphite is also the only non-metallic mineral capable of conducting electricity. Graphite is a good conductor of electricity which makes it useful in batteries and solar panels, graphite has mechanical properties such as metal, is light and cheap which makes it suitable as an electrode in the electrolysis process [11]

Whereas, copper is a metal element in the form of crystals with a reddish color. The chemical name Cupprum is denoted by Cu. Copper is a transition metal for group IB which has an atomic number of 29 and an atomic weight of 63.55 g / mol. Copper in nature is found mostly in compound form or as a solid compound in mineral form (Palar, 2004). Copper has physical properties such as: a reddish-yellow color; malleable so that it is easily formed into pipes, thin sheets, wire; a good conductor of heat and electricity for the flow of electrons; hard when it is not pure; and it has a melting point of 1,084.62 ° C, while a boiling point of 2,562 ° C. While, it chemical properties include: a relatively unreactive element so it is resistant to corrosion; in humid air, the surface of the copper is covered by a layer that is green in color which attracts the basic copper carbonate, Cu (OH) 2CO3; at about 300 ° C, copper can react with oxygen to form black CuO. Whereas at higher temperatures, around 1,000 ° C, red copper (I) oxide (Cu2O) will be formed; it is not attacked by water or water vapor and aqueous, non acids such as dilute HCl and dilute H2SO4 but concentrated and boiling HCl attacks the copper metal and liberates hydrogen gas; peer does not react with alkalis but dissolves in ammonia in the presence of air to form a blue solution of the Cu (NH3) 4+ complex; and a hot copper can react with sulfur and halogen vapors. Reacting with sulfur to form copper (I) sulfide and copper (II) sulfide and for reaction with halogens to form copper (I) chloride

[14]. Copper is known as a good raw material for electrodes according to [10] this is because copper material has good heat resistance, very high electrical conductivity, and high tensile strength even though at high temperatures besides copper also has a relatively cheap price so that making them suitable as electrodes in the electrolysis process.

1.4 Hydrogen Gas

Hydrogen gas is an element that is so abundant that hydrogen alone makes up about three-quarters of the mass of the universe. Hydrogen is present in the water, which covers 70% of the earth's surface and in all organic matter. Hydrogen itself is very simple, consisting of only one proton and an electron. Hydrogen is very environmentally friendly because when "released" hydrogen gas will immediately diffuse into the air, and does not pollute soil or groundwater because hydrogen is colorless, smelly, and non-toxic and does not cause acid rain, hollow out ozone or produce harmful emissions while the weight of hydrogen itself is 14 times lighter than air.

Hydrogen is a good and efficient choice for fuel because it has the highest combustion energy per kilogram compared to other fuels, meaning it is more efficient for the same weight than the fuels used today. Hydrogen generally offers 2-3 times more energy than most common fuels. It combines easily with oxygen, releasing a large amount of energy in the form of heat. Currently, hydrogen can be produced by electrolysis using an electric current to reduce water to hydrogen and oxygen.

Hydrogen can be used as an alternative fuel because the hydrogen molecules that have been released from water can produce energy. This energy will be captured by the combustion engine or a device called the Fuel Cell. In Fuel Cell, hydrogen energy can be converted into electrical energy with a high level of efficiency. Fuel Cell is a promising technology to be used as a source of electricity for housing, and as a source of electricity to drive or run vehicles [12].

2. Method

The method used in this research is an experimental method with descriptive quantitative analysis. In this research, a study was conducted on the production of hydrogen gas through the electrolysis of seawater. After the electrolysis reactor has been designed and built, the seawater electrolysis process can further be carried out. The electrolysis process in seawater is carried out to analyze how the influence of time variations and types of electrodes in the production of hydrogen. The main tools and raw materials used are seawater which can also function as an electrolyte solution and a set of electrolysis equipment that has a test tube that acts as a storage of the produced hydrogen gas [13]

The volume of hydrogen produced by the electrolysis of seawater is measured by determining the amount of water present in the test tube before and after the electrolysis process occurs within a certain measurable time, namely the "water displacement" method which will also be used to calculate the rate of hydrogen gas production. An AC-DC adapter was connected using a crocodile cable to the positive pole (anode) and negative pole (cathode) and also connected to the ammeter to determine electric current that was used. A pair of electrodes made of copper and graphite was used interchange and inserted into the test tube placed up-side down in a large container box filled with 400 ml of seawater for hydrogen gas produced through the electrolysis process. The use of test tubes in the electrolysis process in this study is due to the small volume of hydrogen gas produced.

Data collection of the hydrogen volume that was done by measuring and recording the water level within the test tube before and after of the electrolysis process at a certain time. The measurement time of electrolysis process varied from 10 – 55 minutes with an increment of 5 minutes and a constant voltage of 12 volts. Therefore, the rate and volume of hydrogen gas produced by the electrolysis of seawater using a graphite and copper electrode will be compared to analyze which electrode is more effective and efficient in producing hydrogen gas within a certain measurement time of the seawater electrolysis process. The following diagram shows the setup of the electrolysis equipment (adapted from [16]

Journal of Physics: Conference Series **2019** (2021) 012095 doi:10.1088/1742-6596/2019/1/012095

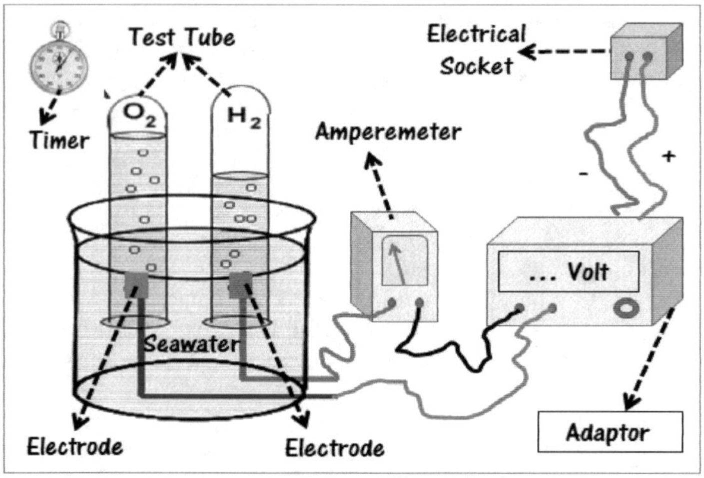

Figure 1. Process Electrolysis

3. Results and Discussion.

The table and graph below shows the measurement results of the volume of hydrogen gas produced from the electrolysis of seawater.

Table 1. Total Time of Electrolysis Process against the Production of Hydrogen Gas using Graphite and Copper Electrodes

No	Time of the Electrolysis Process (minute)	Graphite Electrode			Copper Electrode		
		Current (A)	Volume	Rate of Production	Current (A)	Volume	Rate of Production
1	10	0.41	6.0	0.60	0.25	1.5	0.15
2	15	0.35	6.5	0.43	0.35	2.5	0.17
3	20	0.33	7.0	0.35	0.38	3.0	0.15
4	25	0.31	7.5	0.30	0.40	4.5	0.18
5	30	0.29	7.5	0.25	0.42	6.0	0.20
6	35	0.29	8.5	0.24	0.48	8.0	0.23
7	40	0.23	8.5	0.21	0.48	11.0	0.28
8	45	0.20	9.0	0.20	0.50	14.5	0.32
9	50	0.17	9.0	0.18	0.49	17.0	0.34
10	55	0.17	9.5	0.17	0.44	16.5	0.30
Total	325		79.0			84.5	

The table above and following graphs and show the difference in the volume and rate of production of hydrogen gas produced through the electrolysis of seawater with a constant voltage of 12 volts, and with different times and types of electrodes as mentioned above. As shown in the table above that at first, of 35 minutes the volume of hydrogen produced by electrolysis process using graphite electrode was higher than using copper electrodes. When the time of the electrolysis process reached 40 minutes, the volume of hydrogen produced by electrolysis process of seawater with copper electrode starting to increase higher than graphite electrode. Accordingly, the production rate of hydrogen that was produced by electrolysis process using copper electrode was also increased during the time range of process. It was contrary different with the production rate of hydrogen using graphite that was decreased as shown in the table and graph.

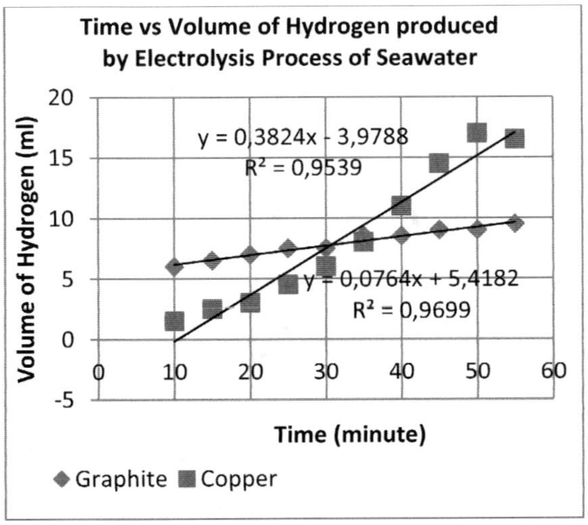

Figure 2. Graph of the volume of hydrogen gas (ml) vs time variations (minutes) of electrolysis process of seawater

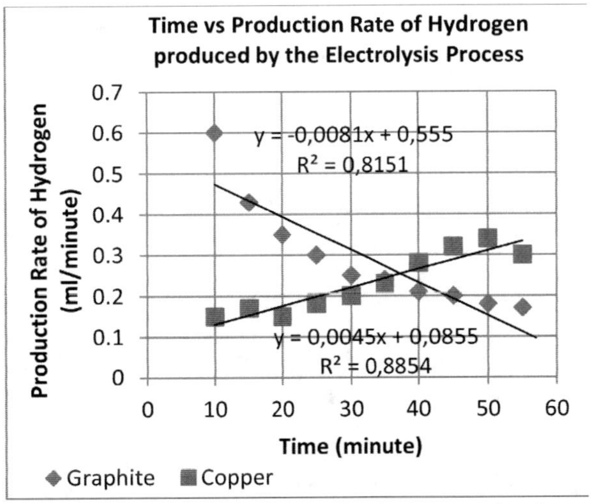

Figure 3. Graph of the rate of hydrogen production gas (ml) vs time variations (minutes) of electrolysis process of seawater

Interestingly, at time process of 55 minutes (total of 325 minutes or 5 hours 25 minutes into using electrodes), the volume of hydrogen that was produced by using a copper electrode indicated decreasing; while for graphite electrode was increased. The longest time of the electrolysis process of seawater caused a trend in decreasing the electric current through the equipment. And further it will cause decreased in volume and rate production of hydrogen. The decreasing of electric current through the equipment of electrolysis process may be assumed because of corrosion process of electrodes, especially for copper electrode. It was different with graphite which showed that the current was decreased after the first 10 minutes. Accordingly, [9] indicated that the greater the electric current supplied, the faster the decomposition reaction so that there is also an increase in the production of hydrogen gas that occurs from the cathode poles. While, [7] described that decreasing in current occurs because the longer the electrode is used, it will experience oxidation due to corrosion of chlorine in seawater; so the effectiveness of electrodes will be decreased and this is what causes a decrease in current. It was supported by the results of research by [2] which explained that the result of the corrosion process on the voltaic cell electrodes will reduce the service life of the electrodes. This of course has an impact on the reduction of hydrogen gas produced in the electrolysis of seawater. However, after that total time process of 5 hours 25 minutes, the electrolysis process of using graphite and copper electrode produced 79 ml and 84.5 ml of hydrogen gas, respectively. It was in agreement with the research conducted by [17] where after a few hours of electrolysis there was a decrease in the production of hydrogen gas. The results obtained in this study indicate that copper electrodes are better than graphite electrode which was in accordance with the results of research conducted by [17] which concluded that copper electrodes are superior to graphite electrodes in terms of efficiency.

4. Conclusion

From the data analysis obtained in this study, it can be concluded that the total volume of hydrogen gas produced in the electrolysis process of seawater by using copper electrodes was much greater than the total volume of hydrogen gas produced by using graphite electrodes. However, the lifetime of copper electrodes was longer than graphite electrodes during the electrolysis process of seawater. The suggestions need to be considered and made for further research is to further wide the range time of process to clearly identify the service life of the electrodes. Another method such stoichiometry can be utilized to accurately determine the volume of hydrogen gas produced by the electrolysis of seawater. The

Journal of Physics: Conference Series **2019** (2021) 012095 doi:10.1088/1742-6596/2019/1/012095

sensor which is able to accurately measure the water displacement was also suggested to be used for next research of hydrogen production based on the electrolysis seawater.

References

[1] Addinia, E. M. (2018). *Sifat-Sifat Fisik Air Laut.* Surabaya: Universitas Islam Negeri Sunan Ampel Surabaya.

[2] Afiah, S. (2017). *Studi Karakteristik Daya Listrik Air Laut dengan Prinsip Sel Volta dan Efek Korosi Elektroda.* Makassar: Universitas Hasanudin.

[3] Anggaretno, G., Rochani, I., & Supomo, H. (2012). *JURNAL TEKNIK ITS Vol. 1, No. 1*, 124-128.

[4] Djakaria, M. N. (2012). *AIR LAUT.* Retrieved from file.upi: http://file.upi.edu/Direktori/FPIPS/JUR._PEND._ GEOGRAFI/194902051978031-DJAKARIA_M_NUR/AIR__LAUT

[5] Fukuzumi, S., Lee, Y.-M., & Nam, W. (2017). *ChemSusChem*, 4264-4276.

[6] Higa, M., Watanabe, T., Yasukawa, M., Endo, N., Kakihana, Y., Futamura, H., Miyake, H. (2019). *Water Practice & Technology Vol 14 No 3*, 645-651.

[7] Ibrahim, J. M., & Moussab, H. (2020). *KeAi Chinese Roots Global Impact Materials Science for Energy Technologies 3*, 780-807.

[8] Isana, S. (2010). Perilaku sel elektrolisis air. *prosiding seminar nasional.*

[9] Jumiati, Sampurno, J., & Faryuni, I. D. (2013). *Jurnal POSITRON, Vol. III, No. 1*, 6-11.

[10] Junaidi, A. (2019). Pengembangan Metode Pembuatan Elektroda Tembaga-Karbon Dengan Metalurgi Serbuk. *Jurnal Sains dan Teknologi Indonesia.*

[11] Luo, L. (2020, Maret 9). *Four Characteristic of Carbon.* Retrieved from Sciencing: sciencing.com

[12] NASA. (2018). *renewableenergyworld.com/hydrogen/tech.* Retrieved from renewableenergyworld.com: https://www.renewableenergyworld.com/hydrogen/tech.html

[13] Rahmi, H. I. (2019). *Desain Sistem Reaktor Dan Pengukuran H2 Berbasis Elektrolisis Dengan Katalis CO2.* Jakarta: UNJ.

[14] Riadi, M. (2020, September 2). *Tembaga (Definisi, Karakteristik, Sifat, Penggunaan dan Dampak Keracunan Limbah).* Retrieved from kajianpustaka.com: https://www.kajianpustaka.com/2020/09/tembaga.

[15] Rustana, C. E., Kharudini, D. S., & Rahmi, H. I. (2019). *ICESE*, 129.

[16] Rustana, dkk. (2021). The effect of voltage and electrode types on hydrogen production from the seawater electrolysis process. SNF UNJ 2021.

[17] Slama, R. B. (2013). *Natural Resources*, 1-7.

[18] Song, Z., & Xie, Z. H. (2018). *Micron 112*, 69-83.

The Effect of Voltage and Electrode Types on Hydrogen Production from The Seawater Electrolysis Process

C E Rustana[1], Sunaryo[2], S J Muchtar[2], I Sugihartono[2], W Sasmitaningsihhiadayah[1], A D R Madjid[1] and F S Hananto[1]
[1] Physics Department, Faculty of Science and Technology, State Islamic University of Maulana Malik Ibrahim Malang-East Java, Indonesia (65144)
[2] Physics Departement, Faculty of Mathematics and Science State University of Jakarta East Jakarta, Indonesia (13220)
E-mail: CE.Rustana@yahoo.com, Sunaryo@unj.ac.id, sxjxm1904@gmail.com, wiwis_hidayah_87@fis.uin-malang.ac.id, armedia@uin-malang.ac.id, faridsamsu@fis.uin-malang.ac.id

Abstract Besides the limited supply, turns out the fossil fuels also causes carbon dioxide (CO2) pollution in the atmosphere and causing global warming. Therefore, renewable energy that environmentally friendly are needed, which is hydrogen gas considered capable of being an alternative to replace fossil fuels. One of many simple ways and effective to produce hydrogen gas is by electrolysis. Seawater was chosen in this research because of its abundant availability, have high efficiency and low cost to produce hydrogen gas. By varying the voltage from 1.5 volts to 24 volts with an increase of 1.5 volts, this research was conducted to determine the effect of the difference in voltage and type of electrode on the volume and rate of hydrogen production through electrolysis of seawater with constant time 10 minutes each. The type of electrode that used in this research are copper and graphite. The result showed that the productivity coefficient of copper electrodes was 0.41 ml with determination coefficient of 0.97; while, productivity coefficient of graphite electrodes was 0.32 ml with determination coefficient of 0.93. This research also shows that a maximum of volume hydrogen gas of 8.5 ml was produced through the electrolysis of seawater using copper electrodes at a voltage of 21 volts. This result is much greater than using the graphite electrodes that only produced a maximum volume of hydrogen gas of 7.1 ml at a voltage of 22.5 volts.

1. Introduction

Currently, the use of energy continues to increase due to an increasing population accompanied by global developments. The increase in global energy consumption was recorded at around 1.5%, always increasing every year starting from 2010 [1]. However, current energy use throughout the world is mainly supplied from fossil energy. In addition to their limited availability, the use of fossil fuels also produces Carbon Dioxide (CO2) pollution in the atmosphere so that it has an impact on global warming. Therefore, environmentally friendly renewable energy resources are needed. Renewable energy is energy that comes from nature without having to deplete the existing resources on earth.

Among the available renewable energy sources, hydrogen gas is one that is considered capable of being an alternative to replacing fossil energy, especially in transportation because the environmental effects produced by hydrogen gas are very minimal. Hydrogen gas also has many advantages, namely, it has a large combustion energy, which is around 286 kJ/mol, it also has very little effect on the environment because the combustion emissions produced are in the form of water [2]. One of the simple and effective ways to produce hydrogen gas is by means of the electrolysis process. The water electrolysis method is widely used because besides being a simple method of producing hydrogen gas, it is also in order to obtain high purity hydrogen [3].

The electrolysis method generally uses an alkaline solution because it is easy to do, but this technology is quite expensive. In addition, the alkaline solution has many disadvantages, namely the high overvoltage and the amount of resistance it has. In addition, the selection of electrodes in the electrolysis process must also be based on the capabilities they have because it will also affect the large or small volume of hydrogen gas that will be produced [4]. A good electrode for use in the electrolysis of seawater is, of course, an electrode that is a good conductor of electricity, the price is economical and is also resistant to seawater corrosion. The use of seawater which is like electrolyte was chosen to produce hydrogen gas through the electrolysis process, because of its abundant availability and promising to become a renewable energy source by producing hydrogen gas with high efficiency and low cost [5]. Understanding the abundant availability of sea water and its potential which can be used to produce hydrogen gas through the electrolysis process, this paper will discuss and analyze the effect of variations in the voltage and type of electrodes used in the electrolysis of seawater on the production of hydrogen gas produced.

2. Production of hydrogen gas through electrolysis of seawater

Hydrogen gas is one of the chemical elements on the periodic table with the symbol H and atomic number 1. Hydrogen gas is a gas that is odorless, non-toxic and colorless. Hydrogen gas is the lightest element in the world because it only has an atomic mass of 1.00794 smu. Hydrogen gas is one of the most abundant elements because there are about 75% of the total mass in the universe [6]. Hydrogen gas can be used as an alternative fuel because it can replace fossil energy, especially in transportation because the environmental effects produced by hydrogen gas are very minimal. The advantage of hydrogen gas is that it has large combustion system energy, around 286 kJ/mol, and has very little effect on the environment because the resulting combustion system is water [7].

One of the processes to produce hydrogen gas is water electrolysis which is carried out by utilizing unidirectional electrical energy to break up water into oxygen and hydrogen. Electrolysis is the process of decomposing compounds by direct electric current through an electrolyte solution. Electrolyte solutions can conduct electricity because they contain ions that can move freely. Electrical conductivity occurs because the electrolyte solution is flowed from the current source and will give a different charge to the two electrodes used. The resulting charge will be divided into negative ions (anions) and positive ions (cations). Hydrogen gas will form at the cathode and oxygen will form at the anode. The factors that influence electrolysis are variations in the amount of voltage applied, the metal properties of each type of electrode and the factor of the electrolyte solution used.

In the electrolysis process, the greater the value of the voltage difference given, the greater the production of hydrogen gas produced. This can occur due to variations in the voltage difference used, because the greater the voltage difference used, the greater the current obtained in the reaction, the faster the decomposition reaction, the formation of hydrogen gas that occurs from the cathode poles will also be greater [8]. The amount of current and voltage generated by electrolysis cells depends on several factors, namely the size of the electrodes and the content of the electrolyte solution (salt) [9]. The electrochemical equation for electrolysis of water is described as follows:

At the cathode: $4H^+(aq) + 4e^- \rightarrow 2H_2\,(g)$ (1)

At the anode: $2H_2O\,(l) \rightarrow O_2(g) + 4H^+(aq) + 4e^-$ (2)

Total reaction: $2H_2O(l) \rightarrow 2H_2\,(g) + O_2(g)$ (3)

The energy used comes from the electricity that is flowing that leads to Gibbs energy (ΔG) as follows:

$$\Delta G = \Delta H - T\Delta S \quad (4)$$

The reverse voltage required in the electrolysis process to occur can be obtained by the ΔG equation [10]

$$V_{rev} = \frac{\Delta G}{nF} \quad (5)$$

With n = 4 and F which is 96,485 J in the standard state (298.15 K and 1 atm), the reactions that usually occur in the electrolysis process are as follows:

$$2H_2O(l) \rightarrow 2H_2(g) + O_2(g)\ \Delta H^\circ = 572\ kJ \quad (6)$$

With the value $\Delta G = 474.74$ kJ. mol^{-1} and $\Delta S = 0.3262$ kJ. mol^{-1}, produces a reversible voltage value of V_{rev}= 1.23 Volt

However, the electrolysis of mostly abundant sea water rather than the purified or fresh water is a more promising way to generate clean hydrogen energy. Sea water is water that comes from the sea, has a high salt content (salinity) and has a salty taste. The world's oceans have an average salinity of 3.5%. The salt content in seawater is 55% chloride, 31% sodium, 8% sulfate, 4% magnesium, 1% calcium, 1% potassium and bicarbonate, bromide, boric acid, strontium, and fluoride which only less than 1%. Seawater is electrolyte due to its salt composition. Electrolyte solution is a compound that can conduct electric current. The electrolyte solution can function as a conductor of electricity, where the electric current is carried by the movement of ions [11]. NaCl dissolved in water will break down into Na+ ions and Cl- ions.

The chlorine content in seawater can cause corrosion that is the degradation of materials (metals) due to electrochemical reactions with their environment. Basically, it is a reaction of metals into ions on a metal surface which is in direct contact with aqueous environment, in this research it is sea water and oxygen. The chlorine content of salt in sea water causes corrosion to the metal to be faster, because it has electrolyte properties that will provide a good atmosphere for the oxidation-reduction reaction [12]. The rate at which corrosion occurs is greatly influenced by the salinity level of seawater. Corrosion rate will increase with increasing salinity of sea water due to the influence of chloride [13].
The production of hydrogen by electrolysis of sea water that contains Sodium chloride (NaCl) make seawater as a natural catalyst [14] will occur a cation reduction reaction at cathode as follows:

$$Na^{+}\ (aq) + e^{-} \rightarrow Na\ (s)\ E^{0}red = -2.71\ volt \quad (7)$$

$$2H_2O(l)+2e^{-}\rightarrow H_2(g) + 2OH^{-}(aq)\ E^{0}red= -0.828\ volt \quad (8)$$

The more positive "the oxidation is, the easier it is for the substance to oxidize. Therefore, chlorine gas will be oxidized at the anode. So that the overall reaction that occurs in the electrolysis of seawater is:

$$2Cl^{-}(aq)\rightarrow Cl_2(aq)+2e^{-}\ E^{0}oks = +1.36\ volt \quad (9)$$

$$2H_2O\rightarrow O^{2}(g)+4H^{+}(aq)+4e^{-}\ E^{0}oks = +1.23\ volt \quad (10)$$

From the total reaction it can be seen that using seawater electrolyte, hydrogen is formed at the cathode; while chlorine gas is formed at anode [15].

3. Electrode

An electrode is a conductor used to pass an electric current from a power source to a material. There are various forms of electrodes, namely in the form of wire, plate and usually that are made of metal such as zinc or copper, but can also be non-metal with an electrical conducting material such as graphite. Copper electrodes were chosen in this study because this material has high electrical conductivity, excellent heat resistance, high temperature resistance and low price, making it suitable as an electrode in this electrolysis research [16]. The physical properties of the copper electrode include a thermal conductivity of 380.7 W /m-k, a melting point of 1083°C, a specific heat of 0.092 Cal /g°C, and a coefficient of thermal expansion of 17×10^{-6}/°C. The chemical properties of the copper electrode are 99% copper in composition, with a density of 8.9 g/cm^3.

However, graphite electrode was used since it has good electrical conductivity and cheap. Graphite also has a diffusion coefficient that tends to be closer to the true value when compared to other electrodes such as stainless steel and brass, so that the graphite electrode is suitable for use in the electrolysis process of seawater which contains chloride [17]. The physical properties of graphite are solid phase, density 2.267 g/cm^3, melting point 4300-4700 K, boiling point 4000 K, heat of melting 100 kJ / mol, heat of evaporation of 355.8 kJ/mol and heat capacity 8,517 J / mol K. Its chemical properties are very unreactive at ordinary temperatures, reacts directly with fluorine, if it burns in limited air it will produce carbon monoxide, if excess air will form carbon dioxide, when heated in air it will react with oxygen forms carbon dioxide and reacts with water to form carbonic acid [18].

4. Research methods

This research was conducted using the experimental method. The electrolysis process of seawater is implemented by varying the voltage and type of electrode to determine the volume of hydrogen gas produced. The electrolysis process occurs when the electrodes in the electrolyte solution are given a voltage difference. This study uses copper and graphite electrodes. The variation of the voltage difference given from the 1.5 volt battery and AC-DC adapter that converts the AC source into a DC voltage source, there are sixteen kinds of voltage variations, 1.5, 3, 4.5, 6, 7.5, 9, 10.5, 12, 13.5, 15, 16.5, 18, 19.5, 21, 22.5, 24 volts. The following diagram shows the setup of the electrolysis equipment.

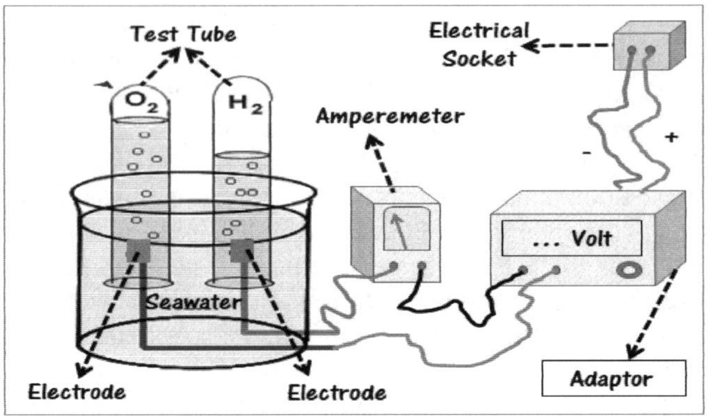

Figure 1. Process electrolysis

The data collection technique in this research is the "water displacement" method to obtain the volume of hydrogen gas, which will also be used to calculate the rate of hydrogen gas production. The "water displacement" method is to measure the volume of gas produced by a chemical reaction based on the reduced volume of water in the test tube used. The technique of analyzing the data in this research is descriptive quantitative, that is, after the volume of hydrogen gas is known from the water displacement method, then the calculation of the hydrogen gas production rate is carried out by dividing the resulting volume of hydrogen gas obtained by the time required for electrolysis [19].

5. Result and Discussion

Table 1. The Measurement result of the volume and rate of hydrogen gas production using copper electrodes in the electrolysis of seawater

No	Voltage (Volt)	Current (A)	Volume of Hydrogen (ml)	Production Rate of Hydrogen (ml/min)
1	1.5	0.05	0.3	0.03
2	3	0.08	0.5	0.05
3	4.5	0.12	0.7	0.07
4	6	0.15	1.3	0.13
5	7.5	0.18	2.1	0.21
6	9	0.21	2.7	0.27
7	10.5	0.25	3.5	0.35
8	12	0.27	4.3	0.43
9	13.5	0.31	5.1	0.51
10	15	0.36	5.8	0.58
11	16.5	0.42	6.4	0.64
12	18	0.51	7	0.7
13	19.5	0.6	7.8	0.78
14	21	0.7	8.5	0.85
15	22.5	0.55	8	0.8
16	24	0.53	7.7	0.77

Table 2. The results of measurement of the volume and rate of hydrogen gas production using graphite electrodes in the electrolysis of seawater

No	Voltage (Volt)	Current (A)	Volume of Hydrogen (ml)	Production Rate of Hydrogen (ml/min)
1	1.5	0.03	0.2	0.02
2	3	0.06	0.4	0.04
3	4.5	0.09	0.6	0.06
4	6	0.01	0.9	0.09
5	7.5	0.13	1.8	0.18
6	9	0.15	2.5	0.25
7	10.5	0.18	3.3	0.33
8	12	0.2	3.8	0.38
9	13.5	0.25	4.2	0.42
10	15	0.31	4.6	0.46
11	16.5	0.35	5.1	0.51
12	18	0.42	5.8	0.58
13	19.5	0.5	6.3	0.63
14	21	0.53	6.5	0.65
15	22.5	0.6	7.1	0.71
16	24	0.38	5.2	0.52

Figure 2 below shows the results of the volume of hydrogen gas produced by the electrolysis of seawater at various voltages for each type of electrode in a constant processing time of 10 minutes. When using copper electrodes, the volume of hydrogen gas produces a hydrogen gas productivity coefficient of 0.41 ml / minute (R^2 = 0.97). Meanwhile, the use of a graphite electrode resulted in a volume productivity coefficient of 0.32 ml / minute (R^2= 0.93). These results show that the use of copper electrodes can produce a larger volume of hydrogen gas compared to using graphite electrodes in the electrolysis of seawater.

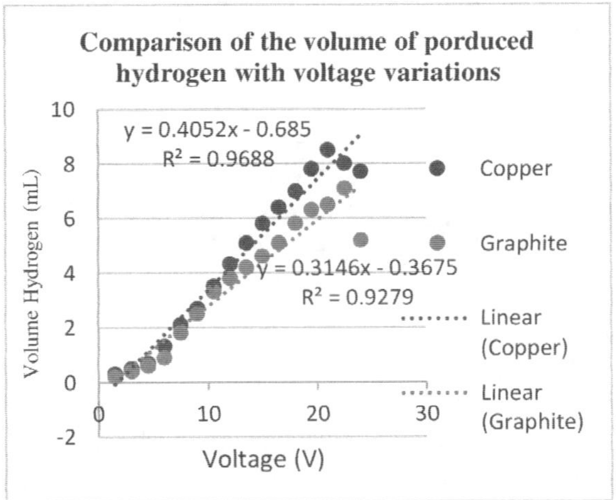

Figure 2. Comparison of the volume of produced Hydrogen with Voltage variations

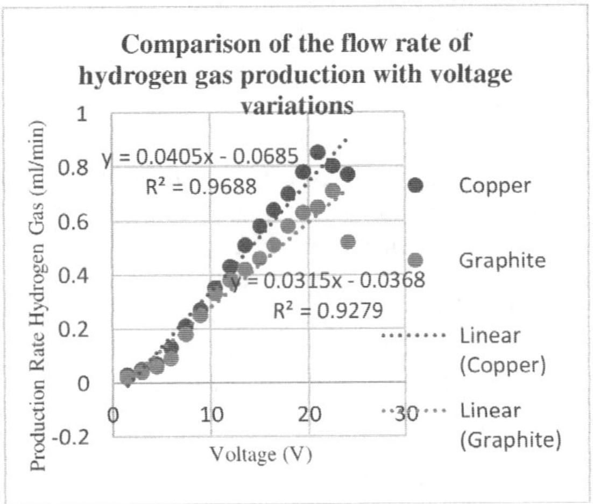

Figure 3. Comparison of the flow rate of hydrogen gas production with voltage variations

Figure 2 shows that the higher the voltage used in the electrolysis of seawater for each type of electrode has an impact on the increase in the volume of hydrogen gas produced through the process. The value of R2 > 0.7 represents the strong relationship between the independent variable X (the voltage variation) and the dependent variable Y (volume of hydrogen gas). This is understandable because the greater the applied voltage, the faster the reaction of decomposition of seawater will be. Thus, the volume of hydrogen gas produced will increase. Whereas, the value of b in the simple linear regression equation serves to show the magnitude of the change in the value of Y (the production of hydrogen gas) due to changes in the value of X (voltage variation) given to the electrolysis of seawater. From these values of determination coefficient, it can also be seen that the type of electrode that has a more optimal performance in producing the production rate and volume of hydrogen gas at various voltages used. However, at voltage variations ranging from 22.5 to 24 volts, the volume of hydrogen gas produced through the electrolysis process using the two types of electrodes also decreased, this can be understood to happen because of the chlorine content in seawater that causes a decrease in function, thereby accelerating the damage to the metal electrodes [12] and of course it will decrease in the resulting volume of hydrogen gas which was produced.

Figure 3 shows the rate of hydrogen gas production against voltage variations for each type of electrode used in the electrolysis of seawater. The use of copper electrodes in the seawater electrolysis process resulted in a hydrogen gas production rate coefficient of 0.041 ml / minute (R^2= 0.97). The graphite electrode produces a hydrogen gas production rate coefficient of 0.032 ml / minute (R^2= 0.93). These results show that the use of copper electrodes can produce a higher hydrogen gas production rate than using graphite electrodes in the electrolysis of seawater. The results of this study as shown in Figure 3 indicated that the greater the applied voltage, the greater the production rate of hydrogen gas produced through the electrolysis of seawater. This is in accordance with the research results of Jumiati et al. (2013) [20] who explains that the greater the voltage causes, the electric current is also greater, and the reaction of decomposition of sea water that occurs will also be greater and gas formation of will also get faster and bigger. In addition, the results of this study are also in line with the results of research by N Bellel and M Sahli (2011) [21] which explains that the variation of the flow of hydrogen gas produced will continue to increase with increasing variations in the voltage used.

From graph 3 it can also be seen that at the variation of the voltage from 22.5 to 24 volts, the production rate of hydrogen gas produced in the electrolysis process using copper electrodes has decreased. Meanwhile, the use of graphite electrodes in the electrolysis of seawater shows that the rate of hydrogen gas production only decreases when the voltage reaches 24 volts. This decrease in the rate of hydrogen gas production is understandable because copper electrodes have less resistance to chlorine than graphite electrodes. Due to the presence of chlorine in salt in seawater, the copper electrodes are corroded, causing a decrease in the effectiveness of the copper electrodes. This is in accordance with the research by NH Saputra, et al. (2020) [22] which indicated the electrodes that experience corrosion due to sea water can cause a decrease in the effectiveness of such electrodes in conducting electricity. However, the results of the research show that the use of copper electrodes is far better than the use of graphite electrodes in producing hydrogen gas through the electrolysis of seawater.

In this case, the use of copper electrodes produces a larger volume of hydrogen gas compared to the use of graphite electrodes in the electrolysis of seawater. The results of this study are in accordance with the results of research by Slama (2013) [23] who conducted research to test various types of electrode materials in producing hydrogen gas in the electrolysis process. Slama's research using NaCl solution in the electrolysis process concluded that the copper electrode is first, bronze is second, stainless steel is third and graphite is last. The results of the study using copper electrodes show that the largest volume of hydrogen gas produced through the electrolysis of seawater occurs at a voltage of 21 volts. At this 21 volt voltage difference, the volume of hydrogen gas produced through the electrolysis of seawater using copper electrodes is 8.5 ml with a production rate of 0.85 ml / minute. This result is much greater than the volume of hydrogen gas produced in the electrolysis of seawater using a graphite electrode at a voltage difference of only 7.1 ml of 22.5 volts and a production rate of 0.71 ml / minute. The results of this study as described above indicate that the productivity rate of hydrogen gas produced through the electrolysis of seawater using copper or graphite electrodes has different results compared to the results of research by N Bellel and M Sahli (2011). As it is known that N. Bellel, and M. Sahli showed that the best voltage to produce the largest hydrogen gas production rate up to 3.5 ml / min was obtained through the electrolysis of brine using a graphite electrode at a voltage of 18 volts. This can be understood because there are differences in voltage and salinity variations in the electrolyte solution used.

6. Conclusion

Based on the analysis of the results of the study, it is concluded that the variations in the voltage and type of electrode have an effect on the volume and rate of hydrogen gas production through the electrolysis of seawater. Copper electrodes in the electrolysis of seawater have a greater impact on the productivity of hydrogen gas. The volume and the rate production of hydrogen production was respectively 8.5 ml (R2 = 0.97) and 0.85 ml / min (R2 = 0.97) occurred when using copper electrodes with a variation of 21 volts.

Reference

[1] BP Statistical Review of World Energy 2019 *BP* from https://www.bp.com/content/dam/bp/business-sites/en/global/corporate/pdfs/energyeconomics/statistical-review/bp-stats-review-2019-full-report.pdf diakses tanggal 12 oktober 2020.
[2] Dincer, I 2012 International *Journal of Hydrogen Energy* https://doi.org/10.1016/j.ijhydene.2011.03.173.
[3] Richard, SP 2004 *A Techo-Economic Analysis Of Decentralized Electrolysis Hydrogen Production for Fuel Cell Vehicles* (Departement Of Mechanical Engineering: University of Victoria Australia).
[4] Hasan, A. 2007 *Aplikasi Sistem Fuel Cell Sebagai Energi Ramah Lingkungan di Sektor Transportasi dan Pembangkit* Teknik Lingkungan.
[5] Nikolaidis, P., & Poullikkas, A. 2017 *In Renewable and Sustainable Energy Reviews* https://doi.org/10.1016/j.rser.2016.09.044.
[6] Nikolaidis, P., & Poullikkas, A. 2017 *In Renewable and Sustainable Energy Reviews* https://doi.org/10.1016/j.rser.2016.09.044.
[7] Dincer, I. 2012 International Journal of Hydrogen Energy https://doi.org/10.1016/j.ijhydene.2011.03.173
[8] Irfana Diah Faryuni, J. J. S., Sampurno, J., & Faryuni, I. D. 2013 *POSITRON* https://doi.org/10.26418/positron.v3i1.4757.
[9] Sani A 2018 *Analisa Baterai Air Asin dengan Elektroda Tembaga dan Aluminium. Jurusan Teknik Elektro* (Fakultas Teknik Universitas Muhammadiyah: Surakarta).
[10] Ursúa, A., Gandía, L. M., & Sanchis, P. 2012 *Proceedings of the IEEE* https://doi.org/10.1109/JPROC.2011.2156750.
[11] Lowrie,P.E.W. 2005 *Mitzubishi Cyclon* (Proceeding of Electrolytic Gas: USA).
[12] Utomo S. 2015 *Jurnal Teknologi* 7(2): 93-103.
[13] Sasono, E. J. 2010 *Efektivitas Penggunaan Anoda Korban Paduan Aluminium pada Pelat Baja Kapal AISI 2512 terhadap Laju Korosi di dalam Media Air Laut* (Universitas Diponegoro: Semarang).
[14] Yilmaz, A. C., Uludamar, E., & Aydin, K. 2010 *International Journal of Hydrogen Energy* https://doi.org/10.1016/j.ijhydene.2010.07.040.
[15] Brady J E, 2008 *Kimia Universitas Asas dan Struktur* (Jakarta: Binarupa Aksara).
[16] Junaidi, A. 2019 *Jurnal Sains Dan Teknologi Indonesia* https://doi.org/10.29122/jsti.v15i2.941.
[17] Zhang, W. M., & Ba, H. J. 2012 *Corrosion Engineering Science and Technology* https://doi.org/10.1179/1743278212Y.0000000017.
[18] Cotton & Wilkinson 2007 *Kimia Anorganik Dasar* (Jakarta: UI-Press).
[19] Rahmi, H. I. 2019 *DESAIN SISTEM REAKTOR DAN PENGUKURAN H2 BERBASIS ELEKTROLISIS DENGAN KATALIS* CO_2 (Jakarta: UNJ).
[20] Jumiati, Joko Sampurno, Irfana Diah Faryuni 2013 *Jurnal POSITRON* Vol. III, No. 1 (2013), Hal. 06-11,Pontianak.
[21] N. Bellel, M. Sahli 2011 *International Review of Physics (I.R.E.PHY.)* Vol. 5, N. 4 August 2011.
[22] Nofrizal Hayadi Saputra, Sugeng Hari Wisudo, Mochammad Riyanto, and Adi Susanto 2019 *Jurnal Teknologi Perikanan dan Kelautan* Vol. 10 No. 2 November 2019: 135-147.
[23] Slama, R. B. 2013 *Natural Resources* 1-7.

Journal of Physics: Conference Series **2019** (2021) 012097 doi:10.1088/1742-6596/2019/1/012097

Watt Peak Meter of Solar Panel

M Manfaluthy[1], A Pangestu[1], R Arif and L A Sanjaya[2]
[1] Department of Electrical Engineering, Jakarta Global University
Jl. Boulevard Grand Depok City, Tirtajaya, Depok 16412, Indonesia
[2] Department of Physics, Faculty of Mathematics and Natural Science, Universitas Negeri Jakarta, Indonesia
Email: mauludi@jgu.ac.id

Abstract. The design of a watt peak meter of solar panel has been carried out using an LDR sensor to obtain the strongest light as the sun moves. The electronic circuit used includes LDR KY018, WEMOS ESP32, servo motor, Solar Charger Controller, INA219 Power Sensor, Battery and 10 A / 5V Relay. The mechanical structure uses angle iron construction which is assembled in such a way as to support solar panels with the SUNLITE Model 156P-20 type. Solar panels are positioned on 2 poles and attach the bolts as the rotary axis. LDR that gets sunlight will provide input to the WEMOS ESP32 to drive the servo motor in line with the direction of the sun in the east - west direction. The result led to an average 93,7% close to number as stated in the data specification of solar panel.

1. Introduction

Electricity has become the most important part that cannot be separated from human life today. Almost all human activities, both at home, offices and industry depend on electricity. Electricity can be generated with an electric generator and more than 99% of the electrical energy used today is generated by electric generators in the form of alternating current which is easily distributed over long distances [1].

The kinetic energy to drive a turbine generator is obtained from energy or steam power generated from the combustion of fossil energy sources (petroleum, coal, and natural gas). Most of the sources of electrical energy in Indonesia are supplied from fossil energy, but in the next 20 years, this energy source will run out and other energy sources are needed to continue to produce electrical energy. Currently there is one form of unlimited energy (non-conventional), namely energy produced from sunlight. This energy is an alternative source that can be converted into environmentally friendly electrical energy which is processed using solar cell panels or photovoltaic [2].

The sun has more than enough energy to supply the world's energy requirements, and unlike fossil fuels, it will not run out anytime soon. Solar power's main constraint as a renewable energy source is our capacity to convert sunlight into electricity in an efficient and cost-effective manner.

A typical question is how to calculate the solar panel output. Given the influence solar panel manufacture has on the overall system, it makes logical. There are a few variables and stages to calculating the power output of your solar panels. Of course, it's hard to anticipate how much electricity a photovoltaic system will generate in advance. There are a lot of things to consider. There are many elements involved such as the location, panel orientation, the direction of shade, the weather, the presence of clouds and the temperature on the roof. The watt-peak (WP) is therefore an indication based on a standard. It corresponds to the maximum electrical power that can be supplied by a photovoltaic panel under standard temperature and sunlight conditions,

The solar panel efficiency number indicates how many watts your solar panel can generate under ideal conditions. These ideal conditions, known as Standard Test Conditions (STC), are simulated in a laboratory/Manufacture where solar panels are tested. Standard test conditions for solar panel wattage would imply that your solar panel is operating at 77 degrees Fahrenheit with 1000 watts of sunlight per square foot [3].

In these ideal conditions, a 200-watt solar panel will generate 200 watts of electricity. This standard is a good way to ensure that when solar panels are manufactured, they meet specific criteria. Efficiency, how much sunlight shines on the solar panels, and which direction your solar panels are facing are the main factors in determining solar panel output. The WP listed by manufacturers makes it possible to compare different photovoltaic panels. The main issue for implementor is they cannot sure how many WP can be in the same surface area, the higher the WP, the better the panel performs.

They are several research on new energy alternative especially in solar Panel or photovoltaic, the trend is to optimize the power generated by the photovoltaic itself and to measure the characteristic of the photovoltaic. The following paragraph, we will summarize what we have been done before.
The real-time solar energy monitoring system was one of the papers we reviewed. The proposed work employs a 125-watt Poly Crystalline silicon solar panel as a monitoring system. The voltage and current sensors are used to measure the voltage and current from the panel, respectively. The temperature sensor is installed on the solar PV module to measure the current temperature, which greatly affects the efficiency of the photovoltaic system. A pyranometer measures the amount of solar irradiance in a planar surface in terms of W/m2. The Microcontroller is critical in processing the measured data and forwarding it to the cloud platform via the Wi-Fi module for concurrent observation and decision making.[3]. We found that it was done in this paper and result can be accessed from smartphone or PC. They way to develop IoT using microcontroller A CC3200 system-on-chip (SoC) with inbuilt Wi-Fi connectivity. This is the efficient way to make real time result.

A virtually reliable Solar PV monitoring system [4] is developed with LABVIEW software for computing the performance of a 5-Watt Solar Module. The electrical parameters like voltage, current, temperature, humidity and irradiance are measured using sensors and store the data in the DAQ (Data Acquisition) unit, which provide an interface to the PC. LABVIEW tool plot the I-V and P-V graph based on the data acquired and compute the Maximum voltage, Maximum current, Fill factor and efficiency of the solar panel. The watt peak calculation is still not measure on this research. [4].

Other papers we referred to is the solar tracking system. Solar tracker is a solar cell propulsion device that is made to automatically move at an angle of 0-180 degree and vice versa. Where the driving work is carried out by motors and sensors that function to track the receipt of solar energy by the solar cell panels. [5]. This paper is solving the issue about how much sunlight shines on the solar panels and Which direction your solar panels are facing.

It was expected that system can reduce the power load generated to drive the solar tracker motor and solar panels can produce maximum power. This research [6] found that solar tracker it has problems with the power generated to drive the motor has a large power to rotate the solar panel to always follow the movement of the sun. From the results of the study, the minimum power consumption occurs at 17:00 at 690 8 kg weight 8.12 voltage 0.11 when the solar panel is cross sectioned.

The developed prototype [7] was composed of current sensor type ACS-712 5A, voltage sensor, LM35 temperature sensor, LDR sensor, servo motor, and 11cmx11cm sized solar panel. The data of solar horizon coordinate system will be connected to the Arduino Uno software; therefore, the microcontroller works as the servo motor's controlling to drive the solar panel. Two pieces of servo motor allow the solar panel to rotate horizontally and vertically in the direction of the sun. Physical parameters, i.e., voltage, electric current, light intensity, and temperature, are measured using sensor modules, then stored in a personal computer (PC) using the PLX program – DAQ. This paper is still used PC and DAQ to collect data.

Other paper presents how to improve the photovoltaic (PV) system efficiency by using a method to integrate the solar tracker and the maximum power point tracking (MPPT). The integrated system provides a closed-loop solar tracker without the sensors. Instead of using the solar sensor, the output power of the MPPT is employed as the feedback signal to the solar tracker. The solar tracker estimates the solar azimuth and elevation angles using an astronomical algorithm based on the latitude, longitude, and the date-time of the local site. To improve the solar tracking accuracy, the fuzzy logic controller is employed to adjust the angle according to the power slope of maximum power with respect to the solar tracker angle. From the simulation results, the proposed method increases the PV energy by 23.23% compared to the fixed PV panel. It improves the efficiency of the existing integrated solar tracker by 0.25% based on the simulation models[8].

The last researcher presents real-time measurement of grid connected photovoltaic system using Zigbee sensor networks. The measurements were performed using sensors that are chosen based on their simplicity and lower price. [9]. The advantage of the solution of this paper had developed the network

by installing the sensor on every PV while they can collect all the information about the weather parameters were measured using OPT101 photodiode for irradiance sensors and DHT22 for temperature and humidity sensors[9]. However, there is no watt peak calculation involve on this solution.
Having said above explanation form several papers reviewed, Therefore, in writing this paper the author tries to provide new innovations by designing a prototype circuit for solar tracker which combine with Internet of things, called Watt Peak Meter of Solar Panel. So that it is expected that the system that will be ensure the value of watt peak compared to datasheet of manufacture, aside it will make easy to have real time measurement without facing or sitting in front of the solar panel. In this research we would cover the three main factor that influence the solar panel. This watt peak meter is based on a solar panel tracker which can solve the issue about direction or facing of solar panel and to absorb the optimum sun shines. Another concern is giving an easy way to measure wattage peak as a real time measurement.

2. Method

The SUNLITE Photovoltaic was selected to be measured in this research. The nameplate from the manufacture state is the 20-Watt Peak, please refer to Figure-1 for further detail technical specification. Solar panel I-V Characteristic Curves are used to give a visual representation of the current and voltage (I-V) characteristics of a particular photovoltaic panel (cell or array) giving a detailed description of its solar energy conversion ability. Knowing the electrical I-V characteristics of a solar panel is critical in determining the output performance and therefore its efficiency. Measuring the output power of a solar panel, for example the 20W – SUNLITE Solar Panel is not that difficult if we apply some simple steps.[10].

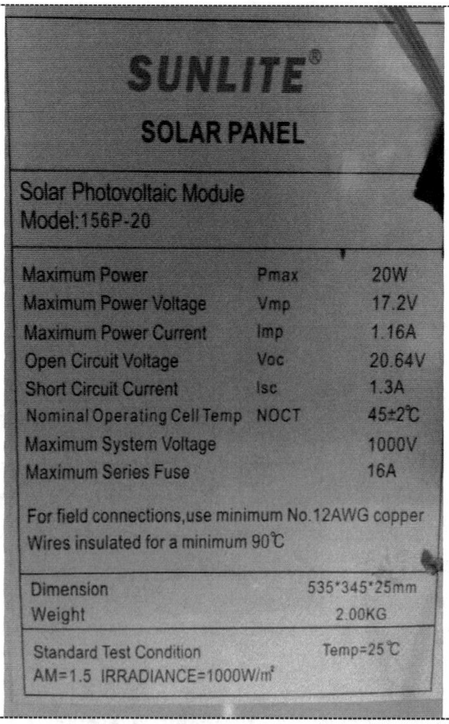

Figure 1. SUNLITE Photovoltaic 20 WP

The product label of the SUNLITE 20-Watt 12 Volt Polycrystalline Solar Panel gives us the electrical characteristics of the panel, according to the manufacture SUNLITE when it is exposed to an irradiance of 1000 W/m^2.

The electrical data label states that the panels Open Circuit Voltage, (V_{OC}) is 20.64 volts and that its Short Circuit Current, (I_{SC}) is 1.3 amps. We can use Ohm's Law then to find the output power of the solar panel, right. But V times I gives us 20.64*1.3 = 26.83 watts, which is a lot more than the 20 watts quoted by SUNLITE.

Further explanation will refer to Figure-2.The open-circuited voltage, V_{OC} means that the PV panel is not connected to any load, so its terminals are therefore open (infinite resistance) resulting in maximum voltage, in this case 20.64 volts, at its terminals. As its terminals are open there will be no current flow (I = 0) because there is no electrical circuit or load for the current to circulate through. Then the output power of the solar panel in this instance is P = V*I = 20.64*0 = 0 watts. No generated electrical power.

Likewise, the short-circuited current, I_{SC} means that the PV panel terminals are shorted or connected (zero resistance) creating a fully closed electrical circuit allowing the maximum panel current, in this case 1.3 amps, to flow. However, as the terminals are shorted together there will be no output voltage drop (V = 0), so the output power of the solar panel will be P = V*I = 0*1.3 = 0 watts. No generated electrical power.

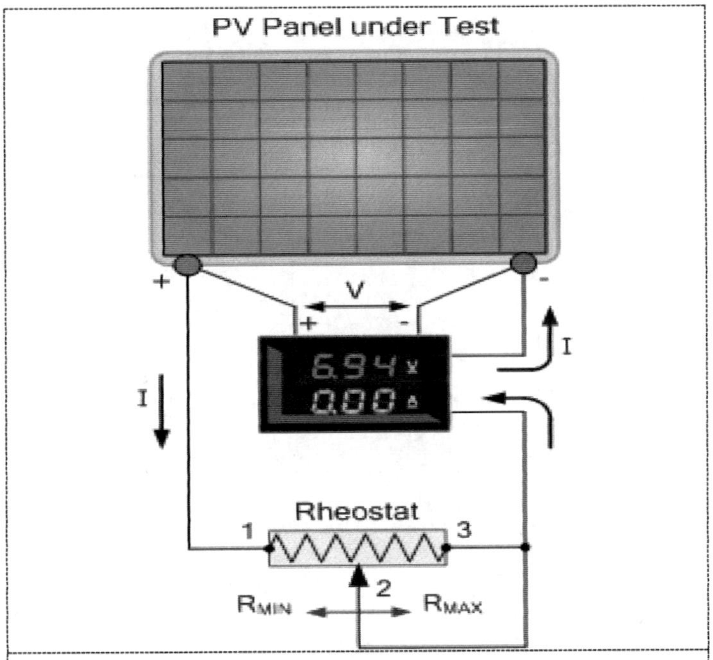

Figure 2. Conventional Method to measure watt peak

We said previously that the output power of a solar panel mainly depends on the electrical load connected to it. This load can vary from an infinite resistance, ($\infty\Omega$) to a zero resistance, (0Ω) value thus producing an open-circuit voltage, V_{OC} at one end and a short-circuit current, I_{SC} respectively, at the other. Then we need to be able to find an external resistive value somewhere in between these two extremes.

As the theoretical maximum power, P_{max} was shown to be 26.83 watts, and the maximum open-circuit voltage, V_{OC} as 20.64 volts. If we assumed the panel has a maximum wattage of 150 watts and a maximum terminal voltage of 30 volts, this would give us the panels dynamic resistive value of:
The way to achieve this R= V.V/P = (20)*(20)/30 = 13.3 Ohm value using rheostats resistors. Thus giving us a fully variable resistance between a minimum value of zero Ohms (0Ω) and the maximum value of 13.3Ω. We know that our solar panel has a manufacturer rating of 20 watts, so this would be the minimum power rating of our rheostat.

As we do internet of things, to measure the voltage across the terminals of the PV panel we would require a voltage sensor and current sensor or Power Sensor (INA219) as input for WEMOS ESP32 to calculate.

2.1. System Concept

The overall block diagram of a Watt Peak meter on the Solar Panel system is shown in Figure 3. The working principle of the system can be seen as follows:

1. The component of the LDR # 1 module detects the sun's rays,
2. LDR # 1 then provides input to the WEMOS ESP32 microcontroller,
3. WEMOS ESP32 drives the servo motor towards the strongest beam of sunlight,
4. When LDR # 2 gets sunlight, it will then provide input to the WEMOS ESP32 microcontroller to stop so that the position of the strong sunlight is between LDR # 1 and LDR # 2.

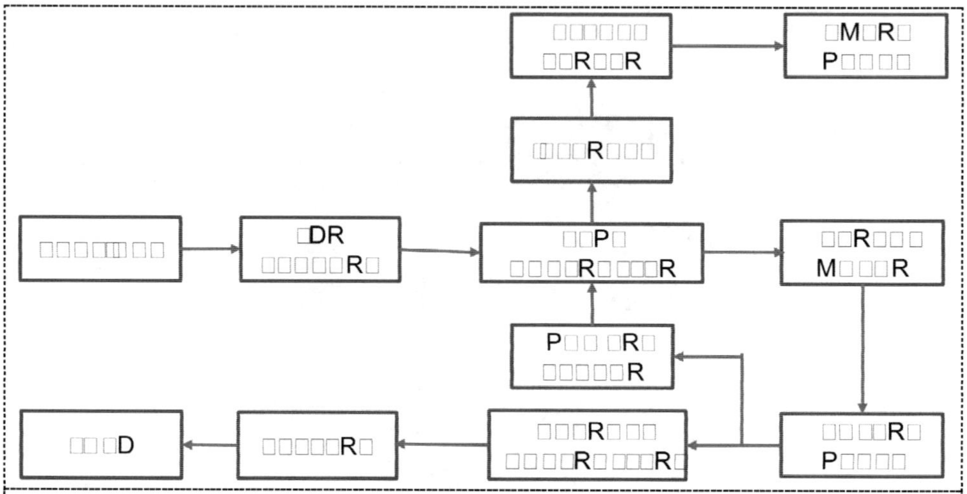

Figure 3. Block diagram of the proposed system

5. After knowing the position or direction where the strong sunrays are, the servo motor mechanic will move the solar cell panels,
6. The sun rays detected by the LDR will be forwarded to the solar charge controller module.
7. Voltage, current and power measurement data is generated by the INA219 sensor which is then sent via an internet connection to the Blink server where the data was previously received by the WEMOS ESP32 microcontroller.
8. Solar Charge Controller is used for charging the battery which is used as backup power when sunlight can no longer be obtained.
9. The output of the Solar Charge Controller is also used to supply LED light power using a 10 A / 5 V relay which then turns ON and OFF via Blink for operation.

2.2. Software Design

The Watt Peak Meter of Solar Panel system based on the internet of things, mainly includes transmitter software and receiver software. The transmitting terminal program mainly includes that the WEMOS ESP32 receives the data sent through the Wi-Fi port, furthermore the program flow chart of the sending end is shown as in Fig.4. The main issue in case of delay due to the Wi-Fi limited connection only, it really depends stability on internet connection, however the delay is still can be tolerated since the data also only showing the result not have some action to drive any actuator.

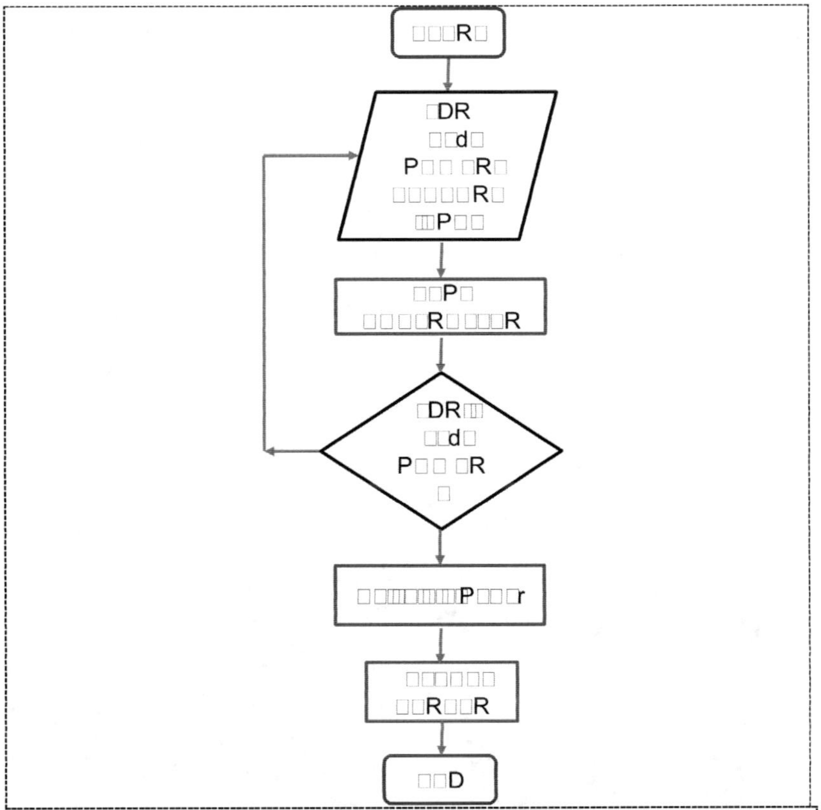

Figure 4. Flowchart Watt Peak Meter

2.3. API Setting
An internet connection-based remote-control system is made by requiring an intermediary software that connects the microcontroller and the controlled device or known as the API (Application Programming Interface). Of the many APIs, researchers chose to use Blink. Blink functions as a medium for connecting internet connections with microcontroller devices such as Arduino, MCU Nodes or Mini CPUs such as the Raspberry Pi. The value of the Blink application is the ease of use and the source code for each command in the application.

Each API includes an entry access in the form of a token key which we then copy to our source code via Sketch Arduino. This application is used to control relay 1 (one) and relay 2 (two) as well as to monitor sensor data, visualization, and others. In the Blink application there are 3 (three) main components, namely applications, servers, and libraries.

Blink server functions to handle all communication between hardware and smartphone. Blink is not tied to several types of microcontrollers, but must still be supported by the selected hardware.

3. Result and Discussion
In this section, we divide in two subsection namely electronic and mechanic part and measurement output.

3.1. Electronic and Mechanical Part
Solar panels are components that convert light energy into electrical energy. Electrical energy generated by a panel depends on the intensity of the light produced by the solar panels used are 6 volts per cell or panel. While Maximum output power is 20 watts. In the figure 5 and 6, we may see the schematic and a prototype of the proposed watt peak meter.

The Watt peak solar panel consists of electronic parts, and mechanical part. The electronic part aside of software, we implemented the WEMOS ESP32 controller, Relay, Motor Servo, Voltage and Current sensors, LDR sensor and SCC (solar charge controller). The whole system diagram can be seen in Figure 5. The sensor used in this design is a sensitive sensor light or so-called light sensitive resistance (LDR). The components are a component that is a variable resistance where the resistance These components depend on the intensity of the light. On Generally, the stronger the infrared light, the smaller the resistance of the resistor and vice versa the less the intensity of the light the larger resistance resistor. In this design the LDR sensor is used to distinguish the intensity of light in the two cardinal directions, namely east and west. By using these two light sensors, we can compare to the angle of light. The sensor output is connected to controller input, wherein the light sensitive resistance (LDR) is coupled with one fixed resistance to form a divider resistor voltage. Thus, the output of the circuit is a voltage magnitude which depends on the light intensity. The controller reads voltage the two sensors and compare them. To find useful R for motor movement.

The controller used is Arduino WEMOS ESP32. The controller functions to control the motor in the direction of the beam of light the sun is by reading the light sensor then compare it to get the difference or error and use to drive the motor or panel. When the sensor voltage the east is smaller than the west the controller will move the motor and panel to the west until the position is balanced, namely the east sensor same as western sensor.

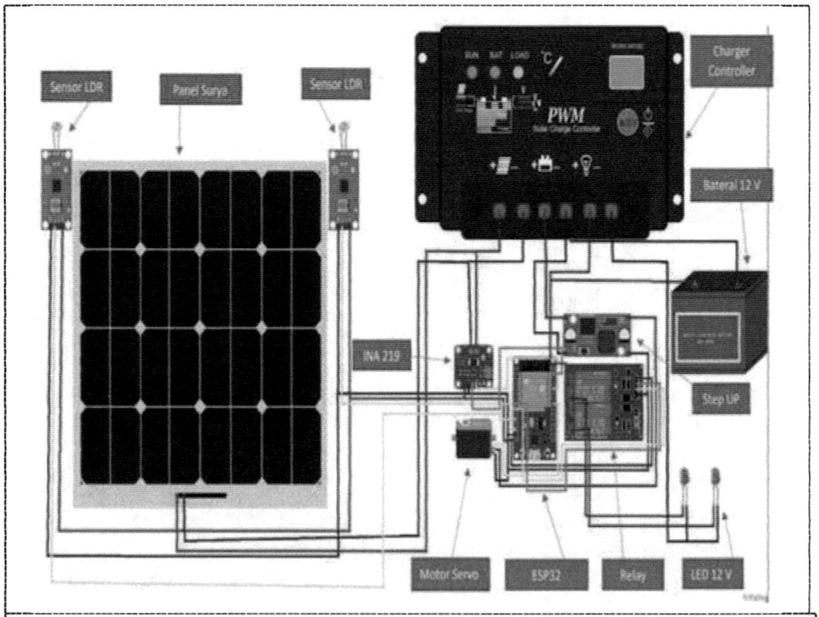

Figure 5 . Mechanical and Electronic System Schematic

The motor driver is a circuit that serves to amplify the current and drive the motor. The amplifier used is a type of amplifier H bridge is IC L293D. The advantages of the H bridge amplifier can reverse the motor current so that the motor can reverse direction. The amplifier can amplify a current of 200 mA. Direction of motion and direction determined by the microcontroller. There are 2 inputs from the amplifier to motor for setting on/off and direction of rotation.

Motor is a component that converts electrical energy into mechanical energy. The type of motor used is a gear motor permanent magnet. The function of the motor is to move the panel to the position in the direction of the sun, the motor is controlled by the controller through the current amplifier.

Journal of Physics: Conference Series **2019** (2021) 012097 doi:10.1088/1742-6596/2019/1/012097

Figure 6. Implementation of Mechanical and Electronic System

3.2. Measurement Output

The measurement output can be seen in the smartphone or personal computer. Figure 7 shown the value result of Voc and my search along together we can see the watt peak meter result.

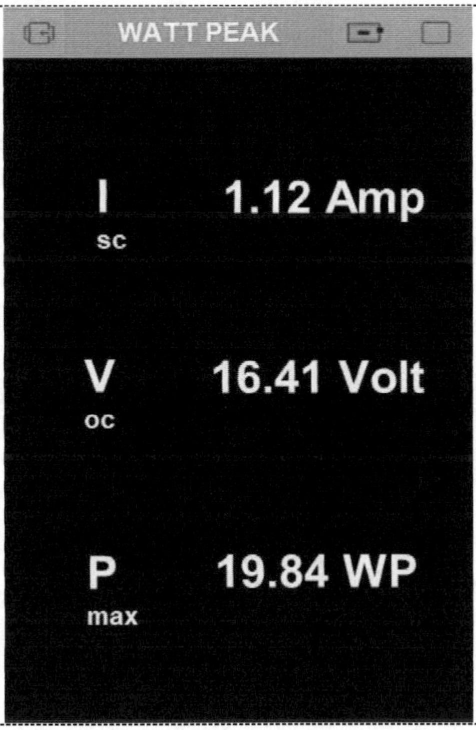

Figure 7. Illustration on Dashboard

Journal of Physics: Conference Series **2019** (2021) 012097 doi:10.1088/1742-6596/2019/1/012097

Using the data from the table-1 below, we can plot a graph of the measured voltage against the panels current as shown. The high voltage and current are popping up around 13.00 PM. The Voltage open circuit is 16.41 Volt and Current short circuit is 1.12 A

Table 1. Voc, Isc conventional measurement during best weather condition

Time	Voc (Volt)	Isc (Amp)	*Rated Current SSC (Amp)*
08.00	13.36	0.98	10.0
09.00	13.86	1.01	10.0
10.00	14.24	1.02	10.01
11.00	15.13	1.05	10.02
12.00	16.35	1.09	10.02
13.00	16.41	1.12	10.02
14.00	15.74	1.06	10.02
15.00	14.59	1.02	10.01
16.00	13.62	1.00	10.00

Having taken our readings and tabulated the results in the above table 1, we can clearly see that the maximum power occurs when a panel voltage of about 16.41 volts producing 1.12 amperes thus giving a calculated output power of 18.38 watts.

This value closely matches the manufactures data label for an operating voltage (V_{mp}) and operating current (I_{mp}) of 17.2V and 1.16A respectively, giving a dynamic panel resistance at maximum power on $V_{mp}/I_{mp} = 17.2/1.16 = 14.8$ ohms and we could if so, wished fine-tuned our measurements to get even closer to the 20-watt, 14.8 ohms target.

3.3. Analysis

Figure 8 present the voltage vs time. We can see from the measurement that the 16.41 is the high voltage around 13.00 PM in the noon. While the highest current is 1.12 Ampere at same time. This value we obtain from the blink dashboard without sitting in front of the solar panel itself. With integrated the IoT will help PV manufacture to be faster to testing the PV. We might conclude that the peak time would be around 12.00 – 13.00 PM with clear sky condition (best weather day).

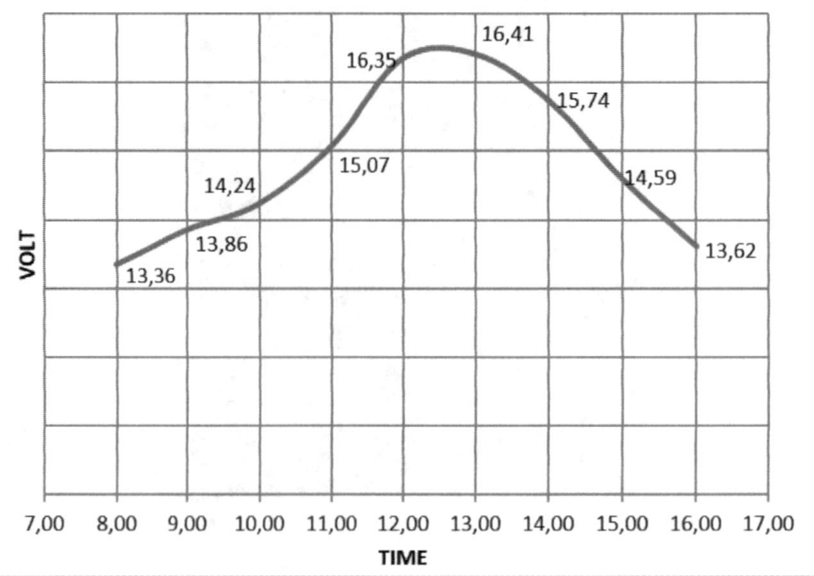

Figure 8. Voltage Peak at 13.00 PM

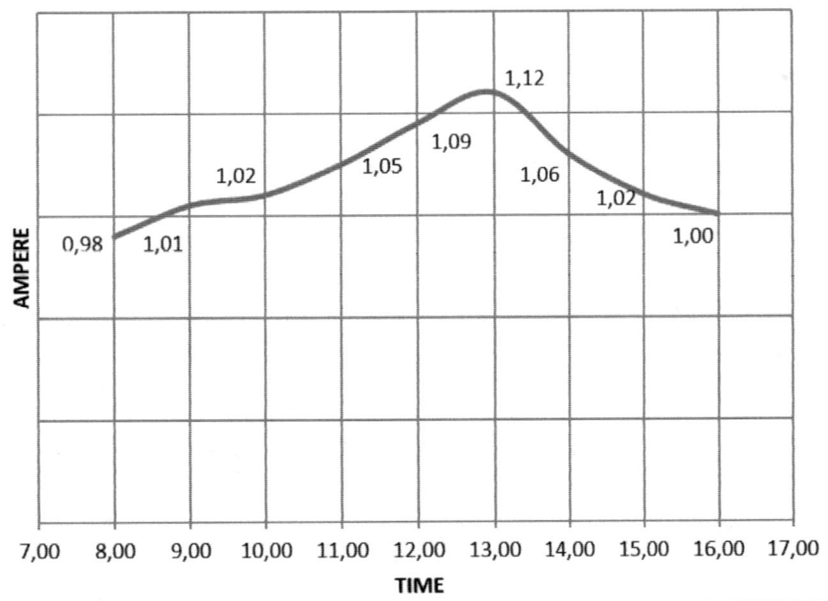

Figure 9. Current Peak at 13.00 PM

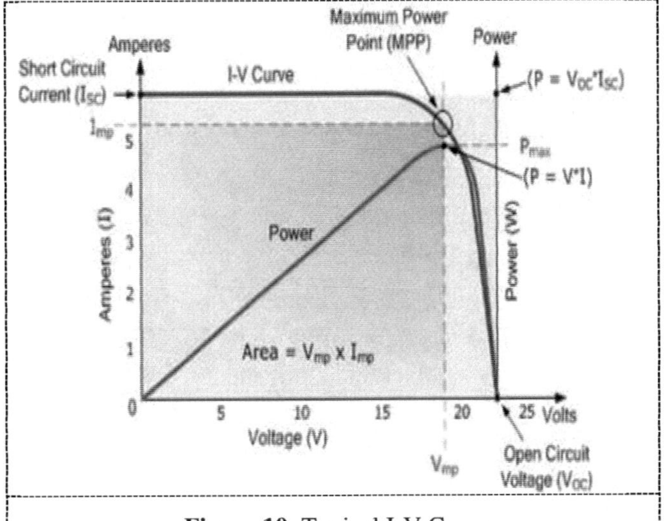

Figure 10. Typical I-V Curve

The I-V characteristics graph for all typical PV solar panels can be seen in Figure 10. Measuring the power of other types and ratings of solar panels will yield similar results; the only difference could be in the voltages and current values. Also, for an open-circuit (zero current condition) and a short-circuit, power is zero (zero voltage condition).

The maximum power output of a PV panel is defined as its peak DC output, which is calculated by multiplying the voltage and current. The optimum operating point for our photovoltaic panels has been identified here at the midpoint of the bend in the character traits curve (or knee). In other words, the maximum power point, or MPP, is the point at which a solar panel produces more power.

Fill factor (FF) is the ratio of actual maximum attainable power (represented by the green, blue box) to the product of short circuit current Is/c and open circuit voltage Vo/c (represented by the light blue box).

Journal of Physics: Conference Series **2019** (2021) 012097 doi:10.1088/1742-6596/2019/1/012097

The Fill Factor is essentially a measure of a PV module's efficiency, with the theoretical maximum value depending on factors such as the type of silicon used to build the module. However, deviation from the expected value or changes in Fill Factor can provide an indication that a fault is present.
The I-V characteristic curve demonstrates an important property of a photovoltaic solar panel, or cell, in that it demonstrates that it is a current source device rather than a voltage source device, such as a battery. Unlike a battery, which has a fixed terminal voltage (12V, 24V, etc.) And delivers variable amounts of current to a connected load, For a given amount of solar insolation, a photovoltaic cell or panel provides a constant supply of current over a wide voltage range.

Measuring the power of a solar panel is no longer difficult, as it no longer necessitates a slew of multimeters, power resistors, or a single rheostat capable. Thank you to IoT technology for making life easier.

The maximum power or maximum production power of solar panels that can be obtained is 18.37-Watt peak when direct sunlight hits the solar panel without any obstructions in the form of shadows of trees, houses, or sunlight obscured by a cloud collection. If a continuous solar panel receives sunlight for 5 hours and has a productive power of 18.37 Watt, it will produce 91.85 Watt-hour. This stored energy of 91.85 Watt-hour can be used to turn on a 10-Watt lamp for 9 hours or other devices. It is necessary to add solar panels if you need more power according to operational needs.
In the figure 8 and 9, some data related to the best time and best power result during testing in sunny weather conditions. The highest power analysis obtained from solar tracker test data can be calculated by the following formula:

$$Pmax = Vmp\,.Imp \quad (1)$$

Where :

Pmax	: *Maximum Power*	(Watt)
Vmp	: *Maximum Power Voltage*	(Volt)
Imp	: *Maximum Power Current*	(Ampere)

Then the *Maximum Power* of the Solar Panel is (datasheet source):

Pmax = Vmp . Imp
= 17,2 Volt . 1,16 Ampere
= 19,95 Wattpeak round up to 20 WP.

However, to produce maximum power can be obtained from:

$$Pmax = Voc.Isc.FF \quad (2)$$

Where FF (*Fill Factor*) formula:

$$FF = \frac{Vmp\,.\,Imp}{Voc\,.\,Isc} \quad (3)$$

So at the peak of the highest measurement at 13.00 with Voc equal to 16.41 Volt and Isc equal to 1.12 A then the results is:

$$FF = \frac{Vmp\,.\,Imp}{Voc\,.\,Isc} = \frac{17{,}2\,.\,1{,}16}{20{,}64\,.\,1{,}3}$$

$$= \frac{19{,}95}{26.83} = 0.74$$

The maximum power generated based on the datasheet is:

$$Pmax = Voc.Isc.FF$$

$$= 20{,}64\, Volt\,.\; 1.3\, Ampere\,.\; 0.74$$
$$= 19{,}86\, Watt$$

So that the maximum power generated at 13.00 hours is:

$$Pmax = Voc.Isc.FF$$
$$= 16{,}41\, Volt\,.\; 1{,}12\, Ampere\,.\; 0.74$$
$$= 13{,}6\, Watt$$

Thus, by involving the Fill Factor (FF) the maximum of Power would be 13.6 Watt.

The use of Blink as an application for power monitoring is also quite helpful in the data collection process, but the data obtained is only real time data, because the database contained in the application platform is stored directly on the Blink server and researchers cannot access the database making it difficult for researchers to create. Further analysis of the peaks in the voltage and current obtained during the test period.

4. Conclusion

On this basis, to sum up, everything that has been stated so far, our design is successful in measuring watt peak result which is very close to datasheet form PV manufacture. The prototype of watt peak meter has been successfully made including the Mechanical and Electrical with a maximum power of 19.84 Watt from the maximum power of 20 Watt-peak PV. By using the WEMOS ESP32, we tested the performance of our circuit and proved its portability. The modification of the Blink mobile application was successfully carried out and functioning properly but is limited to a-Wi-Fi connection only.

The overall performance fit the aim of this work which is to design a watt peak meter portable. Future research should be devoted to the development to add another several sensors to measure more characters of PV. The future research should focus to enhance and play with data measured which can be shared as part of big data. With the help of Computer Engineering, the watt peak meter collected can be better analysed to provide practical help for the PV manufacture.

References

[1] Wildian, Roni, Syafrialdi., 2015, Rancang Bangun *Solar Tracker* Berbasis Mikrokontroller ATmega8535 dengan Sensor LDR dan Penampil LCD, Universitas Andalas, Padang, Indonesia.

[2] Rezkyanzah, Jeneiro. Et al. (2016) Perancangan *Solar Tracker* Berbasis Arduino Sebagai penunjang Sistem Kerja *Solar Cell* dalam Penyerapan Energi Matahari, UPN Veteran , Indonesia.

[3] https://www.energuide.be/

[4] Kavitha, V., & Malathi, V. 2019. A Smart Solar PV Monitoring System Using IoT.Amit Kumar Rohit, Amit Tomar, Anurag Kumar, Saroj Rangnekar2017, “Virtual lab-based realtime data acquisition, measurement and monitoring platform for solar photovoltaic module", International Journal of Resource-Efficient Technologies,pp.1-6.

[5] Rinaldi, Rizal. et al 2020 *Design of Open Loop Single Axis Solar Tracker System,* Telkom University, Bandung, Indonesia.

[6] Putra Aldony, Reco, Putra. et al. 2019 *Minimum Power of Solar Panel Movement in Solar Tracker System Prototype*. Politeknik Negeri Sriwijaya, Palembang, Indonesia

[7] Indrasari , Widyaningrum. et al. 2018 *Active Solar Tracker Based on The Horizon Coordinate System , The 8th International Conference on Theoretical and Applied Phisycs.*

[8] Soetedjo, Aryuanto et al. 2020 *Intergration of Solar Tracker and Maximum Power Point Tracking for Improving Photovoltaic (PV) System Efficiency,* Institute Tekknologi Nasional, Malang, Indonesia.

[9] Syafii, M. I. Rusydi, R. Putra and M. H. Putra, "Real-time measurement of grid connected solar panels based on wireless sensors network," 2016 International Conference on Sustainable Energy Engineering and Application (ICSEEA), 2016, pp. 95-99, doi: 10.1109/ICSEEA.2016.7873574.

[10] sunlightwell.net

Journal of Physics: Conference Series **2019** (2021) 012098 doi:10.1088/1742-6596/2019/1/012098

Magnetic Properties of Silicon Dioxide (SiO_2) from Reed Biomass

N P Har[1,a], D A P Wardani[2,b], B Hariyanto[3,c], N Kurniawati[3,d], N Darmawan[4,e] and Irzaman[1 f]
[1]Physics Departement, FMIPA, IPB University
IPB Dramaga Bogor, West Java, Indonesia, 16680
[2]Chemistry Program, FMIPA, Palangka Raya University
Palangka Raya,Central Kalimantan, Indonesia, 73111
[3]Physics Program, FMIPA, Palangka Raya University
Palangka Raya, Central Kalimantan, Indonesia, 73111
[4]Chemistry Departement, FMIPA, IPB University
IPB Dramaga Bogor, West Java, Indonesia, 16680
Email: [1,a]nazopatul_biofis@apps.ipb.ac.id

Abstract. Silicon dioxide (SiO2) from reed biomass has successfully synthesized by annealing temperature variation at 500 oC, 600 oC, and 700 oC. The electrical properties of silicon dioxide from reed biomass have been analyzed by LCR-meter. The data collection point of each sample was 200 points. The frequency range of each sample was collected at 50 Hz to 5 MHz, alternating current voltage at 1 V, and alternating current at 50 mA. Then, the magnetic properties have calculated in Vicinity Magnetic Induction (VMI) and magnetic flux. The VMI value result of each sample has calculated at the order of 10-7 – 10-6 tesla and magnetic flux value at the order of 102 - 104 weber. Application of magnetic properties in the form of Vicinity Magnetic Induction (VMI) and magnetic flux of silicon dioxide from reed biomass was the forerunner as a magnetic sensor.

1. Introduction

Reed (Imperata cylindrica) is one of the natural vegetation that is very abundant in the tropics and subtropics. The potential development of silicon dioxide (SiO_2) from reed (Imperata cylindrica) is very interesting to study. The potential development of silicon dioxide (SiO_2) is based on the widespread use of SiO_2-based materials in industry. Silicon dioxide is widely used for various purposes with various sizes depending on the application needed, such as in the tire, rubber, glass, cement, concrete, ceramic, textile, paper, cosmetic, electronics, paint, film, toothpaste and adsorbent industries [1,2] . In addition to processed form, silicon dioxide is also widely used directly for refining vegetable oil, as an additive in pharmaceutical products and detergents [3].

Besides reeds biomass, silicon dioxide can also be extracted from rice straw biomass[4] bamboo leaves [5,6], rice husks [7-9], bagasse [10] and coconut husks [11].

This study aims to study and analyze the ignition temperatures on the silicon dioxide content of reeds and measure its magnetic properties including of Vicinity Magnetic Induction (VMI) and magnetic flux in the frequency range at 50 Hz to 5 MHz.

Journal of Physics: Conference Series **2019** (2021) 012098 doi:10.1088/1742-6596/2019/1/012098

2. Methods

Silicon dioxide from reed biomass was extracted in several steps including: (i) alang-alang leaves 1 g soaked in 12 ml of 3% technical HCl at 200°C for 3 hours (ii) reed samples rinsed with aquabides until neutral. (pH = 7), (iii) reed samples were dried and burned in an open space, (iv) reed charcoal is burned in a furnace 500°C, 600°C, 700°C, (v) extraction of SiO_2 from ash samples is dried in a furnace at a temperature of 800°C at a temperature of 100°C/hour (calcination process), (vi) magnetic properties analysis test including Vicinity Magnetic Induction (VMI) and magnetic flux.

3. Result and Discussion

Samples of silicon dioxide from the reeds were characterized using an LCR meter in the frequency range of 50 Hz to 5 MHz. Measurements were carried out as many as 200 points on each sample. The characterization data will display several values, such as the value of frequency, charge, and inductance. These values can be used to calculate and analyze the magnetic properties of the sample. The frequency value (f) can be used to calculate the value of the angular velocity (ω) using equation 1:

$$\omega = 2\pi f \quad (1)$$

Next, the charge value (Q) is used to calculate the internal current with equation 2:

$$I = \frac{Q}{t} \quad (2)$$

Where t is the time required to obtain 200 data points from the LCR meter. The following is a graph between the angular velocity value and the current value.

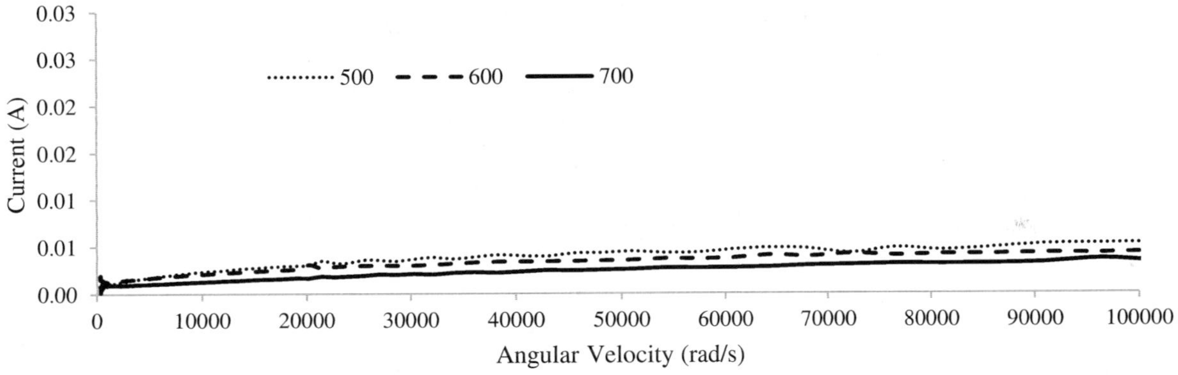

Figure 1. Electrical current as function of angular velocity

Figure 1 shows the electric current data against the angular velocity of 0 rad/s – 100000 rad/s SiO_2 at annealing temperatures of 500 °C, 600 °C and 700 °C of reeds. Based on the resulting pattern, it shows that the higher the angular velocity, the higher the value of the electric current. Samples treated with annealing temperature of 500 °C had the highest electric current value than samples treated with annealing temperatures of 600 °C and 700 °C.

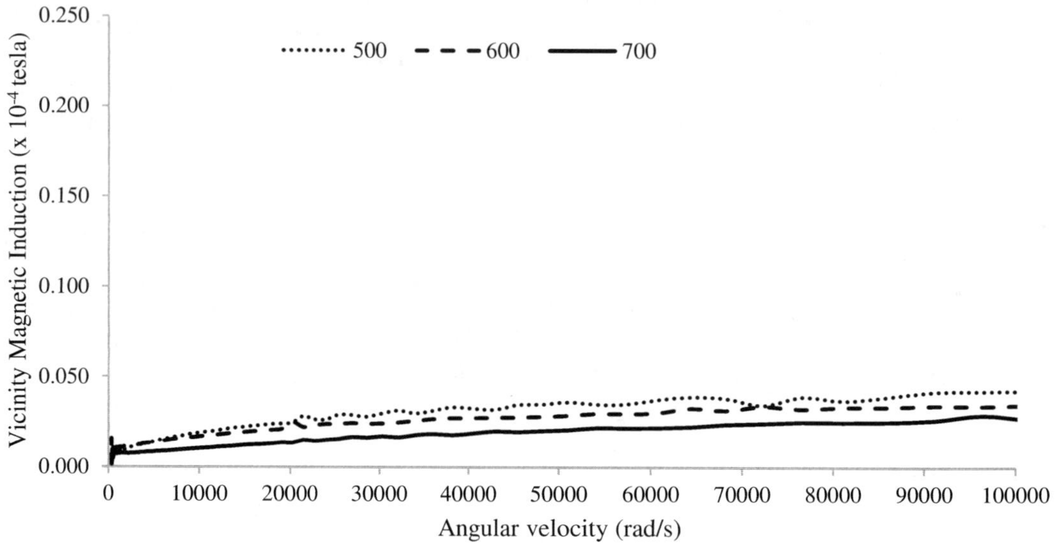

Figure 2. Vicinity Magnetic Induction as a function of angular velocity

The value of the electric current can be used to calculate the value of the surrounding magnetic induction in the sample [12]. The resulting ambient magnetic field is known as Vicinity Magnetic Induction (VMI). Based on Biot Savart's law, we can calculate the VMI value using equation 3:

$$B = \frac{\mu_0 I}{2\pi r} \tag{3}$$

The relations between the angular velocity value and the VMI value is shown in Figure 2. The VMI graph pattern is directly proportional to the current graph pattern. Figure 2 shows that the VMI value of each sample increases as the angular velocity value increases. This may indicate that the sample circuit is capacitive. The VMI values of the annealing temperature treatment samples of 500 °C, 600 °C and 700 °C are on the order of 10-7 – 10-6 tesla. The highest VMI value is shown in the annealing temperature treatment sample of 500 °C.

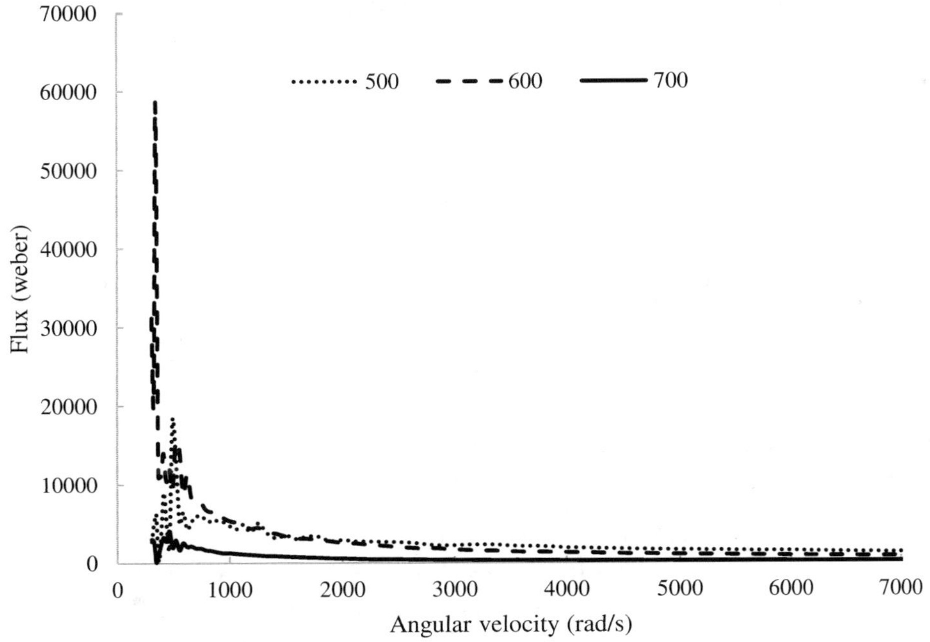

Figure 3. Magnetic flux as function of angular velocity

In addition to current and VMI data, the inductance value is also measured on the LCR meter which can be used to calculate the magnetic flux value in each sample. The value of magnetic flux is shown in figure 3 in the angular velocity range of 0rad/s – 7000 rad/s. The results obtained show that the graph fluctuates at angular velocities below 1000 rad/s, while at angular velocities above 1000 rad/s the magnetic flux values tend to be constant.

4. Conclusion

Based on the results of the analysis of magnetic properties show that the value of Vicinity Magnetic Induction (VMI) calculated on silicon dioxide samples from rice straw with order 10^{-7} - 10^{-6} tesla and magnetic flux values calculated on samples with order 10^{2} - 10^{4} weber on all annealing temperature variation at 500 °C, 600 °C, and 700 °C. Application of magnetic properties in the form of Vicinity Magnetic Induction (VMI) and magnetic flux of silicon dioxide from rice straw is the forerunner as a magnetic sensor.

Acknowledgments

This research was funded by Hibah Penelitian Terapan Unggulan Perguruan Tinggi (PTUPT) from Deputy of Research and Development, Ministry of Research and Technology/National Research and Innovation Agency of Republic Indonesia with contract number: 1/E1/KP.PTNBH/2021 dated March 8th, 2021.

References

[1] Askwar, H, Jong-Kil, K, Pradip, BS, Dang, VQ, Godlisten, NS, Gideon, E, & Hee, TK. Synthesis of mesoporous silica with superior properties suitable for green tire , *Journal of Insudtrial and Engineering Chemistry,* 18, 1841-1844.2012.

[2] Wanvimon, A, Nuchanat, N, & Garry, LR.2004. Application of rice husk ash as fillers in the natural rubber industry. *Journal of Applied Polymer Science,* 98, 34-41. 2005.

[3] Suka G I, Simanjuntak W , Sembiring S, Trisnawati E. Karakteristik silika sekam padi dari provinsi Lampung yang diperoleh dengan metode ekstraksi. *Jurnal MIPA*. 37 (1):47-52.2008.

[4] Nazopatul, PH, Irzaman, & Irmansyah. Crystallinity and electrical properties of silicon dioxide

(SiO_2) from rice straw. *AIP Conference Proceedings*, 2202, 020028, 2019.

[5] Aminullah, E. Rohaeti, B. Yuliarto, & Irzaman. Reduction of silicon dioxide from bamboo leaves and its analysis using energy dispersive x-ray and fourier transform-infrared. *IOP Conference Series: Earth and Environmntal Science*, 209, 012048, 2018.

[6] Irzaman, Novi, O, & Irmansyah. Ampel bamboo leaves silicon dioxide (SiO_2) extraction. *IOP Conference Series: Earth and Environmntal Science*. 141 012014, 2018.

[7] Sintha, I, Dahrul, M, Ismawati, SS., Kurniati, M., Irmansyah, & Irzaman, Optimalization of silicon extraction from husk ashes by excessive magnesium addition on increasing rate of temperature reduction. *Earth and Environmental Science,* 65 012032, 2017.

[8] Irzaman, Indah, DC, Aminullah, Akhirudin, M, Brian, Y, & Ulfah, S. Biosilica properties from rice husk using various HCl concentrations and frequency sources. *Egyptian Journal of Chemistry In press.*

[9] Casnan, Erliza, N, Hartrisari, H, Irzaman, & Eti, R. Scalling up of the phyrolysis to produce silica from rice husk*, Journal Engineering Technology,* 51 (6), 747-761, 2019.

[10] M. Z. Adli, Y. W. Sari, & Irzaman, Extraction silicon dioxide (SiO_2) from charcoal of baggase (Saccharum officinarum L), *IOP Conference Series: Earth and Environmntal Science*, 187, 012004, 2018.

[11] MF Anuar, YW Fen, MHM Zaid, & NAS Omar. Optical studies of crystalline ZnO-SiO_2 developed from phyrolysis of coconut husk, *Material Research Expres in press.*

[12] Vania, R, Endah, KP, Nazopatul, P, Tony, S,& Irzaman. 2020. *Engineering Materials (pp 197-201).*Trans Tech Publication.

Simulation of atmospheric dynamics during heavy rain events on Nias Island using WRF model and Himawari-8 satellite data

R Siburian[1,2], I J A Saragih[1,2], M Situmorang[1]*, K Tarigan[1], K Sembiring[1], M Sinambela[1,3], and S Humaidi[1]

[1]Department of Physics, Faculty of Mathematics and Natural Sciences, Universitas Sumatera Utara, Medan – Jl. Bioteknologi No. 1, Padang Bulan, Medan, Sumatera Utara 20155, Indonesia
[2]BMKG – Kualanamu Meteorological Station – Jl. Tengku Heran No. 119, Desa Pasar V Kebun Kelapa, Deli Serdang, Sumatera Utara 20552, Indonesia
[3]BMKG – Balai Besar MKG Wilayah I – Jl. Ngumban Surbakti II No. 15, Sempakata, Medan, Sumatera Utara 20131, Indonesia

*Email: marhaposan@usu.ac.id

Abstract. Nias is the largest island in the Indian Ocean west of Sumatra. The geographical condition of Nias Island which is a small island surrounded by waters causes strong weather dynamics and high intensity of rainfall. This study was conducted to determine changes in weather dynamics conditions during high rainfall on Nias Island, including the parameters of the vertical profile of air humidity, vorticity, divergence, vertical velocity, reflectivity; and cloud top temperatures. The simulation was carried out on 4 days of heavy rain in 2020, each representing each season period, namely 7 February 2020 for December-January-February (DJF), 30 April 2020 for March-April-May (MAM), 3 August 2020 for June-July-August (JJA), and 8 October 2020 for September-October-November (SON). The data used are Final Analysis Data (FNL) with a spatial resolution of 1°x1° as input data for the Weather Research and Forecast (WRF) model, IR1 data (band#13 – 10.4μm) Himawari-8 satellite, and observational rainfall data from the Binaka and Global Meteorological Stations. Precipitation Measurement – The Integrated Multi-Satellite Retrievals for GPM (GPM IMERG) with a spatial resolution of 0.1°x0.1°. The results showed that the presence of Cumulonimbus convective clouds caused heavy rain, with cloud top temperatures reaching -60°C to -80°C. The relatively humid atmosphere, accompanied by the convection mechanism that occurs, causes the convective activity on Nias Island to be quite intense.

1. Introduction

Nias Island is the largest island among a group of islands on the west coast of Sumatra, which is administratively within the province of North Sumatra. As an area surrounded by waters, Nias Island has a high potential for marine tourism, including Sail Nias activities. In addition, the topographical conditions cause the weather conditions on Nias Island to be very local and dynamic and have very intense convective activity [1]. Intense convective activity causes high rainfall on Nias Island. Cumulonimbus convective clouds can cause extreme weather events, such as heavy rain, strong winds, thunderstorms, and high waves [2]. Extreme weather is one thing that needs attention to support safety in tourism and transportation on Nias Island.

As part of disaster mitigation to support the safety of community activities, accurate weather forecasts are needed to anticipate the occurrence of extreme weather. One method that can be used to improve the accuracy of weather forecasts is to use Numerical Weather Prediction (NWP) [3]. The

Weather Research and Forecast Model (WRF) is a mesoscale NWP that is open source and flexible so that it can be used for operational weather forecasts or developed for research needs [4–7].

This research was conducted to simulate the weather conditions on Nias Island and its surroundings during heavy rains using the WRF model as NWP product and Himawari-8 as the remote-sensing product. Time-series analysis was carried out on weather parameters from the WRF model output, including surface wind, humidity, vorticity, divergence, and vertical velocity, as well as cloud top temperature (IR1) data from the Himawari-8 satellite.

2. Literature Review

2.1. Rain

Precipitation is a meteorological phenomenon in the form of deposits that fall from the atmosphere to the earth's surface [2,8]. Precipitation can occur in several forms, namely rain, drizzle, snow, and hail [9]. Rain is a hydrometeor in the form of water droplets that fall to the earth's surface and has a diameter of more than or equal to 0.5 mm. Rain occurs when water vapor in the atmosphere condenses into droplets that cannot be held in the air (often referred to as saturated).

The following are the criteria for rain events based on rainfall intensity according to the BMKG:

Table 1. Rainfall intensity classified according to the rate of precipitation (BMKG)

	Rainfall per hour	Accumulated daily rainfall
Very light	0.1-1 mm	< 5 mm
Light	1-5 mm	5-20 mm
Moderate	5-10 mm	20-50 mm
Heavy	10-20 mm	50-100 mm
Extreme (violent rain)	> 20 mm	> 100 mm

2.2. Convective activity

Convective activity or convective cloud growth can be observed using the interpretation of Himawari-8 satellite imagery [10,11]. Interpretation of satellite imagery using cloud top temperature thresholds is used to determine cloud types. The lower the cloud top temperature value indicates the colder the cloud top and the higher the position. Clouds of this type can be identified as high clouds and/or convective clouds. Low clouds generally have warmer cloud tops and their positions are relatively close to the earth's surface. Clouds with low cloud top temperatures can cause rain [12]. Convective clouds in the mature stage with cloud top temperatures around 195-241 K can cause heavy rain [13]. The decrease in the value of the cloud top temperature becoming cooler can be used as an indicator of increasing rainfall intensity, such as -45°C for light rain, -47°C for moderate rain, and -50°C for heavy rain [14].

Cumulonimbus clouds are a type of convective cloud that causes extreme weather events, such as heavy rain, strong winds, and hail. The stages of Cumulonimbus cloud development can be identified based on the cloud top temperature value using IR1 data from the Himawari-8 satellite, consist of the growth stage (Cumulus Stage) with a cloud top temperature value of -30°C to -50°C; mature stage (Mature Stage) with a cloud top temperature value of -60°C to -80°C; and a decaying or extinct stage (Dissipating Stage) with a cloud top temperature value of -50°C to -55°C [15].

2.3. Atmospheric dynamic

Atmospheric dynamics are changes in weather conditions that can be seen using several parameters, including wind, humidity, vortices, convergence/divergence, and vertical velocity. Atmospheric dynamics analysis can be done by looking at changes in the vertical profile of weather parameters from time to time (time-series).

a) Vorticity is a vector field (curl) velocity of fluid rotation in microscopic size which has vector dimensions [16,17]. In the Northern Hemisphere (BBU), if the relative vorticity is greater than zero, it indicates a cyclonic movement that causes the air mass to move upward (convection). Circulation and vortices have a relationship and are calculated with the approach that vortices are defined as circulation divided by the area covered [17] so that it can be written in the following equation:

$$\zeta = \frac{\eth v}{\eth x} + \frac{\eth u}{\eth y} \qquad (1)$$

b) Convergence is the meeting of several different wind patterns where the wind speed is getting weaker. Shady wind conditions support the updraft mechanism which is the main prerequisite for the formation of convective clouds. Divergence is an area of wind spread where the wind speed is getting stronger [9,17,18]. Divergence values can be calculated using the data of the zonal wind (east-west wind; N) and meridional wind (north-south wind; V) using the following equation:

$$Div\, V = \nabla_{H.\acute{v}} = \frac{\eth u}{\eth v} + \frac{\eth v}{\eth v} \qquad (2)$$

c) Vertical velocity is a parameter to measure an air mass movement in a vertical plane. There are two methods that can be used to determine vertical or omega motion, namely the kinematic method which is based on the continuity equation, and the adiabatic method which is based on the thermodynamic energy equation [17,18]. In the calculation used in this study, the positive vertical velocity value indicates an increase in the value of pressure per second in a vertical plane or there is a downward movement of an air mass. Here's the equation for the vertical velocity in (x, y, z) coordinates:

$$\omega = -\rho\, g\, w \qquad (3)$$

3. Methodology

3.1. Reseach data

The research was conducted on several case studies of heavy rains on Nias Island for the 2020 period, namely 7 February 2020, 30 April 2020, 3 August 2020, and 8 October 2020. The following are some of the data used in this study:

a. Rainfall measurement data per 3 hours from the Binaka Meteorological Station;
b. Daily rainfall accumulation data from GPM IMERG final precipitation with a spatial resolution of 0.1°x0.1° downloaded from NASA's Earth Data webpage at https://disc/gsfc.nasa.gov/;
c. IR1 data (band#13 with a wavelength of 10.4μm) Himawari-8 satellite from BMKG;
d. FNL (Final Analysis) data with a spatial resolution of 1°x1° downloaded from the NCEP FNL Operational Model Global Tropospheric Analyses web page at https://rda.ucar.edu/datasets/ds083.2/. In each case study, 36 hours of FNL data were used, with the first 12 hours on the previous day being used as a spin-up of the WRF model.

3.2. Reseach Methods

The following steps are conducted out in this research:

1) Determining the time of the case study based on rainfall measurement data for the 2020 period from the Binaka Meteorological Station. The case study selected was one day of heavy-very heavy rain (rainfall > 50 mm/day) to represent the four seasons. Based on these criteria, case studies were selected on 7 February 2020 for December-January-February (DJF), 30 April 2020 for

March-April-May (MAM), 3 August 2020 for June-July-August (JJA), and 8 October 2020 for September-October-November (SON).

2) Processing the IR1 data using SATAID software, which displays a time-series graph of cloud top temperatures at coordinates 97.70°E 1.16°N (Binaka Meteorological Station) and displays the spatial map of cloud top temperatures in the Nias Island and surrounding areas.
3) WRF model simulation using FNL data in four case studies. This stage consists of pre-processing in the WRF Pre-processing System (WPS) including geogrid.exe on static geographical data; ungrib.exe on gridded meteorological data; and metgrid.exe, numerical processes in WRF-ARW (real.exe and wrf.exe); and WRF Post-processing using ARWpost.exe.
4) Plotting the WRF model output data using the GrADS software.
5) Analysed the convective evolution and changes in atmospheric conditions in each case study.

The following is a workflow diagram in research:

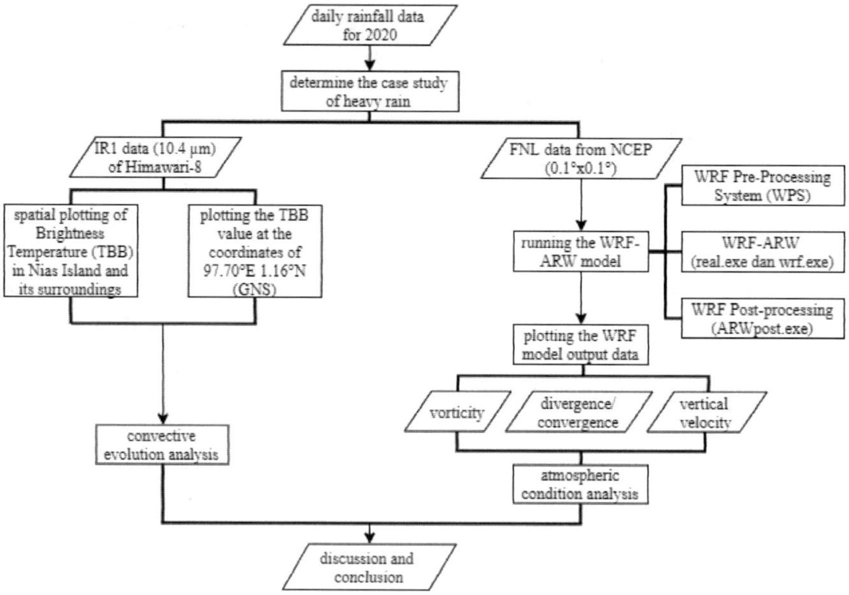

Figure 1. Research flow chart

3.3. WRF-ARW model configuration

Numerical simulations were conducted using the WRF Version 4.1.2 model with three-domain nesting (Figure 2) applying horizontal resolutions of 27, 9, and 3 km and the configuration as shown in Table 2.

Table 2. Configuration of WRF scheme used in research

WRF configuration	**D01**	**D02**	**D03**
Run Hours		36	
Time Step		18	
Center Lattitude		1.16°N	
Center Longitude		97.70°E	
geog_data_res	10 m	5 m	5 m
dx	27 km	9 km	3 km
dy	27 km	9 km	3 km
e_we and e_sn	100	88	76
e-vert		34	
Parameterization Scheme		*TROPICAL physics scheme*	

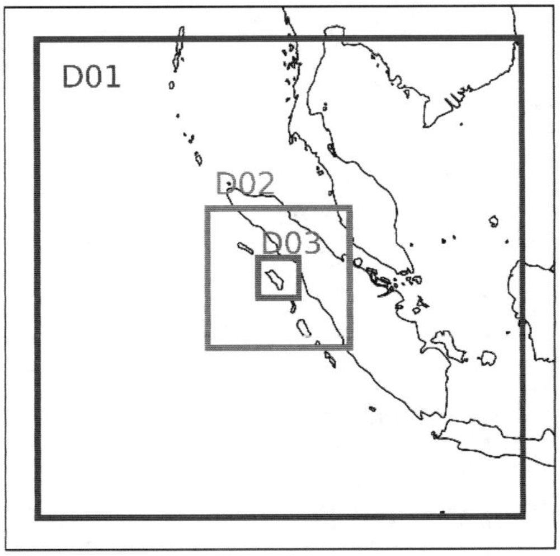

Figure 2. WRF model domain map; D01 = first domain, D02 = second domain, and D03 = third domain

4. Results and Discussions

The following is the time-series rainfall data based on observational data from the Binaka Meteorological Station in each case study:

Table 3. Rainfall measurement data (mm) in the case study of heavy rain

Observation Time (UTC)	7 Feb 2020	30 Apr 2020	3 Agu 2020	8 Okt 2020
00.00-03.00	0.0	TTU	0.0	1.0
03.00-06.00	0.0	16.0	0.0	8.2
06.00-09.00	0.0	16.5	55.5	0.2
09.00-12.00	0.0	0.5	16.5	16.3
12.00-15.00	0.9	TTU	0.0	33.1
15.00-18.00	32.0	TTU	0.0	1.4
18.00-21.00	20.0	10.4	0.0	1.2
21.00-24.00	1.0	44.4	0.0	2.9
00.00-24.00	53.9	87.8	72.0	67.5

Time-series data of 3-hour rainfall shows variations in the period of heavy rain. On February 7, 2020, heavy rain occurred at 15-18 UTC hour intervals; 30 April 2020 at 21-24 UTC; 3 August 2020 at 06-09 UTC; and 8 October 2020 at 12-15 UTC. Heavy rains on 30 April 30, 2020, and 8 October 8, 2020, started with light-moderate rain, in contrast to heavy rains on February 7, 2020, and August 3, 2020, which did not start with light rain. This indicates that intense convective activity occurs in a relatively short time.

4.1. Analysis for February 7, 2020

The spatial distribution of rainfall on 7 February 2020 (Figure 3) shows that there were heavy rain areas on Nias Island at 06-09 UTC, but not observed at the Binaka Meteorological Station (GNS). At 15.00 UTC, rain areas were seen forming in the eastern waters of Nias Island which moved westward. Rainfall at 15-18 UTC in GNS is about 15-25 mm. IMERG GPM data shows the accumulation of daily rainfall in the GNS is around 25-35 mm. This shows that the WRF model is able to capture the time of the rain event on February 7, 2020, but the rainfall value is lower (under-estimate) than the observation data.

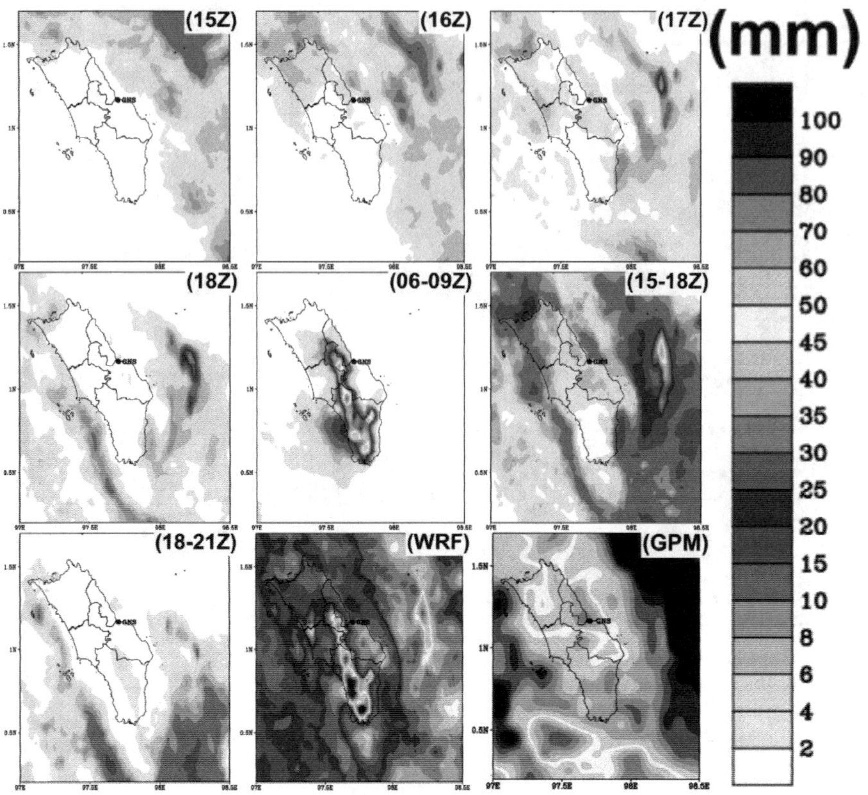

Figure 3. Hourly, 3-hour and daily accumulation rainfall distribution maps from IMERG's WRF and GPM models on February 7, 2020

The time-series graph of cloud top temperature (TBB) on 7 February 2020 (Figure 4) shows two periods of convective cloud growth around the GNS. The first period, starting at around 02.00 UTC, reaches the mature phase at around 03.00 UTC, and undergoes a decay phase until around 05.00 UTC. The lowest cloud top temperature in the first period was around -55°C. The second period, starting at around 09:00 UTC, reaches the mature phase at about 14:30 UTC, and decays until around 24:00 UTC. The lowest cloud top temperature in the second period reached -84°C.

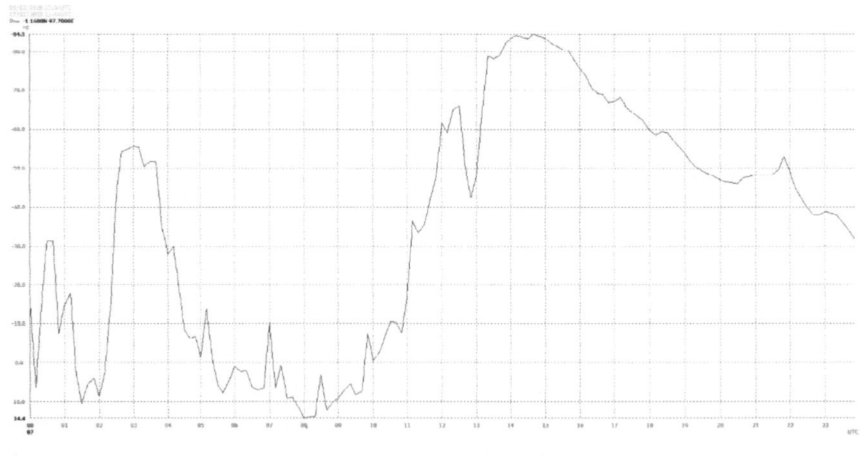

Figure 4. Time-series graph of cloud top temperature at coordinates 97.70°E 1.16°N on February 7, 2020

There is a match between convective activity that occurs with the distribution of rain output from the WRF model on February 7, 2020. Heavy rain that occurs at intervals of 06-09 UTC is caused by convective activity that occurs at intervals of 02-05 UTC. Heavy rains that occur at intervals of 15-18 UTC are caused by convective activity that occurs at intervals of 09-24 UTC.

Based on the time-series analysis of cloud top temperatures, cloud top temperatures were mapped in the Nias Island and surrounding areas (Figure 5). Seen at 08.00 UTC there are seeds of convective clouds in the waters west of Nias Island. This convective cloud cluster is developing, and at 14.00 UTC it reaches the mature phase with the widest cloud coverage area (red color), even covering the entire area of Nias Island. The fading of the red color indicates an increase in the temperature of the cloud top, which means that the convective cloud is undergoing a decay phase.

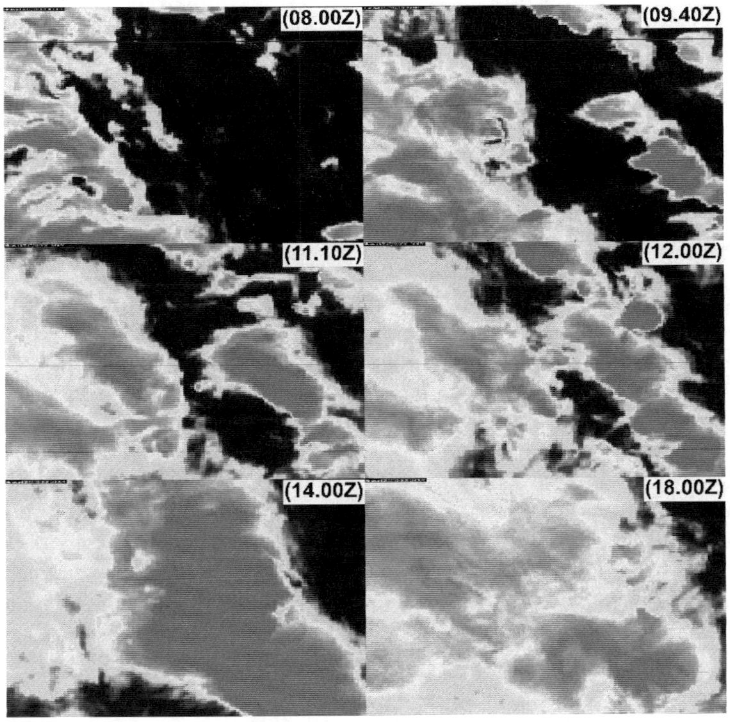

Figure 5. IR1 image of Himawari-8 satellite in Nias Island and surrounding areas on February 7, 2020

The vertical profile of air humidity (Figure 6.a) shows humid air conditions, namely in the lower layer (1000-800 hPa) with a humidity value of around 75-95%, in the middle layer (700-400 hPa) with a value of around 50-85 %, and in the upper layer (300-100 hPa) with a value of about 40-60%. The vertical velocity profile (Figure 6.b) shows that in the interval of 03-06 UTC there is a positive vertical velocity value in the lower layer, which means that there is an upward vertical movement of air (updraft). The updraft phenomenon is part of the convective cloud growth phase. At intervals of 12-15 UTC, it is seen that there is a negative vertical velocity value in all layers, which means that there is a downward vertical movement of air. The downdraft phenomenon is an indication that convective clouds have reached a mature phase. The vertical vorticity profile (Figure 6.c) shows that at intervals of 03-24 UTC there is a positive vorticity value in the lower layer, which means that there is a lifting of air masses to the top layer (lifting). This phenomenon means that there is intense convection activity on February 7, 2020. At the 12-18 UTC interval, there is a negative vorticity value in the middle layer, which means that the air mass is moving downwards. This phenomenon shows the mass of water vapor that falls into precipitation or rain. The vertical divergence profile (Figure 6.d) shows that at intervals of 12-15 UTC there is a negative divergence value in the lower layer and a positive divergence value in the upper layer. This means that there is a convergence at the bottom layer along with divergence at the top layer. This pattern of air circulation indicates the occurrence of convective activity.

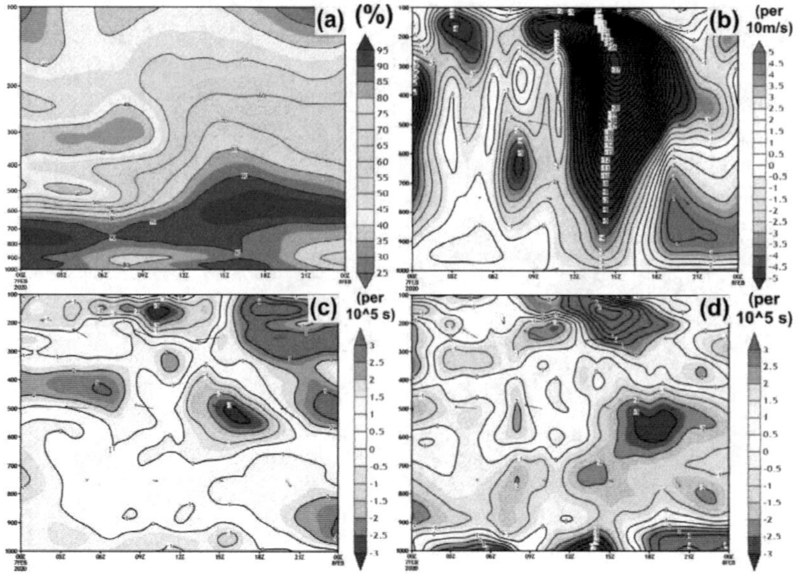

Figure 6. Vertical profile of atmospheric parameters on 7 February 2020; (a) humidity, (b) vertical velocity, (c) vorticity, (d) divergence

4.2. Analysis for April 30, 2020

The spatial distribution of rainfall on April 30, 2020 (Figure 7) shows that there are rain areas in the waters east-south of Nias Island in the morning, at intervals of 00-03 UTC. This rain was measured in GNS at intervals of 03-06 UTC with a rainfall value of 16.0 mm. At night, at intervals of 12-15 UTC, there is a rainy area in the waters west of Nias Island. At intervals of 18-21 UTC, rain areas were seen forming in the waters east of Nias Island, around the GNS. The daily accumulated rainfall value based on the output of the WRF model in GNS is in the range 4-8 mm, while based on IMERG GPM data it is in the range 35-45 mm. This shows that the WRF rainfall model is not able to capture the time of the heavy rain that occurred on April 30, 2020.

The time-series graph of cloud top temperature (TBB) on April 30, 2020 (Figure 8) shows intense convective activity around the GNS. The first convective cloud growth begins at around 00.30 UTC, reaches the mature phase with a cloud top temperature of -60°C at 01.00 UTC, and decays until 02.30 UTC. The growth of the second convective cloud begins at 02.40 UTC, reaches the mature phase with a cloud top temperature value of around -65°C at 04.10 UTC to 06.00 UTC, and decays to 08.00 UTC. At 11.00 UTC there was a third convective cloud growth, reaching the mature phase with a cloud top temperature value reaching a value -65°C at 17.00 UTC to 18.00 UTC, and decaying up to 20.00 UTC. This convective cloud decay was followed by further convective cloud growth which reached a cloud top temperature of -77°C at 21.40 UTC. There is a match between the convective activity that occurs with the distribution of rain on April 30, 2020. Intense convective activity occurs followed by rain that occurs throughout the day.

The convective activity that occurred on April 30, 2020, is mapped spatially (Figure 9). At 01.30 UTC, the seeds of convective clouds were seen in the waters east of Nias Island. This convective cloud cluster developed and reached the widest coverage area of convective clouds (red color) 02.30 UTC and then decayed (fading red color) until 04.00 UTC. At 06.00 UTC, convective cloud seeds were seen growing in the south of Nias Island until 08.00 UTC. At 14.30 UTC, convective cloud seeds were seen growing in the waters east of Nias Island and moving westward. The process of developing convective clouds is very fast and is followed by the growth of several seeds of convective clouds in the vicinity. At 17.30 UTC it was seen that some of the convective cloud seeds had merged to form a convective cloud cluster with a relatively wide cloud area coverage. The decay of this convective cloud cluster was followed by the growth of several convective cloud seeds which then formed a convective cloud cluster that covered the northern part of Nias Island.

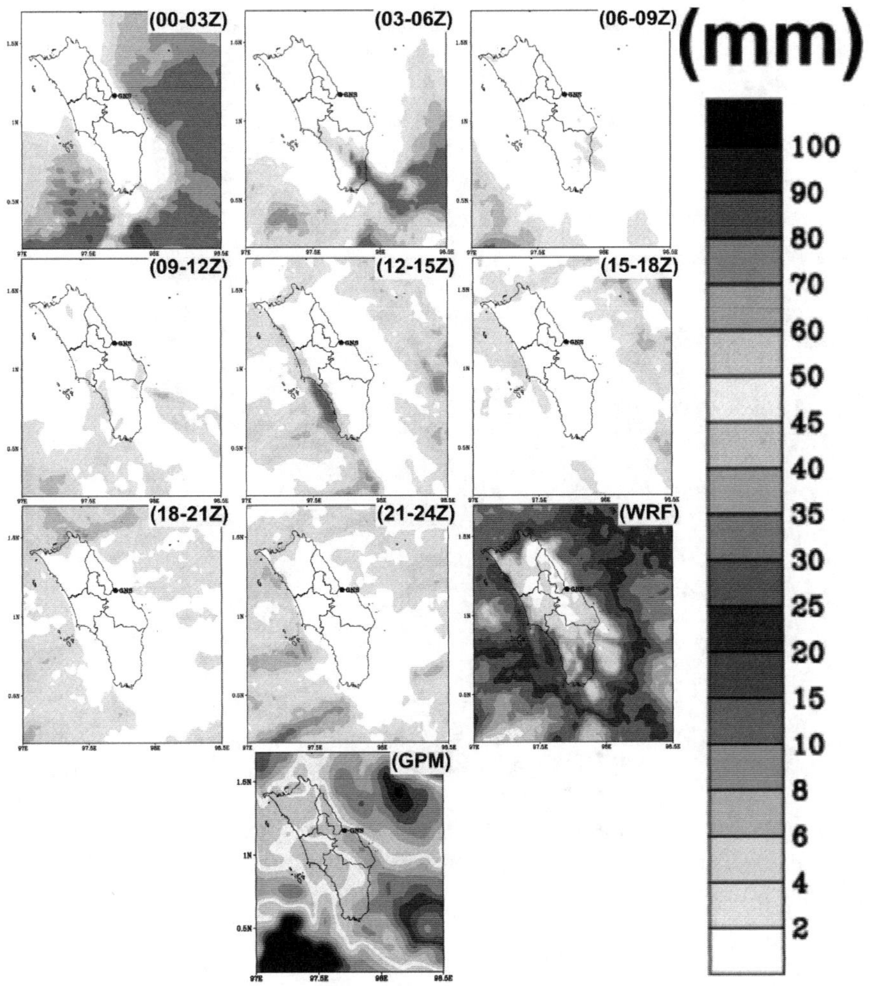

Figure 7. 3-hour rainfall distribution map and daily accumulation from IMERG WRF and GPM models on April 30, 2020

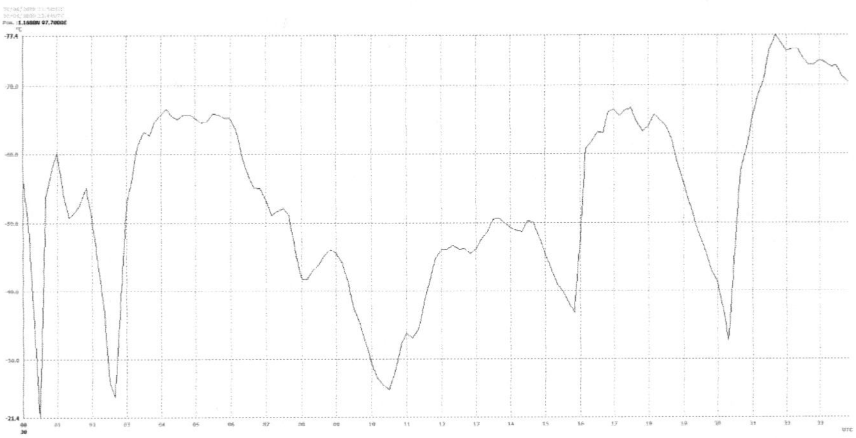

Figure 8. Time-series graph of cloud top temperature at coordinates 97.70°E 1.16°N on April 30, 2020

Journal of Physics: Conference Series **2019** (2021) 012099 doi:10.1088/1742-6596/2019/1/012099

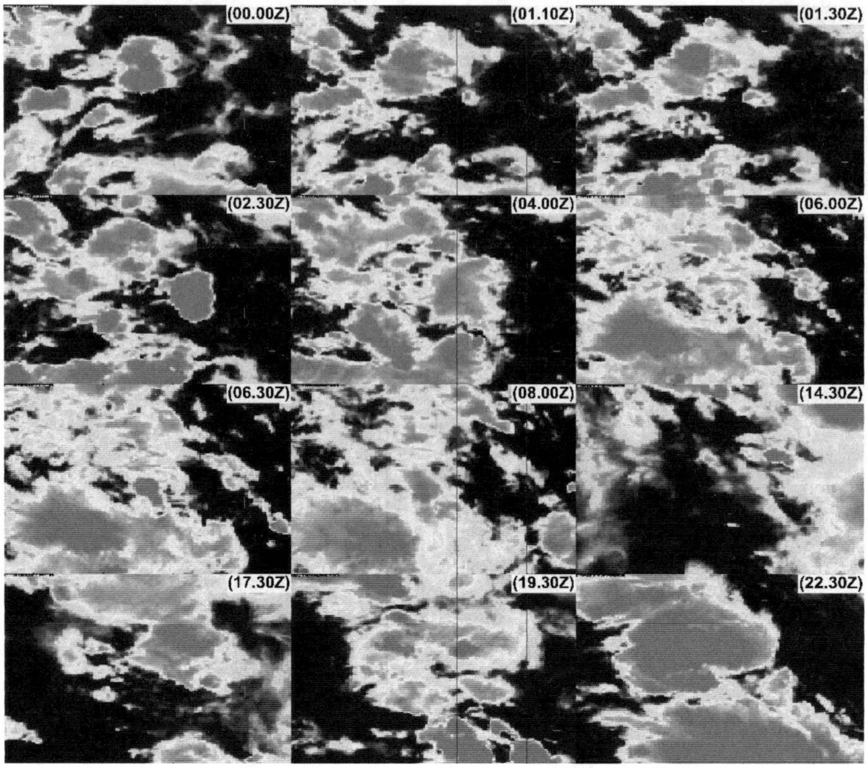

Figure 9. IR1 image of Himawari-8 satellite in Nias Island and surrounding areas on February 7, 2020

The vertical profile of air humidity (Figure 10.a) shows humid air conditions, namely in the lower layer (1000-800 hPa) with a humidity value of 75-95%, in the middle layer (700-400 hPa) with a value of 50-90 %, and in the upper layer (300-100 hPa) with a value of 30-70%. The vertical velocity profile (Figure 10.b) shows that at intervals of 03-06 UTC there are positive vertical velocity values in all layers, which means that there is an upward vertical movement of air. The updraft phenomenon indicates the growth phase of convective clouds. At intervals of 09-24 UTC, it can be seen that there is a negative vertical velocity value in all layers, which means that there is a vertical downward movement of air (downdraft). This negative vertical velocity value is still accompanied by a positive value, which means that new convective cloud growth occurs when the previous convective cloud decays or convective continuity activities. The vertical vorticity profile (Figure 10.c) shows the values of positive and negative vortices spread across layers throughout the day. This indicates that there is a process of moving air masses up and down simultaneously. This is following the convective activities and rain that occurs throughout the day. The vertical divergence profile (Figure 10.d) shows that at intervals of 09-24 UTC there is a negative divergence value in the lower layer and a positive divergence value in the upper layer. This means that there is a convergence at the bottom layer along with divergence at the top layer. This pattern of air circulation indicates the occurrence of convective activities.

Journal of Physics: Conference Series **2019** (2021) 012099 doi:10.1088/1742-6596/2019/1/012099

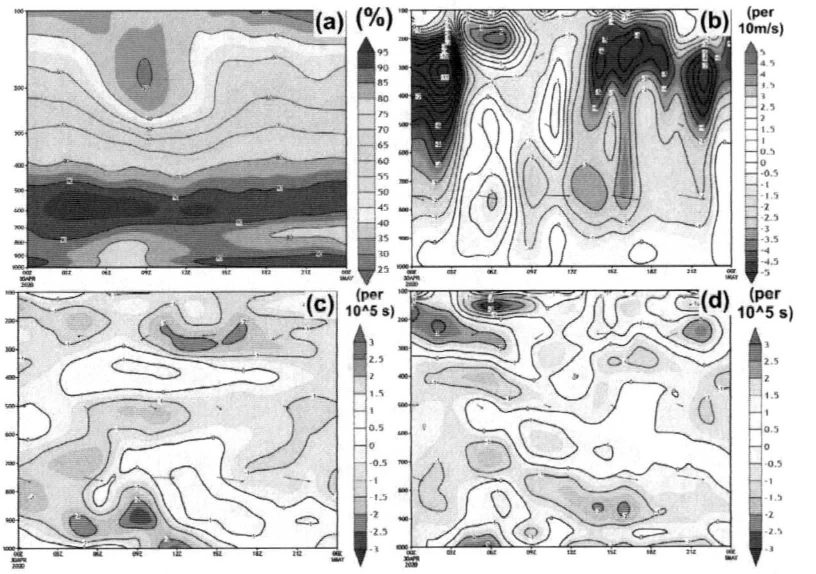

Figure 10. Vertical profile of atmospheric parameters as of April 30, 2020; (a) humidity, (b) vertical velocity, (c) vorticity, (d) divergence

4.3. Analysis for August 3, 2020

The spatial distribution of rainfall on August 3, 2020 (Figure 11) shows that there are heavy rain areas in the eastern part of Nias Island at 06-09 UTC, with rainfall values in GNS in the range of 10-25 mm. The highest rainfall in this period occurred at 08.00 UTC. The daily accumulated rainfall value around the GNS based on the output of the WRF model is in the range of 20-35 mm, while based on IMERG GPM data it is in the range of 20-45 mm. This shows that the WRF model can capture the time of heavy rain on August 3, 2020, even though the rainfall value is lower (under-estimate) than the observation data.

The time-series graph of cloud top temperature (TBB) dated August 3, 2020 (Figure 12) shows the growth of convective clouds occurring from 05.00 UTC to 08.00 UTC with cloud top temperatures in the range of -10°C to -20°C. At 09.00 UTC, the cloud top temperature value decreased and reached a relatively stable condition from 11.00 UTC to 24.00 UTC, with values in the range of -30°C to -50°C. This cloud cluster is indicated not to be a convective cloud, but a stratiform cloud with a high cloud top (Cirrus cloud). Heavy rains that occurred on August 3, 2020, were indicated to be caused by convective clouds growing at intervals of 05-09 UTC.

The spatial map of cloud top temperatures (Figure 13) shows that at 06.30 UTC there were convective cloud seeds on the west coast of Nias Island, and moving east. This convective cloud cluster has developed until it reaches the mature phase at 08.10 UTC with the coverage of the convective cloud area (red color) is not too wide. The limited coverage of the convective cloud area formed indicates that this convective cloud is a single cell Cumulonimbus cloud.

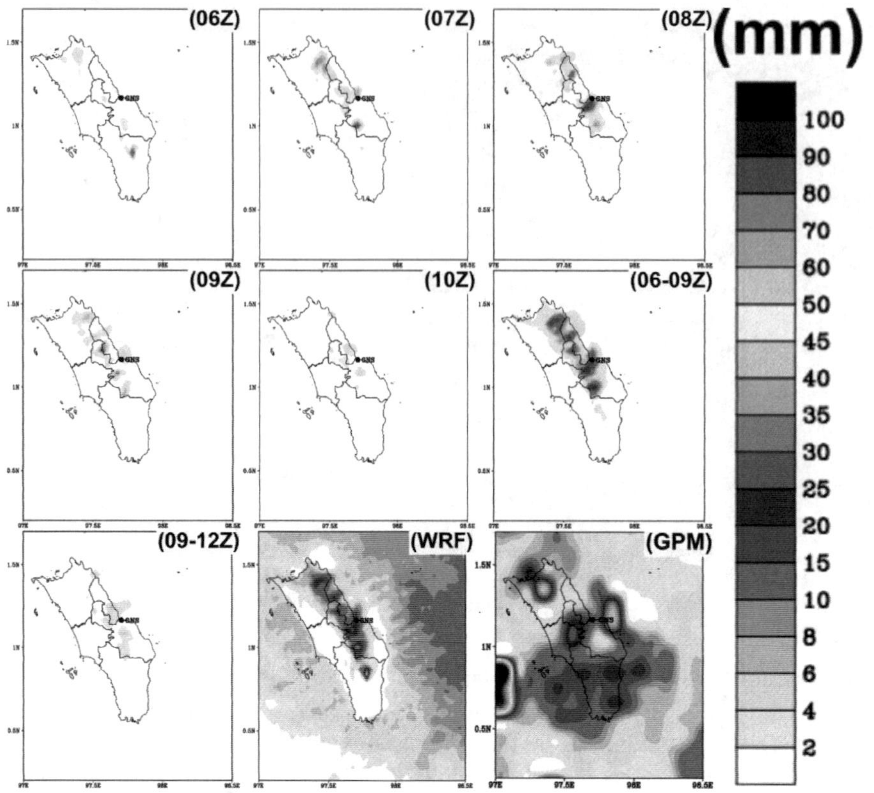

Figure 11. Hourly, 3-hour and daily accumulation rainfall distribution maps from IMERG's WRF and GPM models dated August 3, 2020

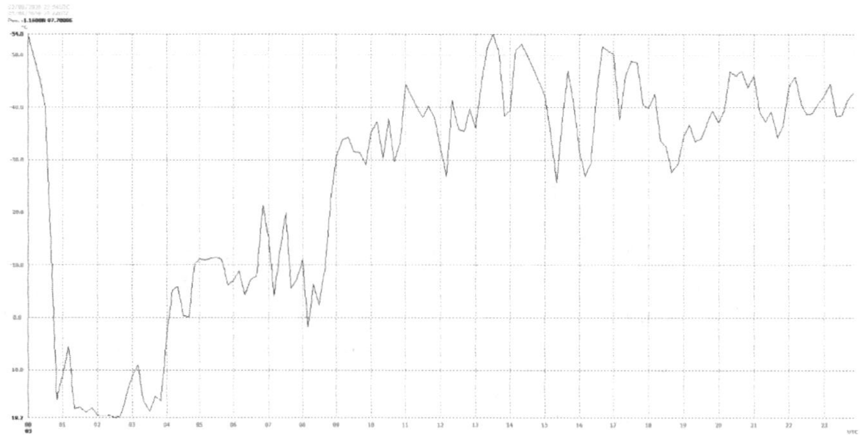

Figure 12. Time-series graph of cloud top temperature at coordinates 97.70°E 1.16°N on August 3, 2020

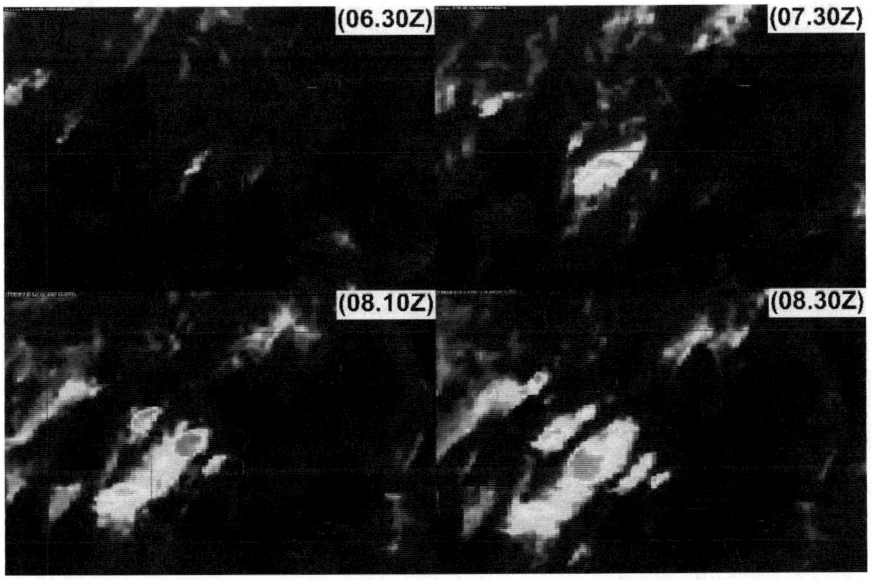

Figure 13. IR1 image of Himawari-8 satellite in Nias Island and surrounding areas on August 3, 2020

The vertical profile of air humidity (Figure 14.a) shows that in moderately humid air conditions, namely in the lower layer (1000-800 hPa) with a humidity value of 70-90%, in the middle layer (700-400 hPa) with a value of 50-70%, and in the upper layer (300-100 hPa) with a value of 30-60%. The vertical velocity profile (Figure 14.b) shows that at intervals of 03-06 UTC there is a positive vertical velocity value in all layers, which means that there is an upward vertical movement of air (updraft). The updraft phenomenon is part of the convective cloud growth phase. This updraft phenomenon is then followed by a downdraft phenomenon (negative vertical velocity value) which is not too strong because the convective clouds formed have relatively small cloud area coverage. The vertical vorticity profile (Figure 14.c) shows negative vorticity values in the lower layers throughout the day. The negative vorticity value indicates that the air is difficult to lift so that convective activity is difficult to occur. The vertical divergence profile (Figure 14.d) shows the dominant positive divergence values occurring in all layers throughout the day. This divergence condition causes the dissolution of air masses so that convective activity is difficult to occur.

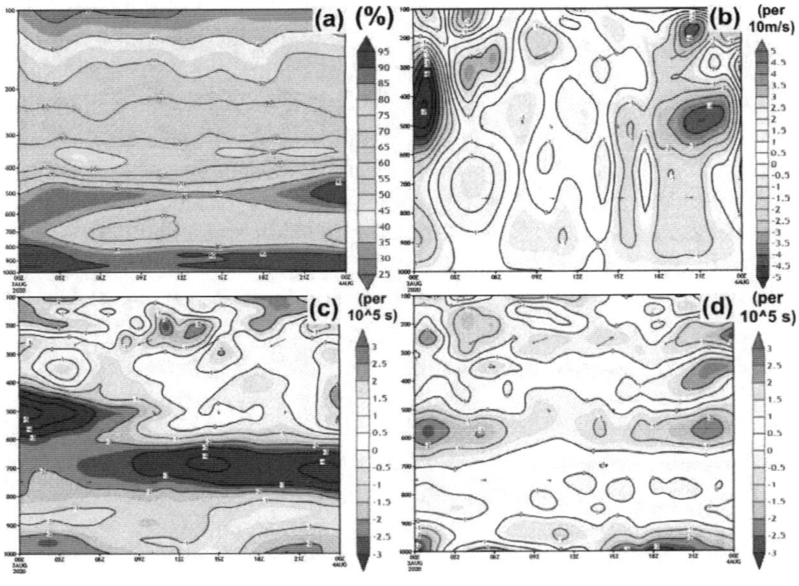

Figure 14. Vertical profile of atmospheric parameters on August 3, 2020; (a) humidity, (b) vertical velocity, (c) vorticity, (d) divergence

4.4. Analysis for October 8, 2020

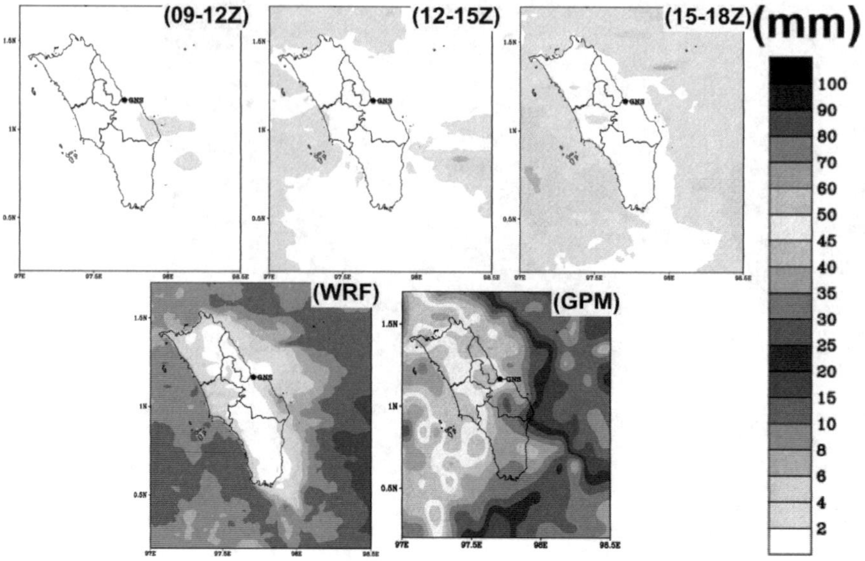

Figure 15. 3-hour rainfall distribution map and daily accumulation from IMERG WRF and GPM models on October 8, 2020

The spatial distribution of rainfall on October 8, 2020 (Figure 15) shows that at intervals of 09-12 UTC, rain areas formed on the east coast of Nias Island around the GNS, with rainfall values of 6-10 mm. At intervals of 12-15 UTC, rain areas formed in the waters west of Nias Island, while at intervals of 15-18 UTC also formed in the western and eastern waters of Nias Island. The daily accumulated rainfall value in the area around the GNS based on the output of the WRF model is around 6-20 mm, while based on IMERG GPM data it is in the range of 45-60 mm. This shows that the WRF model is able to capture the time of the rain event on October 8, 2020, but the rainfall value is lower (under-estimate) than the observation data.

The time-series graph of cloud top temperature (TBB) on October 8, 2020 (Figure 16) shows that at 03.00 UTC there was a convective cloud growth, reaching the mature phase at around 04.20 UTC with a cloud peak temperature reaching -72ºC, and then decaying until 06.00 UTC. The decay of the first convective cloud continued with the growth of the second convective cloud, which lasted until around 09.20 UTC, reached the mature phase at 10.00 UTC with the cloud top temperature reaching -60ºC, and then decayed until 15.00 UTC. At 22.00 UTC, there was a growth of convective clouds that reached the mature phase at around 23.40 UTC with a cloud peak temperature reaching -60ºC.

There is a correspondence between convective activity that occurs with the distribution of rainfall on October 8, 2020. Rain measured at intervals of 03-06 UTC is indicated to be caused by the first convective cloud, which develops at intervals of 03-06 UTC. A second convective cloud that developed at intervals of 09-15 UTC resulted in rain measured at intervals of 09-15 UTC, with the highest value reaching 33.1 mm. Convective activity that still occurs causes rain to still occur until 24.00 UTC, although with low rainfall intensity.

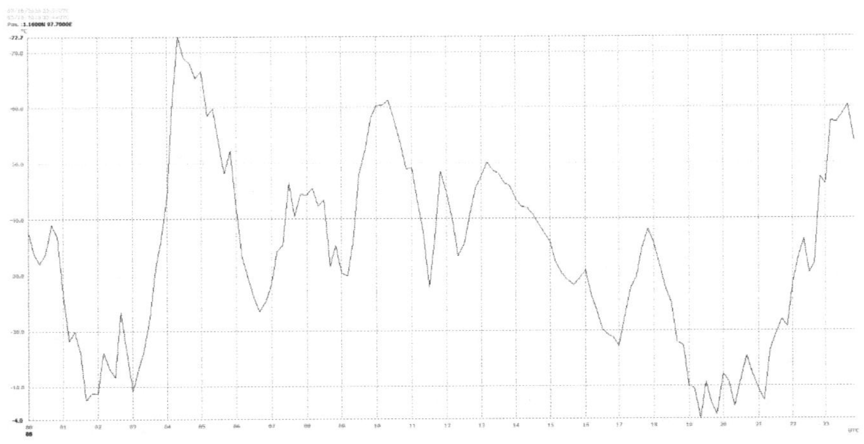

Figure 16. Time-series graph of cloud top temperature at coordinates 97.70°E 1.16°N on October 8, 2020

The spatial map of cloud top temperatures on October 8, 2020 (Figure 17) shows that at 03.00 UTC there were convective cloud seeds around Nias Island. Convective cloud clusters continue to develop, although the cloud area coverage is relatively small.

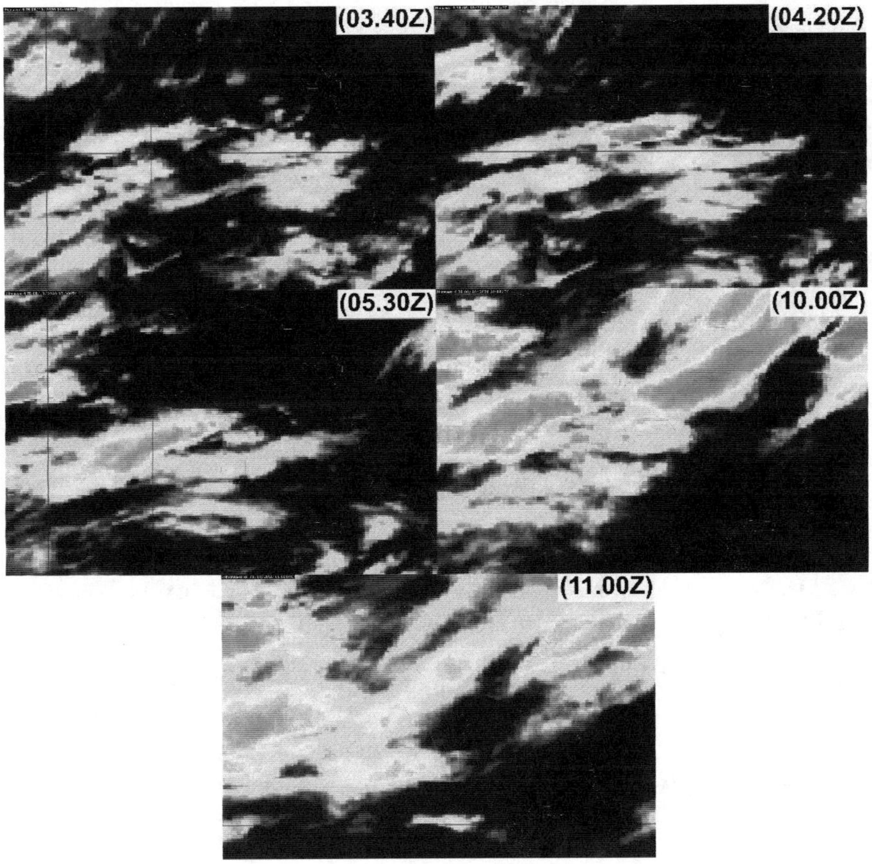

Figure 17. IR1 image of Himawari-8 satellite in Nias Island and surrounding areas on October 8, 2020

The vertical profile of air humidity (Figure 18.a) shows humid air conditions, consisted of in the lower layer (1000-800 hPa) with a humidity value of 80-95%, in the middle layer (700-400 hPa) with a value of 70-90%, and in the upper layer (300-100 hPa) with a value of 40-70%. The vertical velocity profile (Figure 18.b) shows that at intervals of 03-08 UTC there is a positive vertical velocity value in the lower layer, which means that there is an upward vertical movement of air. The updraft phenomenon is part of the convective cloud growth phase. At intervals of 12-15 UTC, it is seen that there is a negative vertical velocity value in all layers, which means that there is a downward vertical movement of air. The downdraft phenomenon is an indication that convective clouds have reached a mature phase. The vertical vorticity profile (Figure 18.c) shows that in the interval of 06-09 UTC there is a positive vorticity value in the middle layer, which means that there is a lifting of air masses to the top layer (lifting). The positive vorticity value that is not too dominant indicates that the convective activity that occurs is not too strong so that the convective cloud formed only has a relatively small cloud area coverage. The vertical divergence profile (Figure 18.d) shows that at intervals of 09-18 UTC, there are negative divergence values in the lower layer and positive divergence values in the middle to upper layers. This means that there is a convergence in the lower layer along with a divergence in the middle layer. This pattern of air circulation indicates the occurrence of convective activity. The divergence that occurs in the middle layer indicates that the convective clouds that are formed have a cloud top height that is not too high, only reaching the height of 800-700 hPa.

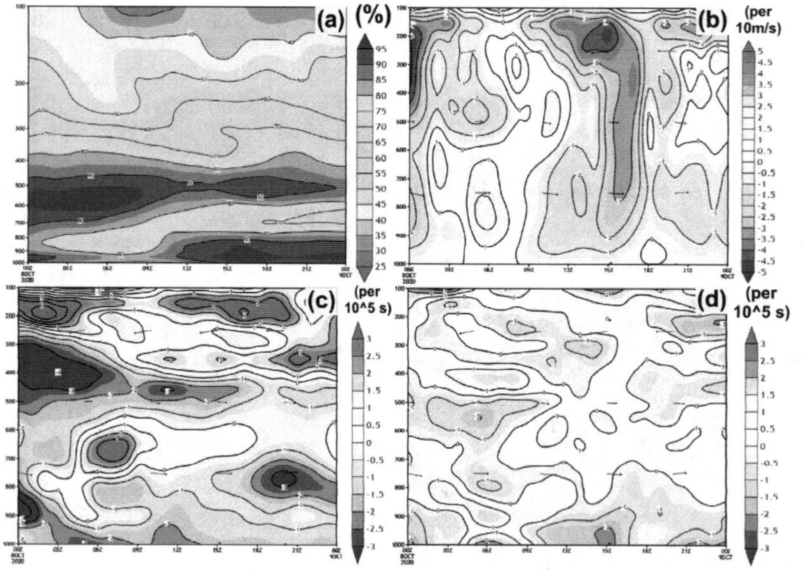

Figure 18. Vertical profile of atmospheric parameters on October 8, 2020; (a) humidity, (b) vertical velocity, (c) vorticity, (d) divergence

5. Conclusions

Convective activities that occur on Nias Island and its surroundings causes rain with varying intensity. Heavy rains are generally caused by convective clouds that have relatively wide cloud coverage areas or are included in the category of multi-cell convective. Heavy rain caused by convective clouds needs to be anticipated because it can occur in a relatively fast and sudden time. In general, the growth phase of convective clouds on Nias Island can reach a duration of 2-3 hours, then reach a mature phase of about 1-2 hours. The relatively humid atmosphere, accompanied by the convection mechanism that occurs, causes the convective activity on Nias Island to be quite intense. Convective cloud growth is characterized by positive vertical velocity values and positive vorticity values in the lower layers (1000-800 hPa), as well as circulation patterns of convergence in the lower layers and divergence in the upper layers (300-100 hPa) and/or intermediate layers (700-500 hPa). Several case studies in this study show the occurrence of several cycles of convective cloud growth in one day or the so-called continuity convective system. The mature phase of convective clouds that form generally reaches a cloud top temperature in the range of -60ºC to -80ºC. This convective cloud is a Cumulonimbus cloud. In general, from the four case studies, the WRF model is able to capture the timing of the occurrence of rain on Nias Island, but the rainfall intensity is lower (under-estimate) than the observation data.

References

[1] Tjasyono B H 2012 *Meteorologi Indonesia Volume I* vol I, ed S Suratno, W Fitria and D L Sari (Jakarta: BMKG)

[2] Tjasyono B H K and Harijono S W B 2017 *Meteorologi Indonesia 2: Awan & Hujan Monsun* (Jakarta: BMKG)

[3] Coiffier J 2011 *Fundamentals of numerical weather prediction* (Cambridge University Press)

[4] Skamarock W C, Klemp J B, Dudhia J, Gill D O, Barker D M, Duda M G, Huang X-Y, Wang W and Powers J G 2008 *A Description of the Advanced Research WRF Version 3* (Colorado: NCAR)

[5] Powers J G, Klemp J B, Skamarock W C, Davis C A, Dudhia J, Gill D O, Coen J L, Gochis D J, Ahmadov R, Peckham S E, Grell G A, Michalakes J, Trahan S, Benjamin S G, Alexander C R, Dimego G J, Wang W, Schwartz C S, Romine G S, Liu Z, Snyder C, Chen F, Barlage M J, Yu W and Duda M G 2017 The weather research and forecasting model: Overview, system efforts, and future directions *Bull. Am. Meteorol. Soc.* **98** 1717–37

[6] Skamarock W C, Klemp J B and Dudhia J 2014 Prototypes for the WRF (Weather Research and Forecasting) Model *Natl. Cent. Atmos. Res.*

[7] Skamarock W C, Klemp J B, Dudhia J, Gill D O, Zhiquan L, Berner J, Wang W, Powers J G, Duda M G, Barker D M and Huang X-Y 2019 *A Description of the Advanced Research WRF Model Version 4* (Colorado: NCAR)

[8] Tjasyono B H and Harijono S W B 2012 *Meteorologi Indonesia Volume II Awan dan Hujan Monsun* (Jakarta: BMKG)

[9] Tjasyono B H 2012 *Mikrofisika Awan dan Hujan* ed H Widiyatmoko, R Setyaningsih and W Fitria (Jakarta: BMKG)

[10] Panjaitan A 2012 Pemanfaatan Citra Satelit untuk Informasi Meteorologi Penerbangan *Materi Diklat Penerbangan*

[11] Pandjaitan B S, Panjaitan A, Indarto S and Susilawati A 2018 Utilization of Infrared and Water Vapor Channel Combination in Making the Modification of RGB Composite for Nighttime Convective Cloud Monitoring (Case Study of Tropical Cyclone Cempaka and Dahlia)

[12] Handoko 1994 *Klimatologi Dasar* (Jakarta: Pustaka Jaya)

[13] Parwati, Suwarsono, Yulianto F and Suprapto T 2008 Penentuan Nilai Ambang Batas Untuk Potensi Rawan Banjir Dari Data MTSAT Dan QMORPH (Studi Kasus: Banjir Bengawan Solo 2007) *J. Penginderaan Jauh* **5** 56–63

[14] Hanna J W, Schultz D M and Irving A R 2008 Cloud-top temperatures for precipitating winter clouds *J. Appl. Meteorol. Climatol.* **47** 351–9

[15] Saragih I J A I J A, Kristianto A, Silitonga A K and Paski J A I A I 2017 Kajian Dinamika Atmosfer saat Kejadian Hujan Lebat di Wilayah Pesisir Timur Sumatera Utara Menggunakan Model WRF-ARW dan Citra Satelit Himawari-8 *Unnes Phys. J.* **6** 25–30

[16] Wirjohamidjojo S and Swarinoto Y S 2013 *Meteorologi Sinoptik* (Jakarta: Puslitbang BMKG)

[17] Holton J R 2004 *An Introduction to Dynamic Meteorology: Fourth Edition* ed R Dmowska, J R Holton and H T Rossby (California: Elsevier Academic Press)

[18] Tjasyono B H 2008 *Sains Atmosfer* (Jakarta: BMKG)

Effect of pH on the Morphology and Microstructure of ZnO synthesized using Ananas comosus Peel Extract

A S Rini[1], S D Rahayu[1], Y Hamzah[1], T M Linda[2], Y Rati[1]
[1]Jurusan Fisika FMIPA, Universitas Riau, Kampus Bina Widya
Jl. HR. Soebrantas, KM 12,5, Tampan, Pekanbaru 28293, Indonesia
[2]Jurusan Biologi FMIPA, Universitas Riau, Kampus Bina Widya
Jl. HR. Soebrantas, KM 12,5, Tampan, Pekanbaru 28293, Indonesia
Email: ari.sulistyo.rini@gmail.com, sucirahayu629@gmail.com, yanuar.hamzah@gmail.com, tetty.martalinda@yahoo.com, yolandarati@gmail.com

Abstract. This research investigated the effect of synthesis pH on the microstructure and morphology of the zinc oxide (ZnO) particles prepared using pineapple (Ananas comosus) peel extract. In this study, ZnO powder were synthesized at different pH, i.e. 8, 9, 10, 11, and 12. ZnO samples were characterized using UV-Vis spectroscopy, Fourier Transform InfraRed spectroscopy, scanning electron microscopy (SEM), and X-ray diffraction (XRD). The UV-Vis absorbances spectroscopy shows the optical absorption peak of the ZnO sample occurred in a wavelength range of 300-360 nm, with bandgap energy of ~3.22 eV. The FT-IR spectrum shows the peak of Zn-O absorption at the wavenumber of 437.55 cm-1 and the reduction of aromatic compounds with increasing of pH. The micrograph of ZnO particles synthesized at different pH shows that the pH affected the size and shape of ZnO. Micro-sized particles with a granular shape have been found at pH 8, and 9, a spherical shape have been found at pH 10, while micro-sized flower-shaped particles have been found at pH 11 and 12. The XRD pattern reveals a wurzite hexagonal ZnO crystal phase with the hkl plane of (101) as the strongest peak, as well as the purity of the sample increasing as the pH value rises. As the conclusion, synthesis pH has a significant impact on the optical, structural, and morphological properties of ZnO biosynthetic powder.

1. Introduction

In recent years, biosynthetic nanoparticles have gained tremendous attention with numerous applications in waste water treatment, food industry and medicine. The biosynthetic method employs materials derived from organisms such as microbes and plant extracts that are environmentally friendly [1]. Biosynthetic techniques provide a way to overcome the drawbacks of physical and chemical synthesis techniques, which use hazardous chemicals, consume a lot of energy, and produce waste that can harm the environment [2]. The use of natural ingredients in synthesizing nanoparticles makes these materials safer to be applied in the fields of agriculture, medicine and the food industry. Those natural ingredients act as a reducing agent, stabilizing agent as well as a capping agent because it contains secondary metabolite organic compounds (phytochemicals) such as flavonoids, alkaloids, tannins and saponins [3].

Zinc oxide (ZnO), II-VI compound n-type semiconductor, has a wide energy band gap of 3.37 eV and a high exciton binding energy of 60 meV at room temperature. Its unique properties and diverse applications in transparent electronics, ultraviolet (UV) light emitters, piezoelectric devices, chemical sensors, and spintronics have attracted a lot of research attention [4]. ZnO is also categorized as a safe material and has antibacterial properties and has the potential to be used in medicine.

ZnO biosynthesis has been previously reported using extracts of watermelon peel [5], red clover flower [6], alang-alang leaf [7], moringa root [8], and cherry fruit [9]. Pineapple peel contains a lot of antioxidants which are composed of phytochemical compounds. Pineapple peel extract shows potential as a bioreductant, stabilizer and capping agent in the formation of ZnO. The ZnO formation process is influenced by several supply conditions such as temperature, precursor concentration, solvent and acidity level. In this study, ZnO biosynthesis was carried out using pineapple peel extract and Zinc Nitrate Hexahydrate as precursors by varying the pH values of 8, 9, 10, 11 and 12. Optical absorption properties, functional group bonds, morphology and crystal structure of ZnO samples were analyzed based on spectroscopic characterization. UV-Vis, Fourier Transform InfraRed (FT-IR) spectroscopy, scanning electron microscopy (SEM) and X-ray diffraction (XRD).

2. Method

2.1. Preperation of pineapple peel extract
The pineapple fruit peel was removed and thoroughly washed with distilled water, then dried under the sun for a week. The peels were thoroughly crushed in a blender and stored in a sealed container for future use. 5 grams of Ananas comosus peel powder were extracted with 500 mL distilled water and heated for 1 hour at 80°C while stirring using a hot plate. Finally, the extract is filtered using Whatman filter paper no. 1.

2.2. Synthesis of ZnO
1.47 grams of Zinc Nitrate Hexahydrate ($Zn(NO_3)_2.6H_2O$) was dissolved in Aqua DM and put into a beaker containing pineapple peel extract. In this study, samples were made with varying pH, namely, pH 8, 9, 10, 11, and pH 12. To get the desired pH, NaOH solution was added into the extract solution mixture. The solution was heated at 40°C until a white precipitate is formed. The mixture was then allowed to cool to room temperature and centrifuged at 4000 rpm for 10 minutes. Finally, the ZnO sample was oven-dried at 60°C for a day.

2.3. Characterization
The optical absorption properties of ZnO were characterized using a SHIMADZU UV-Vis spectrophotometer with a wavelength range of 300-800 nm. The functional groups of the sample were detected using SHIMADZU IR Prestige-21 FT-IR spectroscopy. The shape and size of ZnO particles were characterized using Hitachi Flexsem 1000. The structure and crystal size of the samples were studied using an X-ray diffractometer (Rigaku MiniFlex) with CuKα radiation ($\lambda = 0.154$ nm).

3. Result and discussion
ZnO powder with various synthetic pH values shows different color for each particular ZnO sample (Figure 1). pH 8 powder is a brownish yellow color. The color of the powder lightens as the pH rises. The brownish yellow color was probably originated from the pineapple peel extract due to low quantity of NaOH as reaction catalyst. The color change from pH 8 to pH 12 also indicates an increasing reduction rate [9]. Meanwhile, at pH 12, the ZnO powder obtained was white, this indicates the higher the purity of ZnO. These results will be explained further in the XRD analysis.

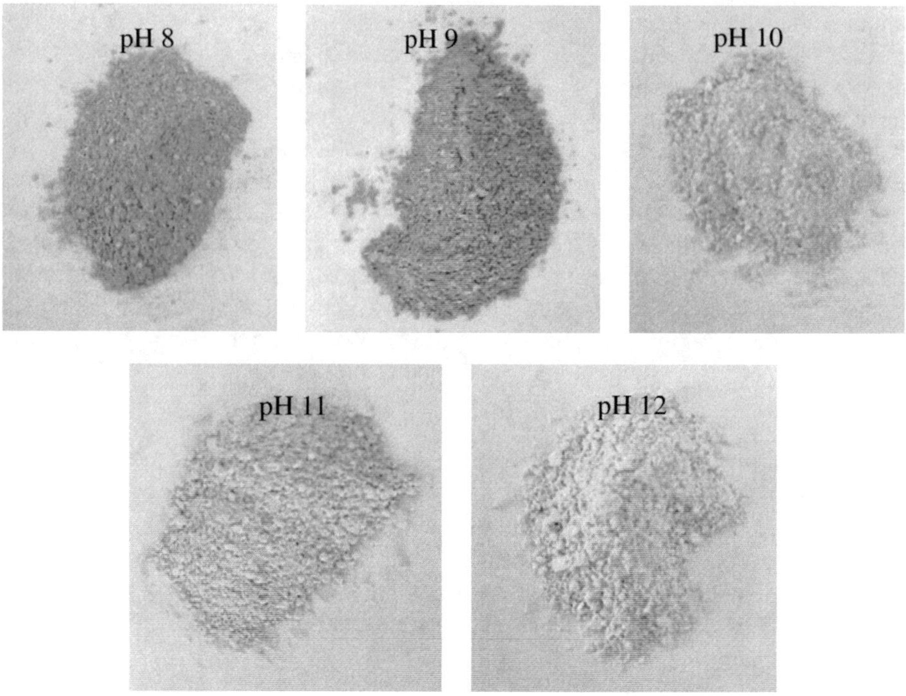

Figure 1. ZnO biosynthetic powder.

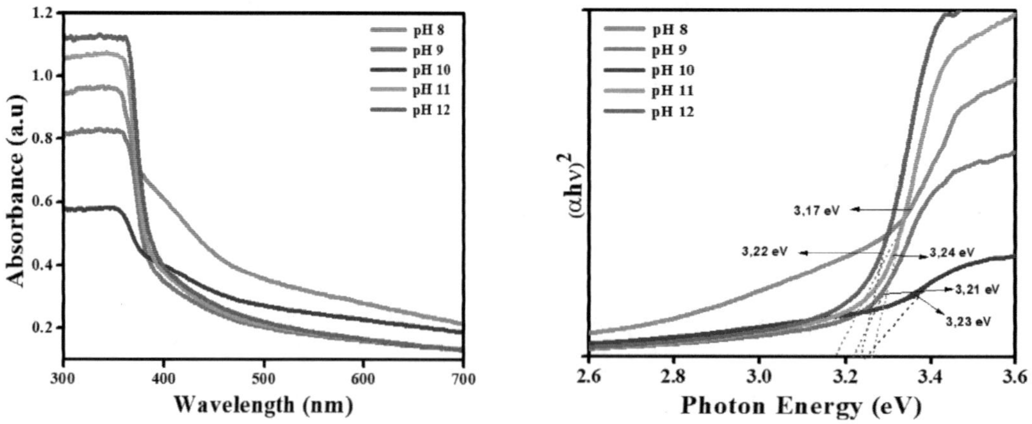

(a) (b)

Figure 2. (a) The UV-Vis absorption spectrum of the ZnO sample and (b) the bandgap energy curve of the ZnO sample synthesized using pineapple peel extract with varying acidity levels.

The absorption UV-Vis spectrum in Figure 2 of the ZnO sample has a characteristic absorption peak at a wavelength of less than 400 nm. The wide absorption peak at ~376 nm (~3.37 eV) is the characteristic peak of ZnO [10]. The pH 8 sample is unique in that there is absorption in the visible light region. This is related to the brownish yellow color of the pH 8 sample from the extract, so it will absorb light in the complementary color spectrum, namely purple-blue (λ = 380-450 nm) [11]. The highest optical absorption value was found in the pH 12 sample and the lowest in the pH 10 sample.

Bandgap energy (Figure 2) was determined from the Tauc Plot method of photon vs (αhυ) 2 energy curves. The bandgap energy value of the ZnO sample obtained is in the range of 3.17-3.23 eV. The difference in the band gap energy value is caused by differences in the particle size of the ZnO sample where the band gap energy will increase as the particle size of the ZnO sample decreases [12].The value of the ZnO band gap energy is reported to vary based on the shape of the particle. Ashar et al [13] reported the band gap energy of ZnO material is 3.21 eV for flower-shaped ZnO and 3.24 eV for pseudo-sphere ZnO obtained from the precipitation method. The difference in the band gap energy value comes from the difference in oxygen vacancy levels in ZnO nanoparticles due to different shapes [14].

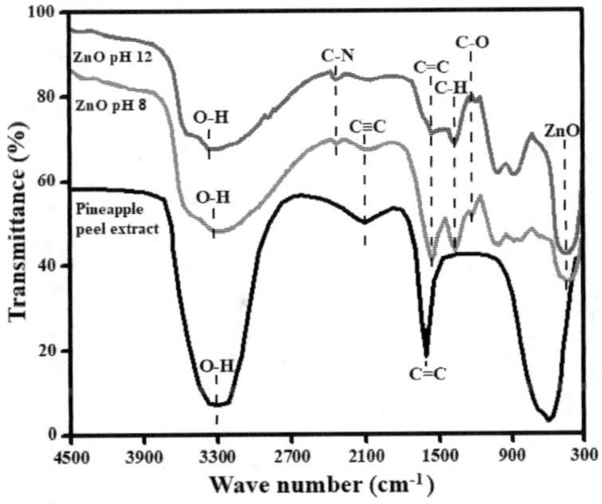

Figure 3. FT-IR spectrum of ZnO samples and pineapple peel extract.

The FT-IR characterization aims to determine the functional groups involved in the synthesis and confirm the presence of Zn-O bonds in the sample. The characteristics of the Zn-O bond based on the FT-IR spectrum (Figure 3) are found at the wavenumber of 437.55 cm-1. This position was also found in the biosynthesis of ZnO using pomegranate peel extract [15]. The absorption peak of the Zn-O bond appears to increase with increasing pH value. These results are in accordance with the research conducted by Thi et al (2020) using orange peel extract.

The functional group bonds found at wave numbers 3303.53 cm-1-3356.97 cm-1, 2335.67 cm-1, 2104.09 cm-1, 1610.27 cm-1, 1562.77 cm-1, 1372.76 cm-1 and 125.20 cm-1 represent the functional groups OH, CN, C≡C, C=C, C=C, CH and CO, respectively. The C≡C functional group at pH 12 looks reduced, while at pH 8 it is still clearly visible. This is thought to be lacking in the addition of NaOH in helping the reduction process so that the pineapple peel extract is still dominant. While the C=C functional group at pH 8 and pH 12 compared to pineapple peel extract shifted to a lower wavenumber. This wave number shift according to Malek and Nahid [9] proves the performance of the reducing agent and the formation of ZnO.

Figure 4 shows the surface morphology (SEM) of ZnO at a magnification of 30,000 times. At each pH value, the particle shape of the ZnO sample changes. Particles were micro-sized, like a ganular, and had indistinct grain boundaries in pH 8 and 9 samples. Meanwhile, pH 10 has micro-sized spherical shape, but it is not uniform. Some of the particles in the pH 11 sample have a micro-sized flower shape, but it is not fully formed. When compared to other samples, the pH 12 particles were in the form of uniform micro-sized flowers with the size (307,6 ± 107,18). The different shape of the particles is due to the increase in the pH value affecting changes in the reaction between H+ ions and OH- ions so that the hydrolysis and condensation processes of particles are not controlled, this causes the formation of interconnected particles and has different shapes [17]. Table 1 shows the average particle size of the ZnO sample. The relationship between acidity level and particle size is in line with the results of research by Ashar et al [13] who reported that changes in size and morphology also affect the band gap energy of ZnO.

Figure 4. SEM photos of ZnO samples with a magnification of 30,000 times.

Table 1. Average sizes of ZnO particles

Sample	Morphology	Average sizes (nm)	Photon energy (eV)
pH 8	microgranular	410,8 ± 157,22	3,17
pH 9	microgranular	336,8 ± 153,1	3,21
pH 10	microspherical	244,18 ± 136,21	3,23
pH 11	microflowers	219,68 ± 72,41	3,24
pH 12	microflowers	307,6 ± 107,18	3,22

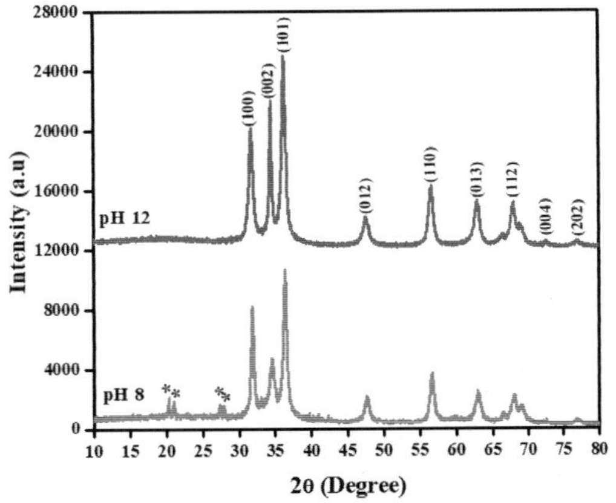

Figure 5. XRD pattern of the ZnO sample.

Figure 5 shows the X-ray diffraction pattern of the pH 8 and pH 12 samples. The diffraction peak of the pH 12 sample has a higher diffraction intensity than pH 8. This indicates that the pH 12 sample has high crystallinity properties. ZnO diffraction peaks correspond to Crystallography Open Database (COD) No. 96-900-4182 at an angle of 2θ = 31,77o; 34,42o; 36,25o; 47,55o; 56,60o; 62,88o; 67,96o; 72,68o dan76,95o represent the hexagonal wurtzite ZnO crystal planes of (100), (002), (101), (012), (110), (013), (112), (004) and (202) respectively. These results are in line with the research conducted by [16]. The ZnO sample pH 8 exhibits another crystalline phase in addition of ZnO wurtzite phase at diffraction angle of 20,29o; 21,01o; 21,36o dan 27,92o. This is due to the presence of another crystalline phase originated from the pineapple peel extract [18]. It can be concluded that the pH value affects the purity of ZnO, the higher the pH value, the higher the sample purity.

Table 2. Crystal size and lattice parameters of ZnO samples at the highest diffraction peaks (101)

Sample	2θ (degree)	Crystal size (nm)	a=b (Å)	d_{hkl} (Å)	c (Å)
pH 8	36,363	15,687	3,243	2,469	5,183
pH 12	36,268	14,81	3,250	2,475	5,206

The crystalline size of the ZnO sample was calculated using the Debye Scherrer equation. The sample at pH 12 had a smaller crystal size than the sample at pH 8 (Table 2). The lattice parameters a and c increased with increasing pH value. The increase in the value of lattice parameters a and c on increasing the pH value with decreasing crystal size occurs due to changes in microstructure, size and shape as well as defects in particles (microstrain) [19]. The lattice parameter values a and c match the lattice parameters of the hexagonal wurtzite ZnO structure with a c/a ratio of 1.6. It can be concluded that the higher the pH of the sample, the smaller the crystal size. These results have also been obtained in a study conducted by Aziz et al [20]. According to Aziz et al [20] different pH values will affect the crystal size, because pH affects the reduction in tensile strain in ZnO with an increase in basicity when the reaction occurs.

4. Conclusions

Biosynthesis of ZnO powder using pineapple peel extract with various pH values has been obtained and analyzed. Increasing the synthetic pH value provides sample with maximum UV-Vis absorption and loss functional groups from pineapple peel extract derived from secondary metabolites. With increasing pH, the morphology of ZnO changes from micro-sized like a granular, micro-sized spherical to micro-sized flowers. Furthermore, increasing the pH results in pure ZnO crystal phase without any other impurities phases. As a result, the optical properties, functional groups, morphology, and structure of ZnO were found to be affected by the synthetic pH.

Acknowledgments
Acknowledgements are addressed to Lembaga Penelitian dan Pengabdian Masyarakat, Universitas Riau for facilitating this research.

References

[1] S. Ahmed, Annu, S. A. Chaudhry, and S. Ikram *J. Photochem. Photobiol. B Biol.* vol. 166, pp. 272–284, 2017, doi: 10.1016/j.jphotobiol.2016.12.011.
[2] S. Sawan, R. Maalouf, A. Errachid, and N. Jaffrezic-Renault *TrAC - Trends Anal. Chem.* vol. 131, p. 116014, 2020, doi: 10.1016/j.trac.2020.116014.
[3] K. Vijayaraghavan and T. Ashokkumar *J. Environ. Chem. Eng.* vol. 5, no. 5, pp. 4866–4883, 2017, doi: 10.1016/j.jece.2017.09.026.
[4] Ü. Ozgur, D. Hofstetter, and H. Morkoç *Proc. IEEE* vol. 98, no. 7, pp. 1255–1268, 2010, doi: 10.1109/JPROC.2010.2044550.
[5] R. Fadillah, Y. Rati, R. Dewi, R. Farma, and A. S. Rini *J. Phys. Conf. Ser.* vol. 1816, no. 1, 2021, doi: 10.1088/1742-6596/1816/1/012019.
[6] R. Dobrucka and J. Długaszewska *Saudi J. Biol. Sci.* vol. 23, no. 4, pp. 517–523, 2016, doi: 10.1016/j.sjbs.2015.05.016.
[7] S. N. Kane, A. Mishra, and A. K. Dutta *J. Phys. Conf. Ser.* vol. 755, no. 1, 2016, doi: 10.1088/1742-6596/755/1/011001.
[8] C. S. Espenti, A. G. Rama Krishna, and Y. V. Rami Reddy *SN Appl. Sci.* vol. 2, no. 8, pp. 1–11, 2020, doi: 10.1007/s42452-020-2945-3.
[9] F. Malek and M. Nahid *J. Nanostructure Chem.* vol. 8, no. 1, pp. 93–102, 2018, doi: 10.1007/s40097-018-0257-6.
[10] B. N. Singh, A. K. S. Rawat, W. Khan, A. H. Naqvi, and B. R. Singh *PLoS One* vol. 9, no. 9, 2014, doi: 10.1371/journal.pone.0106937.
[11] V. Hoseinpour, M. Souri, N. Ghaemi, and A. Shakeri *Heal. Biotechnol. Biopharma* vol. 1, no. 2, pp. 39–49, 2017, doi: 10.22034/HBB.2017.10.
[12] M. Singh, M. Goyal, and K. Devlal *J. Taibah Univ. Sci.* vol. 12, no. 4, pp. 470–475, 2018, doi: 10.1080/16583655.2018.1473946.
[13] A. Ashar et al. *J. Alloys Compd.* vol. 678, pp. 126–136, 2016, doi: 10.1016/j.jallcom.2016.03.251.
[14] A. Alkauskas and A. Pasquarello *Phys. Rev. B - Condens. Matter Mater. Phys.* vol. 84, no. 12, pp. 1–11, 2011, doi: 10.1103/PhysRevB.84.125206.
[15] S. N. A. Mohamad Sukri, K. Shameli, M. Mei-Theng Wong, S. Y. Teow, J. Chew, and N. A. Ismail *J. Mol. Struct.* vol. 1189, pp. 57–65, 2019, doi: 10.1016/j.molstruc.2019.04.026.
[16] T. U. Doan Thi, T. T. Nguyen, Y. D. Thi, K. H. Ta Thi, B. T. Phan, and K. N. Pham *RSC Adv.* vol. 10, no. 40, pp. 23899–23907, 2020, doi: 10.1039/d0ra04926c.
[17] R. Wahab, S. G. Ansari, Y. S. Kim, M. Song, and H. S. Shin *Appl. Surf. Sci.* vol. 255, no. 9, pp. 4891–4896, 2009, doi: 10.1016/j.apsusc.2008.12.037.
[18] R. Adawiyah and Belinda Gabriella Anak Bronny Int. *J. Technol. Res. Eng. ISSN (Online, no. July* pp. 26–40, 2019, [Online]. Available: www.ijtre.com.
[19] M. J. Chithra, M. Sathya, and K. Pushpanathan *Acta Metall. Sin. (English Lett., vol. 28, no. 3)* pp. 394–404, 2015, doi: 10.1007/s40195-015-0218-8.
[20] R. A. Abdol Aziz, S. F. Abd Karim, and N. A. Rosli *Key Eng. Mater.* vol. 797, pp. 271–279, 2019, doi: 10.4028/www.scientific.net/KEM.797.271.

Polyvinyl Alcohol-based Hydrogel: A Systematic Literature Review on Thermal Properties by Differential Scanning Calorimetry

P A Putro*, A S Sulaeman and A Maddu*
Department of Physics, Faculty of Mathematics and Natural Sciences, IPB University
Jl. Meranti, Kampus IPB Dramaga, Bogor, Jawa Barat 13220, Indonesia
Email: permanenputro@apps.ipb.ac.id; akhiruddin@apps.ipb.ac.id

Abstract. The purpose of this systematic literature review (SLR) study is to assess the importance of thermal properties and the depth of analysis on Polyvinyl alcohol (PVA) -based hydrogel. Google Scholar is used as a search for database sources using Publish or Perish software, which is related to PVA-based hydrogel's thermal properties from 2018 to 2020. Thermal properties are focused on the DSC thermogram characteristic. These results strongly indicate that PVA-based hydrogel's thermal properties are very important and considered for their potential application.

1. Introduction

PVA has been considered as one of the attractive synthetic polymers over a few decades as it possesses desirable properties such as non-toxicity, biodegradability, biocompatibility, water solubility, and chemical and thermal stability [1–7]. It has prepared by chemical modification like partial or complete hydrolysis of polyvinyl acetate [6, 7]. PVA (**Figure 1.**) holds an enormous number of hydroxyl (–OH) groups encourage a various reaction, like with new functional groups [8–10], cross-linking between the -OH groups and other polar molecules to form entanglements and aggregates [11–13]; Hence, It has highly hydrophilic and dispersible in water [11,12] . A semicrystalline hydrophilic, odorless, tasteless, and translucent white color granular, as PVA physical characteristics. Moreover, It has an excellent film-forming, emulsifying and adhesive property [14, 15].

In recent years, physical and mechanical properties of PVA has been improved by blending with various polymers [16-20], incorporation natural materials [21–24], organic molecules [25, 26] and nanoparticles [27]. Based on its excellent properties, PVA have led to various biomedical [28] and pharmaceutical [29, 30] applications including but not limited to: fibrous meshes and membranes for wound dressings [31,32], drug delivery systems [33–35], interventional radiology embolization particles [36], and hydrogels used in orthopedic medicine [37], contact lenses [38]. The wide range of applications for PVA has resulted in many investigations over the past four decades into both its fundamental thermal [39–46] and structural [47–49] properties as well as the crystallization behavior of PVA hydrogels, PVA nanofibers [50, 51], and PVA blends and composites [52, 53].

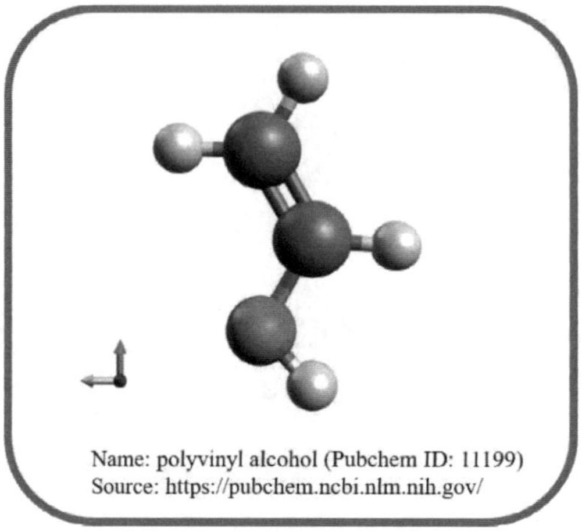

Figure 1. Polyvinyl alcohol structure

The techniques of thermal analysis provide important tools in order to characterize thermal properties of these materials. When a polymeric system absorbs heat, many physical processes and phase transitions take place [54]. In order to understand these physical and macromolecular changes, the tools of thermal analysis, namely thermogravimetric analysis (TGA) and differential scanning calorimetry (DSC), have been used by researchers [55, 56]. The techniques of thermal analysis have been used for the determination of the crystallinity of copolymers, which varies with incorporation of some additives [57]. The incorporation of nanoparticles in polymer-based hybrid materials is found to affect the thermal properties of this material [58, 59].

From thermal analysis, relaxation processes that occur in the polymeric material, such as α-relaxation and β-relaxation can be understood. The α-relaxation (glass transition) phenomenon is associated with side (functional) groups of polymers, whereas β-relaxation is associated with main chain molecular motion. Compatibility (complete mixing) of polymer blends can be determined using DSC curves. Miscible blend films exhibit one phase and a single glass transition temperature (T_g). Immiscible blends are phase-separated and exhibit multiple T_g's, one for each component [59, 60]. The microstructural modifications can also be studied by estimating thermal parameters like glass transition temperature (T_g), and melting temperature (T_m) using DSC [61].

A systematic literature review (SLR) is defined as systematic and explicit methods to classify, select, and critically evaluate relevant primary research, and to extract and analyze data from several previous studies' [62]. An SLR can provide useful insights to both academics and practitioners [63]. An SLR should sensitive; because it aims to seek as many studies as possible hence reduced bias and to be effective for under review [64]. Moreover, SLR an important tool for synthesizing scientific evidence, enhancing the validity of the results of studies, and identifying areas of uncertainty where the study is needed [65].

This study aims to carry out a SLR analysis on thermal properties of PVA-based hydrogel by DSC, following the Preferred Reporting Items for Systematic Reviews and Meta-Analyses (PRISMA) methodology [66], DSC as a tool or instrumentation to deepen the study of its thermal properties. DSC has an important role in observing thermal properties as well as its excellent performance (namely T_m and T_g). However, in certain circumstances, the thermogram results cannot provide strong and accurate information regarding the T_g value. The dynamic mechanical-thermal analysis (DMTA) is here as an alternative solution for measuring the value of T_g. Thus, the sample preparation process here also greatly determines the quality of the thermogram data used as primary data in future research and report.

2. Method

Through a holistic and organized specification that adheres to standard protocols, SLRs provide readers with detailed knowledge of literature in a region [67–70]. SLRs also help to clarify current information gaps and to find areas for future study [71, 72]. Although, different authors propose different approaches for conducting the systematic literature review process, there are three main steps found to be shared in all of them [63]. This research covers these measures following Ortiz-Martínez et al. [62] suggested guidelines.

2.1. Protocol

Under the PRISMA process, the latest systematic review has been carried out. It consists of well-defined stages of analysis and the creation and definition of information source eligibility requirements, literature search strategy, literature selection process, and data synthesis based on selected literature [62, 66]. According to DSC results, this systematic approach involves defining and discussing PVA-based hydrogel of thermal properties through their parameters.

2.2. Eligibility criteria

To determine and limit the form of literature in this study, the determination of eligibility requirements will be chosen. Eligibility criteria consist of criteria for inclusion and exclusion. The inclusion criteria in the present review were: (1) articles that documented PVA-based hydrogel thermal properties; (2) articles published in 2018 to 2020; (3) reports that have detailed relevant details. In addition, the exclusion requirements were: (1) unquoted articles; (2) review articles; (3) chapter books and books; (4) patents; (5) works written in a language other than English; (6) PVA-based hydrogel articles without thermal analysis.

2.3. Search strategy and information source

In order to obtain the data generated by published papers in 2018-2020, a bibliographic search was carried out. In a systemic analysis with PRISMA, the database quest can be done with applications such as Publish or Perish (PoP). The PoP search technique uses keywords so information that is accurate and applicable to the topics agreed upon by all writers can be comprehensively contained. For the format of the keywords, the Boolean algorithm is used and we used "polyvinyl alcohol" AND "thermal properties" AND "thermal analysis" AND "differential scanning calorimetry" AND hydrogel while Google Scholar is the chosen data source and we obtain the meta-data type details ($n = 641$).

2.4. Study selection

Selection of data based on the PRISMA technique. The review team consisted of all writers who screened names, abstracts, and full texts of the works for possible inclusion separately. According to the inclusion and exclusion criteria, they were assessed by all authors in the section on the eligibility criteria. To handle the 641 works accumulated by one database, the Mendeley Version 1.19.4 program was crossed over. We chose the subject of the remaining works and the works associated with them. When the review team was unclear about whether or not a given work met the inclusion requirements, the complete works were downloaded for further examination. Besides, in search of potentially appropriate articles, the selected works' reference list was also examined; however, no additional works were found in this step. Finally, after the full-text eligibility assessment, a total of 13 papers were found to be suitable for the present systematic review. The article selection process has been explained in **Figure 2.**

2.5. Data collection

Because of this review phase, data extraction by the review team must be precise and unbiased. From the 13 papers included in the present systematic review, qualitative and quantitative data were extracted using a data extraction form built for use in the present analysis. Table 1 presents a portion of this data extraction form.

2.6. Data items
The review team performed data extraction on the basis of DSC data (namely T_m and T_g) of the 13 selected article variables for which data relevant to the publication of the article were searched, such as the first author, year, title, materials, and thermal parameters.

2.7. Synthesis of results
In this current systematic review, a synthesis of the findings based on the discussion of the empirical patterns in the data provided without meta-analysis lacked adequate statistical details. The search method with PoP resulted from some variable data such as patent, book, website, etc. It does not support the meta-analysis performed.

3. Results and Discussions

3.1. Article selection
Several stages of selection were passed based on the illustration in **Figure 2**. Of the 641 papers initially retrieved, 61 were chosen based on the title and abstract. Of these, 13 were included in the study, along with an additional study found by manually reviewing the reference lists of the articles chosen. Table 1.a to 1.b lists the 13 papers included. The dates of release range from 2018 to 2020. In the sample populations, there was no overlap.

3.2. PVA characteristic
The chemical stability and hydrophilic nature of PVA are great [73, 74]; It caused undergoes a rearrangement of chemical bonds when PVA is heated above 300°C, followed by the establishment of some form of linkages by hydrogen bonding with water molecules [75, 76]. Highly conjugated aromatic structures are formed during the heat treatment of PVA, transforming them into cross-linking agents at high temperatures [77].

Generally, PVA is known to decompose in two stages [59, 78]. In the first stage, which begins at 245°C, the formation of volatile products is followed by dehydration. The water molecule (H_2O) is removed from the polymer chain, and polyene structures are formed. These residues are mainly polymers with unsaturated structures in conjugation. In the second stage, the remaining residues of polyene are further degraded from 430 to 450°C to produce carbon and hydrocarbons.

3.3. Thermal properties by DSC thermogram
Measurements of differential scanning calorimetry can be used, in particular, to consider the thermal background and stability of polymer samples [79]. Two endothermic peaks related to the T_g and T_m transition are observed based on the thermograms [80]. As estimated from the extensions of the pre-and post-transition baselines, the T_g is calculated as a mid-point in the thermograms [20], [81]. The degree of polarity and essence of the material is explained by the T_g [80].

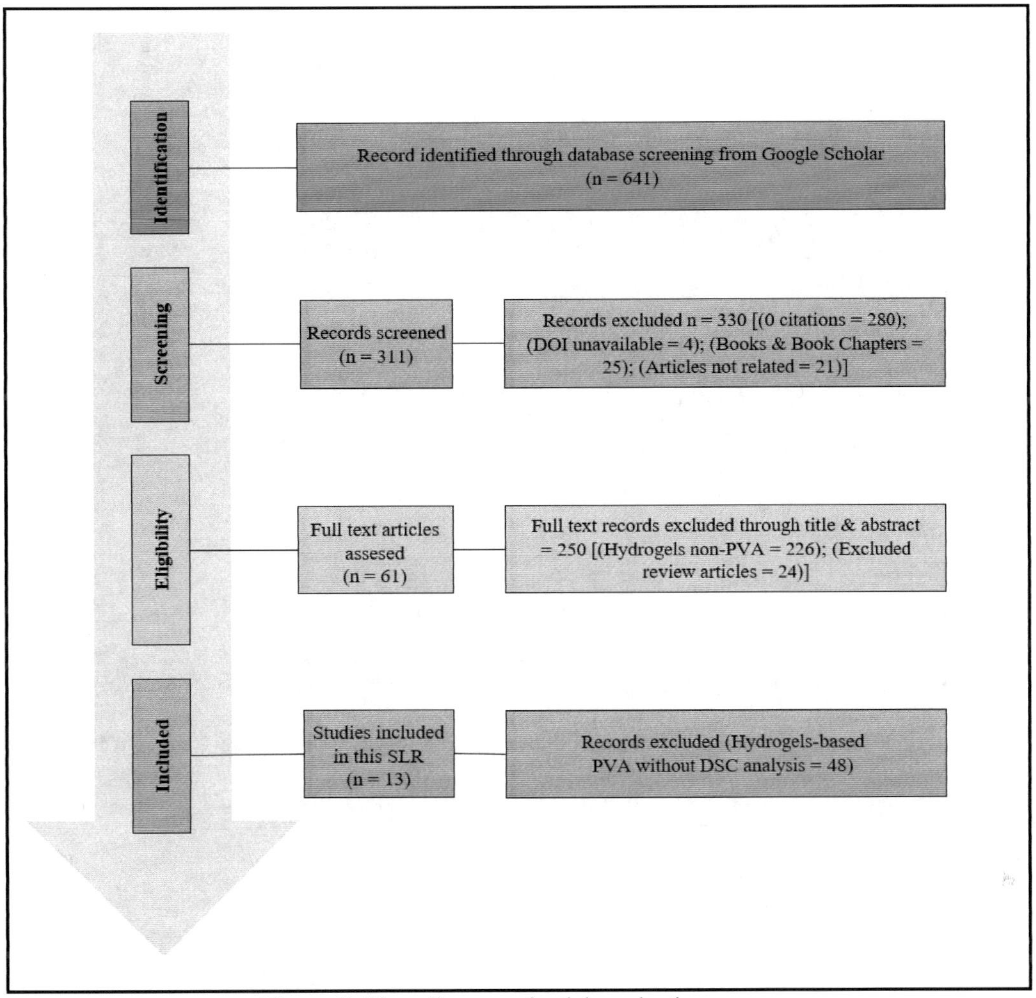

Figure 2. Flow diagram of articles selection process

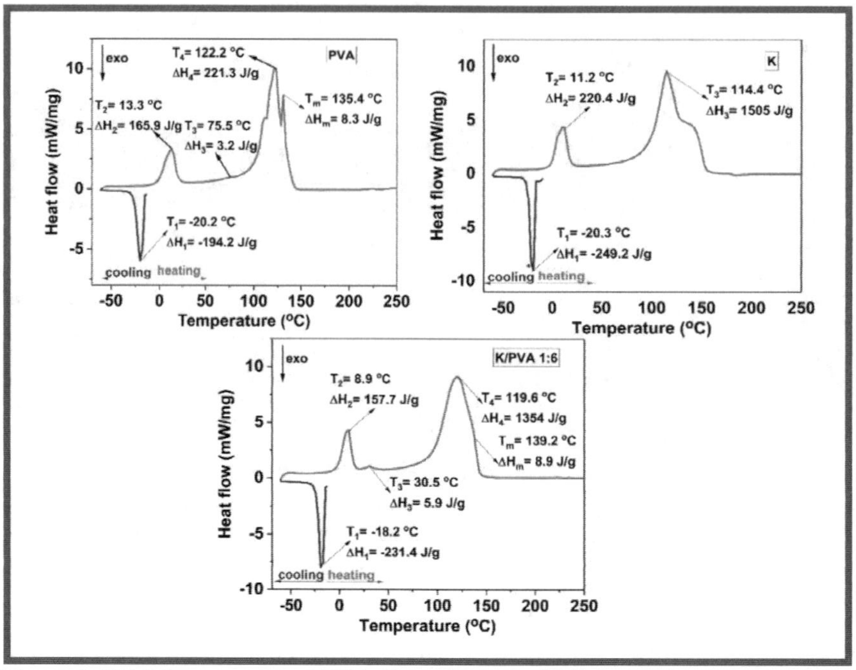

Figure 3. DSC thermograms of PVA, K, and K/PVA [82]. Note, K is k-carrageenan.

The thermograms' cooling curve mentioned by Croitoru et al. [82] contains one characteristic, namely the large exothermic peak attributed to hydrogel network water crystallization (**Figure 3.**). The melting temperature of the non-associated (free) water in the hydrogel is indicated by an endothermic peak for PVA at ~13°C and K at ~11°C. The sample with the highest degree of crystallinity and the highest thermal stability was K/PVA (1:6) at ~9°C [83].

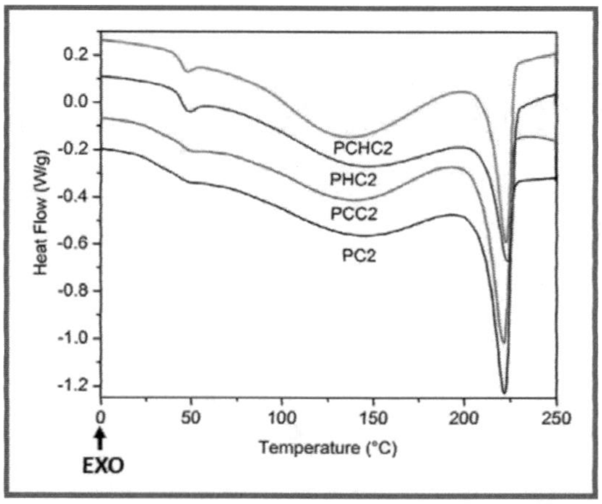

Figure 4. DSC thermograms of PC2, PCC2, PCH2, and PCHC2 [84]

Table 1.a Summary of article results obtained with SLR method

Authors	Years	Titles	Materials	T_g	T_m
Khaleghi, *et al*. [84]	2018	Synthesis and characterization of new honey incorporated double-network hydrogels based on poly(vinyl alcohol) and acylated chitosan	PVA/ CaCl2/ Chit-MA/ Honey	*	*
Kamoun, *et al*. [20]	2018	In-situ UV-photopolymerized PVA-g-GMA hydrogels for biomedical applications: I. Synthesis, characterizations and grafting optimization	PVA-g-GMA	*	
Martínez-Gómez, *et al*. [85]	2018	Characterization of poly-D-mannuronate and poly-L-guluronate block fractions from sodium alginate and preparation of hydrogels with poly(vinylalcohol)	MM/PVA and GG/PVA	*	
Yang, *et al*. [27]	2018	Polyvinyl alcohol/chitosan hydrogels with enhanced antioxidant and antibacterial properties induced by lignin nanoparticles	PVA/Ch/LNPs	*	*
Chee, *et al*. [86]	2018	Investigation of the effects of orientation on freeze/thawed polyvinyl alcohol hydrogel properties	PVA/CAF	*	*
Tanriverdi, *et al*. [87]	2018	Preparation and In-vitro evaluation of melatonin loaded HA/PVA gel formulation	HA/PVA/M	*	
Long, *et al*. [88]	2019	Development of a long-term drug delivery system with levonorgestrel-loaded chitosan microspheres embedded in poly(vinyl alcohol) hydrogel	PVA/CS-LNG		*
Ullah, *et al*. [89]	2019	Natural and synthetic materials based CMCh/PVA hydrogels for oxaliplatin delivery: Fabrication, characterization, In-vitro and In-vivo safety profiling	CMCh/PVA	*	*
Freitas, *et al*. [90]	2020	Investigation on miscibility, thermal, crystallographic diffraction and dynamic-mechanical properties of poly(vinyl alcohol)/poly(vinylpyrrolidone)/zirconium phosphate nanocomposites	PVA/PVP/ZrP	*	*

Table 1.b Summary of article results obtained with SLR method

Authors	Years	Titles	Materials	T_g	T_m
Zulfiqar, *et al.* [91]	2020	Efficient removal of Pb(II) from aqueous solutions by using oil palm-bio-waste/MWCNTs reinforced PVA hydrogel composites: Kinetic, isotherm and thermodynamic modeling	OPB/PVA/MWCNTs		*
Reena, *et al.* [92]	2020	Synthesis and characterization of cross-linked hydrogels using polyvinyl alcohol and polyvinyl pyrrolidone and their blend for water shut-off treatments	PVA/PVP		*
Croitoru, *et al.* [82]	2020	Physically croslinked poly (vinyl alcohol)/Kappa-carrageenan hydrogels: Structure and applications	PVA/CAR	*	*
Sirousazar and Khodamoradi. [93]	2020	Freeze-thawed humic acid/polyvinyl alcohol supramolecular hydrogel	HA/PVA		*

In general, the presence of a single peak T_g for a polymeric blend in the DSC thermograms can be connected to good impermissibility and contact between the constituents [94]. However, in this case, a single value for T_g has not been associated with the formation by hydrogen bonding of inter-polymer complexes since k-carrageenan shows no presence of this type of transformation. The T_g may be attributed to a peak at ~75 ° C for PVA hydrogel, following the literature's values [95]. The combined plasticizing effect of k-carrageenan and water induces a decrease in T_g values for the K/PVA mixture relative to PVA. While in the 100-150 °C interval, broad endotherm peaks appear, which could be due to water vaporization [82].

Table 2. Formulation of samples for PC2, PCC2, PCH2, and PCHC2 [84]

Formulation	PVA (%)	$CaCl_2$ (%)	Chit-Ma (%)	Honey (%)
PCHC2	98.50	0.10	0.70	0.70
PHC2	99.20	0.10	0.00	0.70
PCC2	99.20	0.10	0.70	0.00
PC2	99.90	0.10	0.00	0.00

Khaleghi et al. [84] successfully compared the effect of Chit-MA with honey using parameters of T_m and degree of crystallinity. The DSC curves of PC2, PCC2, PHC2, and PCHC2 are shown in **Figure 4.**; and, the formulation of the sample can be seen in Table 2. According to them, honey added to PVA increased the melting temperature and decreased the degree of crystallinity significantly compared to the addition of Chit-MA. This difference is possibly due to the mutual affinity of hydroxyl-rich surfaces between PVA and honey, leading to increased hydrogen bonds affecting the molecular mobility of the PVA chains and the matrix's crystallinity behaviour [84], [96].

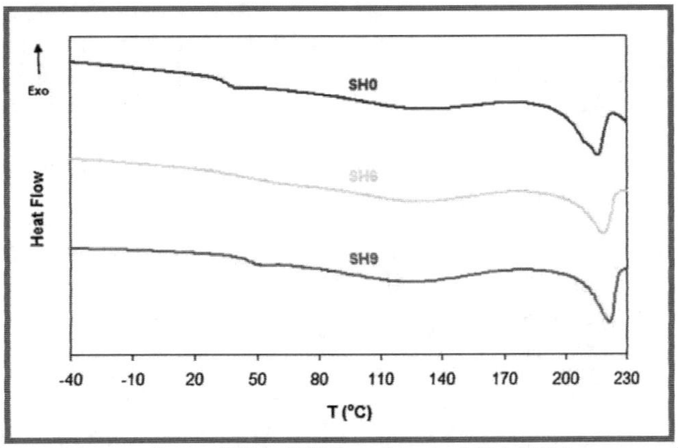

Figure 5. DSC thermograms of pure PVA (SH0) and typical HA/PVA supramolecular hydrogels loaded with 6 and 9 wt.% HA (i.e., SH6 and SH9) [93]

T_g is correlated with changes in the PVA amorphous region and is dependent on the mobility and free rotation of its chains, and T_m is determined by the crystalline region transition [96]. In comparison with the PC2 study, the T_g of both systems increased towards higher temperatures. Strong interactions between honey and Chit-MA with the PVA matrices that limited the free rotation of the PVA chains in the interfacial zone could be due to this. PCHC2 showed the highest T_g compared to the other samples that could be due to Chit-MA and honey's synergic effect. Compared with the other samples, PCHC2 displayed the highest T_g, which may be attributed to the synergistic effect of Chit-MA and honey together.

In the latest report, Sirousazar and Khodamoradi [93] succeeded in observing the effect of humic nanoparticles (HA) loaded with PVA supramolecular hydrogels based on the T_g value. The increased melting point of PVA supramolecular hydrogels loaded with humic acid nanoparticles (HA) could be attributed to their increased gel fraction values and, because of the presence of HA, formed more crosslinking zones within the supramolecular hydrogels. Besides, samples showed broad and shoulder-like endothermic peaks around 40-55 °C, which may be due to the transition of samples to the glassy-rubbery phase. As the peaks were very weak and broad, the T_g obtained from DSC thermograms (**Figure 5.**) would not be accurate and consistent for the samples examined. Therefore, they propose that the exact values of T_g should be recorded based on the results of the dynamic mechanical-thermal analysis (DMTA) as their solution.

4. Conclusion

PVA-based hydrogel has several thermal properties obtained from DSC measurements. In this SLR study, we found thermal properties parameters usually analyzed on PVA-based hydrogel, which is more focused on T_m and T_g values. However, not all of the data obtained have a good thermogram curve and can be analyzed. Finally, we recommend taking measurements using DMTA; it has a higher degree of accuracy for analyzing T_g. We hope that this study will provide challenges and opportunities to deepen the analysis of thermal properties using DSC measurements, and the thermogram curve can be used as one of the primary data in future research.

References

[1] Sedaghat E Rostami A A Ghaemy M and Rostami A, 2019 *J. Therm. Anal. Calorim.* **136**, 2 p. 759–769.
[2] Freitas D F S Mattos G C and Mendes L C, 2020 *J. Therm. Anal. Calorim.*
[3] Narasagoudr S S Hegde V G Vanjeri V N Chougale R B and Masti S P, 2020 *Chem. Data Collect.* **27** p. 100370.
[4] Chiellini E Corti A D'Antone S and Solaro R, 2003 *Prog. Polym. Sci.* **28**, 6 p. 963–1014.
[5] Alexandre N *et al.*, 2014 *J. Biomed. Mater. Res. Part A* **102**, 12 p. 4262–4275.
[6] Buraidah M H and Arof A K, 2011 *J. Non. Cryst. Solids* **357**, 16–17 p. 3261–3266.
[7] Saadiah M A Zhang D Nagao Y Muzakir S K and Samsudin A S, 2019 *J. Non. Cryst. Solids* **511** p. 201–211.
[8] Hamzah M Ndimba R M Khenfouch M and Srinivasu V V., 2017 *J. Mater. Sci. Mater. Electron.* **28**, 16 p. 11915–11920.
[9] Nandimath M and Bhajantri R F, 2020 *Opt. Mater.* **105** p. 109892.
[10] Putro P A Yudasari N and Maddu A, 2020 *J. Phys. Conf. Ser.* **1491** p. 012033.
[11] Bercea M Morariu S and Teodorescu M, 2016 *J. Polym. Res.* **23**, 7 p. 142.
[12] Giteru S G Ali M A and Oey I, 2019 *Food Hydrocoll.* **87** p. 270–286.
[13] Roza L Putro P A and Isnaeni I, 2019 *AIP Conf. Proc.* **2169** p. 060008-1-060008–13.
[14] Reis E F dos *et al.*, 2006 *Mater. Res.* **9**, 2 p. 185–191.
[15] Kasai D *et al.*, 2019 *J. Polym. Environ.* **27**, 3 p. 472–488.
[16] Shahbazi M Rajabzadeh G Rafe A Ettelaie R and Ahmadi S J, 2016 *Food Hydrocoll.* **60** p. 393–404.
[17] Zhou X Y Cui Y F Jia D M and Xie D, 2009 *Polym. Plast. Technol. Eng.* **48**, 5 p. 489–495.
[18] Zou G-X Jin P-Q and Xin L-Z, 2008 *J. Elastomers Plast.* **40**, 4 p. 303–316.
[19] Olkhov A A Vlasov S V. Iordanskii A L Zaikov G E and Lobo V M M, 2003 *J. Appl. Polym. Sci.* **90**, 6 p. 1471–1476.
[20] Kamoun E A Omer A M Khattab S N Ahmed H M and Elbardan A A, 2018 *J. Appl. Pharm. Sci.* **8**, 1 p. 034–042.
[21] Ma Q Du L Yang Y and Wang L, 2017 *Food Hydrocoll.* **63** p. 677–684.
[22] Su J-F Huang Z Zhao Y-H Yuan X-Y Wang X-Y and Li M, 2010 *Ind. Crops Prod.* **31**, 2 p. 266–276.
[23] Su J-F Huang Z Liu K Fu L-L and Liu H-R, 2007 *Polym. Bull.* **58**, 5–6 p. 913–921.
[24] Gaikwad K K Lee J Y and Lee Y S, 2016 *J. Food Sci. Technol.* **53**, 3 p. 1608–1619.

[25] Song P Xu Z and Guo Q, 2013 *ACS Macro Lett.* **2**, 12 p. 1100–1104.
[26] Chen Y-N Jiao C Zhao Y Zhang J and Wang H, 2018 *ACS Omega* **3**, 9 p. 11788–11795.
[27] Yang W *et al.*, 2018 *Carbohydr. Polym.* **181** p. 275–284.
[28] Costa-Júnior E S Barbosa-Stancioli E F Mansur A A P Vasconcelos W L and Mansur H S, 2009 *Carbohydr. Polym.* **76**, 3 p. 472–481.
[29] Mc Gann M J Higginbotham C L Geever L M and Nugent M J D, 2009 *Int. J. Pharm.* **372**, 1–2 p. 154–161.
[30] Nugent M J D and Higginbotham C L, 2007 *Eur. J. Pharm. Biopharm.* **67**, 2 p. 377–386.
[31] Hong K H, 2007 *Polym. Eng. Sci.* **47**, 1 p. 43–49.
[32] Zhou Y Yang D Chen X Xu Q Lu F and Nie J, 2008 *Biomacromolecules* **9**, 1 p. 349–354.
[33] Kenawy E-R Abdel-Hay F I El-Newehy M H and Wnek G E, 2007 *Mater. Sci. Eng. A* **459**, 1–2 p. 390–396.
[34] Yoon S-D Chough S-H and Park H-R, 2007 *J. Appl. Polym. Sci.* **106**, 4 p. 2485–2493.
[35] Zeng J Aigner A Czubayko F Kissel T Wendorff J H and Greiner A, 2005 *Biomacromolecules* **6**, 3 p. 1484–1488.
[36] Vaidya S Tozer K and Chen J, 2008 *Semin. Intervent. Radiol.* **25**, 03 p. 204–215.
[37] Baker M I Walsh S P Schwartz Z and Boyan B D, 2012 *J. Biomed. Mater. Res. Part B Appl. Biomater.* **100B**, 5 p. 1451–1457.
[38] Hyon S-H Cha W-I Ikada Y Kita M Ogura Y and Honda Y, 1994 *J. Biomater. Sci. Polym. Ed.* **5**, 5 p. 397–406.
[39] Tubbs R K, 1965 *J. Polym. Sci. Part A Gen. Pap.* **3**, 12 p. 4181–4189.
[40] Peppas N A and Merrill E W, 1976 *J. Appl. Polym. Sci.* **20**, 6 p. 1457–1465.
[41] Gaur U Wunderlich B B and Wunderlich B, 1983 *J. Phys. Chem. Ref. Data* **12**, 1 p. 29–63.
[42] Bu H S Aycock W Cheng S Z D and Wunderlich B, 1988 *Polymer (Guildf).* **29**, 8 p. 1485–1494.
[43] Holland B J and Hay J N, 2001 *Polymer (Guildf).* **42**, 16 p. 6775–6783.
[44] El-Zaher N A and Osiris W G, 2005 *J. Appl. Polym. Sci.* **96**, 5 p. 1914–1923.
[45] Li C Vongsvivut J She X Li Y She F and Kong L, 2014 *Phys. Chem. Chem. Phys.* **16**, 40 p. 22145–22158.
[46] Li C *et al.*, 2015 *Polym. Degrad. Stab.* **119** p. 178–189.
[47] Tsuboi K, 1968 *J. Macromol. Sci. Part B* **2**, 4 p. 603–622.
[48] Mallapragada S K and Peppas N A, 1996 *J. Polym. Sci. Part B Polym. Phys.* **34**, 7 p. 1339–1346.
[49] Tadokoro H Seki S and Nitta I, 1955 *Bull. Chem. Soc. Jpn.* **28**, 8 p. 559–564.
[50] Naebe M Lin T Staiger M P Dai L and Wang X, 2008 *Nanotechnology* **19**, 30 p. 305702.
[51] Koski A Yim K and Shivkumar S, 2004 *Mater. Lett.* **58**, 3–4 p. 493–497.
[52] Arvanitoyannis I Kolokuris I Nakayama A Yamamoto N and Aiba S, 1997 *Carbohydr. Polym.* **34**, 1–2 p. 9–19.
[53] Jang J and Lee D K, 2003 *Polymer (Guildf).* **44**, 26 p. 8139–8146.
[54] Chung D D L, 2001 *J. Therm. Anal. Calorim.* **65**, 153–165.
[55] Menczel J D and Prime R B, 2009 *Thermal analysis of polymers: Fundamentals and applications* John Wiley & Sons, Inc.
[56] Groenewoud W M, 2001 *Characterization of polymers by thermal analysis* 1st ed. Amsterdam, the Netherlands: Elsevier Science B. V.
[57] Bertolino V Cavallaro G Lazzara G Milioto S and Parisi F, 2018 *J. Therm. Anal. Calorim.* **132**, 1 p. 191–196.
[58] Majka T M Raftopoulos K N and Pielichowski K, 2018 *J. Therm. Anal. Calorim.* **133**, 1 p. 289–301.
[59] Baraker B M and Lobo B, 2018 *J. Therm. Anal. Calorim.* **134**, 2 p. 865–878.
[60] Mudigoudra B S Masti S P and Chougale R B, 2012 *Res. J. Recent Sci.* **1**, 9 p. 83–86.
[61] Baraker B M and Lobo B, 2018 *J. Therm. Anal. Calorim.* **134** p. 865–878.
[62] Ortiz-Martínez V M Andreo-Martínez P García-Martínez N Pérez de los Ríos A Hernández-Fernández F J and Quesada-Medina J, 2019 *Fuel Process. Technol.* **191** p. 211–222.

[63] Kushwah S Dhir A Sagar M and Gupta B, 2019 *Appetite* **143** p. 104402.
[64] Moher D Liberati A Tetzlaff J Altman D G and Group T P, 2009 *PloS Med.* **6**, 7 p. e1000097.
[65] Ferreira González I Urrútia G and Alonso-Coello P, 2011 *Rev. Española Cardiol. (English Ed.* **64**, 8 p. 688–696.
[66] Moher D *et al.*, 2015 *Syst. Rev.* **4**, 1 p. 1.
[67] Afrooz S and Navimipour N J, 2017 *J. Circuits, Syst. Comput.* **26**, 12 p. 1730004.
[68] Aznoli F and Navimipour N J, 2017 *J. Netw. Comput. Appl.* **77** p. 73–86.
[69] Ahmad M O Dennehy D Conboy K and Oivo M, 2018 *J. Syst. Softw.* **137** p. 96–113.
[70] Mehta N and Pandit A, 2018 *Int. J. Med. Inform.* **114** p. 57–65.
[71] Ganeshkumar P and Gopalakrishnan S, 2013 *J. Fam. Med. Prim. Care* **2**, 1 p. 9.
[72] Tandon A Dhir A Islam A K M N and Mäntymäki M, 2020 *Comput. Ind.* **122** p. 103290.
[73] Mukherjee G S, 2009 *J. Therm. Anal. Calorim.* **96**, 1 p. 21–25.
[74] Krumova M López D Benavente R Mijangos C and Pereña J ., 2000 *Polymer (Guildf).* **41**, 26 p. 9265–9272.
[75] Xu J *et al.*, 2001 *Polym. Degrad. Stab.* **73**, 1 p. 29–31.
[76] Park J-S Park J-W and Ruckenstein E, 2001 *Polymer (Guildf).* **42**, 9 p. 4271–4280.
[77] Wiśniewska M Chibowski S Urban T and Sternik D, 2011 *J. Therm. Anal. Calorim.* **103**, 329–337.
[78] Lakouraj M M Tajbakhsh M and Mokhtary M, 2005 *Iran. Polym. J.* **14**, 12 p. 1022–1030.
[79] Tellamekala S Ojha P Gopal N O Nagabhushana H and Ch R, 2020 *Mater. Res. Innov.* p. 1–13.
[80] Guirguis O W and Moselhey M T H, 2012 *Nat. Sci.* **04**, 01 p. 57–67.
[81] Eldin M S M Kamoun E A Sofan M A and Elbayomi S M, 2015 *Arab. J. Chem.* **8**, 3 p. 355–365.
[82] Croitoru C Pop M A Bedo T Cosnita M Roata I C and Hulka I, 2020 *Polymer (Guildf).* **12**, 560.
[83] Cha W-I Hyon S-H and Ikada Y, 1993 *Die Makromol. Chemie* **194**, 9 p. 2433–2441.
[84] Khaleghi M *et al.*, 2018 *Adv. Polym. Technol.* **37**, 8 p. 3596–3606.
[85] Martínez-Gómez F Guerrero J Matsuhiro B and Pavez J, 2018 *Int. J. Biol. Macromol.* **111** p. 935–946.
[86] Chee B S Goetten de Lima G Devine D M and Nugent M J D, 2018 *Mater. Today Commun.* **17** p. 82–93.
[87] Tuncay Tanrıverdi S Cheaburu-Yilmaz C N Carbone S and Özer Ö, 2018 *Pharm. Dev. Technol.* **23**, 8 p. 815–825.
[88] Long J *et al.*, 2019 *ACS Appl. Bio Mater.* **2**, 7 p. 2766–2779.
[89] Ullah K Sohail M Murtaza G and Khan S A, 2019 *Int. J. Biol. Macromol.* **122** p. 538–548.
[90] Freitas D F S Mattos G C and Mendes L C, 2020 *J. Therm. Anal. Calorim.*
[91] Zulfiqar M Lee S Y Mafize A A Kahar N A M A Johari K and Rabat N E, 2020 *Polymers (Basel).* **12**, 2 p. 430.
[92] Reena Kumar A Mahto V and Choubey A K, 2020 *J. Mol. Liq.* **301** p. 112472.
[93] Sirousazar M and Khodamoradi P, 2020 *Mater. Today Commun.* **22** p. 100719.
[94] Kantoğlu Ö Çaykara T and Güven O, 2013 *J. Appl. Polym. Sci.* **127**, 1 p. 500–507.
[95] Stavropoulou A Papadokostaki K G and Sanopoulou M, 2004 *J. Appl. Polym. Sci.* **93**, 3 p. 1151–1156.
[96] Li C She M She X Dai J and Kong L, 2014 *J. Appl. Polym. Sci.* **131**, 3 p. 39872.

Investigation the Vulnerability Potential of Landslide in Jayagiri Area Using the Geoelectric Method

R Rusdiana[1], Nurhasan[1], S Viridi[1], D Fitriani[2]
[1]Department of Physics, Faculty of Mathematics and Natural Sciences, Bandung Institute of Technology, Indonesia
[2]Department of Geophysics, Faculty of Mathematics and Natural Sciences, University of Padjadjaran, Indonesia
Email: randirusdiana18@gmail.com

Abstract. Landslides are the third-largest natural disaster and have resulted in many victims, both property and casualties. One area where landslides occurred with casualties and heavily damaged houses was in Jayagiri, Lembang, West Java, Indonesia.
One essential parameter in researching landslide-prone areas is the sliding area, where rocks or soil above the plane can move, resulting in landslides. Research on landslides continues to be developed from various fields of science. This study uses the Geoelectric method to determine and analyze the sliding area carried out in the Jayagiri area, Lembang. A reasonably significant resistivity value characterizes this field compared to the surrounding rocks. The measurement result on three lines shows that the study area is divided into two parts: the dominance of high resistivity and low resistivity. 2D inversion modeling shows that the slip plane has a resistivity value of 500 ohms.m at a depth between 9 m at the top of the slope and 13 meters at the bottom of the slope. The existence of a relatively shallow sliding field indicates a depth of water infiltration with a resistivity value of 10 Ohm.m, which is shallow enough so that it can be assumed that this area has the potential for landslides.

1. Introduction

Landslides ranked third in Indonesia throughout 2020. This disaster caused a lot of losses, both property and loss of life. Landslides can occur, one of which is caused by climatic conditions. One of the areas where many landslides occur is West Java Province. The climate in West Java has changed compared to the previous year. Air humidity increased from 74 percent in 2019 to 77 percent in 2020, triggering an increase in precipitation with a high level of intensity. In 2020, precipitation ranged from 2000-4000 mm/year, higher than the previous year. The highest precipitation occurs in November and December 2020. One area where landslides occurred in West Java with casualties and heavily damaged houses was in Jayagiri Village, Lembang District, West Bandung Regency, which occurred in December 2020.

In addition to climatic conditions, other parameters that can trigger natural disasters (landslides) are disturbances in soil stability, slip planes, and gravity. The triggering factors for landslides can be in the form of hydrological and geological influences [2],[3],[4]. Landslide identification is generally done through a macroscopic geological approach supported by hydrological data [5],[6],[7]. The physical and mechanical properties of rocks have a significant influence on landslides [8]. The type and composition of slope-forming soil affect soil parameters that contribute to soil shear strength and slope stability [8]. In this study, another approach was used, identifying rock structures that make up the landslide zone based on the physical properties of the rock. This study aims to obtain the distribution of resistivity value variations, the structure of the slip plane layer, the depth of the water catchment area, and the estimation of areas that are most prone to landslides. One method that can be used is geoelectric. The Geoelectric method measures the electrical properties of rocks below the surface. Electrical properties of rocks below the earth's surface have been widely used for various purposes, including for the analysis of volcanic structures [9,10,11]. The Wenner configuration can be used to map the subsurface in two dimensions. This method is non-destructive to the environment, relatively inexpensive, and can detect the subsurface layer structure of the soil and a slip plane. The slip plane in the landslide area is characterized by very contrasting variations in resistivity values between the two soil/rock layers [12,13].

2. Method

The research was conducted in Lebak Cihideung, Jayagiri Village, Lembang sub-district, Bandung, West Java, as shown in Figure 1. The research area has a slightly steep to steep slope and is included in the zone of moderate to high ground movement potential. In addition, the research area is an area with high-intensity precipitation. Land movement in this area can occur if precipitation exceeds the average, especially in bordering river valleys, cliffs, road cliffs, and slopes [14].

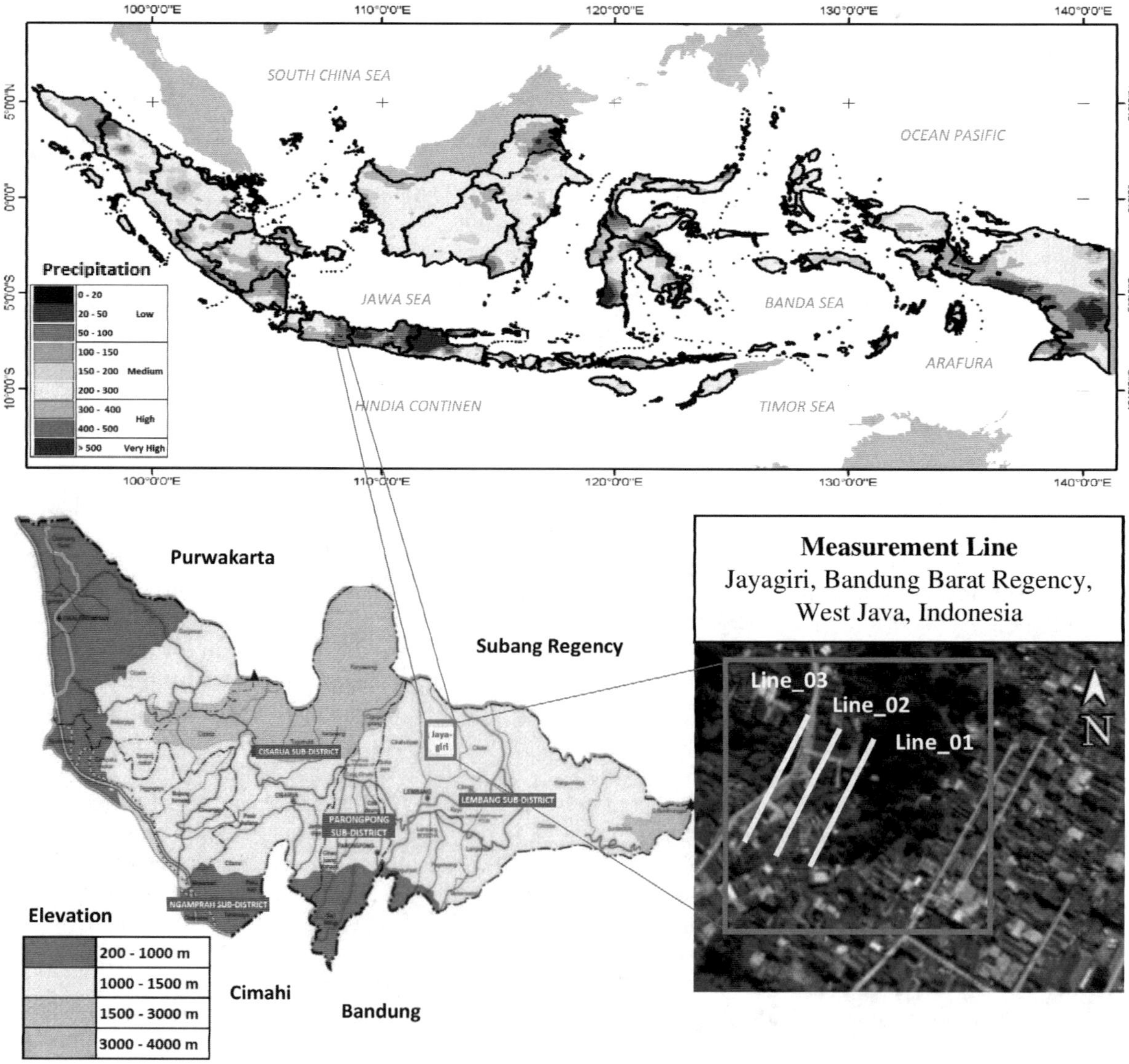

Figure 1. Research Site Map

Journal of Physics: Conference Series **2019** (2021) 012102 doi:10.1088/1742-6596/2019/1/012102

Based on the geological structure, the Lembang sub-district comprises three rock formations: Quaternary volcanic rocks, Neogene volcanic rocks, and Plio-Pleistocene volcanic rocks [15]. Quaternary volcanic rock formations are the most influential on soil movement based on the three rock formations. In contrast, Plio-Pleistocene Volcanic Rocks and Neogene Volcanic Rocks have less influence on soil movement. Quaternary volcanic rock formations are young volcanic deposits consisting of sand tuff, pumice tuff, lapilli, solid andesite fragments, and breccias. These rocks are the result of the eruption of Mount Tangkuban Parahu.

Based on the soil texture, Lembang District consists of clay soil, clay, and sand [14]. The texture of clay comes from the weathering of rocks and other chemical elements. This soil is the result of clay deposits that are above it. The location of the clay soil texture is close to the Lembang Fault. Coarse textured sand soil sourced from the eruption of Mount Tangkuban Parahu. This type of soil is close to the Tangkuban Parahu volcanic area, which is in the northwest area of the Lembang District. In contrast, the clay texture consists of volcanic ash from Mount Tangkuban Parahu, deposited so that its location is in the northern part of the Lembang District. The topographical condition of the Lembang Sub-district has the characteristics of getting to the south the lower elevation. The location of the clay texture is to the south of the sandy soil.

This research was conducted on three parallel profiles with measurements in a relatively north-south direction. The length of each profile is 80 meters with a space between the electrodes of 5 meters using the Wenner configuration as shown in figure 3. In the Wenner configuration, the distance between the current electrode and the potential electrode is constant [17]. The Wenner configuration has an excellent vertical resolution, high sensitivity to lateral changes but weak current penetration with depth, The geometric correction factor K for this configuration is 2πa. The following equation can obtain the apparent resistivity value.

$$\rho_a = 2\pi a \left(\frac{\Delta V}{I}\right)$$

where ρ_a is the apparent resistivity (ohm.m), I is the current (Amperes), a is the electrode distance (m), and V is the potential difference (volts).

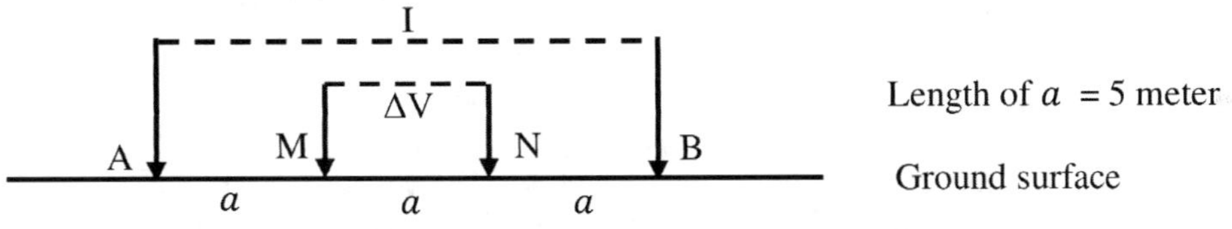

Figure 2. Wenner configuration geoelectrical measurement scheme

The measurement data obtained are variations in apparent resistivity values, electric current strength, potential difference, electrode spacing, and recording time.

3. Results

3.1 Line_01

The 2D inversion modeling was carried out using the Res2DInv software. Figure 3 below shows the cross-section of the 2D inversion in the form of resistivity distribution on Line_01. The altitude (elevation) of line_01 ranges from 1248 m to 1250 m above sea level. The percentage of 2D inversion error is 2.4% for 15 iterations.

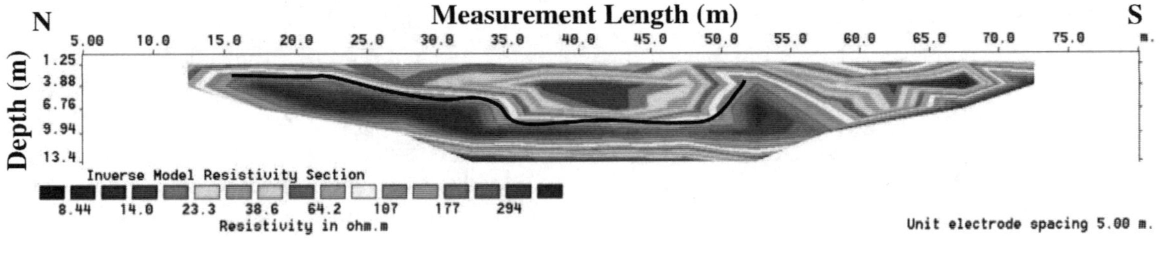

Figure 3. Cross-section of the 2D model of the resistivity distribution for Line_01

Line_01 is located at 06°, 48' 30.6732" South latitude and 107°, 36' 52.0920" East longitude until 06°, 48' 28.8046" South latitude and 107°, 36' 53.6933" east longitude. The distribution of resistivity values of the Line_01 ranges from 7.43 m-553.61 m and is dominated by areas of high resistivity. The maximum depth that can be detected is 13.4 m below the surface.

3.2 Line_02

Line_02 is located at 06°, 48' 30.3403" South latitude and 107°, 36' 53.0117" East longitude until 06°, 48' 29.6935" South latitude and 107°, 36' 54.6957" east longitude. The results of measuring the Line_02 profile are shown in figure 4. The elevation ranges from 1229 m to 1232 m above sea level. Variations in resistivity values ranged from 5.97 m-530.07 m. The maximum depth detected is 13.4 m, with an error percentage of 3.2% for 15 iterations.

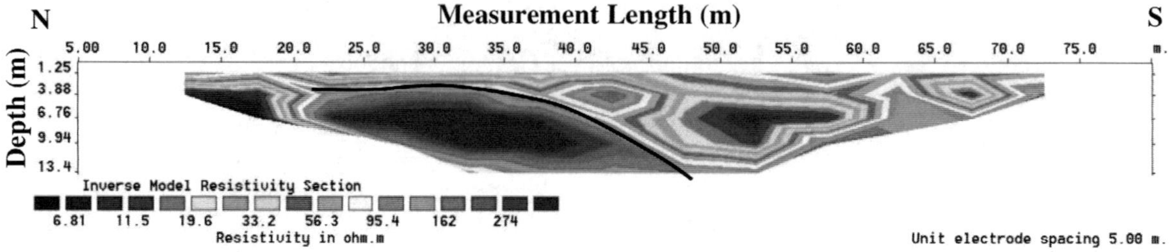

Figure 4. Cross-section of the 2D model of the resistivity distribution for Line_02

3.3 Line_03

Line_03 is located at 06°, 48' 31.2918" South latitude and 107°, 36' 53.0168" East longitude until 06°, 48' 29.7969" South latitude and 107°, 36' 55.4921" east longitude. The elevation ranges from 1226 m to 1232 m above sea level, and the variation of resistivity value of the Line_03 ranges from 8.61 m-932.31 m. The maximum depth that can be detected is 13.4 m. the percentage of error obtained is 2.3% for 15 iterations.

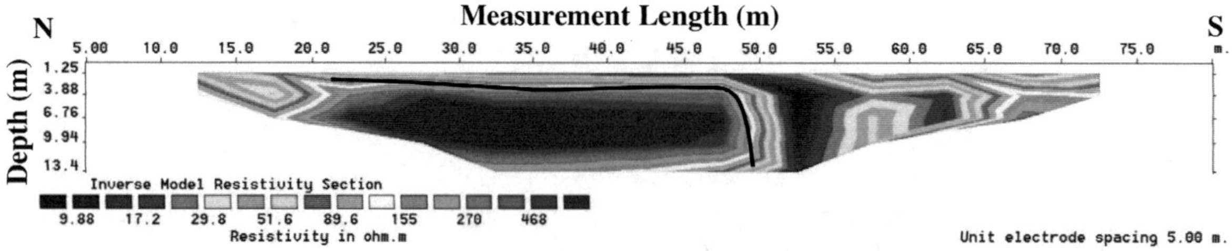

Figure 5. Cross-section of the 2D model of the resistivity distribution value for Line_03

4. Discussion

The 2D resistivity cross-sectional profile of the study area is shown in figure 6. The cross-sectional model shows a pattern that corresponds from one profile to another—the variation of resistivity values from tens of ohmmeters to the order of hundreds of ohmmeters. The high resistivity structure indicates an impermeable layer, while low resistivity indicates infiltration that fills the rock pores. The water that fills the pores of the rock will continually add to the mass of the slope that could result in a relatively high level of vulnerability to landslides, especially if the degree of slope is large.

The 2D resistivity cross-section of Line_01 in figure 6a shows the high resistivity dominant part is at a measurement distance of 10 m - 60 m with a depth of 2 m - 10 m below the surface. In contrast, the dominance of low resistivity is at a measurement distance of 25 m - 80 m below the surface. The high resistivity layer can be assumed as an area/zone where water infiltration occurs. Meanwhile, layers with low resistivity values are zones with high water saturation that contribute to slope mass. A water-saturated zone that adds to the mass of the slope is a trigger for soil movement [18].

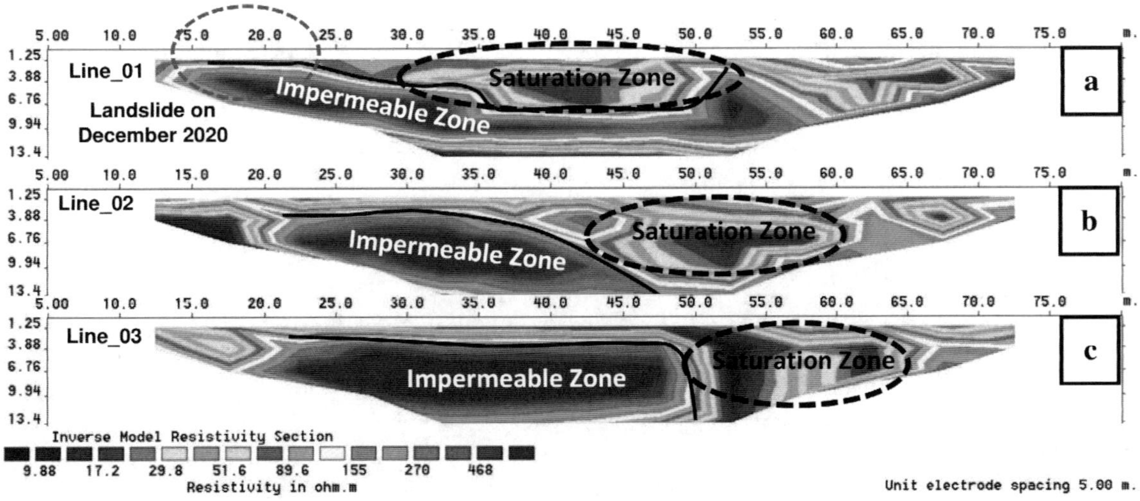

Figure 6. 2D model correlation of resistivity value variation of measurement profile (a) Line_01, (b) Line_02 and (c) Line_03

On the Line_01 profile, towards the relative south, the seepage zone is getting thicker up to a measurement distance of 50 meters. The permeable layer above the non-permeable layer at a distance of 25 m to 55 m which forms a basin, is an area of water seepage which increases slope mass. The water catchment area on the Line_01 profile ranges between the surface to ± 9 m, with the center of the measurement point 35 m – 50 m being the deepest area. At the same time, the shallowest layer is at the measurement point of ±18 m to ±25 m.

Figure 6b shows the profile of Line_02. In this profile, the high resistivity dominant part is at a measurement distance of 20 m – 45 m, and the low resistivity dominant part is at a measurement distance of 45 m – 80 m. Layers with high resistivity values (order hundreds) show a relatively complex structure to absorb water to be indicated as an avalanche field. The depth of the infiltration area ranges from the surface to ±13 m, with the deepest layer being at the measurement point of 50 – 65 m, while the shallowest layer is at the measurement point of 20 m – 45 m with a depth of only about 4 m.

In the Line_03 profile in figure 6c, the dominant part of the high resistivity is at a measurement distance of 20 m – 50 m. While the dominance of low resistivity at a measurement distance of 50 m - 80 m. The cross-sectional model of the Line_03 profile shows a more precise structure between the seepage zone in the shallow layer and the water seepage zone in the deeper layer. In this profile, the depth of the infiltration area ranges from 0 to ±13 m, with the deepest area being at the measurement point of 45 – 60 m, while the shallowest seepage area is at the measurement point of 25 m – 45 m, which is around 3.5 m.

The potential for landslides in the research area can be minimized by taking mitigation measures, including carrying out a terracing system and reforestation with plants with deep root systems and proper spacing. In addition, mitigation can be done by reducing the slopes' steepness, carrying out soil compaction, and closing fractures on the slopes to quickly prevent water from entering the soil.

5. Conclusion

Based on the resistivity distribution, the research area is divided into two parts: the high resistivity domination section and the low resistivity domination section. The distribution of resistivity values in the study area ranges from 5.97 m-932.31 m. The resistivity values in the Line_01 vary between 7.43 m-553.61 m, the Line_02 between 5.97 m-530.07 m, and the Line_03 between 8.61 m-932.31 m. The water catchment area on Line_01 reaches a depth of ± 9 m, with the center of the measurement being the deepest area. On Line_02 and Line_03, the depth of the infiltration area reaches ±13 m at each measurement point between 50 – 65 m and 45 – 60 m. The 2D inversion model shows the slip plane has a resistivity value of about 500 ohms.m is at a depth between 9 m at the top of the slope and 13 meters at the bottom of the slope. Based on these results, the measurement position of 0 to 50 m has a reasonably shallow slip plane that shows the depth of water absorption with a resistivity value of 10 Ohm.m, which is relatively shallow. Thus, it can be assumed that this area has a high potential for landslides compared to other positions. Meanwhile, the measurement area at a position of ±50 m - 80 m with the characteristics of a reasonably thick water catchment area takes longer to reach its saturation point.

Acknowledgments
LLPM ITB funded the research through the 2021 Community Service Program. The author would like to thank all those who have supported and assisted the research process, especially LPPM ITB, ITB's earth physics friends, Head of Lembang District, and Jayagiri Village Head.

References

[1] *Provinsi Jawa Barat dalam angka 2021*. Badan Pusat Statistik Jawa Barat. ISSN: 0215-2169 No. Publikasi/Publication Number: 32000.2103 Katalog /Catalog: 1102001.32
[2] Highland, L.M., dan Bobrowsky, P., The landslide handbook: *A guide to understanding landslides*: Reston, Virginia, U.S. Geological Survey Circular 1325. 2008: 129-130.
[3] Varnes, D.J., *Landslide Hazard Zonation: a review of principles and practices*. Paris: The United Nations Educational, Scientific and Cultural Organization. ISBN 92-3-10195-7,1984.
[4] Karnawati, D., *Geologi Umum dan Teknik*. Yogyakarta: Program Studi S2 Teknik Sipil UGM. 2005.
[5] Goranson, G., Larson, M., Bendz, D., dan Akesson, M. 2012 *Mass transport of contaminated soil released into surface water by landslides (Gota River, SW Sweden) (*Sweden: Hydrol Earth Syst. Sci. 16)
[6] Gundogdu, B., *Relations between pore water pressure, stability and movements in reactivated landslides*. Thesis. Civil Engineering, Graduate School of Natural and Applied Sciences, Middle East Technical University. 2011.
[7] Igwe, O., Sassa, K dan Fukuoka, H., Excess pore water pressure: a major factor for catastrophic. IAEG 2006, *Paper Number* 159. 2006.
[8] Irayani, Zaroh, dkk. 2016 Investigasi bidang gelincir tanah longsor dengan metode tahanan jenis dan pengujian sifat plastisitas tanah (Studi kasus di bukit Pawinhan, Sijeruk, Banjarmangu, Banjarnegara). Vol. 12 No. 2 (2016) Hal. 53-57. p-ISSN 1858-3075 | e-ISSN 2527-6131.
[9] Nurhasan, D. Sutarno, W. Srigutomo, Sparisoma Viridi, D. Fitriani. Integrated Geophysical Measurements for Subsurface Mapping at Papandayan Volcano, Garut, Indonesia (Preliminary Result), *American Institute of Physics Conference Proceeding*, Vol. 1454, 154;DOI: 10.1063/1.4730710. 2012.
[10] Hashimoto, T., T. Mogi, Y. Nishida, Y. Ogawa, N. Ujihara, M. Oikawa, M. Saito, Nurhasan, S. Mizuhashi, T. Wakabayashi, R. Yoshimura, A. W. Hurst, M. Utsugi, Y. Tanaka. Self-potential studies in volcanic areas (5) -Rishiri, Kusatsu-Shirane, and White Island-, *J. Fac. Sci. Hokkaido Univ*., Ser. VII, 12, 2, 97-113. 2004.
[11] Nurhasan, Y. Ogawa, N. Ujihara, S.B. Tank, Y. Honkura, S. Onizawa, T. Mori, and M. Makino. Two electrical conductors beneath Kusatsu-Shirane volcano, Japan, imaged by audio magnetotellurics and their implications for the hydrothermal system, *Earth Planets Space*, vol. 53, 1053 - 1059. 2006.
[12] Sri, C. W., Tris, A. H., Pariadi, Putri, H., Resty, F. N., Raisa, K. D., dan Ori, M. Aplikasi Metode Tahanan Jenis 2D Untuk Mengidentifikasi Potensi Daerah Rawan Longsor di Gunung Kupang, Banjarbaru, *Jurnal Fisika FLUK*, Vol. 8 No.2. hal 95-103. 2011.
[13] VSI (Vulcanology Survey of Indonesia). (2005). *Pengenalan Gerakan Tanah*. Departemen Energi dan Sumber Daya Mineral. *http://ESDM.ac.id/pengenalan_gerakan_tanah/VSI*.
[14] Widodo, Tri dkk. Aplikasi sistem informasi geografis dan penginderaan jauh untuk zonasi

kerawanan bencana gempa bumi sesar lembang, *Jurnal Dialog Penanggulangan Bencana Vol. 8.* 2017

[15] Badan geologi, Kementerian energi dan sumber daya mineral. *Tanggapan gerakan tanah di kecamatan lembang kab. Bandung Barat, Jawa Barat (12-2-2015).* 2015. https://vsi.esdm.go.id/index.php/gerakan-tanah/kejadian-gerakan-tanah/708-tanggapan-gerakan tanah-di-kec-lembang-kab-bandung-barat-jawa-barat-12-2-2015

[16] Prasetyaningtyas, kukuh. *Analisis curah hujan dan sifat hujan Desember 2020*. Badan Meteorologi dan Klimatologi dan Geofisika https://www.bmkg.go.id/iklim/informasi-hujan-bulanan.bmkg?p=analisis-curah-hujan-dan-sifat-hujan-desember-2020&lang=ID

[17] Telford, W.M., Geldart, L.P. and Sheriff, R.E. 1990 *Applied Geophysics: Second Editon* (USA: Cambridge University Press)

[18] Ludman, A., and Coch, N.K., 1982, *Physical Geology*, Mc Graw-Hill

Journal of Physics: Conference Series **2019** (2021) 012103 doi:10.1088/1742-6596/2019/1/012103

Analysis of the Structure of Ceramic Materials of Clay, Mount Kelud Volcanic Ash and Sea Water

N J F Nasution[1], K Sembiring, Z Sitorus
Department of Physics, Universitas Sumatera Utara
Jl. Dr. Mansyur , Medan 20155, Indonesia
Email : njeihan@gmail.com

Abstract. Based on the durability of ancient roman buildings that are 2000 years old and are always exposed to sea waves. In which there is a mixed reaction of volcanic rock with seawater. Indonesia has many active volcanoes, this is because Indonesia is located at the confluence of the Eurasian and Indo-Australian tectonic plates. One of the mountains formed in the area where the plates meet is still active, the mountain is Mount Kelud. Mount Kelud is a volcano in the province of East Java, Indonesia, which is classified as active. The existence of volcanic ash resulting from the eruption of Mount Kelud is quite potential as a ceramic material. So the purpose of this research is to make a ceramic construction made from clay, volcanic ash from Mount Kelud and sea water by the method of die pressing and sintering at a temperature of 1000°C. So it can be known how the physical properties, mechanics and especially the crystal structure formed from ceramic samples. The results of the research that has been carried out are that the highest density value is in the composition of a mixture of 50% volcanic ash and 50% seawater, the highest compressive strength value is also found in the composition of a mixture of 50% volcanic ash and 50% seawater, and from the XRD results, the structure The crystal formed is hexagonal with the phase formed is SiO2.

1. Introduction

Ceramics are basically very close to everyday life. Generally, in the manufacture of ceramics, it can be done by varying the ceramic raw materials and the shape of the granules added with water or without water, pressing and sintering. One of the building materials that can be used as an absorption wall is ceramic. Ceramic comes from the Greek "keramos" which means something that is burned from clay minerals that are dried in the sun and hardened by burning at high temperatures. The material for making ceramics in the form of clay is one of the materials whose uses are very beneficial for humans because the materials are easy to obtain and the finished products made of clay are very wide. So that in this study a building material was made in the form of construction ceramics. The mixed ingredients in the manufacture of this construction ceramic are Kelud mountain dust and sea water. In this study, a ceramic material was made as a raw material for clay soil by mixing Kelud mountain ash and sea water with conventional compression molding techniques. So that the mixture is expected to produce a ceramic material that has good physical and mechanical properties, characterization tests for construction ceramics include: physical properties (density, porosity and shrinkage), mechanical properties (compressive strength and hardness), surface morphology using SEM- EDX (Scanning Elecron Microscopy-Energy Dispersive X-Ray Spectrometer) and XRD (X-Ray Diffraction). The final results of this research are expected to be applied as construction ceramics or porous ceramics that are beneficial to the community and have economic value.

Journal of Physics: Conference Series **2019** (2021) 012103 doi:10.1088/1742-6596/2019/1/012103

2. Method

First, the materials were collected (clay, Kelud mountain ash and sea water), then the clay was dried under the hot sun to remove its moisture content for 3 days. Then, the clay was ground or crushed using a mortar and mortar to make it smoother and then filtered the clay and Kelud mountain ash with a size of 200 mesh. Then the material is weighed with a total weight of 20 grams and then homogenized. After that it is printed on a cylinder size with a diameter of 5 cm and a thickness of 1 cm, the Die Pressing method. Then the sample was dried for 3 days to remove the moisture content and then burned at a sintering temperature of 10000C with a holding time of 3 hours. After that the samples were cooled and tested on ceramic samples.

3. Result

3.1 Volume Decrease

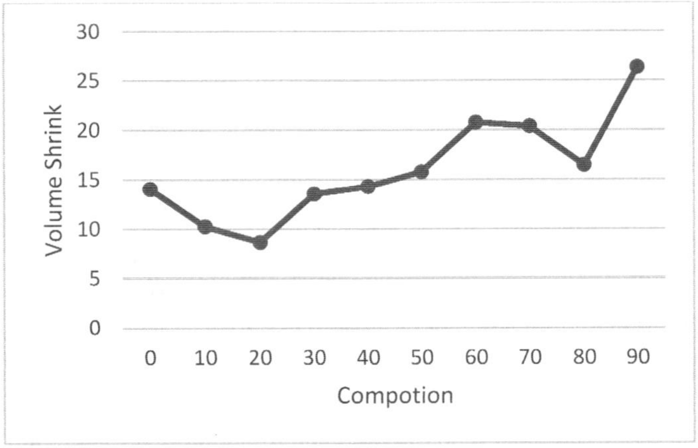

Figure 1. *Volume Shirnkage vs Compotion Chart*

The volume loss graph shows that the burn loss value for each variation of the construction ceramic mixture has increased and decreased. The higher the sintering temperature, the ceramic will undergo a high compaction process so that it will produce a relatively high volume loss value. The lowest burn loss value was owned by sample L20 of 8,64%, while the highest value of burn loss was owned by sample L90 of 26,30%.

3.2 Porosity

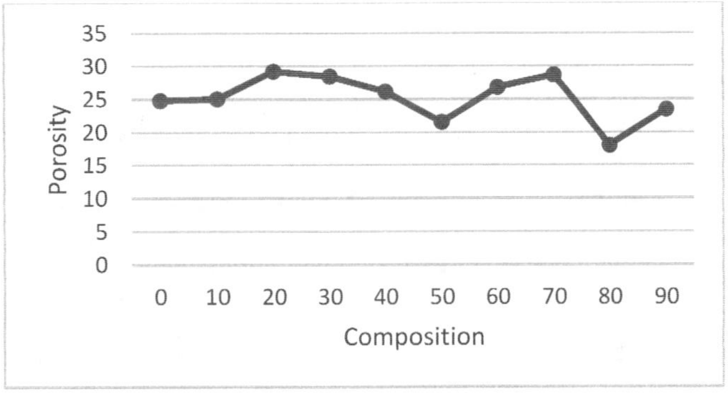

Figure 2. Porosity vs Composition Chart

Low porosity indicates that the material has a solid alloy because the pores in the material are relatively lower, the L20 sample has the most optimum level of porosity.

3.3 Density

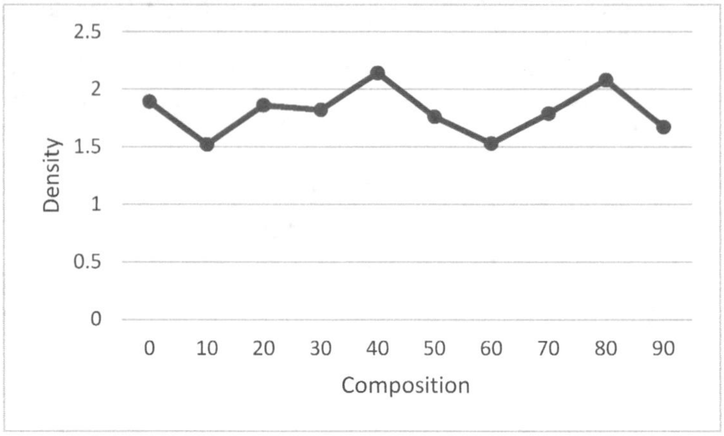

Figure 3. Density vs Composition Graph

Density test data on samples using seawater solvents show that the L50 sample has the highest density. an increase in density indicates that the mass density of the sample is getting higher in each volume, or in other words the L50 sample has the strongest bond between particles because the mass density reaches its optimum point.

3.4 Hardness

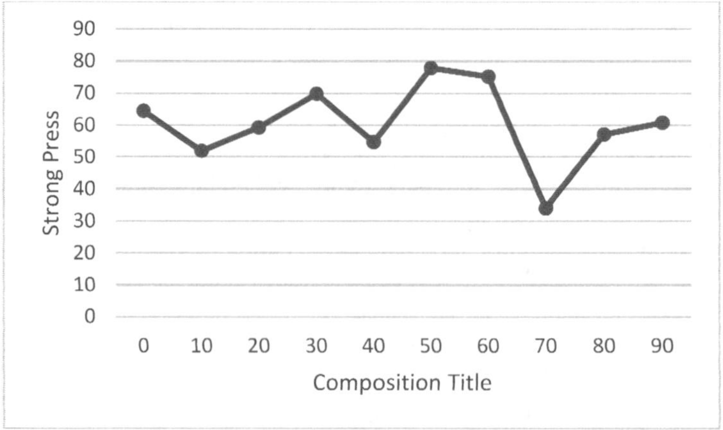

Figure 4. Compressive Strength Graphics vs Composition

From the test data on the compressive strength of the sample, the lowest compressive strength was found in the L70 sample with a value of 33.95 MPa, while the highest compressive strength was found in the L50 sample with a value of 77.83 MPa.

3.5 XRD

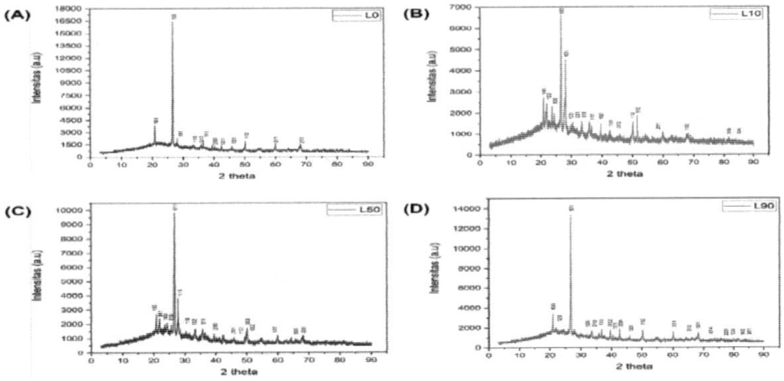

Figure 5.Displays the XRD diffraction pattern

In the picture above, which displays the XRD diffraction pattern in the sample using seawater, it shows the same crystalline phase, namely SiO2, according to the peaks formed. However, there are some differences between the diffraction patterns formed, where the L10 sample produces several new peaks when compared to the L0 sample. Based on the data sheet obtained, the new peaks that appear still show the SiO2 phase. As the concentration of Mount Kelud ash decreases to 50% at L50 to 90% in the L90 sample, the peaks formed have decreased in intensity, this indicates that a reduction in the ash concentration of Mount Kelud by more than 10% will reduce the level of crystallinity in the sample.

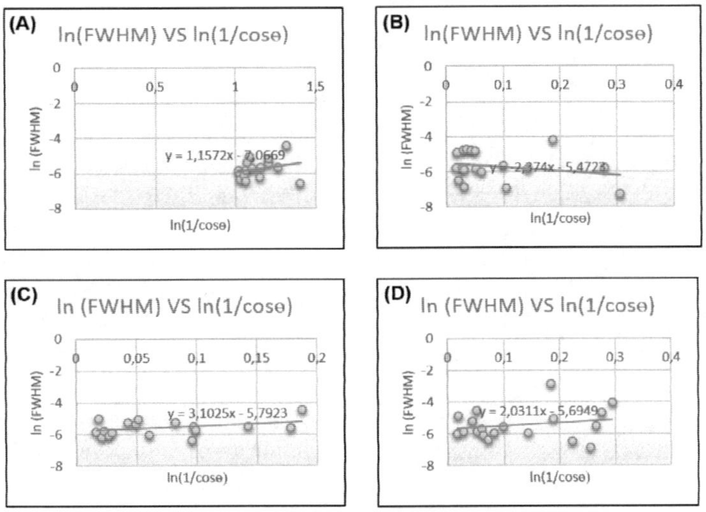

Figure 6. The crystal size

As well as the crystal size can be determined using the Debye Scherrer equation. The largest crystal size was found in the samepl L0 which was 162.57 nm, while the L10 sample had a significant decrease in crystal size to 33 nm. In sample L50, the crystal size increased again with the decrease in the concentration of Mount Kelud ash and along with the increase in the concentration of seawater. The crystal size again decreased in sample L90 with a relatively low decrease.

Table 1. Crystal size in ceramic samples using seawater

Sampel	Crystal size (nm)
L0	162,57
L10	33
L50	45,44
L90	41,23

4. Conclusion

The size of the crystal atoms in clay-based ceramics has been investigated with variations in the composition of Mount Kelud ash with seawater or aquadest of L0: 162.57 nm; L10: 33 nm ; L50:45,44 nm and L90:41,23 nm.

The composition of Mount Kelud ash with seawater or aquadest in the sample can affect the physical and mechanical properties of ceramics. The optimum physical properties are indicated by the lowest porosity value, namely 18,03% in sample L80, while the optimum density is obtained at 2.14 x 103 Kg/m3 which is the highest density possessed by sample L50. The mechanical properties of the L50 sample also showed an optimum value of 77,83 Mpa.

The crystal structure and morphology of the sample have a significant influence on the mechanical strength of the ceramic. The crystal structure with hexagonal shape has relatively more optimal mechanical strength when compared to the monoclinic structure, this is because the crystal bonds in the hexagonal structure are stronger so that it has an impact on increasing the morphological structure formed.

References

[1] Amri, Sjafei. 2005. Teknologi Beton A-Z. Jakarta: Yayasan john Hi-Tech Idetama
[2] Apriyanti, Eny. 2017. Pengaruh Karakterisasi Pada Pembuatan Membran Keramik
[3] Komposit Abu Vulkanik Untuk Pengolahan Air Bersih. *Prosiding Seminar Nasional Kimia UNY*, Yogyakarta : 17 Oktober 2017. Hal 305 – 313. Ariswan. 2016. *Hand Out Struktur Kristal.* Yogyakarta: Universitas Negeri Yogyakarta
[4] Astuti, Ambar. 1997. "Pengetahuan Keramik". Cetakan pertama, Universitas Gajah Mada : Bogor Das Braja, M. 1988. Mekanika Tanah (Prinsip-Prinsip Rekayasa), Erlangga : Jakarta
[5] Baiq, Asma. 2014. Aktivasi Tanah Liat. Universitas Negeri Surabaya : Surabaya. Beiser, Arthur . 1992. *Konsep Fisika Moderen Edisi ke Empat.* (alih bahasa: DR.The Houw Liong). Jakarta : Penerbit Erlangga
[6] Djakaria, M Nur. 2008. Air Laut. Universitas Pendidikan Indonesia : Bandung. Ginting, Eva Marlina dan Nurdin Bukit. 2014. Karakterisasi Material. Medan. Unimed Press.
[7] Jiwani, Atmita. 2007. *Studi Pengaruh Alur Pemanasan Pada Preparasi Bahan Semikonduktor CuInSe2 Massif dengan Teknik Bridgman.* Skripsi. Yogyakarta: Universitas Negeri Yogyakarta
[8] Karleskint,G. 2009. *Introduction to Marine Biology*. Boston: Cengage Learning. Page : 47
[9] Khairunnisa, Suri. 2018. Karakterisasi Dan Pembuatan Keramik Berpori Berbahan Dasar Tanah Lempung Dan Abu Kayu Damar. [SKRIPSI]. Medan : Universitas Sumatera Utara.
[10] Laili, Carol, M., dan Parsons, T. 1993. *Biological Oceanography Pergamon. University of British* Coloumbia Vencover: Canada.
[11] Leni, Ariani (2010). Peristiwa di laut. Jakarta: CV Graha Ilmu Mulia. hlm. 9. ISBN 9789797560140.

[12] Pond, S. 1978. Introductory Dynamic Oceanography. Pergamon Press. Page 5. ISBN 0750624965.

[13] Rizki, Faisal & Nadia. 2015. Studi Kuat Tekan Beton Normal Dengan Abu Gunung Kelud Sebagai Bahan Aditif Pengganti Semen. Jurnal Konstruksia, 6 (2), 41 – 47

[14] Sembiring, A. D. 2010. Pemanfaatan Limbah Padat Pulp Untuk Bahan Baku Pembuatan Keramik Berpori Yang Diaplikasikan Sebagai Filter Gas Buang Kendaraan Bermotor Dengan Bahan Bakar Premium. Disertasi S-3. Medan: Universitas Sumatera Utara.

[15] Smallman, R. E & Bishop, R. J. 2000. *Metalurgi Fisika Modern & Rekayasa Material Edisi Keenam.* (alih bahasa: Ir. Sriati Djaprie, M. Met). Jakarta: Erlangga

[16] Van Vlack . 2004. *Elemen-elemen & Ilmu Rekayasa Material*. Jakarta: Erlangga Wiyatmo, Yusman. 2003. *Fisika Modern.* Yogyakarta :Pustaka Pelaja Muis

Permeability and Permittivity Measurements of Teflon in Millimeter Wave

E Handoko[1*], M A Marpaung[1], R Fahdiran[1], Z Jalil[2], and M Alaydrus[3]
[1]Dept. of Physics, Universitas Negeri Jakarta
Jalan Rawamangun Muka 13220, Jakarta, Indonesia
[2]Dept. of Physics, Syiah Kuala University
Banda Aceh 23111, Indonesia
[3]Dept. of Electrical Engineering, Universitas Mercu Buana
Jalan Meruya Selatan No.1, Jakarta, Indonesia
Email: erfan@unj.ac.id

Abstract. In order to expand a new millimeter wave absorbing material over the V-band frequency range, teflon were prepared with thickness of 1.4 mm, 1.5 mm, and 1.6 mm. The variation of electromagnetic variables (complex permeability $\mu=\mu'-j\mu''$ and complex permittivity $\varepsilon=\varepsilon'-j\varepsilon''$) properties were studied over the V-band frequency range with varying sample thickness. Both dielectric loss (ε'') and magnetic loss (μ'') were leading to strong fluctuations, whereas the dielectric constant (ε') was found to decrease and magnetic permeability (μ') increased in the frequency range of 50–67 GHz.

1. Introduction

Recently, millimeter wave (mmWave) frequencies are being used for many applications in the modern world, including telecommunications, remote sensing, defense and military applications [1–5]. The mmWaves are electromagnetic (EM) waves defined within the frequency range of 30–300 GHz. According to the formula which is given by the wavelength $\lambda= c/f$, where f is the frequency (in hertz) and $c = 3 \times 10^8$ *m/s* is the speed of light, the mmWave band corresponds to a wavelength range of 1 mm at 300 GHz increasing to 10 mm at 30 GHz [1,3,5]. Thus, effective mmWave absorbers are a serious matter in the world for many applications in the gigahertz frequency. It is well-known that the EM wave absorption abilities of absorbers are crucially determined by the complex permeability and permittivity [6,7]. There are numerous research results about mmWave absorbers with its application.

For example, barium hexaferrite $BaFe_{12-x}Al_xO_{19}$ ceramics [8], polymers [9], $BaTiO_3/Co_3O_4$ composite powders [1], have been successfully exhibited mmWave absorption properties. This work is to investigate the permeability and permittivity of teflon in the mmWave band 50–67 GHz. We evaluated the reflection S_{11} and transmission S_{21} values with a thickness d.

2. Experimental Method

Teflon was obtained from the Indonesia materials market and cutted with dimensions 3.76 mm×1.88 mm×1.6 mm (±0.05) for the mmWave measurements. Teflon was measured with a Rohde Schwarz ZVA 67 vector network analyzer (VNA), which was used to find the scattering parameters (S_{11} and S_{21}) with the WR15 sample holder in the frequency of 50–67 GHz. Base on Nicolson-Ross-Weir (NRW) method, the complex permeability (μ) and permittivity (ε) of teflon were calculated by using S-parameters (S_{11} and S_{21} values) [10–12]. The VNA measurement mechanisms of teflon is illustrated in Fig. 1.

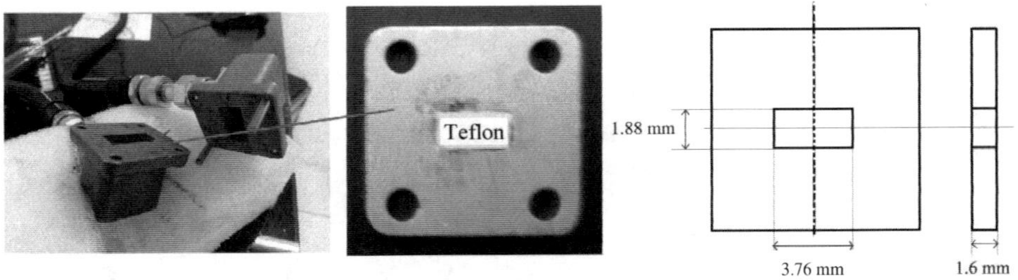

Figure 1. Schematic illustration of the measurement of S_{11} and S_{21} values of teflon using the WR15 sample holder.

3. Results and discussion

The complex reflection (S11) and transmission (S21) results were measured with VNA in the frequency range of 50–67 GHz as shown in Fig. 2. The S_{11} and S_{21} data were used to find the complex permeability and the permittivity of teflon with thickness of 1.4 mm, 1.5 mm, and 1.6 mm through the NRW method, which is then formulated using the following steps.

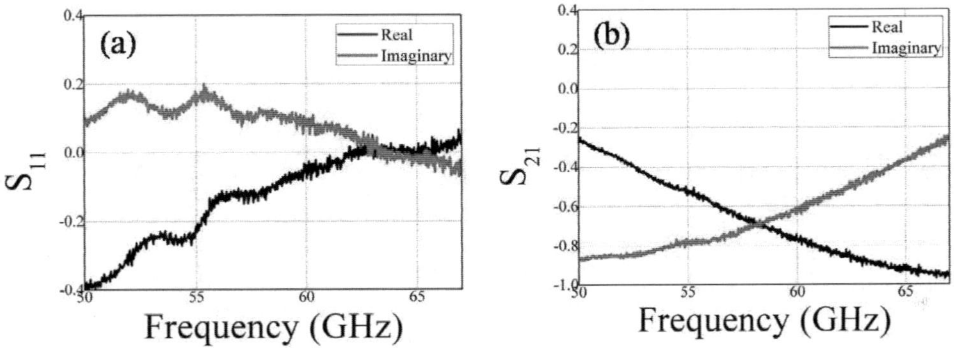

Figure 2. Scattering parameters (a) S_{11} and (b) S_{21} of teflon

$$K = \frac{(S11)^2-(S21)^2+1}{2(S11)} \tag{1}$$

$$\Gamma = K \pm\sqrt{K^2-1}\,, \quad |\Gamma| \leq 1 \tag{2}$$

$$T = \frac{(S11)+(S21)+G}{1-(S11+S21)G} \tag{3}$$

$$\frac{1}{L^2} = \left(-\left[\frac{1}{2\pi d} ln\left(\frac{1}{T}\right)\right]^2\right) \tag{4}$$

finally complex permeability and permittivity of teflon can be formulated by

$$\mu = \frac{1+G}{L(1-G)\sqrt{\frac{1}{l_o^2}-\frac{1}{l_c^2}}} \tag{5}$$

$$\varepsilon = \frac{l_o^2}{\mu}\left(\frac{1}{l_c^2} - \left[\frac{1}{2\pi d} ln\left(\frac{1}{T}\right)\right]^2\right) \tag{6}$$

where Γ, T, λ_o and λ_c are reflection coefficient, transmission coefficient, free space and the cut off wavelength, respectively.

By using the equation (5) and (6), The complex permeability (μ) and permittivity (ε) values can be calculated as shown in Fig. 3.

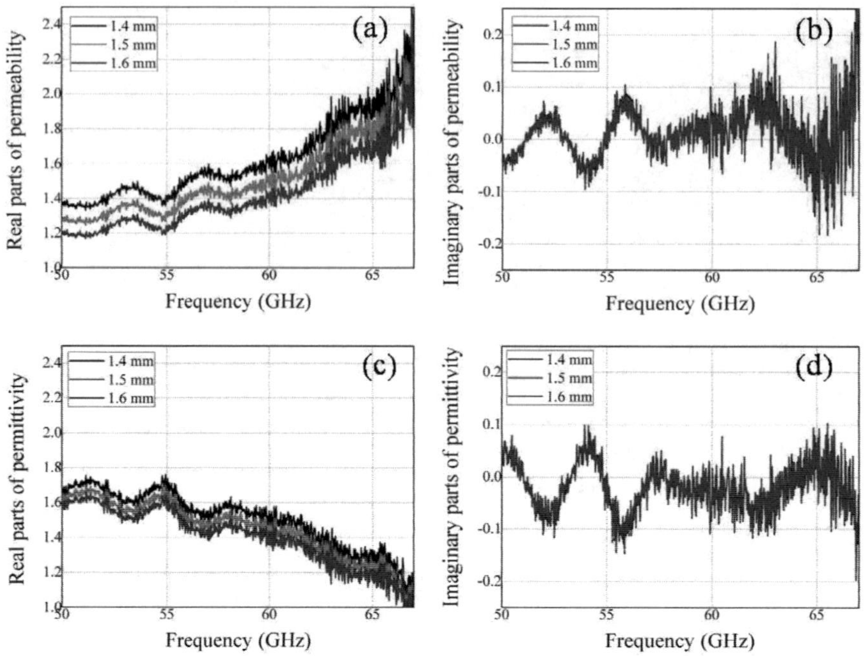

Figure 3. The complex permeability (a and b) and permittivity (c and d) values of teflon with thickness of 1.4 mm, 1.5 mm, and 1.6 mm in the mmWave band 50–67 GHz.

The complex permeability ($\mu = \mu' - j\mu''$) and permittivity ($\varepsilon = \varepsilon' - j\varepsilon''$) that were calculated by using the NRW method with the S-parameters (S_{11} and S_{21}), which were simultaneously characterized by VNA. It is the well known that The real (μ' and ε') and imaginary (μ'' and ε'') parts reflect energy storage and dissipation, respectively [13–16]. The values of the μ' presents the trend of increases with increasing frequency and decreasing thickness. The increase in the μ' leads to decrease in the ε' that the dominant contribution of magnetic energy storage come from this sample in the mmWave band 50–67 GHz. Meanwhile, both the ε'' and μ'' were leading to strong fluctuations.

4 Conclusions

In summary, teflon with 1.4 mm, 1.5 mm, and 1.6 mm of thickness have been characterized by VNA and calculated by the NRW method to find the complex permeability and permittivity in the mmWave band 50–67 GHz. The increase in the μ' with decreasing thickness leads to decrease in the ε' with increasing thickness. Both the ε'' and μ'' show to strong fluctuations.

Acknowledgment

This work was financially supported by the Penelitian Terapan Unggulan Perguruan Tinggi (PTUPT) 2021 (grant no. 281/E4.1/AK.04.PT/2021) of Kementerian Pendidikan, Kebudayaan, Riset, dan Teknologi, Republik Indonesia.

References

[1] W. Tian, R. Ma, J. Gu, Z. Wang, N. Ma, and P. Du. 2018, J. Mater. Chem. C. 6. pp. 12965–12975.
[2] C. Choi, K. Ko. 2010, Journal of Navigation and Port Research International Edition Vol. 34, pp. 111-115.
[3] F. C. Commission, N. Technology, and D. Division, "FEDERAL COMMUNICATIONS COMMISSION OFFICE OF ENGINEERING AND Millimeter Wave Propagation : Bulletin Number

70 July, 1997.
[4] J. S. Mccloy, K. Korolev, J. V Crum, and M. N. Afsar. 2013, IEEE TRANSACTIONS ON MAGNETICS, VOL. 49, NO. 1, JANUARY 2013, pp. 546–551.
[5] maxim zhadobov, nacer chahat, ronan sauleau, catherine quement and yves drean. 2011 International Journal of Microwave and Wireless Technologies, 3(2), 237–247, doi:10.1017/S1759078711000122.
[6] B. Zhao, B. Fan, G. Shao, W. Zhao, and R. Zhang. 2015, ACS Appl. Mater. Interfaces, 7, 18815−18823. DOI: 10.1021/acsami.5b05482.
[7] B. Zhao, B. Fan, Y. Xu, G. Shao, X. Wang, W. Zhao, and R. Zhang. 2015 "ACS Appl. Mater. Interfaces, 7, 26217−26225. DOI: 10.1021/acsami.5b08383.
[8] M.G. Vakhitov, D.S. Klygach, D.A. Vinnik, V.E. Zhivulin, N.S. Knyazev. 2019, *J. Alloys Compd.*, 816, 152682. doi.org/10.1016/j.jallcom.2019.152682
[9] J. Krupka, 2016, IEEE Microwave and Wireless Components Letters, vol. 26, no. 6, pp. 464-466, June, doi: 10.1109/LMWC.2016.2562640.
[10] E. Handoko, I. Sugihartono, and M. A. Marpaung, Maulana Randa, Mudrik Alaydrus and Nofrijon Sofyan. 2018, Materials Science Forum, 929, 109–115.
[11] E. Handoko, A. M. Mangasi, S. Iwan, M. Randa and M. Alaydrus. 2016, *International Conference on Radar, Antenna, Microwave, Electronics, and Telecommunications (ICRAMET)*, Jakarta, 2016, pp. 28-30, doi: 10.1109/ICRAMET.2016.7849576.
[12] E. Handoko *et al* 2018 *J. Phys.: Conf. Ser.* 1080 012002.
[13] X. Qiu and L. Wang, 2017, *Nanoscale*, 9, 7408-7418.
[14] Y.C. Qing ·W.C. Zhou · S. Jia · F. Luo · D.M. Zhu. 2010, Appl Phys A (2010) 100: 1177–1181 DOI 10.1007/s00339-010-5738-5.
[15] Handoko, Erfan; Budi, Setia; Sugihartono, Iwan; Marpaung, Mangasi Alion; Jalil, Zulkarnain; Taufiq, Ahmad; Alaydrus, Mudrik. 2020, Materials Express, 10, pp. 1328-1336(9). DOI: https://doi.org/10.1166/mex.2020.1811.
[16] Chunhong Gong, Jiwei Zhang, Chao Yan, Xiaoqiang Cheng, Jingwei Zhang, Laigui Yu, Zhensheng Jin and Zhijun Zhang. 2012, J. Mater. Chem., 22, 3370.

Journal of Physics: Conference Series **2019** (2021) 012105 doi:10.1088/1742-6596/2019/1/012105

Millimeter Wave Absorption Properties of Teflon at Frequency Range from 50 GHz to 67 GHz

E Handoko[1*], M A Marpaung[1], R Fahdiran[1], Z Jalil[2] and M Alaydrus[3]
[1]Dept. of Physics, Universitas Negeri Jakarta
Jalan RawamangunMuka 13220, Jakarta, Indonesia
[2]Dept. of Physics, Syiah Kuala University
Banda Aceh 23111, Indonesia
[3]Dept. of Electrical Engineering, Universitas Mercu Buana
Jalan Meruya Selatan No.1, Jakarta, Indonesia
Email: erfan@unj.ac.id

Abstract. In this paper, teflon material with dimensions 3.76 mm×1.88 mm×1.6 mm was prepared from the Indonesia materials market for the millimeter wave characteristics. The permeability, permittivity, and millimeter wave absorption properties were studied using a vector network analyzer in the frequency range of 50−67 GHz. The results indicated that teflon material have a good reflection loss and broadband absorption and a appropriate absorber in the millimeter wave frequency range.

1. Introduction

Millimeter radio waves, which are also called the millimeter band, are electromagnetic waves with wavelength between 1−10 millimeters and frequencies of 30−300 GHz [1,2]. The millimeter wave (milliwave) technology is applied mainly in Radio Detection And Ranging (Radar) systems to determine the range, location, or velocity of objects [3,4]. It can be used to detect aircraft, ships, cars and missiles. It is the well known that radar system consists of a transmitter and a receiver with the same antenna arranged in such a way as to quantify the electromagnetic resonance produced by a distant target [5]. Then, various signal processing techniques are employed to retrieve the desired information about the target. Fortunately, millimeter wave propagation has its own rarities including effects of the various physical factors on propagation and free space propagation [6,7]. One of the most significant issues in millimeter wave technology is that absorption by the solid materials throughout the band increases with frequency. Designing for millimeter wave spectrum use must take into the propagation characteristics of this frequency range. While lower frequency signals can penetrate more easily through buildings and propagate for many miles, millimeter wave signals do not penetrate solid materials very well and can travel only a few miles or less. [8–10].

Some research results on solid materials as millimeter wave absorbers have been published in many papers. Barium hexaferrite $BaFe_{12-x}Al_xO_{19}$ ceramics obtained the optimum reflection losses (RL) of –30 dB at frequencies above 39–42 GHz [11], polymers produced an optimal dielectric loss tangents at frequency of 30–50 GHz [2], and $BaTiO_3/Co_3O_4$ composite powders reached the RL of –40 to –50 dB with sample thickness of 1.8 mm and the bandwidth of 5 GHz [1]. Therefore, this work is to characterize teflon material with using vector network analyzer (VNA) in the frequency of 50–67 GHz. The millimeter wave absorption properties of teflon material are studied in detail.

2. Method

The cutted teflon material with dimensions 3.76 mm×1.88 mm×1.6 mm was obtained from the Indonesia materials market. For the millimeter wave measurements, teflon material was measured with a Rohde Schwarz ZVA 67 vector network analyzer (VNA) with the WR15 sample holder in the frequency of 50–67 GHz. The scattering parameters (S_{11} and S_{21}) data were recorded by VNA. The complex permeability (μ) and permittivity (ε) values were calculated by Nicolson-Ross-Weir (NRW) method using scattering parameters [12–14].

Journal of Physics: Conference Series **2019** (2021) 012105 doi:10.1088/1742-6596/2019/1/012105

The reflection loss (RL) of the prepared absorbers versus the frequency is studied by calculating with various thicknesses of 1 to 5 mm, through transmission line theory using the equations $\mathrm{RL\ (dB)} = 20\log\left[\frac{(Z_{in}-Z_o)}{(Z_{in}+Z_o)}\right]$ and $Z_{in} = Z_o\sqrt{\frac{\mu}{\varepsilon}}\tanh\left[\left(\frac{-j2\pi f d}{c}\right)\sqrt{\mu\varepsilon}\right]$, where Z_{in}, Z_o, μ, ε f, c, and d are the input impedance, impedance of free space, complex permeability, complex permittivity, millimeter wave frequency, velocity of electromagnetic waves in free space, and thickness, respectively [15]. The VNA measurement mechanisms is illustrated in Fig. 1.

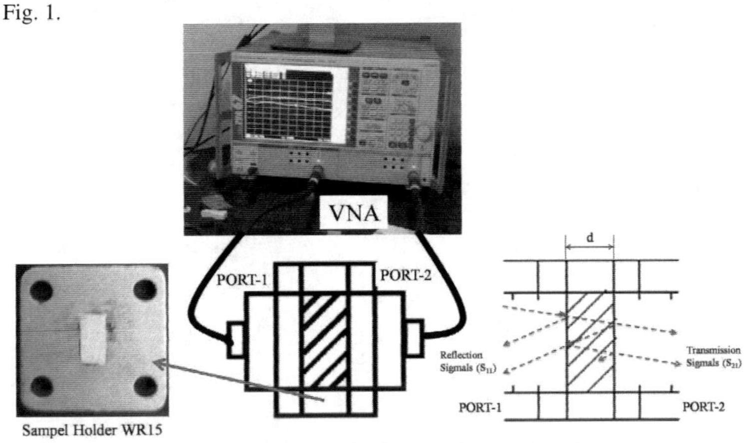

Figure 1. Measuring equipment, sample holder, and schematic illustration of the measurement of S_{11} and S_{21}.

3. Results and discussion

The permeability ($\mu = \mu' - j\mu''$) and permittivity ($\varepsilon = \varepsilon' - j\varepsilon''$) in the range from 50 GHz to 67 GHz are shown in Fig. 1. The real part of permeability (μ') slightly increased over the whole frequency range (Fig. 1a). The real part of permittivity (ε') remains decreased in the frequency range of 55 – 67 GHz (Fig. 2b). While for the imaginary part of permeability (μ'') and permittivity (ε'') tend to fluctuations in the frequency range of 50 – 57.5 GHz. Furthermore the range values for the μ' and ε' of the sample is observed between 1.0 and 2.5. The measured μ'' and ε'' were small and approximately constant in the 50–67 GHz range.

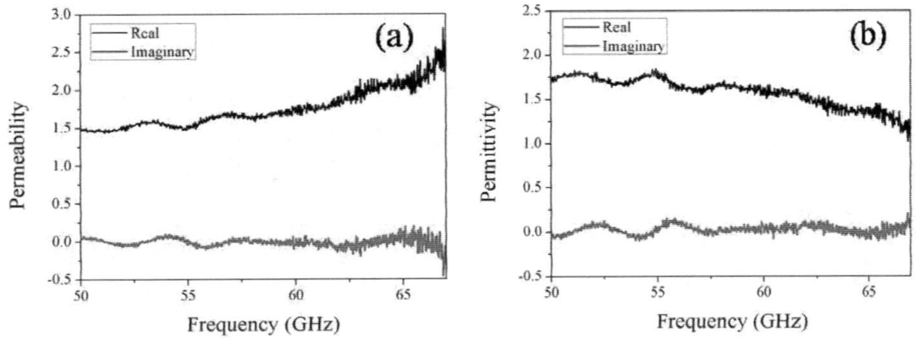

Figure 2. The complex permeability and permittivity as a function of frequency for the teflon material.

In order to find the millimeter wave absorption properties, based on a transmission line theory, the RL values of the teflon material backed by a conductor plate under the normal incidence electromagnetic field were calculated using some variables of the Z_{in}, Z_o, μ, ε f, c, and d [16].

Fig. 3 shows the calculated RL results of the teflon material with various thicknesses. It can be observed in Fig. 3(a-d) that the resulted RL enhance with increasing the sample thickness up to 5 mm. The RL values were obtained less than −10 dB (90% absorption) in the frequency range of 52.19−52.55 GHz, 55.14−55.98

GHz and 52.17−53.44 GHz, 54.55−54.71 GHz for 4 mm and 5 mm thickness, respectively. A minimum RL of −12.73 dB was observed at 52.43 GHz with an absorber thickness of 5 mm. The results also obtained a maximum broadband with RL values less than −5 dB (67% absorption) in the whole frequency range with a thickness of 5 mm (see Fig. 3d and 3f).

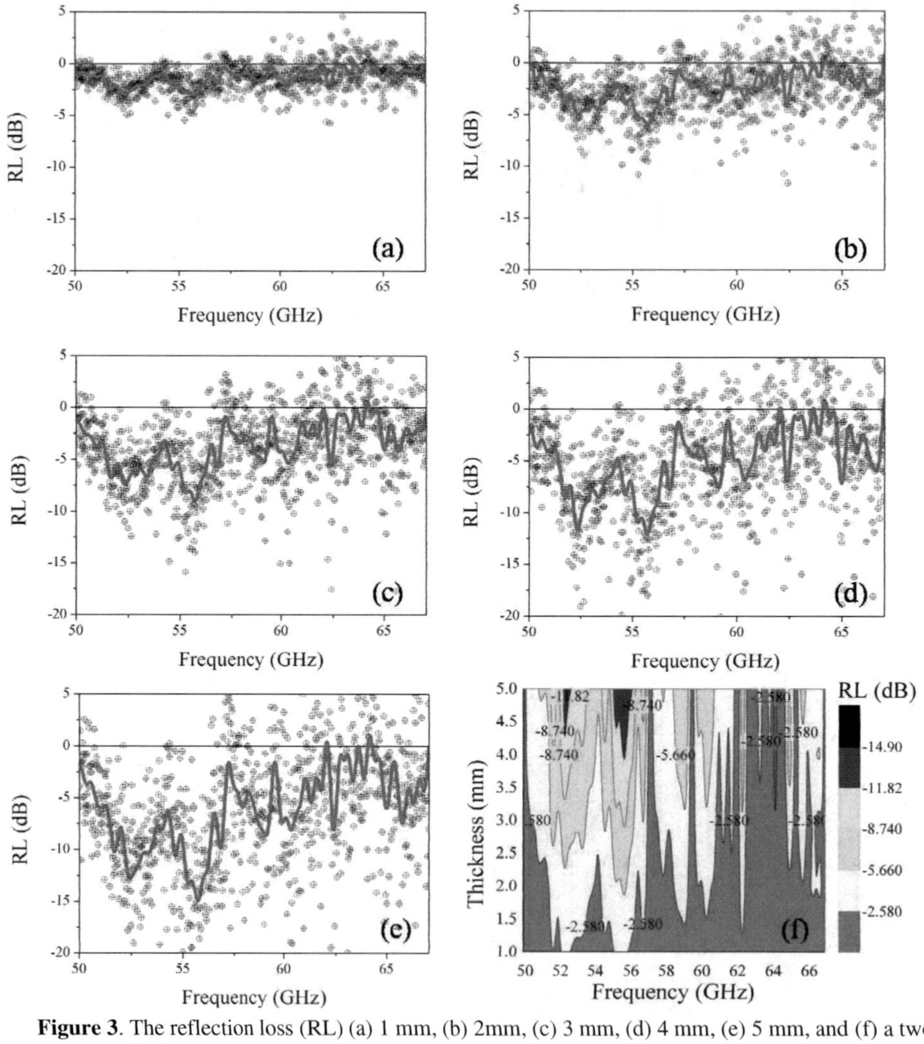

Figure 3. The reflection loss (RL) (a) 1 mm, (b) 2mm, (c) 3 mm, (d) 4 mm, (e) 5 mm, and (f) a two-dimensional (2D) RL image map of the teflon material with different thicknesses in the frequency range of 50–67 GHz.

4. Conclusions

The teflon material was successfully prepared for measuring millimeter band in the frequency of 50–67 GHz. The millimeter wave absorption properties had higher magnetic energy storage (μ') than dielectric dissipation (ε') in millimeter wave frequency range. The resulted RL values showed a good broadband absorption (>90% absorption, RL < −10dB) in the frequency range of 52.17−53.44 GHz and a minimum RL of −12.73 at 52.43 GHz with an absorber thickness of 5 mm. The results particularly indicated a optimum broadband with RL values less than −5 dB (67% absorption) in the whole frequency range with a thickness of 5 mm.

Acknowledgment
The authors wish to thank Prof. Mudrik Alaydrus (Universitas Mercu Buana) for VNA measurements of the scattering parameters (S_{11} and S_{21}) for sample. This work was financially supported by the BLU Program 2021 (grant no. 335/UN39/KU.00.01/2021 and 4/KI/LPPM/IV/2021) of Universitas Negeri Jakarta.

References
[1] W. Tian, R. Ma, J. Gu, Z. Wang, N. Ma, and P. Du. 2018, J. Mater. Chem. C. **6**. pp. 12965–12975.
[2] J. Krupka, 2016, IEEE Microwave and Wireless Components Letters, vol. **26**, no. 6, pp. 464-466, June, doi: 10.1109/LMWC.2016.2562640.
[3] Chang-Mook Choi, Kwang-Soob Ko. 2010, Journal of Navigation and Port Research International Edition, vol. **34**, no. 2, pp. 111–115.
[4] V. A. Zhuravlev, V. I. Suslyaev, O. A. Dotsenko, and A. N. Babinovich. 2012, Russian Physics Journal, Vol. **53**, No. 8, pp. 96–97.
[5] Paul Dixon, 2012, "Theory and Applications of RF/Microwave Absorbers," Director of Technology Emerson & Cuming Microwave Products, Inc 28 York Avenue Randolph, MA 02368. www.eccosorb.com pp. 1–19.
[6] S.W. Harmer1 N. Rezgui2 N. Bowring2 Z. Luklinska1 G. Ren. 2008, IET Microw. Antennas Propag., Vol. **2**, No. 6, pp. 606–614 doi: 10.1049/iet-map:20070235.
[7] E. Handoko, I. Sugihartono, and M. A. Marpaung, Maulana Randa, Mudrik Alaydrus and Nofrijon Sofyan. 2018, Materials Science Forum, **929**, 109–115.
[8] U. Topal, 2011, *Mater. Sci. Eng. B Solid-State Mater. Adv. Technol.*, vol. **176**, no. 18, pp. 1531–1536.
[9] R. S. Meena, S. Bhattachrya, and R. Chatterjee, 2010, *Mater. Sci. Eng. B Solid-State Mater. Adv. Technol.*, vol. **171**, no. 1–3, pp. 133–138.
[10] A. Srinivas, T. Karthik, R. Gopalan, and V. Chandrasekaran, 2010," *Mater. Sci. Eng. B Solid-State Mater. Adv. Technol.*, vol. **172**, no. 3, pp. 289–293.
[11] M.G. Vakhitov, D.S. Klygach, D.A. Vinnik, V.E. Zhivulin, N.S. Knyazev. 2019, *J. Alloys Compd.*, **816**, 152682. doi.org/10.1016/j.jallcom.2019.152682.
[12] E. Handoko *et al* 2018 *J. Phys.: Conf. Ser.* **1080** 012002.
[13] Wahyu Widanarto, Ananda Iqbal Ekaputra, Mukhtar Effendi, Wahyu Tri Cahyanto, Sib Krishna Ghoshal, Candra Kurniawan, Erfan Handoko, Mudrik Alaydrus, 2020. *Composites Communications,* **19,** pp. 51–55.
[14] W. Widanarto, M. Effendi, S. Krishna, C. Kurniawan, E. Handoko, and M. Alaydrus, 2020, *Curr. Appl. Phys.*, vol. **20**, no. 5, pp. 638–642, 2020.
[15] E. Handoko, S. Budi, I. Sugihartono, and M. A. Marpaung, "Materials Express," vol. **10**, no. 8, pp. 1328–1336, 2020.
[16] Yusmaniar, Erdawati, Harini Sosiati, Setia Budi, Mudrik Alaydrus and Erfan Handoko, 2021, Mater. Res. Express **8** (2021) 046101.

X-Ray Peak Profile Analysis of Silica by Williamson–Hall and Size-Strain Plot Methods

B Hariyanto[1], D A P Wardani[2], N Kurniawati[1], N P Har[3], N Darmawan[4] and Irzaman[3]
[1]Department of Physics, Faculty of Mathematics and Natural Science, Universitas Palangka Raya, Palangka Raya 73111, Indonesia
[2]Department of Chemistry, Faculty of Mathematics and Natural Science, Universitas Palangka Raya, Palangka Raya 73111, Indonesia
[3]Department of Physics, Faculty of Mathematics and Natural Sciences, IPB University, Bogor 16680, Indonesia
[4]Department of Chemistry, Faculty of Mathematics and Natural Sciences, IPB University, Bogor 16680, Indonesia
Email: budi.hariyanto@mipa.upr.ac.id, irzaman@apps.ipb.ac.id

Abstract. Annealed silica has been prepared by various annealing temperatures at 800 °C and 1000 °C. The crystallite size and lattice strain of silica were estimated by x-ray diffraction spectra analysis using various calculation methods; Modified Scherrer, Williamson-Hall (W-H), and Size-Strain Plot. Qualitative analysis confirms that the XRD patterns were hexagonal quartz. Annealing temperature cause a change in crystal size and lattice strain and all methods showed a decrease in the value of the crystal size with increasing annealing temperature from 800 °C to 1000 °C.

1. Introduction

Silica (SiO_2) has occupied the prime position for nano-sized materials and devices due to their facile synthesis, high surface area, suitable dielectric and thermal properties, low toxicity, and controllable properties such as optoelectronic, mechanical, and chemical stability [1, 2]. Silica is one of the most abundant families of materials, naturally having major crystalline phases such as quartz, cristobalite, and tridymite as well as several other polymorphs. In addition, amorphous silica can be transformed into the crystalline phase by thermal treatment [1].

Thermal treatment influences the structure and gives impacts for the intensity and 2θ peak position. Because X-ray diffraction peak profile analysis provides information about the crystalline size and lattice strain, it would change the crystallite size and the lattice strain value [2–4]. The crystallite size is a single crystal inside the particles or grains, and it is not generally the same as the particle size due to the formation of polycrystalline aggregates [5,6]. Lattice strain is a measure of the distribution of lattice constants arising from crystal imperfections. The most common sources of the lattice strain are dislocation density, point defects, grain boundary junction, contact or sinter stress, and stacking faults [7]. Crystallite size and lattice strain affect the Bragg peak in different ways. Both these effects increase the peak width and intensity and shift the 2θ peak position accordingly.

X-ray profile analysis is a simple tool for estimating crystal size and lattice strain. There are several methods for this analysis, including modified Scherrer [8], Williamson-Hall method [9], and Size-Strain Plot [5–7]. Previous studies [7,10] reported that the crystal sizes determined from XRD data using these methods were in good agreement with those obtained by Transmission Electron

Microscopy. So, this study focuses on the x-ray peak profile of annealed silica to obtain crystallite size, lattice strain, and its annealing effect.

2. Materials and Method

The material in this study was SiO_2 powder (Sigma-Aldrich 99%, USA). Furthermore, the SiO_2 powder was annealed at 800 °C and 1000 °C for 1 hour. We annealed at that temperature according to previous research [11,12]. Annealed SiO_2 was characterized using a Rigaku Miniflex II x-ray diffractometer with Cu-Kα (λ = 1,54056 angstrom). XRD patterns were collected with a step size of 0.02° and a range of 2θ from 15 to 60°. The XRD data analysis provides information such as the phase of the sample, purity, crystallite size, and lattice strain [13,7]. Crystallite size and lattice strain were analyzed using several methods: modified Scherrer, Williamson-Hall, and Size-Strain Plot.

3. Results and Discussion

3.1. XRD analysis

Fig. 1 shows XRD spectra of SiO_2 annealed at 800 °C and 1000 °C for 1 hour (denoted S800 and S1000). Qualitative analysis using *Match!* software confirms that the XRD patterns were hexagonal quartz (ICDD#00-046-1045). Furthermore, the inset presents selected Bragg reflections (102) and (111), where it is clear that there is a slight shift in peak position due to the increase in annealing temperature. The shift in peak position indicates that there is a change in the structure, in this case, the crystal size and lattice strain [2,4].

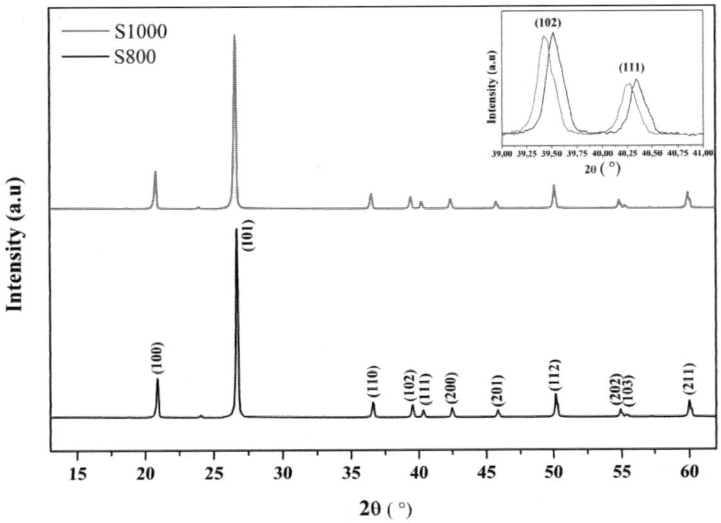

Figure 1. XRD pattern of SiO_2. Inset: zoom of the peak corresponding to the plane (102) and (111)

3.2. Crystallite size and strain estimation

3.2.1. Modified Scherrer method

Derived from the Scherrer equation (Eq. (1)), the modified Scherrer formula aims to reduce errors and obtain the average value of crystallite size through the least-squares method diffraction peaks [8]. After that, by making logarithms on both sides, the modified Scherrer equation can be written as Eq. (2).

$$D = \frac{k\lambda}{\beta \cos\theta} \quad (1)$$

$$\ln\beta = \ln\left(\frac{k\lambda}{D}\right) + \ln\left(\frac{1}{\cos\theta}\right) \quad (2)$$

Where D is the average crystal size, λ is the X-ray wavelength used in nanometers (nm), β is the integral breadth of the full width at half maximum (FWHM) resulting from the small crystal size in radians, θ is the Bragg angle, and k is a constant associated with the crystalline form. By plotting $\ln(1/\cos\theta)$ on the x-axis and $\ln\beta$ along the y-axis, the crystallite size can be determined from the intercept of the y-axis, see Fig. 2.

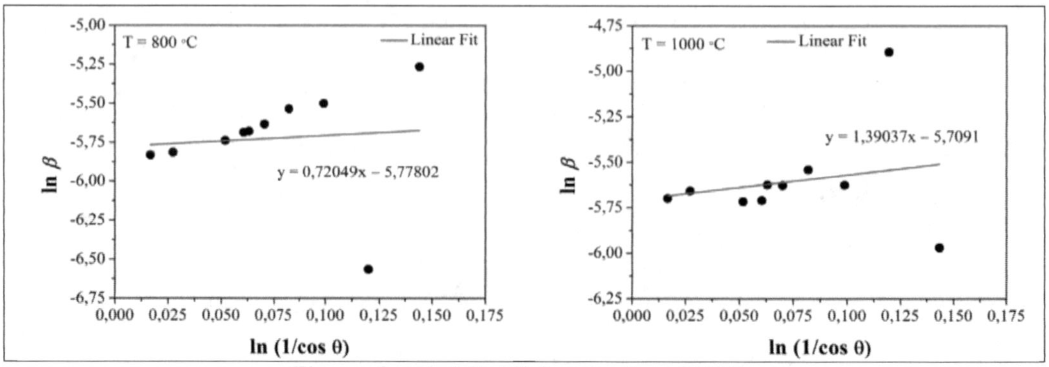

Figure 2. Modified Scherrer plot of SiO_2.

3.2.2. Williamson-Hall (W-H) method

Crystal imperfections and distortions induce strain resulting in the broadening of the peak, given by the relation [6,7]:

$$\beta_{strain} = 4\varepsilon \tan\theta \tag{3}$$

But

$$\beta_{hkl} = \beta_{crystallite\ size} + \beta_{strain} \tag{4}$$

Therefore

$$\beta_{hkl} = \frac{k\lambda}{D\cos\theta} + 4\varepsilon \tan\theta \tag{5}$$

Rearranging this equation,

$$\beta_{hkl}\cos\theta = \frac{k\lambda}{D} + 4\varepsilon\sin\theta \tag{6}$$

The above equations are W-H equations [9]. A plot is drawn with $4\sin\theta$ along the x-axis and $\beta_{hkl}\cos\theta$ along the y-axis (Fig.3). The crystallite size was estimated from the y-intercept and the lattice strain (ε) of the slope via a linear fit of the data.

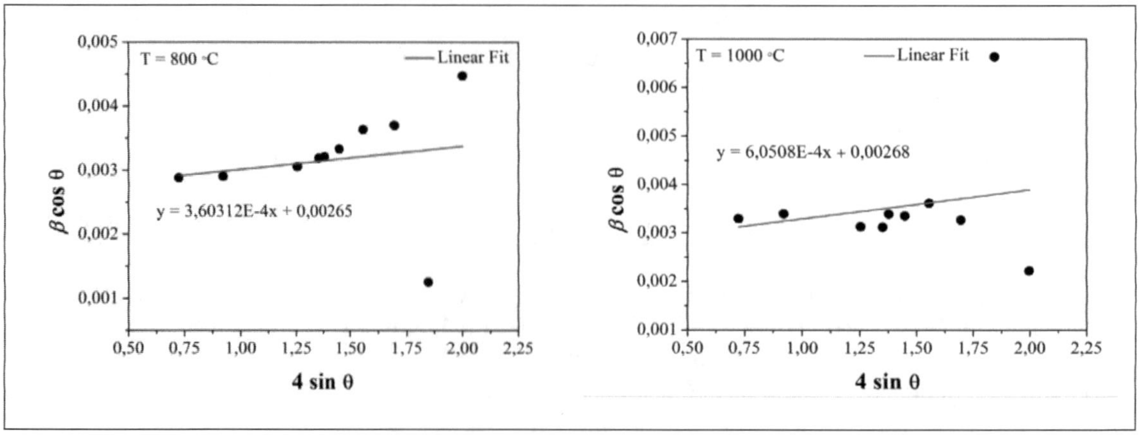

Figure 3. Williamson-Hall plot of SiO_2.

3.2.3. Size-strain plot (SSP) method

W-H method shows that peak profile broadening was essentially isotropic, so that the diffraction domains were isotropic, and there was also a microstrain contribution. SSP method was expanding the isotropic basis for evaluated the calculation of the W–H method. In this approach, the "crystallite size" profile was assumed to be described by a Lorentz function and the "strain profile" of a Gaussian function as follows [5–7]:

$$(d_{hkl}\beta_{hkl}\cos\theta)^2 = \frac{K}{D}\left(d_{hkl}^2\,\beta_{hkl}\cos\theta\right) + \left(\frac{\varepsilon}{2}\right)^2 \quad (7)$$

Where K is a constant that depends on the shape of the particles; for spherical particles, it is given as 3/4, and d_{hkl} is the lattice distance between the (*hkl*) planes [10]. The term $(d_{hkl}\beta_{hkl}\cos\theta)^2$ was plotted with respect to $\left(d_{hkl}^2\,\beta_{hkl}\cos\theta\right)$, shown in Fig.4. The crystallite size was determined from the slope, and the *y*-intercept gives the strain.

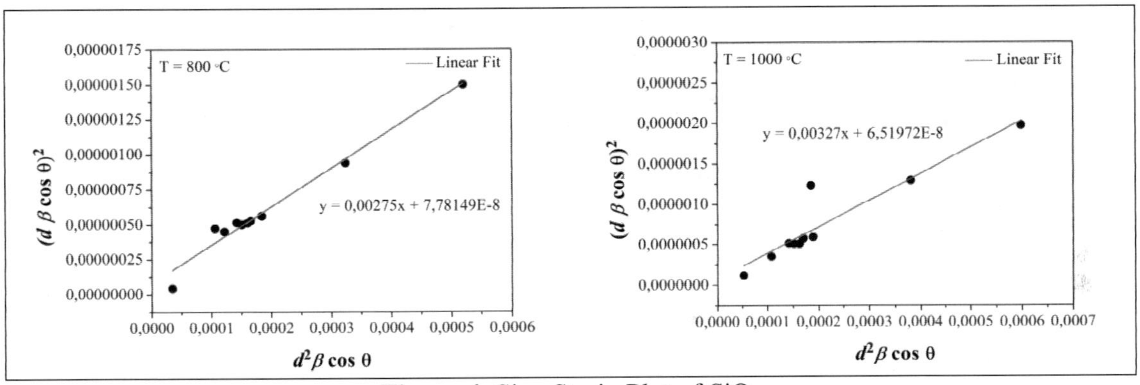

Figure 4. Size-Strain Plot of SiO_2.

The results obtained from the modified Scherrer, Williamson-Hall method, and SSP are summarized in Table 1. Increasing annealing temperature causes a change in crystal size and lattice strain [2–4], although all methods show different results. In addition, the results of all methods showed a decrease in the value of the crystal size with increasing annealing temperature. These results confirm the previous XRD pattern analysis regarding the shift in diffraction peak position.

Table 1. Structural parameters (crystallite size and lattice strain) annealed SiO_2

Sample	Modified Scherrer	Williamson-Hall method		Size-Strain Method	
	D (nm)	*D* (nm)	ε (10^{-4})	*D* (nm)	ε (10^{-4})
SiO_2 800	46,79	54,64	3,60	42,01	5,58
SiO_2 1000	43,67	54,03	6,05	35,33	5,11

4. Conclusion

The XRD analysis was successfully carried out on SiO_2 annealed at 800 °C and 1000 °C. The qualitative analysis confirmed that SiO_2 was in the quartz phase. Furthermore, increasing annealing temperature caused crystal size and lattice strain changes, although the modified Scherrer method, Williamson-Hall method, and SSP showed different results. The crystal size decreases as the annealing temperature increases from 800 °C to 1000 °C.

Acknowledgments

This research was funded by the PKPT Grant (No: 264/UN24.13/PL/2021) by the Ministry of Education and Culture, Research, and Technology of the Republic of Indonesia (Kemendikbud-ristek RI).

Journal of Physics: Conference Series **2019** (2021) 012106 doi:10.1088/1742-6596/2019/1/012106

References

[1] Prabha S, Durgalakshmi D, Rajendran S and Lichtfouse E 2021 *Environ Chem Lett* 19 1667–91
[2] Oufakir A, Khouchaf L, Afqir M, Elaatmani M, Zegzouti A and El H DIRECT DEPENDENCE BETWEEN TEMPERATURE, MICRO STRAINS AND STRUCTURAL DISTORTIONS IN NATURAL SiO2 QUARTZ 12
[3] Haq S, Rehman W, Waseem M, Javed R, Mahfooz-ur-Rehman and Shahid M 2018 *Appl Nanosci* 8 11–8
[4] Kumar D, Singh M and Singh A K 2018 *2ND INTERNATIONAL CONFERENCE ON CONDENSED MATTER AND APPLIED PHYSICS (ICC 2017)* (Bikaner, India) p 030185
[5] Irfan H, Racik K. M and Anand S 2018 *Journal of Asian Ceramic Societies* 6 54–62
[6] Augustin M and Balu T 2017 *Int. J. Nanosci.* 16 1650035
[7] Khorsand Zak A, Abd. Majid W H, Abrishami M E and Yousefi R 2011 *Solid State Sciences* 13 251–6
[8] Monshi A, Foroughi M R and Monshi M R 2012 *WJNSE* 02 154–60
[9] Williamson G K and Hall W H 1953 *Acta Metallurgica* 1 22–31
[10] Nath D, Singh F and Das R 2020 *Materials Chemistry and Physics* 239 122021
[11] M. H. Shahrokh Abadi 2015 *J. Ceram. Sci. Tech.*
[12] Hapsari I L, Irzaman and Indro M N 2020 *IOP Conf. Ser.: Earth Environ. Sci.* 460 012024
[13] Holder C F and Schaak R E 2019 *ACS Nano* 13 7359–65

Journal of Physics: Conference Series **2019** (2021) 012107 doi:10.1088/1742-6596/2019/1/012107

Effect of Pulse Electromagnetic Field Exposure on the Expression of Lipo Protein Lipase (LPL) on the Differentiation of Mesenchymal Stem Cell

Umiatin[1], W Indrasari[1], Taryudi[2], P Sari[3] , [4]N F Mazfufah and [5]I Rosadi

[1]Department of Physics, Faculty of Mathematics and Natural Science, Universitas Negeri Jakarta, Indonesia

[2]Department of Electronic Engineering, Faculty of Engineering, Universitas Negeri Jakarta, Indonesia

[3]Department of Biology, Faculty of Medicine, Universitas Indonesia, Indonesia, Jl. Rawamangun Muka, Jakarta 13220, Indonesia

[4]Stem Cells and Tissue Engineering Research Center, IMERI, Faculty of Medicine, Universitas Indonesia, Indonesia

[5]Department of Biology, Faculty of Mathematics and Natural Sciences, Mulawarman University, Samarinda, Indonesia

Email: [1]umiatin@unj.ac.id , [3]puji_sari@yahoo.com

Abstract. Pulsed electromagnetic fields (PEMFs) have an important role in cell differentiation. Previous study reported that PEMFs had positive and negative effect towards cell differentiation that depends on their frequencies applied to the cells. Human adipose-derived stem cells (ASCs) are mesenchymal stem cells that have an ability to differentiate into several types of cell including adipocytes, chondrocytes and osteocytes. This study aimed to evaluate the effect of human ASCs towards their adipogenic differentiation during PEMFs exposure. Human ASCs were isolated from adipose tissue. The cells then cultured in specific medium of adipocyte that induced ASCs differentiation along with PEMFs exposure. The maximum magnetic field used is 2 mT with a frequency of 75 Hz. To confirm the effect of PEMFs exposure towards adipogenic differentiation, mRNA expression of lipo protein lipase (LPL) was measured in mRNA expression level. The results showed that ASCs cultured on adipogenic differentiation without PEMFs exposure gradually increased LPL expression until day 14 of observation, while ASCs with PEMFs exposure significantly decreased LPL expression from day 2 to day 14. Based on the results, we concluded that PEMFs exposure can inhibit LPL expression that suppressed adipogenic differentiation.

1. Introduction

Adipose-derived stem cells (ASCs) are mesenchymal stem cells that exist in adipose tissue [1]. ASCs have ability to multi-directional differentiation into bone cells, cartilage cells, or fat cells. Their differentiation ability depends on the supplementation in the microenvirontment [1,2]. However, recent study showed that electromagnetic also promote stem cells differentiation. Pulsed electromagnetic fields (PEMFs) enabled to generate magnetic field effects with pulsed intervals that changes microenvironment of physiological activities [3]. Due to the change microenvironment of cells, PEMFs can promote bone marrow mesenchymal stem cells (BMSCs) into osteogenic and inhibit other lineage of differentiation. The technique is safe, environmentally friendly, non-invasive, and simple [4].

Obese mice lose their weight and fat after treated by electromagnetic fields. Extremely low frequency magnetic fields (ELF-MF) also is capable to inhibit the effect of obesity. Inhibitory effect of ELF-MF on obesity may be related to the differentiation of MSCs to adipocytes [5]. Adipogenesis of MSCs expressed lipoprotein lipase (LPL). LPL is an enzyme that hydrolyzes triglycerides and releases free fatty acids in cardiac muscle as an energy source [6]. Furthermore, previous study showed that PEMFs can promote the expression of ALP and bone morphogenetic protein-2 (BMP-2) in BMSCs, and have the effect of promoting the differentiation of BMSCs into osteogenic tissue [7] suggesting the PEMF suppressed adipogenesis due to their negative feedback mechanism.

In this study, we manage to induce ASCs to differentiate into adipocytes using specific media. The induced cells then exposured using PEMFs. To evaluate the adipogenesis, the expression of LPL was measured. We suggest that PEMFs regulate the expression of adipogenesis.

2. Material and Method

2.1. Culture of Mesenchymal Stem Cell

MSCs were obtained from Stem Cell and Tissue Engineering Research Center (SCTE-RC) Indonesian Medical Education and Research Institute (IMERI), Faculty of Medicine, University of Indonesia. MSCs were isolated from adipose tissue [8]. The cells were cultured in the complete medium containing of 1% antibiotic, 1% anti-mycotic, 1% glutamine, 1% heparin, 10% serum platelet-rich plasma (PRP), and αMEM. The expansion of cells was conducted using flasks 25 and incubated at 37°C, 5% CO_2. To maintain the cells, the medium was changed every three days. After the cells reach 100% confluency, the cells were harvested using Triple Select and seeded in the six well culture flask.

To induce MSCs differentiation, the complete medium was replaced to specific medium of adipogenesis (StemPro® Adipogenesis Differentiation Kit). This induction medium contains 1% antibiotic, 1% anti-mycotic, 10% supplement, and basal medium. The adipogenic induction medium was replaced every three days for 14 days.

2.2. PEMFs Exposure

The construction of PEMFs generator has been described previously [9-11]. This PEMFs generator consisted of a signal generator and a pair of 40-cm diameter Helmholtz coil. The coil were connected to a signal generator that delivered repetitive square-wave pulses with pulse duration 1 s and frequency 75 Hz. (Figure 1).

The maximum magnetic field used is 2 mT with a frequency of 75 Hz. The cells were placed in flask between the coil. MSCs from the experimental group were exposed to PEMFs for 10 minutes daily for 14 days. The control group was not exposed to PEMF. On days 2, 4 and 14, most of the MSCs were harvested for qRT PCR analysis.

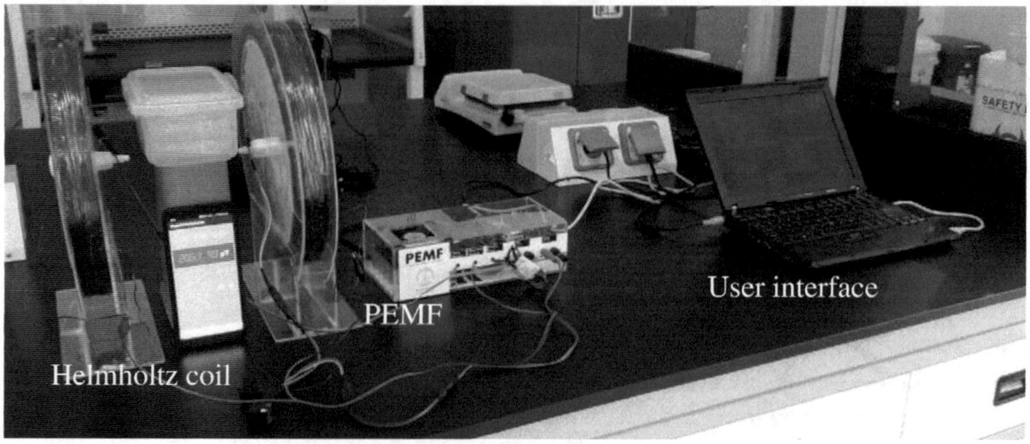

Figure 1. Schematic representation of the PEMF exposure system: (a) PEMF consisting of Helmholtz coil, pulse generator, user interface and DC power supply.

2.3. RNA Extraction and qRT-PCR Analysis
The total RNA was isolated from harvested MSCs on day 0 (as calibrator), 2, 4 and 14 using TRIzol TM LS Reagent (Invitrogen, Carlsbad, CA, USA) according to the manufacturer's instruction. The RNA concentration and purity was performed using a spectrophotometer. Primer sequences are described in Table 1.

Table 1. Polymerase chain reaction primer sequence

No.	Gene	Nucleotides (5'-3')
1	LPL	F: CAGCCCTACCCTTGTTAGTTATT
		R: ACGTTGGAGGATGTGCTATTT
2	GAPDH	F: CCCTTCATTGACCTCAACTACA
		R: ATGACAAGCTTCCCGTTCTC

LPL : Lipo Protein Lipase

RT-qPCR was performed using a standard protocol from the MIQE (minimum information for publication of quantitative real-time PCR experiments) guide [12]. The mRNA expression of GAPDH and LDL was analyzed using Sensi FAST ™ SYBR® Lo-ROX One-Step Kit material, which allowed detection of the PCR products by measuring the increase in SYBR green fluorescence caused by the binding of SYBR green to double -stranded DNA. The qRT-PCR thermal cyling program was as follow: 10 min at 45°C for activation of the reverse transcription enzyme, 2min at 95°C for activation of the polymerase enzyme. Then followed by 40 cycles of 15 s at 95°C, 30s at 60°C and 30 s at 72°C. The level of expression was calculated based on the PCR cycle number (Ct) and the relative gene expression level was determined using the Livak method [13].

2.4. Statistical Analysis
Statistical analysis was performed using SPSS 21 for Windows. Statistical analysis was began with normality and homogeneity testing using the Shapiro Wilk and Lavene test. The data presented as means ± SEM. One-way analysis of variance (ANOVA) with a post hoc Tukey test was used to compare means. The differences were analyzed ($p < 0.05$).

3. Results and Discussion

Mesenchymal stem cells (MSCs) are multi-potent cells that can be differentiated into several types of cells in the specific microenvirontment including adipocytes, chondrocytes and osteocytes **[2]**. Their abilities lead MSCs to be used as a model for cell differentiation study **[14]**. Differentiation of MSCs depends on signaling pathway, transcription factor, growth factors and physical factors as well **[15]**. Furthermore, recent study showed that the use of electromagnetic field also promotes osteogenesis and increases proliferation rate of stem cells in *in vitro* **[7-16]**.

Those studies provided an insight to break down the effect of electromagnetic field toward stem cells differentiation. Hu *et al.,* reported that during adipogenic differentiation of hASCs, LPL is gradually up-regulated but tended to become stable 14 days after induction **[17]**. In this study, we investigate the effect of PEMFs on the adipogenic differentiation of MSCs towards LDL expression of adipogenesis for 14 days. MSCs were cultured in an adipogenic medium that also has an effect to induced the cells to undergo adipogenesis. LPL that regulate human adipose-derived mesenchymal stem cells (hASCs) into adipogenic differentiation was measured by qRT-PCR. The data obtained on day 0 (calibrator), 2, 4, and 14 from each group. We observed that the control group expressed LDL and gradually increased until day 14. However, the MSCs that exposured by PEMF tended to become stable on day 2 and 4 and significantly decreased on day 14 (Figure 2).

Our previous study showed that PEMF exposure may have potent effect to inhibit adipogenic differentiation indicated by PPARγ and ADIPOQ mRNA expression **[9-10]**. Extremely low frequency magnetic fields (ELF-MF) also inhibited adipogenic differentiation of ASCs showed by decreasing proliferator-activated receptor 2 (PPARγ2) expression but had no effect on osteogenic differentiation **[5]**. ELF-MF regulates JNK-dependent intracellular signaling in MSCs **[18]**. JNK signaling is essential in the bone formation **[5]**. However, no data of osteogenesis were recorded in this study. Moderate intensity Static Magnetic Field (SMF) also inhibited adipogenesis of MSCs, which was indicated by reducing LPL and PPAR-γ and by decreasing the number of lipid droplets **[19]**.

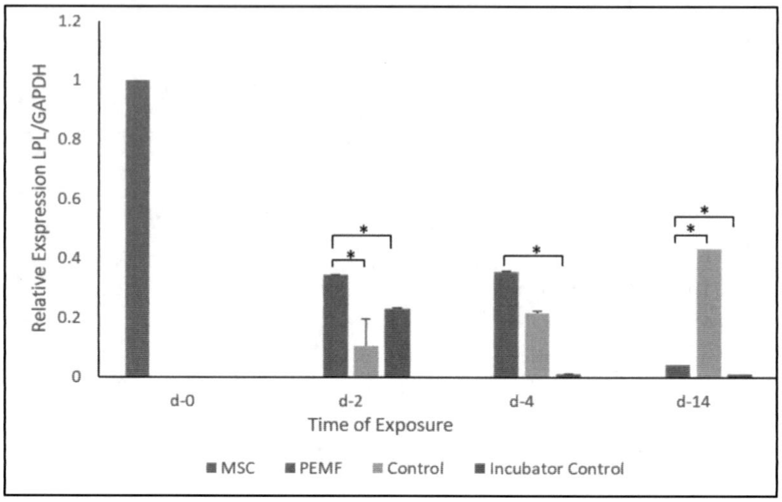

Figure 2. Expression level of LPL indices stimulated by pulsed electromagnetic filed (PEMF).

Based on these results there are some limitations to our study. First, to evaluate adipogenesis, osteogenesis marker should be measured as well because their negative feedback mechanism. It means that if adipogenesis was promoted, osteogenesis will be inhibited. In another study we have found that PEMF exposure can promote fracture healing. This was indicated by an increase in the expression of osteoblast activity marker genes including Wnt 5a, Wnt10b and β-catenin [20]. Second, the mechanism of LPL inhibition after PEMF exposure was not clear yet. Further study should be conducted to evaluate and confirm our results.

4. Conclusion

Adipogenesis of MSCs could be inhibited by PEMF suggesting the exposure of PEMF may provide a potential approach for the treatment of obesity.

Acknowledgments

The research was carried out and financially supported by research grant from Universitas Negeri Jakarta, No 16/KOMP-UNJ/LPPM/V/2020.

References

[1] P. A. Zuk, "The Adipose-derived Stem Cell: Looking Back and Looking Ahead," *https://doi.org/10.1091/mbc.e09-07-0589*, vol. 21, no. 11, pp. 1783–1787, Apr. 2010, doi: 10.1091/MBC.E09-07-0589.

[2] I. Rosadi *et al.*, "The Effect of Human Platelet-Rich Plasma and L-Ascorbic Acid on Morphology, Proliferation, and Chondrogenesis Ability towards Human Adipose-Derived Stem Cells," *Molecular and Cellular Biomedical Sciences*, vol. 3, no. 1, pp. 26–33, Mar. 2019, doi: 10.21705/MCBS.V3I1.43.

[3] G. M. Marcheggiani Muccioli *et al.*, "Conservative treatment of spontaneous osteonecrosis of the knee in the early stage: Pulsed electromagnetic fields therapy," *European Journal of Radiology*, vol. 82, no. 3, pp. 530–537, Mar. 2013, doi: 10.1016/J.EJRAD.2012.11.011.

[4] T. Lu, Y. Huang, C. Zhang, M. Chai, and J. Zhang, "Effect of pulsed electromagnetic field therapy on the osteogenic and adipogenic differentiation of bone marrow mesenchymal stem cells," *Genetics and Molecular Research*, vol. 14, no. 3, pp. 11535–11542, 2015, doi: 10.4238/2015.September.28.5.

[5] L. Du, H. Fan, H. Miao, G. Zhao, and Y. Hou, "Extremely low frequency magnetic fields inhibit adipogenesis of human mesenchymal stem cells," *Bioelectromagnetics*, vol. 35, no. 7, pp. 519–530, Oct. 2014, doi: 10.1002/BEM.21873.

[6] H. Yagyu *et al.*, "Lipoprotein lipase (LpL) on the surface of cardiomyocytes increases lipid uptake and produces a cardiomyopathy," *The Journal of Clinical Investigation*, vol. 111, no. 3, pp. 419–426, Feb. 2003, doi: 10.1172/JCI16751.

[7] M. Zhang *et al.*, "Effects of low frequency electromagnetic field on proliferation of human epidermal stem cells: An in vitro study," *Bioelectromagnetics*, vol. 34, no. 1, pp. 74–80, Jan. 2013, doi: 10.1002/BEM.21747.

[8] I. Liem, J. A. Pawitan, I. K. Liem, D. Suryani, A. Bustami, and R. Y. Purwoko, "Simple lipoaspirate washing using a coffee filter," *Asian Biomedicine*, vol. 7, no. 3, pp. 333–338, 2013, doi: 10.5372/1905-7415.0703.184.

[9] U. Umiatin, W. Indrasari, and N. F. Mazfufah, "Journal of Global Pharma Technology Expression of Adipocyte Binding Protein-2 (AP-2) on the Adipogenic Differentiation of Mesenchymal Stem Cell after Pulsed Electromagnetic Field (PEMF) Exposure," vol. 2, pp. 157–162, 2014.

[10] P. Sari, Umiatin, L. Yunaini, E. Luviah, and N. F. Mazfufah, "Relative expression of PPARγ and ADIPOQ mRNA as markers of inhibition adipogenesis on mesenchymal stem cells after Pulsed Electromagnetic Field (PEMF) exposure," *Journal of Global Pharma Technology*, vol. 12, no. 1, pp. 307–317, 2020.

[11] Umiatin, I. H. Dilogo, S. K. Wijaya, P. Sari, and A. D. Djaja, "Design and development of pulse electromagnetic fields (PEMF) as adjuvant therapy for fracture healing: A preliminary study on rats," *AIP Conference Proceedings*, vol. 2092, no. 1, p. 020028, Apr. 2019, doi: 10.1063/1.5096696.

[12] S. A. Bustin *et al.*, "The MIQE guidelines: Minimum information for publication of quantitative real-time PCR experiments," *Clinical Chemistry*, vol. 55, no. 4, pp. 611–622, 2009, doi: 10.1373/clinchem.2008.112797.

[13] K. J. Livak and T. D. Schmittgen, "Analysis of Relative Gene Expression Data Using Real-Time Quantitative PCR and the 2−ΔΔCT Method," *Methods*, vol. 25, no. 4, pp. 402–408, Dec. 2001, doi: 10.1006/METH.2001.1262.

[14] I. Rosadi *et al.*, "In vitro study of cartilage tissue engineering using human adipose-derived stem cells induced by platelet-rich plasma and cultured on silk fibroin scaffold," *Stem Cell Research & Therapy 2019 10:1*, vol. 10, no. 1, pp. 1–15, Dec. 2019, doi: 10.1186/S13287-019-1443-2.

[15] B. Birru, N. Kumar, P. Sreenivasa Rao, and R. Raju Baadhe, "Role of Signaling Pathways in Mesenchymal Stem Cell Differentiation," 2014, doi: 10.2174/1574888X09666140812112002.

[16] Z. Bao, M. Fan, L. Ma, Q. Duan, and W. Jiang, "The effects of pulsed electromagnetic fields combined with a static magnetic intramedullary implant on the repair of bone defects: A preliminary study," *https://doi.org/10.1080/15368378.2019.1625785*, vol. 38, no. 3, pp. 210–217, Jul. 2019, doi: 10.1080/15368378.2019.1625785.

[17] X. Hu *et al.*, "MiR-27b Impairs Adipocyte Differentiation of Human Adipose Tissue-Derived Mesenchymal Stem Cells by Targeting LPL," *Cellular Physiology and Biochemistry*, vol. 47, no. 2, pp. 545–555, Jun. 2018, doi: 10.1159/000489988.

[18] L. Ling, V. Nurcombe, and S. M. Cool, "Wnt signaling controls the fate of mesenchymal stem cells," *Gene*, vol. 433, no. 1–2, pp. 1–7, Mar. 2009, doi: 10.1016/J.GENE.2008.12.008.

[19] J. Wang, B. Xiang, J. Deng, D. H. Freed, R. C. Arora, and G. Tian, "Inhibition of Viability, Proliferation, Cytokines Secretion, Surface Antigen Expression, and Adipogenic and Osteogenic Differentiation of Adipose-Derived Stem Cells by Seven-Day Exposure to 0.5 T Static Magnetic Fields," *Stem Cells International*, vol. 2016, 2016, doi: 10.1155/2016/7168175.

[20] U. U, I. H. Dilogo, P. Sari, and S. K. Wijaya, "The Effect of Pulsed Electromagnetic Field Exposure on Fracture Healing Through the Wnt Signal Pathway," *OnLine Journal of Biological Sciences*, vol. 20, no. 4, pp. 239–249, Apr. 2020, doi: 10.3844/ojbsci.2020.239.249.

Digital Text Security with Steganography Least Significant Bit and Audio Feature Extraction

M Delina[1], H Nasbey[1], A Gunawan[1] and S Muhasyah[2]
[1]Physics Department, Faculty of Mathematics and Natural Sciences, Universitas Negeri Jakarta. Jl. Rawamangun Muka No1, Jakarta Timur 13220, Indonesia.
[2]Ministry of Foreign Affairs of the Republic of Indonesia. Jl. Taman Pejambon No 6. Jakarta Pusat 10110, Indonesia.
Email: mutia_delina@unj.ac.id

Abstract. The study developed a digital text security system by inserting the classified message or text into digital audio (stego audio) with the least significant bit (LSB) steganography and audio feature extraction. The text bits were inserted into each cover audio frame based on the value of feature extraction energy and zero-crossing rate (ZCR). The study has demonstrated that the least significant bit and audio feature extraction can cover the digital text in stego audio and then be extracted back to the original digital text. There is a small difference in feature extraction energy value between stego audio and cover audio. While there is no difference in feature extraction zero-crossing rate between stego audio and cover audio. The study also measured signal-to-noise ratio (SNR); the detected noise in stego audio because of the change in the least significant bit. The SNR is 111.97 – 130.00 dB.

1. Introduction

Nowadays, digital transmission through the internet increases rapidly [1]. Therefore, the information exchange easily worldwide every second. However, for a classified file that is only accessed by an authorized person, security and authenticity are still the primary concern [2]. During transmission, the classified information has risked being attacked by an eavesdropper [3] such as interruption, interception, modification, and fabrication[4]. As prevention, this classified information can be protected by steganography [5] which covers the information with various media[6], e.g., text [7], images [8], audio [9], and video [10]. The cover is called the cover image. Meanwhile, the text with the cover is a stego image. With the aid of cover image, the classified information is not visible directly and looks natural. Human vision cannot recognize the different between cover image and Stego image [11].

Steganography has been applied in various fields. In pharmacy, steganography present to secure the electronic prescription and patient data from the eavesdropper [12]. In medical, steganography employs to protect the medical image. Therefore, no theft can steal the data nor modified it. An illegal modification of medical images will lead to a false diagnose[13]. In politics, the steganography identified cheaters in an election[14]. In transportation, steganography is practiced for a ticket barcode to protect the passenger data such as transaction, bank account, and traveling history[15]. In Information and Communication Technology (ICT), steganography secures the Internet of Things (IoT) data [16].

This study employed the steganography least significant bit [17] with audio feature extraction [18] to protect the classified text. Here, the information is covered by a 16-bit double cannel digital audio with various genres. By combining audio feature extraction and steganography least significant bit, the study obtained a better security level result.

Steganography employs a cover to hide digital information. This cover is like a wrapping box for the information during the transmission from sender to receiver. The digital information may in bits or a block. The cover itself has various types like text, image, audio, video, sound, or other media [19].

Least Significant Bits (LSB) is a steganography method in the spatial domain. This technique is simple because the cover LBS's can be replaced by message bits directly. For example, in steganography with cover image, the LSB in each image pixels are changed by the message bits.

Feature extraction is essential for audio analysis, such as pattern recognition and machine learning. With this feature, the audio characteristic such as human voice, music, and explosion can be recognized [20]. This feature can also distinguish music genres such as pop, rock, jazz, and dangdut. The feature extraction consists of two main parameters, time and frequency based. The time-based itself composed of energy and Zero-Crossing Rate (ZCR) feature. The energy feature $E(i)$ can be denoted as follow.

$$E(i) = \frac{1}{W_L} \sum_{n=1}^{W_L} |x_i(n)|^2, \tag{1}$$

where W_L and x_i are the frame number at the $-i$th audio sample, respectively. W_L is obtained from ratio between number of samples in the audio cover reference with number of text character. Meanwhile, the ZCR to count the noise signal is denoted as follows.

$$Z(i) = \frac{1}{2W_L} \sum_{n=1}^{W_L} |sgn\,[x_i(n)] - sgn\,[x_i(n-1)]| \tag{2}$$

where

$$sgn[x_i(n)] = \begin{cases} x_i(n) \geq 0, \\ x_i(n) < 0. \end{cases} \tag{3}$$

The difference expected signal with the noise signal can be obtained by measuring the Signal-to-Noise Ratio (SNR) on the following equation.

$$SNR = 10 \log_{10} \frac{\sum_{i=1}^{n} x(n)^2}{\sum_{i=1}^{n} [x(n) - y(n)]^2} \tag{4}$$

2. Method

The research started with the development of audio feature extraction program. Here, there were three audio files as the cover audio in the .wav format (audio 1, audio 2, and audio 3), and a (classified) digital text in the .txt format.

Each text was then inserted into the audio cover by least significant bit steganography technique. The process was started by extracted and inserted the (classified) text into the cover audio, between the initial and final refrain (reff). Furthermore, the initial and final reff time was the stego key. The audio file details and time interval of reff are shown in table 1 and table 2, respectively.

Table 1. Cover audio

Audio Stego	Frequency (Hz)	Data
Audio 1	44,100	10,601,328
Audio 2	44,100	9,532,272
Audio 3	48,000	12,206,064

Table 2. Time interval of the first reff.

Cover Audio	Initial Reff (s)	Final Reff (s)
Audio 1	59	90
Audio 2	96	149
Audio 3	62	89

The text character's number determined the energy feature and ZNR frame. Therefore the character's number was similar to the frame lenght (W_L). The initial and final reff was also obtained to extract the energy feature and ZNR. Here, the energy feature influenced the bit arrangement in the text, where an 8 bits character was inserted in each frame. Once the feature extraction in a frame was above the average energy, the 8 bits character was inserted in the initial frame. Meanwhile, if the energy feature extraction was below the average energy, the 8 bits character was inserted into the middle of the frame.

Finally, the SNR was obtained to analyze the stego audio difference before and after the text was inserted. Thus, the SNR determined the success of the steganography method application. As a comparison, SNR from previous research was employed, the details is shown in table 3.

Table 3. SNR from previous research

Research	SNR (dB)
Wakiyama [21]	45.2 – 50.4
Divya [22]	55.37
Meligy [23]	78.11 – 86.09
Meligy [24]	94.76

3. Results and Discussion

This research obtained three audio files as the audio cover and four texts. Yet, to concise the graph presentation, this paper only showed the figure of audio1.wav. The energy and ZNR feature of audio 1 before the text was inserted was shown by figure 1.

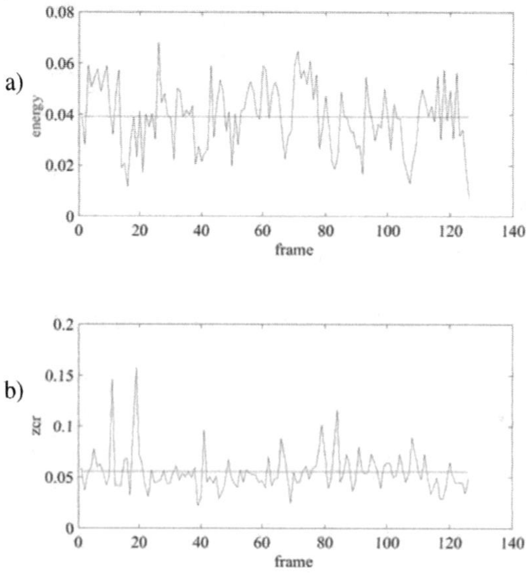

Figure 1. (a) Energy feature of audio1.wav before inserted by text. (b) ZNR of audio1.wav before inserted by text.

Figure 1(a) showed the energy feature of audio1.wav or stego audio 1 before inserted by (classified) text. Meanwhile figure 1(b) was the ZNR of audio1.wav before the audio was inserted by the text.

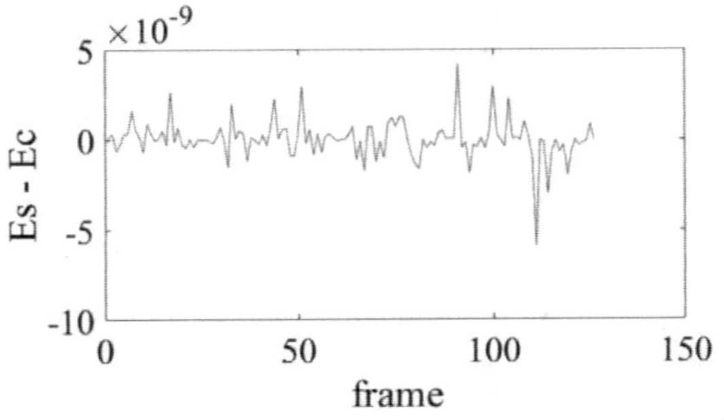

Figure 2. The energy feature diference of stego audio Es with cover audio Ec after the text was inserted to audio1.wav.

Figure 2 showed the energy feature difference of stego audio Es and cover audio Ec after the text was inserted to the audio1.wav. The graph showed that the energy feature difference was small. Therefore, it was not influenced the text position in the stego audio.

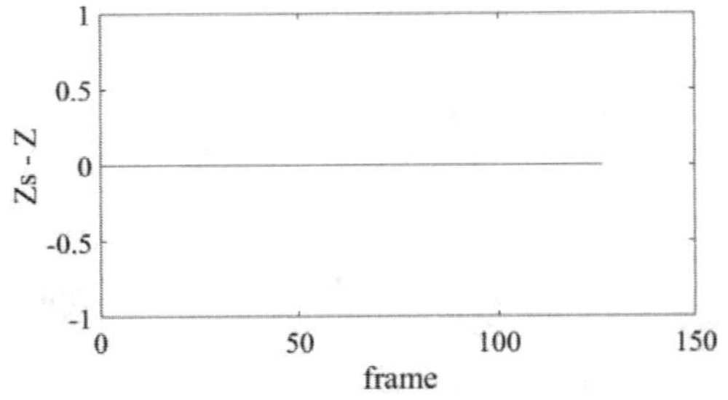

Figure 3. The ZCR feature extration difference of stego audio Es with cover audio Ec after the text was inserted to audio1.wav.

Figure 3 showed that the ZCR feature of stego audio and cover audio has similar value. The comparison of stego audio before and after the text was the inserted obtained from SNR. The SNR of three audio stego and four text were shown in table 4.

Table 4. The SNR of four text and three cover audio

Digital Information	Audio Stego (dB)		
	Audio 1	**Audio 2**	**Audio 3**
Text 1	116.84	125.97	121.03
Text 2	115.37	124.65	119.64
Text 3	121.00	125.77	120.56
Text 4	111.97	121.22	116.30

The SNR was 111.97 – 130.99 dB. This result was more significant than SNR from previous research in table 3. The SNR value larger than 40 dB was good because the stego audio's noise was smaller than cover audio.

4. Conclusion

This study show that the combination of steganography least significant bits with the audio extraction feature give a better security result. It shown that this combination method was able to cover the digital information which is shown by the SNR value 119.95 – 136.03 (dB).

References

[1] S. Dhawan and R. Gupta, "Analysis of various data security techniques of steganography: A survey," *Inf. Secur. J.*, vol. 30, no. 2, pp. 63–87, 2021, doi: 10.1080/19393555.2020.1801911.

[2] M. R. Islam, T. R. Tanni, S. Parvin, M. J. Sultana, and A. Siddiqa, "A modified LSB image steganography method using filtering algorithm and stream of password," *Inf. Secur. J.*, vol. 00, no. 00, pp. 1–12, 2020, doi: 10.1080/19393555.2020.1854902.

[3] R. Ranjith Kumar, S. Jayasudha, and S. Pradeep, "Efficient and secure data hiding in encrypted images: A new approach using chaos," *Inf. Secur. J.*, vol. 25, no. 4–6, pp. 235–246, 2016, doi: 10.1080/19393555.2016.1248582.

[4] M. Delina, B. H. Iswanto, H. Permana, and S. Muhasyah, "The simulation of a symmetric quantum key distribution," *IOP Conf. Ser. Mater. Sci. Eng.*, vol. 434, no. 1, 2018, doi: 10.1088/1757-899X/434/1/012023.

[5] M. Fateh and M. Rezvani, "An email-based high capacity text steganography using repeating characters," *Int. J. Comput. Appl.*, vol. 0, no. 0, pp. 1–7, 2018, doi: 10.1080/1206212X.2018.1517713.

[6] H. Dutta, R. K. Das, S. Nandi, and S. R. M. Prasanna, "An Overview of Digital Audio Steganography," *IETE Tech. Rev. (Institution Electron. Telecommun. Eng. India)*, vol. 37, no. 6, pp. 632–650, 2020, doi: 10.1080/02564602.2019.1699454.

[7] Z. Jalil and A. M. Mirza, "Innovations in Computing Sciences and Software Engineering," *Innov. Comput. Sci. Softw. Eng.*, 2010, doi: 10.1007/978-90-481-9112-3.

[8] T. H. Park, J. G. Han, Y. H. Moon, and I. K. Eom, "Performance improvement of LSB-based steganalysis using bit-plane decomposition of images," *Imaging Sci. J.*, vol. 64, no. 5, pp. 262–266, 2016, doi: 10.1080/13682199.2016.1171452.

[9] Y. Xiang, I. Natgunanathan, D. Peng, W. Zhou, and S. Yu, "A dual-channel time-spread echo method for audio watermarking," *IEEE Trans. Inf. Forensics Secur.*, vol. 7, no. 2, pp. 383–392, 2012, doi: 10.1109/TIFS.2011.2173678.

[10] K. Srinivasan, T. Gowthaman, and J. Kanakaraj, "A novel copyright marking approach using steganography and robust RSA asymmetric-key cryptographic technique in audio files," *J. Discret. Math. Sci. Cryptogr.*, vol. 20, no. 8, pp. 1563–1571, 2017, doi: 10.1080/09720529.2017.1402575.

[11] F. A. P. Petitcolas, R. J. Anderson, and M. G. Kuhn, "Information Hiding — A Survey," vol. 87, no. 7, 1999.

[12] A. Omotosho, J. Emuoyibofarhe, and C. Meinel, "Securing E-Prescription from Medical Identity Theft Using Steganography and Antiphishing Techniques," *J. Appl. Secur. Res.*, vol. 12, no. 3, pp. 447–461, 2017, doi: 10.1080/19361610.2017.1315788.

[13] S. K. Patel, C. Saravanan, and V. K. Patel, "Cloud-based Reversible Dynamic Secure Steganography Model for embedding pathological report in medical images," *Int. J. Comput. Appl.*, vol. 0, no. 0, pp. 1–9, 2019, doi: 10.1080/1206212x.2019.1641273.

[14] B. Lakshmi Sirisha, S. Srinivas Kumar, and B. Chandra Mohan, "Identification of cheaters in elections using steganography," *J. Inf. Optim. Sci.*, vol. 37, no. 2, pp. 271–278, 2016, doi: 10.1080/02522667.2015.1130891.

[15] Alma, R. W. Wardhani, S. Muhasyah, and M. Delina, "Least significant bit steganography method for the digital data protection in the barcode," *AIP Conf. Proc.*, vol. 2169, no. November, 2019, doi: 10.1063/1.5132672.

[16] A. K. Bairagi, R. Khondoker, and R. Islam, "An efficient steganographic approach for protecting communication in the Internet of Things (IoT) critical infrastructures," *Inf. Secur. J.*, vol. 25, no. 4–6, pp. 197–212, 2016, doi: 10.1080/19393555.2016.1206640.

[17] M. Baritha Begum and Y. Venkataramani, "LSB based audio steganography based on text compression," *Procedia Eng.*, vol. 30, no. 2011, pp. 703–710, 2012, doi: 10.1016/j.proeng.2012.01.917.

[18] A. G. Chiţu, L. J. M. Rothkrantz, P. Wiggers, and J. C. Wojdel, "Comparison between different feature extraction techniques for audio-visual speech recognition," *J. Multimodal User Interfaces*, vol. 1, no. 1, pp. 7–20, 2007, doi: 10.1007/bf02884428.

[19] S. Khan, N. Ahmad, and M. Wahid, "Varying index varying bits substitution algorithm for the implementation of VLSB steganography," *J. Chinese Inst. Eng. Trans. Chinese Inst. Eng. A*, vol. 39, no. 1, pp. 101–109, 2016, doi: 10.1080/02533839.2015.1082933.

[20] E. Keogh, "Automatic Feature Extraction for Classifying," *Artif. Intell.*, pp. 127–149, 2005.

[21] M. Wakiyama, Y. Hidaka, and K. Nozaki, "An audio steganography by a low-bit coding method with wave files," *Proc. - 2010 6th Int. Conf. Intell. Inf. Hiding Multimed. Signal Process. IIHMSP 2010*, pp. 530–533, 2010, doi: 10.1109/IIHMSP.2010.135.

[22] S. S. Divya and M. R. M. Reddy, "Hiding Text In Audio Using Multiple LSB Steganography And Provide Security Using Cryptography," *Int. J. Sci. Technol. Res.*, vol. 1, no. 6, pp. 68–70, 2012.

[23] A. M. Meligy, M. M. Nasef, and F. T. Eid, "An Efficient Method to Audio Steganography based on Modification of Least Significant Bit Technique using Random Keys," *Int. J. Comput. Netw. Inf. Secur.*, vol. 7, no. 7, pp. 24–29, 2015, doi: 10.5815/ijcnis.2015.07.03.

[24] A. M. Meligy, M. M. Nasef, and F. T. Eid, "A Hybrid Technique for Enhancing the Efficiency of Audio Steganography," *Int. J. Image, Graph. Signal Process.*, vol. 8, no. 1, pp. 36–42, 2016, doi: 10.5815/ijigsp.2016.01.04.

AUTHOR INDEX